THE LIFE AND LEGEND OF JAY GOULD

The Life and Legend of

JAY GOULD

by Maury Klein

THE JOHNS HOPKINS UNIVERSITY PRESS
Baltimore and London

Printed in the United States of America on acid-free paper

The Johns Hopkins University Press
2715 North Charles Street
Baltimore, Maryland 21218-4319
The Johns Hopkins Press Ltd., London

Johns Hopkins Paperbacks edition, 1997
06 05 04 03 02 01 00 99 98 97   5 4 3 2 1

Library of Congress Cataloging-in-Publication Data

Klein, Maury, 1939–
The life and legend of Jay Gould.

Bibliography: p.
Includes index.
1. Gould, Jay, 1836–1892. 2. Businessmen—United States—Biography. 3. Capitalists and financiers—United States—Biography. 4. Railroads—United States—History. I. Title.
HC102.5.G68K55 1986 332'.092'4 [B] 85-24107
ISBN 0-8018-2880-5 (alk. paper)

ISBN 0-8018-5771-6 (pbk.)

A catalog record for this book is available from the British Library.

FOR DIANA WITH LOVE

Contents

III: REACHING HIGH, 1879–1881

IV: HOLDING ON, 1882–1887

V: LETTING GO, 1888–1892

Illustrations

PHOTOGRAPHS

MAPS

Preface

Years ago, when my interest in Jay Gould first tempted me to do a full biography of him, several colleagues assured me that I was foolish to attempt such a project on a subject with no known papers. Their warning was sound but the sort that no scholar can heed unless he cares only to play safe. I was prepared to find papers or not to find them, and to make do with what I found. My plan was to write a smaller volume, an interpretive biography, and to use it as prelude to a novel on the life of Gould.

But projects assume a life of their own, and often have a way of goading one to seek more than was originally intended. Diligence and a healthy dose of good luck enabled me to uncover far more material than I ever expected to find, and the more I found, the more I was driven to extend the search. In the end I had almost as much material as any biographer could want, much of it never before used or even known to exist. The discoveries pleased but did not surprise me, for I have always believed that the material one needs is out there somewhere if you can but locate it. What did surprise me was the extent to which these materials revised the conventional wisdom about Gould and the events associated with his life. Gradually it became clear that I had not only a fatter biography on my hands but also one that painted a sharply different picture of its subject. Julius Grodinsky gestured at this difference in his fine study of Gould, but he lacked the sources to document his case and made no effort to explore the character of the man.

It is always tempting to preface a biography by explaining why the subject's life is important enough to warrant a lengthy book. Certainly I consider Gould one of the half-dozen most important and influential Americans of his age, but there is more at stake here. Every public figure has a public image, a legend of sorts, surrounding his life. In this case, however, the gap between the facts of Gould's life and the myth that has endured for a century is so striking that it forces other questions to the surface. The most obvious one concerns the role of newspapers in shaping our impression of the late nineteenth century. The role of media has become a major question in our own time but we have been slow to realize how far back the issue runs. There is no doubt in my mind that the Gould of legend was created almost entirely by the press and perpetuated by later writers who relied on these earlier accounts without verifying or adding to them. That is how legend is born and how it flourishes, especially when little is known about the subject.

Ironically, even those wary of old newspaper accounts used the versions of these later writers without realizing where they had originated. The same process held true for characterizing other businessmen of the era until all were labeled with what is perhaps the most persistent and misleading cliché in American history, the "robber baron." Nothing better exemplifies the American genius for coining a phrase without having the foggiest notion of what it actually means. It was the newspapers that forged the image of the robber baron, thereby creating material for later writers such as Gustavus Myers or Matthew Josephson or a host of hacks seeking to capitalize on it. The result has been a severe distortion of history that a generation of scholars has labored in vain to correct. One need only pick up a current textbook to see how tenaciously the old myth survives despite all efforts to revise the story.

That is not surprising. It is far easier to perpetuate error than to correct it, especially when the error concerns a cherished myth. We are always reluctant to surrender myths because they tell a story uncluttered by ambiguities and complexities. In them the roles are clearly defined; we know who the good guys and the bad guys are. In reality the distinction is not so obvious. Gould poses a unique problem in this respect. He remains the arch villain in the Age of Robber Barons, the most hated man in America. Textbook authors, whose task is to convey the most conventional of wisdom, would feel naked without his services in that role.

More important, his career brought into collision two myths dear to the American heart: Horatio Alger and the plain but virtuous citizen vital to the life of a democratic society. Alger was that public-spirited citizen, but he was also looking for every possible means to advance himself. Americans have never found a comfortable way to resolve the clash between private ambition and public interest. The climb to success invariably creates moral dilemmas that have no place in the clean lines of the Alger myth. In just this way Gould served as a sort of id to the American ego: he became the embodiment of what men aspired to and shrank from in horror at the consequences. The dilemma was simply how to achieve what Gould did without using the same methods or paying the same price for success. However, if it turns out that Gould's methods were not so much different from those of his peers as they were more original, the question assumes a very different cast.

Because the gap between legend and reality is so astounding in Gould's case, I have tried to present both in tandem so the reader can compare them. I have also emphasized the influence of Gould's personal life on his business career. Business historians tend to ignore this aspect entirely for reasons I have never fathomed. I do not expect this reinterpretation of Gould to be accepted readily. We do not surrender our myths easily, and scholars no less than generals have the urge to slay the messenger bearing bad news. However, I did not invent this interpretation of Gould; I have merely gone where the evidence pointed.

Some years ago I gave a paper suggesting the need for a wholesale revision of our views on Gould. Joseph Frazier Wall, a member of the panel and the author of a splendidly detailed biography of Andrew Carnegie, commented that my interpretation was intriguing but required more solid evidence to be convincing. I hope this book satisfies that need.

Acknowledgments

As always I incurred in the writing of this book more debts than I can possibly repay or even acknowledge. In libraries and archives everywhere I found specialists willing to share their expertise and expedite my work. Their number included Donald D. Snoddy of the Nebraska State Museum and Archives, Florence Bartoshesky and Marjorie Kierstad in the Corporate Records section of Baker Library, Ardie L. Kelly of the Mariners Museum, Henry E. Brown of the Pennsylvania Historical and Museum Commission, John Aubrey, Diana Haskell, and James Wells of the Newberry Library, Alice C. Dalligan and Joseph F. Oldenburg of the Burton Historical Collection, Phyllis E. McLaughlin of the Iowa State Department of Archives and History, Robert L. Sponsler of the Missouri Pacific Railroad Archives, Tom Dunnings of the New-York Historical Society, Priscilla Sutcliffe of the Cooper Library, Anthony Bruno of the Mugar Library, Clark L. Beck, Jr., of the Alexander Library, Robert A. McCown of the University of Iowa, and John D. Cushing of the Massachusetts Historical Society.

Mimi Keefe and the entire staff of the University of Rhode Island Library provided indefatigable cooperation despite the constant demands I placed on their limited resources. Special thanks are due Vicki Burnett, Robert E. Doran, and Sylvia C. Krausse of the Interlibrary Loan Department, Judy MacDonald in Microform, and Marie Rudd. Mary Chatfield, librarian of Baker Library, was helpful in innumerable ways and procured for me access to the original diaries of Oliver Ames.

A number of friends and colleagues provided help and encouragement along the way, among them Alfred D. Chandler, Jr., Lucius Ellsworth, Lewis Gould (no relation), Robert S. Grodinsky, Albro Martin, Richard C. Overton, Martha Parker, Phillip E. Reeves, and Judith Swift. Edwin Gould granted me access to documents in his possession, and Kingdon Gould, Jr., lent me his collection of reproductions of political cartoons from *Puck*. Harry E. Hammer smoothed my path into the Missouri Pacific Archives. The Union Pacific System, Omaha, Nebraska, provided the railway system maps. John Frisbee, administrator at Lyndhurst, allowed me to examine important letters and pictures, and Adna Watson provided a most informative tour of the house. Dr. Peter M. Small furnished medical information. The burden of typing fell on the willing and capable shoulders of Bonnie Bosworth and Elise Otis, who also performed a variety of secretarial tasks that greatly expedited my work. Mary-Ellen Toscano and Robert Galisa cheerfully undertook a number of thankless duties. At the Johns Hopkins University Press the manuscript benefited greatly from the prudent counsel and staunch encouragement of my editor, Henry Tom, and from the intelligent and sensitive copy editing of Carol Ehrlich.

I am especially grateful to Kingdon Gould, Jr., and his wife, Mary, for their help and encouragement. They provided many of the documents crucial to this study with no restriction on their use. In one of those moments about which historians dream but rarely experience, Kingdon and I discovered the large batch of Gould letters to Clark one evening at Furlow Lodge while looking through some old folders tucked away in a bookcase. Clark had obviously returned the letters to George after Jay's death, and they had remained in the bookcase undisturbed ever since. Although Kingdon had found the two letterbooks, he did not know the Clark letters existed until we happened on them. Since that evening he has taken a keen interest in the project and had in fact long been gathering materials for a biography. From the first he has imposed no editorial restraints, insisting always that his only desire was for someone to get Jay's story straight for a change. I hope this book has fulfilled that desire.

To those named and many others unnamed belongs much of the credit for what is worthwhile in this study; the errors and omissions are entirely my responsibility. Finally, all other contributions pale before that of my wife, Diana, who added much and endured more in enabling me to complete this work. Without her loving support it would have been a much poorer piece.

Railroad Reference Key

A & N	Atchison & Nebraska
ALBANY	Albany & Susquehanna
ALTON	Chicago & Alton
ATLANTIC	Atlantic & Great Western
B & M	Burlington & Missouri River in Nebraska
BOSTON	Boston, Hartford & Erie
BRADFORD	Buffalo, Bradford & Pittsburgh
BURLINGTON	Chicago, Burlington & Quincy
BURLINGTON & NORTHERN	Chicago, Burlington & Northern
BURLINGTON IOWA	Burlington & Missouri River in Iowa
CENTRAL	Colorado Central
CLEVELAND	Cleveland & Pittsburgh
COTTON BELT	St. Louis, Arkansas & Texas
COUNCIL BLUFFS	Kansas City, St. Joseph & Council Bluffs
FORT SMITH	Little Rock & Fort Smith
FORT WAYNE	Pittsburgh, Fort Wayne & Chicago
FRISCO	St. Louis & San Francisco
GULF	Kansas City, Ft. Scott & Gulf
HAMILTON	Cincinnati, Hamilton & Dayton
HANNIBAL	Hannibal & St. Joseph
INDIANA CENTRAL	Columbus, Chicago & Indiana Central
JERSEY CENTRAL	Central of New Jersey
KANSAS CITY	St. Louis, Kansas City & Northern
KATY	Missouri, Kansas & Texas
LACKAWANNA	Delaware, Lackawanna & Western
LAKE SHORE	Lake Shore & Michigan Southern
MANHATTAN	Manhattan Elevated
MEMPHIS	Memphis & Little Rock
METROPOLITAN EL	Metropolitan Elevated
MICHIGAN SOUTHERN	Michigan Southern & Northern Indiana
NEW ENGLAND	New York & New England
NEW YORK EL	New York Elevated
NICKEL PLATE	New York, Chicago & St. Louis
NORTHWESTERN	Chicago & Northwestern
PANHANDLE	Fort Worth & Denver City
PEORIA	Toledo, Peoria & Warsaw
READING	Philadelphia & Reading

RIO GRANDE	Denver & Rio Grande
RIO GRANDE WESTERN	Denver & Rio Grande Western
ROCK ISLAND	Chicago, Rock Island & Pacific
ROCKFORD	Rockford, Rock Island & St. Louis
RUTLAND	Rutland & Washington
SANTA FE	Atchison, Topeka & Santa Fe
SIOUX CITY	Sioux City & Pacific
ST. JOSEPH	St. Joseph & Denver City (Western)
ST. PAUL	Chicago, Milwaukee & St. Paul
WABASH	Toledo, Wabash & Western
WABASH	Wabash, St. Louis & Pacific
WICHITA	St. Louis, Ft. Scott & Wichita
WILMINGTON	Philadelphia, Wilmington & Baltimore
WYANDOTTE	Kansas City, Wyandotte & Northwestern

THE LIFE AND LEGEND OF JAY GOULD

PROLOGUE

The Legend

Few properties on which this man laid his hand escaped ruin in the end. . . . He was not a builder, he was a destroyer, and the truth of this statement may be easily demonstrated by tracing out the subsequent history of the corporations which he got into his clutches.

—Alexander D. Noyes, *Forty Years of American Finance* (1909)

Gould was impeached as one of the most audacious and successful buccaneers of modern times. Without doubt he was so; a freebooter who, if he could not appropriate millions, would filch thousands; a pitiless human carnivore, glutting on the blood of his numberless victims; a gambler destitute of the usual gambler's code of fairness in abiding by the rules; an incarnate fiend of a Machiavelli in his calculations, his schemes and ambushes, his plots and counterplots.

—Gustavus Myers, *History of the Great American Fortunes* (1909)

The whole interest of Gould lay in the manipulation of the securities of his various companies. The development of the roads was an entirely minor concern. In all cases the property was used to aid his financial transactions. . . . Gould made a fortune, but the roads he touched never quite recovered from his lack of knowledge and interest in sound railroading.

—Robert E. Riegel, *The Story of the Western Railroads* (1926)

In his expeditions he used courts and governments; his allies were many, but none his friends; at one time or another in his life he broke almost every man who worked with him, and to the last he remained what by nature he was, a lone worker.

—Robert I. Warshow, *Jay Gould: The Story of a Fortune* (1928)

Where certain of his rivals, intoxicated with power, learned to crave glory too, Jay Gould seemed to place himself above such human vanities. Nor did any social interest, or any sentimental consideration . . . deflect him for a moment from his marvelously logical line of movement. No human instinct of justice or patriotism or pity caused him to deceive himself or to waver in any perceptible degree from the steadfast pursuit of strategic power and liquid assets. . . . self contained, impassive to all pleas or reproaches, he seemed content with his loneliness.

—Matthew Josephson, *The Robber Barons* (1934)

Furtive and deadly as a spider, . . . Jay Gould fed on the betrayal of friends, fattened on the ruin of stockholders, lied and bribed his way to a power that raised him above the law.

—Robert V. Bruce, *1877: Year of Violence* (1959)

His life and career . . . became the ultimate perversion of the Alger legend. He was a certifiable crook by the time he reached voting age. . . . None of his contemporaries quite approached his genius for trickery and thimblerigging, his boldness in corruption and subornation, his talent for strategic betrayal, his mastery over stock and bond rigging, his daring in looting a company and defrauding its stockholders.

—Richard O'Connor, *Gould's Millions* (1962)

Gould was first, last, and always a trader. His special talent, amounting to a compulsion, was for detecting opportunities to seize control. . . . In railroading Gould was a wrecker. . . . The fact that he led an exemplary family life simply brought into sharper focus his utterly ruthless business dealings.

—Richard C. Overton, *Burlington Route* (1965)

Gould's career encompassed almost every known variety of chicanery, from stock watering and industrial blackmail to bribery, market manipulation, and union busting.

—John A. Garraty, *The New Commonwealth, 1877–1890* (1968)

For Gould, the meaning of life could be expressed in a simple formula: wealth equals success. There were no qualifying plus or minus values on either side of the equation.

—Joseph Frazier Wall, *Andrew Carnegie* (1970)

Many of the railroad promoters of the age were neither as civic-minded nor as creative as Hill. Most notorious of all was Jay Gould. Gould was a railroad man, but he built no railroads.

—Irwin Unger, *These United States*, volume 2 (1978)

In his own time Jay Gould became known as the most hated man in America. It is not an honor most men seek, let alone relish, nor is it easily cast off once bestowed. For most of Gould's life the taunting curses of his critics followed him like the tinkle of the leper's bell. Two generations of operators, businessmen, bankers, lawyers, coupon clippers, stockholders, politicians, reformers, journalists, ministers, and other guardians of the public morals reviled him as the supreme villain of his age. Even those who admired Gould's genius often did so with a mixture of awe and revulsion, as if fascinated by the movements of some exotic but deadly predator.

It is difficult to exaggerate the depth of vituperation heaped on Gould by his contemporaries. Daniel Drew said of him simply, "His touch is death." James R. Keene, another operator, denounced Gould as "the worst man on earth since the beginning of the Christian era. He is treacherous, false, cowardly and a despicable worm incapable of a generous nature." Editors reserved their choicest insults for Gould. Joseph Pulitzer called him "one of the most sinister figures that have ever flitted bat-like across the vision of the American people." The *New York Herald* suggested that "he should be called the Skunk of Wall Street, not one of its ubiquitous Wolves and Wizards," while the *New York Times* complained that "the work of reform is but half done when the insidious poison of an influence like that of JAY GOULD can be detected in politics, in finance, in society, and when people claiming to be respectable are not ashamed of being associated with a man such as he."[1]

Alongside the portrait that emerges from these descriptions, Dorian Gray seems but a choir boy in comparison. To his fellows Gould loomed as a dark, sinister figure of almost supernatural powers. A cold, calculating loner, his eye ever on the main chance, he crushed rivals, betrayed friends, fleeced unwary investors, corrupted judges and legislatures, tyrannized working men, and flouted every standard of business morality. At the same time there is something vague and fleeting, even mysterious about this image, as if he were some sort of financial vampire who struck with such consummate skill that victims never understood just how their blood had been drained.

Unquestionably Gould enjoyed the worst press of any American of his era. Reporters disliked him because he spoke little and said less. When he did speak it was either to advance some purpose of his own or to trick reporters into thinking he had answered their questions when in fact he had evaded them. Still they flocked after him. Gould's exploits aroused great excitement among a public eager to feed on the derring-do of business titans. That so much of Gould's legerdemain was cloaked in mystery only heightened its appeal and encouraged journalists to embellish their accounts in any manner they chose.[2]

From this outpouring of abuse emerged the popular image of Gould as a cold, delicate genius with a devious mind, an insinuating manner, and an instinct for the jugular. As caricatured in print and cartoon he betrayed little feeling for anyone or anything. He performed his crimes with the detached air of a surgeon, dispensing his ruthless precision upon friends, foes, and innocent bystanders alike. He purred like a cat and struck like a cobra. The clinical, almost inhuman figure that stalks the dailies resembles nothing less than the ultimate financial technician.

To a surprising extent the press manufactured Gould's reputation as the most hated man in America. That reputation evolved long before his death to a point where it assumed the proportions of legend. Notoriety came early and stayed late in Gould's career. He burst into public prominence with two spectacular episodes, the Erie War and the Gold Corner. Other episodes on Wall Street added to his reputation, but these two escapades implanted Gould's ignominy in the public mind and colored appraisals of everything else he did. To the end of his life he wore them like a scarlet letter, the imprint of which did not fade even after his death.

The sensational publicity surrounding his two scandals gave birth to the legend of Jay Gould, which for two decades spread like a lethal cloud across every enterprise he touched. No corner of his life escaped its corrosive effect. Critics retrieved from oblivion every chapter of his early career and, with scant regard for the facts, paraded them before an eager public as the seedbed of his treachery and deceit. Even his acts of charity, numerous but unpublicized, were twisted into expiations of a guilty conscience to the point where Gould felt compelled to give them up rather than defend them. Had he failed in business, the public would have drowned him in applause. Instead the attacks on his character grew in proportion to the scale of his operations.

The legend persisted because so much of Gould's life was the stuff of legend, and because his style encouraged its growth. In business his secretiveness was so complete that even his closest associates often confessed ignorance of his true designs. He mystified adversaries because they could not catch his drift or depth. Like a chess master Gould usually grasped the whole position well enough to stay a move or two ahead. He found combinations that duller minds overlooked. Recognizing that a straight line was the shortest distance between two points but not always the best route to travel, he seldom took it. In his personal life too he frustrated the curiosity of reporters, associates, critics, and public alike. Few persons of such prominence guarded their privacy as jealously as Gould. In every respect he was a man of mystery.

This aura of mystery may be the most revealing clue to the legend of Jay Gould. Even now the most remarkable thing about the man is how little we actually know about him. His contemporaries knew him no better than we do, and one wonders whether they despised the mystery about the man more than the man himself. To them he was a figure of shadows whose substance they fleshed out from their own imaginations. This was the image handed down to historians and, on the whole, the image historians have preserved intact. The invective hurled at Gould in his own time has been matched by later students of his wrongdoings.

The historiography on Gould resembles a hall of mirrors in which contemporary and modern accounts reflect one another without adding depth to the original image. Unaware that Gould left behind any personal papers, later writers drew their information from the same meager pool of sources: newspapers and financial journals, Gould's testimony before legislative and congressional committees, the memoirs of his contemporaries, the celebrated *Chapters of Erie*, and a spate of hack biographies that appeared immediately after Gould's death in 1892.[3]

Among these sources the newspapers were by far the most comprehensive but also the most suspect. Compelled to operate in the dark, they traded heavily in rumor, innuendo, and speculation spiced generously with sensation. Nor were they the least bit detached or objective in their views. Many dailies considered Gould not merely evil but a direct threat to their existence. Gould's testimony has the virtue of being in his own words. It is reliable but guarded, offered on occasions when he was inclined to conceal more than he revealed. Surprisingly few memoirs by financiers mention Gould, and fewer still avoid inaccuracies. None is quoted more often than that of Henry Clews, the incessant blabbermouth of Wall Street, and none is more riddled with errors of fact. Drawing extensively on newspaper accounts, the hack biographers rehashed the familiar episodes of Gould's life and wrapped them in homilies. None scratched the surface of his career or gleaned fresh insights into his character. The Adams brothers attempted a serious study of the Erie War and the Gold Corner, but their analysis sank beneath the weight of righteous indignation. In the long run, however, they achieved their purpose. Later generations owe more than they know to the Adams brothers for the image of Gould as villain.[4]

If nothing else, the hack biographies served as the first clearing house for the ragtag mélange of source materials on Gould's life. In the ninety years since their appearance little else has been added to the collection. The same tired anecdotes, unflattering caricatures, and blatant inaccuracies have been handed down from one generation of writers to the next like family heirlooms, worn and faded from overuse, their authenticity validated by age if nothing else. Writers seemed content to repeat past clichés without bothering to inquire whether or not there was any truth in them.

Not until 1957 was Gould's career dignified by a serious scholarly study. Julius Grodinsky's massive book was not a biography but a searching analysis of Gould's business activities. While denying none of Gould's faults or misdeeds, Grodinsky showed little interest in flogging him for his villainies. His concern was rather to portray the economic environment in which Gould operated and to demonstrate how Gould grasped that environment more thoroughly than most men and utilized it in ways that were breathtaking in their originality and breadth of vision. In effect he performed a scholarly autopsy on Gould's career and declined to keep a scorecard of moral judgments. Yet even Grodinsky failed to penetrate many of the mysteries surrounding Gould. Handicapped by the lack of primary source materials, he relied on his own shrewd guesswork to fit the available and often disparate pieces of the puzzle together.[5]

Historians received Grodinsky's study with faint enthusiasm, as if an unwanted relative had come to stay. They praised Grodinsky's analysis as vigorously as they rejected his conclusions, apparently willing to be persuaded that Gould was an even more clever rascal than they suspected but unwilling to be convinced that he was anything other than a moral leper. During the 1960s two biographies of Gould revived the hacker's art. Both quoted Grodinsky on occasion but left the clear impression that they understood neither the intricacies of Gould's career nor Grodinsky's explanation of them. Baffled by the complexities of business and finance, they confined their efforts to reruns of the usual hoary anecdotes. Not until 1977, when Alfred D. Chandler, Jr., published *The Visible Hand,* did a prominent historian take Grodinsky's version of Gould's contributions seriously.[6]

Controversial figures expect unpleasant things to be said about them, many of which are untrue, but seldom do these distortions stand unchallenged for a century. At some point historians usually appear on the scene to tidy up the premises and rummage through the debris, setting aside what is valuable and discarding the rubbish. Except for Grodinsky's impressive but limited effort, this process has yet to occur with Gould. He remains a man about whom much has been written and little said, a victim of wrong images and wrong metaphors. His reputation is grounded not in sober analysis but in moral judgments derived from incredibly meager evidence, much of it hearsay. If it is true that legend begins at the point where perceptions depart from reality, then the problem is one of pulling our perceptions of Gould back to a solid mooring of evidence. It is not a matter of substituting one set of moral judgments for another but of simply getting the facts straight.

History needs villains no less than it does heroes, and the process by which it acquires both is remarkably similar. The American hero is not a thinker but a man of action whose deeds cut through the haze of ordinary life. His acts reflect qualities to which other men aspire but which they seldom attain. As Robert Penn Warren once remarked, "The hero is the embodiment of our ideals, the fulfillment of our secret needs, and the image of the daydream self." More than that, the hero must be destiny's child, the right man in the right place at the right time. So too with villains. Although they reflect qualities at the other extreme, they stand apart from the common run of humanity in the same way and for similar reasons. They are despised with no less passion than heroes are admired. The heroes and villains of any age reflect the polar opposites of a value system that is itself constantly in flux. At any given moment the difference between them is never more than what people think about what they have done.[7]

In effect they are the principal characters in history's endless procession of morality plays. As such they are no more than caricatures, one-eyed jacks seldom seen in full face. That is the way it must be, if only because it is far more difficult to dramatize a lifetime than to capture its moments of highest achievement. The life of even the noblest hero betrays mundane, even seedy aspects when probed too closely. No man remains a hero or a villain to his biographer. Stereotypes are the first casualty of character study, and only when an individual's depth and complexity of character are revealed can the extent of his heroism or villainy be appraised.

This is precisely what Gould requires. It is not necessary or desirable to contend that Gould was an angel to dispute his legend. Nor is it necessary even to protest that history has done him a disservice. The point is that history knows next to nothing about him, even though he was unquestionably one of its most significant business figures. Hack biographers and scholars alike, including Grodinsky, have neglected virtually everything about Gould except his business career. Little has been written about his personality, attitudes, or psychological and emotional makeup, or the interaction between his personal life and his business career.

He is not alone in suffering this neglect. Historians and even biographers too often ignore the whole person. Those dealing in business and economic history have been especially prone to this shortcoming. They have rigorously investigated how business affects life but rarely explored how life affects business. An artist is more than his technique, and cannot succeed through technique alone. So too with a businessman. His work cannot be properly understood without putting it in the broader context of his life. Like all men, Gould was not the same in later years as he was when a young man. Just as experience hardened him as a youth, so did it soften and mature him in middle age. To ignore such changes and their effect on a man's behavior is to misunderstand him entirely.

One objective of this book is to set Gould's career in this broader perspective. Another is to square the legend with the facts as we now know them, to determine what he actually did or did not do, what powers he did or did not possess, what the effects of his actions were, and why he did what he did. The truth of any man's life is never a simple thing to unravel, and the task is unusually difficult in Gould's case. As a reporter once observed, "Gould is not the sort of man to leave his tracks clearly marked in the dirt." If anything that is an understatement for a man who practiced secretiveness as if it were a religion.[8]

The tracks are faint but they are there, enough so to piece together a trail that connects Gould's early years to his later career with surprising clarity. The imprint of Gould's youth, the lessons learned and the habits formed, has always been a part of the legend in the most superficial way. When looked at carefully it reveals the spawning ground not only of Gould's character but also of his methods of operation. The discovery of new material, much of it in the possession of descendants, provides new insights into Gould's business career. In short, there is more evidence than anyone has suspected, more than enough to give the legend a decent burial.

If it rises again from the dead, you'll know who to blame.

I

STRUGGLING UPWARD 1836–1860

1

Routes

Gould . . . was small and slight in person, dark, sallow, reticent, and stealthy, with a trace of Jewish origin.

—Henry Adams, *Chapters of Erie*

Many who knew Mr. Gould intimately are in the habit of asserting that his origin must have been Hebraic. . . . His habits of thought and his extraordinary intellect were both Jewish, these people assert, with how much or how little basis in the actual fact of his origin, no one can ever decide.

—Trumbull White, *The Wizard of Wall Street*

You might have seen in Jay Gould's Jewish look, bright scholarship, and pride of manners some promise of an unusual career.

—John Burroughs, *My Boyhood*

Jay Gould was reputed to be a Jew, but he had traits strongly indicative of the American Indian.

—John Torrey Morse, Jr., *Massachusetts Historical Society Proceedings*, 1927

THE MYSTERY surrounding Jay Gould begins with the name itself. It pleased some who disliked him to insinuate that he was Jewish and to dismiss his peculations as typical of his race, as if this were the ultimate calumny to be heaped on him. To their eyes the traits were all there: he looked and acted the perfect Shylock, full of smiling treachery to friends and enemies alike, secretive and cunning in all his dealings, wielding his power with the cruelty and vindictiveness of an oriental despot. Then there was the name, which several generations back had been Gold. No one knows exactly when or why it became Gould, but that simple change, coupled with a family tree crowded with Nathans, Abrahams, Davids, and the like, was evidence enough for those who wished to be convinced.

This sort of epithet was a common form of malice for a generation devoted to ethnic and racial stereotypes. Aside from its obvious bigotry, the charge illustrates a curious short-sightedness on the part of those who hurled it. In their obsession with anti-Semitism they went after the wrong name. Few of them noticed or had the wit to explore the irony of Gould's bearing the name of the man who pursued the golden fleece. For gentlemen who trumpeted the virtues of a classical education this was perhaps the more unforgiveable oversight.

The Gould family tree was in fact an old and modestly distinguished one, planted in America two centuries and six generations before him. Its progenitor, Major Nathan Gold, left England in 1647 to settle in Connecticut, where he was one of nineteen colonists who petitioned King Charles in 1674 to grant Connecticut its celebrated charter. By that time he had served as judge, senator to the General Court, and chief military officer. He had also done well enough to be called Fairfield's richest inhabitant. The Golds drew their fortune from land and shipping, their reputations from a legacy of public service. Nathan Gold, Jr., outshone his father in the holding of public offices, including twenty-two years as town clerk of Fairfield. His grandson, Abraham Gold, married Elizabeth Burr in 1754, a connection that was to make Jay Gould the distant relation of another American with a tainted reputation, Aaron Burr.[1]

During the Revolutionary War Colonel Abraham Gold died a hero's death at the battle of Ridgefield in 1777. Two years later his widow and nine children were burned out of their home when the British set fire to Fairfield. Undaunted, Elizabeth Gold rebuilt the house and reared her children to carry on the family's tradition of prominence in the town. That house survived until her grandson, John Gould, replaced it in 1840 with a mansion adorned by a portico of huge Corinthian columns. A master of clipper ships that worked the China trade, John Gould did everything expected of him except produce an heir. When the last of his two daughters died in 1908, the Gould line in Fairfield had played itself out.

It was the sons of Abraham and Elizabeth Gold who began using the name Gould. One of them, young Abraham, married Anna Osborne in 1788 and seemed destined to follow his forebears to the sea. But his bride noted that three of Abraham's uncles had drowned and rebelled against a life spent wringing her hands atop a widow's walk. As a child she had lived for some years in a place called Beaverdam in the Catskill Mountains. Hearing that a few families were about to take up land there, she persuaded her husband to join them.[2]

They settled in a place called West Settlement, later renamed Roxbury. For Abraham, as for all pioneers, the future was everything. To realize its promise one needed above all else patience and tenacity, two qualities that were imprinted on the Goulds like a coat of arms. Abraham put up a small house near a stream in the meadow. As the first wave of their ten children arrived, he dutifully trudged up the hill and built another house above the road, this one with two floors and an attic. It was a house much like Abraham himself: simple yet sturdy, ambitious yet painstaking in its attention to detail, built not to impress but to endure.

When Abraham died in 1823, the homestead passed into the hands of his eldest son, John Burr Gould. Already thirty-one years old, John was remarkably like his father. He was a man of strong convictions, a staunch Democrat, and a devotee of local affairs. Limited in education, he nourished his intellect on newspapers and whatever books he could find. In the sixty-eight years he lived in Roxbury, John's ambitions never went beyond serving the town as assessor and performing whatever duties were asked of him. His younger brother, Jason, having no inheritance to keep him at home, pulled up stakes and departed for Canada, leaving behind only a name John would later put to use.[3]

Not until the farm was his did John Gould even think of marriage. When the time came for him to seek a wife, he looked no farther than the other side of the mountain. The village of Moresville had first been settled by John More, a Scotsman obsessed by the vision of America, who had bundled his wife and two small children aboard a ship to New York in 1772. From there he sailed to Catskill, bought three horses, some cattle, tools, utensils, and seed, and plunged into mountain country. Following Indian trails when he could find them, More led his family fifty-five miles through the forest to a site near what is now Hobart. The place offered rich soil, a spring, a brook for the stock, and grass for the animals in a nearby marsh. Satisfied, the Mores went to work planting crops, erecting a log cabin, crafting furniture out of felled trees. Whenever More needed ammunition, flour, or other supplies, he had to ride back to Catskill for them. Except for an occasional trapper, they never saw the face of another white man. Their only companions were roving bands of Mohawks, whose friendship they were careful to cultivate.[4]

For all the hardships, John and Betty More prospered in the wilderness. The birth of a third son, Alexander, added to their contentment. Had fate left them undisturbed, they would doubtless have lived out their days in that same cabin. Their granddaughter would not have met John Gould, and one can only wonder what that missed connection would have done to the railroad map of America. But the outbreak of war cut their idyll short. After the British induced many of the tribes to join forces with them, mixed bands of Indians and Tories began terrorizing frontier settlements in New York. More was burned out by a band of marauders but managed to escape with his family. Using what little money he had, More bought some land near Catskill and for nine hard years scratched a living from the thin, rocky soil. Three more children came along, and the war had long since ended. Seeing no reason to toil his life away in Catskill, More sold off his land and started back to the old homestead. On the way he met a man named Clark, who owned some land a few miles east of More. For reasons nobody knows, they decided to swap claims. Once again More cleared the land and built a cabin on the site of what later became "The Square" in Moresville (now Grand Gorge).

This time the Mores stayed put. Their cabin sat at the intersection of several well-traveled trails. More opened a tavern and, as other settlers arrived, served as magistrate and postmaster. On Sundays the tavern turned meeting house as More held services, reading from the Bible and a book of sermons. Betty gave birth to an eighth child. Prosperous and content, they watched their children come of age, marry, and take up farms of their own. Around them Moresville blossomed from an isolated settlement into a thriving village, and the Mores lived long enough to become its patriarchs. Betty died in 1823 at the age of eighty-five; John lingered on to the ripe old age of ninety-five.

Most of their children settled in Beaverdam except for Alexander, who remained in Moresville. It was his daughter Mary who caught the eye of John Gould. She was a pert, comely girl, deeply religious without being dour, simple yet strong-willed in her tastes and habits. There was nothing rushed in their courtship. Not until 1827, four years after inheriting the Roxbury farm, did John Gould wed Mary More. He was already thirty-five years old, his wife six years younger.

By these roundabout routes the Gould and More families were united. In both of them could be found traits that formed a rich legacy for Jay Gould: the pioneers' independence of spirit, a willingness to carve one's own path in life, fierce pride coupled with unswerving loyalty to family, and, above all, an indomitable will. What is lacking on either side is the slightest hint of any Jewish connection. Even if it had been there, those who insinuated its existence might have spared themselves the trouble. After all, what better legacy for acquisitiveness could a man have than a Scottish mother and a Puritan father?

The town of Roxbury nestles in the eastern corner of Delaware County, near the center of the Catskill Mountains. It is an insular world of hollows and hills, forests and fields, walled in by mountains through which only an occasional gap offers access. Two small rivers, the East Branch of the Delaware and a tributary of the Schoharie, transformed the red clay soil into rich farmland and pasture. Dense stands of oak, maple, cherry, hickory, beech, pine, spruce, elm, and chestnut crowded the slopes, and thick clusters of hemlock, with their blue-green foliage, shaded the valleys in an air of perpetual twilight. For most settlers a farm consisted of a few acres wrested grudgingly from these stands of timber and some bottomland wedged between undulating slopes. The fields tilted at such steep angles that houses, cattle, and crops seemed to cling to them as if hanging on for dear life. "The trouble with the Catskills is they got too much land," one sage quipped. "Had to set it all on edge before they could put it on the map."[5]

John Gould's frame house and the barn opposite it leaned into a ridge overlooking a long, narrow valley. The sitting room left of the entrance hall dominated the downstairs with its huge fireplace. Upstairs Mary Gould installed her loom in one of the five large bedrooms. From the windows she could look down upon the meadow through which a brook curled its graceful path to the distant mountains. Every spring she reveled in the beauty of the grass splashed blue with violets and forget-me-nots. Not content to rely on nature, she ringed the house with flower beds. While she bequeathed this love of flowers to all her children, none of the girls rivaled Jay in their passion for them.[6]

As a farmer Gould did well by local standards, which is to say he was less poor than most. He kept a herd of cows and sold butter or cheese to get what cash he needed. Meat came from the cattle, pigs, and chickens, vegetables from the garden, sugar from the maple trees, and fruit from the orchards Gould planted around the house. Mary made most of the family's clothing, and few were the possessions that did not come from their own handiwork. When Mary chanced to acquire an imported tea set, she designed a cupboard with glass doors in which to display it. There it remained for years, safe from harm, seldom used but greatly admired, as if satisfying the need for some glimpse of the larger world beyond the mountains.[7]

When the day's work was done, Gould sat for hours by the dim light of flickering pine knots, poring over his books and newspapers. His passion for education matched that of his wife for religion. It was not the passion of one desperate to escape his station in life but of a man bent on self-improvement. If Gould harbored any larger ambition than following in his father's footsteps, he left no inkling of it. His horizon never stretched beyond the bounds of family and farm. Ironically, the matter of family proved to be Gould's thorniest problem. What a farmer needed most was strong, sturdy sons to help with the work and inherit the homestead, but the empty bedrooms filled up with one daughter after another until Gould despaired of ever having the son he craved. Five girls arrived in quick succession: Sarah (1828), Anna (1829), Nancy (1831), Mary (1832), and Elizabeth (1834).

It would have been easy for John Gould to imagine that fate had played a cruel joke on him, for when Mary at last presented him with a son, it seemed a clear case of too little too late.

The son born May 27, 1836, was not what John Gould had in mind: a frail, scrawny infant with dark eyes and a wreath of dark hair. For more than a week he lacked even the dignity of a name, perhaps because he was so undersized John was debating whether or not to throw him back. Finally he called the boy Jason after the brother who had gone to Canada. He was not deemed worth a middle name, and early in his life the family took to calling him Jay as well as Jason.[8]

Armchair psychologists could have a field day with the spectacle of an undersized boy raised with five older sisters. Certainly the girls doted on him. As Sarah, the eldest, recalled, "He was the pet and the idol of the household. . . . Being frail and delicate he was all the more petted, and generally had his own way." Moreover, Jay lost his mother at the age of four. Long afterward he confided that his only memory of her was being summoned to her bedside as she lay dying. He never forgot how cold her lips were when she kissed him good-bye.[9]

The death of Mary Gould in January 1841 ushered in a period of grief and instability for the family. With his eldest daughter not yet thirteen and no one to run the household or care for the children, John could not afford the luxury of prolonged mourning. That same summer he married again only to be widowed again by Christmas. A weaker man might have resigned himself to the will of the Almighty, but submission was never a Gould trait. The following May John wed a neighbor named Mary Ann Corbin. At fifty he seemed finally to have repaired his broken family; six months later Nancy, the middle daughter, died suddenly. In

1843 Mary Ann presented John with a new son, Abram. The presence of another boy helped dispel the gloom left by Nancy's death, at least for a time. Two years later Mary Ann was dead. Distraught and bewildered, John buried her alongside his daughter and two wives in the little graveyard by the Yellow Meeting House. Whatever bitterness he felt remained locked in his heart. Shorn of all hope concerning marriage, he had no choice but to entrust the older girls, who were still in school, with the enormous burden of cooking, cleaning, mending, making clothes, and caring for the smaller children.[10]

These events had a profound effect on young Jay. In the short space of four years he lost a sister and gained a brother, watched his mother die, and embraced two stepmothers only to have death claim them as well. The sisters who had always coddled him drew even closer to him, becoming in effect surrogate mothers. Necessity huddled the family together, forcing each one to depend more than ever on the others. It compelled Jay to accept new responsibilities, demanded that he grow up before his time. He seems to have grasped these imperatives at a remarkably early age. Even as a boy he impressed others as one who looked years younger and acted a decade older than he was.

In the midst of these sorrows the Goulds endured still another trial. For decades small farmers in a dozen counties had worked their land as tenants to the patroons who had acquired enormous tracts in colonial times. While the rents charged under these leases were nominal, resentment against them mounted until 1839, when a revolt known as the antirent movement broke out. Protesting farmers, disguised as Indians, roamed the countryside in armed bands, intimidating landlords, pressuring other tenants to join them in withholding rents, and punishing dissenters with tar and feathers. In Roxbury tenants organized their own bands of Indians who, clad in crude disguises of sheepskin and calico, assembled on signal from a horn normally used to call farmhands to their meals. To protect their signal, the Roxbury antirenters passed a law prohibiting the custom of blowing the dinner horn. At that point they ran afoul of John Gould.[11]

From the first Gould had antagonized the antirenters by condemning their movement as illegal. Despite repeated warnings he persisted stubbornly in blowing his dinner horn as usual. One day in July 1844 he blew the horn at noontime and was soon confronted by five armed Indians. After a heated exchange the Indians retreated behind a volley of threats. A week later fifteen Indians crept around the house and, upon hearing the bell rung, rushed to surround Gould. Jay was standing in the yard with his father when the raiders appeared. His dark eyes bulged with terror as their leader brandished a sword above his father's head and angrily ordered Gould to stop blowing his horn. Defiant, unmoved, Gould stood his ground and yelled for his hired man to bring out the muskets. Taken aback, the Indians hesitated long enough for a few neighbors to reinforce Gould. After a hasty powwow the braves filed sullenly off the premises and never bothered John Gould again. Years later his daughter Elizabeth, or Bettie as she was called, observed of this incident that her father "would have been shot a dozen times rather than give in to a principle he did not believe." It was not the first time a

Gould was found on the unpopular side of a controversy, and it would by no means be the last.[12]

A year later, when the governor proclaimed Delaware County in a state of insurrection, John Gould rode with the militia to restore order. For the first time he put a lock on his door. Tensions gripped even the public school at Meeker's Hollow. After someone threatened to drown Jay because of his father's stand, John Gould vowed that his children would never again set foot in the place. He and two of his brothers-in-law erected a schoolhouse on the land between their farms, called it Beechwood Seminary, and imported a young teacher from Albany.[13]

During these years of sorrow and turmoil John Gould displayed the attributes characteristic of his ancestors: perseverance, an indomitable will, the determination to see things through despite all obstacles, love of learning, and a passionate, often blind loyalty to family. These were precisely the traits for which Jay would later become legendary, and they describe the child no less than the man. Nor should the influence of his mother be overlooked, for it ran much deeper than a love of flowers. From her, insisted Sarah Gould, Jay inherited his driving ambition, the even temperament beneath his remarkable self-control, and his gift for turning everything to his own profit. Both parents implanted in the boy a simplicity of taste, an aversion to vice or pomp, and a shrewd disposition to the practical.[14]

The legacy bequeathed to Jay was an imposing one. Some thought it too weighty for his frail shoulders to bear, just as they thought his puny constitution inadequate to sustain him through life's sterner tests. Childhood impressed upon Jay the harsh fact that for him nothing would come easily, that whatever path he carved for himself would be a steep, uphill climb. Even to those who loved him or wished him well, he seemed ill-equipped for so rugged a journey.

2

Manchild

He was not what was generally termed a manly boy. He kept out of the rough good-natured games. . . . If in banter the boys attempted to force him to join them, he would make a great outcry, and . . . would sit and mope until the school was called to order. Then he would go to the master's chair and enter a tearful complaint against his enemies. The master would thrash the other fellows, and little Gould would be tickled.

—Trumbull White, *The Wizard of Wall Street*

Frail and undersized, he dreaded the cold darkness; he would recall, thirty years later . . . how the thistles hurt his bare feet; he spoke with vibrant bitterness of his boyhood. He had to plead with his father to be permitted even to attend the village school.

—Matthew Josephson, *The Robber Barons*

Until he was fourteen, Jay attended the nearby Beechwood Seminary, where he was an unwilling, unpopular and occasionally rebellious student. The boy had a passion for acquiring knowledge which he regarded as necessary to his future career, such as mathematics, but he rebelled at learning by rote what a mere country schoolmaster decided was proper.

—Richard O'Connor, *Gould's Millions*

THE MOST STRIKING fact about Jay's childhood is not that he survived it but that he absorbed its lessons so thoroughly. By any measure he seemed a poor bet for the future. The environment favored muscle, size, and stamina, none of which he possessed, and punished weakness with indifference. In that rugged, insular upcountry world, where heritage bound eldest sons to the land, he was a farmboy unsuited for farming. Few other prospects beckoned in so remote a region, and larger ambitions seemed beyond the reach of one cast as the runt of the litter. It would have been easy, even predictable, for Jay to fail in life. His vulnerability went beyond a lack of size and strength. In an age that measured good health by the pound, Jay always looked the part of a sickly scarecrow. His delicate constitution was a dangerous liability in a family plagued by a long history of what John Gould called "the Old Gould Disease, the Consumption that they most all Die with." To these physical weaknesses might be added the emotional turmoil of the constant turnover within his own family and the prolonged tensions of the antirent dispute. Taken together, these forces were enough to crush most boys so ill-equipped to face life's pressures.[1]

Yet they did not crush Jay. On the contrary, they seem to have been the fiery furnace in which his character was forged and from which his genius emerged. His was an intellect honed on adversity and disadvantage. The bitter struggle for survival taught him early how close to the line life is lived, and he did not shrink from the lesson. Instead of bemoaning his limitations or allowing them to beat him down, he responded to their challenge with a superhuman effort. What he lacked in size or strength he compensated for with intelligence and a fierce desire to succeed.

Vulnerability bred in Jay an edge of desperation. All his life he pushed relentlessly against the limits of his endurance, as if he knew he had much to do and little time in which to do it. Because time was always his enemy, he developed an impressive economy of style. There were no wasted motions, no whimpers or complaints when obstacles did not budge. His mind raced to the next task, which he attacked with awesome concentration and staying power. "I knew him once to work at times for three weeks on a difficult problem in logarithms," Sarah Gould observed. These attributes are impressive enough in a grown man. What is astonishing is how fully developed they were in Jay as a mere boy. No one familiar with his childhood would be surprised that he worked himself to death at the age of fifty-six. Even as a youth he possessed a drive and single-mindedness rarely found in adults. Like many gifted people he cultivated his own distinctive way of doing things, habits that emerged early and remained with him all his life. In that sense he was truly a prodigy, talented and wise beyond his years.[2]

Those who knew Jay as a young boy describe him as reserved, serious to the point of solemnity, studious, well-mannered, unusually polite for one his age. In company he seldom spoke, and when he did his words were few and carefully chosen. He listened well and seemed to remember everything, as if storing it away for future use. His school work was neat and precise, even fastidious, the product of a logical, orderly mind.

Jay showed little enthusiasm for sports or games because he was not very good at them. As James Oliver, Jay's favorite teacher at Beechwood, put it, "He was not a boy given to play very much. He was never rude and boisterous, shouting, jumping and all that sort of thing." Studying, however, became a passion with him, possibly because books provided his first glimpse of a larger world, nourished his ambition for some higher destiny than the mountains offered.[3]

Not surprisingly, these traits set Jay apart from other boys. Much has been made of his being a loner who shunned the company of his fellows and was in turn despised by them. At least half the claim is true: if ever a boy marched to the beat of his own drummer it was Jay Gould. A former schoolmate said he was "proud and exclusive, and would not put himself on an equal with the other boys." But being a loner does not imply that one is also a pariah. The charge that Jay moped in a corner and tattled on those who tormented him has no evidence to support it and compelling testimony to refute it. Oliver insisted that "his mental and moral fibre were such that it would have been impossible for him to appeal to a teacher against a school-fellow. His self-reliance and self-respect would have revolted against such a proceeding."[4]

Nothing in Jay's character impressed people more than his self-reliance. "He would never accept assistance in working out hard problems," Sarah Gould declared. No one knew that better than Oliver, who recalled that "if he was sent to the blackboard to work a sum he would stay there the entire recitation rather than ask for a solution." Nor did he crave the approval of others. His sister Anna thought it peculiar that he "never wanted praise or to be noticed if he had accomplished what older boys could not. His satisfaction was in the fact that he had overcome a difficulty."[5]

The portrait that emerges is startling in its maturity at so tender an age. Even as a boy Jay was his own man. He went his own way and set his own standards for measuring himself. Strong-willed, serious, ambitious in his quiet way, he worked hard to better himself and let nothing deflect him from that purpose. Shrewd and watchful, he learned from those around him, soaking up their experience like a sponge. Little escaped his eye and still less his lips. His fertile brain laid deep plans which he kept to himself until the right moment came to spring them. Once convinced of their soundness, he pursued them with dogged tenacity. Neither flattery nor censure could sway his determination or addle his judgment.

While the ordeal of childhood scarred Jay with its tragedies and disruptions, it also left him hardened in spirit and, above all, resilient. Despite his frail, unassuming appearance, there was nothing soft or weak in him. Life had taught him its harsh rules and he accepted them without bitterness or complaint. Those who later denounced him as insensitive or indifferent to human suffering missed the point entirely. He knew what pain was, knew firsthand the full breadth of misery, but his response was that of the stoic: to accept misfortunes for what they were, a part of the game.

Struggling upward. No Horatio Alger hero ever faced a tougher climb to success than Jay Gould, and none managed it with more efficiency or self-effacement. By the age of ten he had already shown himself a survivor; at fifteen he was on his way to becoming a businessman.

It did not take Jay long to discover that farming was not for him. Every morning and evening he brought the cows in for milking and afterward drove them back to the pasture. Sometimes he had to help the girls with the milking. He rode the horse while his father raked hay and did a hundred other chores, each more monotonous than the last to his restless mind. Thistles cut his bare feet in the pasture; the summer sun blistered his legs and gave him headaches. His studies interested him far more, but there was never enough time for them. Often, when his father came looking for him, he grabbed his book and slate, swore his sisters to silence, and ducked into some hiding place to work. "It is too bad," he shrugged apologetically, "but I must study you know."[6]

Not until his early teens did Jay glimpse a part of the world beyond his native valley. His father had business in New York and took the boy with him. When they returned a week later the girls crowded around Jay, eager to hear what he had seen and done. But Jay ignored their pleas until his mind had sorted through the experience to suit himself; only then did he pour out his wondrous tale of adventure in minute detail. It is impossible to know what effect this journey had on Jay, on his vision of what the future held for him. If nothing else it opened his eyes to new possibilities, strengthened his desire to cast his ambitions beyond the mountains.[7]

In the spring of 1849, when he was only thirteen, Jay concluded that neither the farm nor the local school had anything more to offer him. He conceived the idea of attending a private school in Hobart, nine miles away, and broached the question to his father. John Gould dismissed the request as a boyish whim, saying he was too young and was needed at home. The girls understood that when their father said no, the verdict was final. But not Jay. Instead of submitting meekly he marshaled new arguments, asked Anna and Sarah to intercede for him, and found ways to keep the question before his father. His tenacity and ingenuity in pressing the case must have surprised John Gould. Gradually it dawned on him that Jay meant to have his way, that there was nothing boyish or whimsical in his plan. To his credit he did not shrink from either the decision or its broader implications, disappointing as they must have been for him.[8]

"All right," he said at last. "I do not know but you might as well go, for it is certain you will never make a farmer."

Horatio Alger would have blushed at writing the scenes that followed, yet the evidence corroborates their every detail. Early one Monday John Gould hitched up his old bay mare, drove Jay to Hobart, entered him in Mr. Hanford's school, bade him be a good boy, and went home, leaving Jay with a bundle of clothes and fifty cents. Jay found a blacksmith who offered him board in exchange for keeping his books. He knew nothing about bookkeeping, but he had a neat hand and a good head for figures. Always eager to learn, he began here a practice that became a lifelong habit: the willingness to incur responsibilities for which he had no training or experience and, by dint of application, to master the skills needed while actually performing them.[9]

Six days passed. At dusk on Saturday Jay appeared unexpectedly at the farmhouse, tired, hungry, footsore, his eyes brimming with tears of gladness at being home again. The girls hugged him joyfully and stuffed him with his favorite fried cakes and sausage. John Gould offered a consoling smile. Like the girls, he assumed Jay had learned his lesson and had come home to stay. He could not have been more wrong. On Monday Jay rose at dawn to start back; his father drove him far enough over the mountain that he could walk the rest without being late for school. During the five months Jay studied at Hobart, he never once missed trudging home on Saturday.

While Jay was at Hobart, a new teacher came to the Beechwood Seminary. The presence of James Oliver brought Jay back home to school that fall. Fresh out of the state normal school at Albany, Oliver possessed a rare mixture of enthusiasm, intelligence, and discipline. Two of his pupils in that tiny rural school were to leave their mark on the world: Jay Gould and naturalist John Burroughs. Both regarded Oliver as the teacher who made the deepest imprint on their lives. "We all got a real start in that school," Burroughs recalled fondly, "for under Mr. Oliver we acquired a genuine love of learning."[10]

The two boys could not have been more unlike. Jay was quick of mind and motion, precise in everything, devoted to his studies, alert to everything around him. Burroughs was slow (some said dull witted), slovenly, an indifferent student forever distracted by the antics of an insect or some other minor spectacle of nature. Yet somehow they became close friends who played together, swapped knives and marbles, helped each other out of scrapes. When wrestling became the rage at recess, Jay would have no opponent except Burroughs. They went at it by the hour, and for an unathletic boy Jay acquitted himself well. "He was very plucky and hard to beat," Burroughs admitted. "He was made of steel and rubber."[11]

That spring Jay wrote an essay destined to achieve notoriety in later years when critics flung it in his face, delighted by what they took to be its savage irony. Dated April 9, 1850, it was the last composition he wrote at Beechwood that year:

"HONESTY IS THE BEST POLICY"

> By this proposition we mean that to be honest; to think honest; and to have all our actions honestly performed, is the best way, and most accords with precepts of reason. Honesty is of a self-denying nature, to become honest it requires self-denial; it requires that we should not acquaint ourselves too much with the world; that we should not associate with those of vulgar habits; also, that we should obey the warnings of conscience.

> If we are about to perform a dishonest act, the warnings of conscience exert their utmost influence to persuade us that it is wrong and we should not do it; and, after we have performed the act this faithful agent upbraids us for it; this voice of conscience is not the voice of thunder; but a voice gentle and impressive; it does not force us to comply with its requests; while at the same time it reasons with us, and brings forth arguments in favor of right; . . .
>
> And again we find numerous passages in the scriptures which have an immediate connection to this, and summing up the whole we cannot but say—Honesty is the best policy.

Ironies aside, the essay reveals much about Jay few critics have bothered to notice. At fourteen he possessed a vocabulary, grasp of language, and literary style remarkable for his age. During the next few years he cultivated the flowery prose and conceits typical of his time, which may be seen in his *History of Delaware County.* His letters betray a self-conscious, often imitative, attempt at literary expression. In his search for a career he seems to have flirted with the notion of becoming a writer.[12]

The essay also suggests a youthful idealism that expressed itself in earnest, prolonged ruminations over life's larger questions. This passion for weighty abstractions was a favorite playground for nineteenth-century minds, and Jay brooded over them with no less intensity than he devoted to other pursuits. Unlike many other people, his concern did not spring from religious fervor. When the revivalist fever swept through Roxbury during the winter of 1848, it caught Jay in its frenzy only for a brief time. The girls were all devout Methodists, but sometime around 1850 Jay switched to his mother's church, the Presbyterian. For the rest of his life he remained, like his father, a perfunctory church goer.[13]

Jay's absorption with such matters was part of the familiar ritual of passage from adolescence to manhood in which the lofty ideals of youth run hard against the cutting edge of reality. Like most men, Jay would recast his ideals in more practical molds; he was hardly the first to find the world more complex and unyielding as a man than he had imagined as a boy. He saw with his usual clarity the tension between his passion for education and the necessity of seeking his fortune. At age sixteen, barely launched into affairs of business, he confided wistfully to Oliver, "The plain truth is I am growing old too fast; my years are getting the advance of what of all things I value most, an education."[14]

Years of immersion in business activities doused that flame but never extinguished it entirely. Even at the peak of his career he found comfort in his books. While it was common practice for magnates to furnish their mansions with well-stocked libraries, Jay was one of the select few who actually read the books that lined his walls.

Practical concerns occupied John Gould no less than they did his son. As the children came of age, he found himself facing hard choices about the future. Sarah and Bettie had become school teachers, leaving only Anna and Mary (Polly) at home to keep house. It was only a matter of time before he began losing them to marriage; Polly was already on the verge of becoming engaged to James Oliver. Jay was heir to the family farm but had already rejected that career, and Abram was still a small boy. John Gould was pushing sixty, partly lame, unable to manage the farm without help. He took on a hired hand, Peter Van Amburgh, who proved a good worker, but hired help didn't solve the problem of what to do with the farm when Gould grew too old or infirm to run it.

The dilemma was a painful one for John. His parents had lived and died in that farmhouse; he had been born there and so had his children. He basked in the shade of trees he had planted as a boy. Memories crowded every acre, but Gould was not one to let sentiment blur his vision. The harsh truth was that the farm's future depended on Jay, who wanted no part of it. "There is no use of my trying to make a farmer of Jay," he told Sarah. "I cannot do it. . . . I think, if I make this change, that probably he will be satisfied." The change involved swapping the farm for the house and business of Hamilton Burhans in Roxbury. Burhans dealt in tin, sheet-iron, and stoves; his brother Edward, the most prosperous merchant in the village, operated a dry goods store just down the block on Main Street. It was an extraordinary decision for a conservative farmer with no experience in the tin business, yet Gould made it without hesitation. He signed the agreement in July 1851 and agreed to move the following April.[15]

Gould assumed his son would come into business with him. In September Jay was sent to board with the Burhans family and work as a clerk in the store until his father took over. During the next seven months he swept out, ran errands, made himself useful, and learned the business. His duties occupied him from six in the morning until ten at night. Unwilling to drop his lessons, he got Burhans's permission to rise at three, light a fire, and study until six. Beneath the dim glow of a tallow candle he began turning his grasp of mathematics toward the more practical bent of engineering and surveying, using instruments borrowed from Edward Burhans.[16]

Here began Jay's lifelong habit of pushing his stamina to its limit and beyond, of wringing the last ounce of energy from mind and body alike. He forced himself not only to work but to concentrate past the point where fatigue dulled the senses of most men. Whatever he learned about engineering or surveying was entirely self-taught. His curiosity was as insatiable as his appetite for new skills. He impressed everyone with his drive and determination. Anna Gould described her brother as being "very quick motioned in those days."

As spring approached and the Goulds busied themselves with the move to Roxbury, Jay looked about for new opportunities, preferably in surveying. He made no secret of his ambition. "Jason does not intend to stay in B[eaverdam]," Sarah reported. "I don't know where he will go."[17]

When an opening appeared, Jay told no one until his plans were perfected. A close friend, Abel Crosby, told him of a surveyor named Snyder who needed an assistant for preparing a map of Ulster County. An exchange of letters brought Jay an offer of "$20 a month and found." Without telling his father, Jay accepted the position and enlisted his sisters to help plead his cause. When he informed them of the offer, the girls tried frantically to dissuade him. Ulster was rough country, they warned, filled with rattlesnakes and other dangers. "We begged and pleaded with him not to go," Sarah declared. "We thought he was too young. I never expected to see him again."[18]

But Jay's persistence won out. John Gould swallowed his disappointment and gave permission. On an April morning in 1852, a month before his sixteenth birthday, Jay left home with five dollars in his pocket. Behind remained the last vestiges of what little childhood he had enjoyed. Manchild had come of age.

3

Rungs

In some of his first business ventures, can be read the same disposition for silent intrigue, the same secretiveness touching his intentions, the same subtlety and elaborateness of plan, and the same indifference to the feelings or comments of others, as marked the tremendous operations which are the climaxes of his purely speculative career.

—Trumbull White, *The Wizard of Wall Street*

He carried with him a showy mahogany case, containing an invention which the boy hoped would bring him fame and fortune. . . . The mouse-trap was a success, but its inventor laid traps and caught speculative mice all his life.

—Henry D. Northrop, *Life and Achievements of Jay Gould*

When he was seventeen, in his efforts to gain wealth, he had tried invention . . . an ingenious contraption which reflected the mind of the growing boy. There is something characteristic in the device Jay chose for his inventive energies. Cruel, ingenious, and practical, the little trap mirrored Jay Gould as the world later came to know him; but there the comparison ends, for the trap failed to bring him success and riches.

—Robert I. Warshow, *Jay Gould: The Story of a Fortune*

EVERYTHING SEEMED SIMPLE the way Snyder explained it. He handed Jay a small passbook and said, "As you go along you will get trusted for your little bills, what you will eat, and so on, and I will come round afterwards and pay the bills." Jay nodded and set out with his compass and gnodometer, a wheelbarrow-like device with an attachment that measured distances. Each day he sketched a crude panorama of houses, shops, churches, schools, public houses, tanneries, roads, and streams. At night he plotted his notes into a map of the region covered. On his third day out Jay stopped overnight with a farmer. As he entered the charge in his book, the man swore and exclaimed, "Why, you don't know this man! He has failed three times. He owes everybody in the county, and you have got money and I know it, and I want the bill paid."[1]

Jay looked at him aghast. Turning his pockets out, he said meekly, "You can see that *I* tell the truth."

The farmer eyed him dubiously. "I'll trust *you*," he said finally, "but I won't trust that man."

Stunned and disheartened, Jay went all day without food rather than risk another scene. Late in the afternoon, faint with hunger, he found some secluded woods and sat down to cry. "It seemed to me as though the world had come to an end," he said years later. "I debated with myself whether I should give up and go home, or whether I should go ahead. . . . Finally I thought I would try my sister's remedy—a prayer. So I got down and prayed, and felt better about it, and I then made up my mind to go ahead."

He asked for food at the first house he reached and was treated kindly. When he said, "I will enter it down," the woman agreed. A few minutes after leaving, however, Jay heard her husband calling after him. Quaking with fear, he turned and braced for another outburst.

"I want you to take your compass back," the man said, "and make me a noon-mark."*

Relieved, Jay made the mark and started to leave. The man insisted on paying him, saying, "Our surveyor always charges a dollar for these jobs."

"Very well," Jay replied. "Take out a shilling for my dinner."

"That was the first money I made in that business," he recalled, "and it opened up a new field to me, so that I went on from that time and completed the surveys and paid my expenses all that summer by making noon-marks at different places." In one day he had gone from apparent failure to climbing the first rung on the ladder of success. It is doubtful that the lesson hardened his heart against all men. More likely it reinforced his belief that perseverance brings rewards.

Of this incident Jay told his family nothing. He admitted only that he was "a little homesick" and that "the only thing I fear are the snakes." He liked the

*A noon-mark is a north-south line run through a window in the house and marked so that farmers could regulate their clocks by it. When the sun struck the line, it was twelve o'clock.

work and the skills it taught him, declaring it to be "a great field *for a person to engage in* if the object of school is to improve the memory or discipline the mind." A few weeks later Snyder confessed himself unable to pay the money due Jay and the project verged on collapse. Jay proposed to Peter Brink, another surveyor who had worked for Snyder, that they finish the map themselves. Brink introduced Jay to Oliver J. Tillson, who agreed to join them as a partner. Since Brink and Tillson had some money and Jay did not, Jay hired himself to them for thirty dollars a month and board.[2]

Each Sunday the three lads met to divide the country before them, then went their separate ways to work until the following Sunday. The older partners soon discovered that Jay's boyish looks belied a demonic energy. "He was all business in those days, as he is now," Tillson recalled. "Why, even at meal times he was always talking map. He was a worker, and my father used to say: 'Look at Gould; isn't he a driver?' " When the partners settled accounts on the day after Christmas, Jay realized $500 and the odometer he had been using. It was his first business stake. He went home and asked the girls to make him a suit of winter clothes while he pondered his next step. Already he had decided to undertake surveying projects on his own. That summer he had met the most prominent citizen of Greene County, Zadock Pratt; on New Year's Day he wrote Pratt offering, with his assistance, to do a map of Greene County. Although the project never materialized, the acquaintance with Pratt was to prove momentous for Gould.[3]

When the girls had replenished his wardrobe, Jay left home to survey the plank road between Albany and Shakersville. Once in Albany, the old yearning to resume his education seized him again. The conflict between business and school had in fact nagged at him throughout the past year. "But to speak of school seems to fire every feeling in my soul," he lamented to Oliver. "It tells me that while My Schoolmates are boldly advancing step by step up the ladder of learning I have to hold both hands fast to keep myself upon the same round where I stood over fourteen months ago." He resolved the dilemma by attempting to do both at the same time. Jay enrolled in the Albany Academy along with his cousin, Iram More, who lived in that city. The two boys bunked together in More's house and became close friends. On Saturdays and after school, More assisted Jay with the road survey. Their friendship greatly troubled Jay's sisters. A good-natured country boy, slow of wit and fond of pranks, More was an indifferent student. He had just been expelled from the Richmondville school, and the girls feared his influence on Jay.[4]

They need not have worried. In later years people would wonder how Gould could remain friends with a man so utterly different from himself as Jim Fisk. As his relationship with More suggests, Jay learned early not to judge his associates by his own values. He seemed not only to tolerate but to enjoy contrasting personalities, as if, in some vicarious way, they satisfied hidden longings of his own. At the same time he was strong enough, even at sixteen, not to let such companions divert him.

On the road survey Jay discovered that his employers expected him to do the engineering as well. Despite his inexperience, he set about doing the preliminary work. Then one of the directors presented him with a large theodolite, a complicated instrument he didn't know how to use. "I could not for a good while even unloose the needle much less adjust the instrument," he recalled with a smile. With a show of bravura he set up the theodolite, barked orders to the line men, and called out numbers as the directors watched admiringly, all the while tinkering and twiddling desperately in search of a clue as to how it worked. Fortunately the rain that was falling changed into snow, postponing work for two days. Jay took the theodolite home and, with the help of a book, learned its intricacies. In the same way he mastered each new task just before he came to it. He finished the project without his employers ever learning that he had no engineering experience.[5]

When not surveying Jay spent hours at the state library, not only studying but lobbying. A bill had been introduced to appropriate funds for a complete survey of the state, with separate maps of each county. "If this bill passes," he wrote Oliver, "I think I will realize enough to see me through Yale College and that is the extent of my hopes." It is impossible to know how deeply Jay felt his tug of opposing ambitions, but there is no doubt the conflict was genuine. In one outburst of florid rhetoric he unburdened his frustrations to Oliver:

> Is it not an exception to a general rule but that developed manhood shows many a marked streak in youth? It is when trivial sports of youth have lost their wonted sweetness, it is when its wild romance is succeeded by the realities of maturity when he is called upon to fit a station where his utmost exertions and all he knows must be fearlessly put forth in the contest for success. It is then that the remorse of idle moments comes in spectres before the imagination.[6]

While the survey bill languished in committee, Jay quit the academy after the winter term and decided to visit four colleges in which he was interested. It happened too that his grandfather More wanted a favor that fit Jay's travel plans nicely.

The first thing to be said about the celebrated mousetrap is that it was invented not by Jay but by his grandfather More. The old man, nearly eighty and growing senile, thought it a wondrous device and wanted it exhibited at the World's Fair in New York. When neither of his sons would oblige him, More cursed their disloyalty and offered Jay a free ticket if he would register the trap. When Jay agreed, the old man joyfully packed the trap in a handsome box and sent it off with him.[7]

Iram More accompanied Jay to New York. As they boarded a streetcar for the Crystal Palace, Jay set the box on the platform. Within moments a man snatched the box and took off down the street. The boys jumped off the car and ran down the street until they spotted the man with the box tucked under his arm. Without thinking they tackled the man and hung on for dear life, crying, "Stop, thief!" while the culprit thrashed about wildly, trying to shake loose. A crowd gathered to watch the melee until a policeman arrived and whisked all three of them off to the station, where Jay filed a complaint. At the hearing the magistrate ordered the box opened. When the mousetrap was revealed in all its brightly painted glory, the thief gasped in bewilderment. The justice burst out laughing and declared the prisoner to be the largest rat ever caught in a mouse-trap. It was a good enough story to make the *New York Herald*, the first time Jay Gould's name appeared in the metropolitan dailies that were later to transform him into a legend.

When they had done the Fair, Iram returned home and Jay went to visit four colleges in which he was interested: Harvard, Yale, Rutgers, and Brown. Having wandered amid the wealth and wonders of the big city and inspected four venerable temples of learning, he traveled to Roxbury for a few days with the family. The contrast between that simple village and the larger world he had just glimpsed must have jarred the already confused images swirling in his head. He found sister Polly convalescing from an illness that had afflicted her lungs. Anna Gould had become smitten with the Methodist preacher, Asahel Hough, and was glum because he was about to leave for another church. His friend Rice Bouton lay desperately ill with typhoid fever, and sickness had also prostrated Edward Burhans's wife and daughter Maria.[8]

Into this gloomy atmosphere Jay plunged with determined cheerfulness. He drove his sisters to church to hear Hough's last sermon, then insisted on cooking Sunday dinner. When Jay learned that no one would venture near Rice Bouton, he hitched the wagon and went with Bettie to visit him. He did some bookkeeping for his father and in the evenings sat near the fire with him, discussing future prospects. All the while he sorted patiently through his recent experiences, searching for the right thread, the direction he wanted to take. One image seems to have dominated the rest, though he did not yet know what it meant. He revealed it only to his friend Abe Crosby, who listened wide-eyed as Jay described New York and the World's Fair. When he had finished, Jay paused, then said, "Crosby, I'm going to be rich. I've seen enough to realize what can be accomplished by means of riches, and I tell you I'm going to be rich."[9]

"What's your plan?" Crosby asked eagerly.

"I have no immediate plan," Jay replied. "I only see the goal. Plans must be formed along the way."

All hope for the survey bill vanished when its sponsor died unexpectedly. Jay decided to undertake a survey of Albany County and contemplated doing his native Delaware County as well. Only a year earlier he had been the raw recruit learning the ropes as Snyder's assistant; this summer, at age seventeen, he took charge of the project, hiring assistants of his own. For the Albany survey he engaged Iram More and a farmboy named Isaac Wilcox. Jay first had to teach his assistants how to survey. Once they got the hang of it, he retired to a room in his uncle's house to plot their notes into a map. In addition, he had to solicit a clientele for the product, which offered two sources of income: subscriptions for copies and the sale of space around the border for engravings of individual homes or businesses. To get customers he had to canvass every township and visit the more prosperous outlying farmers. He also advertised in local newspapers to publicize the map. The surveyor had become not only mapmaker but promoter as well.[10]

The Albany survey occupied Jay through the summer of 1853. When his funds ran short, he borrowed a few dollars from his sisters, taking care always to give them his note. Somehow he found time to give his father occasional help at the tin shop, even though he had decided to go ahead with the Delaware County survey. During the summer he sent out a prospectus, solicited advice from his relatives scattered throughout the county, persuaded "Champ" Champlin to help with the survey work, and began his long and controversial relationship with the press. For some time Jay had been sending items of interest to another friend, Simon D. Champion, editor of the *Bloomsville Mirror*. In August he unabashedly asked Champion to help promote the use of maps in county schools:

> In Delaware County the Supervisors ought to encourage it by buying maps for each of the School Districts. I want you to give me an editorial to this effect. . . . I want such a notice inserted as a sort of a title page to an advertisement. . . . You must model the editorial over to suit yourself but it must be as strong as it can be made and come direct from you.[11]

The Delaware County survey commenced in mid-September. When More and Wilcox finished their field work in Albany, Jay sent them to join Champlin on the new project. The responsibilities piling up on him were stretching his endurance to the limit. The Albany map still required close attention, and he had to oversee the work in Delaware County as well as doing the canvassing there. His father had suffered a bad fall in the tin shop and could only get around on crutches. "I think I have learned one thing this winter from actual observation," Jay philosophized to James Oliver, ". . . that Happiness consists not so much in indulgence as in self denial." Earlier he had declared, "My youth or inexperience are no excuse to sanction my falling into degraded habits or for yielding to delusive temptations when I have such friends and such good advice before me."[12]

The abstemious habits that characterized Gould later in life were already present in his teens. They were part of his family's tradition, a legacy which, like all tradition, had been shaped by both good and bad example. In recent generations the women seem to have furnished the good examples and the men the bad ones. Especially was this true with alcohol. No subject aroused fiercer emotions among the Goulds. Jay's mother had seen her life made miserable because her father (he of the mousetrap) was addicted to drink. On her deathbed she had pleaded in vain with him to give up liquor. Until his death early in 1854 the old man doted on his grandchildren as once he had on Mary, but they were not fooled. "Poor old man, he little thinks how many sad hours he caused my mother," Polly said savagely, ". . . how quick the tears would roll down her face when she saw him coming intoxicated."[13]

Nor was the problem limited to Grandfather More. In recent years John Gould, worn down and disheartened by his struggle with life, had taken to drinking heavily. That was one reason why the girls had been so reluctant to see Jay leave home, and why Jay was compelled to devote ever more of his precious time to the tin shop. The girls fretted endlessly over their father's condition. "Father has one fault—you know what it is as well as I," Polly lamented to Oliver, "but with that exception a kinder, a more affectionate father children never had." A few months later she poured out her heart to her financé:

> Trials and afflictions of the severest kind have made him such a man as he once was not but he has taken a poor way to drown his trouble. I can but hope he will yet reform, although I know a man of his age seldom changes his habits, they seem so fixed that they cannot be altered. . . . I do hope it may be a good lesson for Jason and I believe it will.[14]

The lesson impressed itself not only on Jay but on all the children. Every one of them embraced temperance with a fervor. Sarah became so strict that she would not allow her children to drink cider or even root beer, fearing that they might acquire the "habit of going to a keg for a drink." Both Anna and Bettie married men who later became ardent crusaders for prohibition. Although Jay sipped an occasional glass of claret or champagne in later years, his distaste for alcohol was obvious to everyone. His boyhood taught Jay to be temperate if not spartan in all his personal habits. He avoided drink, ate sparingly, did not use tobacco in any form, shunned games of chance, and was never known to utter a curse. His clothes were more expensive but no less plain as a business magnate than as a surveyor. Sarah remembered that as a young man "he took pains to look neat, but he did not spend very much time; never carried very much baggage." That habit too was a lesson from childhood. His mother despised vanity as she did alcohol.[15]

Religion also separated the Gould men from their women folk. Neither Jay nor his father succumbed to the evangelical fever that had swept up Mary Gould and all her daughters. Jay attended the Presbyterian church but never professed his faith; John Gould didn't even bother with the ritual. The girls pleaded with them like missionaries among the heathen, but in vain. Somehow Jay managed to develop what his sisters considered a strong Christian character without the trappings of church membership, as if he had privately sorted through what was useful to him and what was not.[16]

Hamilton Harris never forgot his first encounter with Jay Gould. The directors of the Shakersville plank road visited his law office one August day in 1853, accompanied by "what appeared to me a small boy, dark eyed and dark haired, standing not much higher than this table." Harris paid him no attention, thinking he was one of the directors' sons. The directors complained that opponents of the road were seeking to enjoin its construction. While they were wading tortuously through the legal complications, the "boy" asked quietly whether anything prevented the company from going ahead with the work until an injunction had actually been obtained. Harris looked at him in surprise, then replied that nothing did. "Thereupon this little fellow commenced to figure at a table, [and] after figuring away a little time, he asked me if they went ahead with the building of the road I would protect them from any punishment. I told him yes. I would protect them until an injunction was served, and that that when served must be obeyed."[17]

Jay had originally done the survey, then acquired part interest in the contract to build it. After leaving Harris's office, he rounded up every laborer and dray he could find, hauled the lumber, and put the men to work laying the road day and night. On the day the injunction was issued, the disputed portion of the road stood complete. In that brief episode Jay revealed a talent for which he would become legendary: the ability to cut through legal thickets to the heart of the matter, to walk the fine line of the law like a man on a tightrope. In the process he also gained an association that lasted a lifetime. Hamilton Harris was impressed enough to become Jay's friend and one of his lawyers for nearly forty years.

When the Delaware County survey commenced, Jay moved back to Roxbury in a room above the tin shop, where he busied himself with plotting the map, soliciting subscriptions, and repairing or replacing worn equipment. The tin shop also demanded his attention. Gradually Jay relieved his father of the responsibility until he found himself in full charge of the business. Eager to complete the survey that fall, he hired another man, Colonel Zawadille, but was forced to discharge Iram More for sloppy and careless work. Despite strenuous efforts, the survey was still unfinished in mid-December when foul weather shut the work down. Only Zawadille stayed on to help Jay with the plotting; Wilcox took his pay and went home, and Champlin enrolled in school for the winter term. Champ's decision intrigued Jay, not only because Champlin was a good friend but because it spoke directly to his own dilemma. Writing to Champlin in January 1854, Jay revealed that his own indecision had at last been resolved:

> I dont know for my part how you can content yourself to sit and study when there is so wide a world before us. I am sure it would require more presence of mind than I possess. I might have a stock of Geometries, Chemistries, Algebrays, Philosophies and the whole catalogue of studies that make up the routine of a finished education piled shoulder high on either side—all, I am fearfull, could not hinder me from dreamy delusive visits into the world of rat traps and maps.[18]

4

Slips

Gould always made it a point to prove himself valuable. He made his employer's interests his own.

—Trumbull White, *The Wizard of Wall Street*

Jay did very little of the actual surveying. . . . As a matter of fact, Jay was never a great success as a surveyor.

—Robert I. Warshow, *Jay Gould: The Story of a Fortune*

In these obscure years of struggle . . . the young Jay Gould consummated his first important business "deal," an operation . . . which could not have succeeded without the collaboration of his father. While employed as a clerk by the village storekeeper, he had learned that his master was negotiating for a good property in the neighborhood . . . and had offered $2,000 for it. The boy of sixteen quickly made some investigations of his own, then went to his father and by the most urgent pleading got a loan of $2,500 toward purchasing the property himself. In two weeks, thanks to the connivance of his parent, he had been able to sell it out for $4,000. But his employer, it appears, was highly incensed at what he saw as trickery or duplicity in his assistant and summarily dismissed him.

—Matthew Josephson, *The Robber Barons*

THE WORK CONTINUED to pile up. The Albany map was not yet completed, that of Delaware County barely begun. New assistants had to be found and trained for the Delaware County survey and for the Newburgh & Syracuse Railroad, which hired Gould to survey its route through Delaware County. Jay had neither the time nor the energy to take on another assignment, particularly one of such magnitude, yet he could not refuse. The Delaware County survey took on a new dimension when the state agricultural society approached Jay with the idea of supplementing his map with a history of the county based on interviews with its oldest inhabitants and whatever documents he could find. It was an extraordinary project for a boy his age to take up; that he did so at a time when he already had more work than he could do makes it more incredible.[1]

All winter long Jay buried himself in work. When spring blossomed, he plunged into the railroad survey. It was difficult, exasperating work, made all the more so for his having too little time to give it. Family and friends warned him to slow down, to put off some of his projects. "He worked himself too close," Champion observed, ". . . and the load he was carrying was too heavy for him." Jay ignored these warnings; he had learned all there was to know about work except how not to do it. He understood everything about drive and nothing about pace. His first lesson was to cost him dearly in the one coin he could least afford to pay.[2]

In June Jay completed the railroad survey and came home to plot the map. Pale and haggard, he collapsed a day or two later and was put to bed. The doctor diagnosed his illness as typhoid fever and allowed him no food except boiled buttermilk, barley coffee, and trout soup, or what Polly called "water off of trout." The fever lasted nearly three weeks, during which time Jay lost twenty-three pounds from his already scrawny frame. Anna and Polly stayed with him constantly, fretting, praying, seeing to his needs, unmoved by any fear of catching the disease themselves. Choking back her tears, Polly described his pitiful state to Oliver:

> His mind has been as weak as his body and he seems now more like a child than like *Jay.* He has been very nervous, so much so that we have been very careful about doing anything or saying anything that would in the least excite him. It would make him tremble sometimes just having the Doctor come in unexpectedly.[3]

For days he tossed in fitful delirium, mumbling incoherently about grades and routes, stakes and stations, his mind tortured by the work waiting for him and by Sarah's wedding. She was supposed to marry George W. Northrop, a widower with five children, on June 28, but Jay's illness forced a two-week postponement. Once the fever passed, Jay mended slowly, admitting that "the sun wilts me down pretty quick in the middle of the day." His face was so gaunt, his eyes so dark and sunken, the neighbors did not recognize him. His weight returned so slowly that he seemed to swim in his clothes. Yet no sooner was he back on his feet than he resumed work, plotting his maps, sending his assistants into the field, pausing only long enough to attend Sarah's wedding. He submitted to his weakened condition only to the extent of selling his interest in the Delaware County map to his engraver for $1,000, and even then he had to complete the work.[4]

After the wedding Sarah returned to Lackawack and Bettie to Margaretville, where she taught school. John Gould was laid up with a bad back and could scarcely walk, let alone work. "Father grows old fast," Polly lamented, "and I can see that both mind and body are failing." Jay had no choice but to pick up the slack at the tin shop. He spent the rest of July alternating between the store downstairs and the office upstairs, where he spent long hours poring over his maps. The strain was too much too soon. On July 31 he suffered a relapse in the form of a severe bowel inflammation. For two or three days his life hung in the balance. The doctor relieved his agony with morphine, bled him three times, and finally called in two other doctors for consultation. Once again Anna and Polly hovered over him night and day until fatigue wore them down. The other girls were called home and told to prepare for the worst. "I fear he will *never* recover but we are hoping for the best," Polly warned them. "You know this disease terminates one way or the other, *soon*."[5]

That Jay survived his second crisis despite his weakened condition was a tribute not only to his resilience but also to his strength of will, the same inner drive that would not let him quit at anything. Ten days later he was out of bed and downstairs again, weak, unsteady on his feet, but already thinking about the work to be done. Champlin and Zawadille helped with the Delaware County map while Peter Van Amburgh assisted him in the tin shop. With their help Jay was operating at full speed by September, a fact that worried his sisters. "I shall be glad when it is done," Sarah said of the Delaware County map, "so that the child will have less to think about or I dont know as he ever will get entirely well."[6]

But Jay paid little heed to such warnings. In September he attended the county Democratic convention as a delegate from Roxbury. The convention was a handy vehicle for meeting people who might be useful for his projects. Simon Champion was also there as a delegate. The two of them had grown close during the past year. Like most of Jay's friends, Champion was several years older than he was; impressed by Jay's precocity, he helped with the history and did other small favors for him. When Jay came to visit his office, they walked and talked together for hours, discussing politics, making plans, spinning out vast dreams for the future. One subject in particular fascinated them: the possibility of a transcontinental railroad. It is not known what aroused Jay's interest in this project. Proposals to construct such a road had recently been introduced in Congress only to bog down in sectional acrimony. The Bloomville survey had given Jay his first taste of railroads and may have kindled some vague ambition in him. Whatever the reason, he and Champion spent hours plotting routes, building bridges, discussing ways to raise money.[7]

It would be wrong to make too much of these musings. They were no more than boyish fantasies, the pleasant fillers of idle hours when the future could be made to assume any shape one chose.

As autumn dwindled away Jay scratched doggedly at his projects with limited strength. Tired and underweight, impatient with the slowness of his recovery, he seemed more shadow than substance. All three doctors advised him to put business aside for a time but, as Polly mourned, "the idea of giving up all employment seemed to make him worse than ever." She wanted to visit Sarah at Lackawack but hesitated to leave, fearing that "if his health continued to fail as fast as it had done . . . my stay in L/ would be short. . . . If his days on earth are to be short, I must spend those few with him, for than him I have on earth but *one* dearer friend."[8]

Early in December Jay started on a business trip. He stopped at Moresville to visit the doctor and there fell desperately ill again, this time with pneumonia. Anna hurried to Moresville to nurse him. Two weeks passed before he was well enough to be brought home, a month before he was able to leave his bed. Even then his hands shook so violently he could not so much as hold a pen. When finally he mustered the strength to write, he reported his condition to Champion with startling good cheer and utter lack of self-pity:

> The Dr. stands over my shoulder and criticises every movement as an alarming symptom; his orders are for the present live on soups made of shadows. To say the word map requires a portion of castor oil, and to think or transact any kind of business equal to jumping into a mill pond in winter time; but I have dismissed their sympathies and regulate my own diet. I find health and strength to improve in consequence.

Only in closing did Jay permit himself the plaintive admission that "I havent hardly raised a smile for five weeks."[9]

Jay's condition was far more serious than he let on. Even after he was out of danger, the fear persisted that he would not survive the rigors of a Catskill winter. Grave doubts remained as to whether he could regain enough strength to work, let alone drive himself at his accustomed pace. At eighteen he faced the real prospect of invalidism, which preyed heavily on his mind. He could not even bring himself to eat, and only Anna's diligent efforts coaxed his appetite back to solid food.[10]

For several months these doubts lingered unresolved. During that time fate inflicted a perverse and tragic irony on the Goulds. Since June the failing health of both Jay and his father had pressed all four sisters into service as nurses, with Anna and Polly bearing the brunt of the long hours. Jay pulled through and his father was to live another dozen years, but in March Polly, gone at last to visit Sarah in Lackawack, was stricken with consumption and died two weeks later. "Sweet Polly Gould," wrote John Burroughs, who as a schoolboy once had the most desperate crush on her, "was the flower of the family—a very sweet girl." She was twenty-two years old.[11]

Months passed before Jay regained enough strength to resume work. During that time he stayed close to home, did what work he could, and allowed his mind to dwell at length on life's larger mysteries. These months of recovery were unique in Gould's life in that they permitted him a degree of introspection for which he would never again have the time or the leisure.

Events had certainly bent his thoughts toward the subject of death. Along with his family and the disconsolate James Oliver, he brooded over the loss of Polly, to whom he had been especially close. The twist of fate that claimed her while sparing him was very much on his mind through that forlorn winter. Apart from Polly, the wife of Edward Burhans had been bedridden with consumption for months. The Goulds took turns visiting her, none more than Jay, who was a particular favorite of hers. That she was a cousin of Jay's mother only strengthened the bond between them, as did the fact that her daughter, Maria, had been close to Jay since early childhood. The hours Jay spent at Mrs. Burhans's bedside, coupled with the frequency of Maria's visits to Jay while he was ill, set village tongues to wagging about a romance in bloom.[12]

It was not romance Jay discussed with Mrs. Burhans but religion, the enigma of life after death. The question had come to obsess him; he pored over theological treatises on the subject without finding satisfaction. He debated the issue with J. W. McLany, who had come to town to take charge of the Roxbury Academy. One night the two of them went to the cabin of a young man dying of consumption to watch dispassionately as he wheezed toward his final breath. "We viewed the dying process from a physical and psychic standpoint," McLany recalled. "We longed to fathom what we knew to be impossible, the hidden mysteries of life and death; we watched with interest the changes of the body while the soul was departing; we yearned to behold with spiritual vision the immortal spirit as it passed away and to follow it in its flight within the portals of the spirit world." The spirit departed without obliging their curiosity. From their hours of discussion Gould concluded that "as regards the future world, except what the Bible reveals, I am unable to fathom its mysteries, but as to the present, I am determined to use all my best energies to accomplish this life's highest possibilities."[13]

These words proved no idle boast. The past six months had worked a subtle but profound change in Jay. The long siege of illness and the close brush with death impressed on him a sense of his own vulnerability, an acute awareness of his physical limitations. For some people this insight might trigger a sort of spiritual paralysis, an air of resignation and a retreat from life's harsher demands. Others might find solace in a determination to enjoy to the fullest whatever comforts and pleasures remained to them. In Jay discovery seems to have bred a renewed sense of urgency, a desperate need to do all he could as quickly as possible, before he died. He despised wasting time or effort because he had none to waste; his ear was always tuned to the relentless ticking of mortality's clock. Those who saw him as a driven man never understood the urgency behind that drive. He was not, as many claimed, lacking in emotion or affection, but he measured out his feelings in the same spare portions as the food he ate, the letters he wrote, or the advice he offered.

He was, in short, a man running scared, and Jay Gould's way of running scared was to run as hard as he could for as long as he could.

For eighteen months after his bout with pneumonia Jay lingered in Roxbury, convalescing, busying himself with various projects, mulling over his future. He finished the Delaware County survey and took up the history in earnest. For a time he worked in the general store of Edward Burhans, which gave rise to a pair of episodes that later became part of the Gould legend: a controversial land deal and Jay's supposed romance with Maria Burhans.[14]

The land deal involved some estates in Roxbury which had been divided into small parcels. Some of the heirs were anxious to sell and offered their land to Edward Burhans. According to legend, Jay heard of the offer and persuaded his father to make the purchase before Burhans responded. Enraged over what he considered an act of treachery, Burhans summarily fired Jay. It appears that the transaction did occur, but there is no hard evidence to support the charge that Jay betrayed Burhans or lost his position because of the transaction. Anna dismissed the allegation as *"a vile slander and . . . a lie on the face of it."* The original version was apparently derived from Hamilton Burhans, who was a business rival of his brother. His account is maddeningly vague but hardly provides grounds for charges of duplicity or disloyalty:

> Jay Gould, knowing of some of these lots that were for sale and knowing who owned them, prevailed on his father to make the purchase as they had been offered for sale to the man that he had been clerking for. As soon as he bought them, his employer was dissatisfied with his purchase, as he was with every one that bought any lot out of this divided tract of land, but after Gould bought the lots the farmers who owned land adjoining it, soon came in and bought it and he realized a handsome profit.[15]

Similar ambiguities cloud Jay's relationship with Maria Burhans, whom legend elevated to a position as the Ann Rutledge in Jay's life. His niece, Alice Northrop Snow, did as much as anyone to perpetuate the legend, declaring flatly that "this puppy love, nevertheless, in time waxed into a romance that led to formal engagement." No evidence exists for this charming tale. Every one of Jay's friends shared the observation of Simon Champion that Jay "was not given to running with girls much or at all when he was with me. . . . I was a single man too then, and we never talked over the question of marriage." No mention of the relationship appears in the surviving family correspondence, which abounds in gossip on the romantic liaisons of others. Hamilton Burhans, who was in a position to know, rejected the story outright: "As gossip had it, he was to marry her; but it was never so; Jay Gould was too bashful; he never talked matrimony to anybody at that time; he never conceived any such idea." If further testimony is needed, it can be taken from Jay himself. In December 1857, hearing that Peter Van Amburgh had married, he wrote asking his friend for "a chapter or two on matrimonial felicity, that strange uneven sea of human existence upon which I never expect to embark myself."[16]

During the spring of 1855 Jay taught surveying at the Roxbury Academy. By winter his history of Delaware County was ready for publication. It was a remarkable achievement for a boy not yet twenty. He followed the usual format of the genre, blending accounts furnished by oldtimers, bits of lore culled from old newspapers, biographical sketches, and excerpts from yellowing letters and documents with his own narrative. The 426-page history was no slapdash job. In his preface Jay observed the judicious caution of the scholar:

> I do not claim that this work is free from error; perfection, in a history of this character, where much of the information to be relied upon is of an oral and indefinite nature, is an impossibility. I have been careful to weigh

> all the statements presented—to discriminate between truth and fiction—and have suppressed much apparently interesting matter, which lacked the proper authenticity or conflicted with truth; still, doubtless there is room for improvement.[17]

The narrative was vigorous and enthusiastic, the prose flowery and self-conscious. Most impressive of all, Jay completed the entire project in only about two years of part-time work. He hoped the book would bring him a little money and sufficient reputation to do similar histories of other counties, but a cruel accident disrupted his plans. Jay had contracted with a Philadelphia printer to publish the history. While the first edition was being readied, a fire in the shop destroyed the manuscript and most of the engraving plates. It was a devastating loss, yet Jay took the news with astonishing calm. "I am under the unpleasant necessity of informing you that [*sic*] the total destruction by fire of my History of Delaware County," he wrote to Oliver. "I shall leave for Philadelphia in the morning to ascertain the exact state of my affairs."[18]

The printer had no insurance, although he managed, by sifting through the ruins, to recover some portions of the work. Jay saw at once that if he wanted the history, he would have to write it again. It was a difficult and disheartening task, but Jay plunged into it without hesitation. "As you know I am not in the habit of backing out of what I undertake," he informed Oliver, "and shall write night and day until it is completed." That is exactly what he did. Scarcely pausing for sleep, he wrote at a furious pace. On his twentieth birthday he acknowledged the event as "almost the only leisure moment I have taken for a month." Sometime before mid-summer he finished the book, which a Roxbury printer published in September 1856. By that time Jay had left town to embark upon a new and unexpected career. He would never return to his native village except as an infrequent visitor.[19]

It was a departure made with scarcely a backward glance. Unlike some men who clawed their way to success, Jay made no attempt to transform his youth into anything other than what it had been. He never tried to make a shrine of his hometown or portray himself as the poor boy made good. Such pretensions offended a man who never traded in the market of self-delusion. There is no doubt that the struggles of boyhood left a deep imprint on Jay, but he rarely talked about them in later years and adamantly refused to capitalize on them. Reporters hounded him on the subject but failed to crack his reticence. When one of them persisted with questions about his early days, Jay silenced him by snapping, "Do you know that my father's poverty was never worth a single thousand dollars to me?"[20]

Before leaving Roxbury to its quiet insularity, it is worth a brief glimpse at the destinies of those who were close to Gould during these early years.

In May 1855 Anna Gould finally married the Reverend Asahel Hough and embarked on the life of a missionary in the West. Five years later, Bettie Gould married Gilbert Palen, by whom she had four children. They moved to Philadelphia and devoted much of their energy to good works, especially the temperance cause. Sarah bore ten children in addition to the five she inherited from George Northrop and went to live in Canadensis, Pennsylvania, where George operated a tannery. Shortly after Jay left home for good, John Gould sold the tin shop but remained in Roxbury until 1864, when he went to live with Sarah. He died two years later.

Among Jay's circle of friends the most successful was John Burroughs, who gained national prominence as a naturalist. The studious John Champlin went to Michigan, where he became a lawyer and eventually chief justice of the state supreme court. After losing his beloved Polly, James Oliver migrated to Kansas, where he abandoned schoolteaching in favor of a mercantile business. Isaac Wilcox also tried his luck in Kansas, going there in 1863.

Rice Bouton donned the collar of a Methodist minister and finished his days at the celebrated Five Points Mission in New York City. Abe Crosby became a jobber of mill supplies in Rondout, New York, and Andrew Corbin a merchant in Bloomville. Simon Champion continued to publish the *Bloomville Mirror* until 1870, when he moved to Stamford, New York, and started the *Stamford Mirror.* The irrepressible Iram More went out west in 1862 but came home seven years later, never to leave again. Peter Van Amburgh spent the rest of his days in Roxbury, as did Maria Burhans, who eventually married George Lauren, the man who succeeded Jay as clerk in her father's store.

Every one of these old friends outlived Jay, but none outshone him. In later years some of their paths chanced to cross his, and there would be a few hours of reminiscing about the old days. For all of them he became in time a legendary figure, a source of tales to tell family and friends. John Burroughs for one could not shake free of Jay, even though he never again spoke to him:

> It is a curious psychological fact that the two men outside my own family of whom I have oftenest dreamed in my sleep are Emerson and Jay Gould; one to whom I owe so much, the other to whom I owe nothing; one whose name I revere, the other whose name I associate, as does the world, with the dark way of speculative finance.[21]

5

Opportunity's Knock

But after a while Mr. Pratt became dissatisfied with the condition of affairs. Apparently a rushing business was being done, from which there was no adequate return. The books seemed to be so mixed that it was quite impossible to ascertain just how the firm stood. . . . Gould . . . found Pratt looking over the books and puzzled by their intricacies. He discovered that Gould had started a private bank at Stroudsburg in his own name, and he became suspicious that the firm's funds were being used in the bank. Pratt then demanded an explanation and finally threatened to close up the tannery and dissolve the partnership. Gould protested that this would ruin him, when Pratt said that he must buy or sell.

—Henry D. Northrop, *Life and Achievements of Jay Gould*
—Trumbull White, *The Wizard of Wall Street*

Pratt, it seems, was impressed by young Gould's energy, skill and smooth talk, and supplied the necessary capital of $120,000. Gould, as the phrase goes, was an excellent bluff; and so dextrously did he manipulate and hoodwink the old man that it was quite some time before Pratt realized what was being done.

—Gustavus Myers, *History of the Great American Fortunes*

The tannery was doing a rushing business, and was always at capacity. Profits, however, were very small. Then Pratt came on one of his rare visits, and went over the books. The result of the honest old tanner's investigations was a demand upon Gould to buy Pratt out or let Pratt purchase his share in the tannery. There was apparently no answer to Pratt's charges of very original bookkeeping.

—Robert I. Warshow, *Jay Gould: The Story of a Fortune*

OF ALL THE TANGLED episodes in Gould's career, none contains more perverse twists of irony than his venture into tannery. It was his first major business operation, the springboard of his career, and in retrospect a remarkable achievement for a youth with no prior experience in the field. Unfortunately, his accomplishments have been overshadowed by controversy over their outcome. In time the tannery affair assumed its place as a cornerstone of the Gould legend, the first ugly chapter in what his critics portrayed as a sordid career.

Although Jay was to be involved in much larger and more sensational schemes, none have been more carelessly or vindictively handled by later generations than that of the tannery. As a cornerstone of legend it crumbles at the touch, one of those rare instances where the story as written is not merely disputable but simply wrong. In their zeal to blacken Gould's name his critics did more than misinterpret the facts; they got nearly all the facts wrong. There is no better measure of Gould's unpopularity than the perseverance of this fable for more than a century when even a cursory glance at the facts would have betrayed the "weight of historical evidence" as nothing more than the gossamer of hearsay repeated.[1]

As Jay well knew, surveying paid an important fringe benefit: it brought him into contact with prominent people in the region. Men of standing noticed and admired his drive, liked his snap and spunk. Because they could be useful to him, Jay cultivated their favor assiduously. Early in life he revealed a talent for charming people. To some extent he merely ingratiated himself with those he wanted as friends, but there was something more, something compelling in his personality that drew people to him.

It was as easy to like Jay as it was to underestimate him. The undersized body with its small hands and feet that moved with feline grace; the boyish looks with their eager, earnest expression; the shy, friendly smile; the huge, dark, liquid eyes, piercing in their intensity, ceaselessly probing without revealing; the soft, musical voice that seldom rose above *piano*; the reserved, diffident manner, ever respectful, unfailingly courteous, never shouting for attention or challenging others for the spotlight. All these qualities impressed people who met Jay. Those who came to know him soon learned to appreciate the less obvious elements of his character: his demonic energy, fierce determination, and intense concentration. There was in him a tensile strength, a resiliency that seemed incongruous in so frail a vessel. Until they knew better, men tended to dismiss Jay as a boy attempting a man's work. Jay understood this attitude and turned it to his own advantage. He grasped early the value of possessing a reservoir of capabilities known only to himself.

Whatever the alchemy of his peculiar charm, Jay applied it unstintingly on anyone who might be of use to him. During the summer of 1856 he reaped his first harvest in the form of Zadock Pratt. When the two men first met, Pratt was already in his sixties. A distinguished if eccentric career as tanner, banker, farmer, and politician had made him one of New York's richest and most prominent citizens. The son of poor parents, Pratt had moved to the village of Schohariekill and created the largest tanning operation in the country. A shrewd, careful businessman, he practiced small economies, kept detailed statistics, personally supervised every aspect of the work, and introduced a host of technical improvements. In the twilight of his long career Pratt boasted that he had tanned more than 1.5 million sides of sole leather, cleared 10,000 acres of land, used 250,000 cords of hemlock bark, employed 40,000 men, formed more than thirty partnerships, and, he noted proudly, "closed them all in peace."[2]

Unlike most tanners, Pratt cared deeply for the land he stripped, for his employees, and for the community they shared. As the trees came down, the land was converted into farms and pasture. By the 1850s, when the tannery had exhausted its bark supply and closed down, Pratt had become the state's leading agricultural reformer. He turned his own farm into a showplace and Schohariekill into a model town, building houses for his workers, churches, schools, even fishponds. He operated the town bank and financed a variety of small industries to sustain the local economy once the tannery was gone. For people accustomed to seeing tanneries turn overnight into ghost towns, it was nothing less than a new wonder of the world. In gratitude the villagers renamed the town Prattsville in his honor. After growing rich from tanning and banking, Pratt entered politics. An ardent Democrat, he served as state senator and presidential elector. Twice he was elected to Congress, where he introduced the bill creating the Bureau of Statistics and fought for agricultural reform, cheap postage, and increased expenditures for public buildings and monuments.

Above all else Pratt was an eccentric, an outlandish figure in the mold of Paul Bunyan or Mike Fink. No one knew what to make of him. Tall, with chiseled features, athletic, loud, horny-handed, bristling with energy, he threw himself into everything with a passionate disdain for convention. Where other tanners clutched their trade secrets like family jewels, Pratt cheerfully offered his to anyone who wanted them. When a stranger applied to his bank for a loan, Pratt examined not his collateral but his face and hands. If they showed evidence of hard work, the man got his loan. In his pleasures Pratt was notorious for his queer pranks and for his domestic arrangements, which puzzled more orthodox souls. Three wives died young and a fourth divorced him. At age seventy-nine he took a fifth bride, a twenty-year-old employee of the *Shoe and Leather Reporter.* Of this liaison the journal's editor noted solemnly, "She had, no doubt acquired that amiability and flavor of the Swamp that made her attractive to the old tanner."[3]

For all his shrewdness, there was about Pratt an air of ingenuousness. His vanity was childlike and colossal. The most impressive monument to it still adorns the road to Prattsville. After donating a park to the town in 1840, he hired an itinerant sculptor to emblazon some nearby rocks with a melange of images dear to him. Lacking the ability to write, Pratt vowed to "let the rocks tell the story" of his life. As different as Pratt and Gould were from each other, they shared some important values, notably a burning ambition and an appetite for hard work. In one of his notebooks Pratt pasted a motto that suited Gould no less than himself: "There is no greater obstacle in the way of success than trusting for something to turn up, instead of going to work and turning up something."[4]

J. W. McLany remembered a cold, rainy spring day when Jay was visiting him at the hotel. A man splattered with mud rode up, asked loudly for Jay, and was shown to McLany's room. McLany was struck by his appearance: "An old gray haired man, tall, erect, booted and spurred, his boot tops extending above his knees, mud bespattered over his clothing from head to feet. He introduced himself as Col. Pratt . . . wanted Mr. Gould to make a survey of his farm." Jay had met Pratt the previous year and taken care to cultivate his friendship. At one point, knowing the old man's vanity, he offered to do a biography of Pratt. Instead, the colonel employed him to write a speech on one of his favorite topics, "The Horse." It was not Jay's first venture as a ghost writer. For a time he helped an illiterate Irishman by writing love letters to his distant sweetheart for him. The Irishman was delighted with Jay's work and so was Pratt, who paid him $100 for the speech.[5]

The farm survey brought Gould into Pratt's company for several weeks. His quick mind, self-reliance, and go-getting eagerness impressed Pratt deeply, perhaps because it reminded the old man of his own boyhood struggles. From these weeks evolved a plan to go partners in a tannery to be built in eastern Pennsylvania. The precise origins of this scheme are unclear. Both men knew the future of tanning lay in the virgin forest of eastern Pennsylvania. In 1849 Pratt had opened a large tannery at Aldenville. Jay had talked extensively with George Northrop, Sarah's husband, who had gone in with Gilbert and Edward Palen to build a tannery at Canadensis in Monroe County. From Northrop Jay gleaned information about the tanning business and about the vast tracts of hemlock stretching southward between the Lehigh and Delaware rivers. Although the region was still wilderness, the recently completed Delaware & Lackawanna Railroad could provide transportation for a large enterprise.[6]

Intrigued by the possibilities, Pratt and Gould formed a partnership in August 1856. Jay went at once to explore the region. With the aid of a compass he struggled through the dense forest until he found massive stands of hemlock and alder. Bursting with excitement, he hurried back to New York and reported to Pratt. Together they returned to the site, where the abundant supply of bark convinced Pratt on the spot. He went back home, leaving Jay to buy land for the tannery and make contracts with nearby landowners for taking bark from their trees. Numerous assertions have been made about how much money each man put into the partnership, but there is no evidence to sustain them. Doubtless Pratt furnished most of the capital and Gould was expected to do the work, with advice from his mentor. From the first their relationship was that of master and apprentice. Jay understood his role as obsequious disciple and played it perfectly, seldom missing a chance to praise or flatter the old man. But he also had to produce; the responsibility for creating the enterprise was his.[7]

By late August Jay had acquired a site and a batch of bark contracts. On September 1 he led the fifty or so workmen he had recruited to a location that was, in his own words, "fifteen miles from any place." Jay personally chopped down the first tree; before dusk it was sliced up with a portable sawmill and the lumber used to erect a blacksmith's shop. That night Jay slept in the shop on a bed of hemlock boughs. After a two-story boardinghouse went up in four days, the men and a few of the locals honored him with a round of cheers and a resolution naming the village-to-be "Gouldsboro." Not many twenty-year-olds could boast of having a settlement named after them, yet Jay took care to share the tribute with his partner. "Three hearty cheers were then proposed for the 'Hon. Zadock Pratt the *world renowned Great American Farmer,*' " he wrote Pratt, "and a more hearty response I am certain this valley never before witnessed."[8]

The work went forward at a furious pace. By November Jay had put up a second boardinghouse, a barn, a wagon house, four dwelling houses, a post office, and a large water race. He had arranged for stagecoach service to Tobyhanna, contracted for a general store, the first in the county, begun construction of a plank road to reach the railroad, and started work on the vats, hide mills, and other tanning facilities. He seemed to be everywhere at once, driving the work forward, allowing no detail to escape his eye. At night he did the paperwork, pushing himself at the old pace. By February 1857 most of the tanning facilities were ready. An experienced tanner was hired to oversee preparation of the hides in the sweat pits and beamhouse. Jay was busy measuring bark, devising an inventory system, and organizing his accounts so as to provide Pratt with detailed monthly statements. On this last point Jay was insistent, adding, "in the end it will prove a source of economy to us." Already he grasped a principle that would serve him well in later years: "I have almost always noticed that the most successful men are invariably the most careful about small things."[9]

During these months Pratt peppered his partner with advice, to which Jay responded with smarmy deference, saying on one occasion, "I am much obliged to you for all your suggestions. I find them a *good dictionary.*" He understood only too well his utter dependence on Pratt's good will and took pains to feed the old man's insatiable appetite for flattery. In that vein he dangled again before Pratt the prospect of a biography. Nothing was more important, he gushed, than that the world know how "Col. Pratt has arisen by his own untiring industry from comparative obscurity to become the most eminent and *most useful* man in the world in his profession."[10]

This tireless stroking of Pratt's ego should not obscure Jay's genuine achievement. Within a few months he had transformed a patch of wilderness into a flourishing business enterprise and village. As the settlement grew larger, he donated land for a cemetery, raised money to build a church, erected a schoolhouse at his own expense, paid the teacher out of his own pocket, and required his employees to enroll their children. He knew little about tanning, and nothing he had done before even remotely approached the scale of the tannery project. Nor was inexperience his only handicap. He was an undersized young man attempting to boss gangs of older, tougher workmen who must have enjoyed a few snickers at his expense. To get anything done required not only talent but hardnosed leadership, the ability to control men who could physically intimidate him.[11]

The founding of Gouldsboro displayed Jay's extraordinary ability to get things done; to step into a new game, learn its rules on the run, and become a superior player with breathtaking speed. Already one could detect glimmers of what would later become a familiar pattern: even as he took hold of a new business and mastered its intricacies, he searched for ways to expand his interests outward from the original source. Not even the immense challenge of the tannery was enough to occupy his full attention. What others regarded as a main chance was for him merely a point of departure to explore connecting possibilities which, in time, might prove more profitable to him than the original enterprise.

At Gouldsboro land provided his first opportunity. At first Jay was content to contract with landowners for bark from their trees. That winter he began buying land, not only for its hemlock but for its speculative value as well. "I was out to Wilkes Barre yesterday to attend to some business," he blithely informed Pratt in February, "& I accidentally fell in with a bargain in the way of land." Then there was the Stroudsburg bank. According to legend, Jay used Pratt & Gould funds to start a bank at Stroudsburg without Pratt's knowledge and made it a vehicle for his private speculations. The evidence for this charge is sketchy at best. There *was* a bank at Stroudsburg in which the partnership kept its accounts for a time, but it was not the only bank used by the firm. The notion that Jay started the bank apparently comes from Hamilton Burhans, but nothing in his version attests to misuse of funds for speculative purposes or clarifies the question of whether Pratt knew about Gould's involvement in the bank. It is clear, however, that the firm's capital was being consumed by expenditures for construction, equipment, and wages, and that no income could be expected until the tannery was ready to process hides. Since Jay had to scramble constantly to meet notes falling due, and often drew on Pratt for large sums, it seems doubtful that he had much spare cash for peripheral ventures.[12]

By the time Gouldsboro was founded, the leather industry had long since separated the manufacturing process from marketing. Since 1790 the industry had been clustered along Jacob, Pearl, Frankfort, and Ferry streets in New York City, a section known as the Swamp. At first it was the tanners who dominated the Swamp, the stench of their vats adding one more odor to an already fetid region. Gradually, however, the tanners were replaced by (or became) leather merchants, who transformed the Swamp into their commercial headquarters. The most powerful merchants, such as Gideon Lee and Jacob Lorillard, forged a system whereby they supplied inland tanners with hides and sold the finished leather. They also provided tanners with short-term financing, trade information, and market advice. By 1840 the merchants sat astride the most profitable aspects of the leather industry and had reduced the tanner to a mere processor of hides, dependent for his livelihood upon them.[13]

It did not take Jay long to grasp this essential fact about the trade. The Gouldsboro tannery obtained its hides from the firm of Corse & Pratt. Israel Corse's father had been one of the Swamp's pioneer merchants; George Pratt was Zadock's only son. Jay depended on them not only for hides but occasionally for money as well. Never one to overlook an opportunity, he used his dealings with Corse & Pratt to learn something about the industry in general and the Swamp in particular—how it operated, where the sources of profit lay. He also tried to widen his circle of acquaintances among the other leather merchants.

His immediate task was to produce salable hides. Bad luck and inexperience hampered his efforts. Tanning is not a process that can be done by written formula; it is an empirical craft, requiring constant supervision by a practiced eye. Pratt had no peer at the work, but he was not there to do it. As he well knew, Jay was no tanner and therefore had to retain one as foreman. This arrangement left Jay dependent on someone else's judgment in the most vital area of operation. From the first the partners decided to adopt a new technique known as the wet-spent tan bark process, which enabled them to burn as fuel the vast amount of wet refuse bark normally discarded as a by-product. The tannery could employ steam instead of water power if a reliable method of converting the wet bark into fuel could be found. Since the process was still unproven, Jay had to experiment with various methods of conversion. In the midst of this work the steam engine he had purchased proved defective, and the supplier failed before the replacement parts could be obtained.[14]

Jay struggled with these problems through the summer of 1857. Painfully aware that large sums of money had been spent with no return, he begged Pratt to be patient and insisted that all was going well despite the delays. At this critical juncture the Panic of 1857 sent ripples of contraction through the economy. Jay weathered the early storm and seemed safely home when the tannery began at last to turn out hides. Then a rumor got abroad in October that Pratt & Gould had been discredited in New York and had suspended operations. Jay first got wind of trouble when several suppliers besieged him for immediate payment. "Everybody seems frightened to death," he reported, including himself among their ranks. "I could have managed to have got along very well had not this report got abroad," he added, "& as we have a note due 27th . . . I did not sleep a wink last night for fear we cannot meat [*sic*] it."[15]

It took Gould several weeks to extricate the firm from danger. In December he succeeded at last in implementing the wet-spent tan bark process. The rumor of suspension had been laid to rest, and the number of hides going to market increased rapidly. Nevertheless, Pratt's confidence had been shaken. He showered Jay with criticism, suggested changes in his bookkeeping methods, and demanded that the firm withdraw its account from the Stroudsburg bank. Caught in a difficult position, Jay swallowed these reprimands with all the grace he could muster. "I am under many obligations for your good advice & useful suggestions," he replied meekly. "There is an old saying in my scrapbook, 'that it is our best friends that tell us of our faults.' . . . I often read over your letters rainy days and always I think I learn something new from their perusal."[16]

Improved business conditions eased the strain, but a cloud of doubt had entered their relationship. Jay knew that Pratt was unpredictable in his behavior, and now he was becoming more difficult to please. It must have rankled him to be so dependent upon Pratt's favor, to realize that a sudden whim or change in mood might undo all he had worked to build up. For his part Pratt found Gould increasingly difficult to understand. It became apparent to him that Jay was his own man, and that his way of doing things often did not conform to what Pratt regarded as proper or standard practice. During 1858 their relationship cooled until, by the year's end, the differences between them had become irresolvable.

What caused this breach? There is scant evidence to support the myth of the artful swindler hoodwinking a gullible old man. Pratt may have been eccentric but he was no fool, especially in business matters. The lurid tale of cooked accounts, misused funds, and a stormy confrontation has no basis in the surviving records. Nor is it needed to explain the falling out. There is no doubt that Pratt grew dissatisfied with Gould's handling of the tannery. It is possible he distrusted Gould but more likely that he simply could not fathom Gould's way of doing things. A clash of styles and values had emerged between an old man set in the ways that had made him successful and a young man on the make, still feeling his way, alert to opportunities, willing to take risks and employ practices repugnant to his more conservative partner. Pratt had taken Jay under his wing, made him his protégé, and given him a fine opportunity. In return he expected loyalty, gratitude, and the deference due a master from his disciple. Doubtless Pratt assumed their relationship would continue along these lines, but he did not reckon with Gould's phenomenal capacity for growth. After two years at Gouldsboro the protégé had come of age, soaking up experience as a plant draws nutrients from the soil, until he possessed knowledge and acumen far beyond his years. Pratt was slow to grasp this change in Jay, and the discovery must have come as a rude shock to him.

As Jay's knowledge grew, he became impatient with the restraints imposed by Pratt. The ritual of heaping flattery on Pratt became tiresome to him. Gradually the honeyed passages disappeared from his letters; "dear friend" gave way to "dear sir." In his search for fresh opportunities Jay revealed again his compulsion to undertake several ventures at once, to utilize his position in one as leverage for entering others. It was this tendency that disturbed Pratt, who could not help believing that such activities distracted Jay from his primary concern, the tannery.

A different explanation is offered by Hamilton Burhans, who claimed that some of Pratt's old tanning cronies in Ulster County exhausted the local bark supply and went to Pratt in hopes of securing work from him. According to Burhans:

> They had been to see the Gouldsboro tannery—they reported to Col. Pratt that young Gould's management of the large tannery was bad, that he was nothing but a boy, and that he had better buy him out or get him out in some way, or he would ruin him. They so worked on Col. Pratt that he was

> anxious to get "the boy" (as they called him) out and—they would go and take charge of it.

Burhans also thought the rupture was precipitated by a disagreement between the partners over "what house in New York should furnish the hides."[17]

In fact the partners did sever their ties with Corse & Pratt and arranged to obtain hides from Charles M. Leupp & Company, one of the Swamp's most prominent houses. It is not known who instigated the switch or why. Apparently the change occurred in the spring of 1858, a time when Jay had difficulty keeping a foreman. Early in June Pratt was negotiating with a tanner to replace a foreman who had left after only five days on the job. Despite this turmoil, the tannery did a thriving business. In September Jay reported "hides going in at the rate of 300 sides per day." In all he produced 60,000 hides during 1858, an impressive figure for one accused of bungled management. This output satisfied Pratt to the extent that he continued to supply money when called upon by Jay to do so. That summer Jay began buying land, this time on his own and probably without Pratt's knowledge.[18]

At the year's end, when the tannery had weathered all obstacles and was a going concern, Pratt made his move. He informed Gould that he wanted the partnership terminated, and that one of them must buy out the other. So much has been written about Pratt's motives that it is easier to disprove existing versions than to reconstruct what actually happened. Pratt may have grown alarmed at Gould's expansionist tendencies or balked at the increasing demands on him for money. No known evidence supports the charge that he suspected Jay of using company funds for his own speculations. On the contrary, two accounts argue that Pratt wanted Jay out for quite different reasons. Burhans insists that Pratt "had arranged with one of his old partners in Ulster County to take possession of the tannery." John Gardner, who worked at Gouldsboro, goes even farther:

> Pratt having the largest interest—said to Gould—you must either sell or buy—thinking it impossible for Gould to buy—and secretly hoping to force Gould to sell his share—and thereby enriching himself at Mr Gould's expense.

Whatever his motives, Pratt forced the issue with an offer to buy Jay's share for $10,000 or sell his own share for $60,000. He gave Jay ten days to decide, and declared he would allow him time to pay provided good security was given. He knew Jay did not have that kind of money or the influence to borrow it. If his true object was to oust Gould, he appeared to hold an unbeatable hand.[19]

But Pratt was not the first man to underestimate Gould, and he would by no means be the last. Jay went at once to see Charles M. Leupp and his partner/brother-in-law, David W. Lee, who agreed to buy Pratt's share of the tannery. Since both Leupp and Lee were experienced businessmen, it must be assumed that they inquired into the state of the tannery's financial affairs. Moreover, they had been supplying Pratt & Gould with hides for nearly a year, which suggests a firsthand knowledge of the operation. If the working relationship had been unsatisfactory to them, it is unlikely they would have invested in the firm.

Pratt must have been stunned at the news when Jay broke it to him. Since Leupp & Company was among the Swamp's most reputable firms, he had no choice but to accept the surety. On January 27, 1859, Pratt sold his interest to Gould for $60,000. The next day Jay conveyed a one-third share of the property each to Leupp and Lee for the sum of $30,000 apiece. The transaction was a milestone for the merchants in that it marked their first direct entry into manufacturing.[20]

The parting of Pratt and Gould appears to have been amicable, yet it cast long and mysterious shadows over the years. Of this episode the usually loquacious Pratt later made only scant mention to his biographer. In his *Chronological Biography* Pratt filled page after page with the individuals and events important to his lengthy career. Not once did he mention Jay Gould or the Gouldsboro tannery. Long after Pratt's death hard feelings still lingered in the family. In 1892 Pratt's daughter, Julia Ingersoll, scribbled these embittered lines in her journal:

> Jay Gould will have been dead a week tomorrow. What is he to me, not my brother. Maude had two such curious dreams about him, but I will not copy them here. He leaves 75 millions, he still owes my father a few thousands. Will he be sorry now that he owed anything in this world to anyone?
>
> Why did my darling Father say once, when someone called Gould such fearful names, "hush! do not say it loud." Because he suffered ingratitude at the man's hands uncomplainingly, is that the reason why I feel strangely as if I could never speak unkindly of him. What has he put in the upper treasury to draw on, where he has gone?[21]

6

First Blood

Meanwhile the Panic of 1857 had swept over the country and unsettled all business operations, and when Leupp discovered the extent in which he had been involved in Gould's speculations he thought he was ruined. He went to his magnificent home one night and, in a fit of despondency, shot himself dead. . . . it is a fact that Leupp's partners and heirs have always felt very bitter against Gould, and could not help believing that he was indirectly the cause of Leupp's sad and untimely end.

—Henry D. Northrop, *Life and Achievements of Jay Gould*

In a fit of despondency, after a stormy interview with Gould, he shot himself. In the excitement of Black Friday, years later, when thousands thought themselves ruined by Gould, the crowds around the Gold Exchange took up the cry, "Who killed Leupp?"

—Robert I. Warshow, *Jay Gould: The Story of a Fortune*

On leaving the train, Charles Leupp took a cab to his Madison Avenue mansion, walked in the door, marched into his library, locked the panelled doors behind him, picked a revolver out of the desk drawer, put the barrel in his mouth, and pulled the trigger, sending a bullet through his brain.

—Edwin P. Hoyt, *The Goulds*

THE CONTRAST BETWEEN Zadock Pratt and Charles M. Leupp was startling: The one an eccentric, semi-literate rustic, loud and crude, proud of his horny-handed ways; the other suave and urbane, fond of art and literature, reserved and dignified in manner. Where Pratt advertised himself with carved rocks, Leupp built a mansion on Madison Avenue and filled it with fine furniture, books, and paintings. Yet Pratt and Leupp shared a common legacy as pioneers in the leather trade, Pratt as tanner and Leupp as successor to the most prominent house in the Swamp, Gideon Lee & Company.

Leupp went to work for Lee as a boy of fifteen, earned himself a partnership, and married Lee's daughter. In 1843 the firm was changed to Charles M. Leupp & Company, with Leupp and David Lee, one of Gideon's sons, as partners. It suffered serious reverses in 1846–47 but weathered the Panic of 1857 unharmed. Through the years Leupp developed a reputation as a sound, conservative businessman whose integrity was unquestioned. Among the denizens of the Swamp he represented the Establishment. Given this reputation, it would be intriguing to know what prompted the partners to venture directly into tanning. Leupp gave two reasons for wanting the agreement kept secret: "First, to make two-name paper; and second, that they feared a knowledge that they were directly interested in manufacturing would affect the standing of their paper in market." At the very least the arrangement with Gould was a departure from Leupp's usual business practices.[1]

On February 1, 1859, Jay signed a working agreement with Leupp & Company in which the latter was to provide the tannery with hides for the next fourteen months in return for a 5 percent commission and another 6 percent commission for selling the leather. One serious flaw marred the arrangement: the dual role of Leupp and Lee as partners in both enterprises. They controlled the leather house but left Jay in charge of the tannery so as to avoid the appearance of doing business with themselves. Conflicting views over this relationship soon emerged. Although a minority owner in the tannery, Jay understood himself to be the "acting and sole known partner," for which the merchants paid him a fixed sum. He assumed the tannery was his to run as he saw fit; the February agreement merely obliged him to tan hides for Leupp & Company. By contrast the merchants considered themselves entitled to dictate policy as majority owners.[2]

The February agreement required Leupp & Company to furnish at most 1,800 hides per month, well below the 5,000 per month Jay had tanned the previous year. For reasons of their own, the merchants might elect to restrict the tannery's output, while a policy of running the tannery at low volume ran counter to Jay's interests. For one thing, as he later asserted, he had endorsed notes for Leupp & Company amounting to more than $100,000 on the assurance that he would always have sufficient hides and leather from the merchants to cover his obligation. Moreover, the firm held Jay's own notes in payment for hides delivered. For these reasons, and because an expansionist policy suited his temperament as well, Jay moved to increase the tannery's output. In June he informed Leupp & Company that he would ship 21,120 finished sides during the next three months, and that he would begin working 8,000 sides per month "if you consider it safe & prudent to do so—if not—not." He argued that "we can tan 90,000 sides cheaper pro rata than we can 60,000 as the same tannery & machinery does it & the same men oversee it, so that the cost of supervision is not increased."[3]

Jay's logic was impeccable but the merchants, ever prudent, worried about the size of their commitment and the ability of the market to absorb those quantities. They were also bothered by indications that Jay was not confining his attention to the Gouldsboro tannery. This last concern was real enough. Long before his association with Leupp & Company, Jay realized that better opportunities awaited him as a dealer in the Swamp than as a mere technician toiling in the wilds of Pennsylvania. A few weeks after signing the February agreement he entered a partnership with a man named C. B. Freeman to operate a tannery in the latter's name. Gould's sole responsibility was to "attend & take charge of the money matters in New York and to take care of the Hides and Leather." It is likely he made similar arrangements with other tanners as well. By December he had opened an office at 39 Spruce Street in New York as "Jay Gould, Leather Merchant."[4]

As Jay expanded his interests, the merchants discovered that two-name paper could be a two-edged sword. Concerned that Gould was using the tannery to aid his private ventures, they ordered him not to issue drafts unless "advised specifically how & when to do so." Friction also developed over responsibility for the outstanding debts of Pratt & Gould and for some money owed Pratt himself. A period of polite bickering ensued, during which the merchants tried to curtail their liabilities for Gould's activities. These difficulties with Gould must have come as a surprise to Leupp and Lee. They were, after all, reputable businessmen steeped in the best traditions of the Swamp and set in their ways of doing things. To them Jay was an ambitious young man, promising but raw. Like all novices he was expected to learn his place in the pecking order, accept the conventional wisdom, and conform to traditional practices. He was, in short, an outsider, new to the game, for whom acceptance depended on his learning the rules and abiding by them.[5]

To their dismay, Leupp and Lee discovered that Jay had no intention of conforming to traditional norms or practices. Like Pratt they underestimated not only his audacity but his fertile intellect as well. There was in Gould a streak of originality that escaped or baffled conventional minds. He was far too precocious to play the part of the apprentice for very long. Like a good general, he mastered the rules not to follow them blindly but to know when and how to violate them if it suited his needs. That is what others found unsettling about him: he was unpredictable and he was always at it, always concocting fresh schemes or new variations on old ones. What Pratt, Leupp, and Lee all discovered about Jay was not that he was dishonest but that they had a tiger by the tail.

The growing schism between Gould and the merchants owed much to this clash of styles and values. Although Jay was spending more time in New

York, he did not neglect the tannery; in his absence two capable subordinates, J. A. Dubois and Abraham Steers, supervised the work. But the merchants continued to oppose expansion of the tannery's output and to fret over the extent of their liability in Jay's ventures. In their anxiety to resolve the dispute over the February agreement, they offered to buy Jay's share of the tannery. Jay countered with an offer to buy them out instead, and declared emphatically that the tanning contract was a separate issue altogether from ownership of the tannery.[6]

Unable to budge Gould, the merchants in August 1859 arranged for John B. Alley, a Boston leather dealer, to replace Leupp & Company as supplier for the tannery. The fact that Jay consented to this arrangement suggests that he wanted the connection with Leupp & Company dissolved as much as the merchants did. But as long as the merchants remained his co-partners in the Gouldsboro works, they were liable for whatever transactions he undertook in the company's name. Moreover, the tannery had on hand a large supply of hides furnished by Leupp & Company before the agreement terminated. The merchants wished their interest in these hides protected; for his part Jay wanted assurance that the hides in his possession were sufficient to cover the paper he had endorsed for Leupp & Company. In effect he considered the hides as collateral.[7]

The differences between the partners were genuine and difficult to resolve under the best of circumstances. Jay felt as uneasy about the merchants as they did about him. If they regarded him as unpredictable, he was finding them impossible to deal with for reasons that had nothing to do with business. The critical element in their relationship, the factor that proved decisive in shaping its outcome, was a matter wholly ignored in past accounts: the tragic descent of Charles Leupp into insanity.

Leupp's mental stability began deteriorating long before he met Gould. Some said his decline traced back to his wife's death sixteen years earlier, and to the anxieties he felt in raising three daughters alone; others attributed it to the prolonged strain of business. Whatever the cause, Lee noticed a marked change in Leupp as early as 1853, when his disposition grew nervous and subject to wild swings of mood between elation and depression. As his condition worsened, Leupp pulled back from close attention to business, leaving the everyday details in Lee's hands. For a time travel seemed to revive him, as did his release from the daily grind of business. But a series of reverses within the firm forced him to resume his former role. Reluctantly Leupp took up the reins again, going daily to his Ferry Street office and immersing himself in the detail work.[8]

The effort proved too much for him. In October 1858 violent shifts of mood again seized him. As the months passed, the periods of elation grew shorter and the fits of depression deeper, until their intensity became unrelieved. By the summer of 1859 Leupp had fallen prey to hallucinations. He accused close friends of spying on him, insisted that the police were following him, imagined that he had heart disease, and demanded that Lee feel his pulse several times a day to see if it was still beating.

Alarmed at Leupp's condition, the family debated anxiously what to do. "I saw that he was absolutely insane," Lee admitted later. "He had a fear of insanity, and said so often." Lee consulted a physician about obtaining the necessary certificate to put Leupp under restraint, but for some reason this was not done. It was a mistake that haunted the family for years. Possibly Lee found it difficult to convince outsiders of the true gravity of Leupp's state, which he described as "a continuous depression of spirits during the last six or eight months, although a casual observer might have thought him cheerful." To an intimate friend he confided ruefully:

> That Charles was unsound in mind I have seen for a good while—that he was positively insane, I have seen for some time, but I had reason to believe that he would never commit suicide. How terribly in this have we all (for several of us—friends & physicians were conscious of his unsoundness and watching him carefully) been deceived by the wonderful cunning of lunacy.[9]

This final stage of Leupp's decline coincides with the period of his partnership with Gould and raises the intriguing question of the extent to which they actually dealt with each other. Leupp's condition must have affected his dealings with Jay; it is also quite possible that Lee handled most of the business with Gould. The surviving correspondence contains only the firm's signature and, on rare occasion, that of Lee. Jay provides no insights, for in later years he was reticent on the subject of Leupp. As a result, both the nature and extent of their relationship remains an unsolved mystery.[10]

It is, however, possible to state accurately what did *not* happen between them. No bookkeeper rushed to Gouldsboro to investigate affairs there or returned with alarming reports about cooked accounts. Nor did Leupp personally undertake a journey to the tannery to confront Gould with his suspicions. There was no stormy interview between them, no shattering revelation that drove Leupp to end his life. His dealings with Gould, whatever they were, played no part in his tragic decision, and none of his family or friends ever suggested anything of the sort. On the contrary, Lee testified that "there was nothing in the facts of Mr. Leupp's personal, family, social, property or commercial condition, as far as I am aware, to justify any apprehension or distress on his part."[11]

On Wednesday, October 5, 1859, Leupp spent the entire day at his office attending to business, then went home and had dinner with two of his daughters. At about 5:30 William F. Cook, a close friend, called and was invited to sit down with them. Given a glass of ale, Cook lifted it and said, "Your health, Mr. Leupp." To his surprise Leupp sighed loudly, clapped his hand against his brow, and stared at the table, his face wreathed in what Cook described as an "unnatural expression." Later, when Cook started to leave, Leupp implored him to stay the night. Cook saw that his friend was agitated and agreed to remain. He followed Laura and Maggie Leupp to an upstairs sitting room, where the girls took up their sewing. Suddenly Leupp burst into the room and looked about wild-eyed. His frenzied manner startled but did not frighten the girls, who had grown accustomed to their father's peculiar behavior. Without uttering a word he kissed each girl fervently and rushed out the door, leaving his daughters to exchange puzzled glances.[12]

Returning to his room, Leupp fetched a double-barreled pistol, seated himself in a chair, opened his vest, pressed the gun against his heart, and pulled the trigger. Next day Laura Leupp, numb with grief, wrote a close family friend, "Poor father shot himself last night it was not his fault he was insane and has been so some time on that point—Maggie and I were with him he has intended it a long while don't blame him."[13]

Leupp's suicide dissolved the firm of Leupp & Company and left Lee struggling to hold the business together until it could be reorganized on a new basis. The task of sorting out Leupp's estate fell to his executor, William M. Evarts, and to Lee as representative of the family. For Lee this burden came at the worst possible time, when he still brooded over the tragedy and his part in it. At his best he may not have been equal to the work before him, and in the months that followed he was not at his best.

Jay found himself in no less a quandary. With the house of Leupp dissolved, no new hides would be forthcoming except from Alley, which meant that collateral for the paper endorsed by him was reduced to the leather currently in his possession. But title to that leather belonged in part to Leupp's estate along with his share of the tannery. The operation at Gouldsboro had become mired in a maze of legal intricacies from which Jay found it difficult to extricate himself. After being refused a statement on the condition of Leupp & Company's affairs, he offered to buy the tannery outright. On December 19, 1859, an agreement was signed that appeared to resolve all differences. Lee sold his and Leupp's share of the tannery for $60,000, payable at a rate of $10,000 per year. He allowed Jay $10,000 for tanning out stock credited to him. Sale of the tannery was made subject to the approval of Evarts.[14]

Far from settling the dispute, the December 19 agreement touched off another round of bickering in which the mood grew progressively uglier and the genre shifted abruptly from tragedy to farce. During his business career Jay's roles covered a wide repertory; here he was about to make his debut in opera buffa, a genre he would perfect in his years with the Erie Railroad.

Evarts raised objections not to the tannery sale but to the provisions treating the hides still at Gouldsboro. Apparently Lee too had reservations about the terms he had accepted only days earlier, for he informed Gould that he considered the agreement nullified by Evarts's objections. Jay reminded Lee coldly that the agreement gave Evarts the right to approve only the tannery sale, which, he insisted, was "an entirely separate matter from the tanning out of the stock." The squabble led to the signing of another agreement on December 28 which confirmed the original terms of the tannery sale and obliged Gould to pay for two-thirds of the "personal property" involved, the amount to be deducted from the $60,000 purchase price. Curiously, the new agreement did not even mention the hides at Gouldsboro.[15]

Neither agreement clarified the crucial question of who actually controlled the tannery, and therefore the disposition of its leather, until the sale was formalized. Lee claimed that he represented two-thirds of the ownership on behalf of Leupp & Company and so was entitled to dictate policy. Not so, Jay countered in his precise way. The tannery belonged to a special partnership between himself and Leupp & Company formed under the name of Jay Gould. Leupp's death had dissolved the partnership, leaving Jay to settle up the tannery just as Lee was doing with the leather firm. Until settlement was made, Jay argued, he remained in charge of the tannery's business.[16]

It was a nice legal point, but neither side was in the mood for legal niceties. The tannery continued to ship finished leather to Leupp & Company, but at a much reduced rate. Uneasy about the amount of his paper held by the firm, Jay pressed Lee to settle their account once and for all. Lee put him off with one excuse after another before promising at last to ready the account by Saturday, February 25, 1860. Jay summoned J. A. Dubois, his supervisor in Gouldsboro, to New York for the meeting. On Tuesday the 28th they met with Lee, who informed them that he had no intention of preparing the account but wished to see them again the following day. Perplexed, they returned on Wednesday only to be told that Lee was ill.[17]

In fact Lee had left town to execute what he regarded as a decisive stroke. His anxiety over the leather at Gouldsboro was fast boiling up to the edge of obsession, for reasons that are hard to fathom. Terms for the sale of the tannery had been arranged, and the value of the leather in dispute, though substantial, was small change on the overall balance sheet of Leupp & Company. Perhaps Lee needed all the change he could muster, for Leupp's death came at a time when the firm's affairs were in a bad way. When it became obvious that liquidating the leather house would consume much of Leupp's fortune, Lee grew concerned about salvaging what he could for the three daughters.[18]

From Gouldsboro came reports that leather was being mistreated, that Gould had been absent for several months, and that the employees had not been paid and were growing mutinous. Alarmed, Lee consulted a Scranton lawyer named Willard and together they decided to occupy the tannery. On February 29, while Gould and Dubois waited in his office, Lee slipped out of town and hurried to Gouldsboro, where he found young Abraham Steers in charge. That evening Lee called the employees together to hear a statement denouncing Gould and explaining why he, as rightful owner, had come to take possession. After accusing Gould of mismanagement and misappropriating $25,000 sent from the leather house to pay tannery debts, Lee invited anyone in the audience to answer his charges.[19]

As naïve as he was loyal to Gould, Steers refuted Lee's assertions and for his pains was pitched unceremoniously out the door. When the meeting broke up, only about fifteen men sided with Lee. If the employees were mutinous, they did not exactly flock to his banner. Lee stroked his whiskers nervously. With so few men to defend the tannery, he dispatched Willard to Scranton to recruit "armed constables," a polite term for hired thugs. Early on the morning of March 6 Willard returned with ten men and the promise that more were on the way. Lee ordered his men to barricade the tannery and deployed them inside the building. At thirty-two the gentleman merchant braced himself for his first taste of combat.

After learning what Lee had done, Jay consulted his lawyer, Andrew H. Reeder, a man who knew something about frontier warfare from his days as governor of "Bleeding Kansas." Reeder advised him to withhold hides from Leupp & Company until the account was settled and to reclaim possession of the tannery from Lee. Jay reached Gouldsboro on the evening of March 5 and went at once to the tannery, where he was met by Lee and his men. Lee announced that he had taken lawful possession of the works and told Gould to leave. Later Jay claimed that his life had been threatened if he did not withdraw. Next morning a crowd of more than two hundred employees and residents from the area gathered outside the tannery office. Mounting the steps so he could be heard, Jay pleaded his case and asked for their help. To his surprise most of the men in the crowd volunteered to join him in ousting Lee. The remarkable show of support left little doubt as to where the loyalty of the employees lay. They didn't know Lee but they knew Gould and evidently trusted him. He hadn't hired a single Hessian or even asked anyone to come to the meeting.[20]

Lee and his men spent the night huddled inside the tannery, surrounded by hides hanging from beams and stacked on the floor. Jay sent an envoy urging surrender of the premises to avoid bloodshed. When Lee refused, Jay divided fifty men into two companies, sent one to the rear of the tannery, and led the other to the front entrance. He was not armed and admitted to "never having owned, loaded or fired a pistol in my life." But excitement flashed in his dark eyes and a martial spirit stirred in him, as if he were acting out a boyhood fantasy, a scene similar to those described so vividly in his Delaware County history. Not since the antirent war had he experienced such a moment.

While Jay's men ripped away the boards covering the front door, the rear unit demolished the padlock on the back door. Inside, Lee's troops braced for the assault. Suddenly both doors gave way and Gould's men rushed into the building. As both sides fired wildly in every direction, most of the balls glanced off hides or embedded themselves with dull plunks. Amid the screams and howls, the swirling smoke and stench of vat liquor, the defenders broke and ran, flinging themselves out the doorway or through windows with the attackers in hot pursuit. On the second floor the Scranton toughs swallowed hard and leaped out of windows to make their escape. The battle was over within minutes. Only the hides had taken heavy casualties; three men were wounded, one seriously, and Lee caught a piece of buckshot in the finger. He was taken prisoner and freed on parole with instructions never to return. The rout complete, Jay ordered his men back to the humdrum business of tanning. A short time later the press arrived, surveyed the battlefield, and reveled in the spectacle. One local paper emblazoned its story with these headlines:

CIVIL WAR AND THE LEATHER TRADE
ITALIAN WAR ECLIPSED
GREAT FIGHT AT GOULDSBOROUGH
GEN. GOULD VICTORIOUS AND MARSHALL LEE A PRISONER OF WAR

For Luzerne County residents, the Gouldsboro war soon became the stuff of folklore. For Jay, it wasn't much of a war, but it was the only one he had. When the real item came along a year or so later, he showed no disposition to resume his military career.

When the shooting stopped, the conflict moved into the courts. Both sides slapped suits on the other, charging trespass and illegal seizure of property. Convinced that Gould would never surrender the leather, an embittered Lee decided to sue for it. Somehow Jay got wind of his plan and shipped twelve hundred hides to Naglesville, where he hid them in a barn. Gould's action only confirmed every low opinion Lee held of him, as well as his worst fears about the leather. He brought suit to replevy, then asked the court for an injunction forbidding Gould from disposing of any property belonging to the tannery and for a receiver for the tannery.[21]

The court granted the injunction but refused to appoint a receiver. Not until August 21 did Lee and Gould agree to a procedure for settling up the tannery. Jay relinquished all claim to the leather and was released from his earlier agreement to buy Lee's share of the tannery. Lee won a judgment of $3,500 but only after the case plodded through the courts for seven years. By that time the tannery itself had long since been abandoned. In December 1868 Jay sold his interest in the property to Lee for one dollar. For Lee it proved a Pyrrhic victory. He failed to restore the leather house to its former prominence, and the firm passed into the hands of his younger brother. The long fight also blasted Jay's chances of succeeding in the leather business. Given the venerable standing of Leupp & Company and the tragic circumstances of Leupp's death, it was inevitable that the Swamp's Establishment would take Lee's side in his struggle with a brash outsider. Jay saw quickly that he must look elsewhere for his future.[22]

On the face of it, Jay appears to have failed at his first major enterprise. The tannery was defunct and he had lost his toehold in the industry. Yet failure is a slippery thing to measure. While no records are available, there is little doubt that Jay came out of the episode in better financial shape than when he began it. The tannery may have lost money but Jay did not, largely because the Gouldsboro experience opened his eyes to investments in coal, land, timber, and other resources. For the first time Jay gleaned an understanding of the importance and profitability of peripheral investments. It was a lesson he would put to good use.

His education did not stop there. Just as in later years Jay was careful never to be blinded by success, so did he possess the ability to learn from failure, from ventures gone awry. The long struggle with Lee, for example, taught him the utility of lawsuits, an insight he developed to the point where he became, in the words of Julius Grodinsky, "probably the most successful litigant in American history." In every aspect of business Gouldsboro served Jay as a training school, its curriculum grounded in practical knowledge. As always he showed himself to be an attentive student. The growth he achieved during these years enabled him to matriculate to more advanced realms of finance.[23]

But the cost of tuition had been high. Much has been written about what Jay did to Pratt and Leupp, nothing about what effect these relationships had on him. A strain of peculiarity complicated both partnerships: the one with an unpredictable if well-meaning eccentric and the other with a man caught in the throes of insanity. The difficulties of working with such men as partners would have taxed the skills of anyone, let alone a young man in the formative stage of his career. To some extent the outcome must have scarred and hardened Jay, left him wary and skeptical beyond his years. He emerged from the experience without bitterness but shorn of still more of his youthful idealism. The school of hard knocks had impressed its lessons on him, and they would serve him well in the business jungle he was about to enter.

Above all, his tanning career put Jay in the one setting where his highest ambitions could be realized. At the tender age of twenty-four Jay was about to leave the leather industry, but he was in New York to stay.

II

GRASPING 1861–1878

7

Fierce Extremes

The impression is that after the tannery episode Mr. Gould was pretty thoroughly impoverished. However that may be, his marriage put him on his feet again, for his father-in-law was a wealthy merchant, and though it is said that he opposed his daughter's marriage with the hero of Gouldsboro, yet he soon became reconciled.

—Henry D. Northrop, *Life and Achievement of Jay Gould*

Mr. Gould . . . ingratiated himself in the favorable esteem of one of the grocery merchants with whom he had done business. The merchant took him to his house to board and Mr. Gould fell in love with his handsome daughter.

—Henry Clews, *Fifty Years in Wall Street*

The nimble Gould was busily engaged in buying gold and selling the dollar since the opening of the Civil War. . . . He was reputed to have set up machinery whereby information hurried news of victories or defeats to him by telegraph almost a day ahead of his rivals.

—Matthew Josephson, *The Robber Barons*

TWO CONCERNS dominated the rest of Gould's life, business and devotion to family. While these twin passions often conflicted with and cannibalized each other, no other interest even remotely approached their claim on his attention. It was common for businessmen to separate their lives into two spheres, the commercial and the domestic, but few drew their nourishment so deeply from both or clung so tenaciously to them as did Gould.

Both these passions trace their origins to Jay's early years in New York, when he mastered the intricacies of finance and married the only woman he ever loved. Unfortunately it is the period of Gould's life about which we know least, the point at which obscurity simply swallows him. Between the time he settled in New York and October 1867, when he first appeared on the board of the Erie Railroad, Jay left few footprints on the public record. His splashy debut in the financial arena was made all the more sensational by the fact that he seemed to have burst onto the scene from nowhere.

Of course that was not the case. Like many an overnight celebrity he attained notoriety after years of toil and struggle, during which he was as invisible as any apprentice. In coming to New York he removed himself from the eyes of local chroniclers and vanished into the horde of ambitious young men seeking their fortune. During that time his name was absent from those newspapers which a few years later would flog him unmercifully for his sins. None of the financiers who in old age reminisced about their days in Wall Street spoke of Gould's early career from first-hand knowledge for the simple reason that he was unknown to them.[1]

To the casual observer Jay must have seemed no more than another mullet venturing rashly into shark-infested waters. He was the smallest fish in the largest of seas, an outsider in a place where being on the inside counted for everything. He had little money and no connections, the two surest avenues to Wall Street's inner circle. What reputation he possessed had been tarnished by the imbroglio with Leupp & Company. Lacking these assets, he was forced to rely on his wits, his intellect with its marvelous capacity for growth, and his fierce desire to succeed. Between 1860 and 1867 Jay acted out the role later enshrined by Horatio Alger. Driving himself to the limits of endurance, he worked hard, ingratiated himself with those who might advance his interests, probed everywhere for new opportunities. Ever diligent, always hustling, he willingly paid whatever price was necessary to push himself along. As before he absorbed every experience, storing its lessons away for future use.

Like any good Alger hero, Jay was rewarded for his perseverance. By 1867 he had become a consummate financial technician, the likes of which Wall Street had rarely seen. Nobody knows how he did it or at whose elbow he learned; most likely he taught himself as he had done with surveying. The two had more in common than one might suppose, for mathematics lay at the heart of both. Finance posed the kinds of problems that appealed to his intellect and to which he brought awesome powers of concentration. Those who worked closely with Gould came to understand this quality in him. None described it better than the eminent lawyer Thomas G. Shearman:

> When intensely interested in any matter, he devoted his whole concentration of thought upon that one thing, and would seem to lose interest in things, often of greater pecuniary importance but of not so much commercial fascination. He loved the intricacies and perplexities of financial problems.[2]

This phenomenal grasp of mechanics was a talent born of necessity. Since Jay possessed large ambitions and meager resources, he cultivated the art of controlling huge enterprises with minimum holdings, utilizing not only equity control but funded debt, the proxy market, floating debt, contractual flaws, receiverships, and especially legal technicalities. Technical mastery serves most men as a useful tool; in the hands of one as inventive as Gould it became a formidable weapon. And he left nothing to chance. Where others were content to rely on native genius or instinct, Jay did his homework. His ability to extract an advantage, to spring a trap or exploit a weakness, stemmed from his habit of studying every venture in exhaustive detail.

In his approach to business Jay was not only thorough but imaginative. From this combination evolved his knack for doing the unexpected. Unorthodox responses to apparently straightforward situations helped mask his true intentions, bred confusion and uncertainty from which he might extract some advantage. He learned too how to judge the character and motivations of other men. In an age when most business was done through personal contact and the rules were at best slippery, nothing was more important than knowing who could be trusted how far. Jay revealed an uncanny instinct for discerning a man's true colors, for knowing who would fight and who would not, and adapting his tactics to each temperament. Here as elsewhere Jay was a realist. If he grew adept at deceiving others, he was careful never to deceive himself. He did whatever he felt had to be done and did not flinch, however difficult or unpleasant it might be. No cravings of ego, no desire to parade himself, his appetites, or his possessions before others, deflected him from the business at hand. Where others sought the limelight he was content to dwell in shadows, a keen student of the madding crowd but never part of it. The frenzied rough and tumble of Wall Street did not sweep him up in its violent, irrational pulsations; on the contrary, it taught him the virtue of patience, the value of steady nerves under fire. "I avoid bad luck by being patient," he once declared. "Whenever I am obliged to get into a fight, I always wait and let the other fellow get tired first."[3]

The business style that emerged during these years was an astringent one, as lean and compact as Gould himself. There was about it a strong flavor of secretiveness, the silent running of a loner forging his own way upstream. Early in his career Jay formed the habit of telling no one about an operation except those directly involved, and then only as much as he wanted them to know. His passion for secrecy, what critics called his sly and furtive ways, applied no less to confederates than to rivals. Even his lawyers were left in the dark until the need for their services arose. A chronicler of the firm that handled most of Gould's business complained that "he never disclosed to his lawyers, or to anyone else except the particular ally chosen for the particular venture, what he was about to undertake—and never, to anyone, the full extent of what he had in mind."[4]

Gould's secretiveness became one aspect of the darker side of his business style. During these years his attitudes underwent as radical a change as his methods. Qualities that were to blacken his reputation began pushing to the fore, especially a view of ethical and legal niceties that bordered on amorality. He showed himself willing to employ bribery and other shady tactics to gain his ends. It was as if his urgent desire to succeed had blended with the harsh lessons of recent experience to glaze his ambition with a sense of desperation. He was struggling to survive in an arena where no mercy was wasted on those who failed. If he faltered, the thundering herd would trample him underfoot without a backward glance. As an outsider, a sort of Wall Street urchin, he could not afford the luxury of moral propriety enjoyed by those who belonged to the establishment. Nor did he much care about such things. He recognized the game of business for what it was and played with few illusions and fewer pretenses.

Gould's behavior resulted from the interplay between his own needs and the shaping influences of an environment caught in the throes of transition. His willingness to bend the rules when it suited him derived in part from the fact that he entered Wall Street at a time when the rules had become unsettled and malleable. It is no accident that his formative years were those of the Civil War, or that the most sensational and controversial episodes of his career belonged to the immediate postwar period.

As Henry Adams discovered, education has a way of transforming one's character in ways that cannot be predicted. In that process much depends on the school itself, on the values it espouses and the concessions it demands. Much depends too on one's capacity for adapting to an environment in which the value system clashes with past experience. Some are bruised or outraged by that confrontation and recoil from it in disgust. Others shrug and make their peace with things as they are. That was Jay's approach. It was not that he lacked principles but rather that he had long since learned the knack of doing what the Romans did. In this instance his set of Romans turned out to be a very rough crowd.

Wall Street was never much to look at, a narrow strip that began east of Broadway, opposite the site of Trinity Church, and wound its crooked way down to the East River. Cynics of a later age found it altogether fitting that this main artery of speculative finance should start at the doorstep of a church and plunge downhill to a watery grave.

During the early nineteenth century New York replaced Philadelphia as the banking center of the nation and Wall Street emerged as its financial heart. In 1817 a group of thirteen brokers and seven firms founded what became the precursor of the New York Stock Exchange. The Exchange operated as a club of the elite, guarding its prerogatives jealously and admitting few new members. Because of its limited membership and lack of continuous trading, the real action took place on the street, where anyone could deal in shares all day long. Unlike the Exchange's roster of dignified, well-connected gentlemen, the curb brokers were a motley assortment ranging from the genteel to sharpsters in search of a quick killing.[5]

The decade of the 1850s wrought a striking transformation in Wall Street. Two innovations, the telegraph and the rise of intercity express service, gave other major eastern cities access to the New York market. Newspapers throughout the country began featuring business news, including price quotations from Wall Street. The discovery of gold in California, coupled with a boom in railroad construction, flooded the street with mining and rail issues. For a few ebullient years the financial district caught the prosperity fever that infected most of the nation. The Panic of 1857 shattered this speculative mania. So many of the brokers dealing in mining shares failed that the Mining Exchange closed down. William Street, the crowded, bustling haunt of the curb brokers, wore the look of a ghost town.

Amid the debris a generation of older men were swept out of Wall Street, never to return. Their departure ushered in a new era dominated by younger brokers, brash, aggressive men who flocked into New York from every corner of the nation after cutting their teeth on ventures at home. Eager to parlay their stake into a fortune, they played a less polite game of cutthroat. Where their predecessors bet thousands, they risked millions with the verve of hardened gamblers. In combat they asked little quarter and gave less. The traditions of the Street meant nothing to them. Where the older generation at least paid lip service to rules they often violated, the new breed simply ignored them.

These were the men who dominated Wall Street when Gould arrived on the scene. Even in ordinary times they would have created profound changes in the tone of business, but their rise to prominence occurred at a most extraordinary time. The outbreak of war plunged the country into a period of abnormality that proved an ideal spawning ground for businessmen and speculators alike. A host of enterprising young men ignored the call to arms and devoted their energies to supplying the government with uniforms, weapons, blankets, shovels, wagons, foodstuffs, transportation, and other goods and services. In the bargain they earned for themselves not only fortunes but experience that would establish them as leaders of the postwar era.[6]

War bred uncertainty, and uncertainty was to Wall Street what ambrosia was to the gods. "Along with ordinary happenings, we fellows in Wall Street had the fortunes of war to speculate about and that always makes great doings on a stock exchange. It's good fishing in troubled waters." So claimed Daniel Drew, the most lethal shark cruising those waters. The greatest uncertainty of all was the course of the war itself. Every market rose or fell with the tide of news from the front, inspiring some crafty manipulators to find ways of obtaining news before it reached the Street. A favorite ploy was to bribe a telegraph operator or war office clerk. Gould was credited with devising such an arrangement but there is no evidence that he actually managed it.[7]

The task of financing the war opened giddy new realms of speculation. When the government resorted to issuing greenbacks that were not convertible into gold, it set the stage for rampant speculation on the relative value of the two currencies. Since greenbacks were only as good as the government that printed them, they fluctuated with every twist of fortune on the battlefield. Their worth was reflected in the price of gold, which soon became a barometer for the Union war effort. In October 1864 the New York Gold Exchange was formed and within a year occupied a building on New Street. The Gold Room acquired its own peculiar madness as a hippodrome where tumultuous spectacles played themselves out daily. Observers cringed before the turmoil in what one broker described as "a cavern, full of dank and noisome vapors, and the deadly carbonic acid was blended with the fumes of stale smoke and vinous breaths." Journalist Horace White was even less flattering:

> Imagine a rat-pit in full blast, with twenty or thirty men ranged around the rat tragedy, each with a canine under his arm, yelling and howling at once, and you have as good a comparison as can be found. . . . The furniture of the room is extremely simple. It consists of two iron railings and an indicator. The first railing is a circle about four feet high and ten feet in diameter, placed exactly in the centre of the room. In the interior, which represents the space devoted to rat killing in other establishments, is a marble cupid throwing up a jet of pure croton water. The artistic conception is not appropriate. Instead of a cupid throwing a pearly fountain into the air, there should have been a hungry Midas turning everything to gold and starving from sheer inability to eat.[8]

On the stock market the same frantic excitement prevailed. The wartime demand for transportation thrust railroad stocks to the forefront of the market, a position they would hold for decades. Already the Street was reeling beneath what one observer called "the struggle of a few strong men for railroad supremacy." From the heat of battle over rail securities emerged those speculators whose exploits became the stuff of legend: Cornelius Vanderbilt, Daniel Drew, the Jerome brothers, Anthony Morse, Henry Keep, William H. Marston, and a dozen others.[9]

"The war, which made us a great people, made us also a nation in whom speculative ideas are predominant," observed James K. Medbery. On Wall Street the anything-goes, get-rich-quick fever raged in epidemic form. From dawn to dusk the curb swarmed with brokers feeding on dreams like locusts devouring a prairie. At nightfall they descended on the rooms and corridors of the Fifth Avenue Hotel until the disgruntled proprietor evicted them as a nuisance. Despite the swollen volume of trading, the Board persisted in limiting its membership and confining its auctions to two sessions a day. As a result the bulk of transactions occurred among nonmember brokers on the curb or in some place established to permit continuous trading, such as Goodwin's Room. By 1864 the volume of business prompted the creation of a new organization, the Open Board of Stock Brokers.

The Open Board leased space on Broad Street and opened what became known as the Long Room, a cockpit 145 feet long and 45 feet wide in which trading ran nonstop from 8:30 A.M. to 5 P.M. The Open Board charged members a $500 fee and, unlike the regular Board, permitted the public to witness the spectacle on the floor. In short order the volume of business in the Long Room overwhelmed that done by the regular Board. According to Medbery, "In the Regular Board an average of seven million dollars' worth of shares are sold in a day. In the Long Room, brokers roughly estimate that ten times that amount is bought or sold between sunrise and nightfall." The Long Room continued to be used after the two boards consolidated in 1869; it was there that Gould's brokers would stage the great Erie imbroglio.[10]

"This life of the Wall Street man," declared William W. Fowler, "is . . . a life of fierce extremes. One day he is . . . agitated by all the hopes and fears, and torn by all the emotions and passions, engendered by the thirst for greedy riches;

the next day the struggle is over, and he is plunged, at once, into the opposite extreme of stillness and repose."[11]

For the small broker, the newcomer or outsider, these fierce extremes wore an edge of desperation. Success demanded from them a dedication worthy of the priesthood. The routine of such a broker varied little from day to day. He rose early, gulped breakfast, and rode downtown while scanning market reports in the morning papers. He traded until noon, swallowed a hurried lunch, and plunged back into the maelstrom until five or six o'clock, when he rode uptown for dinner, discussing stocks with his fellows up to the moment of their parting. After dinner he might haunt the evening market until sleep plucked him from the treadmill for a few short hours.

Those who stalked opportunity as lone scouts dared not ease their pace lest they lose the trail. In the Wall Street jungle mere survival required all the wits and energy a man could summon. It was a jungle belonging to natives who resented the intrusion of outsiders. If the tribal hierarchy treated lone adventurers (or "guttersnipes," as they were called) with disdain, it regarded successful intruders with undisguised hostility. The day would come when Jay Gould felt the lash of that resentment as no other outsider on Wall Street ever had.

Amid this teeming multitude of fortune seekers Jay made his way with the stealth of a cat, learning what he could, sometimes falling but always landing on his feet, inhaling the coarse, opportunistic atmosphere of wartime until he could breathe its foul air as easily as any other. Like a newcomer to a slaughterhouse he grew inured to the sight of blood, the butchering of carcasses. Wall Street had odors of its own, more subtle and insidious, but after awhile they offended Jay no more than had the stench of the tannery.

Jay's early years on the Street were rugged going even for one of his determination. He needed not only education but direction, a clear focus for his energies. While learning the intricacies of finance he was also searching for the line of business best suited to his talents and ambitions. Of all the misconceptions that persist about Gould, perhaps the worst is the notion that he was primarily a speculator. Although his exploits in the market were the stuff of legend, speculation was for him not an end but the means by which he obtained funds for other ventures. The market was in fact his only source of capital during the years when he was an outsider on the Street with no money of his own, no family ties, no prominent associates or links with the financial establishment. Later he developed other sources of capital, but as the scale of his enterprises increased he was compelled to return again and again to the market for the funds he needed. When at last he declared himself out of the market forever in the mid-1880s, the notoriety of his operations had forged an indelible impression of him as a speculator.

What little we know of Gould's early career in New York confirms the notion that he was searching for the right endeavor to take hold of and develop. It was only natural that he began with the leather industry. His first major venture was a partnership with three other men under the name Wilson, Price & Company. The firm owned half the stock of another company through which it bought hides and sold leather. Gould's partners were businessmen of considerable means, probably acquaintances made during his forays in the Swamp. The partnership was formed in January 1859, the same month Leupp and Lee bought into the Gouldsboro tannery. The controversy that later arose over Gould's outside activities probably revolved around his interest in Wilson, Price & Company.[12]

When Jay entered the new partnership, he was estimated to be worth about $80,000. Besides his share in the Gouldsboro tannery, he owned nine thousand acres of land in Pennsylvania and held title to the bark on another thirty thousand acres. Although his standing as a businessman was still unproven, it was said of Jay as late as February 1860 that "he is a smart enterprising young man of good character and habits, reliable in his statements." During the troubled months after Leupp's death Jay devised a plan that suggests how high his ambitions had begun to soar. To buy the tannery back from Lee, he organized a joint stock company with an authorized capital of $150,000 and applied to the legislature for a special charter. Apparently he intended to use the Gouldsboro Manufacturing Company to expand and consolidate his tanning interests, with Wilson, Price as his agent for procuring hides and selling the leather. It was an impressive scheme, but the tannery war destroyed whatever chance it had to succeed.[13]

There is no doubt that the confrontation with Lee was decisive in forcing Jay out of the leather business. Shortly after the fracas a credit reporter warned of "unfavorable reports in circulation respecting Gould which have greatly impaired if not destroyed his credit in the Swamp." The tannery war also delivered a fatal blow to Wilson, Price, which dissolved a few months later with a loss of $60,000. Jay found himself with little capital and no business prospects. The next year or so marked the nadir of his career. In March 1861 it was reported that he "has not settled his affairs & has no particular location. Is not known to do any business, nor is it ascertained whether or not he is worth anything."[14]

In casting about for prospects that dark winter, Jay happened on one that was to be of enormous significance to his future. His former partner, D. M. Wilson, owned $50,000 worth of first mortgage bonds in the Rutland & Washington Railroad, a small line that had never done very well. In the gloom pervading Wall Street after secession the value of its securities shrank to the point where Wilson was willing to sell out for ten cents on the dollar. Years later Jay told a Senate committee what he did about it:

> I left everything else and went into railroading. That was in 1860. I took entire charge of that road. I learned the business and I was president and treasurer and general superintendent, and I owned the controlling interest in the road. . . . I gradually brought the road up and I kept at work, and finally we made the Rensselaer & Saratoga consolidation, which still exists. In the mean time my bonds had become good, and the stock also; so that I sold my stock for about 120.[15]

Here, as in all his public testimony, Jay omitted more than he revealed and reduced complex matters to beguiling simplicities. The Rutland did mark his entry into railroading and provided him a classroom in which to study every aspect of the business. He did not manage the road in absentia but spent long weeks in Rutland. By the time he leased the road in 1865, he had acquired a solid education as well as a handsome profit. But it is highly unlikely he "left everything else" during the war years. Whatever the Rutland taught him about railroading and finance, it could not tutor him in the mysteries of speculation. That subtle art had to be learned on the curb, where Gould doubtless spent more time than he cared to admit after his reputation as a speculator became a sore point with him.[16]

From the Rutland he gleaned the vast potential of railroads for making money. His experience on the Street taught him a variety of ways to gain control of a rail company on a shoestring. He emerged from the Civil War period a seasoned Wall Street warrior, wise beyond his years, fired with ambitions that dwarfed those of ordinary men and possessed of the audacity to attain them. His timing could not have been more perfect. The nation stood on the threshold of a boom era in railway expansion that would continue for the rest of his lifetime. Like others of his generation, Gould had the good fortune to be in the right place at the right time.

None of this was apparent in 1865. To the inner circle of Wall Street Jay remained an obscure outsider. In his struggle to establish himself Gould made the acquaintance of prominent men but was not accepted by them. The lone exception was a merchant who not only assisted Gould's career but also gave him, with some reluctance, his daughter as a wife.

To those around her, Helen Day Miller embodied the best of Murray Hill society. She had been reared in the strict, conservative tradition, molded by its conventions into a staid, well-mannered girl who wore propriety like a corset and knew little of the world beyond the narrow confines of her own set. Life for her, as for other young women in her circle, was a ritual of duties imposed by obligations to family and friends, church and good works. Her days were details of change fluttering against a landscape as constant and timeless as the seasons.

Nothing in Helen's looks or manner shouted for attention. A small, slender woman of rather plain appearance, her large brown eyes had a gentle, timid cast that offset a small chin and a somewhat prominent nose. She wore her long brown hair piled atop her head in ringlets or pulled tight and coiled into a bun. Her mouth was large and winsome, with full lips and sharply etched corners on which there hovered a faint, tentative smile. Around other people she was reserved and soft-spoken, with the shy, nervous manner of a bird on the verge of taking flight. Her kind and modest disposition made her the most dutiful of daughters in her parents' eyes. She had never done anything extraordinary and no one expected that she ever would.[17]

In many respects Helen's nature was remarkably similar to Jay's, yet they lived in vastly different worlds. Perhaps the most intriguing mystery in Gould's life is how, as a young man who knew nothing of Murray Hill society, he managed even to meet, let alone court, a woman as painfully reserved and inexperienced as himself. None of the accounts written on the subject amount to anything more than romantic blather. Since the Millers lived on East Seventeenth Street, not far from the Everett House where Jay stayed, it is at least possible that they met by accident. More likely they were introduced by Helen's father, whose activities on Wall Street brought him into contact with Jay.[18]

The redoubtable Daniel Miller had been, like his father before him, one of New York's most prominent merchants. Early in 1853 he retired from his produce firm, took an office in Wall Street, and devoted his time to managing his various interests. A credit reporter described him as "more than ordinarily bent on making money," a quality that gave him at least something in common with Gould. Miller had long dabbled in railroads, especially the Harlem and the Erie, and it is possible he held some interest in the Rutland. However they met, Jay apparently impressed Miller with his business acumen if not his qualifications as a son-in-law.[19]

The marriage of Jay and Helen is no less intriguing a mystery than their meeting. Early biographers, scrounging desperately for human interest angles in Gould's life, came up with a fanciful romance. In their version Miller adamantly opposed the marriage, whereupon the couple went to the home of a nearby minister and were wed without fanfare. The notion of a secret marriage despite parental objections has persisted ever since and even drawn a counter assertion that the wedding took place in Miller's home before some fifty people and was followed by a reception for four hundred guests.[20]

The choice here is between one fairy tale and another. In reality Gould as a suitor confronted Miller with a nice dilemma. No Murray Hill patriarch wished to see his daughter marry outside her own set, and certainly not to a social nobody. However, Jay was a bright young man, full of spunk and ambition, with a promising future. There is reason to believe that Miller not only admired Jay's talents but also liked his quiet, unassuming manner and his utter lack of the vices and excesses common to many men in Wall Street. For his part Jay had a way of ingratiating himself with anyone who could be useful to him. In the end Miller eschewed pedigree in favor of more practical concerns. Besides making Helen a good husband and provider, Jay would be a valuable asset to Miller's business interests.

The marriage took place on January 22, 1863. A surviving printed invitation to the ceremony demolishes the tale of a secret elopement. No report of the affair appeared in the newspapers, but it was not the custom of Murray Hill society to parade its ceremonies before the public eye. In 1897 one of Gould's sisters confirmed under oath that she had received an invitation to the wedding, and that the Gould family had known of the event "for months ahead." After living for awhile with Helen's parents, the Goulds eventually settled into a large brownstone at the corner of Fifth Avenue and Forty-seventh Street in 1870. No sooner had the Goulds occupied their new home at 578 Fifth Avenue than the Millers joined them as neighbors by taking up residence at 518 Fifth Avenue.[21]

By that time Helen had already given birth to four of the six children she would bear: George Jay (1864), Edwin (1866), Helen (1870), and Howard (1871). As the eldest son little George became the heir apparent, to be pampered and spoiled even more than the others. Jay doted on his children as he did on Helen. For the rest of his life he devoted himself to wife and children with a regularity that put even Victorian convention to shame. Family had always been important to Jay and grew more so as the years passed, as if he were determined to shield his own from the privations and hardships he had endured as a boy. He protected them and took refuge with them as the world grew more bitterly hostile toward him. Because they were the joy of his life, he did everything in his power to fill their lives with joy.

The one thing he could not give them was more of himself, because there was nothing left to give. During the postwar years Jay's absorption in business was maniacal in its intensity. Every day he marched downtown to wage his wars of finance, and every evening he returned with the life sucked out of him only to turn to his correspondence or other work after the children had gone to bed. As his successes piled up and the scale of his operations increased, the drain on his time and energy grew even more demanding. His constitution had never been the same since his near fatal illness, and all his life he would have to battle frailties of health. As a young man he had learned to get the most from his limited strength, to drive himself to the edge of exhaustion, and to harness everything he did with remarkable self-control. It helped to strip his life of wasted motion, but still he could not find enough time or energy to go around.

Business and family. They were not only the twin pillars but also the fierce extremes of his life, and it was his frustration always to be spread too thinly between them.

8

Ascendance

It is scarcely necessary to say that he had not a conception of a moral principle.

—Henry Adams, *Chapters of Erie and Other Essays*

He became invested with a sinister distinction as the most cold-blooded corruptionist, spoliator, and financial pirate of his time; and so thoroughly did he earn this reputation that to the end of his days it confronted him at every step, and survived to become the standing reproach and terror of his descendants. For nearly half a century the very name of Jay Gould was a persisting jeer and byword, an object of popular contumely and hatred, the signification of every foul and base crime by which greed triumphs.

—Gustavus Myers, *History of the Great American Fortunes*

AT THIRTY-ONE Jay had shed his boyish appearance. The hairline had already begun to recede, making his large forehead seem even more prominent. His strong jawline had vanished beneath a thick, wiry beard which, like many Civil War officers, he may have grown to make himself look older. His dark, cavernous eyes were as piercing as ever; they took in everything and reflected nothing. Often their gaze fixed on some point in space, as if that emptiness held the key to whatever question tumbled in his mind. Their power contrasted strangely with his musical voice, delicate frame, and tiny, almost effeminate hands and feet.

By 1867 he was on his own in every sense of the word. His father had died the previous year, his uncle Jason two years before that. His sisters had all married and scattered. Sarah was in Canadensis with her growing brood of children. Bettie had married Gilbert Palen and gone to live in Tunkhannock. The Reverend Asahel Hough had taken Anna to the wilds of Montana, where they endured incredible hardships as missionaries among the miners and Indians. Young Abram was struggling to make a start in business. At one time or another each of them would require Jay's assistance, and never did he disappoint them. He had become head of the larger family as well as of his own.[1]

The Erie Railroad had a checkered history long before it felt Gould's touch. Chartered in 1832 as a grand project to connect ocean traffic at New York City with the Great Lakes, it floundered for two decades as a political football among local interests. When the road was finally completed in 1851, it ran from Pierpont to the obscure village of Dunkirk on Lake Erie, without access to Buffalo at one end or New York City at the other. The interplay of local politics had forced the company to occupy a route through southern New York that had poor grades, was expensive to maintain, and could not tap rich sources of traffic in northern Pennsylvania. Botched management obscured whatever prospects the Erie had during the 1850s. The road was in such poor shape that, as Edward Mott observed, it "became notorious for the insecurity of travel upon it." Thirty serious accidents occurred in 1852 alone. A succession of financial blunders saddled the road with a persistent floating debt. By August 1859 Erie stock had plunged from 33 to 8 and the company was in receivership.[2]

The road reorganized as the Erie Railway under the presidency of Nathaniel Marsh, who in two years tripled the road's traffic, acquired a short line to Buffalo, developed the valuable Long Dock facility at Jersey City, gained access to the Pennsylvania coal fields, paid off the floating debt, and reduced the mortgage by several million dollars. The Erie also profited from a large oil traffic through connection with the newly opened Atlantic & Great Western Railroad. It seemed assured of a bright future unless one noticed that prosperity had been achieved through the abnormalities of wartime. In July 1864 Marsh's sudden death removed his restraining influence from a speculator who had been skulking through the Erie's history for a decade.

Daniel Drew started out as a drover running herds of cattle across the Alleghenies to New York. Some said it was he who inspired the phrase "watered stock" by his practice of feeding salt to his steers to make them drink prodigious amounts of water and swell in size. During the 1830s he went into steamboating on the Hudson River, where his People's Line dazzled travelers with its floating palaces. Later he drifted into Wall Street and became the most notorious bear since Jacob Little.[3]

A gaunt, cadaverous figure, coarse and sly of manner, Drew laid traps, switched sides, and betrayed allies until his raids became the stuff of legend. When cornered he did not hesitate to resort to what the Street called "squatting" or dishonoring his contracts and taking refuge behind a lawsuit. His demeanor was pious and austere, the product of his early conversion to Methodism. Although fond of quoting Scripture, he took care to shield his business dealings from the influence of Christian ethics. This habit, coupled with his ruthless depredations in the market, led others to regard him as a hypocrite. "Mr. Drew," noted one observer delicately, "is singularly lacking in popularity."[4]

Drew had been preying on the Erie since 1854, when he became its treasurer and promptly secured a mortgage on all its property by endorsing some of its notes. Through every financial upheaval Drew clung leechlike to his hold on the road. At the same time he milked the stock dry, working it up then down, using his inside position to swell his personal fortune with such brazen success that he became known as the "speculative director."[5]

In 1859 Drew persuaded Cornelius Vanderbilt to come into Erie as a director. The two men had undertaken ventures together since their days as rival steamboat entrepreneurs, and their backgrounds had much in common. Vanderbilt was born in 1794, Drew three years later, both of dirt-poor families. After clawing their way to wealth, neither bothered to glaze his crude, uncouth ways with a veneer of social refinement. Vanderbilt had a brutal, tyrannical streak that showed itself no less to his family than to business rivals, yet he was also subject to outbursts of generosity. His emotions veered from one extreme to the other. He was foul-mouthed, possessed an outlandish sense of humor, and showed a zest for life utterly lacking in Drew.[6]

While Drew in all his rail ventures never rose above the level of speculator, Vanderbilt took hold of them as enterprises. At seventy, his powers undiminished, he launched a new career in railroading devoted to forging a trunk line between New York and Chicago. His strategy was as elemental as it was effective: buy a road, put in honest management, improve its operation, consolidate it with other roads when they can be run together economically, water the stock, and still make it pay dividends. By 1865, after a bitter struggle, he had acquired the Harlem and Hudson River roads and put them in the hands of his son William. He then went after the New York Central, which would complete his line from New York City to Buffalo. Since the chief competitor of this new line would be the Erie, Vanderbilt was eager to have it under his thumb as well. That desire brought him into conflict with the wily Drew.[7]

At the war's end deflation and rate cutting reduced the Erie's earnings sharply and produced a large floating debt just as its competitive position was crumbling into ruins. Besides Vanderbilt's line the Erie was threatened with the loss of its lucrative oil business to the newly opened Pennsylvania Railroad. The Atlantic, its key link to the oil fields, grew disenchanted with the Erie and decided to create its own route to the seaboard. The Erie had incurred heavy obligations to purchase new equipment, much of which now sat idle, and the roadbed was in wretched condition. Amid this sea of woes Drew cruised with lethal skill. He resumed his favorite game of loaning money in exchange for stock, which he then dumped on the market to drive down the price and reap fat profits from his short sales. While Drew filled his purse, the Erie scrounged for cash enough to pay its bills.[8]

Both Drew and the Erie had become thorns in Vanderbilt's side. After finally gaining control of the New York Central in 1867, he decided to capture the Erie and oust Drew from its management. His scheme ran afoul of another group intent on bending the Erie to its purposes. Led by John H. Eldridge, a Boston banker whose venal instincts smacked more of Wall Street than State Street, this faction represented the Boston, Hartford & Erie Railroad, a proposed line between Boston and the Hudson River. Since 1865 Eldridge had been soliciting financial aid from the Erie. Dangling the prospect of a connection to southern New England as bait, he wanted the Erie to guarantee interest on $6 million of the Boston's bonds. Even if the Erie could have afforded the commitment, which it could not, the cost of building the Boston far exceeded its value as a feeder. Nevertheless, the Erie board voted in June 1867 to guarantee $4 million of the Boston's bonds, though it attached several conditions to its action.[9]

From this conflict emerged a complex struggle to control the Erie election of 1867. Unhappy with the Erie's offer, Eldridge decided to acquire enough stock to join its board and force its hand. Vanderbilt too was corralling stock for his campaign against Drew, a fact that did not escape Uncle Dan'l's attention. There followed a bizarre three-cornered hat dance in which partners changed so often it was difficult to know who was odd man out.[10]

The scramble for Erie stock revolved largely around the proxy market. Much of the stock belonged to English holders who simply left it in the name of a broker for the sake of convenience. This allowed the brokers, unless otherwise directed, to vote the stock as they saw fit or to sell their proxies to the highest bidder. While Vanderbilt combed the Street for proxies, Drew formed an alliance with Eldridge's agent against their common foe. The agent soon defected to Vanderbilt, leaving Drew to play the part of wallflower.

Vanderbilt's bargain with Eldridge did not take into account the curious bond that had formed between him and Drew over the years. Two days before the election Drew, recognizing that his position was hopeless, went to the Commodore and begged forgiveness. If allowed to keep his seat he promised to mend his ways and serve only Vanderbilt. It was neither the first nor the last time this lachrymose scene would be enacted. Time and again Drew had bamboozled Vanderbilt in some outrageous way and then, brought to bay, had pleaded tearfully for mercy amid a torrent of references to bygone days. The Commodore was far too shrewd to believe that so incorrigible a predator as Drew would actually change his ways. Yet each time he relented, perhaps because he relished the spectacle of watching Drew squirm.

The Boston crowd was flabbergasted at the news but agreed to go along. On October 8 the board was elected and promptly installed Eldridge as president. The next day, however, one of its members resigned and was replaced by Drew, who also reclaimed the office of treasurer. The newspapers buzzed with rumors of an alliance between the two archrivals. In their excited speculations they passed over the board's lesser lights, which the *Herald* dismissed as "a batch of nobodies." One of these nobodies was a broker named James Fisk, Jr., a man sufficiently unknown that some papers recorded his name as "Fiske" or "Fish." Another was a broker so obscure that he was listed only as "J. Gould."[11]

It was through Eldridge that Jay wangled his place on the Erie board. Eldridge had sent Fisk and lawyer Frederick A. Lane to secure Erie proxies. Apparently Jay held enough stock to attract their interest; possibly he held title to a large amount through the brokerage firm he joined sometime during 1867, Smith, Gould & Martin. Whatever his source, a pivotal moment in his career had arrived: for the first time he became an insider.[12]

No one regarded his presence as anything more than a device to aid his speculations. The house in which he had become a partner was a small, respectable firm. Henry N. Smith had been a broker and Henry Martin a banker in Buffalo before coming to Wall Street in 1865. Martin tended the office while Smith and Gould worked the street. A short, dapper man with blue eyes, strawberry hair, and full beard, Smith maintained an impassive expression that belied the bustling gait that made him seem always in a hurry. For two decades his shadow would dog Gould's steps as both friend and foe.[13]

The quest for Erie proxies had still another fateful consequence for Gould: it introduced him to the man with whom he would form the oddest couple ever to grace Wall Street. Not even the genius of Dickens could have invented two more unlike characters than Jay Gould and Jim Fisk. They were the same age, and there the resemblance ended. Short and rotund, his twinkling eyes spread on a broad face, his reddish-yellow hair fringed with curls and his lip masked by a formidable moustache described by one wag as "the color of a Jersey cow," Fisk looked the part of Falstaff and played it to the hilt. His excesses outmatched Jay's abstemiousness by half again. He loved the best of wine, women, and song, not always in that order. For every occasion he dressed and strutted like a peacock.[14]

Above all Fisk craved the limelight. In all his outrageous antics he sought publicity as a moth seeks the flame with what a biographer called "a love of notoriety so extravagant that he preferred to be insulted than ignored." He started out as a peddler in his native Vermont and graduated to jobbing for the Boston drygoods house of Jordan, Marsh & Company. During the war he ran cotton out of the South to feed his employer's mills, and developed a taste for the good life. At nineteen he married Lucy Moore, to whom he remained devoted in his own bizarre way. When Fisk set out to conquer New York in 1864, Lucy stayed behind in Boston. That arrangement allowed Fisk to indulge his taste for "actresses," especially the overripe Josie Mansfield.[15]

On Wall Street Fisk was a whirlwind in both success and failure. He fell in with Drew and opened a brokerage in partnership with William Belden, the son of an old friend of Drew's. Notorious for being a sharpster on the Street and a fast liver off it, Belden's ability fell far short of his extravagant tastes. Board members and curb brokers alike regarded Fisk & Belden as a house of ill repute, partly because they could not separate its bold and risky ventures from Fisk's own splashy displays.[16]

It was easy to dismiss Fisk as a garish, vulgar buffoon, and for some it proved a fatal mistake. Behind the clown's costume loomed a native genius, a cunning, imaginative intellect that made him a formidable presence in business. His boisterous good nature made him easy to ridicule, impossible to dislike, and hard to take seriously. Gould was alert enough not to underestimate his talents; so was an observer who described Fisk as "a shrewd speculator & very careful to take care of No. 1." The Erie provided Fisk a perfect stage for his outsized personality. Within months after landing on its board he assumed the role of ringmaster for the grandest circus ever to play Wall Street.[17]

A rude awakening that the Erie was not his to run came quickly to Vanderbilt. He permitted Drew to manage a pool in Erie stock on behalf of his clique only to discover that Drew was lining his own pockets by secretly working against the pool. In December the Erie board opened negotiations with the Michigan Southern & Northern Indiana for a line to Chicago that threatened to drain much of the New York Central's traffic onto Erie. A fierce, prideful man accustomed to having his own way, Vanderbilt realized the folly of trusting Drew to do his bidding. The only recourse, he concluded grimly, was to buy control of the road. In 1867 the Erie had outstanding about $16.5 million in common and $8.5 million in preferred stock. Vanderbilt began secretly to buy Erie while laying legal snares to prevent the company from increasing its capitalization. In February 1868 his nephew, Frank Work, secured from Judge George G. Barnard, a notorious Tammany Hall wheelhorse who had been elevated to the state supreme court, a series of injunctions against Drew and the Erie board. Without knowing it, Vanderbilt had fired the first salvo in the Great Erie War.[18]

A provision of the General Railroad Act of 1850 prohibited rail companies from increasing their stock but allowed them to issue bonds for construction or equipment and then to convert the bonds into stock. Since Drew had used this "convertible" clause to burn the Commodore in 1866, Vanderbilt knew the danger it posed to his plans. The injunctions asked that Drew be removed as treasurer and that he return 58,000 shares acquired as collateral in 1866. Both Drew and the Erie board were enjoined from selling stock or convertible bonds, but the suit overlooked the executive committee, which had the power to act during the intervals between board meetings. Drew was a member of that committee, as were Eldridge and Gould. Unbeknown to Vanderbilt, Drew allied himself with the two "nobodies," Gould and Fisk, to resist the Commodore's takeover of Erie. Eldridge was willing to follow anyone who promised to give the Boston what it wanted.[19]

Politics never created stranger bedfellows than the trio who united against Vanderbilt. Their first step was to have the executive committee authorize an issue of $5 million in bonds for "improvements" on the road. In one meeting the bonds were approved, reported sold to Drew's broker, and the sale ratified. The bonds were then converted into stock and thrown onto the market along with another 10,000 shares manufactured in December 1867 when the Erie board leased the Buffalo, Bradford & Pittsburgh Railroad, converted its bonds to stock, and exchanged the stock for Erie shares.[20]

Suddenly the market was awash with Erie stock. Vanderbilt's brokers, ordered to buy all the stock offered, dutifully absorbed the new issues. The price of Erie held firm, yet Drew and his associates kept selling short with a serenity that puzzled the Commodore. At his request Barnard extended his injunctions to include the Bradford shares, the guarantee of Boston bonds, and the Michigan Southern connection. The executive committee blithely ignored these mandates and recruited its own judge. It was a peculiarity of New York's judicial system that all thirty-three of its supreme court justices enjoyed equal authority throughout the state in certain equity actions. Drew found a judge in Broome County who obligingly stayed all of Barnard's orders and suspended Frank Work from the Erie board. An outraged Vanderbilt launched a new suit in Richard Schell's name staying the stay and forbidding the Erie board to meet or transact business without Work.[21]

There followed a bewildering crossfire of injunctions. While the lawyers traded volleys, the executive committee converted a second $5 million issue of bonds and dumped them on the market. It also created a $500,000 fund for "legal expenses" and an elaborate report justifying the road's urgent need for improvements. Attached to the report were two letters from the superintendent bewailing the road's sorry physical condition even though a year earlier he had reported it "in better condition and better equipped than at any period during the past ten years."[22]

In publishing this report the Erie clique cleverly portrayed themselves as defending the public from Vanderbilt's grasping attempt to erect a railroad monopoly. To cover their action they applied to Judge Gilbert in Brooklyn, an impulsive jurist who astounded everyone by restraining all parties to all other suits from further proceedings and ordering all the Erie directors except Work to discharge their proper duties. A perfect legal stalemate had been achieved. "One magistrate had forbidden them to move, and another magistrate had ordered them not to stand still," observed Charles Francis Adams, Jr. "If the Erie board held meetings and transacted business, it violated one injunction; if it abstained from doing so, it violated another."[23]

Meanwhile Vanderbilt's unsuspecting brokers snatched the newly offered bait. Since the shares were dispersed through several brokers, Vanderbilt did not grasp what was happening until fresh certificates bearing Fisk's name began to appear on the street. The light dawned too late; the Commodore had already strained even his immense fortune trying to sustain the price of Erie. Pandemonium shook Wall Street as the stock's quotations reeled back and forth. Drew precipitated a spasm in the money market by withdrawing the nearly $7 million proceeds of stock sales from banks and locking the greenbacks in the Erie safe. If Vanderbilt faltered at this critical hour, his failure would launch a panic of grand proportions. But the old man stuck grimly to his task of buying Erie even though he had to borrow on collateral of falling value.[24]

The Erie clique had little time to savor their triumph. On March 11 word came that Barnard had issued contempt orders and intended to clap them all in the Ludlow Street jail. Alarmed, they packed up files and ledgers, bundled stacks of greenbacks, stuffed their pockets with securities and papers, and dismantled the offices. Drew, Eldridge, and some other directors, along with a few clerks, dashed for the ferry. Three directors stayed behind as rearguard and later bore the brunt of arrest. Fisk and Gould were not among them. They felt bold enough to dine at Delmonico's that evening until news that the law was approaching sent them scurrying to Canal Street. Hiring a small steamer, they plunged into a fog so thick that they cruised in a circle for a time and narrowly escaped being rammed by a ferry. Like poor Eliza in *Uncle Tom's Cabin* they crossed the river, seeking on the Jersey shore freedom from the bloodhounds sniffing at their heels, clutching to their bosoms not an innocent child but $7 million in greenbacks and the corporate majesty of the Erie Railway Company.[25]

In Jersey City they ensconced themselves in Taylor's Hotel near the Erie terminal, quickly dubbed "Fort Taylor" by newspapermen who rushed to view the novel spectacle of a corporation in exile. The ebullient Fisk accommodated them with cigars, liquor, and an endless stream of blather about how the struggle against Vanderbilt's monopoly was "in the interest of the poorer classes especially." Alone among the clique Fisk took readily to his new surroundings. Where Drew was out of his element and Gould missed his family, Fisk gleefully installed Josie Mansfield in the next room and reveled in the attention showered on him.[26]

A state of siege suited neither side. Vanderbilt still shouldered his heavy burden of Erie stock which, if not maintained, might destroy his personal fortune and engulf Wall Street in panic. His lawyers got Judge Barnard to unleash a fresh barrage of injunctions, to which the Erie judge responded in kind, but no damage could be done so long as the target hovered out of range. In Albany the state senate created a committee to investigate the Erie's affairs, a move interpreted by cynics as legislative resentment that the judiciary was hogging the spoils of war.[27]

Meanwhile Fisk and Gould managed surprisingly well at operating the Erie from Fort Taylor. As part of their antimonopoly campaign they opened a rate war with the New York Central even though it meant running the road at a loss. When Jay declared that the Erie had come to stay, the New Jersey legislature gratefully rushed through a bill incorporating the company. Possession of legal status in New Jersey would allow the Erie to issue more stock, a fact Vanderbilt must have found unsettling. The financial community followed these events with dumbfounded amazement. So unprecedented was the situation that even respectable journals did not know which way to turn.[28]

Despite the conferral of legitimacy by New Jersey, the Erie clique could not afford to remain in exile. The railroad did its business in New York and so did its managers who, away from Wall Street, were fish out of water. Drew pined for home and so did Gould; both felt uncomfortable as neighbors to Fisk's private saturnalia. The press castigated them as outlaws fled with their loot. New York law forbade arrests in civil cases on Sundays, which allowed them to return home for that day; otherwise they could not set foot in Manhattan without facing the threat of jail. The impasse could be ended by negotiating a compromise with Vanderbilt or by inducing the legislature to pass a bill legalizing the convertible bond issue. The Erie clique decided to pursue both possibilities.

From the start Vanderbilt and the press alike had focused their attention on Drew as the Erie ringleader. If that assumption was true at first, it did not long remain so. Once in exile Drew lost all heart for combat; weary and homesick, he undertook in typical fashion to negotiate his own settlement with little regard for his cohorts. Gradually Fisk and Gould emerged as Vanderbilt's true adversaries, a fact the Commodore was slow to grasp. The odd couple made an effective if improbable team, their strengths masked by Fisk's flamboyant style and Gould's quiet, retiring manner. Where Drew was adept at laying an ambush and Fisk knew how to lead a charge, Jay alone had the talent for planning a campaign. It was his subtle mind that formulated strategy in the weeks to come.

Sensing that Vanderbilt was not yet ready to compromise, Gould concentrated on Albany. An Erie agent was already trying to obtain a bill by bribing that notorious element of the legislature known as the Black Horse Cavalry. Vanderbilt's agents were also on the scene and apparently employed a more generous fee schedule, for on March 27 the assembly crushed the Erie bill by a vote of 83 to 32. Meanwhile the senate investigative committee was preparing its report. Two members supported the charges against Erie and two opposed them, leaving the decision in the hands of A. C. Mattoon, whose personal research included visits to both the Vanderbilt and Drew camps. With fine impartiality he sold his support to both sides. When the report appeared on April 1, Mattoon was found on the side of those denouncing the Erie even though he had declared the opposite intention only the day before.[29]

By that time Jay had already decided to risk a personal visit to Albany. A deal was struck whereby he agreed to appear before Judge Barnard on April 4 provided he was left undisturbed until that date. On March 30 he left Jersey City with a suitcase full of greenbacks and a ready reserve of checkbooks. For three days he wooed legislators with a liberal supply of food, drink, and greenbacks. Vanderbilt's agents set up shop on another floor of the Delavan House, enabling legislators to shuttle back and forth in search of top dollar for their vote. On April 4 Jay dutifully went to New York to answer charges. After a lengthy hearing Barnard deferred proceedings until the 8th but ordered Gould to remain in the custody of a court officer. When Gould declared his intention of returning to Albany, the officer protested, "You are in my custody."

"But you may go with me," Jay replied blandly. "I will still be in your custody."[30]

Back in Albany Jay tried to resurrect the Erie bill. When time came to return to New York, he pleaded illness and asked for a postponement. While the lawyers wrangled, Jay emptied his satchel of greenbacks. "Jay Gould is still at the Delavan House, Albany, with his keeper," noted a bemused reporter. "Many legislative friends are calling on him." Charles Francis Adams, not in the least amused, described Gould as devoting "the tedious hours of convalescence to the task of cultivating a thorough understanding between himself and the members of the legislature."[31]

The inside story of the Albany debacle will never be known. If the rumors that flew and the stories told about the amounts exchanged are even half true, they are enough to satisfy any connoisseur of avarice. Legislators tumbled over each other trying to bid one side against the other. Awed reporters spoke of a trunk filled with thousand-dollar bills, but there was no shortage of exaggeration at Albany. The situation looked anything but promising for passage of a bill nearly identical to the one defeated so resoundingly a few weeks earlier, yet on April 18 the senate passed it by a vote of 18 to 12, with the ubiquitous Mattoon switching sides again. The fight moved to the assembly where, on the eve of what loomed as a desperate battle, Vanderbilt abruptly quit the field. His withdrawal threw the spoilsmen into dismay; in their wrath the assembly turned on Vanderbilt with a vengeance. On April 20 the bill passed by a landslide vote of 101 to 5 and the assembly ransacked its docket for other bills hostile to the Commodore.[32]

The battle of Albany ended with Gould the victor. Despite his prominent role, the press persisted in describing the struggle as one between Drew and Vanderbilt. In the public eye Jay was merely Drew's henchman, a misapprehension he made no effort to dispel. Nor would he ever admit how much money he had dispensed. His testimony before an investigating committee was plagued with so many lapses of memory that Adams, his severest critic, was driven to observe:

> Mr. Gould [in Albany] underwent a curious psychological metamorphosis and suddenly became the veriest simpleton in money matters. . . . This strange and expensive hallucination lasted until about the middle of April, when Mr. Gould was happily restored to his normal condition of a shrewd, acute, energetic man of business; nor is it known that he has since experienced any relapse into financial idiotcy.[33]

Far from being commander, Drew had all but defected to the enemy. His timid, whining soul yearned for peace and responded eagerly to secret overtures from Vanderbilt, who was receptive to talk of compromise. A provision of the new Erie bill forbade the same parties from controlling both the New York Central and the Erie, which rendered Erie stock a useless burden to Vanderbilt. When terms had been hammered out, Drew stealthily enlisted the support of Eldridge and the Boston directors. Aware that Gould and Fisk were not yet willing to compromise, Drew took pains to leave them in the dark.[34]

As early as April 25 rumors flew that the Erie War had ended "through a settlement between the chief belligerents." The exiles abandoned Fort Taylor and returned home to prolonged negotiations with Vanderbilt. Gould and Fisk realized at last what was happening and demanded stiffer terms, but Drew now had a majority of directors on his side. Early in June Drew arranged a secret meeting between Vanderbilt and the Erie directors to finalize an agreement. Gould and Fisk, following up a rumor, located the meeting and burst into the proceedings. Their cries of treachery and protests against terms that all but ignored them fell on deaf ears. Drew had the votes, and Vanderbilt still wielded contempt charges over their heads. After salvaging what they could, they agreed to go along.[35]

The settlement, completed during July, seemed to reward everyone except Gould and Fisk. Vanderbilt agreed to withdraw his suits and the Erie bought back from him 50,000 shares of its stock at 70, or $3.5 million. In addition, Vanderbilt received another $1 million "without consideration" and two seats on the Erie board. Eldridge finally unloaded $5 million in Boston bonds on the Erie at 80. Drew left the Erie management and paid the company $540,000 for release of his disputed contract of 1866. Vanderbilt's cohorts, Work and Schell, were awarded $464,250 to reimburse them for losses suffered while speculating in Erie stock. If charging a corporation for personal market losses seemed a curious practice, so did the payment of $150,000 to Peter B. Sweeny, the chamberlain of the Tweed Ring. Sweeny had been appointed receiver in one of Barnard's court orders, but there had been nothing for him to receive. He was awarded this large sum, in Mott's words, "as a balm for his not having a chance at the Erie treasury."[36]

The settlement added $9 million more to the Erie's burden, sinking that hapless road deeper into the mire of insolvency. Having scraped the oyster clean, the participants were content to toss the empty shell to Gould and Fisk, for whom no other provision had been made. Eldridge resigned the presidency to Gould, who appointed Fisk and Frederick Lane, the Erie counsel, to the executive committee. To wrap up affairs, Barnard purged six directors for contempt by fining them ten dollars apiece. Judgment on Drew, Fisk, and Gould was deferred, then forgotten altogether. In return the Erie dropped a suit it had initiated against Barnard.[37]

The consensus on the Street was that the Erie War had ended with Gould inheriting a sinking and abandoned ship. At thirty-two Jay emerged from obscurity as president of a major railroad, yet this feat went unnoticed. Most observers regarded the two nobodies as scavengers feeding on the carcass left by Drew and Vanderbilt. Even Gould's hold on the presidency was tenuous; he had been elevated by the board and controlled nowhere near enough stock to insure his election by the stockholders in October.

"There ain't nothin' in Ary no more, C'neel," Drew assured his friend Vanderbilt, and Wall Street agreed with him.[38]

It took Gould only a few months to disabuse them of this notion. He strengthened his position by adding Tammany chieftain William M. Tweed to the board. During the Albany debacle the porcine Tweed had sided with Vanderbilt but he cheerfully switched partners in return for some Erie stock, giving Gould a powerful political ally. Tweed included in his baggage the amiable Judge Barnard, who could supply injunctions as easily for Gould as he had for Vanderbilt. To control the election, Jay managed through some clever if underhanded maneuvers to disenfranchise a large number of shares while securing for himself the stock manufactured out of convertible bonds, which had been left in the names of brokers friendly to him. These tactics, which aroused little protest from a press already surfeited with Erie affairs, enabled Gould to win reelection handily.[39]

Once elected, Jay dispensed with board meetings. Power stayed in the hands of the executive committee, which consisted of Gould, Fisk, and Lane. Jay also served as treasurer and Fisk as comptroller. The auditor was sent on leave for a year and his duties assumed by the assistant comptroller, Giovanni P. Morosini, a bear of a man who in 1872 became Gould's confidential secretary and bodyguard. Venetian by birth, Morosini had migrated to the United States in 1850 and worked as a sailor until 1855, when he rescued the son of an Erie official from assailants and was rewarded with a job. His ability won him promotion to clerk in the auditor's office, where Gould found him. Quick to discern Morosini's intelligence and loyalty, Jay advanced him to the auditor's post in 1869. Thus began a close relationship between the two men that lasted the rest of Gould's life.[40]

The Erie was now Gould's to run, at least for a time. His task seemed at best a hopeless one. The road was in wretched physical condition and groaned beneath a heavy burden of fixed charges. The amount of its stock had swollen $21 million since 1866 with little in the way of improvements or assets to show for it. A long history of scandal had blackened the company's reputation beyond redemption, and an imposing array of enemies waited in the wings.

9

Notoriety

Within the past few days we have seen the most gigantic swindling operations carried on in Wall Street that have as yet disgraced our financial centre. A great railway . . . has been tossed about like a football, its real stockholders have seen their property abused by men to whom they have entrusted its interests, and who, in the betrayal of that trust, have committed crimes which in parallel cases on a smaller scale would have deservedly sent them to Sing Sing.

—*New York Herald*, November 22, 1868

At the time that the Gould-Fisk ring was sucking the life-blood of the Erie, the Tweed-Sweeny ring was plundering the city of New York. The two were really one.

—Trumbull White, *The Wizard of Wall Street*

While secretly bribing, Gould constantly gave out for public consumption a plausible string of arguments, in which act by the way, he was always fertile. He represented himself as the champion of the middle and working classes in seeking to prevent Vanderbilt from getting a monopoly of many railroads. He played adroitly upon the fears, the envy and the powerful main-springs of the self interest of the middle class. . . .

—Gustavus Myers, *History of the Great American Fortunes*

IN THE SUMMER of 1868 Andrew Johnson had barely won acquittal from impeachment charges and was serving out his term a hostage to congressional reconstruction. Federal troops still occupied parts of the South to help insure that democratic government did not mean government by Democrats. The nation was gearing up for an election campaign between Horatio Seymour and the Hero of Appomattox. In the West the Union and Central Pacific roads eyed each other from afar, while in Washington the House grudgingly approved funds for the purchase of Alaska. The two favorite novels of genteel ladies that season were Louisa May Alcott's *Little Women* and Elizabeth Stuart Phelps's *The Gates Ajar.*

Everywhere the country seethed with movement as men scrambled feverishly in search of new opportunities. As the war receded and the West opened up, the prospects seemed boundless for those with ambition and a willingness to work. There were railroads to build, factories to erect, mines to dig, farms to wrest from the prairie, business enterprises to undertake, new technologies to exploit, fortunes to be made in a thousand lines of endeavor. The euphoria of boom times seized the nation, a buoyant optimism so seductive as to tempt even prudent heads into bouts of gullibility. Later this generation would recoil from these years with the pain and embarrassment of revelers contemplating their behavior of the previous night. How different the party seemed in the cold, sober light of the morning after!

In this respect at least Jay differed little from others of his generation. Although he never bothered to defend his behavior during these years, it is revealing that when he recounted his career to a Senate committee in 1885 he neglected even to mention his association with Erie. In later years he maintained an inscrutable silence about the events that first brought him notoriety. His reticence encouraged others to believe what Mott called "the storm of distorted statements and positive falsehood with which he was assailed."[1]

None of Gould's actions can be understood apart from the era and circumstances in which they occurred. It is true that Jay wrung a large personal fortune from the Erie; it is also true that he wanted desperately to make something of the road. That was how he differed from a vulture like Drew. Jay stayed with the Erie not to pick the carcass but to make his first attempt at forging a major rail system.

As always, the most pressing need was money. To get it Jay resorted again to convertible bonds, an issue of $20 million to complete the work for which the previous bond issues were intended. Gould put them to other work in a market operation that displayed his grasp not only of the stock market but of the banking system as well.[2]

The national banking system had become notorious for certain weaknesses, most notably its inability to provide an elastic currency supply. During the winter and summer bank reserves tended to be abundant, enabling speculators to obtain money at cheap rates. In the fall, however, large sums were withdrawn from New York to finance the movement of crops. Since both specie and legal tender were already in short supply, this demand for funds tightened the money market. Shrewd operators learned to exploit this weakness by a practice known as the "lockup." Using securities or certified checks as collateral, they borrowed huge sums of greenbacks and simply locked them in a safe. This artificial contraction induced a shortage of money just when demand was heaviest. The money market tightened, interest rates rose, and the price of stocks declined. Banks dared not call in their demand loans lest they force customers to sell stocks in a falling market and precipitate a panic that might crash on their own heads.[3]

In August 1869 Jay sold short some borrowed Erie shares along with those reacquired from Vanderbilt and locked up the proceeds. After reaching 70 in July, Erie tumbled to 46. The market then firmed until shortly after Gould's election on October 13, when rumors that Erie had issued $10 million in convertible bonds swept Wall Street. Jay confirmed the rumors, admitted that half the bonds had already been converted into stock, and warned that Erie would be hard pressed to meet its January obligations. News of this secret issue triggered another fall in stock prices. Then came the revelation that Gould and his cohorts had locked up $16 million in greenbacks. A wave of panic selling demoralized the market and plunged the financial community into "the greatest difficulty . . . in regulating the operation of the finances since the organization of the banking system."[4]

The excitement surrounding Erie attracted the attention of Drew. No longer an insider, the grizzled speculator sniffed a familiar bear scent and decided to sell short 70,000 shares of Erie at around 40. On Friday, November 13, Erie sank to 35, money remained tight, and the bears seemed to have the entire market at their mercy. The stage was set, but for quite another play than actors and audience alike expected. The next day, to everyone's astonishment, Erie suddenly began advancing until it reached 52$^{1}/_{2}$. At the same time a flood of greenbacks returned to the market, easing the stringency of money. The climax came on Monday when, "amid one of the wildest scenes ever witnessed on the Stock Exchange," Erie climbed to 61 in half an hour. The startled bears fought savagely but failed to lower the price by closing hour. Gould had sprung his trap and caught Drew firmly in its jaws.

Unable to fill his contracts, Drew realized at last that he was up against an intelligence far surpassing his own. In the past he had escaped ruin by pleading for mercy or reneging on his contracts. This time he begged Fisk for enough stock to cover. Amid his pleas Drew admitted being party to a scheme against Gould's management and warned of a suit about to be brought in August Belmont's name. Rebuffed by both Fisk and Gould, Drew resorted to bluster. A conspiracy was brewing against them, he hissed, and unless Fisk accommodated him he would give his affidavit to the enemy. "If you put up this stock," he wailed, "I am a ruined man."[5]

These interviews took place on Sunday, November 15, the day before Erie leaped from 52 to 61. The veil of tears that worked so well with Vanderbilt left Gould and Fisk unmoved. Drew left in dejection, unaware that he had given Jay a vital piece of information. On Monday Belmont, representing the foreign holders, asked Judge Sutherland of the supreme court to enjoin the Erie board from issuing new stock and to appoint a receiver. Although his petitions were granted the next day, they came too late. Gould had sent an Erie employee to Judge Barnard that same Monday morning to apply for a receiver. Barnard obliged by appointing Gould himself as receiver and by enjoining all parties in the Belmont suit from proceeding under Sutherland's injunction.[6]

Two receivers had been appointed under conflicting injunctions; once again the lawyers rose to battle like palace guards. The tortuous litigation dragged on into February, enlisting fresh judges who picked at the Gordian knot Gould had devised and countermanded each other's orders until even they wearied of the sport. While the lawyers piled up fees, New York paused to watch the Erie circus come to town again. The executive committee put a special litigation fund at Gould's disposal and directed him to sue Vanderbilt to recover the $1 million bonus and force the Commodore to buy back the $5 million in Erie stock he had unloaded in the July settlement. Caught by surprise, Vanderbilt broke his vow of silence and issued a lame denial that he had ever accepted a bonus or sold any stock to the Erie. The gleeful Fisk exposed this lie by giving reporters details of the July settlement and showing them two checks totaling $1 million endorsed by Vanderbilt, whereupon the scowling Commodore retreated to lick his wounds in silence.[7]

The suits all vanished into legal limbo, leaving Jay in firm command of Erie. In December he issued a report to the stockholders listing his accomplishments since becoming president. It was a prototype of what would become the classic Gould statement. The facts presented were accurate yet misleading, a series of truthful observations that did not convey a true picture of the company's affairs. The items listed accounted for only a fraction of the capital raised through new stock issues. Against the huge sums lost through stock speculations, the July agreement, and legal fees, Jay dangled the promise of recovery from the suits instituted against Drew and Vanderbilt. In the end nothing came of these suits, but in December hope still sprang eternal. "Whenever the facts are fully known and the public becomes aware of what is being done to make the Erie Railway the most magnificent and perfect railway line in the country," Gould concluded, ". . . then the acts of the present managers will be appreciated."[8]

By the year's end it was evident that Gould and Fisk had become figures to reckon with on Wall Street. In the space of a few months they had embarrassed both Vanderbilt and Drew in operations executed with a skill and audacity that astounded even veteran operators. The cornering of Drew gave them a reputation as enfants terribles who had beaten the master at his own game. With Drew gone from Erie, editors discarded at last the fiction of his dominance and directed their indignation at a new generation of villains. The two nobodies had become somebodies, a fact that delighted Fisk as much as it discomfited Gould.[9]

As for Drew, the Erie corner finished him. In November he was forced to settle with Gould at 57 for stock he had sold at 40. He lost over $1 million and joined that pathetic circle of once prominent speculators who haunted the fringe of Wall Street. A series of reverses drove him into bankruptcy in March 1876 and onto the charity of his family until his death in 1879. In flush times he had donated large sums to Wesleyan University and to the seminary that would bear his name, but the gifts were in the form of notes that became worthless after his bankruptcy. Nothing wounded him more than "the fact that I could not continue to pay the interest on the notes I gave to the schools and churches."[10]

But that was Drew's way. To the bitter end, even for God's work, he could not resist selling short.

To bewildered observers the Erie's new management showed two faces of contrasting and improbable complexion. One was the mask of theatricality imparted by Fisk, the other a sober demeanor, the reserve of a gambler intent on wringing the utmost from his cards. That face belonged to Gould, who demonstrated quickly that his bag of surprises was not confined to the financial arena. Where Fisk made Erie into New York's longest running opera buffa, Gould transformed it into the most disturbing element ever to hit the railroad industry.

In December 1868 Fisk began to indulge his fantasy of corporate amenities by persuading Gould to purchase Pike's Opera House as the new headquarters for Erie. Under his watchful eye an army of workmen refashioned the edifice into a baroque palace. Visitors gaped in awe at the "carved woodwork, the stained and cut glass of the partitions, the gilded balustrades, the splendid gas fixtures, and, above all, the artistic frescoes upon the walls and ceilings." The building also contained several lavishly appointed apartments, ground-floor offices rented to business firms, and of course a theater capable of mounting the sort of extravaganzas Fisk adored. Soon afterward he installed Josie Mansfield in a house on West Twenty-third Street, half a block from the Opera House.[11]

For his endless escapades and for the resplendent castle from which he sallied forth, the most errant of knights, Fisk was dubbed the "Prince of Erie." It was all too much for Gould and other staid souls of Erie, who trod the rich carpets of the Opera House with the disquiet of monks in a bordello. They cringed at Fisk's unabashed pleasure seeking, at the proximity of Josie Mansfield and the champagne and poker soirees she hosted for Tweed, Barnard, and other politicos. What bothered them was the kind of attention these roisterings drew to the company. Fisk's flamboyant personality left its imprint on the whole of Erie, lent its affairs a tone of unreality that made it difficult to treat its activities in a serious vein.

Distasteful as it must have been to him, Jay bore Fisk's excesses with remarkable patience. He was never known to censure Fisk publicly; the bond of loyalty formed during the Erie War remained unshaken until Fisk's death. In blunt terms he needed Fisk, who for all his bluster was a man of kind and generous impulses. His thick hide willingly absorbed every volley of criticism. He handled reporters with a deft affability that permitted Gould to retire to the anonymity he craved. His sunny disposition and small charities made him immensely popular with the Erie's employees. Only a man with Fisk's magnetism could send a gang of toughs to disperse striking brakemen, order them to shoot if necessary, and then be deafened by cheers from the strikers when he reached the scene. Jay needed Fisk as a front man in the office and on the Street. He was a shrewd, forceful businessman, as reliable in work as he was erratic at play. His loyalty was steadfast; he once said of Gould, "He knows I'd go my bottom dollar on him." A genuine affection bound them together.[12]

While Fisk trampled merrily on genteel sensibilities, Jay unfurled a scheme to give Erie a competitive edge over its chief rivals, the Pennsylvania and the New York Central. His actions flabbergasted Wall Street, which knew him only as a speculator and did not suspect him capable of actually doing something with Erie.

For three decades after the Civil War the chessboard of railroad strategy was to remain intensely volatile. As the rail network expanded from 36,801 to 182,777 miles, the building of new roads or the merging of old ones into unified lines radically altered competitive relationships in a single stroke. To master the intricacies of so fluid a game required a breadth of vision possessed by few men. Even at the outset of his Erie career Gould revealed a genius for strategic possibilities and the boldness to pursue grand designs even though his resources were never adequate for the vision they served. The lack of resources compelled him to devise methods as novel as the conceptions from which they sprang. Through this process Gould inadvertently made a supreme contribution to his age: he imposed upon rival businessmen an education in the new realities of transportation strategy. For many of them it was a painful and unwelcome lesson.

In 1863 the Erie was still only a New York road dependent, like the other trunk lines, on connecting roads for its western traffic. Four routes dominated the business between New York, Philadelphia, and the West: (1) the lake shore route, a string of independent roads linking Buffalo and Chicago; (2) the Atlantic, which ran from its connection with Erie to Dayton and thence to St. Louis via two other lines; (3) the Pittsburgh, Fort Wayne & Chicago, which dominated traffic between Pittsburgh and Chicago; and (4) the Panhandle road, a recent amalgamation of lines between Pittsburgh and Columbus which reached Chicago via the Columbus, Chicago & Indiana Central.[13]

Although the trunk line managers appreciated the growing importance of western traffic, they had been content to rely on agreements to secure their connections. The Pennsylvania in fact had once owned large amounts of Fort Wayne securities but sold them off as part of a conservative policy to keep the company's debt down by avoiding expansion. It controlled the Panhandle road but not the Indiana Central. In studying these connections, Jay hit upon the idea of trying to capture all four routes for the Erie. His plan threatened to shut the other trunk lines out of western markets.[14]

In December 1868 Jay disposed of the Atlantic's receivership and leased the road to the Erie for twelve years. A month later he negotiated a lease of the Indiana Central and struck a blow at the Cleveland & Pittsburgh, an important road that provided the Fort Wayne access to the oil refineries in Cleveland. Through purchases of stock and proxies Gould obtained enough voting power to control the company's annual meeting, where he rammed through a series of resolutions revamping the company's by-laws and financial procedures in a manner amenable to his control. At the same time Jay quietly scoured the market for Fort Wayne shares and proxies until he secured enough of both in January to control that company's election on March 17.[15]

This succession of moves, accomplished in less than two months, fell like a bombshell among railroad men. Here was a young upstart, scarcely come to office, launching a policy of aggressive expansion against the two eastern giants. Jolted out of their complacency by Gould's raids, galvanized by threats that had not even occurred to them before, both abandoned traditional policies in favor of a hastily improvised counterattack.

The Pennsylvania's response was swift and decisive. Before the Indiana Central's stockholders met to approve the Erie lease, they accepted a higher bid from the Pennsylvania. Afterward the normally prudent Pennsylvania management discovered the road to be "in an unfinished and dilapidated condition, deficient in depot accommodations, rolling stock and shops." In Ohio the Pennsylvania enjoined Jay's high-handed actions at the Cleveland's annual meeting and forced a compromise giving him influence but not control of the road. As for the Fort Wayne, Jay controlled the stock but the Pennsylvania dominated the state legislature. The Fort Wayne's and the Pennsylvania's managers did not want the road's traffic diverted to New York via the Erie, and neither did the state's businessmen. They rushed to Harrisburg and, in only thirty-four minutes, obtained a classification act staggering the election of Fort Wayne directors over a four-year period. Later the Pennsylvania locked up the road with a perpetual lease.[16]

To the north the roads between Buffalo and Chicago had already begun the negotiations that would lead to their consolidation that summer as the Lake Shore & Michigan Southern. This was the route Vanderbilt relied on for his connection to Chicago, but he did not control the several companies. The Erie had access to Chicago over the Michigan Southern, and Jay had become friends with LeGrand Lockwood, who owned a large amount of its stock. Through an alliance with Lockwood Jay hoped to control the newly merged Lake Shore route. He had already tweaked Vanderbilt by gaining entrance to Albany over the Albany & Susquehanna, a road that would soon rise to plague Gould.[17]

By the summer of 1869 Jay had forced the country's two most powerful rail interests into startling reversals of policy. The Pennsylvania had responded with the fury of a maddened bull and used its superior resources to defeat him on every front. In the process it junked the policy of gradual, prudent growth for one of precipitate expansion. Besides securing its western connections, the Pennsylvania built new lines, gained access to New York, New Jersey, and Washington, and acquired a large network of roads in the South. Vanderbilt's turnabout was more subtle but no less complete. He understood the importance of the lake shore route but did not attempt to seize direct control of it until confronted by Gould's maneuvers. Even then he was slow to respond, partly because he assumed Lockwood to be an ally rather than a rival.[18]

Although the Lake Shore battle still hung fire, the war had plainly been lost. Accordingly, Jay undertook to lighten the burden of the Atlantic lease. His attempt to change the terms touched off another round of legal skirmishes that left the Erie in control of the Atlantic pending the latter's foreclosure and sale. The Atlantic might be a derelict but it remained the Erie's only western outlet.[19]

The Atlantic served Jay in another way, one that revealed his gift for dovetailing his personal interest with that of the companies he managed. Since 1866 the Atlantic had been steadily losing its oil business to the Pennsylvania. Although the traffic in oil was increasing, the competition threatened to become even more savage when the Lake Shore completed its line into the oil region. This struggle proved a godsend to Cleveland refiners, who needed special rates to offset their greater distance from eastern markets. In 1867 the refiners commenced the tactic of obtaining rebates by playing the railroads off against one another. The following year three of Cleveland's major refineries, including that of John D. Rockefeller, entered into an elaborate arrangement with the Atlantic, the Erie, and the Allegheny Transportation Company, a pipeline company organized in the fall of 1867 by Henry Harley and W. C. Abbott, two veterans of the oil region who had friends on the Erie board. The series of agreements signed in the spring of 1868 gave Rockefeller and his associates an enormous advantage over other Cleveland refiners; indeed, some historians have used them to account for the spectacular growth in refining capacity by Rockefeller's firm during these years.[20]

What role Gould played in these agreements does not emerge from the scant evidence. Shortly after becoming the Erie's president, however, Jay purchased a majority of the Allegheny's stock and installed Harley as General Oil Agent for both the Erie and the Atlantic. In August 1868 he signed a contract extending the original agreement to four years and increasing the drawback paid by the railroads to the Allegheny on crude shipped to the East Coast from $12 to $15 per car. Since the Erie had not yet leased the Atlantic, Jay induced that road to swallow the new terms through the simple device of selling a quarter of his shares in the Allegheny to the Atlantic's receiver, Robert B. Potter.[21]

Although these agreements were significant, too much must not be made of them. The historian who first made the documents public proclaimed shrilly that Gould educated Rockefeller in "the piratical tactics and amoral business methods of the Erie ring" and that the Standard Oil Trust "must be regarded as the gigantic offspring of the Erie ring." To regard this episode as the origins of a "concert of interest" between Gould and the Rockefellers is to do violence to the facts. The contracts did bring together for the first time two men who in later years would vie for the title of Most Hated Man in America, but no enduring alliance came of it. When their paths crossed in later years, it was often as antagonists rather than as allies.[22]

What the episode does reveal is the prototype of one method used by Gould to extract personal profits from his corporate interests. The increased drawback given the Allegheny in the August agreement enriched that company (and therefore himself) at the expense of the railroads. At the same time he could defend the arrangement as necessary to improve the competitive position of the

Erie and the Atlantic in the struggle for oil business. Jay's involvement in pipelines was no hit-and-run affair. In 1871 he became a stockholder and director in the Pennsylvania Transportation Company, a successor to the Allegheny organized by Harley to control nearly five hundred miles of pipeline.[23]

The tannery experience had taught Jay the value of properties ancillary to the main enterprise. He followed this pattern with the Erie, investing in coal lands, real estate, and other peripheral enterprises, some of which he later sold to his own company for a handsome profit. Gould's position as both buyer and seller strikes modern eyes as a glaring case of conflict of interest. In his own day, however, it was a common practice among businessmen who seldom bothered to make distinctions between their personal interests and their corporate responsibilities. While business moralists frowned on such behavior, those operators below the rank of saint unabashedly sought corporate positions for the purpose of trading on their inside information. Gould may have been more imaginative than most, but he was no pioneer in the art.[24]

Jay utilized his corporate position for personal gain in other ways as well, most notably in market operations. The most obvious device was to release information or spread rumors that made the price of Erie dance and then profit from the stock's rise or fall. This was an ancient Wall Street tactic but few operators matched Gould's skill of execution. On one occasion the executive committee authorized him to negotiate a more favorable contract with the United States Express Company. He and Fisk promptly informed the express company that its annual rent would be $500,000 higher than the previous year. When the express company balked, Jay terminated its contract and declared that Erie would organize its own express business.[25]

It was a typical Gould stroke. Because the express company had been instrumental in touching off a rate war, even a harsh critic of Erie applauded the action as inflicting "condign punishment" on the culprit. United States Express stock plummeted from 59 to 43. Jay bought the stock at low prices, negotiated a new contract with the express company, and sold out as the stock climbed into the sixties. At the same time he could parade the new contract before the Erie stockholders as a triumph with its generous increase of rent.[26]

On another front Jay managed to turn even the bitter rate wars among the trunk lines to his own advantage. These wars followed the customary cycle of international relations: distrust led to the outbreak of hostilities which lasted until one or both sides sued for peace; a settlement would then be negotiated, only to collapse beneath the weight of broken promises. Such was the case early in 1870 when the Erie aroused the wrath of its rivals by offering drawbacks on westbound traffic. An agreement reached in February muzzled the dispute only until June, when the war broke out with renewed ferocity.[27]

The eastbound livestock traffic soon emerged as the most conspicuous battleground. The usual rate from Buffalo to New York was $125 a carload. When Vanderbilt knocked the Central's rate down to $100, Gould put the Erie's at $75. The Commodore went to $50 only to have Gould drop to $25. Vanderbilt then decided to ruin the Erie's livestock traffic by setting his rate at the absurd figure of

$1 per carload. At the same time hogs and sheep were being carried for a penny apiece. Sure enough, the Central filled up with cattle while the Erie's cars ran empty. Vanderbilt cackled with glee until he discovered the reason for his easy victory. Unbeknown to him, Gould and Fisk had bought every steer in Buffalo and shipped them into New York via the Central.[28]

"When the Old Commodore found out that he was carrying the cattle of his enemies at great cost to himself and great profits to Fisk & Gould, he very nearly lost his reason," Morosini laughed. "I am told the air was very blue in Vanderbiltdom." The chastened Commodore understood at last what Drew had meant when he said of Gould, "His touch is death." In 1872 he went so far as to publish a notice declaring that he would have nothing more to do with Gould "unless it be to defend myself. I have always advised all my friends to have nothing to do with him in any business transaction."[29]

In the spring of 1869 Gould seemed well entrenched in the Erie presidency. By one pretext or another he had disenfranchised most of Erie's foreign stockholders or voted their stock on his own behalf. From the other directors he had secured written pledges to support his policies or resign; small wonder that he never bothered to call a board meeting. The executive committee consisted of himself, Fisk, Lane, Tweed, and Sweeny. Tammany supplied him with a stable of judges, political henchmen, police protection, and gangs of toughs when needed. Beyond these he could draw upon his battalion of lawyers which numbered in its ranks some of the era's finest legal talent.[30]

To bolster his position Jay spent freely for political favors and to insure the election of candidates friendly to Erie. In this work he never allowed questions of affiliation to sway him. "It was the custom when men received nominations to come to me for contributions," he told an investigating committee in 1873, "and I made them and considered them good paying dividends for the company; in a republican district I was a strong republican, in a democratic district I was democratic, and in doubtful districts I was doubtful; in politics I was an Erie railroad man every time." When pressed for details on how much was paid to whom, the reply was vintage Gould:

> There has been so much of it; it has been so extensive that I have no details now to refresh my mind; when I went over a transaction, and completed it, that was the end of it; and I went at something else; you might as well go back and ask me how many cars of freight were moved on a particular day, and whether the trains were on time or late; I could not charge my mind with details; I can only tell you what my general rule was; my general rule of action.[31]

It was a formidable citadel Jay had erected, but not an invulnerable one. The Erie's physical and financial weaknesses, like Banquo's ghost, would not down. Its capital stock jumped from $46.3 million in 1868 to $57.8 million by March 1869. The disposition of $11.5 million of new stock in only a few months struck one financial journal as "one of the most extraordinary transactions ever witnessed in financial circles." Company reports listed the stock as paid up, but where had the money gone? Erie stockholders still smarted from the methods used by Gould to rig the election of 1868. Since Jay had taken office the stock had dropped from 71 to 37 beneath the weight of new issues. The English holders were organizing to secure a greater voice in Erie's management; if domestic holders joined the mutiny, there could be trouble in the 1869 election. To counter this threat, Gould borrowed a tactic from his enemies. In May he obtained from the legislature a classification act staggering the election of Erie directors over a five-year period.[32]

Once again Jay managed to vote large amounts of stock owned abroad but left in the hands of American brokers with blank power of attorney. After a decisive victory he used the Classification Act to fortify his position. Bloodied but unbowed, the stockholders dug in for what promised to be a lengthy siege. Under normal circumstances a fight to oust Gould would have drawn lavish attention from the press, but in this instance the campaign received scant notice. Newspapers across the nation blared the name of Gould for quite another reason. To their dismay, the stockholders found their efforts upstaged by what proved to be the most spectacular episode in Gould's career.[33]

The young businessman with the first blush of the celebrated beard (courtesy Lyndhurst Archives).

(a) *Facing page, left,* a young and ambitious Jay Gould of Erie days, wearing his usual unrevealing expression and an unusually natty outfit (Brown Brothers); (b) *right,* Helen Gould as the frail and lovely young lady with whom Gould fell in love (courtesy Lyndhurst Archives); (c) *bottom,* Lyndhurst as seen from the front lawn (courtesy Lyndhurst Archives). (d) *Top,* Helen Gould with young Anna and Frank on the lawn at Lyndhurst (courtesy Lyndhurst Archives); (e) *above,* the greenhouse at Lyndhurst as rebuilt by Gould (courtesy Lyndhurst Archives).

Immediate left, Helen Gould at middle age (courtesy New-York Historical Society, New York City); *below, left,* the crown prince as young dandy: George Gould coming of age (courtesy Lyndhurst Archives); *below, right,* Edith Kingdon Gould, radiating the charms that so enchanted George (courtesy Lyndhurst Archives).

Right, Jay Gould at middle age, when his hairline and health were deserting him and toil and care had flecked his beard with gray (courtesy New-York Historical Society, New York City); *below*, *Atalanta*, Jay Gould's majestic yacht (from the author's collection).

Immediate left, the crown prince uneasy upon the throne: George Gould at middle age as head of his father's empire (courtesy Union Pacific System); *below, left,* an aged and ill Jay Gould recuperating at El Paso during the last year of his life (courtesy Lyndhurst Archives); *below, right,* Benjamin Constant's portrait of a pensive Jay Gould (courtesy Kingdon Gould, Jr.).

10

Infamy

Of all the financial operations, cornering gold is the most brilliant and most dangerous, and possibly the very hazard and splendor of the attempt were the reasons for its fascination to Mr. Jay Gould's fancy. He dwelt upon it for months, and played with it like a pet toy. His fertile mind even went so far as to discover that it would prove a blessing to the community, and on this ingenious theory, half honest and half fraudulent, he stretched the widely extended fabric of the web in which all mankind was to be caught.

—Henry Adams, *Chapters of Erie*

The general belief was that Gould was irretrievably ruined. . . . As a matter of fact, his underhand sales had brought him eleven or twelve million dollars profit.

—Gustavus Myers, *History of the Great American Fortunes*

Like an inspired fiend, Jay Gould had ridden out the storm to safety. . . . Opinion differed afterward as to whether he had gained nothing, lost all he possessed, or garnered eleven millions of dollars at one coup.

—Matthew Josephson, *The Robber Barons*

It required only a year to thrust Gould's name into public consciousness and less than another year to embellish it with trappings of notoriety. In 1869 it required little more than a summer to blacken his name with infamy in the most audacious speculation yet witnessed in Wall Street: an attempt to corner not a stock or group of stocks but the entire gold supply of the nation. Even an age jaded by tales of spectacular coups gasped in wonder at an endeavor of such immensity. So indelible seemed his imprint as author that in later years the day became known as "Jay Gould's Black Friday."[1]

That imprint is misleading. Because the Erie War and the Gold Corner episode were truly the stuff of legend, they fixed Gould's image in the public mind once and for all. Especially was this true after the Adams brothers cast Gould's villainy in bronze. In the process Jay received more credit for orchestrating the corner than he deserved. His critics, now and then, portrayed him as masterminding the whole scheme with cool deliberation, yet Jay protested that he "had no idea of cornering it" and was "forced into it by the bears selling out. They were bound to put it down. I got into the contest. All these other fellows deserted me like rats from a ship."[2]

The precise truth cannot be extracted from the evidence, which is abundant but riddled with contradictions. Most likely Gould started out with less sweeping intentions and encountered complications that plunged him into deeper waters than he had originally cared to test. He chose to extricate himself not by scurrying to shore but by expanding his scheme to breathtaking proportions. The gold episode has about it less an air of grand design than of grand improvisation.

It is important to remember that Jay's venture into gold did not take place in a vacuum. He first bought gold in the spring of 1869, when he was still struggling to secure a western connection for Erie. By then he had lost every outlet except the Lake Shore, where his tie with Lockwood still offered hope of victory over Vanderbilt. Jay also bought into the Toledo, Wabash & Western, a road running from Toledo to St. Louis, with the notion of uniting it with the Lake Shore as a through line to Chicago and the West. That summer he tried to acquire the Albany & Susquehanna, a line from Albany to Binghamton that formed a bridge between the Erie and the New York Central. Its guiding spirit, a hardnosed Scotsman named Joseph H. Ramsey, proved a worthy foe for Gould.[3]

Ramsey thwarted Gould's attempt to control the Albany's election by borrowing a page from his book. He dipped into the road's treasury for some unissued stock to vote for his slate. When Jay took refuge in the courts, the ensuing battle produced seven injunctions before almost as many judges. Gould had Barnard appoint Fisk and another crony as receivers for the Albany; Ramsey countered with a receiver of his own. Fisk led a platoon of Erie irregulars to seize the road's office by force and was bounced rudely out the door. Never one to mourn so trifling a thing as loss of dignity, he offered to play Ramsey a game of seven-up "to see who runs this railroad." There followed a pitched battle between supporters of Fisk and Ramsey who packed themselves aboard two trains that found each other in the Long Tunnel near Binghamton.[4]

The melee forced Governor "Toots" Hoffman, a Tammany stalwart, to intervene. He ordered the militia to operate the Albany until the company's election on September 7. By then the number of suits had climbed past twenty, with more waiting. The election produced two separate boards, one for each side. Wearily the attorney general put the contest in limbo by filing suit against both sides. If nothing else the affair added another chapter to the swelling legend of "Jubilee Jim." It was one that, like so many others, Gould would doubtless have preferred to do without.

This was the landscape against which Jay's gold operation took place. It seems doubtful he contemplated anything like a gold corner in the spring of 1869.

His financial load was heavy and increasing steadily. At the moment Erie had nothing left to give. Most of his own holdings were needed as collateral for loans to buy more stock, which left him vulnerable to any drop in the price of securities. The money market underwent one of its periodic spasms in April, and money was tight. Moreover, the Grant administration had just taken office and no one knew what its financial policy would be.[5]

As usual Jay's most pressing need was to raise cash for his many projects. The gold market caught his eye for two reasons. The most obvious was the hope of realizing speculative profits; the other concerned a theory borrowed from James McHenry, an English financier and dominant figure in the Atlantic road, that if the price of gold went up, more American wheat could be sold abroad and railroads would profit from long hauls of eastbound grain.[6]

McHenry's argument was based on complications arising from the use of both gold and greenbacks as currency. At home business was transacted in greenbacks, but foreign merchants accepted only gold as a medium of exchange. The disparity in value between the two currencies forced merchants in foreign trade to contract for domestic goods in greenbacks and sell them overseas for gold, or vice versa in the case of imports. Most transactions required several weeks to complete. Any change in the price of gold during that interval upset the original terms and might produce serious losses.[7]

To protect themselves, merchants paid a premium to borrow gold and sold it for the greenbacks needed to make their purchases. After a foreign exchange house discounted his bill, the merchant took the gold paid him and returned it to the loaner. This was in theory the legitimate business function of the Gold Exchange, but it also offered choice opportunities for speculation. Merchants who borrowed in this manner were in effect short of gold. A fall in the price of gold could wipe out their profits on business transactions. However, a sharp rise in gold required merchants to put up fresh margins (in greenbacks) against what they had borrowed. In the process some might go bankrupt before their bills were discounted abroad. If speculators could control the available supply of gold, they could use the frantic buying by merchants who were short to help run the price up.

Farmers too were affected by changes in the price of gold. The prices of commodities such as wheat and corn were fixed in the London market and based on gold. Since farmers received greenbacks for their crops, a rise in gold meant higher prices for commodities. In effect the weakened value of greenbacks against gold stimulated exports by making wheat or corn cheaper for foreigners to buy and more profitable for American farmers to sell.* It enabled American producers of breadstuffs to compete with producers in Mediterranean countries for the European market.

What did all this have to do with Gould's plans? By 1869 the center of

*To clarify this point, it may help to recall that quotations on the Gold Exchange expressed the value of gold in terms of greenbacks. Thus, if gold rose from 130 to 135, the amount of greenbacks needed to purchase $100 in gold increased from $130 to $135.

American wheat production had shifted to the Midwest. Increased exports meant large shipments of grain eastward on the Wabash, the Lake Shore, and the Erie. Improved earnings would benefit the roads and enhance the worth of their stocks, which in turn protected their value to Gould as collateral. Herein lay all the elements for another masterful demonstration of Gould's ability to intertwine his personal and corporate interests. If he could advance the price of gold, the Erie would prosper; so would other railroads, farmers, export merchants, and others involved in overseas trade. While wearing the mantle of corporate and public benefactor he could at the same time realize large speculative profits.[8]

It was surprisingly easy to corner the available supply of gold in New York. The amount, usually ranging between $15 million and $20 million, could be purchased on credit with a modest investment and then loaned to merchants who were short. As Jay later asserted, "A man with $100,000 of money and with credit can transact a business of $20,000,000." The chief obstacle to such an operation was the federal treasury, which maintained a large supply of gold. At any time the secretary of the treasury could thwart an attempt to make gold tight by selling part of its holdings. For any corner to succeed, therefore, it was essential to keep the government's gold out of the open market.[9]

When Jay first tested the waters that spring, the new administration had not yet formed a policy in regard to gold. He began buying in April and advanced gold from 130 to 142 before selling out. Other speculators ran the price up to 145 by late May, whereupon Secretary of the Treasury George S. Boutwell announced that the government would sell twice the usual amount of gold each week until further notice. Although this action sent the price tumbling, the government's intention remained unclear. Early in June, reports spread that Boutwell intended to reduce gold sales again.[10]

These mixed signals convinced Jay that nothing could be done in gold without inside information. Although he did not know Grant, he was acquainted with his brother-in-law, Abel R. Corbin. Sly, garrulous, whining, Corbin was a hot-air balloon, puffed up with pretension that collapsed at the first pricking. At sixty-seven he was long retired from a checkered career as lawyer, editor, lobbyist, and speculator. Like Drew he was devoted to Methodist pieties and took care not to let religion intrude on his avarice. As Gould was quick to discern, age had in no way diminished Corbin's appetite for money. Although married to Grant's sister only a year, he liked to leave the impression that he was the president's confidant and adviser. On that point he may have bamboozled Gould, who later dignified him as "a very shrewd old gentleman, much more far seeing than the newspapers give him credit for." Those who knew Corbin, however, shared the opinion of a journalist who called him "the worst and most consummate old hypocrite I ever saw."[11]

That spring Jay impressed on Corbin the McHenry argument and its importance for the national economy. Corbin snapped at the bait. Declaring that the president should hear Gould's views, he promised an introduction. The meeting took place on June 15 aboard one of Fisk's steamboats carrying Grant to the Peace Jubilee in Boston. During supper talk turned to the economy, crops,

Boutwell's gold policy, and what the future held. While everyone exchanged views, Grant listened in his usual sphinxlike silence. Asked his opinion, the president replied that there was an element of fictitiousness in the current prosperity, and that the bubble might as well be tapped one way as another. His words cast a pall of gloom across his audience; as Gould noted, "We supposed from that conversation that the President was a contractionist."[12]

The meeting with Grant convinced Jay that the time was not ripe for bulling gold. By early July gold was at 136 and declining. Money remained tight and a financial editor complained that "the money market has been more completely in the hands of speculators during the last six months than ever before in this country." As the summer wore on, however, it became evident that the country would produce bumper crops. Although Boutwell reduced government gold sales in July, the price continued to fall, as did that of wheat. A sharp decline in grain receipts at ports indicated that little wheat was being moved. The deeper Jay got into his railroad campaigns, the more imperative it became to secure traffic for the roads. His need for funds grew more desperate as his financial load increased. Remarkably, the stringency in money had not yet produced a fall in stock prices, but he could not rely on that situation continuing.[13]

After mulling the matter over, Gould decided to try again. He calculated that if gold could be nudged above 140, wheat would move and his own speculative profits would be substantial. Since everything depended on what the government did, he searched for ways to gain inside information. During the spring the position of assistant treasurer in New York had fallen vacant. As the officer who executed the government's transactions in gold, the right person in this post could be invaluable. Corbin proposed General Daniel Butterfield. Jay was not enthused at the choice but deferred to Corbin's judgment. The fact that Butterfield received the appointment helped persuade Gould that Corbin must have influence in Washington. From that moment Jay fell into the fatal error of relying on Corbin's ability to sway Grant's views. He also lost no time in making himself useful to Butterfield. Shortly after the general assumed his office on July 1, Gould tendered him a loan of $10,000.[14]

To wage a gold campaign Jay needed more credit. On August 5 he bought a majority interest in the Tenth National Bank. Several days later he transferred most of the stock to his Erie and Tammany associates but did not disturb the bank's management or its policies. All he wanted was a friendly bank that would extend him credit on the most tolerant of terms and in a pinch certify checks drawn against assets not yet deposited. In addition to credit, Jay needed allies. In August Fisk bowed out of the scheme after another futile attempt to divine Grant's intentions, declaring later that "the thing began to look scary to me." In his place Jay enlisted two other speculators, William S. Woodward and Arthur Kimber.[15]

Jay could not have been pleased with the way matters stood, if only because there were too many loose ends for his precise intellect. Gold was still tending downward, and there was no way of knowing how fiercely the bears would retaliate once he started buying. Grant and Boutwell remained enigmas to him, capable of devastating his plans with a single stroke. Apart from his dependence on so frail a reed as Corbin, he was about to launch a major operation without his most reliable ally, Fisk. The pact with Kimber and Woodward was no more than a loose agreement to buy $3 million of gold apiece in tandem, with each man free to do what he pleased afterward. Fisk was right in describing the outlook as scary.[16]

In mid-August Gould began to buy gold again. He also tried a letter to Boutwell reiterating his theory on the relationship between grain exports and the price of gold. It was a clever ploy to draw Boutwell out, but the secretary declined the bait with a noncommittal reply. The price of gold advanced only slightly, even though Jay and his allies bought more than they originally intended. Just when the scheme seemed on the verge of foundering, however, a fresh ray of hope appeared. Early in September the president stopped at Corbin's house on his way to the funeral of General John A. Rawlins. Corbin arranged a meeting with Gould, who discovered to his surprise that Grant had changed his views entirely. Reports of a bountiful harvest had convinced him of the need to sell foodstuffs abroad, and the government would not impede the process by putting gold down or making money tight. A short time later Grant met Gould again briefly and repeated these views.[17]

So Gould later testified. Although his version has been disputed, there is strong evidence to sustain it. On September 4 Grant wrote Boutwell that it was "undesirable to force down the price of gold" because of its effect on the movement of crops. Despite assurances that any decision on the matter was his to make, Boutwell took Grant's opinion seriously enough to countermand an order to sell gold. A week later Grant surprised Boutwell with another letter warning that "a desperate struggle is now taking place, and each party wants the government to help them out. . . . I think, from the lights before me, I would move on, without change, until the present struggle is over."[18]

Of this letter Jay knew nothing. Fortified by Grant's apparent change of heart, he stepped up his gold purchases. Already he had offered to put $1.5 million worth in Corbin's name and carry it for him at no charge. Corbin delicately refused, then consented to have the gold placed in his wife's name. It was, as Henry Adams observed, a transaction "worthy of the French stage." Jay also made a similar arrangement for Butterfield, who later denied it.[19]

By the first week in September Gould found himself enmeshed in an undertaking of gigantic dimensions. The web of intrigue spun out over the summer had caught up in its fragile threads half a dozen railroads, the gold market, the stock market, the grain export trade, and the fiscal policy of the United States. The enormity of the stakes must have staggered even Gould. Gradually it dawned on him that he was in too deep to get out, and that his best, if not his only, chance lay in plunging bravely ahead.

Gold rose steadily until it touched 138 on Wednesday, September 8, but Jay had little cause for cheer. Already Corbin was at his door demanding the profits on the gold purchased for him. Gould gave him a check for $25,000 thoughtfully drawn to himself so that Corbin's name need not appear on it. The advance also induced Kimber to sell his holdings and switch to the short side. Woodward looked at his accounts, discovered he had bought upwards of $10 million worth, and complained to Gould that he did not care to hold that amount. He agreed to keep $4 million after Jay offered to take the rest off his hands. Before Jay could catch his breath, the bears launched a savage counterattack. Amid frenzied trading gold slid back to 135 and hovered within a point of that figure for nearly two weeks. Jay found himself dangling in limbo, unable to push gold up again with the resources at hand and uncertain as to whether to stay in or get out.[20]

Help was needed. Fisk was the obvious choice, but Jay hesitated. Perhaps there was a touch of pride in his reluctance; Fisk had smelled trouble from the outset, and Jay didn't care to admit he had been right. Moreover, Fisk was a bull in a china shop, difficult to control in a delicate operation. Jay wanted only to boost gold to 140 or 145 long enough to move the crops and take his profits, but how to restrain his partner's ebullience once the excitement started? Day after day Gould sat brooding at his desk in the Opera House, his eyes fixed on the indicators that told him the latest prices on the Street and his ear tuned to the private telegraph wire that brought him information from the financial district and all parts of the country. Intelligence was plentiful, alternatives scant. For some days Fisk had noticed an edge of coolness between himself and Gould, born probably of Jay's urgent need for help and his reluctance to ask for it. At last Jay broached the subject by asking, "Don't you think gold has got to the bottom?" Fisk replied that it was futile to buy gold unless the market could be controlled, but after further discussion he agreed to come back in.[21]

Much confusion and apparently conflicting testimony followed because it was assumed that Gould and Fisk were acting in perfect concert. Normally they did, but not this time. Although the overall effort was coordinated, each man executed his transactions independently of the other through separate networks of brokers. As Jay put it, "Our interests were entirely separate. He had his own gold and I had mine." Fisk noted that it was their usual custom to "make up a settlement and divide the results, whatever they are. This, I think, is the only case in which that was not done."[22]

To mask so large a movement it was necessary to employ an army of brokers. During the next ten days only one firm, William Heath & Company, handled gold transactions for both Gould and Fisk. Jay entrusted most of his business to Henry Smith, who in turn recruited Edward Willard, Benjamin Carver, Edwin Chapin, and Charles Osborn, among others, to execute orders. Asked later how many brokers he employed, Smith replied, "Probably fifty or sixty; it is impossible to give a correct estimate. . . . I would employ one broker and he in turn would employ several others." Fisk relied on his former partner, William Belden, described by a fellow broker as "a very timid man who was easily frightened." Belden managed to put together a network of brokers similar to that of Smith but on a smaller scale.[23]

Although the separateness of these networks may seem a mere nicety, it was in fact genuine. Many of Gould's brokers disliked or distrusted Fisk and Belden. Willard described Fisk as "an erratic sort of genius" who "never could do business with Smith, Gould, Martin & Co. very comfortably. They would not do business for him. . . . None of us who knew him cared to do business with him. I would not have taken an order from him or have had anything to do with him." Charles Osborn declared afterward that "Fisk was the main man in the whole thing, so far as illegitimate proceedings were concerned. . . . There seemed to have been two separate rings in this matter."[24]

It was crucial not to inform all these brokers about the true scale of the operation or its authors, lest they ruin attempts to orchestrate the rise by buying heavily on their own account. On this point Gould was as clear and precise as Fisk was loose and careless. The arrangement also suited Jay because it enabled him to pursue his objective regardless of what the more impulsive Fisk might decide to do in the heat of battle. What troubled him most was a nagging uncertainty over the government's intentions. He began to visit Corbin every morning and evening, but the old man's humbug did not reassure him. Fearing that he had done too little, Jay let anxiety push him into attempting too much.[25]

Grant stopped at Corbin's again before going on to Washington, Pennsylvania, for a brief vacation. During this visit Jay managed a third interview with Grant. Before Grant's departure Jay approached his private secretary, General Horace Porter, with an offer to buy gold for him. Porter declined, saying "it would be a manifest impropriety for me to do it." After his arrival in Pennsylvania Porter received from New York a sheet of note paper, unaddressed and signed by Gould, noting that $500,000 in gold had been purchased in his name. Annoyed, Porter wrote Gould at once to repudiate this and any other transaction in gold on his behalf.[26]

Even before Porter's reply reached him, Jay decided on another step. On the morning of September 16 he asked Corbin to urge Grant not to change the government's policy until the crops had moved. Corbin dutifully prepared a long letter rehashing Gould's argument and hinting ominously at rumors of impending gold sales which, he insisted, would be ruinous to the national interest. Since Washington, Pennsylvania, lay southwest of Pittsburgh, thirty miles from any railroad or telegraph station, Fisk procured a trusted Erie employee, William Chapin, to deliver the letter by hand.[27]

Chapin dutifully took an early train, reached Pittsburgh at one in the morning, hired horses, and rode all night to Washington. When he arrived next morning, Grant was on the lawn playing croquet with Porter. Chapin waited in the house until the game had finished, then handed Grant the letter. The president read the letter twice, left the room for a time, and returned, lost in thought. Chapin asked if there was any reply. "No, nothing," Grant answered, whereupon Chapin returned to Pittsburgh and telegraphed Fisk, "Delivered all right." Somehow his message was garbled in transmission and received in New York as "Delivered. All Right."[28]

On Friday, September 17, gold closed at 136 5/8, up only a point for the week. So far the bears had staved off Gould's efforts, although the gold supply had dwindled to a point where borrowers were growing uneasy. The situation was volatile, the Street charged with anticipation. The more the bears resisted, the harder the push needed to dislodge them; the harder the push, the greater the chance of precipitating a panic that would engulf them all. Too fast a rise in the price posed as grave a danger as no rise because it might force the government's hand. Chapin's telegram was reassuring but did not entirely dispel Jay's misgivings. What others later described as a gigantic combination felt to him more like a house of cards liable to collapse at the first hard blow.[29]

Smith and Belden's army of brokers attacked the market with renewed fury, but at the close of trading on Tuesday, September 21, had gained only another point to 137 5/8. Although the amounts held were already huge, Jay realized gloomily that much larger purchases would be required. On Wednesday Fisk personally led the charge into the Gold Room. His magnetic presence, with its mixture of banter and bullying, swept like a riptide across the floor, bellowing orders, braying offers to bet $50,000 that gold would touch 145, venting rumors that the bulls could not fail: the government was with them, the president was with them, no gold would be sold, Corbin was about to be made president of the Tenth National Bank.[30]

While these tactics forced the price upward, the bears refused to crumble. Soaring money rates spread a ripple of contraction into the stock and bond markets as well. New York Central plummeted 25 percent in a few hours, and other stocks followed suit. The tumbling stock prices encouraged some gold operators to renew their short selling. If bulls were paying half a percent a day for the currency needed to carry gold, it was tempting for bear operators to loan currency and use the proceeds to borrow gold which could then be sold short. Despite their efforts, however, gold closed at 141 1/2 on Wednesday. To accomplish this required enormous purchases later estimated at between $50 and $60 million.[31]

While the din of battle raged in the Gold Room, Jay sat at his desk in the Opera House, penning another letter to Boutwell. "There is a panic in Wall Street," he began, "engineered by a bear combination. They have withdrawn currency to such an extent that it is impossible to do ordinary business." He urged Boutwell to increase the currency supply by loaning banks some of the government's reserve. Needless to say, he did not mention the bulls' need for currency to promote their gold operation. Once again Boutwell refused the bait.[32]

That evening Jay went as usual to visit Corbin, who greeted him in a state of wild-eyed agitation. His wife had received a letter from her sister, Julia Grant, stating that the president was distressed at Corbin's gold speculations and that he must close them up as quickly as possible. He showed the letter to Gould, who grasped at once what had happened. Corbin's letter to Grant, instead of pacifying him, had aroused suspicions that the old scalawag was dabbling in gold. The message sent through his wife was plain to Corbin: "I must get out instantly—instantly!" For Jay the broader implication was as obvious as it was ominous: Grant would not hesitate to sell gold if the price rose high enough to threaten a panic. Corbin had not an ounce of influence with the president and stood revealed at last for the fraud he was.[33]

A nasty dilemma impaled Jay. He could not prevent Grant from selling to avert a panic, and he could not prevent Fisk from trying to bull the price through the roof without precipitating a slide that would ruin them all. As his mind searched for a way out Corbin hovered at his elbow, nervous and fitful, his watery eyes blinded less by fear of disgrace than by the dollar signs dancing before them. Since he must get out, he whined, why didn't Jay simply close his account and pay over the profits accrued, which amounted to about $100,000? Jay's dark, inscrutable eyes fixed him with a stare. He asked Corbin to say or do nothing until the following morning.

Early on Thursday morning Jay went again to Corbin's house. He refused to buy Corbin's gold but offered him $100,000 on account if he would leave his gold in the market. Before leaving he extracted from Corbin a promise not to reveal the contents of Julia Grant's letter to anyone, adding, "I am undone, if that letter gets out." Corbin agreed, knowing full well that he too would be ruined if the market broke sharply. When Jay arrived at the Opera House he told Fisk only that Corbin was nervous and wanted money. Smith was already there, and Belden soon appeared. Fisk swaggered about in high spirits, his blood up for battle. He too sensed an edge of panic on the Street, as if the bears were struggling to hold back a dam that, once burst, would sweep them all away. He ordered Belden to put gold up to 144 and keep it there. Across the room, in his soft voice, Jay gave quite another message to Smith: sell gold, buying only enough to conceal sales.[34]

Already Jay had determined in his cool, methodical way what must be done. Ironically, gold stood at just the price he would like to have frozen for some months, but he knew it could not stay put. The speculative forces tugging at the price would wrench it violently up or down. Convinced now that the government would intervene if gold went too high, he abandoned without a flicker of hesitation the bulling scheme that had cost him months of difficult work. There was no point in telling Fisk about the letter and a good reason to conceal it from him. They were in too deep and the market was too volatile for them to unload without huge losses. The only hope lay in Jay selling out his holdings in a rising market, which could best be done if Fisk continued to play his role of bull to the hilt. When the price collapsed, Fisk and his brokers would suffer heavy losses but Jay would emerge relatively unscathed. The task then would be to extricate Fisk from the wreckage. There were ways of doing this, using the lawyers and judges at their disposal.

Later it would be charged that Jay saved himself by ruthlessly betraying his closest partner. This accusation not only misreads Gould's character but overlooks the ingenuity of his hastily improvised plan. If both men were caught in the crash, it would be difficult to salvage anything. If word about the letter leaked out, even to Fisk, it might weaken the resolve to play out the bull movement so crucial to Jay's plan. Certainly Jay never regarded his actions as a betrayal of his partner, and neither did Fisk. The most telling evidence on this point is the fact that the episode did not ruffle their friendship in the slightest. Jay was playing a dangerous game, but he could find no better one.[35]

The first sale in the Gold Room that morning was made at 141 5/8. Within minutes pandemonium engulfed the floor as the bears, realizing their danger, fought desperately to stave off disaster. In the center of the room the gilded dolphin spouted its stream of water with placid regularity while around it a raging sea of brokers shoved and shouted at each other like clashing armies broken into the confusion of a hundred separate battles. "As the roar of battle and the scream of the victims resounded through New Street," a reporter observed, "it seemed as though human nature were undergoing torments worse than any that Dante ever witnessed in hell."[36]

The bears fought fiercely but in vain. As the indicator climbed inexorably higher, the smaller fry rushed frantically to cover their shorts at any price. Fisk was there again, exhorting his troops, shouting oaths, offering wagers that gold would touch 145. By the day's end the bears had been routed and a few swept into bankruptcy. When the last exhausted traders slouched from the Gold Room, the indicator stood at 143 1/4 and no one doubted it would go higher the next day. Total clearings exceeded an incredible $325 million, more than three times that of the previous day.[37]

Fisk's brokers had amassed an impressive pile of what he liked to call "phantom gold." That day alone Albert Speyers had bought nearly $6 million, Stimson $7.25 million, Carver $3 million, Russell Hills $7 million, Samuel Boocock $1.2 million, and Heath $3 million. Belden boasted that the clique held calls for $110 million in gold, of which about $7 million had been called in to foment a demand for the next day, when they expected to force the bears and hapless merchants to their knees. Heath had also purchased $3.4 million for Gould. He had been made an unwitting smokescreen by Smith, who took pains to entrust his sell orders to brokers that did no business for Fisk. One of them, Edwin Chapin, sold more than $8 million that day, most of it to Belden or his men.[38]

That evening Gould, Fisk, Smith, Belden, and Willard gathered at the Opera House to plot strategy for the next day. What transpired was less a meeting than a series of separate and overlapping conversations. Jay sat at his desk talking with Smith while Fisk and Belden huddled in a far corner. The strange mixture of moods illustrated again the sharp contrast between the partners. While Fisk cackled gleefully and brayed instructions, Gould sat in thoughtful silence, his voice barely above a murmur when he spoke, calculations tumbling through his mind, his mask of calm betrayed only by his nervous habit of tearing paper into bits or fiddling with objects on his desk. They talked, he listened, saying little and phrasing what he said with exquisite care. "I . . . had my own fish to fry," he recalled, indulging his weakness for homilies, "and I would listen to everything that everybody said . . . but it went in one ear and out of the other. I was all alone, so to speak, in what I did, and I did not let any of those people know exactly how I stood."[39]

Fisk was jubilant. Inspired by the day's triumphs, he was ready to bid gold up from the start next morning and delegated some brokers to terrify bears into private settlements. Belden leaped at the opportunity with the flushed enthusiasm of a nonentity whose chance of glory had at last arrived. Apart from his purchases for Fisk, he gave the latter carte blanche to buy gold in his name.[40]

That same evening in Washington Boutwell went to see Grant, just returned from Pennsylvania, about the crisis boiling up on the Gold Exchange. Merchants and bankers were deluging the secretary with telegrams imploring him to relieve their distress by selling gold. Other reports warned him that the Tenth National Bank was locking up greenbacks and certifying checks to a huge amount for parties involved in the gold movement. Boutwell at once sent a team of examiners to investigate these charges, ordering them to be at the bank before it opened for business on Friday morning. When he related all this to Grant, the president dismissed the high price of gold as artificial and injurious to the country but left to Boutwell the decision of whether or not the government should sell.[41]

The day that would become infamous as Black Friday dawned bright and clear in New York. Long before the Gold Room's opening hour of ten o'clock excited crowds jammed New Street and the alleyway from Broad Street, surging and jostling into doorways while a fortunate few packed themselves into the Gold Room's cramped galleries. Their ranks included genteel businessmen and merchants who ordinarily avoided the Exchange as they would a brothel but who, staring ruin in the face, could not resist coming to watch their fate unravel. Outside speculators, who normally entrusted everything to their brokers, flocked there in droves for the same reason. The idle and the curious, having sniffed in newspaper accounts the scent of impending disaster, mobbed New Street eager for the spectacle. Through their midst streamed knots of brokers, bleary-eyed and haggard from trading sessions that raged until midnight at the Fifth Avenue Hotel, their faces sagging with weariness and apprehension, their courage fortified by a fresh collar, a dose of coffee or brandy. Reporters were everywhere, ferreting out every lead or rumor, searching the crowd for the leading actors in the drama about to unfold.[42]

Early that morning, while the crowds were still gathering, Gould and Fisk decided that the Opera House was too far from the scene of battle to get intelligence quickly enough even by telegraph. They went to Heath's offices at 15 Broad Street and installed themselves in a back room flanked by bully boys. On the way downtown Jay stopped at the Treasury Building to see Butterfield, who told him no news had come from Washington. Once arrived at Heath's, Gould and Fisk remained there the entire morning. Heath popped in to ask Gould for more margin; Jay scribbled a note instructing the Tenth National Bank to give Heath a loan or certify his checks for the amount needed. Belden came in with Albert Speyers, a short, wiry broker described by one observer as "an elderly man of small intelligence." Fisk dispatched him with orders to buy gold at 145, after which Belden went off to rally his brokers, telling them in his loud, exhilarated manner that Gould and Fisk were behind him and gold would reach 200 that day. Smith too was issuing orders: "Sell, sell, sell," he confided to Willard, "do nothing but sell."[43]

In the Gold Room the heaving mob of brokers, unwilling to wait for the opening gavel, were already shouting out bids. Gasps and shrieks greeted the first bid of 145 as the bears, blind with desperation, tried to stave off ruin. Both sides sent bids whizzing like bullets in hopes of inflicting mortal wounds. A messenger handed Speyers a note from Fisk ordering him to put gold at 150. By the time the president's gavel fell at ten o'clock Speyers had accomplished his task: the opening price was listed as 150. Already Smith and Belden had their men pressuring the weaker bears to put up more margin or make private settlements. The room had become a stifling, frantic cockpit in which threats, insults, and cries for relief were interspersed with bids. Nearly three hundred brokers pushed against the iron railing that protected the statue of Cupid, who clutched his gilded dolphin as if fearful it might be snatched away and thrown into the day's auction. All eyes were on the gold indicator, which paused briefly at 150 1/2, then moved inexorably upward again. Speyers hurried to report the price to Fisk, who said, "go back and take all that you can get at 150.[44]

The scene was no less hectic in Heath's back room where Fisk, clutching a large cane like a marshal's baton, barked orders while Jay sat in a corner reading or crumbling bits of paper. Heath dashed in red-faced to report startling news: Dickinson, the president of the Tenth National, had said he could no longer certify checks because bank examiners had unexpectedly arrived. Even worse, rumors were spreading that the bank was in trouble and Dickinson feared a run. Another bank president, Henry Benedict of the Gold Exchange Bank, arrived to protest the staggering volume of clearings thrust upon it. Apprehensive about the bank's ability to withstand the load, he demanded to know if all contracts were to be settled that day. Fisk hesitated, then declared, "I might as well tell you that this is the day."[45]

It was approaching eleven o'clock. Gold stood just above 150, where it stuck for a time. Jay continued to sell in ever-increasing quantities. "I did not mean that anybody should say that I had opened my mouth that day," he noted later, "and I did not." Then came a piece of news that confirmed his worst fears: Joseph Seligman was selling gold. Seligman was Butterfield's broker and would not sell unless he knew something. Apprehensive, Jay dispatched a messenger to Butterfield, who reported that nothing had come from Washington. Something in his response struck Jay as evasive. Putting this together with other bits of intelligence, he deduced that time was short before the government entered the market. He must sell out at once, before it was too late.[46]

Of this decision Jay later observed, "A man who is liable to rapid thinking very often arrives at conclusions without being able to tell the process, and yet he is satisfied the conclusions are correct." When pressed on the point he added,

> I can only say that it is one of those conclusions that a man sometimes arrives at intuitively, that are correct in themselves, and yet if you undertake to give the evidences by which they are reached you could not tell how it was done.

Later critics of Gould ignored this explanation, which is in fact a revealing insight into the peculiar blend of logic and intuition that enabled him to make decisions with lightning speed throughout his career.[47]

Without betraying a hint of his anxiety, Jay told Smith to hurry the selling along. His reticence amid so much excitement confused even Smith, who later insisted that Gould had instructed him not to sell but that he did so anyway on his own initiative because "I was protecting myself, regardless of orders or anything else." Going back to his office, Smith found the place jammed with brokers and had to shout them down before he could give orders. "I was half crazy that day," he admitted.[48]

In the Gold Room the milling herd was fast turning into an ugly stampede. The gold indicator, stalled at 150 1/2, leaped to 155 in only six minutes. A deafening roar went up. Transactions flew so thick and fast that telegraph wires melted or burned under the frantic efforts of operators to keep up with sales. Glassy-eyed brokers shrieked hoarse cries of triumph or wails of doom. The shock waves battered the Stock Exchange next door, where prices were tumbling fast. A stockbroker named Horton rushed into the Gold Room and warned Speyers that he would be shot if he continued his mad bidding. The excitable Speyers dashed over to the Stock Exchange, fought his way onto the platform, and cried angrily that he was a member and wanted to face the cowards who had threatened him; that he intended to keep bidding gold up and offered himself now as a target to anyone who wished to shoot him. When no one obliged, he rushed back to Fisk who ordered him to "go and bid gold up to 160. Take all you can get at 160."[49]

Speyers did as he was told. Bedlam swept the Gold Room as the dial inched toward 160. More brokers and merchants crumbled beneath the pressure and went to settle with Smith or Willard at ruinous prices. Then suddenly Speyers found customers at 160. Gould was selling, of course, and so was James Brown, a banker who represented some merchants short of exchange. A fierce tug of war kept the price around 160 until shortly before noon, when word came that Boutwell had ordered the sale of $4 million in government gold. News of the sell order shattered the floor like a bolt of lightning. Amid thunderous cheers the bears toppled the price, which had climbed above 162, like a deposed tyrant. Fisk's brokers milled about in bewilderment, unable to comprehend so swift a turnabout. Dazed and exhausted, Speyers kept shouting orders for gold at 160 after the price had plummeted to 135. Several observers thought he had gone insane; Fisk called him "crazy as a loon."[50]

Shortly before the collapse Jay sent another runner to Butterfield. By the time he returned with a copy of Boutwell's order, gold had already dropped to 135. Although the news was of no help to Gould, the good general had taken care to sell his own gold well before the collapse.[51]

The chaos in the Gold Room rushed through the financial district like a storm. Gold brokers flapped about the floor like headless chickens, bewildered as to who owed how much to whom, searching for ways to enforce or repudiate contracts made in the last feverish stage of bidding. The Stock Exchange was no less convulsed and the Gold Exchange Bank, still groaning beneath the weight of Thursday's transactions, was utterly unable to handle the mass of clearings thrust upon it. An angry mob gathered outside the office of Smith, Gould & Martin howling for vengeance. There was turmoil at Heath's office as well and grave trouble at the Tenth National Bank, where news of the examiners' presence had triggered a run. The mood on Wall Street was ugly, bordering on violence. A company of militia was hurriedly ordered into readiness.[52]

The situation had deteriorated to the point where, in Fisk's choice phrase, "it was each man drag out his own corpse." He and Gould slipped out Heath's private entrance and took refuge in the Opera House for the weekend, transacting business behind a shield of burly palace guards. For two days the plush Erie offices were awash with lost souls wailing cries of doom and failure. Like flotsam on an incoming tide this procession of broken spirits swirled around Gould and Fisk, who poked gingerly among the debris to see what could be salvaged. A distraught Heath and his partner demanded protection for the purchases they made. Belden was beside himself, as were other brokers staring bankruptcy in the face.[53]

Corbin came around, too, on Friday evening and again the next morning to survey the ruins firsthand. Jay ushered him into another room for a private chat. Corbin thought Gould quite depressed but admitted, "It is difficult to read a man reticent as he." Before much was said, Fisk burst into the room like a wounded bull and showered Corbin with abuse. Shaken and abject, Corbin fled the premises. On his return the next day he encountered Fisk alone and asked cautiously how Gould was taking the calamity.

"Oh, he has no courage at all," Fisk snorted. "He has sunk right down. There is nothing left of him but a heap of clothes and a pair of eyes."

The bells of Trinity Church pealed a funereal dirge across Wall Street that weekend. Everywhere men worked frantically to unsnarl the tangle that paralyzed the financial district. Friday's melee shut the Gold Room down, leaving hordes of unfulfilled contracts and no way to determine prices or enforce settlements. Many firms stood in a technical state of failure because they were unable to submit statements or effect clearings as required by the rules. The hapless Gold Exchange still had not completed Thursday's clearings, let alone taken up Friday's massive load. The inability to make clearings or balance accounts meant that nothing could be resolved except through private settlements. Men who had sold gold at high prices to Fisk's brokers clamored loudly for their contracts to be honored. Scores of smaller dealers had already failed and one distracted broker, Solomon Mahler, shot himself to death Saturday morning.The run on the Tenth National Bank continued through Saturday. Failure was averted only through the intervention of the national bank examiner, who arranged a loan with another bank.[54]

Although Jay had sold off his gold, his position remained precarious. The fall in stock prices left him owing large sums on margin loans. His gold accounts were muddled by the clearing debacle, as were those of Smith and his brokers. Fisk and his brokers, who still owed millions in outstanding contracts, had to be rescued. Faced with myriad difficulties, Jay resorted to a familiar tactic: he utilized the courts to attack and defend, and above all to obfuscate and delay.

Amid the debris of Black Friday it was inevitable that the lawyers would be summoned to sweep up. On Monday Jay and his allies obtained no less than twelve injunctions from friendly judges. One put the Gold Exchange Bank into the hands of a receiver; others prevented the Gold Exchange from selling any gold bought by Gould's or Fisk's brokers under its rules, from taking any action to protect its members except through the courts, and from expelling Smith. "These contracts were made to be settled through the clearing-house," Jay explained later, "and as the bank had failed, they could not be settled through the clearing-house. Then the only way would be for these brokers to go out and get the gold and deliver it; and in this way, unless they were restrained by some legal process, a great many of those brokers would have failed. . . . I wanted to save as many as possible from failure. That was the point; to keep firms from failing."[55]

Including his own. Snarls of protest greeted the injunctions on the Street, where Gould's motives were regarded as something less than charitable. Efforts to push the clearing process along had failed miserably. When the Gold Exchange shut down, another bank undertook the task but was overwhelmed by the sheer bulk of transactions. The Gold Room then appointed a Committee of Twenty, which requested a statement of contracts from every firm or broker. Smith, Gould & Martin outraged the Street by being the only firm refusing to submit such a statement. Using figures provided by other dealers, the committee constructed a sheet for Smith, Gould & Martin that showed outstanding contracts for $7.5 million in sales and $20.6 million in purchases. To avoid having to take this balance of $13.1 million at ruinous prices, Smith sought refuge in the courts.[56]

Amid so much turmoil and confusion the committee made little headway. Faced with a derangement "more serious than anything experienced since 1857," the Gold Exchange proclaimed on October 1 that members must settle their contracts as best they could in private. That was precisely what Gould wanted. It enabled Smith to effect settlements at about 135, the closing quotation on Black Friday. Those who balked at these terms could go to court. In short order Gould's lawyers found themselves handling a flood of "gold" cases, between two and three hundred of them.[57]

If the injunctions extricated Gould's friends from danger, they also helped others caught in the messy tangle. So long as the clearing process remained muddled, brokers could not learn their balances or fulfill obligations because their funds were tied up. In effect the injunctions bought time to stem the panic, calm the market, and apply rational measures. That they suited the needs of Gould and his allies so well should not obscure their broader usefulness.[58]

The gold panic struck Wall Street like a tornado and departed as quickly as it had come, strewing in its wake the rubble of broken houses and individuals. The casualty list included few of those who had unleashed the storm. Smith, Gould & Martin settled its accounts and lived to speculate another day. Fisk escaped ruin simply by repudiating his contracts and hiding behind friendly judges. The Tenth National Bank survived its run and was found to have violated no laws; nor could the examiners uncover evidence that its wholesale certification of checks had been illegal or improper, though they condemned it as unsound policy.[59]

Of the lesser lights Heath and Belden failed, but both returned to the Street, the latter as partner in a firm associated with Gould because, it was rumored, Belden was "custodian of certain secrets of his principals." In 1879, however, the last tatters of his reputation were shredded by scandal and he left the Street. Apart from gold dealers fourteen firms on the stock exchange failed, the most prominent being Lockwood & Company. His failure proved a calamity for Gould. When Lockwood's firm went under, its large block of Lake Shore had to be thrown onto the market. Vanderbilt leaped at the opportunity to buy 70,000 shares "at a price thoroughly satisfactory to himself." Two of his men moved into the Lake Shore's management and doomed the proposal to consolidate the Wabash and the Lake Shore. Instead the Lake Shore became part of Vanderbilt's line to Chicago.[60]

Jay had thus come up empty not only in the gold scheme but in his quest for a western connection as well. The Albany too slipped from his grasp. Although the litigation dragged on until 1871, Ramsey won every decision. In February 1870 he leased the Albany to the powerful Delaware & Hudson Canal Company, an adversary with ample means to fight the Erie. As for the gold suits, Jay orchestrated them over a period of weeks, months, sometimes years, never intending that any should go to trial on their merits. Some he settled without even consulting his lawyers; others he allowed to dangle until his antagonists wearied. One case lingered on until 1877. If nothing else these cases revealed how thoroughly he understood the value of time as a weapon against the eager, and how well he had mastered the art of delay to wear down the most determined opponent.[61]

The dispute over whether Jay made or lost money from his gold transactions will probably never be resolved. It is in fact the wrong question to ask. More to the point, he lost in the broader sense that he failed to attain any of his larger objectives. The gold scheme had collapsed, his personal finances were in disarray, the Erie still lacked a western outlet, and Vanderbilt had grown stronger in the process. A season of difficult and frustrating work had accomplished little more than laying the cornerstone of the Gould legend. The disappointment at gaining so little from so much exertion might discourage or break some men, but not Gould. By the time he testified on the gold panic in January 1870, he had put his losses behind him. "My recollection about these transactions is very indistinct," he declared. "The thing was over, and I banished it from my mind. I have been full of other things since."[62]

This remark was of course scorned as another case of convenient amnesia, and to some extent it was. Yet it was also perfectly true, for it was Jay's habit to empty his mind of past failures. The present gave him more than enough problems to occupy his attention.

11

Ouster

The Erie record of Mr. JAY GOULD *should have sufficed to banish him from decent business society. The perpetrators of the frauds . . . should have no place among reputable people. Whatever they touch they defile.*

—*New York Times,* March 21, 1877

Those who had looked most closely into the matter felt sure that the hour and the man had both come that would severely test the most accomplished of all modern criminals, Jay Gould.

—*The Financier,* March 16, 1872

With all the power they had so suddenly acquired, these bold men were afraid of the man they had stripped of all his consequence in the Erie Railway Company. . . . They dared not remove him from his place as Director, for they feared him yet, and knew he could make them trouble. . . . Jay Gould, even, did not know then the extent of the duplicity and treachery of those he had trusted, and how deep-laid the plot for his dethronement was among them, in complicity with his enemies.

—Edward H. Mott, *Between the Ocean and the Lakes: The Story of Erie*

AFTER HIS REELECTION in October 1869 Jay moved quickly to fortify his position. The Classification Act enabled Fisk, Lane, Tweed, and himself to avoid standing for reelection until 1874. Along with Abram Gould they served as the executive committee which, like the board, met only four times the entire year. Half the Erie's bylaws were deleted to expedite the Gould style of management. Jay remained as president with Fisk as vice-president and Morosini as auditor. None of this deceived Gould into believing that his enemies were routed or his control secure. He tried to appease their wrath by dangling the prospect of a dividend on the preferred stock, but nothing could ease the Erie's debt or arrest the slide of its stock, which tumbled to 21 by the year's end.[1]

By 1870 opposition to Gould was forming on three fronts: the English holders, American holders, and a group headed by James McHenry of the Atlantic. About $45 million of the $78 million in Erie stock belonged to English investors, and a large amount was held on the Continent. Despite their huge investment, foreign holders had been too distant and scattered to influence the Erie's affairs. But the fall in Erie caused them heavy losses, the rush of unsavory publicity alarmed them, and Gould's high-handed appropriation of their proxies outraged them.[2]

After the election of 1869 the English holders formed a protective committee representing 190,000 of the shares left in the hands of American brokers that Jay had used for himself. The committee authorized a New York banker to register these shares in the names of two of its members, Robert A. Heath and Henry L. Raphael, and also sent an agent to Albany to seek repeal of the Classification Act. The legislature debated the question at length, but in the end Tweed's power prevailed. An assembly member informed the disillusioned agent that he might have made better progress had he "brought $20,000 to smooth the way." Sadder but wiser in the intricacies of democratic process, the committee looked elsewhere for relief.[3]

They found none in the Erie's office. When the first batch of 16,770 shares was presented, Erie officials declined to record the transfer. This brazen refusal to perform a routine transaction utterly confounded the committee's representative. He returned with another 60,000 shares to be registered and certified in accordance with the rules of the New York Stock Exchange. In response Jay persuaded an obscure Erie stockholder to file suit against Fisk and the Erie's treasurer, charging them with conspiring to transfer the stock on behalf of the English holders in a manner detrimental to the company's best interests.[4]

The complaint found its way to Judge Barnard, who promptly ordered the 60,000 shares placed in the hands of a receiver and enjoined the English holders from further attempts to transfer them. The receiver in turn delivered the shares to Gould in exchange for a certificate. In this way Jay thwarted efforts to transfer the stock with a stroke that also allowed him to vote the shares yet again. While the crestfallen Englishmen sought redress in the courts, three pro-Gould directors were reelected in October 1870 by a huge margin. The committee protested the election as "unfair, irregular, illegal and fraudulent," but to no avail. At Gould's bidding the stockholders passed resolutions ratifying all actions taken by the Erie management since August 1869. The outcome moved one disgruntled journal to dismiss the proceedings as "such a farce as hardly to merit any notice."[5]

This startling maneuver, as imaginative as it was unscrupulous, displayed Gould's talent for unearthing opportunities and exploiting them before anyone realized what had happened. The smallest opening caught his alert eye, whereupon he would proceed to bend, twist, or ignore the rules of the game and take advantage of the confusion that followed. Although the pattern had long since become familiar, it remained effective because no one could divine where the next blow would fall. The committee was helpless against so original and elusive a foe. Efforts to join forces with unhappy American holders bogged down in mutual distrust. The only hope was to pursue the fight in the courts, which meant attacking Gould in his citadel. Through one device after another Jay delayed final action on the disputed stock until December 1871. By that time a more formidable opponent had emerged in the form of James McHenry.[6]

McHenry's interest in the Atlantic at first allied him with Gould. Together they had in 1868 rescued the road from receivership and leased it to the Erie. When Jay's ambitious scheme to control western outlets fell through, however, he antagonized McHenry by changing the terms of the lease without notice. A protracted legal fight ensued, from which Gould obtained a new lease pending sale of the road at foreclosure. This outcome enabled Jay to operate the Atlantic for the Erie's benefit while McHenry struggled to reorganize the company. Not until October 1871 did McHenry succeed in regaining control. At first he determined to make the Atlantic a strong road independent of its Erie connection, but efforts proved futile. The hard fact was that the Atlantic was burdened with debt, starved for traffic, and reached no major outlets except Cleveland. Its line was scissored on one side by the Pennsylvania and on the other by Vanderbilt's New York Central-Lake Shore system. Without the Erie, the Atlantic could not move oil between Cleveland and the eastern seaboard.[7]

It took McHenry only a few months to concede that, like it or not, the Atlantic could not survive without the Erie. At that point revelation gave way to inexorable logic: If the Erie connection was vital, there must be a lease on favorable terms; to obtain such a lease, he must acquire a voice in the Erie's management; to attain that voice, Gould must be deposed. Accordingly, McHenry organized the "Erie Protective Committee," allied himself with the English banking house of Bischoffscheim & Goldschmidt, and enlisted the support of Francis C. Barlow, who had recently become state attorney general. In London McHenry recruited General Daniel E. Sickles, then serving as minister to Spain. Sickles was a man whose life read like episodes from a tawdry novel. During the war he had lost a leg and most of his military reputation at Gettysburg. He took leave from his post and, for a fee of $100,000, set about the task of unseating Gould.[8]

Barlow already had charge of a suit filed by the Heath-Raphael group to remove the Gould management. In December 1871 Sickles initiated a similar suit while the Protective Committee issued a circular informing Erie stockholders of Barlow's intention "to break up the whole combination of the Erie ring, without respect to persons." The crusade against Gould had been launched, and it could not have caught him at a worse time.[9]

Gould's downfall did not result from the actions of any individual or movement but rather from a complex interplay of events. His control of Erie rested on a variety of elements: the political alliance with Tweed, the protection afforded by Tweed's judges, the support of friendly directors, the inability of hostile stockholders to unite against him, and the steadfast friendship of Fisk in a place where treachery was the rule. By December 1871 all these elements were in disarray and the foundation once thought impregnable seemed on the verge of collapse.

While his enemies gathered, Jay struggled to revive the failing Erie. He continued the search for a western outlet, tried to drain traffic from rival lines by cutting rates, and instituted reforms to reduce salaries and other expenses. To offset the allegations of the Heath-Raphael suit, the Erie board appointed a committee to investigate the charges. When its report absolved Gould, Fisk, and Lane of any "evil intent" or wrongdoing, critics dismissed it as a whitewash. Wall Street buzzed with rumors that the Heath-Raphael suit would topple Gould, that the Erie was about to go bankrupt, and that some of its directors were about to resign in favor of Vanderbilt men.[10]

In these difficult times Jay needed all the help he could muster. Especially did he need Fisk, whose appetite for publicity and affable manner with reporters made him an ideal lightning rod for Erie. In his own outlandish way Fisk was a master of public relations who deflected attention from matters of substance by the simple act of piling one spectacular antic on top of another. Moreover, it was Fisk who hobnobbed with Tammany politicos in the backslapping style so repugnant to Gould's nature. He was one of the boys as Jay could never be.

Unfortunately Fisk had problems of his own, which were fast transforming him from an asset into a liability. During 1871 the fat man's excesses, always an embarrassment to Gould, embroiled him in an ugly scandal. His affair with Josie Mansfield had become a love triangle embracing the tempestuous Ned Stokes. Josie relished her role as hypotenuse and thought nothing of taking Stokes as a lover while still living off Fisk's largesse. The two of them concocted a scheme to blackmail Fisk, using as bait certain indiscreet letters in Josie's possession. A rash of litigation followed and threatened to splash lurid details about the Fisk-Mansfield affair through the dailies. The matter went beyond Fisk's private life, for the lovers hinted repeatedly that the letters contained incriminating revelations about the Erie and Tammany Hall as well. Fisk knew the letters were comparatively harmless and told his friends so in private, but he spurned all advice that he disarm his enemies by publishing the letters himself.[11]

Grave as were Fisk's problems, they paled before those of Tweed. Early in 1871 the genial boss seemed as secure as ever on his throne. Although there were critics who howled for his scalp, notably George Jones of the *Times* and Thomas Nast, the poison-pen cartoonist for *Harper's Weekly*, none of them possessed concrete evidence. Then, unexpectedly, Jones got hold of detailed figures purportedly copied from the Tweed Ring's own account books. On July 22 the *Times* launched a crusade against the Tweed regime and hammered relentlessly for four months. To the amazement of everyone, the attack served as catalyst for a reform movement that handed Tammany a stunning defeat in the November elections. Indictments were prepared that sent ring members fleeing in all directions. Within a few months Tweed's regime was utterly routed. In the aftermath Tweed himself was clapped in Ludlow Street jail, Barnard impeached and convicted, and Cardozo forced to resign.[12]

The collapse of the Tweed Ring had not yet occurred when the Erie held its annual election in October 1871, but the storm warnings were evident. Criticism of the Erie grew so harsh that even the normally hostile *Chronicle* felt obliged to defend it. To counter these threats Jay had only meager resources. The Erie board still supported him, but its loyalty might melt away in the heat of battle. Some directors had lucrative contracts with the company and did not care who was president so long as their private fiefs remained undisturbed.[13]

Jay thought he had found a valuable ally in George Crouch, a journalist who aspired to more than merely writing about high finance. Through careful study Crouch had mastered the inner workings of both the Erie and the Atlantic. He displayed his expertise in a report that challenged prevailing views on Wall Street by praising Gould's management and the Erie's prospects as better than anyone realized. In 1870 Crouch helped Gould put down the English rebellion, but the following year he grew disenchanted with Gould and went to England to assist the foreign holders.[14]

Amid this sea of woes Jay tried to put the best possible face on the Erie's management. At the board meeting just before the election on October 10 he, Fisk, and Lane abruptly resigned, declaring they would not return "unless freely restored . . . by the vote of the stockholders." It was of course a charade staged to refute charges that the trio owed their positions to the Classification Act. The resignations were solemnly accepted and the names submitted for reelection at the annual meeting, where they received a nearly unanimous vote. The Heath-Raphael agent ignored the election. On the heels of this melodrama came the annual report, a surprisingly full document that earned grudging praise even from the *Chronicle.* The Erie had done well enough to resume dividends on preferred stock, Jay announced, and the board dutifully approved a 7 percent scrip dividend. Mindful that the court was about to rule on the disputed 60,000 shares, Lane introduced new bylaws to expedite the transfer of stock. The beleaguered Tweed resigned as director amid rumors of sweeping changes in the management.[15]

The Erie's financial troubles led Gould to conclude that his only hope lay in donning the mantle of reformer. Earlier in the year the company had decided to issue a new $22 million consolidated mortgage, but investors would not touch the bonds. On the advice of a prominent banker, William B. Duncan, Jay devised a reorganization plan of breathtaking scope. It called for the current directors to resign in favor of a coalition board representing the coal and railway companies that fed Erie and the banking houses with large interests in the road. Only Gould and Eldridge would remain from the old board. To avoid what Jay called "merely speculative control," he proposed that J. S. Morgan, Bischoffscheim, and Sir John Rose be made a committee to procure irrevocable proxies for a majority of Erie stock and vote it as trustees. Gould pledged the $24 million worth of stock in his control and offered to help secure repeal of the Classification Act. Once again Gould displayed his gift for advancing his private interest while playing the part of statesman. At the same time he took care to obtain from the Erie board a release absolving himself, Fisk, and Lane from liability for past actions. To encourage support for his plan, he resigned as treasurer and persuaded Fisk to resign as vice-president.[16]

It is intriguing to speculate on what the Erie's history might have been had Jay's plan been implemented. An original and magisterial solution to an impossible problem, it came too late. Before anything could be done, the plan was shunted into the corridors of intrigue and never heard of again.

On November 25, 1871, the Fisk-Mansfield affair burst into headlines as hearings began on the first of their suits. Spectators choked the courtroom along with reporters eager to capture every lurid detail for their readers. They were not disappointed; enough dirty linen was paraded to titillate the public for weeks. The issues at stake were soon buried beneath the sensationalism of what the puritanical George Templeton Strong called "a special stinkpot." The dispute was complicated, far easier to relish than to understand, but the dailies took little interest in the finer points. They pounced on the juicier details to produce their own version of the Fat Man's Follies which roasted not only Fisk but Gould and the Erie as well.[17]

The Erie's lightning rod had lost his magic. Instead of deflecting bolts he now drew them down on the company. Yet Fisk was doing well in court despite the squalid publicity. He blocked publication of the controversial letters and won an indictment against the lovers on charges of blackmail. Stokes had spent a fortune in legal fees with nothing to show for it. The press shredded his reputation and humiliated his proud family until his volatile temperament snapped. On January 6, 1872, Stokes hurried to the Grand Central Hotel ahead of Fisk, waited for him atop the stairway to the second floor, and shot him twice at point-blank range. Fisk bore his wounds with unaccustomed courage and even managed a joke or two. Put in a nearby room, he lingered until the next morning, long enough to dictate a will, greet the friends who gathered for the death watch, and bid farewell to his beloved Lucy, who had been more mother than wife to him.[18]

Jay mourned in an adjoining room with Tweed, Field, Shearman, and others close to Fisk. For some time he sat in perfect silence, as was his way; then, a reporter noted, "Everyone was suddenly startled by seeing him bow his head upon his hands and weep unrestrainedly with deep, audible sobs." There is no doubt that his grief was genuine and deeply personal. With Fisk's passing Gould lost not only a friend but his alter ego. Although he would have other close friends and business associates, none provided him with so outsized a shadow in which to dwell.[19]

The manner of Fisk's death emblazoned the newspapers with copy as sensational as his life had furnished. One financial journal, in a perceptive editorial, castigated him for his amoral excesses while admitting him to be "an extraordinary person" whose life had demonstrated "that any vice is only a virtue exaggerated and overdone." It even offered grudging praise for the Erie's management, the guiding spirits of which "were fitly mated; the audacity, recklessness, and dash in Fisk well complemented the secretiveness, the wiles, and the cool scheming of the quiet Mephistopheles, Jay Gould."[20]

Apparently the image of Gould as Mephistopheles struck a responsive chord, for in later years it became an overworked epithet. In later years, too, when his name had become anathema, he would be charged with shabby treatment of Lucy Fisk after her husband's death. These attacks prompted the widow in 1881 to set the record straight. In a letter to the *New York Herald* she defended "my great friend Mr. Gould" as "the only friend of Mr. Fisk who has responded to my actual needs and wants since his death."[21]

A week after Fisk's death the *New York Herald* published the celebrated Fisk-Mansfield letters in full. None contained references damaging to the Erie or Tammany or anyone else except the three lovebirds, a fact that embarrassed editors who had promised shocking revelations. The fizzling of the letters was about Gould's only consolation that bleak winter. By February 1872 the rebellion against him was hurrying toward a sordid and spectacular climax that he was helpless to oppose. The Heath-Raphael group had finally recovered its 60,000 shares. Sickles had arrived in Albany to secure repeal of the Classification Act and assist Barlow in pressing the suits. McHenry, still in London, had dispatched Crouch to New York to woo Erie directors away from Gould. Before Crouch sailed, a mysterious new element surfaced in the form of a cable to McHenry from W. Archdall O'Doherty, former receiver of the Atlantic. It read in part:

> Will you lodge one and a half million dollars in a trust company here or in Philadelphia, payable only on condition that you shall nominate a majority of Erie board, and have them elected within a week of deposit? I guarantee success, and know that your present plans will fail.[22]

An exchange of cables soon enlightened the puzzled McHenry. O'Doherty was relaying a scheme hatched by Fred Lane, who had concluded that Gould's position was hopeless. Eager to run with the hounds instead of the hare, Lane saw that the fastest way to unseat Gould was not by pressure from without but through subversion from within. As Lane delicately phrased it, "The majority of the Board might, for a small consideration, be induced to resign." O'Doherty informed McHenry that the $1.5 million was a "bonus to be paid for control" and could "only be got back by profit on stock." He suspected correctly that McHenry and his friends were buying Erie shares in anticipation of the rise that would accompany a change in management.[23]

McHenry hesitated. Having entrusted the mission to Sickles and sent Crouch to assist him, he was reluctant to muddle the situation by injecting a new element. When Crouch arrived Lane went after him at once, offering "'any terms' if I would work with him," but Crouch preferred to do his own spadework. On February 15 he cabled McHenry, "Progressing well. Buy all you want and hold confidently." Five days later he wrote his own assessment of the Erie board: Lane could be bought, as could two other Gould loyalists, Henry Thompson and John Hilton, and two of Fisk's cronies, M. R. Simons and George Hall; another Gould man, Charles Sisson, was seriously ill; still another, Justin White, would flow with the tide; and Homer Ramsdell, former president and current director, agreed to support a revolt free of charge. H. N. Otis, the secretary, also threw in with the rebels.[24]

The board was ripe for plucking, Crouch reported: "It's no use letting Barlow and Sickles waste money on the law and legislature." Reluctantly he divulged his scheme to Sickles, who approved it and bent his efforts toward recruiting Erie directors. While Crouch maneuvered, Lane and O'Doherty fidgeted. On February 16 McHenry replied that he would pay the sum requested only if the entire board resigned or the Erie went into the hands of a friendly receiver. This response forced O'Doherty to consult S. L. M. Barlow, the Atlantic's counsel. Dubious at first, Barlow entered the fray by cabling McHenry on the 19th:

> Receivership, through action of the Board, impolitic, useless. Lease now equally so. In future this may be arranged. What is offered is ten members of the present board named by you. Immediate withdrawal of Gould's power. His resignation to be forced afterward.[25]

McHenry agreed but wanted Gould out entirely. A few days later O'Doherty discovered that Sickles and Crouch were working the Erie board on their own. Angrily he warned McHenry, "Beware. Present plans will result in loss of money and disgraceful failure. Save yourself." It took McHenry until March 9 to bring O'Doherty to terms, by which time Crouch had put the finishing touches on his scheme. One by one the directors named their price: Lane and Thompson got $67,500 apiece, Simons and O. H. P. Archer $40,000, and Otis, Hilton, and White $25,000 each. Otis kept Crouch informed of Gould's every move, and Lane had to be watched as well. Crouch was in his element. As early as February 27 he had cabled McHenry, "Majority of Erie Directory with us. Gould powerless. Have loaded up in this market at thirty." Two days later he added, "Failure impossible. Have you bought all you want?" His communiqués resounded with the clang of arms:

> We have moved against the enemy in three columns. One, headed by Sickles, has been diverting him in the Legislature; another, under the Attorney-General, has been threatening a flank movement in the courts; and the third, under yours truly (composed principally of sappers and miners), has succeeded in undermining the very citadel of Erie. In order to cover my mining operations, I kept up an incessant bombardment through the press. . . . Tomorrow the mine will be fired, and the forlorn hope will mount the breach.[26]

The "mine" was a plan forcing Gould to hold a meeting of the Erie board at which the conspirators would move to fill two existing vacancies with Atlantic men, giving them a majority. The retiring directors would then resign one by one and be replaced by McHenry designees. On March 8 nine of the directors sent

Gould a letter requesting a meeting for Monday, March 11. Bristling with confidence, Crouch readied his troops. "Eve of battle," he cabled on March 10. "Victory certain."[27]

Jay sniffed a plot but could not ferret it out. He tried to see Crouch but the latter avoided him. Then, late in February, the whole plot was disclosed by none other than Lane, who had not yet sealed his bargain with Crouch. "They are counting on me," Lane added after ticking off the names of directors who were to resign for a price, "but I shall remain true to you, and they have not money enough to change me."

"All of which," Gould observed dryly, "I took with a slight discount, knowing Lane very well."[28]

So far Jay had dodged all efforts to force a meeting of the board; the March 8 letter lay on his desk unopened. When the conspirators got the vice-president to call a meeting for March 11, Jay secured an injunction. Undaunted, the conspirators assembled in the boardroom shortly before noon for their meeting. Motions flew in rapid succession to elect a new board, officers, and executive committee.[29]

When Jay reached the Opera House, he ordered the conspirators locked inside and the door guarded by Erie henchmen. After notifying employees that he was still in command, he barricaded himself in the president's office with Thomas Shearman, Morosini, and a handful of bodyguards. Outside an excited crowd gathered to watch the spectacle. Sickles, who had left the meeting early, returned with a force of United States marshals to liberate the conspirators. The new management followed the marshals to the president's office in hopes of serving Gould with notice of his removal in favor of General John A. Dix. A marshal pounded on the door, then applied a crowbar. The door was forced, allowing General Dix's irregulars to storm the office. During a brief scuffle Shearman tried to escape and was pitched on his duff by opposing honorable counsel. A deputy marshal, clutching the papers, tried to corner Gould, who dodged nimbly behind chairs and over tables until he found refuge in the law office. Morosini and Shearman, along with their Hessians, managed to reach the same room and locked the door behind them.[30]

As the impasse stretched into the wee hours, more police and more Erie toughs friendly to Gould arrived. Shearman was for routing the enemy with a frontal assault, arguing that the law was on their side, but Jay demurred. Convinced that the time had come to parley, he allowed the dreaded papers to be served by a boy hoisted up to the transom. Sickles took the cue and offered to negotiate if Gould would surrender the premises. Jay agreed and left the building early Tuesday morning, a beaten man to all appearances. On Friday the 15th he resigned as director without explanation. That same week the legislature repealed the Classification Act. In all the victory cost McHenry $300,000 instead of the $1.5 million Lane had proposed.[31]

The victors, plying one another with congratulations, took charge of the Erie. The press whooped with joy at the ouster of the Erie Ring. Although ignorant of the details and puzzled as to exactly what group or groups now controlled Erie, editors and financiers alike rejoiced that the tyrant Gould had been overthrown by "honest men, who have wrested stolen property from the hands of thieves."[32]

They could not have been more wrong. The struggle for the Erie, far from being over, had only entered yet another dismal phase. Once again Gould's adversaries were about to learn that he could be more dangerous in defeat than in victory.

Sickles knew what would entice Gould to resign. Being more interested in victory than retribution, he was willing to offer terms others regarded as overly generous if not ruinous. The Erie would release Gould, Lane, and Fisk's estate from all claims, settle outstanding accounts between Gould and the company, reimburse him for all advances made on the company's behalf, and repay loans estimated at nearly $2 million carried by him for the Erie. As a final sop he reminded Gould that "if you resign, it will send the price of Erie up fifteen points. You can make a million dollars."[33]

Jay hardly needed instruction on this point. Like McHenry's crowd he bought heavily as Erie soared upward on the tide of victory from 35 in mid-March to 67 by the month's end. The victors too fattened on the rise; Crouch left orders to sell at 60 and sailed back to Europe. The boom in Erie persisted despite a raft of bad news and dark suspicions. It was discovered that the road had a floating debt of nearly $5 million, much of it owed to friends of Gould who pressed for immediate payment. An emergency loan from Bischoffscheim's firm staved off disaster, but the threat of receivership lingered. With most of its assets pledged as collateral, the Erie was living hand-to-mouth.[34]

Gradually it became clear that Gould's successors were not entirely robed in virtue. Both the Heath-Raphael group and the American holders distrusted McHenry's crowd, as did some financial editors who regarded the Atlantic as a white elephant and already suspected that McHenry's movement "is speculative and was so from the beginning." Since McHenry's board had seized power without an election, a struggle for control among its factions was inevitable. To woo support the new directors declared a dividend on preferred stock and, ignoring the release given Gould, authorized proceedings to recover 650,000 shares of Erie "fraudulently issued by him." In London McHenry formed a new committee, which requested English holders to deposit their stock for the forthcoming election. This move impaled the Heath-Raphael group upon the dilemma of bowing to McHenry's leadership or commencing an open fight that might ruin both interests. Reluctantly they tendered their proxies to McHenry, enabling him to control the election in July. After some confusion the new board chose as president not Dix but Peter H. Watson, a capable lawyer who had founded the South Improvement Company. At the election McHenry presented the Erie with a bill for $750,000 to cover his expenses in ousting Gould.[35]

On one point the opposing Erie factions were agreed: Gould must be forced to make restitution. Suits were flung against him but languished for want of evidence. The Erie's books were in such miserable shape as to be useless in a court of law. Most of Gould's Erie transactions had been handled by Smith, Gould & Martin, but Watson had no access to the firm's records. Prospects for convicting Gould seemed doomed until November, when an unexpected development revived them in spectacular fashion.

12

Entr'acte

There is but one man in Wall Street today whom men watch, and whose name, built upon ruins, carries with it a certain whisper of ruin. He is . . . one whose nature is best described by the record of what he has done, and by the burden of hatred and dread that, loaded upon him for two and one-half years, has not turned him one hair from any place that promised him gain and the most bitter ruin for his chance opponents. They that curse him do not do it blindly, but as cursing one who massacres after victory.

—*New York World*, September 1, 1873

In those days Mr. GOULD's editors were known to the public as stool pigeons. It was their function to entice the incautious investor or speculator within reach of his ammunition; to "hammer" securities that he wished to acquire, and to exalt by artful misrepresentations the quotations of those he desired to "unload."

—*New York Times*, March 23, 1891

From the facts, Jay Gould was never anybody's friend. Not even his own. Sage was primarily concerned with him as an interest-paying customer and "dummy." . . . it was obvious that Gould would never have accumulated his fortune without assistance. Sage, the accomplished artist in the magic of money, was always ready to help—that is, for a price. . . . Gould was so deeply mired in debt to Russell Sage that he would never be able to get out. Nor did he, until death mercifully released him from his obligations.

—Paul Sarnoff, *Russell Sage: The Money King*

GOULD'S OUSTER from the Erie put him back on the Street, a man without a company. An enormous change had occurred in five years. He had entered the Erie a nonentity and left it a public figure, one of the most notorious financiers in the country. The newspapers portrayed him as a national villain, ever mindful that his misdeeds made good copy. When the Tweed Ring collapsed, George Jones seized upon Gould as the blackguard most suitable for public flogging. Although other papers cried for Gould's scalp, none did so longer or louder than the *Times*. In later years, when Gould acquired or influenced newspapers, even laid the foundation for a communications empire, the howls against him went from shrill to hysterical.

His image suffered too from the type of enterprises he took up. With few exceptions they were properties in trouble, weak financially and possessed of a history scarred by fraud or mismanagement. Proper men regarded him as a sort of speculative roué who haunted the financial red light district, flirting with properties no sound investor would touch and struggling to elevate them into the company of respectable society. Gould's genius lay in his ability to dress and present his trollops in so attractive a manner as to tempt even the most virtuous of gentlemen with their charms. That he succeeded so often at arousing the lust for profit in men of propriety only caused them to despise him all the more.

Those who condemned Gould for specializing in weak properties never understood why he did so. The most obvious reason was that marginal or tainted companies offered the widest latitude for manipulation. They were also easier to gain control of than companies in good standing, an important consideration for someone like Gould, who did not belong to the financial establishment. The better properties were either tightly held or dominated by some imposing figure. Even if accessible, they were often so large as to require more funds than Gould could muster. As a renegade in the eyes of the Wall Street establishment he was shunned by the "old boys' network" that dominated the conservative banking houses. His reputation among such men was neatly capsuled by a financial reporter in 1873: "Not considered high-toned; not at all liked except by those who operate and make commissions out of him." As a result, his chief source of funds remained the profits extracted from market operations.[1]

Apart from practical considerations, the fact was that stable enterprises held little appeal for Gould. His early years had given him a taste for uphill battles. He relished the challenge of breathing life into moribund lines; in time he came to care less for the money than for the game itself. "I didn't care about the money I made," he testified in 1883. "I took the road more as a plaything to see what I could do with it; I have passed the time when I cared about mere money-making. My object in taking the road . . . was more to show that I could make a combination and make it a success."[2]

Scorn and derision greeted remarks like the one above, yet it is absurd to suppose that money alone motivated so complex a figure as Gould. Wealth and power were the markers of success, but it was the contest itself that animated him, drove him on with demonic fury until he literally worked himself to death. Like a chess master he thrived on the challenge of intricate problems, the battle of wits, the devising and unraveling of complex machinations. There was in Gould's approach to business an intellectuality, a quality of abstraction often mistaken for coldness. It was this quality others saw when they described him as distant, aloof, calculating, meditative, or simply quiet.

His very personality seemed abstract to many who dealt with him. Some found his company discomfiting because his presence left so soft an imprint and his features revealed so little of what he thought or felt. "His face always wears a contemplative expression," a reporter noted. In conversation his eyes seldom stayed with the talker, not to avoid looking at him but because his mind churned ceaselessly even then. He seldom spoke and then expressed himself in the fewest possible words in a soft, low voice without inflection or gesture.[3]

Here was a curious figure to be thrust into public prominence. Other men generated legends about themselves through sheer force of personality. In Gould's case the lack of presence created a vacuum that legend rushed to fill.

The Erie served Jay as a springboard for operating in the stocks of other roads. In these operations he had not been overly scrupulous about separating what was done on his own account from transactions done on behalf of Erie. Several auxiliary enterprises were organized and developed with Erie funds but their stock remained in the hands of Gould and his associates. In one instance Jay bought control of two small roads and leased them to Erie. When they proved their value as feeders, he defaulted on the rental payments, forfeited the leases, and sold both roads to the rival Northern Central at a fat profit to himself.[4]

Once evicted from Erie, Jay lacked the inside position to capitalize on such opportunities. His place on the Street had changed in another respect when the firm of Smith, Gould & Martin dissolved in August 1870. Although Jay remained on Wall Street another two decades, his name was never again attached to a brokerage house. Instead he operated as a special partner in a variety of firms, which preserved his anonymity and enabled him to disperse his business among a network of brokers. Whatever his brokers thought of him personally, they prized him as a customer; he was reputed to have paid $100,000 worth of commissions in a single month.[5]

Gradually there emerged the profile of Gould as the deadly and elusive predator, the lone wolf who might run with the pack but was not of the pack. He was without corporate office, lent his name to no firm, and did not even belong to the Exchange. The lack of affiliations allowed him to dwell among the shadows, where he was at his best. After leaving Erie he became one of the leading traders on the Street. For a time he stalked prey with his fellow renegade, Henry Smith. In their bear operations Smith and Gould singled out two stocks for special attention: Erie and Pacific Mail. The latter was a steamship line plying the coastal and Far Eastern trade out of San Francisco. Once a prosperous company, it had fallen on hard times. The transcontinental railroad (completed in 1869) cut into its coastal business while English competitors threatened its overseas trade. Years of complacency had saddled the company with an obsolete fleet of steamers. At a time when Pacific Mail needed strong, vigorous leadership, its management fell into the hands of a stock operator, Alden B. Stockwell.[6]

A short, chunky, red-headed westerner, Stockwell was a newcomer on the Street. When the low price of Pacific Mail caught his eye, he determined to gain control of both that company and its connector, the Panama Railroad. At first his efforts met with striking success. Panama, which had languished in the 50s during 1871, touched 113 in April 1872 and was still climbing. As president of Pacific Mail Stockwell bulled the stock by obtaining an enlarged subsidy from Congress and securing agreements with the Pacific railroads. The price moved upward, but sluggishly, its progress slowed by bears who seemed to divine Stockwell's every move.[7]

The light did not dawn on Stockwell until rumors began flying that the mysterious bears were Smith and Gould. In a brilliant stroke he sent to Albany for a bill authorizing Pacific Mail to reduce its capital stock by half. The bill was defeated, then mysteriously revived and passed. A fierce struggle ensued, during which 315,600 shares of Pacific Mail changed hands in four days. Smith and Gould resorted to their old friend, the injunction, but this time it did not save them. The new bill enabled Stockwell's friends to sell to the company the shares they bought while reducing the supply of stock needed by bears to meet their contracts. In May Pacific Mail reached 87, a gain of 23 points since March. Smith suffered the indignity of losing first on the rise and then, buying at high prices to fill contracts, on the subsequent fall.[8]

As Stockwell carried the laurels of victory off on a Pacific tour, Smith and Gould clung to the bear side of the market. In September they got burned again, this time in Erie by their old mentor, Drew. For some reason Smith blamed Gould for this disaster and demanded that Jay reimburse him for his losses. When Jay declined, Smith vowed angrily, "Then I'll get good and even with you before another year!"[9]

This outburst capped a rift that was already under way. Despite repeated drubbings, Smith clung to the bear position with a tenacity that did not appeal to Gould. Unwilling to lash himself to the mast against all weather, Jay boldly changed front by covering his shorts and joining Stockwell to bull Pacific Mail. The change became evident in October when, amid frantic trading, Pacific Mail shot up from 73 to 103. Smith again took heavy losses and Stockwell received credit for the coup.[10]

While the Street buzzed with excitement over that melee Jay launched a new line of attack. He knew Smith was short in Chicago & Northwestern, as were several of the road's own directors. At the same time Vanderbilt's son-in-law, Horace Clark, and Augustus Schell were attempting to gain control of the Northwestern on the Commodore's behalf. Gould was quick to grasp the possibilities offered by an alliance. That Clark and Schell had been his foes in Erie days mattered not at all to him; Wall Street men could not afford the luxury of permanent enmities. Having already changed front, he now proceeded to change partners as well.[11]

Despite heavy buying Clark and Schell made little headway until Gould informed them of the heavy short interest in Northwestern. Sensing the chance for a corner, they joined forces to bull the stock. As the stock began to rise, a puzzled observer described its climb as "a conundrum to the street." Alarmed, Smith spread rumors that Northwestern was about to issue $10 million in new stock. The price dropped briefly, then bounded upward again. Fate intervened on November 11 when a large fire ravaged Boston and caused the market to plummet. A year earlier Smith had reaped a fortune by selling short after the disastrous Chicago fire. Inspired by that success, he increased his line of shorts, including Northwestern, by large amounts. To his dismay the market recovered quickly. With Gould and his allies buying all the Northwestern the market offered, the price marched steadily upward.[12]

On November 20 Northwestern reached 95. Smith, Drew, Travers, and other bears had contracts to fill at 75, but not a share could be found and the price was still climbing. Only then did they discover that Gould had laid the trap that snared them. Boiling with rage, Smith looked for drastic measures to relieve his predicament. The Erie's suits against Gould leaped into his mind. They had languished for want of evidence. The books of Smith, Gould & Martin contained all the evidence needed to revive the suits. Samuel Barlow, the Erie's counsel, did not have access to the books, but Smith did. He went at once to see Gould.[13]

"You must let me have Northwestern," he demanded, "and let me have it so I can get out of this fix whole." When Jay refused, Smith played his trump card. "If you don't help me out of this, I will turn over the Smith, Gould & Martin books to Barlow—and you know what that means!"

"Turn them over," Gould replied coolly. "I have no objection."

Northwestern touched 100 on Thursday, November 21, the day Smith delivered the books to Barlow. After feasting on their contents Barlow and Watson filed suit at once, demanding from Gould $9,726,541.26 "fraudulently appropriated by him." The complaint itemized at length Gould's wrongdoings as president of the Erie. An arrest order was procured but not served until late Friday afternoon. By the time the deputy sheriff had located Gould, Northwestern had reached 105. News of his arrest, with bail set at a staggering $1 million, threw Wall Street into a state of frenzy. Just as the Northwestern corner neared fruition, its guiding hand was carted off to jail on serious charges.[14]

Smith hoped that Gould's removal at a critical moment would smash the corner, but he made the fatal error of tipping his hand. Clark and Schell provided bail and Jay was back on the Street in half an hour. Meanwhile, amid intense excitement, frantic bears rushed to cover their shorts as sales of Northwestern were recorded at 110, 111, 112, 116, 125, 130, 140, 150, 152, 155, 160, 165; the last bid on Friday was for 200, with no takers. A reporter asked Clark to explain the wild gyrations. "The only reason I can see," he replied solemnly, "is that people have arrived at a just appreciation of the stock."[15]

Saturday was settling day for the bears who had failed to deliver on their contracts. Gould let the smaller operators off at 150 or less to save them from failing; the larger fish paid above 200, the highest being 230. In desperation Smith threatened to renege on his contracts but eventually swallowed his loss. By Wednesday Northwestern had dropped back into the 80s. Financial editors denounced the whole affair and its participants. The presence of Clark and Schell caused the victors to be labeled "the Vanderbilt party." This reference distressed the Commodore, who liked to preserve the fiction that he never indulged in speculation. His indignation over this slander prompted him to publish the celebrated notice denying that he had transacted any business with Gould since July 1868.[16]

Brilliant as the campaign had been, the Northwestern corner left serious problems in its wake for Gould. The Erie suit had to be confronted and a large amount of Northwestern stock disposed of somehow. The embittered Smith severed all business connections with Gould. In time he would come back to haunt his old partner, but for now Jay had the last word. There had been a moment during the excitement when Smith, purple with rage, shook his finger in Gould's face and sputtered, "I will live to see the day, sir, when you have to earn a living by going around this street with a hand organ and a monkey."

"Maybe you will, Henry, maybe you will," Gould cooed softly. "And when I want a monkey, Henry, I'll send for you."[17]

The Erie suit threatened Gould with ruin, even imprisonment. He avoided both by grasping the situation more thoroughly than did his opponents, and by doing what was least expected of him. The obvious response would have been to delay the suit in court, to fight until his adversaries wearied, but Jay understood that the Erie's managers were willing, even eager, for compromise. The road was floundering, stockholders were impatient for results, and management could not afford a long siege in court. Instead of fighting, therefore, Jay astonished everyone by offering to settle at once. "I do this for the sake of peace," he wrote Watson, "because any litigation of such questions is more annoying to me than the loss of the money involved, and because I am sincerely anxious for the success of the Erie Company, in which I have a large pecuniary interest."[18]

To discharge his obligation he offered the Erie a package of stocks, bonds, properties, and real estate worth by his reckoning more than the $9 million demanded by the suit. On December 19 a special committee of Erie directors confirmed Gould's estimates; by unanimous vote the board accepted the package and waived all further claims against Gould. News of the compromise rocked Wall Street. Flabbergasted observers, still reeling from the Northwestern affair, heaped praise on the settlement. Amid great fanfare the Erie board opened a special "Reclamation Account" with a balance of $9 million. Excited trading drove the price of Erie from 54 to 62. Jay added to the clamor by confiding to a reporter that the compromise terms gave him the privilege of buying 200,000 shares of Erie at about 50.[19]

The Gould package was in fact a gift horse without teeth, but that revelation did not emerge until the summer of 1874. By then fresh controversies plagued the Erie management and the road was about to enter receivership. Much bickering arose over why the special committee had declared the package worth $9 million, and why it had taken eighteen months to discover its true value. New suits were filed against Gould and dragged on until 1876, when he agreed to a new settlement. This one cost him money, but by then he was well able to pay. Once again the Erie dropped all claims against him, leaving unresolved only the mystery of the Smith, Gould & Martin books. After the first debacle Smith stored the books at his farm in New Jersey. Early in the spring of 1874 parties unknown appeared at the farm, intimidated the hired man, seized the books, and vanished. They were never seen or heard of again.[20]

For the remainder of the century Erie floundered haplessly through one crisis after another. That solvency and stability continued to elude the road could hardly be blamed on Gould. His influence on Erie's jaded history had been significant but not decisive. He had neither cost the Scarlet Woman of Wall Street her virtue nor rescued her from a life of sin and degradation.

Without a company like Erie to manage, Gould's operations lacked focus and cohesion. For a time he was content to watch and wait, to let his speculations spread like tendrils (some would say tentacles) in all directions until circumstances attached them to the field of transcontinental transportation. It was an arena uniquely suited to Jay's genius. The development of the West posed challenges that fascinated and ultimately obsessed Gould. It became the core of his life's work and his most enduring legacy.

Late in 1872 Pacific Mail underwent another convulsion as Stockwell proved more adept at managing stocks than steamships. In February 1873 the stock plummeted to 55 from its October high of 110 amid rumors that Stockwell was unable to meet his personal obligations. In the struggle for control of the company Jay allied himself with other interests against Stockwell, who in May was deposed as president and carried the fight into the courts. The sparring dragged on until December, when a compromise was arranged that left Pacific Mail in the hands of Gould's friends.[21]

Stockwell took his fall philosophically. "When I first came to Wall Street I had $10,000 and the brokers called me 'Stockwell,'" he noted dryly. "I scooped some profits, and it was 'Mr. Stockwell.' I got to dealing in a thousand shares at a time, and they hailed me as 'Captain Stockwell.' I went heavily into Pacific Mail, and folks lifted their hats to 'Commodore Stockwell.' Then one day Jay Gould came along. Smash went Pacific Mail, and I went with it. They did not call me 'Commodore Stockwell' after that. Then it was the 'red-headed son of a bitch from Ohio.'"[22]

Pacific Mail marked Gould's entry into transcontinental transportation. He came in as a speculator and did not even hold a seat on the board; nevertheless, he had found the doorway to his destiny. It also brought him together with the man who would become his lifelong friend and business partner, Russell Sage. It is hard to imagine two beings more unlike than Fisk and Sage, yet their relationships to Gould were rooted in common soil. It was the perfect mating of misfits, men who had few other friends and cared nothing for the ordinary rituals of society or the approval of its arbiters.

Twenty years older than Gould, Sage was born in upstate New York aboard a wagon heading west. Like Gould he was a poor country boy who clawed his way to wealth, in his case as merchant and banker. Prior to the Civil War he dabbled extensively in politics and served in Congress before turning his shrewd intellect to Wall Street, where he pioneered the techniques of puts and calls, spreads and straddles. Sage speculated in stocks but also involved himself in corporate enterprises. He discovered early that the surest investment was to lend money at high rates on sound collateral. For this purpose, and because money itself held a peculiar fascination for him, Sage always maintained a large cash reserve.[23]

On the Street Sage was regarded as a sharp if eccentric operator who took risks and, like Gould, preferred dealing in low-priced securities. His passion for holding cash gave him a trapdoor that others misinterpreted as fiscal conservatism. By 1872 Sage was renowned as a skinflint who wore second-hand clothes and parted with a dollar only with the greatest reluctance. At fifty-three he made a loveless marriage with Olivia Slocum, a plain, birdlike spinster of forty-one. Later Olivia became a close friend of Helen Gould and her children, especially daughter Helen, a timid, painfully shy creature in whom Olivia must have seen something of herself. The Sages resided at 506 Fifth Avenue, a block below the Goulds.

Sage's lean, gaunt frame made him seem taller than he was. At five feet ten inches he towered over Gould, which enhanced their image as an odd couple. Possessed of a generation more of business experience, he was not content to follow Gould's bidding as Fisk had done. They joined forces in some enterprises and went their separate ways in others. What bound them together was a deep sense of loyalty rooted in friendship.

None of this had occurred at the time of Sage's election to the presidency of Pacific Mail in December 1873. By then the company had assumed new significance for Gould, for he had purchased his first shares of Union Pacific stock on the advice of his partners in the Northwestern corner. Both sat on the road's board and Clark was its president. Whatever Gould's original motives for purchasing the stock, an unexpected chain of events transformed his interest in Union Pacific into one of the longest and most extensive commitments of his career.[24]

On April 29, 1873, Joseph J. Marrin, a lawyer representing certain parties in the Black Friday suits, spotted Gould across the dining room at Delmonico's. He strode to the table, accused Gould of reneging on his promise to compromise the suits, and punched him in the nose. When Jay filed complaint, Marrin paid a fine of $200 and declared his satisfaction worth every penny of it. The press had a field day with the incident, especially the *Times*, which all but applauded Marrin's action.[25]

The *Times* had another reason to rejoice at Marrin's assault. Since December 1872 it had blasted Gould for his growing influence in the rival *Tribune*. After the death of Horace Greeley on November 25 his assistant, Whitelaw Reid, bought control of the paper with the help of a loan from Gould. On Christmas day the *Times* opened its campaign against the Gould interest with a warning that he was "not the sort of man to leave his tracks clearly marked in the dirt." Four months later it was assailing him as the "Dictator of 4 states." When the *Tribune* proposed to erect a new tower, the *Times* suggested that "JAY GOULD, under the mask of *Mephistofeles*, with an armful of *Tribune* shares and an admiring crowd of the purchased legislators of 4 states, would make a good central figure."[26]

Although details of the transactions have never come to light, it appears that Jay took some *Tribune* shares as collateral for the loan. He made no attempt to influence editorial policy, but criticism of his activities vanished from the paper's columns. That fact, coupled with his unsavory reputation, was evidence enough for rival editors that Gould's sinister shadow darkened the *Tribune*'s newsroom. The *Sun* echoed the *Times* by labeling the *Tribune* a "stock-jobbing organ" and denouncing Reid as "Jay Gould's stool pigeon." For years the connection with Gould hung like a millstone about Reid's neck. On one occasion the *Tribune*'s financial editor was pummeled on the floor of the Stock Exchange for being Gould's stooge.[27]

The *Tribune* affair offers an early glimpse of a significant yet curiously neglected dimension of Gould's activities: his role in the communications industry. Much is known about his dominance in the telegraph field, far less about the nature and extent of his influence on newspapers. Throughout his career Gould was accused of buying or controlling newspapers to use as mouthpieces for his operations. The charges were grounded more in suspicion than in concrete evidence, yet they persisted stubbornly.

Gould's relationship to the press was not only complex but paradoxical. His aversion to publicity about his private life was well known. The notoriety thrust on him by his business activities, which made him a favorite whipping boy among editors, only reinforced his passion for privacy. Jay's instinctive response to the abuse heaped on him was to retreat deeper into the shadows, to evade reporters and shun interviews. "Uncle Jay's attitude toward the press and its personnel was strictly passive," his niece recalled, "a mixture of aloofness and disgust."[28]

At the same time Jay understood thoroughly the importance of publicity to his operations. The market danced to rumor no less than to real news; enterprises in which he was interested had to be puffed and rival ones attacked, which meant active promotion, giving out interviews, planting stories. Jay disliked doing these things, but he did them because they served crucial functions. He mastered the art of saying much while disclosing little. If a reporter printed what Jay wanted said, his next request for an interview would be warmly received. It startled reporters to discover that Gould could be as frank as he was evasive, but always for a purpose. Most found it impossible to get anything out of him unless he wanted some bit of information spread. "The public heard from him," grumbled one newsman, "only when he, not the public, would profit by the utterance."[29]

There were other ways of influencing the news. It was unnecessary to buy newspapers or even editors when one needed their cooperation only on a few select matters. A loan, a market tip, or some other favor was usually enough to obtain puffs and plants. Jay used reporters as he did brokers, both to reveal and to conceal movements. A journalist named William Ward offers a typical example. Although Ward's affiliation is not known, his function was to write financial articles on topics suggested by Gould from a slant provided by Gould. Sometimes Gould handed Ward an entire article or letter written by himself, to be published anonymously or under a pseudonym.[30]

For his services Ward received occasional market tips. "A little W Union won't hurt you," Jay advised in January 1874. "I think it is the next big card. The stockholders are petitioning . . . for a 50 per cent bond dividend representing the surplus of the company . . . & I would like to have you write it up strong." Gould took care not only to prime Ward with material but to orchestrate its flow as well. "You need not use that information I gave you about Lake Shore till Thursday," he suggested in July 1874, "but talk about the Redemption, the new law & the Granger suits which have gone against the Railroads."[31]

Like other aspects of his career, Gould's involvement in newspapers grew more controversial as it became more extensive. His appearance on the scene occurred at a time of transition for the major New York dailies. Charles A. Dana had taken over the *Sun* in 1868. A year later the *Times* lost its founding father, Henry J. Raymond, and slipped into a long if gradual decline. James Gordon Bennett, Jr., replaced his father at the *Herald* in 1872, the same year Reid acquired the *Tribune*. The able Manton Marble still dominated the *World* but with diminished enthusiasm, and William Cullen Bryant of the *Evening Post* had gone into virtual retirement in 1870. A new generation of publishers was emerging, one eager to cut old party ties, launch crusades, and build circulation through a peculiar blend of news, sensationalism, and entertainment.[32]

In most cases they resented the presence of a shadowy figure like Gould flitting among their ranks. Over the years their cries against Gould grew ever more shrill, as did those of the financial journals such as the *Commercial and Financial Chronicle*, the *Financier* (later the *Public*), and *Bradstreet's*. Some publishers came to regard Gould as a threat to their very existence because his expanding interests caught up not only newspapers but control of the major telegraph and cable companies as well. Although many observers noticed the fact of Gould's entry into both the transportation and the communications field, few grasped the significance of their relationship.

Jay was guilty of no such oversight. During the next decade he pursued two parallel careers, one in railroads and the other in communications, ever aware of the vital interplay between them.

13

On Track

So at the same time Jay Gould roved through the West eyeing the ruined husks of transcontinental railroads.

—Matthew Josephson, *The Robber Barons*

Getting out of Erie in a strong cash position on the eve of the panic of 1873, he was ready to attack the Union Pacific. Running down its stock with ingenious bearish tactics, he bought a controlling interest for very little.

—Thomas C. Cochran and William Miller, *The Age of Enterprise*

The entrance of Gould into the field of western railroads was in large part accidental. He saw the speculative opportunities in the depression of Union Pacific securities in the period shortly after the road's completion. In his subsequent manipulations he became interested in a large number of western lines, which he then proceeded to use for his own advantage.

—Robert E. Riegel, *The Story of the Western Railroads*

It is impossible to determine the extent to which some vision of a transcontinental railroad possessed Jay in his youth, but there is no doubt that such a dream seized him in midlife and did not release its grip until his death. His fascination for the West went beyond the profit motive or even the challenges the West posed for his voracious intellect. Perhaps it was the sheer scale of the undertaking that fired his imagination. To fill so vast a canvas with life, to shape the intricacies of so grand a design, required consummate artistry and boldness of execution. This was the stuff of empire. How could it not appeal to one eager to leave his imprint on history?

In 1873 the Union Pacific-Central Pacific route still constituted the only rail line between the Missouri River and the West Coast, though other projects were busily laying plans if not rails. The Union Pacific had been a financial white elephant since its organization in 1863, its history shorter but no less checkered than that of Erie. To build eleven hundred miles of road across wild, uninhabited country was too vast a project to attempt without government aid. In the Pacific Railroad acts of 1862 and 1864 Congress provided a land grant of twenty alternate sections per mile and a loan in the form of subsidy bonds given at the completion of twenty-mile sections. The government bonds were made a second mortgage, enabling the Union Pacific to issue its own first mortgage bonds as well. The acts also gave the road title to coal and iron found on its land and required a board of twenty directors, five of whom were to represent the government.[1]

In effect the Pacific acts defined the Union Pacific as a mixed enterprise between private and government investors. This status was to plague the road for decades because its ramifications were so imperfectly understood. Public demand for a transcontinental road had prodded Congress into offering financial support which, despite radically changing circumstances, Congress persisted in regarding as an obligation to be repaid in full. While the government assumed it had provided both generous aid and protection for investors, the men in charge of the Union Pacific protested that the terms made it impossible to build the road as a profitable enterprise. Certainly businessmen regarded it as a high-risk venture. "I would not have put a dollar in the enterprise," declared Horace Clark in 1873, "because it occurred to me that it was a wild waste of money to think of doing such a thing."[2]

Those who took it up included Thomas C. Durant, a promoter whose early brush with medicine led everyone to call him the Doctor; Oakes and Oliver Ames, proprietors of the most noted shovel works in America; Sidney Dillon; and a coterie of New Englanders, most of them from Boston. To minimize risk they chose the familiar course of extracting their profits from construction of the road. In November 1864 Durant let the first contract to an agent, H. M. Hoxie, who in turn assigned it to a corporation the Doctor had picked up a year earlier and renamed the Credit Mobilier of America. The new company became the vehicle for limiting the partners' liability as well as siphoning profits from work on the road.[3]

The saga of building the first transcontinental railroad, with its heroic struggles against nature and the Indians, has become part of American folklore. Another myth, less savory but no less spectacular, evolved from the constant infighting among its managers, who clawed and scratched like alley cats from the first mile to the last. By May 1869, when the Golden Spike was driven at Promontory, Utah, the promoters were balkanized into factions who had blustered and bungled their way through three construction contracts strung together by feeble compromises until their affairs had become hopelessly snarled in litigation. The road had always lacked strong, effective leadership. Durant, its most forceful personality, had the energy of a hurricane and was as unpredictable; he sowed dissension everywhere before resigning from the board just after the Golden Spike ceremonies. The president, Oliver Ames, was earnest but ineffectual, a lamb of indecision among wolves seeking spoils.[4]

As the company limped from crisis to crisis, the price of its securities languished. By 1871 the Union Pacific was an afflicted enterprise clutching at straws. Despite issues of land grant and income bonds, it never managed to get out from under a huge floating debt estimated at nearly $13 million in 1869. So wretched was its credit that one short-term loan cost the company 17½ percent interest. Most of the road's loans came from its own directors, who often charged 12 or 14 percent interest. In December 1870 the stock fell to 9 amid rumors of impending default. The crisis was averted but in January Oakes Ames declared insolvency and two months later his brother left the presidency. For years the Ames brothers had been the staunchest supporters of the Union Pacific. Although both remained influential in the road, they never again dominated its affairs. Just as Oakes climbed back from failure, a new blow struck from an unexpected quarter.[5]

For several years rumors had circulated about the corrupt practices of the Credit Mobilier. As early as 1869 Charles Francis Adams, Jr., referred ominously to a "Pacific Railroad Ring." In September 1872 the publication of some indiscreet letters elevated the rumors into a full-blown scandal. Two congressional committees pursued at length the charges that Credit Mobilier stock had been used to bribe certain congressmen, and that exorbitant profits had been reaped from construction of the road. The storm of public excitement far exceeded any actual proof of wrongdoing and transformed the Credit Mobilier into a vivid symbol for the growing revulsion against railroad overcapitalization and corruption. The Union Pacific would never live down its association with the most notorious scandal of its age. This new role as a public target further depressed the company's securities at a time when it was struggling to avert bankruptcy.[6]

The departure of Oliver Ames as president in 1871 marked the beginning of a competition among eastern trunk lines to capture the Union Pacific for their own use. Tom Scott of the expansion-minded Pennsylvania was the first to gain control, but his hold lasted only a year. Vanderbilt, coveting the Union Pacific as a feeder for the Lake Shore, managed to acquire enough stock in 1872 to install Horace Clark as president. When Clark was reelected a year later, the Union Pacific seemed firmly wedded to the Vanderbilt system. Within three months, however, a chain of unexpected developments relegated that assumption to the dustbin.[7]

Jay's version of how he got into Union Pacific has about it a flavor of ingenuousness as deceptive as it is beguiling. Clark and Schell recommended the stock, he told two different congressional committees, so he issued some orders to buy down before going off to the White Mountains. During his absence Clark fell mortally ill and his Union Pacific was dumped on the market. Jay discovered that his order had caught 100,000 shares and began to inquire seriously into the company's affairs. Only then did he learn of the huge floating debt, the $10 million issue of income bonds about to fall due, the large amount of paper endorsed by directors at ruinous interest rates, and the threat of impending receivership. Alarmed at the falling value of the stock, he scanned the Union Pacific board for reliable men and found two: Oliver Ames and Sidney Dillon.

Together they staved off immediate disaster and, by dint of hard work, transformed the road into a paying property.[8]

Jay always had a knack for not taking a story far enough. Clark did die suddenly in June 1873, but it is unlikely that the brokers of Vanderbilt's son-in-law would dump his stock on the market. While the facts about the wretched state of the road's finances are perfectly true, it is hard to swallow the notion that so careful and calculating an operator as Gould learned these things belatedly or got saddled with a load of Union Pacific by accident. It is doubtful he would leave an open order to buy down without knowing something of the company's affairs. A more palatable version was revealed years later by David B. Sickels, a banker who claimed to have bought control of the road for Gould during the winter of 1873–74 "by a secret combination with Messrs. Dillon & Atkins, after we had purchased in the open market all the stock available." Other evidence confirms Sickels's account that Gould struck a deal with some of the men long associated with the Union Pacific sometime in December 1873 or January 1874. General Grenville M. Dodge told his brother that "Jay Gould came in at the last moment." Gould assured Dodge that he intended to stick to the road and "make it a big thing."[9]

In fact the Union Pacific ideally suited Jay's needs and talents. Here was another grandiose, oversold property that had yet to realize its promise. Its tarnished past, tainted image, and history of internecine strife made any change of management look good. The dismal state of its finances offered choice opportunities for creative manipulation. Although the board possessed strong figures, it lacked a dominating personality. The Credit Mobilier scandal, coupled with the deaths of Oakes Ames and Clark, had left the management demoralized and rudderless. By any measure the Union Pacific had nowhere to go but up. Besides the vacuum at the top, it suffered from arrested development at all levels.

The staid Bostonians who had long dominated the Union Pacific did not welcome Gould with open arms. The reputation that preceded him was not one to cheer the proprietors of a moribund railroad, but their plight was too desperate for them to refuse help from any quarter. In time he won the support and friendship of Oliver Ames and his son Frederick, Elisha Atkins, Ezra Baker, F. Gordon Dexter, and especially Sidney Dillon, whom he met for the first time. Dillon joined that circle of men who remained Gould's closest friends and associates for the remainder of his life.

Already sixty-one years old when he met Gould, Dillon too had come up the hard way. The son of a poor farmer in Montgomery County, New York, he took up contracting in the late 1830s and virtually grew up with the railroad. He was reputed to have built thirty roads and once declared that he would not take a construction contract "unless I could make 20 percent on it." Dillon entered the Union Pacific early in its construction and tacked an evasive course through its inner struggles. Apart from similar roots, he appeared to have little in common with Gould. He was an imposing figure: tall, handsome, rugged, with a constitution as indestructible as Gould's was fragile. A leonine mane of white hair crowned chiseled features fringed with snowy sideburns. Though courteous and affable, Dillon's approach to business was, like Gould's, brisk, blunt, and all-consuming. His rapport with Gould was grounded in a mutual admiration and trust that survived their periodic disagreements. After his partner's death in 1875 Dillon spent the rest of his career as a Gould stalwart in a host of enterprises. The surest sign that Jay had taken hold of a company was the accession to its board of the triumvirate of Gould, Sage, and Dillon.[10]

In March 1874 Jay was elected to the Union Pacific board with four of his brokers. Dillon was made president, an office he would hold for a decade, and Atkins vice-president. With the help of his new allies, Gould rescued the company's finances from the brink of disaster. During the next year he orchestrated a refunding of the income bonds and cleared up the floating debt. It was a long and grueling campaign, brilliantly conceived and executed. Those acquainted with Gould's career sensed a familiar scenario unfolding. Once into the road's management, he did everything critics expected of him. He took command of its finances, helped formulate its strategy, operated extensively in its stock, issued glowing reports about its future, reiterated his determination to stay with the company until that future was realized, and bulled the stock at every turn. The logical progression would be for him to profit from the stock on the rise, milk the company's finances dry, destroy the road's standing through the same devices he had used earlier to build it up, sell the stock short, and exit with large bear profits, leaving a ruined property in his wake.[11]

But Jay departed entirely from this script. He stunned Wall Street by staying with the Union Pacific as the driving force behind its development. His goal was to make the road into what it had never been, a stable, profitable enterprise. To astonished observers this behavior amounted to nothing less than a metamorphosis. Hardly anyone realized that Jay had in fact tried to do the same with the Erie only to have his efforts swallowed by the financial and legal pyrotechnics of those years. Among his peers Gould was regarded as no more than a cunning operator, the "grizzliest of bears." With few exceptions the financial community clung to this belief despite the contrary evidence offered by Jay's role in the Union Pacific and later in other enterprises as well.[12]

No one scrutinized Gould's movements with a closer eye than Collis P. Huntington. As the dominant figure in the Central Pacific he shared the transcontinental route with Union Pacific and admitted to being the uneasiest of bedfellows. Gould impressed him as "a clever fellow, but . . . the most reckless speculator in the world, I think." By contrast Huntington was a builder, a developer whose own brand of recklessness was expanding projects faster than he could raise funds to sustain them. For more than a decade he had labored mightily on Wall Street to raise capital for the Central Pacific, the Southern Pacific, and an eastern venture, the Chesapeake & Ohio.[13]

The ordeal had worn him down and left him impatient with men who merely raided properties instead of developing them. Even the physical differences between Huntington and Gould suggested their contrast in styles. Where Jay was small and lithe, a dark, elusive figure, Huntington was large and impassive, a bear of a man with granitic features and piercing eyes. At fifty-three his full beard was marching toward snow white and his thinning hair toward extinction. Their progress was accelerated by Huntington's encounter with Gould, which marked the beginning of a long and sometimes bitter rivalry. Like so many others, he distrusted Gould at first and wanted nothing to do with him. Unlike many others, he came to know Gould well, to trust and even to like him despite their business differences.[14]

The Pacific roads were a curious spectacle: two independent companies bound together in a symbiotic relationship that forced them to cooperate in everything from schedules to supplies to setting rates. Their only competition for through traffic came from Pacific Mail, a company in which Gould was also prominent. Huntington assumed that Jay would follow his personal interests and could not be counted on to do right by the railroads. Gould's position in Pacific Mail permitted him to play one against the other, jiggling their stock prices in whatever direction suited him, and ultimately to scuttle one or the other or both.

During 1874 relations between the rail and steamship lines deteriorated into a rate war despite Jay's efforts to renew a traffic agreement that had lapsed the previous fall. Within Pacific Mail one element, associated with the Panama Railroad, wanted the steamers to compete vigorously with the Pacific roads and resisted Gould's efforts to make peace. After a heated contest Sage was elected president of both Pacific Mail and Panama that spring, but his interests did not coincide with Gould's. Through the summer Dillon struggled feverishly to end the rate war that enabled Pacific Mail, in Jay's words, "to divert nearly all the China trade via the Isthmus." Huntington too was furious with Pacific Mail and explored with Gould and Dillon the idea of a rival steamship line owned jointly by the Pacific roads. In September they sent Captain George Bradbury to charter five steamers in England, where the depression had left ships lying idle.[15]

The cautious Huntington reminded Leland Stanford that Gould was "rather fast in such matters," and urged him to organize the steamship company in California. "If we go into this," he warned, "it must not get out of our hands. . . . The more I see of these U.P. people the more I am convinced . . . that this steamer company must be controlled by us." What bothered him most was his inability to fathom Gould's true intentions. How deep was his commitment to Union Pacific? What was his interest in Pacific Mail? Did he really want a new steamship line or were the negotiations a ploy to set up a bear raid in Pacific Mail?[16]

By November Bradbury had found the ships but Huntington's California partners had not yet chartered the new company. Meanwhile, Pacific Mail was fast sinking beneath its floating debt of about $4 million and a management engulfed in strife. Sage invited a new proposal for peace from Huntington, then rejected the terms offered. His rebuff convinced Huntington that the new steamer line was needed, but he did not want to move as rapidly as Gould, who was pressing him to charter the steamers at once. In giving the reason for Jay's haste, Huntington revealed what he took to be the difference in their approaches to business:

> I have been disposed to go slow so as to be sure that we have the best ships at low rates if we charter any, all the time hoping that we could make some satisfactory arrangement with the P.M.; while Gould, who has been largely short of P.M., to make money out of his short interest, would make almost any contract, as he would make money in any way.

For these reasons Huntington concluded it would be best if "Gould were not in, and I will endeavor to find the opportunity to tell him so, of course, in a friendly way."[17]

But events soon extinguished that hope. The showdown in Pacific Mail came on December 3, when a stormy confrontation forced Sage out as president of both Pacific Mail and Panama. "They have had a flare-up in P.M., and have not got settled down yet," Huntington reported gloomily. Stanford finally forwarded a charter for the new Occidental & Oriental Steamship Company, but just when Huntington was ready to move, Gould and Dillon balked. The capitalization of $10 million was too high, they protested, twice what they expected. Huntington could not deny the validity of their complaint; the amount surprised him, too. "Just why you made the capital so large as $10,000,000 I do not know," he growled at Stanford, "I suppose you do." Yet he clung to the suspicion that the delays were part of some "bold stock speculation" on Gould's part.[18]

"I think . . . they are not our kind of people," Huntington concluded of the Union Pacific crowd. The stronger this belief grew in him, the more convinced he became of the proper solution. "The fact is," he admitted, "I see no other way to have peace and control the China trade than for us to control U.P."

Jay was in fact playing a different and far deeper game than Huntington realized. Like many complex men he possessed an uncanny ability to find courses of action that served a variety of motives at the same time. In the case of the steamship companies his strongest desire was to protect his interest in Union Pacific. To raise the value of its stock he had to promote the development of business along its line, protect its strategic position, and keep earnings high. Each of these matters received his close attention during 1874–75, and in each instance he achieved striking success.

How did Pacific Mail fit into his plans? He was undoubtedly operating as a bear in its stock, but he had compelling reasons for driving the price down. The rate war with Pacific Mail hurt Union Pacific earnings at a critical time. Peace could be obtained by negotiating an agreement, by forcing one through repeated attacks on Pacific Mail's weakened finances and/or formation of a rival steamship line, or by capturing Pacific Mail and imposing terms. Gould pursued all these options at once, shifting emphasis from one to another as circumstances changed. Sage's ouster was part of an effort to batter Pacific Mail into submission. It succeeded in knocking the stock down, which not only helped Jay's speculation but made it cheaper to buy control. The formation of Occidental added competitive pressure from without. For that reason Jay was eager to get the new line in operation.[19]

On one point Huntington was wrong. The delays came not from Gould but from his Boston friends, who were advised by counsel against subscribing so large an amount. Dillon hesitated to move without approval of the executive committee until he learned that a decision had to be made at once on leasing steamers and that a new charter would take a month to procure. Huntington offered to place part of the stock in California, but Gould was adamant that half remain in hands friendly to Union Pacific. He could not afford to let the matter drag, for Occidental was crucial to his plans. In January 1875 he broke the impasse by asking the Bostonians to take $500,000 if he and Dillon would take the rest. Oliver Ames took exactly that amount, leaving Gould to absorb $2.5 million and Dillon the other $2 million. The start of the new company, coupled with defeat of a subsidy increase bill, dropped Pacific Mail below 31, its lowest price since late 1873.[20]

Having set the stage, Gould did not miss his cue. While the bears gleefully unloaded Pacific Mail, he switched sides and bought heavily until the price climbed back to 41. In one stroke he caught the short sellers by surprise and acquired control of Pacific Mail. On March 3 he, Dillon, and Oliver Ames went on the board and Dillon assumed the presidency. Peace between the railroads and the steamship line was restored and rates were advanced sharply. Jay assured Huntington that the two steamship companies would be run in harmony and offered him a seat on Pacific Mail's board.[21]

Huntington hesitated, then declined. The abrupt turn of events puzzled and alarmed him. Gould's recent moves gave him power over through rates and put him in a position to damage the Central Pacific in a variety of ways. So far all had gone smoothly; Gould remained favorably disposed toward Occidental, and his presence in Pacific Mail improved relations with that company. There were occasional squabbles, one of which sent Dillon storming into Huntington's office shouting that "he would be God damned if he would allow the U.P. to be trampled underfoot by the C.P. as she had always been." But Dillon's outbursts cooled quickly and were for Huntington a welcome relief from Gould's inscrutability. Never had he encountered a man so fastidious in his dealings, so adept at piling complications onto the most straightforward of situations. So precisely did he walk the tightrope of obligation that Huntington was moved to confess in exasperation, "I think him the most difficult man to do anything with that I ever knew."[22]

If only he could discern Gould's ulterior motives! While these remained a mystery to him, Huntington continued to distrust Gould's friendliness. During the spring of 1875 both Union Pacific and Pacific Mail surged upward. Huntington was convinced that Jay was bulling the stocks as a prelude to selling out and going short. As long as Gould remained in the catbird seat, Huntington saw no choice except to "be prepared, as far as we can, for his doing almost any outrageous thing" and, above all, to "avoid a quarrel with him, and watch for the time when we are ready to control the U.P., and then go in and get control of it."[23]

Yet, for all the warning flags he hoisted, Huntington agonized over whether he judged Gould fairly. "I am afraid they will play us false," he muttered uneasily, "although I am not sure that I have any good reason for thinking so."[24]

Huntington was correct in attributing the spring market movement to Gould. It was in fact the culmination of a long and difficult campaign to extricate the Union Pacific from its financial morass. The thorniest problem was the road's poor credit, which forced Jay and other directors to carry much of its floating debt themselves. It could not sell bonds or paper at reasonable rates until investors had confidence in its future. The obvious way to win confidence was to demonstrate the road's ability to earn a profit. No amount of stock jobbing could accomplish this task. It required close attention to detail, a rigorous effort to expand the volume of business, keep rates high and expenses low, and cultivate new sources of traffic. To undertake so ambitious a program in the teeth of a depression seemed foolhardy at best, yet Jay plunged into it without a flicker of hesitation.

In this campaign Gould regarded the price of Union Pacific stock as the key barometer of progress. An upward movement would help the company's credit, provided it was grounded in performance rather than manipulation. It would be difficult, for the stock had tumbled below 15 in November 1873 and had crawled back only into the thirties a year later. Nor would it be easy for a man with Gould's reputation to convince anyone that he was managing the road for some larger purpose than a rise or fall in its securities.[25]

"I am satisfied Union Pacific is the best piece of railroad property in the country," Jay proclaimed in September 1874. Nine months later the stock stood above 78, the company's credit was secure, and the board had declared its first dividend ever. Controversy would arise over whether the dividend had actually been earned, but on one point opinion was unanimous: An astonishing turnaround had occurred in Union Pacific, and the credit belonged to Gould. Even then much of the praise was grudging, for few observers were prepared to believe that he had done anything more than offer another display of his market wizardry.[26]

14

Developer

His control was always exercised from the East, and it is probable that he never saw some of his properties, owing to his infrequent western trips.

—Robert E. Riegel, *The Story of the Western Railroads*

Gould was a speculator [who] cared little for the quality of his railroads as transportation machines, and even less for building up the territory through which they passed; his eye was continually out for quick profits.

—Richard C. Overton, *Gulf to Rockies*

There was no such thing as corporate loyalty in Gould's book . . . no adventure in opening the vast new territories to settlement. If he helped to change the face of the country, it was only inadvertently; his whole attention was focused on a company's books and its fluctuations on the stock market. A railroad to him was a set of books, a safe full of securities.

—Richard O'Connor, *Gould's Millions*

THE STUNNING RECOVERY of Union Pacific offers a revealing insight into the persistence of Gould's reputation as a predator and wrecker of properties. It convinced observers not that he had changed stripes but only that he had grown more artful in his machinations. They watched Union Pacific float upward and waited confidently for the balloon to burst. When to their amazement it did not, critics reacted as if the law of gravity had somehow been defied. Only a few ventured to explore the possibility that Jay regarded the road as something more than a pawn in his speculative combinations.

Reticent as always, Jay offered skeptics scant evidence on which to modify their views of him. Although he bent any willing ear on the potential of the Union Pacific and his determination to see it through, these pronouncements were dismissed as rhetoric intended to puff the stock. By 1874 the possibility of a dialogue between Gould and the public had already become nonexistent. He was not one to bare his soul or disclose his practices, and journalists refused to believe what he said anyway. As a result, few men outside the company understood his actual role.

The evidence leaves no doubt that Gould shaped the destiny of the Union Pacific to a far greater degree than anyone has suspected. He revised its financial structure, waged its competitive struggles, captained its political battles, revamped its administration, formulated its rate policies, and promoted the development of resources along its lines. He did these things by immersing himself in every aspect of the road's operations until his knowledge of its affairs was encyclopedic in its breadth and detail. No rail magnate of the nineteenth century knew his road more thoroughly, and few mastered their enterprise in so short a time. There was no shortcut to this achievement; it was the product of hard work and intense concentration.

What drove Jay to undertake this task remains an intriguing mystery. It could not have been simply the desire for gain, for there were quicker, less taxing ways of making money. Perhaps the immensity of the challenge appealed to him. Possibly, too, his tarnished reputation had begun to bother him. Although he would never admit to such a concern, he may have seen in the Union Pacific a chance to erase the stigma attached to his name. Whatever his motives, Jay identified himself with the road in a way that transcended his usual ability to weave corporate and self-interest inextricably together. A new phase in his career had opened. He did not leave the market or cease his speculative forays; nor did he confine his activities to the Union Pacific. But he had taken hold of the road, and he did not mean to let it go.

The Union Pacific suffered from all the weaknesses of a pioneer enterprise. The grandest construction project of its time, it had been built too rapidly at enormous expense through country too empty to support it. By 1874 the company had issued nearly $37.6 million in stock and $49.5 million in bonds on which it paid about $3.3 million interest. The prolonged dispute over repayment of the government loans loomed as a threat to an already strained financial structure. Handicapped by what Charles Francis Adams called a "remarkably narrow charter," the company lacked a free hand in such crucial areas as rate making and branch building. By any measure the Union Pacific was a vulnerable enterprise facing an uncertain future.[1]

These problems were not of Gould's making. He inherited them as liabilities to be settled in the coin of policy. His solutions were neither original nor unique; the basic premises already belonged to the conventional wisdom of shrewd industrialists. Earnings must be kept high and expenses low, which required close monitoring of costs. New sources of business must be developed and old ones protected from competition. Obligations must be refunded at lower rates of interest. A favorable political climate must be cultivated to ward off hostile legislation at both state and federal levels. The government claims against the company must be contested to the bitter end. Clear channels of authority for administering the road must be established.

The last point was crucial, for Jay preferred to function as a de facto ruler. He held no position except member of the board and the executive committee. If anything he resembled a party boss who remained in the background while exerting his authority through officeholders and party faithful. This arrangement depended on having reliable men in key positions. A cadre of brokers served Gould's interests on Wall Street while Dillon acted as his lieutenant in New York. The Boston directors handled routine matters and worked closely with Gould on financial and policy questions. What Jay lacked most was a capable operating man whose intelligence and judgment could be trusted far from the scene of action. During 1874 such a man emerged in the person of Silas H. H. Clark, who soon became Gould's western adjutant and close friend.

A few months younger than Gould, Clark was a New Jersey farmboy who started in railroading at the bottom. He had worked his way up to conductor when his zeal caught the eye of Sidney Dillon, who in 1867 sent him to Omaha as freight agent for the Union Pacific. The move began Clark's long association with the railroad. In April 1874 he succeeded T. E. Sickels as general superintendent and commenced what proved a lifelong relationship with Gould. Along with their Algerlike climb from humble origins the two men shared a love of books, an appetite for work, and delicate health. Sensitive, alert, dedicated, shrewd, a tough infighter prone to bouts of insecurity about the loyalty of those above and below him, Clark served Gould well for eighteen years and earned for himself wealth and status in the process. A man of striking appearance, his deep-set, lustrous eyes could assume the haunted distance of a poet one moment and the fierce glare of a prophet the next. Beneath a prominent nose he wore a long, scraggly beard that obliterated not only his lower face but tie and collar as well.[2]

As superintendent Clark found himself in an awkward position. His formal orders came from Dillon, yet he also received a steady stream of instructions from Gould by private letter. Clark learned early which had priority. The dangers inherent in serving two masters was averted by the close rapport between Gould and Dillon. Their ability to administer the road's affairs through this triangular arrangement for nearly a decade reflects a close bond of friendship and trust among the three men. Clark also found himself in dual roles at a more subtle level. He was the servant of the Union Pacific and the trusted agent for Gould's interests in the West. So long as the two were compatible Clark could fill both comfortably, but the day would come when they clashed and forced him into making difficult choices.

Once Jay took the measure of Clark's abilities, he relied heavily on him for information and advice. The key to Gould's mastery of Union Pacific affairs lay in his extraordinary store of knowledge, much of it gleaned from personal inspection tours along the road. The first took place in September 1874; thereafter he made it a point to go west once or twice every year. On these trips he was no eminent visitor whisked through crowds of fawning dignitaries. Traveling only in daylight, his eye scrutinized every detail of the road, farms, ranches, mines, industries, towns, crops, even the weather. At every stop he pumped employees for information; the officers soon learned to pack their heads like students when Gould came to call. In villages and towns he talked to bankers, merchants, manufacturers, editors, politicians, anybody who could provide data useful to him. He wanted to know about local needs, potential markets, what service could be provided by the road or himself to develop business where little or none existed, what problems or obstacles existed, who were the important men to know.

Gould supplemented this fund of information with a network of other sources. In conversation he picked the brains of speculators, railroad men, politicians, engineers, inventors, anyone with something of value to him. He fired off letters in all directions, asking questions, seeking information or data. The newspapers of major cities and important towns along the road, as well as trade and financial journals, were culled assiduously and relevant items clipped for reference or dispatch to Clark or some other officer. A flood of inquiries poured from his pen, many of them scribbled late at night from his desk at home, on his personal blue monogrammed stationery with the initials intertwined like snakes crawling up a staff.[3]

Prospects looked bleak in September 1874. Business in the East remained depressed, tumbling farm prices curbed immigration to the plains, and in Nebraska an invasion of grasshoppers cut the corn crop to half of what it had been in 1873. Like most railroads the Union Pacific relied heavily on local traffic, where lack of competition allowed it to charge high rates. Any decline in local traffic made the road's earnings more dependent on through traffic, the rates for which had been slashed by competition with Pacific Mail. It was this dilemma that spurred Gould's efforts to resolve that conflict. When bumper crops in California helped keep earnings high despite the war, Gould augmented this bit of good fortune with a concerted attempt to hammer down expenditures. Dillon wanted to spend generously for improvements on the road, but Jay's demand for economy overruled him. Clark got the same message, and the performance of one officer drew a stiff reminder from Gould that "money is made by saving as well as earning."[4]

Instead of publicizing the road's improved performance Jay decided to withhold the figures entirely. "It not only embarrasses you in keeping down expenses," he explained to Clark, "but it gives the eastern roads the idea that we are making too much money." There was another, more subtle reason: Jay wanted the news kept in reserve for the campaign to bull Union Pacific securities. The opportune moment came in February 1875. In a statement to the bondholders Dillon declared the refunding of income bonds a success, then announced net earnings of $5.9 million for 1874 along with glowing statistics on the increase in ore and bullion traffic. Before Wall Street could absorb this sensation, Gould unleashed the market operation by which he seized Pacific Mail and imposed a settlement of the rate war. Early in March Union Pacific rose above 40 before the bears, caught unawares by these events, struck back furiously.[5]

For Gould the critical hour had arrived. Some dissidents on the board were scheming to control the forthcoming election; everything depended on the willingness of the Boston directors to trust him. "Tell our friends in Boston not to sell a share but to place what they can off this market," he urged treasurer E. H. Rollins. "The result will be that when the market turns there will be no stock for sale . . . I shall not sell a share of my stock at any price . . . I shall stand by it." If they would stand fast, he promised, "we can plant the price to 50 or above & keep it there."

Boston held fast and Jay delivered his pledge. The forces amassed behind his bull campaign proved irresistible. Besides the pact with Pacific Mail, the Supreme Court rendered a verdict that was in Gould's opinion "worth a couple of hundred thousand dollars a year to the company." He unveiled a new plan to settle the dispute with the government. On the road Clark kept trains moving despite horrendous weather. Traffic remained heavy at rates advanced in March by 40 to 100 percent. The company, Dillon crowed happily, was "at high tide." At the annual meeting on March 10 the public learned of the road's fat earnings for the first time. After reciting a long list of favorable developments the board declared a quarterly dividend of 1½ percent, the first ever for Union Pacific. Gould's ticket won reelection easily. The opposition never had a chance; Jay held 140,000 shares, Oliver Ames 30,000, and Dillon 10,000.[6]

News of the road's earnings and dividend threw Wall Street into pandemonium. As the entire market bounded upward, nearly 389,000 shares of Union Pacific changed hands in a single week. While Gould was credited with orchestrating the bull movement, skeptics wondered whether the dividend had been earned or whether it was "not a measure adopted at the instance of a very prominent speculator." On March 30 the stock reached 67 and knowing observers waited expectantly for the inevitable fall that accompanied Gould's operations. They were doomed to disappointment. Earnings remained strong and Jay improved them with another rate hike. The upward surge continued until June, when Union Pacific touched 79. A month later the company paid the first dividend. As earnings swelled Jay continued to preach the gospel of economy. In 1874 expenses amounted to 44.5 percent of earnings; Gould wanted them cut to 40 percent. It was a message he would repeat incessantly over the coming months.[7]

By the summer of 1875 even Gould's severest critics had to concede that he had done something more substantial with Union Pacific than manipulate its stock. Exactly what role he played in its affairs remained no less a mystery to them than did his ultimate intentions for the company.

From the outset Gould regarded coal as the key to the development of the Union Pacific. The road offered a variety of resources for local traffic, among them coal and mineral ores in the mountains, cattle and corn on the plains. In Gould's mind none of it rivaled the potential of coal, which could provide cheap fuel for trains and supply a region starved for energy. The importance of coal to the company's future impressed itself on Jay during his first western tour. He came home full of praise for Clark's work, then added, "The only thing I have any solicitude about is the coal business. I would like to see the coal business developed to its fullest extent."[8]

Here too the road's managers had already amassed a short but sordid history. In July 1868 the company had leased its coal lands to two Missouri dealers, Cyrus O. Godfrey and Thomas Wardell, who were to locate and operate mines. The Union Pacific agreed to buy all coal produced at prices on a sliding scale ranging from $6 to $3 a ton over the life of a fifteen-year contract. It also gave the contractors a 25 percent rebate on all shipments to other customers. Godfrey transferred his interest to Wardell, who on April 1, 1869, assigned the contract to a Nebraska corporation called Wyoming Coal and Mining Company. From the first the two men acted as agents for several Union Pacific directors, who owned 90 percent of Wyoming's stock and held it "for the use and benefit of the Union Pacific," but the stock remained in their hands and so did the profits from Wyoming's contract.[9]

The result was an absurd situation in which the Union Pacific paid exorbitant prices for coal to a company it was supposed to own but did not, and assumed all risks for a business over which it had no control or influence. For nearly five years the board fumbled about ineptly trying to annul the contract or get control of Wyoming. The government directors condemned the contract as unwise, but still the impasse dragged on. Gould gave it short shrift. One day after the election he cut through the morass with the kind of blow for which he would become notorious. Rather than debate the issue any longer, the company simply abrogated the contract and seized the mines to work as its own property. A startled Wardell did not resist the takeover but sought refuge in the courts, where his suit lingered for six years before finally losing.[10]

Once in control, Jay resolved to "take hold of this department . . . and get the mining and selling of the coal upon the most economical & efficient basis." It was not an easy task. In 1874 the Union Pacific was new at the business of operating mines. Production was low, costs were high, and labor was scarce. No one had created an effective organization or systems for marketing, monitoring production, or setting prices. The company operated seven mines in Wyoming: four at Rock Springs, two at Carbon, and one at Almy, near Evanston. All were beset with labor unrest. When hard times and low wages drove some miners to strike in 1871, they were fired and replaced with Scandinavian labor at lower wages. By 1874 their disgruntlement matched that of the older hands. An attempt to organize the miners brought rumblings of another strike. The threat confronted Gould with his first major labor crisis, and he did not underestimate its importance. "Our coal business will never work satisfactorily," he declared, "till we master the labor question."[11]

Jay showed no sympathy for the miners' plight. Like most businessmen he dealt with the labor issue from the narrow perspective of control and cost. Unionization threatened his control of the property and higher wages thwarted his efforts to reduce costs. The best way to lower costs and eliminate discontent, he concluded, was to fill the mines with Chinese. The road already used Chinese laborers as section hands despite protests from local editors and displaced workers. They received $32.50 a month compared to the $52 paid white men, and Jay thought the figure could be driven even lower. Whatever tensions the presence of Chinese might spawn, their use would enable Jay to reduce costs. For that task he had a model in some bituminous mines he owned at Blossburg, Pennsylvania. Although those mines were also embroiled in labor strife, they managed to produce coal at a cost of $1.25 a ton. Jay pegged the Union Pacific's goal for the coming year at $1.30, a reduction of more than 70 cents a ton. That fall the first Chinese went into the Almy mines; in November the white miners retaliated by going out on strike.[12]

Although the strike occurred at a time when Jay was desperate to improve the Union Pacific's earnings, he urged a hard line. Dillon feared the strike would spread to other mines and warned that "it would be very hard to have a general strike of miners this winter." For Gould, however, the paramount issue was bludgeoning down the cost of labor. "I would not make any concessions to the miners," he told Clark, "but fight it out with them even at the cost you mention . . . & in the spring I would let Mr Serat fill up his mines with Chinese labor and thus settle the status permanently."[13]

The strike ended less in a settlement than a truce. In the spring of 1875 Jay instructed Clark to fill a new section at Almy and all other vacancies with Chinese labor. At Almy the cost per ton fell 25 cents, the company saved about $40,000 a year, and the union movement was broken. Gould wanted the figure cut another 25 cents but conceded "it is better to come down gradually." In the East a wage cut had just been imposed on anthracite miners after the failure of their prolonged strike. With Jay's approval Clark promptly reduced wages at both Carbon and Rock Springs. "With Chinese at Almy & native miners at the other point," Jay observed, "you can play one against the other & thus keep master of the situation." The miners thought the cost of provisions and clothing would also be lowered to offset the cut in wages. When that did not occur, they went out on strike.[14]

"I am not at all disappointed at the action of the miners," Gould admitted as Clark moved vigorously to crush the strike. He replaced the strikers with more Chinese, whipped up public sentiment against the miners, and asked Governor John M. Thayer for troops to protect the company's property. Thayer, former senator from Nebraska and an old Union Pacific hand, appealed to General George Crook, who sent two companies of federal troops to each mine. There was no trouble; the miners were beaten from the start. Gould and Dillon showered Clark with congratulations on routing a "dangerous labor combination." Before long Chinese would outnumber whites in the company's mines, nurturing a legacy of bitterness and prejudice that would one day explode into violence, but such matters were no concern of Jay's. With Chinese workers installed in all the mines, he unfurled an ambitious new goal: "I shall be disappointed if we do not reduce the cost of our coal to not much over $1 per ton."[15]

The broad objective was simple. Jay wanted to monopolize the entire coal trade between Ogden and Council Bluffs, Iowa. If rival operators could be driven out, the Union Pacific could charge high prices for its coal, fill empty cars running in both directions, and use the profits to "give the company its own fuel free." To do that he needed an organization capable of running the mines efficiently at full capacity and snatching business from competitors through aggressive marketing tactics. "I am anxious to sell all the coal we can to local consumers," he declared. "What I want is to gain so large a local trade that we can keep our mines & rolling stock busy all the time."[16]

Part of that trade would have to be wrested from Huntington and his associates, who had similar ambitions for their Rocky Mountain Coal Company. Friction between them was increased by the fact that Rocky Mountain had to ship from Evanston via the Union Pacific. In June 1874 an ugly squabble erupted when the Union Pacific raised rates sharply. After conceding Huntington a lower rate, Gould startled him with a bold proposal that the Pacific roads form a new company to control all the mines at Evanston for their joint benefit. Huntington was tempted by the idea, but the old fears held him back. "Much could be made out of such an arrangement," he wrote Colton, "if we could, in some way, keep control; but I do not like to be mixed up with Gould in anything of which it is possible for him to get control."[17]

While negotiations played themselves out, Gould pushed Clark to whip the coal department into shape, like an army girding for battle. "To mine coal cheaply in the long run you want to have the best of machinery," he advised. The cost at Almy and Rock Springs should be driven down to $1.25. If the Central Pacific insisted on competing, "I think we should regulate the rates of transportation & the price of the coal delivered as to monopolise the coal trade of Utah." Like many a general, however, Jay soon learned that it was easier to devise a battle plan than to implement it. He tended to set ambitious goals on the assumption that Clark would somehow find ways of realizing them. Not surprisingly, the results often fell short of expectations.[18]

Production lagged, costs dropped slower than Jay hoped, and efforts to seize commercial markets fell far short of his goals. The situation at Council Bluffs exemplified his frustrations. The Union Pacific reached Council Bluffs via its own bridge, on which it charged tolls. As early as October 1874 Gould suggested selling coal in the Bluffs at Omaha prices, which meant waiving the toll. For fifteen months he urged this policy but deemed it more important to maintain Clark's trust by deferring to his judgment than to impose policy by command.[19]

If orders were gloved in velvet, advice flowed freely. Jay deluged Clark with suggestions on every aspect of the coal business—marketing and pricing tactics, how best to utilize machinery, where to locate storage sheds, how to insure quality ("in competition this is necessary to retain and satisfy our customers"), how to speed delivery, drive out competitors, and avoid false weighing. The possibility of producing coke from company coal especially intrigued him. Convinced that availability of coke would revolutionize the silver industry in Utah, he hired specialists to experiment with different processes and sustained their research for years.[20]

"You know this coal business is a sort of hobby of mine," Jay demurred. "I will hardly be satisfied until we get the cost of mining reduced to $1.00 per ton & mine & ship about 1,000,000 tons per year." The receipt of some unsatisfactory coal returns ten days later moved the hobbyist to blister Clark's ear on the need to lower costs and push sales. Gould's close attention to coal development lessened noticeably after 1880. By then the coal department had become a profitable and growing operation. Much of the credit for its success belongs to Gould. He never realized his goals of the dollar ton or the million-ton output, and commercial sales never fulfilled his expectations. Yet between 1875 and 1880 production more than doubled and the cost per ton dropped 65 cents. The foundation for future expansion had been laid.[21]

What Gould did with coal he tried also to do with soda. In this case the results did not bear out the effort. The discovery of vast soda deposits southwest of Laramie sent Gould to investigate the site on his first western tour and prompted Dillon to explore the market for soda products and the feasibility of erecting a refinery. The Union Pacific planned a spur line to Soda Lakes and perfected its land titles there. Jay wanted other parties to develop the business so that "our hands are left free for other matters," but when the project lagged, he hired a geologist to survey the lake deposits and, as he had done with coking, discussed with scientists every known conversion process. The company's chief engineer was sent to Europe to scout techniques, machinery, and skilled workmen. Unfortunately, the best conversion process required salt water and seaweed, neither of which were abundant in Wyoming. Still Gould pushed the project, but delays and problems stalled it until 1883, when another attempt was made to revive it. Nothing came of the venture in Jay's lifetime, but a century later the Union Pacific would profit greatly from the mining and refining of trona in that region.[22]

Occasional setbacks never dampened Gould's enthusiasm. His sharp eye roved the line seeking new enterprises beneficial to the company and the towns along its route. This last point was crucial to him, for thriving towns meant new customers and increased business. In Laramie, for instance, the company built a rolling mill that became the town's largest employer. Here was Gould's notion of local development at its best. The town got a major plant at a time when its economy was stagnant and its prospects bleak. In return it offered generous tax concessions that helped make the mill profitable. The railroad got cheaper rails and some traffic.[23]

No town received more attention than Omaha. Its strategic location impressed Jay with the need to promote industries that drew their raw materials from the West, so that the Union Pacific would get the long haul and fill its empty eastbound cars. Smelting offered a prime opportunity. Since the road carried ore and bullion from western mines, Jay encouraged the owners of the Omaha and Grant Smelting Company to extend their works. To the disgust of Colorado smelters the Union Pacific aided the Omaha company with loans and rebates.

But poor management kept the company flirting with failure until a disappointed Gould was driven to proclaim in 1877, "I am half inclined to think we better take up the Omaha works & run them for our own a/c or build new works for the purpose." In the end he did neither. Consolidation with a Colorado company in 1882 transformed Omaha and Grant into a prosperous enterprise. Jay also suggested that the company mine its own ore "to supply the market the same as we do with coal." He found in Baltimore a market for the large amounts of low-grade copper ore in Utah that could be captured with a rate concession.[24]

If mining was vital to the Union Pacific, so were ranching and farming. To fill the vast stretches of empty land along the route with settlers the West had to be promoted vigorously, the sale of company lands improved, and a policy devised to resolve the conflicting interests of cattlemen and farmers. In all these areas Jay took an active role. Promotion was old hat to him; he had used his influence with the press as freely to hype new ore bonanzas and the virtues of grazing or farm land as he did to boost the price of securities. Since 1869 the Union Pacific, like other land-grant roads, had dispatched a flood of promotional literature and agents to lure immigrants to its territory. Despite these efforts, the company's land department made a poor showing during the 1870s. Its head, Oscar Davis, attributed its sorry record to difficulties beyond his control. Wyoming was thought too dry for cultivation, and Nebraska had been ravaged by drought and grasshoppers for several seasons. There had been prolonged disputes with the federal government over securing titles and with local communities over the tax status of company lands. Attempts to sell land in large blocks had been thwarted by the checkerboard pattern of alternating government and company sections. Davis had attacked the problem with energy and imagination but could not budge these obstacles.[25]

In Jay's view Davis had spent large amounts of money with little to show for it. After being warned in 1874 to "come down to hard pan all around," Davis trimmed his operation but the results still disappointed. "This is the only department of our business that is not satisfactory to me," Gould informed Clark. "This one needs a thorough overhauling or a *clean wipe out*." In February 1878 Davis was fired and replaced by Leavitt Burnham, who was ordered to pump new life into the sagging department. Jay warned Clark that competition for immigrants would be fierce and added wistfully, "I wish every foot of our lands east of Kearney was sold & in possession of an actual settler—it would soon give that division a large local business."[26]

Apart from promotion and sales, there was the thorny question of how the land should be used. At first the Union Pacific sold land with little concern for the peculiar needs of cattlemen and farmers. By the time Gould arrived it was painfully apparent that lands in western Nebraska and Wyoming held little appeal for farmers. Years of drought reinforced the old myth of the Great American Desert, and a host of authorities reasserted doubts about the arability of land west of the hundredth meridian. This debate interested Gould because of its practical ramifications. If lands west of the hundredth meridian were not suitable for farming, the land department should confine its promotional efforts to stockmen. In July

1875 Grenville M. Dodge advised Gould to do just that, arguing that "unless the elements change they cannot west of Grand Island get more than one crop out of five. It is too dry." Instead men of small capital should be encouraged to put modest herds on the land and build their holdings up. The road would benefit, in Dodge's view, because "the man who raises stock, ships three times as much over the road as the man who raises grain."[27]

Gould's western tour that fall convinced him that Dodge's policy was sound. He first tried to protect stock growers by not selling land west of North Platte to settlers. When, to his surprise, stockmen petitioned the company to reconsider the move, he suggested dividing the land into ranches so as "to sell the bad with the good & and thus locate the cattle more permanently along our road." Here was another chance to secure a long haul if large herds could be located at the western end of the line and stockyards established at Omaha. "We cannot pay too much attention to stock development on the west end of the road," he reminded Clark. "It is the only business the country is adapted to & it requires some encouragement & inducements to get it started."[28]

Gould was prepared to offer both. In June 1877 he paused during his western tour to visit W. A. Carter, whom an English correspondent described as "the great man of Fort Bridger, and all western Wyoming for that matter." Earlier, in New York, Carter had unfolded to Gould his vision of a cattle shipping complex near Fort Bridger. Yards could be built to receive herds from Montana, Idaho, Utah, Nevada, even Oregon and Washington territory. Jay saw that the creation of a shipping center near Fort Bridger would give the Union Pacific a long haul to Omaha. Impressed by what Carter showed him, he reduced the rates on stock shipments at all points and gave Carter special rates on all cattle shipped. In addition, he agreed to build extensive yards adjacent to the road and promised Carter assistance in erecting his own facilities. Carter promptly expanded his yards, put up a hotel for buyers, and constructed other facilities. Jay reminded Clark that "we ought to have plenty of buyers on hand to buy & ship. We ought not to let any cattle be *driven* farther east for shipment. The more of a center we can make Judge Carter's yards the more buyers will be attracted to it & the more cattle will be brought in for sale." By November Carter had already shipped 250 carloads and had more cattle than the Union Pacific could find cars for.[29]

A similar pattern was followed in Nebraska, where the Union Pacific built cattle pens at Schuyler, Kearney, and other points to capture northward drives. The business transformed Ogallala from an obscure way station into a full-blown cowboy capital. In each case Gould searched for the right man to take hold as Carter had done. At the Omaha stockyards William A. Paxton and his associates emerged as the chosen few who received special rates and other concessions, as one embittered rival complained.[30]

Jay understood that rates served a variety of needs and circumstances. An inflexible tariff discouraged business and hampered development. The road had to take the long view and charge not only what the traffic would bear but also what it would bring. Gould prodded E. P. Vining, the freight agent, with suggestions for cultivating new sources of business. In a typical instance he urged Vining to "bring out the Utah barley . . . at a *concession* if necessary as the cars will go back loaded with corn." The basic formula was to charge high local rates for captive traffic, match or beat rates on competitive traffic, offer special rates to attract new business, and strive to keep cars filled in both directions.[31]

When new opportunities arose, Jay swooped hawklike to exploit them. The discovery of gold in the Black Hills excited him with prospects of heavy traffic and the opening of new territory. An evening's conversation with the geologist who had surveyed the Black Hills for the government informed him that the region was suitable for grazing as well as mining. The mining district could best be reached from the southwest, which made Cheyenne the logical point of departure. Clark was ordered to open a stage line from there to the mining district, but Gould wanted confirmation that the gold rush was more than a flash in the pan before committing himself to building a road. There was also trouble with the Indians, who resented the growing white invasion of lands guaranteed them by treaty. In 1876 their anger erupted into an uprising that culminated with the massacre at Little Big Horn. News of that affair shocked Americans and set Clark to fretting about its effect on immigration and commerce.[32]

Jay knew better. "The ultimate result," he assured Clark, "will be to annihilate the Indians & open up the Big Horn & Black Hills to development & settlement & in this way greatly benefit us."[33]

The message drummed into Clark was simple and incessant: find new sources of income and new ways to cut expenses. In this quest Jay left no corner unswept. Government mail service offered a choice target. The volume of mail carried increased sharply; the compensation received from the government did not. On advice of counsel Jay advanced the proposition that the Union Pacific had the right to charge the same rates for mail as paid by private parties for express and baggage. Services provided by outside agencies also bothered him. Rather than pay someone else to handle the road's express business, the Union Pacific organized its own service. Once in operation, Jay extended the service to the Kansas Pacific and connecting roads in Utah and tried to induce the lines east of Omaha to accept it as well. Even the contract for Pullman sleeping cars rankled him because he did not want the company "deprived of so large a source of revenue for the benefit of outsiders."[34]

On the other side of the ledger the problem as always was to strike a balance between paring expenses and maintaining the road. Most operating costs could be lumped into three categories: labor, materials, and equipment. Like most of his peers, Jay remained tight-fisted about wages. By 1877 he had squeezed his Chinese labor down to $27 a month, which in his opinion was "still $2 per mo too high." That same year the Union Pacific joined the eastern lines in new wage cuts, which Jay hoped would be "acquiesced in as one of the necessities of the times." Wherever possible he shopped the bargain basement; in one instance he found twenty locomotives for sale at half price by the bankrupt Northern Pacific. In buying rails he would secure several quotations and compare notes with Clark before committing himself.[35]

To keep net earnings high, which in turn enhanced the value of the road's securities, Gould wanted to run the Union Pacific at 40 percent. Between 1875 and 1879 the figure averaged 41.4 percent. During those years the tonnage hauled more than doubled while the net increase in rolling stock amounted to only 26 locomotives and 146 freight cars. It is impossible to calculate the extent to which these gains were achieved by filling hitherto empty cars, by paying mileage on the cars of other roads, or by simply running equipment into the ground. The government directors expressed little concern about the road's physical condition until 1878, and even then their reservations were mild.[36]

Whatever its effects on the road, Gould's policy restored the Union Pacific's standing on Wall Street. Investors smiled on a company that paid out nearly $12 million in dividends in only four and a half years. Some carped that the money might be better spent on improvements or equipment or put into a sinking fund against the government debt, but the grumblers made few converts. That a real transformation had occurred could not be denied even by Cassandras who continued to utter gloomy prophecies for the company. At the same time a more subtle change was taking place: in making the Union Pacific respectable Jay had also earned new respect, however grudging, for himself.[37]

The struggle to develop the Union Pacific took its toll on Gould and Clark even as it forged a bond between them. Part of that bond arose from their shared habit of overwork. They were both men of ragged health accustomed to driving themselves past the edge of endurance. More than once during the decade they dropped in their traces, victims of illness induced by strain or exhaustion. Clark fell sick in the fall of 1874 and did not recover until the following spring. Because it was a pivotal time in the company's affairs, he tended his duties anyway despite solicitous advice from the home office. "It is of the first importance you should look after your health," Gould urged, while Dillon, hardy as a bull despite his age, admonished Clark not to "go out on the line this winter to expose yourself. Give the orders and have it done by others."[38]

But Clark shrank from Dillon's advice. He was still new in his position and insecure in his relationship with those above and beneath him. Fearful that some part of his authority might be snatched away, he was reluctant to delegate responsibility. At the same time he hesitated to make decisions on his own initiative. Instead he tried to oversee everything while taking no action on important matters without instructions from Gould or Dillon, even though the extent of his authority had been made explicit. "Mr Clark is now really General Manager," Dillon said on one occasion, "and has . . . full charge of the entire road and all its departments." Gould assured Clark repeatedly that he had the power to act. When Clark still took refuge in caution, he was told bluntly in 1877 that "you must not hesitate to take *responsibility* of doing what in your judgment is best for the interest of the road on the *spot. Your acts will be sustained.*" He renewed the suggestion that Clark find a capable man to take charge of the operating department so as to "relieve yourself of the ugly wear & tear of details."[39]

Relief from the ugly wear and tear of details was an even more urgent need for Gould himself. At forty he drove himself with the same intensity as at twenty, and the pace was beginning to tell. In December 1875 he fell ill and was confined for nearly a month. Once recovered, he could do little reading or writing because of "neuralgia in my eyes." His eyes troubled him for months afterward; attacks of facial neuralgia would plague him the rest of his life. That spring he took Helen to California at the end of his western tour.[40]

Excursions to Florida or California were no more than palliatives. The disease was overwork and the condition had become chronic, but for Jay the remedy was always worse than the ailment. He was fast approaching the crossroads where the scope of his affairs expanded just as his physical capacities began gradually to diminish. He was a man seated hard upon a tiger, and he did not pause to ask why he was there. He would not let go unless the tiger stopped, and the tiger moved ever faster. How then to make the ride bearable? There was something at once poignant and pathetic in the suggestion Jay tendered Clark shortly after his own illness: "In regard to your health let me advise you to do no work or think of business after 4 PM & get some one to read you to sleep nights. I do that and get a good nights sleep in that way."[41]

15

Politico

If we were to judge from outward manifestations and from results, Gould's policy was to obtain the largest immediate return possible, and then to withdraw from the company while there was still time.

—Nelson M. Trottman, *History of the Union Pacific*

The great . . . railroad companies, are often through their land grants and otherwise brought into relations with the Federal Government. Bills are presented in Congress which purport to withdraw some of the privileges of these corporations, or to establish or favor rival enterprises, but whose real object is to levy blackmail on these wealthy bodies, since it is often cheaper for a corporation to buy off its enemy than to defeat him either by the illegitimate influence of the lobby, or by the strength of its case in open combat.

—James Bryce, *The American Commonwealth*

The present opinion of this stupendous work [Union Pacific], derives its tone and character from developments subsequent to its construction. The "Credit Mobilier" has tinged everything.

—*Report of the Government Directors,* 1874

IT WAS INEVITABLE that Gould's efforts to develop the Union Pacific would carry him into the political arena. No railroad could avoid involvement in state or local governments to protect its interests. However, the peculiar status of the Union Pacific as a mixed enterprise produced a long, bitter relationship with the federal government that ultimately brought the company to its knees and drove Gould from its management. This perpetual battle at so many levels of government squandered resources and vitality the Union Pacific could ill afford to lose.

The political role of railroads has long been misunderstood as part of a morality play in which selfish, powerful corporations trampled the public interest until their depredations were curbed by regulation. If there exists a "folklore of capitalism," to use Thurman Arnold's phrase, so too is there a folklore of reform in which the struggle between the railroads and "the people" occupies a prominent niche. Like most myths it contains enough truth to mislead and enough falsehood to cloud understanding. However ruthlessly the Union Pacific wielded its power, even its worst abuses were often born of weakness rather than strength. Outsiders tended to regard the company as invincible, a corporate leviathan of tyrannical disposition. In fact size made the monster not only a force but also a target. Its enemies might be smaller and weaker, but their number was legion and their demands incessant. The "people" were, after all, not disinterested bystanders but a collage of individuals, firms, interest groups, rival lines, speculators, and communities, each seeking some advantage.

Within its territory the Union Pacific functioned as queen on the chessboard of economic development. Its strategic position made it the source of opportunity and the scapegoat for failure. No policy, however enlightened, could have satisfied the whole range of interests clamoring for its services. Choices that favored one group disappointed others, and the disenchanted were quick to blame their misfortunes on the greed and malevolence of an impersonal corporation. Farmers could attribute their distress to high rates and unfair classification policies, businessmen to rebates or other discriminatory practices that gave some rival an advantage over them. Communities not located on the road's line condemned the railroad for sentencing them to economic stagnation, while those on the line complained bitterly if they did not receive rate structures favoring their growth at the expense of rival towns.

The political arena emerged as the battlefield on which these conflicts were played out. Here too size made the Union Pacific vulnerable though it loomed like a juggernaut in the eyes of a given adversary. The company could exert muscle but not everywhere all the time. It had to contend with three state or territorial legislatures as well as town councils and county courthouses. To transact its business the road needed friends at every level: governors, congressmen, judges, legislators, mayors, aldermen, sheriffs, agents, editors, and merchants. Contrary to myth, the road seldom got as good as it gave. For every favor granted a dozen were demanded in return—free passes, outright payment, securities or market tips, contracts, agencies, a job (often for a relative), a political contribution, an investment in some enterprise, preferential rates or treatment.

"I took a contract eleven years ago to educate the people of this state to keep their hands off the Union Pacific railroad," scowled George L. Miller, editor of the *Omaha Herald*, who wanted a printing contract. "The corporation . . . would not miss this pittance which I ask for on a purely business basis. Let them let me have this thing for $4000 a year, no matter what change may occur. No management can complain of it and if they do, who the devil cares? I am kicked and knocked around here deprived of business and opposed in my political influence . . . and I am entitled to the only protection you can give."[1]

Here was an attitude shared by friends and those predators who swarmed the company like flies tormenting a bull. The railroad might not have a soul to damn or a body to kick, but it had a pocket to pick and the art of venality was not confined to corporate officers or speculators. For the Union Pacific politics became an exercise in self-defense. It worked hard to elect friends, harder still to defeat foes. More effort went to resisting hostile legislation than to obtaining favorable bills. Ever sensitive to publicity, the road's officers showed more vigor in punishing unfriendly editors than in rewarding sympathetic ones. In the courts the company had to defend against many more suits than it brought. Local juries rarely sided with the railroad, and most judges were beyond its influence. Appealing lost decisions was effective but expensive.

All this made rich pickings for the lawyers who feasted on railroad business. Even Gould, the master of litigation, sometimes cringed at its cost. "There are a swarm of lawyers besides that have been living off the property," he once said of the Kansas Pacific, "but I dont propose to pay them a cent—unless we have to." Much of the time they had to. As resentment against the railroad swelled, the clashes multiplied and the price of victory soared. Other roads faced this same dilemma but did not have to grapple with the federal government as well. Like some ancient family feud the dispute with Washington raged on into the closing years of the century, a ritual of acrimony that had long since lost touch with the realities of changing times. Aware that the company's future hinged on the outcome, Jay made a dogged effort to resolve the conflict. His attempt failed, as did those of his successors; neither Gould nor Dillon nor Fred Ames lived to see the issue settled.[2]

In Nebraska Edward Rosewater understood the political clout of the Union Pacific better than most. A maverick Republican, he founded the *Omaha Bee* in 1871 and used its columns to attack standpatters, corruptionists, and vested interests. Where Miller's *Herald* cozied up to the Union Pacific, Rosewater blasted the railroad for meddling in state politics. His smooth, round face with its walrus moustache flushed with indignation when he described the spectacle of "droves of men taken from the shop to vote." When the Union Pacific wanted someone elected or a bill defeated, it set up "oil rooms" in an Omaha hotel to lubricate public servants with a liberal supply of liquor, cigars, and good times.[3]

While some of Rosewater's charges were true, he exaggerated the railroad's power. Certainly its officers regarded their influence as something less than a steamroller. By 1875 the Union Pacific had already lost a court battle with Nebraska over tax exemption and was fighting an attempt to tax its land grant. It had done little in state politics other than support candidates and make a feeble gesture toward relocating the state capital along the line. When a new state constitution emerged in 1875, the Union Pacific was slow to react because opinion was divided over whether its regulatory provisions applied to a road chartered by Congress. Dodge warned Gould that "the whole article appears to be aimed directly at us" and thought it could be beaten, but only if lobbying was kept as covert as possible. "It would never do," he observed, "for us to undertake to beat it and fail."[4]

Jay doubted the provision could be used against the Union Pacific, but he disliked it as an unhealthy precedent and so requested Clark to "lay the pipes for its defeat." The constitution won overwhelming approval. In its fight against taxation the company fared little better. After losing the court battle, Gould struck a bargain with one lobbyist who used his influence to reduce the company's taxes in exchange for a fee of half the amount saved. The bargain smacked more of desperation than of a corporate leviathan flexing its muscles. The harsh truth was that having friends in high places cost a lot of money and yielded uncertain returns. State politics mattered little to the Union Pacific beyond the need to protect its interests.[5]

Except for the congressional delegation. They were important allies for the Thirty Years' War with the federal government, yet in the election of helpmates the Union Pacific gained only limited success. John M. Thayer was a case in point. Territorial pioneer, Indian fighter, Civil War hero, and longtime friend of the Union Pacific, he served as one of Nebraska's first senators until defeated in a bid for reelection in 1871. Three years later he tried again with help from the railroad, but the old general had evidently lost his appreciation for the element of surprise. "Mr. Thayer came on here [New York] to borrow money with half Nebraska well informed as to his mission," Gould related to Clark, "and while we all feel friendly to him we thought it indiscreet to advance money to aid him further than had already been done." Thayer lost and was consoled with an appointment as territorial governor of Wyoming, in which position he remained useful to the Union Pacific.[6]

Nor was Union Pacific influence strong enough to prolong the career of Phineas W. Hitchcock, one of its best friends in the Senate. A veteran of state and national politics, author of the Timber Culture Act, Hitchcock was considerably more than the senator from Union Pacific. Nevertheless, he lost his seat in 1876 to former governor Alvin Saunders. Gould regretted Hitchcock's ouster but figured Saunders would do as well. Finding friends seemed less important to Jay than eliminating enemies. Lorenzo Crounse, a Nebraska congressman, had long been a thorn in the Union Pacific's side. As convention time neared in the summer of 1876 Jay told Clark, "I depend on you to beat Crounse—defeat him at all hazard." He and Dillon went so far as to pause for a week in Omaha during an inspection tour to orchestrate the attack that denied Crounse the nomination.[7]

Then there was Rosewater. He was not only a hostile editor but an influential voice among Nebraska Republicans. The Union Pacific had no more implacable foe. His stridence annoyed Jay, then infuriated him to the point of retaliation. If Rosewater believed the worst about the railroad he had good reason. On Jay's orders Clark denied the *Bee* company business, such as land advertisements, and refused to carry the paper on its trains. The Union Pacific tried to build up the rival *Republican* in hopes of reducing the *Bee*'s circulation. When that did not suffice, Jay was ready to deny Rosewater access to Associated Press dispatches. "Rosewater ought to be *squelched*," he snapped, but all his power and ingenuity proved unavailing. The *Bee* remained Omaha's most prominent newspaper, rivaled only by Miller's *Herald*, and Rosewater kept hammering away at the

Union Pacific. What he neglected to mention was that he too had tried without success to deal with the railroad before becoming its ardent opponent.[8]

At the local level the company had to contend with a welter of conflicting interests. It was an axiom of the age that towns without a railroad wanted one, and those with a railroad wanted another one. Promoters hounded county officials for bond issues to support construction. If a county through which the Union Pacific ran toyed with an issue for some competing line, Jay's reaction was swift and unsparing. In 1875 a narrow gauge project called the Nebraska Central asked Douglas County for $125,000 in bonds. Gould fired off a telegram to Miller of the *Herald* asking, "Is it true that all our Omaha friends are going in to build a competing road to us West? If so, I will vote for the immediate removal of our shops and all our works to Council Bluffs." The warning was heeded and the proposition defeated, but new projects hatched as fast as old ones were laid to rest. Here as elsewhere the Union Pacific found itself constantly on the defensive.[9]

The pattern of politics in Nebraska was repeated elsewhere. In most cases the Union Pacific used its influence not to get but to avoid or prevent. Time and again it demonstrated that a powerful corporation could be as inept as the littlest lobbyist. The railroad meddled in Wyoming politics for years without much success beyond arousing public sentiment against it. No adequate solution was found to the problem of "translating its economic power into political effectiveness." Ham-handed influence could backfire into hostile legislation. Bond subsidies for branch lines were nice to have, but not if they were also given to competing roads or if the issue was so profligate as to make them worthless in the market. Apart from bonds, the company's chief interest was to protect itself from taxation or regulation.[10]

Jay understood these limitations well. In contemplating a branch toward the Black Hills, he saw that any legislation obtained in Wyoming might cost the road more than it was worth. "I would prefer to defeat any bonds at Cheyenne," he concluded, "unless they were given specifically to us & the legislation was such as we could live under. . . . We dont want to leave a *loop hole* for the KP [Kansas Pacific] to get north." In 1878, seeking legislation for an extension into Montana and Idaho, he told Clark flatly that "exemption from taxation & county bonds are all I would ask for." Montana balked at giving even that assistance, and in the end Jay built without it.[11]

Along with its other effects, expansion swelled the ranks of the enemy. Within a few years the Union Pacific would penetrate five more states or territories. That meant five more legislatures and governors, five sets of officials, interest groups, judges, lawyers, editors, local politicians, and the whole swarm of "friends" seeking a free ride of one kind or another. By that time the clash of interests had grown fiercer because of mounting public hostility against the railroad. Like other railroads, the Union Pacific soon discovered that victory was seldom worth the cost of war, especially when the battle had to be waged on two fronts.

For the right bard, capable of translating legalisms into poetry, the clash between the Pacific railroads and the government might have been stuff worthy of Homerian epic. Instead it came more to resemble Theatre of the Absurd, a tragicomedy entitled *Waiting for the Dough.*

The basic issue could not have been more simple. The government considered it had loaned the Union Pacific money and wanted it back, or at least demanded to know when and how it would be paid. For thirty years the government clung tenaciously to this proposition, which seemed reasonable enough until one realized how drastically circumstances changed during that period. Nineteenth-century minds were innocent of any notion of social overhead, the larger benefits bestowed on the country by the presence of a transcontinental railroad. In their canon a debt was a debt and must always be repaid. Nor were they moved by the argument that changed conditions made repayment impractical and unrealistic, even though the Pacific Railway Commission, after hearing exhaustive testimony, concluded in 1888, "It is universally conceded by every person of intelligence . . . that it is and will be absolutely impossible for the Union Pacific . . . to pay the indebtedness to the United States at its maturity."[12]

There are several reasons why this impasse persisted for so many years. The original Pacific Railroad acts were hastily and carelessly drawn. Legitimate disputes arose over the thicket of ambiguities housed in their language. In stipulating the terms of repayment Congress presumed a set of conditions that ceased to exist by the time the road opened. Under the best of circumstances the terms of repayment demanded revision to fit new economic realities, but by that time the whole issue had become tainted by the furor over the Credit Mobilier, which blew the issue out of all proportion.

The damage wrought by this stigma can hardly be exaggerated. At a time when dispassionate examination of the debt question was needed, the process was subverted by irrationality, intrigue, and political infighting. Congress made little effort to distinguish between the plight of the road and the profits reaped by those who had built it. Nor was it sensitive to changes in the economic environment that threatened the road's survival as a paying enterprise. There was scant evidence that the government understood the Union Pacific's function, let alone its needs. It must have disheartened Gould, who had toiled long and hard to develop local business as the line's backbone, to hear the government directors as late as 1878 characterize the Union Pacific as "essentially a through line."[13]

The two sides could not even agree on where the road began. For several years there existed a controversy as to whether Omaha or Council Bluffs was the eastern terminus. The law required the road to be operated as one continuous line from a point on the western boundary of Iowa. When the Union Pacific bridged the Missouri River in 1871, it interpreted the span as an extension of the Iowa boundary and therefore not part of the line. Having Omaha as terminus enabled the company to issue separate bonds for the bridge and to charge tolls for its use. It also subjected customers to the inconvenience of dummy transfers. Passengers and freight were unloaded at Council Bluffs, run across the bridge on bogus trains, and reloaded at Omaha. Inevitably suit was brought to challenge this practice. Dillon's own nephew, Judge John F. Dillon, ruled Council Bluffs to be the proper terminus but allowed the company to charge reasonable tolls on the bridge.[14]

The terminus spat was one of several sideshows to the main event, repayment of the government loan. The original act stipulated that once the road was completed, its federal obligation was to be discharged by applying half the earnings on government transportation and at least 5 percent of its net earnings. This apparently straightforward provision hatched a brood of controversies, three of which eventually wound their way to the Supreme Court: When were the interest payments due? When was the road actually completed? How should net earnings be defined?

Originally Congress assumed that earnings on government transportation would more than cover the interest on government bonds. For various reasons this did not occur; by July 1870 the government had paid out $3.7 million in interest against less than $1.3 million in retained earnings. As the deficit mounted, debate began over how to repay it. The legislation also left unclear whether the interest must be paid to the government as it fell due every six months or whether it could be deferred until the bonds matured. As the interest account crept past $5.5 million, Congress passed a bill withholding *all* earnings on government traffic. The legislation allowed the Pacific roads to file suit for recovery of the disputed sum, which the Union Pacific did immediately.[15]

By that time the argument had taken on some new wrinkles concerning the so-called 5 percent clause. Interest was to accrue from the date of the road's completion, but when was that? The government argued that it was July 15, 1869, when the last subsidy bonds had been issued. But in 1869 the government had withheld patents for half the land grants of both Pacific roads until they corrected certain deficiencies in construction. The work was not certified as completed until October 1, 1874, which, the company insisted, ought to be the proper date.[16]

There was merit and inconsistency in the position of both sides, but the confusion ran still deeper. The clause required the company to apply at least 5 percent of net earnings to the interest account, but no standard definition of net earnings existed. Was it the amount that remained after deducting only operating expenses, or after deducting operating expenses and fixed charges such as interest on other bonds? How did expenditures for new construction and equipment enter the calculations? Baffled by the law's sloppy language, the government filed suit asking the Court of Claims to resolve both the net earnings and date of completion questions.[17]

All three issues were pending in the courts in 1874, when Gould was struggling to put the Union Pacific back on its feet. Huge sums were at stake. Favorable decisions would boost the company's securities; unfavorable ones might devastate them. Any legislation passed by Congress would have similar effects, and the halls of the Capitol that winter swarmed with lobbyists eager to gain their ends by waving the bloody shirt of Credit Mobilier at the Union Pacific. The infighting grew fierce, and as the pressure mounted so did the cost. Dodge, the company's chief lobbyist, had already turned in $25,000 worth of "Washington drafts," and no relief was in sight.[18]

So much depended on which way the political winds blew in Washington. In Jay's view the key to stabilizing Union Pacific finances lay in restoring investor confidence. For that reason it was essential to maintain or improve the price of its securities, but this was impossible so long as the road's destiny was tied to the vagaries of governmental action. Every suit or court decision, every threat of unfriendly legislation, every departmental order disturbed the market, often forcing Gould and his associates to sustain the stock with heavy purchases. The bears pounced gleefully on every rumor and supplied plenty of their own invention. On Wall Street a large short interest gathered expectantly that winter, waiting for the crash to come. When it did not, they attacked furiously on the market and in Washington.

The perils facing Jay convinced him that a settlement with the government was imperative. Every controversy sapped the road's vitality while feeding its enemies. He did not want the government as a partner because it cramped his style, deprived him of a free hand. To break the impasse, Gould decided to push for a compromise even though it meant negotiating with a hostile Congress.

Prospects in Washington were discouraging. Dodge was there lobbying against another attempt by Treasury Secretary Benjamin Bristow to compel the 5 percent payments. "It will be impossible to get any legislation from Congress this winter for Union Pacific," Dodge warned. "The best we can expect to do is to prevent any adverse legislation from going through." Nevertheless, Jay devised a plan to offer the government. He proposed annual payments of $500,000 to be treated as a sinking fund with interest compounded semiannually and continued until the amount equaled principal and simple interest on the government debt. In return the government would hand over all disputed funds previously withheld and pay for its future transportation in cash.[19]

Dodge took the proposal to President Grant, who responded favorably but suggested that the payments increase to $750,000 after twenty years. The change was made and Grant asked Bristow to prepare a bill incorporating the proposal. Then Bristow fell ill and the matter hung fire. Dodge was sick himself but managed to sound out some members of the Appropriations Committee. "Everyone seems to see the importance of it," he wrote Gould, "but most of them seem to doubt that it can be done this session." He added, "The matter is of so much benefit to the Government that . . . they ought to accept it but you know how chary they are of everything now."[20]

Pressure might be brought by making the offer public, but Dodge hesitated. "If we put it in the papers and began to run up stock on it," he noted, "they would be suspicious of its being a stock jobbing operation and the moment that was so they would keep hands off." Wall Street already thought as much, but Dodge knew nothing of the sort was intended because Gould had told him so explicitly and Dodge took pains to verify the point: "*From your letter, I understand that you don't care anything about any temporary effect; that if anything is done, you wish it to be something to base permanent operations upon and I am acting upon that supposition.*"[21]

Dodge had a bill drafted and incorporated it into a letter sent Bristow over Dillon's signature on February 10. Bristow brought the proposal before the cabinet, which demanded several changes favorable to the government. Although Gould was the toughest of bargainers, he agreed readily because the compromise was too important to be lost by quibbling over terms. The entire cabinet except one approved the revised proposal, whereupon Grant drafted a message to Congress recommending passage of the bill. During these negotiations the Supreme Court, in a decision unrelated to Union Pacific, defined net earnings in terms that were, as Jay beamed, "as full & complete as though we had had it prepared by our lawyers to meet our own case." Hopes for a settlement never burned brighter, yet within a week they were snuffed out with the abruptness of a vanished mirage. Grant's message never materialized, and Congress adjourned without acting on the matter.[22]

What had happened? It was a case of bad timing, Grant told Dodge. The Senate Republicans had informed the president that any new legislation would force an extra session of Congress, which no one wanted. Later, however, he came to believe that certain bear interests were responsible for Grant's change of heart by convincing him that the proposal was a mere stock-jobbing scheme. If Grant needed reason to hesitate, he need only think back a few years when he was embarrassed by another grand scheme in which Gould was prominent. Dodge took the setback hard. "No one can be more disappointed than I am," he wrote gloomily, "as I think I never worked anything up with so much satisfaction." Well might he mourn, for the two sides would never again come so close to an amicable settlement. A decade later even Charles Francis Adams would look back on Gould's proposal as "the most beneficial, the most business-like and the most financially sound of all the plans to meet the obligation to the Government."[23]

Despite this lost opportunity Jay managed to rout the bears and plant Union Pacific at high figures. That summer the company began paying dividends, which made it vulnerable to charges of distributing to stockholders funds that ought properly to be applied toward the government obligation. In November the Supreme Court ruled that interest on the government bonds did not have to be paid until their date of maturity. Still the government refused to release the sums withheld since 1873. As the new session of Congress drew near, the position of both sides hardened.[24]

So too was the strategic chessboard bristling with new combinations, each with its own lobby in Washington. The Kansas Pacific joined forces with the Burlington & Missouri River in seeking legislation to compel the Union Pacific to prorate through rates. Congress obliged with the "Equal Advantages" Act in 1874, but Union Pacific had so far managed to evade its provisions. Tom Scott was still trying to coax a reluctant Congress to subsidize his southern transcontinental line, the Texas & Pacific. Gould and Huntington labored to thwart Scott's efforts, but their collaboration was a delicate one. Huntington wanted to build his own road to New Orleans, a threat no more palatable to Jay than that posed by Scott. To complicate matters, Scott's chief engineer and lobbyist was none other than Dodge.[25]

This welter of railroad intrigue descended upon a Congress reeling from years of scandal and mindful of an approaching election. A movement for retrenchment and reform was already under way, and its converts were not likely to embrace any bill soiled by the names of Jay Gould or Union Pacific. No interest was shown in reviving the proposed settlement from the previous session although the clamor for repayment of the debt rang as loud as ever. Freed by the court decision from paying the government interest, Jay wanted the other disputed issues left in limbo until the courts resolved them as well. Dodge agreed that the proper strategy was to seek relief in the courts and to stop any legislation. The Kansas Pacific's pro rata campaign got nowhere, but in March 1876 the House Judiciary Committee recommended a new sinking fund with harsher payments than those already required by law.[26]

Some intricate maneuvering followed. Gould approached Huntington with a bold proposition to merge the Pacific roads and develop the Southern Pacific jointly. The startled Huntington chewed the idea over and concluded that "a better way than for the C.P. and U.P. to divide would be to consolidate on almost any terms." He joined Dillon in renewing the latter's earlier offer to the government and added a wrinkle of his own by offering to sell most of the land grant back to Washington at $2.50 an acre. Confident of success, Huntington persuaded a dubious Gould to let him take charge of the fight. On March 27 Dillon joined in the land offer. Oliver Ames was summoned to a conference at which it was agreed not to oppose Huntington's bill in return for a share of the Southern Pacific and Huntington's help in getting a settlement. However, it quickly became clear that Huntington was too obsessed with his own bill to be of much aid. As the fight between him and Scott grew nastier, Gould found himself pushed into the role of intermediary. On one side Scott was slinging the mud of past scandals at both Pacific roads; on the other Huntington pressed relentlessly the argument that he could build a southern transcontinental without government aid.[27]

Fearful that Huntington might get a line independent of Union Pacific, Gould lent support to a narrow-gauge project with potential as the nucleus of a rival line from Utah to California. Using this threat as leverage, he tried to force Huntington into seeing the folly of his tactics. Too late Huntington grasped the danger in his course; not until the end of April did he agree reluctantly to drop his bill for the session if Scott did likewise and to throw his weight behind the sinking fund bill. The House was unimpressed and, despite intense lobbying, passed its punitive bill by a landslide vote. The severity of its terms drew criticism even from conservative financial editors, and the Senate quickly buried it in committee. However, in July 1876 Senator Allen G. Thurman of Ohio reported out a substitute bill with virtually identical provisions and an even harsher repayment plan.[28]

Thurman's bill did not pass but it loomed like a spectre over the forthcoming session, as did renewal of the clash between Scott and Huntington. The presidential campaign was under way and clouded matters with the usual outpouring of emotionalism. Politically there was bad blood abroad as stump speakers grappled with the legacy of Grant's scandal-plagued administration and the bitter controversies surrounding Reconstruction policy. Internal rifts within both parties hampered efforts by the railroad lobbies to line up support for their interests. Gould sensed that a major storm was brewing, that revival of the Thurman bill, which in his view "repudiates the existing contracts between the Govt & the Company," would trigger a desperate battle at a time when the mood of the country was hostile to railroads. In this fear he was entirely correct. What he could not foresee was that the battle would be fought in the turbulent aftermath of the most controversial election in American history.[29]

The outcome of the contest between Tilden and Hayes hinged on disputed returns from three southern states. No tradition or formula existed for resolving the matter. Congress convened on December 4 and plunged at once into an acrimonious, fiercely partisan debate. Tension and uncertainty gripped the country as it had not since the secession crisis. Amid charges of fraud and coercion there was ugly talk of violence, even of secession.[30]

While the tempest raged, lobbyists prowled their haunts with their usual intensity. Scott was busy trying to melt the glacial Hayes with his charm and Dodge had taken up his post. Their efforts forced Huntington to admit that he was "having the roughest fight with Scott that I ever had." In desperation he turned to Gould for help, offering a half interest in the construction company that would build the proposed Southern Pacific line. Despite the past year's failure Jay again joined with Huntington, hoping that an alliance might forge a compromise that would unite all sides in common cause. In less than a week the Scott steamroller clanked to a halt. After two futile attempts to force his bill out of committee he went to Huntington's room and struck a bargain. Overnight the two foes became allies pushing for a compromise bill described by Huntington as "nothing like what we want, but as good as we can get."[31]

This volte-face was too blatant to help Gould much. Editors pounced savagely on the spectacle of a warm embrace by men who for months had been alerting the Republic to the venality of the other. Reformers were outraged that so sordid a business should brazen its way into the midst of a national crisis. Time was running out; if the election dispute was not resolved by Inauguration Day (March 4), the country would find itself without a president. The congressional calendar was jammed to the bursting point, and lobbyists swarmed the floors of both houses like clouds of locusts. Despite these storm warnings, Gould and Huntington had Senator John B. Gordon of Georgia introduce another sinking fund bill with the land sale provision intact. Huntington was certain his friends could get it through safely but got a rude awakening when the bill's enemies reduced it to shambles. By late February the Gordon bill was dead and so was the Scott-Huntington grab-bag subsidy bill. A version of the Thurman bill came onto the Senate floor and was narrowly beaten back after an impassioned debate.[32]

The election crisis played itself out, leaving Gould's hopes for a settlement in ruins. Embittered Democrats charged wildly that "the House was controlled from the start by Tom Scott & Jay Gould." Tilden himself was heard to blame Gould for his defeat. The reform-minded Hayes took office along with an unusually large number of new faces in both chambers, which meant restructured committees. Scott and Huntington ended their truce and girded for renewed combat. Prospects for defeating the Thurman bill looked bleak, and no one believed the new Congress would even glance at an alternative. Depression still gripped the country and exploded that summer in a wave of strikes and violence against the railroads.[33]

Events during the winter of 1877–78 confirmed Jay's worst fears. While a renewed effort by the Kansas Pacific to get a prorate bill was beaten back after a fierce struggle, Dodge could not even manage to weaken the Thurman bill with amendments. It passed the Senate 40–19, swept through the House unaltered with only two negative votes, and was signed by Hayes in May 1878. From the railroads' point of view, the terms amounted to unconditional surrender. Virtually every point contested by them was incorporated in the act. Net earnings were defined as the sum left after deducting operating expenses and interest on the first mortgage bonds. All earnings on government transportation were to be retained, with half the sum applied to current interest and the rest to the sinking fund. The Treasury Department would establish the sinking fund and invest its balance in United States bonds. The two Pacific roads would pay annually into this fund 5 percent of their net earnings up to a specified amount. In addition, they would pay annually a sum equal to the difference between the amounts required above and 25 percent of their net earnings. No dividends could be paid until these obligations were met.[34]

Reaction to the Thurman Act veered between rage and resignation. "I think we can manage to get along under it if compelled to," sighed Charles Crocker. A livid Huntington scorned Thurman as a demagogue and declared savagely, "Lying was his best forte." Gould too was deeply upset, and subsequent events did nothing to ease his distress. That summer Carl Schurz, the new Secretary of the Interior, ruled that all unsold acreage in the Pacific land grants was subject to preemption under the Homestead Act. Congress created a new bureau to audit the Pacific roads and a board of commissioners to poke into a number of questions that troubled it.[35]

Routed by two branches of government, the Union Pacific hastily sought refuge in the third. So far the courts had been sympathetic to the company's position. Jay assumed that the Supreme Court's decision in 1875 laid to rest the disputes over transportation earnings and the interest due date. But Congress and the Treasury Department squelched that hope by simply ignoring the decision in their later enactments. After passage of the Thurman Act, Jay's only chance for a favorable settlement rested with the courts. As late as February 1878 Charles E.

Perkins grumbled that "the Union Pacific has so far had Congress and the Courts upon its side." By summer the company had neither. In June the Court of Claims fixed 1869 as the road's completion date from which 5 percent of net earnings was owed. Six months later the Supreme Court defined net earnings more strictly than the Thurman Act, then in May 1879 upheld the Thurman Act by a 6–3 vote.[36]

While much has been written about the effects of the Thurman Act, one repercussion has gone unnoticed: it drove Gould from the Union Pacific. Although he did not leave the board until 1884, his thinking underwent a profound shift. The act undid years of hard work building up the company and blasted his plans for its future. The justness or equity of the law mattered less to him than its practical effect, which was to saddle the company with an obligation it could never discharge at a time when competition had begun to drive rates down and money was needed to build branch lines. Furthermore, the act insured that bickering between the company and the government would continue unabated for years to come.

The Thurman Act hurt Gould financially by impairing the value of his Union Pacific stock, but this loss had little to do with his change of heart. If profits had been his main concern, Jay could have sold out at any time after the spring of 1875, when he first ran the stock up. Of course he wished to make money, but there was something more. He had taken hold of a transcontinental line, had breathed life into a moribund property, but he did not control the entire line and his own part had to be shared with Boston and the government. By 1878 his interests had expanded greatly in other directions. His ambitions grew even faster than his assets, and the larger they became the more he chafed at his imperfect hold on the Union Pacific.

What Jay wanted above all was a business empire of his own making and under his own control. The Supreme Court decisions merely confirmed what the Thurman Act had already told him: the Union Pacific would not do as the nucleus of the rail system he longed to create. Having reached that conclusion, he began looking elsewhere. Ezra Baker recalled Gould being so upset by the Thurman Act that he threatened to sell his Union Pacific shares and build a competing road from Omaha to Oregon. "I did not like to be in partnership with the Government," Jay admitted in more circumspect language. "It was constant turmoil and quarrelling, and the moment you took this property and made a dividend, paying which we did by our hard work, working night and day, then the Government came in and began attacking us. . . . That tired me out."[37]

It was not only the government that tired him out but some of his Boston associates as well. Part of Jay's program to obtain a settlement involved a vigorous campaign to clear the air of lingering antagonisms between the company and the government. The most prominent of these were the Credit Mobilier suits. The corporation still existed, its affairs hopelessly tangled by a decade of internecine warfare. Some of its stockholders, like Dillon and Oliver Ames, were still in Union Pacific, but most had sold out or been driven into exile. All wished to realize something on their stock by winding up Credit Mobilier and dividing its few remaining assets, especially a controversial $2 million note given it by the

Union Pacific in 1869. Nothing could be done because a blizzard of suits and countersuits, the legacy of years of bickering among the warring factions, tied the company in knots.[38]

Doubtless Gould might have preferred to avoid this swamp altogether, but he could not. The $2 million note hung like a sword above the Union Pacific's head. Jay noticed it even before his election to the board and was bothered enough to demand assurance from Oliver Ames that no attempt at payment would be made. But in 1875 Credit Mobilier filed suit to collect on the note. Gould was furious at this action because it thrust the contaminated name of Credit Mobilier back into the headlines amid his negotiations with the government. He reminded Ames coldly that the suit was "stirring up a dirty cesspool just at a time when we want to stand well at Washington." It was, he warned, "a serious blunder" and should be withdrawn. When it was not, Gould resolved to slay the Credit Mobilier dragon once and for all.[39]

Twice he asked the Union Pacific to file countersuits. The executive committee rejected his request and instead appointed him head of a committee to settle all outstanding claims between the companies. Gould had other ideas. He knew that those with interests in both companies were anxious to resolve the dispute; the trick was to devise a plan that separated friends from foes. In his brilliant, imaginative way Jay devised just such a solution. When the Union Pacific declined to bring suit, he did so personally as a stockholder. At the same time he offered releases to Credit Mobilier holders who agreed to turn their shares over to Union Pacific. If the railroad held most of Credit Mobilier's stock, it would render a judgment harmless. This clever tactic enabled Gould to protect his friends while pressing the suit against those who refused to sign.[40]

The Boston directors were loath to antagonize a man who had not only revived Union Pacific but was making them money in other enterprises as well. Most surrendered their stock in December 1875; a handful refused and were sued by Gould. The most stubborn holdouts were Rowland G. Hazard and Oliver Ames 2nd, known as Governor Oliver. As son and executor of Oakes Ames he wanted to extract maximum value for his father's estate which, he explained later, had "lost its Union Pacific stock and we had nothing but the Credit Mobilier stock." He had never approved his uncle Oliver's conciliation of Gould and did what he could to block him. Jay countered by having Governor Oliver dropped from the Union Pacific board in 1877. One entry in Governor Oliver's diary in March 1879 capsuled the lengthy test of wills: "UP elections. Had talk with Gould. He called me his adversary. Told him I should go for him soon." As often happens, the turn of fortune's wheel would give him a golden opportunity.[41]

A lengthy legal crossfire ensued and dragged on for years. Although an adverse decision in 1883 forced Gould to buy for Union Pacific the shares of Governor Oliver and some others, he accomplished his main objective. By 1878 the Union Pacific owned a majority of Credit Mobilier shares at no cost. Nevertheless the trouble caused him by this ordeal doubtless reinforced Jay's conviction that his future lay elsewhere than in Union Pacific. The series of brilliant maneuvers by which Jay realigned his holdings also provided him with a choice occasion to repay certain people for the aggravation inflicted on him. That was Gould's way; he would not permit revenge, if such it was, to intrude on his plans unless it fit the inexorable symmetry of his logic.[42]

Apart from its effect on his business dealings, the long ordeal in Washington thrust Gould into a position of influence within the Republican party. Although Jay had always been active among politicians, he took little part in politics except as a lobbyist. In his youth he had been a fiercely partisan Democrat until the demands of business shunted his attention elsewhere. Since then he had kept politics on the short leash of self-interest. The vast, tumultuous issues of war and reconstruction touched him only insofar as they provided troubled waters for him to fish. Through their turbulences and cross-currents he sailed an unswerving course in pursuit of his own ambitions. The stormy days of Erie and the Gold Corner plunged him into politics as never before, but always as consumer rather than participant. Some politicians, like Tweed or Thomas C. Platt, became his friends; others he regarded as mere peddlers of commodities vital to his business. He bought influence as ironmakers bought ore. To get it he willingly courted the favor of politicians, paid them deference, and met their price if they could deliver what he wanted.[43]

Gould's experience with the Union Pacific did not change this pattern so much as add new dimensions to it. The once-ardent Democrat became a staunch Republican. It was a natural conversion for one so prominent in business affairs, and one rooted in practical necessity. The Republicans were in power and likely to remain so, not only in Washington but also in the states traversed by the Union Pacific. The party's economic policies suited Gould, who never strayed far from orthodoxy on such matters. Although he was not issue oriented, his grasp of their complexities was thorough and sure, especially if they affected his own affairs.

The Union Pacific provided Jay with direct connection to the inner circle of the Republican party. Edward H. Rollins, the road's secretary-treasurer, was a prominent New Hampshire Republican who had served in the House and won election to the Senate in 1877. Rollins was in turn close to William E. Chandler, one of the most astute political managers in Washington, who had served Union Pacific as a lobbyist for years. Chandler was close to Dodge and had managed Grant's campaign in 1872.[44]

In extending his connections to men of leadership in the party, Gould discounted past differences. He cultivated the friendship of two men he admired greatly, Grant and Garfield. The former he would assist in business, the latter in his drive for the presidency. Another Republican heavyweight, James G. Blaine, became a Gould favorite, especially after standing in against the Thurman bill so vehemently that his opponents tarred him as "Jay Gould's errand boy." In 1876 hostile papers like the *Sun* denounced Gould's support of Blaine's run at the nomination in what amounted to a preview of 1884.[45]

As his network of friends in high places grew, Jay emerged as a reliable source of campaign funds at both the state and national levels. It is easy to dismiss this largesse as an investment in influence, but one suspects something more was involved. Gould had joined that elite known as "men of large affairs." Like most other members he viewed the Republican party as a rudder of stability for an erratic ship of state. It befitted men like himself to tender it support, and his support in turn confirmed his arrival among the ranks of those who counted. He had, in short, become something of a power broker, a man to be admired or feared but never ignored.

To have achieved so prominent a place fed his craving for respectability, a need he never acknowledged with anything more than a mournful smile or weary shrug, but one that must have gnawed at him for years. He did not care to be liked so much as respected, but once branded a pariah he found the scar impossible to remove. Although he came to accept his fate, he never ceased resenting the inequity of the sentence passed on him. Nor could he miss the pain it inflicted on his wife, that shy, eminently proper Murray Hill lady who was not accustomed to having doors closed or noses tilted against her, let alone enduring a husband hounded by gangs of police and lawyers. For her and for the children who meant everything to him, Gould pursued respectability as a fringe benefit of more elemental rewards.

But there were limits to what he could accomplish. In politics as in business he remained a shadowy figure, a shy, furtive creature feeding at twilight at the forest's edge, allowing his enemies only the most fleeting of glimpses. His movements continued to baffle those who feared his growing stable of railroads, telegraph lines, newspapers, and other enterprises. For all their efforts at following his tracks they saw little, fathomed less, and therefore assumed the worst.

16

Chess Player

Gould was the type of man who would not have been content in the development of his properties and the waiting of dividends. He wanted more action and larger returns.

—Robert E. Riegel, *The Story of the Western Railroads*

Gould was first, last, and always a trader. His special talent, amounting to a compulsion, was for detecting opportunities to seize control. . . . His countless deals involving the Union Pacific are legendary; he nearly ruined the road.

—Richard C. Overton, *Burlington Route*

To serve a territory with no railroad competition was the ambition of every railroad operator. . . . A monopoly of this kind was perhaps the most important strategic advantage of a railroad, provided, of course, the monopolized area either originated valuable traffic or served as a market for goods produced in other areas. Territory thus controlled was looked upon as "natural" territory. It belonged to the road that first reached the area. The construction of a line by a competitor was an "invasion." Such a construction, even by a business friend of the "possessing" road, was considered an unfriendly act. The former business friend became an enemy.

—Julius Grodinsky, *Transcontinental Railway Strategy, 1869–1893*

BETWEEN 1870 and 1890 railroad strategy was a chess game played on a board so volatile as to keep the players in a state of constant anxiety. The number of pieces exploded and arranged themselves into combinations of such infinite complexity that even grandmasters despaired of grasping the whole board, let alone staying a move or two ahead of their rivals. There was no sure path to victory. Some played aggressively and others passively; some feasted on complex positions while others hewed to clean and simple tactics. Changing circumstances made the rules fluid and often short-lived, which frustrated efforts to impose stability or create lasting alliances.

In the beginning certain constraints of geography and resources fixed the rules of the game. Each road had its own territory consisting of the region tributary to it. The most basic rule of all was that of one road for one territory. As Grodinsky noted, every road regarded its tributary region as "natural" territory to be dominated like a feudal barony. Any attempt by another line to enter the territory was denounced as an "invasion" and usually led to war.[1]

The territorial rule resembled a vow of chastity in being easiest to keep where the temptation to violate it did not exist. In the railroad-starved West no sane promoter would build a second line where there was not yet traffic enough to support one. Early diplomacy dealt mainly with problems of connecting rather than competing. To move traffic beyond their own lines the major roads had to cooperate in such matters as physical connections, transfers, creating rate and classification structures, devising uniform operating and accounting procedures, standardizing technology, and coordinating schedules. These questions dominated the early relationship between the two Pacific roads. Gradually there emerged a rail network composed of independent lines integrated by a series of cooperative agreements reached through a process of continuous negotiation.[2]

These arrangements expedited the flow of through traffic, which at first provided western roads with most of their income. Since through rates were at the mercy of competitive pressures, rival roads often formed alliances to avoid rate slashing. In 1870 the three carriers with direct routes from Chicago to Council Bluffs created the Iowa Pool for sharing eastbound traffic from the Union Pacific. It worked well enough to inspire similar efforts elsewhere and to incur Gould's enmity. The alternative to cooperation could be seen in the savage rate war between the Pacific roads and Pacific Mail after their failure to maintain agreements.[3]

While striving to increase through traffic, railroad managers also worked hard at cultivating local business. Since local rates were immune from competition, they could be kept at high levels. Early in the game Gould and his peers recognized that local business must form the backbone of a road's earnings and pursued its development vigorously. This made defense and expansion of the territory imperative, but here fresh difficulties arose. Expansion involved staking claim to untapped regions by building branch or feeder lines into them. Branches were costly and risky ventures at best, but if the parent road did not build them someone else might. Every hamlet in the West hungered for a rail connection and listened eagerly to any promoter with plans in his pocket. These local projects, called "sucker roads" by Gould, preempted a piece of the territory and loomed as potential menaces if they should fall into the hands of rival roads or evolve into larger lines.[4]

During the 1870s changing conditions steadily eroded territorial integrity and the cooperative agreements that sustained it. Competition reared its ugly head as new roads crowded onto the board, slicing up domain, multiplying points of dispute, and tangling the calculus of diplomacy. Every new road not only clamored for its share of traffic but also posed the threat of becoming a beachhead for invasion. Five years of depression intensified the struggle and exposed certain grim truths about the brand of competition practiced by railroads. For the first time Americans witnessed the spectacle of a few large enterprises vying for the same business. Every road needed a steady flow of traffic to remain solvent because fixed costs constituted so large a proportion of its total costs. Given this fact, it made sense to fill empty cars with freight obtained at reduced rates even if doing so violated rate agreements. Once a road succumbed to this temptation, war usually followed.[5]

The fragile diplomacy of cooperation crumbled beneath the relentless pressure of costs and the proliferation of roads. It could not satisfy the needs of players anxious to protect their territory and expand their share of business. Squabbles over rates and division of traffic grew more frequent and difficult to resolve. If war broke out, the combatants resorted to a limited arsenal of weapons. They could cut rates or withhold business if the enemy was a connecting line. Sometimes alliances were formed with other roads on which the enemy depended. They could defend their own territory by building branches and threaten the enemy with retaliatory invasion. Unfortunately, all these weapons inflicted serious damage on both sides and rarely gained either side a decisive victory. To hammer a rival into bankruptcy only made it a more formidable enemy, since a bankrupt road need not pay interest on its bonds and could therefore slash rates with impunity.

It was a curious game, the diplomacy and war of railroads, one that baffled the logic of ordinary minds. Conflicts persisted even though nothing more than Pyrrhic victories seemed possible. Rate wars cost both sides money and so did invasions. Defeated enemies did not vanish but simply reappeared in new guise, ready to fight again. The weak were less a victim of the strong than a menace to them as nuisances or pawns for some stronger foe. They could be eliminated only by being absorbed, which meant saddling solvent roads with insolvent ones. Buying sucker roads made no sense financially, yet it was done again and again, always in the name of defense, to prevent stronger rivals from getting them. Defense became the supreme rationale for rate wars, branch building, and construction of new lines or acquisition of existing ones. No aggression was intended; it was a matter of protecting the territory. Of course, what constituted the territory had a way of shifting as needs changed. The players were no more modest than monarchs in their assertion of eminent domain.

Gould emerged as one of the game's grandmasters, famous for his dazzling and unexpected combinations. Contrary to myth, he did not earn that reputation at once. His resurrection of the Union Pacific caused investors to perceive him as one who, by some financial alchemy, breathed life into moribund properties. However, he had yet to demonstrate his talent as a strategist. In taking up the game he did not question its rules or challenge the territorial orthodoxy. His brilliance lay not in innovative ideas but in originality of method. He simply played the game with such ingenuity as to force wholesale revision of the rules by exposing every weakness or loophole in them. It was ironic that circumstances pushed him into the role of catalyst for revamping the game itself; his original objective had been only to secure a dominant position for Union Pacific.

In 1874 the Union Pacific still owned the middle of the continent. The bankrupt Northern Pacific lay sleeping on the upper Missouri and the southern routes were but glimmers in the eyes of Scott and Huntington. Pacific Mail remained the only competitor for through traffic, while on land budding rivals included the Kansas Pacific and a cluster of small roads south of Omaha, most of them dominated by the Burlington. Since all these roads connected with the Union Pacific, they could not compete for through traffic so long as that road imposed discriminatory rates. The bitter pro rata controversy sprang from efforts to break this stranglehold.[6]

Gould's strategic objectives for the Union Pacific were clear and simple. Through rates must be kept high, which required either an agreement with Pacific Mail or control of that company. On eastbound traffic he wanted a larger share of the through rate, a goal that brought him into perpetual conflict with the Iowa Pool roads. Above all he wanted to protect the road's territory, and like all railroad men he took an expansive view of what constituted its "natural" territory.

Jay thought Pacific Mail had been laid to rest when he assumed control in March 1875, but he reckoned without the Panama Railroad, the tiny but vital rail link across the isthmus. In April Trenor Park, a fiery gnome of a speculator, surprised Gould by capturing Panama and ousting its Pacific Mail board members. Park staged his coup at a time when he and Senator John P. Jones of Nevada were seeking Gould's help in building a railroad between Los Angeles and Owens Valley, where they owned some silver mines. The project appealed to Jay as part of a through line from Utah to the Pacific coast which could serve as a counter to Huntington's Southern Pacific. Park and Jones suspected that Gould was using them as pawns in his negotiations with Huntington and bought Panama to exert some pressure of their own.[7]

Park's first move was to abrogate Panama's contract with Pacific Mail. When Jay threatened to move oriental traffic overland, Park announced plans for a rival steamship line. Through the summer and fall of 1875 the two men sparred relentlessly, negotiating agreements, proclaiming them with great fanfare, and repudiating them before the ink had dried. Pacific Mail became the most active stock on the market as its price twitched in tune with every new development. Park knew Pacific Mail was hurting for money and had to meet $1.35 million in payments, of which $500,000 was owed to Panama. He also understood that a rate war with Panama would transform joint ownership of Union Pacific and Pacific Mail into a liability. Freight carried at low rates would have to be divided between the two routes; since Gould was strongest in Union Pacific, he would favor it at the expense of Pacific Mail. Accordingly, Park organized a steamship company to begin operations in April 1876, when the contract with Pacific Mail expired. Pacific Mail, which sold at 39 in January, dropped to 18 amid rumors that Gould was unloading. When the company defaulted on its note to Panama, Park twisted the knife by attaching its steamers.[8]

None of Park's moves surprised Gould, but he was helpless to prevent them. When the rate war commenced, Jay moved to curb his losses by the familiar tactic of reversing his position. He sold Park's group 25,000 shares of Pacific Mail, giving them control of the company, and the Union Pacific men retired from its management. This apparent surrender enabled Gould to effect a brilliant change of front. The point of controlling Pacific Mail had been to eliminate competition. By dumping the bankrupt company into Park's lap, Jay made continuation of a rate war unpalatable to the new owners. They were, in Dillon's words, "so loaded up with Pac Mail & Panama stocks that they fear a fight as much as we do." After selling out, Gould resorted again to curbing competition through negotiated agreements. In Washington Dodge lobbied as tirelessly against Pacific Mail's government subsidy as he had for it only months earlier. Jay obtained a new contract from the steamship company that, except for minor squabbles, maintained rates for two years. Although Pacific Mail remained a thorn in Union Pacific's side, it rarely seized center stage in the competitive drama.[9]

Defense of the territory was for Gould a two-front campaign. At the eastern end Union Pacific fought to preserve its agricultural and livestock traffic from invasion by the Iowa Pool roads or local poachers. In the Rocky Mountains it wanted to dominate the mineral and ore business. The trick was how to protect and develop so vast a region without straining the company's resources or ability to pay dividends. Three tactical options were available: cooperation, construction, or consolidation. The first was cheap but unreliable, the others expensive and uncertain in their results. Like his peers, Jay resolved the dilemma of whether to bargain, build, or buy by attempting each in turn. He exhausted all three in dealing with the Kansas Pacific.

The Kansas Pacific traced its origins to the Pacific railroad acts, which provided for several branches to supplement the transcontinental line. One of these connected Sioux City with Fremont, Nebraska, on the Union Pacific line. Another, known as the Central Branch, extended a hundred miles west of Atchison and connected with nothing except prairie grass. A third branch, which became the Kansas Pacific, linked Kansas City and Denver. Through a subsidiary,

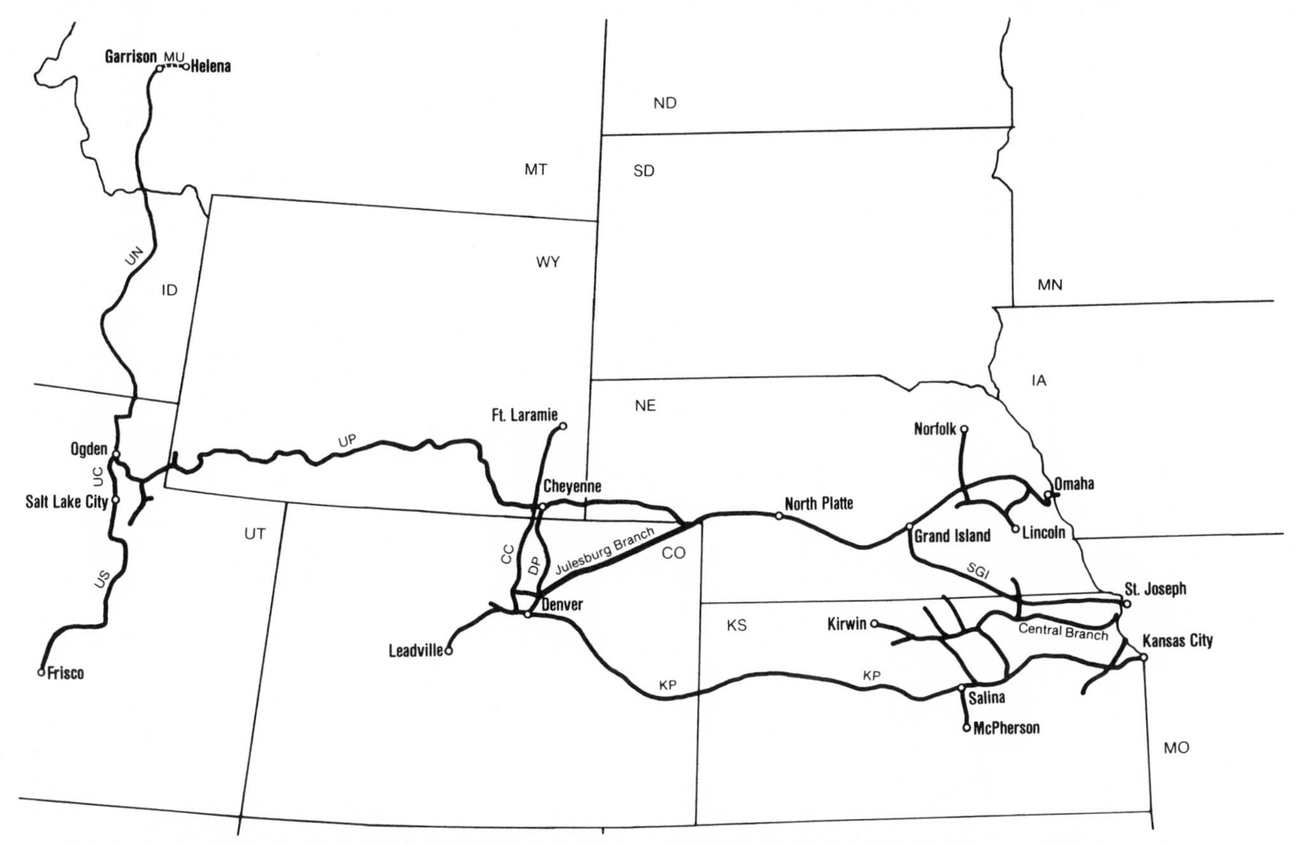

The Union Pacific Railroad with connecting branches, the Kansas Pacific, and the Central Branch.
UP—Union Pacific; KP—Kansas Pacific; SGI—the St. Joseph & Grand Island (originally Grand Island branch and St. Joseph & Denver City); CC—Colorado Central; DP—Denver Pacific; UC—Utah Central; US—Utah Southern; UN—Utah Northern; MU—Montana Union.

the Denver Pacific, it reached the Union Pacific at Cheyenne. The act of 1862 stipulated that the branches be run as "one continuous line" with the transcontinental route, but circumstances transformed the Kansas Pacific into a fierce competitor for through traffic. The Union Pacific kept it at bay by imposing rate discriminations.

Predictably, the Kansas Pacific sought relief by invoking the "one continuous line" clause and demanding legislation to compel the Union Pacific to prorate through rates. Although discrimination hurt the Kansas Pacific, there were deeper reasons for its financial woes. The road had always been mismanaged, and its capitalization rivaled that of the Union Pacific even though it had been built across prairie free of mountains or other engineering obstacles. Nevertheless, the aftertaste of Credit Mobilier induced Congress in 1874 to pass an act ending rate and other forms of discrimination. When the Union Pacific ignored the act, the Kansas Pacific launched a rate war, filed suit to force compliance, and commenced a vigorous pro rata lobby in Washington.[10]

Gould's dealings with the Kansas Pacific involved not only the issues of competition and pro rata but also access to Colorado. Some years earlier the Union Pacific had let slip an opportunity to dominate the Denver Pacific; now it faced the choice of depending on a line controlled by its rival or building one of its own. The stump of an alternative route already existed in the form of the Colorado Central, which ran between Denver and Longmont. It could be extended to parallel the Denver Pacific northward to Cheyenne or eastward across the cattle country to Julesburg on the Union Pacific line. The Central's promoter, William A. H. Loveland, had worked closely if not always smoothly with the Union Pacific for years. His eagerness to build was matched by Jay's reluctance to embark on so costly a venture. Wielding the Central as a club, Gould tried first to bargain for what he wanted. Little did he suspect that the matter would occupy him for half the decade and culminate in one of the most controversial transactions of his career.[11]

Months of negotiating with Robert E. Carr, a St. Louis banker and president of the Kansas Pacific, boiled the dispute down to essentials. Obliged to protect the interest of his Boston associates in the Central, Jay proposed consolidating that road with the Denver Pacific. Carr replied that if Gould wanted the Denver Pacific he must buy control of the Kansas Pacific, which owned 75 percent of its stock. Jay did not want the debt-riddled Kansas Pacific any more than Carr wanted the debt-riddled Central. He wrinkled his nose at Carr's price tag and told Clark in January 1875, "I dont think we ought to lose a moment's time in building into Colorado."[12]

Still he nursed the negotiations along. In April he got Carr to accept a compromise plan uniting the Kansas Pacific with both Colorado roads in a new company. The Union Pacific would own half the stock and manage them in tandem with its own line. Carr was delighted with the arrangement and Jay agreed it "settles forever the question of *pro rate.*" Within a few months, however, the agreement came unraveled and the situation in Colorado went from fluid to volatile. As the hapless Central veered toward bankruptcy, Dillon rebuffed a demand by Loveland for more money and tried to force the Central into receivership. Loveland returned the courtesy by inflaming local opinion against "eastern moguls" and kidnapping the judge sent to hear the request. Eventually Dillon got his receiver, but Loveland's agitation prevented the Union Pacific from taking possession. "Loveland controls the Colorado side of this controversy," a company attorney conceded, and hostility ran high against both Dillon and Gould.[13]

Amid this uproar the competitive situation changed abruptly. William J. Palmer completed his Denver & Rio Grande from Denver to Pueblo and was looking south and west. In March 1876 the Atchison, Topeka & Santa Fe reached Pueblo, which meant that Rio Grande traffic could move to Kansas City over either the Kansas Pacific or the Santa Fe. Anxious to avoid a two-front war, Carr agreed to pool Colorado business with the Santa Fe while he dickered with Gould. West of Denver the indomitable John Evans, a physician and former governor of Colorado, wanted Union Pacific help in extending his Denver, South Park & Pacific through the mining region to Leadville. Dodge took his case to Gould and Dillon in April 1876 but got only a lukewarm response.[14]

The Kansas Pacific lay at the center of this labyrinth. Jay was certain of that, but how to get at it? The road wallowed in a sea of debt which in November 1876 plunged it into receivership. As a bankrupt it could hurt the Union Pacific in two ways, by pressing the pro rata fight in Washington and by carrying through freight at ruinous rates. Gould's efforts at negotiation were hamstrung by the division within the Kansas Pacific's management. Carr spoke for the stockholders, while the receivers, Carlos S. Greeley and Henry Villard, represented two different groups of bondholders, mostly foreign, who had by far the largest investment in the property.[15]

While Carr clung to the hope of an agreement, Jay saw its futility and quietly switched tactics. He knew the Kansas Pacific could not survive without its Denver line and so made it his prime target by making peace with Loveland and completing the Central from Longmont to Cheyenne in November 1877. This put the Union Pacific into Denver but did not eliminate the Kansas Pacific as a competitor. To do that, Jay observed, "We must so effectively clean them out in Colorado that they will sue for terms."[16]

At the same time Gould thrust himself into the Kansas Pacific's financial affairs through several oblique maneuvers. Aware that the bondholders were divided among themselves, he tried to lure those with holdings in the Denver line into a separate deal. He also induced the road's directors to relieve its floating debt with a funding mortgage underwritten by himself, Oliver Ames, and the Union Pacific. Since the collateral for their mortgage included the Denver Pacific stock, any default would place control of that road in Gould's hands. In addition, he bought some Kansas Pacific income bonds at bargain prices and loaned the company $85,000 on notes maturing in February 1877.[17]

These disparate holdings became weapons in Gould's hands. He struck first by suing to collect the expired February notes, a move that caught Carr by complete surprise. Earlier, as a gesture of good will, Carr had suspended two of his pro rata suits against the Union Pacific. Now, however, he concluded that Gould was "at the bottom of the Denver Pacific troubles in Denver" and agreed with Villard that "we must attack Mr. Gould not only in the Courts but at Washington." A rate war intensified the conflict. Once the Central reached Cheyenne, Jay joined the Iowa Pool roads in an effort to drain eastbound traffic from the Kansas Pacific.[18]

As the fight warmed, Villard emerged as the most dangerous of Gould's opponents. A journalist turned financier, Villard had the confidence of the German bondholders and defended their interest unflinchingly. In July 1877 he broke off negotiations with Gould and devised a reorganization plan to eliminate the leverage exerted by Jay through his junior securities. At the same time Villard lobbied vigorously in Washington for the pro rata bill sponsored by his ally, Senator Jerome B. Chaffee of Colorado. "Gould is there with a barrel of money," Greeley warned, and "unless we spend some money in Washington we will not succeed in what we have undertaken."[19]

To Villard's chagrin Gould and Dodge thwarted the Chaffee bill, leaving them only the courts to worry about. Both sides tried to wangle the case into the arms of a sympathetic judge. Here too the Union Pacific emerged victorious. "The U.P. seems to be now making war on us in earnest," Villard lamented. In desperation he tried to enlist Forbes of the Burlington and Thomas H. Nickerson of the Santa Fe as partners in reorganizing the Kansas Pacific. Both offered moral support but little more.[20]

Villard's position seemed hopeless, but Jay knew better. Events in Washington had turned against the Union Pacific. While the Thurman bill marched relentlessly toward passage, pro rata revived unexpectedly. The hostile mood of Congress prompted Gould to change tactics while his bargaining position was still strong. First he induced other holders of Kansas Pacific junior securities to protect their interests by forming a pool in April 1878. As holder of the largest interest Gould was in a position to dominate the pool; other members included the St. Louis parties who owned most of the stock and some of the bondholders represented by Greeley. Jay persuaded the directors to deposit their holdings by assuming their share of the road's floating debt. On May 2 he joined the Kansas Pacific board along with Dillon and Fred Ames.[21]

In effect the pool gave Jay equity control of the Kansas Pacific. More important, it drove a wedge between the two receivers by wedding the interests of Greeley and the St. Louis directors to his own. Having isolated Villard, Jay proceeded to offer him a reorganization plan favorable to the bondholders. As a token of good faith Jay terminated the rate war in Colorado. Villard accepted the plan, and on June 1, 1878, a traffic agreement embracing the Union Pacific, the Kansas Pacific, and the Central was signed. On June 20 Villard turned the road over to Dillon, heaved a sigh of relief, and prepared to depart for Europe.[22]

That same day, however, the new Kansas Pacific board met in Lawrence, Kansas, and decided to offer the bondholders a different plan with less favorable terms. Steps were taken to effect foreclosure under one of the junior mortgages before the first mortgage holders could act on their prior lien. When news of these actions reached New York, Villard was furious. Europe would have to wait, he informed his wife, because "the scamp Gould" and "the rascally St. Louis people, including my colleagues Carr & Usher, had formed a regular conspiracy in the West to break the contract and cheat the bond holders."[23]

The bondholders were quick to rally behind Villard and a bitter struggle ensued. While Villard pressed for foreclosure, Jay filed suit to oust Villard as receiver. In October the court removed both receivers and appointed a new one, but Villard remained implacable. The bondholders disdained negotiations on terms short of the original agreement; they were, in the words of one editor, tired of following Gould "through endless mazes." Villard prepared a new plan of reorganization and talked of extending the Kansas Pacific to Ogden. As autumn waned Gould was content to wait, to scan the board for openings that might have been overlooked. By 1878 other developments in his affairs added new dimensions to the struggle. Like all good players, he understood the principle of never moving one piece in isolation from the others.[24]

The situation at the eastern terminus differed radically from that on the western front. At the Omaha gateway traffic passed into the hands of the three Iowa Pool roads, the Burlington, the Northwestern, and the Chicago, Rock Island & Pacific. The relationship between the Pool roads and the Union Pacific had been difficult long before Gould arrived on the scene. Profits on through traffic depended not merely on volume of business but also on what percentage of the through rate each side received. Both the Union Pacific and the Iowa roads demanded a higher proportion of the rate; the Pool enabled its members to maintain a united front in this dispute.

The dominant road among the Pool members was the Burlington, which soon emerged as Gould's chief antagonist. Like the Union Pacific it was controlled by a group of Boston capitalists, led by John Murray Forbes. An offspring of the China trade, Forbes had ventured into railroads during the 1840s and counted the Michigan Central, the Burlington, and the Hannibal & St. Joseph among his achievements. Well connected and possessed of inherited wealth, he was in many ways the antithesis of Gould as a businessman. In raising capital or devising policy for his enterprises he preached the gospel of sound construction, conservative financing, and careful management. Although not above an occasional plunge, his contempt for speculators in general and Gould in particular knew no bounds. Sixty years old in 1873, Forbes had lost none of his acumen, energy, or acerbic wit.[25]

An upheaval during the mid-1870s secured Forbes's dominance of the Burlington management at the expense of another Burlington patriarch, James F. Joy. In 1876 Robert Harris was elevated to the presidency and Charles E. Perkins to the vice-presidency. Perkins was the man to watch. An intelligent, forceful executive, he also served as vice-president of the Burlington & Missouri River (Nebraska) and had the advantage of being Forbes's cousin. Not only blood but mutual respect bound the two men together. Perkins shared Forbes's conservative philosophy but not his cautious approach. Squat, thick-necked, his face clean-shaven beneath a walrus moustache, Perkins looked the part of bulldog that he played so well. Four years younger than Gould, he drove himself no less relentlessly.[26]

Like other roads, the Burlington had been forced to hunker down during the lean years after 1873. Except for absorbing its Iowa subsidiary and leasing two small lines, it did not build or acquire any new road between 1872 and 1875. Meanwhile, competition at the Omaha gateway intensified as an assortment of smaller roads enabled the Union Pacific to route eastbound traffic away from the Pool roads. A list of the possibilities reveals how complex the board had grown by 1874: (1) to Kansas City via the Kansas Pacific; (2) to Kansas City via the Council Bluffs; (3) to Hannibal (connecting there with the Wabash) via the Council Bluffs; (4) to Council Bluffs, St. Joseph, or Kansas City via the B & M and the Council Bluffs; (5) to Hannibal via the B & M and the St. Joseph; (6) to St. Joseph or Kansas City via the A & N and the B & M.[27]

The history of these smaller roads was interwoven with that of the Burlington. Where Forbes insisted that the Burlington's destiny lay westward, Joy was enticed by the prospects of north-south connections along the Missouri River. Joy controlled both the Council Bluffs and the A & N, which rendered them harmless so long as he remained in command of Burlington. The B & M extended from East Plattsmouth (just below Council Bluffs) to Kearney, where it connected with the Union Pacific. Although Burlington directors owned a majority of its stock, Perkins insisted on managing the road in the interest of all the stockholders. He favored the Burlington in traffic arrangements but refused to serve the Iowa Pool at the expense of the B & M's own prospects.[28]

All these factors Gould scrutinized with a careful eye. In trying to force a lower percentage on the Pool roads he shifted deftly from one tactic to another. When sporadic rate cutting failed to impress the Pool, Jay approached Perkins with a proposition to throw business over the B & M via Kearney. This was Perkins's first encounter with Gould, and it provided him an education. He wanted the business for B & M and objected vigorously when Joy restrained him. "I am unable to see," Perkins growled, "why the interests of the B. & M. in Nebraska should be made to suffer in order that the Pool lines may be victorious in the present controversy with the U.P." The dispute worsened Joy's already difficult position in the Burlington's management. He could not find a policy to reconcile the Pool's needs with those of the B & M, the Council Bluffs, and the A & N, all of which wanted more business from the Union Pacific.[29]

Gould did what he could to exploit these internal rifts. A new eastern connection emerged in the Chicago & Alton, which extended from Chicago to St. Louis and could reach Kansas City via the St. Louis, Kansas City & Northern. In April 1874 Gould began diverting eastbound traffic onto a circuitous route via the B & M and the Alton. The Pool roads got the message; by the end of May a new agreement was reached and traffic resumed its normal flow.[30]

Nevertheless, Gould had injected another element of discord into the competitive muddle between the Missouri and Mississippi rivers. During the next two years savage rate wars plagued the region as alliances formed and dissolved in rapid succession. The Pool roads maintained their unity but could neither impose harmony on the smaller roads nor curb encroachments by new rivals. Joy's departure from the Burlington management in 1875 placed the Council Bluffs and the A & N in hands no longer friendly to the Pool even though he failed utterly to make the two roads work together. The Hannibal dangled in limbo, a financial albatross dangerous only if it should fall into the wrong hands. The Burlington acquired its own line to St. Louis but still found its position in the West eroding. "We must do something at all Missouri points," moaned one official, "or lose our business."[31]

Prolonged negotiations in 1876 brought a stability of sorts in the form of a new pool on southwestern business. However, the pool was less cohesive than its Iowa counterpart and too fragile a thing to deflect Gould from his course. The B & M posed a more serious problem. By 1876 Perkins concluded that Gould had been using him as a wedge to split the Iowa Pool. That winter he joined the Kansas Pacific in an effort to push a pro rata bill through Congress. Jay doubted any bill would pass, but he was quick to accuse the Burlington of standing behind Perkins's action. The Burlington pleaded innocence while Perkins kept grimly at the task of obtaining a bill.[32]

Gould pinned his hopes on the Alton. No arrangement with the Pool roads should be made, he told Clark, "that would exclude the Alton Line. I think we ought to strengthen that line & if possible break the Pool of the Iowa Lines." He also asked Clark to send "some discreet & competent person" over the St. Joseph and report on its condition. Admitting that "the road can be got for a song & may be dear at that," Jay still thought it "might be worth while to buy it to keep it out of the maws of B & M."[33]

By the autumn of 1876 the Union Pacific and the Burlington were edging toward open war. Disturbed by rumors of a merger between the Pacific roads, John N. A. Griswold, chairman of the Burlington board, talked of joining the Santa Fe in building a line to Ogden if the worst should occur. Gould chafed at his inability to crack the solidarity of the Pool roads. The key lay with the B & M and its ties to the Burlington, the only Pool road with a Nebraska affiliate. Traffic moving over the Union Pacific to Omaha had to be shared among all three Pool roads, while that shipped via the B & M went entirely to the Burlington in return for a drawback. To the extent that pro rata would increase the flow of traffic via this route, it would profit the Burlington at the expense of the other Pool roads and Union Pacific. Although the Northwestern and the Rock Island nursed other grievances against the Burlington, the B & M stood out as the lightning rod of conflict west of the Missouri River. For that reason Gould undertook to remove it as a source of friction. In December 1876 he suggested that "the three Iowa Roads and the Union Pacific lease the B & M in Nebraska in perpituity [*sic*] and have the B & M business all go to Omaha and let all the roads have a share of the business."[34]

Previous accounts of this episode have distorted both Gould's motivations and the sequence of events. Even Grodinsky dismissed the idea as "a specious proposal which had all the elements of fairness." There is no evidence for this assertion and no reason to believe the offer was not sincere. To be sure, Jay realized that such a lease would threaten the Pool's unity by promising the other two roads part of the B & M's traffic now going exclusively to the Burlington. To that extent the proposal was a brilliant gambit against the solidarity of the Pool. More important, however, the lease would solve several other problems for Gould. A harmony of interests would eliminate the threat of costly wars and retaliatory invasions on both sides of the Missouri. It would also remove B & M as an ally of Kansas Pacific in the pro rata struggle. This last was a crucial point early in 1877, when the disputed election imbroglio coated everything in Washington with unpredictability. The lease would, in short, buy Gould peace on good terms and free his hands for other pressing affairs.[35]

Predictably, the offer split the Burlington management. Perkins wanted no part of it, and Forbes saw "great difficulties" in any such management. However, the vision of peace in the Missouri Valley appealed to President Harris and other Burlington directors. In both temperament and philosophy Harris differed sharply from Perkins and Forbes. Unlike Perkins, he saw nothing amiss in seeking peace through compromise, and unlike Forbes, he did not regard Gould as the devil incarnate. In March 1877 he and Griswold met with Gould and representatives of the other Pool roads to discuss the whole matter of Nebraska business. Gould's proposal was favorably received and all sides agreed to meet again with Forbes and Perkins present.[36]

Having split the Burlington management, Jay also took steps to isolate it. He bought enough stock in Northwestern and Rock Island to gain a seat on both boards, while representatives of those roads joined the Union Pacific directorate. Although analysts interpreted the move as hostile to the Burlington, Harris scoffed at mobilizing for war until Gould had disclosed his intentions or negotiations had failed. "As to 'throat cutting' as applied to operating Railroads," he declared, "count me out first, last and always. I object to murder in all forms and especially to suicide."[37]

Gradually the negotiations blossomed into a wholesale settlement of all outstanding issues between the Pool roads, the Union Pacific, and the B & M. Forbes and Perkins fought doggedly against this so-called Quintuple Contract but remained a minority within their board. By the end of March a basic agreement had been thrashed out. "It is a good thing for UP & a wise arrangement," Jay enthused to Clark. "All parties agree to build no more railroads in Neb." Dillon instructed Clark to halt all construction on branch lines the moment agreement was reached. Wisely Clark did not hold his breath. The negotiations dragged on for three months while Perkins interposed one objection after another and Forbes railed against what he called "the grasping monopoly of the U.P." A conference to arrange details of the accord broke up when Perkins demanded so many concessions that the other roads developed second thoughts about the overall contract.[38]

Doggedly Harris tried to nurse negotiations along with fresh proposals, but in June the stillborn Quintuple Contract was quietly buried, with credit or blame for its demise going largely to Perkins. The Burlington and the B & M concluded a new agreement for interchanging traffic at East Plattsmouth. Gould tried to arrange a pact with the Northwestern and the Rock Island. When that failed, Clark negotiated an informal agreement by which the Union Pacific agreed to build branches north of the Platte River while the B & M remained south of it.[39]

The peace that followed, later described as the "Golden Age" of the Iowa Pool, proved little more than an interlude. Jay had failed either to break the Pool or to negotiate an agreement with it. In the process he had aroused the enmity of Forbes and Perkins, whose future dealings with him would be strongly colored by personal dislike. As Forbes admitted privately, "My objection to him [Gould] is chiefly personal, but goes deeper to the manner in which he is likely to use any power he gets." In May 1878 Forbes replaced Harris as president, with Perkins as his right hand in the West. Their ascendance meant that Gould would face tough, hostile bargainers in future conflicts with the Burlington.[40]

Given his later reputation for aggressive expansion, it is surprising to discover how gingerly Gould approached the matter of branch building. "I have a dread of branches," he admitted in 1875. The reasons for his reluctance are not hard to find. He was already carrying a heavy load for Union Pacific and frequently had to sustain its stock in the market along with whatever other irons he had in the speculative fire. Moreover, the road's charter contained no provision for building branches, which obliged Gould and his associates to undertake construction on their own as personal investments of high risk.[41]

However much Jay deplored the "expense of building & maintaining branches," he viewed them as indispensable for tapping local business and discouraging intruders. As offensive or defensive weapons they were costly; once employed, they could not be withdrawn. They had to be planned with great care because a branch in the wrong place created suckers instead of feeders and dragged the system down in debt. The cost was high, the risks great, and the results uncertain, but to do otherwise invited stagnation or invasion. That harsh lesson impressed itself on Jay at both ends of the Union Pacific line.

If branches must be built, the trick was to minimize risk as much as possible. Gould's favorite method was to encourage local interests to undertake the project with their own resources. Where possible he sought local subsidies in the form of bonds, tax relief, or other concessions. The Union Pacific could loan equipment, rails, and supplies as well as provide transportation. In most cases Jay anticipated selling the finished branch to the parent company, but that outcome was never automatic. Contrary to legend, he did not exploit branch building as a device for lining his pockets. Rather, he became personally involved in response to competitive pressures, and then only when local efforts proved inadequate.

Aside from Colorado, branch building in the West revolved around the Ogden terminus and partnership with the Mormons, whose hunger for railroads often dragged a hesitant Union Pacific along with them. In January 1870 Brigham

Young completed the Utah Central between Ogden and Salt Lake City. Long before the last rail went down he eyed extensions in both directions. To the north lay Mormon farm communities huddled in the Cache Valley, and beyond them the untapped mineral wealth of Idaho and Montana. More rich mineral land lay to the south, where the potential for extractive industries seemed unlimited if transportation could be found. By 1871 two narrow-gauge projects, Utah Northern and Utah Southern, were organized and hard at work.[42]

Despite its financial woes, the Union Pacific provided help because it dared not lose its stake in the mineral trade. As the Utah Southern crept southward, the growth of business impressed even hardened skeptics. By 1873 Utah boasted thirty ore-reduction furnaces, eleven of them within a dozen miles of Salt Lake City. A large traffic in coal and coke moved over the road, and mining interests in eastern Nevada waited impatiently for the railroad to reach them. Although construction languished after the onset of depression, the Mormons pleaded with Dillon to keep aid flowing toward the Utah Southern, the Utah Northern, and the Summit County, a small coal road. Gould responded with what proved to be a decisive influence in Utah affairs. Here as elsewhere his role evolved gradually and his presence was felt long before it made headlines. Confronted by "an epidemic of railroad building among the Mormons," he realized the Union Pacific had invested too much in Utah to back down. Acordingly, in October 1874 he had the company loan the Mormons $200,000 to extend the Utah Southern twenty-five miles in return for half the road's stock.[43]

In February 1875 the Utah Southern crawled the last of its seventy-five miles to the Juab mines and a clamor arose for further extension. Jay encouraged rumors to that effect because they served his effort to enlist Huntington as an ally in Washington. The Utah Southern was part of the scheme concocted by Gould, Park, and some Nevada parties to create a new through line between California and Utah. Huntington didn't know what to make of it, and John Young added to his puzzlement by pressing the Central Pacific to invest in Utah Southern and Northern. When Young persisted in offering his Utah Southern stock to Huntington and then marching to the Union Pacific office with a similar proposal, Jay moved to end the uncertainties in Utah. He did not like the Mormons as partners any more than he liked the federal government. It had nothing to do with Mormonism but rather with his distaste for shared control. By January 1876 the Union Pacific controlled the Utah Southern board and put Bishop John Sharp in as president. Later in the year Jay also took the little Summit County road off Brigham Young's hands. In railroad-hungry Utah word spread that Gould had taken hold and expansion would surely follow. Jay neither discouraged such talk nor rushed to fulfill it. As he confided to Clark, "I don't think we want to extend the U.S. Rd till we know whether it will *hurt* or help UP."[44]

Although the Utah Northern posed a different problem, it forced Gould into a similar response. There John Young had recruited an "eastern capitalist" in the form of Joseph Richardson, a Connecticut manufacturer, who took up the work in 1871. By February 1874 Richardson had built seventy-seven miles from Ogden to the hamlet of Franklin, but thrifty Mormon farmers gave the road little business. Richardson saw that his investment could be salvaged only by extending the road northward to Montana, where railroad mania had reached fever pitch. When Young resigned as president of the road in October 1875, Richardson applied to the Union Pacific for aid.[45]

Gould listened willingly because he knew Montana was starved for railroads. Farmers and stockmen lacked access to markets, and the mining industry could do nothing without better transportation. Investment money had dried up after the Panic of 1873 and the Northern Pacific, which had been Montana's great hope, had gone bankrupt before reaching the territorial border. In January 1876 the legislature approved a $3 million bond issue for Northern Pacific, an outright subsidy of $1.15 million for Utah Northern, and $750,000 for a small local line. To Montana's astonishment Richardson and the Union Pacific declined the offer; they wanted subsidies, but so profligate an issue would glut eastern markets. By October Jay had formed a syndicate to extend the Utah Northern and had reached agreement with Richardson to act as contractor for the work. In February 1877 the legislature approved a new subsidy, and Montanans waited eagerly for the work crews to begin.[46]

Gould had other ideas, however. He did not like the Utah Northern's route or the fact that it was a narrow-gauge road. It made more sense to him to build a new standard-gauge line northward from Evanston, where coal was plentiful and grades easier. His syndicate again surprised Montana by rejecting the new subsidy, ostensibly because of an unfavorable tax provision but also because, as Jay informed Clark, it "requires us to use the Utah Northern Rd." Then plans for the new road were disrupted in March 1877 by the death of Oliver Ames. Jay mourned his passing as "a serious loss to us—he was always ready to do his full share & . . . just such a friend as the Union Pacific Rd needs for the next few years in order to plant herself thoroughly master of her tributary territory." The lament was sincere, for Ames and Dillon were his closest associates and the directors most supportive of an aggressive policy.[47]

The growth of Montana business convinced Jay that further delay would prove fatal. Dillon learned that "nearly 100 Steam Boats went up the Missouri this past summer with freight for the territory," and no one doubted its enormous mineral wealth. The moribund Northern Pacific was desperate to keep rival lines away until it could resume construction. Unable to gain support for his Evanston route, Gould turned his attention back to extending the Utah Northern. In October he ordered contracts let, only to be stymied by a dispute with Richardson over choice of routes north from Franklin. A perplexed engineer wired Dillon, "Gould says let work Richardson says dont let it."[48]

Here again was the problem of shared control that Gould found so repugnant. The obvious solution was to buy Richardson out, but Ames was dead and neither Dillon nor the Boston directors wanted any part of it. When his friends demurred, Jay assumed the contract himself. In April 1878 he foreclosed the road and reorganized it as the Utah & Northern. The crucial question was how far north through Idaho must Gould build to secure the Montana market and coax another subsidy offer out of the legislature. During 1878 "blue jays," as some wag dubbed the letters hastily scrawled in blue ink on his private blue stationery, flew thick and fast on the subject. Fifty miles of track would intersect the Corinne wagon road; another fifty miles would reach the Snake River at Blackfoot. Jay decided to build only the first portion, then changed his mind when prospects encouraged him to try for "a clean sweep of Montana business."[49]

By late summer Gould was already looking beyond Blackfoot to Montana. "The more I hear of & from Montana & Northern Idaho," he enthused, "the more I am impressed with the wonderful mineral wealth in gold & silver of that region & when the U & N is once extended so as to give them a chance to mine the lower grades of ore. . . . It will be Colorado over again."[50]

The branch-line fever raging in the Rocky Mountain states was slow to infect Nebraska, where the Union Pacific showed little enthusiasm for construction. This lukewarm attitude had partly to do with the restraints imposed by the presence of the Iowa Pool, the B & M, and other roads. Within their fluid and delicate relationships branch building was an extension of diplomacy, an act tantamount to mobilization. Moreover, Gould hesitated because of his conviction that "as a general rule railroads that depend for support on agriculture never pay." Not until 1876 did the Union Pacific commence building its first branch, the Omaha & Republican Valley, which left the main line thirty-six miles west of Omaha and penetrated the region south of the Platte River. Jay approved this project for two reasons: he got county bonds to help finance construction, and he wanted the road as a weapon in dealing with Perkins. Even then he had to take most of the bonds himself, and it took more than a year of concerted effort to dispose of them.[51]

Nebraska interested Jay less than the Black Hills, where the gold craze had drawn twenty thousand people by 1876. A line built north from Nebraska or northeast from Cheyenne would open rich stock and farm country as well as tap the mining boom, yet the Union Pacific hesitated. In October 1876 Dillon said flatly, "We should build 100 miles next year," but indecision lingered on the question of whether to start from Nebraska or Wyoming. In March 1878 Jay informed Clark that 120 miles would be constructed that year. Work began from a point 100 miles west of Omaha but stopped at the village of Norfolk and never resumed. The Union Pacific would not commit funds, and Jay had his hands full with other projects. By then, too, the Thurman Act had forced him into wholesale revision of his plans. In February he had promised that "if we get our matters in Washington settled I shall feel like going ahead pretty vigorously in the way of constructing feeders," but his hopes in that direction were blasted. Instead he began work on a spur from Hastings to Grand Island, completing the connection to the St. Joseph.[52]

Spread as he was across so vast a battlefield, Jay simply could not do everything at once. It was an obvious lesson he was slow to learn and reluctant to accept. The temporary peace with the Pool was for him no more than breathing space for plotting his next gambit. At the western end Gould had put together an impressive system of roads in the mountain states, but he was immersed in plans for further construction, acquisitions, consolidations. His eyes still looked longingly toward the Black Hills and Montana even while his mind raced with other possibilities.

Montana held a peculiar appeal for Jay because it bore an imprint of family. Some years earlier his sister Anna and her husband, the Reverend Asahel Hough, had gone there to do missionary work. Anna had been the first white woman in the territory aside from some camp followers whose notion of the missionary position was quite another one. Living in a one-room mud hut that barely deflected the biting winter wind, enduring extreme hardships, she helped preach the gospel among prospectors and Indians until her health broke.[53]

Their Montana was one Jay never saw, yet brother and sister shared a common bond of dedication that drove them to the brink of collapse. Where Anna's simplicity of purpose sustained her, Jay's complexity of vision consumed and ultimately devoured him. To the righteous, ever in search of useful parables, this was but a just reward when one served God and the other Mammon.

17

Cross-pollination

But . . . the wily little man, with the advice of a technician, a certain General Eckert, set about building a telegraph line of his own along the tracks of his railroads, which he named the "Atlantic and Pacific Company."

—Matthew Josephson, *The Robber Barons*

The company's board of directors was a miniature social directory, with the names of Astor and Vanderbilt leading all the rest. During the panicky summer of 1877, to their dismay, Gould began buying into Western Union, and suggested that a place be made for him on its august board. He had already been black-balled by the New York Yacht Club, and the Western Union board considered it sheer impertinence for him to seek entry to its even more exclusive precincts. For Gould, to the Vanderbilts and Astors, was an out-and-out bounder, a cad, a villainous outsider. Worse yet, it was whispered, a Jew.

—Richard O'Connor, *Gould's Millions*

Beginning with the friends of his youth, Gould had defrauded all who trusted and befriended him. After driving one benefactor to suicide by his treachery, he swindled the dead man's daughters, a circumstance recalled as a mob milled through Wall Street chanting, "Who killed Leupp!" and responding with the name of Gould as they hunted him during the Black Friday panic. . . . Selover probably recalled the tragic fate of the noble Leupp as he manhandled Gould.

—Denis T. Lynch, *The Wild Seventies*

IF IN THE YEARS after 1873 Gould had been content to develop only the Union Pacific system, his legacy would have been full and impressive. If he had confined his career to railroads alone, the results would have placed him among the giants of his age. The fact that he played a dominant role in other industries while pursuing his rail intrigues elevated him into a truly awesome figure, one of those elemental forces that defy the bounds of ordinary mortals.

That Gould was engaged in erecting a business empire eventually became clear, although its nature and extent eluded the closest observer. Obviously railroads constituted its core, but how far did his ambitions extend? Jay was not one to disclose his dreams, and it is possible that even he did not comprehend them all of a piece. Like other men driven by demons, he may well have become prisoner of his own momentum, assembling an empire less by blueprint than by the process of accretion. But if he did not design the whole, he understood thoroughly how the parts fit together.

Gould's involvement in steamship lines reflected one extension of this broader vision; his role in the telegraph industry revealed another. Where the first always remained subsidiary to other interests, the latter ultimately became Jay's largest commitment outside of railroads. Critics have never known quite what to make of this aspect of Gould's career. They are unanimous in depicting it as unscrupulous stock jobbing but fall strangely mute on the larger question of why he acquired the dominant firm in the industry and controlled it for the rest of his life. That sort of commitment is not common in mere speculators. Nor have critics bothered to explore how the telegraph episodes related to Gould's other activities.[1]

To view Gould's telegraph ventures as disjointed grabs for profits is a distortion that obscures his subtle genius for accomplishing several objectives at once. His combined holdings in railroads and the telegraph placed Gould astride two of the most vital industries of his time, crucial arteries from which leverage could be exerted in many directions. The telegraph was indispensable not only to the railroads but to the flow of information throughout the country. It carried dispatches to newspapers, quotations to brokers, messages for businessmen and government officials, and commercial information of all kinds. The ability to monitor or control the flow of such information, or to influence its dissemination in ways profitable to oneself, created the potential for abuses of the gravest sort.

The storm of controversy aroused by Jay's telegraph ventures derived mostly from this concern. To his contemporaries, stock jobbing was small change compared to the immense power a man might wield atop a telegraph monopoly. Control of the wires was to them an ominous extension of Gould's influence over the press, which many already regarded with a suspicion bordering on hysteria. Here were the ingredients for what ultimately became a conspiratorial melodrama of national proportions: railroads, newspapers, the telegraph, armies of Wall Street brokers, judges, and politicians, all under the sinister thumb of Jay Gould and interlocked by his Mephistophelian genius for the single purpose of piling up riches. As early as 1875 the *New York Times* felt obliged to mock the power invoked by his name:

> But straightaway we are assured that "JAY GOULD" is at the bottom of the whole affair, as he is said to be at the bottom of everything that goes on nowadays. We strongly suspect that he will yet be found to have been the origin of the "Beecher" scandal, and to have had something to do with the hard Winter, frozen water-pipes, and plumbers' extravagant bills. He doubtless formed a "ring" with the plumbers sometime last Summer, and then produced the recent severe cold, so as to get all his machinery to work.[2]

The uproar over his power must have surprised Gould, who spent so much of his time and energy swimming against the tide. He could on occasion be ruthless, cunning, vindictive, or inscrutable, but he hardly regarded himself as Alexander or Napoleon. Many of his campaigns were sieges won by patience and maneuver rather than lightning strokes, and a surprising number ended in stalemate or defeat. The telegraph wars were typical of the pattern. They went on intermittently for six years, during which time Gould skated more than once to the edge of defeat before finally prevailing.

The telegraph revolutionized communications in the nineteenth century by enabling messages to be sent over long distances faster than man or beast could carry them. Its ability to transmit information at astounding speeds radically altered business practices and the railroad system, which used the wires for dispatching trains. Early in its history the industry tended toward consolidation. The immense amount of business generated by the Civil War, coupled with the upheavals and dislocations it produced, permitted one firm, Western Union, to emerge as the undisputed leader by absorbing its nearest rivals in 1866.[3]

The blossoming of Western Union into the industry giant did not proceed without controversy. Some said the plant grew fast because it was so well watered. Between 1863 and 1869, when Cornelius Vanderbilt bought control of the company, the stock mushroomed from \$3 million to \$41 million. Vanderbilt reversed this trend by paying no dividends for four years and using surplus earnings to buy \$11 million of Western Union stock, which he deposited in the company's treasury. Turning a deaf ear to the wails of disgruntled stockholders, the Commodore ran his usual tight ship, bought up some small competing lines, and erected a massive new office building on Broadway.[4]

The key to Western Union's success lay in its symbiotic relationship with the railroads, most of which chose to contract the business to telegraph companies. Usually they agreed to furnish transportation and construction materials, maintain the lines along their tracks, operate the offices located in train depots, and provide their own operators to handle dispatching. In return they received unlimited free service on their road and a limited amount off the road. By 1870

Western Union possessed most of these contracts and had nearly 9,000 of its 12,600 offices in depots. Roads having a contract with Western Union could not make one with any other telegraph company.[5]

The Western Union stable did not include the Pacific roads, which were mandated by their charters to build a telegraph line. Union Pacific rejected terms from Western Union and in 1869 turned its line over to a small firm, the Atlantic & Pacific Company, in exchange for 24,000 shares of A & P stock. A year later A & P corraled the Central Pacific but in 1873 still trailed Western Union in miles of poles by 65,757 to 3,065, in miles of wire by 154,472 to 7,460, and in number of offices by 5,740 to 250, not counting the Franklin Telegraph Company, which it controlled. Peace between the two companies had been assured by an agreement in 1872.[6]

Until the winter of 1873–74, A & P stock attracted no attention. Union Pacific had deposited its shares with Morton, Bliss & Company as collateral for a loan and thought so little of the stock's worth that it gave the bankers an option to purchase 12,000 shares at 12. Evidently that option inspired in Levi P. Morton the notion of forming an independent line around A & P to challenge Western Union. He helped strengthen A & P's hold on Franklin and in February had the stock listed on the New York Exchange. A month later Gould entered Union Pacific, of which Morton was also a director. At once he pounced on the telegraph stock as too valuable an asset to let go.[7]

It is impossible to divine Gould's motives at the outset of his telegraph adventure. Given the load he had just shouldered in Union Pacific, it is doubtful he aspired to swallow a Vanderbilt property the size of Western Union. More likely he saw, like Morton, an opportunity for market profits and a toehold in the telegraph industry from which might evolve the sale of A & P to Western Union at a tidy profit. All the ingredients were in place for that most classic of Gould operations, the attack on a large, established company through the instrument of a small, obscure competitor into which he breathed new life.

After some tough bargaining Jay induced a reluctant Morton in April 1874 to waive his option. A week later Dillon canceled A & P's agreement with Western Union, but he and Gould were far from ready for war. Jay realized that nothing could be done without a thorough overhaul of A & P's decrepit physical and financial condition. To whip it into fighting trim he arranged a connection with Direct United States Cable Company, a new rival to the Anglo-American cable, forced a lease of Franklin Telegraph to A & P, and took steps to string A & P's wires along the Colorado Central. In Washington Dodge was busy promoting bills on the patent question, lathering up antimonopoly sentiment against Western Union, trying even to get its contract with the War Department annulled.[8]

Two elements were lacking in Jay's campaign to transform A & P into a serious competitor: strong executive leadership and a superior technology to offset the size advantage enjoyed by Western Union. The first he obtained from General Thomas T. Eckert, the second from Thomas A. Edison. In December 1874 Gould approached Eckert, then general superintendent of Western Union, with an offer to become president of A & P. It was their first meeting, and Eckert came away deeply impressed. Years later he recalled what proved to be the beginning of a lifelong friendship:

> Those negotiations were characterized by a directness and a frankness that at once compelled my unreserved confidence, as well as an appreciation of his breadth of view in surveying the telegraphic field of the country and its possibilities, *although it was a subject to which he was almost entirely new.*[9]

This last point is important for the light it sheds on a later controversy. Contrary to the assertions of Edison's biographer, Matthew Josephson, Jay appears to have taken no interest in the telegraph industry prior to his involvement in Union Pacific. By that time Edison had already favored telegraphy with his genius for invention and for muddling business arrangements. In 1870 Edison devised an automatic, or high-speed, telegraph with funds advanced by the Automatic Telegraph Company, a new firm headed by former diplomat George Harrington and Josiah C. Reiff, financial agent for the Kansas Pacific. Harrington's relationship with Edison was a stormy one, broken by repeated quarrels over money only to be patched up again. Amid one such upheaval in 1871 Edison assigned his partner a two-thirds interest in all his inventions and patents for automatic telegraphy for five years. This agreement would later become vital to Gould.[10]

Quixotic and careless in financial matters, Edison blotted up money like a sponge for his work. During 1872 he churned out one invention after another in a frantic attempt to stave off his creditors. In November he approached William Orton, president of Western Union, for funds to develop one of his brainstorms, the duplex, which could transmit two messages simultaneously on one wire. Edison labored not only on the duplex but also on a quadruplex capable of sending four messages on one wire, two in each direction. His relationship with Orton followed the familiar pattern of ruptures and reconciliations. At one point he gained use of Western Union's experimental shop by agreeing to list the company's electrician, George Prescott, as co-inventor on any patents even though this arrangement violated his earlier agreement with Harrington. Meanwhile Harrington and Reiff tried to peddle the automatic system. The Panic of 1873 so crippled their finances that in June 1874 Reiff sounded Orton on the possibility of selling Automatic to Western Union. Nothing about Automatic interested Orton except the rights to Edison's patents, which was the one commodity Reiff was not prepared to sell.

Orton could afford to be imperious. He knew Automatic faced a dismal future if Edison perfected the quadruplex and sold it to Western Union. In July Edison was ready to demonstrate the instrument to Orton; he also needed $10,000 to avoid foreclosure on his shop. Orton advanced him $3,000 and suggested he apply to Harrington or Reiff for the balance. This was shabby treatment for a man whose services were valued. Orton considered Edison "a very ingenious man, but very erratic," who, left to his own devices, swallowed huge sums with little to show for it. Moreover, Orton was himself besieged from all sides. Vanderbilt wanted Western Union run efficiently, with expenses pared to the bone. When Edison tested the quadruplex successfully in July, Orton saw that Prescott was recorded as co-inventor but did not bother negotiating an agreement to acquire rights to the instrument.[11]

While Edison dangled between his conflicting obligations, Western Union installed the quadruplex on several of its lines with superb results. At the stockholders' meeting in November Orton paid lavish tribute to the "invention that will solve satisfactorily the most difficult problem which has ever been presented: how to provide for the rapidly increasing volume of business without an annual expenditure for the erection of additional lines and wires that would prevent the payment of reasonable dividends." A technical journal agreed that "the automatic invention fades into insignificance alongside of the great 'quadruplex invention,' which is bound at once to revolutionize telegraphy."[12]

Edison was not impressed. Orton had paid him nothing and his debts continued to mount. Not until December 10 did the two men meet to negotiate terms. After Orton agreed to advance him $5,000, Edison submitted three offers, each one lower than the last. Orton countered with an even cheaper proposal. Compelled to leave at once for Chicago on business, Orton promised to resolve the difference when he returned. It was a departure he would soon regret. He liked Edison and fully appreciated the value of the quadruplex, but he also believed he could get it cheaply because Edison had nowhere else to go. Nothing would please his directors more than to acquire such a prize at a bargain price. In his eagerness to wring the best possible terms from Edison, Orton blundered into the costliest mistake of his career.

"Our night of suspense is over," Reiff exulted to Edison on September 11. "*We will shake the foundation of things.* The *money* and satisfaction is from another direction." Since his failure with Orton, the former cavalry officer had been scouting for someone to take Automatic off his hands. He found his man in Gould, who agreed to buy the company if the deal included Edison's patents. Jay then went after Eckert. A brusque, punctilious man of forty-eight, Eckert possessed "the powerful build and moustachioed face of a German brewmaster" and the manner of a Prussian officer. He had expected to be president of Western Union until Vanderbilt put in Orton, with whom he did not get along well. Jay learned of his discontent and resolved not only to acquire his services for A & P but also to use him as an intermediary in reaching Edison. Years later Edison gave his version of what followed:

> One day Eckert called me into his office and made inquiries about money matters. I told him Mr. Orton had gone off and left me without means, and I was in straits. He told me I would never get another cent, but that he knew a man who would buy it. I told him of my arrangement with [Prescott] and said I could not sell it as a whole to anybody; but if I could get enough for it, I would sell all my interest in any share I might have. He seemed to think his party would agree to this.[13]

On December 20 Gould, with Eckert at his elbow, crossed the Hudson to Edison's shop in Newark, where he examined the quadruplex, listened intently to the inventor's explanation, poked around a bit, and departed as abruptly as he had come. During the next few days, while both Prescott and Orton were out of town, Reiff and Gould came to terms on the sale. When Jay asked about Prescott's share of the quadruplex, Reiff assured him that Edison's 1871 agreement with Harrington invalidated the later contract. Apparently that was good enough for Jay. On the evening of January 4, 1875, Eckert brought Edison to Gould's home, where he accepted an offer of $30,000 plus 3,000 shares of A & P stock and, to his delight, the position of electrician for the company. A week later Eckert resigned from Western Union and was quickly named president of A & P. Aghast at this news, Orton rushed back to New York and prodded his executive committee into approving one of Edison's earlier proposals. The inventor informed Orton curtly that he had rescinded the agreement with Prescott because the rights had already been assigned to Harrington in 1871.[14]

The twin blows of Eckert's defection and the loss of Edison's patents hardly toppled Western Union. The stock dropped only four points, partly because it had not yet lost the quadruplex, which A & P could not use until the dispute over ownership was resolved. Jay used that controversy as an excuse to withhold delivery of the A & P shares pledged as payment for the Automatic company. Instead he persuaded the ailing Harrington to sell his interest in the Edison patents. On April 16 Harrington assigned his rights to Gould for $106,000 and fled to the gentler climes of Europe. With these rights in hand, Gould demanded that the purchase of Automatic be reopened because Reiff had misled him on the Prescott agreement. Too late Reiff and his fellow stockholders in Automatic realized that Harrington had sold them out. When he insisted that Gould return all deeds and assignments to Automatic, Jay transferred them instead to A & P and declared the matter out of his hands. The inevitable lawsuits dragged on for years.[15]

Edison watched in bewilderment as these events unfolded. He refused to take up his duties as electrician for A & P because of a personality clash with Eckert. Aware that the automatic system needed further work by Edison, Gould tried in vain to lure him back to the shop by offering him the 3,000 shares of A & P. Edison refused to accept the shares unless Reiff got his as well, or enter the shop so long as Eckert had any authority over him.[16]

Ever the eccentric himself, Edison never knew what to make of Gould, who to his mind belonged to a tribe of eccentrics utterly foreign to his own. His visits to Gould's brownstone produced conversations filled with short circuits. Jay's efforts to initiate him into the mysteries of railroads and finance fell as flat as Edison's attempt to regale his host with droll stories. Three decades later Edison recalled that "Gould had a peculiar eye" and decided "there was a strain of insanity" in his nature. It bothered Edison that Gould "had no sense of humor. I tried several times to get off what seemed to me a funny story, but he failed to see any humor in them. I was very fond of stories and had a choice lot . . . with which I could usually throw a man into convulsions." He conceded, even admired,

Gould's ability to absorb piles of data on the ventures he took up but concluded that "Gould took no pride in building up an enterprise. He was after money and money only. Whether the company was a success or failure mattered little to him."[17]

Fortunately, posterity does not revere Edison for his ability to read character. He missed entirely an instructive parallel between himself and Gould. While condemning Jay's lust for cash, Edison insisted that "the money with me was a secondary consideration." No doubt he was sincere in this belief, yet the pursuit of funds consumed an inordinate amount of time and energy and led him into business relationships that were at best questionable. For Edison money was a necessary tool to keep body, soul, and laboratory together. What he failed to grasp was the extent to which Gould required even larger sums to practice his own peculiar brand of creativity.

Later it would be charged that Gould not only swindled Edison out of the quadruplex but also consigned the automatic system to a premature death. The evidence suggests quite another conclusion. Gould's desire to upgrade A & P spurred the company to acquire in one swoop Edison's automatic system, the Wheatstone automatic system, and the duplex and quadruplex. Within a short time A & P established twenty-two automatic stations that for two years handled its through business between the seaboard and the Middle West. It soon became clear that automatic functioned best when combined with the quadruplex, and that the system's limitations were prohibitive without more work on Edison's part. A & P switched to another system because, in the words of one authority, "as a system, automatic telegraphy has never been successful." Far from stifling technological innovation, Jay embraced it as a weapon against Western Union. The problem was his inability to deploy it effectively. The rights to the quadruplex were tied up in court, and Edison had simply closed his mind to the floundering automatic system. For months Jay tried in vain to rekindle his interest.[18]

In February 1875 Gould opened a rate war only to have Western Union match the cuts. Three months later he persuaded Tom Scott to string A & P wires along the entire Pennsylvania Railroad system. Rumors of an agreement between the telegraph companies began to surface. Gould's Boston associates lent support by buying some stock and approving the sale of Union Pacific's 24,000 shares of A & P at 25, the current market price. Jay took 16,000 of these shares and the Bostonians divided up the remainder. As the contest heated up, Eckert fell ill with pneumonia and was hors de combat for the entire summer. Much as Jay needed the general, he insisted that Eckert not rush his convalescence. With Eckert ill and Edison disenchanted, Gould could do little to improve A & P's position as a competitor. His best hope lay in inducing Western Union to buy him out, but the rate war had made no impression on Vanderbilt or Orton. Any approach to them would be interpreted as a sign of weakness.[19]

Nevertheless, in July 1875 Jay undertook to negotiate a sale or lease to Western Union. At the same time he bought Western Union heavily to give himself a bargaining chip. "I don't think an amalgamation of the WU and A & P. T. Co's far off," he told the Boston crowd hopefully. He persuaded them to buy some Western Union by agreeing to guarantee them against any loss. "We shall carry through the consolidation of the two telegraph companies," he promised, "in such a way as to make Western Union an active ally of the Union Pacific." Rumors flew that Vanderbilt was selling out, and that in his wake Gould and Orton were conspiring to oust his friends from the board. The market danced eagerly to every new tune until the negotiations were pronounced officially dead in September. Eckert recovered from his illness, but in December Jay fell prostrate for nearly a month and suffered from neuralgia of the eyes for weeks afterward. Weak and haggard, he crawled out of bed in January to resume the fight because, he warned Rollins, "The W.U. are buying [A & P] stock and it is necessary to look after it."[20]

Unable to improve the bleak position of A & P, Gould could find no better option than to launch another rate war. This time the fight was long and bitter, a relentless siege to hammer down Western Union earnings and with them the price of its stock. "The A. & P. are whipping the W.U. badly," Jay reported in his optimistic way. "I don't see why A & P can't be made worth 50." He promoted the rate cuts as a blow against monopoly, but financial editors sneered at "the professed reduction of rates *pro bono publico.*" Feeling the pinch, Western Union skipped its April dividend and made a clumsy swipe at buying A & P out. By June the stock had declined ten points since the first of the year.[21]

Gould had embarked on a siege from which there could be no retreat without heavy losses. He had the advantage of a market tending downward in hard times, but he was fighting a powerful company whose aged leader made no bones about despising him. Despite its reverses, Western Union resumed dividends in July because it dared not do otherwise. To push the attack Jay needed reinforcements. A rate war in the east had thrown Vanderbilt's New York Central and Garrett's B & O into opposing camps. Gould turned this dispute to his own advantage by approaching Garrett with a plan designed to serve both his railroad and telegraph interests. On one hand they organized a new through line to combat the Iowa Pool and the eastern trunk lines; on the other, Gould persuaded Garrett that the B & O's contract with Western Union was flawed and could be canceled in favor of one with A & P. While perfecting these moves, Jay startled Western Union with a new round of rate cuts and by signing contracts to displace the larger company on some midwestern roads.[22]

Then fate intervened. On January 4, 1877, Commodore Vanderbilt died and the scepter passed to his capable but cautious son, William.* At fifty-five William had dwelled too long in the Commodore's shadow to possess his steely resolve. Inclined by temperament more to preserve than to attack, his instincts were those of the investor rather than the entrepreneur. Moreover, he inherited the mantle at a difficult time. The country was in the worst throes of the depression, rate wars plagued the eastern trunk lines, and Western Union was besieged by Gould on one side and antimonopoly critics on the other. These circumstances fed William's conservative instinct, led him to prefer compromise to

*Unless otherwise indicated, all future references to "Vanderbilt" are to William.

conflict, and left him vulnerable to those who understood this aspect of his character.[23]

Gould took the Commodore's death as a signal to unleash his most concerted assault on Western Union. The A & P board was reshuffled to include a formidable array of railroad talent and James R. Keene, an Englishman just arrived in the East after wringing a fortune from the rough-and-tumble San Francisco Bourse. Keene resembled Gould in that his slight frame, sallow face with its fringe of beard, sedate expression, and quiet demeanor belied a speculator daring to the edge of recklessness. Keene's presence lured his friend Russell Sage into the A & P venture, thereby reuniting him with Gould for the first time since their falling out over Pacific Mail.[24]

Utilizing the money and influence provided by his new associates, Jay attacked on several fronts. Eckert announced that ten thousand miles of new lines would be constructed, "connecting every important city and town in the Union." The rate war was expanded to the point where both sides incurred losses despite increased business. Then Jay struck at the citadel of Western Union's power, its railroad contracts. Already the Pennsylvania system and a few midwestern roads had broken their exclusive ties with Western Union. The B & O followed suit, the Missouri Pacific announced its intention to do likewise, the Utah roads pledged themselves to abandon Western Union, and Gould let it be known that he was negotiating similar arrangements with Erie, Northwestern, and Rock Island.[25]

Aware that any expansion of service by A & P required perfection of the automatic system, Jay pleaded again with Edison to discharge his duties as company electrician. A testy exchange made it painfully clear that Edison was lost to the cause, and without him A & P lacked the technical capacity to compete with Western Union. That left the stock market as the only front on which to do battle. Gould used his maneuvers on the business front as cannon fodder for a prolonged bear attack on Western Union. As early as February the old merger rumor reared its well-worn head. "There has been no year for the past three or four when the same game was not played," noted a financial editor acidly, "and it may be well . . . that the *canard* has been used and finished so early in the year." Within a few months the editor would discover that his canard was anything but a dead duck.[26]

Against these blows Vanderbilt buckled but did not break. As Western Union's earnings dropped, so did the price of its stock. During the first half of 1877 the company paid dividends without earning them. It met every attempt to cancel a railroad contract with an injunction; later the courts sustained virtually all the original contracts. A more serious problem threatened Vanderbilt in the form of the eastern trunk-line war, which depressed the earnings and stock prices of his extensive rail holdings. While standing firm against A & P, Vanderbilt terminated the rate war in April by making concessions that, in one opinion, surrendered a vital principle his father had labored long and hard to establish.[27]

Gould pounced on this compromise as an insight into the character of his new adversary. Vanderbilt's need to protect his rail empire above all else led Gould to probe for a weakness that might be turned to advantage in the telegraph war. If Western Union would not yield to frontal assault, perhaps its flank could be turned. From this line of reasoning evolved one of Gould's most complex and ingenious combinations. So subtly did he weave its disparate elements together that its ultimate purpose eluded Vanderbilt and most of Jay's own associates until the very end.[28]

The key lay with certain rivals of Vanderbilt's Lake Shore system. The Michigan Central paralleled Vanderbilt's Michigan Southern from Chicago to Detroit, where it could deliver traffic to the Lake Shore, the Canada Southern, or the Great Western of Canada. For years James F. Joy had been the dominant figure in Michigan Central, which enjoyed good relations with both Canadian roads. It was also friendly with the Union Pacific, a coziness enhanced by the fact that Dillon held a large interest in the Canada Southern. Vanderbilt too had some holdings in Canada Southern and wanted representation on the Michigan Central board to tighten his grasp on business east of Detroit. To secure part of the Union Pacific's eastbound traffic, he also needed friendly relations with at least one of the Iowa Pool roads.[29]

Aware of these complex interests, Gould seized upon control of the Michigan Central as a bargaining chip for the telegraph war. Once again he demonstrated his gift for turning a single maneuver to several purposes. His accession to the Northwestern and Rock Island boards in April 1877 advanced his campaign against the Iowa Pool while also alerting Vanderbilt to potential danger west of Chicago. The threat was heightened by rumors that Gould was also buying heavily into Michigan Central and Canada Southern. In May an open fight for control of the Michigan Central erupted, with the opposition ticket headed by none other than Dillon.[30]

In June 1876 Joy was replaced as president of the Michigan Central by Samuel Sloan, the capable chief of the Delaware, Lackawanna & Western and a man soon to become a close ally of Gould. Both Sloan and Joy were associated with Moses Taylor of the National City Bank. Faced with the Dillon ticket on one side and pressure from Vanderbilt on the other, Taylor worked hard at gathering proxies. The balance of power was a delicate one: Taylor's faction controlled 80,000 shares, Gould 40,000, Dillon 25,000, and Vanderbilt 30,000. On two occasions Vanderbilt called at Taylor's home to plead his case, but Taylor would promise him nothing. Unable to exact concessions, he could use his proxies to support Taylor's man or Gould's man, which for Vanderbilt was no choice at all. "There are the proxies," he growled at Taylor after their second meeting; "you can use them." Gould and Dillon followed suit, and Sloan was reelected without opposition.[31]

The election was irrelevant to Jay, whose object was simply to impress Vanderbilt with the fact that control of the Michigan Central could be had by acquiring the Gould-Dillon shares and thereby to set the stage for a little horse trading. A month later rumors of a telegraph merger surfaced anew. Officers of both companies issued denials, but Western Union soared from 56 to 79 when word leaked out that talks were in progress. Late in August it was announced that Western Union would purchase from Keene and Sage 72,502 shares of A & P at 25, somewhat higher than the current market price. Within days the telegraph war vanished and rates advanced sharply.[32]

To the very end Wall Street dismissed the merger rumors as no more than "the annual skirmish between the directors of the two companies." When Orton hailed the acquisition of A & P as an end to "wasteful competition," editors demanded to know what had become of "the great professions of 'opposition to monopoly' which have been urged with so much persistency by the Atlantic and Pacific officers for years past?" No one detected the connection between the sale and the fact that in June 1878 Vanderbilt controlled the Michigan Central election and assumed its presidency.[33]

The sale of A & P climaxed a campaign lasting nearly three years with a victory that surpassed the fondest hopes of Gould's associates. Having failed to make A & P an efficient, viable rival to Western Union, Jay found himself saddled with a weak property unable even to earn its expenses. Despite bleak prospects, he not only managed to dispose of it on lucrative terms but also to leave Vanderbilt with the impression that he had made a good deal. The war was over, rates were restored, and Western Union, which had dropped as low as 56, nudged past 84 during the fall. For Gould the whole affair had turned out well after "a laborious and anxious commercial fight," to use Eckert's delicate phrase. To most observers it was a stunning and complete triumph, but Jay knew better. He had walked away with profits but without a telegraph company. Few people realized how important the latter was to him; he would soon correct their ignorance. After all, the device of unloading a company on Western Union had worked so well that it might be worth another try.[34]

Every victory has its casualties, though rarely did the Wall Street wars reach the level of violence inflicted on Gould during the telegraph fight. On August 2, shortly before noon, Jay left Broad Street and headed down Exchange Place. As he approached New Street Major A. A. Selover crossed the street from the opposite side and engaged him in conversation. Selover was another of the westerners anxious to make their mark on the Street, but so far he had done little. A brawny six-footer, he was a friend of Keene and knew Gould, who had helped him earlier with some small loans.[35]

After chatting briefly Jay turned to leave. Suddenly Selover grabbed his lapels and shook him violently. "I'll teach you what it is to tell me lies!" he bellowed, and swung at Gould, who struggled frantically to get free. Selover pushed him backward until Jay stumbled and sprawled onto the stoop near the drop area in front of a basement barber shop. He struggled to his feet and tried to fend off his adversary, but his strength was no match for Selover's. The Major hoisted him over the railing and let him dangle above the eight-foot drop, his left hand clutching Gould by the collar while the right rained blows on his head. "Gould, you are a damned liar!" he roared.

"I am not a liar," Jay protested weakly. Unable to find footing, he hung limply and buried his face in the coat that all but swallowed his head.

"Gould, you *are* a liar!" Selover repeated, showering him with blows and profanities. A bystander tried to intercede and was shoved aside. Finally the barber, watching from below, sent out his black helper, whose shouts startled Selover into letting go of Gould. Jay dropped into the basement area and crumpled in exhaustion. "It is characteristic of Mr. Gould," smirked a broker later, "that he landed on his feet." The barber flung open the window and Jay crawled inside. Selover glowered at the drop area for a time, then departed. Pallid, trembling, Jay let the barber tend him as best he could, then walked up the stairs and recovered his battered hat while the barber's helper gathered up his pen, pencil, and broken watch chain. Aside from bruises, he suffered no injury. He refused all interviews, issued no statements, and pressed no charges.

Selover, basking in victory at Long Branch, was more talkative. He did it because Gould had lied and cheated both him and Keene in a transaction involving Western Union. "It would exhaust the capacity of the English language to fittingly characterize the meanness, the duplicity, and the treachery with which this scoundrel has treated me," Selover sputtered. "For weeks and months he has lied to me . . . all the time pretended to be my friend, and yet in secret he has been constantly plotting my overthrow." He did not mean to damage Gould but merely "to disgrace him publicly, as he deserved."

The facts behind the dispute never emerged clearly. There was much confused talk about Gould betraying Keene and Selover in a pool involving Western Union, about quarrels between Gould and Keene on market matters, even about Gould wanting revenge because Keene and Sage had scooped control of A & P from him. Keene dismissed the tales as "too preposterous for notice" and assured reporters that "I have lost nothing by Mr. Gould's maneuvering and have no apprehension of doing so."[36]

Whatever the truth, Selover knew less of it than he thought. He was a small-time roller who may have picked up the wrong scent from both Gould and Keene. His pummeling of Gould remained his only contribution to Wall Street, and it was not well received. The Long Branch crowd objected to Selover's brand of frontier justice on the grounds that he was so large and Gould so small. The *Chronicle* noted with dignity that "the practice of appealing to blows as a remedy for disagreement in a stock operation will hardly be accepted as a satisfactory method of settlement." The *New York Times,* which relished any discomfort afforded Gould, proclaimed with savage glee, "If reprisals like this are to be countenanced, Mr. JAY GOULD will be hung by the nape of the neck and pummeled by indignant stock operators from January to December."[37]

It is a measure of Gould's unpopularity that later generations would find in this episode a convincing example of his perfidy. Yet the market was a game Jay played better than most in an arena where the rules were at best malleable. The deal that enraged Selover appears to have been a commonplace maneuver that only a naïve or reckless operator would overlook. If Selover was duped, he need look no further than his mirror for retribution. As a reporter later observed of Gould, "It was his business to utilize for personal ends the follies and passions of his fellow-men. . . . And yet, though Mr. Gould was widely accused of failing to keep the faith, there were many among his harshest critics who were ready to admit that he seldom, if ever, turned upon an associate without having had previous reason to believe that the associate was turning upon him."[38]

Gould's response to the Selover affair was typical. He issued no threats, pressed no charges. The revenge he exacted came in the market, where Selover's strength was no match for his. However, the scars of that experience never again allowed him to walk the streets freely, as he once did. Thereafter he traveled in the company of Morosini, his husky secretary, or some other bodyguard. To his sorrow he discovered that his way of doing business had hidden costs of a kind for which he had never bargained.

III

REACHING HIGH
1879–1881

18

Homebody

Perhaps his devotion to family was due in part to the fact that his public career placed him apart from other men, and made him an object of fear and hatred. He was an exile from the sympathies of his fellow-men. . . . The members of his family were his only intimate friends.

—Henry D. Northrop, *Life and Achievements of Jay Gould*

It is strange that this appreciation of pure and poetical things should exist in the soul of a man of such financial grimness. But it was doubtless Mr. Gould's nature before his life took on its acquired thirst for gold. When that thirst was in a measure satiated he turned again to his fundamental instincts and his great conservatory was the result.

—Trumbull White, *The Wizard of Wall Street*

I shall always believe that with Uncle Jay money was secondary to his intense desire to win, his burning will to succeed, in life, in business.

—Alice Northrop Snow, *The Story of Helen Gould*

AT FORTY-THREE Gould had already fought the wars of Wall Street for nearly two decades, and his body had begun its retreat toward the hollow shell it would soon become. There seemed barely enough flesh to cover even his frail frame. His thinning black hair had receded to expose a high, cerebral forehead and make his large ears even more prominent. The wiry thicket that masked his lower face contrasted with his complexion, swarthy yet tinged with the pallor of one who seldom saw sunlight. There was about his appearance a quality of abstraction, of intellectuality, as if brain had sucked body dry to fuel its own incessant needs. Even his eyes, those dark pools of mystery, confirmed this impression, whether glowing softly or fixed with laserlike intensity or glazed with preoccupation, as if his mind had gone to dwell in some remote place beyond the reach of ordinary mortals.

Those who dealt with Jay found him conspicuous for his inconspicuousness. His clothes were of excellent cut and quality, usually in black or deep blue, with ties of matching or equally subdued colors. He wore no jewelry except watch and chain. In warm weather his felt top hat gave way to a Panama. If certain early photographs hint at elegance, the fling was brief. His manner was as quiet as his attire, unassuming but not inattentive. "His concentration was so intense that you noticed it," recalled Alice Northrop Snow, a niece who visited the Goulds often as a child. "When he spoke, he became perfectly oblivious of everything around him. When he listened, his eyes would never leave the speaker."[1]

Gould's reputation for being cold and aloof owed much to the fact that he was a shy, reserved man whose emotions registered on so small a scale, such as tearing bits of paper or tapping a pencil, that only initiates recognized them. Every act, every gesture embodied a sort of conservation of energy. In middle age as in youth he drove his emaciated body to the brink of exhaustion by sheer force of will, but the demands on him had increased while his reserves had diminished. The result was a perpetual energy debt that required more frequent repayment as he grew older. One suspects that his icy self-control was in part a facade to conceal the ravages of fatigue.

Fatigue followed him home, where he returned each evening a spent warrior. At dinner he sat with bowed head, sometimes too exhausted or preoccupied to utter a single word, picking indifferently at the plate before him. "Watching him," Alice Snow wrote, "I used to wonder if he knew what he was eating." Helen enforced strict silence at the table and saw to it that servants "melted from view as though through mysterious trap-doors when their functions had been performed." In the evening, after kissing the children good-night when they filed in at bedtime, he puttered with his flowers or curled up with a book in his library. Yet leisure could not hold him even in the evening. There were always "blue jays" to write, associates to visit or receive, projects to contemplate, decisions to make, a hundred details pressing his weary brain for attention before the wars resumed next morning. Even the most adroit of jugglers, tossing so many balls into the air, dared not remove his eye lest they crash down on his head. If nothing else, the night offered him silence and respite while his mind ticked with fresh possibilities.[2]

Home was both haven and fortress for the Goulds, shutting the noisy, clamorous world out and themselves in. Except for the Millers and intimate friends like the Russell Sages, they entertained little and went out less. That society snubbed the Goulds did not bother Jay at all. He despised the glitter and grind of social functions, or "unescapables," as he called them. If resentment burned inside him he gave no outward sign. Apart from family, books and flowers were his passion and solace; he preferred their company to that of sports or hostesses. He also enjoyed fine paintings but not music, having inherited the tin ear of his mother and other Goulds. In this sense his pleasures were no less private than his business practices.[3]

The self-contained existence that satisfied Jay imposed a peculiar burden on his wife. So faint is the imprint left on history by Helen Gould that it is possible to recover only a trace of it. She had never aspired to be a grande dame, but neither did she wish to be an outcast, especially when she fretted about the children's future. She disapproved of fashionable society with its vulgar pursuit of fads and snippets of malice exchanged at elegant soirees; the pretensions of nouveau hostesses were for her both tasteless and tiresome. But she had also been cut off from the world of her upbringing, the polite society of Murray Hill, in which dignity, genteel manners, and staid conventions counted for more than the size of the family fortune. Cast adrift from her heritage and unable to find mooring in the brazen glitter around her, she sought refuge in the harbor of her family.[4]

The bitter truth was that Helen had no choice. Murray Hill was a place where good citizens could be outraged by the excessive ringing of a church bell. It must have shocked her beyond belief to find herself with a husband who was vilified in the press, hounded by sheriffs and summons servers, assaulted in the streets, linked with the most famous and infamous men of the time. How it must have puzzled her to connect the affectionate soul she knew at home with the nefarious monster that stalked the pages of the dailies. She had married a man different from other men, and those differences forced her quiet, gentle nature to withdraw ever more into the shelter of her family. There at least her sense of duty could find fulfillment. With a strict eye she supervised the household in such a way that no detail intruded on Jay unless he wished it, which he rarely did. Her skill and dedication prompted Sarah Gould Northrop to observe shrewdly that her brother had married "just the right woman," one who relieved him of every care and distraction at home. A subdued, dignified tone suited Helen, who detested ostentation no less than did Jay. Like him she dressed well but in drab grays or darker colors.[5]

To admiring outsiders, including even Jay's harshest critics, the Goulds embodied ideals of domestic bliss to a degree that put Victorian convention to shame. There were no false notes in the image. Whenever Jay was away on business, which was often, he longed most to be home again or at least to have Helen with him, and he did not hesitate to tell her so. *"I wish I had you here with me now,"* he wrote mournfully from Geneva in July 1879. *"I would give all this beautiful scenery for a few kisses & embraces*—my only consolation is that when I get home I shall have all I want of them, which will be not a few, so you must not disappoint me."[6]

But if it was a home filled with love, so too did it bristle with inner tensions that were not so apparent. For Jay they were the product of his obsession with work; for Helen they were the price exacted by her devotion to duty. Where Jay was drained by exhaustion, Helen suffered from "nervous disposition," a fashionable term for the stress of anxieties and frustrations from which propriety allowed her no release. As her health began to fail, she came to rely ever more on her housekeeper, Miss Terry. True to her code, Helen languished and wasted away in dutiful silence, concerned more about her family than about her own decline.

The insular world shared by Helen and Jay extended to their children. Unlike many women of her class, Helen took personal charge of rearing the children. There is no doubt she was the dominant figure in their childhood. Although Jay was an affectionate, even doting father, the pressures of business limited his presence. The children learned early to obey him at once and not to disturb him unless invited, but it is misleading to view Gould as a stern Victorian patriarch. The trials of his own childhood had instilled in him a devotion to family that was his true religion. He had no other creed and did not pretend otherwise, an honesty that moved even the *New York Times* to declare, "We do not like Mr. GOULD. We do not think he is a good man to have around. But it is much to his credit that he is wholly free from hypocrisy in the matters of religion."[7]

The gospel of family was one Jay preached and practiced fervently. It is ironic that Gould's most appealing virtue was also his greatest weakness, perhaps his Achilles heel as a businessman. That was his unswerving loyalty to his children and his passionate desire to see them, especially George, succeed him in the office. From the first everyone knew that George was being groomed as the heir apparent, and no one knew it better than George himself. Jay doted on all the children but none more than George, who even as a child was imperious as only a spoiled regent can be. Of the three older children he was the most personable and outgoing. Edwin was quiet and studious, young Helen almost morbidly shy and timid. Small and plump, with an oval face dominated by dark, sensitive eyes, she inherited her mother's Murray Hill reserve and conservatism. Even as a girl her attitudes struck others as prissy and old-fashioned. Until she was twelve Jay called her "Nellie" and his wife "Ellie" to distinguish between them.[8]

While Jay was busy transforming his business holdings, he changed homes as well. The Goulds lived at 578 Fifth Avenue until 1882, when they moved across the street to an unpretentious four-story brownstone at 579. The furnishings reflected Helen's taste for muted elegance leavened by Jay's fondness for fine paintings, a well-stocked library, and a conservatory. Pleasant as it was, 579 did not satisfy Jay's longing for a sanctuary from the city. In 1877 he leased an estate at Irvington-on-Hudson from the widow of George Merritt. Three summers at Lyndhurst so enchanted the Goulds that in 1880 Jay bought the property for $250,000. It was the most ostentatious purchase he would ever make, and probably the happiest as well.[9]

Lyndhurst was among other things a national architectural treasure. General William Paulding built its first section in 1838, employing as his architect Alexander Jackson Davis, the country's foremost advocate of Gothic Revival style. When Merritt bought the house in 1865, he summoned Davis back to transform his original villa into a boldly romanticized Gothic castle by adding a new wing and a great tower. From the massive stone gate the road meandered through vast carpets of lawn sprinkled with stands of linden and elm, beech, birch, and pine, most of them huge and ancient, with billowing canopies and enormous limbs. The bark of the copper beeches resembled the hides of old elephants, rubbed, gnarled, wrinkled, and imposing. Suddenly the house loomed above the trees like Camelot sprung to life, a sprawling asymmetrical cruciform busy with turrets, bays, finials, buttresses, trefoils, stone traceries, and crenellations, its white façade of Sing Sing marble gleaming in the sunlight. On the other side the ground sloped toward the river, leaving the mansion to dominate the skyline.[10]

The spacious interior drew upward to ceilings with rib vaulting of haunched beams carried on corbels. Davis personally designed every interior detail and had his cabinetmaker build furniture to harmonize with the character of the rooms. His decor employed the current vogue of using one material to simulate the appearance of another. The massive stonework of the vestibule was actually plaster grooved and etched; the stonelike ceilings and marbled walls were plaster painted or feathered by dipping quills in hot linseed oil. The ornate dining room had simulated marble columns and plaster walls painted to resemble leather. In the hallway canvas walls with a plaster base gave the illusion of stone, with pine treated to look like oak.[11]

What more perfect setting for a dark, mysterious figure like Gould than a Gothic castle where things were never what they seemed? The image is tempting but wrong, for there was nothing dank or gloomy about Lyndhurst. Its cheerful, airy interior danced with light and colors flowing from a subtle blend of textures. For all the family it was an enchanted house: quiet, isolated, remote from the heat and turmoil of the city, surrounded by breathtaking scenery. There Helen could preside over the modest teas for genteel neighbors that were to her taste. The eighty-five-foot bell tower never saw a bell, but the children loved to play in its glass-enclosed rooms. The elaborate carriage house and stables allowed Jay and his sons to indulge their fondness for horses. He transformed the great hall on the second floor into an art gallery, crowding the walls with paintings by Rousseau, Daubigny, Corot, Diaz, Rico, Leroux, Messonier, and others of the Barbizon school then in vogue. For the spacious, two-room library he purchased a large collection of books and devised his own catalog for them.[12]

Apart from these pleasures, Lyndhurst offered Jay an even more sublime treat: the largest greenhouse in America. It was a true colossus, 380 feet long and 37 feet wide, with 60-foot wings at either end. A huge onion dome in the Moorish style rose 100 feet above the central portion and offered a spectacular view of the Hudson Valley from the Palisades to the northern highlands. Merritt's collection of plants had been sold off after his death, which obliged Jay to restock the greenhouse. He couldn't wait to get started. That 1880 was one of the busiest and most portentous years of his career did not matter. Scarcely had he closed the sale on Lyndhurst when, despite the rush of other business, he plunged eagerly into the task. Merritt's head gardener, Ferdinand Mangold, was retained to supervise the work. By the time Gould returned to New York in the autumn he had acquired a collection worth an estimated $40,000.[13]

Then disaster struck. Early on the morning of December 11 fire demolished the greenhouse and its exotic collection. The heartbroken Jay responded as he had years earlier when fire destroyed his history of Delaware County, by moving at once to replace his loss without a whimper. In the spring of 1882 the firm of Lord and Burnham, working with an architect hired by Gould, completed the first metal-framed greenhouse built in America. The new design covered the same area but suited Jay's taste by dispensing with the Moorish observation tower. The west section and wing was devoted to a grapery and cold house with rhododendron, camellias, hyacinths, and various bulbs. In the central portion, an extended semicircle, Jay surrounded the lovely fountain with a huge collection of palms. The east section and wing contained a fern house, carnation house, croton house, mixed planter, rose house, and, most impressive of all, an orchid house. Nothing fascinated Jay more than the delicate beauty of the orchid. Importing species from all over the world, he put together the most impressive collection to be found anywhere in the country.[14]

Jay approached his pleasures with the same intensity he gave to business. Botany became a passion with him; he pored over books on the subject until his knowledge was broad and expert. The grounds on either side of the greenhouse were planted in flower and vegetable gardens. Like Merritt before him, Jay spent large sums draining swampy areas and planting them with trees, shrubs, and orchards. He bought pedigreed cattle and put them to pasture on the land. If there was something incongruous in the image of Gould as country gentleman, he did not trouble himself about it. From Lyndhurst he commuted to the city by train or on his yacht. In 1880 he had a small steel bridge erected above the tracks of the New York Central that ran between Lyndhurst's sloping lawn and the dock. Those associates who were also his neighbors sometimes shared the ride with him. Cyrus Field turned up often enough to become a standing joke.[15]

A ride with the boys or game of croquet on the lawn. A stroll about the ground at twilight to watch the sun setting behind the Palisades, its last rays shimmering in the treetops. Quiet evenings in the library with a book, or an hour's contemplation of beauty in the art gallery. Simple pleasures enjoyed on a grand scale, most of them the tastes of his youth, refined and packaged in luxury. None delighted him more than his morning pilgrimages to the greenhouse. A small, solitary figure in felt slippers, he was transported for a brief time from the jungles of finance to this lush, fragrant Eden of his own creation. What communion or consolation did these plants offer a man whose native language was silence? Perhaps their delicate beauty nurtured his spirit as religion nourished the souls of other people.

None of these tastes satisfied the curiosity of a public starved for gossip about the titans whose clashes in the business arena had already become the stuff of legend. There was nothing heroic about Gould. His bland personality and inconspicuousness seemed wholly at odds with the brilliance and daring of his exploits. He was neither a sport nor a peacock, had no charisma, and kept mostly to himself. He did not fit anybody's notion of manhood, yet some mysterious power enabled him to outsmart and ruin men who physically could crush him underfoot. Those puzzled by these shattered stereotypes viewed him as something alien and despised him for it. His appearance and manner, his habits and tastes were effeminate, they sneered, his character timid if not cowardly. He was a dark, furtive creature operating in shadows, using methods filled with deceit and treachery to achieve unsavory objectives. What kind of specimen was Gould?

A few advanced intellects of the age came up with an insight of their own: perhaps he was Jewish.

Lyndhurst was not merely a fairytale castle for Gould but also a fortress against a hostile world. Millionaires were always choice prey for cranks and extortionists, and Gould's unpopularity made him a prime target. The two earlier assaults left Jay fearful for his safety and that of his family. Threatening letters became as commonplace as those requesting money and were ignored until October 1881, when one signed "An Old Victim" warned Gould that he would be shot dead within six days. Jay turned the letter over to Chief Inspector Thomas Byrnes, the man he always relied on when cranks or threats besieged him. Further letters offered to spare Gould if he would furnish stock tips through newspaper notices in cipher. For nearly a month Jay played along until Byrnes's detectives snared their man. To their surprise he was J. Howard Welles, the son of a prominent lawyer. The incident cemented the relationship between Gould and Byrnes, who became a frequent caller at 579. He provided Jay with police protection and was rewarded with market tips that did more for his standard of living than a detective's salary ever could.[16]

The Welles episode was no isolated affair. On the contrary, such incidents grew more frequent as Jay vaulted up the ladder of success. Industrialization had caught the country up in the throes of furious and often violent change. The mad rush of economic growth had unleashed forces of social dislocation that bordered on chaos in the minds of those who had most to lose from any upheaval. As a primary agent of change Gould became a target for those maddened by its effects. Only a few months earlier the lesson had been impressed on him in shocking fashion. One summer afternoon at Lyndhurst Alice Northrop came around the porch and found Jay walking unsteadily toward her, clutching a telegram, his face drained of color and his hands trembling violently. "Alice," he whispered hoarsely, "Garfield has been shot! President Garfield has been shot!" She led him to a wicker chair on the porch where, after a few minutes, he regained his composure. Two things about the incident impressed Alice in later years. Jay's mask of self-control, which she had never seen slip, vanished completely, revealing a man of deep if private feelings. His head sagged back, then snapped forward again. Within moments the mask had returned and he went inside. "At dinner," she marveled, "no one could have told how that telegram had shaken him."[17]

What was it in the shooting that shattered Gould's mask of imperturbability? Certainly it shook the financial markets at a most inopportune time for Jay's operations, yet he seldom flinched in the face of adverse business news. Over the years he had grown fond of Garfield, whom he considered a sound and able president, but perhaps he saw something that cut even closer to the bone. Here was a grim reminder that even men of prominence were vulnerable to the capricious acts of fanatics or lunatics. For all the precautions he took, never again would it be possible for Jay to move about without peering over his shoulder. The death of Garfield, followed so closely by the Welles incident, pushed the normally reclusive Goulds ever more toward a garrison lifestyle.[18]

The more Jay retreated into isolation, the wider grew the chasm between him and those who knew him only through business or the newspapers. To his inner circle he was one person, to the public quite another. He demanded much of his associates and paid them well for their services. Those who served him found their loyalty and consideration returned in full measure. They were not deceived by his image but were helpless to counteract it, and eventually Jay himself gave up the attempt. "If I denied all the lies circulated about me," he told a reporter wearily, "I should have no time to attend to business." He shunned the usual solvents for a tarnished reputation: society, religion, philanthropy. His acts of public charity were responses to local needs and as idiosyncratic as Gould himself. When yellow fever ravaged Memphis in September 1879, Jay sent the Howard Association $10,000 and exhorted its president to "keep on at your noble work till I tell you to stop and I will foot the bill. What are your expenses?" A few months later he gave $5,000 to relieve settlers along the Kansas Pacific hurt by a year of crop failures.[19]

If these were gestures intended to enhance Gould's image they were curious choices, wholly lacking in the publicity value of a church, a seminary, an orphanage, or a school on which his name could be emblazoned for posterity. What the two incidents have in common is a direct response to immediate needs, made without frills or fanfare. Gould's acts of charity bore the same economy and simplicity of style as his business methods.

The true source of Jay's philanthropy, like so much else in his life, was his family. He saw to it that his half-brother Abram was trained in business and employed on the roads he owned. Anna Hough received help when her health broke, and large donations went to support the good works of Bettie Palen and her husband. No one needed Jay's help more than Sarah Northrop and her brood of fifteen children. When George's tannery failed, Jay set Northrop up as a storekeeper in a Pennsylvania town where they could be near the Palens. It was to no avail. Past sixty and in declining health, his spirit crushed by the collapse of the tannery, Northrop lingered only a few months before taking his own life.[20]

Sarah's grief was edged with desperation, for her husband had left behind little beyond some debts. Jay was quick to set her up with quarterly checks every year and a fifth check at Christmas along with detailed instructions on closing up the store. All debts would be paid, he promised, and the children were to continue in school. The remarkable thing is not that Jay undertook the support of his sister's large family but that he did so with characteristic thoroughness. Not content merely to sign checks, he took a close interest in the children and followed their progress closely. When Ida, the oldest girl, finished Vassar and wished to open a preparatory school of her own, Jay moved the family first to Hackettstown, New Jersey, and then to Camden. During Sarah's frequent visits to 579 she was obliged to furnish her brother with detailed accounts of each child. Alice appeared regularly at 579, too, and soon became young Helen's dearest friend.[21]

What Alice remembered best through these difficult times was her first Christmas at 579. On that magic morning the children tumbled downstairs, hurried through breakfast, and marched single file into the sitting room. Aunt Helen was there, beaming with delight, and so was Jay, his weary face aglow with twinkling eyes and a broad grin. At home the Northrops wrung the most from their few gifts by opening them one at a time, but here each child had a chair or table piled high with packages. With yelps of glee they all pitched in at once. Sarah had warned Alice not to expect too much, but Alice quivered with excitement at the treasures she found: a gold Swiss watch from her aunt, an exquisite brooch from Edwin, a check from Jay, and so many others she could hardly contain herself.

When the mountain of presents had been leveled, everyone trooped into the dining room for Christmas dinner. A huge vase of scarlet anthuria and poinsettias adorned the center of the table, with smaller vases at each end. Gleaming silver compotes filled with hothouse fruit, candied fruit, and bon bons were scattered everywhere. They sat down to a feast of terrapin, oysters, turkey with stuffing, cranberry jelly, vegetables, and plum pudding. That evening, full and drowsy, they gathered in the library before the huge Christmas tree ablaze with candles and topped with a smiling angel. Everyone received a box of candy and one last gift before going off to bed.[22]

Christmas at 579 was a fairyland of delight much like Lyndhurst, and for the same reason. For both Jay and Helen family was a balm that soothed the pain of isolation, a respite from lives of unremitting pressure. The shadow of Jay's childhood loomed over these fierce loyalties as it did over so much of what he did. The ordeals of those years spawned a curious paradox in his nature. In scratching and clawing his way to the top, experience had imbued him with self-reliance, hardened him into the most private of men. At the same time his physical frailty, his bouts with sickness and exhaustion, impressed him deeply with the importance of caring and being cared for. Eventually the scars imposed by repeated clashes with a hostile world taught him that a man of large ambitions and limited strength cannot be all things to all people. Family and friends counted for most; the rest could be let go. Perhaps he even smiled at the irony that, through a quirk of fate, the runt of the litter had become the one on whom the others depended.

In choosing these loyalties Jay resolved only a part of the paradox. A sense of balance between the demands of his work and the needs of his family continued to elude him. As his vision of empire hurtled toward realization, the fierce extremes of his life tugged even harder at his ebbing strength. Unable to satisfy both, he fused them into one grand design. His business empire would be his legacy to the children, and his fortune the mortar that would bind them together as a family.

19

Gambit

In his mind it is mapped out into a series of chessboards, set with curves and parabolas, as well as squares and corners,—chessboards which run into each other curiously, although a separate game goes on upon each. Pawns, knights, castles slip deftly from one to another in kaleidoscopic confusions, out of which only one pair of eyes in the world evolves orderly and coherent plan.

—*Stockholder,* August 20, 1878

The collapse of Pacific Mail after his retirement in the spring of 1876, his lack of faith toward his railroad associates in the directorate of the Atlantic & Pacific, his chicanery in undermining the Hannibal, and his faithlessness in dealing with Villard . . . gave him and his controlled Union Pacific a bad name.

—Julius Grodinsky, *Jay Gould: His Business Career, 1867–1892*

Gould has always quarreled with his closest friends. Charley Osborne was with him for years, and so was S. M. Mills and finally Belden. But where are they now. One of them he ruined, and the others would have been ruined if they had staid long enough. . . . One of his earliest business associates was Leupp, who was ruined and ended his life by suicide. Next comes Fisk, murdered while carrying out Gould's schemes. Tweed, his co-Director, died in prison. Judge Barnard, his social friend and legal ally, deposed from the Bench and permanently dishonored.

—*New York Times,* January 30, 1879

IF IN THE AUTUMN of 1878 Gould had struck a balance sheet of his activities, he would not have been pleased at the outcome. On the positive side he had wrung a profit from the telegraph venture and left the industry. Apart from speculative holdings his chief asset remained a block of about 170,000 shares of Union Pacific which, he insisted repeatedly, "I hold for a long pull." But the pull had grown more onerous in recent months. The rise in earnings slowed while fixed costs soared, forcing the company to skip a dividend. The contract with Pacific Mail was about to expire, which meant another round of thorny negotiations. Although peace with the Iowa Pool still reigned, intrigue and discontent raged beneath the surface. The emergence of Forbes and Perkins as the dominant voices in Burlington assured the hostility of that company toward Gould. Clearly Union Pacific was scudding into rough weather at a time when Jay's personal and financial commitment to the company had never been greater.[1]

In effect Gould's position reflected the sum of past failures. He had neither broken the Iowa Pool nor found a way to protect the territory from invasion. His toehold in Northwestern and Rock Island had brought meager results. In Kansas Pacific the struggle dragged on, tying up more resources. The protracted fight over Credit Mobilier still occupied his attention. Above all, a settlement with the government continued to elude him, crippling attempts to win investor confidence. There seemed no way to free Union Pacific from the shackles of its past. It was too expensive to compete against cheaper modern roads, and management had to be shared with Boston and Washington.

The conclusion was inescapable: if Union Pacific had become a losing position, the chess board had to be shaken up with fresh combinations, perhaps even a sacrifice. The bears sensed the road's vulnerability and launched an attack in August. Their assault unmasked another failure on Jay's part: he misread the market and found himself on the wrong side at a critical moment. The dangers arising from his miscalculation occupied Gould for months and forced him to reassess his entire business position. From this long night of the soul emerged a series of stunning combinations that revamped his holdings, rearranged virtually every piece on the board, and shook the business community to its foundations. Although he did not yet realize it, Gould had reached the bank of his Rubicon. The gambit he essayed to salvage a bad position proved the turning point in his career.

Within the archives of Wall Street mythology can be found the hoary saga of Gould's narrow escape from disaster in the fall of 1878. As legend has it, Gould's short position in a rising market enabled the bulls, led by Keene, to force the calling of loans secured by Union Pacific shares. Keene summoned Gould to a conference and demanded immediate settlement. A distraught Jay sold 40,000 shares of Union Pacific but still could not meet his obligations. At a second meeting Keene insisted that Gould leave Wall Street and New York forever or face ruin. Broken and bowed, "in the shape of a dish-rag," Gould agreed. However, one of his creditors was Sage, who balked at Keene's tactics and furnished Gould a check for $2 million to pay off his tormentors.[2]

Here was the sort of tale on which the dailies thrived: a colorful, action-packed confrontation between two titans of Wall Street. This one made its debut three years after the fact in the *Times* under the usual guise of revelations from an "insider." More likely it was a fable spun from rumors kicking around the Street. Although Gould did run into financial trouble, the evidence departs radically from the tale told by the *Times*.

By 1878 the burden of Union Pacific had worn Gould down. For five years he had defied critics by working to strengthen the company; as late as 1876 he summarily rejected an opportunity to sell 100,000 of his shares at a handsome profit. Much had been accomplished, yet it rankled him to own a majority of the stock and still be deprived of absolute control. He was tired of endless battles with Boston, Congress, cabinet members, lobbyists, local politicians, and editors who sneered at "Jay Gould's road." Early in 1879 he would sound Dodge and Clark on a plan for "relieving himself as much as possible from the load." Circumstances had changed dramatically since the past summer. When the board decided reluctantly to pass the July dividend, the bears picked up the scent at once and sent Union Pacific tumbling six points. Gould was forced to buy heavily and to borrow $200,000 from the company to meet an expiring note.[3]

At this critical juncture the market offered Jay mixed signals. The gloom of the depression years had enabled him to execute one bear raid after another with brilliant success, but how long could he stay with the role of Ursus Major? The downward trend could not last forever, and the need to maintain Union Pacific left him vulnerable to the predations of the Twenty-third Street crowd, who had already bared their claws. On the other hand, two factors struck Gould as drags on the market. The resumption of specie payments, due to begin in January 1879, would in his opinion tighten the money supply and depress prices. Moreover, early forecasts for bumper harvests in the Northwest soon gave way to reports of severe crop damage.[4]

If the market declined that autumn, a line of shorts could provide some badly needed cash. Gould's eye fell on Northwestern which, along with its granger twin, Chicago, Milwaukee & St. Paul, had been bulled nearly twenty points before slumping back in July. As Northwestern's largest stockholder Gould could not ignore this sudden drop engineered by Cammack, who was also leading the bear charge against Union Pacific. A new pool was formed to lift the grangers again; given Jay's position in Union Pacific and Northwestern, he seemed a logical member. Once again, however, he did the unexpected. Convinced that the market would fall, he sustained Union Pacific while selling the granger roads short.[5]

In the bitter struggle that followed, the bears seized on Union Pacific as Gould's Achilles heel, hoping to ruin his credit by destroying its value as collateral. Jay was forced to borrow another $250,000 from the company and promised the treasurer that "after the storm blows over I can reborrow & let you have what you want." A few days later he impressed Clark with the urgency of keeping net earnings high to pay dividends the next two quarters. "The recent decision of Secy Schurz has frightened some of our stockholders & I have had to become a large buyer," he confided, "but if I can have my next two dividends I will run out of debt."[6]

As autumn waned and the reports of crop damage proved exaggerated, it became apparent that Jay had miscalculated. Rumors flew that a "leading operator" was in trouble. W. L. Scott, one of Jay's opponents in Northwestern, told the road's president that Gould's losses had put him in a tight position. In October Gould informed M. L. Sykes, the Northwestern's secretary, that he had received certain proposals "looking to a change of ownership in whole or in part, of his Union Pacific property." The fact that Jay approached a man who had long opposed him suggested that he needed cash and could only get it by peddling some of his Union Pacific shares. Sykes hastened to sound the other Iowa Pool members on the possibility of buying into Union Pacific. The Rock Island was willing but Forbes, ever cautious, hung back. Rumors that Gould was hurting buzzed furiously up and down Wall Street, but with Gould things had a way of never quite being what they seemed on the surface.[7]

The business was all too murky for Forbes. As the beat grew faster, it became impossible to know who was dancing with whom. He was bothered by fresh tales of a Gould-Vanderbilt alliance and of negotiations for merging the two Pacific roads. While he pondered, Jay acted. In November he sold 30,000 shares of Union Pacific to Sage and Keene, who joined him in talking consolidation with Huntington. The move relaxed the pressure on Union Pacific, but the bulls sent Northwestern soaring despite all Gould could do. From an August low of 32 it climbed into the 50s by December. Even worse, the market greeted the approach of resumption with a smile instead of a frown, beginning a rise that continued for thirty months.[8]

This second miscalculation deepened Jay's predicament, as did another matter that has gone unnoticed. Since March 1877 Gould had been special partner in the firm of William Belden & Company, which included two of Belden's brothers and Washington E. Connor as partners. "Wash" Connor had been a loyal Gould associate for more than a decade. A credit reporter described him as "smart & cunning & likely to take care of himself." Where Jay had always found Connor steadfast, time had done little to improve Belden's character. Morosini later asserted that Jay was in the firm only because the brothers, "having in their possession claims arising from the gold panic against Mr. Gould forced him to form a brokerage firm with them."[9]

During 1878 trouble developed. Although the details were never disclosed, Morosini claimed that "Wm Belden and his brother after robbing Mr. Gould right and left wound up by stealing all the securities they could lay their hands on. Something like 3 millions of dollars and came very near ruining Mr. Gould entirely." In January 1879 the matter burst into the open. Connor charged the brothers Belden with transferring assets to their own name and threw the firm into receivership. In the legal skirmishes that followed, the Beldens were arrested on a charge of converting $30,000 worth of securities to their own use. Connor opened a new house under his own name, with Morosini as partner and Gould as "special." It served only one customer, who kept the partners well supplied with work.[10]

Gould's losses in this affair came at the worst possible time. In January the market battle boiled toward a climax in which, as Dodge put it, "the entire street is fighting Mr. Gould who was a bear on the Northwest stocks." Amid furious trading Northwestern spurted to 65 and St. Paul to 48 while the bears regrouped for another go at Union Pacific. Belden roamed the Street telling anyone who would listen that Gould had ruined him and was now on the rocks himself. Banker Patrick Geddes picked up the tale and informed Forbes that "Goulds cash means are, I think, very much reduced. I am told that he is no longer feared or followed."[11]

The newspapers too smelled blood. For months they had been wringing all they could from the rivalry between Gould and Keene. Since the latter was as quiet and unassuming a citizen as Gould in private life, the effort amounted to the bland leading the bland. During January 1879 the *Times* virtually ignored Gould until the 30th, when it splashed a lurid account of "JAY GOULD'S DIFFICULTIES" across its front page. On the editorial page the *Times* moralized that Gould was "one of the very few Americans whose misfortunes will be generally regarded as a public gain."[12]

During January the *Times* also lambasted the Pacific roads for their role in politics and demanded a settlement with the government. On the 28th news of the Supreme Court's ruling in the net earnings case enabled the bears to hammer Union Pacific down 10 percent in five minutes. Gould and Sage managed to push it back up, but Jay had granger shorts to cover and was fast running out of cash. He could not afford another attack on Union Pacific that might imperil its value as collateral. The time had come for hard decisions, not merely about his market position but about the future of his railroad interests. In either case it made no sense to maintain his dominant holdings in Union Pacific. How could he extricate himself from danger so long as that block of stock remained his chief asset? It hamstrung his efforts in the market, invited attack from his foes, left him without flexibility to respond. Selling more of the stock meant surrendering control of the company, a move he had been contemplating for months.[13]

A revealing step in that direction had been taken as part of the November deal with Sage and Keene. They regarded the Utah & Northern as crucial to the Union Pacific's future and insisted that Gould sell it to the company. So did Dillon, who had reversed his opinion of the branch. "Our directors seemed to think the U & N ought to belong to the U.P.," Jay told Clark, "& I have acquiesced." The Utah & Northern proved one of the best bargains Union Pacific ever got, becoming within a few years its most profitable branch. Jay's forced sale of the road lightened his load but did not solve his financial dilemma. The obvious solution was to sell more Union Pacific, but how to do it in a way that transformed difficulty into opportunity? Once again Gould displayed his gift for conceiving inspired moves in tight corners. Borrowing the tactics employed earlier with Pacific Mail, he disarmed his opponents by bringing them into the company they were attacking. In mid-February he agreed to sell another 70,000 shares to a syndicate that included Sage, Cammack, Keene, Osborn, and W. L. Scott. The price was about 60, which gave Gould a sizable profit on his original investment. As part of the deal Sage, Keene, Scott, and Cammack took seats on the Union Pacific board.[14]

News of the transaction shot Union Pacific up to 81 amid rumors that Gould was about to leave the company. Not everyone regarded the sale as the last act of a desperate man. The *Stockholder* ridiculed the *Times*'s exposé and offered the parable of an old rat who, weary of the world, retired from its travails and was later found "in the middle of an enormous Cheshire cheese." Gould's own explanation of the sale was so straightforward that no one believed it. On February 22 he penned three brief letters to Clark, who was prostrate with anxiety over the news. Ever sensitive to his friend's insecurities, Jay reassured Clark that he had sold because shared ownership would leave Union Pacific "less open to attack than when it was all 'Jay Goulds.'" The buyers were "all entirely friendly" to the company and his "personal friends." Their presence would "enable us to get any amt of fresh capital for the perfection of our system of branches." As for retiring, "Of course I should like rest & would like to put some good man in my place. Still I doubt if it will be thought advisable." Morosini reinforced this point in a letter of his own. "Mr. Gould has no idea of severing his connection with the Union Pacific," he reassured Clark, "for her prosperity is identified with his own. It was for strategic reasons better known here in Wall Street than any where else that he has let others in."[15]

It is worth noting that the explanation given Clark in private was identical to that offered the public and repeated eight years later before the Pacific Railway Commission. There is no more reason to doubt the truth of what Gould said than there is to believe it was the whole truth. He did not care to lie; neither did he wish to reveal his hand entirely. The sale was in fact a brilliant gambit played for deeper considerations than those he admitted publicly. He sacrificed control of Union Pacific in exchange for an improved position and the resources to effect new combinations. The proceeds from the sale gave him cash to cover his granger shorts and to push other operations.

Few of Gould's contemporaries understood that he had played a gambit or knew what to make of it, and historians have fared no better. Blinded by the rush of events during the next year, they have overlooked his continuing presence in Union Pacific. He had surrendered majority control but remained the largest stockholder, voting 123,700 shares in his own name at the annual meeting in March 1879. Although he began quietly to dispose of more stock that spring, he remained a director until 1884. His friends still constituted a majority of the board and Dillon and Clark, the president and chief operations officer, were both loyal allies.[16]

The gambit provided Jay with the freedom and resources to move in other directions, but it did not show him where to travel. The course he pursued was so subtle as to deceive friends and foes alike. He did not abandon Union Pacific and had no intention of doing so; his financial and emotional investment in the company was too large. On the contrary, every step he took on its behalf was genuine and useful. To all appearances the sale had not affected his role. He began again to loan the company money and continued pushing expansion on every front. The day after announcing the sale he joined the subscription list for extending the Utah Southern toward Frisco, where the Horn Silver Mine promised a lucrative traffic.[17]

All seemed as before, yet with every act, every gesture, Gould was at the same time preparing the ground for a radical shift in his own position. Until events disclosed to him what that position should be, he was content to conceal his personal ambitions inside the Trojan Horse of service to Union Pacific. To appreciate the subtlety of this dual role one need only follow his handling of the Kansas Pacific affair. Already Jay sensed that the outcome of this struggle was as pivotal to his own plans as to the future of Union Pacific. Victory offered three major prizes: elimination of a dangerous competitor, control of the Colorado gateway, and greater leverage against the Iowa Pool. Moreover, a settlement would boost Kansas Pacific securities, in which Gould had a large investment.

"Gould has made himself by buying into KP & NW," Forbes grumbled in May 1878, but by year's end it looked more as if he had unmade himself. Northwestern hung about his neck like a noose, and the Kansas Pacific had him stymied. In December Villard won a court decision that thwarted all attempts by the junior holders to foreclose. On behalf of the Denver Extension holders he intended to foreclose, smash the pool created by Gould, and seize control of the property. Privately he admitted that "it may take years to bring the foreclosure proceedings to a final determination."[18]

Jay could not afford to wait years. In January he offered to pay back interest to August on the Denver Extension bonds if their holders would exchange them for new bonds with interest reduced from 7 to 5 percent. The bondholders responded favorably but hesitated when Villard opposed the settlement. In mid-February Jay agreed to raise the interest rate to 6 percent and to pay the back interest in cash. These concessions came just as news of the Union Pacific stock sale hit the Street and were enough to satisfy Villard. The effect of the gambit on these negotiations was affirmed by Dodge, who noted that "the K.P. people are working hard to make a deal with us since these new developments in U.P."[19]

As usual there was more to the settlement than met the eye. On the morning of March 2 Villard met with John D. Perry, who had come to New York with Carlos Greeley to represent the St. Louis interests in the Kansas Pacific pool. Villard startled Perry by expressing doubt that the bondholders would accept Gould's latest proposal. An alarmed Perry conveyed the news to Greeley, who confronted Gould with it the next morning. Jay shrugged wearily; he had made his last offer and was tired of the whole affair. But if the bondholders refused to yield, Greeley protested, it would be ruin to the property and to their St. Louis friends. Why don't you sell out? Jay suggested. Greeley replied that he couldn't sell unless all sold. The dark, inscrutable eyes gleamed suddenly. In that case, Jay said blandly, why don't I buy you all out? I will get some of my Union Pacific friends to help me.[20]

After some brisk negotiating Gould agreed to buy out the other pool members at the valuations fixed in the original pool agreement. Sage, Dillon, and Fred Ames took small shares, giving the purchase an appearance of being on behalf of Union Pacific. For weeks Gould and his Boston friends had been buying all the Kansas Pacific stock outside the pool they could obtain at prices ranging between 11 and 22. By these moves Jay acquired a majority of the road's junior securities, putting him in position to control its reorganization if he could settle with the bondholders. His dealings with both Villard and the pool representatives had been carried off in strict secrecy. The pool members did not dream that a settlement with the bondholders was near, and Wall Street was still trying to digest the implications of the Union Pacific sales. In that sense the gambit also served as a smokescreen concealing Gould's activities in Kansas Pacific.[21]

A few days after the Union Pacific annual meeting on March 5 the story broke that Gould had settled with the bondholders, acquired control of the Kansas Pacific, and would reorganize it as "a tributary to the Union Pacific." No one knew how to square these transactions with the sale of the Union Pacific stock except to conclude that Jay had not left the company after all. Praise was heaped on Villard when the Denver Extension bonds, which had sold at 35 two years earlier, leaped to 102½. This sensation overshadowed the rise in Kansas Pacific from 17 to 22, where it stood at the end of March. Since Jay paid 12½ for the pool's stock, he already had a nice profit, but the best was yet to come.[22]

It was crucial to Jay's plans that reorganization of the Kansas Pacific proceed smoothly. On April 1 he assured Villard of his "entire good faith" and declared his willingness "to advance *at once* the requisite amt to pay the back interest & also the $125,000 to the committee." This last was another sop to the bondholders' committee, which Gould had originally offered $100,000 for expenses. It is tantalizing to speculate on just how cozy Gould and Villard grew during these months. Villard was absorbed in his project to build a line eastward from Oregon to meet the Northern Pacific. Jay subscribed to his construction company and agreed to extend the Utah & Northern northward to connect, perhaps even to consolidate with Villard's road. The subject occupied them through the spring and early summer of 1879, the same period in which the organization of Kansas Pacific took place.[23]

While curious eyes followed Gould's handling of Kansas Pacific, he was busy elsewhere. If his motives were shifting, his objectives remained constant. He wanted to control the Colorado gateway and smash the Iowa Pool. The confused situation in Colorado seemed incapable of swift resolution. The Colorado Central had been leased to Union Pacific, but the Denver Pacific remained in receivership and Jay owned none of its securities. The Kansas Pacific owned three-fourths of its stock but most of the bonds were held in Amsterdam. There was trouble too with the Santa Fe and the Rio Grande. In October 1878 the Santa Fe leased the Rio Grande, but so many disputes arose that by spring Palmer was fighting to reclaim his road. Meanwhile, John Evans was doggedly trying to push his South Park road toward Leadville, where a mining boom was luring hordes of prospectors.[24]

The Rio Grande lease blocked Union Pacific access to southern Colorado and New Mexico. Leadville lay at the tip of a triangle southwest from Denver and northwest from Pueblo. Evans's line snaked toward it from Denver; Palmer and William B. Strong, the Santa Fe's general manager, both wanted to build toward Leadville from Pueblo. Jay did not like being shut out, but he recognized that "so long as the D & RG is under Strong's control & we have no line of our own into that country it is better to submit than fight." He had Clark arrange a pool with the Santa Fe for Colorado business while he figured a way out of the stranglehold. One obvious solution was to buy into the South Park. Strong talked of leasing the road, which enabled the canny Evans to play him off against Gould. Jay held no illusions about Evans; as he reminded Clark, "The Gov is on record in the past as selling us out whenever he gets a good chance." He would bargain, but at the same time he ordered a line surveyed west of Colorado Springs. The important thing, he stressed, was to "own our own line."[25]

The struggle in Colorado keenly interested the Iowa Pool roads. Both the Northwestern and the Rock Island wished to see the Rio Grande allied with the Union Pacific, since all their Colorado business came via that route. The Burlington, however, had grown friendly with the Santa Fe; Strong had in fact come to the latter road from the Burlington. Forbes was anxious to keep the Santa Fe and the Rio Grande in harmony but feared that personal antagonisms between Strong and Palmer might drive them to "renew their alliance with Jay Gould, which I think is not their natural alliance."[26]

Forbes would have done better to look closer to home, where Gould was searching relentlessly for the right combination to undermine the Iowa Pool. The peace imposed in 1877 was eroding beneath new competitive pressures. An uneasy coexistence prevailed in southeastern Nebraska, where Gould and Perkins faced off during the summer of 1878. As the Union Pacific's Republican Valley branch crawled westward, the B & M organized its own line to tap the Republican Valley. After Perkins and Clark met to exchange the obligatory threats, F. Gordon Dexter sounded his Boston friends on the B & M board and assured Gould that "they desire a fair & equitable division of Nebraska business with U.P." Jay was willing to concede Perkins the region along the Republican River. "If the B & M & ourselves can work in harmony," he wrote Clark, "it is money in both our pockets."[27]

Gould wanted to avoid a branch war because he was after bigger game, the Iowa Pool itself. The territory east of the Missouri River was a tinderbox of roads that were financially weak but dangerous as potential links in a new through line to Chicago. Most of them were bankrupt, nearly bankrupt, or fresh from bankruptcy, which meant their securities could be acquired cheaply. Jay regarded them as the key to freeing Union Pacific from dependence on the Pool roads. To do that Union Pacific needed its own line to St. Joseph and/or Kansas City so as to route traffic away from Omaha without depending on the B & M. The trick was to find the best combination and put it together without arousing suspicion among the Pool roads.

Throughout 1878 and early 1879 Gould labored at this task, campaigning on several fronts at once, disguising his intentions by moving secretly, using one move to deflect attention from another, and keeping them so disparate that no clear pattern emerged. While the financial community followed his market maneuvers and amused itself with tales of his downfall, Jay was busy securing the lines he needed. If he was in deep trouble, his activities on this front do not reveal it.

He wanted the St. Joseph and the A & N because "if we should lose one or the other of these connectors it might place us in the power of the Iowa Pool Lines—this we could not afford." But the A & N had converted two-thirds of its bonds into stock, making it difficult to acquire. The St. Joseph was in fact two properties, a railroad and a bridge company, and Jay did not want one without the other. The Burlington too was interested but Forbes, as always, moved cautiously. In January 1879, at the very time he was supposed to be without cash, Gould bought the first mortgage bonds of the St. Joseph and picked up the bridge company as well. The public knew nothing of the transaction or Gould's plans. "With this line in our hands," he exulted, "we will be independent of the Pool lines." Traffic could now flow over the Union Pacific's Hastings branch to St. Joseph for connection with the Hannibal. Another possibility loomed: control of the A & N, along with some construction, would create a line through Atchison to Kansas City, from which St. Louis could be reached by the Missouri Pacific or Kansas City roads and Chicago via the Alton. The Pool roads would be outflanked.[28]

Gould was aware that rates between Chicago and both St. Louis and Kansas City had been demoralized since March 1878, when the Southwestern Pool had collapsed. The Kansas City startled the Pool roads by declaring its intention to build to Council Bluffs and sounded Union Pacific on the use of its bridge into Omaha. Jay welcomed the news but warned Clark to "take & maintain a position of strict impartiality." He did not want the Pool roads antagonized until his own arrangements were perfected. Already he had in mind securing a line to St. Louis by acquisition of the Kansas City.[29]

Silently, deftly, Gould shoved his pieces into place for the coming attack, scattering them across the board in such a way as to veil his intentions. Had the financial community been privy to these maneuvers, they would have found it strange behavior for a drowning man. The gambit had worked to perfection. It had cost him little and given him much—time, money, room to maneuver, and the sort of blurred attachment on which he thrived. Did he still control Union Pacific? Were his activities on behalf of the company or himself? The Street always found it difficult to determine whose colors Gould was flying, and never more so than after the gambit.

"Since the Kansas Pacific surrender, we are about the only fighting enemies left against him," Perkins warned in 1878. "In short, he is more dangerous than ever." The coming months would fulfill that prophecy on a scale beyond even Perkins's wildest imaginings.[30]

20

Blitzkrieg

Jay Gould . . . does not go gossipping about the streets; he does not discuss the affairs in which he is interested with the crowd. He is a silent, earnest, self-contained, and, it may be, an unscrupulous man, who not only knows what he wants, but how to get it, and straightaway proceeds to do it. His ability in organization is profound. His capacity to grasp, to analyse, to reach conclusions through complicated and knotted accounts, is perhaps unequalled. His depth has never been fathomed.

—Anglo-American, quoted in *Stockholder*, July 20, 1875

All his tactics, therefore, hinged upon what Thorstein Veblen has defined as "disturbance of the industrial system." He would pursue a deliberate policy of mismanagement "as a matter of principle," deriving his gains from the discrepancies between the real value of the affair and its supposed or transient value in the security markets. In good times he would give an appearance of gauntness and misery to his enterprises; in bad times he would pretend affluence.

—Matthew Josephson, *The Robber Barons*

From whatever point of view you regard it, the consolidation of 1880 seems to me, judging by the light of experience, to have worked nothing but benefit. I cannot see that it has worked injury anywhere . . . the transaction was far more favorable than anything which I have been able to effect during more recent years. In other words, the consolidation long since more than justified the price paid for it.

—Charles Francis Adams, Jr., *United States Pacific Railway Commission Testimony*

THERE WAS NO transformation, no change of character, simply a change of front. Just as Gould had always been a trader, so had he always been a businessman who used the market to finance larger ambitions. His image as a predator feeding on the carcasses of lame corporations was a legacy of the Erie War and Gold Corner years, long obsolete but frozen in the public mind. For two decades he had speculated but had never been content to operate merely as a speculator. All his life he had wanted his own enterprise to run in his own way. He thought he had found it in Union Pacific, but events had proven otherwise.

During the fateful year of 1879 Gould perceived that circumstances had changed radically and responded with a series of bold, unexpected moves. He did not become, as Grodinsky argued, "the competitive bull thrown into the stabilized china shops, overturning rate compacts from the Rockies to the seaboard." The bull was already loose and crashing into crockery, but most men shut their ears to the din. Gould kept his ears and eyes open, as did his astute rival Perkins, who grasped what the tumult was about: "I have long been of the opinion that sooner or later the railroads of the country would group themselves into systems and that each system would be self-sustaining—or . . . cease to exist & be absorbed by those systems near at hand & strong enough to live alone."[1]

Perkins persuaded himself (and later historians) that Gould had been the one to untether the bull. "If Gould's combinations turn out to be solid—& more than mere stock operations," he predicted, "the general tendency in the direction of consolidations will be hastened. The crystallizing process is going on faster and faster." He was wrong on one point: Jay was not the cause of change but simply its catalyst. The forces had already been unleashed; he was merely the first to harness them to his own purposes. Jay understood, as did Perkins, that competition had rendered the territorial concept extinct. Too many roads, large and small, crowded against one another and more were coming. Fearful of isolation, they eyed each other warily and tried to calculate the threat posed by each new line, harmless in itself but lethal as a link in some fresh combination. The permutations seemed endless, a mathematics of frustration and despair.[2]

In so unsettled a state bold strategists pointed to self-contained systems as the surest refuge. If the territory could not be defended then it must be expanded, the points of battle multiplied, key outlets secured without reliance on outsiders. Weak adversaries must be neutralized, strong ones cowed into submission or that higher form of cooperation practiced by monarchs who divide neighboring realms on an amicable basis. An age newly introduced to Darwin could appreciate the spectacle of strong roads devouring weak ones until only the fittest survived. There were dangers in a policy of indiscriminate gorging, but these were as yet only dimly foreseen.

Years earlier, during his Erie tenure, Jay had grasped the essence of these forces and tried to implement them. The attempt failed for lack of resources, but now the gambit had given him the funds and freedom to act. Perkins shared his insight but not his freedom; Vanderbilt possessed ample resources, but without the vision or temperament to use them boldly he remained the sleeping giant, passive and docile until aroused. Though others aspired, no one else was in a position to assume the role Gould claimed for himself. The setting suited his needs perfectly. The securities of impoverished roads were cheap and easily obtained. He could boost their value by combining old derelicts into new lines, new systems. The market would follow his lead as a trader, convinced that he was up to his old tricks and sure to sell out at the opportune moment.

As usual the market would be wrong. The profits gleaned from rising securities would enable Gould to pyramid his holdings, would provide cash or collateral for more acquisitions. Traders often pyramided but for speculative purposes; Jay did it to erect a business empire. It helped immensely that his name worked magic on the Street. Prior to 1879 the dominance given Gould in speculative circles was astonishing to behold. Scarcely a week passed without some rumor, combination, maneuver, or attack being attached to his name. But all that had been paled in comparison with what was to be.

Clark wished to go to Europe that summer, so Jay obliged by booking him passage. Perhaps they would go together, if Jay could spare the time. Early in April the last obstacle to reorganization of the Kansas Pacific was cleared. News of the settlement stunned Wall Street. The stock, languishing at 22, rose from the dead to touch 60 at the month's end. While speculators reveled in the coup, Jay was busy with larger visions. "You doubtless notice we have been investing in Wabash," he informed Clark. "In fact we have secured through the Wabash a direct line from Kansas City & St. Jo to Toledo *independent of the Pool lines. . . .* When our Grand Island connection is completed we can take care of ourselves."[3]

Wabash. Here was another road with a past. An amalgamation of eighteen early corporations, the Wabash ran between St. Louis and Toledo. As the only St. Louis road reaching the Great Lakes east of Chicago it did a thriving grain business, but not enough to stave off receivership after the panic of 1873. In a region notorious for savage competition the Wabash emerged as a leader in cutting rates, which left it financially weak despite a large traffic. Reorganization in 1877 solved none of the road's chronic problems; it was physically and financially run down, and title disputes clouded its future. The stock became the plaything of speculators until November 1878, when Commodore C. K. Garrison scooped up the road.[4]

A shrewd, affable man of sixty-nine, Garrison amassed a fortune in banking, shipping, and utilities before venturing into railroads. As owner of both the Missouri Pacific and the Kansas City he controlled traffic between St. Louis and Kansas City. When the Alton shut him out of Chicago, Garrison acquired the Wabash and announced plans to build to that city. With the Kansas City extending toward Omaha, Garrison seemed on the verge of preempting Gould's fondest dream of a line from Omaha to Chicago independent of the Pool.[5]

The threat posed by Garrison galvanized Gould into action. His greatest advantage was the element of surprise. No one expected a tough, aggressive

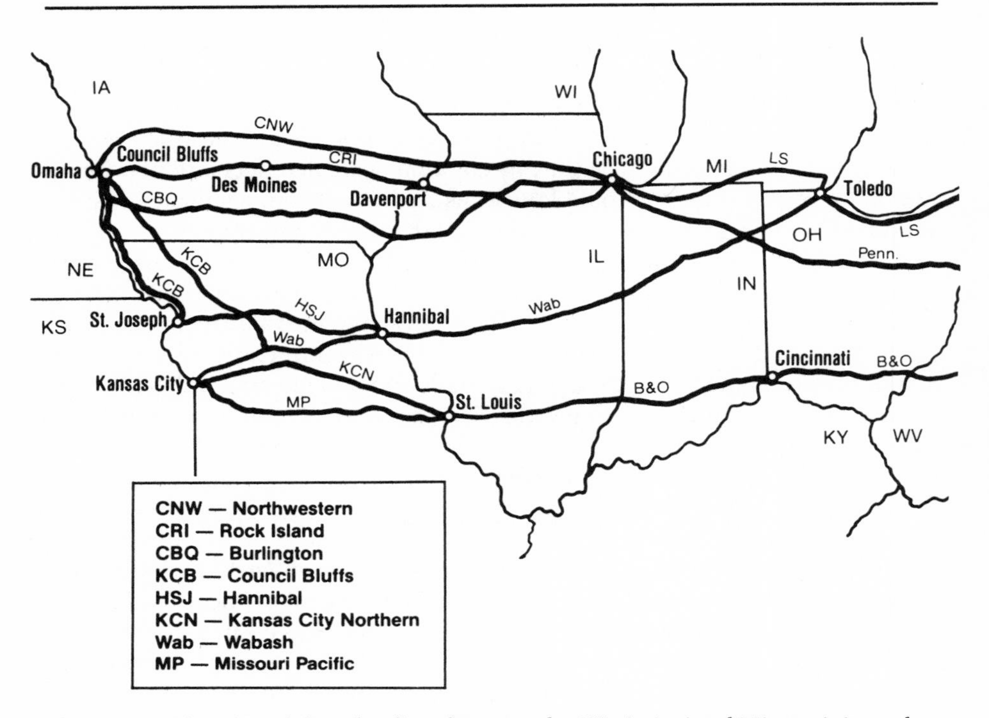

The Iowa Pool roads and the other lines between the Mississippi and Missouri rivers that became pawns in Gould's blitzkrieg.

campaigner like Garrison to sell out quickly, and no one expected Gould to turn up east of the Mississippi River. During the winter of 1879 Jay secretly bought stock at bargain basement prices. Early in April he induced Garrison to part with his holdings along with an agreement for use of the Kansas City line. Not until April 22, when Garrison resigned the Wabash presidency, was Gould's presence disclosed. The stock, which floundered below 19 as late as April 12, soared into the 30s, another speculative corpse sprung to life. Astonished financial editors, having predicted vast new combinations under Garrison, belatedly swung their trumpets toward Gould. The burning question was how he would handle connections eastward. Would he fight Vanderbilt or strike an alliance with him?[6]

All eyes were on Vanderbilt, who had within the past year threatened to withdraw from all rate pools after capturing the Michigan Central and the Canada Southern. In May 1879 he toured the West amid rumors that he was after both the Alton and the Northwestern. A month later his friends strengthened their position on the Northwestern board, moving Perkins to suggest that the Burlington seek closer ties with the Pennsylvania Railroad. Forbes scoffed that he was "not much scared at Vanderbilt's big schemes" and dismissed the Wabash as a mere "stockjobbing venture."[7]

While Vanderbilt traveled, Gould struck again on an unexpected front by organizing a new telegraph company, American Union, heralded as "part of a vast scheme that may extend from the Atlantic to the Pacific Coast." Later he claimed he did it because Western Union had reneged on a promise to make Eckert its general manager, "and I made up my mind that I would put this man at the head of as big a company as I had taken him from." As usual there was a trifle more to the story. One suspects more than coincidence in the fact that Jay chose to help his friend at precisely the time he acquired Wabash. Hard bargaining with Vanderbilt lay ahead, and useful leverage might be gained from a new competitor nipping at the heels of Western Union. To push expansion vigorously, Jay organized a construction company with Connor at its head and began exploring ways to break existing contracts between railroads and Western Union.[8]

Gould's entry into Wabash brought him new associates who were to remain with him for years, notably banker Solon Humphreys and A. L. Hopkins, a first-rate railroad man. The presidency was given to Cyrus W. Field, promoter of the first Atlantic cable, whose strong following in speculative circles and past association with Wabash made him a useful ally. Sage and Dillon also joined the board along with the venerable Joy, whose presence would attract English investors to Wabash securities. The creation of so formidable a management again refuted the image of Gould as a friendless predator. As the Wabash coup demonstrated anew, Jay relied on associates in everything he did.[9]

Joy was a case in point. He still championed the interests of Detroit and resented Vanderbilt's monopoly of traffic to that city. Gould lured him to the Wabash board by promising to extend the road from Toledo to Detroit "as rapidly as men & money can do it." Joy also controlled the Eel River, one of five small roads forming a short route from Detroit to Indianapolis. Vanderbilt's Lake Shore depended on this route as an outlet for southwestern traffic. An alliance with Joy, therefore, offered Gould a chance to bait Vanderbilt in his own lair. The Detroit extension would enable Wabash to connect with three roads that paralleled the Lake Shore. Two of them, the Canada Southern and the Great Western, reached Buffalo; the Grand Trunk ran to Portland, Maine. Vanderbilt controlled the Canada Southern but not the Great Western, and he had blocked efforts by the Grand Trunk to reach Chicago. He also owned both roads between Detroit and Toledo.[10]

Gould was quick to exploit the possibilities in this strategic maze. He invited the Great Western and the Grand Trunk to join him in building the Detroit extension and also approached Vanderbilt with an offer to lease one of his Detroit lines, arguing that it would be wasteful to construct a third line. Puzzled by Gould's recent activities and uncertain of his intentions, Vanderbilt rejected the idea of a lease but agreed to discuss a contract for use of trackage by all three companies. He had more than Detroit to worry about. The new Wabash board included an Erie representative, which sparked rumors that Gould planned to use his alma mater as an outlet to the seaboard. The Erie held nominal control of the Bee Line, a road on which the Lake Shore depended for connections southwest of Cleveland. For years the Bee Line had been the Wabash's mortal enemy; in hands friendly to Gould it could wreak havoc with the Lake Shore. Then there was Chicago. Jay appreciated the value of a line into that city as a weapon against both

Vanderbilt and the Iowa Pool, but how to get it? One possibility lay in a motley assortment of roads scattered about Illinois and northern Missouri. In June Gould acquired two of them for the Wabash. One provided another connection to Omaha, while the other could be extended to Omaha with only ninety miles of new construction.[11]

Slowly, subtly, Jay spun a web large and delicate enough to embrace not only Vanderbilt but the Iowa Pool as well. His snatching of small roads caught the Burlington by surprise, as did the revelation that he possessed the St. Joseph. In June he let contracts for building the Hastings extension that would link Union Pacific with the St. Joseph. That same month he finally gained control of the Kansas Pacific. By putting up $1.5 million of his own money he had become the only creditor outside the bondholders. To facilitate reorganization he created a new $30 million consolidated mortgage for which he and Sage served as trustees. The new bonds were exchanged for all outstanding Kansas Pacific issues and enough set aside to build a branch to Leadville if negotiations with Evans fell through. Some additional securities were deposited as collateral for the new bonds, including the nearly 30,000 shares of Denver Pacific stock. At the time it seemed an unimportant detail, an impression Jay was careful to leave undisturbed.[12]

The parade of sensations unfolded so rapidly the keenest observer could not digest their mysteries. Financial editors put together combinations faster than railroad men and ascribed most of them to Gould. Values on the stock market had advanced 20 percent since the first of the year, with Gould the acknowledged leader of the bull movement. Everyone agreed that further surprises were in store, but no one had the faintest hint of Gould's intentions. Jay did nothing to enlighten them. In the midst of the frenzy he decided after all to go to Europe with Clark, but he took pains to conceal even this plan from the Street until the last possible moment. He did not tell Clark that business no less than pleasure lured him abroad. Instead he merely observed, "I guess we have cut out about all the work our Secretaries can do while we are away."[13]

What was he up to? In Gould's absence friends and foes alike pondered the question of his intentions. Gould's scoop of the Wabash, carried off in complete secrecy, had caught Perkins looking west, where more danger was brewing. Palmer seemed on the verge of winning his fight to free the Rio Grande from its lease to the Santa Fe, which meant a renewal of the fight between them for the Royal Gorge, the gateway to Leadville. Gould was after the South Park and would get it unless someone intervened. At bottom it came down to whether the Burlington wished to rid itself of uncertainty by extending the B & M to Denver. "I doubt if we can long depend on any business not controlled by our own lines," Perkins argued, "& if that is what it is coming to we ought perhaps to go through to the mountains."[14]

Forbes disagreed. He conceded the "alarming activity in broken down roads" but preferred letting others have "the lame ducks and put good money into mending them, than to try and outbid them beyond any moderate limits." A better plan was to cement Burlington's alliance with Rock Island, which would allow them to "safely set at defiance any of Jay Gould's schemes of attack for he is still largely concerned in Union Pacific notwithstanding his real or pretended sale of large blocks of stock and this is *his* vulnerable point." As for Colorado, Forbes wanted to extend the B & M perhaps seventy-five miles up the Republican Valley, leaving a gap toward Denver that could easily be filled later.[15]

Perkins shook his head. Forbes's logic was impeccable, but his tactics were dubious. Lightning had departed and lightning would return; it was dangerous to sit still while Gould moved so rapidly. Too many roads like the Hannibal were still ripe for plucking. In Perkins's judgment the Burlington should not only build to Denver but "since all my efforts for 18 months to make an *alliance* with the Atchison have failed *I should favor a consolidation there also.*" Perkins realized the Santa Fe was too much for conservative men to swallow, but he was distressed at their refusal to nibble at his suggestions. He could not convince even Forbes.[16]

It would have been instructive to Forbes to know that Gould had in fact sold off most of his Union Pacific stock by June. His motives baffled even Dillon, who assumed like everyone else that Gould's recent acquisitions had been made on behalf of Union Pacific. If so, why was he selling, and if not, why had they been made? The only man blessed with some certainty was Villard, who knew at least what Gould was not doing. In June Gould dropped out of Villard's Oregon project. The ostensible reason was to avoid a clash with the Central Pacific, but in fact his decision offered a clue to the mystery that perplexed Dillon: while the Union Pacific looked northward, Gould had turned his eyes to the south and west.[17]

Jay took none of the family to Europe with him except George, because there was business to be done and it would do George well to learn something of it. At fifteen the heir had already commenced his apprenticeship, struggling to absorb lessons that taxed minds far more alert than his. In one exquisite transaction Jay provided him a textbook on economy of motion. He was after the bonds of the Denver Pacific and sent his offer to the owners in Amsterdam. "I supposed they would take a day or two smoking before they would make up their minds," he recounted later, ". . . and it would take more time than I had. Finally, I went over there and saw them. I got in there in the morning at 10 o'clock and washed and got my breakfast, and let them know that I was there, and they met me at 11 o'clock, and 12 o'clock I bought them out and paid them." No one at home or abroad could divine his motive for the purchase beyond the obvious fact that it enabled him to control the foreclosure recently ordered by the court. Months were to pass before events unraveled the enigma to an astonished Street.[18]

The rest of the journey was not so hurried. Gould paused in Ireland to admire the sparkling fresh countryside and endured a horseback ride through the wilds to visit the lakes of Killarney. Apart from the spectacular scenery, his eye was drawn to every ruin; he could hardly wait to get home and read up on Irish history. "The only drawback," he concluded of Ireland, "is the squalid poverty of the poor." After exploring Dublin and Belfast, which Clark especially wanted to see, they went first to Scotland, then to London. The rain drove them quickly to Paris, which Jay found light, cheerful, and full of Americans. He was the most conventional of tourists, roaming the monuments and galleries until they tired him out, and eschewing the night lights. Switzerland appealed to him more than any other place he visited. He relished its quiet beauty even though he thought the Alps inferior to the Rocky Mountains. "I am sure you would enjoy a summer in Switzerland," he told Helen. "I trust it is in store for us."[19]

Jay remained abroad nearly two months but his absence did not slow the pace of developments at home. Early in July the Wabash announced plans to merge with the Kansas City, creating a line from the latter city to Toledo with extensions to Omaha, Detroit, and Chicago. Coupled with the Kansas Pacific, the new Wabash outflanked the Iowa Pool roads and could wreak havoc with the existing flow of traffic. In acquiring control of the Wabash and the Kansas City Gould had defanged Garrison as a challenger. Before his departure he had blunted efforts by the owners of the Council Bluffs and the A & N to procure local aid for extending to Columbus on the Union Pacific line. Clark was told in no uncertain terms to make life *"red hot"* for the interlopers: "You cannot go for that concern too strong & I hope you will show them that it is not profitable to get into our territory. I think you should use them up before you get through with them. Please consider yourself ordered to do so."[20]

It was Gould's way to stress the urgency of a matter by repetition, as if words would flail his foes into submission. His ferocity on paper contrasts sharply with his quiet, almost meek manner of doing business in person. On the question of invaders he was adamant. Some days later he reminded Clark to "leave no stone unturned to kill the A & N prospective for bonds. . . . We have built up Columbus & can destroy it & build up a rival town (as we did at Lone Tree or started to do)."[21]

Vanderbilt received the same hardnosed treatment. After stringing out negotiations for weeks, he took refuge in a technicality as grounds for not leasing one of his Detroit lines. Gould responded with a cable ordering the Detroit extension built at once. In short order the Wabash leased the strategic Eel River and agreed to pool earnings with the Bee Line. Loss of the Eel River made it impossible for Vanderbilt to invade Wabash territory without constructing a new line. Wabash then demanded and received from the Trunk Line pool a Toledo rate on California traffic proportional to that given Chicago and St. Louis. One observer saw in this "recognition of the obvious fact that the Union Pacific now extends to Toledo." The Bee Line pool ended a long and bitter rivalry and brought the Wabash into friendly relations with the Pennsylvania.[22]

While Vanderbilt licked his wounds, Forbes groped for a policy to counter "this carnival of bankrupt Roads." For months rumors had circulated that Gould was after the Hannibal; Forbes did not want him to get it but could not bring himself to bid for the road. The truce between Union Pacific and B & M had become a powder keg awaiting a match. Perkins talked peace with Dillon, then allowed his traffic manager to launch a rate war on cattle shipments. Dillon was furious at what he deemed Perkins's treachery but hesitated to negotiate a settlement until Gould and Clark returned.[23]

Home Jay came late in August to a dithyramb of rumor. Dodge heard he had not only bought the Denver Pacific but had arranged for "a through line to the Atlantic." As the Omaha extension neared completion skeptics examined the Wabash-Kansas City consolidation and wondered "whether Mr. Gould intends to use it as he did Atlantic & Pacific Telegraph, for breaking rates on competing lines, until they are compelled to buy up the opposition." Jay did not let their curiosity linger; when the gangplank lowered, he hit the ground running. While events in eastern territory held center stage, he struck suddenly in the West. During his absence the court had invalidated the Rio Grande lease and appointed a receiver for the road. A disgruntled Santa Fe appealed the decision and began work to parallel the Rio Grande into Denver. Before it could act, however, Gould and Sage bought from Palmer a half-interest in the Rio Grande, loaned the road money, and induced Palmer and Evans to join in building toward Leadville. Gould increased his holdings in South Park and the construction company organized by Evans to extend the road.[24]

His timing was immaculate. Jay had deferred action in Colorado until the lease dispute was resolved, then moved in time to shut the Santa Fe out of both Leadville routes. When the mining boom reached its peak that fall, the securities of both roads soared in value. The Santa Fe was forced on the defensive in both Colorado and Kansas, where Gould added some small lines to the Kansas Pacific. His influence in the Colorado roads promised a growing volume of traffic for the Kansas Pacific just when he needed that road as a bargaining chip.[25]

Every transaction heightened the mystery concerning Gould's ultimate intentions. The pieces for a colossal system were being gathered, but they did not fit together well and no one understood Gould's relationship to each of them. Most observers assumed that Union Pacific must form the heart of any Gould system, although in September one concluded that "Mr. Gould's recent acquisitions in the Southwest are believed to have made him comparatively indifferent about the control of the Union Pacific, that no longer being a prime necessity for his transcontinental scheme."[26]

This remarkable prophecy missed the mark only in assuming that Jay would utilize the Kansas Pacific as his link between Colorado and the Missouri River. In fact Gould conceived a much deeper plan involving nothing less than a radical transformation of his position. Few military campaigns rival the scale and audacity of his vision or the decisiveness with which he executed it. So skillfully did his scheme unfold that no one grasped its design until the last piece fell into place. The first step lay not in using the Kansas Pacific but in merging it with the Union Pacific.

Few episodes in Gould's career have been more mangled by myth than this consolidation. Two generations of muckrakers howled indignantly at what they deemed the apotheosis of fraud; later historians, while conceding the technical brilliance of the maneuver, dismissed it as another predatory grab for profits. Certainly Gould made money from the deal (though not the huge sums claimed by his critics), but his most important returns came in less tangible form. The obsession with profits has blinded historians to Gould's motives and prevented them from exploring a more fundamental question: What role did the consolidation play in the broader scheme of his affairs?[27]

The notion of a consolidation had been abroad since 1875, when Gould first tried to merge the roads and their Colorado branches. For years his twin objectives had been to secure for Union Pacific a line east of Omaha independent of the Iowa Pool and to eliminate the Kansas Pacific as a competitor. In the newly merged Wabash, which connected with the Kansas Pacific at Kansas City, he had forged a line from Toledo to Denver that outflanked not only the Iowa Pool but part of the Union Pacific itself. One salient fact unlocks the riddle of all that followed: Gould wanted a rail system of his own creation, free from the government or contrary boards of directors or shared control of any kind. A financial killing would do Gould no good unless it advanced him toward this goal as well. However the world viewed him, Jay saw himself as a builder, a pioneer of western development, blazing the trail of progress not with the plow but with rails, mines, commerce. The empire he fashioned was a monument to himself, like Zadock Pratt's carved rocks, and a legacy for the sons who were to follow in his steps.

But what was that system to be? He had already dismissed Union Pacific as a possibility for all the reasons discussed earlier. Although the Kansas Pacific's prospects had improved, its legacy of debt and mismanagement would remain a handicap for years. Moreover, a system built around the Kansas Pacific would throw Gould into direct competition with the men who had been his closest associates for years. Sentiment aside, it was neither sound policy nor shrewd strategy to alienate his closest allies. A wiser course would be to merge the two roads and use the profits to jump into another competitive arena altogether, disguising every step in that direction to resemble part of some other scheme. This way he might retain not only his friends but his influence in Union Pacific, which could prove useful.

In September 1879 Jay proposed merging Union Pacific, Kansas Pacific, and Denver Pacific on the basis of an equal exchange of stock. The Union Pacific board still wanted the consolidation but balked at the terms. When the meeting broke up, Jay said softly, "Gentlemen, you are making a mistake." He was about to go west; Dexter recalled that he "had his war paint on and his trunk in hand." On the trip Gould inspected not only the Kansas Pacific but the Central Branch and the Missouri Pacific as well. When he returned in October negotiations resumed, but neither side would budge. Humphreys and Dodge were then appointed a committee to "prepare a plan for the equitable consolidation."[28]

To Gould's surprise the opposition included Dillon, who told Sage that the only fair rate of exchange was two shares of Union Pacific for three of Kansas Pacific. Sage repeated these views to Gould, who conveyed them to Clark in a tone at once sorrowful and ominous:

> I really think the value comparatively is the other way . . . & if there is any difference of opinion I shall withdraw the offer to consolidate & shall close my connection with UP & let the future demonstrate whether my offer & my plan was a wise one. I feel very much hurt that Mr D should not have frankly expressed his views to me rather than behind my back.

Clark had just applied to Gould for reassurance after a misunderstanding with two subordinates, and he must have felt uneasy at being forced to choose between his protector and his company. In many respects his position was the strangest of all. Although Jay insisted he was "entirely satisfied with all you have done & no one can give me an erroneous impression as to your acts," Clark played ostrich. As a Union Pacific director he refused to act on any question pertaining to the consolidation. Tired and distraught, he fell ill again in December and fled to California for a long rest.[29]

The situation was a delicate one, yet the rift between Gould and his associates was never as great as it appeared in retrospect. Everyone agreed the consolidation was desirable; the dispute was over terms, not policy. Even Dillon stressed the need to settle the matter quickly. "I think the longer it stands," he told Fred Ames, "the more complicated it may get." The Boston directors hesitated partly because they were preoccupied with other threats. Dexter feared the Burlington might build to Denver, and the Santa Fe was still making noise in the same direction. Garrison owned some small Kansas roads that could be used to extend his Missouri Pacific toward Denver. Governor Oliver Ames, who controlled the Central Branch, was also building westward. Harassed by these enemies, the Boston directors assumed that the Kansas Pacific would neither expand nor become unfriendly while in Gould's hands.[30]

Jay exposed the fallacy of this thinking in another series of lightning strokes. He seized first on an improbable target, the Central Branch. Its stock had sold as low as fifty cents a share prior to a reorganization in 1877 under R. M. Pomeroy. By 1879 the road belonged to Governor Oliver, who bought control at prices ranging from 1 to 125. He and Pomeroy conceived the idea of building toward Denver in hopes of attracting a buyer. After starting construction Ames flirted with Garrison, Perkins, and the Rock Island before approaching Gould.[31]

Much has been made of the way in which Ames and Pomeroy snookered Gould by holding back freight trains until his arrival to create the illusion of an immense traffic. Gould saw trains nosing out of every station and later admitted to being "wonderfully impressed with the amount of business." The tale is amusing but beside the point. Jay would have bought the Central Branch if he had seen ghost track because the road was crucial to his plans. On October 16 Ames visited Gould in New York and suggested it was better "to buy us out than it was to fight us." He offered a majority of the stock at the outlandish price of $300 a share. They met again on November 6 for some hard bargaining and the next morning came to terms in half an hour. Ames noted in his diary that "Gould & Sage were delighted with their trade & so was I." He should have been ecstatic, for ultimately Gould bought 7,616 shares at an average price of 240.[32]

"Just where this Gould combination is going to stop it is not easy to see," lamented Perkins on November 8. That same day he was jolted by news that Gould and Sage had been elected to the Hannibal board. Five days later Gould dropped another bombshell by purchasing the Missouri Pacific after trading tough talk with Garrison at what Jay later described as a "stormy interview or two." Garrison threatened to push his Kansas Central westward; Gould replied that he would extend the Kansas Pacific to St. Louis. To this Garrison had no response. Having lost the Wabash, the Kansas City, and the Central Branch to Gould, he was in no position to fight. "You had better buy me out," he scowled at last; "that is the cheapest way." Garrison exacted a steep price: $3.8 million for 4,000 shares (half the stock) or $950 per share. The purchase included the Kansas Central and another small Kansas road.[33]

The Missouri Pacific deal triggered a rumor that Gould was about to buy half a dozen other roads as well as merge Union Pacific with Kansas Pacific, but a week later it vanished behind the shadow of the most sensational transaction yet witnessed on Wall Street. On November 21 the *Tribune* broke the story that Vanderbilt planned to sell 200,000 shares of his New York Central to a syndicate headed by Drexel, Morgan and the Wabash trio of Gould, Sage, and Field. Vanderbilt hastened to assure reporters that no such negotiations were in progress, and Field issued a denial. Panic swept the market as the rumor hovered like a dark cloud, refusing to pass or unleash its storm, until the sale was officially announced on the 26th. Vanderbilt gave the syndicate 150,000 shares at 120, with an option for another 100,000 shares at the same price. Behind the London firm of J. S. Morgan & Company, which subscribed for 50,000 shares, Gould was next in line with 20,000 shares. Among his friends Sage took 15,000, Field 10,000, Fred Ames 5,000, Dillon 5,000, and Humphreys 5,000.[34]

Speculation ran rampant on why Vanderbilt had sold, and why he sold to the "Wabash syndicate." One broker noted shrewdly, "I suppose he wanted peace." Certainly that was his reason for allowing the Wabash crowd so prominent a place. Gould was pressing him on two fronts, the railroads and the telegraph. Where the Commodore would have met Gould's tactics in kind, his son snatched eagerly at the olive branch. Vanderbilt did not want Gould to participate but relented when Jay threatened to divert eastbound Wabash traffic onto the Baltimore & Ohio. "It was a choice," he explained, "between continuing the competition for western connections and making its members my friends. I thought it wise to do the latter." Financial editors united in the belief that the deal signaled an alliance between Vanderbilt and Gould; one went so far as to declare it a "practical consolidation of their interests." The *Tribune* called it "the most powerful railway combination ever known."[35]

That impression was precisely the one Gould wished to convey. Participation in the syndicate brought him prestige, profits, and power of a sort he needed most at that moment. It did not hurt his reputation to travel in the company of such distinguished bankers; men who had shunned him in the past might be more amenable to extending him credit in the future. Early in 1880 the impression of a Gould-Vanderbilt rapprochement was strengthened by a traffic agreement between the Wabash and the New York Central.

At the same time Gould struck a decisive blow in Colorado at the Santa Fe's threat to parallel the Rio Grande into Denver. He announced plans for a new road to parallel the Sante Fe from Pueblo to Fort Dodge, Kansas, where short extensions would connect it with the Kansas Pacific. This surprise attack on an exposed flank brought the Santa Fe's management to heel; in February it signed a treaty resolving all differences with the Rio Grande.[36]

The road was never built, but its shadow was sufficient to subdue the Santa Fe and swell the value of Jay's Rio Grande holdings. The certificates he had purchased at 22 in the fall jumped to 75 in February 1880. There was yet another benefit, for Gould had in his usual way made one move serve several purposes at once. The shadow that threatened to parallel the Santa Fe would, if constructed, also parallel the Union Pacific.

Perkins watched in helpless consternation as Gould closed the ring around the Burlington. The Wabash had opened its Omaha extension and would soon enter Chicago. Gould had snatched the Hannibal and was pursuing the Council Bluffs and the A & N as well. Rumor had him buying into the St. Louis & San Francisco, the Santa Fe, even the Texas & Pacific, where Tom Scott needed all the help he could get. It alarmed Perkins that Gould had "drawn in Vanderbilt including probably the Chicago & Northwestern. I am afraid he may go further & get the Rock Island or the Alton or both into his combination." Rumor could be discounted, but Perkins had done that before only to be proven wrong. "The acquisition of so much Railroad property ought to make Gould conservative," he mused, "but how far he will try to use his power to whip the B. & M. & the Atchison into subjection I do not feel sure."[37]

The immediate danger, he warned Forbes, "is that the rapidity & brilliancy of Gould's movements will frighten all of the New York Roads into surrender. The Rock Island people have slept with him before & while perhaps they did not like it still they have shown that they can stand it!" Perkins knew what had to be done: "There ought to be a counter combination." Since Gould could control the business over his own roads, the Burlington must beat him into the Rock Island's bed and invite the Santa Fe in as well. Above all the Burlington should acquire feeders to help itself and prevent Gould from getting them. Already Forbes was negotiating for the bankrupt Missouri, Kansas & Texas, but the Dutch bondholders wanted too high a price for him. At the very least, Perkins argued, they should grab the Council Bluffs, but Forbes considered it less a feeder than a sucker.[38]

Perkins had good reason to feel frustrated. He could not convert even Forbes to his views, let alone the Burlington and B & M boards, which had dawdled for months before bringing themselves to approve consolidation of the two roads. Old habits and conservative views do not change easily—Perkins understood that, but he also knew that Gould could not be fought with obsolete weapons and hoary tactics. In his dogged manner he prodded Forbes with the reminder that "Gould moves so rapidly it is impossible to keep up with him with Boards of Directors."[39]

Perkins was not alone in feeling the noose tighten. By January 1880 the Union Pacific directors realized that Gould had them cornered. Governor Oliver Ames, having pocketed his profits, watched the spectacle with disinterested relish. "I saw Sidney Dillon and Dexter and Fred Ames as gloomy and unhappy a set of men as I ever saw," he recalled. "Mr. Gould had them in his power." Jay knew they had no choice but to accept his terms, which were precisely those recommended by the Dodge-Humphreys report. Later Gould insisted that he did not influence their conclusion. No influence was necessary, for by then he was in a position to dictate terms regardless of what the report said.[40]

The timing of the extension threat was perfect. Earnings on Gould's western roads remained high, thanks to the mining boom and bumper crops coupled with short harvests abroad. On a rising market led by what had come to be called "southwestern fancies," Kansas Pacific soared to 97½ on January 9, eleven points higher than Union Pacific. As the gap widened, it grew harder to dispute Gould's argument for an equal exchange. Even the Burlington played into Gould's hands by completing its merger with the B & M and threatening extension to Denver. Forbes's hope of curbing Gould succeeded only in alarming Dillon, who summoned Fred Ames and Dexter to New York. The Burlington's action, he declared, "makes it necessary for us to do something immediately and to determine upon some plan to meet them." The "something" was consolidation with the Kansas Pacific, which now seemed imperative; otherwise the Union Pacific might find itself with two formidable competitors south of its line. The Kansas Pacific held another trump card in the dormant pro rata suits, which could be revived at any time.[41]

"It was always a bugbear, this Kansas Pacific," sighed Elisha Atkins, and now it had them. The Boston directors understood clearly what seems to have escaped later critics, that the danger confronting them was very real. They knew Gould was not bluffing; if the consolidation failed, he would complete his own line and, regrettable as it might be, compete with the Union Pacific. Whatever the directors thought of Gould as a friend, they shrank in terror from the prospect of him as a rival. "A man of Mr. Gould's ability," exclaimed Dexter, "—for he is the first man of the country on that subject—with such a weapon as the Kansas Pacific, let alone his Missouri Pacific, could have built branches and cut rates and cut us all to pieces."[42]

In their anxiety the directors did not wait for overtures from Gould but went to him and insisted that the merger be consummated. On the evening of January 14 Sage, Dillon, Fred Ames, Baker, and Dexter came to Gould's home for a meeting that lasted until midnight. Although their recollections differ somewhat on details, they agree that the negotiations were protracted but not rancorous. Dexter found that Gould's earlier irritability had given way to a "pleasanter frame of mind" and that he was more "open to reason." Jay demurred that the consolidation was no longer in his best interest. His holdings in Union Pacific were reduced to 27,000 shares, a sizable amount but small compared to his investment in the other lines. He had resigned as a Union Pacific director four days earlier and was intent on erecting his own system, for which he needed the Kansas roads. The Bostonians reminded him of his long association with Union Pacific and insisted he "was bound by previous conversations" to go through with the consolidation. Sage sided with Gould, that being where his own interest lay.[43]

How Jay must have relished watching the Bostonians beg for the very terms they had shrugged off earlier; how he must have savored getting precisely what he wanted in the guise of yielding generously to the importunings of his friends. When at last he yielded, Dexter scribbled the agreement on a sheet of paper. It stipulated that Union Pacific, Kansas Pacific, and Denver Pacific stock be exchanged *at par* for shares in the new Union Pacific *Railway* Company. The $4 million in Denver Pacific stock, once converted into Union Pacific shares, would be used to buy from Gould the St. Joseph railroad and bridge at par for the bonds and 20 for the stock. When the Bostonians insisted on having the Central Branch and the Kansas Central, Jay agreed to sell them at cost provided the roads were then leased back to the Missouri Pacific. This arrangement allowed him to use the roads but prevented the building of extensions or branches.[44]

At long last the light dawned on why Gould had gone to such lengths in acquiring Denver Pacific securities. In one stroke the agreement gave value to what had seemed worthless stocks and bonds if certain legal hurdles could be jumped. The Denver Pacific stock was still deposited as collateral for the Kansas Pacific mortgage, the trustees for which were Gould and Sage. It could not be used in the consolidation without first being released from trust. Normally such proceedings moved at a glacial pace, but in this instance suit was brought at once against the trustees and somehow "the entire machinery of complaint, answer, trial, decree, and execution was carried to a finish" in only ten days. When it was discovered that the Union Pacific could not acquire the Denver Pacific while holding the Colorado Central under lease, the lease was promptly canceled.[45]

The way was now clear for formal action. On Saturday afternoon, January 24, while Jay sat waiting in his own office down the hall, the Union Pacific board accepted Gould's resignation, then approved the consolidation. On Wall Street, already inflamed by rumors that something big was about to break, Kansas Pacific touched 98 and Union Pacific 94½ that day. Those with long memories recalled that a year earlier Kansas Pacific stood at 9⅛, a fact that moved one editor to dust off his Latin: "*ex nihilo nihil fit.*" For his various holdings Jay received about 73,300 shares of new Union Pacific worth about $6.7 million at current prices.[46]

A century of criticism has been heaped on this transaction. The tone was set in 1887 by the Pacific Railway Commissioners, who concluded "that Gould and his associates had planned the consolidation primarily for profits in securities." In the process, later historians charged, he saddled the Union Pacific with a swollen capitalization and a motley collection of unprofitable railroads. In time these assertions became frozen in myth although they were never proven and rest largely on inference or circumstantial evidence. For that reason they are worth a closer look.[47]

Henry Villard gave later generations their benchmark for profits by asserting that Gould "cleared more than ten millions of dollars by the operation." Curiously, no one has attempted even a rough calculation of what Gould actually made. There are considerable data, not enough for a precise figure but more than enough to correct the unfounded numbers that have passed down through the years. Gould made nothing on the Central Branch or his $1 million in Denver Pacific stock, both of which he turned in at the cost to him. His profits in Denver Pacific came from the bonds, of which he bought about $2 million worth in Amsterdam at 74. These were exchanged, presumably at full value, for the new Kansas Pacific consolidated bonds. The difference in values constituted Gould's profit, but he did not keep the entire $2 million for himself. For example, he let Sage have $353,000 of the bonds at 74, and he may have sold more to other friends. That was his usual way of conducting operations, contrary to the image of him as a lone wolf.[48]

The other properties "unloaded" on Union Pacific were the Kansas Central, the St. Joseph, the St. Joseph bridge, and the Hastings & Grand Island. The two St. Joseph interests he shared with Sage, Clark, Dillon, and his Boston friends, who took a total of 30 percent of his holdings. For his 70 percent and the other two roads, Gould's own accounts show he paid $1.55 million and received in the consolidation 24,502 shares of Union Pacific worth $2.45 million at par. But Union Pacific was not selling at par on February 16, when Gould obtained the stock. At current market prices the shares were worth about $2.25 million, which left Jay a profit on the branch roads of approximately $700,000, a tidy sum but hardly the stuff of legend.[49]

What of the Kansas Pacific itself? Gould acquired 43,393 shares at 12½, or $542,416, and exchanged them for 40,382 shares of Union Pacific worth $3,715,144 market value. In addition, he owned $1,254,130 in Kansas Pacific securities (his share of the 1878 pool) and bought $1,351,867 in junior bonds from pool members for $598,284. These were later exchanged for Kansas Pacific consols, presumably at full value. Jay thus cleared nearly $3.2 million on the stock and an unknown but probably large amount on the bonds. Here too the profits were shared; Sage had a sixth interest in the pool securities and others held lesser amounts. Nevertheless, Gould reaped an enormous return on his Kansas Pacific holdings, possibly as much as $5 million. Yet he insisted years later, "I did not make any [money] out of the consolidation, because these securities were worth as much before the consolidation as they were after."[50]

Gould spoke the truth in his usual careful way, but of course no one believed him. Blinded by their obsession with the question of profits, critics missed the whole point of the transaction. Jay derived his profits not from the consolidation but from the fabulous rise in value of Kansas Pacific securities during the months prior to the consolidation. The agreement served merely as the vehicle by which Gould sold his holdings at peak prices. A more relevant question, then, is who or what created the rise in value. By any measure the credit belongs to Gould. He bought Kansas Pacific securities from men who would never have sold so cheaply had they believed them to be worth anything more. By reorganizing the road and acquiring other properties he transformed the Kansas Pacific into the formidable nucleus of a new transcontinental line. He had the good fortune to do this at a time when business was good and prospects better. The revitalized Kansas Pacific was still financially weak but, in Grodinsky's words, "as a threat to its competitor it had the strength of Samson."[51]

Jay understood this perfectly, and so did other investors. Kansas Pacific soared in value not for what it was but for what it might become in the hands of a man with Gould's ability. Those who attribute the rise to the merger rumors overlook the fact that Kansas Pacific had already reached 85 by October 30, well before the rumors surfaced. Gould the market magician nudged the stock upward, but it was Gould the railroad man who made the road a force to be reckoned with. In bringing these dead properties to life he created a system of his own, but it was not the one he wanted. For that reason, and because the price was right, he was willing to sell.[52]

To a man the Boston directors insisted afterward that they had acted to protect the Union Pacific, that the price paid was reasonable, and that, in Atkins's words, "without the consolidation, the road to-day would not have been worth a great deal of money." They seconded Baker's view that the transaction was "not forced upon us by Mr. Gould." Indeed Baker went so far as to defend Gould's conduct as "very proper" because "the rest of us were trying to bring about exactly what we persuaded him to do." When Gould was later asked what the effect of his system on the Union Pacific would have been, he replied simply, "It would have destroyed it." The Bostonians were convinced of Gould's ability to do just that; at least they were unwilling to gamble at high stakes that he could not. One may question their judgment but not their motives. In retrospect it is hard to argue against the decision, for everything in Gould's later career suggests that he would have done precisely what they feared he would do.[53]

While the consolidation served Gould well, it proved a mixed blessing for Union Pacific. A disastrous competition had been averted, but at heavy cost. The burden of an enlarged capitalization plagued the company during the competitive wars of the next decade, and critics were quick to seize on the consolidation as the cause. In fact the financial woes of the Union Pacific had deeper, more complex roots. The road had always been overcapitalized, and it suffered more harm from the endless dispute with the government than from any acquisition it ever made. The policy of buying rival lines in self-defense would soon become the standard survival tactic of American railroads; Union Pacific got in on the ground floor and did not do too badly with its acquisitions. The Kansas Pacific and the Denver Pacific remained an integral part of the system. After a slow start the St. Joseph properties proved valuable as part of a short line to St. Louis, the Central Branch and the Kansas Central less so as local agricultural branches.[54]

As matters stood in January 1880, however, the decision seemed both rational and sound. The Union Pacific had freed itself from the tyranny of the Iowa Pool and could deliver its traffic to Gould's Missouri Pacific-Wabash system at Kansas City. Two days after the consolidation the new company held its first meeting and restored Gould, a large stockholder once again, to the board and executive committee. On the surface little had changed, but in fact much had changed. Those who thought that Jay might ease up after the consolidation could not have been more wrong. The Gould whirlwind had not slowed but merely shifted direction. The first clue to its new thrust came a few days later with the announcement that he had become president of the Missouri, Kansas & Texas.[55]

Young men of ambition: Jay Gould (*right*) with his Roxbury neighbor, Hamilton Burhans (courtesy Lyndhurst Archives).

Above, left, the kindly visage of the "speculative director," Daniel Drew (Brown Brothers); *above, right,* the philosopher king at rest: Charles Francis Adams, Jr., long after his writing of *Chapters of Erie* and his miserable experience as president of the Union Pacific (courtesy Union Pacific System); *immediate left,* the Commodore, Cornelius Vanderbilt, wearing an expression that Gould doubtless saw on more than one occasion (Brown Brothers). *Facing page, top,* Sidney Dillon and General Samuel Thomas strolling together at Jekyll Island (courtesy Lyndhurst Archives); *bottom, left,* the outsized Jim Fisk in a moment of rare repose (Brown Brothers); *bottom, right,* Russell Sage follows his fortune on an early ticker tape (Brown Brothers).

Above, left, Silas H. H. Clark, Gould's most trusted lieutenant, who served his friend loyally for eighteen years and was enriched by the relationship in every sense (courtesy Union Pacific System); *above, right,* H. M. "Hub" Hoxie, who served as Gould's chief operations man on the Missouri Pacific and handled the strike of 1886 (courtesy Iowa State Department of Archives and History); *immediate left,* Giovanni P. Morosini, Gould's faithful secretary and bodyguard (courtesy Lyndhurst Archives).

The intractable Henry Villard and friends in Oregon, 1883 (courtesy Union Pacific System).

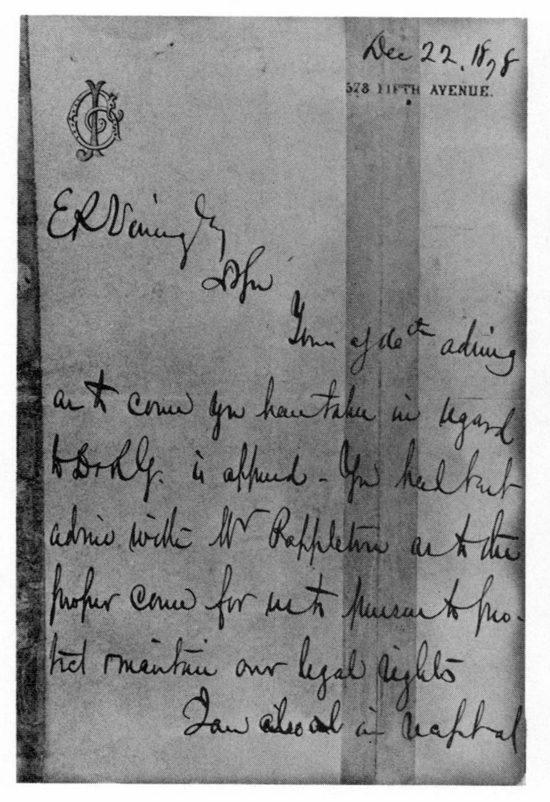

Dec 22. 1878
578 FIFTH AVENUE.

A "blue jay" reaffirming Gould's emphasis on cultivating local business (courtesy Union Pacific System).

578 Fifth Avenue

December 22, 1878

E. P. Vining, Esq.
Dear Sir:

Yours of 16th advising as to course you have taken in regard to D. & R. G. is approved. You had best advise with Mr. Poppleton as to the proper course for us to pursue to protect and maintain our legal rights.

I am also in receipt of the classified statement of freight earnings—the local growth of our company is a very gratifying feature. Our policy must be to build up and develop a local business which cannot be diverted or taken away.

Yours truly,
JAY GOULD

Wall Street looking west from Broad Street, with Trinity Church looming in the background, 1895. Note the elaborate street clock (*left*) and the Commercial Cable office on the same corner. Note also the spider's web of telegraph wires (courtesy of the Museum of the City of New York).

An Erie business car pauses during an inspection tour. The small, bearded figure on the platform appears to be Jay Gould (courtesy Smithsonian Institution).

An interior shot of the *Atalanta*, Gould's private car (courtesy Smithsonian Institution).

Some families pose with their favorite rolling stock, a pay car, at Hays, Kansas, 1875 (courtesy Union Pacific System).

An engine and crew on the Kansas Pacific, 1887. The pride of the men in their machine is obvious (courtesy Union Pacific System).

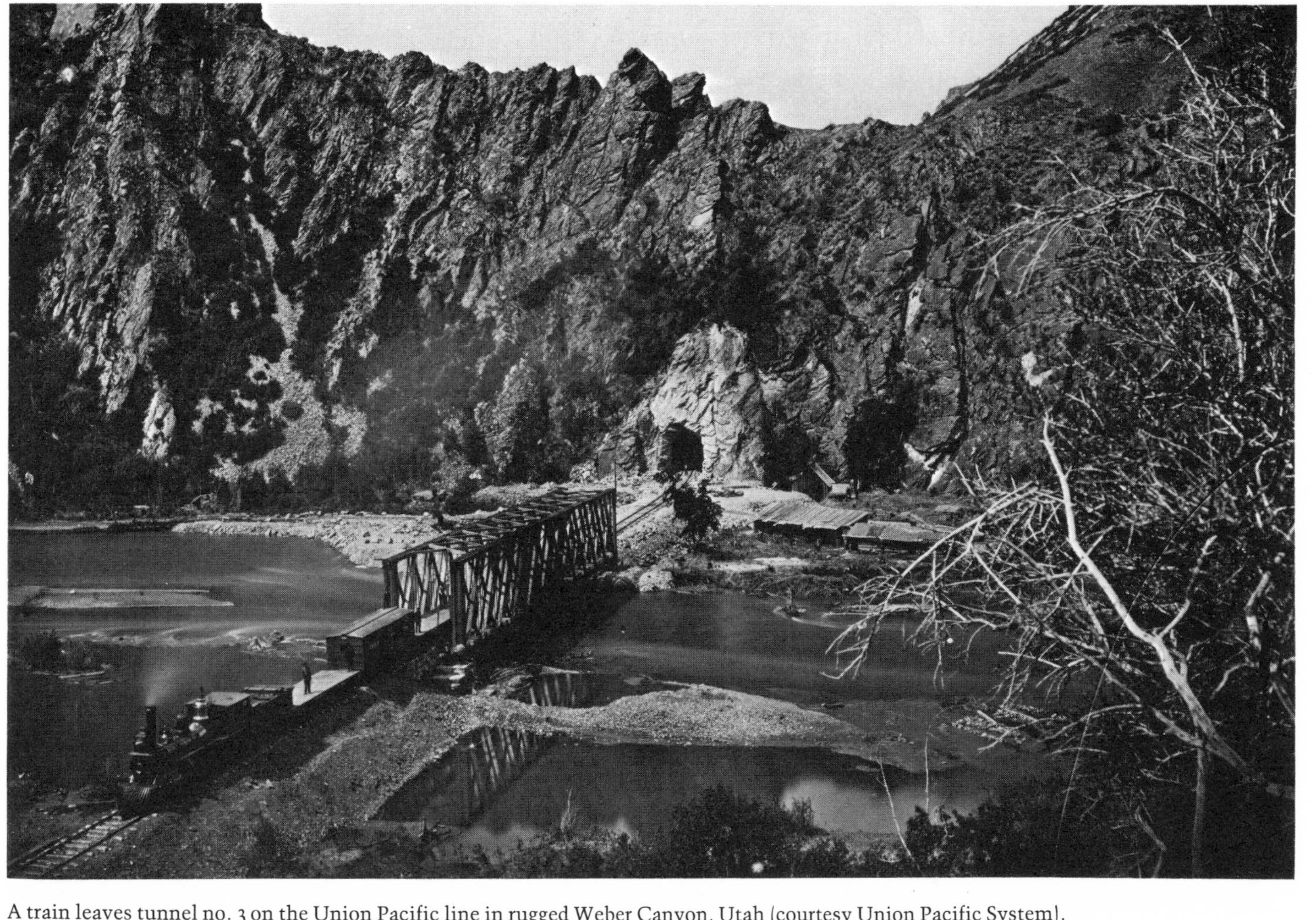

A train leaves tunnel no. 3 on the Union Pacific line in rugged Weber Canyon, Utah (courtesy Union Pacific System).

Rails south: an Iron Mountain construction train at work before a crowd of curious onlookers (courtesy Union Pacific System).

21

Phantasmagoria

The greater part of his empire, even at the peak of his influence in 1881, was kept together by minority holdings, and occasionally by no holdings at all.

—Julius Grodinsky, *Jay Gould: His Business Career, 1867–1892*

It is not six months since one of the three greatest operators in Wall Street . . . said contemptuously: "Nothing that Gould undertakes will ever amount to anything." . . . we have from the first assumed Mr. Gould's ruling idea to be constructive instead of destructive.

—*The Public*, April 15, 1880

The essential difference between our methods and those of the Union Pacific management is that we are straightforward and say what we mean and do what we say we will, while they are not the one and do not the other.

—Charles E. Perkins, letter to John Murray Forbes, March 7, 1880

By the winter of 1880 Gould had emerged as one of the leading railroad men in the country. Financial editors tumbled over one another in paying homage to the new "railroad king" who by one calculation directly controlled 8,168 miles of road, "the largest combination of roads in the control of any one individual or corporation in the world." For all their musings, however, the grand design eluded them. Wiser heads like Perkins and Huntington, who had long since learned to take Gould seriously, did no better than the editors in divining his true purpose. Gould's holdings were too diverse and sprawling for anyone to know where the heart of his system lay. He had long been identified with the Union Pacific but the consolidation had blurred that connection. His interest in Wabash was shared with others, and few traders believed he was in the road to stay. The Missouri Pacific, which he owned outright, was too small to be more than a link in some larger system, and other roads appeared to be little more than pawns in the game.[1]

Whatever their other interests, most railroad men were identified with some primary road: Vanderbilt with the New York Central and the Lake Shore,

Forbes with the Burlington, Huntington with the Central Pacific and the Southern Pacific, Boston groups with the Union Pacific and the Santa Fe. For several years this was true of Gould as well, but no longer. Did he still wear the badge of Union Pacific or had he discarded it for something else? If the latter, what was the something else? While Wall Street and railroad men alike agonized over these questions, Gould compounded their confusion by popping up on two new and unexpected fronts: the Katy and Texas & Pacific roads.

The Katy extended from Hannibal through the Indian Territory to Denison, Texas. Its main line crossed both the Wabash and the Missouri Pacific with a branch to the Kansas Pacific, giving the road a peculiar *Y* shape. Despite a promising business, the Katy lapsed into receivership in 1874 and languished through the depression years. Like most southwestern roads it was saddled with a heavy debt. Control of the bonds rested with a committee of Dutch bankers whose cooperation held the key to reorganization. Most likely Gould visited them during his trip to Amsterdam that summer, but no suspicions were aroused. The Burlington had long interchanged traffic with the Katy and was expected to dominate its reorganization. However, Forbes and the Dutchmen came to loggerheads over the rate of return on the bonds. The bankers wanted a guarantee of 5 percent, which Forbes thought high. In his usual cautious manner he wished to dissect every angle of possibility before acting. When the Wabash entered a higher bid, Forbes reasoned that the bankers would prefer 4 percent from Burlington to 5 percent from the Wabash. He was wrong. The bankers rejected his offer, whereupon the Wabash offered 6 percent.[2]

While Forbes fiddled, Gould burned him with another flanking raid. Quietly he bought Katy stock until it soared from a low of 6 to 35 in November. Puzzled observers attributed the rise to the bidding war for Katy bonds. "It would require a very powerful microscope," snorted the *Tribune*, "to discover one-tenth of that price in the present value of the stock." The light dawned in January 1880 when Jay controlled the Katy election and became its president before any deal had been made with the bondholders. The Dutch bankers still threatened to foreclose, but that would take time. Meanwhile, Gould announced plans to connect the Katy with yet another moribund property, the Texas & Pacific.[3]

Tom Scott's vision of a southern transcontinental line had collapsed with his failure to wring a subsidy from Congress. In 1879 the Texas & Pacific extended from Texarkana to Fort Worth, with a branch from Texarkana to Sherman, Texas, just below the Katy terminus at Denison. A short branch reached Shreveport, beyond which lay a gap to New Orleans, the unfinished business of another derelict called the New Orleans & Pacific. Without eastern connections or funds to build westward, the Texas & Pacific faced a bleak future as a local road. Then, in December 1879, Gould organized a syndicate to extend the road to El Paso in return for $20,000 in Texas & Pacific bonds and a like amount of stock per mile. The syndicate included Sage, Dillon, Humphreys, Dodge, George M. Pullman, W. L. Scott, and Charles F. Woerishoffer, one of the most astute brokers in Wall Street.[4]

A construction company was chartered and plans were announced to extend the Katy to Dallas on the Texas & Pacific line. Suddenly Jay possessed the skeleton of yet another transcontinental route, far afield from his other holdings but connected to them via the Katy. Again the timing was superb, the motives cloaked in the guise of serving other purposes. Huntington was pushing his Southern Pacific toward El Paso, while the Santa Fe allied with the Frisco in hopes of creating a new transcontinental line. Both roads, if completed, threatened to drain traffic from the Union Pacific system. Gould's moves could thus be interpreted as an attempt to check the expansion of formidable rivals.[5]

Into this cauldron of confusion Jay poured another ingredient. Late in January he joined a syndicate to dispose of 50,000 shares of Central Pacific stock. The California partners, having clutched their holdings tightly for years, followed the lead of Gould and Vanderbilt in dispersing part of them among the public. Wall Street, dazed by one sensation after another, didn't know what to make of the sale. Did Gould's presence mean that Huntington was about to leave the Central Pacific, or did it portend the long-awaited merger of the Pacific roads? Unruffled by mystery, buyers snatched up the entire 50,000 shares in fifteen minutes.[6]

If nothing else these transactions confirmed Gould's reputation as the phantom of Wall Street. No one knew where he would strike next, in what property or on which side of the market for what purpose. For that matter no one knew where he stood in the lines he already owned or was supposed to own. Businessmen and traders, who would have sold their souls to know his innermost thoughts, sifted his every move for clues but came up empty. Knowing so little of the man, they found no way to take his measure and so despised him the more for making their lives difficult. Of course, he might have revealed everything and no one would have believed a word of it, so profound was their distrust of him. Minds less subtle than his own dealt more in impressions than precision. For men whom unorthodoxy baffled like a menu in French, it was easier to denounce Gould than to understand him.

"He veils his movements in a mystery as profound as that of an African sorceror," grumbled the *Times*. "When he condescends to speak, men listen as they would listen to the Sphynx which looks across the lifeless deserts of Egypt." Editors no less than traders assumed that whatever Gould did was governed by one principle: to extract maximum profits. That explained his activity in debt-riddled southwestern roads. He had, in the *Sun*'s words, used the proceeds from his Union Pacific sale to buy "a heap of rubbishy Southwestern stocks." During the past year Jay had worked wonders in creating a market for them where none had existed before. The mood of the Street was bullish and likely to remain so. As one broker sneered, "It is only necessary now for a stock or bond to be 'rotten and southwestern,' to meet with a ready market."[7]

In February 1880 a fierce storm unroofed Gould's cottage at Long Branch just as Jay was ripping the lid off traffic arrangements on both sides of the Missouri River. Aware that the Burlington did not share its B & M business with the

Iowa Pool, he followed suit for traffic moving along the Union Pacific-Wabash axis, then demanded that all business west of the Missouri be divided and the Wabash made a full member of the Iowa Pool. Perkins realized Gould had scraped a raw nerve in Pool relations; unless agreement could be reached, the spectre loomed of unbridled rate cutting and expansion between the river and Denver. Perkins believed ardently that "the law of Railroad nature" dictated "that each line must own its feeders," but he also saw that the process of acquiring them would transform friends into bitter rivals. Gould's aggressive policy forced him to confront this ugly dilemma.[8]

Convinced that "the present attitude of the Wabash is a part simply of Gould's plan to trade with us on the basis of our not going to Denver," Perkins negotiated an agreement with the Union Pacific in which he promised not to build to Denver provided that "existing Missouri River pools . . . continue undisturbed." The Burlington would keep its monopoly on B & M traffic while Union Pacific, Katy, and Hannibal would not discriminate in favor of the Wabash. This could be done, Perkins presumed, because Gould controlled all these roads. However, he did not take into account a crucial distinction: his agreement was with the Union Pacific, not with Jay Gould.[9]

Gould was quick to enlighten him. The Wabash continued to cut rates on its own and, through some brilliant tactical maneuvers by Gould, captured two small roads in the Burlington's own territory. Plans were announced to extend one westward to Council Bluffs and the other eastward to connect with the Wabash, thereby creating a line parallel to the Burlington. Perkins had wanted to buy the roads defensively but as usual could not get his board to invest in suckers. Now the Wabash had them and could strike at the heart of Burlington prosperity, its lucrative local business. Forbes muttered that "we all exaggerate Goulds desire to fight us," but events overrode even his caution. Gould scooped some small Illinois feeders and was rumored to be after the Northwestern along with another line north of the Burlington. Field left the Wabash presidency and was replaced by Humphreys, who pressed the demand for admission to the Iowa Pool while Gould and Dillon pledged that Union Pacific would live up to its agreement.[10]

Slowly, sullenly, Forbes came to recognize that invective and indignation were inadequate weapons to counter the dangers posed by Gould. In the spring of 1880 the Burlington began a momentous shift in strategy: it abandoned the old conservatism in favor of defensive expansion. Forbes blamed Gould for forcing his hand and never forgave him. For a decade his bitter, almost irrational dislike of Gould would complicate efforts to resolve disputes between their roads. The merger with the B & M, which had itself leased the A & N, marked a first step in this new direction. Soon afterward the Burlington embarked on the practice Forbes feared and despised most, buying plug roads at inflated prices. After much waffling it acquired the Council Bluffs at $125 a share and picked up a small road as a connector to the Santa Fe. Forbes also explored the possibility of alliances with the Rock Island and the Santa Fe. His cause improved greatly in May when T. J. Coolidge, a Burlington director, became president of the Santa Fe.[11]

These moves served notice that the Burlington's leaders would not be cowed. That same spring Vanderbilt rose from his lethargy to enjoin the Wabash's construction of terminal facilities in Chicago. In responding to these challenges Gould played to perfection his role of corporate chameleon. The Wabash resumed cutting rates on southwestern business and the Hannibal announced plans to build to Chicago in retaliation for the Burlington's purchase of the Council Bluffs. Gould, Sage, and Dillon were among the incorporators of the new line, yet Jay took care to warn Perkins privately of the move in advance and suggested that Wabash and Burlington combine to "protect this business fairly in the future." Baffled as to which hat Gould was wearing, Perkins replied that if the Hannibal went to Chicago he would build to Kansas City. By August the threat of war loomed large. A rate war engulfed the territory southwest of Chicago, and the truce west of the Missouri River remained wobbly. For the Burlington one issue dominated all others: the Wabash must be kept out of its Iowa territory. The Wabash wanted a shorter, straighter line and the Hannibal an outlet to Chicago. However, the most pressing question was, what did Gould want?[12]

With his usual agility Jay deepened the mystery. He and Dillon met with Perkins to resolve all differences between the Burlington and the Union Pacific. This was accomplished, Gould told Coolidge, "in a very friendly spirit." He agreed with Coolidge that the Union Pacific and the Santa Fe should treat lines east of the Missouri River "all fairly and they should stay on their own side of the river." An invasion would trigger "a free fight that would in the end ruin every line west of Chicago." In this dialogue Gould assumed the voice of Union Pacific. At his urging the Hannibal had made what he deemed "a perfectly fair adjustment" with the Burlington. He had also advised the Wabash *"to make a strong alliance . . . such as Mr. Perkins intimated would be satisfactory."* But he did not control the Hannibal, and the "Wabash people" insisted on their Iowa extension. "I explained to Mr. Perkins," he added blandly, "that I was not likely to have so much influence with the Wabash for the reason that I have sold my stock in that Company."[13]

Prolonged negotiations merely confirmed what Gould had said. The Hannibal management told Perkins it could not accept Gould's advice and suggested the Burlington buy control of the road. The Wabash finally entered Chicago and promptly slashed passenger rates. Complaints arose that the Union Pacific still diverted most of its business to the Wabash. In Iowa both sides took a hard line. Perkins declared that the Burlington would parallel any new road and build on to Toledo. In that case, Humphreys replied defiantly, the Wabash would extend into Nebraska.[14]

By September the war drums had grown loud enough to frighten not only the combatants but their neighbors as well. Forbes was at his wit's end, his sarcasm long since turned to acid rain. Haunted by the spectre of the crash following the orgy of construction in the early 1870s, he longed for peace but could not negotiate with the enemy. Dillon he dismissed rudely as "so stupid," and Gould was beneath contempt. Instead he appealed to his old friend Fred Ames to intervene with Gould. "I, of course, can do nothing with Gould," Forbes protested, "and the last time we met only mischief came of it. I know he dont like me and I certainly dont like him."[15]

Ames agreed to do anything in his power to prevent a disturbance, but got from Gould the same response Perkins received earlier. "I am doing all I can to secure peace and good fellowship between these large interests," Jay insisted, "but if I fail I object to having the C.B. & Q. inflict a punishment on the innocent Bond holders of the U.P. *Two wrongs will not make one right.*" This remark betrayed Gould's unease at the renewed threat of a Burlington extension to Denver. Perkins detected this weak point and argued, "The only way to get peace with the Wabash is to make Gould understand that war means injury to his Pacific Roads."[16]

But how to corner so elusive an opponent? To Perkins the question was how "to tie Gould East of the Missouri where he operates under so many aliases." He thought Gould "personally should also be party to such a contract if practicable," but Jay was far too shrewd to blunder in that way. He signed contracts or agreements only when it served purposes of his own; otherwise he operated through the instrumentality of bankers, brokers, promoters, and corporations or their officers.[17]

Those who concluded from this that Gould shirked responsibility simply missed the point. Few entrepreneurs assumed more risks or responsibilities than Gould, and none were more meticulous in assuming them. As Dodge observed, he defined obligations precisely and never reneged on those he incurred. But he also thrived on leaving sloppy minds with the impression that he had committed himself when in fact he had not. To Jay this was part of the game in which alertness and perception counted for points. So too with contracts, which could restrict his movement, his ability to shift course without warning. Operating through others enabled him to roam freely, a dealer in the unexpected. It was impossible for others to know what he actually controlled, let alone discover his intentions. He was the consummate one-eyed jack, an enigma, a phantasmagoria.

Perkins didn't know what to believe. After reading Jay's response to Ames he concluded that "Gould either cant or wont make peace—I dont believe he cant." The trick was to find a compromise that would lure Gould to the bargaining table. Perkins could find only one: let the Wabash and the Burlington build the Iowa line jointly and divide the local business equally, with a provision barring branches. The Wabash would not invade Nebraska and the Burlington would not extend to Toledo or Denver. The proposal did not thrill Perkins, who snarled that "to let the Wabash come into our Iowa Country even on joint ownership *without our building any thing in retaliation is a dangerous precedent—& a backdown.*" But the Union Pacific, Santa Fe, and Pool roads were nervous about the prospect of a war that threatened to shatter the last pretense of the territorial principle. A "general treaty" might restore harmony, prevent further construction, and, most important, stabilize rates. It would also buy time to find out what Gould really intended.[18]

Curiously, Jay proposed a similar compromise to Coolidge at about the same time. This discovery of common ground led to a series of meetings in October that hammered out an agreement. The Iowa road would be built jointly; the Wabash was allowed some minor branches in Iowa but no main lines. It agreed, as did the Union Pacific, to leave southern Nebraska to the Burlington, which pledged not to extend to Denver. In a separate agreement the Southwestern Pool curbed the passenger war by admitting the Wabash as a member and giving it an equal share of the business.[19]

Reasonable as this compromise was, it met a peculiar reception. One editor, not known as Gould's friend, labeled it "a complete surrender of essential points in dispute" by the Burlington. The passenger rate agreement collapsed almost before the negotiators left the room and war raged for nearly two years. By November 1880 the fare between Chicago and St. Louis had dropped to one dollar. Despite this imbroglio, the main agreement held firm. Railroad executives breathed a sigh of relief. They had produced peace with honor—for now.[20]

"I believe Gould must have some . . . plan in his head for eventually putting all of his lines together," Perkins surmised that November, but he could not locate its mainspring. Gould roads sprawled across the map like great rivers and tributaries, but where was the source into which they all flowed? Perkins searched for clues and found only contradictions. Ever nimble, Jay maintained his relentless pace in the West, Midwest, Southwest, and, of all places, the East. At first Gould's eastern forays seemed no more than skirmishes in his battle with Vanderbilt. The illusion of an alliance between them palliated but did not solve their conflicts over railroad and telegraph matters. Neither cared to depend on the other for connections out of Chicago. In Gould's search for an eastern outlet the obvious tack was to ally with one of the other trunk lines. Although he renewed his flirtation with Garrett of the B & O, their liaison had more to do with telegraph than railroad wars. Rumors also flew of his buying into Erie, which Jay used to divert attention from his true target, the Lackawanna.[21]

It seemed the most improbable of choices. A profitable coal carrier, the Lackawanna extended from Hoboken through northeastern Pennsylvania to Syracuse. Jay knew the region from both his Erie and his Gouldsboro days. In 1880 it offered him two commodities in which he traded best: personalities and possibilities. The road was controlled by Moses Taylor and his close associates, Samuel Sloan and Percy Pyne. Gould and Sloan had clashed in the Erie War and more recently in Michigan Central, but they joined forces readily against Vanderbilt. Sloan had long itched for a Chicago connection; Jay understood this aspiration and saw how to harmonize it with his own. In May 1880 Lackawanna tumbled more than twenty points and Gould bought heavily. Late in August the Lackawanna startled everyone by announcing plans for an extension to Buffalo. The incorporators of the new line included Gould, Sage, Dillon, Fred Ames, Pyne, and Humphreys, a mixture that spoke volumes. Skeptics dismissed the venture as a

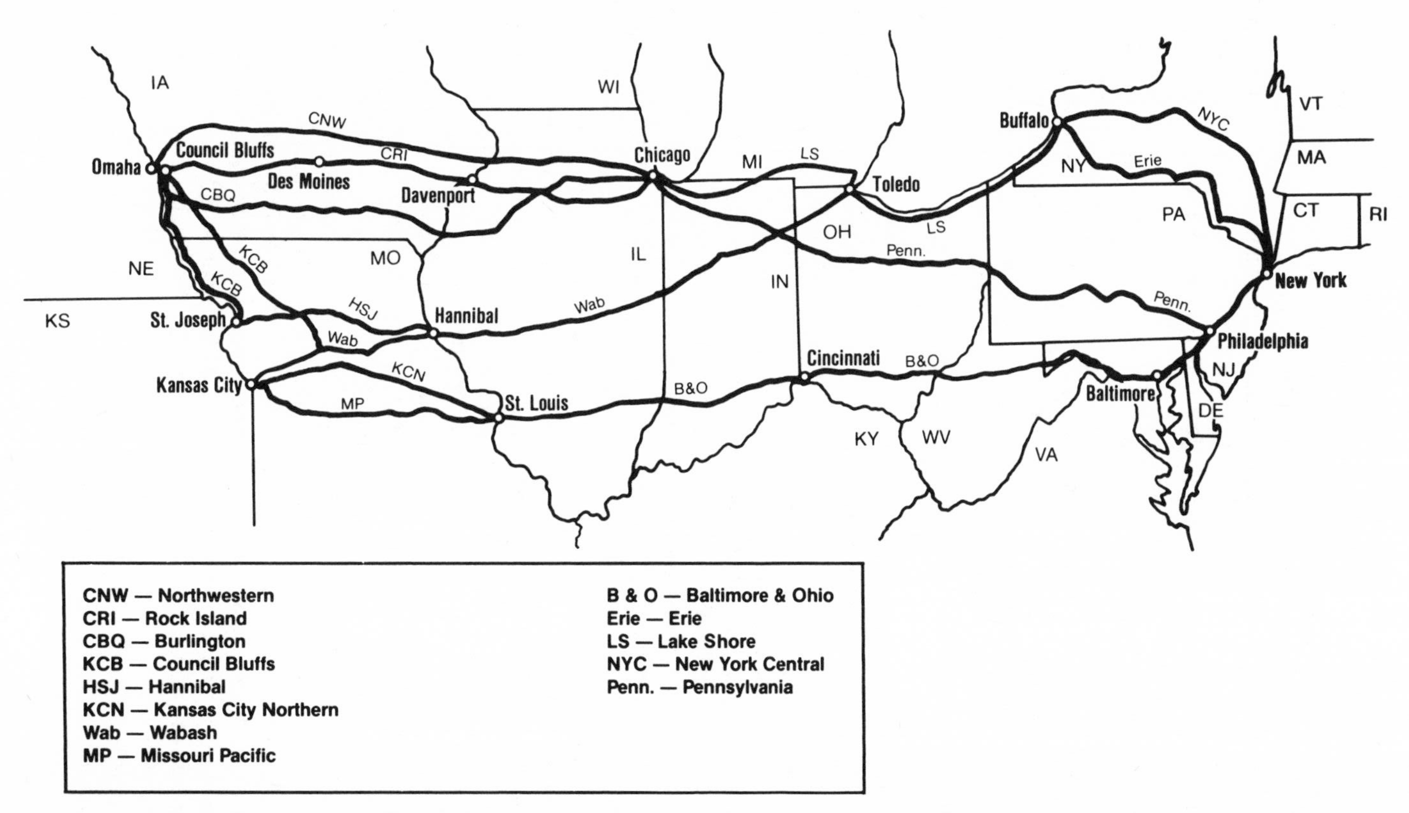

The battlefield between the eastern seaboard and the Missouri River, showing the eastern trunk lines, the Iowa Pool roads, and some of their connectors and antagonists.

signal that "Mr. Gould has a large line of Lackawanna stock which he would like to dispose of" and mocked the building of "Paper railroads."[22]

Those who thought it a game of bluff were soon undeceived. The work went steadily forward and Wabash negotiated traffic agreements with Great Western and Lackawanna. Suddenly there loomed the prospect of another trunk line through territory in which, Vanderbilt complained, "there were five trunk lines and only business for two." The Wabash served notice that it would "parallel the Lake Shore in every direction that may be deemed desirable, Mr. Vanderbilt having refused the Wabash business at Chicago on any terms." Lackawanna stock climbed from its May low of 68½ to 110 by the year's end.[23]

In the West Gould seemed intent on protecting the Union Pacific. Another upheaval in Pacific Mail forced him to reenter that hapless company. Park had been ousted in 1878 by Henry Hart, who presided over two years of agreements made and broken with the Pacific railroads. Sage and Gould tried to reconcile all sides to the pooling arrangement whereby the Pacific roads paid the steamship company a subsidy in return for the right to fix rates, but to no avail. In January 1880 Pacific Mail launched a rate war just as Gould was perfecting the consolidation with Kansas Pacific. He and Sage responded by buying enough stock to dump Hart in favor of a board representing all interests, and a new pooling contract soon followed.[24]

In November 1880, shortly after closing the Burlington agreement, Jay tried to find a home for the South Park by merging it with the Rio Grande. Palmer made a suitable offer, but a snag developed when Evans demanded payment in cash. When the Rio Grande balked, Jay turned to his Union Pacific friends. Evans put forth a knotty proposition and was treated to a display of Gould's distaste for wasted motion. "The proposed contract is too complicated," he wired Evans. "Suppose, in lieu of everything, we make a cash sale at 90 up, and you remain as president?" Evans hesitated, then demanded par in cash. To his surprise Gould accepted at once; the transaction required only two days. The entire stock of the South Park, 35,000 shares, had been distributed as a bonus among three construction companies that built the road. Gould had already received 5,716 shares as a dividend on his stock in one of those companies. He now bought from Evans's group another 25,908 shares for about $2.6 million. Two months later he sold 30,993 of these shares to the Union Pacific at par, or about $3.1 million.[25]

Long afterwards critics charged that Gould sensed impending doom and unloaded his holdings on the Union Pacific. The evidence does not sustain this tribute to his foresight. Not even soothsayers could have anticipated how abruptly the Leadville bubble would burst. "There never was such a collapse as that," mourned Charles Francis Adams. For two years the South Park did well enough to pay its interest and a dividend to boot. Far from being a reluctant buyer, the Union Pacific coveted the road to fend off the Rio Grande and perfect its hold on every line north, east, and west of Denver. In fact the company spent another $6 million in unwise extensions which, coupled with poor management and the falloff in mining, did more to make the road a burden than the original purchase.

In later years Adams deplored the acquisition but admitted that at the time it "could have been sold by the Union Pacific for a large actual profit."[26]

The South Park transaction attracted little attention because Gould upstaged it with efforts to promote a merger between the Pacific roads. The idea had a long history and undeniable inner logic. Huntington favored it and so did the government auditor, but Jay could not convince his own associates. The attempt revealed just how apprehensive the merger mania had made everyone. Perkins feared that "as soon as Gould has consumated the consolidation with Huntington he will be on hand for a trade with Coolidge." Gould's activities in the Southwest did little to alleviate his anxieties. While one hand did service for Union Pacific, the other seemed to be fashioning a rival system out of the Katy and the Texas & Pacific. Why was he puttering in the desert, so far afield from his other interests? The answer unfolded gradually over a period of months like pieces of some elaborate puzzle of which only Gould knew the ultimate design. In fact the southwestern roads were crucial to Jay's plans, but his intentions were so artfully disguised that few managed to solve the riddle until it stood revealed for all to see.[27]

Jay was no stranger to the Southwest, having intervened in the subsidy wars between Scott and Huntington. The Union Pacific feared a southern rival built by either man, but at least it had a shared interest with the Southern Pacific crowd. David Colton argued that, "disagreeable as the medicine is, it is better for us to have Gould buy Scott out, and get rid of him and all of these double contests, and then try to take from Gould the West End of the T. & P. Road." His reasoning cut to the heart of the difference between Scott and Gould. "I never had much respect for Tom Scott's ability to *accomplish* any great undertaking," he sneered. "He can give everybody a Pass, and get them to say he is a 'big Injun' and good fellow—but he is not the man to lay down a Hundred or Two Hundred Thousand Dollars Cash, to carry a scheme of his own."[28]

Gould, however, was to Colton quite another breed, "the reverse of Scott; he is a one man power; consults no one, advises with no one, confides in no one, has no friends, wants none—is bold. Can always lay down Two or Three Hundred Thousand Dollars to accomplish his plans and *will* do it if he thinks it will pay." Time proved Colton a prophet. What Scott failed to do with all his charm and flamboyance, Gould accomplished through tenacity and hard work. As always he relied upon a trusted associate to oversee the work. In the Southwest that man was the ubiquitous Dodge, who served as chief engineer for the Texas & Pacific. A tough, imaginative workhorse, Dodge was neither as fragile nor as sensitive as Clark. He respected Gould and served him well while never losing sight of his own interests.

The challenge confronting Gould was formidable. He had not only to extend the Texas & Pacific 600 miles across uninhabited desert to El Paso, but also to fill the gaps northward to connection with the Katy and southeastward to New Orleans. In typical fashion he undertook all three tasks at once. While construction proceeded westward, Jay let contracts for the northern link and bought out the creditors of the moribund New Orleans & Pacific. In Louisiana he unearthed a land grant belonging to a predecessor road. Scott had overlooked its value but Jay was not so careless; through some intricate maneuvers he managed to have it transferred to the New Orleans & Pacific by the year's end. Another construction company was formed to build the Louisiana road. Through it Gould cleverly funneled the line's securities into the hands of Missouri Pacific, Katy, and Texas & Pacific stockholders, thereby making them investors in the Louisiana venture. Those who accepted the offer eventually realized large profits, which enhanced Gould's reputation among followers and foes alike.[29]

In all this work Dodge did yeoman service—surveying, bargaining, lobbying, buttonholing anyone with useful information, which he promptly relayed to Gould. On Wall Street the puffing of west Texas had begun. "Already the effect is magical," gushed one correspondent. "New counties, new towns, new farms are springing into existence by hundreds on the line of the road, in advance of it." Scott's annual report glowed with optimism and was followed by news that construction crews would attempt to reach El Paso by January 1882, a year ahead of schedule. The completion date was advanced because Huntington was marching his Southern Pacific toward El Paso. By allying with another Texas road and the owners of the Sunset route, he hoped to forge a through line from El Paso to New Orleans without the Texas & Pacific. Gould wanted to reach El Paso before Huntington or his friends, but torrential rains stalled work in both Texas and Louisiana. Heavy cotton shipments clogged the line west of Fort Worth, forcing carloads of building material to sit idle on sidings.[30]

While Gould struggled to beat Huntington to El Paso and perfect his connections, he also explored ways to reach the Rio Grande and the Gulf. The Katy could be extended to Laredo or Galveston by new construction or buying shadow roads and finishing them. One company, the International & Great Northern, occupied the Laredo route but had withered in the depression after some disjointed building. Dodge reckoned it would take 525 miles of new road to reach Laredo unless Gould acquired the International and another company building toward Fort Worth from Galveston.[31]

As a first step Jay perfected his hold on the Katy. His methods were deft, if not delicate, and showered him with criticism. In April 1880 a furor arose when he borrowed Katy stock, transferred it to his own name, and then closed the transfer books "without reasonable notice." In response Gould pointed out that the Katy was a Kansas corporation, and that due notice of the closing had been given there. The episode typified what Jay's critics detested most about his way of doing things. He had obeyed the rules precisely but with a fine disregard for custom. Nothing obliged him to educate speculators who did not know the Katy was chartered in Kansas. Why should he freely surrender an advantage born of information available to anyone who cared to obtain it? To him such fine points were a crucial part of the game; to others they smacked of deviousness.[32]

The fight for Katy drew heavily on Jay's bag of tricks. In June he startled the bondholders by leasing the Katy to the Missouri Pacific without consulting them. Before the bondholders could react, Jay raised enough cash to pay back interest and demanded that the court return the company to the stockholders. The court complied readily, whereupon Gould increased the Katy's stock by $25 million and implemented a consolidated mortgage to provide funds for expansion. Plans were announced to build 600 miles of road, including extensions to Fort Smith, Arkansas, and Camargo on the Rio Grande where connections could be made to Mexico City. Before traders and pundits could digest these moves, Jay astounded them with another coup by acquiring the St. Louis, Iron Mountain & Southern.[33]

The Iron Mountain owned the short line from St. Louis to Texarkana. Its 685-mile system included 195 miles of branches, the most recent acquisition being the Memphis & Little Rock. Originally a mineral road, it had weathered the depression years and climbed toward financial health under the leadership of Thomas Allen. Gould's scrutiny of the road disclosed certain key facts: it did not actually reach St. Louis but relied on a small Illinois line to enter East St. Louis; it drew upon the Texas & Pacific for most of its long-haul traffic; its superior route forced the Katy to cut rates or give rebates to compete for business; and it lacked a southern connection unless the International completed its line.[34]

Each of these points served Gould well. In February 1880 the Wabash bought the road used by the Iron Mountain for access to St. Louis. On the southern flank Gould's position in Texas & Pacific enabled him to squeeze Allen by shunting traffic onto the Katy, while the New Orleans & Pacific threatened to drain cotton business from the Iron Mountain. By December Allen found himself surrounded and besieged. Recently elected to Congress, he was in no mood to squander his fortune in a fight against the massed legions of Gould, Sage, and friends. He sold Gould 40,000 shares for about $2 million. Purchases from other sources increased Jay's holdings to 70,000 shares, slightly less than a third of the company's stock. Two weeks later Gould dropped the second shoe by acquiring control of the International.[35]

Suddenly the clouds of mystery began to dissolve. In January 1880 Gould had put together the skeletal outline of a southwestern system which few observers took seriously. During a year of hectic trading he had planted his flag on so many fronts that no one could discern his true direction of march. Amid the intricate maze of his maneuvers, however, ran the one consistent thread of expanding and linking his holdings in the Southwest. By January 1881 there emerged a system with the potential of stretching from El Paso to the Gulf. It was a vast territory, largely uninhabited and crying for railroads to hasten its development. "Immigration from the north is beginning to pour into this country," Dodge advised. " . . . Now is the time to act, and the quicker you move in the matter the better it will be for all your interests in the south."[36]

Such messages were music to Gould's ears. Here at last was what he had been searching for, the task worthy of his talents. His career had been a persistent search for the right challenge to engage his genius to the fullest, as a general seeks the war, the battlefield, that will fulfill his destiny. Jay had found his field and for the rest of his life never let go of it, whatever else he did. It was there for the taking, and above all it was his to take.

The headquarters of his operations had also become clear. Logic dictated that his southwestern roads funnel their traffic toward St. Louis, where Jay controlled nearly all the railroads entering from the West. This fact was to elevate an unlikely candidate from the status of prosperous local road to the exalted role of parent company for Gould's southwestern empire.

22

Imminent Domain

In nearly all his railroad operations he repeated, to a greater or less extent, his career in Erie.

—Trumbull White, *The Wizard of Wall Street*

My relations with Mr. Gould were entirely different from the views that Wall Street had of him. He never gave me an order that he did not live right up to it. There are hundreds of times where he had gone into enterprises that he could have avoided taking the responsibility . . . it would not do to go to Mr. Gould and talk to him on a matter and have him speak favorably of it and then go off and consider that he was behind it. You had to have him say to you, "Go ahead; I think this ought to be done." You needed something definite. When he said that it was all you needed. I used to often see men who criticised Mr. Gould very severely, whom I found trying to get out of things that they had approved or adopted because they seemed to be unfortunate.

—Grenville M. Dodge, *Dodge Record*

Mr. Gould is busy, in distant Texas, as in the Pompeii of an extinct civilization excavating for buried lines which as soon as found are to be placed in the Borbonean museum of the stock-exchange, labelled at millions, and made lively at prices in the fullness of time to be provided for them.

—*Stockholder*, March 9, 1880

DURING THE HECTIC summer of 1880 Jay paused long enough to erect an obelisk above the family graves at the Yellow Meeting House in Roxbury. It was his lone gesture to a past that had taught him well but held little appeal for him. Sentiment for Gould seldom faded into nostalgia. He had no longing to go home again. There were no triumphant processions to Roxbury, not because of his unsavory reputation but because he simply lacked interest. The loyalties that once sustained him he repaid to the living rather than the dead, not only to family but to the associates on whom he relied so heavily.[1]

The roster of key executives swelled with each new acquisition. A. L. Hopkins came with the Wabash, E. B. Wheelock with the New Orleans & Pacific, and three of his most able lieutenants with the forlorn International: H. M. Hoxie, R. S. Hayes, and A. A. Talmage. Drafted into the Missouri Pacific, all three displayed an immense capacity for work. Hoxie was round-shouldered with large eyes and taut lips, walked with a limp, and spoke in a staccato style that compressed more in a sentence than most men put into a paragraph. A tough, capable manager, he demanded much of his men and repaid them with unswerving loyalty. Hayes was a thoughtful, competent officer whose deliberate manner provoked Talmage into exclaiming, "Why, damn it, get rid of your work; do not let it remain unsettled . . . a man cannot last long who lives in an atmosphere of put-off and worry." Talmage was right; over the years Hayes grew gray and careworn until neurasthenia forced his retirement. Of the three only Talmage seemed to bear his burden without strain. A handsome six-footer with sparkling eyes and a magnetic personality, sharp tongued, quick tempered but warmhearted, steadfastly loyal to those above and below him, Talmage ruled his men with the ease of a beloved despot. An associate considered him "nearer of absolutely being an ideal Manager than any one I ever knew."[2]

Gould was quick to appreciate their talents. Their rapid promotion reflected his need for capable operating men to run the system he was putting together. Those who wondered what Jay was up to, or who dismissed him as a mere stock jobber, might have learned much from the attention he gave personnel matters. The larger his empire grew, the more he relied on the advice of key subordinates to guide his decisions. Because their intelligence carried so much weight with him, he took care to treat them well. Earlier Gould had strengthened another crucial area, the legal department, by recruiting the formidable talents of Judge John F. Dillon.* In 1879 the erudite judge, tired of starving on the federal bench, retired to a professorship at Columbia. A few months later he signed on as counsel for the Union Pacific and thereafter devoted his legal talents to it and to Gould's enterprises.[3]

Important as these men were to Gould, none supplanted Clark in his confidence and affection. Time strengthened the bond between them even as it

*Hereafter, to avoid confusion, "Judge Dillon" refers to John F. Dillon and "Dillon" to Sidney Dillon.

emphasized their differences. The diffident, soft-spoken Clark moved at only one speed: slow. Thoughtful and deliberate, he was fond of saying that "65 percent of things would naturally settle themselves if left alone—the remaining 35 percent now and then demanded prompt and determined action." This fatalistic attitude clashed with Clark's insatiable ambition, which prompted an acquaintance to label him "an anomalous mixture of weakness and efficiency." Perhaps this contradiction spawned the periodic attacks of neurasthenia which plagued Clark all his life. During the spring of 1881 he fell ill again and required an extended rest. When he returned, Dodge warned Hoxie that Clark was "pressing very strongly for control in the southwest. You know he is an old favorite, but I do not think his health is equal to the work, and I have so stated very frankly." Whatever friction Clark's position as favorite may have aroused among other officers, it assured his personal fortune. By 1880 Gould had taken charge of Clark's portfolio. On one transaction alone, involving Denver Pacific bonds and Rio Grande stock, Clark earned $50,000 without lifting a finger.[4]

There was a poignant quality in Gould's devotion to Clark. He fussed much more over his friend's health than over his own. "We are glad to learn that you return to Omaha with restored health," he wrote in the spring of 1881. "The important thing now is for you to avoid overloading yourself with cares & details." If one wonders why Gould cared so deeply for Clark, he provided the answer in his usual succinct way. "I appreciate your friendship very highly," he told Clark, "because I know it is the real stuff."[5]

The reasons behind the emergence of the Missouri Pacific as Gould's parent road were easy to find only if one knew where to look. Apart from its strategic location at the gateway to the West and the Southwest, it was the only major road Jay controlled by absolute stock ownership. His holdings in Wabash were minimal, and he had sold the Union Pacific stock received in the consolidation. As the link between two great rivers, connecting two cities of vast commercial potential, the Missouri Pacific could unify a system sprawled across a huge domain from the Rocky Mountains to the Louisiana Delta.

Observers were slow to grasp Gould's design because conventional wisdom dictated that any large system must include a major trunk line, which the Missouri Pacific was not. Jay pursued an original course rooted in two assumptions: that profits could be had by developing untapped territory where competition was minimal, and that St. Louis could be elevated into a transportation entrepot rivaling Chicago. He had long been impressed with the potential of St. Louis. On his western swing in 1879 he reminded a gathering of local businessmen that "we have some 3,000 miles of railroad, all of which centre here, and I could not do otherwise than strive for the prosperity of this city."[6]

To utilize the Missouri Pacific as the nexus of his system Jay had to rearrange its lopsided financial structure, which had $15 million in bonds but only $800,000 in stock. During the summer of 1880 he merged the road with a handful of subsidiaries into a new company, the Missouri Pacific *Railway*, with a capital stock of $12 million. The move attracted little attention because few observers caught its larger significance. An expanded stock base shrank the price per share, enabling Gould to distribute it widely, put it on 'Change for public trading, and use it for acquisitions. It provided the instrument he needed to weld his disparate southwestern holdings together. In March 1881 Gould made a whirlwind tour of the southwestern roads accompanied by Sloan and Eckert. Impressed by what he saw, he told reporters that "in a short time certain developments will be made of general interest to the Northern and Southern sections of the country." This bland teaser hardly prepared the public for the blitz that followed.[7]

Control of the Texas & Pacific was assured in April when Tom Scott, now an invalid, sold his holdings to Gould. A month later Scott was dead, never to see Gould complete in short order the line he had fumbled for years to realize. Suddenly rumors flew that Jay planned to combine all the roads into a southwestern company. The Katy, already leased to the Missouri Pacific, absorbed the International by an exchange of stock. Negotiations began for consolidation of the Texas & Pacific with the Iron Mountain, which had increased its capital stock from $21.5 to $35 million for just such a purpose. When no agreement could be reached, Gould changed tactics and merged the Iron Mountain with the Missouri Pacific. A few weeks later the Texas & Pacific absorbed the New Orleans & Pacific. The swiftness and immensity of these transactions fired the Street's imagination. Despite warnings from conservative journals, the securities of all Gould's southwestern roads soared to record heights. Between January and June 1881 Missouri Pacific leaped from 85 to 110, Texas & Pacific from 41 to 73, Iron Mountain from 52 to 86, International from 50 to 92, and Katy from 40 to 53. This virtuoso performance inspired one western editor to the phrase "Great G——D."[8]

By June 1881 Gould had transformed the shadow of a southwestern system into a reality. The excitement generated on Wall Street created a market for southwestern securities where none had existed before. Had he been content merely with gain, Jay might have realized huge profits. Instead he sprang another surprise. Having forged a system, he proceeded to enlarge it in all directions by a program of new construction. To finance his expansion Jay relied on techniques long perfected. He utilized local aid to build branch lines and had subsidiary roads issue vast amounts of new securities. The Katy, for example, authorized a stock increase of $47 million to build into Mexico. A favorite device was the consolidated mortgage, which Gould employed on most of the roads under his control. This blanket mortgage (or "monster mortgage," as one journal called it), while refunding a mass of underlying obligations at lower rates, invariably contained a large additional sum for acquisitions, improvements, or construction.[9]

The lavish outpouring of new securities obliged Gould to create a market for them. There is no mystery in Gould's preoccupation with the securities market; it was his fountainhead of capital. His projects, indeed his business career, lived or died on his ability to boom new issues, sustain old ones, and manipulate prices to his advantage. In the end, of course, the company had to perform well enough to attract investors, but Jay understood the role of promotion. He knew that investors fed on a curious mixture of performance and prospects, that they flocked eagerly to the right promises dangled in the right way. His grasp of human behavior enabled him to feed this appetite of dreams with brilliant success.

Apart from the issue of new securities, Jay relied on that old standby, the construction company, for bringing associates into a venture at minimal risk. The modus operandi was simple but effective. Gould and friends as directors of a road let contracts for work to a construction company owned by themselves. Jay did not hold shares directly but listed them under the name of Morosini or Guy Phillips, his secretary. The company agreed to build the road in exchange for bonds at a rate of, say, $25,000 per mile, with the stock thrown in as a bonus. Invariably the securities received far exceeded the actual cost of construction. One critic asserted that Gould's southwestern roads were "bonded at from $30,000 to $40,000 per mile. It is fair to presume that their actual cost of construction was less than half that sum."[10]

But how else to attract capital to high-risk ventures? The construction company lured funds to projects that no one would otherwise touch by reducing the risk involved. The road got built, investors profited from the work, and a harmony of interests was created. The road groaned beneath a heavy mortgage, but in flush times that seemed a trifling price to pay. Five construction companies were created for Gould's southwestern projects alone, and a sixth built the Lackawanna extension. None would earn profits unless the road's securities climbed in value. Prospects and promises were needed to fire the imagination of traders. Gould had to make both the road and its securities perform well, a task that chained him ever more to the market. The larger his empire grew, the more vulnerable it was to attack by traders who suspected Jay had overextended himself. Never one to play safe, Gould had always skated on thin ice by attempting too much with too little. Increased wealth did not cure the habit but merely enlarged the scale. In choosing to develop raw country Jay assumed enormous risks from which less venturesome souls shrank. As Dodge observed, boldness was among Gould's most impressive qualities, but it exposed him to constant danger.[11]

Peace on the prairie was always a fragile thing, and in 1881 it came unraveled again. The passenger war raged on, and the Southwestern Pool dissolved in June. Sporadic bouts of rate slashing, especially by the Wabash and the Rock Island, kept levels of bickering and distrust high on all sides. The opening of new lines or merging of old ones buffeted existing agreements like shock waves, especially in the Southwest where the Gould roads, Southern Pacific, Santa Fe, Frisco, and Rio Grande were all jockeying for connections in their race across the continent. The region was in upheaval, its tensions spilling onto the roads on both sides of the Missouri River. Nevertheless, optimists clung to the hope that the status quo might be maintained. As late as April 1881 an analyst of the

Burlington's system concluded that "there is little probability, though, that Denver will be reached in the immediate future."[12]

As always, the unknown factor was Gould. For all his activity in the Southwest, opinion remained divided over where the heart of his empire lay. In one bold stroke Jay revealed his priorities and upset the delicate balance of power west of the Missouri. He decided to extend the Missouri Pacific from Atchison to Omaha on the Nebraska side of the river. The Wabash owned a line to Omaha but Gould wanted one for his own system to tap directly the flow of traffic from the Union Pacific. At the same time he explored the possibility of building an extension to San Francisco to serve Texas & Pacific, Union Pacific, and Atlantic & Pacific. That Jay could conceive two such explosive plans in one breath suggests the scale and audacity of his ambitions.[13]

An Omaha extension meant in effect that the Missouri Pacific would parallel the Wabash from Omaha to St. Louis. So much, then, for the primacy of the Wabash in Gould's grand design. Aside from the question of how Jay proposed to square relations between the two roads, the extension violated the spirit, if not the letter, of the October 1880 agreement between Burlington, Union Pacific, and Wabash. The pact had stabilized competition in eastern Nebraska, but Jay proposed to build smack through the counties claimed by the Burlington. Early in June Gould went to Omaha, examined the chosen route, and incorporated a company to build the road. Perkins got wind of it a week later. In a careful letter he told Gould that the rumor had stirred up "excitement among our Nebraska Managers, and the revival of various schemes of extensions into Kansas &c—Hence I am tempted to apply at headquarters for the facts if you are willing to give them to me—?"[14]

The report was true, Jay replied. The Missouri Pacific needed a line into Omaha because the Burlington had absorbed its former connectors, the Council Bluffs and the A & N. "Since then," Gould added, "our business has gradually shrunk to small proportions." There could be no objection, he assured Perkins, since nothing more was intended. This response threw Perkins into a quandary. Here was Gould up to his old tricks, taking advantage, shifting roles, tightroping the letter of an agreement while trampling its spirit. Neither Gould nor the Missouri Pacific had signed the October agreement and were therefore not bound by it, yet Perkins regarded Gould as the real power behind two of the signees, Union Pacific and Wabash. The ambiguities Jay had exploited so brilliantly now came back to haunt him. If Perkins had looked closely, the mystery of Gould's position might have been revealed to him. A careful observer could have seen that Missouri Pacific meant far more to Gould than Union Pacific.[15]

But Perkins did not look closely. The legacy of past dealings with Gould narrowed his vision no less than it steeled his resolve. He had considered Union Pacific of primary importance to Gould for so long that he could not free his mind from that assumption. "Gould is building his Missouri Pacific into Nebraska," he charged, "notwithstanding his agreement last summer as head of the Union Pacific not to do so." The extension posed so many threats to the Burlington that

Perkins determined to retaliate wherever he could. Unable to get at the Missouri Pacific, he lashed out at the Union Pacific instead. On July 20 the Burlington board formally abrogated the October agreement. Three days later Perkins informed T. J. Potter, "It looks a little now as if we should go to Denver at an early day."[16]

An extraordinary situation had arisen in which Perkins proposed to flog one company for the sins of another on the grounds that both were puppets guided by the same hand. Dillon protested the abrogation bitterly and sputtered that it was "highly unjust to the Union Pacific to hold it responsible for action taken which it could not and can not prevent, and which it did not and does not favor." From Saratoga Vanderbilt, as holder of a "very large interest in both companies," pressed similar views on Forbes and offered his services as mediator.[17]

The pleas for peace fell on deaf ears. To Forbes and Perkins the crux of the matter was the slipperiness of Gould and his friends. Legal niceties aside, Gould had promoted and accepted the October agreement, which his new line violated regardless of what alias it assumed. Gould, Sage, and Dillon sat on the boards of Missouri Pacific, Union Pacific, and Wabash; Clark was an officer in the first two. Who did they represent at a given negotiation? If Union Pacific directors could ignore the agreement simply by changing hats, why should the Burlington honor it? Forbes lectured Vanderbilt that agreements must depend "not upon their legal or technical validity, but upon the honest purpose and determination of all persons participating in their formation to live up to them." In short, Perkins and Forbes insisted on regarding the agreement as one made between individuals, not corporations.[18]

Jay hoped to slip through this loophole but when Perkins denied him, the only recourses were to make concessions or war. In a tough letter Gould denounced Perkins's action as "unprecedented among Railroad men." The Missouri Pacific alone was responsible for the extension, he reiterated, "and I take this responsibility." If Perkins built to Denver, he warned, the Missouri Pacific would extend to Chicago, throw out lateral branches in Nebraska, and transfer its business between Hannibal and Chicago to the Burlington's rivals. "We want peace but are ready for war if you insist on making it," Gould concluded. "Carrying out your menaces or extending your line to Denver means war." Dillon added a letter of his own, but Perkins was not about to be cowed. He rebutted Gould's arguments and declared that the Burlington would not "remain with our hands tied while the other party insists on being free." The time for talk had passed. On August 19 the Burlington board authorized the Denver extension; nine months later the new route opened for business. Girding for war, the Burlington elevated Perkins to the presidency and restored Forbes as chairman of the board.[19]

In precipitating this clash Gould damaged both the Union Pacific and the Wabash, neither of which could compete with the Burlington's more efficient route and service. While trying to stave off competition between Omaha and Denver, the Union Pacific also found itself embroiled west of Denver, where Palmer of the Rio Grande organized a new company, the Denver & Rio Grande

Western, to extend his road to Ogden. Already there was talk of an alliance between the Rio Grande and the Burlington to form a through line from Chicago to Ogden. The transcontinental wars had begun, and the chessboard would never look the same again. Gould did not cause these wars but he clearly triggered them. By 1881 railroad men all over the country were shaping and adjusting their policies in response to his moves. Jay surely knew that his Omaha extension invited retaliation by the Burlington and that the consequences of such a fight would be irreparable. Why, then, did he do it? Probably he miscalculated the Burlington's determination and allowed his desire for a system of his own to override all other considerations. As he insisted all along, the Missouri Pacific required a connection to the Union Pacific and he resolved to have it at any cost. He had committed himself to the Missouri Pacific as his instrument of empire, and from that premise all else flowed.[20]

Those who doubted this premise, and there were many, received another clue that autumn. Through a series of intricate maneuvers Jay secured control of the St. Louis Bridge Company, which owned a bridge over the Mississippi, a tunnel to connect with the railroads, and valuable terminal facilities. Since the Missouri Pacific reached one end of the bridge and the Wabash reached the other, Gould leased it jointly to the two roads. By adjusting the toll structure he transformed a deficit-riddled enterprise into a profitable one. The company's stock grew in value, and as a bonus Jay held an important terminal area. He also bought some barge companies and reorganized them into a new enterprise. He was in St. Louis to stay.[21]

No sooner had war erupted on the plains than showdowns loomed in the Rockies and the Southwest. The same forces were at work there: ambition, mistrust, insecurity, the fear that if one did not expand the enemy would steal a march. In this atmosphere even local clashes got caught up in the larger machinations of transcontinental strategy. Trouble arose in Colorado when the Union Pacific invaded Rio Grande territory by extending the South Park, which was not included in the March 1880 agreement. Palmer countered by throwing out branches and pushing his new line toward Ogden.[22]

The battle in Texas pitted Gould against his old antagonist, Huntington, who was as unyielding as Perkins and much more aggressive. Like Perkins, Huntington had the support of a staunch, determined partner, Charles Crocker. Less urbane and witty than Forbes but no less choleric, Crocker said of Gould, "I am very suspicious of, and want as little to do with him, as possible." Huntington respected Gould but was clever enough neither to fear nor to underestimate him. While the Texas & Pacific laid a mile of track a day, Huntington rushed the Southern Pacific toward El Paso and planted himself squarely in Texas by buying out his Texas ally and picking up a small Louisiana road. By July 1881 he posed the threat of undercutting the Gould system with a line from San Francisco to New Orleans, from which he might build northward to Kansas City or St. Louis.[23]

Jay was quick to recognize danger, and he knew his adversary's rugged character well enough to waste no time on threats or bluffs. In the spring of 1881 he discovered a flaw in Huntington's armor. The Southern Pacific had built part of its road on land awarded the Texas & Pacific by a congressional grant. Jay visited the office of Judge Dillon and, twisting a piece of paper between his thumb and finger as he often did when something serious was on his mind, asked casually, "If one man builds a house on the lot of another, without the consent of that other and especially against his protest, to whom does the house now belong?"

"It becomes annexed to the land and belongs to the landowner," Judge Dillon replied.

"Well, I supposed that was the law," Jay continued, "and if it is the law I want a suit brought at once."[24]

Two battalions of lawyers converged on an old adobe building in New Mexico where, after several days' battle, Judge Dillon secured an injunction against the Southern Pacific. While the legal fight raged, Gould pursued his plan for a new line to San Francisco. The Union Pacific owned a Utah branch that curled southwest from Lehi into a mining district. To the west lay a forlorn road called the Nevada Central. It was a worthless property running from Austin, Nevada, to a point on the Central Pacific, yet Gould induced the Union Pacific to buy it and commence extending the branch from Lehi westward. At the same time he organized a syndicate with Santa Fe, Frisco, and Atlantic & Pacific for constructing the line to San Francisco. Reports went out of a grand coalition gathering against Huntington's group.[25]

Gould and Dillon were later criticized for wasting money on a derelict that did not even connect with the Union Pacific. While this hindsight is true enough, it misses the point. The purchase was a two-edged threat to wield against both Huntington and Palmer. Further, it enabled Gould to attack Huntington in the guise of protecting the Union Pacific when in fact his true interest was the Texas & Pacific. Once again he had made one action serve a multitude of needs. His threat aroused Crocker, who urged Huntington to "give Mr. Dillon pretty square talk and tell him that all this stuff and talk about building parallel to us will not be tolerated." Crocker had very definite views about how to deal with Gould and Dillon. "If Dillon thought for a moment we would connect with the C.B. & Q.," he snorted, "it would scare him out of his boots." Huntington dutifully sounded the threats but was more intent on finding a peaceful solution. He knew Gould better and thought he would respond to a compromise if it served his interest. The trick, as always, was to discover just what Gould's true interest was.[26]

Not until October did Huntington realize that, for all the sound and fury in the Rockies, the crucial battlefield lay in Texas, where rival crews marched relentlessly toward each other while the lawyers haggled. If Huntington completed his road to El Paso, it would parallel the Texas & Pacific for ninety miles through a region where, Crocker admitted, "there is no local business." If Gould won his suit, he would get that portion of the Southern Pacific built on the Texas

& Pacific land grant. The situation was ripe for a compromise, and overtures were soon forthcoming. The flinty Crocker warned Huntington against "that little fellow in Broadway" and advised him to "do more *watching* than 'praying,' when you come in contact with him." Huntington protested that Gould and Dillon had been very friendly. "Their friendship," Crocker replied acidly, "is evidenced by the petitions which they filed . . . to steal what little road we have built in Texas. I assure you that I shall never go to sleep on their smiles, and I shall go on & organize the Companies to build from Corinne eastwardly."[27]

Here was the familiar lament of the righteous, condemning unscrupulous rivals for turning a situation to their own advantage, a practice which they themselves, as high-toned men, would never dream of doing. Businessmen of this ilk displayed a gift for moral solipcism that compared favorably with such advanced practitioners as politicians and the clergy. Huntington was not of that breed. Like Gould he was a shrewd pragmatist who assumed any man would honor an agreement that served his interests. Like Gould, too, he wanted a compromise because he was stretched thin. Even more, he saw how the right kind of agreement would enable him to take advantage.

Accordingly, in a series of meetings on Thanksgiving Day, Gould and Huntington hammered out one of the most successful compromises in railroad history. They agreed to share the ninety miles of track east of El Paso and divide equally the earnings on through business from the Pacific coast. The Texas & Pacific dropped its injunction, transferred its land grant to the Southern Pacific, relinquished all claims west of El Paso, and promised not to extend west of that town. Huntington agreed not to parallel the Texas & Pacific east of El Paso or build competing roads to the north and east. Other provisions covered rate and pooling arrangements on traffic.[28]

Even Crocker, impressed by the terms Huntington had extracted, switched from war chants to warblings of peace. The Gould-Huntington agreement enabled both sides not only to protect their interests but also to stabilize competition in the Southwest, at least for a time. Two months later, in January 1882, the new allies joined forces to buy a half interest in the Frisco, which itself owned a half interest in the Atlantic & Pacific. By this move they hoped to curb expansion by the Santa Fe and (most important to Huntington) prevent construction of another road to California.[29]

The Gould-Huntington agreement endured far longer than anyone expected, requiring only four amendments between 1881 and 1927. It kept the peace but did not succeed in keeping Huntington at bay. While Jay lived up to the bargain, Huntington pushed and probed until he managed to get a through line to New Orleans. Alone among railroad men his ambitions equaled Gould's for a transcontinental system entirely under his command, and he would move heaven and earth to create it. Perhaps that common vision was what enabled the two men to understand each other so well.[30]

Only a month earlier, in another corner of the Southwest, the Earp and Clanton brothers had shot their differences out at the O.K. Corral. Gould and

Huntington were too smart to come out with guns blazing. Both had long since imbibed the maxim given young Collis by his father: "Do not be afraid to do business with a rascal, only watch him; but avoid a fool."[31]

In the East Gould's plans were screened behind a haze of mystery and confusion. He was into the Lackawanna, had loaned money to the Philadelphia & Reading, and in February 1881 turned up on the board of the Jersey Central. This flirtation with three of the five major anthracite carriers excited speculation that Gould was about to invade the coal industry.[32]

Eastern territory posed difficult problems for Gould. Of the four trunk lines dominating traffic flow, the Pennsylvania and the New York Central were too big to swallow and the Erie promised only financial indigestion. The fourth and smallest line, the B & O, offered the shortest route to St. Louis via the Ohio & Mississippi but did not own a line to Philadelphia or New York. It reached the former city via the Philadelphia, Wilmington & Baltimore and relied on the Pennsylvania for access to New York. When the Reading and the Jersey Central jointly opened a new line between Philadelphia and New York in 1880, the Pennsylvania responded by obstructing the B & O's use of the new Bound Brook route in every way possible.[33]

Here lay the ingredients for another of Gould's elaborate intrigues. He could not buy the B & O because John Garrett and his son Robert had no intention of selling. The telegraph fight had made him friendly with both the Garretts and the Pennsylvania, which gave him an element of surprise in that he had no apparent stake in the dispute between them. He entered the management of the Jersey Central on the pretext of linking that road to the Wabash through some construction and the use of certain roads owned by the Pennsylvania. Negotiations among the officers of all three companies began in late February 1881.[34]

Meanwhile, Jay secretly formed a syndicate with John Garrett, Sage, Dillon, and some bankers to buy control of the Wilmington. Bostonians held most of the road's stock with one large block belonging to Nathaniel Thayer, a longtime investor in western railroads. Although Thayer owned only 6,000 shares of Wilmington, he blithely contracted to deliver Gould 120,000 shares at 70 on March 15. Gould closed the deal and headed west thinking he had acquired a majority of the stock. Word spread that the Wilmington had been scooped by a syndicate eager to challenge the Pennsylvania on its home turf by uniting the interests of B & O, Jersey Central, and Reading. However, on March 1 Gould learned that Thayer could not fulfill the contract. He had not consulted his fellow stockholders on the Gould agreement but had simply assumed they would follow his lead. Instead they formed a committee that negotiated a sale to the Pennsylvania at 80. Left high and dry, Thayer found himself short on the contract and obliged to pay a penalty of $500,000 to the syndicate. Forbes thought he got off cheaply at that price.[35]

Undaunted, Jay resumed negotiations with the Pennsylvania as if nothing had occurred, keeping as a club his rumored interest in the Reading. When agreement was finally reached in June, one critic dismissed the new Jersey Central-Wabash route as one of Gould's "unions against nature." Before Gould could improve his position, fresh disturbances unsettled eastern territory. The dissonant element was, of all people, Vanderbilt. After the New York Central lost considerable business to its rivals early in 1881, Vanderbilt concluded that the other trunk lines were secretly cutting rates and retaliated with a devastating rate war. The fight lasted nine months and made a shambles of Gould's recent contracts, routes, and plans.[36]

Nothing puzzled outsiders more than trying to figure out who was fighting whom. Vanderbilt and Garrett were clearly at loggerheads, but Garrett, embittered over the telegraph war, had also broken with Gould and become his enemy. Where, then, did that leave Vanderbilt and Gould? Their telegraph differences had been resolved, but the Lackawanna extension posed a threat Vanderbilt could not ignore. Had these ancient enemies renewed their battles or had Vanderbilt suddenly become Gould's "wicked partner"? They huddled frequently that summer, not only in town but at Saratoga, prompting suspicion that they had "common interests in too many railroads to be engaging just at present in any other than a sham fight."[37]

It is likely that neither man could have stated in simple terms where he stood. Both were maneuvering for advantage and groping for a coherent policy to deal with a complex situation. Vanderbilt responded to the Lackawanna threat by injecting himself into the Reading muddle, thereby preempting the role some observers had assigned to Gould. Jay charted a more original course. In October he tried to snatch the B & O's connector to St. Louis, the Ohio & Mississippi, in an election coup. His tactics were clever but rested on a technicality that the court quickly swept aside.[38]

Having failed to dent Garrett, Jay switched fronts and struck Vanderbilt on a most improbable flank, New England. He had not touched that region since his fling with the Boston during Erie days, and it was the Boston, now reorganized as the New York & New England, that drew him back. After years of futility the New England was about to reach the Hudson River, but it still lacked an entry into New York. One possibility involved using the tracks of Metropolitan Elevated, which had recently been acquired by Gould, Sage, and Cyrus Field. The New England's president agreed to sell Gould and his friends 15,000 shares at 60, about 20 below the market price. They were elected directors at the annual meeting in December along with the presidents of both the Pennsylvania and the Erie, which gave the ring of a grand coalition against Vanderbilt.[39]

Reporters promptly pronounced the New England a Gould road and embellished every promise of future expansion. Sugarplums of earnings and dividends danced in the rhetoric of every speech at the stockholders' meeting. Amid this sea of euphoria Gould alone sounded the proper note of understatement. Called upon to make a speech, he remarked that he would reserve comment until the New England paid a dividend.[40]

Then there was Mexico.

Little has been written about Gould's foray into Mexico, yet for a time it loomed large in his thinking. Tension between the two countries eased in 1880 when General Porfirio Díaz yielded the presidency to Manuel González and granted concessions to two railroads, the Mexican Central and the Mexican National. As the flow of capital southward picked up and promoters swarmed the capital in search of other concessions, American newspapers rhapsodized over "the vast commerce which development of the resources of Mexico will build up" between the countries. The selling of Mexico had begun.[41]

The Central was to run from El Paso and the National from Laredo to Mexico City. During the 1870s General Palmer of the Rio Grande had pushed the Central, but by 1880 he had been cut off from El Paso by his pact with the Santa Fe and the march of Huntington's road. Undaunted, he abandoned the El Paso route to his rivals and took the concession to build the Laredo line. During the autumn of 1880 Matías Romero, the strongest advocate of American investment in the Díaz ministry, visited New York to secure capital for a railroad of his own and to promote harmony among American interests intent on building in Mexico. He enlisted the support of Grant, who had lost his bid to regain the White House and was looking for work. On November 11 Romero hosted a banquet at Delmonico's for Grant and about twenty leading railroad men.[42]

For Gould the Romero visit served as an entering wedge. A month later he acquired the International and pushed completion of its line to Laredo, which prompted him to look farther south. Clashes at home with Huntington and Palmer loomed on the horizon, and Palmer owned the concession most useful to Gould in Mexico. Like all superior chess players, Jay understood the value of creating a complicated position from which he might grasp combinations that escaped less astute opponents. Quietly he bought some outstanding rights in the concession for the Laredo route as a club to wield against Palmer. Then, in June 1881, Gould secured a concession for another line between Laredo and Mexico City. Organized as the Mexican Oriental Interoceanic & International Railroad Company, the new route impressed Dodge as "the best line in the republic, both in an engineering and a commercial point of view." By December Jay had made peace with Huntington and was fighting Palmer, who had laid track from Laredo to Corpus Christi and was spending large sums to prepare his invasion of Mexico. Dodge was wrapping up work on both the International and the New Orleans roads, which meant a large force of men and material could be moved into Mexico.[43]

All efforts to consolidate interests with the Central failed, which obliged Gould to build first and strike an alliance afterward. Then Fred Ames and his Boston friends notified Gould that their investment in the Central was too great a burden for them to subscribe to another Mexican road. Although stretched thin himself, Jay decided to proceed without them. The Oriental Construction Company was formed with Dodge as president and its shares offered to Missouri

Pacific stockholders. Construction of this 680-mile road would be an enormous undertaking. Dodge admitted privately that "I am in something of a quandary as to how I shall work it." An engineer sent to examine prospects for business along the Mexican line did not report for six months. Reminded of his mission, the surprised engineer assumed that Dodge "knew about Mexico, that so far as he could see, there was no business there for a railroad." Apparently Gould never got this message; instead Dodge provided glowing accounts of the resources and potential for business along the line. As always Jay depended heavily on the eyes and ears of his lieutenants on the scene. Dodge was one of the best. He had never failed Gould before, and there was no reason to think he would now.[44]

23

Well-Wrought Earns

Gould . . . has paid a good deal of money into the enterprise himself or he has induced other parties to do so. . . . By many it is looked upon as a very speculative scheme, got up for the purpose of depressing Western Union and forcing them to buy, or absorb it, or at least . . . as a handle by which to operate in the Stock Market. Others, however, look upon it as a legitimate enterprise.

—R. G. Dun Report on American Union, July 10, 1879

In some properties, as with the Western Union Telegraph, he forced a reputable concern to admit him to partnership through the shrewd and daring use of a species of corporation blackmail, in which he was always an adept.

—Alexander D. Noyes, *Forty Years of American Finance*

Jay Gould was not an originator of systems. Others with ideas secured charters, began railroads and other schemes, and then, when money was needed, Gould would step in and profit by their energies by purchase at low figures. This was never more forcibly illustrated than by his connection with the elevated railroad system of New York City.

—Trumbull White, *The Wizard of Wall Street*

From the beginning . . . the elevated railroad business has been a robbery of the public and of individuals. The bills were not honestly passed; the roads were not honestly built; the rights of adjacent property owners were not honestly considered; the lease of the roads to the Manhattan was a fraud; the litigation relating to the roads has debauched and disgraced the judiciary, and the manipulation of the stocks has been a swindle.

—*The Public*, November 9, 1882

AFTER ABSORBING A & P the Western Union wallowed in prosperity. In October 1878 the stock touched par where, except for periodic fluctuations, it remained for two years. It paid regular dividends and earned praise as a "very good stock for anybody to buy who could afford to own it and hold it." The death of William Orton in April 1878 scarcely ruffled the company because his successor, Dr. Norvin Green, was even more capable. One nuance of the transition, unnoticed at the time, was to have fateful consequences. Eckert was not promoted to general manager as promised but remained president of Atlantic & Pacific, a position that nurtured his bitterness at being slighted. In March 1879, two months before Gould formed American Union, Eckert resigned his office, but, "by bluff, bullying and persuasion," was induced to remain until the year's end.[1]

In that post Eckert served Gould no less effectively than if he had already come on board. The new venture posed formidable challenges for Jay. Unlike A & P, the American Union had to create a system from scratch. The key to success lay in invading that bastion of Western Union strength, the railroad contracts. As early as November 1878 Gould recruited an eager ally in John Garrett, who had already ousted Adams Express and the Pullman Company from the B & O and was anxious to rid himself of Western Union as well. That winter Eckert, untroubled by qualms over conflicting loyalties, did secret reconnaissance for Garrett's campaign to dump Western Union. When Garrett reclaimed his lines, he maintained ownership but allowed them to be operated as part of the American Union system, which had acquired some small companies and built mileage through a construction company. Gould moved slowly during 1879 because of his preoccupation with railroad matters. His attempt to put lines along the Wabash system goaded Western Union into a futile injunction. Elsewhere the larger company resorted not only to legal action but also to cutting down poles; the head-knocking that followed gained Western Union little besides bad press. By the year's end American Union owned 12,000 miles of wire, operated 4,000 miles of wire along the B & O, leased another 12,000 miles in Canada, connected with one overseas cable, and was about to reach a new French cable coming ashore on Cape Cod.[2]

Stung by the failure of its strong-arm tactics, Western Union retreated to the magisterial view that the new rival would "very soon be broken up." By 1880 Gould was ready to strike. The law, backed by a Supreme Court decision, allowed railroads to share their business with other telegraph companies if they so inclined. Eckert assumed the presidency of American Union and was ready to push expansion vigorously. Once Gould launched his blitzkrieg of railroad acquisitions and consolidations, every move was designed to serve not only his rail but his telegraph interests as well. For a long time this dual motive escaped the notice of nearly all his business rivals. Awed by the scale of his railroad activities, startled by the myriad of threats he posed, they were kept busy reacting to one unexpected thrust after another. It never occurred to them that he was intent on erecting not one but two empires. To their eyes the telegraph venture was a speculation in the same mold as the A & P coup.[3]

Gould was careful not to disturb this impression; indeed, his tactics followed the pattern of his earlier campaign. In January 1880 American Union gained access to the Pennsylvania system and won a legal battle for use of the Great Western's lines, which thrust it into the heart of the Midwest. Jay's merger of Union Pacific and Kansas Pacific in January enabled him to compete with A & P on those systems. Judge Dillon found a loophole in the 1869 lease between Union Pacific and A & P which, in his view, obliged the railroad to operate its telegraph for all comers. A & P denied American Union the use of its lines, and Western Union did the same on the Kansas Pacific. Bolstered by the opinion of counsel, Jay laid secret plans to seize the telegraph lines along both roads. His troops struck on February 26 with complete success; the next day an American Union officer reported, "Last night I took possession of all lines on the Union Pacific." Garrett timed the seizure of lines on his western system to coincide with this raid.[4]

Injunctions flew like hail as Western Union scrambled not only to recover its lost lines but to protect those on the Missouri Pacific and other Gould roads. The struggle moved to the courts, which was fine with Gould; the legal battles were diversionary moves to maintain pressure at one point while he attacked elsewhere. Over the next few months American Union won more legal skirmishes than it lost as Gould relentlessly expanded the system. He cut rates on a small scale but was careful not to trigger a rate war. The A & P experience had taught him that in telegraph wars victory went to the biggest battalions. By autumn Jay had established American Union as a serious competitor. The construction company declared a 50 percent dividend and attracted some curious bedfellows as investors, including Sage, Dillon, Connor, Fred Ames, Robert Garrett, Charles Woerishoffer, Thurlow Weed, Thomas C. Platt, and Thomas Nast, the cartoonist who had skewered Boss Tweed so unmercifully. Late in September American Union went on the New York Stock Exchange, where it traded at about 60.[5]

From the outset Jay understood that the battle would be won or lost in the market. While the pundits assumed he was angling to sell out to Western Union, he was actually trying to buy control of that company. His tactics blinded observers because they viewed his holdings in Western Union as mere leverage to force a sale. If the cognoscenti thought they knew Gould's game, they were at a loss to explain Vanderbilt's. In the winter of 1880, following the sale of his New York Central stock, he was reported at peace with Gould; by spring they had clashed over railroad matters and were supposedly at war. Rumor asserted that Vanderbilt had sold Gould 100,000 shares of Western Union and was "clearing out." Both men issued vigorous denials, and a month later the stock baffled the Street with a sharp decline.[6]

The stock fell because Gould had formed a bear pool to hammer it down. Convinced that competition with American Union would hurt Western Union's earnings and depress its price, he decided to build a position under the guise of later covering a large short interest. Where his allies sniffed market profits, Jay was after the stock itself, and to get it he had to sell it. As the Western Union annual meeting approached, public reports confirmed what Gould already knew, that management was divided over how to deal with American Union. One faction, led by Green, wanted to mobilize for war to the knife; another, headed by Vanderbilt and banker E. D. Morgan, a large stockholder, represented investors who craved peace and uninterrupted dividends. Gould muddied the waters by entering the proxy market to secure a large block of Western Union votes, thereby sparking rumors that he was about to seize control and force a consolidation. On cue Sage denied that American Union wanted a merger. "The people," he proclaimed solemnly, "will no longer submit to a monopoly of the telegraph business."[7]

The election passed without incident, but early in December Western Union's report revealed that earnings had dropped and a surplus of $1.6 million had been wiped out. The stock plunged below 78 amid rumors that dividends would soon cease and that a rate war loomed on the horizon. Legal battles with both Garrett and the Pennsylvania still raged with scant hope of resolution. To this dark outlook Gould added fresh gloom by cutting some rates, which Western Union promptly met, and announcing that American Union would construct two overseas cables of its own.[8]

The cable announcement revealed again Gould's genius for striking the right blow at the right place at the right time. The telegraph companies had working arrangements with different cable companies: Western Union with Anglo-American Cable, A & P with Direct Cable, and American Union with a French cable firm. None owned its cable link, all of which were controlled by foreign capital. After several rate wars the cable companies agreed in September 1880 to pool receipts and fix minimum rates. From these ingredients Jay concocted a familiar dish. He denounced the foreign cable monopoly and promised to smash the artificially maintained rates with his own American cables. That he would also impair Western Union's earnings and insure himself a cable connection in the bargain received less publicity. When the existing cables refused his demands, Gould organized his company, capitalized it at $10 million, called for 70 percent of the subscription at once, and got it within 40 hours.[9]

As the squeeze on Western Union tightened, the bears rushed in for the kill. To their amazement the price turned upward again until it passed 89 on January 5. While confused and angry traders groped for explanations, Gould had already executed his plans with quiet precision by selling off most of his Union Pacific at high prices and buying Western Union at low prices. "As an exhibition of superb manipulation," marveled one analyst, "this price movement has never been surpassed." Few traders outside Gould's own circle knew who had managed the coup or suspected that it was not an end in itself but merely a tactical maneuver in the campaign for a much larger prize. Rumors swept the Street like violent gusts that Vanderbilt had cornered Gould, that a telegraph merger was imminent, and a dozen other tales. Only Forbes, watching the melee from Boston, leaped early to the conclusion that Gould would emerge from the debris in possession of Western Union.[10]

In all Gould bought 90,200 shares, making him the largest holder of Western Union. That Vanderbilt and Morgan were ready for peace he surely knew. He and Vanderbilt had met several times to discuss railroad matters since the latter's return from Europe in November, and on at least one occasion chewed the telegraph situation over at length. Now, through Dillon, he let Vanderbilt know he was amenable to talk and waited for him to make the first move. As with the Union Pacific merger a year earlier, Jay put the other side in the position of approaching him with the very proposal he wanted most. Instead of forcing the issue he could assume the role of mediator, anxious to please all sides.[11]

The summons came on Sunday morning, January 9. Jay brought Sage, Dillon, and Fred Ames with him to Vanderbilt's mansion. He recalled the meeting as "very stormy and threatened in the end to break up in a row." He parried Vanderbilt's thrusts about American Union in his coy, careful way by insisting that he did not control the company. Only its directors and stockholders could make an arrangement, but he would, in the interest of peace, use all his "influence with the persons holding the controlling interest." This favorite conceit of Gould's confounded Vanderbilt as it had Perkins; the usual way of potentates was to display the keys to their kingdom with pride rather than to deny ownership. Little was accomplished beyond an agreement to meet again the following evening. Next morning Vanderbilt summoned Green to his house and growled, "I have seen the Great Mogul." Thus did Green learn for the first time that negotiations had begun.[12]

On January 10 Gould, Sage, and Dillon met with Vanderbilt, Green, and two Western Union directors. The bargaining was tough but friendly. Gould and Sage played their familiar Mutt and Jeff routine to perfection, with Gould inclined to be lenient while Sage demanded a higher price for American Union. The final agreement contained something for everyone. Western Union issued $15 million in stock to exchange for American Union shares, a price Green estimated at twice the latter's original cost. It issued another $8.4 million in stock to take up outstanding A & P shares at 60. Gould had owned much of this stock since 1877, when his associates happily sold their holdings to Western Union at 25. Finally, Western Union appeased its shareholders with a stock dividend of $15.5 million representing earnings invested in the company but never capitalized. Therein lay a superb irony. The value reflected in this dividend was the fruit of Commodore Vanderbilt's reinvestment policy. What the father had accumulated by sound policy the son now handed over in large measure to the Commodore's mortal foe.[13]

Although the terms were kept secret for more than two weeks, word of the negotiations leaked out the next day. Western Union shot past 116 and American Union past 94 as frenzied traders rode the roller coaster of every rumor. The prospect of a telegraph monolith capitalized at $80 million aroused a storm of protest. Embittered voices asked what had come of all the promises to smash

Western Union's monopoly. The Board of Trade condemned the merger, the Cotton and Produce exchanges talked of organizing an independent telegraph, and Chicago merchants subscribed $500,000 to do just that. Hamilton Ward, New York's attorney general, was petitioned to intervene but concluded the state had no case. The Anti-Monopoly League met to vent its wrath. In Albany the assembly rushed through a bill to prevent the merger only to have the senate bury it in committee. Congress grumbled but did little.[14]

Financial journals united in denouncing the consolidation as "another immense stock-watering upon which the people must pay dividends." The *Tribune*, supposedly Gould's mouthpiece, predicted it would "tend strongly toward . . . control of the telegraphic system by the Government as complete as its control of the mails." Everyone was slow to grasp the most significant aspect of the merger: Gould was not merely selling off another company but was about to take control of Western Union itself. Jay kept a low profile, fending off reporters with the disingenuous comment that "I am an outsider in this matter." As early as January 15 the *Tribune* hinted at his new role, but the light did not dawn until early February, when the actual consolidation took place. Gould, Sage, and Eckert joined the Western Union board and Dillon soon followed. Although Green had fought him hardest, Jay admired his ability enough to keep him on as president. True to promise, however, Eckert was made vice-president and general manager.[15]

"The country finds itself this morning at the feet of a telegraphic monopoly," mourned the *Herald*, sounding a theme it would reiterate with the monotony of Cato's "*delenda est Carthago.*" Events would disclose the *Herald* and its publisher to be somewhat less than disinterested in the matter about which they raised such fierce howls. *The Public*, which loved not Gould more but Vanderbilt less, scolded the *Herald* that "the country stands precisely where it did two years ago, except that its telegraph wires were then controlled by W. H. Vanderbilt, and are now controlled by Jay Gould. The change may not be for the better; it certainly could not be for the worse."[16]

Once in the saddle Gould abruptly canceled the American Union and A & P cable contracts. The French and Direct Cable companies resorted to suits but were beaten decisively in court. Gould's American Telegraph opened its first cable in September 1881 but broke down the next day. Nevertheless, it delivered an unmistakable message to the other companies: Gould's domination of the land wires along with an overseas cable signaled disaster unless they came to terms with him. In the spring of 1882 American Telegraph and the three established cable companies formed a pool to fix rates and divide earnings. Rates were increased and harmony restored.[17]

Acquiring Western Union enabled Gould to exploit the symbiotic relationship between railroads and the telegraph. Thereafter an ironclad contract with Western Union became a regular feature of all his rail ventures. During the summer of 1881 he obtained a new contract with Union Pacific. A year earlier, Bartlett and Judge Dillon had found a loophole enabling Union Pacific to do business with other telegraph companies; now they cheerfully helped Gould nail the loophole shut. As part of his agreement with Gould that November, Huntington extended the contracts between his roads and Western Union an additional ten years. The larger issues of rail strategy caused analysts to overlook this apparently minor proviso but it was vital to Gould. Later he would utilize rail maneuvers as blinds to conceal his true objective, the telegraph contract. Even his most astute business rivals were slow to grasp this tactic.[18]

A long and difficult campaign, brilliantly executed, finally brought Gould the communications empire he had long coveted. It never again slipped from his hands. The annual election in October confirmed not only Vanderbilt's departure but Jay's resolve to have a strong board. Solid management was needed to counter a fast-rising wave of opposition. The path of victory was strewn with embittered casualties already gathering their hosts for a fresh assault. Foremost among them was Garrett, who had twice followed Gould to the ramparts only to be left dangling there. After a nasty exchange of letters with Eckert, Garrett withdrew the B & O's wires from American Union and turned them over to a small company, Mutual Union.[19]

The telegraph wars were far from over. Like the railroad wars, they had merely entered a new phase.

The third and most incongruous pillar of Gould's business empire consisted of a rail system in his own backyard. His venture into elevated railroads has always seemed an anomaly, a sideshow to the main events of his career. The els were not a field in which he had past involvement or experience. Nor did Jay need fresh interests to occupy idle hands; he was already carrying responsibilities that would have staggered a dozen lesser men. Why, then, did Gould, at a time when his energy and resources were absorbed in projects stretching from coast to coast, decide to plunge into a new endeavor so far removed from his other activities?

The answer lies in a curious mixture of timing and circumstances. The elevated railroads fit the classic mold of enterprises that attracted Gould. They were properties of enormous potential with checkered pasts and futures rendered uncertain by legal snarls and corporate infighting. Their history had been brief but turbulent. Two early efforts to construct el lines floundered in the depression after 1873. The city's growing need for mass transit induced the legislature in 1875 to create a Rapid Transit Commission with authority to choose the form of transportation, select routes, and plan construction. The commission could either utilize the routes of the existing companies and allow them to extend their lines or it could create a new corporation for private capital to build under its supervision.[20]

In effect the commission did both. It authorized New York Elevated to extend its line up Ninth Avenue to Harlem, down to South Ferry, and up Third Avenue. The Gilbert Elevated was awarded comparable routes on Sixth and Second avenues. At the same time the commission created a new company, Manhattan Elevated, to build any line not completed by the other companies within specified dates. Manhattan issued $2 million in stock, most of which was taken by men interested in the other two roads, but it had nothing to do unless one of the companies faltered in its mission. The fact that Manhattan began life as an anomaly was to loom large in the convolutions that followed.[21]

The men who dominated the existing companies were familiar to Gould. The promoters of New York Elevated included W. L. Scott and David Dows, whom Jay knew from the railroad wars, and Samuel J. Tilden, then governor and presidential aspirant. Until someone penetrates the murky waters of Tilden's business dealings, his role will remain unclear. Apparently he persuaded Cyrus Field to buy into New York Elevated in May 1877 and become its president. The Gilbert line, which in June 1878 became Metropolitan Elevated, was dominated by Commodore Garrison and his son William, George M. Pullman, Horace Porter (whose path had crossed Gould's during the Gold Corner episode), and José F. de Navarro, a Spaniard with interests in banking, insurance, and shipping.[22]

Under Field's vigorous leadership New York Elevated expanded from six to thirty-one miles in less than two years. Metropolitan got bogged down in litigation with property owners along its route but managed to open its Sixth Avenue line in June 1878. To do the work, its owners formed a construction company that built the $9.7 million road for $21.5 million in securities. When friction developed between the two companies, the legislature suggested helpfully that they settle the matter between themselves. They did so in a way that caught everyone by surprise. Since the law prohibited consolidation, the two companies resurrected the ghost of Manhattan and leased their roads to it.[23]

In return for a 999-year lease of both lines, Manhattan issued each company $6.5 million in its own stock and agreed to guarantee interest on their bonds and pay a 10 percent dividend on their stock. The companies provided Manhattan with $9 million to complete construction of their lines, which was done in 1880. By this transaction Manhattan emerged unexpectedly as a holding company with the lease as its sole asset. It owned no securities in the two companies but controlled them through the lease, on which it had to pay stiff obligations. It possessed a monopoly of the city's elevated lines but at a price that ruled out dividends on its own stock.[24]

Investors greeted the lease with skepticism. For all the clamor about the need for mass transit, the els were still an unproven commodity. Commodore Vanderbilt had once spurned the chance to invest in a subway project by remarking brusquely, "I shall be underground a damned sight sooner than this thing." His son shunned the elevated because he believed people would never consent to tromping up and down stairs to reach the railroad. One analyst described Metropolitan as "a venture by men who undertook it upon a risk, much as the Union Pacific road was undertaken by a party of adventurous men."[25]

Others admired the "prolonged iron bridge—the first of its kind in the world," insisting the els were "well-equipped and have been, on the whole, well managed and have served the public well." By the spring of 1880 these brave words were drowned in a sea of troubles. A bill to compel five-cent fares passed the state assembly; although the senate crushed it, the attempt was a portent of things to come. Field and his friends grew dissatisfied because New York Elevated's earnings far exceeded those of Metropolitan. The Metropolitan crowd exacerbated the differences between the two factions by selling off all its Manhattan stock. Field responded by disposing of his 13,000 shares and resigning from the Manhattan board. Tilden went further by liquidating his interests in both Manhattan and New York Elevated, thereby violating a pledge to Field. Manhattan sank 30 points amid rumors that it would be unable to meet its lease obligations and that Field wanted the lease annulled. The court added to these miseries by ruling that property owners could sue the els for damages resulting from smoke, stench, cinders, or other nuisances.[26]

Although Manhattan avoided default, it remained in a state of siege for the next year as the contending factions struggled to reconcile their differences. New York Elevated's earnings provided Manhattan with a surplus that went to cover part of Metropolitan's inevitable shortfall. Moreover, the lease had equalized the capital stock and funded debt of the two operating companies, but in 1880 Metropolitan upset the balance by borrowing money to complete its Second Avenue line. The parent company stayed afloat only by drawing on a surplus accumulated before its first payments were due. Unless the earnings of its subsidiaries climbed sharply, Manhattan would default once the surplus dried up.[27]

In July 1880 the factions tried to break the deadlock through a full merger but could not agree on terms of exchange. Prolonged negotiations accomplished nothing. Meanwhile, one element of Metropolitan headed by Navarro and the Garrisons bought enough Manhattan to control that company. Amid these maneuvers the court of appeals delivered a stunning blow by ruling that the elevated's structures were taxable as real estate. Despite this muddled situation, all three stocks held firm until April 1881, when Manhattan led a downward plunge. The Garrisons sold out and announced they were "glad to do it." An appeal for relief from the newly imposed tax was denied by the city and jeered down by the press. *Bradstreet's* dismissed Manhattan as "water pure and simple," and denounced the "enormous mass of water injected" into Metropolitan. As Manhattan bottomed at 21 amid rumors of impending default, editors sounded the call for men strong enough to "take Manhattan at its present price and make it pay." A candidate stepped forward in the person of Sage, who bought the stock cheap and offered a plan to insure that Manhattan would honor its lease obligations for two years. Officers of both operating companies pledged fidelity to the lease.[28]

Here matters stood in May 1881. The next six months witnessed a string of events so bizarre and tangled that no one has yet unraveled their inner history. The *Times* thought it did in December 1881 by splashing across its first two pages an exposé of how Gould, Sage, and Field masterminded the swindle that gained them control of the elevated roads. "There is no more disgraceful chapter in the history of stock-jobbing," thundered an accompanying editorial. The *Times*'s version has shaped subsequent interpretations of the Manhattan episode for nearly a century. It is possible that events and motives were just as the *Times* claimed them to be, but the evidence is scant, circumstantial, often contradictory, and filled with puzzling holes. The Manhattan legend remains a tantalizing mystery, one not easily accepted or rejected with certainty on the evidence available.[29]

The precise nature of Gould's role is especially difficult to fathom. Later he testified that he had bought no elevated stock until September 1881. Certainly he had other fish to fry that spring. The telegraph merger had just been consummated, he was about to unite the Missouri Pacific and the Iron Mountain, and he had just bought out Tom Scott's holdings in the Texas & Pacific. In the Scott transaction Jay also acquired the *New York World*, which he later insisted was "really a mere accident." If so, it proved a fateful one, for about the time the paper changed hands it opened an attack against Manhattan that continued until early October. The purpose was to hammer down the price of Manhattan, asserted the *Times*, which called the campaign "wholly unprecedented in the history of journalism." The attack came primarily from a column called "Wall Street Gossip," where the tidbits were always attributed to a "prominent broker" or "leading operator." In fact, the *Times* charged, "The views really stated are those of Gould himself as filtered through his brokers and dependents."[30]

Thanks largely to the *Times*, the significance of the *World*'s campaign has been exaggerated by later students of this episode. It is likely (but not definite) that Gould did orchestrate the attack on Manhattan. He had done that sort of thing throughout his career, as had other traders blessed with access to print. But it is one thing to suggest that he used the *World* as a tool in his market operations and quite another to assert that from the start he directed a concerted bear campaign against Manhattan. Given the muddled state of elevated affairs, it presumes much to suppose that Gould knew in May what his goals were to be and how best to reach them. Or that the campaign was effective in producing concrete or significant results.

Apart from devoting considerable attention to the elevated fracas, the *World* did little that differed from its rivals. Far from being "wholly unprecedented," the puffing or pelting of stocks was standard practice. Rumor was the nectar of Wall Street on which traders and journalists alike fed, the former to nourish their schemes and the latter to spice their columns. The reporters of most New York papers and financial journals routinely larded their accounts with observations from "prominent brokers" or "leading operators" when they could not get direct information from some market kingpin. The *World*'s campaign persisted over several months, but it was hardly unique in either duration or choice of target.[31]

The emphasis on Gould's role has obscured the more important point of whether the campaign produced any tangible results. Given the crowded, fiercely competitive arena of New York journalism, it is absurd to suppose that the views of any single publication, let alone one of the city's least influential dailies, could exert a decisive effect on the price of Manhattan. Nor was the *World* alone in attacking Manhattan, which had been roundly abused by the press for over a year. During 1881 the two organs most hostile to Gould, the *Herald* and the *Times*, blasted Manhattan as harshly if not as regularly as did the *World*. The kindest word anyone could muster about the company was that it possessed leases of great potential value. Although the dailies thrived on ridiculing each other's crusades, none attacked the *World* at the time. Only in retrospect did the campaign seem unusual, and even then it might have gone unnoticed had it not been triggered by the first of several startling developments.

On May 18 Hamilton Ward, the state attorney general, suddenly filed suit to vacate Manhattan's charter and appoint a receiver for the company. No one inside or outside the els expected the action or had requested it. Why, then, did he intervene? Pressed by reporters, Ward explained that the company's request for tax relief had called his attention to its affairs. It was not only insolvent, he discovered, but had forfeited its franchise and had no legal existence or right to lease property. That same day an obscure bondholder sued to restrain Manhattan from paying dividends to the operating companies. The effect of these actions on the market was to leave Metropolitan "badly demoralized" and Manhattan "more than ever a football of speculation." One analyst concluded that "there is certainly some cunning hand pulling the wires against these companies, but whose it is can only be suspected at present." The most likely candidate was Field, just returned from Europe, but he vigorously denied the charge.[32]

Some Manhattan stockholders, girding for battle, asked Ward to discontinue his suit. The *Chronicle*, no friend of Manhattan, cautioned Ward not to put the state in the position of acting as an "irresistible bear" and warned that he "could not have moved better for speculative interests." In June it was revealed that large blocks of Manhattan and Metropolitan had been acquired by "Sage, Dillon and others of the Gould party." Field was alleged to be supporting elevated stock and "breathing out threatenings and slaughter against all enemies, his eye being evidently most steadily fixed on Mr. Gould." But where did Gould stand in the affair? Although his closest associates had entered the fray, his own role remained obscure. He and Field were neighbors and had been friends since working together in Wabash two years earlier.[33]

The most plausible explanation is that Gould did not enter the elevated fracas until late June, and then came in at the behest of Sage. Where normally he took the lead in ventures, this time he was drawn in by a close associate who was already there, knew the ground better, and needed his help. It did not take much to recruit Gould despite his heavy work load. He would not refuse an appeal from Sage, and he could never resist anything that promised both large profits and the challenge of solving so intricate a problem of corporate finance. The elevateds posed the sort of Gordian knot on which his intellect thrived. For years it had been said they would yield immense profits to anyone capable of putting their house in order. Gould and Sage knew Field well enough to realize he was not the man for the job, and no one else was in a position to accomplish it or oppose a concerted effort on their part.

Gould's presence in the contest was revealed early in July, just as the nation went into shock over the shooting of President Garfield. For the rest of the summer Garfield's fight for life upstaged all else in the press, including the elevated imbroglio. Two suits and a countersuit barred Manhattan from making July payments; in response New York Elevated's directors declared the lease forfeit and demanded return on their property. A week later Gould, Sage, and Dillon were elected to the Metropolitan board along with Dodge, Connor, and Sloan. Only William Garrison, Navarro, and Porter remained from the old board.[34]

Apparently Sage induced a reluctant Jay to come on board after negotiating terms with Metropolitan's largest holder, Sylvester H. Kneeland. Playing upon Kneeland's fears that New York Elevated might break the lease and compete with Metropolitan, Sage argued that a strong board could "build up the company." A powerful figure like Gould was needed to handle Field, he added, not only for his genius but for his following as well. Kneeland agreed to give Sage, Gould, and friends control of the Metropolitan board even though they owned a paltry amount of stock. Gould consented to join the board under those conditions. Some reports said the new directors were prepared "to treat the Manhattan lease as its notorious illegality deserves."[35]

Gould's entry into Metropolitan coincided with another bizarre and unexpected development. Ward abruptly dropped his suit in New York City and brought a new complaint before a court in Kingston presided over by Judge Theodore R. Westbrook. The new suit charged insolvency but not fraud and omitted the request that Manhattan's charter be forfeited. Ward's complaint was drawn by one of Connor's lawyers, who had also filed one of the suits to enjoin Manhattan from making its payments. On July 13 Westbrook heard arguments on the appointment of a receiver for Manhattan; two days later he came to New York City and named as receivers none other than Judge Dillon and A. L. Hopkins. All sides approved Judge Dillon but a co-receiver was needed because the Judge was not a practical railroad man. Ward asked the attorneys of both operating companies for a list of candidates even though the companies were not actually parties to the receivership suit. He also asked Westbrook to delay selection twenty-four hours while he ascertained "whether or not any of the gentlemen named had relations of a professional, official, or business character with the elevated railways."[36]

Not even the golden age of Tweed could boast a more curious legal proceeding. Gould's needs had been served to perfection. Manhattan had saved its charter and was entrusted to men in his employ. He had acquired influence in Manhattan without spending a dime on its stock and had made it more difficult for Field to break the lease. In this good work his best agents, unwittingly or not, had been Ward and Westbrook. How fittingly ironic that Westbrook had once sat in Congress with Tweed and later presided over his civil trial. How convenient, too, that another member of that same Congress was Russell Sage, who in this instance probably knew the judge better than Gould did. If coincidence be the stuff of conspiracy, here surely were the ingredients for a public outcry of grand proportions.[37]

But none occurred, at least not until months afterward. No pattern had yet emerged, and no one suspected even remotely the significance of these actions or foresaw the spectacular climax for which they prepared the ground. Later the *Times* would thunder indignation, but at the time it said little even though Judge

Dillon and Hopkins were widely known to be Gould's men and the *Times* seldom missed a chance to flog Gould for a villainy in progress. The choice of receivers not only escaped criticism but was hailed as a signal that all sides were about to resolve their differences. Field's position was simple: he wanted the lease honored or annulled. In mid-July he sued to have New York Elevated returned to its owners. Instead Westbrook issued an order merely requiring Manhattan to show cause why New York Elevated should not be returned.[38]

Both the Ward and Field suits dragged on through the scorching summer. Some Manhattan stockholders sued to recover the $13 million in stock issued to the operating companies in 1879, claiming that Manhattan had received nothing in return. As the legal quagmire deepened, Gould promised Field "enough lawsuits to last him the rest of his life" unless all sides reached a settlement. Time was running out on Manhattan, which had until September 30 to meet its lease obligations. The receivers' report confirmed that earnings would not provide enough funds and that New York Elevated still earned more than the guaranteed dividend, while Metropolitan fell short of its interest. The *World* labeled their report a "financial coroner's inquest."[39]

A sense of demoralization pervaded the elevated battlefield at a time when the business outlook was uneasy. The market fluttered with every change in Garfield's condition, and the bears were remorseless in milking that advantage. Garfield's death on September 19 brought apprehensions about his successor, who was hardly esteemed as a statesman. Against this backdrop of uncertainty it should have been easy to hammer Manhattan down. The deadline for default was approaching, hopes for a settlement had all but vanished, and most financial reporters (including the *World*) were singing a requiem to the company. Yet the stock held surprisingly firm. After touching a low of 15 1/4 on August 5 Manhattan recovered enough to range between 19 1/2 and 25 during September.[40]

Those who anticipated Manhattan's demise reckoned without the quixotic Judge Westbrook, whose behavior in this matter has yet to be explained. Possibly he wished to resolve once and for all a messy situation that had plagued financial circles, the courts, and the city of New York for years. Whatever his motives, Westbrook pursued a course breathtaking in its naïveté and indiscretion. While pondering the suits before him, Westbrook entered into private correspondence with Wager Swayne, the counsel for Manhattan. Swayne was Judge Dillon's law partner and widely known to devote most of his practice to enterprises controlled by Gould. These facts merely compounded the impropriety of a judge consulting privately and at length with the attorney for one party to the suits before him.[41]

For whatever reason, Westbrook seemed bent on preventing a default that would prove the insolvency charged in the Field and Ward suits. He showered Swayne with questions, asked his views on key points, and on one occasion invited him to Kingston to present them in person. If that were not enough, he suggested to Swayne several lines of argument that might be used on Manhattan's behalf. As the September 30 deadline neared, Westbrook declared himself "unprepared to say what the effect of the failure to pay . . . would be. If it can be paid in any way it would have a most happy effect." Who would raise the necessary funds? Westbrook's thoughts on that subject are worth quoting:

> It occurred to me it might be Mr. Gould. I do not understand the mysteries of Wall Street, having never speculated either there or elsewhere; but I can see with Mr. Gould's great interests how such an act by him to save the property would be a good financial operation as one affecting all his interests. To accomplish this result I am willing to go to the very verge of judicial discretion.

He was willing, the judge added, "because we have great practical questions before us in which the arms-length etiquette of courts is useless."

In this extraordinary correspondence the well-meaning but gullible judge had long since passed the bounds of judicial discretion. Whatever he revealed to Swayne was passed along to Judge Dillon, often at Westbrook's request, and probably to Gould as well. A grand opportunity had been dumped in Jay's lap without his lifting a finger. He had access to intelligence unavailable to Field or Ward. The judge's zeal to protect Manhattan and its stockholders put that company into a different light for Gould. He surmised that Westbrook would go to any lengths to avoid the default, which pumped life back into Manhattan at a time when press, public, and traders alike had given it up for dead. Manhattan's board at once proposed raising the cash needed by selling receiver's certificates. To hear this request Westbrook came to New York on September 29 and held court in a room adjacent to Gould's office in the Western Union Building. He approved the request despite a chorus of protests. Sage, in his role as Metropolitan's president, said the certificates "would not be worth a cent apiece." Field growled that he "wouldn't give a dollar for as many as a jackass could draw downhill."[42]

While the receivers tried to float their certificates, the deadline passed. Field and Sage petitioned Westbrook for return of their properties. Playing his part to the hilt, Sage called Manhattan's insolvency "hopeless and irretrievable" and denounced the certificates as "a futile and foolish attempt to put life into a corpse." Privately he and Gould told Kneeland that Manhattan had reached the end of its rope, a belief that already permeated the entire Street. Even more privately, while the dirge swelled, Jay bought Manhattan. At first he acquired equal amounts of Manhattan and the operating companies, reasoning that "whichever side won I would have the same interest at stake." On September 26 Westbrook wrote Swayne the letter cited above; that same week Gould began buying heavily into Manhattan. The company's transfer books were to close on Saturday, October 8, thirty days before the annual meeting. Late that afternoon, after trading had ended, it was revealed that Gould controlled 48,000 of Manhattan's 180,000 shares. His associates owned enough Manhattan to give them dominance of the company.[43]

Wall Street reeled in astonishment at the news. The *Times* quoted one of Gould's "closest friends" as saying Jay's purpose was to press vigorously Manhattan's claim to recover the $13 million worth of stock issued to the operating companies. Although this was an old threat, it carried new weight in Gould's hands. Field continued to sputter war chants even as he rushed to sound Gould on terms of peace. "I got thoroughly disgusted with the eternal fight," he claimed afterward. "There was nothing but litigation. I went to Mr. Gould . . . to see if we couldn't make one more effort to get the property out of the hands of the lawyers and the court." Committees from the three companies were appointed to thrash out a compromise. They met secretly in Gould's office and in a private room at Delmonico's. The series of conferences led observers to believe that "the close of the elevated railroad war is at hand." Signs of harmony were everywhere, the most conspicuous being Field's election to the Western Union board on October 12. Amid this glow of optimism Manhattan soared to 45 by the week's end.[44]

If Field still harbored hopes of getting New York Elevated out of the Manhattan lease, he was soon undeceived. Jay would entertain no agreement that did not keep the elevated roads together under Manhattan. In this bargaining Field's hole cards were his and Ward's suits, but these were easily trumped with help from Judge Westbrook. On October 19 Westbrook informed Swayne and Judge Dillon privately that he had decided against Field's application and wished to confer with them about it. Two days later he made the decision public and suggested further that Manhattan's claim for repayment of its $13 million in stock might have validity. Judge Dillon, who five months earlier had assured Field that such a claim was untenable, promptly asked the court's permission to sue for recovery of the stock, and Westbrook promptly complied.[45]

Months later Field would call the decision "a surprise and a disappointment," while Gould, with his best poker face, declared himself "thunderstruck." Beaten on every front, Field came hastily to terms. The lease was modified to give New York Elevated a guaranteed dividend of 6 percent; Metropolitan would receive the same amount only if earned, and Manhattan would get any surplus above the 12 percent required for the others. All back taxes and bond interest would be paid but not the six months of overdue dividends, and all suits were to be dropped. After the three boards ratified the new lease, it remained only to deliver Manhattan from receivership. The obliging Judge Westbrook performed this service a few days later without even bothering to notify Ward. What about the attorney general's suit? a reporter asked Gould. "The Attorney-General has informed us," came the reply, "that he would not stand in the way of our project, our settlement."[46]

It was not quite that simple. Ward was furious at having been ignored repeatedly during the settlement process. Privately he complained to Swayne that Westbrook had kept him in the dark and then had asked him to approve a fait accompli. But Ward was about to leave office and had no heart for further combat. He had begun the contest as Don Quixote only to find himself in the end playing Sancho Panza to the judge's knight errant. There was nothing left for him, he concluded, "except to submit to what seems to have been disposed of substantially in my absence."[47]

With the Ward suit eliminated and the new lease approved, Manhattan climbed into the low 50s. A later investigation exonerated Westbrook from all charges of wrongdoing, if not from gross impropriety and terminal naïveté.

Although it was widely suspected that Gould had gotten to one or both men, not a shred of evidence was produced, most likely because none existed. The episode merely proved again that Gould's shrewd grasp of human nature enabled him to know a judge without having to own him. Apparently no one thought to ask the intriguing question of what role, if any, the relationship between Sage and Westbrook played in this legal opera buffa. Whatever the answer, the cloud of suspicion followed Westbrook to his grave in 1885.[48]

On November 9 the Manhattan board reorganized and elected Gould president. As with Western Union, his accession to power marked not the end of struggle but the opening of a new phase in a prolonged and bitter contest. The opposition centered around Kneeland, who denounced the settlement as a betrayal of Metropolitan. The next few years would fulfill the *Times*'s prophecy that the elevateds "are likely to prove a mine of wealth to the lawyers."[49]

By the end of 1881 Gould sat firmly astride what would be the three pillars of his business empire: Missouri Pacific, Western Union, and Manhattan Elevated. No one yet recognized this fact or even suspected that he was there to stay. Like other men of boldness and ambition who had reached high, he was about to learn that the getting of empire was easier than the holding of it. He had at last climbed the mountain of success that had been his life's dream and had driven him relentlessly since boyhood. Finding himself at the top, he discovered that the only way down required the most precipitous of falls.

IV

HOLDING ON
1882–1887

24

War

In general, during the time that Mr. Gould was engaged in building up and consolidating great enterprises, it was a satisfaction to record the progress of the work, and its success was predicted. When he began to wreck railroads again—to squeeze the juice out of his oranges, and offer the empty skins to the public at an advanced price, we called attention to the fact that he was swallowing the juice, and advised investors not to buy empty skins at twice the price of good oranges.

—*The Public*, May 24, 1883

The telegraph has doubtless acquainted you before this with the enormous contract recently signed by Mr. Jay Gould, under the terms of which he is to "control every orange grove in Florida for thirty years," and has also, no doubt, given you some small notion of the danger that is to be apprehended from the corner in "alligators" which, in conjunction with his manipulation of oranges, Mr. Gould seems bent upon securing. . . . It may be, however . . . that Mr. Gould really intends to build a new line of railroad from the Gulf to the Atlantic, crossing the most productive part of the alligator fields . . . and that, having established this connection, he intends to transfer the entire wheat production of Minnesota each year to Florida with a view to fattening and increasing the growth of his alligators.

—*New York World*, March 27, 1881

IF ANY SCENE presaged the decade to come, it was the moment when Alice Snow glimpsed her uncle's reaction to the news that Garfield had been shot. Guiteau's bullet not only slew a president but also shot a hole through the fragile fabric of Gould's empire. Although the assassination in no way caused the rush of events that followed, it served to unleash forces long welling up behind the façade of prosperity—much as the shooting of Ferdinand would plunge a later generation into a war long gathering. A new era in railway strategy, for which Gould had served as catalyst, surged forward with a fury that swept him no less than others along in its momentum.

Although railroad men still paid lip service to the territorial doctrine in 1881, they were fast discarding it in favor of self-contained, interterritorial systems. Bitter experience taught them that the only sure connections were their own. Having lost faith in cooperation as a means for resolving disputes, they resorted increasingly to construction and consolidation. The result was an orgy of railroad building unparalleled in the nation's history—40,962 miles of new track during the 1870s and a staggering 71,212 miles the following decade. In the dozen years prior to 1890 railroad mileage more than doubled from 80,832 to 163,359. While much of this was branch line into untapped regions, a significant portion opened new through lines to places served by existing roads.[1]

System building dictated that each company have its own line to every point it wished to serve. Time eventually revealed the tea-party logic of this notion. The object of a self-contained system was to immunize the company from uncertainty by assuring connections; expediting the flow of traffic; maintaining rates; obtaining longer hauls, which brought a larger proportion of through rates; and penetrating new local territory. But the presence of more roads vying for business at the same points intensified competition and triggered a series of wars that drove rates inexorably downward. To their dismay railroad men discovered that the gains of expansion, on which they counted so heavily, were undercut by the need to haul larger amounts of traffic at lower rates than ever before. They invested heavily in new mileage only to find the growth of business offset by declining tariffs.[2]

Too late railroad managers realized that system building did not immunize them at all but merely enlarged the battlefield. Ironically, the unremitting rate wars of the 1880s drove them to a desperate search for fresh ways to harmonize their interests through cooperation. Moreover, the cost of building new lines and buying old ones imposed financial strains on every company. The strong were forced to cut or eliminate dividends; the weak flirted with bankruptcy. Expansion also flooded the market with masses of new securities that ultimately dragged it downward by sheer weight. As the market grew saturated it became more difficult to float new issues, which left several roads burdened with a large floating debt.

All these forces worked against Gould. He had invested three years of hard, sustained work to erect his empire, and the strain had taken its toll. In the spring of 1882 a reporter found him looking "somewhat careworn and weary." He needed rest or at least respite, but the worst was yet to come. For the next three years he would be locked into the grim task of holding his empire together. He had ridden the crest of expansion, prosperity, and optimism; now the tide was turning and he was obliged to swim against falling markets, pessimism, and constant bickering and war among the railroads. Moreover, success had cost him the advantages of obscurity and surprise. The possession of empire changed his role from hunter to hunted. His need was no longer to attack properties but to defend them, not to break the market but to sustain it at all costs. The enormity of his holdings rendered him vulnerable to all the enemies he had incurred along the way. There may have been railroad or Wall Street men as imposing as Gould, but none fought as many battles on as many fronts with as many formidable adversaries at the same time. The German high command might have learned at his elbow the art of fighting multifront wars with minimum forces. Moreover, Jay fought with a reduced arsenal in that foes were now familiar with his methods and no longer underestimated his ability. Every struggle forced him to reach deeper into his bag of tricks.[3]

The switch from offense to defense was for Gould a joyless transition. The pressure was unrelenting and compounded by the refusal of the financial community to take his new role seriously. Despite overwhelming evidence to the contrary, the Street persisted in regarding him as a trader rather than a businessman. His enterprises continued to be viewed as vehicles for his market operations even though Gould had long since reversed his priorities. Nothing revealed this misunderstanding more clearly than the widespread belief that Gould was operating chiefly on the bear side between 1882 and 1884. That a falling market did not suit his larger business interests did not impress analysts. They excoriated Gould because he was a bear and bears were wreckers of property, which Gould, to their thinking, had always been. The brilliant Charles Woerishoffer was the fiercest of bears, but he was a trader whose operations didn't undermine the value of companies he controlled. Ironically, it was Jay's dual role as businessman and trader that brought the harshest criticism down on his head.

For the rest of his life Gould struggled to escape the shadow of his past. The reputation of wrecker had followed him throughout his career and would chase him into the grave. Eventually the tinkle of the leper's bell wore him down, drove him to acts born of sheer exasperation. These resentments were but part of the price paid for success. Having risen to prominence against the grain, Gould accumulated an eager audience for his expected fall. The wisest of men, after all, understood that what goes up must eventually come down—in one way or another.

The heart of the matter was that Gould had too much to do and, for all his wealth and power, too little to do it with. His rail empire was still a work in progress, an inchoate world struggling to be born. Some parts had been acquired for the long haul, others merely to keep them out of enemy hands or as chips for a game never played. They lay scattered over the map like pieces of an unfinished puzzle. It would be intriguing to know what design Gould would have imposed had he the leisure to contemplate the problem. Instead, adversity and struggle dictated the outcome.

Nowhere was his hold more tenuous than in the East, and no road illustrated his dilemma better than the Wabash. By 1883 it had grown from 1,578 to 3,518 miles but the funded debt had swollen from $35 to $70 million. It was not a coherent system but a patchwork of "every neighboring piece of road lying around loose." In a region of strong, efficient roads it survived only by cutting rates, a policy that doomed it in the long run. Despite impressive gross earnings, the Wabash ran a deficit because of its large debt and high operating costs. What the Wabash needed above all was time to mend its weaknesses and impose order on its chaotic system. But it got no peace, thanks largely to the truculence of Vanderbilt, whose behavior mystified contemporaries no less than historians. The prolonged rate war of 1881 was but prelude to a course of intractable aggression Vanderbilt was to pursue for two years. It sapped the Wabash's earnings to the point where the company was flirting with default by the year's end. Wabash common sank from a high of 60 to 33¼ and preferred from 96¼ to 64¼.[4]

Gould responded to the crisis by assuming the Wabash presidency in December 1881. This was for him a highly unusual step and reflected the road's importance in his larger plans. Together with the Missouri Pacific it held the St. Louis terminal facilities. By controlling both roads, Gould could arrange the division of rates to favor one or the other. The Wabash, like the Katy, was important for the benefits it bestowed on Gould's primary interest, the Missouri Pacific system. To serve the one it sometimes became necessary to sacrifice the other, yet Gould dared not let the Wabash deteriorate to the point of threatening his control. Like many satellite roads it was a weak friend but a dangerous enemy, capable in the wrong hands of wreaking incalculable harm. The trick, therefore, was to find the cheapest possible way to keep the Wabash in tow.[5]

Jay gave out no figures on Wabash earnings until April, when his "ghastly exhibit," as one analyst called it, revealed the toll taken on Wabash earnings by the rate war of 1881. By that time Vanderbilt had surprised Gould with several moves. During the fall of 1881 he managed to gain control of the strategic Bee Line, which gave him a through route to St. Louis and enabled him to fight the Wabash directly. In the East he bought enough Reading stock to restore the peripatetic Franklin B. Gowen to the presidency in place of a man friendly to Gould. With Vanderbilt's support Gowen launched an ambitious extension program that included connecting the Reading with the New York Central.[6]

These moves forced an abrupt shuffling of partners in the East. Gowen struck an alliance with Garrett in hopes of obtaining a line into New York City for both their roads. To do that they needed to secure the Jersey Central, still under Gould's influence. Meanwhile the Pennsylvania, alarmed by Vanderbilt's presence in Reading, joined hands with Gould to oppose it. In February Gowen and Garrett, claiming to hold 92,000 of the Jersey Central's 185,000 shares, invited the current management to step down. Judge Francis S. Lathrop, the road's receiver, declined the offer. Amid this skirmishing a "brief and apparently unimportant bill" was quietly rushed through the legislature authorizing corporations to issue stock at par for taking up bonded debt, of which the Jersey Central had $8 million eligible for quick conversion. An outpouring of fresh stock would drown Gowen and Garrett as it had the Commodore during Erie days.[7]

When the bill passed over the governor's veto, an injunction sent the squabble into the courts, where it lingered for months. Lathrop died early in March, depriving Gould of a crucial ally. In April Jay fell ill and went to Florida to gather strength. After many delays the Gowen-Garrett faction forced an election on June 23 and swept Gould and his friends off the board. By this victory Vanderbilt eliminated the shadow threat of Gould's through route via the Jersey Central and the Pennsylvania. His fight with the Wabash raged on undiminished, with both sides resorting to fictitious billing practices so outrageous that the trunk line commissioner denounced them as "the most utterly scandalous of the many scandalous tricks . . . played in the traffic departments of the railroads." Part of the larger problem lay in the fact that the flow of traffic from the West had yet to fulfill the extravagant predictions made for it. By 1882 traffic from west of the Mississippi constituted less than 11 percent of eastbound and 14 percent of westbound trunk line freight. Chicago delivered more freight to the trunk lines in five days than California did in an entire year.[8]

East of the Mississippi the Wabash had all the traffic it could handle but at rates too low to earn a profit. Moreover, it still lacked reliable connections between Detroit and the Lackawanna extension at Buffalo. George Seney's Nickel Plate road paralleled Vanderbilt's system from Buffalo to Chicago and St. Louis. From the start rumor insisted that the Nickel Plate was a Gould project, his final solution to the problem of the missing link. In the fall of 1882 the road was nearing completion and no one doubted that the promoters would unload it for a hefty profit. Vanderbilt dismissed the Nickel Plate as "a poor piece of work" and warned that "no railroad can parallel us that will not starve to death."[9]

Speculation was rampant over who would grab the dubious prize, with Gould considered the most likely candidate. He started west with Hopkins on October 21, two days before the Nickel Plate opened for business. On October 26 Wall Street learned that the Nickel Plate had been sold, but the buyer was not disclosed for another two months. When the truth was at last revealed, the purchaser proved to be none other than Vanderbilt, who paid an exorbitant ransom for the road he had disparaged so freely. Why did he acquire an overpriced, poorly constructed road for which he had no use? Aside from ridding himself of a competitor, his aim was apparently to keep it out of Gould's hands and to use it in the fight against Wabash. It may have been a Pyrrhic purchase but it delivered another blow to the Wabash's sagging fortunes. The road was staying afloat only through notes endorsed by Gould, Sage, and Dillon; as early as July rumors flew that Jay was about to leave the Wabash presidency. In a November interview he confirmed the reports, adding that his faith was "pinned to the Missouri Pacific" and that the Wabash was "one more care than he bargained for."[10]

Leaving Wabash was one thing, abandoning it quite another. As the threat of receivership loomed, Gould conceived the idea of attaching the Wabash to his system by leasing it to the Iron Mountain. "Why, it will be another Kansas & Texas lease—that's it," cackled Sage to a reporter. "We shall give the Wabash its net earnings but nothing more. We shall guarantee nothing." The lease was signed on April 10, 1883; ten days later Jay announced a $10 million issue of collateral trust bonds to clean up the Wabash's floating debt. The lease kept the Wabash in the family but the fight with Vanderbilt soon deteriorated into a rate war that dwarfed its predecessor. By the end of 1883 an ironic reversal of roles had taken place. To recover lost traffic Vanderbilt began cutting rates ruthlessly, while Gould, thrown on the defensive, emerged as peacemaker eager to promote stability but helpless to deflect Vanderbilt from his intransigent position. Despite Gould's efforts the war raged through most of the year and into the next, leaving chaos in its wake. Against this violent storm Jay stood a grim and weary vigil like the captain of a sinking ship, clutching at straws of hope while the sea lapped ever nearer his feet.[11]

Things were no better in the West, where new transcontinental routes and alliances were fast transforming the railroad map. The Burlington's Denver line reached both Chicago and Kansas City. Palmer was doggedly pushing his Rio Grande Western toward Ogden where, in tandem with the Burlington, it would create a line independent of the Union Pacific. To the north Villard had revived the Northern Pacific and was driving it westward to connect with his Oregon roads. To the south the Santa Fe and the Frisco were struggling to piece together a transcontinental line. Even the Union Pacific entered the lists by commencing work on its Oregon Short Line to Portland.[12]

The growth of new rivalries raised the spectre of rate wars from which the Union Pacific had the most to lose. It was least prepared to endure a siege of low rates, and traffic gained by other roads would come at its expense. High rates, coupled with a monopoly on through traffic and lucrative local traffic, had enabled the Union Pacific to support its bloated financial structure, but high rates and a traffic monopoly would be the first casualties of a competitive struggle. The problem of meeting this crisis was compounded by the ambiguity of Gould's role in Union Pacific affairs. Although holding little stock, he was widely believed to exert a dominant influence in its policies.

In 1882 Jay still needed the Union Pacific to feed the controversial Omaha branch, which he predicted would "prove one of the most valuable acquisitions the MoPac has." Something else bound him to Union Pacific, something few outside his circle ever sensed or would have believed him capable of. Gould's long association with the road, his awareness of the role it played in his career, evoked in him a loyalty much like the nostalgic affection for one's first love. Fate would take him elsewhere and to other loves but never weaken this bond to Union Pacific.[13]

For his part Perkins believed the Union Pacific still danced to Gould's tune and acted accordingly. Before entering Denver in May 1882 the Burlington agreed to pool Colorado traffic but changed its mind when the amount of business exceeded its expectations. Despite this turnabout, Union Pacific, Santa Fe, and Burlington managed to negotiate a new agreement in mid-July. Clark, again ailing, praised the terms one day only to insist the next that the pool include a pledge that the Burlington would build no new mileage in Colorado. Perkins and his negotiator, Potter, adamantly refused to mix the pooling and territorial questions.[14]

The hitch was a familiar one. Perkins and Potter agreed that any territorial clause would be worthless unless Gould was a direct party to it. Three months of hard bargaining followed, during which time the Rio Grande began delivering all its eastbound traffic to the Burlington. "This the U.P. feel very *keenly*," Potter reported, but retaliatory moves by the Union Pacific pushed the roads nearer the brink of war. Both sides were having trouble keeping their freight agents from, in Potter's words, " 'getting over the fence' and making trouble." Then, in October, the Union Pacific abruptly withdrew its territorial demand and agreed to a pool on terms less favorable than those originally proposed. The Union Pacific caved in because it had other troubles that made the prospect of war intolerable. "Mysterious rumors" swirled that "the corporation was rotten," that bankruptcy loomed, and that Vanderbilt was about to gain control. The floating debt mounted alarmingly as Dillon's effort to pay off the debt with an issue of collateral trust bonds was blocked by a lawsuit. Investor confidence was shaken and no one knew where Gould stood in regard to Union Pacific. The Colorado negotiations shed no light on his position, and Jay did nothing to enlighten the curious.[15]

During the Colorado parlays Perkins and Gould tangled on another front, the Hannibal. Once again Perkins's inability to read Gould's true position left him baffled. The Hannibal was, as Forbes put it, an *"orphan"* going downhill fast, but the Burlington relied on it for connection to Kansas City. Gould and Sage sat on its board but so far had exerted little influence. For months Forbes made desultory swipes at buying the road at bargain prices. Then, in September 1881, John R. Duff suddenly cornered the stock. As the price soared to 359, Gould and Sage were reported among those caught short. After squeezing his victims Duff faced the task of unloading his stock in a market that regarded it "much like a pack of marked cards." The road had no apparent value to anyone except the Burlington; even the suspicious Perkins could not conceive how Gould would profit from grabbing it. On that premise Forbes and Perkins decided to play a waiting game to force the price down. Gould encouraged this policy by intimating to Perkins that he didn't want the Hannibal and would be happy to see the Burlington acquire it. Perkins made a careful investigation of the road and was content to string the negotiations out for nearly a year. By September 1882 he was ready to close with Duff at $42 per share only to learn that Duff had just sold the road to Gould, Sage, Dillon, and Fred Ames.[16]

Perkins was thunderstruck, his methodical mind utterly at a loss to comprehend why Gould had bought the road or what he would do with it. For the next month the lights in the Burlington chancery burned late as Perkins and his officers toiled at unearthing Gould's intentions. Every possibility was explored, dissected, and fitted with alternate scenarios for counterattack. They agonized at length over whether Gould would play the Hannibal "as a Wabash, or Missouri Pacific, or Union Pacific card." The Hannibal's president told Perkins that Gould planned to unite the road with Missouri Pacific and build to Chicago—but not for awhile. Perkins took it seriously enough to devise a wide range of reprisals.[17]

Impressive as this orgy of preparation was, it did Perkins no good because once again Gould had him looking in the wrong direction. He let Perkins dangle in suspense until January 1883, when he sent John B. Carson to propose a meeting. Carson told Perkins that Gould was ready to sell the Hannibal and had also mentioned "something about our telegraph relations" which Carson did not understand. Ten days later he reported that Gould and Dillon were willing to sell at their cost plus six percent interest but that Sage, in his usual role, objected. After a February meeting the negotiations recessed while Gould took one of his Florida vacations. In his absence rumors of the Chicago extension sprouted anew, prompting a complaint from Forbes that "Gould is building Roads in the newspapers from Quincy to Chicago." He suggested that Perkins build some paper roads of his own but added dryly, "I suppose you have several other things to do besides playing a game of Bluff with the great Jay Hawker and Chief Thief."[18]

Early in April Gould informed Perkins that a revised telegraph contract for the Burlington lines was a prerequisite to the Hannibal sale. Suddenly the light dawned; Perkins saw that the Hannibal served not Gould's railroad but his telegraph interest. There followed sixteen hectic days of negotiation in which letters and telegrams flew furiously between Boston and New York. Gould was at his cagiest, shifting ground from one disputed point to another, trying to wring every last advantage from Perkins by being conciliatory one moment and hardnosed the next. When an impasse developed over the telegraph contract, T. J. Coolidge intervened with Gould, apologizing that "I should not do so but I think the negotiations ended without it." Jay welcomed Coolidge's gesture, which led to a compromise settlement. As late as April 26 Jay feared the deal would fall through, but another flurry of telegrams produced an agreement the next day. In September 1882 Forbes had declared, "I don't think *today* I would give Gould what he paid for H & St Jo"; seven months later he agreed to a sum above that amount plus a new telegraph contract. Most financial journals reporting the transaction did not even note that a telegraph agreement was involved.[19]

The Colorado pool and the Hannibal trade should have eased tensions between Burlington and Union Pacific, but their effect was short-lived. The forces unleashed by expansion had altered the strategic situation so profoundly as to render old assumptions obsolete and conflict inevitable until order was established on some new basis. Like it or not, the competitive map had been redrawn and was changing almost daily. The territorial concept had gone the way of the dodo, and with it the stability that enabled roads to maintain rates effectively through pools. Agreements rose and fell like card houses, blown down by every shift in the competitive winds.

On this sea of change the Union Pacific pitched like a rudderless ship. Low rates caused a sharp decline in its net earnings, and in 1883 even gross earnings dropped as the Colorado mining boom collapsed and the Horn silver mines in Utah gave out. As the stock began to slip, both Dillon and the government directors asserted bravely that prospects looked favorable. Charles Francis Adams, Jr., toured the system, liked what he saw, and published a letter urging New England investors to buy. Apparently they did, for by February 1883 Bostonians held about a third of the company's stock. A month later Adams was elected to the board. None of this euphoria deceived Gould. He knew the Union Pacific was in deep trouble and doubtless suspected who would be blamed for its downfall, but there was nothing left for him to do. Having reduced his holdings to a token, Jay took little interest in the management. The presence of Adams, who had the confidence of the growing army of New England investors, made Gould's withdrawal easier.[20]

In April 1883 the Rio Grande Western reached Ogden. A war on rates between Utah and the Missouri River erupted despite several attempts to resolve differences. By October a $4 rate had been knocked down to 50 cents and then to 25 cents a month later. That autumn the storm burst in full fury from Utah to the Atlantic seaboard. The eastern trunk lines were engulfed in war even though the Lackawanna had at last been admitted to their pool. The chain of events triggered by the Burlington extension to Denver had reduced the Iowa Pool to a shell of its former self, and the widening war in the West renewed tensions between Burlington and Union Pacific. The Southwestern Pool was foundering in a sea of strife with no relief in sight.[21]

With brutal swiftness there followed a series of changes that confirmed the demise of the old order. Palmer's costly expansion program left Rio Grande helpless to withstand a rate war. In August he was forced out as president and replaced by Frederick Lovejoy. To the north Villard resigned from his Oregon companies and was rumored about to leave Northern Pacific as well. Another severe drop in earnings sent Union Pacific stock tumbling. Already the word was out on Wall Street that Dillon would soon surrender the presidency to Adams. The toll of battle soon claimed another victim, the Iowa Pool. Earlier it had admitted Wabash, Missouri Pacific, and St. Paul to membership in hopes of reconciling differences, but to no avail. The Burlington's Denver extension drove a wedge between its interests and those of the other members that could not be reconciled. In November 1883 the St. Paul announced its intention to withdraw from the pool at the year's end. The Union Pacific, desperate for traffic and allies, joined with the St. Paul and the Rock Island in a new pact aimed at the Burlington. On December 5 they consummated the Tripartite Agreement, described by one analyst as an alliance "thought before to be incredible." The Iowa Pool was dead and the Burlington isolated.[22]

How ironic that the Iowa Pool should perish in the end without any effort on the part of the man who had worked so long and hard to destroy it. Gould had not lifted a finger, had even opposed Tripartite at first. However, when E. P. Vining, the Union Pacific's controversial traffic manager, was named commissioner of the new pool, Jay had Missouri Pacific and Wabash apply for admission. The Northwestern grumbled but joined, leaving the Burlington as the lone outsider. Gould hoped the new combination would "result in a long and permanent peace and . . . help all our stocks." It did not. War continued to rage on both sides of the Rockies. In desperation both Perkins and Lovejoy of the Rio Grande appealed for help to Adams, whom they perceived as the rising star in Union Pacific. If they expected detachment from this renowned theorist on railroad questions, Adams soon undeceived them. He jolted Lovejoy with a truculent opinion that, while war was distasteful, his policy would be to "force the fighting until some results of a permanent nature were reached" and "one party or the other was thoroughly worsted."[23]

Having rebuked Lovejoy, Adams told Perkins that the Burlington had forced Union Pacific into its new alliance by joining hands with the Rio Grande. Perkins denied this charge heatedly, insisting that the Union Pacific "was deliberately the aggressor in the Utah fight," and that Clark regarded "the building of a Railroad anywhere near the Union Pacific as a high crime for which somebody ought to be punished." Yet Perkins held precisely that view and acted on it in building the Denver extension as retaliation against Gould. Railroad men, like political leaders, were adept at justifying the same actions for which they condemned others, and were no less wedded to obsolete ideas. Still, the tone of Adams's response steeped all sides in gloom. If philosophers opted for war, what chance had mere mortals of finding the path to peace?[24]

In the Southwest Gould's primary concern was to control the flow of traffic between that region and the twin gateways of St. Louis and Kansas City linked by his Missouri Pacific. By 1882 the "southwestern system of roads," as the press called it, consisted of one east-west line and three north-south lines. The Texas & Pacific stretched from El Paso to Texarkana and intersected all three north-south lines. At its eastern end the New Orleans & Pacific (now part of the Texas & Pacific) and the Iron Mountain formed a route from New Orleans to St. Louis. To the south the International extended from Longview to Laredo, with a branch to Galveston. North of Texas & Pacific the Katy reached its twin arms into Kansas and the Mississippi Valley at Hannibal. By monopolizing the north-south lines Gould dominated traffic between St. Louis and eastern Texas. Smaller roads in this area, dependent on Gould lines for connections, retaliated with rate wars and attempts at expansion. Amid the fighting Jay succeeded through some intricate maneuvers in picking up the Galveston, Houston & Henderson, a small road linking the International to the Gulf, and managed to keep a firm grip on the north-south artery.[25]

The weak link in the Southwest lay on the east-west line, where the Texas & Pacific proved no match for Huntington. Where Vanderbilt brayed and blustered, Huntington bored and clawed like a mole toward completion of his "Sunset Route" from El Paso to New Orleans. For a time he toyed with the idea of securing the Frisco and the Atlantic & Pacific, which Crocker thought "the best road to do business over between New York & San Francisco." In August 1883 those roads completed their line to a connection with the Southern Pacific on the Colorado River. By then Huntington had leaped at a chance to acquire the late Charles Morgan's system of steamships and railroads in Louisiana and Texas, giving him the road he needed from Houston to New Orleans as well as a formidable north-south line (the Houston & Texas Central), some branch roads, and a fleet of steamers.[26]

Huntington could now move traffic from San Francisco to New Orleans by his own rails, and from New Orleans to New York by his own steamers. With fine disregard for the agreement with Gould he began shifting eastbound business from the Texas & Pacific to his own line. At the same time he bled transcontinental business from the Union Pacific and the Northern Pacific by slashing rates. A potent new challenger for through traffic had appeared and would not be denied. Gould saw what was happening but was helpless to prevent it. The El Paso end of Texas & Pacific had already become a financial albatross. Apart from being poorly built, it had little local traffic and Huntington was siphoning away its through business. By 1883 it had become the Wabash of the West, a load Gould could not afford to carry but dared not drop.[27]

Like it or not, Jay had to concede the territory west of Fort Worth to Huntington. He concentrated, therefore, on the region served by the Iron Mountain and the Texas & Pacific's eastern end, a country rich in cotton, sugar, and timber. In a small way at first he began the tactic he would follow for a decade of constructing small branches and feeders wherever business and/or a subsidy could be had. Contrary to legend, Gould always appreciated the importance of local traffic as the backbone of a road's income. He implemented on his southwestern roads the same development policy he had emphasized with the Union Pacific. The great transcontinental wars underscored this need by intensifying the competition for through traffic. Development was a quieter, less spectacular course to pursue, but it paid better in the long run.[28]

During 1883 Gould was reported about to gain control of the East Tennessee, Virginia & Georgia, a large but financially weak system owned by the same clique that had built the Nickel Plate. A few months later he and Sage turned up on the board of the powerful Louisville & Nashville, which kindled fresh rumors of a Gould incursion into the Southeast. It was all smoke without fire. Jay lacked the resources to expand and was too embattled to open new fronts. Like other conquerors he had marched too far too fast and was now absorbed in defending and consolidating his gains. It was all he could do to hold on.[29]

In Mexico he could not even do that. There Jay's plans died aborning, a victim of circumstances and the pressures assailing him elsewhere. Without the help of Ames and the Boston group Gould could not raise the funds to proceed. Although he announced in September 1882 that enough money to build the road had been subscribed, the construction company did just enough work to keep the concession alive. By March 1883 Dodge conceded that there was "no prospect of our going ahead with the Oriental road," whereupon Gould negotiated a consolidation of his project with the Mexican Southern headed by Grant. Although his interest in Mexico remained keen, he built no railroads there. In the end it was the stubborn Palmer who completed the road between Laredo and Mexico City.[30]

25

Siege

I sometimes think I should like to give up business entirely. The care and worriment attending large business interests are very great, but besides that fact the manner in which motives are impugned and characters assailed is very unpleasant.

—Jay Gould, *Commercial and Financial Chronicle*, March 25, 1882

The yacht of Mr. JAY GOULD, *it appears, ran through a tug yesterday for the purpose of hitting a schooner on the other side. The natural conjecture that Mr.* GOULD *had "gone short" of both the injured vessels will, we trust, prove to be baseless . . . but he ought not to prey upon our useful merchant marine.*

—*New York Times*, September 5, 1883

I am so fully convinced that Gould . . . read[s] all messages that look like R.R. messages that I dare not trust the wires except with a cypher which I change from day to day. . . . George Jones . . . brought Tweed down to his marrow bones—why may he not win much credit by getting Gould into his proper quarters, the P—— a place I will not mention for fear of the law of libel!

—John Murray Forbes, letter to John W. Garrett, December 15, 1882

There appears to be a turn in the tide of that success with courts and judges which hitherto has marked Mr. Gould's career. . . . Indeed, some observers have remarked that there seemed to be a race between the members of the judiciary to see who could most speedily render a decision against the once all-powerful ruler of courts and the stock market.

—*Bradstreet's*, January 6, 1883

WHILE GOULD's railroads embroiled him in perpetual warfare, the telegraph and elevated companies subjected him to prolonged siege. No sooner had he captured these properties than he was obliged to defend them from attack by erstwhile allies. Much of the fighting took place in the courts and on Wall Street, two arenas in which Gould had long reigned supreme. But times were changing and so had Jay's position. As the holder of large properties he found himself on the defensive and compelled to adopt new tactics. Moreover, the telegraph and elevated contests invoked other issues that ultimately proved more damaging to Gould than the fights themselves. Both resurrected the spectre of his pervasive influence in politics and communications.

These charges were not new, but in the heat of battle they assumed new dimensions. In 1881 Gould was regarded as one of the nation's leading financiers, the ruler of an imposing business domain. By 1884 there emerged the image of a sinister figure who controlled not only a transportation empire but a communications monopoly embracing the national telegraph system, the overseas cables, several leading newspapers, and the Associated Press. Independent newspapers depended on Gould's telegraph, as did businessmen, bankers, brokers, and stock exchanges. His command over the flow of information enabled him to rig the market, confound business adversaries, promote his enterprises, tear down rivals, and punish organs that opposed him. So too in politics, where his wealth and power earned him the deference of presidents, congressmen, cabinet members, governors, legislators, and, of course, the judges who served him so faithfully. His dark, elfin figure joined the standard repertory of political cartoonists, who relished his high forehead, pointed ears, and dark eyes with their mischievous glint. From their pens the Mephistopheles of Wall Street sprang vividly to life.[1]

This image of Gould, in the form it assumed during the 1880s, was shaped not by disinterested critics but by men he had worsted in battle and who feared the effect of his power on their own fortunes. It rested not on evidence but on suggestion and inference, the ancient premise that smoke always signaled fire. The question of how Gould managed to accomplish the vast schemes attributed to him, or where he found time to get them all in, went unaddressed. If intelligent businessmen like Forbes believed Gould spent his day poring over intercepted telegraph messages, it was easy enough to persuade ordinary minds that he was the devil incarnate.

By Jay's own estimate it took "twenty or thirty meetings" to hammer out the elevated agreement. Everyone approved it except Kneeland, who sued to have it set aside. His lone opposition threw a wrench into the consolidation. The plan called for Manhattan to issue first preferred shares to New York Elevated and second preferred to Metropolitan in exchange for their stock, but Kneeland's group refused to accept the new shares. Gould then divided the new stock into "stamped" and "unstamped" shares. This attempt to force Kneeland's hand instead forced that of the governing committee of the Exchange which, after much waffling, decided not to list the new shares. Furious at this rebuff, Jay disarmed their objections by preparing a new consolidation plan. In December the state engineer approved the new stock issue and two judges in separate cases upheld the validity of the October 22 agreement. The supreme court denied an appeal and the new attorney general refused Kneeland's request for a new suit to annul Manhattan's charter. The governing committee agreed finally to list stamped certificates for those companies obtaining stockholder approval. Only Metropolitan failed to do so, thanks to Kneeland's influence.[2]

Beaten on every front, Kneeland fought doggedly on, assuming in a curious way Jay's familiar role of patient gadfly biding his time while new troubles assailed his opponent. The *Times*'s exposé of the elevated coup triggered an angry war of words among rival papers. The nickel fare and taxation controversies flared anew, forcing Gould into a testy exchange of letters with the mayor. In the spring of 1882 hearings on the conduct of Ward and Westbrook gave the elevateds more bad press even though both men were exonerated. Threatened by a legal siege, Manhattan obtained an injunction barring Metropolitan holders from bringing more suits against the October agreement. Meanwhile, Kneeland confidently awaited Metropolitan's annual meeting in July, at which he expected to oust the Gould-Sage puppet board.[3]

The nickel fare campaign stalled and Jay arranged a legislative compromise on the taxation issue. In June he had the Metropolitan board pass two resolutions that caught Kneeland by surprise. One shifted the annual meeting date to November; the other prohibited stockholders from voting unless they were registered on the books, which required them to obtain stamped certificates. An outraged Kneeland persuaded the attorney general to bring suit removing the Gould clique from the board. A temporary injunction permitted the unstamped shares to vote and produced a stormy meeting. As tempers flared, the features of Sage and Field twitched nervously while Gould, his expression "enigmatical as ever," flitted restlessly from place to place. He left during the counting, which gave Kneeland's ticket a decisive victory. "We have lost the election," Sage told a *Herald* reporter, "but this will not be the end of it. There will be more litigation or—a compromise."[4]

Beset with other problems, Gould preferred compromise and delegated the negotiations to Field. Victory had hardened Kneeland's position, however, and Field gave up in disgust after several of his proposals were rejected. The prospect of an open competition between the elevated lines loomed, but both sides were content instead to lunge and parry in the courts for another eighteen months. In November 1883 hearings on Kneeland's suit to invalidate the October agreement finally commenced. Public interest centered on the appearance of Gould, who by this time had launched a busy career testifying before courts, commissions, and committees. As always he sat twiddling his spectacles, imperturbable except when "his eyes would light up with a merry twinkle as he spoke of his Wall street business." The crowds who jammed the courtroom strained to hear his testimony, given in a voice so low it barely reached the stenographer.[5]

A sensation arose over the whereabouts of the books listing Gould's transactions in Manhattan during 1881. The court demanded them as evidence but Jay did not know where they were. Connor was in Europe amid dark hints that he had taken the books as traveling companions. Of the partners that left only Morosini, who was ordered to produce the books or face contempt charges. Day after day he appeared before the judge to protest that he had hunted high and low, consulted Sage, even ransacked an obscure storeroom at Connor's office to no avail. The *Times* heaped ridicule on the hapless Morosini and demanded his incarceration as an example. After two weeks the judge denied the contempt motion, but the trial dragged on into the new year. Through these months Gould remained curiously impassive, content to await the verdict or preoccupied with battles raging elsewhere. His inactivity was matched by Kneeland's intransigence. A speculator unaccustomed to leading crusades, Kneeland found himself in the unlikely role of hero among editors and businessmen who despised Gould.[6]

Like Kneeland, the enemies of Western Union concluded that the best way to beat Gould was to borrow his tactics. They launched not only a barrage of lawsuits but a battalion of telegraph companies old and new: B & O, Mutual Union, Bankers & Merchants, American Rapid, and Postal. Bankers and Postal were both organized within months after the consolidation, the offspring of promoters eager to catch the wave of antimonopoly indignation. The guiding hand of Bankers was a smooth operator named A. W. Dimock, that of Postal none other than Gould's longtime market rival James R. Keene. American Rapid had begun the previous summer but announced expansion plans a few weeks after the merger. Mutual Union, incorporated in October 1880, followed suit by vowing to acquire "nine-tenths of the profitable telegraph business of the country."[7]

Surveying this host of potential allies, Garrett decided in September 1881 to lease the B & O's lines to Mutual Union. The dominant figure behind Mutual was George F. Baker, president of the First National Bank. Destined to become one of the nation's most powerful bankers, Baker was, like Gould, a man of deep silences, as inscrutable as he was capable. Jay resorted to his favorite tactic of having several suits filed challenging the legality of Mutual Union's policies and status. He suspected that Baker, like most bankers, preferred negotiated settlements to long and costly wars; the suits were his way of inviting Baker to the conference table. Baker took his cue promptly and proposed a community of interests. In return for dropping the suits, Gould agreed to buy about 30 percent of Mutual Union's stock and place it along with Baker's holdings in joint trusteeship. Analysts leaped to the conclusion that Western Union was about to swallow another rival only to discover, as one admitted, that "the relations between these two telegraph companies are the puzzle of the street." Scarcely a week passed without some development "strangely incongruous" with their supposed pact.[8]

While the Street puzzled, Gould tried to seize Mutual Union through one of his stealthy maneuvers. He increased his holdings to 40 percent and, leaving the stock in his own name, sold it all to Western Union at cost. Baker countered by recruiting other stockholders to place slightly more than 50 percent of Mutual Union's stock in trust with himself and two allies as trustees, making it impossible for Gould to gain control. Undaunted, Jay enjoined the new trust, then revived the old elevated gambit by persuading the attorney general to bring suit invalidating Mutual Union's charter. In several editorials the *Herald* dismissed Gould's suit as "novel and ludicrous." All this was familiar if irritating ground for Gould, but an ominous trend was emerging: one of his most reliable weapons was losing its potency and being turned against him. A close observer noted that "litigation for stock-jobbing purposes" had infected Wall Street with "an epidemic of lawsuits, many of them directed at Gould properties in general and Western Union in particular." In New York the attorney general, with fine impartiality, also sued to invalidate Western Union's charter. On the heels of this action the court dismissed Jay's suit against Baker.[9]

Mutual Union was also elbowing its way into railroad contracts with anti-Gould interests, notably Vanderbilt. Jay recognized the danger of allowing it to become the rallying center for his telegraph enemies and came to terms with Baker. Western Union agreed to lease Mutual Union at a stiff price, thereby eliminating another rival. But more waited in the wings, and the lease triggered a rash of suits challenging its legality. The telegraph siege did not end because, as one observer quipped, "the Western Union lion has just lain down with the Mutual Union lamb, and is now digesting it."[10]

Late in 1882 Garrett's search for allies brought him to the Burlington, which had long been eager to escape its contract with Western Union. The suit to invalidate Western Union's charter inspired Forbes to a Machiavellian scheme worthy of the archenemy whose tactics he deplored so ferociously. His idea was to "put some powder behind the present attack" by covertly supplying a third person with some shares of Western Union so that he might intervene in the suit on behalf of the public interest. Gould had used this tactic with great success in the past, but for what Forbes considered evil purposes. That Forbes embraced it for the cause of righteousness only attests to the extent his judgment had been warped by his hatred of Gould.[11]

Morality aside, Forbes lacked Gould's skill at execution. In seeking the right stalking horse, he invited George Jones of the *Times* to add Gould's scalp to that of Tweed. "You go in to *crusade,*" Forbes argued. ". . . He cannot buy you out." Forbes could not intervene personally because he had "a truce with Western Union and with Gould and am not ready to go to war." He could offer financial and legal help but must "keep my hands free to stop fighting or meddling whenever business requires it." When Jones declined, Forbes asked Richard T. Olney, a corporation lawyer, to scour the list of Western Union stockholders for a suitable replacement. At the same time he handed Olney a copy of the Burlington's telegraph contract to study, explaining, "Our hope is to break it on some flaw." Although Olney found a small loophole in the contract, the Burlington concluded the risk wasn't worth the effort and Forbes's brief fling at intrigue died aborning.[12]

While Forbes puttered, a more formidable coalition gathered against Gould. Once again the catalyst was Garrett. While supporting Mutual Union in 1882 he decided to open a second front by challenging Gould's cable monopoly. When the new pooling agreement commenced that spring, many businessmen and newspapers were unhappy because rates went up and all overseas messages had to be sent or received over the wires of Western Union. No one liked the increased costs, especially the newspapers that were utterly dependent on the telegraph and cable systems. Nor was Forbes alone in his dark suspicion of "Gould and his detectives on the Western Union deadhead lines." The threat posed by Gould's control of the cables drove some new allies into Garrett's camp, notably James Gordon Bennett, Jr.[13]

A force field of emotional loose ends charged with flashes of brilliance, Jamie Bennett had already established himself as the black sheep of journalism. His father had made the *Herald* New York's leading paper in advertising and circulation. For a decade young Bennett maintained its position without contributing anything new beyond some extravagant hoaxes and the financing of highly publicized expeditions such as sending Stanley in search of Livingstone and the ill-fated *Jeannette* polar mission. Possessed of an able staff, he decimated it with outbursts of explosive temper and a tyrannical arrogance. On one occasion he demanded from his editor a list of men considered indispensable and discharged them all, saying, "I will have no indispensable men in my employ."[14]

Tempestuous, impulsive, self-indulgent, and erratic, Bennett lived life at the extremes like the spoiled child he had been and, some said, still was. Social ostracism made him even more eccentric. His escapades following bouts of drinking became the stuff of legend, but even they paled before the scandal that arose when his engagement to the daughter of a prominent Washington family was abruptly terminated. In January 1877 the lady's brother accosted Bennett outside the Union Club and horsewhipped him. A duel followed in which neither man was hit. It was the incessant gossip that wounded Bennett more, driving him at last to seek refuge in his beloved Paris and reduce his presence in America to flying visits.

In that unpredictable way history has of connecting disparate events, Bennett's exile was to have fateful consequences for Gould. The fact that Bennett ran the *Herald* by cable, and that his paper used more cable dispatches than any other daily, made him far more dependent on that service than other editors. Pugnacious by temperament, Bennett fancied himself a reformer and loved a crusade, especially if it made good copy. His father had battled the telegraph monopoly since the 1860s; an opposition cable suited the son's needs perfectly. It defended his self-interest, promised profit, and provided him an ideal target in the form of Gould, who was already the city's favorite whipping boy. That Gould dominated Western Union, owned rival papers, and supposedly controlled the Associated Press strengthened Bennett's desire until he became Jay's most persistent tormentor.[15]

In August 1882 Robert Garrett reported from Europe that Bennett was "willing to put his money in clean and clear." By October Garrett was openly soliciting subscriptions for a new cable. Gould and Field sent out peace feelers and were rebuffed. An alarmed John Pender, M.P., chairman of the English Direct Cable, crossed the Atlantic to mediate but returned empty-handed. Gould refused to liberalize the cable contract and Garrett shared his son Robert's belief that "Pender Gould and Field are acting in concert and I have no doubt whatever that their alliance is as close as it is possible to make one."[16]

Shortly after Pender's departure the western branch of Associated Press defected to Mutual Union. Bennett chose that moment to fire the opening salvo in what proved a lengthy barrage against Gould, Western Union, and the evil cable monopoly. To astute observers Bennett's change of front was blatant. In May 1881 he had applauded Gould's laying of a rival cable, had even blasted the cliché that "whenever anything goes wrong in Wall street Gould is held responsible. . . . It is about time this foolish gabble with respect to Jay Gould should stop." But that was last year. Since then Gould had united the cables into a pool Bennett now condemned as "the most gigantic system of organized robbery in existence in any civilized country."[17]

The loss of Mutual Union in February 1883 stalled but did not stop Garrett. Aside from raising money for the new cable, his chief problem was breaking Western Union's stranglehold on the railroad business. The best hope lay in joining forces with Gould's enemies, such as Vanderbilt, and with the other telegraph companies. The glib-tongued Dimock was expanding Bankers throughout the East and Southeast in extravagant fashion because, he later testified, "the existing state of things was too oppressive to be borne." He also procured a contract with the newly formed United Press and vowed never to sell out to Western Union. Postal did little under Keene, but in August 1883 it elected a new management dominated by John W. Mackay. A shrewd, energetic Irishman five years Gould's senior, "Bonanza" Mackay had extracted a fortune from his mining empire in the Sierra Nevadas and come east to test his mettle in Wall Street. Together with banker George S. Coe he locked up a majority of Postal stock in a pool for three years and embarked on an expansion program. That same month Postal, Bankers, and American Rapid formed an agreement to exchange business.[18]

Mackay was a formidable opponent: rugged, independent, unfazed by Gould's reputation. His arrival on the scene came at a most inopportune time for Western Union, which faced an ugly labor dispute that summer. The fledgling Brotherhood of Telegraphers demanded not only better wages and hours but equal pay for both sexes. Unmoved, Eckert issued a disingenuous reply and hewed to his usual hard line. On July 19 the Brotherhood struck all the companies. Although Western Union was hardest hit, enough operators stayed on the job to maintain a semblance of service. For a month the Brotherhood held out, but the telegraph companies remained implacable; "the Company will close every office rather than submit," snarled one Western Union official. Starved into submission, their cause distorted by the press and undermined by sporadic acts of violence, the strikers capitulated on August 15. Asked his views on the strike by a congressional committee, Jay asserted that no large corporation treated its employees better than Western Union, that the operators were well paid, and that "the poorest part of your labor" was at the bottom of the strike. "Your best men do not care how many hours they work," he added sanguinely; "they are looking to get higher up; either to own a business of their own and control, or to get higher in the ranks."[19]

As usual Jay was content to recite the canons of orthodoxy held by most businessmen. The law of supply and demand dictated wages and prices; stockholders, not individuals, ruled corporations; government regulation was unwise; monopoly was a myth, competition a reality; the laboring class needed education, not unions; the Republic was threatened less by capital than by "large masses of uneducated, ignorant people"; success went to those with the grit and ambition to seize it. Jay never lapsed into heresy or originality on such questions, and he was aware on this occasion that the matter went beyond theorizing. The strike added fuel to the demand for a competitive telegraph system to reduce rates and ensure *"certain and absolute secrecy"* of transmission. Bills were circulating in Congress to investigate the practices of Western Union and revive the old notion of a postal telegraph run by the government. On Wall Street the opinion prevailed that Gould was angling to unload Western Union on the government at a fat price. The *Herald* reflected a common view by sneering, "What has Western Union got to sell?"[20]

From the beginning critics dismissed Gould's acquisition of Western Union as merely another predatory raid and predicted that he would exit after squeezing the company dry, leaving it with a bloated capitalization on which dividends could never be earned. When dividends continued to be earned, they were attributed to monopoly rates. Undeterred by the fact that no one could agree on what constituted a suitable basis for determining its proper capitalization, critics insisted that disaster loomed on the horizon. That opinion still prevailed at the end of 1883, and for good reason. Garrett, Mackay, and Dimock expanded their lines rapidly and goaded Western Union into cutting rates. Although the company won an important suit, litigation assailed it on every side. The annual report, issued in October, was skimpy enough to arouse suspicion that Gould had something to hide.[21]

The hosts arrayed against Western Union grew stronger when Mackay threw his weight into the cable fight as well. In October he and Bennett signed contracts for the construction of two new Atlantic cables. Pender made another effort to avert a cable war but to no avail. Confident that Gould had been thrown on the defensive, Bennett cranked up the *Herald* for an all-out assault on the cable monopoly. That summer Jay sent George abroad to peddle 50,000 shares of Western Union in London, but the English would not pay his price. Wearily Jay dug in for a long siege, aware that, in the eyes of his enemies, Western Union looked increasingly like a fortress ripe for the fall.[22]

The talk about retirement began in 1883, although there had been hints of it earlier. Jay was finding it ever harder to maintain his mask of indifference. His nerves had begun to rebel against the strain of prolonged concentration. Fatigue and tension not only drained his energy but weakened his resistance to personal attacks. The hounds were always at him, braying incessantly, nipping at his heels, frustrating his attempt to assume with dignity the role of businessman he had tried so desperately to carve for himself. When he talked of retirement in 1883, he meant not from business but from Wall Street. The time seemed right to close his career as an active operator. On New Year's Day George joined Morosini and "Wash" Connor as general partners in the latter's firm; Jay continued as special partner but hoped to turn most of the work over to his son. He had commissioned a magnificent new yacht, the *Atalanta,* which was expected to be ready in June. In January he announced plans to leave that summer for a two-year cruise around the world. It was not ill health that prompted the trip, he assured a reporter. "I am going to try a little play. I did not have an opportunity when I was young, and I must do my playing later in life. If I like it I may keep it up."[23]

No one believed a word of it.

In time the next-month-for-sure voyage became a standing joke among editors, another routine from Gould's incomparable repertory. No one ever knew whether Jay was sincere or merely casting out bait; indeed, how could anyone know? As usual Jay said little and everyone believed the opposite of what he said. Clearly he wanted the Street to accept the decision as a sign that hereafter he was to be regarded as a businessman rather than a trader. That did not happen, because Jay did not leave on his trip or quit the market. He may have wanted to, but his business affairs were in a tangle; as one editor noted astutely, "Mr. Gould's desire to travel and his desire to litigate do not fit each other."[24]

The harsh truth was that Jay could not quit the market because his fortune and his empire were inextricably tied to its vagaries. To finance his vast holdings he relied on profits from trading operations and on bank loans for which securities were pledged as collateral. He depended too on the resources of associates, brokers, and traders who followed his lead and were no less at the mercy of market fluctuations. So long as prices advanced, as they did from 1879 to 1881, Gould had no difficulty financing his expansion program. Buoyed by success, he let his ambitions outrun his trading judgment. When the downturn came, it caught him heavily loaded with stocks run up by his own operations to levels they would never see again. No one could have predicted that Garfield would be shot or foreseen the confluence of forces that would send prices tumbling for three years. Jay was hardly alone in riding the crest of prosperity's wave only to find himself thrashing against its downward plunge. If he had been purely a trader, he might have moved more decisively to cut his losses when the light dawned, but the properties crucial to his empire had to be protected.[25]

By the early 1880s the market itself had changed and so had the folklore concocted by analysts to explain its workings. The orgy of railroad construction swelled both the list of securities and the volume of trading. During flush times this outpouring of new issues was readily absorbed, but when the economic skies darkened, the weaker stocks fell fastest and farthest. Since several of Gould's enterprises belonged to this category, their collapse aggravated his problem of holding the empire together while defending his stronger properties. Embittered investors who had bought freely on Gould's promises charged him with bad faith. Their wrath stemmed in part from the misconception that men like Gould and Vanderbilt controlled the market and were therefore responsible for its behavior.

Wiser heads understood that no person or group could control the market, yet the larger and more impersonal the market grew, the more financial writers embraced the myth of individual dominance. By 1881 they shared a folklore that endured even though it flew in the face of common sense. Their columns depicted the market not as a mechanism governed by vast, impersonal forces but as the plaything of big rollers who toyed with the smaller fish before gobbling them up. Of the men who galvanized 'Change the kingpins were Gould and Vanderbilt with their retinues of brokers and followers. It was generally conceded that together they could do with the market what they pleased, that neither could be trusted, and that their depredations had driven small investors or "the public" from the market. Financial writers ascribed to them extraordinary powers and scrutinized their every movement for its significance. Their views were avidly sought and never taken at face value; their intrigues and intentions were discussed with the breathless intensity columnists now devote to the foibles of film stars and rock singers. The Street's rank-and-file believed "that Mr. Gould and Mr. Vanderbilt are responsible for every rise and fall in the market."[26]

The fact that financial writers knew very little about the motives or operations of Gould and Vanderbilt obliged them to trade heavily in rumor and surmise. While they exaggerated the influence of both men on the market, Jay never deceived himself on the subject. "No man can control Wall street," he told a reporter. "Wall street is like the ocean . . . too big, too vast . . . full of eddies and currents. . . . To attempt its direction and control would be as wild as to try to turn back the Atlantic." He scoffed at the way newspapers "magnify individual men in Wall street. One-half of what you read and three-fourths of what you hear is pure romance." Although this was but Gould's old habit of self-deprecation ("I am a mere passenger in all my financial transactions"), he meant what he said.[27]

By the onset of 1882 Jay felt like a passenger headed for uncharted waters. Despite his enormous holdings, traders and editors alike could not fathom which side of the market he was on; this question was to baffle and tantalize them for years. A bear raid on Rio Grande in December 1881 was widely attributed to Gould even though the stock had been declining for six months. When the market slumped that same month, observers concluded that Gould had "assumed his old and familiar part of Ursa Major." A Philadelphia newspaper was awe-struck by his ability to drag down values:

> It was a striking instance of the immense power which one man can exercise. Jay Gould has done more in a single week, without the aid of a solitary bear event or disaster, than all the Vanderbilts, Mills, Cammacks, Keenes, Sages and Osbornes were able to accomplish with the assistance

of an assassination, a change of administration, a tight money market, a railroad war, a short crop, a drought, and a flood, so long as the little dark man was not in sympathy with them.[28]

This hyperbole might have amused Gould if his situation had been less serious. For more than a year he had been on the bull side, but in recent months the business decline made it difficult even to maintain values. In January 1882 a major Parisian bank failed, prompting European holders to unload their American securities. As talk of a panic filled the Street, Jay declared publicly that he was a bull. Most editors assumed as usual that he meant the opposite; one called it "rather a sudden turn—sudden enough, indeed, to take one's breath away." Cynics declared Gould was selling everything that moved, and even those who dismissed such rumors doubted that he really wanted a rising market. Although Jay reiterated his position, the slide continued. By mid-March those carrying large lines of stock on borrowed money were growing nervous.[29]

The situation had become intolerable for Jay. It was not merely the rumors that he was liquidating or about to fail, or even the Street's refusal to believe what he said. What galled him most was his inability to convince anyone that he was first and foremost a businessman, not a trader. Like a fallen woman he was doomed never to escape the shadow of his past. Times had changed and so had he, but the old reputation lingered, thanks in part to the dogged efforts of his enemies. In the public eye Gould remained the sly, clever, treacherous, enigmatic, elusive villain of Erie and Gold Corner days. If conventional wisdom knew that caterpillars became butterflies, it preferred to believe that leopards never changed their spots.

Exasperated by this dilemma, Jay conceived an extraordinary solution. Sage didn't like it and told him so; that Gould persisted despite this objection suggests how desperate he was to prove himself. On Monday, March 13, Sage, Field, and Frank Work watched as Morosini dumped pile after pile of securities on Gould's desk. The display totaled $53 million par value, most of it Missouri Pacific, Western Union, and elevated shares. Jay called attention to the fact that all the shares were unsigned, meaning they had never been transferred or used as collateral. He offered to show more stock and another $30 million in bonds, but his friends had seen enough.[30]

Traders and analysts alike were astonished that the most secretive man on Wall Street should resort to so public a performance, but they did not budge from their convictions. As one wrote, "No one in the slightest degree acquainted with Wall street imagines in his wildest dreams that Mr. Gould has them to keep as permanent investments, or for any other purpose than to sell." The wily Addison Cammack capsuled the Street's reaction. "My God! fifty-three millions more to be sold?" he exclaimed in mock surprise. "Where *will* they go to?"[31]

Sage was right; it was not a good idea. Apart from being wholly out of character for Gould, the exhibit could not possibly succeed. Although the market responded to his gesture with an upward bound, Gould's strongbox was in itself hardly enough to counter the larger forces depressing prices. Once recovered from their shock, critics pounced mercilessly on the "spring opening" as a confession of weakness. In unison they assailed the attempts by Gould and Vanderbilt to peg the market upward as blatant and futile. "When people exhibit goods," one editor sneered, "the presumption is that they want to sell." Ironically, this confirmed the very belief Gould was anxious to dispel: that he wanted a rising market in order to sell out.[32]

Gould's motives escaped observers because this time they lay too near the surface. The spring opening was a rash, even desperate act intended to relieve emotional as well as financial pressure. It was a plea, so obvious as to be subtle, for recognition as a legitimate businessman, and when it backfired there was little else he could do. Jay had trained his audience too well to believe in his inscrutability. After its initial rise the market sagged downward again and Gould let it go. So did Vanderbilt, who was distracted by the suicide of a brother. Ill and weary from the wars, Jay fled to Florida and then to the West. For the rest of 1882 the market reflected the weakness and uncertainty of the business climate. Jay reverted to his customary veil of secrecy. His roads issued skimpy reports and his own operations remained a mystery to the Street, which as usual interpreted his every bullish pronouncement as a sign he was selling. Nearly all his stocks closed the year at prices near their January quotations despite the downward trend of the market, but this fact went unnoticed.[33]

The *Atalanta* was to be launched on April 7, 1883, so Jay brought the family and friends down by special train to Cramp's Shipyard in Philadelphia. A festive crowd had gathered, the largest to witness a launching there in years, eager to glimpse what some called "the most magnificent private craft afloat." A corps of dignitaries waited on the high platform built around the yacht. The 230-foot ship, her gleaming white hull dressed with bunting, was an awesome sight. Jay had spared no expense to create a floating palace. The dining saloon was paneled in light wood carved with groupings of fruit. Plush strawberry curtains trimmed in gold edged the portholes and a huge oriental rug covered the floor. The chairs were richly upholstered, and one wall held a built-in piano. At Lyndhurst the silver bore Helen's initials, but here they were stamped "J. G." Everything had a nautical flavor including the silver bowls and the china, a rare pattern decorated with sea mosses and shells. There was an opulence about the *Atalanta* that exceeded anything at 579 or Lyndhurst. Aside from his estate the yacht was Jay's grandest act of self-indulgence, a luxurious, self-contained paradise where the world could not get at him or his family. If he planned to retire, no one could doubt that he had provided himself suitable accommodations.[34]

The crowd fell silent as workmen wedged the last fastenings loose. Shy little Nellie, her brown hair tumbling out of a claret bonnet, shanked a bottle of champagne against the shuddering hull and squeaked, "I name thee *Atalanta.*" A week later Connor confirmed that Gould would retire as soon as arrangements for the cruise were made. On April 25 Jay appointed veteran seaman John W. Shackford captain of the *Atalanta* and put him in charge of outfitting the vessel. The stage was set but critics still doubted the play would go on. "Mr. JAY GOULD protests that he . . . will quit the world (which is Wall-street), and will wreck no more railroads, rig no more markets, and buy no more newspapers," scoffed the *Times*. ". . . The wily little man has many devices. The penitential game is one of them."[35]

Early in May, while the Street pondered Gould's declaration, Vanderbilt announced his retirement, leaving his sons in charge. A few days later he boarded ship for a month in Europe. It was not ill health that prompted his decision, Vanderbilt insisted just before sailing. Asked about the million dollars he was reputed to have withdrawn before departing, he replied solemnly, "It is the custom on the other side to fee waiters and attendants." Vanderbilt had his little joke and sailed. In June Gould returned to Philadelphia for the *Atalanta*'s trial run. Everything went smoothly; the yacht was ready to leave, but Jay was not. Month after month the market continued its decline while analysts traded profundities about its behavior. Most pointed to the gloomy state of business. The rate war still blighted earnings, several industries suffered from overproduction as trade slackened, and the number of failures swelled. The weakness of the iron industry especially bothered Gould, who was fond of saying, "It is no time to buy stocks when the price of iron is falling."[36]

The savants of the Street persisted in accusing Gould of leading the downward march. According to one theory the big rollers had misread the market and found themselves overloaded with stocks. Having failed to unload on each other, they were trying to dump them on smaller traders and the "outside public." But no one was buying, brokers were starved for business, and an editor wondered if Gould "still has the sagacity which he once displayed." Against these forces Jay could do little to raise or even maintain prices. Every time he and Sage issued bullish statements, critics contended that "they have been hard at work selling all day." When Union Pacific began to fall, the *Times* labeled Gould, "the most active, persistent and formidable bear on Union Pacific of all the crowd of speculators."[37]

By the month's end the press had enshrined Gould's impending voyage and retirement, "which, by the way, did not occur again last week." Already it had dawned on some that Gould might actually be in trouble, that his peculiar antics were in fact attempts to ease the burden he was carrying. Everyone knew his load was heavy and pointed especially to the large amount of Wabash paper he and Sage had endorsed. By summer the belief persisted that Jay's bulling efforts were confined to his major holdings and that he was hammering the rest of the list down. Even when the telegraph strike forced him to sustain Western Union it was asserted that "he certainly had a big line of shorts out." No one bothered to explain why one of the country's largest holders of securities, who was supposed to be overloaded with stocks, should be leading the bear attack. This contradiction, and the air of mystery surrounding his operations, were glossed over by the old assumption that he was a speculator who would sell anything for a short-term profit with the possible exception of Missouri Pacific or Western Union.[38]

Small wonder that Gould's trading policy baffled analysts, who remained blind to the possibility that his market tactics were dictated less by speculative than by business considerations. Ironically, they reached the right conclusion from the wrong premise. Whatever the extent of Jay's difficulties, they arose less from speculative forays than from the strain of trying to keep his empire intact against a shrinking market. He had not misread the downward trend but was helpless to deflect it. His business dilemma was brutally simple: he dared not let anything go but lacked the resources to keep everything together.[39]

In August the market broke sharply and three brokerages failed in as many days. Gould's southwestern roads, the Northern Pacific, and the Rio Grande were hardest hit. Jay took the unusual step of denying publicly that he had attacked Villard's stocks; his statement was "met with a laugh" on Wall Street. As Wabash, Texas & Pacific, and Katy plunged to new lows, analysts concluded that Gould was supporting Missouri Pacific and Western Union while letting everything else go. To their surprise he was unable to do even that. Missouri Pacific dropped seven points and Western Union, which everyone assumed Gould would never let drop below 80, fell to 72. "The bears have had their innings," Sage growled bravely, "and now it is the turn of the bulls." One editor expressed the more common view that "no one seems to look for anything but a continuance of liquidation. . . . There is no leader on the bull side. Mr. Vanderbilt will do nothing; Mr. Villard's power seems utterly broken; Mr. Gould . . . is plainly in a more uncomfortable position than it is usually his luck to get into."[40]

By contrast the bears had acquired a formidable new leader, although few of the Street's denizens yet realized it. The Villard stocks were being pounded down not by Gould but by Charles F. Woerishoffer, a German broker whose boldness inspired one colleague to remark, "He could have marched upon a cannon's mouth with a jest on his lips." Scholarly in appearance, Woerishoffer came to New York in 1865 at the age of twenty-two, served as clerk for August Rutten, and started his own firm in 1870. For a time he was content to run a "good strong house doing a fair business." By 1877 his three partners had all retired or left the firm and Woerishoffer's career entered a new phase, as if his genius had been unleashed and allowed to roam free. His plunges grew larger and more daring. A credit reporter found him "more speculative than formerly which causes a little shade in his standing with very prudent houses."[41]

It took Woerishoffer only a few years to earn a fortune dealing in railroad securities. Gould tested his mettle in one transaction, found it made of flint, and was careful never to underestimate his ability. The emergence of Woerishoffer as leader of the bears had fateful consequences for Gould. It pitted him against the most brilliant operator on the Street at a time when the prevailing business climate favored the bears. The relentless attack on Northern Pacific helped drag the whole list down. In September Jay tried to spark a rally by joining forces with some of Vanderbilt's brokers. The attempt fizzled when, as one wag quipped, an early frost "killed some corn and a great many bulls." Publication of the Northern Pacific report confirmed the soundness of Woerishoffer's judgment and sent prices tumbling again. Every attempt to bull the market foundered on the rocks of more bad news, and matters did not improve even when the news was favorable. In October Western Union won an important court decision, which Jay hoped would ignite a rally. The stock started upward, but orders dried up so abruptly that one observer sneered, "Seldom, if ever, has there been seen in Wall Street a flatter failure than this long expected and much debated decision."[42]

Reporters found much to ridicule that sorry season. Gould testified before a Senate committee and was drawn into recounting some of his boyhood experiences. The *Times* gave it the headline, "THE PREYING CHRISTIAN." His retirement and the *Atalanta*'s prolonged virginity still provided merriment. After the yacht rammed a tugboat in New York Harbor, Morosini joked that when Gould aimed his yacht at the tug he didn't know it was loaded. In mid-October the market dropped to its lowest level in three years. A combined effort by Gould and Vanderbilt sent the bears briefly to cover, Cammack grumbling that he would not pit his brains against Vanderbilt's money. Vanderbilt's abrupt return to the Street was prompted by more than falling prices; one of the sons now running his enterprises was rumored to have suffered huge speculative losses and gone to his father for help. The senior Vanderbilt denied the story and used the occasion to announce that he was buying stocks again.[43]

Jay must have welcomed Vanderbilt's help and that of Keene, who turned bull on the Villard stocks, but it was not enough. All hopes that the bottom had been reached were dashed by an avalanche of bad news. The Wabash report was a disaster, Texas & Pacific was no better, and Union Pacific was falling "in a way almost equal to the Villards." Rate wars still raged across the continent, the money market was behaving weirdly, and Wall Street was unsettled by the election of John G. Carlisle as speaker, which, Huntington scowled, would prove as disastrous as Garfield's slaying. An investigating committee was poking into the affairs of Oregon & Transcontinental, prompting Villard to resign as president of Northern Pacific. Rumors were flying that Keene was in trouble and about to default.[44]

The same had been whispered for months about Gould, whose position remained Wall Street's favorite guessing game. Every report that he was a "very determined bull" was countered by one insisting he was "undoubtedly selling out his holdings of every kind as fast as he can while still maintaining the appearance of supporting his stocks." Jay was in fact not selling out but holding on. He once said that the only stock a Wall Street man must have was a stock of patience. At the year's end, laid up with a cold, he needed all the patience he could muster. The market was still sinking with no relief in sight. A prominent house advised customers that "prices have gone down in obedience to laws which seem to be still operating in favor of decline." Gould's attempts to raise the market had failed utterly, and his speculative forays had brought no relief from the relentless bear siege.[45]

On the evening of December 1 the usual crowd of brokers gathering in the lobby of the Windsor Hotel were surprised to find a sofa with Cammack perched on one side, Woerishoffer on the other, and Gould between them, "listening most of the time with his usual cool attention." From nine until nearly eleven they talked in low, earnest voices while onlookers tried to divine their topic. Those eager for portents should have used their eyes instead of their ears, for the tableau told the tale: the bears had Gould between them and would not let him go.[46]

26

Immolation

Then came whisperings that important failures were imminent and these quickly turned into a statement that Mr. Gould was in financial straits, and that was followed by damaging assaults upon the Gould specialties.

—*Commercial and Financial Chronicle*, January 26, 1884

Mr. Gould is undeniably the most potent person in the speculative ranks. He frightens the operators into buying or selling as a rule, very much as he pleases. He accomplishes more with this power, probably, than Mr. Vanderbilt does with his money.

—*Bradstreet's*, March 29, 1884

Mr. Gould's god is his pocket, to fill which he has prostituted every attribute with which he is endowed. Like the Ishmaelite of old, his hand is against every man, and every man's hand is against him. Those who have ventured to look upon him in the light of a friend have invariably found, sooner or later, that they had taken unto their bosom a serpent whose sting is death. But to attempt to sketch the character of Mr. Jay Gould in its true colors would be futile, since no language is equal to the task. His cold, heartless villainy stands alone, like the man himself, who has so ruthlessly swindled those whom he regarded as his "friends" that he has now none to swindle.

—*Railway Times* (London), July 12, 1884

THE ROOM IN WHICH Gould worked at 579 was a curious one, neither office nor library in the conventional sense but a chamber thirty feet long and ten or twelve feet wide, arranged, like the man himself, with fine disregard of the decorator's art. Two sides were lined with low bookcases crammed indiscriminately with richly bound and cheap, well-thumbed volumes. A long table covered with green cloth, piled high with letters, telegrams, documents, pamphlets, cards, and notes, filled nearly half the room. A huge map of the United States occupied the wall above a capacious sofa, and elsewhere could be found smaller maps marked with red and blue lines drawn across states, continents, oceans. At one end of the table stood a handsome case containing specimens of rope and cable. On the wall opposite the table lay a broad fireplace, its giant andirons crowned with sphinxlike heads. A plain bronze clock sat atop the mantel with hands frozen at five minutes to nine.[1]

Perhaps Jay did not care to hear the ticking, the inexorable rhythm that pressed him as relentlessly as the stacks on the table. He had passed many long, gray hours in that room, with many more yet to come. The new year looked to be the grimmest Gould had ever faced. Like "Chinese" Gordon, who found himself shut up in Khartoum that same year, Gould was under siege with no relief in sight. Talk of retirement lingered but was not taken seriously. Four years earlier he had told George Miller of the *Omaha Herald*, "I have been in the harness a long time & want rest. My ambition was long since satisfied & I am ready to go into quiet retirement & hope ere long I shall have the opportunity." Instead he had ordered up more harness, enough to break a man half his age and twice his strength. Now it was too late; times had gone sour, the stakes were immense, and the boys not yet ready to step into his place. Having chosen to put on the red shoes of empire, Jay found he must dance until he dropped.[2]

In January Gould hoped to put his affairs in order for a long trip south, but matters grew steadily worse. The stock, grain, cotton, and provisions markets were all in turmoil and being pummeled by bears. The shrinkage of values in stocks and bonds exceeded anything seen since the crisis of 1873. Villard was ruined and prostrate with anxiety. Western rates were still in shreds from the clash between the Burlington and the Tripartite members. Stories persisted that Union Pacific "would not see the end of its trouble until President Dillon went out of it." In the East Sloan and the Lackawanna were at odds with the trunk lines, and the New England stunned everyone by going into receivership. Gould declined to lend the road money and left the board. One embittered holder, recalling the night two years earlier when Gould declined to make a speech until the first dividend was paid, growled that the "chief grief" of the stockholders "was not the want of the dividend but the loss of the speech."[3]

As always the dogs of war barked loudest in Wall Street and forced Gould into some heroic measures to muzzle them. He recruited Dodge to organize a pool for bulling Texas & Pacific, then selling at 19. Armed with favorable data from Clark, he seized on Union Pacific as the vehicle for an upward push. "GENERALISSIMO GOULD IN COMMAND," brayed the *Herald* as the battle raged. The market surged briefly only to sink again beneath the weight of bad news from other quarters. Rumors swirled that Wabash notes totaling $5 million, which Gould had endorsed, were about to fall due, and that he was edging near the brink of failure.[4]

Gamely Jay tried next to lift the market at its weakest point, the Villard stocks. Amid great fanfare he organized a syndicate to lend Oregon Transcontinental enough money to stave off a maturing loan. The bears counted on the company's having to meet this loan by selling the shares it had deposited as collateral, which would force the price down and enable them to cover shorts cheaply. Gould's stroke allowed the company to reclaim the stock and sent the bears scrambling for cover. "A more unexpected, rapid and brilliant *coup* has never been seen on 'Change," marveled one observer. A more astute critic asked why Gould required a syndicate to "raise a sum of money which at one time he could have drawn his check for without trouble."[5]

It was an ingenious maneuver, but again the effect was short-lived and left Gould facing the old dilemma of how to liquidate some holdings on a falling market without jeopardizing his control of vital properties. For months the man of silence had been trying to talk the market up; by the winter of 1884 he must have seemed positively voluble. The persistent rumors that he was financially embarrassed cracked his façade of indifference. He talked irritably of suing those who spread tales of his impending failure, and there were reports of an exhibition similar to the spring showing of 1882. To ordinary eyes this suggested that Gould was faltering under the strain; to alert enemies it must have revealed that he was in grave trouble.[6]

In February the market appeared to improve. Vanderbilt helped by announcing that "the bears . . . have been scaring people, but their doom is sealed." By the month's end Gould felt encouraged enough to risk his long-delayed cruise south. He stayed only a month, probably because the news from home allowed him little rest. On the telegraph front Garrett's B & O Telegraph had captured lines on the Nickel Plate and the New York, West Shore & Buffalo, another new road on the crowded corridor between New York and Chicago. Bankers was expanding, Siemens was preparing to lay the Mackay cables, and Bennett opened a withering fire of publicity with the *Herald*. In Washington a Senate committee began hearings on a postal telegraph bill introduced by an old Gould nemesis, Senator Nathaniel Hill of Colorado. The crusade against the waterlogged telegraph monopoly had begun in earnest.[7]

Western rates remained demoralized despite repeated efforts at compromise. The conflict in Nebraska took an ugly turn when Perkins discovered that Vining, the Tripartite commissioner, had hired an agent "to annoy and demoralize the B. & M. and . . . increase the ill-will." Bristling with indignation, Perkins wanted to "knock the bottom out of Colorado rates" and "give notice of our intention to terminate existing pools east of the Missouri River." He was restrained from these drastic actions and plunged instead into hastily called conferences that yielded uneasy truces.[8]

The Vining uproar came at the worst possible time for Union Pacific, which was under attack from all sides. Congress was looking for ways to enforce the Thurman Act and stiffen its terms. Earnings continued their alarming decline and the floating debt mounted steadily from expenditures on the Oregon Short Line. In March the board declared the quarterly dividend even though it had not been earned, and Dillon took pains to conceal the true state of things by issuing as uninformative a set of figures as possible. The Street was confused but not fooled; the stock nosed relentlessly downward. By April it was clear that a crisis loomed and could not be staved off much longer. Although the Union Pacific was no longer Gould's road, he had been the architect of its system and policies. If it defaulted, the bitterest criticism would be aimed at him. Even worse, a collapse might deal the market (and his own financial position) a lethal blow.[9]

Whatever respite Gould gained from his vacation was soon lost as a break in the price of wheat sent the market tumbling again. Although his stocks led the decline, Jay surprised the Street by doing little to sustain them. Missouri Pacific dropped to 80 and Western Union to 67 even though both were paying healthy dividends. Bewildered analysts, unwilling to believe the bears would dare attack Gould directly, concluded that he must be liquidating and revived the tale of an impending sea voyage. Vanderbilt had already turned cautious and denied that he and Gould had ever combined to support the market. As the slide continued, no one knew what to believe.[10]

Early in May Western Union and Union Pacific both skidded to 61. "Unquestionably," declared one analyst of Western Union, "Mr. Gould is paddling out of the property as fast as he can." Wabash was expected either to fund its June coupons or to default on them. The Street was now convinced that Gould was "turning all his securities into money with characteristic perseverance and boldness." If so, he must have found it hard going. The fall in prices reduced the collateral value of his stocks and exposed him to real financial danger. He needed time to regroup, but there was no time. For months the clouds had been gathering; suddenly, unexpectedly, the storm burst forth with a fury that confirmed the Street's worst fears.[11]

The first hint came on April 30 with the failure of James Keene. His suspension was brief, less than two months; the Street mourned his loss but did not tremble at the news. On the morning of May 6, however, the Marine National Bank abruptly closed its doors and the firm of Grant & Ward suspended. As excited crowds milled about the streets word trickled out that the bank had loaned large sums to a house gone bust. The crash engulfed not only Ulysses Grant, Jr., but also his father, who had entrusted nearly all his money to the firm.[12]

These failures from an unexpected quarter threw Wall Street into turmoil for a week, but the worst was yet to come. On May 14 news leaked out that John C. Eno, president of the Second National Bank, had fled to Canada after squandering $4 million of the bank's funds in stock operations. Later that day the Metropolitan National Bank closed its doors after the brokerage headed by its president, George I. Seney, announced bankruptcy. A prominent banker, Seney had, like Woerishoffer, emerged from the larva of respectable conservatism to become the freest of speculative butterflies until he allowed himself to be caught and crushed by a falling market. By dusk a Brooklyn bank and half a dozen smaller brokerages had gone down in the rubble as panic swept the financial district. The next day A. W. Dimock & Company failed when their pet scheme, Bankers Telegraph, nosedived from 119 to 45. That afternoon the prominent house of Fisk & Hatch suspended. Those with long memories shivered anew at the nightmare of 1873 as prices tumbled, the money market tightened, and margin calls flew like bullets.[13]

"This disturbance to-day is a senseless thing," Gould lamented when the carnage had slowed. "The general calling of loans has been wholly unnecessary and foolish." His remorse was genuine. Although none of the failures touched him directly, the convulsion put him in a desperate situation. Missouri Pacific sank to 63 1/2, Western Union to 49, Union Pacific to 35 1/4, Texas & Pacific to 9 1/2, Wabash to 9. Shrinking collateral values produced calls for more margin on pledged shares; Wabash and the southwesterns had become worthless as collateral. The bears sensed Jay's weakness and hammered his stocks unmercifully. At first he tried to squeeze the bears by borrowing all the stock he could to make it scarce, but they cheerfully paid high premiums for its use and sold short more heavily than before. Cammack declared that Gould had lost his head.[14]

Money was needed to withstand the attack, but where to get it? Borrowing against a shrinking market was not promising, especially with half the Street clamoring for loans and the banks paralyzed with timidity. Vanderbilt had just sailed for Europe; Jay apprised him of the situation by cable and received a noncommittal reply, which the guffawing regulars interpreted as "very much obliged, but I guess I won't buy anything for the present." Nor could Jay draw on his most reliable cashbox. Sage had been caught with too many privileges outstanding at eight or ten points above current prices and was in trouble. On May 16, while a throng of anxious broker's clerks fidgeted in his tiny outer office, Sage received stocks at a glacial pace until the required hour of 1:45 P.M., whereupon he shut his door and turned his deaf ear to the howling mob. That evening rumors swirled through the heaving lobby of the Windsor that Sage would lay down on his contracts, that the prospect of huge losses had shattered the old miser's nerves, broken his health, driven him insane.[15]

The old man was distressed but hardly broken. At sixty-eight he wore his years far more lightly than Gould. "My health is as good as ever and I have no intention of retiring yet awhile," he chirped to a reporter after the siege had lifted. But at last he broke under fire. On Monday, May 19, the crush in his office eased only to resume the next day. Alarmed by complaints that Sage was ignoring the rules, Gould and Connor hurried that evening to his home for a conference. "He has been thrown into a condition of nervous prostration by the excitement," Jay explained afterward, "and by the threats made against him. He is a gentle old man and the turmoil around his office has completely unnerved him." Sage would, he assured reporters, take all stock put to him the next morning.[16]

On Wednesday Sage opened his door to all comers and the crisis passed, but he was in no position to help Gould. As the bear attack intensified, Connor and his brokers stood their ground like an imperial guard, absorbing the barrage of sell orders as fast as they arrived. If any of Gould's friends deserted the ranks at this critical hour, no record exists of the fact. The problem was not lack of loyal troops but of sufficient firepower to repulse so ferocious an assault. Gould was, as Connor later admitted, "in deep water." Rumor asserted he had borrowed $2 million in sterling bills and was seeking loans wherever they could be had. "Is the load which Gould is carrying too heavy for him?" asked the *Herald* gleefully. Next day the *Times* chortled that "the boys had not only got him under but were walking up and down on his prostrate form."[17]

At this point legend merges inseparably with events. Years later the story arose that Gould escaped disaster by boldly seeking from his enemies what his friends could not provide. On a Sunday morning he sailed to Long Branch for a meeting with the bear leaders. Brandishing a copy of an assignment drawn up by his lawyers, he threatened to announce the next morning that he could not meet his engagements. The bears realized that Gould's failure would trigger a convulsion that would sweep them into ruin as well. Since they needed Western Union to cover their short contracts, they agreed to buy 50,000 shares from Gould if he promised not to announce. Convinced that the transaction amounted to little more than a stay of execution, they provided Gould with the cash by which, to their amazement, he extricated himself with a brilliant operation in Missouri Pacific.[18]

This tale is plausible in that it is consistent with Gould's character. It may be an embellished version of what happened; the problem is that not one shred of evidence exists to substantiate it. Even those who disseminated it after Gould's death conceded that "this story, like many others told in Wall Street, probably has a mixture of truth and fiction." None of the editors following Gould's every movement made any reference to it at the time, not even as one of the countless rumors that spiced their columns. The slightest hint of such a showdown would have caused a sensation in the press, yet there was nothing. If the meeting occurred, it remained one of the Street's best-kept secrets. Jay's few surviving letters for 1884 disclose nothing on the subject in content or tone.[19]

While the extent of Gould's difficulties remains a mystery, there is no doubt that the panic strained his already overextended resources. All his properties were in trouble, some for reasons apart from the market collapse. If not already at the brink of failure he must soon approach it unless the spiral of liquidation could be checked. At least one part of the showdown saga is true: Gould responded to the crisis with a stunning coup in Missouri Pacific. It was a logical place to strike. The Street interpreted Gould's inability to defend even the stocks dearest to him as proof that he was in dire straits. If he could start one of them bounding upward, the short traders might aid his cause with panic buying. Since Gould and his friends already held large amounts of Missouri Pacific, it was the easiest stock for them to bull. Moreover, a rise in one of Gould's specialties would ease the pressure on others by striking terror in the hearts of the bears. Traders had long feared to attack Gould's stocks because of his reputation for trapping unwary transgressors. Lately, emboldened by reports of his demise, they had ventured out to feed on the carcass. Punishing the foolhardy in one stock might send the rest scurrying for cover.

Jay launched his campaign in May. "There is never any difficulty in deciding that a Gould movement is really on," smiled one observer. "The real difficulty is in discovering when it is finished." In three weeks Gould trapped the shorts and sent Missouri Pacific from 64 to a high of 100; Western Union recovered ten points in the rise. These advances gave him breathing room but his position remained precarious. Critics still maintained that he was locked "in a sort of death-grapple with the bear combination which has been after him so long." The danger lay not only in the market but in the business arena where, one analyst noted, "Mr. Gould's own properties, stocks and bonds alike, are all going to pieces."[20]

The railroad scene looked bleak. Early in June the Reading defaulted and was soon followed by the West Shore and the Rio Grande. The once proud L & N saw its president resign amid charges he had borrowed treasury stock for his own speculations. Unlike other directors, Gould frankly conceded the crime in hopes of stifling rumors of worse. More of his hours went into prolonged sessions preparing a financial plan that was never used. Among Gould's own roads Union Pacific, Texas & Pacific, Wabash, and Katy were in such deep trouble that collapse of the entire southwestern system appeared imminent. Few observers believed Gould could keep it intact, let alone maintain control. The *Herald* celebrated his discomfort with an editorial captioned, "JAY GOULD MUST GO."[21]

The horrors of June for Gould went beyond the market crisis. During the very weeks he was trying to salvage his position by bulling Missouri Pacific, the full weight of empire crashed down on him. Every one of his major properties except Missouri Pacific reached a flash point in its affairs. There was little time for deliberation or calm maturing of plans; if Draconian measures were needed they were taken without hesitation. It was difficult to play the part of statesman with the barbarians knocking at the gates. For months it remained doubtful whether all or even part of the empire would survive. Gould had carried great loads before, but always on the attack; this time he was on the defensive.

The first blow fell on Wabash, which failed to earn its interest due June 1. The lease gave the Iron Mountain an option to advance the deficit or not. For months Gould and his spokesmen had issued repeated assurances that Wabash would not default. As late as May 13 he insisted that receivership "has not even been considered." A week later the Iron Mountain announced it would not pay the Wabash's charges. If the Wabash defaulted, the bondholders could apply for a receiver and oust Gould from control. Moreover, $308,333 of the $2.2 million in

Wabash notes endorsed by Gould and Sage were about to mature; if the Wabash could not pay, the endorsers would be forced to provide funds at the very time they were strapped for cash.[22]

Gould met these threats with a stroke worthy of his reputation: ingenious, strikingly original, unexpected, technically legal, and ethically dubious. On May 21 Gould, Sage, Dillon, Humphreys, and Hopkins held a special meeting of the Wabash executive committee at which they decided to secure a receiver *prior* to default. Two friendly judges were found in Missouri who consented readily to the application and named as receivers Solon Humphreys and Thomas E. Tutt, a St. Louis banker, without notifying any other creditors. From this decision would spring years of litigation winding all the way to the Supreme Court, where the chief justice would denounce it as "without precedent" and "dangerous in the extreme." The bondholders had been forestalled by receivers appointed to protect the Wabash from holders of the promissory notes falling due.[23]

The receivers took office on May 29 and applied the next day for permission to sell receiver's certificates for taking up the notes endorsed by Gould, Sage, Dillon, and Humphreys himself. In granting the request the court agreed not to make public the names of any endorsers except Humphreys to spare them *"the personal inconvenience and injury which might result to them from the publicity thereby given to their business affairs."* By early June Gould still retained control of the Wabash, his obligations as endorser had been relieved, the amount of those obligations kept secret, and the unsecured notes established as a prior lien to the mortgage bonds, giving Gould a powerful voice in any reorganization. As a precedent this last point opened a legal Pandora's box that would plague courts and investors for years. When the truth was revealed, enraged critics and investors excoriated Gould for his trickery. One analyst asked savagely "whether the federal judiciary . . . is really the law department of Mr. Gould's southwestern railroad system."[24]

The Wabash scheme was unsavory but it did the job. So too with the Texas & Pacific, which Jay controlled not through stock ownership but through an operating contract with the Missouri Pacific. When it became clear that Texas & Pacific could not earn its June interest, Gould avoided default by having the Missouri Pacific advance the funds. The paid coupons were not canceled but kept alive by the Missouri Pacific as security for its advances. By this device Jay and not the bondholders held a claim against the Texas & Pacific and could apply for a receiver if one were needed. When critics denounced the scheme as undermining the integrity of mortgage bonds, Gould shrugged them off by saying it was "nothing new" and "had been done two or three times before." Alarmed bondholders hurriedly organized a committee to investigate the road's affairs. Here, as in Wabash, a long and difficult fight lay ahead.[25]

In one area Jay snatched victory from defeat. For months the battle over Manhattan had been at an impasse, with Kneeland rebuffing every attempt at compromise. His course seemed vindicated in April when the court invalidated the agreement of October 1881. On the surface this decision doomed Gould's control of the elevateds, but again he proved equal to the test. Aware that New York Elevated was by far the more prosperous of the two operating lines, he struck an ingenious bargain with Field that brought Manhattan stockholders a share of New York's profits. The new agreement left Metropolitan to fend for itself on the expectation that it could not alone earn the dividends paid by combined operation.[26]

Here at least the financial panic helped rather than hurt Gould. Metropolitan fell fifteen points in May, forcing Kneeland and his followers to sell off part of their holdings. The buyers proved more amenable to compromise. Early in June Field and a Metropolitan committee agreed on a new plan converting the shares of all three companies into one class of stock. All three boards ratified the plan, with Kneeland doggedly voting in opposition. The new Manhattan Elevated was born with Gould firmly in control. Predictably, the newspapers howled in protest. "It will be a worse thing for the people of New-York," warned the *Times*, "the day that Mr. Gould is confirmed in sole control of the main lines of travel of this city."[27]

Gould had no time to savor victory. No sooner had Manhattan simmered down than the Union Pacific pot boiled over. As earnings dwindled, the company was helpless to effect economies; it tried cutting wages but had to relent when the men walked out. In April the court ruled against Union Pacific in the net earnings case, and the long battle over the Thurman Act took an ominous turn. Since its passage in 1878 the Union Pacific had ignored the payment provisions, preferring to gamble on either a negotiated settlement or victory in the courts. By the spring of 1884 it was evident that the gamble had failed, and Congress was in a vindictive mood. The House passed an amendment to the Thurman Act raising the sinking fund payments from 25 percent to 55 percent. At the same time Senator George F. Edmunds of Vermont prepared a report asking the attorney general to enforce the penalties against Union Pacific for paying unauthorized dividends and against all directors who had voted for those payments.[28]

Where the House bill threatened to bankrupt Union Pacific, the Edmunds report struck directly at Gould and his associates. The crisis also embarrassed Adams, who for two years had been encouraging New Englanders to invest in Union Pacific and had speculated in the stock himself. Since May 1883 he had been a director and had steadily increased his role in the company's affairs. With the stock going to pieces, Forbes feared that Rock Island or Northwestern might start buying and compel the Burlington to do the same. He did not like that any more than the prospect of having a bankrupt road on the Burlington's flank, and hoped Adams and his Boston friends could pull Union Pacific through.[29]

The Union Pacific board shared Perkins's view that Adams "has induced a lot of people to buy Union Pacific, and does not know how to get himself out of the scrape." From this insight emerged a plan that not only relieved Adams's conscience but elevated him into the man on horseback. A man of "unimpeachable moral and business character," the nation's foremost authority on railroads, his voice carried a weight and probity no other spokesman could muster. At the board's urging he hurried to Washington to convince Edmunds and the rest of the

Judiciary Committee that any measure against the Union Pacific would not be in the public interest. By dint of some strenuous pleading Adams obtained a most unique agreement.[30]

Adams was quick to absolve himself of blame. Everything he knew and heard convinced him the property was sound, he told the senators; in voting for dividends the past year he was not aware that the law was being violated. How he could have endorsed the stock to investors for two years without knowing anything of the Thurman Act controversy was a question he chose not to explore. Instead Adams emphasized that adverse legislation would destroy the stock, ruin Gould, precipitate a panic, and sweep away an army of investors who had bought Union Pacific in good faith. Persuaded by this reasoning, the committee agreed to delay action if the Union Pacific pledged to pay no more dividends, made certain financial concessions, and replaced Dillon with none other than Adams himself.[31]

This was an extraordinary arrangement. In effect a Senate committee with no authority to do any such thing went beyond law making to law enforcement, ignoring both the attorney general and the Justice Department in the process. The Union Pacific had no choice but to accept. Its debts were mounting, the road was in poor shape, borrowing power was exhausted, rate wars continued to ravage earnings, and the stock had dropped below 40. On June 18 Dillon resigned on the plea of exhaustion, a plausible excuse for a man of seventy-two who had recently lost both wife and mother-in-law. Adams was elected president and the company's offices returned to Boston. The press hailed the change as a defeat for Gould; the *Herald* celebrated the occasion by reprinting Adams's estimate of Gould in *Chapters of Erie.*[32]

As usual Gould's role in the affair puzzled observers. To all appearances he had been routed from Union Pacific, but his influence there had been minimal in recent years. Although the Boston directors instigated the Adams mission to Washington, Jay approved it for reasons of his own. He had taken Adams's measure and had seen his vanity as a lyre eager to be plucked. Doubtless he agreed with Perkins's astute view of Adams: "An entirely honest man himself, morally speaking, he, nevertheless, suspects everybody less stupid than himself of not being honest." Gould realized that Adams's reputation might accomplish what all his own manipulative skills had failed to achieve, a rally in Union Pacific stock.[33]

Summer's heat emptied out Fifth Avenue as wives and their progeny fled to the resorts, abandoning the city to their sweating spouses. The gloom filling the papers went beyond the shroud cloaking Wall Street. A cholera epidemic ravaged China and had invaded France; worried observers wondered when it would spread to American shores. Khartoum was still under siege, and in July a relief expedition found seven survivors of the Greely expedition, from whom tales of cannibalism and other horrors soon emerged. In August New York was jostled by a mild earthquake lasting thirty seconds, considerably less time than it took to read the *Herald*'s account of it.[34]

Through these sultry weeks Gould stuck grimly at the task of shoring up his damaged financial position. His closest allies were of no help. Dillon still mourned the loss of his wife and Union Pacific while Sage, worn down by his ordeal, secluded himself for six weeks in the Long Island cottage of Dr. John P. Munn. Left alone, Jay had to combat not only his financial dilemma but also persistent reports that he was "following fast in the footsteps of Mr. Villard, and hastening toward . . . a grand collapse." These rumors subjected him to yet another indignity. For some years Gould had served on the board of Mercantile Trust Company and had given the bank much of his corporate business. He owned little stock in the bank, which was dominated by the Equitable Life Insurance Company. Early in the year the Street was titillated by a story that when the bank asked Gould to take up some loans, he refused adamantly and dared the company to sell his collateral. Louis Fitzgerald, Mercantile's president, denied the rumor and produced a loan book showing that Gould's last loan had been paid off in October 1883. Since then, Fitzgerald added, Gould had not borrowed a dollar from Mercantile because he could get money more cheaply in the call market.[35]

The issue revived in July when Gould abruptly retired from the Mercantile board. Pressed for an explanation, he denied having any loans with the bank but declared that the "constant reiteration of the story was not doing it any good and I concluded that I would sell my stock." On the Street it was said that Gould had large loans, which Mercantile decided not to renew, and that the Equitable people had asked him to leave the board. The hostile press applauded his departure; Bennett's *Herald* alleged that Gould's presence on the board had cost the bank $5 million in withdrawn deposits.[36]

Although critics ridiculed Gould's "sacrificing himself for the public," there is no evidence that he was asked to leave. Both Fitzgerald and Henry B. Hyde of Equitable remained Gould's associates for the rest of his life, and Mercantile continued to handle his business. The bank's directory included several of his closest allies, "whose presence," the *Herald* conceded, "serves him nearly as well as if he remained in office." If nothing else the episode reflected the widespread belief that Gould was still flirting with insolvency and that his presence damaged the reputation of financial institutions.[37]

Even as the *Herald* wrote Gould's financial obituary he was laying the groundwork for another market maneuver. To sustain the upward rise sparked by Missouri Pacific he seized upon an improbable leader. Jay had been buying Union Pacific since early June, when he picked up 25,000 shares. "The company has & is passing through a fiery ordeal but will I think come out all right in the end," he told Clark. "It seems to me that with good rates & good crops this property must improve rapidly & the stock again become a good investment." Had these remarks been made in public they would have been dismissed as Gould puffery, but they were confined to private letters. They offer a clue to what would become clear half a dozen years later. Although Gould's interests had shifted elsewhere, he could not bring himself to sever ties with the Union Pacific altogether.[38]

Analysts pronounced his efforts to lift Union Pacific "a fizzle" and labeled the stock "a safe target for bears," but by July's end it had risen from a low of 28 to 48. An extension of his sterling loans improved Gould's credit and helped him renew the attack. During August the entire list marched upward "in a raging fever of advance," and Union Pacific topped 57, or double what it had been in June. Gould engineered the rise even though the company's situation had improved so little that, in the words of one critic, "Nobody now discusses the intrinsic value of the Union Pacific, for the street is fully convinced it is hopelessly bankrupt."[39]

After Union Pacific peaked late in August, Jay permitted himself a few days in the Catskills. Except for a brief cruise earlier in the month it was his only vacation that summer. The strain left his nerves ragged and his body shriveled from anxiety. A reporter who bumped into him was startled at how careworn he looked. His face was thin and drawn, "and except for red rings about his eyes his flesh was colorless." Dillon was seventy-two, Sage sixty-eight, and Gould only forty-eight, yet Jay appeared much older and more weary than the others. He needed rest, urgently needed the long cruise he had talked about for so long, but he could not leave the ramparts while the battle still raged. By sheer force of will he summoned the strength to meet the next challenge, sustaining himself on nervous energy and little else.[40]

A new threat loomed on the telegraph front that summer. Dimock's failure left Bankers insolvent and spawned rumors that Western Union would gobble up the company. Instead Mackay stepped forward with a plan to unite Bankers in a pool with Postal Telegraph. After weeks of negotiation Garrett's B & O Telegraph joined the pool in mid-July, creating what the *Herald* trumpeted as the "most formidable opposition to the Western Union Telegraph monopoly which has ever been organized."[41]

Strong as the combination appeared, it expired in the cradle. Where Gould and his friends repeatedly advanced funds to tide their properties over in lean times, none of the rival leaders were disposed to do so. Mackay brayed that he would rather lose all the money he had put into the telegraph than sell to Gould, but when Bankers' debts proved larger than expected he quickly withdrew his support. On September 24 John W. Garrett died, leaving the B & O in the hands of his erratic son. Two weeks earlier Robert had abruptly pulled B & O out of the pool on the grounds that Bankers could not carry out its part of the bargain. Bennett was content to promote his cable and pummel Western Union and its master in the *Herald*. With no one willing to help, Bankers floundered into receivership.[42]

Confusion in the enemy camp provided Gould with what he needed most—time. Western Union was vulnerable to a concerted attack that season. The strike and rate cuts had reduced net earnings nearly 15 percent and trimmed the dividend from 7 to 6 percent. The Mackay-Bennett cable opened in December and offered rates 20 percent lower than Gould's lines. While his foes regrouped, Jay turned his attention to unfinished business. He left the Northwestern and Rock Island boards, surrendered the Wabash presidency to the venerable James F. Joy, and somehow found time to formulate a reorganization plan for the Wabash. It was drastic, effective, and carefully protective of his interests at the expense of the bondholders. Joy went abroad to woo the English bondholders only to be hooted down with cries of "Gould's man." Journals on both sides of the Atlantic denounced the plan as "a monstrous proposal, monstrous in its audacity."[43]

At home Jay suffered a personal if indirect wound when Adams drummed Clark out of the Union Pacific. Adams had taken an immediate dislike of Clark, whom he described as "a narrow, tricky, dishonest and treacherous promoted railroad conductor, utterly unequal to his position." Gould consoled his friend but had little to offer him; he already had an able operating man in Hoxie. Instead he invited Clark to accompany him on the fall tour. Jay hoped to leave in September but held back because of an obligation larded with the peculiar mixture of loyalty and self-interest so typical of him. "We expected to have gone west before this," he apologized to Clark in mid-October, "but I have concluded to stay here & do what I can to help the election of Blaine."[44]

The political situation was a disturbing force to Wall Street and to Gould personally. He had no influence with Grover Cleveland, while Blaine had been his friend for years. His fondness for Blaine formed a seamless web with his need to have a reliable man in the White House during these critical times. But Blaine's nomination had split the party nationally and in New York. The campaign had gone badly, and even the faithful knew Blaine was in trouble. Jay contributed $5,000 and used his influence to solicit funds from others. As part of this effort he agreed to attend a gala dinner arranged by Cyrus Field and others to honor Blaine. In a year filled with misfortune and bad timing it proved one of the worst decisions he made.[45]

The banquet took place at Delmonico's on the evening of October 29, a week before the election. That same morning Blaine had committed the unwitting blunder of allowing the phrase "rum, Romanism, and rebellion" to fall unchallenged from a minister he later called "an ass in the shape of a preacher." He compounded this mistake by sitting down with a galaxy of two hundred prominent businessmen, financiers, and politicians. The press was excluded except for a reporter from the friendly *Tribune*, and the event produced nothing of importance. What gave it significance was the presence of so glittering an array of wealth and influence to pay homage to the candidate. Hostile newspapers seized on the spectacle and, in the waning days of the campaign, flogged Blaine unmercifully as the tool of privilege.[46]

Although much has been written about the effect of this dinner on the election outcome, its consequences for Gould have gone unnoticed. The *World*, no longer in Gould's control, lampooned the event as "Belshazzar's Feast" and printed a devastating cartoon of Blaine surrounded by Money Kings festooned with diamonds while a tattered, starving family begged for crumbs from the table. The *Herald* took another tack. For months Bennett had been attacking Gould on one front and Blaine on another. Since June, when Gould declared for Blaine, the *Herald* had on occasion pilloried the connection between them. The presence of so notorious a recluse as Gould at the dinner inspired Bennett to new heights of malice. In the last days before the election he fused his two crusades into one savage assault on a hapless creature labeled, thanks to the coincidence of initials, "Jay Gould Blaine."[47]

Gould had no part in organizing the banquet and sat not at the dais but at a remote table with such familiars as Sage, Connor, E. F. Winslow, and R. M. Galloway of Manhattan Elevated. For Bennett's purposes, however, Jay's mere presence was sufficient. Next day the affair became, in the *Herald*'s distorted version, "GOULD'S BANQUET TO BLAINE," with this explanation:

> It was announced that the feast would be given under the auspices of a committee of which Cyrus W. Field was the head. Jay Gould's name [does] not appear on this committee, but Mr. Gould was, as a matter of fact, the most prominent of Mr. Blaine's entertainers, and a glance at the long array of names of Mr. Gould's other friends and associates who were present justifies the strong suspicion that it was Jay Gould who was back of the whole affair, and that he was the real host.

On this premise the *Herald* blitzed readers for the next six days with responses to "GOULD'S DINNER TO BLAINE." Unnamed businessmen and bankers sputtered indignantly over how "that dinner with Gould has killed Blaine." For four days several columns were devoted to reprinting outcries of disgust from newspapers across the country. A series of editorials harangued "Jay Gould Blaine," the tool of monopoly, and asked the question, "Jay Gould or the People?"[48]

Unfortunately, the ordeal did not end once the polls closed. It became clear early that the election hinged on the outcome in New York, where returns dribbled in slowly. Although both sides claimed victory, the vote was close enough to prolong suspense for nearly two weeks pending an official canvass. Inevitably disputes and charges of fraud arose. The *Herald* offered an easy explanation for the delayed returns: they were being withheld, perhaps even falsified, by Western Union, the puppet organization of Jay Gould. The "Jay Gould-Blaine-*Tribune*" conspiracy hoped to steal the election by use of "Jay Gould-Western Union" dispatches. Saloons and hotel lobbies swarmed with angry crowds demanding to know the results; as the delay stretched out, their mood grew nastier and more confused. Excited stories of an "intended demonstration in the vicinity of Jay Gould's residence" prompted the police superintendent to keep all off-duty men on reserve at the station houses. Aware that the city was ready to explode, Gould tried to break the tension. On November 7, before the canvass was completed, he sent Cleveland a telegram of congratulations and avowed his belief that "the vast business interests of the country will be entirely safe in your hands." It was a statesmanlike act and, as usual, one that coincided with his self-interest, if not his preference. If, as Jay now believed, Blaine had lost the election, the best thing was to eliminate the uncertainty and suspense that gripped the country and played havoc with markets.[49]

His enemies viewed the telegram in another light. "Everybody knows that Jay Gould is a coward," sneered a "gentleman" quoted by the *Times*, "and fear of bodily injury undoubtedly caused him to abandon the idea of tampering with the vote of this State, if he had any such idea." The *Herald* agreed. Beneath such headlines as "GOULD IN THE PILLORY" and "A COMPOUND OF DEVIL AND COWARD" it published extracts from other papers denouncing Gould. Two weeks later Norvin Green issued a vehement denial that Western Union had tampered with returns, and no evidence to that effect was produced. In fact the election had been a genuine cliff-hanger; Blaine lost New York, and with it the presidency, by a mere 1,149 votes out of 1,167,169 cast.[50]

The election that cost Blaine his life's dream also sealed the image of Gould as archvillain. In his relentless campaign Bennett, with help from George Jones, *Puck*, and other critics, had brilliantly fused together the disparate elements of Gould's reputation as manipulator, monopolist, corruptionist, sneak, betrayer of friends, and coward. Although his dark influence in politics had long been assailed, the charge that he tried to steal a national election added a new dimension of outrage. The gap between man and legend had been made complete; never again could Gould escape his image as the most hated man in America. Ironically, he had already wreaked exquisite if unwitting revenge on Bennett. In May 1883 he had sold the *World* to Joseph Pulitzer, whose imaginative leadership would soon lift that paper past the *Herald* in circulation and advertising.[51]

In private Jay was less sanguine about Cleveland's election than his telegram suggested. It left the market in such an unsettled state that he still hesitated to leave for the West. After the Union Pacific boom tailed off in September, Gould maintained what he called "a large interest" in the road. In March he would leave its board, yet he clung to the company with a tenacity that transcended his manipulation of its stock.[52]

As this longest and most dismal year of Gould's career crawled to a close, the debate over his solvency was slowly laid to rest. Unlike "Chinese" Gordon Jay would live to fight another day, but the unremitting horrors of battle had left him a changed man—ashen, haggard, stooped, his face lined with care and his expression lifeless. The ordeal seemed to have shrunk him down to fit Jim Fisk's description of him as a pair of eyes and a suit of clothes. Even the eyes, those flashing pools of energy, tended now to stare dully into space, glazed with fatigue until some new problem ignited them. As his strength ebbed, the attacks of neuralgia grew more frequent and severe. Although no one yet suspected, Gould could no longer endure the nervous strain of so constant a grind on so many fronts. To outward appearances he remained active, even frenetic in business, but his roles were reduced, his involvements became more measured, and his flagging strength was revived by longer and more frequent vacations.

As always Jay told the public nothing of his physical debility. He wanted no one's sympathy and wasted no energy on self-pity. With tired eyes he surveyed his besieged empire and found its contours battered but not broken. The market remained shaky; tough battles loomed over reorganization of the Wabash and the Texas & Pacific; the Katy was in revolt; the Missouri Pacific had to be expanded; the telegraph struggle continued; the trunk line rate war still raged; and the transcontinental pools were in shambles.[53]

Nevertheless, the king still lived and the crown prince was coming of age. There was much to be done.

27

Realignment

The fight of Gould against the Burlington and the Atchison was thus partly responsible for the most promiscuous paralleling and duplication of railroad mileage in the country's history. . . . Gould again thought in terms of establishing new through routes. His mind turned to grand competitive schemes, ignoring again, as he had so many times before, the cost elements and the relative importance of through traffic vs. local.

—Julius Grodinsky, *Jay Gould: His Business Career, 1867–1892*

The tribe of Hinckleys, Cranes, Graves, Villards, and other cranks and thieves led by Hopkins and Gould, have built and will build Roads wherever fools with money will follow, and where three Roads, stimulated by contracts, are thus built to do the work of one, it, in the very near future, leads right up to a necessity for the nearest solvent Railroad to buy the other useless ones, and thus load up the said solvent Road . . . with foolish Branches which have been a dead weight ever since.

—John Murray Forbes, letter to C. E. Perkins, August 25, 1885

Mr. Gould . . . mastered not only the general conditions, but every detail. . . . If the matter . . . was one which could be adjusted, and both sides were willing, it was a pleasure to transact the business. He was so clear and direct, so intelligent upon the matter in hand that no time was wasted in useless discussion on irrelevant facts, but the point was arrived at at once and the possible concessions on either side were considered and made. If the subject, however, was one which he did not care to bring to a head and his interests were adverse to the adjustment . . . he could be as vague, as indefinite and unsatisfactory as any man that ever lived.

—Chauncey M. Depew, *Railway Review*, December 17, 1892

ALL HIS LIFE Gould loved maps. As a young man he had made them; now he made the railroads and towns that filled them. The walls of his office swarmed with maps and so did his worktable. On their pliable surfaces his small, lithe hands charted the course of empire. He knew, of course, that no other gimmick was more useful in promoting a railroad, which always looked vastly more enticing on a map than in actuality. Like a good general he used maps to plot campaigns and was careful never to confuse their deceptive symmetry with the realities of battle. Yet he also relished their worth as the stuff of dreams. Jay built and rebuilt systems, rearranged old lines and plotted new ones, on maps. What was his dream of a transcontinental railroad but an unbroken ribbon of one color spread across the length of a map?[1]

This love of maps reflected a basic element of Gould's nature that escaped those blinded by his reputation as a speculator. He was above all an active, not a passive investor. In the wake of his recent travails he could have followed Vanderbilt's lead by gradually selling off his holdings, investing the proceeds in government bonds or other safe securities, and retiring from business to live off his income. Certainly his health would have benefited, but one doubts that he ever considered it. Unlike most wealthy men Gould put little money into real estate, bonds, or other safe investments. Instead his funds went into live enterprises that opened new territory, promised growth, and demanded constant attention. He was not only a creative financier but a creative businessman who thrived on the challenge of working with unformed clay. Dodge was more right than he knew in saying that Gould would "stand in history as having risked and planted his millions in developing a new country while others merely risked and planted their millions in a country . . . where there was no risk as to returns."[2]

This creative impulse, which had driven Gould throughout his career, did not perish in the fiery ordeals of 1884. On the contrary, he emerged wanting more than ever to leave his children the larger inheritance of an ongoing business empire. Ultimately this obsession became the tragic flaw that consumed him, but obsessions are not easily dismissed and to all appearances he remained its willing prisoner. Realizing that he no longer possessed the strength to do everything, he devoted full attention to developing the properties under his control.

The railroads were the trouble area, the one most disturbed by the upheavals of 1884. Jay confronted a radically changed map. He had lost all influence on the Union Pacific and with it his power over connections west of Omaha. The Wabash was in receivership and the Texas & Pacific hurrying there, leaving him secure control of only the Missouri Pacific–Iron Mountain and Katy systems. Both were north-south systems linked together by the Missouri Pacific at one end and the Texas & Pacific at the other. Without the latter road and the Wabash, Gould's rail empire would be reduced to a stable but small domain dependent on other lines for connections in all directions and surrounded by powerful rivals: Huntington and the Santa Fe to the southwest, the Iowa roads and Union Pacific to the northwest.

This state of things forced Gould to rethink his entire strategy. The Missouri Pacific was a healthy property but without east-west lines it could not be a transcontinental system, and Gould still clung to that old ambition. If he stood idle, he would likely be invaded, perhaps even swallowed, by his enemies. To stay in the game he must either keep the floundering portions of his present system intact or build new lines. The original empire had been erected in jerry-built fashion by piecing together a mishmash of existing lines augmented by construction. While this approach was easy to condemn by hindsight, it was a logical response during a period of frantic change and intense competition. For three years Gould had absorbed himself in the building of empire; now he was obliged to probe the structure for flaws, weak points, inadequacies.

They were many and glaring. Too many roads and branches did not do enough business to earn their keep. Those with ample traffic were compelled to haul it at rates too low to pay interest on their swollen capitalization. Roads that got by in an era of high rates were helpless against newer, cheaper roads with shorter lines and more efficient service. Many of the lines in Gould's system were old, dilapidated, and inefficient—motley wholes assembled from ill-fitting parts. So rapidly had the map changed that once-promising roads had become dinosaurs in only half a decade. Gould had made money by starting their stocks low, running them up, selling at the top, and selling short on the way down. That game was finished; many of them could now be bought for loose change and there were no buyers.[3]

The southwestern roads also served Missouri Pacific as feeders and connectors. From the beginning Gould adopted a policy of making subsidiary lines enhance the prosperity of the Missouri Pacific at their own expense. Several techniques were employed, including the manipulation of payments for terminal services, inequitable division of through rates and traffic, and favoritism in the areas of maintenance and equipment. In Gould's priorities the Missouri Pacific–Iron Mountain system was the sun of his rail universe and all other lines merely satellites, important only so long as they served a useful purpose.[4]

By 1885 the feeders were becoming suckers, yet Gould dared not cast them adrift until he had replacement lines. From this dilemma emerged a plan that transformed Gould into one of the foremost railroad builders of his age and made him once again the catalyst for change on the competitive map. He would cling to the satellites as long as possible and try to retain their best parts. At the same time he would construct new lines to eliminate his dependence on them and to tap virgin territory. The new lines would be cheaper, well built, and located precisely where he wanted them. They would be entirely under his control with no legacy of debt or shoddy construction. No construction companies would be needed; he could finance the work by new issues of Missouri Pacific securities. The timing was right: materials were cheap, labor was cheap, and money would soon be cheap again.

Like a man who had lived all his life in other people's clothes, Jay plunged eagerly into planning a wardrobe of his very own. The first time he had tried to forge an empire the quick way, through consolidation. This time he would do it gradually, through construction of lines that fit his own design. A glance at the map showed the western outposts of the Missouri Pacific in Kansas and Nebraska to be the logical starting points. Beyond them Gould looked longingly toward Colorado, San Francisco, and, over his shoulder, toward Chicago.

All the western railroad presidents agreed that rate slashing was ruinous if not suicidal, and all were quick to deplore what Hughitt of the Northwestern called the "needless paralleling of existing systems." After lip service, however, came the dilemma of finding a solution that both resolved differences and satisfied all parties. Diplomacy had failed and war merely escalated the conflict. As Adams astutely observed, no one could escape the old habit in every dispute of overhauling "the long record of the past . . . in order to see whether every one is even with every one else." Tensions ran high, suspicion higher still. Strenuous efforts to impose stability on rates and expansion had produced nothing. The strategic board remained a powderkeg awaiting a match.[5]

Once again Gould stood poised to light the fuse. His calculating mind boiled a complex situation down to its essence. To the north the Burlington and Tripartite roads were struggling to patch up their differences before a war of expansion engulfed them. Even as they talked peace Perkins organized a line to the Twin Cities, Hughitt prepared to invade northern Nebraska, and R. R. Cable of the Rock Island eyed southern Nebraska. Caught between these ambitions, Adams strove to play the part of statesman and came nearer that of fool. "Adams does not know what he wants," scoffed Perkins, "but he has a vague feeling that he wants something." In Gould's opinion Adams was running the Union Pacific *"for the galleries"* and "trying to make out that he is a big man as a railroad manager. He may be. I dont take any stock in him in that score. We will wait & watch results—the growth of the country may bring him out."[6]

The country *was* growing, as Gould well knew. Plentiful rainfall and good crops sent settlers streaming into the plains states. This influx promised local business for areas still untouched by rails, a factor that influenced Gould no less than other presidents. Peace still reigned in Kansas, where the Missouri Pacific and the Santa Fe had agreed not to build into each other's territory. The door to Kansas lay open, and with it a line to Colorado. The Iowa roads were distracted by their own quarrels; the Union Pacific, beset by woes on all sides, had an inexperienced president eager for peace. All these considerations fed a bold plan by Gould to move aggressively in both Kansas and Nebraska.

As early as 1883 Jay began laying the groundwork for expansion. In August 1882 he bought the St. Louis, Fort Scott & Wichita, an obscure, 94-mile road in southern Kansas for which he constructed another 156 miles of track. The Santa Fe protested but did nothing. During 1883 the Missouri Pacific obtained charters for three projects which, if built, would amount to a wholesale invasion of Santa Fe territory. These lay dormant through 1884 but could be revived at any time. The Missouri Pacific also operated by agreement the Central Branch which, although in wretched shape, could be used to spearhead a line to Colorado or an invasion of Nebraska.[7]

All this was but prelude for the coming campaign. In 1885 the Missouri Pacific chartered or acquired eight roads in Kansas alone as Gould's agents scurried across the state seeking local aid. He understood thoroughly what sustained a new line. "I dont think much myself of the through business," he remarked. "It is the local business that is the life of a road." But the new projects also served broader considerations. Together they constituted disparate pieces of a puzzle, apparently unrelated but offering several possibilities for a new through line to Colorado. By May reports asserted that Jay would definitely build to Denver and push extensions into southern Nebraska.[8]

This activity startled Adams. The apostle of peace looked up from his mediation with Perkins to find Gould marching on his flank. While he talked and studied, Jay put crews in the field building northwest from Council Grove, Kansas. Unable to divine Gould's intentions and convinced that the "Kansas mania will not stop while there is a point on the map which has not got a railroad, and which can vote bonds," Adams visited Gould's office. During their talk the usually impassive Gould exploded with anger. "He was mad all the way through," reported a shocked Adams.[9]

Whether the outburst was merely for effect or a symptom of Jay's growing inability to control his emotions is unknown. Convinced that Gould meant to invade Union Pacific territory unless his hold on the Central Branch was assured, Adams gave the Missouri Pacific a new lease for twenty-five years and agreed with Jay to build no new roads competitive with each other, including lines to Denver. In effect the pact gave Gould a free hand to expand in northern Kansas and southern Nebraska while allowing Adams to fight the Burlington without worrying about his southern flank. Now it was Perkins's turn to fulminate. He scolded Adams for the arrangement, while Forbes deplored a contract resting "upon so flimsy a guarantee as Gould's promise" to build no competing lines. "Gould has as many aliases as a London professional thief," Forbes snapped, "and in fifteen minutes will find a substitute to do his bidding if he sees his interest in it."[10]

In Nebraska the Omaha extension reached only to Papillion, from which it leased trackage and terminal rights from the Union Pacific. In September 1883 Clark and some other Union Pacific officials obtained a charter for the Omaha Belt line. Nothing was done until July 1885, when Gould decided not only to construct the Belt but to extend the Missouri Pacific into Omaha and throw out branches to Lincoln, Wahoo, and Nebraska City, which would plant him squarely in the B & M's territory. By then Clark had been ousted from Union Pacific and was eager to serve Gould. After wresting control from the Union Pacific faction, who assumed their road would dominate the project, Clark assumed the Belt presidency and began construction with materials supplied by the Missouri Pacific. To build the extensions Gould suggested a front company of local people "*so that I would not appear on the surface till the road was constructed* . . . in order to cover our tracks." The ruse did not fool his rivals but it left them puzzled over his precise intentions. They might have shivered at his thoughts. On the premise that the Burlington was "invading our natural territory of the Central Branch," Jay wanted to "lay out a system through Lincoln west with spurs to leading towns thus touching their best territory. It might pay us to build to Cheyenne & Denver while material & labor are so cheap."[11]

By the year's end Gould had added 687 miles to the Missouri Pacific system. To finance acquisitions he issued no new bonds but merely increased the company's capital stock from $30 million to $36 million. Impressive as this growth was, it would pale before the figures of the coming year. Led by the Missouri Pacific, the major western roads mobilized for a war of expansion. Fear and suspicion swept all hope of compromise before them. As Grodinsky observed, "In a vast territory all the roads, against their own will, against their repeated desires to avoid the building of new roads, found themselves forced into major building programs." Friend trampled over friend in order to smite foe. Hughitt apologized to Adams for building across his line but insisted he must punish the Burlington even if it meant punishing the Union Pacific in the bargain.[12]

Everyone felt the pressure not only to expand but to do it at once, while the time was ripe. Money and materials were cheap and likely to go higher next year. Perkins got the message when he tried to buy rails and found the mills crowded with orders. By October even Forbes thought it "wise to organize and be prepared for war." A vicious cycle was emerging: the fall in prices and earnings that hurt the railroads also made it inviting for them to construct new lines, which intensified an already savage competition and dragged rates lower still. Gould did not create this cycle, but his aggressive policy did much to unleash its fury.[13]

During the next two years Gould constructed 1,980 miles of new track and bought another 308 miles of existing road. His crews fanned out across six states and the Indian Territory (now Oklahoma). In Kansas he struck at the heart of Santa Fe territory and thereby goaded that company into massive retaliation. Subsidized by county bonds, both companies laid track ahead of demand or, in some cases, regardless of demand. By 1890 Kansas possessed more mileage than New York, Pennsylvania, or the New England states.[14]

Into this conflict plunged the Rock Island which, after first threatening to invade Burlington territory, decided instead to build southwest into Kansas and Colorado. Within three years the Rock Island expanded its system from 1,400 to 3,000 miles, most of it in territory contested by Missouri Pacific and Santa Fe. The latter retaliated by building its own line to Chicago, a step that threw all existing relationships between Chicago and Denver into chaos. Gould responded by pushing his plans for an extension to Colorado. His problem was to find the right route and connections. Since the Central Branch agreement precluded his extending that road or his central Kansas line, Jay picked up the charter for a dormant project known as the Denver, Memphis & Atlantic and used it to build south of the Kansas Pacific.[15]

In May 1886 Perkins learned from an intercepted telegram that Gould was contemplating a line to Pueblo. This was disturbing news because relations in

Nebraska were still explosive and he was agonizing over whether the Burlington should invade Kansas. In fact Gould had long since decided to build, but he had to arrange a suitable connection for through business. Not until the winter of 1887 did he find a satisfactory path through the competitive thicket in Colorado. The Rio Grande offered him good terminal facilities in Pueblo and Denver, from which traffic could move west via the Rio Grande Western. These possibilities induced Jay to build directly across western Kansas, and from the state line to Pueblo.[16]

Gould had assured Perkins in December 1885 that he would build only to Lincoln, not to Nebraska City as well. As Clark built toward Lincoln, however, he began to waver. Hoxie warned that an extension to Nebraska City "would invite open retaliation by the Q into our territory—that if built our hand should not appear until after construction." Gould decided to adopt the tactics used so effectively in Kansas. Clark would persuade some local people to form a company, build the road, issue securities to pay for it, and sell them to another intermediary, who would later transfer them to Gould. Although the Nebraska lines were small, they loomed large in Gould's thinking. He envisioned a large coal and lumber business at Lincoln and was convinced the extension would pay from the start.[17]

While spreading his web of rails across the prairie Gould did not neglect the Southwest, where a region rich in coal, lumber, and other resources lay waiting to be tapped. He conceived the notion of a new through route between Kansas and the Southeast by following the Arkansas River from the Katy to the Iron Mountain at Little Rock and on to the Mississippi River at Arkansas City. Roads built by local promoters already occupied the route between Fort Smith and Arkansas City; one was in receivership, the other floundering. In December 1886 Jay acquired the line between Little Rock and Arkansas City at foreclosure sale.[18]

Possession of this line confronted Gould with a hard choice. He could buy the Little Rock & Fort Smith and construct the link from Fort Smith to the Katy, or he could build toward Kansas from a point farther north on the Iron Mountain. During 1886 Gould commenced his own line to Memphis from Bald Knob, 58 miles north of Little Rock. William Kerrigan, his general superintendent, favored an extension westward from Bald Knob, arguing that the region between Fort Smith and the Katy (or Kansas, if the whole line were built) offered no local business. Kerrigan's route would cross more promising territory and terminate in Memphis rather than Arkansas City. He knew Gould was negotiating an alliance at Memphis with the Richmond Terminal system, which reached most of the major cities in the Southeast.[19]

Kerrigan's argument was sound but did not reckon with Gould's desire to "build up Little Rock as a commercial center and of developing the State of Arkansas." It was not through traffic that attracted him but local resources, particularly coal and lumber. A tour of Arkansas convinced Jay to follow his original plan. In short order he bought the Fort Smith road, constructed a line from Fort Smith to the Katy, and acquired one feeder and built another to secure coal lands. In public he talked up the advantages of the Arkansas Valley route as a through line; privately he followed his practice of developing local resources.[20]

The same policy prevailed elsewhere. In Missouri Gould built short branches wherever the region promised subsidies and coal or other resources. In Texas he added 225 miles to the Katy system. As his Arkansas system grew, Gould turned his attention toward the region south of Texarkana. He wanted a line from the Iron Mountain to Shreveport and Alexandria, not only to develop the territory but also to reach New Orleans should he lose control of the Texas & Pacific. However, the city of Shreveport rejected his subsidy proposal and a rival road, the St. Louis, Arkansas & Texas (known as the Cotton Belt), already occupied the region and was struggling to forge a system parallel to the Iron Mountain. Realizing that war was inevitable, Gould elected to wait for an opportunity to go after the Cotton Belt itself.[21]

By the end of 1887 Gould had erected a formidable rail empire. Since December 1884 the Missouri Pacific–Iron Mountain system had expanded from 1,960 to 4,150 miles. In addition he controlled Katy, Texas & Pacific, International, and Central Branch, another 4,311 miles, giving him a total of 8,461 miles without even including the Wabash. During that three-year period the Missouri Pacific's capital stock increased by about $14 million and its funded debt by about $15.5 million. However, the vast amount of new line enabled Gould to show a decrease in the stock *per mile* from $27,228 to $10,701 and in bonded debt *per mile* from $39,009 to $20,089. These remarkable figures revealed again Gould's ability to present facts in the most favorable light. He did not remind stockholders that most of the new mileage lay in undeveloped regions that did not yet (and might never) have the earning capacity of more established routes.[22]

In less than a decade Jay had transformed the Missouri Pacific from a road between St. Louis and Kansas City to one of the largest systems west of Chicago. He had made it prosperous and powerful but not secure. His aggressive policy merely enlarged the field of battle and galvanized his adversaries on every front. In Nebraska Perkins, whose road enjoyed the reputation of never losing a fight, watched his every move. In the West Huntington and William B. Strong, the Santa Fe's president, whom Adams described as "very belligerent," applied relentless pressure against Gould. Huntington, commanding the dominant system in the Southwest, perfected his control by unifying his western roads within a holding company. Strong responded by building even more road than Gould. Besides the Chicago extension he criss-crossed Kansas with branches and countered Gould's Pueblo line with one of his own. In 1886 he struck at the heart of Gould territory by acquiring the Gulf, Colorado & Santa Fe, which ran from Galveston to Fort Worth. By extending this line northward to a connection in Kansas, Strong created a new through line to Kansas City and also paralleled the Katy. In January 1887 he bought a small Missouri road and announced plans to build into St. Louis.[23]

Where Gould had been aggressive, Strong showed himself positively bellicose. Already the Santa Fe groaned beneath a staggering burden of debt, but no matter. The roads had been constructed and the result could only be an intensi-

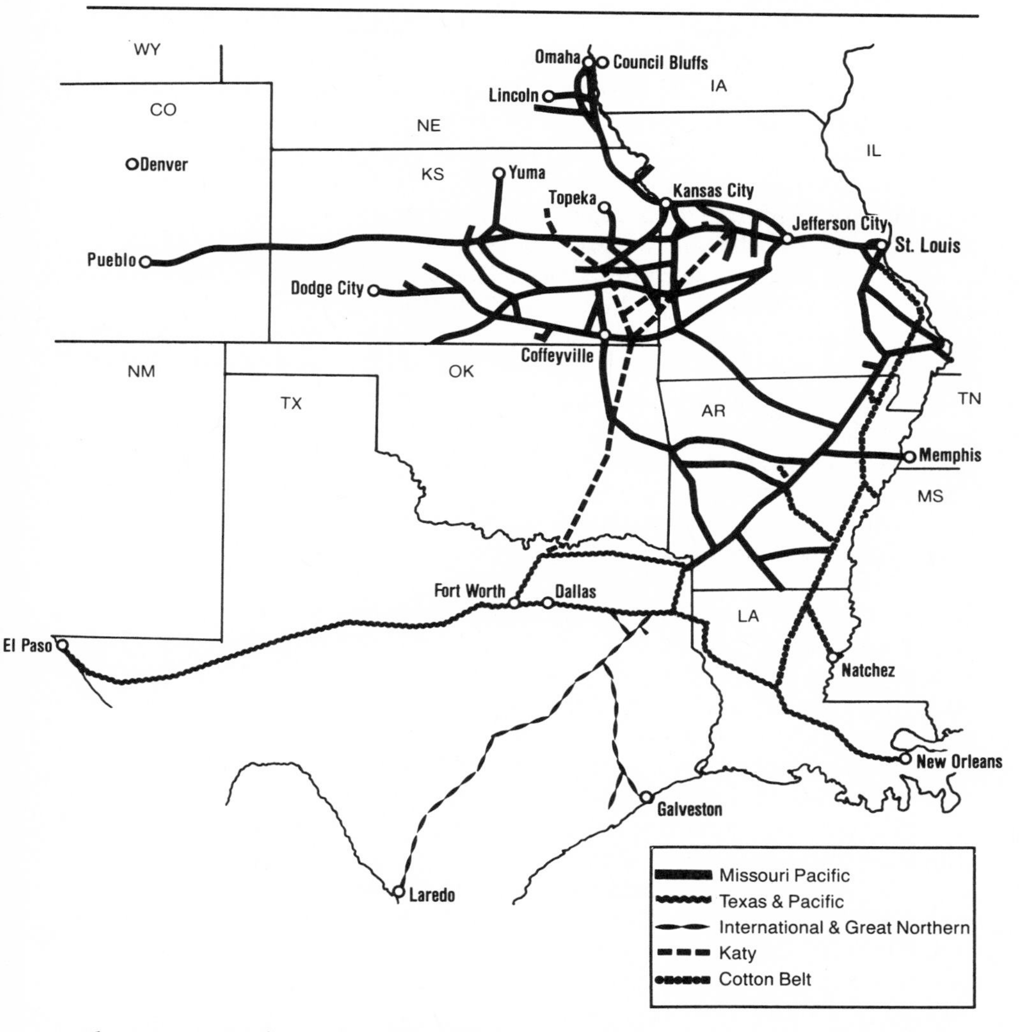

The Missouri Pacific and the Gould "southwestern system," including the Cotton Belt, or St. Louis, Arkansas & Texas Railroad.

fied struggle for traffic and more rate cutting. Amid this escalating fight Gould found himself engaged on a second front: the mounting challenge to his control of properties that only a few years earlier formed the heart of his southwestern system.

Nothing came easy for Gould during these truculent years. The maze of legal and financial arrangements fashioned during the blitzkrieg of 1879–1881 now conspired to haunt him as their tortuous complexities came unraveled. Two of them, the Wabash and the Texas & Pacific, underwent reorganization fights lasting half a decade. In these battles Gould displayed an ingenuity and ruthlessness that drove opponents to revile him in the bitterest of language. His tactics were so novel, so utterly audacious, as to confound bankers, lawyers, judges, and security holders alike. Men steeped in conventional wisdom, who clung to precedent as if it were a life raft, did not appreciate being plunged into uncharted waters where they flailed helplessly while Gould swam with the sureness of a shark.

The strain of these contests taxed Gould's concentration and stamina severely, but nothing deflected him from his purpose. Convinced that the Missouri Pacific needed both roads as friendly satellites, he undertook to dominate their reorganization. The Wabash posed the kind of problem on which Gould's intellect thrived. Its affairs were a nightmare of legal and financial complexities that bewildered most observers. Yet in 1884 Jay produced a reorganization plan within weeks of his receivership coup. As Grodinsky observed, "The bondholders had hardly grasped the essentials of the problem before Gould was ready with a proposal." The plan was speedily rejected, leaving the Wabash, in Perkins's words, "floundering about and trying to make a settlement."[24]

Everyone conceded that the Wabash required drastic financial surgery. Part of its problem involved the large number of leased lines that did not earn their interest. Gould had the receivers ask the court for permission to lop off all nonpaying lines. In a curious decision Judge David J. Brewer, the same magistrate who had appointed the receivers, denied the request but invited the bondholders of these branches to foreclose if interest was not paid. The receivers took the cue and stopped interest payments, forcing the bondholders to reclaim lines of little value except as part of some larger system. The prospect of these roads floating free in a region ravaged by fierce competition horrified the major carriers. Perkins went shopping at once among the castoffs.[25]

Meanwhile, Gould entered protracted negotiations with the committee representing the English bondholders. Joy went back to England in May 1885 with a modified version of the plan. After several months, agreement was reached on a compromise plan for a purchasing committee headed by Joy to buy the road at foreclosure sale. One of the other three members was O. D. Ashley, the Wabash's secretary and a Gould man. "If agreement could be launched while present boom lasts," Gould told Joy in August, "it will go off like hot cakes." Here was a cute twist: Having in the past used agreements to boost stocks, Jay was now using the market to boost an agreement.[26]

The sale took place in April 1886 and the end seemed near for what an observer called "one of the most complicated reorganization movements ever attempted in this country." This bit of naïveté was shattered on June 1 when the purchasing committee issued a pamphlet detailing the harsh realities facing the Wabash. After emphasizing that "there is no point on its lines not subject to competition as intense as competition can be," the committee painted a bleak picture of the road's financial situation and submitted five propositions necessary to complete reorganization. Among other things, senior bondholders were asked to take reduced interest and exchange overdue coupons for new bonds. If they chose instead to foreclose on individual components of the main line, they would simply wreck the system that gave their securities value.[27]

Once again Gould had upset the applecart of precedent by converting the primary strength of a mortgage into a weakness. Despite its gloomy cast, the pamphlet earned praise from financial journals. The *Chronicle*, long a Gould critic, labeled its arguments "sound and their conclusions fairly drawn." Some unimpressed bondholders decided to take their chances in court despite Ashley's warning that "the most complicated litigation ever known in the railroading of this country" would ensue. Their action came before Judge Walter Q. Gresham, who took a dim view of liberal receiverships. His presence spelled trouble for Gould, who had no direct hold on the bondholders.[28]

He did have the purchasing committee, however, and used it freely to squeeze the bondholders. Resorting to a favorite Gould device, the receivers bought back their certificates and, instead of discharging them, turned them over to the purchasing committee to hold as a lien prior to that of the bondholders. Judge Brewer obliged with an order allowing the receivers to pay interest only on those bonds designated by the committee, which chose to pay only those bondholders who accepted the reorganization plan. These high-handed tactics were swept aside in December 1886 when Judge Gresham delivered a landmark decision on what one critic called "the greatest scandal in all recent American railroad history."[29]

In a lengthy review Gresham criticized Judge Brewer's appointment of receivers in 1884 and roundly condemned their practices since that time. The testimony disclosed for the first time some of the reasons why Wabash remained important to Gould. Since 1884 the receivers had paid the Missouri Pacific $3.2 million for labor and supplies incurred during the lease; Gresham thought this money properly belonged to the bondholders. Gould, Sage, Dillon, Humphreys, Hopkins, and Charles Ridgeley, all Wabash directors, owned the Ellsworth Coal Company adjacent to the railroad. Gresham found evidence to suggest that, in selling coal to the Wabash, "the price paid for the coal purchased for use was too high, and that the freight upon the coal shipped was too low." In all, the receivers, one of whom owned stock in Ellsworth, paid that company $80,711 in rebates, more than its entire capital stock. Deeper down, beyond the judge's scrutiny, could be found items such as trackage agreements. In one case the Wabash general agent complained that the amount charged by Missouri Pacific for use of some Katy track was "considerably in excess of what we should pay," but efforts to reduce it proved unavailing.[30]

Gresham granted the bondholders' request that the receivers be ousted and a new one appointed for Wabash lines east of the Mississippi River. On January 1, 1887, when the new receiver took charge, the Wabash reverted to its premerger status of two independent systems separated by the river. The Street was astonished; reorganization seemed light years away, and Joy predicted the Wabash system would be broken up. Gould climbed out of the rubble with another proposal for a reorganized company embracing the lines west of the river, the road from Chicago to St. Louis, and the Detroit division. When Gresham quashed this new plan, Jay carried the case to the Supreme Court. As the months rolled by, the bondholders stubbornly resisted every blandishment proffered by Gould.[31]

During the three long years of this travail Gould was also locked in a dogged battle over reorganization of the Texas & Pacific. There his claim rested on an operating contract fortified by his familiar device of a floating debt held by the Missouri Pacific. The road consisted of three divisions, each with its own territory and mortgage structure. The eastern division from Shreveport to Fort Worth, with a branch to Texarkana, did the largest business. By contrast the Rio Grande division from Fort Worth to El Paso had failed utterly to fulfill the extravagant predictions made about its future. The New Orleans division, between that city and Shreveport, was another road with more promise than performance.[32]

Like the Wabash and so many other roads of the era, the Texas & Pacific was a monument to failed dreams. It had already lost the fight for transcontinental business to Huntington. Local traffic was ample on the eastern division, growing on the New Orleans, and nonexistent on vast stretches of the Rio Grande. Predictably, bondholders of each division held exaggerated views as to the value of their portion relative to the others. They agreed only that Gould was somehow to blame for their woes, and that a primary reason lay in his policy of operating the Texas & Pacific in the interest of the Missouri Pacific.

During 1884 a committee of eastern division bondholders tried to eliminate this conflict of interest by demanding the right to control the board and name the chief operating officers. After lengthy negotiations Gould accepted a compromise plan that preserved his practical control of the road. Early in 1885 he ordered a report prepared on traffic relations among all the roads under his command. This detailed analysis emphasized that if the Texas & Pacific were divorced from the Gould system, most of its through business "could be sent by other channels, and its rates on both through and local traffic would be badly demoralized." It also concluded that neither the Rio Grande nor the New Orleans division could sustain itself if cast adrift.[33]

During 1885 the situation went steadily downhill. The Texas & Pacific continued to run a deficit, its physical condition "rendered it dangerous to life and property," and the funding scheme fell flat. In November the board appointed a committee of five directors, none of them Gould allies, to investigate the road's physical and financial condition. Six weeks later the committee recommended seeking receivership, but before it could act Gould protested some $1.13 million in notes held by the Missouri Pacific and persuaded the court to appoint two receivers friendly to him. The bondholders responded in March 1886 by creating a committee on reorganization headed by General Isaac J. Wistar of Philadelphia.[34]

On the surface the fight resembled the Wabash conflict all over again, but appearances were deceiving. The committee represented eastern division bondholders, most of whom resided in Philadelphia. No sooner were the receivers appointed than Gould and Wistar joined forces to prepare a reorganization plan. By concealing Jay's involvement and letting the plan emanate from Philadelphia they hoped to disarm critics opposed to anything shaped by Gould. Wistar's loud show of independence fooled no one. When an outline of the plan leaked out in January 1886, the *Chronicle* abruptly entered the lists as champion of the junior security holders. In a position paper disguised as an article it urged holders to organize at once "unless they are willing to see their rights trampled upon, and the control of the property put entirely beyond their reach in the hands of Mr. Jay Gould, or his *alter ego*, the Missouri Pacific."[35]

No amount of protest by Wistar could erase the fact that the plan would allow Gould to dominate the reorganized company. Although Jay agreed to some modifications, the plan was assailed on all sides and the Wistar committee denounced as a tool of Jay Gould. The general was in fact no one's tool but rather a shrewd, practical man. Aside from protecting his own interests, he cooperated with Gould on the premise that the Texas & Pacific could not survive without the Missouri Pacific as a friendly connector. The eastern division was tied to Gould roads at both ends; a falling out would trigger not only rate wars but probably an invasion as well.[36]

It is the genius of financiers and lawyers alike to transform the simplest of propositions into a labyrinth and profit from the bewildered groping that follows. Himself a master architect of labyrinths, Gould on this occasion sought a swift, clean resolution in his favor. His attempt succeeded only in unleashing a swarm of eager practitioners of the art. While the Wistar committee pushed its plan, the stockholders and other bondholders rushed to form their own committees. The *Chronicle* insisted that the Texas & Pacific was "prospectively one of the most valuable in the country" and "in a better position to-day than ever before." Security holders asserted similar views and charged Gould with having "purposely allowed it to go to ruin . . . for the purpose of gobbling it up in the interests of the Missouri Pacific."[37]

In June 1886 the Rio Grande division bondholders unveiled a plan designed to split the opposition by giving eastern division bondholders even better terms than Wistar proposed and using junior bonds instead of stock to fund the floating debt. When the eastern division faithful began to waver, Gould warned Wistar that his people would not get "such favorable terms in any future reorganization." There was another screw to turn. Gould had just constructed a branch for the Katy from Dallas to Greenville in the heart of eastern division country. By amending its charter he could extend the road east to Texarkana, southeast to Longview, south to Waco, and west to Jacksonboro, creating a system that would make him independent of the Texas & Pacific. Late in July Wistar heard that the Missouri Pacific was widening a narrow-gauge road westward from Jefferson on this route and demanded an explanation from Gould.[38]

The reply is worth quoting as a vintage example of Jay's style:

> A vote was passed by the stockholders of the M K & T to widen this road on account of the widening of the St. Louis, Arkansas & Texas and their determination to build a road from Mt. Pleasant, through Sulpher [*sic*] springs to Sherman.

This lone sentence illustrates perfectly Gould's ability to confound and infuriate those who dealt with him. His explanation is clear, precise, entirely plausible, and utterly dissembling. The true message lurks in the subtext like a shark beneath placid waters. The work was in fact being done by the Katy, which happened to choose just this time to undertake it. Gould controlled the Katy but not its stockholders and therefore was technically not responsible for the decision. Before Wistar could digest these complexities, Gould wrote the very next day suggesting that "in the interest of Tex & Pac reorganization . . . it might be well to put this road and the Dallas & Greenville into the new company."[39]

By July even the *Chronicle* complained that "there is getting to be too much literature in Texas & Pacific affairs, and the situation becomes less clear." A group of London bondholders agreed; they appointed a new committee headed by Robert Fleming, a hardnosed Scotsman who had played a prominent role in several other reorganization efforts. Here was a foe worthy of Gould. Fleming had a strong committee, support from some bondholders in all divisions, and an underwriting syndicate led by Drexel, Morgan and Kuhn, Loeb. With the latter's help he startled Wistar by offering to buy the eastern division bonds at nearly par. The price enticed the committee but not Wistar or Gould, who realized that the sale would leave them without a voice in the reorganization.[40]

This first gambit led to the bargaining table. Fleming was too shrewd to suppose Gould could be elbowed aside, and he had data showing that in 1885 the Iron Mountain had delivered three times more freight to the Texas & Pacific than it had received. The trick was to find terms acceptable to both the Fleming and Wistar committees and to Gould. After some tough trading with Fleming and Jacob H. Schiff of Kuhn, Loeb, Jay reached agreement on a modified plan. The two committees were merged, with Wistar retaining a majority of one. The Fleming plan was adopted with some changes, the most notable being an increase in the capital stock from $32 to $40 million. Of this amount $6.5 million would be used to fund the floating debt, giving the Missouri Pacific a large interest but far from a majority. The Fleming group would get six seats on the reorganized board. Gould agreed to join the underwriting syndicate and was allotted a share of $930,000, of which he reserved $250,000 for Wistar and his friends.[41]

The *Tribune* noted astutely that the compromise would "secure to Mr. Gould his practical control of the property but yield better terms to security-holders than were originally proposed." The changes impressed the *Chronicle*, which embraced the new plan warmly and warned that security holders "should not be misled by any talk of 'selling out to Gould' or 'surrender of control' &c., but examine for themselves." To its chagrin the very interests it had fought to protect quickly rebuffed the plan. The stockholders' committee trotted out its own plan and vowed a "fight to the end." The Drake committee, representing holders of some third mortgage liens known as land-grant bonds and income bonds, also demanded better terms. Once again hopes for an amicable agreement were snuffed out.[42]

Gould knew what was going on, knew the game well because he had played it so many times before. Apart from the interests they represented, men like S. J. Drake and Henry Clews of the stockholders' committee wanted their own pound of flesh from the carcass. Wistar saw it too but balked at paying "that lying rascal Clews . . . one cent of their 'expenses,' which is what they really want." While he raged, Jay reached an accommodation with the stockholders and placated Drake's committee with better terms spiced with an allowance of $75,000 for its fees and services. Wistar protested this offer in three pages of earnest detail only to have Gould brush them aside as irrelevant. The issue was not financial logic but practical necessity. Drake's committee threatened litigation that could stall settlement indefinitely. If the figure of $75,000 seemed large, Gould added, "after the litigation had been opened it would soon swell to larger proportions." Lest Wistar miss the point, Jay noted with exquisite delicacy that "it is this compensation that will lubricate the committee & lawyers and thus facilitate the settlement."[43]

Jay was no stranger to lubricants or litigation. A few months earlier he had defanged the opposition of John R. Dos Passos with a fee of $50,000 to switch his legal talents to the cause of the Wistar committee. Ultimately Wistar's group approved Gould's offer to the Drake committee, and the road plodded relentlessly toward a foreclosure sale, which finally occurred in November 1887. Even then more months of preparation would be needed to complete the reorganization.[44]

Through it all Gould clung tightly to his hold on the property. As a hedge he kept alive his plan for "making Dallas the future terminus of the Iron Mountain road," but he had no intention of letting the Texas & Pacific slip from his grasp. Here as with the Wabash he had invested three years of patience, tenacity, ingenuity, and hard work only to find the outcome still in doubt. During those years he had transformed the Missouri Pacific into a major system and in the process had triggered the largest and most explosive war of expansion ever witnessed. On several projects the costs exceeded his estimates, and everywhere rate wars slashed into earnings. The reorganization battles dragged on with no end in sight, and yet another struggle loomed on the horizon, this one over the Katy. Nor was this all. The impetus of growth revealed to Gould serious deficiencies in the operating structure of his organization. Like so many men of vision, he was to discover that forging an empire was a simple task compared to that of running it efficiently.[45]

28

Workday

To the operating man accustomed to efficient car and train movements and to the shipper accustomed to expeditious deliveries, the Gould railroads were by no means a blessing. A Gould road in the Southwest was a byword for poor service. . . . the Gould roads were poorly maintained; in fact usually drastically undermaintained.

—Julius Grodinsky, *Jay Gould: His Business Career, 1867–1892*

The simple fact is, as I have at last ascertained to my great cost, S. H. H. Clark did not know what a railroad was. He thought he did, and Mr. Gould thought he did; but the man was so utterly incompetent, and not over honest. His weakness with our employes [sic] *was something incredible. He turned over the road with its force in a wholly demoralized condition and materially in a condition which could not stand six months hard work.*

—Charles Francis Adams, Jr., letter to G. M. Dodge, March 26, 1886

At a special meeting of the Spread the Light Club, composed of pioneer Knights of Labor of New York and Brooklyn . . . Gould is charged in the address with "treason and high crimes against humanity"; with trampling upon every just principle and every ennobling sentiment.

—*New York Herald*, April 16, 1886

On February 6, 1885, George turned twenty-one years old. To herald his coming of age he was elected a member of the New York Stock Exchange and presented with his father's power of attorney. His apprenticeship done, he assumed new responsibilities as an officer and director in most of Gould's enterprises. No one doubted that Jay was grooming his successor. "It is his wish," George explained, "that I can be so thoroughly trained that I can take hold at any time." Jay confirmed that "my son represents my interests in these companies. I can rely upon him. He has not only proven himself a good pupil but an able man."[1]

That, at least, was Jay's fervent belief, and to all appearances it seemed justified. George had shown himself to be the most loyal and dutiful of heirs. After some schooling he had stepped willingly into the harness of his destiny and pulled zealously if not with the fierce determination of his father. At Jay's insistence he had mastered the telegraph, served as clerk at Connor's elbow, and absorbed the lessons taught him in his father's office. Jay spared no effort to give him all the necessary tools; what he could not provide was the will to use them with the same dedication that had characterized his own career. If George possessed unusual gifts or weaknesses they had not yet been revealed. The clay had been molded but not fired by experience.

Certain differences in style and character enabled the son to escape the father's stigma. Where Jay was shy and reticent, George grew affable and gregarious as his confidence bloomed. He lacked Jay's quick, penetrating intellect and perceptive eye; indeed, some thought him rather slow-witted. He once described his workday to a reporter this way:

> My chief business at present . . . is to attend to the affairs of the Missouri Pacific. During the summer I go to my office at a quarter before ten and remain there until three, when I get on board of the Atalanta and come home. At the office . . . I am kept busy watching the reports concerning the construction of upward of one thousand miles of new track in the Missouri Pacific system. You can't imagine the work this involves. Sealed bids and proposals are made directly at our office by contractors and supply agents for everything we see, from a locomotive to pens. . . . Then I have to watch all the reports, statistics and returns. Some days I have to sign my name one thousand times. . . . I have to do that to stock certificates. It will make your arms and your back tired. It is so monotonous. In the evening father and I often work from three to five hours studying the reports of wheat, corn, freights in general, passenger traffic, what one section of the country is doing and what another section should do and don't.

Congeniality served George as ingenuity had his father, with results more telling in society than in business. He fancied fine clothes, wore a trim, finely waxed moustache, and strutted like a dandy. Small and lithe, he loved sports and fast horses. Pleasure occupied him no less than work, his tastes running to clubs, parties, and the theater. Since Jay also enjoyed the theater, they sometimes went together.[2]

One trip to the theater in November 1884 proved eventful for father and son alike. They went to Daly's to see a bit of fluff called *Love on Crutches*. The cast included a winsome creature named Edith Kingdon, whose bewitching eyes and hourglass figure entranced George. Afterward he wangled an introduction and discovered to his delight that Edith was no ordinary actress but a young woman of good family who had turned to the stage to help her impoverished mother. In his earnest, guileless way George opened a proper courtship. A few months later they became engaged. When George broke the news to his parents, he plunged the Gould household into the worst domestic crisis it had yet known.[3]

Helen was aghast at what George had done. No amount of explanation about Edith's respectability could penetrate her Murray Hill conventions. That Edith was charming and beautiful did not matter; she was an actress, a social nobody, and gentlemen did not marry either aberration. Helen had wanted so much for her first-born, and had instead been presented with a mockery of every value she held dear. It was not only the blasting of tradition that tormented her but a fear of scandal as well. Two summers earlier, amid the business horrors of 1884, Morosini's daughter, Victoria, had eloped with a coachman discharged by her father and gone to live with him above a liquor store. Morosini was a social nobody in Helen's eyes, but wealth had instilled in him a fierce desire to give his children the best of everything. The papers feasted on it for weeks, wringing every last ounce of sensationalism from the family's miseries. Convinced that a similar humiliation awaited her, Helen asked Jay to intervene. It must have pained her deeply when he took George's side. He would do nothing to mar George's happiness, and he did not share Helen's social prejudices. His own match had not been one of social equals. As for Edith, Jay found much in her to admire. "She went on the stage to earn her own living and to support her mother," he later told a reporter, "and that, I think, was very much to her credit. I honor her for it." He knew something about hard work and about those with spunk enough to make their own way in life.[4]

As the engagement entered its second year Helen showed no signs of relenting. In June 1886, when the Daly company left to perform in England, George set tongues wagging by following close behind. Gossip mongers breathlessly awaited news of an elopement, but none came. Instead George returned and put the issue squarely to his father. Unable to reconcile all sides, Jay could only arrange an awkward compromise. Helen would not disavow her son but neither would she tender her approval. There would be a marriage without a wedding—no ceremony, no celebration, and no guests. By a quirk of fate even the timing rubbed salt in Helen's wounds. Two weeks before the marriage, Victoria Morosini Schilling again leaped into the headlines. In September 1886 she abruptly left Schilling and vanished. Day after day the papers titillated readers with the latest rumors of her whereabouts and Morosini's efforts to locate her.[5]

Against the backdrop of the second saga of Victoria, George's marriage took place. The arrangements were as furtive as some of Jay's legendary deals. On September 14 Edith and her mother came up to Lyndhurst. Jay startled Miss Terry, the housekeeper, and the other servants by asking them to come down to the parlor because "George is to be married." They assembled dutifully along with the other children. The Reverend Washington Choate of the Irvington Presbyterian Church arrived and performed a simple ceremony steeped in funereal gloom. Through the vows Helen stood in grim, tight-lipped silence. When it was over, Jay stepped forward and kissed the bride. Edith bore the ordeal with dignity until the end, when she dropped her head on George's shoulder and cried softly. After Miss Terry signed the certificate as witness, two carriages pulled up to the entrance. Reverend Choate left in one, the newlyweds in the other. Helen retired to her room without a word.[6]

A few days later Alice Northrop was sitting on the sofa in her aunt's bedroom when Helen rose suddenly and began to pace the floor. Within minutes she burst into tears and cried, "Why has this happened? Just to think of it, George married to an actress! What next? How do we know that Helen won't fall in love with a coachman?" Although she gradually warmed to Edith, her health and spirits had suffered a blow from which they never fully recovered. The prim Miss Terry, who like all housekeepers knew more than her employers, shook her head sadly. "I am sorry myself that George picked out an actress," she ventured to Alice Northrop, "but how can he be blamed? Your uncle and aunt never seemed to have the knack of bringing many young people into their home."[7]

In the long run Edith proved a better wife than George a husband. Radiating warmth and charm, carrying a bit of the stage in her manner, she eased gracefully into her new role. Alice Northrop thought her the most beautiful woman she had ever seen. No one doubted her loyalty to George or the family, and Jay doted on her as he did on George. In August 1887 she presented George with a son, who received the name Kingdon. Jay delighted in his grandson. A few months later, when Edith complained that George spent his evenings studying navigation, Jay assured her that, "judging from my own experience, the caresses & kisses of a lovely wife such as he has are far preferable to navigation or indeed any thing else in this world."[8]

Gould was no stranger to labor troubles but he was no more prepared than anyone else for the upheavals of the 1880s. At every turn the same adversary confronted him. The Knights of Labor had fought the losing battle against Western Union in 1883 and forced Union Pacific to back down from its wage cuts the following year. In March 1885 a wage reduction on the Wabash ignited a strike against the Missouri Pacific and all the Gould southwestern roads. After the governors of Missouri and Kansas intervened as mediators, the Missouri Pacific astonished everyone by restoring the cuts and conceding other demands of the strikers. *Bradstreet's* pronounced it a "complete surrender of the management," while a St. Louis newspaper exclaimed that "no such victory has ever before been secured in this or any other country."[9]

Since Gould was absent on a cruise when the strike occurred, his role in the settlement remains unclear. Shortly after his return, however, another crisis boiled up when the Wabash began laying off shopmen who belonged to the Knights. Finding the Wabash receivers implacable, the Knights ordered another strike against the Gould roads in August. Gould responded by inviting Knights' leader Terence V. Powderly and his executive board to a conference at his home. Observers were dumbfounded at the spectacle of the most formidable railroad man in America deigning even to sit in the same room with the head of a union, but Jay was never one to let convention impede him. He had no intention of recognizing the Knights as a bargaining agent, but neither did he regard its mere existence as anathema. In the meeting with Powderly Gould promised to use his influence with the Wabash and to hear any grievance arising on roads under his control.[10]

The Wabash discontinued its layoffs and Powderly called off the strike. While the conference left Gould with the belief that he had a rapprochement with the Knights, the outcome was widely interpreted as a capitulation on his part and a smashing victory for the union. In one stroke the Knights emerged as the most powerful labor organization in the nation, its image enhanced by besting Jay Gould. New members flocked to its banner in droves. In the mushrooming local assemblies the mood was defiant, brimming with confidence, and spoiling for another confrontation with Gould. The hold on their exuberance exerted by Powderly and his board, never terribly strong, slipped away altogether.[11]

On January 10, 1886, delegates from all the local assemblies on Gould's southwestern system convened in St. Louis and approved resolutions that the executive committee of the district assembly, headed by a man named Martin Irons, construed as investing it with the power to call a strike on its own. The spark came on February 18 when the Texas & Pacific discharged a foreman in one of its car shops. Claiming violation of the agreement, Irons issued an ultimatum, then ordered the shopmen out. The shopmen complied even though many of them did not know why they had been asked to walk. Five days later Irons extended the order to assemblies on all the Gould lines and five thousand men left their shops. In this precipitous fashion the strike that was to seal the Knights' fate was launched.[12]

The strike caught everyone by surprise, none more so than Powderly and Gould. Powderly had not authorized it, did not want it, and found himself in an impossible situation. By extending the strike to the Missouri Pacific system, Irons had violated Powderly's earlier pledge not to take any such action without first consulting the road's officials. When A. L. Hopkins raised this point, Powderly suggested that the foreman be reinstated pending an investigation. Hopkins reminded him that the Missouri Pacific had no control over employees of the Texas & Pacific. "We have carried out the agreement made last spring, in every respect," he emphasized, "and the present strike is unjust to us and unwise for you." The Texas & Pacific receivers, bolstered by a court order, gave the strikers three days to return to work or be discharged. A citizens' committee in Marshall, Texas, where the strike began, tried to mediate but failed and signed off.[13]

Gould had departed on January 5 for a prolonged cruise in the West Indies and did not return until March 23. News of the strike reached him in Havana, where George cabled him for instructions. After a lengthy discussion in cipher, Jay sent Hoxie the following message:

> You shall be fully supported in dealing with the strike. If shop-men at Sedalia, Denison and other places stop work, why not close the shops; and if trains are interfered with so as to deprive the Company of earnings to pay its labor, suspend the whole pay-roll during the continuance of the strike.

Standard versions of this episode cast Gould in his usual diabolical role: he wooed the Knights falsely, made concessions when he had to in 1885, awaited his chance to smash the union, and pounced on the opportunity offered by the strike in 1886. No actual evidence of Gould's motives is offered; as Norman J. Ware so eloquently put it, "Gould was like that."[14]

While Gould may not have been adverse to breaking the union, the evidence suggests a different scenario. Historians tend to assign leaders too much credit as architects of policy while ignoring the influence of actions taken by subordinate officers. It was Gould's practice, dating back at least to Union Pacific days, to delegate responsibility to his operating chiefs and to sustain their actions unless they courted disaster. Clark had received this authority for the Union Pacific, and Hoxie held it for the Missouri Pacific. Gould did not provoke the strike and could not possibly have foreseen it. Earlier he had broken a precedent most businessmen considered an iron rule by dealing directly with Powderly to obtain an agreement which, he said repeatedly, satisfied him. There is no reason to doubt his sincerity and no evidence pointing to darker motives. He appeared to have bought labor harmony through the use of unorthodox tactics. When the agreement was violated, Gould felt betrayed. His response was consistent with what he had done in the past: he reaffirmed Hoxie's authority to deal with the crisis and provided him with guidelines.

At the same time Gould knew his man. "Hub" Hoxie was a loyal, capable official of long experience beginning with the Union Pacific, where his name earned a footnote in history by being attached to an early construction contract that gained notoriety in the Credit Mobilier investigation. Frank W. Taussig, a Harvard political economist, interviewed Hoxie and described him as "not only an able man, but a straight-forward and humane one imbued with a strict sense of duty and discipline, but disposed to just treatment of his subordinates." That appraisal coincided with his reputation, yet the Knights came to despise him even more than they did Gould. He was their direct adversary on the firing line, but there was something more. The strain, the tension, the sheer exasperation of the strike affected Hoxie or at least aroused deeply held sentiments. A dispassionate chronicler spoke of his "grim, unswerving fixedness of purpose" that the time had come to decide "whether he should run his own railroad or have the Knights of Labor run it. He never forgot the issue and never relaxed his determination to 'fight it out on that line.' "[15]

Like Gould, Hoxie regarded the strike as unjust, misguided, and in direct violation of the 1885 agreement. The Knights had struck the Missouri Pacific even though, by their own admission, they had no grievance against it. He refused to negotiate a settlement with union representatives because there was literally nothing to settle and they had already betrayed a pledge the company had honored. Convinced by its flimsy pretext that the strike was less a dispute over issues than a test of strength, he resolved not to budge an inch. On March 8 he presented his views to the employees in a lengthy circular; the next day Superintendent Kerrigan formally discharged all strikers and began recruiting replacements. Irons dismissed the circular as "too insignificant to reply to in detail" and countered with a list of thirteen demands. Powderly asked the Texas & Pacific receivers to submit to arbitration but got nowhere.[16]

As March waned the Knights' position deteriorated steadily. The engineers, conductors, and firemen had agreements of their own and ignored the strike call. Frustrated strikers resorted to killing engines, halting trains, and shutting down yards. In 1885 the strike had remained orderly and won public sympathy to its side. Now, as the mood turned uglier and incidents of violence increased, the Knights lost their tenuous hold on public opinion. Hoxie got an injunction and ran his trains under the protection of United States marshals. Irons and Powderly both tried to bring Hoxie to the table only to be reminded that the grievance concerned another railroad. The governors of Missouri and Kansas attempted to mediate as they had in 1885. Their investigation found no evidence that the Missouri Pacific had violated the March 1885 agreement and recommended that its terms be restored as a basis for settlement. Hoxie accepted the offer but Irons spurned it.[17]

By the time Gould returned on March 23, Hoxie had reduced the strikers to desperation. His mixture of patience and firmness contrasted sharply with the erratic and inept leadership of the Knights. The fight had become for him an opportunity to roll back certain practices that had crept in since the 1885 strike. The Order's interference in the details of management had reached a level Hoxie found intolerable; discipline had grown lax and costs were rising. Irons continued to press his list of extravagant demands, insisting lamely that "Mr. Hoxie wanted trouble. He has provoked it." A passenger train was derailed, some bridges and a roundhouse burned, and more trains blockaded by force. Another dispute led by the Knights engulfed the yards at East St. Louis, and disturbances broke out in several towns along the Gould system. Powderly was at his wit's end. On March 13 he issued a secret circular warning his members that unwise and premature strikes were undermining his work and the principles of the Order. Every day brought news of more unauthorized strikes by local assemblies which, like the uprising against Gould, the executive board could neither harness nor disavow. Unwilling to watch Irons lead his followers over a precipice and aware that the strike was on the verge of collapse, Powderly once again appealed directly to Gould.[18]

On March 27 Gould, asked to submit the "Southwest difficulties" to a board of arbitration, stated the position he would maintain consistently. The

Missouri Pacific board had put the matter in Hoxie's hands and all dealings must therefore be with him. Negotiations with a "law-breaking force" would be fruitless; if the men would return to work, Hoxie would meet them "in the spirit in which he has heretofore successfully avoided rupture and cause for just complaint." After an exchange of letters Jay agreed to a private, informal meeting with Powderly. Later he testified that, having read Powderly's secret circular, "my motive in seeing him was to brace him up, because the sentiments expressed in that circular were so different from the acts of his associates that I thought he needed bracing up."[19]

On a Sunday morning in March Powderly slipped unobtrusively into 579. After talking until 2 P.M., he left and returned that evening with an associate, W. O. McDowell. The discussion was frank and cordial. Powderly declared that Irons and his men were "in rebellion to our order" and talked of removing their charter. Gould then showed him a dispatch he planned to send Hoxie the next morning. By that time trains had resumed operations everywhere except at Fort Worth and Parsons, Kansas. In hiring men Hoxie was instructed to give preference to former employees, "whether they are Knights of Labor or not," with two provisos: he was not to employ anyone who had damaged company property or discharge anyone hired since the strike began. The message closed with the statement, "We see no objection to arbitrating any differences between the employes and the company, past or future." Powderly murmured approval, asked for a copy, and departed.[20]

A colossal misunderstanding arose from the last sentence of the dispatch. That same evening Powderly telegraphed Irons that Gould had "consented to our proposition for arbitration. . . . Order men to resume work at once." Someone leaked both dispatches to the newspapers, which printed them Monday morning. Jay read his message to Hoxie before he had even sent it and was furious. He intended the controversial sentence not to change his position but to confirm that Hoxie remained in charge of arbitration and was free to undertake it with any employees of the road. That same morning Gould notified Powderly of this despite a plea from McDowell not to "split hairs." Powderly was sick that day and could not meet again with Gould until the next morning, March 30.[21]

At that meeting the misunderstanding became glaringly apparent. Had Powderly known Gould better, he would have realized he had made the cardinal mistake of presuming too much from too little. Jay had long been on record as favoring arbitration, but only between individuals and the company. In his absence the board had cloaked Hoxie with the authority to handle any such cases. Gould neglected to mention that the board had in fact approved instructions telegraphed by him; instead he emphasized that Hoxie had been given the responsibility and he as president couldn't "take these matters out of his hands." McDowell was moved to recall Lincoln's lament that he had so little influence with the administration. He and Powderly protested that Hoxie was intractable, and the meeting ended on an inconclusive note.[22]

If nothing else, the session produced a curious irony in the positions staked out. Powderly negotiated as if he had control of the strike, which he did not, while Gould disclaimed having charge of developments when in fact Hoxie had done his bidding to perfection. The outcome left the strikers and the public alike bewildered as to whether the strike was on or off. Hopes soared, then were dashed by the squabble over arbitration, leaving the workmen more demoralized and desperate than ever. The Knights issued a victory statement followed three days later by a protest that returning members were not being taken back by Hoxie. Irons roared that "it was the thriving of such men as Jay Gould that enslaved all countries."[23]

Despair vented itself in renewed violence. Mobs stormed the yards in half a dozen towns, and blood was spilled in Fort Worth. Still the number of trains running continued to grow. The widespread distress spawned by the stoppage had inflamed public opinion against the strikers. On April 3 Powderly's board approached Hoxie, who reiterated the stand he had taken all along: he would hear grievances from actual employees but not outsiders, he needed to hire back only about half the former employees, and he would have the final say on which men were taken back. The board members rejected this as a violation of the agreement made with Gould in New York.[24]

For Powderly the worst had come to pass. Neither the company nor the strikers would relent. The strike could not be won but dared not be lost. To his horror it had become the battlefield on which the Order's survival depended. His hopes for a settlement blighted, Powderly saw no alternative to throwing his board's belated support behind a strike already crumbling. "This will be a fight to the death," vowed one of his board members. The district assemblies issued a manifesto summoning the "Working Men of the World" to arms. "Gould, the giant fiend," it bawled, "Gould, the money monarch . . . must be overthrown; his giant power must be broken, or you and I must be slaves forever."[25]

On April 11 Powderly wrote Gould an angry letter declaring, "You can settle this strike. Its longer continuance rests with you and you alone." He bungled even this attack by giving McDowell, who acted as courier, a second letter instructing him to offer Gould a last chance to settle. McDowell was to keep the contents of the first letter confidential "until after five o'clock of the day you deliver it; then if he makes no reply let it go to the world." In a blistering reply Gould pounced on its language as a vague threat to him personally and deftly shifted blame for prolonging the strike back to Powderly. Although Powderly's letter explicitly disavowed any threat to Gould himself, its tone was ominous and easily misconstrued. He had botched not only the strike but the war of words as well.[26]

As early as April 6 Gould pronounced the strike ended. Trains were running, strikers were being hauled into court by the hundreds, and more men were drifting back to work. When appeals by the Knights for funds and support proved unavailing, it remained only to play out the last dismal scenes. A riot in East St. Louis led nervous deputies to fire into the mob, killing six bystanders. In Washington a House committee opened hearings on the strike; as usual Gould's appearance attracted the most attention. On April 26 a Missouri Pacific train was derailed, killing two crewmen. The hate mail in Hoxie's box mounted steadily.

"You are a God damn sucker," wrote one anonymous admirer. ". . . I would give all the money I am worth to see somebody crack your skull, if you have any, you rotten sucker."[27]

Gould came in for his share of abuse as well. The Knights flayed him in another proclamation. A learned professor rebuked his views on strikes, while the Methodist bishop of St. Louis, in a burst of Christian charity, said of Gould, "I would like to see some one take him by the neck and kick him through New-York, so long as he wasn't killed." A labor meeting in St. Louis celebrated May Day by hanging him in effigy. These were the last gasps of an uproar that ended not with a bang but a whimper. The Knights rejected an appeal by a citizens' committee to call the strike off, then on May 4 declared it officially over. Hoxie announced he would take men back on the terms he had outlined in March. By one estimate only about a fifth of the strikers were rehired.[28]

The spring of 1886 witnessed an epidemic of strikes, of which that against the Gould roads was only the most spectacular. A host of other unions struck that same season, engulfing the country in labor turmoil. All these were upstaged by the fiasco in the Southwest. Apart from lacking solid grievances, the strike was managed so poorly as to insure failure against so formidable an opponent as Gould. The day after the strike was called off, the sensational Haymarket bombing shifted public attention to Chicago. Of all the epitaphs applied to the southwestern strike, the *Tribune*'s hit the mark most succinctly: "In all the history of labor controversies, perhaps there never was another strike more unreasonable in its inception or unwise in its conduct."[29]

Within months the Knights, without lifting a finger, gained their revenge on Hoxie and, indirectly, on Gould as well. Overworked and spent from the ordeal, Hoxie broke down soon afterward. An attack of gallstones sent him to Saratoga, where a reporter found him looking "like a broken down consumptive country merchant." Early in the summer, he underwent an operation. Sage joined him for a few days in August as did Gould, who was enduring another siege of neuralgia. Both must have been shocked at the sight of Hoxie's ashen features. The gray eyes flashed sparks, however, when talk turned to the strike. After returning to New York he submitted to another operation and for several weeks wasted away in agonizing pain. His skin turned dark brown until his friends scarcely recognized him. Somehow he clung to life until he had arranged the marriage of a favorite niece. On November 23 he died and was buried in Des Moines alongside his only child, a boy who had died at three. A resolution by the Missouri Pacific board attributed Hoxie's death "in large measure . . . to his faithful maintenance of private right against anarchic passion during the organized attack upon . . . this Company." The loss of Hoxie did more than cost Gould one of his most capable officers: it opened the door for the return of Silas Clark.[30]

No part of the Gould legend is more entrenched than the reputation of his roads as shoddy, ill-equipped, and indifferent to good service. Contrary to belief, he did not practice neglect or undermaintenance as a policy. Those who insist that Gould ignored operating details need only examine his letter books to be undeceived. He had always made it a point to master every aspect of his roads, but financial problems consumed so much of his energy that he was obliged to rely heavily on trusted subordinates.

With his operations chiefs Gould was demanding but never imperious. He told them what was wanted, gave them the power to accomplish it, took care not to intrude upon their authority, always backed them up, and scrutinized their performance closely. Having entrusted an officer with responsibility, Gould expected him to wield it vigorously. He encouraged initiative at all levels and was more apt to remind a man to exercise his authority than to reprimand him for exceeding it. His instructions were succinct, precise, and filled with suggestions. As always Gould's appetite for information was insatiable; his pen, wasting no strokes on amenities, scribbled off a relentless stream of questions, inquiries, and requests for data or progress reports. These missives flew in flocks until he got a satisfactory response. Not only the road's officers but local bankers, merchants, editors, politicians, and other citizens served Gould as a vast intelligence network to supplement what he garnered on his own through newspapers, financial journals, street rumors, and other sources.[31]

I find there is considerable complaint as to the character of the work turned out by the Missouri Car & Foundry Co. It would be well to put a new & careful inspector there to supervise the work. They cannot afford to slight our work or give us a cheap job.[32]

As in Union Pacific days, the western tour remained Gould's primary method of gathering information for himself. What some magnates turned into a triumphal procession Gould made a study in economy of motion. The itinerary was carefully laid out before his private car, the *Convoy*, left New York and usually covered the entire system: Missouri Pacific, Iron Mountain, Katy, Texas & Pacific, and International. Jay took with him George, Sage, and an assortment of other associates. The first stop was St. Louis, where the general manager provided a briefing and Kerrigan, the general superintendent, joined the tour. The pale, clean-shaven Kerrigan towered over Gould but possessed the same unruffled demeanor and low, grave voice. On the tour he served as guide and front man while Gould remained as inconspicuous as possible.[33]

While his car clattered along, Jay studied the track or pored over a mound of reports. A stenographer waited at his elbow to take down notes or telegrams to be dispatched at the next stop. Occasionally he would pause and gaze out the window, surveying the crops, herds, townships—any detail that offered a clue to the state of things in the area. Down the line officials scrambled frantically to prepare for his arrival. Gould crammed as many stops as possible into each day's schedule. At each one he asked Kerrigan how much time was allotted to their stay, glanced at his pocket watch, and followed the superintendent on his rounds.

After listening quietly and asking some questions, he wandered about on his own, eyes and ears alert for useful bits of information. When time expired he was back in his seat, impatient to start and to hear whatever else Kerrigan had learned.

At every stop the presence of Gould's car attracted a crowd of curious onlookers. Prominent citizens vied for the honor of welcoming him, but Gould was as unassuming abroad as at home. He met willingly with bankers, merchants, farmers, anyone with information useful to him, but shunned social displays as a waste of time and energy. Usually he dined and slept aboard the *Convoy*, tucked away in the quietest corner of the yard. On his way home in October 1886 Gould stayed the night in Pittsburgh's Union Station. Before his departure it was discovered that *Convoy*'s roof was too high to run through one of the Pennsylvania's tunnels and Gould was obliged to take another car to New York. Two weeks later he contracted with George Pullman for a new private car, this one to bear the same name as his yacht, the *Atalanta.*[34]

All right hurry forward all the new equipment. As we have the cash on hand to pay for it the quicker we get it earning money the better. Have you taken up the subject of additional cars for lumber & coal traffic?[35]

Rigorous as the schedule was, the tour offered a break from an even more grueling office routine. From Lyndhurst Jay commuted by yacht or train; at 579 he took a cab downtown only in rough weather. Otherwise Sage sauntered up from his house, often in time to cadge some breakfast, after which they strolled briskly up to Fiftieth Street and then over to the elevated station at Sixth Avenue. There have been odder couples than the lanky Sage and the diminutive Gould, but none less pretentious. Few of Manhattan's stockholders and critics, let alone patrons, recognized or imagined its president as another straphanger in the crush. Gould alighted at Cortlandt Street and walked to the Western Union Building at 195 Broadway. His private office was on the second floor in a corner across from the elevator, marked only by a sign bearing his name in black letters. The rooms were small, the furnishings plain, and Gould's own desk a piece that, in one reporter's opinion, "never ought to have cost more than $25."[36]

There was more variety than elegance in the work that churned across its worn surface. The matters requiring Gould's attention on the Missouri Pacific alone were staggering in volume and diversity. One may ignore his demurral that he served as president on condition that "there should not be any detail allotted to me. I have not the physique to stand it." In fact he immersed himself in detail. On new projects he decided when, where, and how much construction would proceed. He negotiated or approved contracts for rails, ties, locomotives, rolling stock, Pullman cars, and other supplies. The two areas of procurement and construction occupied a major portion of Gould's daily routine. A private wire connecting his office with the road's headquarters in St. Louis expedited the flow of inquiries and information.[37]

I can buy 25 Consolidated Locomotives of Baldwin at $10,400—Aug Sept & Oct deliveries. I am inclined to take them. What do you advise?[38]

Gould also made decisions on whether to build terminals, depots, shops, and warehouses. Several factors entered into the choice of location for these facilities: traffic statistics, the cost and availability of land, local demand, concessions by local officials, and possibilities for future development. These questions required not only extensive study and negotiation but a measure of diplomacy as well. Communities regarded these facilities as plums and did not hesitate to lobby for them. Aware that every decision disappointed more people than it pleased, Jay took care to placate local interests as much as possible. Sometimes he dangled such facilities as bait or wielded them as a club to obtain concessions. He also fielded complaints about service and tried to accommodate requests for additional service. His goal was not only to satisfy customers but to induce more firms to locate along the road.

I would like to have the Sewer Pipe company move their works to the Oak Hill & Carondelet line. Can you arrange this. We want to fill up that road with manufacturers.[39]

Financial matters claimed their share of Gould's time. There were always loans to place or take up, funds to be transferred, subscriptions to honor, new bond issues to float and old ones to refund. The handling of security issues alone involved a huge amount of work ranging from negotiations with bankers to such details as engraving and certificate signing. The acquisition of new properties generated another branch of paperwork, as did Gould's habit of inviting associates to invest in every venture. On another front Jay had a myriad of legal business to oversee. Lawyers in half a dozen states awaited his word on whether to fight or compromise, press or delay, negotiate or obfuscate. For all the legal talent at Gould's disposal, it was he who had to decide what best fit the larger web of his plans.[40]

Despite his punishing work load—one must remember that he had other enterprises to tend besides the Missouri Pacific—Gould never lost sight of the bread and butter aspect of his railroads, operations. Like other executives he had always monitored such vital signs as earnings, expenses, the ratio between them, tonnage, and traffic flow, offering his usual barrage of suggestions. He had always believed that profits flowed from mastery of the smallest details. In the past he had simply been too busy to devote enough attention to operations, but his retirement from the market enabled him to concentrate on managing his properties and training his sons in the business they would inherit. The climate wrought by unbridled expansion, with its fierce competition and declining rates, put a premium on sound, efficient management. In 1886 the strike impaired Missouri Pacific earnings; a year later the economy showed signs of turning sour at a time when Gould's expansion program had increased the system's obligations.[41]

It seems to me we ought to investigate the reason why the Kansas City & Ft Scott Rd should get 35 miles to a ton of coal while the Missouri Pac only get 25 miles. Either our scales are wrong or the quality of the coal we [use] are not so good. I presume the Ft Scott use principally the coal from south of Ft Scott same as our Minden & Cherokee coal. The difference is very great. It may be our engineers & firemen are careless in the use of coal. It might be well to make a premium to the engineer & fireman who during the year got the greatest miles to a ton of coal.[42]

As these pressures intensified, the squeeze on profits demanded that the system be run with strict economy. Gould could not accomplish this by himself; he needed an organization staffed by capable, dedicated officers. Every large railroad struggled with the problem of finding good men and a suitable organization for utilizing them. Growth strained the capacity of administrative structures and left many executives unprepared for dealing with the new competitive order. On the Missouri Pacific Gould had the final word on questions of personnel and organization, and here too circumstances forced him into a more active role after 1886. In these critical areas he experienced his most telling failures, largely for a reason his critics would never have suspected. Those who accepted the image of Gould as a cold, friendless clinician who used men up and discarded them could not have been more wrong. His fatal flaw as a businessman was rather an excess of loyalty to friends whose abilities were not equal to his confidence in them.[43]

Far from being aloof or ruthless, Jay was if anything too loyal to men he trusted and too trusting of those loyal to him. The most obvious example was Clark, for whom Gould's affection could not be shaken. Clark had also been close to George for years, which appealed to Jay's tribal instinct at a time when other longtime associates found the heir apparent overbearing. For a time Clark served Gould as a sort of ambassador without portfolio. After Hoxie's death some regarded Talmage as his logical successor, but they reckoned without Gould's sense of loyalty. Talmage may have been the better choice, but Clark was his closest friend and the man he trusted most. Gould waited only two days before offering him the position.[44]

Although the evidence is mixed, it suggests that Adams may have been more right about Clark than Gould was. The "long-bearded conductor of the monkey head," as Adams described him in his genteel way, was knowledgeable but deliberate, a disciple of masterful inactivity. He had the misfortune to take charge at a time of turbulence ill suited to his methodical style. Rate wars raged on every front; hostile legislation loomed in Congress and state legislatures alike;

the new branches had yet to prove their worth and costs were mounting on those still unfinished; the operating ratio for the system had begun to rise alarmingly; staff and organizational problems plagued several departments. These matters required swift, incisive action of the kind for which Clark had never been known. Moreover, he had a history of wilting under pressure and taking to his couch. Gould certainly knew all this from previous experience, yet without hesitation he delivered the fate of the Missouri Pacific system into the hands of his friend.[45]

Gould was aware of some internal problems but had depended on Hoxie to resolve them. A proposed reorganization of the traffic department in July 1886 was postponed because of Hoxie's illness. His decline left the company without an effective general manager for half the year. Kerrigan assumed some of his duties but was in no position to introduce changes. On his western tour in October Jay saw much that displeased him. Among other things, the operating department struck him as inefficient. In his usual way he said nothing and, after his return, put reporters off with some standard blather about prosperity. He did not yet have the facts and, with Hoxie still disabled, could not easily get at them. He could not ask Kerrigan because it was Kerrigan's management that troubled him most. One item particularly bothered him: the procurement of ties.[46]

I notice also Mr Kerrigans remarks about the condition of the new roads received & the great expenditure of ties & labor immediately *required to bring them up to Mr Kerrigans standard. As all the roads (excepting what we build ourselves) were subject to the sole inspection & acceptance by our engineering department it seems to me Mr Kerrigan & our engineering dept should agree on a common standard. If more ties are required (which I rather doubt) let them be put in in the first instance.*[47]

As early as July 1886 Gould inquired of Hoxie about the ties and suggested it "might be well to hold up for a while." The tour revealed to him that in Hoxie's absence quite the opposite had occurred, but there was little he could do without a trustworthy general manager in St. Louis. These suspicions, coupled with the need for a reliable lieutenant, doubtless influenced Gould to bring in his old friend. Before Clark even set foot in St. Louis he was summoned to New York for lengthy consultations.[48]

Ties posed a special problem for railroads in the West, where hardwood was scarce. For some time the Missouri Pacific had procured them from contractors in Missouri and Arkansas, buying only the amount immediately needed and furnishing trains for their delivery. As Gould's investigation gradually discovered, the expansion boom had opened this practice to certain abuses. Subordinate officers, in collusion with the contractor, bought ties at prices well above the market price and in quantities far exceeding current needs. In effect they spent exorbitant sums stockpiling a perishable item prone to rot in the southwestern climate. Moreover, the contractors were selling ties to other roads as well. If their supply ran short, they could load part of the Missouri Pacific surplus onto Missouri Pacific trains and deliver them elsewhere—in effect selling the same ties twice.[49]

Clark took his time changing the procedures and showed no sign of pursuing broader reforms. In April Jay briefed the executive committee on the investigation. By that time the tie scam had in Gould's mind meshed with his concern about expenses. Each month the figures looked gloomier. In the past the Missouri Pacific–Iron Mountain system had been run at an operating ratio of 50 to 55 percent; during the first quarter of 1887 the ratio soared to 70 percent despite high earnings, and the April showing was even worse. "The Wabash is actually making a better showing than MoP," Gould complained to Clark, "something unheard of in their past histories." His suggestion that Kerrigan "learn a lesson from poor old Wabash" smacked of fine if unwitting irony, for the Wabash's manager was none other than Talmage, the man some thought should have succeeded Hoxie.[50]

In mid-May Gould crumpled under the strain. "I have been sick with neuralgia for past 10 days which unfits me for business," he reported to Clark. "If I get no better soon will have to surrender." That remark was the loudest, indeed the only complaint he is known to have uttered about his health. His suffering must have been great, yet at the onset of his agony he dragged himself on three consecutive days to testify before the Pacific Railway Commission. So remarkable was his self-control that reporters scrutinizing his every move caught no inkling of his illness. Each day Jay arrived precisely on time, doffed his hat, bowed to the commissioners, and took his seat. While an overflow crowd craned their necks to catch his soft, low voice, Gould fielded questions with a candor that surprised everyone.[51]

The toughest inquisitor was E. Ellery Anderson, a lawyer of considerable railroad and political experience. His dense, probing questions reduced Sage to fidgeting nervously over trifles lest he betray something while insisting there was nothing to betray. Dillon took refuge in platitudes and the lapses of memory allowed tired old men. When Anderson tried to work Gould into a corner, however, he found himself sparring with shadows. A reporter noted that "Gould is frankness itself, and that low, persuasive voice always gently turns the current of evidence the way he wants it, no matter how hard the fact it may be damned with." Isaac Bromley, Adams's assistant, watched the proceedings closely and agreed that Gould made an "excellent impression" as a "most willing and, at times, voluble witness."[52]

Jay's dark eyes, empty of curiosity, oblivious of the spectators, met Anderson's directly or dropped to the table leg. The small, graceful hands toyed with his watch chain or rolled a ball of paper or rubbed noiselessly against one another. When a document was handed him, he put on his gold-rimmed spectacles to inspect it, then quickly pulled them off again. If Anderson wanted facts they were provided or others put in their place. Gould's books were demanded and, to everyone's astonishment, offered with pleasure. In tight spots an anecdote relieved the pressure, and where memory failed it was because so full a career did not permit details to be kept on call. Only once, when Anderson tried to trap Gould on the withdrawal of Denver Pacific stock from the trust, did Jay grow animated. His voice rose loud enough to be heard as, with a vigorous shake of his head, he defended the honesty and propriety of the transaction and assumed all responsibility for it. Bromley found it "quite refreshing to find anybody among those concerned in this transaction who acted as though he believed in it and was willing to stand up for it."

It was a virtuoso performance, but it left Gould exhausted. The neuralgia clung leechlike to him, fueling rumors that he was too sick even to take the cruise advised by his doctor. As if to refute these tales, he took Helen and three of the children to Virginia early in June. On the 7th the *Atalanta* docked at Mount Vernon, where Colonel Harrison H. Dodge, the superintendent, personally escorted them on a tour of Washington's home. Jay enjoyed it immensely, yet even at rest his mind was at work. The land caught his eye, as it did on inspection trips. Pointing to a hill northeast of the grounds, he asked why it had not been acquired to protect the estate. Dodge replied that the Mount Vernon Ladies' Association's charter did not permit it to buy land. To Dodge's surprise, Jay offered to purchase the land and donate it to the association, stipulating only that his part in the transaction not be made public. From that brief exchange Mount Vernon gained a valuable piece of land that to this day has rarely been associated with Gould's name.[53]

Jay returned from Virginia much improved. "I am not dead yet in spite of Wall Street rumors," he quipped to a reporter, ". . . although I may never be wholly free from neuralgia." He anticipated an easy summer and did not expect to come down from Lyndhurst every day. George could manage things in the office. But late in July the *Times* broke a story on the tie scandal; a few days later came reports of another scam involving the illicit sale of passes by associates of Cowan, the tie contractor. There was more color than accuracy in the *Times*'s account, but it prompted a scathing editorial denouncing Gould as "not a railroad man but a stock jobber" who had been "making ducks and drakes of his own property" and had "never shown the slightest capacity as a railway manager or any marked interest in the proper work of a railroad manager."[54]

---

*On the Iron Mountain in May while the earnings increased only $37,000 the expenses inc. 146,000. On the single item of fuel for freight trains the increased cost appears to be 37,500 agt 20,500 in 1886 with an inc of only 24,000 train mileage—fuel cost 15¢ per mile—*simply an impossibility. *The MoP same month with a freight train mileage of 442,000 fuel cost $31,194* or *7¢ per mile. This is only one item but extravagance since Mr Kerrigan took the swing has crept into all the depts of the business under his controll.*[55]

Hopes for a restful summer faded quickly as Gould wrestled with the problems of reform and reorganization. He admitted the company had been "terribly swindled on wood also ties & lumber." The more Gould probed, the more convinced he grew that Kerrigan was one of Cowan's secret partners. An analysis of the payrolls, he wrote Clark, would reveal "a systematic increase since Hoxie's death scattered through the different departments—all a part of a deep laid conspiracy to make your first year a failure. A ring headed by Mr Kerrigan with others in it have been moving systematically in this direction."[56]

Whether Gould actually believed this or merely said it to encourage Clark remains unclear. The harsh truth was that he had no choice but to pin his hopes on Clark's ability to take hold, and time was running short. "It looks to me that we are approaching a period of depression," he warned in July, "& those who keep their houses in order & practice the closest economy will be in condition to meet it." But Clark stuck to his habit of refusing to make changes without Gould's approval in each case. His response to Jay's request for a plan of reorganization was so general that Gould devised one of his own. "I have always believed in simple organization," he observed. "I have found in my management of railroads that I got better results from a simple organization than one too large & complicated."[57]

The plan was simplicity itself. Beneath the first vice-president he envisioned two general managers, one each for the Missouri Pacific and the Iron Mountain. Echoing their dialogue of a decade earlier, Gould assured Clark that the general managers "ought to relieve you of the detail work *which it never was my intention you should assume.*" The accounting department under C. G. Warner was fine, and the coal department got good results. In traffic W. H. Newman had been offered a job elsewhere at $12,000 but stayed on for a matching salary and promotion to third vice-president. The finance department was sound but Clark was authorized to reorganize the legal and purchasing departments, and machinery too if necessary. One change was inevitable: Kerrigan was gone by the year's end. Although Cowan had skipped to Europe, Gould ordered Clark to "leave no stone unturned to detect & punish the parties implicated." Other reforms came slowly, and Gould's plan of reorganization was never wholly implemented.[58]

Autumn brought little to relieve Gould's anxieties. Although earnings remained high, expenses ate up most of the increase. Few of the branches were pulling their weight, the Memphis and Pueblo extensions lay unfinished, and costs on the Nebraska City line soared beyond all expectations. The bond market was weak and his favorite barometer, the price of iron and steel, was sagging. Gould saw what was coming. "I think we may prepare for a shrinkage of earnings the coming year," he warned in September, "& it is wise to take time by the forelock & keep our pay rolls & expenses as low as possible." Reluctantly he postponed work on several branches for which subsidies had been promised and preached to Clark the gospel of retrenchment. "I am afraid I have given you a rather blue picture," he confessed, "& I hope it is overdrawn." In fact the picture would turn darker than he had dared imagine. Gould's instincts were sound, his choice of policy correct. The big question was how effectively Clark could execute it.[59]

The trouble with all new enterprises is that we have struck one of those periods when the public distrust the securities of all new enterprises & hence a complete cessation of investment with a result that every thing we do must be done by main strength.[60]

Like all railroad men, Gould tried to strike a balance between economy and starvation in delivering the best service at the cheapest possible cost. He understood the axiom that you spent money on improvements when times were flush and ran lean when earnings dropped. He realized too that railroads in farm country could survive only by cultivating other, more reliable sources of traffic. On this point lessons learned during his Erie and Union Pacific days still served him well. While encouraging all kinds of commerce and industry to locate along his lines, he concentrated on developing staple resources that provided a large traffic and were always in demand. Coal loomed largest in his plans, as it had on the Union Pacific. It was a high volume, heavy tonnage business immune to competition from other lines. In a region starved for wood it was a cheap source of fuel, which meant lower operating costs. By developing mines along his own roads Gould created another lucrative symbiosis.

The relationship between the Wabash and the Ellsworth Coal Company that Judge Gresham had so roundly criticized scarcely ruffled Jay. "I have done precisely the same thing on the Missouri Pacific," he declared, "and shall do it again whenever I have the opportunity." His goal was to mine enough coal to supply the system's entire need and do a large commercial business as well. To do that he purchased vast tracts of land wherever the geologists found promising deposits and built branches to the fields. By 1887 Gould had clusters of mines scattered in Missouri, Kansas, Arkansas, and the Indian Territory. That year the Missouri Pacific system carried 2.38 million tons of coal, or 56 percent more than it hauled in 1884. Of that amount 76 percent came from mines along the road, compared to 66 percent in 1884. In 1887 the system obtained 72 percent of the coal it used from these mines, as opposed to only 62 percent in 1884. At the same time, the proportion of commercial to company coal carried jumped from 39 percent to 52 percent during those three years.[61]

Impressive as this growth was, Gould considered it only a beginning. R. M. McDowell, his man in coal, was kept busy scouting new fields and adding tracts to existing ones. Careful investigation preceded every purchase. In his systematic way Gould pored over maps, reports from McDowell and the geologists, production cost estimates, and potential yields before making an offer. He used coal company earnings not for dividends but to buy more land. When one of the companies rashly declared a dividend in 1886, he wrote icily, "I did not wish any dividend declared . . . I thought that was understood some time ago. How is it that one was paid?" Finding coal along the Missouri Pacific system was not enough for Gould; he wanted the road to own as much of the property as possible. Once fields were acquired, the trick was first to produce enough coal to fill the cars and then find enough cars for the coal. Gould was unsparing in his efforts to keep the coal moving. When the tour of October 1886 disclosed to him the road's inability to utilize its cars efficiently, he peppered his lieutenants with advice and inquiries on the subject for months. The theme never varied: traffic had to move, and coal had priority.[62]

The Gould method of intertwining rail and coal development revealed itself clearly in Arkansas, where his interest in the Little Rock & Fort Smith road had less to do with creating a new through route than with the presence of coal fields southeast of Fort Smith. Although Henry Wood, an agent in Arkansas, assured Gould of their value, Jay instructed McDowell to have them examined by a coal expert. Given a favorable report, he proceeded within days to secure control of the Fort Smith, order a contract for connection to the Katy, authorize Clark to construct a branch "through the coal lands or such point as will hold the territory south east of Ft Smith in our interest," and send Wood to buy up to 15,000 acres of coal land. From these maneuvers Gould acquired the Jenny Lind tract, a field embracing 4,831 acres with estimated reserves of 50 million tons of coal.[63]

Arkansas offered not only coal but another resource Gould wished to develop, lumber. His agents scoured the countryside for bargains on good timberland along the road. Louisiana, too, was rich in timber, and there Gould had the advantage of an obscure land grant he had discovered and contrived to acquire for the New Orleans Pacific in 1880. To supplement these holdings he bought extensive tracts wherever his man in New Orleans found them. On one occasion in 1886 he authorized the purchase of 100,000 acres at the bargain price of five cents an acre.[64]

I will cooperate in making St Louis a model for expedition and cheapness—in fact we are immensely increasing our facilities in St Louis with the ultimate view of taking care of all the traffic of the eastern roads for local distribution.[65]

Every empire required a capital. For the Missouri Pacific system the logical choice was St. Louis, but with Gould logic depended always on his ability to get what he wanted. His most obvious desire, certainly the most publicized one, was to monopolize the bridge, terminal, and transfer business for the lines converging on the city. He needed also to integrate the two wings of his rail system, which still lacked a decent connection. To these Gould added a vision of St. Louis as both entrepôt and industrial center, a hub of enterprise fed by his rail system.

Most observers had conceded Gould's domination of bridge and terminal facilities since 1881, when he secured control of the St. Louis Bridge Company by threatening to build a new span forty miles above the city. For five years the other lines reaching St. Louis grumbled about the excessive charges and transfer delays imposed by Gould's control. In 1884 an ingenious maneuver enabled Jay to thwart plans for a rival bridge, but new projects soon took its place. The persistence of efforts to break his stranglehold induced Gould to shift tactics. While continuing to obstruct proposals for another bridge, he undertook to tie the major eastern roads to St. Louis by improving the facilities.[66]

Gould had in mind letting the eastern roads join in the bridge and tunnel lease then shared by Missouri Pacific and Wabash. They would also become partners in all land, shops, terminals, warehouses, and other property of Union Depot and its transfer road. In return they would agree to deliver all freight and passenger business to St. Louis via the bridge. Aware that rumor associated the Vanderbilt lines with the proposal for a new bridge, Gould broached the plan first to George Roberts of the Pennsylvania. In December they hammered out a draft agreement on which negotiations continued into the winter. As a sweetener Jay ordered plans prepared for a new passenger station, machine shops, and enlarged car shops.[67]

While pursuing this grand design Gould also had to perfect the connections of his own lines in St. Louis. Iron Mountain trains could interchange traffic with the Missouri Pacific only via track along congested Poplar Street. To eliminate this bottleneck Gould launched two projects, a partial belt line across the southern part of the city and an elevated road for passenger traffic between the Iron Mountain and Union Depot. The belt line, known as the St. Louis, Oak Hill & Carondelet, was to be 6.3 miles of double track. Gould conceived it not only as a connector for interchanging freight but also as an industrial park where firms requiring transportation facilities might be induced to locate.[68]

Delays stalled completion of the Oak Hill until October 1887, two years after Gould had hired agents to secure a charter and right-of-way. At that the Oak Hill fared better than the elevated road, which never got off the ground thanks to the mayor, who vetoed its chartering bill in July 1886. Jay was furious at the setback. "I dare say there is not another city in the United States of any commercial consequence where such a needed public improvement would be denied," he complained. His response was vintage Gould. The right hand vowed to make do with the Oak Hill and Poplar Street tracks and "save the expense" of an elevated; the left hand started dickering for a new bill, using as leverage his familiar threat of moving his headquarters elsewhere, in this case Kansas City.[69]

It is doubtful Gould seriously considered a move, if only because his holdings and commitments in St. Louis were so extensive. His threat did not succeed. A new elevated bill was forthcoming, but Gould found its provisions unacceptable. By December he had abandoned the elevated scheme in favor of a surface connection south of Broadway, which was completed in 1889. Never one to nurse grudges or mourn losses, he patched up his quarrel with city officials and resumed work on two projects that would make St. Louis the unmistakable capital of his rail empire: a new Union Depot and what the press called an "immense" office building.[70]

MONOPOLY IN HADES.
How the Place will be Run, Two Years after Jay Gould's Arrival.

IN THE ROBBER'S DEN.

JAY GOULD SURPRISES EVEN THE HARDENED MONOPOLISTS.

SHYLOCK'S BAD BARGAIN.

"THOU STICK'ST A DAGGER IN ME—I SHALL NEVER SEE MY GOLD AGAIN!"

WESTERN UNION
TELEGRAPH COMPANY.
"ALL HOPE ABANDON,
YE WHO ENTER HERE!"
MONOPOLY
STRIKE
F. Opper

THE
MONOPOLY
NEWS
DELIVERY
ASSOCIATED
PRESS
DISPATCHES
gould
ASSOCIATED
PRESS
DISPATCHES
N.Y. WORLD
N.Y. SUN
N.Y. STAR
NO
MONOPOLISTS
NEED
APPLY!

WORLD
N.Y. TRIBUNE
EXPRESS
WESTERN
UNION
MONOPOLY

CONSOLIDATED.

29

Casualties

I hope there was none of that dinner gambling scandal entrusted either to mails or to Jay Gould's Dionysian ear, the Western Union.

—John Murray Forbes, letter to C. E. Perkins, December 29, 1888

Mr. Jay Gould's name has for years been synonymous to most of the people of this country with Wall street speculation. In fact, he has been the greatest speculator and manipulator of prices that the American stock market has ever known. . . . Whether Mr. Gould will refrain from taking part in the future remains to be seen, and it must be added that all the events of the year have only tended to make him more supreme should he again enter the speculative field.

—*Bradstreet's*, May 22, 1886

The great cable rate war begins to-day. It is a war of monopoly and extortion against independence and fair prices—a desperate attempt . . . to force the independent Commercial Cable Company to join the pool and consent to advance the cable tolls to sixty cents a word.

—*New York Herald*, May 5, 1886

Mr. Gould . . . was a man of decided views and strong will, yet he never strongly expressed them. . . . He would sit quietly rubbing his hands while the members of the boards would express their views as to what ought to be done. He would listen to all that was said, and after every one was through he would say: "How would it do to pass such and such resolutions?" naming them over as the case might be. His suggestions would just hit the point, and the members . . . would all declare that he had suggested just the right thing to do.

—Norvin Green, quoted in Northrop, *Life and Achievements of Jay Gould*

IN FEBRUARY 1887 Edwin turned twenty-one and received the same emblems of manhood bestowed on George two years earlier: his father's power of attorney, keys to the strongboxes, and the vault combinations. Hostile papers twitted "Master Eddie" with the same gusto accorded "My son George." The tales told of "Master Eddie's" speculations in his father's stocks were reminiscent of stories about the younger Vanderbilts. In truth he resembled Jay far more than did George. His quiet, retiring manner masked a shrewd mind and strength of character. He wore a beard similar to Jay's and possessed the same dark, fathomless eyes. His sensitive disposition and simple tastes drew him closer to his sister Nellie than to George.[1]

Ed is doing nicely as my secty & likes it so much that he cannot stay at home Saturday. I think he will make a good careful business man & will be a great help to us when he comes to have experience & gets through college.[2]

Jay wanted the younger boys to get the college education he never had, but Edwin had other ideas. He enrolled at Columbia, as did Howard behind him, but his obvious talent for business induced Jay to let him quit school and come downtown, where he proved a steady, reliable hand. Like George he was soon put on the boards of several companies, but the line of succession remained unshaken. Whatever ability Edwin displayed, he could never hope to be more than George's lieutenant in the family business. George was moving up, taking on more responsibilities. In May 1887 he replaced his father on the Frisco and Pacific Mail boards. Not everyone shared Jay's delight at having two of his boys in the office. Already there were whispers about their presence crowding other officers who had been associated with Gould for years.[3]

Nellie was growing up too. The little dumpling of a girl had become a timid, sensitive young lady tortured by her own shyness. She crept in her mother's shadow, embracing her conservatism in values, habits, and dress. Her retiring nature concealed a bright, inquiring mind shackled by convention and dulled by lack of challenge. The two passions that filled Nellie's soul, religion and her father, both allowed a life of service. She joined the West Presbyterian Church of Dr. John R. Paxton on Forty-second Street, where the baptism ceremony so unnerved her that only the presence of Mrs. Sage at her side sustained her through the ordeal.[4]

As her mother's health waned, Nellie assumed more responsibilities for the household. Nothing pleased her more than to serve her father, whom she idolized. Jay welcomed her attention and made her a favorite in quite another way than he did with George. He educated her well with tutors and saw that she, like the boys, learned to use the telegraph. When she approached the age of courtship, however, Jay's protective instincts overrode all else. Where his wife feared the disaster of a wrong marriage, Jay's lack of social ambition left him indifferent to suitors with impeccable credentials. Distrustful of fortune hunters, fearful of cranks, he restricted Nellie's movements and provided security wherever she went. In town he would not permit her to ride in open country or off the main avenues. Jay had reason to be uneasy; there had been a rash of love letters and proposals, along with an occasional rude stranger at the door of 579 demanding to see her. The absence of legitimate suitors did not bother Nellie. Young men terrified her, and none of them could rival her affection for Jay. Compared to her father, she once told Alice Northrop, the others seemed like pygmies.[5]

Jay's protectiveness toward his children extended to the rest of his family as well. Regular checks still went to Sarah Northrop, and large sums found their way to the other sisters. As Sarah's children came of age, Jay helped them get started. Reid Northrop found a position with American Refrigerator Transit, a Missouri Pacific subsidiary. Ida was provided funds for a building to house her preparatory school. Jay saw to it that the structure was large enough to include a twelve-room residence for Sarah and the other children. Alice longed to become a medical missionary and was sent to Wellesley. When an accident left her lame and unable to continue school, she asked Jay to finance a prolonged stay in France. Fifteen months later Alice came home cured of lameness and ready to teach French at Ida's school. When she was steeped in gloom after leaving Wellesley, Jay took time to console her with stories of his own disappointments as a boy.[6]

Jay knew something of failure and how it weighed on youthful spirits. He understood too the importance of a helping hand extended at a critical moment. Doubtless he saw in the faces looking to him the same fears and uncertainties that had gnawed at him as a boy, driven him to claw his way toward success with a fury born of desperation. Perhaps he valued success so highly because he found its opposite impossible to bear. At the very least it enabled him to do for his family all that his own family had not been able to do for him.

They have had a big speculation in [Western Union] since you left carrying the price up to 70. The movement I think was engineered by Newcomb on the story that you Field & Garrett & Mackey [sic] *went to meet in London and settle the telegraph controversy.*[7]

On the first day of 1885 Gould's special partnership in Connor's firm expired. The horrors of the past year made him willing, indeed eager, to leave the market, but the market was in no condition to be left. He could not afford to quit the Street until he completed his liquidation program, which required a strenuous effort to lift values. The partnership was renewed for another year; by the time it expired Gould had repaired his finances and built a strong cash position. Once again he announced his retirement from the market. No one believed him;

he had said the same thing too many times before. The *Times* mocked Gould's "touching farewell to his predatory activities" by scoffing that he had "made as many last appearances as any actress on the boards."[8]

Credibility aside, Gould's declaration had the misfortune to coincide with the passing of a generation. In November 1885 Charley Osborn died of Bright's disease at age forty-eight. Scarcely was Osborn in his grave when Vanderbilt, transacting business in his study with Robert Garrett, dropped dead of a cerebral hemorrhage. Although sixty-four years old, he had enjoyed a reign of only nine years outside the shadow of his father. A few weeks later heart disease forced the boldest of the bold, Charles Woerishoffer, to quit the Street; in May 1886 death claimed him at the young age of forty-three.[9]

With brutal swiftness the Street seemed emptied of giants. Pundits were quick to suggest that "even if Mr. Gould did contemplate retiring Mr. Vanderbilt's death may have caused considerable change in his plans." Woerishoffer's demise reinforced this argument. Of the leaders who had dominated the Street in recent years David Morgan was gone, Keene had lost his bite, and Villard was limping back from ruin. Henry N. Smith had once again lingered on the short side of a rising market and gone bust, carrying Billy Heath's firm down with him. Even the venerable Sage had cut back his dealings in privileges. The small, sickly Gould, all 112 pounds of him, had outlasted them all. The market was his to take if he wanted it, and few believed he would resist so golden a temptation.[10]

They were wrong. Gould had worked too hard at getting out to want back in. For the past year he had toiled incessantly to liquidate his speculations. Aware that the trunk line rate wars remained the chief obstacle to a rising market, he concentrated his effort there. Early in the year he used Lackawanna as a leader to lift the market, but the movement petered out in March. Investors were not buying, and all his ingenuity and skill could not restore their confidence. Further efforts convinced Gould that the market would not budge until the carnage of the rate wars ceased. He tried to intervene personally with the trunk line presidents, but his overtures produced little more than a flurry of correspondence.[11]

Where Gould failed in the role of conciliator, another man succeeded. The summer of 1885 witnessed J. P. Morgan's emergence as a mediator among rail interests. Like Gould he undertook to negotiate peace among the trunk lines. His reputation as a banker gained the confidence of investors who distrusted any scheme advanced by Gould. Although Morgan produced little more than an uneasy truce, his dogged effort sparked hopes of a general settlement and sent the market upward. Without lifting a finger Gould was presented with an opportunity to liquidate the last of his speculative holdings. The boom lasted until Vanderbilt's death in December, by which time Jay had put his portfolio in order and announced his retirement.[12]

Nothing attests more eloquently to the strength of the Gould legend than the chorus of disbelief that greeted this declaration. To the jaded cadre of Gould watchers it was inconceivable that he would mean what he said, let alone say what he meant. One suspects that if he had dropped dead on the street the boys would have deemed it a trick to jiggle the market. Having ridden the crest of

Morgan's wave safely to shore by October 1885, Gould spent years trying to convince the Street that his retirement was genuine. He said as much to a congressional committee the following spring and told a reporter in July 1886, "I am out of the Street, and nothing whatever could induce me to go back into it." Another *World* reporter interviewing him the following year got the same message. "I am wholly out of the market," Gould insisted. "I closed up all my contracts a year ago last January, and nothing could induce me to return to the street."[13]

Still, analysts continued to see his imprint everywhere. During the autumn of 1886 Morgan successfully reorganized the Reading and thereby eliminated a major obstacle to peace among the trunk lines. His feat triggered another frenzy of speculation as heavy trading sent the market bounding upward in what Gould himself called "the wildest speculation" he had ever seen. When the collapse came in December, he professed to be "glad that I am no longer in Wall Street," adding with a twinkle in his eye, "Here in my office I am entirely free from the anxiety and excitement, and besides I can sleep well at night."[14]

The Gould aerie was not so snug as he made it out to be. Tight money and a ragged market carried over into 1887 amid signs that leaner times lay ahead. Outwardly Gould remained hopeful about business prospects, but in February 1887 he noted privately "a blue feeling prevailing. The strikes here—the Inter State Bill & prospects of war in Europe are unsettling matters." It was a good time to be out of the speculative arena. Storms were brewing that would keep him busy with corporate affairs. He would seldom be free from anxiety or sleep well at night, but never again because of his market position.[15]

The jury in the WU $2,000,000 suit after being out all night rendered a verdict of $240,000 which we will of course appeal. . . . We can carry it to the U.S. Court at Washington which will take 7 years before Mr Stokes will see any cash.[16]

The collapse of the rival telegraph pool in the fall of 1884 left the opposition to Western Union in disarray. While Garrett operated the B & O Telegraph as an independent company, Bankers and its subsidiary, American Rapid, soon followed Postal into receivership. In May 1885 the affairs of the three latter companies were reported "in such a mixed condition that nothing can be predicted with certainty as to their future." Gould, however, took nothing for granted. He knew that a coalition of his enemies was intent on reorganizing Bankers into a strong competitor. To all outward appearances he was preoccupied with railroad matters and indifferent to their progress. In fact he watched their every move, gathering information in his secretive, methodical way.[17]

One item in particular caught his eye. In October 1883 Bankers had acquired American Rapid through a contract that allowed each company to string wires on the other's poles. This intermingling made it difficult to distinguish the property of one company from the other. To the men reorganizing Bankers this seemed an unimportant detail. They secured an able receiver, General James G. Farnsworth, and looked toward a foreclosure sale in July 1885. But American Rapid had its own receiver, Edward Harland, who did not fall readily in line with this plan. Early in June he filed suit asking the court to determine which facilities belonged to each company. This investigation would take time and might ultimately free American Rapid from the contract of 1883. When Bankers demanded assurance that American Rapid would not negotiate a contract with another company in the interim, the judge promised that no such contract would be permitted without giving Bankers a chance to protect its interests.[18]

The judge reckoned without Gould's ingenuity. In the past Harland had tried vainly to interest Western Union in American Rapid; suddenly he found Gould receptive to the notion. An intermediary brought the two men together aboard Gould's yacht, where Harland agreed to a contract authorizing Western Union to operate American Rapid's lines and act as its agent. No one knew of this contract, not even the judge who had appointed the receiver. Armed with this secret weapon, a conventional mind might have used it to bargain with the syndicate or obstruct the reorganization plan. Gould wasted no time with such prosaic maneuvers; instead he concocted one of those strikingly original schemes that were his trademark.

Brandishing the new contract, Gould's counsel asked the court for an order directing Farnsworth to hand over the property of American Rapid to its agent, Western Union. That judicial investigation had yet to separate the property of the two companies did not deter Gould; his petition included a list of what belonged to whom. Without inquiring into the inventory shown him, Judge Donohue granted the request. On July 10 Eckert led a platoon of Western Union men to Bankers's offices and presented the order to Farnsworth, who protested that he could not distinguish the wires of one company from those of the other and could not act until he had consulted counsel. Eckert entertained no such doubts. As soon as Farnsworth left, the Western Union men pitched into Bankers's wires with axes until all were severed except those reaching Washington. The dangling ends were dragged away and run into the Western Union Building.[19]

In one bold stroke Eckert crippled Bankers's ability to do business. The stoppage was complete for ten days and full service was not restored until August 15. Amid howls of outrage the chairman of Bankers's reorganization committee denounced it as the most "iniquitous proceeding under the cover of the law since the days of Tweed and Sweeny." Inevitably a rash of suits followed, but Gould was now in position to welcome a game of legal obstructionism. The $2 million damage suit instituted by E. S. Stokes, a major holder of Bankers's securities, did not go to trial until the spring of 1886 and turned quickly into a forensic circus between the opposing counsel, Robert Ingersoll and Joseph Choate. "Jay Gould was the dictator," thundered Ingersoll. "He gave the command. . . . I do not believe that since man was in the habit of living on this planet any one has ever lived possessed of the impudence of Jay Gould."[20]

Ingersoll's sentiments were widely applauded. The Bankers incident rekindled memories of Erie days with its furtive maneuvers, secret documents, a compliant judge, a scheme at once imaginative and outrageous, walking the thin edge of the law to an explosive climax. If Gould hoped to silence the leper's bell he had to avoid behavior that resurrected spectres of the past. But business always counted more with him than image. Experience had taught him to take whatever action he deemed necessary regardless of appearances. The situation at hand, not the opinion of outsiders, determined his course. For Gould the most salient fact about the Bankers episode was its outcome. Whatever the cost to his already stained reputation, the raid accomplished its purpose. Although Bankers reorganized as the United Lines Company, it ceased to be a major competitor of Western Union.[21]

Having devastated Bankers with a knockout blow, Gould waged against his other rivals a war of attrition that to all appearances inflicted its heaviest casualties on Western Union. In 1885 the B & O Telegraph launched a rate war that cost both sides dearly. A year later Mackay reorganized Postal, which then arranged to operate the newly formed United Lines as part of its system. This combination ran in tandem with Commercial Cable, the outfit created by Mackay and Bennett to oppose Gould's cable monopoly. Although Garrett and Mackay did not cooperate with each other, each did enough business to cut deeply into Western Union's income. Between 1884 and 1886 the company's earnings dipped from $19.6 million to $16.3 million even though it carried 1.2 million more messages. Profits fell from $6.6 million to $3.9 million, the lowest figure since the merger with American Union.[22]

As earnings shriveled, dividends were cut, then paid in scrip, and finally passed altogether. Gould did not flinch as his casualties mounted. On the contrary, he extended the war from land to sea in May 1886 by slashing cable rates from 40 cents to 12 cents a word. Commercial Cable countered with the bold decision to drop down only to 25 cents and pleaded with customers to continue their support at a rate more than double that offered by the Gould cables. On the eve of battle Bennett launched another of his impassioned crusades in the *Herald*, urging the public to support Commercial in the "war of monopoly and extortion against independence and fair prices." For a month, blending righteousness with venom in his inimitable style, Bennett hammered daily at the theme that the Gould cable monopoly wanted only to drive Commercial out of business. Enough customers stood by Commercial to insure its survival, while on land B & O kept driving telegraph rates down to unprecedented levels. By the end of 1886 Western Union's losses not only eliminated the dividend but produced a deficit for the year. There was widespread doubt that Western Union, with its heavy load of fixed charges, could withstand a prolonged campaign.[23]

Still Jay did not waver. The toll of battle did not shake his confidence or change his belief that he was pursuing sound business policy. The enemy forces were smaller, weaker, less cohesive, could not afford to bleed as freely, and lacked an effective leader. United they might prove dangerous, but Gould realized that would not happen. Garrett had forfeited his chance by scuttling the pool and forcing what Stokes called "ruinously low" rates on everyone with his war.

"Indeed," Stokes complained bitterly, Postal and United Lines had "suffered a great deal more from this absurd competition with the Baltimore and Ohio than they did by the Western Union, and there was nothing gained by it either." Mackay had shown in 1884 that he was no man on horseback and Bennett, for all his fulminations, confined his interest to the cable.[24]

Above all, Gould had taken the measure of B & O and its president. It is curious how regularly business historians neglect the role of personality, temperament, and personal problems despite the enormous influence they exert on the outcome of events. One need not resort to the excesses of psychobiography or the formal apparatus of analysis to avoid discussing careers and transactions as if they took place in a vacuum. Gould understood these intangible and often subtle influences thoroughly. He was, among other things, a shrewd judge of character and close student of human nature. In all his dealings he took care to know the man as well as the facts.[25]

Long experience with the B & O, coupled with his knowledge of railroad affairs, had given Gould insight into its true condition. The public regarded it as a sound company with unwatered stock and a large surplus on its books. The fact that its stock paid regular dividends and sold at high prices impressed many investors, but Gould knew the B & O was living on its reputation. The weakest of the trunk lines, it had not in recent years expanded aggressively, made satisfactory traffic agreements, or modernized its equipment. Its telegraph and express operations were both losing money; Jay knew it and said so, but no one listened. Some said the property languished in John Garrett's last years because the old man "was afflicted with softening of the brain" and could not carry out his schemes.[26]

The father bequeathed the son his ambitions but not his abilities. Charming, voluble, cultivated of manner, fastidious of dress, and fond of society, Robert Garrett seemed more the country squire than the transportation baron. His broad, boyish face, edged with thick sideburns, was a full moon radiant with smiles and exuberance. He longed to realize his father's dreams only to discover that personal appeal was a poor substitute for business acumen. In seeking to fulfill his destiny Garrett's reach exceeded his grasp. His expansion efforts turned out as badly as his forays into the telegraph and express fields. To get business Garrett's railroad, like his telegraph, cut rates ruthlessly and became the chief disturber of stability among the trunk lines. By 1886 Garrett's policies had accomplished little more than a floating debt that beckoned toward receivership.

As early as August 1885 Garrett apparently conceived the idea of unloading the telegraph operation on Western Union at a fancy price. Perhaps he hoped to repeat Gould's earlier successes with this maneuver; unfortunately, he was no Gould and the man on the other side of the table was. Jay showed no interest in buying at Garrett's price. During 1886 he dismissed every rumor that negotiations with B & O were in progress. The officers of Western Union, he declared, "have fully decided that they will not buy up any competing lines. If we did so it would only encourage another crop of competitors that would want to sell themselves out to us."[27]

There was another reason for Jay's refusal to deal, one he was careful not to mention. Garrett was beginning to show symptoms of mental illness. Later, when his behavior grew too erratic to ignore, concerned friends traced his decline to the trauma of watching Vanderbilt drop dead at his feet. Whatever the cause, Garrett's growing instability clouded his business judgment. He pushed the telegraph war with reckless disregard of its effects on his strapped finances. During the spring of 1886 Mackay approached him with yet another scheme to consolidate the rival telegraph and cable companies against Gould. Garrett negotiated with him for months, agreed to terms, then abruptly sailed to Europe for his health. When he returned the merger plan was scrapped on grounds so vague as to leave Mackay, Stokes, and Bennett utterly bewildered.[28]

By the year's end Garrett could find no alternative to disposing of his interest in the B & O rail system. The decision must have been a painful one, given his family's long association with the road. To make matters worse, his search for a buyer resulted in one of those farces that make Wall Street so entertaining to spectators. Early in 1887 Garrett agreed to sell the B & O to Alfred Sully, a financier made prominent by his role in the Reading settlement. Sully and his associates controlled the Richmond Terminal, a holding company in southern railroads. Rumors flew at once of a gigantic combination between the B & O and Terminal systems.[29]

Unwilling to believe that Sully could manage the deal alone, traders agonized over the identity of his partners. Gould's name was on everyone's lips, but the betting was divided over whether he was with or against Sully. As the days rolled by with no new developments, talk turned to whether there would even be a sale. The *Times*, which had been the first to break the story, back-pedaled to the opinion that "Mr. Sully's scheme . . . has fallen through, because it was too big for them to handle." Garrett, smiling and genial, kept telling reporters there was nothing new. On Wall Street the confusion was fast converting B & O from blue chip stock to laughing stock. Nothing broke right for Sully. He counted on his Terminal associates to join him, but that plan fell victim to a power struggle within the company. In desperation Sully went to Gould, hoping to raise cash by offering the B & O Telegraph to Western Union and the express business to Adams Express. Jay shrugged him off with the remark that he "had no wish to engage in new enterprises."[30]

Gould's position was not the great enigma the Street imagined it to be. For all his denials he wanted the B & O Telegraph, but not at the fanciful price Garrett had put on it. He knew John Garrett and his sons, in their bitterness over the earlier telegraph wars, had long despised him, and Robert Garrett was adamant in his determination never to sell to Gould. This animosity, coupled with Garrett's erratic behavior, ruled out any normal business transaction. Jay could afford to wait because time was on his side. He continued to insist that the B & O Telegraph was losing money; Sage had declared as early as March 10 that the B & O had a floating debt of $7 million. The Street in its usual way dismissed their remarks as bear bait.[31]

Thwarted on every front, Sully neither exercised the option nor admitted that he had let it lapse. On the Street traders, frantic for information, listened eagerly to a dozen contradictory explanations of the Sully stall. While they pondered its mysteries a new element of confusion appeared in the figure of Henry S. Ives, a young broker of overweaning ambition and dubious reputation. Short, pale, his boyish features accented only by gold spectacles and dapper sideburns, Ives looked more the retiring intellectual than the unscrupulous plunger. Heralded as the latest Napoleon of Wall Street, his background was unsavory enough for the *Times* to dub him "the young and wicked Ives." Somehow he persuaded Garrett to give him an option on the block of B & O stock, presumably on the grounds that Sully's had lapsed. Earlier Ives had acquired the Cincinnati, Hamilton & Dayton and was dickering for control of some connecting lines. By uniting these roads with the B & O Ives threatened to forge a system capable of altering the balance of power among the trunk lines. It was a grandiose scheme and wholly beyond Ives's means, but he played a good hand of bluff. To doubters he announced boldly that he could raise whatever amount was needed, "without having to skirmish around in Wall-street either," but more days passed without money changing hands.[32]

For all his bravura Ives didn't have the front money, and no reputable firm on Wall Street would touch him. Desperate for funds, he followed Sully's footsteps to Gould's office. As an enticement he showed Gould information given him by Garrett which confirmed Jay's belief that the B & O was in serious trouble. Gould filed this intelligence away but declined to commit himself. For public consumption he maintained his lack of interest in the B & O Telegraph and asserted that Western Union had grown stronger despite the rate war. "What we intend to do," he said dryly, "is to let them go on alone competing till they compete themselves to death."[33]

The B & O mystery played on into July to an appreciative if uneasy audience as Ives procured one extension after another from Garrett. Then, on July 20, Garrett punctured the Ives balloon by declaring in an open letter that negotiations were "absolutely at an end." Ives countered lamely with a suit to recover his pledge deposit but he was finished. When his firm suspended three weeks later, traders on 'Change greeted the announcement with thunderous cheers. Napoleon had met his Waterloo and was banished to an Elba of legal proceedings. There followed an indictment for grand larceny from which he narrowly escaped and a civil action from which he did not. Reduced at last to a plaything of the press, Ives was humiliated and ridiculed like the village fool.[34]

With Ives disgraced and Garrett abroad, matters seemed more muddled than ever until Morgan intervened to end the suspense. Early in September he organized a syndicate to relieve the B & O's financial distress on condition that its management be placed in hands satisfactory to the bankers. When the bankers treated the public to its first glimpse in years of the company's accounts, the figures verified what Gould and Sage had long been saying. The floating debt was enormous, the book surplus a mirage, and the B & O Telegraph owed its parent company nearly $3.9 million. Overnight the mood on the Street changed. Morgan ended nearly a quarter-century of Garrett rule over the B & O and promised to eliminate that company as a disturber of trunk line and telegraph rates. The quick sale of the B & O's express service convinced operators that the other auxiliary companies would soon follow. Relieved of several burdens at once, the market bounded sharply upward. The bulls, spurred by "the belief that before long Mr. Gould will be found in possession of both the cable company and the Baltimore & Ohio Telegraph," traded furiously in Western Union. On September 9 transactions in Western Union constituted 40 percent of the day's total volume.[35]

Despite repeated assertions that the sale to Western Union was a foregone conclusion, uncertainty lingered for another month. The problem was Garrett, for whom the telegraph issue had become something of a phobia. He consented to the other changes but could not bear surrendering the telegraph to Gould, and certainly not for less than his own exaggerated price. Unable to budge Garrett, Morgan took matters in his own hands. On October 6, in brisk, no-nonsense fashion, he personally negotiated the sale of the telegraph company to Gould for $5 million in Western Union stock. Next morning officers of both companies formally ratified the contract. No one was surprised by the transaction, but the Street, after six months of masterful inactivity, was intrigued by the spectacle of Morgan and Gould "rushing the deal through with very singular haste."[36]

The reason was painfully obvious. Garrett, still sojourning in Europe, got wind of the negotiations and booked passage on the next ship home. He stepped ashore on the morning of October 7, just in time to read the details in the morning papers. Grim, tight-lipped, he refused to be interviewed, but at the Hoffman House that evening he said loudly to Stokes and some friends, "It's no trade, gentlemen." He promised a fight, perhaps an injunction, certainly delaying tactics of some sort. The Street braced expectantly for a battle royal, but two days later Garrett left New York as abruptly as he had arrived and went into seclusion at his country estate near Baltimore. Reports circulated that the ordeal had unhinged his mind.[37]

Gould's patience during the long campaign had been amply rewarded. Western Union absorbed a property "about twice as great as the entire plant of any competing system the company ever had" and in the bargain eliminated the chief obstacle to rate stability. With the B & O removed, Postal and United Lines were quick to strike an agreement with Western Union restoring rates, abolishing the rebate system, and promising cooperation. Only the cable rate war lingered on, largely because of Bennett's intransigence. Contrary to reports, Gould had no desire to swallow the other companies. They posed no threat to Western Union, and their presence helped deflect charges of monopoly.[38]

The notion of a government postal telegraph was bandied about for months but nothing came of it. Western Union's dominance remained unchallenged, and the prosperity lost during its wars quickly returned. Against all odds Gould had made himself master of the telegraph industry in less than a decade. For the rest of his life Western Union remained a sound enterprise with a strong management under Gould's leadership. While critics such as the *Herald* carped periodically about its inefficiency, less biased analysts conceded that "on the whole the service rendered by the Western Union Company is satisfactory in quality." In his handling of Western Union Jay demonstrated again what he wished so ardently to prove and what so many people obstinately refused to accept: that he was not only a successful man of business but a capable and constructive one as well.[39]

It soon became clear that the telegraph battle *had* snapped something inside of Garrett. He resigned the presidency without a fight and was persuaded to travel for his health. At Camden Station in Baltimore he seemed buoyant and relaxed as he bade farewell to a crowd of friends. The mere mention of the telegraph incident, however, caused him to burst out suddenly, "DAMN IT. THEY'VE STOLEN THE TELEGRAPH COMPANY!" A few days earlier he had shocked friends at the Maryland Club by accosting several men engaged in conversation. "You are all friends of Gould," he snarled, quivering with excitement, "and you are glad that I have been sold out to him! . . . you were all saying how glad you were that I have been euchred." Since then Garrett's behavior had alternated between the rational and the bizarre. Now, after boarding his private car, he rushed suddenly to the rear platform and cried out, "DON'T LET JAY GOULD STEAL THIS STATE OF MARYLAND BEFORE I GET BACK."[40]

Garrett went to Mexico and then to Europe until the tragic drowning of his brother in a yachting accident summoned him home in July 1888. He seemed improved in health and spirits, but three weeks after his return the old symptoms returned. He was whisked away from a resort into seclusion from which he never again emerged. Hallucinations tormented him constantly, often driving him to fits of violence. The spectre that haunted him most was that of Gould. At the resort he once offered a newsboy ten dollars to stand in front of the hotels shouting, "Jay Gould ought to be in Sing Sing." He imagined that Gould dogged his steps, accused visitors of being in Gould's pay, pleaded with his attendants to protect him from Gould, and talked of having Gould sent to prison. For eight years he dwelled in the shadows of insanity, the hallucinations growing more frequent and the melancholia deeper. His death in 1896 took place at Deer Park, Maryland, where his father had died twelve years earlier under eerily similar circumstances. There was a fitting if morbid irony in that coincidence. During the telegraph fiasco a broker had observed shrewdly, "Robert Garrett has been largely the victim of his father's ambitions."[41]

And his own. Garrett possessed visions of grandeur without the means to realize them. Like Gould he aspired to empire, to the mastery of large affairs, but he lacked Gould's genius, his sagacity and fierce, monomaniacal dedication, and above all his staying power. The fires of experience that forged Gould bent and crumpled Garrett. Business proved an all-consuming flame for anyone lusting after greatness. Those who dared its challenge and were found wanting paid the price of Icarus. Those who dared and succeeded, like all heroes of morality tales, paid quite another kind of price.

Manhattan continues to show handsome increase of earnings. The 5¢ fare on 2nd car is a great success.[42]

In the midst of the telegraph fray another tragedy played itself out on the side stage of Manhattan Elevated. Here the stricken hero was Field, whose behavior since the elevated merger of 1884 had become a source of concern for both Gould and Sage. Always fond of the limelight, Field had like Garrett aspired to a larger role than he could fill. He had been lionized for his work on the Atlantic cable, become "somewhat familiar with celebrities abroad," as Sage contemptuously put it, and achieved a modest success in business. Apart from Manhattan he owned stock in rail, steel, mining, and utility companies along with a newspaper, the *Mail and Express.* At Ardsley Park, his estate near Lyndhurst, Field played the part of country squire, surrounding his manor with houses for his children. Social and intellectual pretensions were but one of many outlets for an intense, restless personality ever in search of a cause on which to focus his nervous energy.[43]

At sixty-seven Field seemed a prosperous man whose popularity had survived his association with Gould and dabblings in the muck of Wall Street. In fact he was adrift in a sea of troubles. Both his sons were financiers of more ambition than talent. Cyrus, Jr., had been chastised for his recklessness and had pledged to borrow no more money from anyone. Edward suffered from personal difficulties; his doctor advised Field that Edward's "imprudent indulgence in stimulants contributes to his melancholy." Of Field's daughters one was severely ill with gout and another in the throes of schizophrenia. These family sorrows beset him at a time when he had suffered business reverses of his own. In recent years he had plunged into the stock and commodity markets with more vigor than judgment, funneling his speculations through young Edward's brokerage firm.[44]

In particular Field had taken to kiting Manhattan. The company had done well since the consolidation in 1884, paying regular dividends of 6 percent. In two years the number of passengers carried jumped from 96.7 to 115 million and gross earnings from $6.7 to $7.4 million, a modest increase due in part to experiments with the nickel fare. Morgan joined the board in 1885, making him a director in two of Gould's three major properties. Like Western Union, Manhattan had become a sound, stable property. Field helped Gould make it so, promoting among other things such innovations as the nickel fare. Gould had been dubious of the lower fare but agreed to try it and was gratified by the results.[45]

Given Field's long association with the company and the industry, his speculative bender must have puzzled his colleagues as much as it pained them. During the first half of 1886 Manhattan hovered between 120 and 129. In August it shot past 140, paused briefly, then soared to 175 in October. Through his newspaper Field asserted that the stock was worth 200, and no one doubted his intention of putting it there. The sudden rise and Field's injudicious boasts could not have come at a worse time. Manhattan was seeking permission for a Broadway extension, Henry George was campaigning for mayor on a platform that included municipal ownership, and the court of appeals had overturned the company's victory in the crucial property damage suits. Aware of the effect an inflated stock price would have on these issues, Jay asked Field several times to stop his operation in Manhattan.[46]

Here was a delicious irony: the most notorious speculator of his age asking a respected public figure to cease his market operations for the good of the company. In fact time and circumstances had reversed their positions. Gould had become the prudent man of business, while Field seemed bent on proving himself in the speculative arena with a recklessness surpassed only by his own sons. The warnings made no impression on Field. He continued buying Manhattan, much of it on margins below the 20 percent customary on the Street. On June 14, 1887, a Chicago wheat corner collapsed and sent commodity prices tumbling. Some firms failed, a Cincinnati bank closed its doors, and one or two other banks were rumored to be in trouble. The convulsion roused 'Change from its summer doldrums as sell orders poured in from the West. By June 23 the shock wave appeared to have passed with only minor damage. Manhattan stood at 156½, about where it had been all year long.[47]

The next day, however, panic swept without warning through the market like a tornado, flattening prices with unexpected force. Gould's stocks took the worst beating: Missouri Pacific dropped fourteen points, Western Union nine, and Manhattan a ghastly forty-one. Rates on call loans soared as bizarre rumors engulfed the floor: Field had failed! Sage had shut himself in his office! Field and Gould had quarreled! Ives had fled the country! Gould had dropped dead in his office! That same afternoon the whirlwind vanished as quickly as it had struck and prices recovered much of their lost ground. The slide lasted only a few hours, long enough to damage those with large holdings on margin. The most prominent casualty was Field, whose troubles went beyond Manhattan to the overextended position of his son's brokerage.[48]

Once Field's plight became known, the Street leaped to the conclusion that Gould and Sage had engineered the panic to bring about his downfall. Suspicion turned to certainty on June 29 when Field sold 50,000 shares of Manhattan to Gould. The announced price was 120, but several papers insisted that Field had been compelled to accept 90. The next day Field disposed of his remaining 25,000 shares, allegedly at 90, to a syndicate headed by Gould. He had, in the pungent phrase of the *Times*, been "struck down in the very house of his friends." Together the purchases made Gould undisputed master of Manhattan.[49]

Within days a new and enduring chapter in the Gould legend had been forged. Field became a pathetic lamb made to lie down with the lions or wolves (choose one) of Wall Street. "The partnership of the wolf and the lamb has from of old been unhappy for the lamb," moralized the *Times*, "and this was a partnership for the shearing of lambs." The ambush had been too perfect. Gould and Sage not only hammered down Manhattan but knew which banks had Field's loans and where to tighten the money noose. They forced him to seek help from his executioners, then stripped him of the stock at the bargain price of 90. Field was, the *Times* sympathized, "too good for the trade he has been following during these latter years." For Gould it was all in a day's work, another scalp for a crowded belt.[50]

This version of events has endured for nearly a century. Field's most recent biographer not only swallowed it whole but added a suitable motive: "The time was right for sweet revenge on Cyrus Field for lowering the elevated railway fare six years before." It is astonishing how often revenge is assigned as a motive for Gould's actions, as if business was for him a form of personal vendetta. Why he should wait six years to avenge a policy he had come to support is not made clear. Like so much of the Gould legend, this episode has persisted through sheer repetition. Yet there exists a variety of evidence that tells another tale altogether. What is more remarkable, biographers and historians have slighted or ignored that evidence even though it has been available all along.[51]

That Field's market antics disturbed Gould and Sage was no secret on Wall Street. As early as August 1886 a veteran financial writer noticed that they considered Field "over anxious and overreaching." Sage recalled that Gould "had urged Mr. Field at the beginning & repeatedly afterwards not to advance the price of Manhattan stock as by so doing property owners would ask greater damages." Ignoring these pleas, Field rushed impetuously toward disaster. After the collapse on June 24 he appealed to Gould for help not once but several times. Jay's own position was secure. Although his stocks took a beating in the fall, he was out of the speculative market and had ample cash in the bank or out on call loans. None of his Manhattan was pledged as collateral for loans. While Field's misfortune was not his own, his failure could have adverse effects on Manhattan and on the market in general. For that reason, and because Field was his friend, Gould agreed to loan him a million dollars in bonds. It was not a rare gesture; he had advanced large sums to other friends, including Green and Eckert.[52]

Field thought a million would be enough to see him through, but he was wrong. The next afternoon he returned to Lyndhurst with his son Edward and Daniel Lindley, his son-in-law and Edward's partner in the brokerage firm. By chance John Terry called at Lyndhurst that same afternoon and was met in the parlor by Helen Gould. To his surprise he found her terribly upset because Jay was not well and she was afraid "the discussion of business matters would prove fatal to her husband." Terry did what he could to reassure her and departed. The next morning Gould informed him that the Fields had stayed until eleven o'clock and left only after Jay had agreed to take 50,000 shares of Manhattan at 120. In his usual way Jay asked Terry if he wanted some of the shares. "No," Terry replied, "I dont think it worth more than Western Union." Sage took 5,000 of the shares.[53]

Gould's lengthy session with the Fields probably convinced him that the situation was more serious than the Street knew. Field controlled the United States National Bank, which held large amounts of his Manhattan as collateral for loans. Logan C. Murray, the bank's president, was as pliable as he was ambitious. Sage learned that he had recently been overcertifying checks for Field. Moreover, reports were abroad that the bank examiner noticed "a certain bank" held an exceedingly large block of Manhattan as collateral. If Field's loans were called and he could not pay, the collateral would be dumped on the market. The crash of Manhattan on June 24 resulted from the trading of relatively small lots; the day's sales totaled only 20,555 shares. A large block thrown on the market would send Manhattan to the basement and perhaps trigger a real panic. If Field failed, he might well bring down the bank, possibly other banks as well, and certainly some brokerage houses.[54]

Once again Field thought the cash from this sale would bail him out, and once again he miscalculated. Back he went to Lyndhurst with Terry for another interview with Gould. They were ushered into the library, where they found young Nellie and Alice Northrop. When Jay arrived, he led Field and Terry to the other end of the room. He seldom asked the children to leave while he transacted business and did not do so on this occasion. From her chair Alice watched the discussion, caught snatches of conversation about Manhattan and Field's predicament, heard various sums of money mentioned. None of it made sense to her. What impressed her deeply was Field's appearance. While Terry did the talking, the old man slumped in a chair looking pale and utterly spent. On the few occasions he spoke, his words and limp gestures reminded Alice "of a drowning man beseeching a rope from shore. I don't think I have ever seen a picture of more abject despair."[55]

Field had come in hopes of persuading Gould to take his remaining 25,000 shares of Manhattan. Jay was noncommittal after the meeting but a short time later told Terry, "I don't think that those gentlemen can go through and perhaps I had better take the balance of the stock." Terry arranged a meeting the next morning and the sale was concluded at the same price of 120. It was customary to allow thirty days for payment, but Field could not afford to wait that long for the money. Gould understood this and agreed to make immediate payment for both transactions even though it put him to the inconvenience of having to call in some loans. Ironically, these calls were widely interpreted as part of the scheme to squeeze Field when in fact they were made to assist him. Gould stated publicly why he had called the loans, but his explanation was ignored.[56]

In all Jay put more than $10 million in Field's hands, for which he received the notes of a man in deep financial trouble and 75,000 shares in a company he already dominated at a price the stock would never again see in his lifetime. Whatever his personal feelings about Field, he took the stock because to let Field fail or to allow the stock to be dumped on the market invited disaster. If the whole affair was an elaborate charade devised by Gould and Sage, it resulted in the best acting and worst bargain of their careers. Far from taking advantage of Field, Gould gave him extremely generous terms. Apart from the loans and immediate payment, Field would have been hard-pressed to get anything like 120 elsewhere. Manhattan was a good stock but not worth the inflated prices Field had paid for it; during the next five years it would range between 90 and 110. Even the *Times*, which announced that the sale price was "unquestionably 90," thought that figure "a fair and reasonable price."[57]

Throughout their careers Gould and Sage had been roundly abused by the press. Bitter experience had taught them to nurse their resentment in silence, but the charge that they had sheared Field taxed their self-restraint to the limit. Sage was livid after reading the version given in the *Times.* The allegation that Gould had paid only 90 for the stock especially rankled him. Other parties besides himself had taken some of the Manhattan off Gould's hands at 120 and might accuse him of double-dealing. Sage confronted Field with the article and demanded that he do something to correct it. Field laughed off the article, saying no one would believe it, but he agreed to publish a correction. Sage was not mollified. It bothered him that Field would not "acknowledge like a man that Mr. Gould had saved him." On his own Sage issued a fiery statement to the *Times* defending Gould, who had been "treated shabbily all around," and denouncing Field for his silence. The *Times* ridiculed Sage's indignation and thought well enough of its comedic value to amuse readers with a fanciful scene between Sage and Field.[58]

Not until July 8 did Field respond with a statement declaring that he had received 120 for the stock, that Gould had acted "throughout the transaction in a perfectly straightforward manner," and that "the most friendly feelings" existed between them. His brother, the Reverend Henry Field, confirmed these facts in a separate note to the *Times.* These belated admissions did nothing to modify the popular version of an ambush, which had by then spread far and wide. Gould said nothing more to the press on the matter but the issue remained very much alive with him. He gave his version of the facts to his children and even to his sister Sarah, who recalled his "great mortification" at being charged with treating Field unjustly.[59]

It was uncharacteristic of Gould to seek vindication or demand redress for anything he did. That the Field episode drove him to this extreme suggests how deeply his sense of honor had been wounded. He went to his grave believing he had saved Field and prevented what might have been a major panic. Sage and Terry thought so, too, as did others who knew something of what had occurred. In fact, the very *Times* columnist who had done so much to propagate the original version came eventually to this same conclusion. The conversion of this savage, longtime critic of Gould was perhaps the bitterest irony of all in this unhappy affair. Five years later, commenting on Field's death, he had this to say:

> The truth is that Mr. Field's speculation in Manhattan was so wild that a smash was inevitable. It was carried on against the protests of his associates; and when the final and expected end came, Mr. Gould, though he pricked the bubble, took care to do it in such a way that a crash which would have carried down one or two important financial institutions, and would probably have caused another panic, was safely averted. Mr. Field was left with something. Had he gone on to the inevitable catastrophe, he would have been stripped of everything.

Whether Gould drew any consolation from these words or even saw them is not known. Clearly they have gone unnoticed by the generations of writers who continue to pass along the tired old tale of Gould and Sage's perfidy.[60]

Field was left with something, but it brought him little comfort. Having lost his fortune, he spent his remaining years salvaging what he could from the ruins of his finances and family. He bore Gould no grudge and was among those present at his wife's funeral. Personal sorrows haunted him mercilessly. By the time of his death in July 1892 Field had lost his wife to pneumonia and a daughter to tuberculosis. All his sacrifices to save Edward's firm proved futile. Edward failed, was imprisoned for forgery and other charges, pronounced insane, and confined in the same asylum as his sister Alice. Amid the debris of his life Field might have done well to recall the sentiments he had uttered in an interview half a dozen years before that fateful June day in 1887:

> Speculation is making people crazy. Why, when I went to Delmonico's for lunch this afternoon, I saw a throng of pale and anxious men congregated about a stock indicator, watching it as if it had been the pulse of a dying friend. It was a melancholy sight. Speculation is carried to an extreme which is sheer madness.[61]

The election of a democratic Prest has been what we feared and business generally is sadly out of joint. It is I think much worse here in the east than with you in the west. The only way is to grin & bear it.[62]

After Blaine's defeat in 1884, Gould never again took an active part in politics except to protect his business interests or help friends seeking office. Although he followed events closely, he had neither the time nor the energy to do more than look after his own. By 1887 Gould's apprehensions about Cleveland had been relieved. In October he admitted the administration had been "a great success and has grown with me every day. I believe Prest. Cleveland stands today stronger with the business interests of the country than any other person on either political side, and I think we ought all to work to secure his re-election—*This is what I propose to do.*"[63]

Gould's pledge probably had less to do with an active role in the campaign than with merely establishing some rapport with the man who might occupy the White House for another term. He had been burned too badly in the debacle of 1884, to say nothing of the invective heaped on him for his role in the New York gubernatorial campaign two years earlier. "As a general rule I don't like to get into politics," he demurred, but the very nature of his business bent that rule badly. Congress was forever investigating the telegraph and threatening action against Western Union. The elevateds had always been a political issue in New York City, and the railroads still sparred with an endless succession of state and national politicians, legislatures, city councils, county courthouses, state commissions, and, after February 1887, the new Interstate Commerce Commission.[64]

Gould did not flatly oppose the Interstate Commerce Act; like many railroad men he welcomed some of its provisions and expressed reservations about others, notably those outlawing pools and making it illegal to charge more for a short haul than a long haul. The ultimate test of the act, however, would be its practical effects, which only time would reveal. Meanwhile, railroad men and investors alike grappled uncertainly with the problem of adjusting to the new order of things. There was no precedent for federal regulation, no clear clues as to what its impact might be. Aside from the new law, the business of politics remained for him what it had always been: an interminable struggle to obtain desirable legislation and to defeat adverse legislation.[65]

Although Gould's political influence had become more shadow than substance, he remained a favorite whipping boy for the press. The 1880s witnessed a revolution in journalism led by the very man to whom Gould had sold the *World*, Joseph Pulitzer. Among other innovations the new *World* emphasized scandal and sensation, gossip, colorful trivia, interviews real or contrived, and crusades against monopolies, boodlers, and corrupters of the public morals—all written in a light, breezy tone with catchy headlines. Pulitzer's astounding success with the *World* did more than trigger a circulation war among New York's dailies; it prompted newspapers across the nation to imitate his approach. So sweeping was the change that in 1887 a close observer proclaimed that the *World* had "affected the character of the entire daily press of the country."[66]

The new journalism burst onto the scene at the time when Gould's own position was undergoing major changes. Now past fifty and in declining health, he still worked compulsively, but success had made him more conservative. The hungry, grasping years were over, flown like youth itself. He had become one of the richest and most powerful men in America. The empire he had amassed as a legacy to his sons must be preserved and developed. This was the work that had come to occupy his life, that drove him as only obsessions can drive. In one sense he had matured and mellowed; in another he was no less desperate and demonic.

These changes accentuated the already wide gap between the Gould of fact and the Gould of legend, between his current role in the business world and the leper's bell of his reputation, which the press had done so much to create and popularize. By 1887 Gould's public image had undergone nearly two decades of embellishment, and the imperatives of the new journalism offered every motive to perpetuate rather than correct it. Gould as responsible businessman made for dull copy compared with the dark, satanic figure who manipulated an enormous empire in furtive, mysterious ways. Newspapers locked in mortal combat for readers were loath to exchange the possibilities inherent in the popular image of Gould for the bland portrait of a business executive. The *Herald* went so far as to coin a name for the Gould it preferred: "The Little Wizard of Wall Street." In 1887 it began referring to him simply as the "Little Wizard" or the "Wizard."[67]

It is a singular coincidence that the two dailies most hostile to Gould were the very papers hit hardest by the *World*'s rise to dominance. The *Times* clung to its staid ways and sank to the bottom of the circulation ladder. The *Herald* adapted readily to the new style but never managed to recapture its leadership in circulation or advertising. During the 1880s their antagonism toward Gould became a weapon in the struggle for survival. For years the *Times* had been railing against Gould's control over communications. It never let the *Tribune*'s editor forget having borrowed money from Gould or forgave the *World* its part in the elevated consolidation of 1881. All the dailies joined gleefully in denouncing the *Tribune* as Gould's mouthpiece but none so doggedly as the *Times*, which, a decade after Whitelaw Reid acquired the *Tribune*, still delighted in referring to him as Gould's "able lieutenant."[68]

The conviction that Gould was extending his influence to other papers drove the *Times* to the brink of hysteria. In 1878 it reported the rumor that he had secured Denver's *Rocky Mountain News* through W. A. H. Loveland. To the *News* and the *Tribune* he soon added the *World*. Early in 1881 he was rumored trying to buy shares in the *Evening Express* as part of a scheme to control four of the seven papers in the Associated Press. The dailies obtained much of their news from the Associated Press, which in turn used Western Union to transmit its dispatches. If Gould dominated the telegraph and the Associated Press, he could monopolize the flow of news. The ownership of newspapers would enable him to tell the public exactly what he wanted them to hear. In February 1881 the *Times* sounded the alarm in an editorial entitled "HIS MAJESTY JAY GOULD":

> So long as [Gould] was supposed to be merely engaged in making a fortune the public cared little about him, but he has now thrown off the mask, and is seen in the act of seizing not only the supreme power of the Nation, but all our powers, our bodies, and to a large extent our souls. . . . To the control of the telegraph lines Mr. GOULD is now said to be determined to add the control of the Associated Press, and it is popularly believed that by the purchase of one more newspaper he will achieve this end. He will thus, to a large extent, control what has been the free press of America. The newspapers will print only such news as he allows the telegraph wires to carry, and will express his opinions with the same fidelity now exhibited by the *Tribune* and *World*. . . . Undoubtedly he does own a good many miles of railroad, but it may not be true that he buys a new trunk line every day at 10:30 A.M. Undoubtedly he owns the *Tribune* and the *World*, and possibly he is unfortunate enough to own the *Express*. . . . Finally, it may be true that fifty millions of Americans will allow Mr. GOULD's foot to rest on their neck, and then again perhaps they will not.[69]

The *Times* overlooked how small and delicate Gould's foot was, especially when it came to newspapers. He owned the *World* for about two years and apparently made use of its financial columns, although one must remember that nothing definite is known about his relationship to the paper. There is a general consensus that he exerted no real influence on Reid or the *Tribune*. The paper was friendlier to Gould than were the other dailies because of Reid's policy not to roast him editorially, but Jay utilized the *Tribune* only in the sense of making himself more accessible for interviews to its reporters. He had no known personal interest in any other New York daily. It was Field who bought the *Express* and combined it with the *Mail*. If Gould wanted a New York paper, he had the opportunity in 1887 when Field was forced to sell. An Iowa editor and close friend of Dodge was eager to transform the *Mail and Express* into "a vigorous, healthy, virile, agressive [*sic*] Republican paper" if someone would loan him the funds to buy. "Could not our friend who helped us in 1884," he asked Dodge, "find it to his interest to help me into this now?"[70]

Dodge's friend showed no interest and Field's paper passed into other hands. Gould had no need or desire to own newspapers. His goal was not, as the *Times* insisted, to erect a media monopoly; the press interested him only as an outlet for views sympathetic to his properties and market operations. He did this by cultivating the friendship of editors, doing them favors, giving them market or investment tips. On one occasion in 1878 Jay authorized Clark to finance an evening edition of the *Omaha Republican*. He also wanted to strengthen the morning edition. "Its local dept I think should be more thoroughly worked up," he advised. "Every thing that transpires in Omaha & Nebraska ought to appear in the 'Republican' first if possible. It will cost a little more to do it well at first but the results in the end [will] more than repay the outlay."[71]

On Wall Street the calumny of being Gould's organ was bestowed on different publications at different times. The *Wall Street Daily News* was a leading contender, thanks more to its indefatigable editor, Charles D. Keep, than to Gould. From its first issue in May 1879 the *Daily News* earned a reputation for having access to inside information. Keep got his tips straight from headquarters, a habit that "enabled him to amass a snug fortune outside of his professed business." He had known Gould since his days as a *Times* reporter but, far from confining his practice to Gould, he insinuated himself into the confidence of Cammack, Keene, Woerishoffer, and other operators.[72]

There was nothing unusual about speculators renting editors or businessmen owning newspapers. Field had the *Mail and Express* and Villard owned the *Evening Post*, although he carefully separated himself from its management. Where other businessmen drew snide comments for their manipulation of the press, Gould aroused hysteria and outrage. The reason lay mainly in his control of Western Union, which exposed him to a wide variety of accusations. For years business rivals had charged Gould with intercepting and reading their messages; the *Herald* and the *Evening Post* also suggested that Gould used "his monopoly to keep himself informed about all business transactions that pass over the transatlantic cables." The Stock Exchange used lines provided by a subsidiary of Western Union, and many brokerages leased private wires from Western Union. What prevented Gould from pirating advance information from these sources as well?[73]

Congress agonized over these questions for years. One Senate committee concluded in 1884 that Western Union possessed "a practical monopoly control of commercial and financial news." That same year Gould was accused of delaying election returns to profit from the advance information. He was regularly suspected of doing the same with news important enough to register an impact on the market. No evidence was produced to sustain these charges, but they helped keep the issue of a news monopoly alive for years. So did a bitter controversy over the relationship between Western Union and the New York Associated Press. The *Times* and the *Herald* had long railed against Gould for trying to dominate the Associated Press, which in turn drew fire from rival news agencies for receiving special rates from Western Union. The complexities of this struggle escaped most newspaper readers, who found it entertaining to follow accounts of "sly Mephisto's" efforts to "get control of the Associated Press, and through it to influence the press of the country and manipulate the channels of news."[74]

While the enthusiasm for a government telegraph gradually waned, the attacks on Gould and Western Union continued. In July 1886 the *Times* blistered Charles A. Dana of the *Sun* for having "betrayed" the Associated Press and "put it into the control of men in Mr. GOULD's pay or associated with him in his enterprises." By that time the public image of Gould had emerged in full bloom. The age of the celebrity had dawned, ushered in by the new journalism, and caught Gould in the glare of excessive attention. Every paper in New York was as eager to quote Gould as it was to abuse him. Reporters banged on his door, buttonholed him on the street, tried to invade his office, and trampled the shrubbery at Lyndhurst, hounding him for interviews. Alice Northrop recalled two reporters interrupting supper at 579 with a threat to "write something anyway" if Jay did not talk to them. Some editors were reputed to offer a standing bonus of $500 for an interview with Gould or $100 for one with any of the children.[75]

During the late 1880s Gould's relations with the press entered a new phase. Reluctantly he made himself more available for interviews because his business interests demanded it, but he never found a way of dealing comfortably with those lying in wait for him. Where his good friend Chauncey Depew handled gangs of reporters with the affability of a ward heeler, Jay went rigid as a gatepost. Relaxed intimacy was possible for Gould only with old friends; familiarity with strangers was beyond him. He resented the intrusions on his privacy and on his time, of which he had precious little. The more he was maligned and misquoted, the more he retreated into a shell of silence or doubletalk punctuated with long pauses.[76]

He could not escape because the circulation wars held him prisoner in yet another, more subtle way. The papers that were eager to quote Gould were no less quick to denounce rivals who quoted him as his mouthpieces. So potent had Gould's name become that editors tarred each other with it in a vicious form of guilt by association. By 1887 Gould's standoff with the press was already two decades old, its roots a tangle of complexities tracing back to Erie days. It was no longer possible to separate the legend from the man or treat either one without misunderstanding and bitterness on all sides.[77]

I send you herewith my cheque for $1800 to the order of Wm C. Heath. Please send to me . . . the papers showing the transfer of the pew.[78]

The business wars took their toll on Gould's troops as well. In the midst of the Field fray came word that A. A. Talmage, that handsome bull of a man, had died suddenly. He was only forty-three, the father of eight children. His death removed the last of the trio that had come to Gould with the Southwest system. Gould had been fond of all three men, none more than Talmage, who was the most personable. The loss of these experienced lieutenants left Clark unchallenged in the realm of operations. Fate had, with another of its ironic twists, settled the question of whether Talmage might have been a better choice to succeed Clark.[79]

In the New York office one officer suffered wounds severe enough to disable him for a time. Amos Lawrence Hopkins was another of those improbable figures with whom Gould formed a close attachment. He belonged to a distinguished Massachusetts family that had for generations chosen the professions over business. His father, Mark Hopkins, gained renown as a teacher, served Williams college as president for thirty-six years, and wrote extensively on moral theology and ethics. Amos graduated from Williams in 1863 and joined the First Massachusetts Cavalry. Wounded at the Battle of the Wilderness, he lingered in the service until 1866. Two years later he commenced his railroad career as a superintendent on the Housatonic. His association with Gould earned Hopkins a fortune but marked the beginning of his fall from social grace. He was, like his father, diligent and hard working, but to proper eyes he toiled in a dubious business for the most dubious of employers.[80]

Hopkins married well only to find himself a widower at an early age. In 1879, at the age of thirty-five, he wed Minnie Dunlap, the beautiful, accomplished daughter of a noted operator on the Chicago Board of Trade. After five years in Chicago they moved to New York and occupied the house adjoining Gould's at One East Forty-seventh Street. The change of scene soon revealed the marriage to be a disaster. Hopkins was staid and reserved, "all business," as the boys said. His friends bored Minnie, whose restless, vivacious spirit yearned to enter society. She found ready sponsors in the William C. Whitneys and soon gathered a circle of new admirers enchanted by her beauty and gaiety. In November 1886, without warning, Minnie filed for divorce and went to live with her parents in Chicago. Her complaint charged Hopkins with infidelity.[81]

Despite efforts to hush up the incident, the *World* got hold of it and broke the story. The ingredients were in place for a scandal that promised to be as prolonged as it was sensational. Humiliated and outraged, Hopkins denied the charge under oath and prepared a countersuit. The whole affair must have horrified Helen Gould, coming as it did only two months after the trauma of George's marriage. Her proper, unbending spirit withered at this mockery of all she held sacred. Jay was not steeped in Murray Hill tradition but he honored nothing more than family in its fullest Victorian sense. To have one of his closest associates mired in the muck of a divorce scandal could not have been pleasant for him in an already difficult season.

Under the circumstances it would have been easy and certainly expedient for Jay to detach himself from Hopkins until the air cleared. But Hopkins was one of his dearest friends, and not even the taint of scandal could induce him to abandon a friend under fire. Believing Hopkins to be the victim of a great injustice, knowing a little something firsthand about mistreatment at the bar of public opinion, Jay stood firmly behind him. A reporter later observed that Gould "came in for a good share of gossip and abuse for his devotion to Hopkins," but that was nothing new to him. Nor did the continued loyalty of Whitney, then a cabinet member, and his wife to Minnie Hopkins sway Gould. In the fall of 1887 Minnie went to live in San Francisco. Hopkins retired from business for two months, but he would not relent in his determination to obtain a divorce on his own terms.[82]

By October 1887 Gould himself was near collapse from illness and the strain of business. So depleted was his strength that he surrendered at last to the lengthy cruise he had been threatening to take for years. It was no longer a matter of choice. He realized, as did the entire family, how weak and listless Helen had become, how she seemed to languish for days on end. Leaving his affairs in the hands of George and Edwin, he booked passage on the *Umbria* for himself, Helen, the other children, and one of Helen's sisters. The sailing date was October 29; on the 25th the *Atalanta* sailed for Marseilles, where the Goulds would board her.[83]

The papers pounced gleefully on the story. Was it another ruse or would Gould actually go this time? Skeptics remained unconvinced until October 27, when Gould visited the *Umbria* to inspect the cabins he had reserved. By then another complication had turned up in the form of a suit initiated against Gould and Sage for their part in the Denver Pacific stock episode in 1880. The *Herald* talked hopefully of a criminal suit as well and warned that the proceedings might delay Gould's departure. Although the lawyers managed to protect Gould's travel plans, the issue was far from dead. The *Herald* would keep it alive as ammunition for another crusade.[84]

Shortly before 2 P.M. on the day of departure three carriages arrived to take the family to the dock. The reporters had already gathered, two or three dozen of them, oblivious to the gaggle of royalty who had also booked passage, waiting only for Jay. They were impressed chiefly by the unassuming appearance of the Goulds, their quiet dress and modest demeanor. "Mr. Gould," noted the *Times* man, "looked as he generally does, a trifle shabby as to clothes."[85]

They waited expectantly while Gould chatted with Sage and Dillon. Morgan arrived for a brief conference, followed by two Manhattan officials. While Nellie exchanged tearful hugs with some schoolmates, Jay said good-bye to George and Edith. The *Times* man surveyed the remains of Edith's once proud hourglass figure and observed snidely that she "now probably weighs more pounds than her husband." His business done, Jay glanced toward the gang of reporters. To their surprise he joined them and, smiling and relaxed, shook the hands of those he knew. "What will you boys do without me?" he asked softly. They laughed and promised to store up a little energy. For a few minutes he fielded questions with an ease rarely seen in him before. Was he taking a physician? Jay shook his head and quipped that he had no more need of one than the average reporter. The trip was for rest and pleasure; he would transact no business. As he turned to leave, the *Herald* man inquired about the stock market.

"About the stock market?" he replied with a twinkle in his eye. "Oh, I've nothing to say. Some people say that the market goes up when Gould goes away. Goodby."

Then he was gone, leaving the papers to speculate endlessly over the true purpose of his trip. That it was exactly what he said it to be made little impression, perhaps because it made for dull copy. Jay did not care; for once he was free of the rumor mongers. He loved the sea, possibly because it was the only place he could truly shed the cares of business. The *Atalanta* had always been for him a floating haven, a refuge from the constant pressure that assailed him. Once during the summer of 1886 he entered the yacht in a grand race but that was a rare occasion. His competitive spirit did not extend to his pleasures, and he had no greater pleasure than the relaxation afforded by a prolonged cruise. For that interlude at least he was free to rest and reflect, to enjoy his family and bask in the solitude he so cherished.[86]

True to Jay's word, the Goulds stayed abroad nearly five months. They spent a week in London and another in Paris until bad weather chased them to Marseilles. From there the *Atalanta* sailed a leisurely circuit about the Mediterranean, stopping at Nice, San Remo, Florence, Naples, Rome, Sicily, and Greece before crossing to Egypt. Helen disliked Rome but found Egypt enchanting and seemed to come alive there. She longed to sail down the Nile, but Howard fell ill with typhoid fever and they were obliged to hurry back to Alexandria for medical help. Aboard the yacht they endured choppy seas along the north African coast, found a crush of visitors in Tunis, and went instead to Gibraltar and the Canary Islands. At home the boys got their first taste of managing affairs on their own. George found it more work than he bargained for. "I shall be very glad to have papa home again," he wrote his mother. "Although I think I can attend to every thing it is a great responsibility to have him so far away."[87]

But Jay did not hurry back. Settled in a deck chair beneath the hot Mediterranean sun, a book cradled in his lap, he had time to contemplate events of the past year or two. The pace had been frantic and conditions difficult, and he sensed that harder times were coming. The lengthy cruise, although restful, provided Jay a respite from the battlefield but could not heal his wounds. To his dismay Helen did not improve but returned home more spent than ever. Although Jay felt better, he had begun to suspect that something was also wrong with him, something more serious than mere fatigue or the torments of neuralgia.

There was.

V

LETTING GO
1888–1892

30

Deceptions

Since the lamented Captain Kidd landed on the Atlantic coast on a mission of miscalculated benevolence, no incident in the history of piracy is more interesting than the return of the famous American Corsair, of Broadway, and "Wizard of Wall Street."

—*New York Herald*, March 20, 1888

In the newspapers of last Sunday Mr. JAY GOULD *made his appearance as a journalist, and as the author of a controversial letter assailing Mr.* JAMES GORDON BENNETT *of the* Herald. *Mr.* GOULD'S *debut in this new capacity cannot be called successful, and the wonder is that a man so universally believed to be shrewd and keen in his perceptions, should have been willing to put his name to such a foolish performance.*

—*New York Sun*, April 1, 1888

The morning newspapers of New-York reached, perhaps, their lowest deep of degradation yesterday morning when all of them, with the single exception of THE TIMES, *allowed Mr.* JAY GOULD *to use their columns for a virulent personal attack upon Mr.* JAMES GORDON BENNETT.

—*New York Times*, April 2, 1888

The persistent reports concerning the health of Jay Gould culminated yesterday in the publication of rumors that he was insane.

—*New York Tribune*, July 24, 1888

Gould's orders to submerge the identity of the Katy appeared designed with malice aforethought. In that way it became practically impossible, even for expert accountants, to untangle the affairs of the two companies, to say that this or that was wrong, or illegal, or unethical, or unauthorized. Glaring examples . . . were set in the Katy shops, at Sedalia, at Nevada, at Parsons, and at Denison. Overnight these properties became "Missouri Pacific Shops," and weird stories of the wholesale overhauling of Mo. Pac. equipment, the switching of good equipment for bad—all charged to the Katy—are heard to this day.

—V. V. Masterson, *The Katy Railroad and the Last Frontier*

FOR TWO decades Gould had confounded the business and financial worlds with his gift for deception. To friends and foes alike, perhaps even more to the public, he was the grand deceiver, the master of illusion, the wizard. Nothing attests better to his reputation than the fact that even after his retirement he remained the most potent force in the market on the strength of rumor alone. On 'Change he had become a novel version of Adam Smith's invisible hand, invoked by traders and analysts alike to explain mysteries otherwise ineffable, wielded like a magic wand to wave stocks up or down. He may well have died a hundred deaths on the Street, so popular were reports of his illness or demise. Rare was the operation of scale that was not linked to his name.

In the last years of his life Jay perpetrated the greatest deception of his career, one that endured until the day he died. It was in fact a double deception, intended to fool the public on one level and his own family on another. So successful was he that to this day no one knows exactly when the ruse began. What little evidence there is points toward the spring or summer of 1888 as the time when Gould's personal physician, Dr. John P. Munn, told him that he had tuberculosis. The white plague had cursed the Gould family for years and had amounted to a sentence of death in every case. Jay knew this, yet his response was as typical as it was remarkable: he swore Munn to absolute secrecy. The public must not know, must not even suspect the truth, lest it complicate his already difficult business affairs. If the Street fed voraciously on mere rumors of illness, what excesses would the real article provoke? The shock would be too much for Helen in her weakened state, and he wanted the children spared as long as possible.[1]

Thus began the longest and noblest deception Gould ever foisted on an unwary public. The symptoms were not yet severe, but as they grew worse it required incredible strength of will to conceal them. Munn kept his pledge faithfully. He became Gould's inseparable companion on trips for business or pleasure, not only to render aid but to avert the need for relying on other doctors. Eventually he was obliged to accompany Gould in business meetings, even though the presence of an outsider among the inner councils of corporate management was too singular to pass unnoticed. The necessity of having Munn always at his elbow was but one of several practical problems caused by Gould's condition. For Gould the worst of it was struggling to maintain his enormous workload despite the disease and the need to mask any symptom that might betray his true condition. This grim ordeal went on day after day for nearly four years, possibly longer.

It was a magnificent performance, and a successful one. For years the sports on the Street had sneered at Gould as unmanly, cowardly, physically afraid. During these last years Jay showed a brand of courage few men could match, but no one knew of it. Although the fight took its toll on his nerves, he did not complain of pain or discomfort. Indeed, who could he complain to besides Munn? He never used illness as a device to gain sympathy or soften public sentiment toward him, never even made a virtue of his stoicism. He bore the agony of this last ordeal as he had so much else in his life, with strength and dignity, without frills or fanfare. He saw what must be done and did it without flinching at the cost to himself.

I am greatly surprised at the theory you advance viz that private outside investments of officers of Railway Cos innure [*sic*] to and are in fact the real property of the corporations they happen to be President of.[2]

After touring the Mediterranean the *Atalanta* crossed the ocean for a leisurely meander through the West Indies, where Gould lingered long enough to escape the celebrated blizzard of March 1888. He could not avoid a host of other storms brewing at home, inklings of which probably reached him by cable at various ports. All the Gould stocks had been pounded down by repeated raids since his departure. Missouri Pacific had lost twenty points, Manhattan nineteen, Katy thirteen, and Western Union six. The decline created serious problems for the Missouri Pacific. Kuhn, Loeb found it impossible to place Iron Mountain bonds on a falling market. The Street was full of tales that earnings on the Gould system had plummeted and would force a passing of the dividend. Some charged that the brothers Gould were bearing their father's stocks and needed a passed dividend to help them out. Amid these pressures neither George as acting president nor the executive committee dared to act without Jay.[3]

The annual report, released just as the *Atalanta* berthed in Florida on March 19, befuddled analysts with its apparent contradictions. An astounding 2,137 miles of road had been added to the system, swelling the total to 6,974 miles. Gross earnings were up nearly $3.5 million, net earnings down nearly $1 million. A huge increase in expenses raised the operating ratio from 62 to 71½ percent. One figure revealed the ravages wrought by prolonged rate wars: the revenue per ton per mile, which stood at 1.63 cents in 1882, had declined steadily to 1.15 cents in 1888. The Iron Mountain and most of the branches showed deficits for the year, and the Katy was a disaster with a loss of nearly $1.3 million. Most of the new mileage had yet to earn a dime or prove its earning capacity. While analysts divided over what it all meant, the Street rendered its verdict by knocking the stock down half a dozen points even though a decision on the dividend had yet to be announced.[4]

Was the Missouri Pacific in deep trouble? Critics were unsure because the market and railway earnings alike were demoralized by rate wars, the new Interstate Commerce Act, and a strike on the Burlington. The *Times* cautioned that "the causes which have brought down the Missouri Pacific's earnings are equally operative on all the roads in the Western territory." Unless some solution was found, the passing or reducing of dividends would soon become wholesale and the Missouri Pacific "merely the first in the procession."[5]

By the time Gould landed in Florida the situation was fast approaching crisis proportions. George met his father in Florida and briefed him on what had been happening. Four days later the board, with Jay presiding, decided to reduce the quarterly dividend from $1\frac{3}{4}$ to $1\frac{1}{2}$ percent. In an interview he denied that George had gone short of Missouri Pacific and insisted the stock "is worth 150 if it is worth a cent." The market steadied somewhat, but Gould was not deceived. He knew there was much to do, all of it requiring his close attention. What he did not anticipate was an attack from quite another quarter that drained energies he needed elsewhere and provoked him into a rare outburst of temper.[6]

The attack had been building for months, awaiting Gould's return to unleash its full fury. Jay's parting jest to the reporters had been to ask what they would do without him; he soon discovered that the *Herald* at least had done plenty. For whatever reason, the paper would not let go of the Denver Pacific issue. The transaction had first drawn attention in 1885, when a suit filed by a Kansas Pacific bondholder opened the question of Gould and Sage's role in releasing the Denver Pacific shares from trust. The suit was settled a few months later and the question lay dormant until revived in 1887 by the Pacific Railway Commission hearings. That testimony prompted some obscure bankers to file civil and criminal charges in October, but both were dropped a week later on the grounds that the statute of limitations required such suits to be brought within five years.[7]

After Gould departed on his cruise unmolested, there followed another of those bizarre legal olios that Gould seemed to inspire without even trying. On his own the bankers' lawyer tried to revive the criminal suit. For two months nothing was heard from the district attorney. On December 31, 1887, the day his term in office expired, he suddenly announced that the charges against Gould and Sage were indeed criminal and ought to go before a grand jury. The new attorney general, like his predecessor, doubted whether the statute of limitations allowed him to prosecute. The *Herald* had no such qualms. It demanded that indictments be brought, badgered the attorney general to take action, and missed no opportunity to abuse Gould personally.[8]

Sage wasn't worried. "Life is too short," he philosophized, "to allow that matter to bother me." In February the grand jury voted not to indict after the judge informed it that the statute of limitations barred prosecution. An outraged *Herald* cried foul and urged the attorney general to go before the March grand jury. He did and was rebuffed again. Still the *Herald* argued that justice should not be thwarted by a mere technicality and filled its columns with opinions to that effect from American, English, and French lawyers. These pronouncements enlivened the controversy but could not change the awkward fact that the attorney general had no case. Stripped of sensationalism, the charges had too little substance to stand up in court. One financial journal examined them in detail and concluded that "had the issue been made up and tried on its merits," the plaintiffs would not have "found a foot of solid ground on which to stand."[9]

This reasoning made no impression on the *Herald*. On March 20 it launched the most vituperative phase of its campaign, mocking Gould, denouncing his influence on Wall Street, charging him with destroying the credit of

American railroads abroad. The "Little Wizard" now became the "Corsair" as well, an inspiration drawn from his prolonged cruise. The *Herald* commemorated Gould's return with a series of articles interspersed with poetic snatches from Byron's *Corsair* and *Childe Harold*.[10]

How much of this abuse Gould read is unknown. He was familiar enough with the main points to discuss them in an interview the day he returned to New York. His mood was one of cold fury. The *Herald* had roused the stubborn, combative spirit he had shown since childhood but had nearly always suppressed in public forums. Not this time. He had come home with an ailing wife to a tangle of financial crises that only hard, unremitting work could unravel. Confidence in his rail system had been shaken and his integrity as a businessman attacked yet again. He was accustomed to vilification, but this time it embraced his eldest son and chosen successor. The charge that George had betrayed his trust by speculating in Missouri Pacific seemed to weigh heaviest on Jay's mind. Perhaps it was the blow that snapped his cord of self-restraint and drove him to lash out at tormentors he had for so long borne in silence.

Slowly at first, the anger seeped out in a *Tribune* interview. He called the accusations against George "cruel—simply cruel. There is not a morsel of truth in the charge that he has been going short of the market." As for the indictment, he offered to place a detailed account of the Denver Pacific affair before the presidents of leading trust companies for their judgment on whether it was handled according to business principles. What, then, asked the reporter, were the motives behind the attack against Gould and Sage? Gould paused, his dark eyes flashing and his lips compressed.

"The motives behind this assault," he answered slowly, "are—a newspaper, a cable company and a woman."[11]

The reporter gaped in disbelief, amazed at so sensational a statement from one notorious for his reticence. Gould knew what effect his words would have. "I have said a good deal more perhaps than is necessary," he added, "but it may be about time for me to let certain people and interests see that it is not always necessary to keep silent while other people do all the talking." Tongues wagged furiously downtown and in the Windsor lobby that night over the meaning of Gould's words, over the very fact that he had uttered them. Two days later Jay amplified his remark in another *Tribune* interview. He thought he knew why Bennett and the *Herald* were against him. Years earlier Bennett had inherited from his father a block of American Union stock and gained a seat on the board. In that position he demanded "precedence over all other newspapers in the transmission of special reports." When the request was denied, Bennett became a "disgruntled and disagreeable associate" and was not offered a seat on the Western Union board after the merger. Gould thought the *Herald*'s criticism of him and Western Union dated from Bennett's retirement as a director. As for the cable venture and the rate war that still raged, "the unprofitableness of the Mackay-Bennett cables is the chief cause for Mr. Bennett's animus against me."[12]

It was the reference to a woman that most surprised and fascinated the Street. During Gould's trip the Hopkins divorce had at last gone through, arranged so quietly that confusion reigned over whose suit had prevailed. What had Minnie Hopkins to do with the Denver Pacific business? Jay spun an intriguing tale. "In the first place," he declared, "the claim . . . against Mr. Sage and me is blackmail, pure and simple blackmail." The opposition lawyers had first attempted a civil suit in the name of a German banking house only to have the firm refuse to act as plaintiff. Left without a suit, the lawyers were encouraged to seek a criminal indictment not by bondholders but by a "prominent politician and office holder."[13]

No one with a taste for gossip missed Gould's allusion. In his view the hand behind the suit belonged to William C. Whitney, the champion of Minnie Hopkins, who was out to punish Gould for his steadfast support of Amos Hopkins. Whitney wanted an indictment, the lawyers a settlement to line their own pockets. Gould declined all negotiations. "Everybody in this neighborhood knows that I was right in supporting Mr. Hopkins," Gould insisted. "I know that he was an innocent and a greatly abused man. I am a man of peace, but gradually I have been driven by people who have not only injured me, but my friends, to defend both them and myself."[14]

Gould could not have startled the Street more if he had hurled a bomb onto the floor of the Exchange. The city's most silent and enigmatic citizen suddenly could not say enough about matters of great delicacy. A man who for years had shrugged off criticism now defended himself with obvious passion by venting sensational charges. Here was a side of Gould's character never seen before, perhaps because earlier frays had confined themselves to the business arena. This one scraped the raw nerve of those values he cherished most, devotion to family and loyalty to friends. In Hopkins's case the two were intertwined; Gould was not only sustaining a maligned friend but upholding the sanctity of the family itself, which in his eyes had been violated by Minnie Hopkins. No one was surprised at his vehement defense of George, but his fervent support of Hopkins baffled those who did not understand how important these issues were to Gould. He would not compromise them, would not retreat even if their defense forced him to step out of character.

Although Bennett's precise role in this imbroglio is unknown, it is doubtful the *Herald* would have proceeded without his approval. More likely he orchestrated the attack, driven like Gould by emotional dictates that overrode the logic of self-interest. The mounting ferocity of the *Herald*'s campaign, the violent, almost hysterical tone of its language, suggests motives that transcended the cable war or even the desire to sell papers. Bennett's hatred of Gould was as well known as his desire to bring him down. If he hoped to lure Gould into a war of words, he succeeded beyond his wildest dreams. In lashing out at his tormentors Gould committed what was for him a singular mistake, one that can be explained only as a rare instance of allowing his feelings to override his judgment. He chose to fight his adversaries with their own weapons on their own battlefield. It was a fight he could not possibly win. No one knew better than he the folly of trying to beat a newspaper at its own game.[15]

The *Herald* responded with a torrent of abuse that exceeded all past bounds. The "Corsair" (a.k.a. "Little Wizard") was promoted to "the Skunk of Wall Street." An editorial angrily denied all charges by "this ghoul in human form, this Satan of the modern world" who had "done more to disgrace the fair name of the United States and injure American credit than twenty Benedict Arnolds." Lest readers get the wrong impression, another editorial blandly assured them that "if there has ever been any 'bitterness' in the HERALD it has been contained in the facts reported, the cold, hard, undeniable facts which we were bound as journalists to print. . . . We have no feeling in this matter whatever. On the contrary, we are thoroughly impassive." Whitney issued an indignant denial and there was considerable talk about how Gould had stooped to defaming women.[16]

Gould stood this shelling for five days, and there is no doubt that it hit home. His dander was up; the chains of self-control snapped. Throwing discretion aside, he rashly sent the papers an open letter to Bennett. In it Jay quoted a memorandum from a former director of American Union to substantiate the charges he had made against Bennett. Had he confined himself solely to the issues, the letter might have caused only a ripple of excitement. For Gould, however, the central question was not business but character assassination. He was enraged at being slandered by a man whose own character, in his opinion, mocked those values Jay held dearest. After reviewing some of the epithets hurled at him by the *Herald* he flung some mud of his own. For thirty years Bennett's life had been "one of shame . . . a succession of debauches and scandals" to the point where no decent gentleman would allow "Bennett the libertine" to cross a threshold "where virtue and family honor are held sacred." In bitter, unsparing language Gould recalled the more notorious episodes of Bennett's days in New York. He closed with a defense of his actions in the Denver Pacific controversy and accused Bennett of being in league with the opposition lawyers.[17]

All the dailies pounced gleefully on the letter except the *Times*, which made a virtue of refusing to print it. The *Tribune* ducked the controversy editorially and signed off after publishing the letter. Among the others there was general agreement that Gould's debut as a journalist was, in the *Sun*'s words, "a foolish performance," and much bewilderment over why he had done it. In lamenting his assault on a member of their craft, none mentioned the *Herald*'s prolonged campaign of abuse that preceded the letter. The *Herald* was ecstatic that Gould had played into its hands. In one stroke he had wrought the miracle of drawing sympathy for Bennett from the feuding warlords of New York journalism. He had kept the issue alive enough for the *Herald* to milk three more weeks of controversy out of it. Even worse, he had unwittingly shifted the focus of discussion from the Denver Pacific affair to the letter itself.[18]

The *Herald* punished Gould fully for his indiscretion. It howled with virtuous indignation at "the revenge of a coward," filled whole columns with disapproving reactions, and hammered away editorially at Gould and the "growing cancer in our business system" that permitted "a ruffian to wreck railways, subsidize newspapers, buy legislators, own judges, degrade American credit, acknowledge no law in dealing with his fellow men but the law of the highwayman, and in doing so amass a vast fortune." For four days it printed comments from an impressive list of newspapers censuring the letter and its author. Throughout this barrage of defamation the *Herald* had things its own way except on the issue that counted most, the indictment.[19]

Amid the mudslinging the *Herald* had long since tried and convicted Gould of a crime, as had many other newspapers. On April 11, however, the recorder absolved the judge of error in his instructions to the grand jury. On that note the legal contest came to an end, much to the *Herald*'s dismay. Having already consigned the judge to the hell reserved for Barnard, Cardozo, and Westbrook, it blasted the recorder for allowing Gould and Sage to escape on a technicality. In reporting the "cold, hard, undeniable facts" the *Herald* again ignored the inability of the plaintiffs to make the substance of the charges stick. Neither did it quote the opinion of Charles Francis Adams, who could hardly be accused of partiality toward Gould. After investigating the evidence Adams declared himself "unable to see wherein any wrong has been done to the trust, and certainly none has been done to the Union Pacific in the handling of the Gould-Sage trust." In fact, Adams added, conversion of the Denver Pacific stock enabled the Union Pacific to realize "a handsome profit, probably some 30 per cent."[20]

Later Adams would berate himself for not using the situation as a lever against Gould, but he never doubted the honesty or accuracy of the views he expressed. With the indictment a dead issue, the *Herald* exhausted its arsenal of invective and turned its attention elsewhere after a last few swipes at the telegraph monopoly. A chastened Gould said nothing more and crawled gratefully out of the headlines for a time. Early in May, when the smoke of battle had cleared, he and Sage quietly resigned as trustees of the Kansas Pacific mortgage. The *Times* tried to revive interest in Gould's "wholesale prostitution of certain New-York newspapers," but compared to the *Herald*'s onslaught it was pretty tame stuff and went nowhere. The most fitting capstone to the Bennett affair came that summer when Gould and Mackay negotiated an end to the cable rate war on terms the *Herald* chose not to splash across its headlines.[21]

While we controll the MK&T I suggest . . . you spend as little as possible on the property and get all the money we can out of it. All the old rails ought to be picked up & shipped to St Louis.[22]

No amount of brave talk could alter the bleak outlook confronting Gould in the spring of 1888. Rate wars continued to chew up earnings with no end in sight. "Hardly a day passes but that some new reduction is announced," the *Chronicle* mourned, "and it is almost impossible to keep track of them all." The mysterious attacks on Missouri Pacific stock, coupled with an unfavorable annual report, hurt the company's credit. Gould's policy of not issuing monthly earnings figures, long criticized by analysts, allowed the year's data to hit the Street as a surprise. This time it was not a pleasant one. Kuhn, Loeb, which had contracted eighteen months earlier to extend the company's first mortgage bonds at 4 percent, demanded better terms to offset its declining credit.[23]

During Gould's absence Clark laid off 10 percent of the work force (fourteen hundred men) for two months. Construction projects slowed to a crawl except for some coal spurs and the new track in St. Louis. Gould was impatient to begin work on the new depot and office building but lacked funds until other projects were wound up. The company's map glittered with branches that still existed only on paper. Financial pressures also complicated the two pending reorganizations and precipitated a crisis on a third subsidiary, the Katy. Yet another Wabash plan, formulated in November 1887, required Gould's attention, and the Texas & Pacific settlement bogged down in more squabbling over details.[24]

Gould hoped to complete the Texas & Pacific reorganization in time for the company's May meeting at which he planned to surrender the presidency to one of the receivers, John C. Brown, and install George as first vice-president. To accomplish this Wistar had first to buy off the other receiver, whom he castigated as "a low political adventurer, without property, credit, veracity, or character." There followed some strenuous haggling with the mortgage trustees over the fees to be paid them, a dispute over how much new stock was due the Missouri Pacific for its claims, and a disagreement between Gould and the Kuhn, Loeb-Drexel, Morgan syndicate over the terms offered the bankers. Somehow Gould and Wistar resolved these difficulties before the May meeting. The new board, considered in some quarters to be independent of Gould, was in fact friendly if not beholden to him. The election of officers went according to plan, but Jay was taking no chances. Quietly he bought Texas & Pacific stock in large amounts; months would pass before his true position was revealed.[25]

Meanwhile, in another part of Texas, a struggle loomed that soon rivaled the Texas & Pacific contest in bitterness and complexity. Like other pioneer roads, the Katy had fallen victim to the expansion mania of the 1880s. For years it was the only north-south road between Texas and the Mississippi-Missouri valleys except for the Iron Mountain. By 1888, however, the Santa Fe, lashing out at Gould on all fronts, completed a line west of the Katy from Kansas through Indian Territory to Galveston. Another parallel road emerged east of the Katy when the Frisco built to a connection with the Santa Fe's Gulf line at Paris, Texas. By giving the Santa Fe an outlet to St. Louis, this line threatened not only the Katy but Gould's entire southwestern system. Rate wars broke out at once, signaling an end to the stability that had prevailed in Texas since 1885.[26]

Flanked on both sides, the Katy followed the downward spiral of other debt-heavy roads, its ledger a seismograph registering the shocks of rising expenses and falling income. The Gould regime had not been popular in the best of times; one clash with some income bondholders lasted two years before Gould agreed to compromise terms. Complaints turned to cries of alarm in October 1887 as Katy securities took the same battering given Missouri Pacific amid rumors of impending default and receivership. In November William Dowd, a Katy director and president of the Bank of North America, assured doubters that he had examined the company's books "quite thoroughly" and found it had laid considerable steel rail, was in first-class condition, had built 160 miles of new road at only $12,000 per mile, had no floating debt, and had enough cash on hand to meet interest for three months. It is possible that much of this statement was correct, but the securities kept dropping and rumors kept flying.[27]

Among those disturbed was Dodge, who had built part of the road, had induced many people to take its bonds, and held a parcel of its securities himself. Dodge had heard stories of equipment "being stripped to be used on other parts of the system," but he knew this could not be. Missouri Pacific employees had assured him that the Katy "was about the best of its leased properties." Personally he thought the road had "a better plant" and was "better able to do business than any other road in the state." Still, certain questions nagged Dodge. The Katy had cost his Panhandle Road considerable business the past year because it could not deliver coal or provide stock cars promptly. He had heard that expenses had soared. Why? Where was the power? The rolling stock? What was the truth about the road's condition?[28]

This last question became the object of a long and tortuous quest. The stock exchange in Amsterdam, where many Katy securities were held, formed an investigating committee to defend against "the alleged machinations of bear traders." In February the *Chronicle* obtained earnings figures for nine months of 1887 which confirmed rumors of "satisfactory gains in gross, but heavy losses in the net" because of a 26 percent rise in expenses. That same month the Missouri Pacific found it necessary to advance the Katy $520,000 for maturing coupons. On the eve of Gould's return the Amsterdam committee reported that during the past year the road had been deprived of at least $1 million through manipulation of traffic in favor of the Missouri Pacific. At a second meeting on March 26 the group vowed to elect a board free from Missouri Pacific control. There was talk of extending the road to St. Louis and Kansas City to eliminate all dependence on the Gould system. Wall Street braced for another lively fight.[29]

Gould's position was, as usual, more complex than his adversaries knew. The Katy served his southwestern system in several ways. It was part of the through routes between Kansas City and the Southwest and between Kansas and the Southeast. It was also a link in the Texas & Pacific's transcontinental line with southern Kansas and western Missouri. Moreover, the Katy owned the stock of the International, which for months had been rumored on the brink of receivership. Loss of the International would cost Gould's lines the traffic of southern Texas, which would be diverted north over the Katy for delivery to the Burlington at Hannibal. The International was in dreadful shape physically and financially, but Gould could not afford to lose control of it.[30]

At the same time Jay blanched at the cost of retaining control of the Katy. Like so many roads constructed in an earlier era, its fixed charges were too high to be sustained at the low rates imposed by newer, more cheaply built competitors. A reorganization was needed to bring its capital structure into line with prevailing competitive realities. For Gould the Katy's strategic value had been diminished by his new Arkansas Valley route, which extended beyond Fort Smith to a connection with the Katy at Waggoner, Indian Territory. By pushing this line northward to Kansas he could create his own through route between Colorado and the Gulf, provided he could hang onto the International. The trick, then, was to maintain his hold on the Katy as long as possible and to keep the International at all costs.

Gould made his first move in mid-April by summoning the Missouri Pacific board to a special meeting at which he summarized relations with the Katy. In reviewing eight years of operation under the lease he noted that the Katy's main line had been laid with steel rails and large expenditures made for improvements. He attributed the road's current troubles to the failure of the corn crop, the impact of new competitive lines, and the lengthy fight with the income bondholders. The Katy had already borrowed $800,000 from the Missouri Pacific, Jay added, and "additional large advances" would be needed until earnings improved.[31]

In that phrase lay the snare of a claim on which to seek a receiver if one were needed. Elsewhere Gould said more than was necessary, suggesting that the wounds of the *Herald* battle still bled. He tendered the movement to elect an independent board his "cordial approbation" and "cheerfully decided not to make any opposition." But he could not resist indulging one of his favorite conceits, that he had served the Katy eight years "without receiving a dollar of salary" and was glad to be relieved. Nor could he withhold his disappointment that the proposed new board consisted largely of the income bondholders, whose suits had cost the Katy dearly and whose attorney, E. Ellery Anderson, late of the Pacific Railway Commission, was to be the new president. Although Gould had approved payments of $50,000 and $40,000 to two lawyers in the Texas & Pacific reorganization, he felt obliged to mention that Anderson received $30,000 for his services to the income bondholders.

The style and content of these remarks were familiar Gould devices of precisely the sort that infuriated his detractors: a mixture of willful pettiness and virtuous indignation, intended to score points in a business dispute. One suspects they also served as a vent for the frustration Gould felt at being constantly misunderstood and ridiculed. The hurt feelings were as genuine as the points made were calculated, but this was too complex a form of expression for critics to digest. Moreover, Gould had assumed this pose so frequently over the years that it had become his trademark. The *Herald* dismissed his letter as "sublime impudence . . . laughable to everybody but the sufferers by his rascality." The Katy dissidents promptly fired back a letter disputing all of Gould's points. Defending his position in an interview, Jay insisted that the Katy's obligations needed scaling down and suggested a joint committee to investigate the road's affairs. He sent no representatives to the Katy annual meeting on May 16. The dissidents elected their own board, passed resolutions condemning their predecessors, and agreed warily to form a joint committee.[32]

By that time Gould had already moved against the International, using what the *Times* called "methods so peculiarly his own." With his claim as a floating debt creditor in hand, he had the International's lease to the Katy canceled on the grounds that the attorney general of Texas was about to forfeit the former's charter. Before surrendering the Katy he summoned a last board meeting at which resolutions were passed transferring the International's stock to the Missouri Pacific as collateral for the Katy's floating debt. Once this was done, Jay issued an order to cease all interest payments on the Katy's behalf after May 31.[33]

Through these maneuvers Gould hoped to separate the International from the Katy and put himself in a strong position if the International defaulted. He had also negotiated an agreement with S. W. Fordyce of the Cotton Belt by which that road and the Missouri Pacific, "though separately owned are to be operated as though the ownership was the same." This alliance was kept secret, a hole card for Gould to play at the suitable moment. Meanwhile the new Katy directors, in the saddle at last, found themselves astride a dead horse. They enjoined Gould from transferring the International's stock but could do little else. While the court pondered that tangle, they decided reluctantly to seek a receivership for the Katy only to have the application bog down in a legal quagmire, where it floundered for months. The joint committee was appointed and fell at once to bickering over procedures. By that time Gould had left the city, headed west on an inspection tour that proved more than he had bargained for.[34]

I am so glad the baby is a boy and I am to have a namesake. It is especially gratifying that my namesake is your son because I am very proud of you.[35]

The first rumors, vague and contradictory, reached New York on June 2. Somewhere west of Kansas City Gould had fallen ill and the *Atalanta* had been switched to an eastbound train for immediate return home. Spokesmen in both New York and St. Louis issued swift denials. Reports of Gould's homeward journey were scotched the next day when he arrived in Pueblo and allowed himself to be driven about town. The *Times* complained that it was "becoming the weekly fashion" to announce that Gould was "dangerously sick and his death imminent," but even it accepted reports that he had failed physically to a noticeable extent during the past year.[36]

Gradually a story was pieced together. Gould had been stricken with what he insisted was an attack of neuralgia, nothing more. Dr. Munn was aboard the *Atalanta* and had prescribed some mountain air instead of a return journey. Munn admitted that Gould was suffering greatly from insomnia but said the neuralgia had improved. Others weren't so sure. Clark's personal physician, who had also attended Gould, confirmed that he was better but "very weak," and that the insomnia plaguing him was "chronic" and "excruciating." A man who saw Gould in Kansas City described him as "haggard" and "not the man that he was a few months ago."[37]

These gleanings suggest that Gould may have been afflicted not with neuralgia but with symptoms of tuberculosis. Jay himself revealed nothing and shunned reporters. After a few days in Pueblo he started east, taking care to make the usual number of stops. A reporter caught him stretching his legs on the platform at Carondelet and came away shocked by Gould's appearance: the stooped shoulders, the beard mottled with patches of white, the face drawn and pinched in pain. He walked slowly, throwing an occasional listless glance upward, "never once straightening his head in the old defiant state."

"We have had a tiresome journey," Gould explained wearily.[38]

But he would not quit. Instead of going directly home Gould went to Memphis to inspect the new road and its connecting line, the East Tennessee. Shortly afterward he made himself available for interviews to assure reporters he was all right again. On June 16 he came home to what he already knew was a difficult situation. The Manhattan board had just reduced its quarterly dividend from 1½ to 1 percent; Jay had telegraphed the Missouri Pacific executive committee to do likewise. Other roads pared dividends as their reports revealed the same pattern of carrying more for less. The business outlook was anything but encouraging.[39]

Throughout July Gould's health remained the subject of intense speculation. John Terry, who saw Gould often, insisted he was much improved and warned that "people who fool with his properties, under the impression that he is a sick man, are liable to find that he is as clear-headed as ever he was in his life." Reports spread that Gould had cancer because a Malden specialist had been summoned; a few days later the betting switched to Bright's disease. Repeated denials had little effect on the Street, where the latest tale declared Gould insane. Sage vowed that Gould would "show some of these folks down here in Wall street before long that he is very much alive," but after his return Jay did not venture downtown. Instead he secluded himself at Lyndhurst, doing little business, enjoying his family, and puttering with his flowers.[40]

The rumors still troubled him, if only because of the damage they might inflict. For that reason he granted a rare interview at Lyndhurst. A Philadelphia reporter found him emerging from the greenhouse, his arms cradling pots of rose bushes. Despite the heat he wore a blue flannel suit, thick felt slippers, and a wide-brimmed straw hat. The reporter thought him thin and wan, his face a mask of pain. "So they make me out a hopeless case," Gould smiled. "Well, it isn't bad as that. I am more miserable than actually sick." Facial neuralgia bothered him, he admitted, especially in wet weather. That was the source of his misery and of the insomnia that gave him fits. He had been in agony before going west but the trip had helped him. Now he was "a gardener first, last and nearly all the time." He planned to rest all summer, avoid business as much as possible, and await the birth of his second grandchild. "I won't undertake to say that my mind is free of thought about my enterprises—a man can't leave his intellect in his office and bring nothing but his body home—but I am diverting myself as much as possible."[41]

The interview helped, but it was not enough. Late in July Gould turned his convalescence into a public event by taking Helen to Saratoga for a month. All the children except George and little Frank came along, as did Alice Northrop and Dr. Munn. Gould rented two of the best cottages at the United States Hotel, where the presence of so prominent an invalid added luster to what proved to be the resort's most successful season in years. Reporters scrutinized his every move, seeking clues to his true condition. Denied interviews, they gleaned what they could from other guests, waiters, and stablemen to pad their accounts. Gould was aware of the interest stirred by his presence, although outwardly he remained oblivious to it. He was far from being the only millionaire at Saratoga, but he attracted more attention than all the others put together. There were more secluded retreats at which to seek rest, but he chose to put himself on display in a social greenhouse for reasons having more to do with business than health.[42]

True to his word, Jay took life easy at Saratoga. He declined the waters in favor of leisurely strolls and drives through the countryside. For meals he chose a table near the door of the cavernous dining room, where he might be seen but not surrounded. In the evenings he sat placidly on the piazza, chatting with Munn or acquaintances who stopped by to greet him, reading the papers, indifferent to the strands of music wafting across the lawn, watching the waters of the fountains dance and sparkle in the multicolored lights. He looked in on a grand ball (but did not dance) and on the practice session of a billiards champion. When the circus came to town he perched eagerly in a chair on the front piazza of the hotel to watch the parade. Every evening at six he took his constitutional with Morosini, who had brought his wife and daughter to the spa.[43]

The sharpest eyes failed to detect him transacting any business. Frank Work, Cammack, Henry Clews, the Wormser brothers, and other denizens of the Street were in residence, but Gould refused to talk shop. He visited none of the branch offices set up by brokerages at Saratoga and did not even bother to install a private wire to his cottage. "If ever a man were careless of stocks and stock market fluctuations," grumbled a skeptical reporter, "Jay Gould just now is posing as that man." Except for a meeting of the Missouri Pacific executive committee, he was content to bask in his role of man at leisure. The usual brand of camp followers huddled expectantly on hotel verandas, trading gossip and inside information. "What they don't know about Jay Gould," sneered a reporter, "would stock a library."[44]

To all inquiries Gould offered the same response. The country air had greatly improved his health. He was sleeping better, eating better, and feeling better than he had in ages. "I was overworked," he admitted when he finally permitted an interview. Thrusting his delicate hands forward, he added, "You see my hands are as steady as ever they were. Two months ago they trembled visibly, my whole nervous system was unstrung, and I was greatly troubled with insomnia." Now he slept soundly eight hours a night and took a two-hour nap in the afternoon. Several observers were satisfied enough to suggest that he had been "playing possum," that "much of his indisposition had been a sham."[45]

The *Times* man was not convinced. He noticed that Gould never strayed beyond the watchful eye of Munn or one of the children. From a discreet distance he studied Gould's expression as he sat reading the papers on the porch of his cottage and found an air of "strained thoughtfulness" from which the tension never relaxed. The newspaper was thrown down, picked up again, teased and twisted until it lay in tatters. There was no life in his eyes, no animation in his voice, no apparent interest in anything about him. Certainly Jay found nothing in the papers to give him comfort. During his weeks at Saratoga General Phil Sheridan and Charles Crocker died. In Maryland Robert Garrett was losing his battle against insanity and had been isolated from visitors. The business outlook remained bleak as rate wars raged with unabated fury in every part of the country. One item must have aroused Gould's interest. In Paris an international conference of doctors convened to debate the causes and possible cures of tuberculosis. Great excitement had been aroused by the efforts of a German scientist, Robert Koch, to isolate the bacillus that caused the disease.[46]

As the *Times*'s man suspected, Jay was not as well as he let on. Alice Northrop noticed that "he had to lie on his bed and rest before he could go anywhere," and insomnia still tormented him. Nellie took her father for drives, read to him for hours, and helped Munn administer massages to relax him into sleep. The condition of her mother caused Nellie no less anxiety. Helen had endured the trip to Saratoga only because she thought it her duty to go. In her quiet, uncomplaining way she showered concern on Jay while her own strength waned. "I wish that I had your strong constitution," she sighed to her mother, "for I really feel about played out."[47]

Late in August Jay sent Helen back to Lyndhurst and went on to the Catskills for some fishing and sightseeing. He stopped at Furlow Lake, a lovely stretch of water isolated atop a hill eight miles from Arkville. Its beauty and solitude must have appealed to him, for soon afterward George acquired the property and laid plans to construct a lodge. From Furlow Jay journeyed to Roxbury for a rare visit to his birthplace. Nellie, who had dutifully accompanied her mother home, hurried back to meet him there. She found him in good spirits. They dined on trout he had caught and bantered playfully. "I have never seen Father so merry," Nellie wrote her mother.[48]

The next few days were among the most pleasant Jay had ever known. He took the children to enjoy the view at the Grand Hotel and hired wagons to haul them to the summit of Utsayantha Mountain. Amid these wanderings Gould retraced the steps of his boyhood, visiting relatives and old friends he had not seen in years. He introduced the children to Cousin Maria Burhans Lauren, whom he was supposed to have courted in his youth. One stormy evening an old man with a hat pulled down over his eyes climbed aboard the *Atalanta*. A smile brightened Jay's pallid face. "You needn't try to disguise yourself," he laughed, "I know you." It was Peter Van Amburgh, the former hired man. Next day Jay took Edwin and Dr. Munn to Van Amburgh's house, where they feasted on homemade bread, butter, and honey. After walking over the farm Jay insisted that Peter visit him at Lyndhurst. If he would come, Gould added slyly, he could pick a cow from the Lyndhurst herd to take home as a present.[49]

Peter took him at his word. Some months later, outfitted in his "Sunday-go-to-meetin' " clothes, he arrived at Lyndhurst and was treated in regal style.

After disposing of the amenities he asked about the cow. Gould sent him out with Ferdinand Mangold to inspect the herd. He went to the city and returned to find Mangold beside himself. "Picked out our prize Jersey," Mangold grumbled, "that's all. The finest one of the lot." Jay could not stop laughing. "Peter, I always said you were a smart one," he cried, "and now I think Mr. Mangold agrees with me. The cow is yours. I will send you her pedigree papers and pay for transportation."[50]

While at Roxbury Jay received word that Edith had given birth to a second son, who was named after his grandfather. Nothing could have pleased Jay more. Already the doting parent had become a doting grandparent, and it suited him to hear the news in a setting where he was surrounded by his own past. "Father has been so different," Nellie wrote her mother; "the old memories and the old friends have quite brightened him up." But the pain was not gone, and neither were his twin nemeses of fatigue and insomnia. At home the harness awaited. For two months Gould had avoided hard work; he could delay no longer. While the children lingered a few days longer at Roxbury, he went back to New York with Edwin and Munn on September 3.[51]

Summer's ease faded faster than the tan he had acquired while fishing. The full effects of recent expansion programs had crashed down on railroads everywhere. Ruinous competition sent rates tumbling ever lower, shattered existing agreements, and forced even strong roads to curtail dividends. Hostility and suspicion poisoned relations among rail leaders to the point where the most optimistic executives despaired of finding ways to restore harmony. Every road had mobilized for war, determined to fight rivals to the bitter end and willing to incur the cost of battle because it could find no better course.[52]

This bloodletting wreaked havoc with Gould's plans in several ways. By the year's end all his construction projects would be finished except for the Carondelet connection in St. Louis. He had counted on paying for the work by issuing new stock, but the turmoil of recent years had weakened the market for railroad securities of all kinds. During 1886 Missouri Pacific hovered around par; in December 1888 it bottomed at 66¼. Unable to sell stock or bonds at decent prices, Gould and his associates had to carry the floating debt themselves. They could expect no relief until earnings improved, which in turn required a settlement among the warring roads that would enable them to advance rates. Meanwhile losses piled up, especially on the new branch lines. Gould was appalled by the figures crossing his desk. Until peace could be negotiated, the belt, already tightened, must be hitched in another notch. "We have got to meet the low rates by larger cars & larger loads in them," he warned Clark, "or our expenses will drive us sooner or later to the wall." He flooded Clark with detailed suggestions.[53]

To these problems were added the interminable reorganization fights. The Texas & Pacific had proceeded smoothly enough for the court to discharge the receiver in October 1888. The stock Gould had quietly bought over a period of months gave him control when combined with Sage's holdings and the shares tendered the Missouri Pacific in the reorganization plan. At the annual meeting in February 1889 the independent directors were replaced by Gould regulars. John Brown assumed the presidency and Gould became chairman as planned. After years of struggle Gould had managed to retain the Texas & Pacific as part of his southwestern system.[54]

The Wabash enigma continued to tax the patience of saints and the resources of sinners. Since the Gresham decision in December 1886 the company's lines east and west of the Mississippi River had been operated as separate roads, each with its own receiver. Gould had no intention of allowing this separation to become permanent. The reorganization plan he had formulated in November 1887 included concessions designed to reconcile all interests. Most observers hailed the plan as a fair compromise, but a group of bondholders decided to challenge it with a suit and, to everyone's surprise, won a decisive victory in February 1889. Emboldened by success, the bondholders announced a scheme to bid against the purchasing committee at the foreclosure sale in May. The fight seemed far from over.[55]

On the Katy it had begun in earnest. Bickering over procedure scuttled the joint committee's work, as did the insistence by the Katy's new board that its investigators prepare an independent report for the stockholders. Their report rehashed familiar charges that under the lease of 1880 the Missouri Pacific had permitted the Katy to deteriorate physically, diverted traffic belonging to the Katy to its own lines, charged excessive terminal fees, and burdened it with excessive construction costs. Gould denied these allegations and blamed the Katy's woes on other factors. New competitors had sapped business and forced down rates; hostile state legislation had also hurt earnings. The Katy's bloated debt made it difficult even to earn interest at prevailing rates, let alone provide funds for equipment and maintenance.[56]

Gould thought he had done what he could for the Katy under difficult circumstances. In 1886, for example, he had spent a million dollars on new equipment for the road. Although his policy was to route as much traffic as possible over Missouri Pacific lines, Jay was quick to cite figures showing that the Katy received more business from Gould's roads than it delivered to them. The Katy may have suffered from discrimination at Gould's hands, but the root of its trouble lay in the problems endemic to older roads in the changing competitive arena of the late 1880s.[57]

Having failed to reconcile their differences through the investigating committee, both sides renewed the request for a receiver. After prolonged hearings Judge Brewer resolved the muddle by appointing two receivers for the property and the road was formally separated from the Missouri Pacific system. The change presented Gould with two problems: he had to formulate a policy for traffic relations with the Katy, and he had to decide what to do about the International.[58]

Those who imagined that illness had left Gould too weak to conduct business were soon undeceived. A week before the Katy receivers were appointed, he was placidly touring a flower exhibit in White Plains and admiring the dahlias of the Reverend C. W. Bolton. That same afternoon the Street was astonished to learn that he had acquired control of the Cotton Belt. The move was a typical Gould blow, delivered at the moment it was least expected. No one knew of the alliance he had made with Fordyce months earlier or even suspected his influence in the road. In one stroke Gould had eliminated the Iron Mountain's most formidable competitor and realized two crucial objectives. Rate cutting between the two lines ceased at once, as did construction on the Cotton Belt's extension to St. Louis.[59]

Gould well knew what effect the purchase would have on the Katy's new management. It showed him, in the words of one analyst, to be "so careless" of the Katy's position as to willingly "make his new acquisition a substitute for it in his Southwestern system." He was in fact taking quiet steps in that direction. "When the KT is separated from the MoP system," he asked Clark, "should we not send our Ft Scott business from St Louis via Garnett?" None of his moves were overt or hostile; Gould had no wish to antagonize the new Katy management until he saw which way they moved. Meanwhile he was content to watch and wait until his adversaries were ready to trade or an opportunity arose to steal a march.[60]

The waiting game would not do for the International, which remained a puzzle to Gould. In the legal battle over the road's stock, the court gave the Katy's receivers temporary control of the disputed securities but not of the road itself. Uncertain about the road's true value, Jay hesitated over what course to pursue. It lacked adequate rolling stock, many of the locomotives were worn out, and the deficits kept mounting. Gould had loaned the road $250,000 in the spring; unable to borrow elsewhere, the executive committee asked him for a like amount to pay the November interest. "I am at a loss whether to keep on paying or let it go," he wrote Clark. "Is the company gaining any money or likely to?" Finally he agreed to loan the money in return for a lien on the company's property.[61]

Later it would appear that Gould was merely playing his favorite game of creating a claim he could use to advantage in legal proceedings, but in private he sounded more like a man anxious to avoid a large financial loss. "What is Eddy doing with this property?" he asked Clark in January. By that time he had advanced over $700,000 to the International on Eddy's assurance that "he could run the road at 70% & that I would get my money back. Is it possible by stopping every thing in the way of expenses or expenditures to save enough between this & April to pay me up . . . I begin to be worried about the outcome of this property." A month later he would plunge the International into a corporate and legal battle of such novel dimensions as to provoke a legislative investigation.[62]

Did he initiate this fight merely to save his investment, or was he playing a deeper game than he admitted to Clark? Here was Gould's peculiar brand of deception at its best. It was his way to paint objectives in a variety of colors and to withhold the full canvas of his motives even from those nearest him. He told them everything necessary to accomplish their purposes without disclosing everything about his own. The point was not to conceal or trick but simply to reveal no more than was required. Deception was for Gould not only a technique but a style, a way of dealing with a world that had taught him some harsh lessons.

It was a style so obvious as to be subtle, a gallery of intricate gestures, conceits, mannerisms, and poses decorated with wit and lodged within an exterior so drab and unassuming that its presence went unsuspected by the ignorant or unwary passer-by. Those close to Gould learned his style without entirely comprehending it; few outsiders grasped the least bit of it and so regarded him as something beyond their ken. For them Gould would remain an image spread to infinity across a hall of mirrors.

Deception was a useful tool for a frail, undersized man struggling against a hostile world. It could shield him from pain and allow him a degree of privacy not otherwise attainable. What it could not do was alter the realities it cloaked in different guises. Gould could hide his illness from the world but not from himself, and he knew the day would come when he could no longer do even that. Neither could he pretend any longer that Helen's condition would improve. Her decline had been gradual but inexorable. Like Jay she bore her suffering with a silent dignity that allowed the family to hope for her recovery even after a mild stroke late in the summer. On November 6, however, a second stroke left her paralyzed. The doctor offered no hope of recovery.[63]

31

Losses

When the end came it found Mr. Gould sitting at his wife's bedside holding her hand. She did not recover consciousness and passed peacefully away. During her entire illness Mr. Gould was at her side or within call, except when matters of vital importance called him away. The long ordeal had changed him very much. It has added to the slight stoop in his shoulders and increased the careworn look in his face. Mr. Gould himself is not a well man. . . . His wife's death is a great blow to him.

—*New York World*, January 14, 1889

The smooth and plausible story which George Gould put in circulation . . . to the effect that Mr. Hopkins had retired because he wanted to rest, was laughed at. Well-informed men have learned during the past two years something of what manner of man young Gould is. . . . They point out that the young man is animated, just as his father used to be, with a belief that he can run all the diverse and gigantic enterprises . . . and conduct all his speculations without any assistance from able and trusted lieutenants. . . . Mr. Gould, it is pretty well known, discarded G. P. Morosini to please his son. Mr. Morosini felt the treatment very keenly, but he never uttered a word against Mr. Gould.

—*New York World*, February 10 and February 11, 1889

The so-called Clearing House plan seems quite as feasible as other agreements which have been entered into for the same territory. . . . Probably there will be more objections to other plans than to this one. What we should do is to restore rates quickly and agree in a high sense to maintain them without waiting for any permanent arrangement. . . . While we are hesitating rates in many places are continually dropping. How much lower can we see them go before we waive personal views and fears of some inequalities which might attach to a tentative agreement.

—E. F. Winslow, letter to C. E. Perkins, December 4, 1888

The railroad situation to-day is one of simple anarchy. We need law and order. We resemble nothing so much as a body of Highland clans,—each a law under itself,—each jealous of its petty independence, each suspicious of the other, but all uniting in their dread of any outside power which could compel obedience. . . . In other words a railroad Bismarck is needed.

—Charles Francis Adams, Jr., "Memorabilia *1888–1893*," December 23, 1888

AFTER THE stroke Helen was not expected to live the week but she lingered a month and then another, paralyzed, helpless, unable to speak more than the single word "yes." For a quarter-century this woman with a voice as quiet as the clothes she favored, the colors she preferred, the texture of the life she led, had worn duty like a millstone until she sank uncomplaining beneath its weight. In these last weeks she showed something of her husband's will, his tenacity in the face of all odds. She would have lived if only to spare her family the pain of her death, but on a Saturday evening in January 1889 she lapsed into unconsciousness and died the next night. She was not yet fifty years old. Gould did not have to trouble with arrangements; five years earlier he had bought land and erected a mausoleum at Woodlawn Cemetery. It was a magnificent edifice, an Ionic temple of granite and marble set on a generous tract of land overlooking a lake and the plots of some men who would make intriguing company for Gould through eternity: Sidney Dillon, Joseph Pulitzer, and William C. Whitney. The temple was unadorned except for a stained-glass window at the back and a simple nameplate for its only occupant.[1]

Jay had lost the only woman he ever loved, at a time in his life when he most needed her. His bent, forlorn, pallid figure was a pathetic sight during the services, stifling pain he did not know how to release. "I am glad you are so happy in your domestic relations," he later wrote George. "Mine were equally so & I really think I loved your mother a little better every succeeding year during our twenty five years of married life." A void opened in Jay's life that work could not entirely fill. Young Helen saw it and rushed at once into the vacuum, taking charge of the household and of the younger children, assuming domestic duties as her mother had done so that Jay need not think of such things. As she became Jay's companion, he began to educate Helen in the ways and pitfalls of business. He encouraged her charitable and religious work and on Sundays accompanied her to church when his strength permitted.[2]

Less than a month after Helen's death Gould suffered a loss of quite another kind when Hopkins abruptly resigned all his offices in the Gould system. He had been with Gould a dozen years and was widely regarded as his most capable lieutenant. Except for the venerable Sage, he had also become Gould's most trusted friend and adviser. In that capacity he had succeeded Connor and Morosini, whose business association with Gould ended with Jay's retirement from the Street. Ever discreet, Hopkins kept silent on his reasons for the decision. George, who had become his father's spokesman, said the parting was amicable.[3]

The Street gossips told different tales. One version had Hopkins leaving in resentment over a cut in salary; others made George the villain of the piece. "Mr. Gould is pushing George forward all the time," said a broker flatly, "and anyone who stands in his way has got to go." A variation on this theme alleged that George had plotted Hopkins's downfall as he had that of Morosini and other friends of long standing because he was jealous of their influence over Jay. George's elevation to high offices in his father's corporations at so tender an age did not endear him to older hands. The heir apparent had ruffled some feathers with his lack of deference to his elders. One of the stories floating about the Street concerned a clash between George and Sam Sloan at a Missouri Pacific board meeting in which the crusty Sloan barked, "Sit down, young man! You'll know more when you're older!"[4]

The precise reasons for Hopkins's departure never emerged. On one point the record is clear: the salary cuts were not aimed at him or anybody else. The desperate state of Missouri Pacific finances had forced Gould into Draconian measures to slash expenses. He instructed Clark to impose a temporary 10 percent reduction on *all* officers and employees earning $100 per month or more. However, Hopkins felt the pinch in other ways; in January 1889 Jay and George took over certain duties for which Hopkins had received $6,000 a year. While it is unlikely Gould wished to lose Hopkins, let alone to force him out, Hopkins knew as well as anyone the importance Jay attached to the grooming of his son as successor. Most likely the financial loss, coupled with the sense of a diminished role, convinced him the time had come to move elsewhere. The separation appears to have been friendly, and Hopkins remained on the boards of Gould's corporations. Nevertheless, he was an able man whose advice and talents would be missed in difficult times.[5]

Clark did not leave, but he was unhappy. It was not the salary cut that bothered him; Gould had helped him make more money than he would ever see in paychecks. He was feeling overworked and unappreciated because Gould was on the warpath again over economy. The fall figures on the Missouri Pacific system were so bad that no amount of fudging could make the annual report into anything more than a dirge. Alarmed, Gould sent flocks of blue jays winging westward to St. Louis crammed with data, details, suggestions, comparisons with other roads, urgent pleas for retrenchment. "The way to reduce expenses," he hammered, "is to *reduce* them." It was too much for the sensitive Clark to assimilate. Whatever he did never seemed enough; Gould demanded more than he could possibly give. Discouraged, beset with anxiety, Clark poured out his feelings not to Gould but to an old friend, Dodge:

> I cannot allow the physical condition of this property to decline. . . . In the long run this will tell. Meantime there is much fault found with the

> expenses, etc., so much so that I am almost discouraged. True, I feel we had perhaps too much passenger train service, but with all the extra work put upon me in way of making settlements of old disputed matters, is it not reasonable some things should be overlooked? . . . Have tried to look after and develope [*sic*] the local business giving but little of my time to through traffic believing the life of any road must depend upon its local trade. Am I not right?

Still Clark was unconvinced Gould would see things in this light. "I am blue," he confessed, "blue as blue can be."[6]

There was justice in Clark's complaint. He had shouldered enormous responsibilities under difficult circumstances. Several reasons for the system's poor performance, notably the cluster of evils spawned by endless rate wars, were beyond his control. He was expected to keep the road healthy and efficient on rations fast approaching a starvation diet. His plight was far from unique: every road had its own version of this conflict between the chief operations man and those responsible for the company's financial showing. Clark's job was to maintain the system in good physical condition, operate it efficiently, and cultivate business, which required large expenditures for maintenance, improvements, equipment, and facilities—to say nothing of expansion. Gould wanted a first-class road but he also had to pay the bills. His desire to upgrade the property always had to be balanced against financial realities. Several issues of bonds were about to fall due, and Gould wanted them refunded at lower rates of interest as he had done with Missouri Pacific firsts in 1888. He could not do this if bankers thought the company's finances were in disarray.[7]

Far from bleeding the company to pay dividends or line his own pocket, Gould was pursuing what was universally regarded as sound, conservative policy. His plight was symptomatic of a disease raging in epidemic form among railroads. The finances of every major road were deranged by the vicious cycle of expansion and ruinous competition. Nearly all had cut dividends and some were struggling to stave off bankruptcy. Gould's situation was unique only in that he had a past to live down. Years of effort on his part had failed to establish his credibility among investors, analysts, and business rivals. They eyed his every move warily and, doubting his sincerity, shied away from agreements or proposals suggested by him. Distrusted by his peers, isolated from them by his indifference to society and to the pleasures that might have brought him into association with them, Gould went his lonely way in search of the legitimacy that would always be denied him.

Gould fought to preserve the financial integrity of the Missouri Pacific because the stakes involved not only his personal fortune but his credibility as a businessman. Ill, haggard, anxious, heartsick over the loss of Helen, he immersed himself in work because it had to be done and that was all he knew to do. The torrent of letters flowing to St. Louis must have been as wearisome for Jay to write as they were for Clark to read. They were brief but dense, the product of numbing hours hunched over his table with the green cloth, poking and picking among mounds of dispatches, reports, telegrams, letters, and clippings, sifting and searching for nuggets of insight. The shrewd, calculating intellect Gould once pitted against the market now embraced the intricacies of cost accounting with the same mastery of detail.

Figures, like maps, described for Jay the realities of a railroad system that mattered most. The salary reduction for officers showed him an amount saved without a sense of the hardships imposed. His surgeon's eye ran down the payroll and sliced with deft, precise strokes:

> I notice we are paying S. D. Barlow $3,000 per year as secy of Iron Mt Rd. . . . I should think Geo. C. Smith could perform them & save the salary. As I understand . . . you have also dispensed with Mr Haynes as Inspec of Loco & Machinery. H R Holbrook Eng Pueblo will I suppose be dispensed with soon. Chief Surgeon at $6,000 looks high. J H Richards should be dispensed with. Why should we pay 74.50 per mo for Ast Supt of Tel at Marshall Tex. J. M. Moore Atty at Little Rock 416.16 should be dispensed with also Asst Atty at Moberly Mo 99.50 per mo. Engineering force at Hot Springs & Pueblo I suppose will be dispensed with by this time—also 200 per month paid to Atty at Greenwood also Asst Eng Huggins & Hanson at Van Buren Ark 250 & 120 per mo.

He ordered Clark to prune the Cotton Belt payroll of deadwood the same way. "Wherever one salaried officer can do the work of two," he added, "keep the best man & let the other go & the same on our own roads."[8]

The losses on passenger service appalled Gould. Calculating that on the entire system in November 1888 freight earned 78 cents per train mile while passenger service *lost* 30 cents, he sent Clark detailed performance analyses for the worst branches along with lists of routes where mixed trains might be run to advantage. "I hope you will not think my letters a bore," he apologized. "The gravity of the situation must be my excuse." Mixed trains struck Gould as an ideal solution on unprofitable branches; lines that did not pay their way would be abandoned if necessary. "A reduction of $200,000 per month on the system," he observed, "which is entirely practicable will straighten out our finances." The suggestions poured forth relentlessly for three months until Gould broke down again. He hoped to go west that spring but did not feel well enough to attempt it. In March he had Munn put on the Missouri Pacific board, which enabled the doctor to attend meetings without arousing comment. To conserve his strength Gould summoned the executive committee only once a month, sometimes once a quarter.[9]

The annual report proved less grim than Gould had feared, and the depressed state of railroad affairs in general cushioned its impact on the Street, but the May statement roused Gould to another outburst. He had ordered expenses curtailed, yet they continued to mount. Other roads showed large decreases; why not the Missouri Pacific? Why did Clark not clamp down on his superintendents, who apparently had not taken the decree for economy to heart? "It will not do to give our Supts discretion in the matter of expenses," he lectured, "especially as when in this case it is eating into the vitals." Superintendent

Ricker of the Iron Mountain was "not a money maker" and neither was Harding, the superintendent at Wichita. Clark knew from experience that Gould's threats were more bark than bite. Neither Ricker nor Harding, whose management Gould called "a great failure," lost their jobs, but the traffic manager did. W. H. Newman, promoted to third vice-president two years earlier at Clark's urging, had disappointed everyone. In May 1889 he resigned for reasons of health, but the evidence suggests that the parting was neither pleasant nor voluntary. Newman's performance had displeased Gould, soured Clark, and antagonized officers of connecting lines.[10]

In September the salary cuts were restored. A month later Gould finally got off for an inspection trip. He came home in an expansive mood. "I write to say how much pleased I was . . . with the whole appearance of our railroad system," he told Clark. "The physical condition of the property is better than I have ever seen it and the money spent all shows." If Clark took heart from this praise his relief was short-lived. By the year's end Jay had uncovered evidence that someone was obtaining car contracts by paying off either the Missouri Pacific's purchasing agent or its master car builder. Since the purchasing agent was Abram Gould, Jay's reaction was predictable. "I cannot believe my brother would take commissions," he wrote, "hence the matter seems to lie with the other party." A worse scandal was brewing in the traffic department. During his tour traffic agents everywhere complained to Gould about business lost because they were not allowed to offer rebates, yet the July statement showed $90,000 paid in rebates for that month alone. Who was pocketing the money? The freight department displeased Gould as well; he wanted it investigated and reorganized.[11]

In December Clark finally wilted under the pressure and went south for his health. In his absence the Goulds transacted all necessary business through George Smith, the secretary in St. Louis. They could not have done otherwise, but Clark's insecurities rushed to the surface when he found out about it. Hurrying back to St. Louis, he asked Gould for an explanation. Once again Jay was obliged to soothe his friend's anxieties; this time an edge of impatience crept into his voice. "Our object in sending communications to Mr Smith during your absence was our solicitude about your health," he replied. "Once [and] for all nothing can shake our confidence in your [*sic*] or interfere with your authority and I greatly regret our well intended act should have received any other construction."[12]

I could hardly go to our stockholders and ask them to take Income bonds at par for their cash assessments, when the Committee were proposing to take what was left at 37½. I thought when the compensation of the reorganization committee was fixed at $500,000, that it should have been more than ample.[13]

Like other railroad men Gould sensed that the era of free-wheeling expansion had ended. The task now was to integrate his sprawling tangle of lines into a coherent system. To compete with other roads Gould needed efficient service on routes attractive to shippers, but until the pending reorganizations were resolved, he could not be certain which of the lines involved would emerge as friends or foes.

The long battle over the Wabash reached its climax in May 1889 when the purchasing committee outbid the minority bondholders to acquire all the eastern divisions as one entity. Six weeks later these were merged with the lines west of the Mississippi into a new company. Gould's voice remained dominant in the policies of the reunited road. Fixed charges had been scaled down considerably, though not as much as Jay would have liked, and several unprofitable branches had been lopped off. The reorganization was a victory for Gould in that he retained control of his eastern outlet, but it cost him dearly in time, energy, and money.[14]

The Katy fight showed no signs of abating. In mulling over the situation with Clark, Jay reiterated that he would lose his investment in the International "unless it can be scraped up out of the road. It might be well to leave pay rolls & everything else & let Eddy pay me up. The Receiver will take care of the pay rolls & bills." Gould could then use the funds to buy control of the second mortgage bonds. "What the company owes me would buy nearly a million of them," he figured, "& as the principal can be declared due—title can be made to the road & the controll in this way secured to the Iron Mountain where it really belongs." While the battle for possession of the International's stock dragged through a New York court, Gould decided to try an end run in Texas. After the Katy intervened, the judge gave Gould the receivership he wanted—but in a state rather than a federal court.[15]

State receiverships were unfamiliar ground for Gould and his lawyers, but they adjusted readily. The Katy appeared to have won the legal skirmish for control of the International's stock and prepared to replace Gould's board with its own men at the election on April 1. However, Jay had a little April Fool's joke in store for them. James S. Hogg, the state attorney general, enjoined the Katy from voting its International stock on the grounds that control of a competing line violated state law. As Gould well knew, Hogg had won election in 1886 on a platform lambasting the domination of Texas railroads by "foreign" corporations. The legislature was about to pass a bill requiring all carriers to maintain their offices, shops, and accounts within the state. Hogg's crusade was directed primarily against Gould's influence, yet in this instance he unwittingly served Gould's cause. The request for an injunction had come from one of Gould's local attorneys who took pains not to disclose the identity of his client or the fact that his public-spirited action had other motives behind it.[16]

This choice display of Hogg-tying left the International in the hands of Gould's board. There followed months of sparring in which Gould offered one reorganization plan, the stockholders another, and the foreign bondholders a third. The stockholders' committee, headed by banker Frederic P. Olcott, vowed to save the road from dismemberment but could not persuade all sides to accept drastic surgery even though fixed charges were about $2.84 million against net earnings of $1.5 million. Gould wanted no part of the Katy unless charges could be scaled down to the level of earnings. Meanwhile, the Katy acquired a line to Kansas City and talked of building to St. Louis. In November the Olcott committee cut through the thicket with a new plan similar to one Gould had proposed years earlier. When he and Sage pledged support, efforts were made to resolve all outstanding suits between the two companies and agreement was reached on the International. In February 1890 Gould bought the International's stock on the understanding that a half-interest would be transferred to the Katy. The committee's progress encouraged the belief that the fight was over, but technicalities in one form or another would delay final agreement for another two years.[17]

At precisely the time Gould and Olcott joined forces in the Katy reorganization, they were locked in battle over the Cotton Belt. Gould and Sage had acquired control of that road by purchasing $6 million in second mortgage bonds and a block of stock for about $2 million in cash. In searching for ways to scale down the company's obligations, Jay resorted to what had become a familiar tactic. The Central Trust Company, of which Olcott was president, had as trustee for the first mortgage issued coupon-bearing certificates which were obligations against itself as trustee. When the coupons fell due in May 1889, they were duly paid by Mercantile Trust as Gould's fiscal agent. Instead of canceling the coupons, however, Mercantile kept them alive as a lien against the road. About $140,000 out of $490,000 in coupons was handled this way before the Central Trust caught on and demanded that the coupons be canceled or surrendered.[18]

The marvel of Gould's style lay not only in its originality but in his ability to play the same game so many times with fresh comers. "To be fooled once, or even twice, by a certain set of circumstances does not argue perhaps a lack of sagacity," moralized the *Chronicle*, "but to be fooled over and over again . . . subjects one at least to a charge of infatuation." Olcott was not amused. Central Trust applied at once for a receiver and the court appointed Fordyce. The first mortgage bondholders, mostly Germans, formed a reorganization committee headed by Olcott. In the ensuing contest Gould followed his old script of pressuring the bondholders by means sweet and sour into accepting a lower rate of interest. Early in November he and the committee agreed on a plan that left control of the Cotton Belt in Gould's hands. That object secured, Jay offered concessions in the affable manner of one unwilling to quibble over details. He balked only when the committee proposed taking for itself all surplus securities at low prices.[19]

The plan was released in January 1890. Analysts hailed it as eminently fair to all concerned, partly because it appeared to eliminate Gould and the Missouri Pacific from control. Once again they misjudged Gould's role in the proceedings, just as earlier they had welcomed the appointment of Fordyce as an impartial receiver, unaware that he had long since aligned himself with Gould. The presence of a committee of bankers served to reassure investors that the road would be saved from Gould's clutches and protected from his depredations. It seems not to have occurred to them that Gould might defend the road from the bankers, or that he could work as easily with them as with anyone else. He did precisely that in all four great reorganization battles of these years, and in three of them obtained the results he desired.[20]

Few observers suspected how closely Gould worked with Olcott's group to shape the reorganization. He fended off an attempt by the committee to create another voting trust that would interfere with his control. It was not public knowledge that, apart from his large interest in second mortgage bonds, he had acquired 52,000 shares of stock by February 1890. Later he purchased some first mortgage bonds as well. These holdings enabled him to orchestrate the details of reorganization. After a foreclosure sale in October 1890 the road was reorganized as the St. Louis Southwestern Railway. Under the agreement Jay had the right to name six of its nine directors. Fordyce was made president and Edwin Gould vice-president. About the only thing Gould did not get was the name of the new company; he wanted it called the "Cotton Belt" because "the road is generally known by that name now."[21]

In his search for improved routes Gould looked hardest at the region between southern Kansas and the Gulf. Completion of the Arkansas Valley line would connect the Kansas-Colorado roads to the Iron Mountain system without relying on Katy tracks. It would not, however, satisfy Gould's desire for an efficient route between the plains states and the growing markets of the Southeast. His lines were too circuitous to compete for this business with other routes. A new short line was needed between the Arkansas Valley and the Texas & Pacific in Louisiana. The Iron Mountain also required a more direct line to New Orleans.[22]

During 1889 Gould pushed these projects as vigorously as his pinched finances would allow. By spring his pessimism had vanished and he urged work on the Arkansas Valley forward as rapidly as possible. The new line opened in November, giving Gould an unbroken route from Pueblo to the Mississippi River. He also asked Clark to explore alternative routes for a short line between Fort Smith and the southern end of the Iron Mountain. The town of Fort Smith loomed large in Gould's plans for two reasons. It was the logical terminus for the short line south and potentially the heart of his favorite industry. As a developer Jay never forgot the lessons learned during his Erie days about the symbiotic relationship between coal and railroads. In Colorado, Texas, Missouri, Kansas, Arkansas, and the Indian Territory, Gould's agents combed the countryside for coal lands. For all the talk about new through routes, nothing did more to influence the course of his construction program than the location of coal fields. The network of branches around Fort Scott, Kansas, for example, was built to eliminate dependence on Katy track but also to service the mines at Rich Hill.[23]

Fort Smith had abundant coal of excellent quality. In tests ordered by Gould, it gave trains 46 percent more mileage per ton than coal from other mines. "Two tons of Ft Smith are equal to three tons of any of our other coals," he enthused, "and besides the important feature of being smokeless it will largely take the place of anthracite at many western points for domestic use." The proposed short line would expedite not only through traffic but the hauling of coal southward to Texas and other markets. However, instead of commencing work on the short line, Gould moved to shorten the Iron Mountain's route to New Orleans by constructing a new road between McGehee (near Arkansas City) and Alexandria on the Texas & Pacific line. This was a difficult and costly project, largely because three rivers had to be bridged. Construction problems delayed completion of this 190-mile line until the last day of 1891 but did not slow Gould's vision of his company's future in New Orleans. In December 1889 he responded warmly to a proposal that all the roads entering the city combine to build a belt line.[24]

Traffic from the Iron Mountain and Arkansas Valley lines could also connect at Memphis with the Richmond Terminal system, a sprawling conglomerate of lines penetrating most of the Southeast. Gould saw great possibilities in Memphis. He knew the financiers behind the Terminal were desperate to find business for their white elephant Memphis line. Early in 1889, therefore, he negotiated a pact to exchange traffic with Terminal roads at Memphis and Arkansas City. Elsewhere the same relentless urge to extend, connect, improve, perfect, left its imprint. In Omaha the belt line still did not connect with either the stock yards or the Nebraska City road. Another road offered to buy the Missouri Pacific's terminals in Omaha, but Gould could not bear to let anything go that promised a future. "I think we will stand stronger in Omaha to own & control our own terminals," he told Clark. Then it dawned on him that the Union Pacific owned track from the stockyards to the junction toward Nebraska City. Recalling that Adams wanted access to some gravel quarries on the Missouri Pacific line, Gould offered to swap trackage rights. He also asked to lease part of another Union Pacific branch that would solve another connection problem. He could construct the link, Jay added, "but it seems cruel to build an independent line when there is not business enough for the road already there."[25]

Gould knew his man. Adams could not resist the appeal of sweet reason. It took only one meeting between them to reach agreement on the trackage rights, a striking contrast to the stalemate in St. Louis over the proposed terminal association. There Gould found himself in a race against time. At Alton, twenty miles above St. Louis, a new bridge was going up. Critics hailed it as a death blow to Gould's stranglehold on traffic entering St. Louis. For years merchants and railroads alike had denounced the Gould monopoly for its high bridge tolls, excessive terminal charges, and delays. All the roads converging at East St. Louis were obliged to detach their locomotives in favor of the transfer company's engines, which hauled the trains across the river to Union Depot.[26]

The new bridge threatened to lure roads away from the Gould facilities. Jay discovered in September 1889 that its representative was urging other railroads to "make no contracts as their Bridge would furnish them a cheaper outlet." Through a bit of intrigue he learned too that the Merchants Bridge Company had been scheming for months to gain access to Union Depot. His source was John O'Day, general manager of the Frisco, who had quarreled with his president and was about to leave the company. Anxious to ingratiate himself with Gould and Sage, O'Day emptied his bag of confidences about the Frisco: its secret agenda for expansion, financial woes, and plots detrimental to Gould's interests.

As owner of $2 million worth of Frisco stock Gould was attentive to this intelligence.[27]

Gould did not fear the Frisco's expansion plans because he knew the road was in desperate financial straits, but the bridge was another matter. For nearly three years Gould had endeavored to enlist roads in his proposed Terminal Association as joint proprietors of the bridge, the tunnel, Union Depot, and related facilities. The Pennsylvania hung back despite Gould's efforts to persuade its president otherwise. In October 1889 Jay decided to wait no longer. Six roads had agreed to join, and Morgan assured him that the Big Four line would come in. The Terminal Association approved a $5 million bond issue to build a new union station and switching yards.[28]

Perhaps the most astonishing thing about all this work was Gould's ability to generate the same enthusiasm for every project, large or small. Through much of 1889 he had preached economy until the sermon sickened Clark, yet for Gould economy never meant retrenchment in the broadest sense. Like a good general he was incapable of viewing retreat as anything more than a temporary stratagem. In these last years he showed the same streak of optimism, the same determination to achieve despite all obstacles, that had set him apart as a youth. Gould's drive was elemental, a force of nature churning relentlessly onward, unstoppable except by death. It was a quality common to the titans of business, a magnetism that made them at once compelling and repugnant figures. Journalists gestured vaguely at it, but only the novelists—Frank Norris, Robert Herrick, and, above all, Theodore Dreiser—captured their heroic dimensions, the epic proportions of their deeds, the complexity of their personalities, their inexorable drive.

So it was that Gould, depleted by disease, crushed by the loss of his wife, beset with a host of business problems, applied himself all the harder. When his strength faltered, he continued to work flat on his back. Around his desk swirled a world of smaller matters that received no less attention than major concerns. A connecting road wanted faster joint service to compete with a rival line for meat and fruit shipments; the World's Fair Committee in St. Louis requested a contribution and use of the road's facilities; a business acquaintance wished the Missouri Pacific to give orders to his son's firm. An employee's mother fretted about her son being too sick to work and was told by a sympathetic Gould, "We are very sorry to hear of your sons illness. Do not let him come to the office till he is well enough." Reid Northrop, now president of American Refrigerator Transit, ventured to make decisions without consulting the St. Louis office. Jay chastised him gently with the reminder that ART, "if not directly under Mr. Clark's control, should be managed by you subject to his directions . . . and it is my wish that you confer with him freely and accept his directions as law."[29]

In the realm of larger affairs Gould was widely viewed as a gobbler of railroads when actually he was putting his own house in order. Earnings improved enough to maintain a quarterly dividend of 1 percent, which strengthened Gould's hand with the bankers. "We think we have gotten our corporation in a snug financial condition," he told one of them, "and that we had better keep there for the present." Gould was in his inimitable way playing better cards than he held, for there still loomed the great unsolved riddle plaguing railroad men: what to do about the rate wars that were bleeding them into bankruptcy.[30]

I hardly see how any new Association could be formed that could include the Northwestern and Union Pacific, so long as they adhere to their present attitude, and the present agreement can have little vitality left after their withdrawal and it would hardly be worth paying the expenses for keeping it up.[31]

It was a new game. Slowly, grudgingly, the players came to that realization even though they remained baffled by its rules and foibles, mystified by its inner logic. Expansion was no longer feasible, and neither was the exaggerated sense of independence that made railroad men seem more like feudal barons. Indeed, the one had doomed the other. Crowding the board with so many new pieces had produced not victory but a colossal stalemate among the competing forces, a sort of trench warfare in which losses were enormous and gains negligible. The players had discarded diplomacy in favor of war only to discover that no one could win. For nearly a decade the carnage had gone on despite frantic efforts to end it.

By the late 1880s the toll of battle had mounted to a level that threatened catastrophe. Rates dropped precipitously, a boon for shippers but disastrous for railroads. The effect on income accounts made it difficult for roads to carry the added costs of expansion, let alone to provide adequate funds for maintenance, equipment, and improvements. Dividends were curtailed, the value of railroad securities eroded steadily, and new capital grew ever harder to secure. An epidemic of reorganization fights reflected the inability of older roads to stay in the new game without scaling down their bloated capital structures. Everywhere relations among roads were poisoned and demoralized. Unwittingly the men who had created the mightiest rail system in the world had set the stage for its decline and fall.[32]

Clearly war had failed and diplomacy offered the only hope for relief, but what form should it take? The players were neither stupid nor even short-sighted. In retrospect it is easy to see how their dilemma arose, but far less easy to see how it could have been avoided. They did not want war but were helpless to find solutions to complex problems. Again and again they tried without success to adjust their differences through the machinery of diplomacy. For two decades the chief instrument of their efforts had been the pool. Everyone knew the weaknesses of pools: they had no standing in law and could not be enforced if a member refused to keep the faith, which was often. Nor could they prevent cheating in the form of rebates or other clandestine arrangements. As the number of lines proliferated, it became impossible to harmonize differences among so many competing roads. "If everybody who can patch up a roundabout line of railroad communication between two points is justified in cutting the rates," Perkins once complained, "it must end . . . in great injury to *all* the roads.[33]

There were more pools and more roads in every pool than ever before. In five years the Transcontinental Pool jumped from two to thirteen members and split into eastern and western pools. Rate wars made shambles of the arrangement during the middle 1880s as they did elsewhere. The Iowa Pool gave way in 1885 to an organization with more lines and less cohesion, the Western Freight Association, which fell apart a year later and was replaced by the Western Traffic Association. There was a Southwestern Pool and a Northwestern Pool, and one for Colorado-Utah business. Pools for freight business were separate from those for passenger business. The mechanisms were elaborate and often ingenious; the problem was that none of it worked for any length of time. The competition was fierce, the stakes high, and diplomacy among railroads no less righteous than that among nations. Tortuous hours of negotiations produced only lulls between storms and more frustration when agreements broke down faster than they could be patched up. A Burlington director expressed the prevailing mood of exasperation when he growled, "If two or three wildcat roads west and northwest of Chicago can make rates for that whole region, what are we coming to? . . . Can they be boycotted or starved into submission and compelled to do business on business principles?"[34]

Then in 1887 the Interstate Commerce Act put pools out of their misery by prohibiting them. Coming on the heels of a period of unbridled expansion, this action deprived railroad executives of their most familiar vehicle for adjusting differences just at the time when competition had grown most intense. The result was predictable. As the pools folded, wars of unparalleled ferocity broke out everywhere. It soon became apparent that clashes among enlarged systems added a new and ominous complication: the fight could no longer be confined to one region but affected rates elsewhere. By the winter of 1888 observers described the rate situation as "totally demoralized" and in "chaos." The value of railroad securities plummeted again amid another round of dividend cuts. Through long, dreary months the wars raged on with little relief and less hope of settlement. While railroad men despaired, criticism mounted even from friends of the industry. In a sweeping condemnation the *New York Sun* found "nothing in our commercial history so disgraceful" as the record of the northwestern roads during 1888. The conservative *Chronicle* charged that the "controlling power in each corporation has been wholly selfish, bristling all over with hostile purposes towards every other. No right of territory, no settlement of rates, no adjustment of business, stood for a moment as a hindrance to the insatiable craving of getting business."[35]

The desperate state of affairs was hardly news to railroad men. They were not only rushing hellbent toward self-immolation but fast losing control of their own industry. The problem was to find a remedy that was both workable and within the constraints of the Interstate Commerce Act. In vain they ransacked the past for useful models. Gould might observe wistfully that if western roads could only get the rates of six years ago "they would hardly know what to do with the money earned," but he realized the *status quo antebellum* could never be restored. What then? A few years earlier Adams had speculated that if railroad executives all lived on the same street, "a sort of Presidents' Road," and went down to their offices together in the morning, "there would not be one quarrel then where there are five quarrels now." He would soon revise that tribute to the powers of reason downward. The *Chronicle* suggested a more compelling principle on which to proceed. In its view, "Some authority over and above these differing managements strong enough to force a permanent arrangement of present rivalries . . . must come in before lasting order can be brought out of the Western chaos."[36]

Intelligent railroad officers recognized the need for some power larger than themselves but were unwilling to surrender the right of managing their own affairs. The central dilemma was but another version of an old political conundrum: the creation of a greater agency would curb the autonomy each individual guarded so jealously. Some of the best minds among railroad men wrestled with the problem in 1888. Perkins, in a confidential letter to Albert Fink, the trunk line commissioner, proposed resolving "the existing disastrous conditions" by creating "some common agency, representing all the lines, which shall have the sole rate making power." The notion was discussed but nothing concrete emerged. Over the months ideas circulated and plans were drafted. The public knew nothing of these activities until late November, when the first formal proposal was revealed. To the surprise of many, the man behind it appeared to be Gould.[37]

In his quiet way Gould had been brooding over the problem for months. The rate wars were the driving force behind the financial woes of the Missouri Pacific and his incessant pleas for economy. Aware that the problem was industrywide, he saw that old habits and prejudices must be discarded. The solution, whatever it might be, required a departure from past practices, a leap of faith by men not known for keeping the faith. His own region seemed a good place to start. Three personalities dominated the southwestern systems: Huntington, Strong of the Santa Fe, and himself. As the *Chronicle* conceded, "If these three parties therefore should determine to act in harmony, there would be no powerful interest to oppose them." And if they could agree on a plan for common action, it might win broader acceptance.[38]

Thus did Gould, the man labeled for two decades as the most notorious wrecker of values, embark on a new role as railroad statesman, a leader in the quest for stability and harmony. Late in November he hosted a series of closed meetings for Huntington, Strong, Midgley, E. F. Winslow of the Frisco, and representatives from the Iowa roads. Ostensibly they had come together to resolve difficulties in southwestern rates; in fact they were debating the merits of proposals breathtaking in scope. A committee headed by Winslow produced three alternative plans, each more sweeping than the last. The least radical, known as the clearing house plan, apparently originated with Midgley. It proposed replacing all freight and passenger associations with a clearing house managed by an executive board of three. A board composed of one representative from each road west of the Mississippi would prescribe all rates, rules, and regulations for freight and passenger traffic as well as divisions on through business. Disputes would be referred to the executive board.[39]

In effect member roads would surrender all authority over making rates. The plan also decreed that "the bidding for business by means of private concessions shall cease," thereby rendering rebates extinct. The Interstate Commerce Act was supposed to have done that, but most railroad men shared Perkins's view that "there is more cheating going on to-day than ever before, in the way of secret rebates of one kind or another." Part of the problem lay in the slippery relationship between presidents and their traffic officers. Presidents might sign agreements and pledge not to cut rates or accept rebates, but they also reminded traffic managers and agents that their jobs depended on the amount of business they procured. While traffic men used any means foul or fair to get business, presidents protested their innocence of such practices. Often they did not in fact know what was going on, but this was the hypocrisy of men who tightened screws only to profess astonishment that the screws dug deeper. On the other hand, traffic agents had long been notorious for their barnyard ethics, and the desperate scramble for business seemed to have eroded the last vestiges of their scruples. Adams later claimed to have discovered "a depth of railroad morals among freight agents lower than had ever previously existed, and that is saying much."[40]

The clearing house plan struck at the root of this evil by eliminating competition among traffic agents. Indeed, the proposal that presidents delegate all rate-making power to a central body was, in Grodinsky's words, "the most radical step in railroad unity of action ever suggested." However, neither Gould nor Huntington thought it went far enough. Believing "more complete consolidations of interests necessary," Gould offered a second plan based on creation of an "Operating Company" and Huntington a third on the premise of an "Owning Company." The details of these plans were never disclosed because they were, in Winslow's tactful words, "so comprehensive as to require further time . . . to insure their successful presentation to the various companies interested." Strong, for one, refused to entertain anything more comprehensive than the clearing house. Reluctantly Gould and Huntington abandoned their schemes and agreed to try the clearing house in hopes it might prove a stepping stone to something better.[41]

But the clearing house also foundered. Apprised of its provisions, Perkins pronounced the scheme "Quack medicine." The plan "certainly has good features however," he conceded, "& it may be worth trying if all come in. But there is the rub." His perception was accurate. Hughitt thought the plan too strong, the Alton considered it too weak, and Cable of the Rock Island found it "insufficient or lacking in permanency." Hughitt exemplified the difficulties involved in reaching agreement. When the meetings began, he pleaded with Midgley to "do everything possible to bring any kind of agreement about." Ten days later he dismissed all the plans as "chimerical"; all that was needed, he huffed, was for the presidents to agree among themselves to maintain rates. For himself, Hughitt wished "to be left alone, to transact business by legitimate means, without interference of others, or interference with others." The press promptly labeled the plan a "trust" similar to the industrial trusts that had become anathema to the public. Reporters also attributed the plan to Gould, and the impression stuck despite Jay's denials that he had anything to do with it.[42]

Attaching Gould's name to the scheme amounted to a kiss of death. Another meeting was scheduled for December 19 but some western roads hesitated to attend. Hostile papers attacked Midgley as Gould's puppet and suggested that Gould wanted harmony to "cause a boom in stocks so that he can sell out his large holdings at a profit." A Burlington official was convinced the articles emanated from a Rock Island officer "who is doing all he can to prevent any combination of such lines, and especially on any line proposed by Mr. Midgley." The western associations put Midgley on the spot by scheduling their own meeting for December 19. As commissioner of the Southwestern Association Midgley was obliged to attend. To emphasize the point, several officers of western roads were reported to "emphatically declare that they will not be at the [New York] meeting, nor join in any agreement Mr. Gould may propose."[43]

The clearing house plan faded from view but the problem remained. "While we are hesitating rates in many places are continually dropping," Winslow despaired. "How much lower can we see them go before we waive personal views and fears of some inequalities which might attach to a tentative agreement." Perkins wondered, might it "not be best in the end for the strong Roads to let *palliatives* alone & just let the disease run its course?" Huntington took issue with Adams's view that the time for wholesale consolidation had not yet come. "When there is only one railroad company in the United States," he snapped, "it will be better for everybody concerned, and the sooner this takes place the better."[44]

Despite brave talk of impending settlements, a sense of rage and frustration prevailed everywhere. On Gould's own Missouri Pacific Sloan lashed out in his acerbic way, "Some of us will go to prison if we do not pay some attention to the wishes of our stockholders." He demanded and got from the executive committee a resolution instructing Clark to increase rates and "take no business that will not be remunerative; and to invite the cooperation of such other lines as are willing to join him in this end." It was a fine gesture, but individual action was no longer adequate to the crisis. The fact that trunk line rates were also demoralized added urgency to the situation. Gould still wished some sort of joint action but the clearing house debacle had eliminated him from any leadership role. Instead, that position was assumed by the one man in whom most railroad men had confidence: Morgan.[45]

What Gould could not do, Morgan did. On behalf of three banking houses he invited the western presidents to meet at his house on December 20. The response was impressive. Seated around the large table in Morgan's dining room were the host, John Crosby Brown of Brown Brothers, George C. Magoun of Kidder, Peabody, Gould, Perkins, Hughitt, Adams, Cable, Winslow, and Frank Bond of the St. Paul. Strong did not attend but Magoun, a director, looked after the Santa Fe's interests. Cable offered a plan to maintain rates which Adams dismissed as "merely the old story once more . . . to bind the railroads together with a rope of sand." Gould thought the same. Finally the railroad men signed it with what Adams called "expressions of contempt and a good deal of suppressed wrangling, especially between Cable and Gould." The meeting adjourned until the next morning.[46]

Afterward Adams strolled up the avenue with Gould. Shocked by Gould's "dreadfully sick and worn" appearance, he nevertheless poured out his displeasure with what had occurred. A strong organization was needed, he insisted, headed by some outside power. The bankers might fill that role, and perhaps the machinery of the Interstate Commerce Act might be employed as well. Gould jumped on the idea. "Yes!" he replied, "and why not call the Commissioners in now; invite them to meet us, and co-operate with us in developing a scheme." Jay saw in a flash that, among other advantages, involving the commissioners would remove the suspicions of conspiracy that had poisoned the earlier plans. He agreed to present the idea the next morning and they shook hands on it. There was in this meeting of the minds between Gould and the author of *Chapters of Erie* not only irony but a fitting emblem of the bizarre thread between fact and fiction, image and reality, in Gould's career. In recent years Adams had climbed into the bed of agreement with Gould more times than he cared to admit, as once again he did on the matter of what was needed to bring peace.[47]

Unfortunately, Jay could not deliver. He arrived late for the meeting and seemed on the verge of collapse. The Morgan meetings, like those called earlier by Gould, took place during the time Helen lay paralyzed. For ten weeks Jay dashed anxiously between the press of business and his vigil at the bedside of a dying wife, snatching time from one to serve the other. By late December the strain had worn him down. Whether Adams realized any of this is not known, but he saw at once that Gould "was not going to take hold. He suggested the idea, but in so weak and vague a way that it made no impression." Disappointed, Adams made a vigorous plea to involve the bankers and the ICC commissioners in the creation of a new plan. To his surprise the presidents assented except for Perkins. All agreed to maintain rates for sixty days and to reassemble at Morgan's house after the holidays to thrash out a plan.[48]

During the interval the rate agreement held except for a minor squabble between Missouri Pacific and Rock Island. Gould issued orders to Clark that if any employee cut rates in violation of the agreement, "I want him discharged at once and not re-employed." At the meetings held January 8–10 two new faces, Strong and A. B. Stickney of the Chicago, St. Paul & Kansas City, joined the talks, but no one from Southern Pacific, Alton, or Illinois Central attended. Again Gould was in no condition to exert any influence. He had entered the final week of his vigil and looked "worn, reduced and nervous, with a tired film on his eyes." In the discussions he did little more than spar feebly with the belligerent Cable.[49]

Helpless to affect the outcome, Jay watched the proceedings follow a predictable course. Morgan tried vainly to push various plans of organization toward a vote. Adams pressed the argument to involve the ICC commissioners because "everything depended on the way in which whatever we did was brought before the public." Perkins growled that he would not attend a meeting at which the commissioners were present. Bond then offered a proposal for what amounted to another pool. When Morgan asked for a vote, Hughitt protested that Bond's scheme was a pool and therefore illegal. Adams warned again of the dangers in ignoring "the prevailing sensitiveness of public opinion," a view Morgan seconded. After more discussion Adams suggested a committee to prepare a plan in consultation with the commissioners. The presidents approved this idea and appointed Adams, Bond, and Strong as members. They agreed to meet again in two days to consider the committee's plan.[50]

Although the sessions took place under a veil of secrecy, the presence of so many notable figures attracted widespread attention. Interest perked up even more when all the trunk line presidents appeared for the second meeting on January 10 along with Ashley of the Wabash and a coterie of directors, officers, and bankers from companies already represented. The reporters snooping about Morgan's house, frantic for information, snatched eagerly at every rumor and badgered the participants for statements. Their net caught a few foolish fish, including one banker who predicted smugly that "the work is already done so far as the prevention of rate wars is concerned. You will hear of no more rate wars."[51]

Inside 219 Madison Avenue could be found another brand of euphoria. The committee was ready with a plan and an accompanying report, Adams having suppressed his contempt of the commissioners long enough to gain their approval. After some haggling over details the western presidents adjourned until 2 P.M., when they gathered at the Windsor. There, to Adams's delight, the plan was approved to create an organization for "the maintenance of rates, the enforcement of the Interstate Commerce law and the arbitration of disputes among railroads." Later it would be called the Interstate Commerce Railway Association. Some observers hailed the news as the kingdom come at last. The *Sun* called it "the most sweeping reform that ever was instituted in a great commercial system. It is nothing short of a revolution in railroad methods." Even the staid *Chronicle* flirted with the term "revolution."[52]

None of this rhetoric impressed Gould. Convinced that the plan lacked strong support, he watched and waited. Late in January the presidents met in Chicago to perfect the new organization. A week's work produced little more than the election of Aldace F. Walker, one of the ICC commissioners, as chairman of the executive board. Gould, still in mourning, did not attend and neither did Adams, who sensed his grand design being whittled away by expediency. Certain roads refused to sign the agreement; others had never been invited to join. Gould was reluctant to sign unless the Katy and the Kansas City, Fort Scott & Gulf were included. The Gulf in turn demanded a pledge that Gould and his new ally, the Richmond Terminal, would not manipulate rates between Kansas City and the Southeast. Perkins suspected Gould, as "Head Devil in the New York Sun," of being behind that paper's attack on railroads.[53]

While the recriminations flew, rates began to soften. On February 20 the presidents gathered again in Chicago to finalize the agreement. Gould sent only a third vice-president to represent him. Adams attended but left in a huff when some presidents balked at signing unless every other road signed. Roswell Miller of the St. Paul offered an alternative plan, which Adams thought would "substitute a mere nothing for the weak something which had already been adopted." On the same afternoon that a sulking Adams departed for Boston, the presidents reversed themselves. The proviso requiring every road to sign before the plan became operative was dropped, and all but four roads approved the new association. "The ship is in port," Adams mused privately. "It isn't much of a ship it is true, and those on board are a mutinous set. What next? . . . I propose to take care of the Union Pacific. I am convinced there is going to be trouble."[54]

So was Gould. Although he finally signed the agreement in March, he entertained no illusions about its worth. The railroad situation, he told a reporter, was the worst he had seen in thirty years. Rates held firm until June, when trouble arose in several regions. The Alton became the first road to leave the organization, which, one of its officers sneered, "did not amount to a hill of beans." There was more trouble in the Northwest, a clash in the West between Missouri Pacific and the Rock Island, and bickering among the Kansas City roads. Adams, true to his word, took care of Union Pacific by closing a formal alliance with the Northwestern that made the roads "in all essential through traffic respects one company." Gould protested that the pact violated the Association agreement.[55]

On every front the agreement was coming apart with the usual sulfurous charges and countercharges. By 1890 the hopes vested in the Association had all but vanished. It had failed every test with machinery too weak and cumbersome to handle efficiently the load of disputes thrust upon it. The final blow came in January 1890 when Walker decreed that the Union Pacific–Northwestern pact defied both the letter and the spirit of the agreement. The two roads responded by pulling out of the Association. Walker professed not to be concerned. He called a formal meeting of the Association for February 11 to take up the withdrawals. Jay listened to the talk of compromise with a mixture of weariness and disgust. The Association lingered on, but it was as good as dead.[56]

More than a year had passed since the presidents' meetings with little to show for it. The roads remained without a plan or an organization more effective than the regional associations. They had shown themselves as inept at diplomacy as they were implacable at war. For himself Gould was no longer certain which route to follow in search of stability, so he tried a little of both. After failing to unite the roads around his plan he had yielded the floor to Morgan and the bankers, who fared no better. Tired and ailing as he was, Gould concluded that he must again play some active part in bringing about a settlement. What that role would be he was not yet sure, but he knew it would not involve the Association. He no longer had the time or the energy to waste on shadows.

"Is it worth while for us to be represented at the meeting of the President's Association in Chicago," he asked Clark sardonically, "or shall we simply send flowers for the corpse?"[57]

32

Last Hurrah

I have been interested in railroads ever since I was a boy. I now think a railroad train is one of the grandest sights in the world. I like to see the great driving-wheels fly around. . . .

—Jay Gould, interview in the *New York World*, July 2, 1887

Jay Gould is the axis upon which Wall street now revolves. Never was one man's power so great in speculation as now, and he is the one who wields it.

—*New York World*, November 26, 1890

Gould . . . never in all his history evinced the power and strength he did a month ago when he acquired the domination of the U Pacific and this was recognized amongst the strongest and best people. He had become a necessary good or evil as the case may turn out. You know you and I have talked over the Gould situation and agreed that his reputation in the community took 10% off all the securities where he was in control. Is this to be the situation again?

—W. T. Walters, letter to G. M. Dodge, undated [December 1890]

WHILE THE corpse twitched out its prolonged agony, the rate wars of February 1890 spread relentlessly despite a flow of traffic so heavy that one observer called it "phenomenal." Reluctant to enter the fray, Gould found it impossible to stay out. Midgley found all the Kansas City roads except the St. Paul and the Missouri Pacific guilty of securing freight by illicit means. The most damaging blow came when some roads cut the St. Louis–Kansas City passenger rate, a business Gould regarded as "our bread and butter." He retaliated by cutting fares west of Kansas City with a vengeance. When the western presidents hurriedly called a meeting to discuss some form of agreement, Jay declined even to send a representative. Cable snarled that Gould could "settle the whole Western rate war in five minutes if he wants to," an outburst that was disingenuous at best. Privately he offered to cease the fight west of Kansas City while letting it continue east of that city. "The plan is entirely inadmissable and I would not consider it for a moment," Jay wrote Clark. ". . . Our interest lies in having a complete settlement made." On that point he remained unyielding.[1]

By May rate structures were in shambles and at least one observer insisted that restoration depended "to a great extent, upon the course that Jay Gould will pursue." In fact he was already arranging a peace plan with Walker and a rapprochement with the new management of the Santa Fe. During the past year financial troubles had forced a turnover in which George Magoun became chairman of the board and Allen Manvel replaced Strong as president. To Gould's surprise they adopted the same policy of expansion that had already brought the company to financial grief. Amid the rate wars of May the Santa Fe acquired control of the Frisco.[2]

At another time this acquisition would have roused Gould to vigorous retaliation. Circumstances had changed, however, and he had changed with them. While the Santa Fe might use the road against him, it would also be included in any settlement and could no longer disturb rates on its own. Consolidation had the advantage of reducing the number of participants in an agreement. Instead of donning war paint, therefore, Jay sent George to congratulate Magoun and express his desire for "working in harmony with the Atchison people." This was no mere platitude; already he had wired Clark, "We think well of restoring rates both east & west of Missouri River to full tarif [*sic*] rates. The matter is referred to you with full power." A few days later Gould signed a treaty drawn up by Walker to restore rates.[3]

The agreement amounted to little more than a truce but it gave Gould time to perfect a larger scheme for peace in the Southwest. He persuaded Huntington and Magoun to join him in forming a new association. Unveiled in September, the Southwestern Railroad & Steamship Association embraced the lines south of Kansas City and west of the Missouri River. An experienced reporter hailed the compact as "one of the shortest and strongest documents of the kind every formulated." Those who doubted the sincerity of Gould's desire for peace were astonished at his willingness to bind himself to a stringent agreement. In fact he had been advocating the necessity of strong contracts for two years, but as usual his arguments were dismissed as smokescreens for some other purpose. Gould's feat in creating the new association forced his peers to take his ideas more seriously. If successful it might serve as a model for other regions; at the very least it offered a glimmer of hope that railroad men might be weaned away from past habits to fresh approaches.[4]

It is very important for the Texas & Pacific to hold if possible, the control of the long-leaf pine, as it will furnish us a very large amount of traffic in the future.[5]

In mythology Atalanta was the virgin huntress who, being swift of foot, refused to marry anyone except the suitor who could first defeat her in a race.

None succeeded until Hippomenes dropped along her path three golden apples given him by Aphrodite. Atalanta paused to collect the apples and lost the race.

No one knows what intrigued Gould about this legend or what lesson he drew from it. The source of his fascination died with him, like the secret of *Rosebud.* He gave the name to both his yacht and the new railroad car built for him in 1887. It was a magnificent vehicle, oversized, paneled with carved mahogany and curly maple, handsomely appointed with plush carpeting, silver plumbing fixtures, and stained glass windows above the transoms. Within could be found four staterooms, snug servants' quarters, a galley, a large dining area at the front, and a spacious observation room with desk and bookshelves at the rear. Gould modified the original design by lopping three feet off the dining area and adding it to the observation room, which served as his traveling office. Throughout the decor was elegant but not opulent, arranged to reflect Gould's preference for comfort above display. On lengthy trips the car became a sort of luxurious prison in which he ate, slept, worked, and secluded himself from the eyes of curiosity seekers.[6]

During Gould's last years the car saw far more use than the yacht. His health no longer permitted him to endure the harsh New York winters, but prolonged cruises no longer suited him or his lungs. After Helen's death Jay never again undertook a lengthy voyage or separated himself from the affairs of empire even though his strength continued to decline. He still went south every winter but by rail, traveling overland on long inspection tours of the southwestern system. The tour was a handy device for combining business with rest and masking his need for the dry desert and mountain climate. On these trips Nellie became his constant companion along with George or Eddie, Dr. Munn, and sometimes the younger children.

In 1890 Gould left New York shortly after the annual orchid show at the Eden Musée, to which he had contributed some of his prize specimens. George went along and Clark joined them in St. Louis. For a month the *Atalanta* clacked along the rails of the southwestern system at an easy pace to avoid tiring Jay. As the train rolled south George marveled that his father worked "as hard as ever" despite feeling unwell. George did his share dutifully but his mind wandered constantly to more pleasant matters, especially to Edith. While Jay toiled over his stacks of papers or rested, George penned long, mooning letters to his wife. "I wish I was with you to night in our snug library," he cooed, "you laying on the lounge with one of your pretty legs in its black stocking showing up to the knee, just far enough to be suggestive. Perhaps you may guess what my thoughts are running on in connection with you to night."[7]

Not for George the emotional reticence of his parents or the slavish devotion to work of his father. He was an altogether different personality from Jay, more the good fellow fond of a good time, guileless in the manner of one who preferred not to have his powers of concentration taxed. All this made him more popular than Jay but less equipped to follow in his footsteps. Those who thought there was something of the devil in Jay would eventually discover too much of the world and the flesh in George.

The journey through Texas brought Gould not only sunshine but useful information on a host of problems there. The reclaimed Texas & Pacific had to be rejuvenated and competitive differences with the Huntington system harmonized. Hogg was still crusading against the wickedness of "foreign" corporations and a hostile legislature was about to create a regulatory commission. Nor could Gould formulate plans with any certainty while the outcome of the Katy and International struggles remained in doubt.[8]

During 1890 the Katy ground slowly toward reorganization under the Olcott plan. Two new members of the board were Standard Oil men, and the road's president, H. K. Enos, had been Rockefeller's man all along. The foreign bondholders' spokesman refused to approve the reorganization unless the Katy remained independent. Even the Olcott committee felt obliged to deny that Gould exerted any influence on their plan. Against this array of forces Jay could do little despite his large holdings in the Katy. He had long been operating on the assumption that the Katy would go its own way.[9]

The International was another matter. To keep it Gould had to come up with a plan suitable to the bondholders and the Katy. By September he thrashed out a scheme and tried to enlist the support of Enos. Among other provisions the plan assessed the stock 10 percent and provided new income bonds in return. Jay agreed to underwrite any assessments not paid and to take the stock and bonds, a common provision he would soon put to uncommon use. The plan was a conservative one for political as well as financial reasons. "I think it desirable," Gould advised, "to get through without opening the reorganization to the charge of watering the stock or bonds."[10]

He had reason to be wary. Any reorganization was "predicated on getting consent of Atty Genl of Texas in pending suits," but Hogg was mounting a gubernatorial campaign that summer on his favorite theme of hostility toward railroads. Alarmed by Hogg's agitations, the railroads funneled money toward Hogg's opponent. In August Gould and Adams agreed to build no more mileage in Texas "until such time as capitalists could feel more security in their investment in the State." Huntington and Enos joined them but to no avail. Hogg rode to victory and Texas voters approved the creation of a railway commission. Late in September, while still attorney general, Hogg intervened in the International foreclosure with a suit attacking the company's bonds as watered and therefore illegal. This action kept the foreclosure tied up in court for nearly a year.[11]

Financial matters here are getting very squally and money is likely to be very tight this fall.[12]

The opening of 1890 found the market for railway securities still in ruins, enervated by years of excessive construction, rate wars, blighted earnings, and reduced dividends. Unable to sell bonds at decent prices, many systems accumulated large floating debts. The Missouri Pacific was no exception. Since 1887

Gould had advanced the company nearly $3 million. When the road failed to peddle enough bonds to repay him, Gould agreed to take a parcel of securities and some New Orleans Pacific lands. By May the floating debt approached $7 million. A brief surge in the market that spring encouraged the Missouri Pacific to offer its stockholders a new 5 percent collateral trust bond and one share of stock for $950. Gould and Sage agreed to take any bonds not subscribed by other shareholders. Although the Missouri Pacific ran a deficit for the quarter, the executive committee declared a 1 percent dividend in hopes of boosting the new issue. Privately the directors agreed in writing to accept their dividend in Texas & Pacific stock at 50. Nearly $2.5 million of the $10 million in new bonds went unsubscribed. Gould took two-thirds of the leftovers and Sage the rest.[13]

As Jay predicted, money remained tight all winter. Spring brought some relief, but agitation over the silver issue left Wall Street skittish. In July Congress passed the Sherman Antitrust and Silver Purchase acts. The Supreme Court struck down Gould's favorite tactic by ruling that all forms of debt were subject to the first mortgage bonds prior to receivership. Those who thought the decision aimed at Gould would have been chagrined to learn that he found it useful in reorganizing the Cotton Belt.[14]

In September, after much wringing of hands, the executive committee declared another 1 percent dividend even though it meant running a deficit. On the sun-baked plains the forecast of poor crops and decreased earnings signaled another round of retrenchment. Gould's campaign to impose economy and efficiency on the Missouri Pacific system had never ceased. With Spartan dedication he monitored the constant stream of data that flowed across his desk, analyzing, evaluating, calculating, searching for weakness, puzzled by problems that seemed to defy solution. Neither illness nor fatigue released him from the millstone of detail he had borne since boyhood. Somewhere in that fine grind of facts lay the answers he craved if he could but find them. Of that he was as certain as of any belief cradled in habit.[15]

There was something of the mathematician in all Gould did, something of the theorist striving to impose his models on a stubborn reality. For years he had tried to perfect the coal business, and still success eluded him. In July he complained again that "our arrangements for selling commercial coal are out of joint." What Gould demanded was not merely order but system, grounded in economy and efficiency. The new competitive realities had long since taught him that victory went to those who performed best at the cheapest cost. Apart from commercial sales coal as fuel exerted a major influence on costs. The trick was to get the right coal from the right mines to the right roads. It cost more per ton to mine coal in one place than in another, and some coals gave better mileage than others. Often the savings in these areas offset differences in the cost of transportation, making it cheaper to use coal from more distant mines. Organizing these variables into the most efficient system was precisely the kind of problem that fascinated Gould. He spent hours juggling figures, devising equations only to discard them as soon as he found better ones. The process never ended because the variables changed constantly.[16]

Every aspect of operations had its own formulas, which on paper assumed a transparency never found in shops or yards or anywhere on line. Jay wondered why Clark and the master mechanic could not agree on a standard locomotive. He could not understand why so many freight cars were reported in poor condition despite large expenditures for the past three years or why Reid Northrop could not manage to keep enough refrigerator cars in Omaha and Sioux City, where business went begging, instead of leaving them scattered on eastern lines. Train schedules offered yet another form of mathematics that intrigued Gould. Amid the press of other work he always found time to dash off elaborate suggestions to Clark on how many trains should be run to what points at which times to maximize business. While the left hand preached economy, the right rattled off plans to expand and upgrade passenger service. "These are not orders," Gould demurred after scribbling a detailed proposal (complete with sketch map) to improve service between St. Louis and Wichita, "but suggestions."[17]

Here again was the paradox so typical of Gould and so tortuous to Clark. Contrary to what many thought, Jay's fundamental instinct was not to fight or plunder but to build, expand, realign, improve. Where his crusades for retrenchment waxed and waned with the financial climate, the urge for growth was constant. Entrepreneurs have their own variant of the creative impulse, which drove Gould throughout his career. He exemplified what Fritz Redlich called the "daimonic figure" who excelled at the art of "creative destruction." When competitive wars served this drive Gould fought them; when war no longer served its purpose he became an advocate of peace. Quicker than most he sensed changing conditions and adjusted his tactics accordingly, but the goal itself was as inexorable as any force of nature.[18]

Driven as he was by this creative impulse, Gould chafed at the very restraints he imposed on Clark. When money grew tight and crops looked unpromising in the summer of 1890, he reverted to his familiar cry of curbing expenses wherever possible. The August figures confirmed his fears that the system would not earn enough to pay taxes and dividends for the year; of the $7.92 million needed, it had netted only $4.2 million in eight months. A red flag went out to Clark on expenses for the rest of the year. In the next breath, however, Gould authorized Clark to run a preliminary line from Alexandria to the Sabine River and talked of extending it to a connection with the International once he had that road under control. This would penetrate the pine region northwest of Alexandria, where Gould already had agents buying up more acreage and had invested in two sawmills.[19]

Despite the bleak financial outlook, Gould pressed forward with projects to integrate the system, create new routes, and develop new sources of traffic. He still wanted to build the short line from Fort Smith to the Gulf but could not decide on the proper route and so concentrated instead on the cutoff from the Iron Mountain to New Orleans. When Grant reported that a line south from Columbia, Louisiana, would pass through a long leaf pine region, that was all Gould needed to hear. He ordered the route surveyed and organized a new company to construct the road. Visions rose in his head of trains hauling grain to New Orleans and returning with cargoes of lumber. He looked into building an elevator at New Orleans and tried to resolve the dispute that stalled progress on a belt road.[20]

In Missouri and Kansas Gould was no less active. Completion of the line from Rich Hill to Fort Scott eliminated the need to use Katy track between Sedalia and Fort Scott. The belt complex around Fort Scott went forward and a new line began south of Minden toward the coal, lead, and zinc mines near Webb City. "Why cannot we build up Fort Scott as a point for zinc smelting?" Gould asked Clark. Aside from reaching new mines and promoting new industries, he wanted a cutoff between southeastern Kansas and the main line at Jefferson City. To that end he formed a company to build westward from Joplin, resuming the old game of locating the route that combined a superior profile, the best prospects for business, and the most generous local aid. Gould realized too that an efficient system required not only physical but administrative integration. In one of several moves to clarify and simplify structures he swept a dozen Kansas branches into one company.[21]

Aside from construction Jay resorted to that old standby, acquisition. The Kansas City, Wyandotte & Northwestern paralleled the Missouri Pacific from Kansas City to Nebraska. Gould worried about what might happen if "the UP or some active competitor" got hold of the road, and he noticed some appealing features about it: an ample supply of rolling stock, good shop sites, a lucrative trackage rental contract with A. B. Stickney, and a line that, with slight extension, could shorten Gould's Colorado route by thirty miles. In January 1890 Jay secretly bought control of the Wyandotte, let the road default in March, and operated it through a receiver loyal to him. This arrangement allowed Gould to make the Wyandotte a feeder for the Missouri Pacific while concealing his ownership.[22]

In all this work Gould was obliged to lean more heavily than usual on Clark. The course of their relationship never strayed from the familiar channel it had assumed during the Union Pacific years. Clark handled his enormous workload with deliberation, ever careful of his ground and hesitant to make decisions despite Gould's insistence that he assume the authority delegated him. "Where you are having as sharp work as you are at Joplin and other points to secure important rights," Jay prodded on one occasion, "you must act on your own judgment without waiting to telegraph here. We will approve what you do afterwards." But he surely knew Clark could not shed the habits of a lifetime and that too much pressure would only send him to bed. Amid his deluge of inquiries and instructions Gould remained more solicitous of his friend's health than of his own. During a heat wave he fired off a telegram that read simply, "You ought to take Mrs Clark & leave St Louis during this hot weather." Hearing no reply, he urged Clark to "get away from St. Louis immediately for a rest. Why not come East and float around the Catskills and Adirondacks. Please answer."[23]

I enclose $2500 toward the deficiency of the Manhattan Eye & Ear Hospital for the current year.[24]

The Gould of legend is a man unkind and ungenerous by nature, indifferent to worthy causes, insensitive to human needs outside his own family. It is a false image, grounded in the same misunderstandings that spawned so much of the legend. Jay's charitable acts were, if anything, more private than his business transactions; indeed, secrecy was often a condition of the gift as in the case of the Mount Vernon donation. The role of philanthropist, like that of socialite, never appealed to him. He endowed no seminaries or universities, no libraries or hospitals, no museums or art galleries. His manner of giving was as inconspicuous as his personal style, designed to attract as little attention as possible to himself. The evidence reveals no clear pattern to his gifts. He did not seek out causes so much as respond to appeals. Morosini affirmed that Gould tended to act on impulse, and that his aid to needy individuals or friends down on their luck was frequent enough to require a special ledger account.[25]

In later years Helen took charge of the charitable work with her father's blessing. The incessant flow of begging letters was sorted, worthy requests investigated, and contributions sent those passing muster. Besides relieving Jay of considerable work, this experience trained Helen for her own career in philanthropy. She tended naturally to favor religious causes and influenced her father in that direction. Despite attempts to carry this work on quietly, word sometimes leaked to the papers and provided another forum for abuse. On one occasion Gould learned that two churches in Irvington were afraid that a vacant lot between them would be used for a saloon. Without being asked he bought the parcel and deeded half to each church. The *Times* got wind of it and printed a tiny account under the heading, "GOULD SOOTHES HIS CONSCIENCE."[26]

In Roxbury Jay renovated the old Baptist Church, brought a granite monument from Scotland to honor the More family, and pensioned his boyhood minister in a home near Lyndhurst where the old man could enjoy the grounds and flowers. The minister outlived Gould and remained a welcome guest at Lyndhurst to the end of his days. Nevertheless, a *World* reporter learned at a More family reunion in Roxbury that Jay was "not any more popular in his young days than he is now." In the winter of 1892 Gould agreed to host a meeting of the Presbyterian church extension society. After listening to appeals for funds and pledges of support from the laymen present, Jay waited until the session broke up to hand its chairman an envelope with a check for $10,000. It was done without fanfare and probably at Helen's request. Next day, however, the gift was censured by ministers and editors alike. Some thought it hypocritical or sanctimonious, a vulgar act of display; others regarded it as niggardly for so rich a man.[27]

Gould read the accounts in dismay. Slumped in a deep chair, he let the papers drop and sat for a time with his eyes closed, his face a mirror of defeat. He had also just donated $25,000 to help New York University purchase a new site, apparently because he was impressed by the fact that Samuel Morse had developed the telegraph in his laboratory at the university half a century earlier. Now he told Alice Northrop glumly, "I guess I'm through with giving. . . . It seems to cause nothing but trouble, trouble. Everything I say is garbled. Everything I do is purposely misconstrued. I don't care especially about myself, but it all comes back so on my family."[28]

Nothing more was said on the subject, and Gould did not stop giving. Although the press never stopped riding him, no major paper mounted another campaign of malice like those of the *Herald* and the *Times* in earlier years. Most followed the lead of the *World*, which preferred to twit Gould in its deft, satirical way to amuse rather than outrage readers. It had great fun with the adventures of Thomas S. Brady, a janitor renowned as Gould's double. Like so many other phenomena of city life, Jay was relegated to the entertainment columns until some larger issue roused the fury of the press against him. On those occasions the leper's bell tinkled anew, reminding all that the "little wizard" was forever to remain a prisoner of his past.[29]

You are authorized to use the pruning knife in fact we supposed you had a "carte blanche" *to act.*[30]

The papers were full that fall of reports that Professor Robert Koch had perfected a new treatment for tuberculosis. Some heralded it as a cure for the disease despite Koch's caution that he had experimented on human beings less than six months. Scientists, physicians, sufferers, and reporters flocked to Berlin to examine Koch's work or apply for relief. Gould followed the accounts with keen interest, as he did any development connected with the disease. He might have rushed to Germany in hopes of a cure much as the promise of laetrile lured a later generation to Mexico, but he did not. The attempt would have revealed the secret he wished most to keep and would put all his affairs in what he deemed an untenable position. Accordingly he watched from afar and remained in harness, pulling his enormous load with grim determination. As his time drew down and his strength dwindled, he might have been expected to shed more of his responsibilities. Instead he plunged headlong into a new project of awesome dimensions.[31]

Gould's decision to recapture the Union Pacific was reminiscent of so many of his earlier coups in which he accomplished several purposes at one stroke, leaving in his wake confusion as to his true objective. He seems not to have contemplated the move months in advance but rather come to it as a last resort. To some extent events forced his hand, impelling him to shoulder burdens he knew would push him nearer the grave. There may even have been a touch of nostalgia in the prospect of one last campaign to save the road that meant so much to his own past, a last hurrah with the same men who had been with him before. Gould was never one to feed on memories, but he always preferred to surround himself with the familiar and he had reached a stage in life where little else was left him.

However, Jay did not permit nostalgia to stray from inexorable realities.

At the heart of the matter lay his deepest concern: how to resolve the rate wars and promote a form of stability that strengthened the Missouri Pacific's position. That quest brought him into conflict with the Union Pacific on several issues. To the casual observer there seemed little basis for friction between the two systems. They did not compete directly and had exchanged traffic on an amicable basis since 1884. Gould and Adams got on well together despite past differences. Trouble arose not from a clash of personalities but from changes in the competitive structure that illustrated the swiftness with which expansion transformed relationships among systems.

In the new competitive order the Union Pacific's former transcontinental monopoly was now shared by half a dozen lines. Among the Iowa roads the Burlington reached Denver, the Northwestern penetrated Wyoming, the St. Paul entered Kansas City, and the Rock Island latticed Kansas. The Rio Grande Western extended to Ogden and the Sante Fe had built feverishly in Kansas, as had the Missouri Pacific. Gould's roads also vied for a share of the business in Kansas and Colorado and at Omaha. Like other presidents Adams responded by carrying the fight into new territory. The Union Pacific staked out a position in the Northwest by completing the Oregon Short Line and acquiring the Oregon Railway & Navigation Company. It also entered the Southwest by gaining control of the Panhandle road in April 1890. To buy peace in his own territory Adams signed a pact with the Northwestern.[32]

These moves troubled Gould for several reasons. The presence of the Union Pacific in the Northwest intensified the already fierce rate wars in that region. Control of the Panhandle complicated Gould's efforts to secure Colorado business and posed the threat of an alliance between Adams and Huntington. The Northwestern pact dealt the crumbling Interstate Commerce Railway Association a fatal blow. In May 1890 Adams took another momentous step. Unhappy over the Northwestern agreement, the Rock Island and the St. Paul demanded from the Union Pacific use of its Omaha bridge or they would build one of their own. This argument induced Adams to sign contracts granting the roads joint use of the bridge, use of Union Pacific terminal facilities in Omaha, and access over Union Pacific tracks to Lincoln and Beatrice, Nebraska. The arrangement gave the Rock Island its own line to Denver and enabled both roads to compete for transcontinental business at the expense of the Missouri Pacific.[33]

Gould's displeasure with the Union Pacific was also fed from another source. The significance of his position in Pacific Mail has gone unnoticed in later accounts. It was not overlooked at the time, although reporters again misread his motives. When Jay left the Union Pacific he maintained his interest in Pacific Mail. The steamship line continued to receive a monthly subsidy from the railroads and joined in the transcontinental pool. Whatever Pacific Mail's value as an investment, Gould found it useful for exerting leverage on transcontinental rates and for pressuring rail competitors (notably Huntington) he could not reach by other means. During the middle 1880s he shared power in the company with Huntington and Henry Hart. In May 1887, when Jay surrendered his seat to George, Hart assumed the presidency with a ringing declaration that hereafter the company would be run on behalf of the stockholders and not as a pawn of the Pacific railroads.[34]

Gould had other ideas. The transcontinental rate war had slashed Pacific Mail's earnings and eliminated its dividend. By 1887 this fight also affected the through business of Gould's expanded system in Nebraska, Colorado, and Texas. Jay spent much of that summer at Lyndhurst laid low by neuralgia, his mind preoccupied with the internal problems of the Missouri Pacific, the Ives caper, and the vagaries of Cyrus Field. In the aftermath of the market collapse that ruined Field, rumor asserted that Hart and his supporters "had their loans called and had to walk up to Capt. Gould's office and settle much in the same way that Cyrus W. Field had to do." Hart would be ousted, a columnist predicted, and replaced by George Gould. The report proved accurate but premature. Hart did not capitulate until October, when the price of Pacific Mail had been clubbed down from 55 to 32. Gould and Huntington returned to the board and George was installed as president.[35]

For two years George applied earnings to improvements instead of dividends. In trying to increase Pacific Mail's business, however, he ran afoul of Huntington. The company competed for the trans-Pacific trade with lines run by the Northern Pacific and Canadian Pacific roads and with the Occidental & Oriental, the line created jointly by Gould and Huntington more than a decade earlier. O & O still belonged to Central Pacific and Union Pacific. Although the two companies maintained joint agencies in Hong Kong and Yokohama, they did not have a pooling agreement and relations between them were strained. Where the Goulds favored Pacific Mail, Huntington wanted O & O to receive most of the business.[36]

In January 1890 George negotiated a contract with the Northern Pacific to run a monthly steamer between Tacoma and the Far East. The Northern Pacific agreed to drop its service and to divide earnings on through business, and the city of Tacoma offered Pacific Mail title to some waterfront property. Although the agreement was a coup for Pacific Mail, Huntington objected vehemently to any deal that benefited the Northern Pacific at the expense of his own Pacific roads. In the ensuing struggle Huntington prevailed. George reported tersely, "I regret that certain railroad interests in our board prevented the confirmation of the contract." But Huntington did not stop there. In May he joined forces with Senator Calvin S. Brice and General Samuel Thomas to buy control of Pacific Mail. Their victory forced George from the presidency, and both Goulds retired from the board.[37]

The loss of Pacific Mail coincided with Adams's contract for use of the Omaha bridge. Without the steamship line Gould had no indirect way to influence Union Pacific by exerting pressure on transcontinental rates. He had lost this instrument just when he most needed it. These two blows, delivered only weeks after the Union Pacific had absorbed the Panhandle line, alerted Gould to the danger of an alliance between Adams and Huntington. Before him loomed the spectre of Union Pacific, Central Pacific, Southern Pacific, Panhandle, and Pacific Mail in league against him. Adams regarded Huntington as an "old scoundrel,—shrewd, wily, vulgar and unscrupulous," but, as Jay knew from long experience, business made strange bedfellows. He did well to worry, for Adams had in fact explored the possibility.[38]

Friction between Adams and Gould flared up on another point, the Central Branch. The trackage rights agreement negotiated so easily the previous year suddenly came unglued in April over Adams's complaint that the Missouri Pacific had violated the lease of 1885. What appeared to be a minor affair assumed new dimensions in light of events that spring. Adams wanted the branch back and the trackage agreement in Nebraska modified; Gould advised Clark to "keep the matter along without any definite decision." The dispute dragged on into the fall, leaving Adams bewildered. Early in April he had seen indications that "Gould, for reasons best known to himself, proposes to make a drive at the Union Pacific. The essential thing is to find out what he is after. For myself, I am at present utterly at a loss. I cannot believe that it is the Central Branch, as that would be a matter of easy adjustment. He has some other idea in his head."[39]

For his part Gould could not know how completely demoralized Adams was over his own situation. The Union Pacific had become his cross to bear, "the skeleton in my closet," and its affairs a "valley of exasperation to me." The leap from theorist to manager had turned out badly for Adams. His railroad was in trouble and his program in shambles. The fanfares sounded at his accession to office had become a dirge mourning the failure of his policies. After six years he was no nearer a settlement with the government than Gould had been before him. Yet another funding bill was before Congress that spring, inciting the same stale rhetoric and sensational accusations of yesteryear. The *World* blasted the "Union Pacific Barons" and their "rich corporation" in a series of articles and declared that Gould was behind the new funding scheme. Like its predecessors, the bill vanished into the graveyard of dashed hopes, a monument to Adams's unfulfilled pledge.[40]

Adams had also promised to pay off the floating debt he had inherited and to manage the road without the sharp practices of the financiers who had brought it into disrepute. In August 1886 he paid off the last of these notes with a flourish, but his expansion program soon produced an even larger debt. A blunder in financing the purchase of the Oregon lines did not help Adams's cause. Only part of the bonds issued for this purpose were sold before the market turned sour; the rest were withdrawn to await a better day and used as collateral for short-term loans. As the money market grew worse, the floating debt mounted ominously. Adams deemed this sort of work unclean and shrank from contact with "human lepers," as he called financiers. He despised Wall Street and admitted to a "want of natural capacity and ineffective education" in its mysteries. Yet he could hardly avoid an arena that influenced the affairs of his company so decisively.[41]

This burden of short-term notes put Adams at the mercy of current income. The monthly statement became the barometer by which he lived or died, and its figures depended on the amount of traffic, prevailing rates, and efficiency of operations. After taking office Adams made little effort to conceal his disdain for the ineptitude of the previous administration. Contemptuous of "practical" officers who saw "what was immediately before them and nothing beyond," he tried to infuse the management with "an element of new and educated men." The idea was fine but its application proved by his own admission to be the most disastrous of his failures. He learned from bitter experience, as had Gould, that knowing what should be done and finding the right men to do it were two different problems. For all his erudition, Adams was a terrible judge of men. After sacking Clark he went through four replacements in six years without finding what he deemed a capable chief of operations. His last choice, W. H. Holcomb, disillusioned Adams to a point beyond despair.[42]

By 1890 Adams was a beaten man, crushed by the weight of failure and longing for release from his "prison-house." The Omaha office, on which his fate depended, had become a "mass of incompetence swathed in red tape" and a hotbed of intrigue among rival factions. Privately Adams showered "Flabby-guts" Holcomb with a stream of invective that taxed even his vocabulary. Repeated trips to the West did little to improve matters. "I find myself in a peck of trouble from the amount of clashing," he confessed to Dodge. The difficulty went beyond morale; earnings for the winter months were horrendous. Business improved in the spring but only on low-paying traffic. Overwhelmed by tonnage carried at cheap rates, Adams found himself in the same dilemma as Gould and other presidents: business increased but profits languished because of soaring expenses. As net earnings declined, the road was hard-pressed to meet its due notes.[43]

Adams was quick to blame "Flabby-guts" for his misery, but he had not helped his position by contracting a new line from Portland to Seattle and some other branches. The financial skies, cloudy for months, darkened ominously in October. The silver bill, a new tariff, and liquidation by English houses contributed to what Gould labeled "the unsettled state of financial affairs." Despite brave talk the Street was jittery; as money grew tighter Adams found himself snapping awake before dawn "in regular panic,—am fairly caught in the grip of a great catastrophe." On October 7 he headed west in one last attempt to straighten out the nightmare at Omaha. From then on events became for Adams, in Kirkland's choice phrase, "Götterdämmerung—self-consciously arranged."[44]

When Adams returned in early November, the remarkable John Succi was about to launch his record-breaking forty-five-day fast. On the Street Adams found starvation of another kind in the form of "very heady liquidation." As stocks tumbled, some observers detected Gould's "fine Italian hand" behind the bears. The suspicion that he was back in the market galvanized operators. Someone was hammering Union Pacific. A Wall Street paper thought to be Jay's mouthpiece reported that "a general feeling prevails in railroad circles that Union Pacific is managed by Harvard graduates who have big heads and small experience." A banker friendly to Adams heard that the attack came from "men who talk with Gould and follow his leadership, circulating the stories he tells them."[45]

"WHAT IS GOULD UP TO?" cried one headline. As rumors flashed like sheet lightning about the Street, the mood resembled that of Gould's heyday as a manipulator. Only a few months earlier reports had Gould about to lighten his burden by leasing his railroad system to the Santa Fe. Now it was asserted that he had regained his health, that he could "make or break railroad rates in the West" and therefore could "make or break the prices of stocks at will."[46]

Gould followed his habit of visiting the Windsor in the afternoon to browse through the papers and chat with acquaintances, imperturbable as ever, disclosing nothing of his intentions. On Friday, November 7, he excited attention by going downtown to Morgan's and some other financial houses. Next day Adams recorded a "regular financial gale blowing in the street, and, if not a panic, something very like one." On Monday the tempest howled itself into a panic checked only when a broker dropped dead on the floor and the Exchange closed for half an hour. Gould's stocks led the downward slide, which persuaded skeptics that he was the eminence grise behind the fall. Jay blamed the decline on "the trouble in London" and saw no reason for it to continue. Nothing in his words hinted that he had anything to do with the market.[47]

Adams was unconvinced. While assuring bankers that the Union Pacific was unaffected and required no new loans, he alerted Hughitt to a "very formidable combination against the Union Pacific on account of its alliance with the Northwestern road." On Tuesday, November 11, the storm hovering above the Street broke in earnest, smashing the Villard stocks. Three houses and a bank closed their doors amid fears that more would follow. While operators struggled to stem the panic next day, rumors flew that Gould had bought control of Union Pacific and would turn Adams out. Jay waved the reports aside with the cryptic remark, "I am not buying anything I can't pay for." From Boston a shaken Adams wired his associates to "take no stock in talk of change of management here." For months he had been wooing Dodge as his replacement, but at this critical hour the general was abroad; urgent appeals went out asking him to return at once. "One thing is apparent," Adams confided to banker R. S. Grant, "that is, that the whole pack, headed by Gould, are now at work to pull me down."[48]

For three days the Street bailed furiously to keep values afloat. Gould's role in the liquidation continued to puzzle observers, who claimed to see his influence in half a dozen places besides Union Pacific. Unable to get at the facts, Adams admitted that "the mere existence of such a rumor has brought all matters to a standstill at this office." The worst was yet to come. On November 15 Wall Street was devastated by news from London that unwise investments in the Argentine had brought down Baring Brothers. Although the panic was mercifully brief, the Baring failure sealed Adams's doom. He was relying, as he had before in a pinch, on sterling loans from the Barings to see him through the crisis. Without the Baring loans the Union Pacific faced bankruptcy unless it could find funds elsewhere for its due notes. Adams's last hope was the Northwestern, which had Vanderbilt money behind it. Hughitt reached New York on the evening of November 18 and met Adams and Ames at the Union League Club. Although sympathetic, Hughitt offered little encouragement. The Vanderbilts were not enterprising or interested in new ventures, he cautioned, but he would sound them out the next morning. Adams went to bed that night "feeling that the Jay-bird had me."[49]

He was not alone in that feeling. While the Street reeled from the shock of the Baring collapse, Gould launched a blitzkrieg worthy of comparison with those of a decade earlier. On November 17 he instructed Clark to hitch in the belt and defer all new construction except the road to Alexandria. Within days the reason for these precautions became clear: Gould was buying stocks in immense quantities and needed all the cash he could muster. No general struck swifter or more decisive blows than Gould delivered the Street in the campaign. He secured a large block of Santa Fe, added enough Richmond Terminal to give him four seats on its board, and regained control of Pacific Mail, putting George back in as president. Before the impact of these moves even registered, Jay revealed that he had gone back into Union Pacific and was prepared "to do what I can to help its relations with the other roads."[50]

"JAY GOULD AGAIN THE MASTER SPIRIT IN WALL STREET"

"HIS EVERY MOVE WATCHED BY ANXIOUS SPECULATORS"

"A VERITABLE PERIOD OF ONE MAN DOMINATION IN WALL STREET"[51]

To awestruck observers the clock seemed to have rolled back a decade. Here was the Gould of old, springing from the shadows to seize the market by the jugular "like some savage beast emerging from the jungle in which he has lain concealed." No one doubted he had returned to the Street or that he commanded its destiny. In one stroke he convinced the *World* that he looked and felt better than he had for fifteen years. His purchases on so grand a scale, coupled with bullish remarks in interviews, banished fears of further decline. Once again operators racked their brains over what Gould had done or might do. The flood-tide of rumor piled sensation on sensation. The Rockefellers were said to have bought control of Northern Pacific and joined forces with Gould to promote railroad harmony. Huntington appeared to have climbed aboard by helping Gould regain Pacific Mail. If true it meant a gigantic combination embracing Northern Pacific, Union Pacific, Southern Pacific, and Missouri Pacific, a "railroad monster" sitting astride "practically the whole transcontinental business."[52]

Why had Gould done it? The *World* suggested that he "COVETS MORE MILLIONS." That was too simplistic an answer for the *Herald*, which insisted that Gould "wanted these properties much, but he wanted revenge still more." Each of his moves had been a retaliatory strike against someone who had offended him during the past year: Adams, Brice and Thomas, Huntington. Especially was he eager to avenge the wrong done "My son George" in Pacific Mail. "He cares little for what is said of himself," the *Herald* noted, "but he never forgets a slight or an injury to George." So too with Adams, the man who had called Gould a "pirate" in *Chapters of Erie*. Even critics unwilling to believe that Gould would expend millions on behalf of personal vendettas conceded that his actions might be tinged with the desire to punish.[53]

Adams certainly believed it. With his gift for taking everything personally he marched to his doom convinced that Gould's sole purpose was to mortify him. On Monday, November 17, Ames and Morgan met with Gould and got their first inkling of his intentions. He wanted control of the Union Pacific, Ames reported to Adams, "to work it into some enormous, vague scheme he is meditating of a railroad combination which is to solve the problem, do away with competition, make everyone rich and, at one stroke, reduce chaos to order." To that end he was willing to assume the burden of keeping the road out of bankruptcy.[54]

On Wednesday, November 19, Adams spent the day at the Knickerbocker Club, reading, ruminating, even admiring Gould's "skill and the far-reaching, well designed and hidden manipulations through which he had accomplished his end." Ames and Hughitt joined him late that afternoon with the expected news: the Vanderbilts' policy was "fixed, and limited to Chicago,—they would not go west of it." Their refusal snuffed out Adams's last hope of saving the Union Pacific without Gould's assistance. Ames found himself in an awkward position. He and his family had long been intimate with the Adams clan; he had also been Gould's business associate for two decades. He knew that Adams had failed in his mission and desperately wanted out, but not in this way. Nevertheless, the Bostonians had immense sums invested in the Union Pacific, and Gould offered the only hope of survival.[55]

Next morning Ames and Adams went downtown to Gould's office. Finding him out, they returned at noon and found Gould waiting for them, "quiet, small, furtive, inscrutable." The contrast between Gould and Adams could not have been more striking. Gould, worn and haggard, a rumpled scarecrow swallowed by his chair, yet sensitive to Adams's brittle ego, fumbled for a way to let him down gently. He had never been one to crow in victory, if indeed he regarded this as victory. Certainly Adams regarded it as defeat. The heir to presidents sat erect, his patrician countenance stern, aware that his humiliation was complete. With the noblest of intentions he had made a fool of himself; he knew it and Gould knew it. The scholar had been battered by realities immune to his theories or principles. He had been routed in the arena he longed most to conquer, had not merely failed but embarrassed himself before those he wished most to impress. Anxious not to embarrass him further, Gould was in his quiet way kind but diffident. He had never done well dealing with the feelings of others. His standard cure for an awkward moment was silence—attentive, sympathetic silence punctuated by nervous gestures.[56]

He greeted Ames and Adams politely and did everything possible to keep the conversation pleasant. Adams mentioned the press attacks on his administration and the remarks attributed to Gould that a change would be forthcoming. Jay protested that he had said no such thing; the accounts were "outrageous" and he had taken pains to contradict them. Even so, countered Adams, the outcome was the same: his credit and his position had been undermined to the point where he could "no longer carry the load." If he could no longer carry the load, Adams continued, then Gould must take it. Jay nodded assent. A meeting would be called for the following Wednesday to make the transition. Gould asked how much the company owed and, getting a figure from Ames, outlined a plan he thought would work.

Adams had nothing more to say and got up to leave; Ames stayed behind. Gould showed Adams out. As they shook hands at the door Adams found solace in the fact that Gould did not meet his gaze. Stripped of everything but his caste and self-pity, he saw that the whole of his railroad career had boiled down to this final scene. At this pivotal moment he stared a dying man in the face and as usual his thoughts were on himself—how he looked, how well he might play this last scene, how gracefully he might exit. As their hands met he thought Gould looked "smaller, meaner, more haggard and lined in the face, and more shrivelled up and ashamed of himself than usual;—his clothes seemed too big for him, and his eyes did not seek mine, but were fixed on the upper buttons of my waist-coat. I felt as if in the hour of my defeat I was over-awing him,—and as if he felt so, too."[57]

To the very end Adams displayed his inability to read men. Clutching this last shred of dignity with pathetic determination, he surrendered the ship he had boarded so confidently six years earlier. "Gould, Sage and the pirate band were scrambling on deck at 10," read his diary entry that day. Gould, Sage, Henry B. Hyde of Equitable Life, and banker Alexander E. Orr were elected directors and the venerable Sidney Dillon restored to the presidency at age seventy-eight. Resolutions of appreciation to Adams were drafted. The ceremonies done, Adams went home to Quincy to drench his journal in mea culpas for nearly two years. The hurt in him personalized the episode into a plot engineered by Gould, whose "long memory" sought revenge for his ouster in 1885, or perhaps even for the Erie article. This was but the rhetoric of self-abuse. Although certain of the conspiracy against him, Adams recognized in more lucid moments that only one man had caused his downfall, and that man was himself.[58]

Gould's explanation of the affair was simple to the point of deception. Besieged by reporters for the reasons why he had recaptured Union Pacific, he said in his disarming way, "There is nothing strange or mysterious about it. I knew it very intimately when it was a child, and I have merely returned to my first love."[59]

Good will come out of the present depression. Some railroad managers will not feel so big as they did a few months ago. I have noticed a mellowing already on the part of a few of them. The big head is a terrible disease, and if I ever get it I hope my friends will put me in an asylum.[60]

Gould, Sage, Dillon, Ames. They were old men, Gould no less than the others, and tired. Time was slipping away from them and there was much to be done. They had not the strength to do it, yet they would not quit. You might as well have asked them to stop breathing as to quit. The spin of the wheel had brought them together again where they had first fallen in as associates. To complete the reunion Jay summoned Clark as general manager, thereby giving his friend two oversized jobs to fret about at once. If nothing else the new regime wore an air of the familiar. Even Forbes, himself only a year younger than Dillon, conceded that "I guess we can get on quite as well with the new as the old rule there."[61]

Unfortunately, crisis rather than nostalgia had reunited them. The lift given the market by Gould's purchases proved a fleeting burst of sunlight in the deepening financial gloom. Money grew tighter with no relief in sight. The pinch forced several brokers to suspend. Adams, who was personally overextended, found the market "swept cleaner of money loanable on time than I ever knew it to be before." The Union Pacific statement for October showed a decline of nearly $500,000 and sparked rumors that receivership was inevitable. To the crowd in the Windsor lobby Jay argued that things were not as bad as they looked. Privately he realized how dangerous the situation was, and that it was likely to get worse. The floating debt stood at $11.5 million, a huge load to carry in a squeezed market. Gould sent Clark the familiar mandate to "use the pruning knife. . . . I am afraid we have given you a big contract." Dillon had already issued orders to cut expenses on every front. "We are so short of money here," he added, "that it is necessary . . . to postpone the payment of every bill possible, including taxes. Arrange as pleasantly as possible to stand them off for the time being." He persuaded some banks to extend notes and Gould advanced enough cash to meet immediate obligations. These moves bought a reprieve until January, when the mad scramble for cash would begin in earnest.[62]

Hope for the future rested on the larger campaign to secure peace and higher rates. The Interstate Commerce Railway Association was dead if not formally buried, prompting one officer to moan that "the situation in the West is so bad it could hardly be worse. Rates are absolutely demoralized." One road after another reported the same dismal litany of higher tonnage carried at lower rates. Some managers estimated that the "existing widespread demoralization" had reduced net earnings by as much as 30 percent. Gould presented figures showing that a difference of only 1.4 mills per ton mile between current rates and those established by the President's Agreement had cost the roads $22.4 million in revenue.[63]

Although regional associations like the Southwestern offered a ray of hope, the search continued for a broader solution. Some officers demanded the legalization of pools. Walker argued forcibly for Gould's notion of a joint agency to eliminate competitive soliciting. Enos of the Katy stated bluntly that "there are too many railroads in the West" and "the remedy would be found in consolidations." The merger trend among railroads, like the rise of giant industrial combinations, had become a source of fascination as well as a hot political issue. That systems were growing larger by swallowing rival roads was an undisputed fact, but analysts could not decide what to make of it. Some agreed with Enos that consolidation brought harmony by reducing the number of combatants. Walker countered with a reminder that "this is not what the public expects or desires."[64]

All railroad men recognized the gravity of the situation but few were willing or able to abandon old ways of thinking in favor of new departures. They remained prisoners of their pasts. As Joseph Schumpeter observed, "Past economic periods govern the activity of the individual" and have "entangled him in a net of social and economic connections which he cannot easily shake off. They have bequeathed him definite means and methods" which "hold him in iron fetters fast in his tracks." Not so Gould. Throughout his career he had shown a remarkable ability to adapt himself to the shifting winds of change. He had played the competitive game too long and too well to slip gracefully into the role of conciliator, but he had also learned the importance of dancing to necessity's tune.[65]

This sense of urgent necessity lay behind the November blitzkrieg. Unlike his earlier strikes, this one was not intended to expand Gould's empire or even to give his system a competitive edge. The advantages on which Gould once thrived no longer served his purpose. He saw the futility of playing the same old game when it amounted to a brawl among the crew of a sinking ship. Haunted by the failure of past efforts to secure peace, he had moved to strengthen his hand in the bargaining. Control of Union Pacific and Pacific Mail would enable him to wreak havoc with transcontinental rates if other roads preferred war to peace. While they enlarged his capacity for mischief, he viewed them as deterrents to mischief by others. He had not only eliminated disturbing elements but also thrust himself center stage in the fight over transcontinental rates. As for the Santa Fe purchases, he admitted taking that stock "in the interest of general harmony among the Western and transcontinental roads." The *World* surmised that Gould wished "to obtain interests in the principal Western roads large enough to enable him to absolutely command the situation."[66]

After the Southwestern Railroad and Steamship Association became a reality, Gould began work on another, more advanced plan with Huntington and Manvel embracing the transcontinental roads. Significantly, he chose to unveil its details at the board meeting called by Adams to arrange the transfer of power in Union Pacific. The new association would include all roads west of Chicago and St. Louis. All regional associations in that territory would be abolished or made subordinate to the new organization, which would be empowered to set through rates, manage competitive business, operate joint agencies for procuring traffic, and determine routing over member roads. Guarantees would be devised to prevent rebates or deviations from posted rates. A board of arbitration would resolve disputes. The most novel provision, one long advocated by Gould, prohibited member companies from building "new railway lines that might compete in the territory or with the business of another member of the association" without the consent of all parties affected.[67]

The memorandum omitted a more ambitious scheme by Huntington and Manvel of creating a huge "operating company and taking in all the Western roads and dividing the roads, pro rata among all the lines on the basis of earnings of 1890, letting each company pay its own operating expenses out of its proportion." Gould realized this plan would never carry; he knew from experience that the proposed association would be hard enough to sell. "I hardly look for permanent peace among the railroads until the millenium," he quipped to a reporter. "But if peace were secured for only a year that would be a great deal."[68]

Earlier Gould had joined other presidents in requesting that Morgan summon another conference. Through this forum he hoped to use his plan as the basis for an agreement. Huntington and Manvel signed the memorandum but Cable, after indicating approval, hung back. The *World* misread the plan as "the same as the old trunkline pool." Perkins dismissed the scheme as "two years ago repeating itself" and growled that "under existing laws *no* agreement will be kept by all the Roads." Forbes took a different view. "Gould has not shown his usual sense in making himself so prominent in this," he conceded, "but on the other hand . . . it is better perhaps to assent to Goulds proposal than to continue the present state of things."[69]

On the morning of December 15 the presidents gathered at Morgan's home for another conclave. The attendance was even more impressive than before: Gould, Sage, Dillon, Huntington, Hughitt, Perkins, Cable, Roswell Miller of the St. Paul, Manvel, Ashley of the Wabash, Stickney, Stuyvesant Fish of the Illinois Central, George Coppell, chairman of the Rio Grande board, Palmer of the Rio Grande Western, James J. Hill of the Great Northern, T. F. Oakes of the Northern Pacific, Aldace Walker, a sprinkling of directors from several roads, and the corps of bankers. Gould read his memorandum as the first item of business. The discussion quickly revealed little enthusiasm for the proposal. Once again Jay had miscalculated the presidents' willingness to reach beyond their fears. In the terse language of the minutes, "General discussion followed as to the best means of securing the objects sought to be accomplished by this meeting."[70]

Morgan offered a plan he described as "simple but comprehensive," and to his mind effective. After some debate and amending the plan was approved by all the presidents except Stickney, who abstained. The new association would be managed by an advisory board composed of the president and one director from each company. The board was empowered to "establish and maintain uniform rates" and to oversee outside agencies securing traffic at competitive points. Any officer quoting a rate below the official tariff would be discharged. Established rates were binding on all members unless changed by a vote of four-fifths of the board. Arbitrators and commissioners would be appointed and the Alton, the Katy, and the Kansas City, Ft. Scott & Memphis would be invited to join. Miller was delegated to call the first meeting of the advisory board after the directors of each company ratified the plan.[71]

"RAILROAD KINGS FORM A GIGANTIC TRUST" cried the *Herald*. Since the meeting was closed, reporters could only surmise what had happened. The *Herald* described it as the Gould plan modified. The *World* too referred to the "Gould combination," while the *Chronicle* did not even mention the Gould proposal. Everyone agreed the new Western Traffic Association looked suspiciously like the old Interstate Commerce Railway Association. The *Times* sneered that the meeting "fell awfully flat" and produced "merely the reconstruction of the former Morgan Association." The outspoken Stickney was quoted as telling the presidents, "Two years ago you formulated something similar and it was a failure. Railroad men are not built in such a way that they will abide by this contract."[72]

Gould's heart sank at the outcome. Months of work had ended in what he could only regard as failure. Gone from the plan were his two pet provisions, the joint agency and the prohibition against building into rival territories. Perkins had been right; it was two years ago repeating itself. Jay came away with half a loaf and a stale loaf at that. Nevertheless, he swallowed his disappointment and took the lead in bringing the new association to life. The roads under his control quickly ratified the agreement. In public statements he declared himself "well satisfied" with the plan; behind the scenes he was active in setting the stage for the first meeting of the new advisory board. Several of the presidents wanted the session held in Chicago, but Gould remembered the fiasco of two years earlier. It was imperative that the meeting take place in New York, he urged Morgan, "where the influence of stockholders and bondholders will be more plainly felt. If called for Chicago, the public will say it is simply the old farce which killed the former Presidents' agreement." On this point at least Gould's views prevailed. Miller scheduled the meeting for January 5 at the Windsor Hotel in New York.[73]

An old world dying, a new one struggling to be born. Gould's role in that struggle had fallen far short of his hopes. It is impossible to know whether his plan would have produced the desired results, but at least it departed from past practices that had long since demonstrated their impotence. The presidents shared Gould's sense of alarm but not his vision of what was required to stave off disaster, preferring instead to cling desperately to the familiar in hopes of riding out the storm. Time would prove their choice tragically wrong. The financial squalls of recent years were but prelude to a coming whirlwind. Having done what he could to huddle the roads together in mutual protection, Gould was obliged more than ever to look out after his own. He knew the others would do the same.

"You are all gentlemen here," Stickney had twitted the presidents. "In your private capacity as such I would trust any of you with my watch, and I would believe the word of any of you, but in your capacity as railroad presidents I would not believe one of you on oath, and I would not trust one of you with my watch."[74]

Gould must have flicked a sardonic smile at this outburst. Asked to comment, he related the story of the time Daniel Drew went to a tabernacle where sinners were confessing their crimes. One man in particular horrified Drew by thumping his chest and accusing himself of ghastly offenses. Drew turned to the man next to him and asked, "My friend, who is this man who has done these awful things?"

"I don't know who he is," came the reply, "but I guess from his account of himself he must be Daniel Drew."

33

Surrender

The good deeds of this man must have been more than usually unobtrusive to have so completely escaped notice. It is incredible that his life should have been devoid of them, but neither in number nor in kind have they been sufficient to extort admiration or create imitators. Wealth in his hands has been, in RUSKIN'*s phrase, as the net of the spider, entangling and destroying, not as the net of the sacred fisher who gathers souls out of the deep. And the end of it all is execration and contempt, shattered nerves and a ruined digestion. There is nothing in such a life for the votary of the lowest forms of pleasure to envy, nothing for the most indulgent critic of human nature to admire.*

—*New York World*, October 4, 1891

What will happen when the crash comes? Is Gould's head turned? He has gone on in defiance of every immutable law, whether of morals or business management. . . . He is "smart"; but he is not as smart as Napoleon was, and he is playing a mighty dangerous game. Perhaps his quickness may save him. The Union Pacific and the Missouri Pacific are, in my brief, both doomed; the crash will be a resounding one; but, when the fragments fly and the dust rises, will Gould really be under the ruins? I doubt it. It would surprise me far less to find myself under them. He is an infernal scoundrel, a moral monstrosity, but he is astonishingly quick!

—Charles Francis Adams, Jr., "Memorabilia 1888–1893," April 19, 1891

Dear Father, the more I see of the world the more I cling to my home and love it. I do not wonder that some long to find their happiness at home with those they love instead of in social ambition.

—Helen Gould, letter to her father, April 7, 1891

IT TOOK the presidents five days to thrash out the details of the new association. When they had finished, Miller pronounced it "the best agreement we have ever had." Gould thought it was about all that could be done for the present. The questions of a joint agency and prohibition of duplicate building were postponed for later consideration. Although fifteen roads became members, a number of important roads did not. Before adjourning the presidents resolved that all these roads were "necessary to this agreement." Scarcely had the ink dried when minor scraps broke out over southwestern rates. Gould emphasized that "if the Alton road would join the new association it seems to me that all conflicts and friction could be dealt with through that association."[1]

He also jolted some member roads with one of those unexpected moves so typical of his style. When Jay reentered the Union Pacific it was widely assumed that he would move at once to cancel the Northwestern alliance. Instead he attacked the bridge contracts with the Rock Island and the St. Paul. As usual he prepared his ground carefully. In December 1890, shortly after the meeting at Morgan's, he asked Judge Dillon for an opinion on the leases. Earlier the Judge had assured Adams that the contracts were within the bounds of the company's charter; now he informed Gould that they might be ultra vires. Jay then asked Clark if he thought it expedient to take such a position and sounded Fred Ames on his views. "It is my deliberate judgment that this lease if carried out would prove very disastrous to the Union Pacific," he advised Ames. He could not believe that the board, "at the time they passed on this contract, had any idea of how far reaching it was."[2]

Throughout these exchanges Gould maintained that he was unaware of the leases until he returned to the Union Pacific. Of the several objections he raised to the leases, one issue concerned him most: the creation of a new through line to Ogden affected the Missouri Pacific no less than the Union Pacific. Before taking action Gould sent Clark to negotiate with the St. Paul on the basis of "abandoning the right to use our terminals for competing line west of Omaha." Miller rejected any notion of a temporary arrangement. "The contract is legally binding," he asserted, "and I do not see how the Union Pacific can successfully set it aside."[3]

Gould was quick to enlighten him. Using arguments provided by Judge Dillon, calling the leases "the most ridiculous he ever met in his railroad experience," he declared them void and ordered Clark not to honor their provisions. The St. Paul applied for an injunction and the legal battle for the Omaha bridge was on. The affair caught everyone by surprise and caused a sensation on the Street. Ames was deeply upset by Gould's move and warned that it had "absolutely ruined the company's credit in Boston." Trying to fathom Gould's motives, he asked Dodge whether Gould and Clark might possibly regard the leases as detrimental to the Missouri Pacific! Eight days later, however, Ames swung around to Gould's position. At an executive committee meeting he joined a unanimous vote to sustain all actions taken in resisting the leases. "The more the extraordinary character of those contracts was examined," Gould reported to Clark, "the worse they appeared."[4]

As Jay well knew, the legal fight would consume months, during which time no Rock Island or St. Paul trains would run under the leases. Gould used this interlude to negotiate a more suitable arrangement. The Rock Island and the St. Paul retaliated by shifting westbound business to the Burlington. Gould decided to split their alliance by striking a separate agreement with the St. Paul. Apart from finding Miller more reasonable than Cable, he understood that Morgan controlled the St. Paul. Morgan was a friend and an advocate of harmony, Cable a foe prone to belligerence. By November 1891 the St. Paul was working exclusively with the Union Pacific in Omaha, and Gould dispatched Clark to close an agreement "in such a way as not to let the R.I. get any advantage from it."[5]

While the furor over the Omaha bridge made headlines, Gould and his associates were busy scrounging cash for the Union Pacific. The funds for January interest fell short after all, obliging Jay to loan the company another $200,000. Dillon threw himself into the task with a vengeance, toiling fifteen hours a day despite his age and exhorting Clark to "keep the pruning knife at work until every decayed limb is cut off. . . . I rise or fall with you in your management of the Union Pacific." Reporters saw none of this difficult work. Instead they spun fanciful tales about Gould's market operations and his expanding vision of empire. The story that he was about to capture the Santa Fe simply would not die, and he was thought to own a large block of Northern Pacific stock from his attempt to help Villard out of his predicament. In February Gould set tongues wagging by going on an extended tour of the Richmond Terminal system. Dillon went along, as did Helen, the omnipresent Munn, John H. Inman, the Terminal's president, and a covey of officials. The presence of Gould in the South inspired reporters to fantasize on a grand scale. Gould was already master of the Southwest through his alliance with Huntington and his virtual possession of the Santa Fe. By acquiring the 8900-mile Terminal system he could extend his domination over the Southeast as well.[6]

The *World* also thought it detected "Gould's cunning hand" in the Northwest, where vast consolidation schemes were hatching. Even without the Northern Pacific the *World* saw Gould's vision of a true transcontinental system on the verge of realization. Possession of the Terminal and the Santa Fe, coupled with the Union Pacific, would plant Gould's flag on both coasts. He might gain access to New York by acquiring the B & O, and perhaps the Lackawanna, the Reading, and the Jersey Central as an encore. His empire would embrace nearly $800 million worth of railroads stretching almost 30,000 miles, or about 5,000 miles more than could be found in all of Germany. These were impressive numbers even for an age intoxicated with bigness. "This generation," brayed the *World*, "is becoming accustomed to vast figures."[7]

While overheated imaginations toyed with these giddy possibilities, Gould had all he could do simply to survive the rigors of southern hospitality. The convivial Inman, who loved ceremony as much as Gould and Dillon despised it, turned the trip into a triumphal procession. For a week the party marched from

Washington to Atlanta amid an unceasing round of banquets, receptions, luncheons, and sightseeing junkets. Gould and Dillon were obliged at every stop to give a little speech. "If you could have heard the speeches made by Mr. Gould and myself you would have laughed," Dillon wrote Clark. "I think you could do better than either of us." As the train approached Atlanta a weary Helen breathed a sigh of relief. "I sincerely hope that this will be the end of the travelling circus exhibition," she moaned, "and that after that we shall be allowed to travel in peace."[8]

In Atlanta the governor made headlines by refusing to meet Gould, saying, "I object to him personally, because I disapprove of his policy of monopoly and his business methods." Gould ignored the slight, endured a final banquet in Atlanta, and went on to inspect Macon, Brunswick, and Savannah. Only then did he and Dillon escape to St. Augustine for some relaxation. They stayed only two nights because of the heat. Gould was worn down from the grind of receptions, speeches, handshakes, and the crush of wellwishers. Rumors that he had fallen ill hurried north from Savannah. In Florida he suffered from what Dillon called "a slight touch of malaria." Doubtless that fiction was arranged by Munn, who advised Jay to return home.[9]

Gradually the rumors of empire evaporated. Gould made no move to acquire the Santa Fe, and the intrigues within the Terminal ruled out any dealings with that crowd. He wanted stability and sound development; they had failed miserably at both. When the Terminal's rickety financial structure later collapsed like a house of cards, Gould was clear of the debris. After reaching New York on February 14 Jay appeared to recover quickly from his illness. Early in April he took to the road again.[10]

There were plausible reasons for his travels. He needed to inspect the Missouri Pacific and had not been over the Union Pacific since acquiring control. The new bridge at Fort Smith had just opened, and the first meeting of the advisory board of the new association was scheduled for April 14 in Chicago. For all that, it was his health that drove him westward. As Gould's condition worsened, so did the strain on his nerves. The symptoms of pulmonary tuberculosis include fever, excessive fatigue, irritability, depression, night sweats, weight loss, headache, and cough. They are insidious in their resemblance to the symptoms of other ailments and therefore easily palmed off as something else, but Jay was finding them increasingly difficult to control. He had developed a cough that required enormous will to suppress. This ordeal, coupled with the pressures of business, was breaking him down. He worked more at home and required more frequent trips to congenial climates. Living aboard the *Atalanta* also enabled him to orchestrate his contact with other people and afforded some relaxation.[11]

This time George, Edith, and Howard went along. Jay stopped in Washington to call on President Harrison, then spent nearly three weeks touring the Iron Mountain, Missouri Pacific, and Union Pacific lines. When the advisory board convened, Jay was in Fort Smith. He sent a telegram pleading ill health, as did Clark, but no representative from Missouri Pacific, Union Pacific, or Southern Pacific attended. Their absence forced the meeting to adjourn for lack of a quorum. "Why Gould and Huntington did not see fit to attend the meeting when it was arranged as long ago as January, I do not know," Perkins barked. Some accused Gould of refusing to attend a meeting in Chicago.[12]

On his way home from the West Gould paused at Chicago long enough to clear the air. He had missed the meeting simply because he had only three weeks to inspect 7,000 miles of road before urgent business required his presence in New York. He supported the Western Traffic Association and hoped it would soon adopt the joint agency plan. Some thought Gould was dissembling but the evidence suggests otherwise. He had in fact cut short his trip over the Union Pacific to hurry back to New York where several items demanded his attention: the reorganization of the Wyandotte, a possible settlement of the International, final details on the Cotton Belt reorganization, and above all the critical state of Union Pacific finances. "All these matters seem to be piling on us at this time," he wrote Clark, "& are about all I can handle." When the advisory board reconvened on May 6 at the Windsor Hotel, Gould was present. He listened to charges that the Missouri Pacific had snatched most of the traffic in sugar by cutting rates and responded with what the *Times* called "a velvety speech, with just the faintest suspicion of claws in it."[13]

Gould conceded nothing but agreed to abide by whatever decision the advisory board reached after a fair investigation. The sugar business had been left entirely to his traffic manager, J. S. Leeds, who admitted that he cut rates openly after other roads did so secretly and took all responsibility for the action. The board found him guilty by unanimous vote which, under the association's articles, required Gould to discharge him. Jay found himself in a painful dilemma. He loathed firing anyone, yet he had long favored the provision. Leeds had shown himself a loyal if imprudent officer who, cynics sneered, was standing fire in Gould's place. It was not so. "I never could understand," Gould admitted to Clark, "why Mr Leeds by his so called sugar tariff should have placed me in such an embarrassing position." There was no choice but to let Leeds go. "I made the best fight I knew how," Jay explained in confidence, "but when the unanimous vote was against me I felt it was my duty to acquiesce & thus save a general breaking up."[14]

Leeds was sacked because he was guilty and because Gould was determined to show good faith. The board promised to consider the joint agency plan at its July meeting, and Jay would not jeopardize that possibility or the chance to restore harmony. Unfortunately, the good feelings of May petered out by July. At the next meeting some managers complained that the large volume of coal carried by the Missouri Pacific had been secured by cutting rates. The Alton quit the Western Passenger Association and slashed fares, forcing the Missouri Pacific to follow. The advisory board did nothing about the Alton and showed no enthusiasm for the joint agency plan. The *World* took this to demonstrate that Gould "could not run the Association." Once again Huntington ignored the meetings altogether. The board adjourned without accomplishing anything beyond routine business. Opinion was divided on whether it had failed miserably or simply was an idea whose time had not yet come. Good crops and heavy traffic helped keep rates stable, but uncertainty hung about the presidents like a noose.[15]

You must not worry over the situation. Let the younger Directors do the work. I am more impressed with the future of the Union Pacific the more I see of it.[16]

It was not a good season for the boys on the Street. The venerable Frank Work retired after long service to the Vanderbilts. "Deacon" White failed after an unwise attempt at a corner in corn. Victor Newcomb, a neighbor of Gould who had been associated with him in L & N, was pronounced insane and put in an asylum. For a brief time he had been one of the boldest plungers on 'Change. Some old and familiar faces climbed out of financial graves. Keene returned to the floor yet again and Henry N. Smith talked of a comeback after being hospitalized a year for injuries suffered in a carriage accident. Henry Ives put in an appearance, no longer young and wicked but hollow-eyed and sallow, his lungs ravaged by hemorrhages.[17]

This procession of ghosts provided a few flickers of life on a Street grown colorless and lacking in leadership. The generation of speculators dominated by Gould was gone, replaced by a crowd as gray and featureless as the giant corporations that now presided over the economy with their phalanxes of managers, bankers, and lawyers. An air of nostalgia crept into the talk of old-timers who missed the grand spectacles of the past. Gould's return to the market in November had electrified them with a display of the old-time magic. The hope that his return would rejuvenate the market burst like fireworks, lit the sky with dazzling colors, and fizzled again into darkness. The companions of Gould's early years had become relics by playing at the same old game while time passed them by. Jay had gone with the age, sometimes by choice, sometimes swept along by events, to a career in business that dwarfed his achievements as an operator. In this way, as in so many others, he was not one of the boys. His days too were running out. He was more ill than anyone knew and could not find strength to do all that was expected of him. On his shortened list of priorities market operations were nowhere to be found.

It was rather the money market that absorbed Gould's attention. The fight to save the Union Pacific had become a grinding search for funds to meet due notes. Jay, George, Dillon, Hyde, and their subordinates toiled daily at rolling over paper or scrounging cash from reluctant lenders. At the annual meeting in April Jay revealed his long-awaited plan to dispose of the floating debt. He proposed a new collateral trust mortgage for $25 million, of which $10 million would be issued at once, but the market remained anemic and the issue flopped. The mad scramble began anew to raise money for the July interest as well as current notes. What the hunt failed to produce Gould and his associates were obliged to provide from their own pockets. By mid-June Gould had personally loaned the

Union Pacific more than $660,000, some of it without note or collateral. At the month's end he advanced another $650,000 to meet interest payments and arranged a loan of another $100,000 from Manhattan Elevated. Somehow the company weathered its heavy July obligations, but the effort left Gould near collapse.[18]

For the third time that year Jay took to the rails, heading west for a prolonged stay in the mountain country. Munn was aware that current medical thinking emphasized the value of mountain air in treating consumption, and Gould easily masked the journey as Union Pacific business. He visited some mines at Hailey, Idaho, joined the Henry Hydes for a junket into Montana, looked over the grounds in Colorado and Utah, and inspected facilities wherever he went, roaming the line like a railbound Flying Dutchman.

Although Gould was gone nearly two months, the dailies did remarkably little speculating about his condition. "My health is much improved," Jay assured Dillon late in August. "You should not spend more than a day or two in New York till my return." It was sensible advice but impossible to follow, for in his absence another rush of creditors laid siege to the depleted Union Pacific treasury.[19]

The crisis broke with so little warning that afterward a company attorney charged that someone determined "to make a drive on Union Pacific" while Gould was in the West and Ames in Europe. On Friday, July 31, the first round of calls hit the New York office. Alerted to the trouble, George agreed to provide $100,000 to see the company through the weekend. By August 6 the trickle of calls swelled to a flood, touching off another frantic scramble for funds. Rumors spread that receivership was inevitable as heavy trading battered the stock down half a dozen points to 35 7/8, its lowest price since June 1884. Edwin Atkins, who had replaced his father on the board, wired Gould that Dillon was overwrought and would break down unless $1.5 million could be provided at once. Evidently the message caught Jay at a bad time; he replied that his own health was too poor to carry the load and suggested a receiver. Aghast at this response, Atkins got Dexter and Judge Dillon to join him in urging Gould to reconsider. "Unanimous opinion of every director that Receivership should be avoided if possible," Atkins and Dexter pleaded. "Can you not encourage and assist."[20]

The problem was not merely to stave off receivership but to get the company out of the strait jacket imposed by demand loans that kept everyone scurrying constantly for cash. Gould had tried to do this with the collateral trust bonds, but the market was dead. Now he boldly attempted a similar arrangement directly with the creditors, asking them to exchange their demand notes for time loans of two or three years. In return the Union Pacific would deposit a bundle of its securities with each creditor as collateral and pay a premium. Those creditors refusing the offer would be paid in cash from funds provided by a syndicate of the road's friends. "I was half sick, tired and discouraged about Union Pacific until your last suggestion came," a relieved Dillon wrote Gould. "We feel in New York that that can be carried through if all work with a will. Sage is fully alive to it and says it is the true principle to work on." Dodge offered to take half a million,

Dillon the same, and Ames would surely come in. George and Atkins had gone to enlist Morgan's help. "I feel more encouraged to-day than I have for a long time that we will be able to swim through," the old man added, his tired eyes shining. He admitted it "would almost break my heart to see it wrested from us. . . . shall we lay down and see the Union Pacific taken away from us? I say 'No' and hope you say the same."[21]

The stock rebounded at once to 39 but sank back to 32 the following week when, amid reports that the syndicate "appear to have abandoned the idea," more banks called their loans and large blocks of Union Pacific were dumped on the market. From Idaho Gould authorized George to subscribe for $2 million, but some of his associates hesitated. Poor Dillon was at his wit's end. Wilting from the pressure and the summer heat, he lost twenty pounds but would not quit his post despite efforts by George to send him home for rest. "All New York looks upon you as the great barometer in Union Pacific," he appealed to Gould. "I know you are. You will have to give some further encouragement than you have to stimulate the parties."[22]

Dillon was right. The confused situation in New York demanded strong leadership. Gould could not provide it from afar, and he could not return for reasons he could not disclose. Even if he possessed the strength to try, the crisis would not await his return. Instead he sent George to confer with Morgan, who moved at once to fill the vacuum. In a revised version of Gould's plan, the Union Pacific offered its creditors three-year 6 percent notes secured by collateral deposited with Drexel, Morgan. "Again Mr. J. P. Morgan steps in to avert a disaster which hung over Wall Street," proclaimed the *Chronicle*. By the time Gould returned in early September, the situation seemed well in hand.[23]

The air of uncertainty hanging over Union Pacific ignited sensational accounts of a clash between Morgan and Gould. Reports insinuated that the Gould forces were indifferent if not hostile to the plan, that Jay or George or Sage or all three were short of the market and working for lower prices by undermining the agreement. Morgan was said to be "indignant" over Gould's "juggling with the property" and to have told him bluntly to cooperate or be "hustled out" of Union Pacific and Wall Street. The controversy amused readers for a few days without producing any of the predicted fireworks. The hitch actually concerned a disagreement over terms of some Union Pacific notes held by Drexel, Morgan and endorsed by Gould, Ames, and Dillon. Once this difference was resolved, the plan went forward.[24]

"The personal bitterness toward Mr. Gould has probably never been greater than it is to-day," intoned the *Sun*, "and on every side, mingled with oaths and execrations, we hear solemn assertions never to touch any of the Gould properties again." This angry blast was provoked not only by the Union Pacific affair but even more by a furor in Missouri Pacific that broke about the same time. Here truly was the triumph of legend over reality. In Union Pacific Gould was accused of opposing a plan that had originated with him and been executed by the banker with whom he had worked closely in several enterprises for half a dozen years. The situation in Missouri Pacific was even more charged with irony and misunderstanding. Throughout most of his career Gould had been condemned for not pursuing sound, conservative business policies; now he was excoriated for doing just that.[25]

When the Missouri Pacific board did not meet as expected in mid-September to decide on the quarterly dividend, speculation arose about the delay. Normally the question of a quarterly dividend did not excite much interest but the occasion was far from ordinary. After a torpid summer the market sprang abruptly to life in September, thanks in part to record crops and strong buying abroad. The rise encouraged even skeptical operators to hope that "an era of active speculation and higher prices has appeared." Every vital sign was monitored for clues to recovery or relapse, none more closely than those connected with Gould, whose November blitzkrieg had thrust him again to center stage.[26]

"The truth of the matter," noted one astute analyst, "seems to be that Mr. Gould is taking no active interest whatever in the market, though he is not unfavorable to an advance of prices." But this was a minority opinion. Most of the Street clung adamantly to the conviction that Gould was not only still active in the market but the most potent force behind its movements. This belief invested the Union Pacific controversy with a weight beyond its own. The apostles of higher prices viewed the road's stalled financial plan as a drag on an otherwise buoyant market. If Gould was the chief obstacle to prosperity, they reasoned, it must be because he or George or Sage was short of the market. Much of the criticism directed at Gould had less to do with Union Pacific than with the fear his tactics might scuttle the upward movement so long awaited.[27]

These apprehensions were running strong when the dividend question arose. On September 23 the Street got its first hint that Gould thought the Missouri Pacific dividend should be reduced or passed altogether. Reports circulated that the system's performance did not merit a dividend but no one could verify the rumor because of Gould's policy of withholding all earnings figures. Left in the dark, traders knocked Missouri Pacific down nearly ten points in a frenzy that sent the rest of the list tumbling as well. This sudden wave of selling inspired several observers to the same revelation. The whole affair was another of Gould's brilliantly staged coups, revealed the *Times* on Friday, September 25. For half a dozen years he had worked hard to convince the Street that he had left the market and mended his ways. In recent months he had led the chorus of those favoring higher prices. The public swallowed the hook, the market advanced, Gould silently sold, and now he was using uncertainty over the Missouri Pacific dividend to smash prices down again to cover his shorts. The *World* thought Gould had been "on the wrong side of the market during the Summer" or, more likely, George was "enormously short of the market" and Jay was forced to bail him out.[28]

By Saturday critics had pieced the swarm of rumors together into a grand design. The supposed opposition of the Goulds to Morgan's plan fit perfectly. A settlement of Union Pacific's financial troubles would lift the market; continued delay would depress it. The *Chronicle* then jumped on the bandwagon. Labeling the market frenzy "a sample of [Gould's] power and of his methods," it stressed that only a day before the wild trading Gould and Sage admitted having "*personally advanced* a great deal of money 'lately' " to the Missouri Pacific. To the *Chronicle* that statement amounted to a kiss of death, "the epitaph which can be found on the tombstone standing over every distressed property which has fallen into the same management during the past twenty years."[29]

These charges stung Gould deeply. Once again his past was to be overhauled and selected chapters culled from the legend to warn the uninitiated of the perils to be found in dancing with the devil. He must watch his sons being arraigned for the same sins as himself. His every act was to be suspect, his every motive twisted, and his policies scorned as blinds for perfidy. The worst part was the veneer of plausibility about it all. He had indeed been boosting his own properties for months (years would be more accurate) to generate a market for the securities he had to peddle. The reason he and Sage were compelled to lend the Missouri Pacific money was precisely because it could not sell enough bonds to relieve the floating debt. There was nothing new about the floating debt or their practice of providing funds; both had been public knowledge for years. The Union Pacific had the same problem, as did any number of other roads.

Gould was in the same fix as other railroad presidents, but critics trusted them and distrusted him. His policy of not releasing monthly earnings figures or full financial statements on his properties now cost him dearly. He was fiercely proud of the Missouri Pacific's dividend record, had staked his reputation as a businessman on maintaining it as a paying property. With great reluctance he had paid four unearned dividends, hoping each time that improved conditions would ease the pressure before the next quarter. Now he was convinced this practice could no longer continue and that stern measures were needed to insure the road's financial health. To Gould this was sound policy if not common sense, yet the very suggestion of it had unleashed a storm of protest and accusation. In his bleaker moments it must have seemed like the spring of 1888 all over again.

As in 1888 Jay chose to vent his feelings through an interview. This time, however, his response showed less anger than frustration. The fight in him had diminished, leaving an undertone of disgust mixed with pleading. After stating that his responsibility to the stockholders for good management did not include paying unearned dividends, out came Gould's favorite lament: "I give the most of my time, more energy and vitality than I have to spare for such purposes . . . and never receive one dollar as compensation. I accept no salary from any corporation with which I am connected." More than that, he loaned money to his properties at 6 percent when the rate was much higher in Wall Street. The allegations about Union Pacific especially bothered him. He had loaned it large sums, endorsed its paper, and "worked so hard and incessantly to extricate it from bankrupt condition that it pretty nearly finished me . . . yet I am accused of wishing to defeat the very object for which I have been laboring."[30]

So far the entire affair had been nothing more than a guerre de plume waged around mere suppositions. The Union Pacific plan had not failed, and no action had been taken on the Missouri Pacific dividend. When the market steadied on Friday, September 25, the *World* went so far as to proclaim "GOULD'S ATTACK

REPULSED." The bulls held the field, the Union Pacific plan would carry, and Gould would back down on the dividend. Thus, in a few short days the dailies had reported an imaginary war, counted its casualties, announced the victor, and predicted the terms of peace. Events finally caught up with them the following week. On Monday the Morgan committee declared its plan operative and the crisis in Union Pacific passed. Two days later the Missouri Pacific board voted to pass the dividend. In his report Gould reminded stockholders that the company had not missed a dividend during the eleven years of his presidency. He presented figures for the nine months of 1891 showing a deficit of nearly $1.7 million on the system, of which $730,612 represented net deficit and $948,595 dividend payments made in the first two quarters.[31]

The release of Gould's report and the earnings data squelched the controversy over the dividend, but the board meeting itself furnished the papers even more sensational copy. Word got out that Gould had suffered a "hysterical attack" and broken down. The accounts were confused and conflicting. Before the vote Sage delivered a speech in blunt language about the importance of declaring at least a small dividend. Wincing at some of Sage's remarks, Jay rose at once to reply. After a few minutes he hesitated, seemed to lose his way, sat down again, and covered his face with his hands, "his whole frame quivering with . . . great excitement." With Munn's assistance Gould regained his composure but could not continue.[32]

No one on the Street knew what to make of it. The boys had been stung too many times by reports of Gould's condition to swallow them whole. As the *World* observed sardonically, "His bad health and his good health, his nerves and the exact pallor or ruddiness of his complexion have been used so frequently . . . with the sole object of affecting the stock market." George and Edwin issued prompt denials, only to have Sage confirm that Gould was suffering from nervous prostration and had collapsed in exhaustion. For all the digging, however, no one got close to the truth. "Nobody that knows Mr. Gould intimately believes that he has any organic trouble," concluded the *World*. "The whole trouble arises from shattered nerves and this condition of his nervous system affects his stomach."[33]

Gould's secret endured even though the façade had crumbled. His long history of neuralgia and digestive trouble kept attention focused in the wrong direction. One suspects that moralists preferred to have financiers suffer from such fitting ailments as shattered nerves or ruined stomachs than from diseases that might invoke sympathy. Over the next ten days the dailies manufactured fresh rumors. The *World* hinted at an impending showdown between Sage and the younger Goulds, the *Times* at a revolt by Missouri Pacific stockholders aimed at ousting Jay. Sam Sloan was said to be so disgusted by Gould's methods that he would soon resign all his directorships. "He maintains a grim silence as to his intentions," the *Times* added gravely. A year later Sloan still had not spoken or changed chairs.[34]

Try as he might, Gould could not escape the dogged insistence that he was short of the market. Perhaps the most revealing aspect of the affair was the extent to which analysts held Gould personally responsible for the market's failure to respond as they hoped it would. In recent years the financial horizon had been unrelievedly troubled. The market was narrow, the Street edgy and fearful of disaster. For a brief time that autumn the clouds seemed to have lifted. Informed opinion believed that conditions had finally improved and the market would rise. Informed opinion was wrong, as the downward spiral toward depression would later demonstrate. At the time, however, the conviction was so strong, the desire so urgent, that financial analysts regarded any obstacle with violent suspicion.[35]

When the march of events shoved Gould in the path of this euphoria, the result was predictable. He became the roadblock and ultimately the scapegoat for dashed hopes. It was a role assigned him by reputation. George protested that his father had nothing to do with the market's problems but no one listened. The papers could not discuss Gould's activities without invoking selected episodes from his past to clinch their point. Jay could not silence the leper's bell but he learned at least one lesson from this experience. His policy of secrecy had boomeranged disastrously; thereafter the Missouri Pacific published its monthly earnings.[36]

One mystery, the exact state of Gould's health, still had not been solved to anyone's satisfaction by the year's end. In November the story revived that his mind had given way, that he displayed signs of hysteria and was in critical condition. Those with access to him knew something was wrong. A Union Pacific officer told Adams how Gould went nowhere without his physician and described "the way in which he would physically collapse while trying to do business."[37]

I should think the plan of grubbing 200 ft wide for Alexandria extention [sic] *excessive—the grubbing ought to be only wide enough for the road bed.*[38]

Age had chiseled his oversized features down to the weathered contours of a monument, thrust their stark, spare lines into bold relief. His body, loose and limp from weariness, filled clothes like straw. The once mysterious eyes had lost their luster, and the head sagged where once it thrust defiantly upward. Still Jay would not relent. He played out the string of his concerns and found always more string, more knots to unravel, more tangles in which to tether himself. Problems marched toward his desk without regard for his weakened state. It made no difference that he avoided the office; he did as much at home as he would have attempted there. The boys relieved him of as much detail work as he would allow, but there remained an enormous weight of business for him to carry. A sense of urgency drove him onward, not merely the habit of toil but the need to put things in shape for the day when the boys would have to go it alone. If times had been better he might have eased up a bit, although even that is doubtful. As it was, he worked until forced to rest and then rested only until he could work again.

All year long he had waged the war for economy on the Missouri Pacific. "I anticipated last fall a shrinkage of earnings caused by the loss of the corn crop," he reminded Clark in February 1891, "and all my correspondence has been shaped to impress the management with the necessity of corresponding economy." By autumn prospects of bumper crops at home and shortages abroad promised a good export trade. Completion of the Alexandria extension would enable him to funnel a large grain traffic down to New Orleans for shipment overseas, but the work was slow and expensive. Other roads were already combining to keep this traffic flowing eastward to Chicago.[39]

Politics still cast a long shadow across the situation in Texas. In April the newly created railway commission made sweeping rate cuts in the one region where stability had been imposed by agreement. The Katy emerged from its long receivership in the hands of a strong board reinforced by the presence of John D. Rockefeller. Without influence in the new management, Gould still had to resolve the dispute over ownership of the International's stock. Hogg's suit had put the matter in legal limbo for a year, but in June the Texas Supreme Court cleared the air by upholding a decision that Hogg had no right to intervene. Haggling over the reorganization plan proposed earlier by Gould consumed the rest of the year. In February 1892, just when all sides appeared to have reached agreement, Gould sprang one of his patented end runs. It was a clumsy ploy that succeeded only in producing another long day in court. In less than a week both sides agreed to drop all suits and proceed with the original plan. After more than three years of jousting, Gould emerged with the International still in hand.[40]

Elsewhere Gould's empire remained stable and prosperous despite periodic grumblings about monopoly or inefficiency. No new competitors arose to challenge Western Union in the telegraph field. The cable settlement reached during the summer of 1888 was followed a year later by a similar pact for land messages. A battle loomed with Bell Telephone over certain crucial patents, but the dividend record of Western Union continued undisturbed. In the decade of Gould's ownership the company had more than doubled its mileage. With the surplus account standing at $11.4 million in 1891, the board decided to increase the capital stock to $100 million and pay a scrip dividend of 10 percent.[41]

Manhattan also thrived during the decade following the consolidation in 1881. The growth in business was phenomenal, that of earnings merely impressive because of the nickel fare and the expense of handling so enormous a traffic. The critics who carped about the elevated's service paid little attention to the demands thrust on the system by the city's exploding population. Every year the number of passengers surpassed all previous records, exploding from 75.6 million in 1881 to 196.7 million in 1891. For that same period earnings rose from $5.3 to $9.9 million. Nothing Manhattan did escaped criticism, if only because it lived in the fishbowl of the city. Of all Gould's enterprises none touched New Yorkers more directly or kept him more enmeshed in state and local politics. To the dailies and their readers the adventures of faraway railroads, however lurid, were remote, but Manhattan was their daily fare. They rode it, saw it, cringed at its noise, gagged on its smoke, dodged its cinders. And, of course, they complained. The elevateds were inefficient, too crowded, too slow, too dangerous, a blight on the landscape they traversed.[42]

Two issues in particular locked Gould in perpetual battle with the press, the politicians, and a variety of interest groups. The endless string of damage suits by owners of abutting property kept passions inflamed over the road's impact on its environment. Manhattan fought these suits vigorously because every loss cost the company money and established what it considered a dangerous precedent. Critics denounced its tactics as the ruthless, high-handed methods of a grasping monopoly—and Jay Gould's monopoly at that. Even this clash paled before the furor over the question of expansion. For years Gould had been anxious to enlarge the system, which now reached 170th Street with connection to a branch of the New Haven northward and to the Second and Third Avenue els southward. He wanted a third track for running express trains along the entire system. By his own estimate a third track required additional terminal facilities at the Battery, a connecting line from the Brooklyn Bridge to the Sixth Avenue el, a second connector at the upper end of the city, and some new branches.[43]

Manhattan's charter authorized only two tracks and sidings, which compelled Gould to seek an amendment from the legislature. He also needed approval from the Board of Aldermen, the Rapid Transit Commission, and the Park Board. The notion of a third track surfaced as early as 1888 but got nowhere. By the winter of 1891 Gould was ready to push it in earnest. The campaign for an amendment commenced in Albany and a formal plan was presented to the Transit Commission for approval in May. It provoked strenuous objections from property owners, rival transit interests, citizen groups, and the dailies hostile to Gould. The *World* led the charge against what it called Jay Gould's "straddle-bug system," arguing that the third-track plan would destroy Battery Park, ruin Broadway as a boulevard, and "shut out the air and sunlight from thousands of downtown residents." It would also strengthen "a monopoly which is already both dangerously and oppressively large, and . . . which has proved itself inadequate and unsatisfactory."[44]

Battery Park quickly became the emotional battleground of the struggle. The surrender of park grounds for terminals drew fire from many quarters, including the Board of Trade and Transportation. The *World* used this issue effectively but, unlike the *Herald* in 1888, never allowed its abuse of Gould to obscure the larger question at stake. Against this determined resistance Gould made little headway. By the end of 1891 the plan remained in limbo. Although Gould's interest remained keen, there was little more he could do. In November he left the executive committee of Manhattan on the plea of ill health. The year's struggle had accomplished little more than enhancing his unpopularity in the city and providing the dailies with another forum for abusing him. A year later, when the Battery Park issue flared anew, the *Times* viewed it as evidence of Gould's "insatiable appetite for grabbing everything in sight."[45]

Alice, I am not afraid to die. I am not afraid to die. But the younger children—well—I don't like the thought of leaving them.[46]

In the late hours, when insomnia or fits of coughing robbed Gould of sleep, the children were very much on his mind. They too found a place in the public eye, for the same papers that roasted Jay now fawned unabashedly on his children and grandchildren. The older boys were doing well at the office and Howard was enrolled at Columbia. Frank attended the Berkeley Lyceum, Anna a private school near Philadelphia. Dear sweet timid Helen remained his perpetual companion, eager to fuss over him or help in any way she could.

The loss of his wife only heightened Jay's concerns about the children. He was not only a doting father but a fiercely protective one. Once, on a visit home, Anna went to the theater without a proper escort or letting the family know where she was. Even worse, she went with a friend of whom Jay did not approve. "It is just such thoughtless acts that has caused a good deal of my nervous trouble the past year," he admonished, "and I now fear when you return home that you will keep me unstrung." He urged Anna to "entirely drop Miss Freeman. I am sure she is not a girl of good instincts or a proper associate for you." Jay worried even more about Helen. At twenty-three she was little more than a recluse, the willing prisoner of her father's wealth and notoriety, indifferent and inaccessible to suitors, fearful of society with its petty malices. "I feel as though I never cared to see much of some of our neighbors," she said of the Irvington crowd. "They slight so many people and are not really kindly at heart except to people in their own set."[47]

Denied the possibility of a career, Helen buried herself in religion and good works. Too late Jay sensed that she needed something more, that she must be brought into contact with the world in which she would soon have to make her way alone. Without his wife to perform the amenities, Jay undertook the task himself by deciding to hold an afternoon reception during the Christmas holidays in 1891. The occasion was in a real sense not only Helen's debut in society but his own as well. The whole family pitched in to help with the preparations. Delmonico's provided the catering and the Hungarian Orchestra of Munczi Lajos the music. A profusion of ferns, palms, holly, mistletoe, and laurel brought down from the Lyndhurst greenhouse along with potted plants and fresh-cut flowers transformed 579 into a bower of green. In the center of the large table in the drawing room sat a huge basket of crimson orchids from Gould's prized collection. Invitations went out to President Harrison and his cabinet, business associates, relatives, old friends, the whole range of Gould's acquaintances, including many who had fought bitterly with him over the years. Conspicuously absent from the list were the Astors, the Vanderbilts, and other denizens of New York society.[48]

On the afternoon of December 26 Jay took his place at the head of the reception line, pale, nervous, struggling to keep his cough in check. Next to him stood Helen in a gown of blue satin and silver brocade, demurely attractive but overshadowed by the glamorous Edith Kingdon Gould in a canary gown with diamond ornaments and a bouquet of orchids. Reporters hawked the event with great curiosity, expecting "some sensational features, some splurge, some magnificence that would startle and amuse the town with a one day's wonder." Instead they found a quiet but pleasant crush, "planned and conducted in the best of taste and in the simplest manner possible." The success of the affair encouraged Dr. Munn's wife to give an evening reception for Helen two weeks later. In her patient way Helen endured her launch into society and returned safely to port with little inclination for further sailing.[49]

There were reasons beyond lack of interest and timidity for her reluctance. Crank season had come around again. In May a character who styled himself as Vice-President No. 71 of Christ's Followers arrived from Colorado to assassinate Gould unless he paid $5 million. He was nipped and dispatched to an asylum, but three weeks before Christmas a more serious blow was struck, this time at Sage. A stranger walked into the old financier's office and demanded $1.5 million. When Sage refused, the man detonated a bomb that demolished the office. The crank and one of Sage's clerks were blown to bits. Miraculously Sage pulled himself from the rubble unhurt except for damage to his already faulty hearing. The blast caved in part of a wall and rained plaster dust on George Gould, who sat in the adjacent room belonging to Manhattan Elevated. George ran out into the street along with Connor and Morosini, who also had offices in the building. After learning what had happened, George raced back inside to telephone the news to his father. Jay went at once to comfort Mrs. Sage and remained with her until Sage was brought home. He insisted on accompanying Sage to his room while the doctor attended his wounds. "It seemed strange to me at that time," Sage recalled, "that this little man, who was anything but robust himself, could do so much and insist upon doing it."[50]

Then it was Helen's turn. A man claiming to be a German prince insisted he was destined to be Jay Gould's son-in-law and implored a friend to help him kidnap Helen. No sooner was he put away than a broker named Landauer appeared at 579 professing his love for Helen. He too was incarcerated but others took his place. The Sage bombing seemed to trigger crank craze among businessmen in New York. For Gould it was familiar ground, but he removed his name from the door of his office and took other precautions.[51]

The retreat to a garrison lifestyle could not have deprived Gould of much more freedom than illness had already taken away. By 1892 he could no longer keep anything like a regular schedule. He made fewer trips downtown and attended meetings only when unavoidable. The cough grew worse, more difficult to hide, and he was bothered by intermittent fever and bouts of confusion similar to the one that struck him during the Missouri Pacific meeting. At home, when Jay could not work, Helen read to him for hours from his favorite novelists, Twain, Dickens, and Scott. Sometimes for diversion he played bezique with the girls. Never an epicure because of his delicate stomach, he was reduced now to fare a prisoner might have scorned. Fits of insomnia drove him late at night to the small island of sidewalk in front of 579 where, under the drowsy eye of a servant, he paced the hours away like some caged and tormented animal, hawking bloody sputum into his handkerchief. Across the street a cheerful carpet of light spilled out of the Windsor's lobby, where so many of his intrigues had played themselves out to appreciative audiences. Those days were gone, never to return. For his generation of operators the Windsor was fast becoming a memorial chapel.[52]

In February the truth of Gould's condition came as close to being revealed as it ever would before his death. On Tuesday the twenty-third he hosted the meeting of the Church Extension Society at which he presented the controversial check for $10,000. Two days earlier Jay had told the boys in the Windsor lobby that he was about to go south. When the *Atalanta* did not depart, curiosity was aroused. On Saturday word leaked out that Gould had fallen ill after the meeting. A source described as an intimate friend of the Goulds told the *World* that Jay seemed to improve until Friday morning, when he suddenly began to hemorrhage. "Mr. Gould was coughing violently," the source reported, "and his handkerchief was flecked with blood." Three handkerchiefs were soaked with blood before the attack ceased.[53]

Munn was equal to the occasion, assuring everyone that the hemorrhage came not from the lungs but from the stomach and was purely local. The usual denials were put out in all directions and the *Atalanta* made ready again for an extended journey. Within a few days Gould was speeding southward, heading for the dry air of the Southwest. In Texas Jay found the Texas & Pacific and the Southern Pacific at loggerheads over several issues. Huntington learned that Gould was in El Paso and came to call. Wearing a black brigadier's hat, coat buttoned tight despite the heat, clutching a gold cane, the burly Huntington must have winced at the frail, grizzled shadow of his old antagonist. He was all smiles and pleasantries, insisting that Gould visit California before returning east. Reporters tried to make much of their conference but got little help from either man. Gould was feeling better. His temperature went down, the cough dried up, and his appetite improved. Not even a fierce sandstorm interrupted his morning walks in the terminal yard or his trips into town to get his dispatches at the telegraph office. Edith was expecting her fourth child any day, and Jay was anxious for the news. Both the girls were with him, and Howard arrived early in April. He took them on the streetcar to Juarez to view the antiquities. When not on outings with the children he tried to woo local shippers to the Texas & Pacific. And, serving notice that he was not yet gone, he bought a small Texas road formerly leased to the Southern Pacific.[54]

A flood of telegrams kept him informed of developments elsewhere. Rumors were swirling of a change in the Union Pacific management. Dillon was seriously ill with a stomach ailment and expected to resign in favor of Ames. A group of foreign stockholders was gathering proxies to oust Gould at the election on April 27. Their representative, Adolph Boissevain, was confident enough of victory to name Dillon's successor, who was none other than Gould's former officer, R. S. Hayes. Ames, Atkins, Dexter, and Hyde remained loyal to Gould, but in his absence the burden of defense fell on George's shoulders. To the surprise of many, the Crown Prince met the challenge. At the eleventh hour he secured proxies for a large block owned by a London house and carried the election.[55]

Gould was ecstatic, not merely over the victory but because his son had proved himself in battle. He showered George with congratulations. "I had no idea of your strength till I received your telegram the day before the election," he chirped. "I should think Bloodgood & Boissevain a little mortified. They had crowed so loud that they had a walk over. We have now a year leeway to look over this property & can probably buy the stock cheap as earnings I think are likely to be poor." George had also managed to push an amendment for Manhattan's charter through the legislature. He appeared at last to be fulfilling Jay's ardent hopes for a successor capable of holding the vast Gould empire together after his death.[56]

Death was very much on Gould's mind these days. During his stay in the Southwest he lost two old friends, Bishop Sharp, the Mormon patriarch in Union Pacific, and Edward S. Jaffray, a prominent merchant and neighbor whose family had been close to the Goulds for years. At the Union Pacific meeting the board installed Dillon in the newly created position of chairman and elevated Clark to the presidency. It was a fitting and timely gesture, for on June 9 the old lion passed away after a lifetime of service to the road he had helped build. A few weeks later Cyrus Field, worn down by the failures and tragedies of his family, went to his grave.[57]

The weather turned hot in May and Gould moved north toward Pueblo and the mountain air. The better he felt, the more restless he grew. He had never been a good loafer and fishing was not enough to satisfy him. "It would be very nice if you could come out & see me," he wrote George mournfully. "There is very little to do here." George had intimated that the western traffic associations were on the verge of collapse, in which case Jay thought that "general demoralization is likely to ensue. I am glad we have no dividends to look after. I shouldnt wonder if stocks declined to a low figure especially if to a railroad war we add the constant exports of gold which must make money tight later in the season. In the mean time we are in a snug position financially & it is best to keep so. We may be able to buy UP in the 30s before another election comes round if it looks well or we can drop it."[58]

Jay could stay away no longer. If he must die, it would not be of idleness or terminal boredom. Home he came to Lyndhurst, where he resumed a limited schedule. Late in the summer he went back onto the Wabash board, sparking rumors that he had some new scheme for the road. George was right; the traffic associations were going to pieces. One officer, noting that business had increased on the Gould lines while decreasing on other roads, muttered that "Gould is at his old tricks again." The Burlington complained loudly about several decisions of the advisory board and demanded a change in the rules. When that did not occur, it withdrew from the Western Traffic Association. The Southern Pacific followed, along with the Missouri Pacific, which observed that the association had "outlived its usefulness."[59]

By early October the association was defunct. The Transcontinental Association had folded a short time earlier, the Trans-Missouri Association soon followed, and the Western Passenger Association stood at the brink of dissolution. Gould attended the last session of the advisory board. Palmer thought he looked "more chirruppy than ever, and with a tone of injured innocence that was lovely to behold." Jay had every reason to look unhappy. Four years of hard work lay in ruins and the goal of stabilization seemed more distant than ever. In typical fashion he had hoped for the best and prepared for the worst. Another round of rate wars loomed, with results no one cared to contemplate. He would not be there to fight them.[60]

The close of Gould's long and incredible business career had come at last. No one yet realized it, although many saw that he was but a shadow of his former self. Such had been his power that his presence was no longer required; legend alone sufficed to strike fear in the hearts of his adversaries. Jay did not know or accept that his work was finished. He would not quit so long as his heart ticked and his brain calculated. With the last of his strength he dragged himself through each day's duties, relying on the faithful Munn to see him through. On October 11 he attended what proved to be his last meeting of the Missouri Pacific board. He seldom appeared in public any more, but one occasion demanded his presence. Edwin had become engaged in June to Sarah Shrady, the stepdaughter of a physician who had worked with Koch. Although Sarah was only eighteen, the betrothal period was cut unfashionably short. Eddie declared that he did not believe in long engagements; neither did his father, for reasons of his own. The ceremony took place on the evening of October 26. George and Edith had moved out of the house behind 579, which was redecorated and given to the newlyweds.[61]

The chill winds of autumn had begun to blow. It was time to go south, to head for El Paso again, but Jay could not bear the thought. He was not suited to a life in exile and complained that the *Atalanta* was a prison. Gould could no longer live without the dry southwestern climate but neither could he live with it. Week after week the trip was postponed. During Thanksgiving Alice Northrop came to visit Helen. They were in the sitting room one evening when Jay came in and sank into his favorite armchair, his face white as a sheet. "Uncle Jay," Alice exclaimed, "you look so tired!"

"Yes," he replied softly, "I am tired. I stopped in for a little at Madison Square Garden to see the horse show. They had some wonderful animals. It was worth going to—but I think I have taken cold."[62]

It was much worse. His disease had entered the miliary phase, an acute form in which the tubercle bacilli enter a vein, disseminate throughout the body, and establish themselves in every organ. Late that night the girls were awakened by scurrying noises. Peeking out, they saw light coming from Jay's room and a screen being set in place before his door. Next morning they learned that Jay had been hemorrhaging severely and that there was little hope of his recovery. Every effort was made to keep the attack secret, to release the news gradually. The public did not learn of his illness until the morning of December 1, a week after he had been confined to bed. At once the bears circulated rumors that he was dead and sent the Gould stocks tumbling a few points. Denials were quickly issued and the stocks rallied, but for the first time it was evident that Gould lay near death. The grand deception had been a complete success. No newspaper discovered his true ailment until it was revealed in the death certificate.[63]

The family gathered for the death watch. Outside on the sidewalk the reporters waited, passing the long hours in idle chatter along with a handful of curiosity seekers. The boys in the Windsor lobby talked of nothing else, their minds burning less with concern about Gould than with calculations of how the market would react. Oblivious to this ritual, Gould lay semi-conscious in the second-floor bedroom above the conservatory, the same bedroom in which his wife had died. Jay had moved into it shortly afterward. The furniture was comfortable but simple, consisting of little more than bed, bureau, wardrobe, table, and a few chairs. He would die as he had lived, quietly and without frills or fanfare.[64]

He was ready to die, as ready as any man could be. Having held on so fiercely for so long, he surrendered with the same economy of motion that marked everything he did. The will was prepared, and a remarkable will it was. He had provided for the children as best he could. The boys were as ready to assume his place as they would ever be. Nothing remained except to say good-bye to them all, which he did in a voice scarcely quieter than it had been in life. Not an ounce of drama could be wrung from this farewell scene. He was like an old clock worn down to its last tick, which came at 9:10 on the morning of Friday, December 2. There was nothing left for him to do but die, and so he did. He had done enough. Indeed, in the eyes of many he had done far too much.[65]

The body was not embalmed but packed with ice in a large mahogany chest and left in the room where he died. The shades were drawn and a black knot of crepe hung on the front door. Judge Dillon and the Reverend Paxton were summoned to assist in the funeral arrangements. At first it was announced that the services would be public, but this was changed when George feared that a crush of strangers would prevent Jay's older friends from attending. Nevertheless, on Monday, December 5, Fifth Avenue was packed with spectators, milling and swarming toward the door despite efforts by police to keep them back. Someone did a brisk business distributing Edwin Gould's calling cards to those trying to get inside.[66]

Those who managed to gain entry were mostly old friends and associates, their names a summary of Gould's career: Sage, Clark, Huntington, Ames, Atkins, Villard, Sloan, Depew, Morgan, Fordyce, Hopkins, Fitzgerald, Cammack, Connor, Morosini, Judge Dillon, Marquand, William Rockefeller, Isidor Wormser, Cornell, Inman, Clews, Odgen Mills, Cornelius Bliss, Eckert, Austin Corbin, Darius Mills, J. B. Houston, Jesse and James Seligman, Whitelaw Reid, Vining, Norvin Green, Hyde, John G. Moore, and dozens more. "Those who assembled . . . never loved the dead man, and the dead man never loved them," proclaimed a *World* reporter with vast authority, "as he had never loved any of his kind, save those of his blood; so it is the cold truth that there was no sorrow by his bier. There was decent respect—nothing more."

The body lay in a plain casket surrounded by elaborate floral displays, the tribute of those who knew Gould's love of flowers. Helen provided a huge cross made of orchids, Anna a delicate bunch of the rarest orchids. From Houston of Pacific Mail came a ship, fully rigged, inscribed in blossoms, "The voyage ended; safe in port." Beneath the coffin rested a huge square of flowers with the single word "Grandpa" spelled in violets. The service was brief and simple. When the final strains of "Nearer My God to Thee" died away, the guests filed past the casket for a final glimpse at the man they had rarely if ever seen at rest before.

At ten o'clock the next morning the cortege commenced its journey to Woodlawn. Fifth Avenue was again crowded with onlookers, some of whom had been waiting for hours. There was little for them to see. The family alone accompanied the coffin to the cemetery, where Chancellor MacCracken of New York University, an ordained minister, read the Episcopal service and a short prayer. When he was finished, everyone shivered in the cold, raw wind while workmen sealed the casket. The boys on the Street would have said it was done to keep Jay from finding some way to get at them again, but they were a jaded lot. It was rather a precaution inspired by the theft of A. T. Stewart's body some years earlier. The work took only about fifteen minutes, but to those present it seemed an eternity. They watched in horror as hot metal was applied spoonful by spoonful into the cracks on all sides. "There was something indescribably awful about that act," Alice Northrop recalled. "And it was so slow! So unmercifully slow!" George wept openly, and Helen was on the brink of hysteria.[67]

When at last it was done, the men screwed the lid down. Like the marble slab its silver plate was inscribed only with Jay's name and dates. The coffin was placed opposite that of Helen, the opening covered with a marble slab, and the heavy bronze gates closed and locked. No one lingered at the mausoleum. Hurried by the wind, they boarded carriages and were driven back to town. For all of them life would never be the same.

The man was dead, but the legend had only begun. The namesake of a legend had himself become a source of legend almost as fabulous. In the most notable retelling of the story of Jason can be found these concluding words: "What became of the fleece afterwards we do not know, but perhaps it was found after all, like many other golden prizes, not worth the trouble it had cost to procure it."[68]

EPILOGUE

The Legend and the Legacy

He exercised a large influence over the careers of many who had commercial aspirations, and that influence tended to lower the moral tone of business transactions. The example he set is a dangerous one to follow. The methods he adopted are to be avoided. His financial success, judged by the means by which it was attained, is not to be envied. His great wealth was purchased at too high a price.

—*New York Herald*, December 3, 1892

A career in the least comparable with that of JAY GOULD *not only has never been run, but has never been possible before our time. It is in our time that the "operator" has been born, and* JAY GOULD *was an operator pure and simple, although, in a general way of speaking, he was as far as possible from pure and as far as possible from simple. . . . It would be at least very difficult to show that the Nation as a whole is a dollar richer by the existence of* JAY GOULD, *while he himself has become the richer . . . from the expansion of the city and the Nation. He has simply absorbed what would have been made in spite of him and what, if he had not interfered, would have been possessed by somebody else. We cannot say this of* ASTOR *or* VANDERBILT *or* STEWART. *In serving their own ends they were serving public ends, while* GOULD *was a negative quantity in the development of the country where he was not an absolutely retarding and destructive quantity.*

—*New York Times*, December 3, 1892

So far as his life and career made him conspicuous he was an incarnation of cupidity and sordidness—nothing nobler, nothing more. In him the degrading passion of avarice was united with an intellectual capacity which enabled him to gratify the passion by larger processes than most avaricious men are competent for; but it was avarice, nevertheless, as rapacious, as unscrupulous, as remorseless as when exercised in ways which all men agree to call ignoble. His path of wealth was strewn with the wreck of other fortunes and happiness. That he despoiled those who owned stocks and bonds does not make him less a despoiler and a robber. . . . The bane of the social, intellectual and spiritual life of America to-day is the idolatrous homage of the golden calf. . . . Nothing else has contributed so much to promote this evil condition as the apparent worldly success of JAY GOULD. *It has dazzled and deluded multitudes of young men. Jails, insane asylums and almshouses all over the land are peopled with those who aspired to wealth by similar methods, and with their victims. These are but a fraction of those who have been corrupted and morally ruined. The majority are at large, mingling with the community in all the walks of life, excusing, practising and disseminating the vices of which he was the most conspicuous model in modern times.*

—*New York World*, December 4, 1892

THEY WREATHED their laments in hedges, those newspapers that had done so much to create the legend they now deplored. In finding praise or censure for Gould's remarkable career they were commenting more on their own handiwork than on the facts of his life. The legend of Jay Gould lay enshrined in two decades of newspaper files. Within weeks of his death their treasure was plundered and assembled into no less than four biographies by a squadron of alert hack writers, who in this way passed along their misconceptions to future generations of journalists, writers, and other seekers after truth. Thus was the legend perpetuated by sheer repetition.

It should come as no surprise that the legend proved more enduring than the legacy. Gould did all he could to make it otherwise. Having created an empire, he went to extraordinary lengths to preserve it in the hands of his children after his death. Through the instrument of a will composed with the same ingenuity and imagination he displayed in business, he sought not merely to keep his holdings intact but to bind the children together in a common destiny. The net value of the estate was estimated at about $72 million, only $2 million of which was in real estate. Each child received a sixth interest but not outright. The securities were put in trust with the four oldest children as trustees. In effect each would receive the income but could not touch the principal. All stocks in the trust were to be voted as a unit; if the trustees disagreed, George was to have the final voice. George and Helen were appointed guardians of the two younger children, who were to become trustees at the age of twenty-one.[1]

Each share passed intact to descendants, which meant no one could receive more than one-sixth regardless of how many children he or she had. Any child who married without consent of a majority of the trustees forfeited half of his or her share. Jay provided explicitly that if Helen or Anna married, control of their shares could not be transferred to their husbands. These stipulations reflected Gould's wish that his fortune remain entirely within the hands of his children, and that they cooperate in its management. Outsiders were barred from any role in the estate's affairs, partly because of Jay's desire to protect the girls from fortune hunters. He expressed the hope that the younger children would live with Helen until they came of age. To that end he left Helen 579 with all its furnishings and Lyndhurst along with an income of $6,000 per month until the youngest child turned twenty-one.

There were several special bequests. George received $5 million for "having developed a remarkable business ability and having for twelve years devoted himself to my business, and during the past five years taken entire charge of all my difficult interests." Grandson Jay was given $500,000, apparently for being the only namesake. Edwin was tendered the house at 1 East Forty-seventh Street. Gould's three sisters and Abram each received $25,000 and an annual income of $2,000 for life. Jay also left to Sarah Northrop the house and lots in Camden he had bought for her but still held in his name. Not a dime went to anyone outside the family, nothing for servants or charities or public institutions. Jay arranged his fortune to take care of his children for the rest of their lives and for no other purpose. They could provide for anything else as they pleased.

How ironic that a man accused all his life of unloading his holdings had in the end locked them up so tight that no one could pry them from his children. No trust company, lawyer, or executor outside their own circle could intervene in their affairs. All they had to do was get along well enough to work together. In the end this proved the fatal flaw in Jay's plans. Without him there to hold them together, the children soon went their separate ways, a diverse lot with personalities and tastes that not only differed but grated on one another. Within a few years they cleaved into factions and eventually went to war. The trust designed to ensure their unity became instead the cockpit in which their conflicts were fought out with unsparing bitterness.[2]

George proved wholly unequal to the enormous responsibility thrust on him. For a time he held the respect and deference of his fellow titans, but the Street was not a place that allowed one to survive on reputation alone. Too late he discovered the weaknesses concealed by his susceptibility to flattery. George lacked not only his father's genius but his appetite for work. Where Jay learned everything and revealed nothing, George seldom bothered with homework and was careless with details. Fond of society, he and Edith built a palatial mansion in

Lakewood, New Jersey, where they entertained lavishly. George preferred parties and vacations in Europe to the hard grind of work, but the Gould empire was too vast and intricate to be run on a part-time basis. At the century's end George launched an ambitious campaign to create the sort of transcontinental system that had been his father's dream. This gigantic undertaking locked him in mortal combat with E. H. Harriman, who proved a far superior general. George was utterly routed and in 1911 surrendered the presidency of the Missouri Pacific. Two years later he quietly sold the trust's interests in Western Union and Manhattan Elevated. The proceeds were put into safe, conservative investments requiring little attention. The fortune endured but the Gould empire was no more.[3]

Edwin fared better. While continuing to help George with the family investments, he also funneled his own money into a match company, banking, real estate, and other enterprises. By shrewd, careful management he quietly amassed a fortune of his own outside the family trust. His marriage was as happy as his tastes were simple. Where George delighted in champagne suppers, Edwin thrived on the outdoors and the isolation of the woods. At the age of forty-nine he acquired a pilot's license. He and Sally doted on their two children and brooded for months after one died in a freak hunting accident. Later Edwin found solace in establishing the Edwin Gould Foundation, an unusual charity devoted primarily to orphaned and underprivileged children. In both ability and habits he seemed cut more in Jay's image than any of the other boys.

Philanthropy and religion occupied Helen for the rest of her life. What Jay neglected to do in the field of good works was accomplished in good measure by his favorite daughter. Helen achieved renown as a do-gooder of such sweetness and light as to provoke one of Jay's favorite authors, Mark Twain, into complaining from distant Vienna that "we all belong to the nasty stinking human race & of course it is not nice for God's beloved vermin to scoff at each other; but how can I help it when the Abendblatt pukes another mess of Helen-Gould adulation unto me." Reconciled to spinsterhood, she unexpectedly found the ideal husband in Finley J. Shepard, a handsome, athletic gentleman of impeccable manners and congenial tastes. Too old at forty-four to have children, Helen adopted three and became guardian of a fourth. She became the most prim and proper of ladies, devoted to her family, her charities, and a variety of Christian fundamentalist and anticommunist causes. Members of a younger, faster generation, along with all her brothers and sisters except Edwin, found her home suffocating in its piety.[4]

Above all Helen kept alive the shrine of her father's memory. Shortly after his death she had built in Roxbury a lovely stone church still known as the Jay Gould Memorial Reformed Church. She bought the house next door, named it Kirkside, and made it her summer home along with Lyndhurst. At both 579 and Lyndhurst she used her parents' former bedroom for the rest of her life. When the old Gould house in Roxbury came on the market, she purchased it and maintained the place as a public library for the town. Nor did she forget New York University, which received funds for the magnificent Gould Memorial Library along with numerous other gifts.[5]

The three younger children all quarreled with their elders over the marriage proviso in the will and eventually exiled themselves to Europe. Howard was the first to fall away. For a time he remained close to George and helped manage the family properties. Tension arose when Helen scotched his attempt to marry one actress and Edith Gould of all people added her protest to his liaison with another. A furious Howard married the second lady in defiance; it proved a disastrous mistake. In 1909 he obtained a legal separation and fled to England, where he passed the rest of his days in the role of country gentleman. It was a life that suited him, but he could not shake his weakness for actresses. He tried again in 1937, marrying the former wife of actor Oscar Homolka; ten years later they were divorced. Through prudent investments Howard managed to accumulate a large fortune, perhaps the largest of all Gould's sons. Childless, indifferent to his family, he did not even bother to shelter his estate from the tax laws. When he died in 1959, taxes swallowed $50 million of the $62 million he left behind.

Anna made the most spectacular marriage. Small, dark, willful, with black eyes and tiny hands and feet, she seemed to fall in love as easily as Helen fell into worthy charities. There was a fling with an actor and an engagement to Oliver Harriman that terminated abruptly when she met a charming French dandy, the marquis Marie Ernest Paul Boniface de Castellane, or "Boni" to his friends. He found Anna "excessively shy . . . childish, and a trifle malicious; but she possessed charm, and—what is always delightful to a man—possibilities." Boni desperately needed Anna's unique "possibilities." At twenty-seven he was an extravagant rake with a string of insistent creditors and a distinguished family rich only in tradition. Boni's tastes ran to women, gaming, costly and exquisite objets d'art, and spectacular entertainments, none of which he could afford. "I can honestly affirm that Miss Gould's fortune played a secondary part in her attraction for me," he claimed piously. Not a soul believed him. With studied care he pursued a transaction then in vogue: the exchange of an Old World title for New World money.[6]

Their wedding in March 1895 was *the* spring event in New York. For the next decade Boni ran through Anna's income with a zeal that amazed even George, who was no slouch at spending money. Anna stood his extravagances but could not abide his infidelities. One day in January 1905 Boni left the house at three in the afternoon. When he returned at six the place was stripped bare of furniture, carpets, objets d'art, chandeliers, everything; even the electricity and telephone were disconnected. Anna had taken their three children and walked out of his life without a word. After some difficulty her lawyers procured an annulment. Later she married Boni's cousin, the duc de Talleyrand-Perigord, by whom she had two more children. Unlike Boni, the duke had ample wealth of his own. Their marriage endured and kept Anna in France as a member of the titled nobility. She and her children became more French than American.

So too with Frank, the youngest of the children, whose life ran a peculiar course from riches to ruin to resurrection. Bright, charming, debonair, and fun-loving, he graduated from New York University with a degree in engineering and went into the office with George. In 1900 he married a young girl of impeccable social credentials, who bore him two daughters. He did well at business and came to disagree with George's handling of affairs. Gradually he began investing his own money in other enterprises, notably traction and power companies. Unfortunately, his marriage proved a disaster. In 1904 he left his wife, and the children went to live with Helen for nine months every year. After securing a divorce in 1909, Frank wed a showgirl, took back his daughters, turned his business interests over to managers, and exiled himself to a chateau in Normandy. When this marriage too fell apart, Frank sought refuge in orgies of sex and alcohol. By 1917 his dissolution was so complete as to seem hopeless. His wife moved out and an ugly legal battle over the divorce ensued.

During these years Frank saw none of his family except Anna. United by bonds of misery, they nursed a bitter grudge against George and Helen. In May 1919 their resentment flared into the open. Frank filed suit to have George removed as chief executor and trustee of the estate and demanded $25 million for himself and Anna for losses incurred by George's mismanagement. For eight years the suit thrust the Gould name back into the headlines and titillated sensation seekers. Frank's lawyers mercilessly rubbed George's nose in every business mistake he had ever made. They also blistered Edwin for his alleged errors and accused Helen and Howard of compliance. Stung by the charges, George angrily dragged all the family skeletons out of their closets. Back and forth the recriminations flew, each one raising fresh welts of bitterness until the fight triggered a rash of suits and countersuits involving the rights of Jay Gould's grandchildren and great-grandchildren.

A New York Supreme Court justice removed George as executor and divided the estate into six separate trust funds. George appealed but died four years before the final decision was rendered in 1927. The estate was put under the administration of four trust companies. A booming stock market had restored its value to about $66.5 million, far less in real dollars than it had been worth thirty-five years earlier. An army of fifty lawyers consumed $2.7 million in legal fees. Even worse, the fight left scars that never healed. It destroyed the last semblance of family unity. Jay's most ambitious plan, the one that mattered most to him, had been shattered. In its rubble lay his fondest hope for the future; the fortune survived, but in the hands of children who had become strangers to one another.

Amid this debacle Frank rose like a phoenix from the ashes of his own disgrace. He married a third time, to Florence LaCaze, a woman of great character and strength. Their union endured. Frank stopped drinking and turned to business with a vengeance. He bought real estate in the south of France, invested in hotels, built casinos at Nice, Juan-les-Pins, and Bagnoles-de-l'Orne, and introduced the notion of chain ownership to France. By 1930 his real estate holdings on the Riviera alone were worth an estimated $20 million. New York's smart set called him "King of the Croupiers," but he did not mind. After all, his father had been something of a gambler in his day. Like Edwin and Howard, he never touched the capital of his inheritance but used the income to amass a fortune of his own. Before his death in 1956 he gave New York University $2.5 million and an estate at Ardsley-on-Hudson.

Where Frank rose from the dead, George fell from glory. His failures in business were followed by the collapse of what his father had ardently hoped to be his domestic bliss. In later years Edith's weakness for sweets put pounds on her with cruel haste. George took up with a showgirl named Guinevere Sinclair, whom he installed in a house in Rye. Guinevere bore him three children, known in the family as "George's bastards." Everyone knew about the arrangement; George proved as careless at philandering as he was in business. Edith died of a heart attack in November 1921. A few months later George married Guinevere, but he too suffered from high blood pressure and heart trouble. In May 1923 he died in France without a single member of his family at his side. Thus ended the short, unhappy reign of the Crown Prince at the age of fifty-nine.

George was the first to die, Anna the last in 1961. In her later years the duchesse de Talleyrand found herself without country or family. Her three children by Boni had all died young. Of the two born to the duke, the boy had killed himself when his parents objected to his marrying an older woman. The girl wed Count Joseph de Pourtales and remained thoroughly French. After the duke's death in 1937, Anna was left alone with her fortune and her memories. When Helen died in 1938, Anna surprised everyone by purchasing Lyndhurst from the estate. For more than twenty years she came home from France to summer there, staying always in the room she had occupied as a little girl, the small tower bedroom with the oriel window. During those months she dwelled again in the ruins and monuments of her childhood. Only the past held any savor for her, and only at Lyndhurst could she recapture even faint traces of its flavor.

Before her death Anna willed Lyndhurst to the National Trust for Historic Preservation along with an endowment for its maintenance. Although Congress accepted the gift, the mayor and trustees of Tarrytown were reluctant to honor a robber baron like Gould. After Lyndhurst was opened to visitors the guides blathered on about the architecture, the furnishings, the books and paintings, the duchesse—anything but Gould or his career or family. Even this last, lovely corner of the legacy found itself invaded by the legend, which hung like a curse over everything Jay touched or did. All the children knew it, felt it, had to live it down. It was a part of the legacy, not only for them but for later generations as well. In 1966 the *Washington Post* published a six-part exposé on the business activities of Dominic Antonelli and Kingdon Gould, Jr., who was, the paper emphasized, the great-grandson of "the Mephistopheles of Wall Street." Kingdon's response might well have come from Jay. "Why," he asked, "are we Goulds always chosen to play the heavy?"[7]

In arranging his legacy with such fastidious care, Jay foresaw none of the forces that snuffed out his hopes. He could not have, for they belonged to a world foreign to him. Certainly he was not the first parent to miscalculate how his children would turn out. The legacy was, like so many others, the triumph of faith over experience; not so much a wrong guess about the future as a dream that never came true. With the legend he was on firmer ground, if only because he had fought its influence for the last quarter-century of his life. Yet here too his failure was complete, for reasons mostly beyond his grasp. The legend endured despite repeated efforts to correct its errors, growing Hydralike in response to every blow struck against it.

Nothing attests more to its strength or its character than the editorial response to his death. Only the *Tribune* offered a balanced or compassionate appraisal, perhaps because Whitelaw Reid knew Gould better than other editors, but he had always been regarded as Gould's hireling. Tributes poured in from the officers of the Gould corporations, but these were easily dismissed as dutiful memorials. Many of those who knew him as friend or foe insisted that he was quite a different man than the public imagined him to be. "I know there are many people who do not like him," Huntington admitted. "I will say that I always found that he would do just as he agreed to do." Alonzo Cornell, who fought some bitter political battles with Gould, called him "the most misunderstood businessman in the country. . . . He was the soul of honor in his personal integrity." E. Ellery Anderson, Gould's antagonist in the Pacific Railway Commission hearings and in other clashes, declared, "I have always found, even to the most trivial detail, that Mr. Gould lived up to the whole nature of his obligations. Of course, he was always reticent and careful about what he promised, but that promise was invariably fulfilled."[8]

Statements by those closest to Gould shared two themes: that Gould was grossly misunderstood and that, in Connor's words, "the people who spoke unkindly of Jay Gould were those who in many instances had never seen the man and . . . never had any business dealings with him." Dodge consoled George with the assurance that "now when your father has gone those who were closest to him, and knew him best, and gave him credit and honor for many of his deeds that others misconstrued during his life, will now have the satisfaction of seeing them properly appreciated by the country."[9]

It was not to be. The legend only gained ground faster after Gould's death, helped along by certain elements attendant to his passing. Much was made of the fact that the market did not break on hearing of Gould's demise but remained steady and actually advanced. "No harder judgment was ever passed upon a departed millionaire," wrote Burton J. Hendrick. Many observers agreed with the *Times* that the advance showed that "his death was regarded as the removal of a danger even from those corporations." This was an absurd view. Critics seemed to have forgotten that by 1892 the Street was well prepared for Gould's death, and that an army of his associates, brokers, and allies stood ready to counter any bear attack with concerted buying. Nevertheless, the market advance in response to Gould's death became part of the legend. So did the will, with its lack of any bequests for anything or anyone outside the family. That a millionaire could depart this earth without a gesture toward philanthropy did not sit well with public expectations, especially when the millions had been accumulated by means deemed unsavory and unscrupulous.[10]

The Gould legend was firmly in place years before his death and has remained intact for nearly a century. Two questions posed at the beginning still demand answers. Why did the legend arise, and why has it endured?

Clearly the legend has its origins in the two sensational episodes that first brought Gould's name to the public eye, the Erie War and the Gold Corner. The press accounts of these events, and especially the masterfully malign summaries of them by the Adams brothers, fixed Gould's reputation for decades to come. Later exploits on the Street enhanced it to the point of coloring the interpretation of everything else Gould did. Part of the reason lay in Gould's habit of secrecy and his stubborn refusal to cater to public opinion. The fact that he often took pains to defend his actions suggests that he cared deeply about his image, especially in the last decade of his life, but he would not bend a policy to improve it or concede an inch to its polishing. In the end he accepted his reputation reluctantly, even bitterly, as the price of his determination to go his own way.

Of course, secrecy was a way of life on the Street and in corporate corridors. All businessmen had to endure the gap between rumor and reality in discussions of their affairs, but Gould was a genius at the game and played it on a grander scale than anyone else. He was a phenomenon the likes of which the Street and the business world had never seen and will probably never see again. To Americans in the industrial age the lives of all the business titans were the stuff of legend. Boys found their heroes, their models for success, in the men who had started out poor and made it big. It was Horatio Alger's gift not to invent but to articulate the formula that transformed the success story into a morality play. Alger's novels were the most accessible but by no means the only guidebooks for acquiring wealth with honor. Americans had always shown a passion for riches and had long worshiped at the shrine of the self-made man. The "public teachers," as the *World* called them, hammered away at the subject, as did the authors of countless handbooks and manuals on success.[11]

In an age of tumultuous change, however, rules have a way of getting bent, broken, or blurred until they become irrelevant. As I have noted elsewhere, the one event that separates American history into two distinct eras is not the Civil War but the Industrial Revolution. The shock waves of change unleashed by industrialization affected every aspect of American life, leaving in their wake confusion and a sense that everything had been pulled to polar extremes. Progress spawned great wealth and immense poverty, success and scandal, materialism and misery. Growth seemed to pull society apart at the seams, embroiling it in ugly and often violent clashes. Prosperity brought with it problems on a scale never before imagined. Amid this upheaval the old verities no longer seemed sure guides to behavior; often they seemed irrelevant or inapplicable to the new realities.[12]

Like other societies, Victorian America protected itself from the forces of change by drawing the veil of convention around its institutions. One cannot fathom the strictures and attitudes of Victorian culture without realizing how deeply rooted they were in the tumult of industrialization. They served not merely as a rationale for the gospel of progress but as a defense against its excesses and repercussions. They may be likened to a snug, overfurnished parlor, secure and orderly in its proprieties, curtains drawn tight against the storm that rages outside, battering the shutters with rising fury. The revolution wrought by industrialization was not a pretty spectacle. "There is something terrible about it," murmurs a character in Frank Norris's novel *The Pit* uneasily, ". . . something insensate. In a way, it doesn't seem human. It's like a great tidal wave. . . . I suppose it's civilisation in the making, the thing that isn't meant to be seen, as though it were too elemental, too—primordial; like the first verses of Genesis."[13]

Gould was not only part of this revolution but one of its prime movers. His role as an instrument of change, an author of upheaval, was too conspicuous not to attract attention. He emerged as the foremost villain of the age not merely by piling up a fortune; that was acceptable, even desirable, so long as one did it in the proper way with appropriate gestures toward convention. His rise to success followed the classic pattern of the rags-to-riches myth except in the crucial area of method. He did not display the probity and purity of conscience so prominent in Horatio Alger heroes, but neither did the vast majority of businessmen. Elsewhere, however, Gould snubbed convention at every turn. In business he was ruthless and devious, clever and unpredictable, secretive and evasive. Above all he was imaginative, not only brilliant but thoroughly original. In an age that relished flourish, he possessed a stunning economy of motion. Few men matched the cold realism by which Gould conducted his affairs. By realism I mean that he usually grasped the facts of a situation, accepted them for what they were, and acted on that basis. He was unfettered by cant, less cowed or deceived by illusions than any other entrepreneur of his age. He did not cling to shibboleths or axioms. He might invoke them as explanation to the public or in testimony but he was careful not to let them impede or deflect his course.

Therein lay one of the cruelest ironies about Gould. Critics condemned him for his deceit and treachery. Yet the impression lingers that many of them actually loathed Gould because he was too honest for minds used to dealing with reality under wraps. Gould knew what he wanted, went after it, and did not mouth pieties to justify his course. Nor did he ever try to disguise himself in airs of respectability. Religion interested but never absorbed him. Never a strong sectarian, he made no pretense of piety and never coated his activities with a glaze of Christian virtue. By shunning polite society he cut himself off from the surest and safest road to cleansing a tarnished reputation. By keeping his numerous acts of charity behind the scenes he deprived himself of the standard act of atonement expected from men of vast wealth. As the *Atlanta Constitution* observed, "The trouble with Mr. Gould was that he did not make arrangements with the newspapers to herald his deeds of benevolence, and the result was that no one outside of his small circle of intimates and familiars knew the extent of them."[14]

Gould recognized the game of business for what it was and played with few illusions and fewer pretenses. If the markers of success were wealth and power, he meant to win them. In that sense his purpose was as frank and open as his methods were devious and closed. Yet the markers meant less to him than the game itself. There was an intellectuality in his approach to business, the joy of a lawyer handed a juicy brief to write or a Gordian knot to unravel. All these aspects set Gould apart from other men. The role of unsparing realist was not easy in an age that preferred to cushion the harshness of social and economic change with bloated sentimentalism. It was particularly difficult for a man whose habits, tastes, and appearance departed from the conventions of masculinity.

Whatever criteria one chooses for the image of the Victorian male or the entrepreneur, Gould did not fit them. He was not hardy or robust, forceful or domineering, strong or clumsy. He slapped no backs and crunched no bones in a handshake. There was nothing of the peacock in him, no airs of pretension or hint of the autocrat. He competed in no game more rugged than croquet, and his passion for books and flowers was hardly the basis on which to build a reputation for machismo. He was not, and never cared to be, a bon vivant or one of the boys at the club. As he admitted with typical candor:

> I have the disadvantage of not being sociable. Wall Street men are fond of company and sport. A man makes one hundred thousand dollars there and immediately buys a yacht, begins to drive fast horses, and becomes a sport generally. My tastes lie in a different direction. When business hours are over, I go home and spend the remainder of the day with my wife, my children, and my books. Every man has normal inclinations of his own. Mine are domestic. They are not calculated to make me particularly popular in Wall Street, and I cannot help that.[15]

None of this would have mattered had Gould not risen to a position of wealth and power by besting more manly but less talented rivals. Some men resented losing to Gould because he did not fight "fair" or "like a man," not as a Titan but as a Siren, elusive and beguiling. Combat with him was not two rams locking horns but the deadly maneuver of spider and fly. Gould always relished a good fight, but on his own terms and in his own way. It must have infuriated "sports" to be outsmarted, even ruined, by this puny specimen whom physically they could crush underfoot but against whose machinations they were often as helpless as babies. Some regarded Gould as effeminate; more often he was accused of physical cowardice. "Everybody knows that Jay Gould is a coward," sneered an observer in 1884. This charge fit neatly with his "sneaky" business tactics. The image of Gould as a furtive figure operating in the shadows reinforced the impression of a man anxious to avoid not only the light of publicity but any direct confrontation with his rivals. The pummelings by Marrin and Selover helped cement this part of the legend firmly in place.[16]

Here then were all the ingredients for an ideal villain of the new industrial society. Gould was a man of dubious past who had amassed immense wealth and power in secretive fashion by unscrupulous means. He respected no values or conventions, held no loyalties, had no friends, honored no trust. No one could understand or identify with him. He marched to the beat of a drummer whose rhythms baffled and disturbed others for reasons they could not grasp. In an age unsettled by vast, impersonal forces he seemed to personify the evils gnawing at the bosom of society. He was the perfect foil for an age that liked to interpret its nature in moral parables. That is why editorial writers rushed to protest the influence of his example on young men before his body was scarcely cold in its bed. He was the embodiment of the darker side of the American Dream, a reflection of the forces that tainted the gospel of success. There could be no remorse at his passing; one might as well mourn the lifting of a plague.

Or so it appeared to the thousands of Americans who learned of the world by reading newspapers.

If this study has shown anything, it is the extraordinary extent to which the Gould of legend was a creation of the press. For the last decade or so of his life Gould existed more as legend than as man in the minds of most Americans. Whatever they knew of him came from newspapers, and no one looked very closely at how much the newspapers actually knew. The cloak of secrecy Gould drew about himself enabled the clichés and hoary tales about his life and deeds to endure with only token opposition. When he or his friends registered objections, no one believed them. The facts of his business operations or his private life were impossible to obtain, and few reporters had the interest or motive to seek them out. For the New York dailies locked in the circulation wars of the 1880s, the Gould of legend was far more useful and entertaining than the man himself could ever have been. He became a character in an ongoing drama filled with color, action, conflict, passion, intrigue, and occasionally even a particle of truth.

At least three of the New York dailies regarded Gould as their mortal foe, and two of them had fought him implacably since Erie days. The others warred with him intermittently; at his death the harshest critic was the very paper he had once owned. All feared, or professed to fear, his influence over the press and the potential for manipulation of news arising from his control of Western Union. Their dislike of Gould was not only fierce but deeply personal because he appeared to threaten the heart of their own citadel. The Gould legend was not the product of public-spirited editors doing their civic duty with an air of detachment. It was born of animus and fueled with the energy of a blood feud. The financial journals may have been less moved by a sense of vendetta but they found Gould no less inviting a target for a quarter-century. Their audience was confined largely to the business world, however, which limited their influence compared to that of the dailies.

The press, especially the New York papers, did more than create the Gould legend; they passed it on to posterity. It is not difficult to answer the question of why the legend has endured for so long. The major reason is simply that later generations have drawn almost everything they know about Gould from the newspapers, sometimes without even knowing it. Twentieth-century writers relied heavily on the hack biographies that appeared shortly after Gould's death. Without exception these works were compilations of material lifted, often verbatim, from the newspapers. The bulk of their text came from issues for the period December 3–12, 1892, when the dailies carried lengthy accounts of Gould's career, his heirs, his will, and quotations from those who knew him. Even the casual reader dipping into these biographies will discover identical passages in each of them. The clippings were garnished by excerpts from other readily available sources such as Gould's testimony before state and congressional committees, *Chapters of Erie,* and articles culled from periodicals.[17]

This reliance on newspaper accounts should not be surprising given the lack of other sources. There are of course reminiscences by Wall Street men who knew Gould, but most are as inaccurate as they are unflattering, none more so than the often cited memoir of Henry Clews. A close reading suggests that the chroniclers of Wall Street also leaned heavily on the papers for information to refresh their memories. So did Alice Snow, who left the most sympathetic portrait of Gould. It is unfortunate that of the men who worked closely with Gould, only Dodge left behind a memoir. Even if they had erred in the opposite direction, they would at least have presented a different picture.

One other factor helped perpetuate the legend. Changing attitudes made it a useful vehicle for explaining complex historical events in a clear and compelling manner. Even before Gould's death he had become the symbol for an era tarnished and embarrassing to its children. The writers known loosely as muckrakers and debunkers shrank in horror from the excesses and iniquities of industrial society. Unable to put the larger experience into perspective, tormented by the contradictions and paradoxes of what Twain dubbed the Gilded Age, they seized upon its dominant business personalities as sufficient explanation for its aberrations. They were not the first to cope with a revolution by personalizing the vast forces that impelled it, and their chosen targets were certainly vivid and inviting enough. As Justin Kaplan has observed, the stereotypes that resulted made this "the age of the glittering phrase. During a Tragic Era (Claude Bowers) and Age of Negation (Charles Beard), the American people, lulled into a Pragmatic Acquiescence (William James) and betrayed by a bloodless Genteel Tradition (George Santayana), created a Chromo Civilization (E. L. Godkin) and watched complacently as the Robber Barons and the Politicos (both Matthew Josephson), working hand in till, pillaged and plundered the country in a Great Barbecue (Vernon Parrington)."[18]

In this crowded field "Robber Baron" must surely be the most overused and least useful label in all American history. It is not surprising that the popular notion of the industrial era continues to be shaped by the phrase and portrait made famous by Matthew Josephson half a century ago. Even scholars have not been immune to its charms; no less respectable a business historian than Glenn Porter has praised Josephson for his "appreciation of the varieties and complexities of history." The historiographical dogfight over the validity of the concept has scarcely dented its clawlike hold on the popular imagination. Certainly Josephson, like Burton Hendrick, Gustavus Myers, and other muckrakers, is far easier and more exciting to read than the laborious treatises of scholars endeavoring to find more in the era's complexities than picaresque characters. The same may be said for a host of writers since Josephson who have steadfastly maintained the tradition of substituting color for comprehension.[19]

In this work later generations merely built on what had gone before. They tended to rely extensively on newspaper accounts and investigative testimony. Both sources are useful but loaded with pitfalls. In the case of Gould writers took the legend intact and passed it along with a generous glaze of interpretive coloring. It suited perfectly their need to indict the Robber Barons for the excesses and failings of their age. Closer scrutiny would have blurred the focus of their indignation, confused the moral, painted patterns of subtle grays where crisp blacks and whites were wanted. These remarks are, of course, the observations of hindsight. The writers regarded themselves as careful students correcting the profligate errors of a literature that glorified the titans. Myers's publisher emphasized the "incontrovertibility of his findings" and insisted that "no one has yet challenged a single fact in Mr. Myers' work," all of which was "documented with references and direct citations from authentic official records."[20]

Josephson went even further. In 1962, some twenty-eight years after publication of *The Robber Barons,* he warned against the attempt by certain "academic historians" to retouch and restore the image of the titans "like rare pieces of antique furniture. This business of rewriting our history . . . has unpleasant connotations to my mind, recalling the propaganda schemes used in authoritarian societies, and the 'truth factories' in George Orwell's anti-utopian novel *1984*." This incredible statement revealed even more than his books how little Josephson understood about the writing of history. Orwell's "truth factories" rewrote history to conceal the past and serve current needs. Honest revisionist history seeks to discover fresh insights and approaches in several ways. It reinterprets older views in a different light and strives to correct errors on the basis of evidence unavailable to or simply not utilized by previous writers. Above all it recognizes that no student of the past ever has the whole truth or all the evidence at his disposal. History is a more intricate and complex entity than can be embraced by any single point of view. Like other fields of knowledge, our store of data has expanded far more rapidly than our comprehension of its meaning.[21]

No study of the past is more urgently in need of revision than *The Robber Barons,* not only because its interpretation is simplistic but because its handling of the facts is careless to the point of being shoddy. It is in effect a compendium of legend and tall tales shaped by a vision of moral outrage, written in forceful, entertaining style. Yet its influence persists, partly because it is an appealing, even beguiling work, and partly because it says what many readers want to hear. One need not be an apologist for industrial capitalism to suggest that the men and the process were more complex than Josephson depicted. Perhaps the worst fault of his book, and of the "Robber Baron" concept, is that they simply get in the way of understanding what was going on. Both are clear, facile, and misleading, a set of blinders handed down by one generation to the next.

It was Gould's misfortune to be measured in the wrong ways by the wrong instruments for the wrong reasons, but that is often the way with legend. Looked at with a sense of detachment, Gould's business ethics were probably no worse and in some respects better than those of most men. What separated him from others was not the dishonesty but his talent, his daring, his sense of vision. He took great risks and stood behind them. He was quick to change his mind but not at the expense of abandoning an ally or a commitment. No man who was his friend need fear for his wallet; no man who was his foe dared sleep at his post.

By any reckoning Gould must be counted among the two or three most important figures in the development of the American industrial economy. Contrary to legend, he was not a meteor flashing across a turbulent sky before vanishing without a trace. He was not the king of speculators content to amass riches but the prime mover in two industries vital to the Industrial Revolution, transportation and communication. No man did more to make the railway map what it is—through his own work and through the influence he exerted on others responding to his moves. He was that rarest of geniuses, the man of vision who possessed the talent and the tools to realize his vision.

His accomplishments must be seen for what they were, whether or not one likes them or him or the way in which the world he did so much to shape turned out. We see in our past what we want to see, what we need to see at a given time. Perhaps the time has come at last for all of us to let go of the legend and to behold industrialization as something more involved than the antics of a few colorful personalities. Only then will the role of Gould and so many of his peers be fully appreciated. Only then will the legacy become more enduring than the legend.

Source Reference Key

BA	Burlington Archives, Newberry Library, Chicago.
BDC	*Biographical Directory of the American Congress, 1774–1971* (Washington, D.C., 1971).
CFA	Charles Francis Adams, Jr., Papers, Massachusetts Historical Society, Boston.
CLC	William E. Chandler Papers, Library of Congress.
CPH	Collis P. Huntington Papers, microfilm edition, University of Iowa Library, Iowa City.
DAB	*Dictionary of American Biography.*
DP	Grenville M. Dodge Papers, Iowa State Department of Archives and History, Des Moines.
DPL	Grenville M. Dodge Papers, Western Historical Collection, Denver Public Library.
DR	Dodge Record, Iowa State Department of Archives and History, Des Moines.
EG	Material in possession of Edwin Gould.
ER	Erie Railroad Records, Pennsylvania Historical and Museum Commission, Harrisburg.
GLB	Jay Gould Letterbooks in possession of Kingdon Gould, Jr.
GLC	Jay Gould Papers, Library of Congress.
GP	Garrett Family Papers, Library of Congress.
HGS	Helen Gould Shepard Papers, New-York Historical Society, New York City.
HV	Henry Villard Papers, Baker Library, Harvard Graduate School of Business Administration, Boston.
JAG	James A. Garfield Papers, Library of Congress.
JFJ	James F. Joy Papers, Burton Historical Collection, Detroit Public Library.
JGL	Jay Gould Papers at Lyndhurst, Tarrytown, N.Y.
KG	Letters and other materials in possession of Kingdon Gould, Jr.
LDS	Materials in library of Church of Jesus Christ of Latter-day Saints, Salt Lake City.
LFP	Leupp Family Papers, Alexander Library, Rutgers University, New Brunswick, New Jersey.

MM	Huntington, Crocker, Stanford Letters, Mariners Museum, Newport News, Virginia.
MP	Archives of Missouri Pacific Railroad Company, St. Louis.
NCAB	*National Cyclopedia of American Biography.*
OA	Oliver Ames Papers, Stonehill College Library, North Easton, Massachusetts.
PRC	*United States Pacific Railroad Commission Testimony*, 1887.
RGD	R. G. Dun & Co. Collection, Baker Library, Harvard Graduate School of Business Administration, Boston.
RI	Ralph Ingersoll Papers, Mugar Library, Boston University.
RL	Thomas C. Cochran, *Railroad Leaders, 1845–1890* (Cambridge, Mass., 1953).
TGC	Thomas G. Clemson Papers, Cooper Library, Clemson University, Clemson, South Carolina.
TPHR	Texas & Pacific Historical Records, archives of Missouri Pacific Railroad Company, St. Louis.
UP	Union Pacific Railroad Company Collection, Nebraska State Museum and Archives, Lincoln.
WSJ	*Wall Street Journal.*
ZP	Zadock Pratt Papers, New-York Historical Society, New York City.

Notes

Prologue

1. Henry Clews, *Fifty Years in Wall Street* (New York, 1908), 119; Richard O'Connor, *Gould's Millions* (New York, 1962), 191; *New York Times*, Apr. 13, 1873. Parts of this prologue are drawn from Maury Klein, "In Search of Jay Gould," *Business History Review* 52:2 (Summer 1978):166–99.

2. See for example *New York Times*, Aug. 8, 1973; Henry D. Northrop, *Life and Achievements of Jay Gould* (Philadelphia, 1892), 298–316; Murat Halstead and J. Frank Beale, Jr., *Life of Jay Gould: How He Made His Millions* (New York, 1892), 290–92.

3. A more detailed analysis of these sources is in Klein, "In Search of Jay Gould."

4. Clews, *Fifty Years*, 619–58; Northrop, *Life and Achievements*; Halstead and Beale, *Jay Gould*; Trumbull White, *The Wizard of Wall Street* (Chicago, 1892); John S. Ogilvie, *Life and Death of Jay Gould and How He Made His Millions* (New York, 1892); Charles Francis Adams, Jr., and Henry Adams, *Chapters of Erie and Other Essays* (New York, 1886).

5. Julius Grodinsky, *Jay Gould: His Business Career, 1867–1892* (Philadelphia, 1957).

6. O'Connor, *Gould's Millions*; Edwin P. Hoyt, *The Goulds* (New York, 1969); Alfred D. Chandler, Jr., *The Visible Hand; The Managerial Revolution in American Business* (Cambridge, Mass., 1977).

7. This observation appears in Robert Penn Warren's introduction to Dixon Wecter, *The Hero in America* (New York, 1972), xiv.

8. *New York Times*, Dec. 25, 1872.

Chapter 1: Routes

1. The basic source for the Gould family genealogy is Charles Burr Todd, *A General History of the Burr Family* (New York, 1902), 261–77. I am also indebted to Kingdon Gould, Jr., for furnishing me with genealogical data in his possession. See also Katherine Moody Spalding, *The Gould Homestead: A Memorial* (Fairfield, n.d.). A copy of this privately printed pamphlet can be found in the Helen Gould Shepard Papers, New-York Historical Society (hereafter HGS).

2. Alice Northrop Snow, *The Story of Helen Gould* (New York, 1943), 18.

3. Hamilton Burhans, "Early Days of Jay Gould," HGS, 1.

4. Snow, *Helen Gould*, 20–32; Irma Mae Griffin, *The History of the Town of Roxbury* (Roxbury, 1975), 4–6; Jay Gould, *History of Delaware County* (New Orleans, 1977), 68–72. This is a reprint of the 1856 edition.

5. Griffin, *Roxbury*, 1–4; Alf Evers, *The Catskills: From Wilderness to Woodstock* (Garden City, N.Y., 1972), 3.

6. Snow, *Helen Gould*, 32–43.

7. Ibid.; *Report of the Committee of the Senate upon the Relations between Labor and Capital*, Senate Hearings, 41 Cong., vol. 28 (Washington, D.C., 1885), 1:1063; Anna Gould Hough, "Mrs. Hough's Reminiscences," HGS, 1.

8. John B. Gould to "Brother," June 5, 1836, HGS; Sarah Gould Northrop, "Reminiscences," HGS, 1.

9. Northrop, "Reminiscences," 1.

10. Some of the dates and details found in Snow, *Helen Gould*, 32–44, and elsewhere are inaccurate. The death dates given here are taken from the family gravesite, which is marked by an obelisk erected by Jay Gould in 1880.

11. For background on the antirent movement see David M. Ellis, *Landlords and Farmers in the Hudson-Mohawk Region, 1790–1850* (Ithaca, 1946), especially chapters 7–8, and Evers, *The Catskills*, passim.

12. There are several accounts of this scene. The one in Snow, *Helen Gould*, 45–54, contains so many errors of fact as to be unreliable. Jay's version can be found in his *Delaware County*, 260–64. See also Burhans, "Early Days," 2, and Griffin, *Roxbury*, 255–59. For the general situation in Delaware County see Edward P. Cheyney, *Anti-Rent Agitation in the State of New York, 1839–1846* (Philadelphia, 1887), 42–47.

13. Burhans, "Early Days," 2; Northrop, "Reminiscences," 3; *Angell v. Gould* (New York, 1897), 87, 914.

14. Northrop, "Reminiscences," 2–3.

Chapter 2: Manchild

1. John B. Gould to A. C. More, July 11–12, 1865, HGS.

2. Northrop, "Reminiscences" (see Chap. 1, n. 8), 3.

3. For descriptions of young Jay see ibid, passim; Hough, "Reminiscences" (Chap. 1, n. 7), passim; *Angell* (Chap. 1, n. 13), 583–84; Frank Allaben, "Was Jay Gould Misjudged?" *National Magazine* (May–June, 1893), 83–87.

4. Allaben, "Jay Gould," 84–85.

5. Ibid., 86; *Angell*, 152, 584; Northrop, "Reminiscences," 3; Hough, "Reminiscences," 10.

6. Northrop, "Reminiscences," 6; Allaben, "Jay Gould," 86.

7. Northrop, "Reminiscences," 8; Hough, "Reminiscences," 8; *Labor and Capital* (see Chap. 1, n. 7), 1063.

8. Northrop, "Reminiscences," 4–5; *Angell*, 89.

9. The Hobart episode is recounted in various places. The most reliable are Northrop, "Reminiscences," 8–9; Hough, "Reminiscences," 8–10; *Labor and Capital*, 1063.

10. Quoted in Snow, *Helen Gould* (see Chap. 1, n. 2), 66. Elsewhere Burroughs referred to Oliver as "a superior man." John Burroughs, *My Boyhood* (Garden City, N.Y., 1922), 68.

11. Clara Barrus, *John Burroughs: Boy and Man* (Garden City, N.Y., 1922), 143. See also Allaben, "Jay Gould," 85; Burroughs, *My Boyhood*, 32–34.

12. The entire composition can be found in *Angell*, 470, 586; Allaben, "Jay Gould," 90, 110. A generous sampling of Gould's letters during this period can be found in *Angell*.

13. *Angell*, 734; Northrop, "Reminiscences," 7.

14. *Angell*, 168.

15. *Angell*, 90; Burhans, "Early Days" (see Chap. 1, n. 3), 2–3. A copy of the agreement, dated July 17, 1851, is in *Angell*, 472–73.

16. *Angell*, 471–72; Burhans, "Early Days," 2–3; *Angell*, 90–91; *Labor and Capital*, 1063.

17. Hough, "Reminiscences," 10; Sarah Gould to Edmund More, Feb. 29, 1852, HGS.

18. *Angell*, 92–93, 152; Northrop, "Reminiscences," 10–11; *Labor and Capital*, 1063; Abel Crosby to Helen Gould, May 20, 1897, HGS.

Chapter 3: Rungs

1. *Labor and Capital* (see Chap. 1, n. 7), 1063–64; *Angell* (see Chap. 1, n. 13), 161–62; Abel Crosby to Helen Gould, May 20, 1897, HGS.

2. *Angell*, 157, 159–60, 163–67, 604–8, 679–89; *Labor and Capital*, 1065.

3. *Angell*, 166–67, 169–70, 474–75, 685–89, 726–27; *New York Times*, Sept. 30, 1883; *Labor and Capital*, 1065.

4. *Angell*, 168–69, 171–74, 475–79, 552–57, 756–57, 980–81. Jay informed the family of his schooling in typically laconic fashion: "I am going to school now at the Albany Academy. School commences at 9 in the morning and closes at two in the afternoon. One recess of five minutes. Eat twice a day." *Angell*, 177.

5. *Angell*, 632–36. At the end of this lengthy letter Gould remarked, "This is the first time ever that one of my letters has exceeded one sheet." The theodolite story is also recounted in J. W. McLany, "Some Memories of Mr. Jay Gould," HGS.

6. *Angell*, 50–51, 104–5, 185–89, 632–36.

7. Accounts of the mousetrap episode, differing only in detail, can be found in *Angell*, 50–51, 154–56, 557, 564–65, 639–41, 644–45, 1096–97; Northrop, "Reminiscences" (see Chap. 1, n. 8), 9–10; Crosby to Helen Gould, May 20, 1897, HGS; *New York Herald*, July 20, 1853. See also the interview with Gould in the *New York Herald*, Feb. 28, 1881.

8. *Angell*, 50–51, 154, 176–77, 557, 620–25, 762–63, 932.

9. Ibid., 443–44, 735, 932; Crosby to Helen Gould, May 20, 1897, HGS.

10. *Angell*, 50–51, 56–57, 105, 537–41, 558–60, 705–12.

11. Ibid., 55–59, 197–201, 205, 207, 236, 480–82.

12. Ibid., 166–67, 206–7, 639–41, 737.

13. Ibid., 644–45.

14. Ibid., 642–45.

15. Ibid., 145; Snow, *Helen Gould* (see Chap. 1, n. 2), 44, 80; Burhans, "Early Days" (see Chap. 1, n. 3), 4–5.

16. For a reference to neither Jay nor his father being a convert, see *Angell*, 607.

17. Ibid., 776–84.

18. Ibid., 18–25, 60–84, 210–13, 561, 712–19; Jay Gould to Sarah Gould, Dec. 6, 1853, HGS.

Chapter 4: Slips

1. For background detail on the railroad survey see *Angell* (Chap. 1, n. 13), 304–7, 311–12.

2. Ibid., 132, 216–20, 304–5, 313–16, 654–57, 757–58.

3. Ibid., 659–63.

4. Ibid., 132, 490–91, 660–69.

5. Ibid., 226–27, 455–56, 665–69, 742.

6. Ibid., 64–65, 230–34, 316, 493, 669–71.
7. Ibid., 285–87, 294, 304–5, 317–18; McLany, "Some Memories" (see Chap. 3, n. 5).
8. *Angell*, 456, 672–76.
9. Ibid., 319, 677–78, 743.
10. Hough, "Reminiscences" (see Chap. 1, n. 7), 14.
11. *Angell*, 754; Barrus, *John Burroughs: Boy and Man* (see Chap. 2, n. 11), 143.
12. *Angell*, 230–33, 240, 669–71; Burhans, "Early Days" (see Chap. 1, n. 3), 4.
13. McLany, "Some Memories."
14. Jay Gould to Peter Wright, Apr. 17, 1855, Peter Wright Papers, Burton Historical Collection, Detroit Public Library; *Angell*, 240, 320; Burhans "Early Days," 4.
15. Hough, "Reminiscences," 10; Burhans, "Early Days," 4. The standard version of this episode can be found in Matthew Josephson, *The Robber Barons* (New York, 1934), 37–38.
16. Snow, *Helen Gould* (see Chap. 1, n. 2), 64–65, 93; *Angell*, 294, 300, 964, 969; Burhans, "Early Days," 4.
17. Gould, *Delaware County* (see Chap. 1, n. 4), v–vi.
18. Jay Gould to Zadock Pratt, Apr. 10, 1856, Ralph Ingersoll Papers, Mugar Library, Boston University (hereafter RI); *Angell*, 641.
19. *Angell*, 198, 479. The portions escaping the fire consisted only of a few proof sheets. Gould also had at his disposal some extracts published earlier in the *Bloomville Mirror*. Allaben, "Jay Gould" (see Chap. 2, n. 3), 89.
20. Northrop, *Life and Achievements* (see Prologue, n. 2), 322.
21. Burroughs, *My Boyhood* (see Chap. 2, n. 10), 34–35.

Chapter 5: Opportunity's Knock

1. The tannery episode is recounted in Halstead and Beale, *Jay Gould* (see Prologue, n. 2), 29–35; Northrop, *Life and Achievements* (see Prologue, n. 2), 41–52; White, *Wizard* (see Prologue, n. 4), 37–48; Walter R. Houghton, *Kings of Fortune, or The Triumphs and Achievements of Noble, Self-Made Men* (Chicago, 1886), 264–66; Gustavus Myers, *History of the Great American Fortunes* (New York, 1909), 396; Josephson, *Robber Barons* (see Chap. 4, n. 15), 39–41; Robert I. Warshow, *Jay Gould: The Story of a Fortune* (New York, 1928), 44–50; M. J. Martin, *Jay Gould and His Tannery* (Scranton, 1945); O'Connor, *Gould's Millions* (see Prologue, n. 1), 34–45; Hoyt, *The Goulds* (see Prologue, n. 6), 14–26. The only attempt to offer a revised version of the story is Lucius F. Ellsworth, "Jay Gould and the Leather Industry, Success or Failure?" in Allen L. Dickes, ed., *Proceedings of the Business History Conference* (Fort Worth, Tex., 1973), 135–59.
2. Evers, *Catskills* (see Chap. 1, n. 5), 342; Zadock Pratt, "Autobiographical Outline," Zadock Pratt Papers, New-York Historical Society, New York City (hereafter ZP). This outline was apparently done during the 1850s, before Pratt's partnership with Gould. In 1868 his biography appeared in book form. Pratt dictated its contents to a reporter for the leather trade journal, *Shoe and Leather Reporter*. Zadock Pratt, *Chronological Biography of the Hon. Zadock Pratt* (New York, 1868). Biographical information on Pratt can be found in ZP and RI; Evers, *Catskills*, 340–51; Frank W. Norcross, *A History of the New York Swamp* (New York, 1901), 114–17; William Hunt, *The American Biographical Sketch Book* (New York, 1848); *Biographical Directory of the American Congress, 1774–1971* (Washington, D.C., 1971), 1566 (hereafter *BDC*).
3. Pratt, "Autobiographical Outline"; Evers, *Catskills*, 346; Norcross, *New York Swamp*, 116; Pratt, *Chronological Biography*, 8.

4. Burhans, "Early Days" (see Chap. 1, n. 3), 8; Pratt, *Chronological Biography,* 32–34; Norcross, *New York Swamp,* 116–17; Evers, *Catskills,* 348–50. This notebook is in ZP. The remaining images still stand along Route 23 outside Prattsville. Pratt's house has also been restored and serves as a museum.

5. McLany, "Jay Gould" (see Chap. 3, n. 5); Gould to Pratt, Dec. 25, 1855, Apr. 10, 1856, RI; Bettie Palen to A. Palen, Mar. 25, 1893, HGS; *Angell* (see Chap. 1, n. 13), 966; Peter Van Amburgh to Helen Gould, Dec. 19, 1892, HGS.

6. *Labor and Capital* (see Chap. 1, n. 7), 1065; Burhans, "Early Days," 9; Pratt, *Chronological Biography,* 42; *Angell,* 149; Snow, *Helen Gould* (see Chap. 1, n. 2), 93–94. For a summary of Pratt's tanning career see Lucius F. Ellsworth, *Craft to National Industry in the Nineteenth Century: A Case Study of the Transformation of the New York State Tanning Industry* (New York, 1975), chap. 6.

7. Gould to Oliver, Oct. 20, 1865, HGS.

8. Ibid.; *Labor and Capital,* 1065; Gould to Pratt, Sept. 5, 1856, RI. Various contracts for bark can be found in Deed Book 68, 60, 161; Deed Book 69, 145, 147; and Deed Book 70, 217, 267, Luzerne County, Pennsylvania. Several of the contracts are reprinted in Martin, *Jay Gould and His Tannery,* 24–27.

9. Gould to Oliver, Oct. 20, 1856, HGS; Gould to Pratt, Sept. 5 and Dec. 24, 1856, Jan. 31 and Feb. 3, 1857, RI. For background information on the tanning process and its technology see Ellsworth, *Craft to Industry,* especially chaps. 4–6.

10. Gould to Pratt, Dec. 24, 1856, Feb. 3, 1857, RI.

11. McLany, "Jay Gould"; Statement of John Gardner, Oct. 3, 1897, HGS.

12. Gould to Pratt, Feb. 3, Oct. 22, and Nov. 29, 1857, RI; Burhans, "Early Days," 11.

13. This summary of the leather industry is drawn from Ellsworth, *Craft to Industry,* chaps. 1–3; Ellsworth, "Jay Gould and the Leather Industry," 136–37; Norcross, *New York Swamp,* 1–68.

14. Gould to Pratt, July 27, Aug. 8, and Oct. 22, 1857, RI; Ellsworth, "Jay Gould and the Tanning Industry," 141–42; Ellsworth, *Craft to Industry,* 215–21.

15. Gould to Pratt, July 27, Aug. 8, and Oct. 22, 1857, RI.

16. Gould to Pratt, Nov. 27 and Nov. 29, 1857, RI, and Dec. 11, 1857, ZP.

17. Burhans, "Early Days," 11.

18. Ibid.; Norcross, *New York Swamp,* 58; J. B. Kissam to Charles M. Leupp, June 5, 1858, Leupp Family Papers, Rutgers University Library (hereafter LFP); Gould to Pratt, July 23 and Sept. 15, 1858, RI; Gould to C. M. Leupp & Co., June 13, 1859, LFP; Martin, *Jay Gould and His Tannery,* 29.

19. Burhans, "Early Days," 12; Statement of John Gardner, HGS. Of this episode Gould later said only, "We carried on the business for a while, and then I bought Mr. Pratt out. . . ." *Labor and Capital,* 1065. There are numerous versions of this offer, none of which seem to agree on the dollar amount or the terms. The figures given here are taken from Burhans, "Early Days," 12, because they are realistic and fit other evidence.

20. Martin, *Jay Gould and His Tannery,* 29, 36.

21. Pratt, *Chronological Biography,* 52; Julia Pratt Ingersoll, Untitled Journal, RI, Box 33.

Chapter 6: First Blood

1. *New York Times,* Oct. 7, 1859; Norcross, *New York Swamp* (see Chap. 5, n. 2), 51–58; R. G. Dun & Co. Collection, New York City, 365:198, 193:655 (hereafter RGD, NYC). See also the correspondence in Gideon Lee folder, LFP. Gould's statement first appeared in a

Wilkes-Barre newspaper, the *Luzerne Union*, Mar. 21, 1860. It was reprinted in the *New York Herald*, Mar. 23, 1860. Authenticated copies of several articles relevant to the affair can be found in "Copy of Local Accounts of the Gouldsboro War," HGS.

2. Martin, *Jay Gould and His Tannery* (see Chap. 5, n. 1), 37; *New York Herald*, Mar. 23, 1860; C. M. Leupp & Company to Gould, July ? and July 27, 1859, LFP. My account of the tannery dispute follows closely that of Ellsworth, "Jay Gould and the Leather Industry" (see Chap. 5, n. 1).

3. *New York Herald*, Mar. 23, 1860; Gould to Leupp & Company, June 13, 1859, LFP. It was the practice of most leather merchants to require tanners to pay for hides received by issuing notes. See Ellsworth, *Craft to Industry* (Chap. 5, n. 6), 64.

4. The contract with Freeman, dated Feb. 22, 1859, can be found in HGS. It is not known precisely when Gould opened his New York office. The Spruce Street address is given in a letter to Leupp & Company, Dec. 12, 1859, reprinted in the *New York Herald*, Mar. 23, 1860, and Allaben, "Was Jay Gould Misjudged?" (see Chap. 2, n. 3), 98. See also Burhans, "Early Days" (Chap. 1, n. 3), 12.

5. Leupp & Company to Gould, July ?, July 27, and Aug. 26, 1859. Notes drawn by Gould and endorsed by Pratt were going to protest as late as November 1859. See the protested notes dated Mar. 3, Nov. 4, and Nov. 5, 1859, RI.

6. Allaben, "Was Jay Gould Misjudged?" 91; Gould to Leupp & Company, June 13, June 28, July ?, and July 27, 1859, LFP; D. W. Lee to Gould, July 22, 1859, LFP; Burhans, "Early Days," 13.

7. Leupp & Company to Gould, Aug. 26, 1859, LFP; Gould to Leupp & Company, Aug. 26, 1859, LFP; *New York Herald*, Mar. 23, 1860; Allaben, "Was Jay Gould Misjudged?" 95.

8. This information is taken from Lee's testimony at the inquest, which can be found in the *New York Herald* and the *New York Tribune*, Oct. 7, 1859.

9. Ibid.; D. W. Lee to Thomas G. Clemson, Oct. 11, 1859, Thomas G. Clemson Papers, Clemson University Library (hereafter TGC). Some friends privately reproached the family for allowing Leupp to go about in his condition. See, for example, Mrs. Calvert to Mrs. Thomas G. Clemson, Oct. 24, 1859, and "E" to Mrs. Thomas G. Clemson, Nov. 26, 1859, TGC.

10. Gould to Leupp & Company, Dec. 27, 1859, LFP. None of the surviving letters from Leupp & Company to Gould are in Leupp's own hand. Samples of Leupp's handwriting can be found in LFP. See also Charles M. Leupp to Thomas G. Clemson, Sept. 30, 1859, TGC.

11. *New York Herald*, Oct. 7, 1859. See also Allaben, "Was Jay Gould Misjudged?" 99–100.

12. This scene is taken from testimony given at the inquest and reprinted in the *New York Herald* and the *New York Tribune*, Oct. 7, 1859. For an example of how one daughter referred to her father's moods, see Laura Leupp to Mrs. Thomas G. Clemson, Sept. 30, 1859, TGC.

13. Laura Leupp to Mrs. Thomas G. Clemson, Oct. 6, 1859, TGC.

14. *New York Herald*, Mar. 23, 1860; Allaben, "Was Jay Gould Misjudged?" 96; Martin, *Jay Gould and His Tannery*, 32.

15. Gould to Leupp & Company, Dec. 27, 1859, LFP; Martin, *Jay Gould and His Tannery*, 31.

16. *Lee v. Gould*, 47 Pa. 398 & Seq.; Copy of Opinion, Application for an Injunction and Appointment of a Receiver, *D. W. Lee v. Jay Gould et al.*, May 2, 1860, Common Pleas Court, Luzerne County, Pennsylvania, LFP.

17. *New York Herald*, Mar. 23, 1860; Allaben, "Was Jay Gould Misjudged?" 96. This is Gould's version of what happened. It is not contradicted by Lee's statement, which is vague and general on events prior to his arrival in Gouldsboro. See *Luzerne Union*, Mar. 14, 1860; *New York Herald*, Mar. 17, 1860; Allaben, "Was Jay Gould Misjudged?" 94–95.

18. A credit reporter observed, "It is positively asserted that Leupp's means will not exceed 100m, which is a great deal less than he was thought to be worth." RGD, NYC, 193:655. The explanation for Lee's concern given by Ellsworth, "Jay Gould and the Leather Industry," 149, argues that Gould's prolonged absences from the tannery, "coupled with Gould's lack of technical experience prompted Lee to doubt Gould's ability to produce high quality leather." But Gould published a letter written him by Leupp & Company, dated Dec. 29, 1859, attesting to the fact that "since you have been tanning for our house yours has been the quickest tannage which our books record, showing in one instance the unusual fact of a sale of all the leather before the maturity of the hide notes." *Luzerne Union*, Mar. 21, 1860; Allaben, "Was Jay Gould Misjudged?" 98.

19. Unless otherwise noted, the narrative that follows is pieced together from the accounts compiled in "Copy of Local Accounts of the Gouldsboro War" and from Allaben, "Was Jay Gould Misjudged?" 91–103.

20. Lee stated in his version that Gould arrived on Mar. 3 rather than Mar. 5, but their accounts of what followed do not differ in the essential facts.

21. William L. Thomson to D. W. Lee, Mar. 9, 1860, LFP; *Jay Gould* v. *Charles C. Niebuhr et al.*, Apr. 6, 1860, LFP; *Lee* v. *Gould*, May 2, 1860. The replevin suit is reprinted in Martin, *Jay Gould and His Tannery*, 22–24.

22. Martin, *Jay Gould and His Tannery*, 32, 36–49; Norcross, *New York Swamp*, 60; RGD, NYC, 193:655.

23. Grodinsky, *Jay Gould* (see Prologue, n. 5), 450.

Chapter 7: Fierce Extremes

1. A good example is the account of Gould's early years by that incorrigible Wall Street gossip, Henry Clews, in which almost every fact offered is inaccurate. See Clews, *Fifty Years* (Prologue, n. 1), 619–22.

2. Quoted in Halstead and Beale, *Jay Gould* (see Prologue, n. 2), 83.

3. *New York Times*, Dec. 3, 1892.

4. Walter K. Earle, *Mr. Shearman and Mr. Sterling and How They Grew* (n.p., n.d.), 70–71.

5. Robert Sobel, *The Big Board: A History of the New York Stock Market* (New York, 1965), 30–61; E. C. Stedman, *The New York Stock Exchange* (New York, 1905), 57–113.

6. Maury Klein, "The Boys Who Stayed Behind: Northern Industrialists and the Civil War," in James I. Robertson, Jr., and Richard M. McMurry, eds., *Rank and File: Civil War Essays in Honor of Bell Irvin Wiley* (San Rafael, Calif., 1976), 137–56.

7. Bouck White, *The Book of Daniel Drew* (New York, 1910), 160; William W. Fowler, *Twenty Years of Inside Life in Wall Street* (New York, 1880), 156–57, 244–47; Stedman, *Stock Exchange*, 151. The charge that Gould profited from advance information can be found in Halstead and Beale, *Jay Gould*, 73–74, and Josephson, *Robber Barons* (see Chap. 4, n. 15), 63–64.

8. Stedman, *Stock Exchange*, 132–66; Fritz Redlich, *The Molding of American Banking* (Ann Arbor, 1951), 85–98; Henrietta Larson, *Jay Cooke, Private Banker* (Cambridge, Mass., 1936), 96–175; Bray Hammond, *Sovereignty and an Empty Purse: Banks and Politics in the Civil War* (Princeton, 1970), passim; Fowler, *Twenty Years*, 411.

9. Stedman, *Stock Exchange*, 119, 168–95; Fowler, *Twenty Years*, 144–357; James K. Medbery, *Men and Mysteries of Wall Street* (New York, 1870), 92–100, 152–93.

10. Medbery, *Men and Mysteries*, 10–11, 39; Stedman, *Stock Exchange*, 155–65; Fowler, *Twenty Years*, 52–58; Sobel, *Big Board*, 77–78; Francis L. Eames, *The New York Stock Exchange* (New York, 1894), 50–52, 57. See also Robert Sobel, *The Curbstone Brokers* (New York, 1970), 21–49. The Open Board also permitted nonmembers to use its facilities for a fee of $50.

11. Fowler, *Twenty Years*, 238.

12. RGD, NYC, 193:660.

13. Ibid., 193:660, 347:737.

14. Ibid.

15. *Labor and Capital* (see Chap. 1, n. 7), 1065–66; RGD, NYC, 193:660. Most earlier sources assert that Gould got into the Rutland through his father-in-law, Daniel S. Miller. I have found no evidence for that link and it does not fit the chronology well. The fact that Wilson owned the Rutland bonds makes a more plausible connection with Gould.

16. Gould to his wife, May 7, 1864, Jay Gould Papers, Lyndhurst, Tarrytown, N.Y. (hereafter JGL). The Rutland was leased to the Rensselaer & Saratoga in June 1865. See Henry V. Poor, *Manual of the Railroads of the United States for 1877–1878* (New York, 1878), 269.

17. Pictures of Helen and Jay Gould can be found in HGS and at Lyndhurst, the former Gould estate now owned and managed by the National Trust for Historic Preservation.

18. For various accounts of their meeting see Clews, *Fifty Years*, 621; Halstead and Beale, *Jay Gould*, 58–68; Northrop, *Life and Achievements* (Prologue, n. 2), 53; White, *Wizard* (Prologue, n. 4), 49–50. Later biographers simply repeat and embellish these versions.

19. RGD, NYC, 201:420.

20. White, *Wizard*, 50–51; Allaben, "Was Jay Gould Misjudged?" (see Chap. 2, n. 3), 109–10; *Angell* (see Chap. 1, n. 13), 786.

21. *Angell*, 468–69. The printed invitation is in JGL. Gould's residences between 1868 and 1870 can be found in the New York City directories for those years.

Chapter 8: Ascendance

1. For background on the later careers of Gould's sisters see Snow, *Helen Gould* (Chap. 1, n. 2), 126–38. Alice Northrop Snow was the daughter of Sarah Gould Northrop.

2. The best and most comprehensive account of the Erie prior to 1867 is Edward H. Mott, *Between the Ocean and the Lakes: The Story of Erie* (New York, 1908), 1–146. See also Grodinsky, *Jay Gould* (Prologue, n. 5), 27–37.

3. Mott, *Story of Erie*, 487–88; Clews, *Fifty Years* (see Prologue, n. 1), 117–56. White, *Book of Daniel Drew* (see Chap. 7, n. 7), purports to be Drew's own diary, revised and edited by White, but its authenticity has been challenged, most notably by John Brooks, ed., *The Autobiography of American Business* (Garden City, N.Y., 1975) 25–27. My reading of the work supports the view that it is a spurious source, and I have chosen not to use it except where the information can be verified elsewhere.

4. Medbery, *Men and Mysteries* (see Chap. 7, n. 9), 168–69.

5. Fowler, *Twenty Years* (see Chap. 7, n. 7), 438–39; Stedman, *Stock Exchange* (see Chap. 7, n. 5), 170–71, 198–99; *American Railroad Journal*, Oct. 22, 1857.

6. Vanderbilt awaits a definitive biography. Of those available, the most adequate

are William A. Croffut, *The Vanderbilts and the Story of Their Fortune* (New York, 1886), and Wheaton J. Lane, *Commodore Vanderbilt: An Epic of the Steam Age* (New York, 1942).

7. Mott, *Story of Erie*, 139–40; Burton J. Hendrick, "The Vanderbilt Fortune," *McClure's Magazine* 32 (November 1908): 46–62; Stedman, *Stock Exchange*, 173–81; Lane, *Commodore Vanderbilt*, 184–228.

8. Grodinsky, *Jay Gould*, 32–33; *American Railroad Journal*, June 6, 1866; Mott, *Story of Erie*, 140–42; Stedman, *Stock Exchange*, 196–97.

9. Adams, *Chapters of Erie* (see Prologue, n. 4), 13; Erie Exec. Comm. Minutes, June 3, 1867, Erie Directors' Minutes, June 5, 1867, Erie Railroad Records, Pennsylvania Historical and Museum Commission, Harrisburg, Pa. (hereafter ER); Mott, *Story of Erie*, 142–43; Grodinsky, *Jay Gould*, 33–34.

10. Differing versions of the 1867 election fight are in Adams, *Chapters of Erie*, 13–16; Croffut, *The Vanderbilts*, 88–89; Mott, *Story of Erie*, 143–44; Stedman, *Stock Exchange*, 198–99; Lane, *Commodore Vanderbilt*, 237–41; Grodinsky, *Jay Gould*, 34–35.

11. Erie Directors' Minutes, Oct. 8, 1867, ER; Mott, *Story of Erie*, 144; Lane, *Commodore Vanderbilt*, 240.

12. Giovanni P. Morosini, "Reminiscences of Mr. Gould," HGS; Morosini to Helen Gould Shepard, May 25, 1893, HGS.

13. Fowler, *Twenty Years*, 529–32; Clews, *Fifty Years*, 223–28. For the brokerage firm's reputation prior to Erie, see RGD, NYC, 417:176. A fourth partner, Abin Altman, operated the house of Altman & Company in Buffalo. Smith and Martin described their functions within the firm in *Investigation into the Causes of the Gold Panic*, House Report no. 31, 41st Cong., 2d sess., 1870 (Serial 1436), 190–202, 281–82.

14. This description is based in part on William A. Swanberg, *Jim Fisk: The Career of an Improbable Rascal* (New York, 1959), 1–2. Swanberg does a superb job of capturing Fisk as a character but is helpless in the face of his business career. He states explicitly that "there is no pretense at close analysis of his complicated financial maneuvers" (299).

15. Ibid., 4–24; Mott, *Story of Erie*, 488–92; Medbery, *Men and Mysteries*, 165–68.

16. Fowler, *Twenty Years*, 477. For the reputation of Fisk & Belden and its partners see RGD, NYC, 416:D91.

17. RGD, NYC, 412:137.

18. The best general accounts of the Erie War are Adams, *Chapters of Erie*, 15–100; Mott, *Story of Erie*, 147–60; Stedman, *Stock Exchange*, 198–208; Lane, *Commodore Vanderbilt*, 243–59; Grodinsky, *Jay Gould*, 49–53. See also the testimony in *Select Committee to Investigate the Erie Railroad*, New York State Assembly Document no. 98 (1873).

19. Mott, *Story of Erie*, 141–42; Stedman, *Stock Exchange*, 197; Fowler, *Twenty Years*, 481–83; *New York Herald*, Feb. 19, 1868; *New York Post*, Feb. 28, 1868; Erie Exec. Comm. Minutes, Nov. 27, 1867, ER. The latter date marked Gould's first appearance on the executive committee.

20. Erie Exec. Comm. Minutes, Feb. 18, Feb. 19, and Mar. 9, 1868, ER.

21. Fowler, *Twenty Years*, 492–94; Stedman, *Stock Exchange*, 433–35; *Commercial and Financial Chronicle* 6:295; *New York Post*, Mar. 4, 1868; *New York Times*, Mar. 6–9, 1868; Adams, *Chapters of Erie*, 22–23.

22. Erie Exec. Comm. Minutes, Mar. 3–5, 1868, ER. A printed version of the special report can be found in these minutes. For Riddle's earlier report see Mott, *Story of Erie*, 146. See also *New York Herald*, Mar. 10, 1868.

23. Adams, *Chapters of Erie*, 27–29; Erie Directors' Minutes, Mar. 9, 1868, ER.

24. Stedman, *Stock Exchange*, 200–202; Fowler, *Twenty Years*, 494–98; Croffut, *The Vanderbilts*, 90–93; *Chronicle* 6:325–26.

25. Adams, *Chapters of Erie*, 29–30; Mott, *Story of Erie*, 150. Details of the flight can be found in the *New York Herald* and other papers, March 1868.

26. Swanberg, *Jim Fisk*, 49–53.

27. Details of the legal war can be found in the *New York Times*, Mar. 15–20, 1868, and are summarized in Adams, *Chapters of Erie*, 32–42. For a description of the legal battle by one of Gould's lawyers see Earle, *Mr. Shearman and Mr. Sterling* (Chap. 7, n. 4), 15–17. Notice of the senate investigating committee is in the *New York Times*, Mar. 7, 1868. Vanderbilt's dangerous position is best described in Grodinsky, *Jay Gould*, 41–43.

28. The bill incorporating the Erie was signed by the governor on Mar. 30. See *New York Herald* and *New York Times*, Mar. 31, 1868. A convenient summary of press reaction to the situation can be found in Lane, *Commodore Vanderbilt*, 248–49. See also *New York Tribune*, Mar. 28, 1868, and *Chronicle* 6:295, 326–27.

29. *New York Times*, Mar. 25–31, 1868; *New York Post*, Mar. 28, 1868; *New York Herald*, Apr. 2, 1868; Adams, *Chapters of Erie*, 43–49; *Chronicle* 6:325–27.

30. Adams, *Chapters of Erie*, 49–50; Mott, *Story of Erie*, 152–54. The legal convolutions may be followed in the *New York Times*, Apr. 5, Apr. 8, and Apr. 10, 1868; *New York Herald*, Apr. 5, Apr. 8, and Apr. 10, 1868; *New York Tribune*, Apr. 7, 1868.

31. *New York World*, Apr. 7, Apr. 10, Apr. 12, and Apr. 15, 1868; *New York Times*, Apr. 12, 1868; Adams, *Chapters of Erie*, 51.

32. Adams, *Chapters of Erie*, 55. The Albany debacle is best followed in the New York papers for April 1868. The final bill is reprinted in the *New York Tribune*, Apr. 21, 1868.

33. *New York Herald*, Apr. 2, 1868; *Chronicle* 6:422–23; Adams, *Chapters of Erie*, 52–53. See also Mott, *Story of Erie*, 447–51.

34. Differing versions of these negotiations can be found in Adams, *Chapters of Erie*, 56–58; Fowler, *Twenty Years*, 500–502; Mott, *Story of Erie*, 155; Croffut, *The Vanderbilts*, 94–95.

35. *Chronicle* 6:519; Mott, *Story of Erie*, 155.

36. Mott, *Story of Erie*, 155. Details of the settlement are contained in the Erie Directors' Minutes, July 10, 1868, and Erie Exec. Comm. Minutes, June 11, June 13, and Dec. 5, 1868, ER. For Sweeny's appointment see *New York Tribune*, July 1, 1868, and *Chronicle* 7:12.

37. Erie Directors' Minutes, July 10 and July 30, 1868, ER; *New York Times*, July 1 and July 25, 1868.

38. Mott, *Story of Erie*, 161.

39. Erie Exec. Comm. Minutes, Aug. 19 and Sept. 18, 1868; Erie Directors' Minutes, July 30, Aug. 19, and Oct. 13, 1868, ER; *Chronicle* 7:556; Grodinsky, *Jay Gould*, 47–48; Denis T. Lynch, *"Boss" Tweed: The Story of a Grim Generation* (New York, 1927), 297.

40. Mott, *Story of Erie*, 498. After Gould's death Morosini called him "the most kind hearted, generous and forgiving man that ever came to Wall Street." Morosini, "Reminiscences."

Chapter 9: Notoriety

1. Mott, *Story of Erie* (see Chap. 8, n. 2), 467.

2. Erie Exec. Comm. Minutes, Aug. 4 and Aug. 5, 1868, ER.

3. For details on the money problem and the lockup see *Chronicle* 7:453–54, 614–15; Adams, *Chapters of Erie* (Prologue, n. 4), 65–66; Fowler, *Twenty Years* (Chap. 7, n. 7), 504.

4. *Chronicle* 7:174, 205, 238, 397, 550–51, 556, 621; *Congressional Globe*, Feb. 13, 1869, 1179; Mott, *Story of Erie*, 162–63; Adams, *Chapters of Erie*, 66–68.

5. *Chronicle* 7:647–48, 653; Mott, *Story of Erie*, 163–66; *New York World*, Nov. 20, 1868.

6. *Chronicle* 7:647–48, 653; *New York Times*, Nov. 18, 1868; *New York Herald*, Nov. 18, 1868; *New York Sun*, Nov. 19, 1868; Adams, *Chapters of Erie*, 71–72.

7. Erie Exec. Comm. Minutes, Nov. 23 and Dec. 5, 1868, ER; *New York Times*, Nov. 25, Nov. 26, Nov. 28, Dec. 4, Dec. 6, Dec. 12, and Dec. 17, 1868; *New York Herald*, Nov. 22, Nov. 24, Dec. 11, and Dec. 15, 1868; *New York Express*, Nov. 30, 1868; *New York Sun*, Nov. 25, Nov. 30, Dec. 5, and Dec. 11, 1868; *New York Tribune*, Dec. 2 and Dec. 9, 1868; *New York World*, Nov. 30 and Dec. 18, 1868. The convolutions of the legal fight are best followed in the New York dailies. For convenient summaries see Mott, *Story of Erie*, 166–72; Adams, *Chapters of Erie*, 71–95; Stedman, *Stock Exchange* (Chap. 7, n. 5), 211–13.

8. *New York Herald*, Dec. 6, 1868.

9. *New York Herald*, Dec. 4, 1868; *New York Tribune*, Feb. 6, 1869.

10. Clews, *Fifty Years* (see Prologue, n. 1), 151–55; Fowler, *Twenty Years*, 505; Mott, *Story of Erie*, 488.

11. RGD, NYC, 412:137; Mott, *Story of Erie*, 489. The quotation is cited in Swanberg, *Jim Fisk* (see Chap. 8, n. 14), 110. The Erie moved into the Opera House in August 1869.

12. Mott, *Story of Erie*, 489–90; Fowler, *Twenty Years*, 514.

13. For more detail on these routes see Grodinsky, *Jay Gould* (Prologue, n. 5), 56–58.

14. George H. Burgess and Miles C. Kennedy, *Centennial History of the Pennsylvania Railroad Company* (Philadelphia, 1949), 176–98.

15. Erie Exec. Comm. Minutes, Dec. 5, 1868, Mar. 15, 1869, ER; *Chronicle* 7:757, 8:89; Mott, *Story of Erie*, 172–74; Burgess and Kennedy, *Centennial History*, 198–99; *New York Tribune*, Jan. 14, 1869; *United States Railroad & Mining Register*, Feb. 13, 1869.

16. Burgess and Kennedy, *Centennial History*, 198; *Chronicle* 8:89, 153, 172, 204, 249, 267–68, 9:241; Grodinsky, *Jay Gould*, 60–62; Mott, *Story of Erie*, 174.

17. *Chronicle* 8:249; Grodinsky, *Jay Gould*, 65. The lake shore route consisted of three roads: Michigan Southern (Chicago to Toledo), Cleveland & Toledo, and Lake Shore (Cleveland to Buffalo).

18. Lane, *Commodore Vanderbilt* (see Chap. 8, n. 6), 261–66; Burgess and Kennedy, *Centennial History*, 200–288; Maury Klein, *The Great Richmond Terminal* (Charlottesville, Va., 1970), 61–63.

19. Erie Exec. Comm. Minutes, July 21, 1869, ER. For details of this legal fight see Mott, *Story of Erie*, 360; Grodinsky, *Jay Gould*, 67.

20. Harold F. Williamson and Arnold R. Daum, *The Age of Illumination, 1859–1899* (Evanston, Ill., 1959), 295–305. The agreements are reprinted in Chester McArthur Destler, "The Standard Oil, Child of the Erie Ring, 1868–1872," *Mississippi Valley Historical Review* 33 (June 1946):103–10. Varying interpretations of their significance are found in Destler, 89–103, Williamson and Daum, *Age of Illumination*, 304–5, and David Freeman Hawke, *John D.: The Founding Father of the Rockefellers* (New York, 1980), 59–62. For background on Harley and Allegheny see J. T. Henry, *The Early and Later History of Petroleum* (New York, 1970), 526–32. This is a reprint of the 1873 edition.

21. Henry, *History of Petroleum*, 531–32. The August agreement and Potter's acceptance of Gould's offer are reprinted in Destler, "Standard Oil," 110–13.

22. Destler, "Standard Oil," 97, 100–101. The whole tone of Destler's article is one of righteous indignation matching that of the Adams brothers except in wit and felicity of

style. See also the rebuttal by Julius Grodinsky and Destler's reply in *Mississippi Valley Historical Review* 33:617–28.

23. Henry, *History of Petroleum*, 532–33. The evidence does not indicate the extent of Gould's commitment or role in this venture.

24. *Chronicle* 6:38–39; *Heath et al. v. Erie Railway Company et al.*, 11 Federal Cases 976 (C.C.S.D.N.Y. 1871), no. 6, 306, 983. The bill of complaint in this case lists several examples of alleged fraudulent transactions by Gould, Fisk, and Lane. In fact its 101 items form the handiest available outline of the acts and abuses supposed to have been committed by the Erie trio.

25. Erie Exec. Comm. Minutes, Aug. 1, 1869, ER; *Chronicle* 8:301.

26. Erie Exec. Comm. Minutes, Mar. 15, 1869, ER; *Chronicle* 10:7; *New York Tribune*, Feb. 25 and Mar. 1, 1869; Grodinsky, *Jay Gould*, 82.

27. H. E. Sargent to W. H. Vanderbilt, Feb. 21, 1870; W. H. Vanderbilt to Sargent, Feb. 21, 1870; "General Notice," Feb. 21, 1870, all in James F. Joy Papers, Burton Historical Collection, Detroit Public Library (hereafter JFJ).

28. Morosini, "Reminiscences" (see Chap. 8, n. 12); Northrop, *Life and Achievements* (see Prologue, n. 2), 247–48.

29. *Railroad Gazette* 2:294; *Chronicle* 11:236, 719, 818, 853; *New York Times*, June 18, 1883; Clews, *Fifty Years*, 119; Lane, *Commodore Vanderbilt*, 259, 261; Morosini, "Reminiscences."

30. The methods used by Gould to disenfranchise foreign holders, and the written pledge to support his policies, are detailed in *Heath v. Erie*, 977–80, 986.

31. *Erie Railroad* (see Chap. 8, n. 18), 551–52, 556–57.

32. *Chronicle* 8:487; *Heath v. Erie*, 985–86; Mott, *Story of Erie*, 176.

33. *Heath v. Erie*, 986–88. Item 71 of the Heath complaint explains the practice that enabled Gould to vote this stock for himself.

Chapter 10: Infamy

1. See for example Herbert Adams Gibbons, *John Wanamaker* (New York, 1926), 1:118.

2. *Gold Panic* (see Chap. 8, n. 13), 135–36. This volume is the single most valuable source on the Gold Corner.

3. Erie Exec. Comm. Minutes, Sept. 10, 1869, ER; *Chronicle* 3:108, 241, 269; John N. Denison to James F. Joy, Aug. 23, 1869, Burlington Archives, Newberry Library, Chicago, (hereafter BA); Grodinsky, *Jay Gould* (see Prologue, n. 5), 73–74. It was Grodinsky who first placed the Gold Corner episode in the broader context of Gould's affairs.

4. Adams, *Chapters of Erie* (see Prologue, n. 4), 139–93; Swanberg, *Jim Fisk* (see Chap. 8, n. 14), 90–107; Mott, *Story of Erie* (see Chap. 8, n. 2), 370. Ramsey's use of Gouldlike tactics for the honorable purpose of defeating Gould threw Adams into moral somersaults. See *Chapters of Erie*, 149–51.

5. Grodinsky, *Jay Gould*, 74, 90, nn. 9 and 10; *Chronicle* 8:453.

6. *Gold Panic*, 132.

7. *Gold Panic*, 25–27; Stedman, *Stock Exchange* (see Chap. 7, n. 5), 215–16.

8. *Gold Panic*, 132; *Chronicle* 9:39. Even Gould's critics conceded the general validity of his reasoning but dismissed it as mere subterfuge for his speculations. See Adams, *Chapters of Erie*, 114–15; Stedman, *Stock Exchange*, 216–17.

9. *Gold Panic*, 148. The supply of gold consisted of specie and gold certificates. The

amount held in New York's national banks on Sept. 8, 1869, totaled $13.2 million. Ibid., 240–41, 368–75.

10. Ibid., 132; *Chronicle* 8:557, 662, 645, 685, 748. Boutwell was just entering on a policy of selling gold and buying bonds to reduce the national debt. See George S. Boutwell, *Reminiscences of Sixty Years in Public Affairs*, reprint ed. (New York, 1968), 2:138.

11. *Gold Panic*, 152, 429. There is conflicting evidence on how long the two men had known each other. Gould stated in January 1870 that he had known Corbin for "something over a year"; Corbin claimed he first met Gould "three or four years ago at Saratoga." Ibid., 151, 243.

12. Ibid., 152–53, 171–72, 243–44.

13. *Chronicle* 8:69, 9:13–14, 45, 69, 108, 173, 261–62; *New York Tribune*, July 9 and July 21, 1869.

14. *Gold Panic*, 318, 437–44. The facts about Butterfield's appointment do not emerge from the evidence. For some reason the Garfield committee did not question him directly on the subject.

15. Ibid., 61, 84–85, 146–47, 172, 219–20, 316, 393. The rules of the bank did not permit a change of management until the election in January. Stout, the bank's cashier, testified that "there was no change whatever" in policy or management after Gould purchased the stock.

16. For the arrangement with Kimber and Woodward see ibid., 136, 219–20.

17. Ibid., 5, 153–55, 246–49, 372–73; Boutwell, *Sixty Years*, 2:172–73. The testimony is conflicting on when this interview took place.

18. *Gold Panic*, 358–60. William B. Hesseltine, *Ulysses S. Grant: Politician* (New York, 1935), 174–75, discredits Gould's testimony on the interview but omits any mention of the Sept. 4 letter. Possibly this neglect stems from his reliance on Boutwell's memoirs in which Boutwell himself apparently forgot about the letter! See Boutwell, *Sixty Years*, 2:169.

19. *Gold Panic*, 151–52, 154, 160–63, 253–54, 272–73, 314–16; Adams, *Chapters of Erie*, 119. On the Butterfield transaction the two men flatly contradict each other. I tend to accept Gould's version because it is detailed and specific.

20. *Gold Panic*, 136, 138, 168, 220, 255, 259, 288; James A. Garfield to Gould, Feb. 2, 1870, GLC; *Chronicle* 9:301, 326, 331–32, 366; Adams, *Chapters of Erie*, 175–87. For all that has been written about the Gold Corner episode, there is no satisfactory account of the event or Gould's role in it. Even Grodinsky's treatment of it is curiously brief and inadequate, even cursory. See Grodinsky, *Jay Gould*, 74–79.

21. *Gold Panic*, 164, 166, 169, 172–73.

22. Ibid., 164, 169.

23. Ibid., 73–78, 90–91, 108–9, 112, 281–85, 299–300; RGD, NYC, 416:D91.

24. *Gold Panic*, 113–19, 217, 219. The corroborating testimony on the separateness of the networks is overwhelming. See ibid., 63–73, 90–92, 100–101, 108–12, 235–37.

25. Ibid., 139–41, 155.

26. Ibid., 445–47.

27. Ibid., 8–9, 155, 174, 249–51, 358.

28. Ibid., 9–10, 155–56, 174, 230–32, 444–45.

29. *Chronicle* 9:366.

30. Ibid. 9:397–98; *New York Times*, Sept. 24, 1869; *Gold Panic*, 279–80.

31. Ibid.; Stedman, *Stock Exchange*, 222; *Gold Panic*, 8, 279–80.

32. *Gold Panic*, 355, 377. The letter is reprinted in full here.

33. Details of this scene are in ibid., 156–59, 251–57, 448–49.

34. Ibid., 63, 94, 142, 175. Gould testified that "I was selling gold that day. I purchased merely enough to make believe that I was a bull."

35. It must be noted that virtually every account of the gold episode, and Gould's part in it, contains numerous errors of fact, large and small, many of them derived from misreading the testimony and documents. I include those of Henry Adams (on whom the others rely heavily), Clews, Myers, Josephson, Warshow, and O'Connor. At least one contemporary observer suspected the scheme to be similar to the version given here. See *New York Sun*, Oct. 15, 1869, and *Chronicle* 9:486.

36. *New York Times*, Sept. 24, 1869; *Chronicle* 9:398; Stedman, *Stock Exchange*, 222–23.

37. Ibid.

38. *Gold Panic*, 64, 75, 91, 93, 108, 236, 398, 401.

39. Ibid., 141, 168–69, 216, 283–84.

40. Ibid., 141–42, 169–70, 178–79, 216–17, 284.

41. Ibid., 227, 355–56, 360–61, 377, 410–12. There are discrepancies of detail between Boutwell's testimony and the account given in his *Sixty Years*, 2:174–76, written much later. In most cases I have utilized the earlier testimony.

42. Medbery, *Men and Mysteries* (see Chap. 7, n. 9), 262–64. The day's events are most vividly followed in the daily newspapers.

43. *Gold Panic*, 14, 64–65, 79, 109, 140, 161, 170, 213; Stedman, *Stock Exchange*, 223. The "sell, sell, sell" quotation is attributed to Smith by Willard. Adams, *Chapters of Erie*, 128, attributes it to Smith himself and adds the phrase "—only don't sell to Fisk's brokers," which does not appear in the testimony. Over the years the quotation was twisted to the point of being attributed directly to Gould. See for example O'Connor, *Gould's Millions* (Prologue, n. 1), 104, and Swanberg, *Jim Fisk*, 150.

44. *Gold Panic*, 64–65; Medbery, *Men and Mysteries*, 264–65; Stedman, *Stock Exchange*, 224–25; *Chronicle* 9:397–98.

45. *Gold Panic*, 56–57, 65, 143–44, 398.

46. Ibid., 143, 161, 167, 181, 238–41.

47. Ibid., 167.

48. Ibid., 286–88. Given the available evidence, it is extremely doubtful that Gould told Smith not to sell.

49. Ibid., 15, 65; Medbery, *Men and Mysteries*, 265. The Speyers incident is erroneously portrayed in virtually every source that cites it. Most depict it as occurring in the Gold Room after the price of gold had already collapsed, even though Speyers's testimony on these points is detailed and explicit. One exception is Stedman, *Stock Exchange*, 225.

50. *Gold Panic*, 15–16, 65–66, 170–71, 202, 207, 213–14, 303–4, 345–46; *New York Herald*, Sept. 25, 1869; *Chronicle* 9:397–98.

51. *Gold Panic*, 161, 238–39, 320–22.

52. Ibid., 202–4, 302–4, 389–91, 404–8; Medbery, *Men and Mysteries*, 266–68; Stedman, *Stock Exchange*, 230–32.

53. *Gold Panic*, 67–68, 81, 149–50, 257.

54. Stedman, *Stock Exchange*, 232–34; Medbery, *Men and Mysteries*, 268–71; *Chronicle* 9:427–29; *Gold Panic*, 25–49, 228–30, 389–96. The examiner later defended his action as necessary to thwart the plot of another clique who sought to use the gold panic as a vehicle for bringing down other national banks as well: "I did it with a full conviction . . . that the failure of that bank would be followed by raiding on other banks, and that the determination was to effect a general crash and a distrust of our national banking system."

55. *Gold Panic*, 38, 150, 285; Stedman, *Stock Exchange*, 232–33. Some of the injunctions are reprinted in *Gold Panic*, 40–50.

56. *Gold Panic*, 285; *New York Tribune*, Sept. 25, 1869; *Chronicle* 9:427–29; Stedman, *Stock Exchange*, 233. A copy of the statement constructed for Smith, Gould & Martin is in Medbery, *Men and Mysteries*, 271.

57. *Chronicle* 9:428; Stedman, *Stock Exchange*, 233; Earle, *Mr. Shearman and Mr. Sterling* (see Chap. 7, n. 4), 77.

58. For some observations on reasons for the injunctions see *Chronicle* 9:428–29; *Gold Panic*, 105–8, 111–12, 115–20, 215–16.

59. *Gold Panic*, 126, 227–30, 392–96; Stedman, *Stock Exchange*, 233.

60. Stedman, *Stock Exchange*, 233; RGD, NYC, 416:D91; Fowler, *Twenty Years* (see Chap. 7, n. 7), 526; *Gold Panic*, 73, 77; *Financier* 1:177; *Chronicle* 9:460, 492, 10:48, 14:287; Morosini, "Reminiscences" (see Chap. 8, n. 12); *New York Tribune*, Oct. 1, 1869; *New York Commercial Advertiser*, Oct. 4 and Oct. 5, 1869; *Stockholder*, Oct. 5, 1869. For more detail see Grodinsky, *Jay Gould*, 79–80.

61. *Chronicle* 10:48, 337; Adams, *Chapters of Erie*, 189–93; Earle, *Mr. Shearman and Mr. Sterling*, 77–78.

62. *Gold Panic*, 142.

Chapter 11: Ouster

1. Erie Directors' Minutes, Oct. 12 and Dec. 2, 1869, Feb. 23, 1870, ER. Gould was not the only railroad executive to borrow the classification device. See for example *Chronicle* 10:525.

2. Ibid.; Erie Exec. Comm. Minutes, Nov. 1 and Nov. 22, 1869, ER; *New York Tribune*, Mar. 2, 1871; *Chronicle* 9:495, 624, 10:7, 12:624; *Heath v. Erie* (see Chap. 9, n. 24), 981, 989; Mott, *Story of Erie* (see Chap. 8, n. 2), 177–78.

3. *Railroad Gazette* 2:469.

4. *Heath v. Erie*, 989; Mott, *Story of Erie*, 178.

5. Ibid.; Erie Stockholders' Minutes, Oct. 11, 1870, ER; *Chronicle* 10:236, 11:524.

6. For the protracted legal struggle over the 60,000 shares of stock see Grodinsky, *Jay Gould* (Prologue, n. 5), 95–96.

7. Mott, *Story of Erie*, 181–82, 365; Erie Directors' Minutes, Jan. 29, 1870, Jan. 10, 1871, ER; *Chronicle* 10:368, 778; Grodinsky, *Jay Gould*, 67, 96–98.

8. Mott, *Story of Erie*, 181–82; *Chronicle* 14:342.

9. Mott, *Story of Erie*, 182; Stedman, *Stock Exchange* (see Chap. 7, n. 5), 247–48.

10. Gould to James F. Joy, July 2, 1870, JFJ; J. M. Walker to Joy, Dec. 8, 1871, JFJ; Erie Directors' Minutes, Apr. 3, Apr. 11, July 31, and Dec. 30, 1871, Jan. 2, 1872, ER; John H. Devereux to William Vanderbilt, Jan. 23, 1871, and to Horace Clark, June 10, 1871, both in Thomas C. Cochran, *Railroad Leaders, 1845–1890* (Cambridge, Mass., 1953), 312 (hereafter *RL*); Gould to James A. Garfield, Jan. 27, 1871, James A. Garfield Papers, Library of Congress (hereafter JAG); *Chronicle* 11:685, 786, 12:142, 624.

11. Swanberg, *Jim Fisk* (see Chap. 8, n. 14), 201–53.

12. Differing versions of the Tweed Ring's downfall are found in Alexander B. Callow, Jr., *The Tweed Ring* (New York, 1966), 253–300; Leo Hershkowitz, *Tweed's New York: Another Look* (New York, 1977), 167–204; Lynch, *"Boss" Tweed* (see Chap. 8, n. 39), 351–90. For Barnard's impeachment see *Chronicle* 15:239.

13. *Chronicle* 13:262–63. For examples of contracts between directors and the company see Erie Directors' Minutes, Oct. 12, 1869, Sept. 8, 1871, ER.

14. Mott, *Story of Erie*, 182–83, 197.

15. Erie Directors' Minutes, Oct. 10, Dec. 12, and Dec. 28, 1871, ER; Erie Stockholders' Minutes, Oct. 10, 1871, ER; Erie Exec. Comm. Minutes, Dec. 19, 1871, ER; *Chronicle* 13:497, 859–60, 14:85.

16. Mott, *Story of Erie*, 179–81; Erie Directors' Minutes, Dec. 28 and Dec. 30, 1871, ER.

17. Unless otherwise indicated, these last episodes of Fisk's life are drawn primarily from Swanberg, *Jim Fisk*, 241–88. The flavor of the hearings is vividly captured in the New York papers for November and December 1871.

18. Stokes's rage against Fisk derived not only from the Mansfield affair but also from a business dispute. See Swanberg, 241–88, and Stedman, *Stock Exchange*, 248–49.

19. Quoted in Swanberg, *Jim Fisk*, 275.

20. *Financier* 1:25.

21. The letter, dated Nov. 14, 1881, is reprinted in Northrop, *Life and Achievements* (see Prologue, n. 2), 222.

22. Swanberg, *Jim Fisk*, 289–90. No satisfactory account of the machinations behind Gould's overthrow exists, partly because the evidence is confusing and conflicting. The fullest and most reliable is Mott, *Story of Erie*, 182–200, which also reprints some of the most relevant documents and testimony. The best source for details of this complex episode is the testimony in *Erie Railroad* (see Chap. 8, n. 18). Where possible I have for convenience cited Mott's reprinting of this testimony. The cable quoted here is in Mott, *Story of Erie*, 192. For the Heath-Raphael group's final victory see *Chronicle* 14:85.

23. The Lane scheme is outlined in a document prepared by Lane himself in September 1872. O'Doherty possessed one of the only copies and released it publicly in February 1873. The full text, along with the cables quoted here, is reprinted in Mott, *Story of Erie*, 190–96. Unless otherwise indicated, the documents quoted in this account are all taken from Mott, 198–200.

24. George Crouch to James McHenry, Feb. 15 and Feb. 20, 1872.

25. Crouch to McHenry, Feb. 20, 1872; Barlow to McHenry, Feb. 19, 1872.

26. A. W. O'Doherty to McHenry, Mar. 3, 1872; Crouch to McHenry, Feb. 23, Feb. 27, Feb. 28, Feb. 29, and Mar. 10, 1872. The figures are cited in Mott, *Story of Erie*, 195, 199. Hall and Sisson also sided with Crouch's forces, but no payment is listed for them.

27. Crouch to McHenry, Mar. 10, 1872. There is both a letter and a cable bearing this date. The letter of the directors is reprinted in Mott, *Story of Erie*, 191.

28. This scene is taken from Mott, *Story of Erie*, 186–87.

29. Erie Directors' Minutes, Mar. 11, 1872, ER. Only two of the directors, Edwin Eldridge and Henry Sherwood, remained loyal to Gould. See also Stedman, *Stock Exchange*, 251.

30. The testimony in *Erie Railroad*, the versions in Mott, *Story of Erie*, 188–89, and Stedman, *Stock Exchange*, 252–53, and the press accounts contradict one another hopelessly on these events.

31. Mott, *Story of Erie*, 195. For Gould's resignation see Erie Exec. Comm. Minutes, Mar. 15, 1872, ER.

32. *Chronicle* 14:310–11, 342–43; *Financier* 1:211, 221.

33. Mott, *Story of Erie*, 189. The formal releases are reprinted in Mott, 199–200. For examples of Gould's accounts with Erie see Erie Directors' Minutes, Apr. 17, 1872, and Exec. Comm. Minutes, Mar. 21, Apr. 26, and June 3, 1872, ER.

34. Crouch to McHenry, Mar. 23, 1872; *New York World*, Aug. 26, 1872; *Chronicle* 14:406–7, 412, 454; *Financier* 1:221, 232-33, 261; Mott, *Story of Erie*, 201–3; Erie Exec. Comm. Minutes, Mar. 22, 1872, ER; *Erie Railroad*, 144–45.

35. *Financier* 1:221, 232–33, 261; *Chronicle* 14:406–7; *Railroad Gazette* 4:118; Erie Directors' Minutes, May 29, 1872, ER; Erie Stockholders' Minutes, July 9, 1872, ER. For the election and Watson see Mott, *Story of Erie*, 203–7, 469–70.

Chapter 12: Entr'acte

1. RGD, NYC, 347:737.

2. *Labor and Capital* (see Chap. 1, n. 7), 1067. The road in question was the Missouri Pacific.

3. Halstead and Beale, *Jay Gould* (see Prologue, n. 2), 211–12.

4. Details on these transactions are scattered throughout the testimony before the Hepburn Committee. For the committee's report see *Report of Special Committee of the New York State Assembly on Railroads*, Assembly Document no. 38 (Albany, 1880). Gould's transactions are briefly summarized in Grodinsky, *Jay Gould* (see Prologue, n. 5), 88; Mott, *Story of Erie* (see Chap. 8, n. 2), 456–57; White, *Wizard* (see Prologue, n. 4), 82.

5. Fowler, *Twenty Years* (see Chap. 7, n. 7), 62; *Financier* 2:453; Gould to William Ward, Nov. 29, 1873, Jay Gould Papers, Library of Congress (hereafter GLC); RGD, NYC, 348:900MM, 348:900 A42, 417:176. The firm of Willard, Martin & Bach, formed two days after the dissolution of Smith, Gould & Martin, included Gould, Smith, and Tweed as special partners. By November 1872 Willard had opened E. K. Willard & Company and Martin had joined the firm of Garland, Martin & Company. Osborn & Chapin was formed in December 1871, with Gould and Smith becoming special partners in February 1872. Later, when Osborn and Chapin each opened his own house, Gould continued to do business with both.

6. RGD, NYC, 418:290, *Financier* 1:294, 311, 2:328–29.

7. Fowler, *Twenty Years*, 536–37; *Chronicle* 12:208, 13:334, 14:11, 457, 16:11.

8. Fowler, *Twenty Years*, 537–38; *Financier* 1:377, 409, 2:328–29; *Chronicle* 14:683. Grodinsky, *Jay Gould*, 112, states that "Gould began to trade in this stock [Pacific Mail] in the fall of 1872." The evidence cited here indicates that he began in the spring of 1872. It is a small point but crucial to the sequence of events that led to Gould's later change of position.

9. Fowler, *Twenty Years*, 538; *Chronicle* 14:790, 15:411, *Financier* 2:40, 217, 236; Mott, *Story of Erie*, 209.

10. Fowler, *Twenty Years*, 538–39; *Chronicle* 15:455, 518, 547; *Financier* 2:272, 313, 328–29.

11. Fowler, *Twenty Years*, 540–41; Stedman, *Stock Exchange* (see Chap. 7, n. 5), 255–56; *Chronicle*, 15:547, 590, 710–11; J. M. Walker to Joy, Mar. 7, 1872, JFJ.

12. Ibid.; *New York Tribune*, Nov. 1, 1872.

13. The scene that follows is in Mott, *Story of Erie*, 210–11. See also Stedman, *Stock Exchange*, 256–57, and Fowler, *Twenty Years*, 542–43.

14. Stedman, *Stock Exchange*, 256–57; Fowler, *Twenty Years*, 542–43. The complaint against Gould is quoted at length in Mott, *Story of Erie*, 210–11. For summaries see *New York Herald* and *New York Tribune*, Nov. 23, 1872.

15. *New York Tribune*, Nov. 25–27, 1872; *New York Times*, Nov. 27, 1872; *Chronicle* 15:688; Mott, *Story of Erie*, 211; Stedman, *Stock Exchange*, 258.

16. *New York Tribune*, Nov. 25–28, 1872; *New York Times*, Nov. 27, 1872; *Chronicle* 15:688, 710–11, 715; Mott, *Story of Erie*, 211; *Financier* 2:448–49, 452–53; Stedman, *Stock Exchange*, 258; Fowler, *Twenty Years*, 542–43.

17. This widely quoted anecdote appears in the *New York Times*, Dec. 3, 1892. Its general authenticity is verified in *The Financier* 2:449. Some sources mistakenly attribute the scene to the aftermath of Black Friday. For Smith's departure from Osborn & Chapin see RGD, NYC, 348:900 A42.

18. For differing views of the situation see *Chronicle* 15:821–22; Fowler, *Twenty Years*, 543; Mott, *Story of Erie*, 211–12; Stedman, *Stock Exchange*, 259; Grodinsky, *Jay Gould*, 106–7.

19. Erie Directors' Minutes, Dec. 19, 1872, ER; *New York Tribune*, Dec. 20, 1872; *Chronicle* 15:830. The entire agreement between Gould and Erie is reprinted in Mott, *Story of Erie*, 213–15.

20. Grodinsky, *Jay Gould*, 107–9; Mott, *Story of Erie*, 219–30. A detailed analysis of the true worth of Gould's package was presented to the Hepburn Committee in 1879 and reprinted in Mott, *Story of Erie*, 215–18.

21. *Chronicle* 15:320, 451, 799, 16:287, 325, 623, 716, 763, 17:16, 21, 246, 586, 773; RGD, NYC, 418:290. The latter source noted in February 1873 that Pacific Mail was "under the control of 3 or 4 persons only & it is doubtful if the above board has ever met as a body."

22. This widely quoted anecdote is taken here from the *New York Times*, Dec.3, 1892. I have taken the liberty of changing the *Times*'s use of the more delicate phrase, "son of a gun from Ohio," to what was clearly intended as the original language.

23. *Dictionary of American Biography*, 16:292–93 (hereafter *DAB*). Sage is yet another major figure who awaits a biographer. The only full-length study is Paul Sarnoff, *Russell Sage: The Money King* (New York, 1965), an amusing but error-plagued work that misreads most of Sage's business activities. Sarnoff offers the novel interpretation that Gould was a "second-rate stockbroker" who owed his fortune to Sage and served largely as his puppet. This approach leads Sarnoff to some startling conclusions. For example: "The scheme to corner the nation's gold was probably born in the brain of Samuel J. Tilden. . . . Certainly, by no stretch of imagination could Gould or Fisk have had the ingenuity to plan such a scheme" (p. 137). No concrete evidence is offered to support such views. The discussion of Sage in Myers, *American Fortunes* (see Chap. 5, n. 1), 447–77, provides more venom than insight.

24. *Chronicle* 17:773. Gould later declared that his shares were decisive in electing Sage president of Pacific Mail. See his letter in the *New York Tribune*, Dec. 4, 1874.

25. *New York Times*, Apr. 30, May 1, May 11, and May 13, 1873; *New York Tribune*, May 1 and May 12, 1873; Stedman, *Stock Exchange*, 264. Marrin said later he slapped Gould with the back of his hand; Gould maintained it was a closed fist.

26. *New York Times*, Dec. 25, 1872, Apr. 13 and Apr. 15, 1873; Bingham Duncan, *Whitelaw Reid: Journalist, Politician, Diplomat* (Athens, Ga., 1975), 46–49.

27. *New York Sun*, Mar. 29, July 1, July 6, July 19, Aug. 7, Aug. 26, Sept. 10, and Oct. 29, 1875, Jan. 19, 1876; Duncan, *Whitelaw Reid*, 48–49; Candace Stone, *Dana and the Sun* (New York, 1938), 128; *New York Times*, July 16, 1875. Duncan adds that "no clear explanation of Reid's arrangement with Gould has been found. . . . If there is a document describing the agreement in the Reid papers, I do not know of it." Duncan, *Whitelaw Reid*, 260. The deal is not mentioned in Royal Cortissoz, *The Life of Whitelaw Reid* (New York, 1921), 2 vols. See also J. N. A. Griswold to Charles E. Perkins, Dec. 13, 1876, BA. An intriguing interpretation can be found in *Stockholder* 13:664.

28. Snow, *Helen Gould* (see Chap. 1, n. 2), 182.

29. Northrop, *Life and Achievements* (see Prologue, n. 2), 316. See Northrop, 298–316, for several good anecdotes about the experiences of reporters with Gould. For Gould's evasive ability see the interview in the *New York Times*, Aug. 8, 1873.

30. An article analyzing Erie bonds and the prospects of the Lake Shore in gloomy terms, and a letter to the editor of the *Tribune* on Lake Shore's earnings, can be found in GLC. Both are undated but in Gould's handwriting; internal evidence suggests that they were written in the summer of 1874. A diligent search has failed to uncover any biographical information on Ward. His function is clearly revealed in the letters found in this collection.

31. Gould to William Ward, Jan. 9, July 5, and Sept. 2, 1874, GLC.

32. For a succinct description see Frank Luther Mott, *American Journalism: A History, 1690–1960* (New York, 1962), 373–434.

Chapter 13: On Track

1. There is no satisfactory history of the Union Pacific. See Nelson Trottman, *History of the Union Pacific* (New York, 1923), 1–98; Robert E. Riegel, *The Story of the Western Railroads* (New York, 1926), 65–94; Robert W. Fogel, *The Union Pacific Railroad: A Case in Premature Enterprise* (Baltimore, 1960); Charles Edgar Ames, *Pioneering the Union Pacific: A Reappraisal of the Builders of the Railroad* (New York, 1969); Arthur M. Johnson and Barry E. Supple, *Boston Capitalists and Western Railroads, 1869–1893* (Philadelphia, 1962), 1–39.

2. *Report of Select Committee (No. 2) on the Credit Mobilier*, House Report no. 78, 42 Cong., 3d sess., 1873 (Serial 1577), 404. Hereafter cited as *Wilson Report*.

3. *Wilson Report*, iv–v, 64; Trottman, *Union Pacific*, 25–32; Johnson and Supple, *Boston Capitalists*, 200–206.

4. Trottman, *Union Pacific*, 32–46, 55–70; Johnson and Supple, *Boston Capitalists*, 206–13; Ames, *Pioneering*, 123–344; Robert G. Athearn, *Union Pacific Country* (Chicago, 1971), 19–98. Construction of the Central Pacific is described in Oscar Lewis, *The Big Four: The Story of Huntington, Stanford, Hopkins, and Crocker, and of the Building of the Central Pacific* (New York, 1938), and David Lavender, *The Great Persuader* (Garden City, 1970).

5. Trottman, *Union Pacific*, 47–54, 71–98; Johnson and Supple, *Boston Capitalists*, 213–15; Ames, *Pioneering*, 323–28, 356–57, 362–63, 379, 384–95, 402–11; Charles Francis Adams, Jr., "Railroad Inflation," *North American Review* 108 (January 1869):130–64. Fisk apparently undertook the Union Pacific raid on his own; Gould denied having any part in it. See *United States Pacific Railway Commission Testimony*, Senate Exec. Doc. no. 51, 50 Cong., 1st sess., 1887 (Serial 2505), 448 (hereafter *PRC*).

6. For differing viewpoints on the Credit Mobilier scandal see Trottman, *Union Pacific*, 71–98; Ames, *Pioneering*, 431–89; Johnson and Supple, *Boston Capitalists*, 216–21; Fogel, *Union Pacific*, 90–110; J. B. Crawford, *The Credit Mobilier of America* (Boston, 1880); Rowland Hazard, *The Credit Mobilier of America* (Providence, 1881). The most detailed sources are the testimony given in the *Wilson Report* and in *Report of the Select Committee to Investigate the Alleged Credit Mobilier Bribery*, House Report no. 77, 42 Cong., 3d sess., 1873 (Serial 1577).

7. Ames, *Pioneering*, 416–17, 423–25; Trottman, *Union Pacific*, 99–100; Grodinsky, *Jay Gould* (see Prologue, n. 5), 113–17; *Railroad Gazette* 2:4; *Financier* 1:161, 201; *Chronicle* 14:215, 16:319, 325.

8. The two versions, differing only in minor details, are found in *Labor and Capital* (see Chap. 1, n. 7), 1066–67, and *PRC*, 2:446–47.

9. *Chronicle* 16:823, 829, 852; Fowler, *Twenty Years* (see Chap. 7, n. 7), 551–52; Grodinsky, *Jay Gould*, 118; Sickels to George Gould, Dec. 30, 1901, letter in possession of Kingdon Gould, Jr. (hereafter KG); Dodge to Nate Dodge, Mar. 27, 1874, Grenville M. Dodge Papers, Western Historical Collection, Denver Public Library (hereafter DPL).

10. *New York Times*, June 10, 1892; Ames, *Pioneering*, 148–49; *Wilson Report*, 511. Dillon too lacks a biography. See RGD, NYC, 373:1342.

11. *Chronicle* 18:266, 273, 376, 456; UP Stockholders' Records, Transfer Agent Book, 90, UP Stockholders' Minutes, Mar. 11, 1874, both in Union Pacific Railroad Company Collection, Nebraska State Museum and Archives, Lincoln (hereafter UP); Trottman, *Union Pacific*, 101; Grodinsky, *Jay Gould*, 121–22; E. H. Rollins to Gould, Apr. 6, 1874, UP.

12. The phrase is from Fowler, *Twenty Years*, 552. See *Stockholder* 12:320–21.

13. Huntington to Charles Crocker, Dec. 7, 1874, Huntington, Crocker, Stanford Letters, Mariners Museum, Newport News, Virginia (hereafter MM). The only recent biography of Huntington is Lavender, *Great Persuader*. For Huntington's constant struggle to raise capital see Julius Grodinsky, *Transcontinental Railway Strategy, 1869–1893* (Philadelphia, 1962), 20–69.

14. Huntington to David Colton, Dec. 12, 1874, MM; *New York Times*, Dec. 3, 1892.

15. Dillon to Oliver Ames, Apr. 9 and July 30, 1874, UP; *Chronicle* 18:8, 370, 543; *New York Times*, Apr. 9, Apr. 10, Apr. 18, May 5, May 26, May 28, and May 29, 1874; *Stockholder* 12:128–29, 353–54; A. N. Towne to Huntington, Aug. 11 and Oct. 16, 1874, Collis P. Huntington Papers, microfilm edition, University of Iowa Library, Iowa City (hereafter CPH); Gould to Rollins, Aug. 28, 1874, UP; Gould to Ward, Aug. 26, 1874, GLC; Dillon and Huntington to Bradbury, Sept. 17, 1874, MM; Huntington to Stanford, Sept. 17, 1874, MM; Gould to Silas Clark, Sept. 14, 1874, KG; Gould to Huntington, Sept. 18, 1874, CPH. Gould's draft of a circular for the new company, dated Sept. 14, 1874, is in CPH.

16. Huntington to Stanford, Sept. 17, Sept. 22, Sept. 30, Oct. 13, and Oct. 21, 1874, MM.

17. Huntington to Stanford, Nov. 10 and Nov. 28, 1874, MM; Huntington to Crocker, Nov. 6, Nov. 7, Nov. 12, and Nov. 14, 1874, MM; *New York Tribune*, Sept. 26, 1874; *Chronicle* 19:498.

18. *New York Tribune*, Dec. 4, 1874; *New York Times*, Dec. 4, Dec. 5, Dec. 8, Dec. 11, Dec. 19, and Dec. 31, 1874; *Chronicle* 19:577, 584, 617; *Stockholder* 13:104; Huntington to Crocker, Dec. 7, Dec. 8, Dec. 12, and Dec. 17, 1874, MM; Huntington to Stanford, Dec. 16, Dec. 17, and Dec. 18, 1874, MM; Stanford to Huntington, Dec. 16, Dec. 18, and Dec. 24, 1874, CPH; Huntington to Colton, Dec. 1, Dec. 8, and Dec. 12, 1874, MM.

19. Gould to Clark, Dec. 24, 1874, KG; Dillon to Gould, Dec. 10, 1874, UP; Dillon to Oliver Ames, Dec. 31, 1874, UP; *Chronicle* 19:613, 632, 640.

20. Dillon to Oliver Ames, Dec. 31, 1874, Apr. 9, 1875, UP; Huntington to Stanford, Dec. 16 and Dec. 17, 1874, Jan. 25, 1875, MM; *Chronicle* 20:162, 203.

21. Huntington to Colton, Mar. 3 and Mar. 12, 1875, MM; Huntington to Mark Hopkins, Mar. 4, 1875, MM; *Chronicle* 20:236, 242, 267; Dillon to Oliver Ames, Mar. 2, 1875, UP; *Stockholder* 13:312.

22. Huntington to Colton, Mar. 22, Apr. 9, and Apr. 12, 1875, MM; Huntington to Hopkins, Apr. 19, 1875, MM; Huntington to Stanford, Apr. 8 and May 29, 1875, MM.

23. Huntington to Colton, Mar. 18, Apr. 3, Apr. 8, and Apr. 12, 1875, MM; Huntington to Stanford, Mar. 16 and Apr. 23, 1875, MM; Huntington to Hopkins, Mar. 4 and Apr. 19, 1875, MM.

24. Huntington to Colton, May 1, 1875, MM.

25. For the stock prices see *Chronicle* 18:35, 20:37.
26. Gould to Clark, Sept. 14, 1874, KG.

Chapter 14: Developer

1. Poor, *Manual, 1875–1876,* 762; *PRC,* 43, 751; Athearn, *Union Pacific Country* (see Chap. 13, n. 4), 215; Trottman, *Union Pacific* (see Chap. 13, n. 1), 105.

2. Biographical information about Clark is scant. There is a brief sketch in J. Sterling Morton and Albert Watkins, eds., *Illustrated History of Nebraska* (Lincoln, 1907), 2:603–5. For Clark's appointment as general superintendent see UP Exec. Comm. Minutes, Apr. 24, 1874, UP. There is a brief sketch and photograph in the *Railway News Reporter,* Annual Review, 1893–94.

3. An appreciation of Gould's appetite for information is most evident in his letters to Clark, KG, and to E. H. Rollins, UP.

4. *Chronicle* 19:312–13; James C. Olson, *History of Nebraska* (Lincoln, 1955), 181–85; Gould to Clark, Sept. 5, Sept. 14, and Nov. 9, 1874, KG; Dillon to Clark, Sept. 2, 1874, UP; Grodinsky, *Jay Gould* (see Prologue, n. 5), 122.

5. Gould to Clark, Nov. 8 and Nov. 16, 1874, KG; Gould to E. H. Rollins, Feb. 1 and Mar. 11, 1875, UP; *Chronicle* 20:202–3, 237.

6. Gould to Rollins, Feb. 18, Feb. 27, and Mar. 4, 1875, UP; Gould to Clark, Jan. 9, Jan. 18, Feb. 8, Feb. 13, and Feb. 18, 1875, KG; Dillon to Oliver Ames, Mar. 2, 1875, UP; *Chronicle* 19:528, 20:8, 157, 163; *Railway World* 20:92, 102; *Chicago Tribune,* Mar. 30, 1875. The annual report is excerpted at length in *Chronicle* 20:266.

7. *Chronicle* 20:260, 306–7, 331–32, 351, 446–47, 22:35; Huntington to Colton, Apr. 12, 1875, MM; Gould to Clark, Mar. 28, Apr. 4, and Apr. 12, 1875, KG.

8. Gould to Clark, Oct. 13, 1874, KG.

9. Trottman, *Union Pacific,* 42–44; *Wardell v. Union Pacific Railroad Company,* 103 U.S. 651; Ames, *Pioneering* (see Chap. 13, n. 1), 247–48. A copy of the contract, dated July 16, 1868, is in the *Wilson Report* (see Chap. 13, n. 2), 591–93. See also *Report of the Government Directors for 1877,* Senate Exec. Doc. no. 69, 49 Cong., 1st sess., 1886 (Serial 2336), 132–33. All reports of the government directors for the period 1864–1884 are contained here.

10. Trottman, *Union Pacific,* 42–44; *Wardell v. Union Pacific,* 651; Sidney Dillon to James F. Wilson, Mar. 20, 1874, UP; *Wilson Report,* 232–35.

11. Gould to Clark, Oct. 24, Nov. 7, and Nov. 8, 1874, KG; *PRC,* 1447; Trottman, *Union Pacific,* 110; T. A. Larson, *History of Wyoming* (Lincoln, 1965), 110, 114; Dillon to Clark, July 23, 1874, UP.

12. Gould to Clark, Oct. 24, Nov. 7, and Nov. 23, 1874, KG; Larson, *Wyoming,* 110. The production cost figures cited were for coal loaded in cars ready for shipment. Gould owned the Blossburg Coal Company, which he apparently acquired during his Erie tenure. In 1883 he sold the company to Erie. See *Chronicle* 36:209, 221; Gould to Clark, Oct. 24, 1874, July 26 and July 30, 1875, KG.

13. Gould to Clark, Nov. 23, 1874, Feb. 1, 1875, KG; Dillon to John Sharp, Nov. 24, 1874, UP; Dillon to G. M. Dodge, Dec. 5, 1874, UP.

14. Gould to Clark, May 1, June 28, July 26, and July 30 (two letters), 1875, KG; Lavender, *Great Persuader* (see Chap. 13, n. 4), 307; Larson, *Wyoming,* 114.

15. Gould to Clark, Nov. 6, Nov. 13, and Dec. 1, 1875, KG; Dillon to Clark, Nov. 27, 1875, UP; Larson, *Wyoming,* 114–15. Thayer had just failed to retain his seat in the Senate

despite financial help from the Union Pacific. See Gould to Clark, Oct. 29 and Dec. 24, 1874, KG.

16. Gould to Clark, Feb. 1, July 30, Oct. 30, and Dec. 1, 1875, Jan. 19, 1876, KG; Dillon to Clark, Feb. 5, 1875, UP.

17. Huntington to Stanford, June 30, 1874, MM; Dillon to Clark, July 3, July 23, and Dec. 28, 1874, UP; Huntington to Colton, July 10, Nov. 13, and Dec. 1, 1875, MM. Dillon described the proposed company as having "control of all coal business west of Evanston."

18. Huntington to Colton, May 3 and May 8, 1875, MM; Gould to Clark, July 30, 1875, Jan. 19, 1876, KG. For examples of Gould's constant demand for information see Gould to Clark, Oct. 28 and Nov. 14, 1874, Oct. 30, 1875, Jan. 6, Feb. 12, Apr. 26, and Nov. 13, 1876, KG.

19. Gould to Clark, Oct. 24, 1874, Feb. 1, 1875, Jan. 29, Jan. 30, Jan. 31, Feb. 23, and Aug. 22, 1876, KG. A later report blasted the company's organization as "simple even to crudeness and quite lacking in system." *Government Directors*, 1878, 139.

20. Gould to Clark, Jan. 22, Jan. 30, Feb. 23, July 28, and Aug. 22, 1876, KG. For the coking experiments see ibid., Oct. 28, 1875, Aug. 29, 1878, KG.

21. Ibid., Aug. 12, Aug. 22, and Aug. 27, 1876, Jan. 6, 1877, KG; *PRC*, 82.

22. Gould to his wife, Oct. 7, 1874, JGL; Dillon to R. H. Thurston, Nov. 11, 1874, UP; T. E. Sickels to Oliver Ames, Dec. 16, 1875, UP; Gould to Clark, Nov. 6, 1875, Jan. 19, Jan. 26, Jan. 31, and Feb. 2, 1876, KG; Dillon to T. E. Sickels, Feb. 1, 1876, UP; T. E. Sickels to J. P. Kimball, June 20, 1876, UP; *PRC*, 82.

23. Dillon to Oliver Ames, Apr. 9, Apr. 29, and June 25, 1874, Feb. 20, 1875, UP; Dillon to Clark, Dec. 28, 1874, Feb. 2, 1875, UP; Gould to Clark, Dec. 25, 1874, Feb. 9, 1875, KG; Athearn, *Union Pacific Country*, 297–98; Larson, *Wyoming*, 111.

24. Gould to Clark, Nov. 8, 1874, Feb. 9, 1875, Jan. 19 and Aug. 12, 1876, Nov. 19, 1877, KG; *PRC*, 1088–90, 1100, 1749–51; Gould to E. H. Rollins, Jan. 11, 1876, UP.

25. Davis to Gould, Feb. 7, 1877, UP; Davis to Fred Ames, Jan. 30, 1878, UP. For detail on the promotional efforts see Athearn, *Union Pacific Country*, 147–71; David M. Emmons, *Garden in the Grasslands* (Lincoln, 1971), 27, 32–38, 147–49; Olson, *Nebraska*, 170–72. *Government Directors, 1877*, 134, noted that "a man can purchase a full section from the railroad company, but it is surrounded on all sides by Government land, which is only open to homesteaders and pre-emptions."

26. Dillon to J. Proctor Knott, May 4, 1876, UP; E. H. Rollins to Davis, Nov. 24, 1874, UP; Gould to Clark, July 11 and Aug. 13, 1877, Feb. 22, [1878], KG; Athearn, *Union Pacific Country*, 149–64; Davis to Fred Ames, Jan. 30, 1878, UP; Davis to Gould, Feb. 7, 1878, UP; Dillon to Clark, Jan. 31, 1878, UP.

27. Athearn, *Union Pacific Country*, 162–63; Charles Francis Adams, Jr., "The Granger Movement," *North American Review* 120 (April 1875): 396–421; Emmons, *Garden in the Grasslands*, 1–24, 128–61; Dodge to Gould, July 11, 1875, Dodge Record, Iowa State Department of Archives and History, Des Moines (hereafter DR), 9:285.

28. Gould to Clark, Jan. 24, 1876, July 11, 1877, Aug. 5, 1878, KG. For the cattle industry in Wyoming see E. S. Osgood, *Day of the Cattlemen* (Minneapolis, 1929); Larson, *Wyoming*, 163–94; Athearn, *Union Pacific Country*, 291–96.

29. W. A. Carter to John W. Garrett, Nov. 8, 1877, Garrett Family Papers, Library of Congress (hereafter GP); Gould to Clark, Aug. 13, 1877, KG; Larson, *Wyoming*, 165. A copy of the circular distributed to stockmen by Carter, dated July 2, 1877, is in the Western History Research Center, University of Wyoming, Laramie.

30. Athearn, *Union Pacific Country*, 291–94; Olson, *Nebraska*, 191–201; *PRC*, 1104–13.

31. Gould to Clark, Oct. 28, 1875, Aug. 21, 1878, KG.

32. Ibid., Jan. 24, Jan. 31, Feb. 2, Feb. 3, Feb. 7, Feb. 15, Feb. 21, and Mar. 2, 1876, KG; Larson, *Wyoming*, 131–32; Herbert S. Schell, *History of South Dakota* (Lincoln, 1961), 125–54. A brief sketch of the Indian uprising can be found in Michael P. Malone and Richard B. Roeder, *Montana: A History of Two Centuries* (Seattle, 1976), 87–103.

33. Gould to Clark, Aug. 21, 1876, KG; Athearn, *Union Pacific Country*, 233–34.

34. Gould to Clark, Jan. 9, Apr. 27, Sept. 25, and Sept. 26, 1875, May 30, July 10, July 14, July 28, July 30, Aug. 12, Aug. 18, and Aug. 21, 1876, July 14, 1878, KG; Sidney Bartlett to Gould, Apr. 22, 1878, KG.

35. Gould to Clark, Apr. 12, July 26, and Aug. 11, 1875, Jan. 19 and July 11, 1877, Oct. 5, 1878, KG.

36. All data were compiled from Poor's *Manual* for the period 1875–1880. In raw numbers tonnage increased from 482,806 in 1874 to 992,886; locomotives from 152 to 178; and freight cars of all types from 2,969 to 3,115. The data from this period cannot be compared with those for later years because the figures given for 1880 and after embrace the consolidated Union Pacific, Kansas Pacific, and Denver Pacific lines. Trottman, who was generally critical of Gould, concluded that "doubtless the road was fairly well kept up for a western road at that period." Trottman, *Union Pacific*, 108; *Government Directors, 1878*, 137–50.

37. The dividend figure can be found in *PRC*, 4802, 4810, or computed from figures in Poor, *Manual*, for the period 1875–1879. For a critique of Gould's dividend policy see Trottman, *Union Pacific*, 106–8.

38. Gould to Clark, Nov. 9, 1874, Feb. 1 and Apr. 26, 1875, KG; Dillon to Clark, Feb. 4, 1875, UP.

39. Dillon to Dodge, Dec. 5, 1874, UP; Gould to Clark, Dec. 5, 1876, July 8, 1877, KG.

40. Dillon to F. Gordon Dexter, Dec. 16, 1875, UP; Dillon to Clark, Dec. 16, 1875, UP; Gould to E. H. Rollins, Dec. 23, 1875, Jan. 11 and Jan. 19, 1876, UP; Dodge to Gould, Dec. 24, 1875, DR, 9:325; Dillon to Fred Ames, Jan. 3 and May 20, 1876, UP; Dillon to Oliver Ames, Mar. 15, 1876, UP; Gould to Clark, Jan. 19 and May 20, 1876, KG.

41. Gould to Clark, Feb. 7, 1876, KG.

Chapter 15: Politico

1. George L. Miller to Dodge, Mar. 5, 1876, DR, 9:411.

2. Gould to Clark, July 18, 1878, KG; Dillon to J. U. Wilson, Jan. 16, 1879, UP.

3. *PRC*, 1333–42; Olson, *Nebraska* (see Chap. 14, n. 4), 158, 219, 225, 240, 247. Besides owning rival newspapers, Rosewater and Miller belonged to different political parties.

4. Olson, *Nebraska*, 153–60, 184–90; Trottman, *Union Pacific* (see Chap. 13, n. 1), 110–17; Gould to Clark, Dec. 24, 1874, KG; Dillon to Clark, Dec. 24, 1874, UP; Dodge to Gould, July 7, 1875, DR, 9:283.

5. Gould to Clark, July 10, July 14, July 15, July 26, and August 11, 1875, Apr. 23, 1876, KG; *PRC*, 1341–42; Olson, *Nebraska*, 190.

6. Gould to Clark, Oct. 29, Nov. 23, and Dec. 24, 1874, KG; *BDC*, 1802–3.

7. Gould to Clark, Jan. 11, Feb. 15, May 8, July 17, Sept. 8, and Nov. 13, 1876, Jan. 19, 1877, Feb. 6 and Sept. 29, 1878, KG; *BDC*, 805, 1124, 1657; Olson, *Nebraska*, 165; Gould to Oliver Ames, Sept. 28, 1876, UP; *PRC*, 1335–36.

8. Gould to Clark, Nov. 10, 1876, July 11, 1877, Sept. 8, Sept. 29, and Oct. 9, 1878, KG; *PRC*, 1346.

9. Gould to Miller, Nov. 15, 1875, DR, 9:315; Dillon to Clark, Nov. 27, 1875, UP; Miller to Gould Nov. 15, 1875, and Gould to Miller, Nov. 16, 1875, DR, 9:319; Gould to Clark, Aug. 11 and Nov. 17, 1875, Aug. 13, 1877, May 12, 1879, KG; *PRC*, 1342–43; Dodge to Gould, Dec. 24, 1875, DR, 9:325; *Government Directors, 1877* (see Chap. 14, n. 9), 126.

10. Lewis L. Gould, *Wyoming; A Political History, 1868–1896* (New Haven, 1968), 42; Larson, *Wyoming* (see Chap. 14, n. 11), 77.

11. Gould to Clark, Nov. 18 and Nov. 19, 1877, Aug. 21 and Sept. 8, 1878, KG; Malone and Roeder, *Montana* (see Chap. 14, n. 32), 130–31; Clark C. Spence, *Territorial Politics and Government in Montana, 1864–1889* (Urbana, Ill., 1975), 126–28.

12. *PRC*, 18.

13. *Government Directors, 1878,* 145. See also Richard C. Overton, *Gulf to Rockies* (Austin, Tex., 1953), 48n.

14. *Union Pacific Railroad Company v. Hall,* 91 U.S. 343, 23 L.Ed. 428; Trottman, *Union Pacific*, 108–9; Poor, *Manual 1876–1877* (see Chap. 7, n. 16), 802; Athearn, *Union Pacific Country* (see Chap. 13, n. 4), 335–37; Dillon to W. E. Chandler, Dec. 16, 1874, UP; Dillon to Clark, Feb. 23 and Mar. 11, 1876, UP; Dodge to Dillon, Mar. 23, 1876, Grenville M. Dodge Papers, Iowa State Department of Archives and History, Des Moines (hereafter DP); Dodge to Dillon and Gould, July 21, 1876, DP; *Railroad Gazette* 6:211; *Chronicle* 20:491, 23:112, 330.

15. Trottman, Union Pacific, 126–29.

16. Ibid., 131–33; *Chronicle* 19:528, 20:8.

17. Trottman, *Union Pacific*, 133–34. A summary of the railroad's position on these issues, written by Dodge, is in DR, 9:615–16.

18. Dillon to E. H. Rollins, Apr. 14, 1874, UP.

19. Dodge to Gould, Jan. 16 and Jan. 27, 1875, DR, 9:221, 229; Gould to Clark, Feb. 8, 1875, KG; Gould to Rollins, Feb. 7, 1875, UP; *Stockholder* 13:281. Gould calculated that the payments with compound interest would in 40 years total $84 million, an amount equal to the government debt figured as principal and simple interest.

20. Dodge to Gould, Feb. 2, 1875, DR, 9:235; Grenville M. Dodge, *How We Built the Union Pacific Railway and Other Papers and Addresses* (Council Bluffs, Ia., n.d.), 40; Gould to Rollins, Feb. 7, 1875, UP.

21. Dodge to Gould, Feb. 2, 1875, DP. The italics are mine.

22. Ibid., Feb. 4, 1875, DR, 9:241; *Chronicle* 20:157, 163; Dodge, *How We Built*, 40–41; Dillon to Oliver Ames, Feb. 16, 1875, UP; Gould to Clark, Feb. 18, 1875, KG; Gould to Rollins, Feb. 18, 1875, UP. The decision referred to was *St. John v. the Erie Railway Company,* 22 Wallace 136. The amended version of the proposal is in Gould to Rollins, Feb. 20, 1875, UP.

23. Dodge to Gould, Feb. 23, 1875, DR, 9:245; *Government Directors, 1878,* 241–42.

24. Trottman, *Union Pacific*, 129–30; *United States v. Union Pacific Railroad Company,* 91 U.S. 72, 23 L.Ed. 224.

25. Trottman, *Union Pacific*, 120–21; Grodinsky, *Transcontinental* (see Chap. 13, n. 13), 18–32, 56–59; Lavender, *Great Persuader* (see Chap. 13, n. 4), 303–11; Huntington to Hopkins, Feb. 22, 1876, MM; Stanley P. Hirshson, *Grenville M. Dodge: Soldier, Politician, Railroad Pioneer* (Bloomington, Ind., 1967), 198–202.

26. C. Vann Woodward, *Reunion and Reaction* (Boston, 1951), 60–62; Gould to Clark, Jan. 11, 1876, KG; Dodge to Gould, Feb. 15, 1877, DP; Dodge to Gould, Jan. 21, 1876, DR, 9:369; Gould to Rollins, Jan. 11 and Jan. 25, 1876, UP; Trottman, *Union Pacific*, 140–41.

27. John Boyd to Huntington, Jan. 13, Jan. 19, and Feb. 4, 1876, CPH; Gould to Huntington, Feb. 14 and Mar. 25, 1876, CPH; Huntington to I. E. Gates, Mar. 1 and Mar. 10,

1876, CPH; Dillon to George F. Edmunds, Jan. 24, 1876, UP; Dillon to Oliver Ames, Mar. 15 and Mar. 22, 1876, UP; Dillon to J. Rodman West, Mar. 27, 1876, UP; *Chronicle* 22:472; Woodward, *Reunion and Reaction*, 73–92; *New York Times*, Feb. 1, 1876; *New York Tribune*, Feb. 8 and Mar. 14, 1876; *New York Sun*, Mar. 14, 1876; Oliver Ames Diary, Mar. 7 and Mar. 14, 1876; Gould to Oliver Ames, Mar. 15, 1876, Oliver Ames Papers, Stonehill College Library, North Easton, Massachusetts (hereafter OA).

28. Dillon to Oliver Ames, Mar. 22 and Apr. 29, 1876, UP; Gould to Oliver Ames, Apr. 2, Apr. 5, Apr. 20, Apr. 21, Apr. 23, Apr. 27, May 27, and May 29, 1876, OA; Oliver Ames Diary, Apr. 6 and Apr. 7, 1876; Boyd to Huntington, May 15, 1876, CPH; Charles H. Sherrill to Huntington, June 2, 1876, CPH; *New York Times*, Apr. 19, 1876; *New York Sun*, Apr. 28, May 9, May 29, and June 9, 1876; Trottman, *Union Pacific*, 140–41; Lewis H. Haney, *A Congressional History of Railways in the United States, 1850–1887* (Madison, Wis., 1910), 84–97; *Chronicle* 22:521, 23:74–75. The House bill differed in requiring additional payments of $750,000 for ten years and $1 million thereafter instead of 25 percent of net earnings.

29. Gould to John W. Garrett, Nov. 29, 1876, GP.

30. The crisis is summarized in Woodward, *Reunion and Reaction*, 3–21.

31. Scott to Huntington, July 26 and Aug. 23, 1876, CPH; Huntington to Colton, Nov. 15, Dec. 4, Dec. 20, and Dec. 25, 1876, MM; Woodward, *Reunion and Reaction*, 113–14; Lavender, *Great Persuader*, 314–15; Gould to Garrett, Nov. 29, 1876, GP. A glimpse of the committee struggle can be found in Harry James Brown and Frederick D. Williams, eds., *The Diary of James A. Garfield* (Lansing, Mich., 1973), 3:398–414. Descriptions of the compromise bill are in *Railroad Gazette* 9:18, and *Railway World* 3:62.

32. *New York Times*, Jan. 5, 1877; *New York Sun*, Jan. 9, 1877; Woodward, *Reunion and Reaction*, 176–77; Dodge to Gould, Feb. 12, 1877, DR, 9:623; Dodge to Gould, Feb. 15 and Feb. 27, 1877, DP; Haney, *Congressional History*, 96–98; Woodward, *Reunion and Reaction*, 114–77.

33. Woodward, *Reunion and Reaction*, 210; Alexander C. Flick, *Samuel Jones Tilden: A Study in Political Sagacity* (New York, 1939), 396–98; Dodge to Gould, Feb. 20, 1877, DR, 9:633.

34. Gould to Clark, Jan. 2 and Jan. 31, 1878, KG; Dillon to Gould, Mar. 27, 1878, UP; Dodge to Gould, Jan. 17 and Mar. 18, 1878, DP; Finley Anderson to Henry Villard, Mar. 13, 1878, Henry Villard Papers, Baker Library, Harvard Graduate School of Business Administration, Boston (hereafter HV); *New York Sun*, Mar. 7–9, Mar. 11, Mar. 13, Mar. 21, Mar. 29, Apr. 5, Apr. 6, Apr. 10, Apr. 12, and Apr. 25, 1878; *Chronicle* 26:368, 420, 471; *Public* 13:168; Trottman, *Union Pacific*, 141–43; Haney, *Congressional History*, 98–101.

35. Crocker to Huntington, Oct. 26, 1877, Apr. 27, 1878, MM; Lavender, *Great Persuader*, 326; *PRC*, 3817; Gould to Clark, Feb. 26 and Aug. 5, 1878, KG; *Chronicle* 26:653, 27:42, 79–80, 149, 237–38; Haney, *Congressional History*, 105–6.

36. *Chronicle* 20:405, 539, 21:467, 529, 535–36; Trottman, *Union Pacific*, 129–30; Perkins to John Murray Forbes, Feb. 18, 1878, BA; *Chronicle* 26:592, 27:332, 28:121, 477, 29:434; Trottman, *Union Pacific*, 134–44; Haney, *Congressional History*, 101–5.

37. *PRC*, 510, 588, 747.

38. Ames, *Pioneering*, 520; Trottman, *Union Pacific*, 86–87.

39. Dillon to Oliver Ames, Feb. 20 and Apr. 23, 1875, UP; Gould to Oliver Ames, July [16], 1875, UP; Gould to Dillon, July 19, 1875, UP; UP Exec. Comm. Minutes, July 21, Aug. 2, Sept. 15, and Sept. 22, 1875.

40. Gould to Rollins, July 20 and July 26, 1875, UP; Gould to Atkins, July 20, 1875, UP (two items). Atkins held the note as trustee.

41. Gould to Rollins, July 24 and Sept. 2, 1875, Jan. ?, Jan. 2 (two items), Jan. 9, Jan.

12, and Jan. 19, 1876, UP; Dodge to Gould, July 28, 1875, Jan. 4, 1876, DR, 9:291, 349; Dillon to Rollins, Dec. 28, 1875, UP; *PRC*, 541–46, 710–21, 814–15; Ames, *Pioneering*, 521; Gould to Henry McFarland, Jan. 6, 1879, UP. Oliver Ames was the brother of Oakes and the father of Fred Ames; Governor Oliver was the son of Oakes Ames. A list of those who signed their shares over to Union Pacific is in Dillon to Rollins, Apr. 22, 1876, UP.

42. Dillon to Oliver Ames, May 11, 1874, UP; Dillon to Sidney Bartlett, Nov. 14, 1876, UP; *Chronicle* 23:623, 37:268; Trottman, *Union Pacific*, 88–89; Ames, *Pioneering*, 521–22.

43. Gould to Clark, Jan. 21, 1879, KG; Dodge to Thomas A. Scott, Dec. 31, 1875, DR, 9:333; Huntington to Colton, Dec. 17, 1877, MM.

44. *BDC*, 1629; Leon Burr Richardson, *William E. Chandler, Republican* (New York, 1940), 29–31, 144–45, 162–63; Hirshson, *Dodge*, 191, 197–98; Dodge to Chandler, Dec. 26, 1875, William E. Chandler Papers, Library of Congress (hereafter CLC).

45. David S. Muzzey, *James G. Blaine: A Political Idol of Other Days* (New York, 1934), 145–46; *New York Sun*, May 17, June 4, June 7, and June 12, 1876.

Chapter 16: Chess Player

1. Maury Klein, "The Strategy of Southern Railroads," *American Historical Review* 73 (April 1968):1054, and *Great Richmond Terminal* (see Chap. 9, n. 18), 16–26; Grodinsky, *Transcontinental* (see Chap. 13, n. 13), 104–5.

2. For a masterful general summary of these cooperative arrangements see Chandler, *Visible Hand* (Prologue, n. 6), 122–23.

3. For background on the Iowa Pool see Julius Grodinsky, *The Iowa Pool: A Study in Railroad Competition, 1870–1884* (Chicago, 1950), 1–38, 101–11.

4. There is no railroad question more complex or difficult to summarize than that of rates. To oversimplify, through traffic may be defined as that passing entirely across a company's line and local traffic as that originating and terminating on its own line. In reality, of course, the vast bulk of traffic originated or terminated at a point beyond the company's line. For such traffic the charge was determined by combining the local rate for mileage traveled on the company's line with the through rate for mileage traveled beyond its line. For example, goods carried from Chicago to Grand Island paid through rates from Chicago to Omaha and the local rate from Omaha to Grand Island.

5. Chandler, *Visible Hand*, 133–37. Fixed costs are those costs that do not vary with the amount of traffic carried. Chandler estimated that by the 1880s fixed costs averaged two-thirds of total costs.

6. For more detail see Grodinsky, *Transcontinental*, 72–74, 88–89, and Richard C. Overton, *Burlington Route* (New York, 1965), 108–13. The Union Pacific discriminated by charging a lower rate for through traffic moving entirely across its own line than for traffic moving partly over its line and partly over one of the other lines.

7. Huntington to Colton, Apr. 7 and Apr. 22, 1875, MM; Lavender, *Great Persuader* (see Chap. 13, n. 4), 312–13; *Stockholder* 12:53. The proposed California road was known as the Los Angeles & Independence.

8. Huntington to Colton, Apr. 7 (two letters), May 3, May 8, May 18, May 26, and May 27, 1875, February 14, 1876, MM; Gould to Clark, Apr. 4, 1875, KG; *Chronicle* 20:515, 593, 615–16, 21:9, 201, 231, 324, 388, 482, 489, 22:130, 247, 257, 305, 322, 345; *New York Tribune*, June 22 and Nov. 30, 1875; *Stockholder* 13:424, 440, 456, 504, 568–69; Gould to Rollins, Dec. 25,

1875, UP; Dillon to Clark, Jan. 24 and Feb. 26, 1876, UP. Gould's response to the attachment is detailed in Gould and Dillon to Oliver Ames, May 7, 1876, UP.

9. Dillon to Clark, Feb. 26, 1876, UP; Dillon to Oliver Ames, Mar. 15 and May 19, 1876, UP; Gould to Clark, Mar. 3, 1876, KG; *Chronicle* 22:201, 248, 487–88, 513, 515, 543–44, 591, 27:68, 123, 229; Dodge to Gould, Feb. 12 and Feb. 20, 1877, DR, 9:623, 633, Feb. 15 and Feb. 27, 1877, DP; Dillon to William P. Clyde, May 31, 1877, UP; Gould to Clark, Sept. 4, 1878, UP, May 8, 1876, KG; Gould to Marvin Hughitt, Sept. 18, 1878, UP; Dillon to Fred Ames, Dec. 12, 1878, UP; *Public* 14:68.

10. *PRC*, 753; Trottman, *Union Pacific* (see Chap. 13, n. 1), 120–21. Between 1867 and 1879 Kansas Pacific net earnings equaled its fixed charges only twice. For that period interest charges exceeded net earnings by $6.5 million.

11. Background detail on the Colorado roads is found in Athearn, *Union Pacific Country* (see Chap. 13, n. 4), 129–36, and Overton, *Gulf to Rockies* (see Chap. 15, n. 13), 6–18.

12. Dillon to Oliver Ames, Mar. 18 and Apr. 9, 1874, Feb. 16 and Mar. 2, 1875, UP; Dillon to Atkins, July 6, 1874, UP; Dillon to Carr, Aug. 3, 1874, UP; Dillon to Clark, Nov. 2 and Dec. 28, 1874, UP; Dillon to Chandler, Dec. 16, 1874, UP; Carr to L. H. Meyer, July 7, 1877, HV; *Railroad Gazette* 6:398; Gould to Clark, Nov. 8 and Dec. 25, 1874, Jan. 18 and Feb. 1, 1875, KG. Details of Carr's offer are in Gould to Clark, Jan. 26, 1875, KG.

13. Gould to Clark, Apr. 5, Apr. 12, Apr. 26, and Oct. 21, 1875, KG; Dillon to Oliver Ames, Apr. 2, 1875, UP; Dillon to Clark, Apr. 23, 1875, UP; Gould to Rollins, May 8, May 9, and May 16, 1875, UP; Huntington to Colton, Apr. 22 and May 8, 1875, MM; Huntington to Stanford, Apr. 23, 1875, MM; *Chronicle* 20:405, 501, 21:466, 22:614, 23:38; Poor, *Manual 1875–1876,* 603; Dillon to Oliver Ames, [Nov. 23], 1875, UP; Dillon to Gould, Aug. 24, 1876, UP; Athearn, *Union Pacific Country,* 217–23; Andrew J. Poppleton to Sidney Bartlett, Jan. 31, 1877, UP. Two aspects of the UP-KP agreement are worth noting: The UP agreed to give the consolidated roads all Colorado local traffic in return for abandoning its pro rata claims, and Gould and Ames agreed personally to take $600,000 of the KP's $1.9 million floating debt. These provisions are embodied in two documents dated Apr. 22, 1875, both in the archives of the Missouri Pacific Railroad Company, St. Louis (hereafter MP).

14. Grodinsky, *Transcontinental,* 90–91; Overton, *Gulf to Rockies,* 16–21, 41–47; Dodge to Evans, Apr. 28, 1876, DP.

15. Poor, *Manual 1874–1875,* 343–44, and *Manual 1877–1878,* 654. The date of the Kansas Pacific receivership appears to have confused historians. Grodinsky gives it as December 1876 in *Jay Gould* (see Prologue, n. 5), 138, and November 1877 in *Transcontinental,* 91. *RL,* 479, gives Villard's term as receiver as 1879–1883. Athearn, *Union Pacific Country,* 226, dates the receivership from November 1873. The correct date, cited in Poor, is Nov. 3, 1876.

16. Carr to Greeley and Villard, May 16, 1877, HV; Gould to Clark, Nov. 19, 1877, KG; Athearn, *Union Pacific Country,* 233.

17. Carr to Greeley and Villard, May 16, 1877, HV; Carr to Villard, Feb. 19, 1877, HV; Greeley to Villard, Feb. 26, 1877, HV; *Chronicle* 24:444; Peter Geddes to Forbes, Apr. 21, 1878, BA. The funding mortgage is described in an affidavit found in HV. Greeley represented bondholders on the eastern line of Kansas Pacific, Villard those on the western end or Denver extension.

18. Carr to Villard, Feb. 19, Feb. 26, and Apr. 30, 1877, HV; Greeley to Villard, Mar. 2 and May 18, 1877, HV; Dillon to Carr, Mar. 2, 1877, HV; Carr to Dillon, Mar. 15, 1877, HV; John P. Usher to Villard, May 13, 1877, HV; Carr to Greeley and Villard, May 16, 1877, HV; Dillon to Gould, July 3, 1877, UP; T. F. Oakes to Villard, Nov. 30, 1877, HV; Villard to Greeley, Dec. 29, 1877, HV; *Government Directors, 1878,* 148.

19. Carr to L. H. Meyer, July 7, 1877, HV; Greeley to Villard, Jan. 2 and Jan. 28, 1878, HV. There is no biography of Villard. See the sketch in *DAB*, 19:273–75.

20. Gould to Clark, Nov. 11, 1877, Jan. 27, Jan. 31, Feb. 4, and Feb. 26, 1878, KG; Greeley to Villard, Feb. 8, 1878, HV; *Chronicle* 26:166, 470; Villard to Carr, Feb. 5, 1878, HV; Forbes to Perkins, Mar. 18, 1878, BA; Forbes to Villard, Feb. 28, 1878, HV.

21. Gould to Henry McFarland, Apr. 27, 1878, UP; Greeley to Villard, May 11, 1878, HV; *PRC*, 165, 451–55; *Chronicle* 26:391–92, 420, 470.

22. Greeley to Villard, May 15, 1878, HV; E. P. Vining to Gould, May 15, 1878, KG; Gould to Clark, May 15, 1878, KG; Villard to Dillon, June 20, 1878, HV; *Chronicle* 26:470, 575, 592, 27:16; Trottman, *Union Pacific*, 149–50.

23. Villard to his wife, June 25, 1878, HV; Charles F. Southmayd to A. H. Holmes, Sept. 5, 1878, HV; L. H. Meyer to Fred Ames, Oct. 16, 1878, HV; *Chronicle* 27:16, 95. The Denver extension bonds were the KP's first mortgage. Part of Gould's plan was to reduce the KP's fixed charges by converting all the pool securities into stock.

24. D. F. Carmichael to Villard, Aug. 3, 1878, HV; Southmayd to Holmes, Sept. 5, 1878, HV; Meyer to Fred Ames, Oct. 16, 1878, HV; Villard to Forbes, July 21, 1878, HV; Villard to Oakes, Sept. 23, 1878, HV; Gould to Clark, Aug. 29, 1878, UP, Sept. 3, 1878, KG; *Public* 14:148, 264; *Chronicle* 27:121, 251, 383, 409, 435–36; Villard to John Evans, Nov. 19, 1878, HV.

25. Richard C. Overton, *Burlington Route*, 27–28; *DAB*, 6:507–8. There is no satisfactory biography of Forbes.

26. Perkins too lacks a biography. See Richard C. Overton, *Burlington West* (Cambridge, Mass., 1941), 111–12, and *Perkins/Budd: Railway Statesmen of the Burlington* (Westport, Conn., 1982).

27. This list is drawn from Grodinsky, *Transcontinental*, 74–75, and *Iowa Pool*, 53–54.

28. Overton, *Burlington Route*, 114–51; Johnson and Supple, *Boston Capitalists* (see Chap. 13, n. 1), 156–80, 222–40, 263–70.

29. J. F. Barnard to Joy, Feb. 10 and Mar. 11, 1874, JFJ; C. C. Smith to Joy, Feb. 11, 1874, JFJ; Perkins to John W. Brooks, Mar. 2, 1874, BA; Perkins to Robert Harris, Mar. 9 and Mar. 20, 1874, BA; Grodinsky, *Iowa Pool*, 42–45.

30. Grodinsky, *Iowa Pool*, 46–51.

31. Joy to Moses Taylor, Jan. 18, 1876, JFJ; W. B. Strong to Perkins, Aug. 1, 1876, BA; Strong to Barnard, July 18, 1876, BA; Overton, *Burlington Route*, 149–50. The rate wars of 1875–76 are summarized in Grodinsky, *Iowa Pool*, 53–67.

32. Gould to Clark, Apr. 26, 1876, KG; Clark to Strong, Apr. 14, 1876, BA; Grodinsky, *Jay Gould*, 134.

33. Gould to Clark, Feb. 15, Aug. 12, Nov. 22, Dec. 5, and Dec. 13, 1876, KG; *Chronicle* 20:586, 21:607.

34. Griswold to A. N. Towne, Oct. 25, 1876, BA; Towne to Griswold, Nov. 6, 1876, BA; Grodinsky, *Iowa Pool*, 68–72.

35. Dillon to Clark, Dec. 23, 1876, UP; Dillon to James F. Wilson, Dec. 29, 1875, UP; Forbes to Perkins, Feb. 23, 1877, BA; Forbes Memorandum, Feb. 24, 1877, BA; Grodinsky, *Jay Gould*, 134–36; Overton, *Burlington Route*, 154–57. Both Grodinsky and Overton confuse the sequence of events by stating that Gould made this proposal *after* gaining seats on the boards of Northwestern and Rock Island in March. The documents cited above make it clear that the proposal was formulated by late December and that Forbes knew about it in February. Curiously, Grodinsky had the correct sequence in his earlier work, *Iowa Pool*, 74–75.

36. Forbes to Perkins, Feb. 23, 1877, BA; Gould to Clark, Mar. 2, 1877, KG. The portrait of Harris in Grodinsky, *Jay Gould*, 135–37, seems unduly harsh. The tone is that of a cold warrior describing Neville Chamberlain at Munich.

37. Harris to Schuyler Colfax, Mar. 12, 1877, BA; *Chronicle* 24:339, 364; *Stockholder* 13:392, 16:484. Gould, Dillon, and Fred Ames went onto the Northwestern and Rock Island boards, while David Dows, H. H. Porter, and W. L. Scott joined the Union Pacific board.

38. Gould to Clark, Mar. 17, Mar. 21 (two letters), and Mar. 31, 1877, KG; Forbes to David Dows, Mar. 15 and Apr. 15, 1877, BA; Forbes to Perkins, Mar. 23 and Apr. 15, 1877, BA; Griswold to Perkins, Mar. 16, 1877, BA; Dillon to Clark, Mar. 31, 1877, UP; Perkins to Forbes, Mar. 23 and Apr. 21, 1877, BA; Harris to Griswold, Apr. 7 (marked "Not Sent") and Apr. 20, 1877, BA; Overton, *Burlington Route*, 156–57; Grodinsky, *Iowa Pool*, 78–85.

39. Forbes to Tyson, June 10, 1877, BA; Gould to Clark, Aug. 13, 1877, KG; Grodinsky, *Iowa Pool*, 83–87; Overton, *Burlington Route*, 157.

40. Forbes to S. W. Simpson, May 29, 1878, BA; Overton, *Burlington Route*, 157–62; Grodinsky, *Iowa Pool*, 112.

41. Gould to Clark, July 30, 1875, KG; Trottman, *Union Pacific*, 183.

42. Athearn, *Union Pacific Country*, 237–39, 265–71.

43. Ibid., 271–78; *Report of the Commissioner of the General Land Office*, House Exec. Docs., 43 Cong., 2d sess., 1873 (Serial 1639), no. 1, pt. 5; Gould to Clark, Oct. 24 and Oct. 28, 1874, KG; Rollins to Dillon, Oct. 28, 1874, UP.

44. Huntington to Stanford, Mar. 12, 1875, MM; Huntington to Colton, Apr. 22, Oct. 9, and Oct. 29, 1875, MM; Gould to Clark, Oct. 30 and Nov. 18, 1875, Jan. 22 and Dec. 5, 1876, Mar. 9, 1877, KG; Dillon to Brigham Young, Feb. 5, 1876, UP; Dillon to T. E. Sickels, Oct. 22, 1875, UP; Dillon to Oliver Ames, Jan. 6, 1876, UP; Dillon to John Sharp, Oct. 20, 1875, UP.

45. Athearn, *Union Pacific Country*, 237–48.

46. Spence, *Territorial Politics* (see Chap. 15, n. 11), 116–26; Gould to Clark, Dec. 13, 1876, KG; Agreement with Joseph Richardson to Extend the Utah Northern, Jan. 14, 1877, UP; Athearn, *Union Pacific Country*, 249–52.

47. Gould to Clark, Mar. 13 and Mar. 17, 1877, KG; Athearn, *Union Pacific Country*, 252.

48. Dillon to Gould, Oct. 24 and Oct. 25, 1877, UP; Dillon to Elisha Atkins, Oct. 31, 1877, UP; Athearn, *Union Pacific Country*, 252–53.

49. Gould to Clark, Jan. 31, [Feb. 1878], Feb. 4, Feb. 6, Feb. 13, Apr. 20, July 12–15, July 17 (three letters), July 20, and July 21, 1878, KG; Dillon to Gould, Feb. 14, 1878, UP; Dillon to Clark, Mar. 23, 1878, UP; *PRC*, 537, 571–72.

50. Gould to Clark, July 21, Aug. 13, Aug. 22, Sept. 29, and Oct. 5, 1878, KG.

51. Gould to Clark, Feb. 8, Nov. 24, and Dec. 5, 1876, Mar. 3, Mar. 13, Mar. 17, July 8, July 11, Nov. 17, Nov. 28, and Dec. 24, 1877, Feb. 6 and Feb. 9, [1878], KG; Dillon to Clark, Apr. 28 and Aug. 28, 1876, UP; Dillon to Oliver Ames, Oct. 18, 1876, UP; Athearn, *Union Pacific Country*, 233; *PRC*, 1269–70.

52. Gould to Clark, Sept. 2, 1875, Feb. 15 and Dec. 13, 1876, Mar. 21, 1877, Feb. 4, Feb. 22, Feb. 26, Mar. 8, and Dec. 15, 1878, KG; Dillon to Oliver Ames, Oct. 18, 1876, UP; Dillon to John M. Thayer, Apr. 25, 1877, UP; Dillon to Clark, Apr. 9, 1877, UP; Dillon to James A. Evans, Aug. 22, 1877, UP; Athearn, *Union Pacific Country*, 233–34; Schell, *South Dakota* (See Chap. 14, n. 32), 140–57; Dillon to Elisha Atkins, Feb. 15, 1879, UP.

53. Snow, *Helen Gould* (see Chap. 1, n. 2), 135–36.

Chapter 17: Cross-Pollination

1. All the works on Gould cited earlier treat his telegraph ventures as speculative forays of the most predatory kind. Even Grodinsky equates his role with that of "corporate racketeers." *Jay Gould* (see Prologue, n. 5), 159.

2. *New York Times*, Feb. 19, 1875. For a series of editorials blasting Gould as the prime example of existing evils in the corporate system, see ibid., Apr. 4, Apr. 5, Apr. 12, Apr. 14, and Apr. 18, 1877.

3. The early history of the telegraph industry is covered in Robert L. Thompson, *Wiring a Continent* (Princeton, 1947). See also James D. Reid, *The Telegraph in America* (New York, 1886), and Alvin F. Harlow, *Old Wires and New Waves* (New York, 1936).

4. *Labor and Capital* (see Chap. 1, n. 7), 865–69, 943–50; *Report of the Committee on Post-Offices and Post-Roads on the Postal Telegraph*, Senate Reports, 48 Cong., 1st sess., 1884 (Serial 2177), no. 577, 4; *Financier* 2:275; *Chronicle* 17:491–92, 859, 18:41.

5. *Labor and Capital*, 881–82, 933–34; *Postal Telegraph*, 64. In 1879 Norvin Green asserted that "I cannot count a dozen lines of railroad in the United States that have any poles. . . . Nineteen out of twenty use the lines of Western Union Telegraph." *Hearing before the Committee on Railroads*, Senate Reports, 45 Cong., 3d sess., 1879 (Serial 1838), no. 805, 2. See also *Land-Grant Telegraph Lines*, House Reports, 49 Cong., 2d sess., 1886 (Serial 2500), no. 3501, 64–67.

6. *Land-Grant Telegraph*, i–x, 21–24, 213; *Railroad Gazette* 2:102; Gould to Rollins, June 16, 1875, UP; Senate Exec. Docs., 49 Cong., spec. sess., 1885 (Serial 2263), no. 2, 35–36; *Labor and Capital*, 961; *Chronicle* 18:136–37.

7. UP Exec. Comm. Minutes, Dec. 31, 1873, UP; Morton, Bliss & Co. to Rollins, Dec. 20, 1873, Mar. 3, 1874, UP; *Chronicle* 18:136–37; Morton to Atkins, Mar. 14, 1874, UP; Oliver Ames Diary, June 1, 1870, Jan. 16, Apr. 24, and June 14, 1873.

8. Dillon to Oliver Ames, Apr. 9 and June 25, 1874, UP; Dillon to Atlantic & Pacific Telegraph Company, Apr. 20, 1874, UP; Dillon to Atkins, May 15, 1874, UP; *French Cable Company*, House Exec. Docs., 47 Cong., 2d sess., 1883 (Serial 2108), no. 46, pt. 2, 5; Gould to Rollins, Nov. 24, 1874, UP; *Chronicle* 19:502, 638; Dodge to Gould, Jan. 6, Jan. 16, and Jan. 27, 1875, DR, 9:213, 221, 229. Delays prevented consummation of the Franklin lease until June 1876.

9. Eckert to Helen Gould, May 16, 1893, HGS. The italics are mine.

10. Matthew Josephson, *Edison* (New York, 1959), 92–111. There are compelling reasons to reject Josephson's version of Gould's role in telegraph affairs. His account is riddled with errors of fact and chronology, and is based on several incorrect assumptions: (1) that Gould controlled A & P as early as 1870 and 1871; (2) that Gould controlled Kansas Pacific, Missouri Pacific, and Wabash as early as 1873; and (3) that Harrington and Reiff were financial agents of Gould, who remained behind the scenes until 1874. Reiff was in fact Gould's business enemy throughout his career. See for example *PRC*, 3901, and the obituary in the *New York Times*, Mar. 2, 1911. The most satisfactory account of Edison's life, and of these episodes, is Robert Conot, *A Streak of Luck* (New York, 1979), 35–78. Although thinly documented, it incorporates the Edison materials at the Edison National Historic Site, West Orange, N.J., far more thoroughly and coherently than Josephson. For a third version, less detailed but offering a different point of view, see Harlow, *Old Wires*, 405–8.

11. Conot, *Streak of Luck*, 60–62. Biographical information on Orton may be found in *Stockholder* 16:386, and Reid, *Telegraph in America*, 529–36, 559–70.

12. Conot, *Streak of Luck*, 62–65.

13. Ibid., 56, 63–65; Harlow, *Old Wires*, 407. In explaining the link between Gould

and Eckert, Conot repeats the old story of how Gould helped his speculations during the Civil War by getting news in advance on the military telegraph, and asserts that Eckert was the man who provided it. He cites no evidence for this tale, and I have found none. Unless new evidence turns up, I see no reason to doubt Eckert's statement that the two men first met during the winter of 1874–75. A biographical sketch of Eckert is in Reid, *Telegraph in America*, 614–17.

14. Conot, *Streak of Luck*, 65–69; Reid, *Telegraph in America*, 595–96; Josephson, *Edison*, 120–21; Harlow, *Old Wires*, 407–8.

15. *Chronicle* 20:55–56; *Stockholder* 13:201–2; Conot, *Streak of Luck*, 68–70; Josephson, *Edison*, 121–25. Josephson asserts that in January 1875 "there was panic at Western Union, whose stock was falling to its lowest levels in recent times," and that Gould, by selling short, "was believed to have gained in a single day twenty or thirty times that which he paid the young inventor." If Gould was then bearing the stock as he often did, there is no evidence how much he made or lost. Certainly the figures would not remotely approach Josephson's claim on so small a movement in the stock. There is ample evidence on the price of Western Union, which during 1875 never dropped below 70¾. In previous years the stock had reached lows of 68 (1874), 43½ (1873), 67½ (1872), and 44 (1871). During January 1875 the price ranged from 79¾ to 70⅞. All figures from various issues of the *Chronicle*.

16. Conot, *Streak of Luck*, 70.

17. Quoted in Josephson, *Edison*, 126.

18. *Chronicle* 20:428; Reid, *Telegraph in America*, 597; Dodge to Nate Dodge, Apr. 27, 1875, DPL. The charge is in Josephson, *Edison*, 103–4, 125–26. For more detail on the technical limitations of the automatic system see Conot, *Streak of Luck*, 68–69.

19. UP Exec. Comm. Minutes, Apr. 22, 1875, UP; Gould to Rollins, Feb. 9, May 9, May 16, June 8, and June 15, 1875, UP; *Chronicle* 20:178, 240; *Railroad Gazette* 6:211; *Stockholder* 13:442, 617; Eckert to Helen Gould, May 16, 1893, HGS.

20. Gould to Oliver Ames, July 22 and July 24, 1875, OA; Gould to Rollins, Sept. 13 and Sept. [26], 1875, Jan. 26, 1876, UP; UP Exec. Comm. Minutes, July 21, 1875, UP; *Chronicle* 21:106, 201, 225, 298, 317, 436; *New York Tribune*, Aug. 17 and Sept. 6, 1875; *Stockholder* 13:648, 680.

21. Gould to Oliver Ames, Feb. 20, Feb. 21, Feb. 23, and Apr. 9, 1876, OA; Gould to Clark, Feb. 25, 1876, KG; *Chronicle* 22:177, 182, 227, 248, 257, 369–70, 416–17, 440, 464, 488–89.

22. *Chronicle* 22:560, 583, 591, 23:9, 57, 280, 516, 525; Gould to John W. Garrett, Aug. 3, 1876, GP; Gould to Ward, Aug. 26, 1876, GLC; John Young to Brigham Young, Brigham Young Collection, Church of Jesus Christ of Latter-day Saints Library, Salt Lake City (hereafter LDS).

23. Where Cornelius Vanderbilt lacks a definitive biography, William lacks any sort of study at all. For brief sketches see *DAB*, 19:175–76, and *National Cyclopedia of American Biography* (hereafter *NCAB*), 30:14–15.

24. *Chronicle* 24:16, 180, 199; *Railway World* 21:174; *Stockholder* 17:209; Fowler, *Twenty Years* (see Chap. 7, n. 7), 554–56.

25. Gould to Clark, Jan. 17, 1877, KG; *Labor and Capital*, 919–20; *Postal Telegraph*, 228; *Railway World* 21:174; *Chronicle* 24:158, 180–81, 316, 364.

26. Conot, *Streak of Luck*, 84–85; *Chronicle* 24:58, 130; Fowler, *Twenty Years*, 557.

27. *Chronicle* 24:136, 250, 299, 309–10, 495, 559, 589. For background detail on the rate war see Lane, *Commodore Vanderbilt* (Chap. 8, n. 6), 291–99.

28. For more detail on this episode see Grodinsky, *Jay Gould*, 153–57.

29. Ibid., 154; Overton, *Burlington Route* (see Chap. 16, n. 6), 25–31; Johnson and

Supple, *Boston Capitalists* (see Chap. 13, n.1), 88–120; Dillon to Clark, June 29, 1874, UP; Dillon to John Sharp, Aug. 10, 1875, UP; *Chronicle* 26:637–38.

30. Grodinsky, *Jay Gould*, 155.

31. Samuel Sloan to Joy, Oct. 5, 1875, Apr. 22, 1877, JFJ; Elijah Smith to Joy, June 20, 1877, JFJ; Moses Taylor to Joy, June 21 and June 22, 1877, JFJ; *Chronicle* 24:387, 462, 485, 606. For more detail on this episode see Daniel Hodas, *The Business Career of Moses Taylor* (New York, 1976), 216–34.

32. *Chronicle* 25:81, 115, 133, 155–56, 178, 188, 206; *New York Sun*, Aug. 21, 1877. Grodinsky, *Jay Gould*, 156–57, was the first to connect the Michigan Central to the sale of A & P.

33. *Chronicle* 25:155–56, 178, 355–56, 26:36, 637–38, 27:36; *Public* 13:404. The total stock of A & P was 140,000 shares, which gave Western Union majority control. It paid for the A & P shares with Western Union shares at 72 and $912,550 in cash.

34. *Postal Telegraph*, 28; *Chronicle* 25:257, 374, 426, 433, 26:135; Eckert to Helen Gould, May 16, 1893, HGS.

35. *New York Times*, Aug. 3, 1877; *New York Sun*, Aug. 3, 1877. These accounts of the Selover episode are so detailed and lurid as to make one suspect the accuracy of their particulars.

36. For differing accounts see *New York Times*, Aug. 3 and Aug. 6, 1877; *New York Sun*, Aug. 20, 1877; *Chronicle* 25:109; Stedman, *Stock Exchange* (Chap. 7, n. 5), 285; Fowler, *Twenty Years*, 563–64.

37. *New York Times*, Aug. 3, 1877; *New York Sun*, Aug. 3 and Aug. 5, 1877; *Chronicle* 25:109.

38. *New York Times*, Dec. 3, 1892. White, *Wizard* (see Prologue, n. 4), 218, agreed that "his triumphs were, for the most part, over men who would have ruined him if he had not ruined them."

Chapter 18: Homebody

1. Snow, *Helen Gould* (see Chap. 1, n. 2), 115, 157.

2. Ibid., 118.

3. Ibid., 117, 119.

4. Biographical information on Helen Gould is virtually nonexistent. Alice Snow provides some glimpses of her character. There are several of her letters at JGL and one childhood letter, dated 1849, in the possession of Kingdon Gould, Jr. Her obituary in the *New York Times*, Jan. 14, 1889, yields little useful information.

5. *New York Sun*, Dec. 9, 1878; Snow, *Helen Gould*, 118–19.

6. Gould to his wife, July 31, 1879, JGL.

7. *New York Times*, Aug. 8, 1883, Jan. 14, 1889. The latter praised Helen for the fact that her children were "trained by herself and not by nurses."

8. Snow, *Helen Gould*, 140, 152–53.

9. The brownstone at 579 belonged to banker and former mayor George Opdyke, who died in June 1880. See *NCAB*, 11:464–65. Lyndhurst now belongs to the National Trust for Historic Preservation and is open to the public. The complex tale of how it was preserved is told in Frank Kintrea, "The Realms of Gould," *American Heritage* 21 (April 1970): 47–65. Most sources, including Kintrea, assert that Gould first leased the property in 1878.

10. For Lyndhurst's architectural significance see the articles in *Historic Preservation* 17 (March–April 1965). The National Trust reprinted this volume under the title *Lyndhurst*. See also the relevant sections in William H. Pierson, Jr., *American Buildings and*

Their Architects: Technology and the Picturesque: The Corporate and the Early Gothic Styles (New York, 1978).

11. For these details I am grateful to Adna Watson, who gave me a special tour of Lyndhurst in August 1973. The point of this vogue was not economy; often the imitation material cost as much as the real item.

12. Snow, *Helen Gould*, 116, 122–23; Northrop, *Life and Achievements* (see Prologue, n. 2), 210–13; White, *Wizard* (see Prologue, n. 4), 234. There is at Lyndhurst a list of the objets d'art in the house and a copy of the catalog of books devised by Gould. Both the paintings in the great hall and the books in the library remain as they were arranged by Gould.

13. Billie Sherrill Britz, "Lyndhurst Greenhouse: Emblem of a Grand Society," *Historic Preservation* 25 (January–March 1973): 15–21; Gould to Clark, Aug. 4 and Aug. 9, 1880, KG; *New York Times*, Dec. 12, 1880; Northrop, *Life and Achievements*, 202–3.

14. *New York Sun*, Dec. 12, 1880; *New York Times*, Dec. 12, 1880; Britz, "Lyndhurst Greenhouse," 16–19.

15. Snow, *Helen Gould*, 155, 165–66; Gould to Clark, Aug. 4, 1880, KG.

16. George W. Walling, *Recollections of a New York Chief of Police* (Montclair, N.J., 1972), 368–72; *New York Times*, Nov. 14–17 and Dec. 13, 1881; Northrop, *Life and Achievements*, 231–32.

17. Snow, *Helen Gould*, 119–21.

18. See Gould to Garfield, May 16, 1881, JAG.

19. Northrop, *Life and Achievements*, 314; *New York Times*, Sept. 6 and Sept. 13, 1879, Jan. 22, 1880; *New York Sun*, Sept. 12 and Sept. 13, 1879; *Stockholder* 17:354.

20. Snow, *Helen Gould*, 107, 111–14, recounts this episode but omits the fact that her father died by his own hand.

21. Ibid., 113–15, 126–27; Helen D. Gould to Sarah Northrop, Nov. 30, 1886, JGL.

22. Ibid., 123–25.

Chapter 19: Gambit

1. *New York Sun*, June 23 and July 1, 1878; *Stockholder* 16:468; *Chronicle* 27:68; Gould to Ward, undated, GLC.

2. This story is repeated in several places, most notably in Grodinsky, *Jay Gould* (see Prologue, n. 5), 161, 164, who took it from the *Philadelphia Press*, Dec. 3, 1892. While admitting that "it is of course not possible to vouch for the accuracy of the story," Grodinsky gave it credibility because "the correspondent of this paper for almost a decade had revealed an extensive acquaintance with many details of Gould's business career." However, there is evidence that the reporter in question did not get this story firsthand. The episode first appeared in the *New York Times*, Nov. 20, 1881, where it was attributed to an "insider." The two versions are so similar as to leave little doubt that one was taken from the other.

3. Gould to Clark, Aug. 18, 1876, KG; Dodge to Clark, Feb. 18, 1879, DR, 9:881; Gould to Henry McFarland, Feb. 26, Mar. 17, and June 24, 1878, UP; Dillon to Elisha Atkins, Mar. 23, 1878, UP; *Public* 13:411.

4. *Stockholder* 16:468; *Chronicle* 27:10; *Public* 14:107; *New York Times*, Jan. 30, 1879.

5. Fowler, *Twenty Years* (see Chap. 7, n. 7), 566; *New York Sun*, July 31, Aug. 7, and Sept. 13, 1878; *Chronicle* 26:469, 574; *Stockholder* 16:484. "Granger roads" was a generic term applied to the northwestern grain carriers.

6. Gould to McFarland, July 30, 1878, UP; Dillon to Elisha Atkins, Aug. 8, 1878, UP; Gould to Clark, Aug. 5, 1878, KG; *Public* 14:68.

7. *New York Times*, July 30, 1878; W. L. Scott to Albert Keep, Sept. 6, 1878, quoted in Grodinsky, *Jay Gould*, 160–62; *Public* 14:68; *Chronicle* 27:117; M. L. Sykes to Forbes, Oct. 28, 1878, BA; Forbes to Sykes, Oct. 30, 1878, BA; Forbes to Perkins, Oct. 30, 1878, BA. The Scott letter was in the archives of the Chicago & Northwestern Railroad. Grodinsky notes that "probably through some oversight these letters were later destroyed." Although Grodinsky possessed copies, his sudden death occurred shortly after he and his wife had changed residences. In the years since his death the family has tried repeatedly but in vain to discover where he stored his research files during the move. As a result, the only Northwestern materials available are those cited by Grodinsky in his books.

8. Forbes to Griswold, Nov. 10, 1878, BA; Stanford to Huntington, Nov. 13 and Nov. 20, 1878, MM.

9. RGD, NYC, 422:695; Morosini, "Reminiscences" (see Chap. 8, n. 12). A detailed but dubious description of Gould's office at Belden's is in the *New York Sun*, Aug. 5, 1877.

10. Morosini, "Reminiscences"; *PRC*, 320, 323; *New York Times*, Jan. 10, Jan. 15, Jan. 21, Jan. 30, Mar. 23, Mar. 25, Mar. 27, and June 3, 1879; *New York Sun*, Jan. 10, 1879; *Stockholder* 16:660. Fowler, *Twenty Years*, 526, describes Belden as "a man of little ability and less conscience," and agrees that he forced Gould to become a special partner.

11. Dodge to Clark, Feb. 1, 1879, DR, 9:873; Patrick Geddes to Forbes, Feb. 1, 1879, BA; *New York Times*, Jan. 30, 1879.

12. *New York Times*, Jan. 30, 1879; Perkins to Forbes, Feb. 1, 1879, BA.

13. Dodge to Clark, Feb. 1, 1879, DR, 9:873; *New York Times*, Jan. 5, Jan. 6, Jan. 21, and Jan. 28, 1879.

14. Dillon to Fred Ames, Dec. 12, 1878, UP; Dillon to Clark, Dec. 20, and Dec. 28, 1878, Jan. 14, 1879, UP; Dillon to McFarland, Dec. 30, 1878, UP; *PRC*, 400, 537, 571–72; Gould to Clark, Dec. 11, Dec. 12, Dec. 14, and Dec. 22, 1878, Feb. 22, 1879 (three letters), KG; Memorandum of Agreement, UP Exec. Comm. Minutes, Dec. 18, 1878, UP; Dodge to Clark, Feb. 18, 1879, DR, 9:881; *Stockholder* 16:737, 772; *Public* 15:130. Gould and others gave the number of shares sold as 100,000, but this figure included the 30,000 shares sold in November. See *Chronicle* 28:200.

15. *New York Sun*, Feb. 18, 1877; *Stockholder* 16:737; Stedman, *Stock Exchange* (see Chap. 7, n. 5), 287; Gould to Clark, Feb. 22, 1879 (three items), KG; Morosini to Clark, Feb. 24, 1879, KG.

16. Dillon to Fred Ames, Oct. 28, 1878, UP; *Stockholder* 16:772; *Public* 15:176.

17. Gould to McFarland, Mar. 25, 1879, UP; Gould to Clark, Feb. 14, 1879, KG; John Sharp to Dillon, Feb. 18, 1879, UP.

18. Forbes to Perkins, May 9, 1878, BA; Villard to John Evans, Nov. 19, 1878, HV; *New York Times*, Mar. 9, 1879. For background see Chap. 16.

19. Gould to D. M. Edgerton, Jan. 11, 1878 [1879], and Jan. 28, 1879, MP; Gould to Clark, Feb. 14 and Feb. 22, 1879, KG; Dodge to Clark, Feb. 1 and Feb. 18, 1879, DR, 9:873, 881; *Chronicle* 28:199.

20. Greeley to D. M. Edgerton, Mar. 4, 1879, MP. See also the flurry of telegrams from Gould to Edgerton, Jan. 8–13, 1879, MP.

21. Greeley to D.M. Edgerton, Mar. 3 (two items), Mar. 5, and Mar. 6 (two items), 1879, MP; Gould to D. M. Edgerton, Mar. 6, 1879, MP; Myers, Rutherfurd & Co. to D. M. Edgerton, Feb. 21, 1879, MP; E. R. Carter to D. M. Edgerton, Feb. 20, 1879, MP; *PRC*, 117–19, 144–45, 175, 345, 361, 460–62, 497–503, 654, 1695. Copies of the original pool agreement and valuation are in MP. Gould's purchase of the pool securities is itemized in *PRC*, 497–503.

22. *Public* 15:130, 146, 176; *New York Times*, Mar. 9, 1879, *Stockholder* 16:772; *Chronicle* 28:216, 252, 271, 277. All data on prices and amounts of Kansas Pacific stock refer to two half-shares. The company issued $50 par value half-shares; to avoid confusion, all figures are translated into full-share equivalents.

23. Gould to Villard, Apr. 1, 1879, HV; Villard to Gould, June 19, 1879, HV; Villard to H. Thielsen, July 5, 1879, HV; Gould to Villard, June 14, 1879, HV.

24. Trottman, *Union Pacific* (see Chap. 13, n. 1), 155, Grodinsky, *Transcontinental* (see Chap. 13, n. 13), 96–98; Robert G. Athearn, *Rebel of the Rockies: A History of the Denver and Rio Grande Western Railroad* (New Haven, 1962), 42–75; Overton, *Gulf to Rockies* (see Chap. 15, n. 13), 14–21.

25. Gould to Clark, Oct. 31, 1878, Jan. 16, Feb. 18, Apr. 9, and Apr. 19, 1879, KG; *Chronicle* 28:18; Athearn, *Rebel of the Rockies*, 70–71.

26. Forbes to Charles J. Paine, Sept. 27, 1878, BA; Grodinsky, *Transcontinental*, 98; Overton, *Burlington Route* (see Chap. 16, n. 6), 167–68.

27. Perkins to Forbes, May 21, 1878, BA; Gould to Clark, July 13 and Aug. 6, 1878, KG; Dexter to Gould, Aug. 5, 1878, KG; *Railroad Gazette* 10:338; Overton, *Burlington Route*, 167–68; Athearn, *Union Pacific Country* (see Chap. 13, n. 4), 233.

28. J. F. Joy to Gould, July 12, 1878, KG; Gould to Clark, Nov. 30, 1878, Jan. 13 and Jan. 16, 1879 (two letters), KG; W. S. Gurnee to Clark, Dec. 14, 1878, KG.

29. Gould to Clark, Dec. 22, 1878, Feb. 18, 1879 (two letters), KG; Albert Keep to Patrick Geddes, Jan. 27, 1879, BA; Geddes to Forbes, Feb. 1, 1879, BA; *Chronicle* 28:275; Grodinsky, *Iowa Pool* (see Chap. 16, n. 3), 113–15.

30. Perkins to Forbes, May 6, 1878, BA.

Chapter 20: Blitzkrieg

1. Grodinsky, *Jay Gould* (see Prologue, n. 5), 165; Perkins to Geddes, Nov. 24, 1879, BA.

2. Perkins to Forbes, May 25, 1879, BA. An excellent discussion of this change can be found in Chandler, *Visible Hand* (see Prologue, n. 6), 145–87.

3. Gould to Clark, Jan. 16, Apr. 9, and Apr. 25, 1879, KG; *Chronicle* 28:373, 447.

4. Anonymous, "A Chapter of Wabash," *North American Review* 146 (February 1888):178–81; *Chronicle* 17:490; *Railroad Gazette* 10:579; Grodinsky, *Jay Gould*, 189–93.

5. *Public* 14:339; *Railroad Gazette* 10:579; *Chronicle* 27:540, 568, 28:200. Sketches of Garrison are in *NCAB*, 7:262, and *DAB*, 7:167–68.

6. *Public* 15:267; *Chronicle* 28:402, 420, 429.

7. *Chronicle* 26:654; *Public* 13:388; *New York Tribune*, May 3, May 12, and May 20, 1879; *Stockholder* 17:149; Forbes to Perkins, May 6 and May 8, 1879, BA; Gould to Clark, May 12, 1879, KG.

8. *Chronicle* 28:454; Reid, *Telegraph in America* (see Chap. 17, n. 3), 577–79; *Labor and Capital* (see Chap. 1, n. 7), 1069; Eckert to Helen Gould, May 16, 1893, HGS.

9. *Chronicle* 28:429; Grodinsky, *Jay Gould*, 196–97. Samuel Carter III, *Cyrus Field: Man of Two Worlds* (New York, 1968), 318, errs in asserting that Gould was responsible for involving Field in Wabash. Field had long been a prominent figure in the road and served as a director. For Humphreys see *PRC*, 370.

10. Gould to Joy, May 1, 1879, JFJ; *New York Tribune*, Nov. 27, 1879; Grodinsky, *Jay Gould*, 197, 212; J. Hickson to Joy, May 14, 1879, JFJ.

11. Both negotiations may be followed in detail in Gould to Joy, May 1, 1879, JFJ; J. Hickson to Joy, May 29 and June 1, 1879, JFJ; A. L. Hopkins to Joy, May 29, 1879, JFJ; Francis

Grey to Joy, June 2 and June 9, 1879, JFJ; Charles Merriam to Joy, May 2, 1879, JFJ; Wager Swayne to Joy, May 17, 1879, JFJ; Samuel Sloan to Joy, May 22, 1879, JFJ; Solon Humphreys to Joy, May 31, 1879, JFJ; F. Broughton to Joy, June 6, 1879, JFJ; Rutter (?) to Field or Humphreys, July 4, 1879, JFJ; Gould to Clark, June 20, 1879, KG; *Public* 15:285, 316, 333, 348, 380, 397; *Chronicle* 28:453, 600, 617; *Stockholder* 17:148. The Bee Line's proper name was the Cleveland, Cincinnati, Chicago & Indianapolis.

12. Perkins to Forbes, May 30, 1879, BA; *PRC*, 472–73; Gould to Clark, Apr. 19 and June 22, 1879, KG; *Chronicle* 28:503, 578; Grodinsky, *Jay Gould*, 172–73.

13. *Chronicle* 28:513; Fowler, *Twenty Years* (see Chap. 7, n. 7), 566–69; Gould to Clark, June 14, June 20, and June 22, 1879, KG; Dillon to Fred Ames, June 27, 1879, UP.

14. Perkins to Forbes, May 30, 1879, BA. For the battle over Royal Gorge see Athearn, *Rebel of the Rockies* (Chap. 19, n. 24), 55–66, 83–90.

15. Forbes to Perkins, May 22, 1879, BA.

16. Perkins to Forbes, May 25 and June 29, 1879, BA.

17. Dillon to Fred Ames, June 14, 1879, UP; Villard to Gould, June 19, 1879, HV; Villard to H. Thielsen, July 5, 1879, HV.

18. *Stockholder* 17:209; *PRC*, 465; *Chronicle* 29:197; Grodinsky, *Jay Gould*, 173.

19. Gould to his wife, July 15, July 16, July 20, July 26, July 29, and July 31, 1879, JGL.

20. *New York Tribune*, July 9 and Nov. 11, 1879; *Public* 16:21–23, 170; *Chronicle* 29:35, 42, 400, 408; *Railroad Gazette* 11:421–22; Gould to Clark, May 1, 1879, KG. The new company, known as Wabash, St. Louis & Pacific, elected its first board in November 1879.

21. Gould to Clark, May 12 and June 20, 1879, KG.

22. Humphreys to Joy, June 26, 1879, JFJ; Field to Hickson, July 8, 1879, JFJ; Field to Vanderbilt, July 24, 1879, JFJ; Humphreys to Hickson, July 24, 1879, JFJ; Field to Grant, July 24, 1879, JFJ; Hickson to Humphreys, July 25, 1879, JFJ; Agreement between Wabash and Great Western railways, July 16, 1879, JFJ; *New York Tribune*, July 25 and Aug. 26, 1879; *Chronicle* 29:42, 87, 121, 171, 226; *Railroad Gazette* 11:416, 422; *Public* 16:70–71, 132. For the Eel River's importance to the Lake Shore see John Newell to Joy, Sept. 8, 1875, JFJ.

23. Forbes to Griswold, July 4, 1879, BA; Forbes to Geddes, July 10, 1879, BA; Forbes to Perkins, July 12, 1879, BA; *Chronicle* 28:521, 29:95; Dillon to J. T. Clark, July 18, 1879, UP; J. T. Clark to Dillon, Aug. 10, 1879, UP. The version of the rate war given in Grodinsky, *Iowa Pool* (see Chap. 16, n. 3), 116, does not square with the sources cited here.

24. Dodge to Nate Dodge, Aug. 14, 1879, DPL; *Public* 16:132, 148, 201, 203, 228, 234; *Chronicle* 29:301, 329, 331, 382, 407; *Stockholder* 17:387; *Railroad Gazette* 11:516; *PRC*, 535–37, 572–74, 1849–53; Gould to Clark, Apr. 25, 1879, KG; Athearn, *Rebel of the Rockies*, 85–86. Athearn errs in stating that the Rio Grande purchase occurred after the consolidation of Union Pacific and Kansas Pacific. Documents on the South Park agreement, dated Sept. 30 and Oct. 8, 1879, are in MP. See also Evans to Gould, Oct. 3 and Oct. 21, 1879, MP.

25. Gould to Clark, Apr. 25, June 20, and Oct. 13, 1879, KG; Dillon to Gould, Sept. 16, 1879, UP; *Public* 16:73, 170; *Chronicle* 29:407. A series of articles on the Leadville boom appeared in the *New York Tribune* during May and June, 1879. See also *Stockholder* 17:212, 244, 276.

26. *Public* 16:187, 245–46.

27. For varying interpretations of the consolidation see Myers, *American Fortunes* (Chap. 5, n. 1), 479–86; Warshow, *Jay Gould* (Chap. 5, n. 1), 138–51; Josephson, *Robber Barons* (Chap. 4, n. 15), 197–201; Riegel, *Western Railroads* (Chap. 13, n. 1), 162–64; Thomas C. Cochran and William Miller, *The Age of Enterprise* (New York, 1942), 148–49; O'Connor, *Gould's Millions* (Prologue, n. 1), 144–49; Trottman, *Union Pacific* (Chap. 13, n. 1), 156–74;

Grodinsky, *Jay Gould*, 175–80; Johnson and Supple, *Boston Capitalists* (Chap. 13, n. 1), 245–51.

28. *PRC*, 476–80, 657, 702–4; Gould to Clark, Nov. 2, 1879, KG.

29. Gould to Clark, Oct. 13 and Nov. 2, 1879, KG; *Chronicle* 29:375; Dillon to Clark, Oct. 28, Oct. 29, and Dec. 3, 1879, UP; Dillon to Fred Ames, Nov. 7, 1879, UP.

30. Dillon to Fred Ames, Nov. 7, 1879, UP; Dillon to Thomas Nickerson, Nov. 8, 1879, UP; *PRC*, 660–63, 705–6.

31. *PRC*, 525; Ames, *Pioneering* (see Chap. 13, n. 1), 522–23. Ames admitted that "when we took [the Central] it was of no value. It did not earn expenses."

32. *PRC*, 524–26, 806–11; Ames, *Pioneering*, 523–24.

33. Perkins to Forbes, Nov. 8, 1879 (two letters), BA; *PRC*, 477, 505–9, 528–29; *Chronicle* 29:489, 505, 538; *Stockholder* 18:84; *Public* 16:324, 329. It is impossible to reconcile the conflicting figures on the transaction. The price per share given in different sources ranges from $750 to $950.

34. *Chronicle* 29:505–6, 530; *Public* 16:308, 324; *New York Tribune*, Nov. 21, Nov. 22, Nov. 27, and Nov. 29, 1879, Jan. 6, 1880; *New York World*, Nov. 22, and Nov. 24, 1879; *New York Evening Post*, Nov. 21, 1879; *New York Times*, Nov. 22 and Nov. 27, 1879; *New York Sun*, Nov. 29, 1879. Details on the transaction are in Harry H. Pierce, "Anglo-American Investors and Investment in the New York Central Railroad," in Joseph R. Frese, S.J., and Jacob Judd, eds., *An Emerging Independent American Economy, 1815–1875* (Tarrytown, N.Y., 1980), 138–60.

35. *New York Evening Post*, Nov. 28, 1879; Pierce, "New York Central," 148–49; *New York Tribune*, Nov. 27 and Nov. 29, 1879; *New York Times*, Nov. 27, 1879; *Bradstreet's*, Nov. 29, 1879; *Chronicle* 29:453, 564; *Stockholder* 18:130; *American Railroad Journal* 52:1317–18, 1421, 53:57.

36. *Public* 17:41; *Chronicle* 30:67, 357; *New York Tribune*, Jan. 14, 1880; Grodinsky, *Jay Gould*, 180–81; Athearn, *Rebel of the Rockies*, 86–88. A copy of the circular describing the proposed Pueblo & St. Louis road, dated Jan. 15, 1880, is in BA.

37. Perkins to Forbes, Nov. 8, Nov. 12, Nov. 18, and Nov. 28, 1879, BA; *New York Sun*, Dec. 16, 1879; *Chronicle* 29:408, 530; *Public* 16:340.

38. Forbes to George Tyson, Dec. 13, 1879, BA; Perkins to Forbes, Nov. 11, Nov. 22, and Nov. 28, 1879, BA; T. J. Potter to Perkins, Jan. 4, 1880, BA.

39. Perkins to Forbes, Nov. 19, 1879, BA.

40. *PRC*, 482, 808–9, 818. For details on the controversy over the Dodge-Humphreys report see *PRC*, 476–81, 583. The report was dated Jan. 16, 1880, but doubtless its contents were known earlier.

41. *Chronicle* 30:40; *PRC*, 660–63, 705–6; Dillon to Dexter, Jan. 8, 1880, UP. For the Burlington consolidation see Overton, *Burlington Route* (Chap. 16, n. 6), 169–70.

42. *PRC*, 510, 731, 753.

43. Recollections of the meeting are in *PRC*, 354–58, 483, 507–10, 659–67, 705–6, 745–48. Atkins was not at the meeting but signed the agreement later. *PRC*, 776–77. For Gould's holdings in Union Pacific see *PRC*, 474–75, 555. He also owned $1 million in bonds.

44. *PRC*, 340–42, 357, 662, 665, 689, 706. A copy of the agreement, dated Jan. 14, 1880, is in UP. See also *PRC*, 691–94. The standard interpretation that Gould forced these smaller roads on the Union Pacific is simply not supported by the evidence. For examples of this interpretation see Trottman, *Union Pacific*, 162–63, and Grodinsky, *Jay Gould*, 178–79.

45. *PRC*, 63, 487–92, 503–5, 516. Gould, Dillon, and Fred Ames sat on the Colorado Central board and executive committee. See Dillon to J. W. Gannett, Dec. 23, 1879, UP.

46. UP Minutes, Jan. 24, 1880, 336–44; *PRC*, 474–75, 512–20, 531–32; *New York Tribune*, Jan. 25, 1880; *New York Sun*, Jan. 25 and Jan. 30, 1880; *Chronicle* 30:93, 108, 118; *Public* 17:67–68, 71. Gould received his Union Pacific shares on Feb. 16, at which time the price ranged between 90 and 94.

47. Johnson and Supple, *Boston Capitalists*, 247.

48. Henry Villard, *Memoirs of Henry Villard, Journalist and Financier* (Boston, 1904), 2:283. Trottman, *Union Pacific*, 167, concluded that "the amount of his profit cannot be ascertained from the evidence available." For the Denver Pacific see *PRC*, 465, 503–4, 523, 556–58. On the Denver Pacific stock Gould insisted that "no director or individual . . . made a dollar out of the transaction, and I am very glad . . . to put a final nail in that coffin."

49. These calculations are based on data in *PRC*, 466–75, 523–24, 530, 574–75. Commissioner E. Ellery Anderson asserted that Gould received 32,960 shares of Union Pacific for properties other than the Kansas Pacific. The evidence does not explain the discrepancy between this figure and the 24,502 shares shown in Gould's detailed accounts.

50. Calculations are based on statements and data in ibid., 361, 457–63, 475, 497–502, 530, 558–59.

51. Grodinsky, *Jay Gould*, 178.

52. *Chronicle* 29:454; *PRC*, 675.

53. *PRC*, 509, 685, 745–47, 760, 777.

54. Even Trottman, *Union Pacific*, 169, a severe critic of the consolidation, concludes that it "was the less [*sic*] of two evils for the Union Pacific." The other evil was Gould's threat to create a rival system.

55. *Chronicle* 30:93, 108, 118; *Public* 17:67–68, 71; *Stockholder* 18:273.

Chapter 21: Phantasmagoria

1. *Chronicle* 29:278, 30:308; *Public* 16:308, 17:245; *Stockholder* 18:97.

2. *New York Tribune*, Mar. 10 and Aug. 14, 1879; *Stockholder* 17:116; Grodinsky, *Jay Gould* (see Prologue, n. 5), 234–35; Forbes to Geddes, Aug. 28, Sept. 23, and Sept. 30, 1879, BA; Amsterdam Committee to Forbes, Oct. 14, 1879, BA; W. J. Ladd to Forbes, Oct. 24, 1879, BA; Forbes to Perkins, Oct. 28, Nov. 1, and Dec. 3, 1879, BA; Ladd to Perkins, Oct. 14, 1879, BA. The Katy lacks a satisfactory history. V. V. Masterson, *The Katy Railroad and the Last Frontier* (Norman, Okla., 1952), is rich in folklore and anecdotes but belongs to the "cops and robbers" school of corporate history. The factual errors concerning Gould's role in the Katy are too numerous to itemize.

3. *Chronicle* 29:424, 30:117; *New York Tribune*, Oct. 22, 1879, Jan. 28, 1880; *New York Times*, Jan. 28, 1880; *Public* 17:27; *Stockholder* 18:276.

4. *New York Tribune*, June 28, 1879; *Chronicle* 29:650; *Stockholder* 18:167; *Public* 16:392, 17:24; *American Railroad Journal* 52:1421; *New York Times*, Jan. 6, 1880. For the Texas & Pacific see Albert C. Simmonds, Jr., "A History of the Texas and Pacific Railway Company, 1871–1924," Texas & Pacific Historical Records (hereafter TPHR), MP.

5. *New York Times*, Jan. 10, 1885; *New York Sun*, Nov. 21, 1879; *New York Tribune*, Dec. 12, 1879, Feb. 7, 1880; *Public* 17:73; Keith L. Bryant, Jr., *History of the Atchison, Topeka and Santa Fe Railway* (New York, 1974), 84–85; Lavender, *Great Persuader* (see Chap. 13, n. 4), 318–34.

6. *New York Tribune*, Jan. 21 and Jan. 29, 1880; *New York Times*, Jan. 22 and Jan. 31, 1880; *Chronicle* 30:108.

7. *New York Times*, Dec. 3, 1879; *New York Sun*, Feb. 9, 1880; *Chronicle*, Investor's Supplement, May 31, 1875.

8. *New York Tribune*, Feb. 4, 1880; Perkins to Geddes, Nov. 24, 1879, BA; Perkins to Forbes, Dec. 20, 1879, BA; Forbes to Tyson, Dec. 13, 1879, BA; *Chronicle* 29:620–21; Gould to Clark, Oct. 18, 1879, KG; Grodinsky, *Iowa Pool* (see Chap. 16, n. 3), 118–19.

9. Perkins to Forbes, Dec. 20, 1879, Jan. 30, 1880, BA; "Memorandum with Union Pacific R. R. Co." Jan. 30 (two items), 1880, BA.

10. *Railroad Gazette* 12:169; *Public* 17:41, 118; Perkins to Forbes, Feb. 11, 1880, BA; Forbes to Perkins, Jan. 14, Feb. 6, and Feb. 26, 1880, BA; Forbes to Geddes, Feb. 7, 1880, BA; Forbes to George F. Hoar, Mar. 10, 1880, BA; Dodge to Perkins, Apr. 1, 1880, DR, 10:75; Forbes to Lucius Tuckerman, Apr. 14, 1880, BA; *New York Tribune*, Mar. 19 and Mar. 30, 1880; *Stockholder* 18:306, 321, 451; *Chronicle* 30:144, 334–36, 409.

11. T. J. Potter to Perkins, Apr. 3, 1880, BA; Forbes to Tyson, Dec. 13, 1879, Apr. 19, [1880], BA; Forbes to Tuckerman, Feb. 14 and Apr. 14, 1880, BA; Forbes to Charles Merriam, May 7, 1880, BA; Perkins to Forbes, May 11, 1880, BA; Forbes to Perkins, Feb. 11 and May 14, 1880, BA; Forbes to Griswold, May 12, 1880, BA; *Chronicle* 30:334–36, 397–98, 448, 533; Overton, *Burlington Route* (see Chap. 16, n. 6), 170–71. The Santa Fe, of course, was also controlled by a group of Boston capitalists.

12. *Chronicle* 31:21, 70, 83–84, 94–95; *New York Tribune*, Apr. 23 and May 27, 1880; *Public* 17:275, 18:51; Gould to Perkins, July 5, 1880, BA; Perkins to Gould, July 7, 1880, BA; Perkins to Forbes, July 7, 1880, BA; Forbes to William Endicott, July 11, 1880, BA.

13. Gould to Coolidge, Aug. 10, 1880, BA. For details of the meeting with Perkins see Dillon to E. P. Vining, July 29, 1880, UP; Dillon to Vining and Thomas L. Kimball, Aug. 6, 1880, UP.

14. Ladd to Forbes, Aug. 24, 1880, BA; Perkins to Forbes, Aug. 30, 1880, BA; Potter to Perkins, Aug. 30, 1880, BA; Edmund Smith to Potter, Sept. 9 (two items, one marked "Thursday night"), 1880, BA; Humphreys to Perkins, Sept. 13, 1880, BA; *New York Tribune*, Sept. 16, 1880; *Chronicle* 31:171, 303–4, 383.

15. *Chronicle* 31:239–40; Forbes to Perkins, Feb. 26, 1880, BA; Forbes to Fred Ames, Sept. 8, 1880, BA. Forbes added, "You and I like your father and Uncle can always understand and trust each other."

16. Fred Ames to Forbes, Sept. 8, 1880, BA; Gould to Fred Ames, Sept. 13, 1880, BA; Perkins to Forbes, Sept. 4, 1880, BA.

17. Perkins to Forbes, Mar. 7, Sept. 4, and Sept. 26, 1880, BA; Perkins to Coolidge, Sept. 26, 1880, BA.

18. Perkins to Forbes, Sept. 16 and Sept. 26, 1880, BA; Perkins to Coolidge, Sept. 26, 1880, BA. See also Forbes to Fred Ames, Sept. 25, 1880, BA, and Joy's letter in *Public* 18:201–2.

19. Potter to Perkins, Oct. 14, 1880, BA; Perkins to Forbes, Oct. 18, 1880, BA; *New York Tribune*, Oct. 13 and Oct. 19–23, 1880; *Public* 18:244, 248, 277–78; *Chronicle* 31:359.

20. *Public* 18:277–78; William K. Ackerman to William Osborn, Oct. 30, 1880, *RL*; Humphreys to Perkins, Nov. 5, 1880, BA; Gould to J. W. Midgely, Nov. 8, 1880, BA; Minutes of Meeting of General Managers and General Passenger Agents, Nov. 8, 1880, BA. The passenger war may be followed in the *New York Tribune*, Nov. 8, 11, 12, and 16, 1880, and *Chronicle* 31:493–94. See also Grodinsky, *Jay Gould*, 218–19, and the interview with A. L. Hopkins in *Railroad Gazette* 12:567.

21. Perkins to Forbes, Nov. 8, 1880, BA; Gould to J. W. Garrett, Sept. 14, 1876, GP.

22. *Stockholder* 17:240, 18:753–54; Hodas, *Moses Taylor* (see Chap. 17, n. 31), 95–156; *Chronicle* 30:520, 31:216–17, 223, 229; *Public* 17:405–6, 18:4, 131, 150–52; *New York Tribune*, Aug. 26, 1880.

23. *Chronicle* 31:217, 248, 304, 330, 32:33; *New York Tribune*, Sept. 13 and Oct. 19, 1880; *Railroad Gazette* 12:512, 567; *Public* 18:259–60, 297.

24. *Stockholder* 16:469, 17:2, 18:547; *Chronicle* 28:554, 30:565; *Public* 17:347; *New York Tribune*, Nov. 11, 1879; Dillon to Fred Ames, Dec. 12, 1878, UP; Dillon to T. L. Kimball, Feb. 9, 1880, UP; *New York Sun*, May 27, 1880. A fuller account is in Grodinsky, *Jay Gould*, 184–86.

25. *PRC*, 1853.

26. *PRC*, 47–48, 1732–39, 1849–60; Trottman, *Union Pacific* (see Chap. 13, n. 1), 193–96; Athearn, *Union Pacific Country* (see Chap. 13, n. 4), 231–33; *Chronicle* 31:45, 68, 358. The charge was made by Trottman, who cites as his only evidence a letter reprinted in *PRC*, 1859, which he appears to have misread.

27. Gould to Clark, Nov. 27, 1880, KG; Perkins to Forbes, Nov. 14, 1880, BA; *New York Tribune*, Nov. 22, 1880; *Public* 18:340; *Chronicle* 31:552, 558, 32:7.

28. Huntington to Hopkins, Mar. 24 and Mar. 25, 1876, MM; Colton to Huntington, Oct. 15, 1877, MM; DR, 10: 137–209; *Chronicle* 21:229–30, 26:302–3.

29. Dodge to Gould, Feb. 20, Mar. 16, and Mar. 20, 1880, DR, 10:41, 65, 67; Frank J. Bond to Gould and Scott, June 23, 1880, DR, 10:99–102; Dodge to Frank Baldwin, June 24, 1880, DR, 10:103; Dodge to Moses Taylor, Aug. 30, 1880, DR, 10:131; Dodge to Frank Bond, Nov. 5, 1880, DR, 10:184; *Chronicle* 30:358, 434, 31:95, 204; *New York Tribune*, Aug. 11 and Aug. 15, 1880; *Public* 17:121, 264; *New York Times*, Jan. 10, 1885; "Memorandum," DR, 10:193; *Bradstreet's*, Mar. 21, 1885; George C. Smith, comp., *The Missouri Pacific Railway Company, Statistics and Directories* (St. Louis, 1887), 121–22; Grodinsky, *Jay Gould*, 256–58.

30. *Railway World* 24:613; *New York Tribune*, Aug. 12 and Nov. 10, 1880; *Chronicle* 31:177–79, 248; *Public* 18:119, 152; Dodge to Gould, Dec. 26, 1880, DR, 10:229–30. Huntington's Texas ally was Colonel Tom Peirce, owner of the Galveston, Harrisburg & San Antonio. See Lavender, *Great Persuader*, 327.

31. Dodge to Gould, Nov. 17, Dec. 26, and Dec. 28, 1880, DR, 10:195–96, 229–31. The other road was the Gulf, Colorado & Santa Fe.

32. *Chronicle* 30:401; *Stockholder* 18:466.

33. *Public* 17:84, 292, 18:308; *New York Tribune*, Nov. 18 and Dec. 5, 1880; *New York Sun*, Nov. 18, 1880; *Railroad Gazette* 12:310; *Chronicle* 31:510, 528, 535, 588, 606, 32:101; Frank Bond to Gould and Scott, June 23, 1880, DR, 10:102; *Railway World* 24:1113.

34. *Public* 18:217; *Chronicle* 25:309, 359, 406; Grodinsky, *Jay Gould*, 258–60.

35. *Chronicle* 30:193, 31:645, 653; *New York Tribune*, Dec. 8, Dec. 14, and Dec. 15, 1880; *Railway World* 24:157; *Public* 17:100, 18:388, 424. Grodinsky, *Jay Gould*, 261, implies wrongly that the International purchase occurred before that of the Iron Mountain.

36. Dodge to Gould, Dec. 26, 1880, DR, 10:230.

Chapter 22: Imminent Domain

1. The monument is in the graveyard of the Yellow Meeting House in Roxbury.

2. The descriptions of Hoxie, Hayes, and Talmage are drawn from Dr. Warren B. Outten, who knew them personally, and are found in W. L. Burton, "History of the Missouri Pacific," 739–40. A copy of this unpublished manuscript is in MP.

3. *PRC*, 182–83, 483–84; *New York Tribune*, May 27, 1879; *Public* 15:333, 18:166.

4. Burton, "Missouri Pacific," 738–39; Dodge to Hoxie, May 12, 1881, DR, 10:499. For examples of investments on Clark's behalf see Gould to Clark, Mar. 28, 1875, Aug. 4, Aug. 13, and Aug. 15, 1880, Mar. 13, 1881, KG.

5. Gould to Clark, Mar. 13 and May 8, 1881, KG.

6. *Chronicle* 31:638–39, 32:285; *Public* 16:340, 18:405–7; Gould to J. M. Osborn, Feb. 26, 1880, KG; *New York Tribune*, May 16 and Nov. 16, 1879.

7. Grodinsky, *Jay Gould* (see Prologue, n. 5), 262–63; MP Stockholders' Minutes, Sept. 22, 1880, MP; *Chronicle* 31:123, 205, 382–83, 606; *Public* 19:180; Gould to Clark, Mar. 13, 1881, KG; Dodge to Gould, [undated], Jan. 11, Jan. 14, and Jan. 20, 1881, DR, 10:239, 249, 259–60, 267; *New York Times*, Mar. 1, 1881; *New York World*, Mar. 13 and Mar. 14, 1881.

8. DR, 10:235–36; *Bradstreet's*, Apr. 16, 1881; *Chronicle* 32:71, 195–96, 205–6, 266, 285–86, 469, 489, 520, 526–27, 545, 552–53, 613, 34:12–13; *Public* 19:136, 232, 259, 264, 275–76, 281, 293; *New York World*, Apr. 13, 1881. For details on the mergers see Grodinsky, *Jay Gould*, 263–65.

9. *Chronicle* 31:123, 178, 32:545, 552–53.

10. *Bradstreet's*, Aug. 6, 1881.

11. Ibid.; *Chronicle* 33:275, 716.

12. *Chronicle* 32:348.

13. Gould to Clark, June 24 and July 5, 1881, KG. A sketch of the proposed San Francisco route is included in the latter.

14. Perkins to Gould, June 15, 1881, BA; *Chronicle* 32:659, 33:24; *Public* 19:376; *Omaha Bee*, June 7, 1881; *Omaha Republican*, June 10, 1881.

15. Gould to Perkins, June 20, 1881, BA.

16. Perkins to Palmer, June 26, 1881, BA; Resolution of the Board of Directors, July 20, 1881, BA; Perkins to Potter, July 23, 1881, BA.

17. Dillon to Sidney Bartlett, July 28, 1881, BA; Dillon to Perkins, July 28, 1881, BA; Forbes to Vanderbilt, July 21, 1881, BA; Vanderbilt to Forbes, July 26, 1881, BA.

18. Perkins to Dillon, July 30 and Aug. 8, 1881, BA; Forbes to Vanderbilt, July 30, 1881, BA.

19. Gould to Perkins, Aug. 4, 1881, BA; Dillon to Perkins, Aug. 4, 1881, BA; Perkins to Dillon, Aug.8, 1881, BA; Perkins to Gould, Aug. 8, 1881, BA; *Chronicle* 33:224–25; Overton, *Burlington Route* (see Chap. 16, n. 6), 173, 176. Overton inadvertently gives the date of the Burlington resolution as Aug. 19, 1880.

20. *Chronicle* 33:46–47, 201; *Public* 20:89.

21. *New York Tribune*, Mar. 11, 1881; *Omaha Bee*, Aug. 18, 1881. For more detail on the Bridge Company and its acquisition by Gould see *Railroad Gazette* 6:297, 10:629, 13:375; *Chronicle* 32:659–60, 33:386–87; *New York Times*, Mar. 11, 1889; Grodinsky, *Jay Gould*, 338–41.

22. *Public* 20:56.

23. Crocker to Huntington, Oct. 20, 1881, MM; *Chronicle* 33:48, 75.

24. Deposition of John F. Dillon, *The Missouri Pacific Railway Company v. The Texas & Pacific RR Company and Others*, Dec. 29, 1893, MP.

25. *New York Tribune*, June 7, 1881; Gould to Clark, June 24 and July 5, 1881, KG; J. T. Granger to Dodge, Aug. 13, 1881, DR, 10:603–4; Athearn, *Union Pacific Country* (see Chap. 13, n. 4), 280–81; *Public* 20:115.

26. Crocker to Huntington, July 29, Aug. 2, Aug. 5, Aug. 6, Sept. 30, and Oct. 7 (two letters), 1881, MM.

27. Ibid., Oct. ? (no. 436), Oct. 20 (two letters), Oct. 27, Oct. 28, Nov.5, Nov. 11, and Nov. 15, 1881, MM; *Public* 20:244, 293. Crocker described the two roads as lying "within a stones throw of each other" over the ninety-mile stretch.

28. Deposition of J. F. Dillon, MP; *New York World*, Nov. 17, 1881. Copies of the complete agreement, dated Nov. 26, 1881, are in MP. A summary is in *Chronicle* 33:623–24.

29. Crocker to Huntington, Oct. 27, Nov. 5, Nov. 7, Nov. 14, and Nov. 18, 1881, MM; *Chronicle* 34:99–101, 116, 243–45, 313. The Frisco purchase is ably treated in H. Craig Miner, *The St. Louis–San Francisco Transcontinental Railroad* (Lawrence, Kans., 1972), 119–35.

30. Lavender, *Great Persuader* (see Chap. 13, n. 4), 336–39; Grodinsky, *Jay Gould*, 350–51; Grodinsky, *Transcontinental* (see Chap. 13, n. 13), 172–73. Grodinsky errs in declaring the agreement "no more than an armistice." Despite evasions and violations, the terms endured.

31. Miner, *Transcontinental Railroad*, 132.

32. *Public* 19:52–53, 118; *New York Tribune*, Feb. 5, 1881; *New York Sun*, Feb. 18, 1881; *Chronicle* 32:230, 299–301, 430–31.

33. *Chronicle* 32:299–301; Grodinsky, *Jay Gould*, 327–28. Bound Brook was the town where the Reading portion of the line connected with the Jersey Central.

34. *Chronicle* 32:205; *Public* 19:121; *New York Tribune*, Feb. 27, 1881. The *Chicago Tribune*, Jan. 12, 1882, asserted that Gould obtained his Jersey Central in a settlement born of a market duel with Keene.

35. *Philadelphia North American*, Dec. 16, 1880; Forbes to Perkins, Mar. 9, 1881, BA; *Public* 19:137, 147; Dillon to Gould, Mar. 1 and Mar. 8, 1881, UP; *Bradstreet's*, Mar. 12, 1881.

36. *New York Tribune*, Apr. 10 and Apr. 26, 1881; *Chronicle* 32:430–31, 454–55, 468, 33:369–70; *Railroad Gazette* 13:256; *Public* 19:259, 264, 281, 339, 376, 405–6, 20:247; Grodinsky, *Jay Gould*, 330–31.

37. J.W. Garrett to John King, Jr., Nov. 19, 1881, GP; J. T. Granger to Dodge, Aug. 13, 1881, DR, 10:603; *Public* 19:405–6; *Chronicle* 32:578, 33:370–71, 385, 560.

38. King to Robert Garrett, Oct. 15, 1881, GP; J. W. Garrett to King, Nov. 19, 1881, GP; *New York Times*, Oct. 14, 1881; *Chronicle* 33:370–71, 396. For background on the Reading muddle see Jules I. Bogen, *The Anthracite Railroads* (New York, 1927), 19–59.

39. *Chronicle* 33:588, 634, 640; *Public* 20:89; Edward C. Kirkland, *Men, Cities, and Transportation* (Cambridge, Mass., 1948), 2:32–53; *New York Sun*, Nov. 24 and Dec. 6, 1881; *New York Times*, Dec. 7, 1881.

40. *New York Times*, Dec. 7, 1881.

41. *New York Tribune*, July 23 and Oct. 1, 1880; *New York World*, Nov. 10 and Nov. 26, 1881; *Chronicle* 31:85, 638; *Public* 18:405–7; David M. Pletcher, *Rails, Mines, and Progress: Seven American Promoters in Mexico, 1867–1911* (Ithaca, 1958), 1–28.

42. *Chronicle* 31:109–10, 317–18, 672; *New York Tribune*, Nov. 12, 1880; Pletcher, *Rails, Mines, and Progress*, 13–14, 159–62; William S. McFeely, *Grant: A Biography* (New York, 1981), 487.

43. Dodge to Fred Ames, Mar. 26 and Oct. 29, 1881, DR, 10:413, 669; *New York Tribune*, Jan. 16, 1881; *Chronicle* 32:100; Pletcher, *Rails, Mines, and Progress*, 81–82, 101, 162–63, 169–70; McFeely, *Grant*, 487; *New York Times*, July 22, 1881; Brief on Behalf of Defendants, *Eugene Kelly v. Jay Gould et al.* (New York, 1892), 2–3, DP; Dodge to Gould, Nov. 17, 1881, DR, 10:679; *New York World*, Dec. 9 and Dec. 17, 1881.

44. J. T. Granger to Dodge, Nov. 9, 1881, DR, 10:675; Dodge to Frances DeGress, Dec. 9, 1881, DR, 10:691; *New York Tribune*, Apr. 2, 1882; "The Oriental Construction Company," DR, 10:759; *Chronicle* 34:605; Dodge to Hayes, Feb. 11, 1882, DR, 10:729; Dodge to Gould, Nov. 17, 1881, Apr. 4 and Aug. 22, 1882, DR, 10:679, 775, 875; DR, 10:951. Pletcher, *Rails, Mines and Progress*, 171, speculates that Gould never intended to build the line but acquired the concession "only to strengthen his hand against his Mexican rivals." The evidence in the Dodge Record alone refutes this supposition.

Chapter 23: Well-Wrought Earns

1. *Chronicle* 26:410, 592, 27:381; *Stockholder* 16:421; *Public* 14:276, 283; Reid, *Telegraph in America* (see Chap. 17, n. 3), 572–79; Eckert to Helen Gould, May 16, 1893, HGS; Harlow, *Old Wires* (see Chap. 17, n. 3), 410.

2. Gould to J. W. Garrett, Nov. 11, 1878, GLC, and Nov. 15, 1878, GP; J. W. Garrett to Gould, Nov. 15, 1878, GP; *Postal Telegraph* (see Chap. 17, n. 4), 129–32; Eckert to J. W. Garrett, Feb. 20, 1879, GP; RGD, NYC, 191:2210; *New York Tribune*, July 17 and Aug. 2, 1879, Jan. 17, 1880; *New York Sun*, July 27 and Sept. 14, 1879; *Chronicle* 29:17, 66, 461; *Railroad Gazette* 11:503; *Public* 16:324; Reid, *Telegraph in America*, 753–57. Eckert's letter is the historian's delight: it is marked "Personal and Confidential" and twice implores Garrett to "Please *destroy this* and consider the information as strictly confidential."

3. *Stockholder* 16:660, 817–18; *Bradstreet's*, Aug. 8 and Dec. 6, 1879; *New York Times*, Dec. 3, 1879; *New York Tribune*, Jan. 17, 1880; *Chronicle* 29:630.

4. *Land-Grant Telegraph* (see Chap. 17, n. 5), 206–7; *Chronicle* 30:66, 90; *Stockholder* 18:257; *Public* 17:68–71, 148, 150; *New York Tribune*, Feb. 28–Mar. 3 and Mar. 10, 1880; Grodinsky, *Jay Gould* (see Prologue, n. 5), 272–74. The court later sustained Judge Dillon's argument. See *New York Tribune*, Oct. 2, 1880; *Chronicle* 31:359, 383; *Railway World* 24:975–76.

5. Potter to Perkins, May 29 and Aug. 16, 1880, BA; *Railroad Gazette* 12:136; *Public* 17:166, 212, 228, 327, 356; *Chronicle* 30:600, 31:21, 121, 327; *New York Tribune*, Mar. 13, June 17, July 1, Sept. 1, and Sept. 16, 1880; *New York Sun*, Mar. 1, Mar. 4, Mar. 13, and May 31, 1880; *New York Times*, Aug. 27, 1880; RGD, NYC, 191:2400, A4. For a good discussion of the rate war question see Grodinsky, *Jay Gould*, 275, 286. Since Western Union owned many more offices at noncompetitive points than American Union, it would suffer less from a rate war.

6. *New York Times*, Dec. 3, 1879, Apr. 30, 1880; *Public* 17:116, 18:199; *New York Tribune*, May 1, 1880; *Chronicle* 30:455.

7. *Stockholder* 18:562; *New York Times*, Sept. 17, 1880; *Bradstreet's*, Sept. 22, 1880; *Public* 18:100, 198–99, 203, 219; *Chronicle* 31:306, 322, 383; *New York Tribune*, Sept. 23, 1880.

8. *New York Times*, Dec. 7, 1880, June 2, 1881; *Public* 18:292, 388; *Chronicle* 31:398, 482, 560, 645, 651, 653; *Public* 18:295; Grodinsky, *Jay Gould*, 278.

9. *Chronicle* 31:651; Grodinsky, *Jay Gould*, 279; Reid, *Telegraph in America*, 592. Gould's new company was called American Telegraph Cable. To build it he organized another construction company.

10. *Chronicle* 31:545, 579, 625, 32:40; *New York Tribune*, Jan. 4, Jan. 6, and Jan. 14, 1881; *Public* 18:395, 19:36–37; Forbes to Perkins, Dec. 21, 1880, BA; *Bradstreet's*, Jan. 15, 1881.

11. *New York Tribune*, May 20 and June 2, 1881; Reid, *Telegraph in America*, 581.

12. Details on the two meetings are drawn from testimony of Gould, Green, and Vanderbilt reprinted in the *New York Tribune*, May 17, May 20, and June 2, 1881; *New York Herald*, May 17 and May 20, 1881; *New York Sun*, May 20, 1881.

13. Ibid.; *Stockholder* 16:660.The agreement is detailed in the *New York World*, Jan. 28, 1881. Grodinsky, *Jay Gould*, 281, notes the irony of the stock dividend. Gould, of course, benefited from it as the largest holder of Western Union. The two Western Union directors at the meeting were H. M. Twombly and Augustus Schell.

14. *New York Tribune*, Jan. 12, Jan. 15, Jan. 29, Feb. 4, and Feb. 5, 1881; *New York Sun*, Jan. 13–15, Jan. 17, and Feb. 6, 1881; *New York Times*, Jan. 14 and Jan. 21, 1881; *New York Herald*, Jan. 14, 1881; Harlow, *Old Wires*, 414.

15. *New York Tribune*, Jan. 12–15, Feb. 5, Feb. 6 and Mar. 10, 1881; *Chronicle* 32:146, 206; *Public* 19:52, 87. Progress of the suits may be followed in the *New York Tribune*, Jan. 26,

Jan. 29, Mar. 17, May 10, May 17, May 19, May 20, June 2, June 21, and June 22, 1881; *Chronicle* 32:156, 207, 266, 527–28, 35:559; *Public* 19:132.

16. *Public* 19:86–87.

17. Ibid., 19:119; *Chronicle* 32:207; *New York Sun*, Jan. 11, 1882; *New York Times*, May 16, 1882, Dec. 28, 1885; *New York World*, May 17, 1882; *New York Herald*, Dec. 19, 1882. The *Herald* article must be used with caution, since it was part of a concerted campaign against Gould's telegraph and cable holdings by parties who were investors in competing companies.

18. *Land-Grant Telegraph*, 4–5, 138; Gould-Huntington Agreement, Nov. 26, 1881, Article 13, MP. A copy of the Union Pacific contract is in *Land-Grant Telegraph*, 43–47.

19. Eckert to J. W. Garrett, Aug. 17, 1881, GP; *New York Times*, Sept. 1 and Oct. 14, 1881; *Chronicle* 33:410; *Public* 20:244.

20. Little has been written on the early history of elevateds in New York. The most detailed account is James Blaine Walker, *Fifty Years of Mass Transit* (New York, 1918), 1–110. See also William F. Reeves, *The First Elevated Railroads in Manhattan and the Bronx of the City of New York* (New York, 1936), 1–24. For a brief scholarly treatment see Charles W. Cheape, *Moving the Masses* (Cambridge, Mass., 1980), 21–35. Carter, *Cyrus Field* (see Chap. 20, n. 9) is thin and unsatisfactory on the elevated railroads and Field's role in them.

21. Walker, *Mass Transit*, 108–110; Cheape, *Moving the Masses*, 33–35.

22. Walker, *Mass Transit*, 110–14; Carter, *Cyrus Field*, 306–8. Commodore Garrison, Porter, Pullman, Navarro, Scott, and Dows were among the incorporators of Manhattan. Brief sketches of Navarro are in the *New York Times*, Feb. 4, 1909, and *NCAB*, 15:246–47.

23. *Stockholder* 13:792; *New York Tribune*, May 23 and Dec. 10, 1879; *Chronicle* 27:148, 280, 28:42, 69, 302; Walker, *Mass Transit*, 112–13; Cheape, *Moving the Masses*, 35–36.

24. *New York Tribune*, May 23, 1879; *New York Sun*, Mar. 3, 1880; *Chronicle* 28:526, 553–54, 579, 29:459; *Public* 15:368.

25. Walker, *Mass Transit*, 99, 113; *New York Tribune*, June 16, 1879; *Stockholder* 16:529–30; *Public* 15:381. Field himself admitted, "The elevated railroads are an experiment yet." *New York Tribune*, Apr. 3, 1880.

26. *Chronicle* 28:538–39, 30:285–86, 315, 346, 357, 376; *American Railroad Journal* 52:1422; *Stockholder* 18:417; *New York Tribune*, Mar. 9, Mar. 11, Mar. 14, Mar. 18, Mar. 21, Mar. 25, Apr. 3, and Apr. 6, 1880; *New York Sun*, Mar. 17 and Mar. 22, 1880; Carter, *Cyrus Field*, 316–17; *New York Times*, Apr. 2 and Apr. 9, 1880.

27. *Chronicle* 30:357, 384, 544, 669, 31:12, 45, 68, 89; *New York Tribune*, June 20, 1880; *Public* 18:4; *Stockholder* 18:673.

28. *Chronicle* 31:95, 123, 304–6, 358, 382, 406, 509–10, 32:414, 421, 468, 489–90; *Public* 18:72, 75, 200, 217, 248, 328, 19:259; *New York Tribune*, Sept. 16, Sept. 17, Nov. 11, and Nov. 12, 1880, Mar. 23, Apr. 14, Apr. 16, Apr. 17, and Apr. 26,1881; *New York Sun*, Nov. 11 and Nov. 12, 1880; *New York Times*, Apr. 28, 1881; *Bradstreet's*, Apr. 16, 1881.

29. The exposé is found in the *New York Times*, Dec. 27, 1881. See also ibid., Dec. 28–30, 1881.

30. *New York Tribune*, May 7, 1882; *New York World*, May 13, 1883; *New York Times*, Dec. 27, 1881.

31. The best way to get the flavor of the *World* campaign is to read the paper for the period May–October 1881. The *Times* cited above reprinted a generous collection of sample passages, but these are taken out of context and, piled atop one another, distort the tone and frequency of the campaign. "Wall Street Gossip" began at least as early as Feb. 12, 1881, and apparently ended on Mar. 11, 1883. On May 7 the *World* observed in an editorial that the

column "records from day to day not our own views . . . but simply the current talk and opinion of leading and active men in Wall street."

32. *New York Tribune*, May 19 and May 20, 1881; *New York Times*, May 19 and Dec. 27, 1881; *New York World*, May 19 and May 22, 1881; *New York Herald*, May 19 and May 22, 1881; *New York Sun*, May 19, 1881; *Chronicle* 32:522; *Bradstreet's*, May 21, 1881.

33. *Chronicle* 32:601, 647, 659; *Bradstreet's*, June 18 and June 25, 1881; *New York World*, June 18, 1881.

34. *Chronicle* 32:24, 33:41; *New York Tribune*, July 2 and July 7, 1881; *New York World*, July 2, July 3, and July 9, 1881; *New York Times*, July 2 and July 7, 1881; *New York Sun*, July 8, 1881. As early as June 17, "Wall Street Gossip" reported rumors that Gould would enter Metropolitan. *New York World*, June 17 and June 19, 1881.

35. *New York Tribune*, May 6, 1881, Dec. 4, 1883; *Public* 20:8, 20, 23; *Chronicle* 32:40. Grodinsky, *Jay Gould*, 295–96, argues that Gould's reluctance was a sham even though an independent judge later concluded, "upon the basis of all available evidence," that it was not.

36. *Public* 20:8; *New York Tribune*, July 15 and July 16, 1881; *Chronicle* 33:67, 74; *New York Times*, July 15, July 16, and Dec. 27, 1881, Jan. 21, 1882; *New York World*, July 16, 1881.

37. For Westbrook see *New York Times*, Oct. 7, 1885; *BDC*, 1902.

38. *Chronicle* 33:74; *New York Times*, July 17 and July 26, 1881.

39. *New York Times*, July 26, Sept. 3, Sept. 15, and Dec. 27, 1881; *New York Tribune*, Aug. 13, 1881; *New York World*, Aug. 20, Sept. 10, and Sept. 16, 1881; *New York Sun*, Sept. 7 and Sept. 29, 1881; *Public* 20:121; *Chronicle* 32:255, 33:281, 304.

40. *Chronicle* 33:118, 322–23, 34:13; W. J. Ladd to Forbes, Aug. 26, 1881, BA; *Bradstreet's*, Aug. 20, 1881; *Public* 20:148. For Garfield's death see the *New York Times*, Sept. 20, 1881.

41. The correspondence between Westbrook and Swayne is reprinted along with the latter's testimony in the *New York Tribune*, May 7, 1882. Grodinsky's account of the Manhattan episode is one of the weakest chapters of his book. For example, he argues that Gould "went ahead carefully with his plan, selecting first the proper judge," and that he "retained his ability in the art of judicial interference." Grodinsky, *Jay Gould*, 293, 300. It was Ward, not Gould, who brought Westbrook into the elevated fight. No evidence was ever produced that Ward, or Westbrook for that matter, acted on Gould's behest. The testimony indicates that Westbrook initiated the correspondence with Swayne, thereby creating a situation of which Gould was quick to take advantage. Grodinsky also has a habit of reversing the chronology of events. For example, he discusses the Westbrook-Swayne correspondence *after* describing Gould's entering Metropolitan's board and the appointment of Manhattan's receivers.

42. *New York Times*, Sept. 29 and Sept. 30, 1881; *New York World*, Sept. 29 and Sept. 30, 1881; *New York Sun*, Sept. 30, 1881. The *Times* and other papers reported that court was held in Gould's own office. This point raised such a howl that Gould, in later testimony, took pains to set the record straight. See *New York Tribune*, May 7, 1882.

43. *New York Times*, Sept. 30, Oct. 1, and Oct. 9, 1881; *New York Sun*, Oct. 1, 1881; *New York World*, Oct. 1 and Oct. 9, 1881; *Public* 20:217; *Chronicle* 33:347, 385, 396–98; *New York Tribune*, May 7, 1882. Only 20,000 of the shares were in Gould's name; the remainder were in Connor's. Gould testified that he bought the entire 20,000 shares after Sept. 30. See *New York Tribune*, May 9 and May 21, 1882. The price paid for these shares was 30.

44. *New York Times*, Oct. 14, 1881; *New York Sun*, Oct. 14, 1881; *New York Tribune*, Oct. 14, 1881, May 9, 1882; *Chronicle* 33:396–98, 406, 410.

45. *New York Sun*, Oct. 15, 1881; *New York Times*, Oct. 19 and Oct. 23, 1881; *New York Tribune*, Oct. 22, 1881, May 9, 1882; *Chronicle* 33:422.

46. *New York Times*, Oct. 19, Oct. 22, Oct. 23, and Oct. 28, 1881; *Public* 20:248, 265, 283; *New York World*, Oct. 22 and Oct. 29, 1881; *New York Sun*, Oct. 25, 1881; *Chronicle* 33:442, 468–69; *New York Tribune*, May 7 and May 9, 1882.

47. *New York Tribune*, May 7, 1882.

48. For the later inquiry into Westbrook's conduct see ibid., May 7–June 1, 1883. The inquiry included other actions besides Manhattan. In its obituary of Westbrook the *New York Times*, Oct. 7, 1885, stated that the proceedings "practically destroyed his usefulness on the bench, and from that time public confidence in him was at an end."

49. *New York Times*, Oct. 22, 1881; *Chronicle* 33:527; *Public* 20:313. The new Manhattan board included Gould, Sage, Field, Dillon, Connor, Sloan, Galloway, E. M. Field, William Garrison, and George Gould.

Chapter 24: War

1. U.S. Bureau of the Census, *Historical Statistics of the United States, Colonial Times to 1970* (Washington, D.C., 1975), 731.

2. For discussion of this pattern in different regions see Grodinsky, *Transcontinental* (Chap. 13, n. 13); Klein, "Strategy of Southern Railroads" (Chap. 16, n. 1). For a brief overview see Chandler, *Visible Hand* (Prologue, n. 6), 159–87.

3. *New York Tribune*, May 7, 1882.

4. Potter to Perkins, Sept. 2, 1882, BA; *Chronicle* 32:71, 33:612, 709, 36:327–28; Anon., "Chapter of Wabash" (see Chap. 20, n. 4), 183.

5. *Chronicle* 34:20, 21; *Public* 21:10; *New York Sun*, Dec. 29 and Dec. 31, 1881; Grodinsky, *Jay Gould* (see Prologue, n. 5), 369–70.

6. *Chronicle* 32:510; Grodinsky, *Jay Gould*, 360–62; Bogen, *Anthracite Railroads* (see Chap. 22, n. 38), 59–60. The Wabash report showed an increase of about 16 percent in gross earnings but a decrease in net earnings of more than 20 percent, leaving a deficit of $1.4 million after fixed charges.

7. *Chronicle* 34:189–90, 197, 203; *Railroad Gazette* 14:123; *Public* 21:118; *New York Sun*, Feb. 18, 1882.

8. *New York Times*, Feb. 21, 1882; *Chronicle* 34:230–31, 263, 337, 434–35, 459–60, 510, 708; *Public* 21:148, 244, 324; *New York Tribune*, Mar. 24, Mar. 31, and June 24, 1882; *New York World*, Mar. 5, 1882. Explanations of the practices known as "short billing" and "half number billing" are in the *Railroad Gazette* 14:292, 718–19. Statistics on traffic flow are in ibid., 14:209–10. The same figures revealed that the trunk lines received 70 percent of their eastbound and 61 percent of their westbound traffic from the five states of Illinois, Ohio, Indiana, Michigan, and Wisconsin.

9. *Railroad Gazette* 14:304, 308–9. For background on the Nickel Plate, known formally as the New York, Chicago & St. Louis, see Taylor Hampton, *The Nickel Plate Road* (Cleveland, 1947), 1–173. Vanderbilt's comments are drawn from several newspaper interviews quoted in Hampton, 165, 171–73, and Grodinsky, *Jay Gould*, 365.

10. *New York Times*, Oct. 29, 1882; *New York World*, Oct. 29, 1882; *Philadelphia Press*, Oct. 30, 1882; Hampton, *Nickel Plate Road*, 177–91; *Chronicle* 35:45, 479, 506; *Railroad Gazette* 14:744; *Public* 22:296.

11. *Chronicle* 36:296–97, 305, 327–28, 338–39, 400, 411, 439–40, 445, 472; *Railroad Gazette* 15:274; *Public* 23:266; *New York Tribune*, Mar. 15, 1883; *New York Herald*, Apr. 20, 1883; *Bradstreet's*, Apr. 21 and Apr. 28, 1883; Grodinsky, *Jay Gould*, 370–72. A copy of the

lease is in MP. It was made to the Iron Mountain because the law prohibited a lease to the Missouri Pacific.

12. For background on these projects see Overton, *Burlington Route* (Chap. 16, n. 6), 180–98; Athearn, *Rebel of the Rockies* (Chap. 19, n. 24), 113–31; Grodinsky, *Transcontinental*, 135–200; Miner, *Transcontinental Railroad* (Chap. 22, n. 29), 119–35; Bryant, *Santa Fe* (Chap. 21, n. 5), 64–105; Athearn, *Union Pacific Country* (Chap. 13, n. 4), 311–18. For the Burlington see also *Chronicle* 34:419.

13. Gould to Clark, Mar. 3, 1882, KG.

14. Potter to Perkins, Feb. 4, Apr. 14, Apr. 17, July 15, July 16, and July 20, 1882, BA; *Public* 20:345; *Railroad Gazette* 14:242, 447; Perkins to Potter, Oct. 29, 1881, BA. Perkins's view of Gould's role is suggested by his stressing the importance of "getting the ownership of Union Pacific *fixed*." Perkins to Forbes, Aug. 25, 1881, BA.

15. Potter to Perkins, July 31, Aug. 29, Sept. 8, and Sept. 18, 1882, BA; Perkins to Potter, Aug. 2, 1882, BA; Potter to S. F. Pierson, Sept. 18, 1882, BA; Potter to Elijah Smith, Oct. 16, 1882, BA; *Public* 22:57, 233, 23:83–84; *Railroad Gazette* 14:652; *Chronicle* 35:578; *New York World*, Nov. 11, 1882; *New York Herald*, Dec. 7 and Dec. 26, 1882.

16. Perkins to Forbes, Nov. 14, 1880, May 5 and Aug. 19, 1882, BA; Forbes to Perkins, Mar. 21, 1881, BA; Forbes to E. Rollins Morse, Apr. 23, 1881, BA; *Public* 19:406, 22:148, 155; *Bradstreet's*, Sept. 17, 1881; J. S. Cameron to Potter, Aug. 25, 1882, BA; Perkins to Geddes, Sept. 5, 1882, BA; *Chronicle* 32:310–11, 33:298, 559, 35:255, 292, 297; *New York World*, Mar. 28, 1882; *New York Sun*, Sept. 3 and Sept. 5, 1882.

17. Potter to Perkins, Sept. 7 and Sept. 25, 1882, BA; Perkins to Forbes, Oct. 2 and Nov. 9, 1882, BA; Perkins to Geddes, Sept. 16, 1882, BA; "Memorandum," Sept. 13, 1882, BA; untitled document, Sept. 16, 1882, BA; *Public* 22:170; *Chronicle* 35:276, 308–9, 360–61, 575.

18. Gould to Perkins, Jan. 16, 1883, with accompanying notes by Perkins, dated Jan. 20, 1883, BA; Perkins to J. L. Gardner, Jan. 20, 1883, BA; Gould to Perkins, Jan. 26, 1883, BA; Perkins to Gould, Jan. 30, 1883, BA; Perkins to Forbes, Jan. 30, 1883, and Memorandum of same date, BA; Forbes to Perkins, Mar. 3, 1883, BA; *Chronicle* 36:195, 221, 399. Carson was a Hannibal official.

19. Perkins to Forbes, Apr. 5, 1883, BA; Forbes to Perkins, Sept. 22, 1882, BA. The entire negotiation can be followed in detail in the packet of letters and telegrams dated Apr. 11–Apr. 27, 1883, BA. These offer the best and fullest documentation of Gould's negotiating techniques in a specific transaction. Memorandums of agreement for the Hannibal sale and telegraph contract, both dated Apr. 21, 1883, are in BA.

20. Trottman, *Union Pacific* (see Chap. 13, n. 1), 202; Potter to Perkins, Feb. 15, 1883, BA; *Boston Advertiser*, Dec. 19, 1882; *New York Times*, May 30, 1883, Aug. 20, 1884; *Report of Government Directors for 1883*, House Executive Documents, 48 Cong., 1st sess. (Serial 2191), no. 1, pt. 5, 639–64; *Chronicle* 36:282–84, 454.

21. Potter to Perkins, July 6, July 23, Aug. 6, and Nov. 16, 1883, BA; *New York Tribune*, Oct. 26, Nov. 16, Nov. 19, and Nov. 29, 1883; Grodinsky, *Transcontinental*, 226–38, and *Iowa Pool* (see Chap. 16, n. 3), 136–49.

22. *Chronicle* 37:188–89, 342, 667, 687; *Bradstreet's*, Aug. 25, Sept. 1, Dec. 8, Dec. 15, Dec. 22, and Dec. 29, 1883; *Stockholder* 22:68, 131, 133, 150, 151, 163, 178, 181; *New York Tribune*, Nov. 19, Dec. 5, Dec. 6, Dec. 14, Dec. 15, Dec. 18–22, Dec. 30, and Dec. 31, 1883. For details see Grodinsky, *Iowa Pool*, 150–67.

23. *New York Tribune*, Dec. 22, 1883; *Bradstreet's*, Dec. 15 and Dec. 22, 1883; *Stockholder* 22:162, 198; Gould to Clark, Dec. 25 and Dec. 30, 1883, KG; Adams to Lovejoy, Dec. 4, 1883, BA; Grodinsky, *Transcontinental*, 239.

24. Adams to Perkins, Dec. 17, 1883, BA; Perkins to Adams, Dec. 18, 1883, BA;

Perkins to Marvin Hughitt, Dec. 18, 1883, BA; Perkins to Forbes, Dec. 31, 1883, BA; Potter to Perkins, Dec. 28, 1883, BA.

25. *Chronicle* 34:264; *New York Herald*, Dec. 6, 1882; Crocker to Huntington, Sept. 1, 1883, MM. Gould's fight with smaller rivals is recounted in Grodinsky, *Jay Gould*, 396–98. For the Galveston, Houston & Henderson episode see *New York Times*, June 30, 1882; *Chronicle* 34:378, 36:674–75; International Exec. Comm. Minutes, Jan. 1, 1883, 98–105, MP.

26. Crocker to Huntington, Apr. 29, 1882, June 30 and July 17, 1883, MM; *Chronicle* 34:263; Grodinsky, *Transcontinental*, 172–73; Miner, *Transcontinental Railroad*, 137–39; James P. Baughman, *Charles Morgan and the Development of Southern Transportation* (Nashville, 1968), 232–34.

27. Grodinsky, *Transcontinental*, 258–61; *Chronicle* 35:162, 319, 36:70; *Stockholder* 22:19.

28. *Chronicle* 34:20, 35:103, 737, 36:312–13, 422–23; *Stockholder* 22:3.

29. *Public* 23:250, 280; *Chronicle* 37:373; Klein, *Great Richmond Terminal* (Chap. 9, n. 18), 115–31.

30. González to Gould, Nov. 7, 1882, DR, 10:919–21; Gould to González, Jan. 8, 1883, DR, 10:971; Dodge to R. S. Hayes, Mar. 1, 1883, DR, 10:1003; Dodge to Gould, June 5, 1883, DR, 10:1037; *Chronicle* 35:320, 36:623, 47:454–55; *Public* 24:57; DR, 10:759–60, 951, 959, 969; *New York Sun*, Jan. 14 and Apr. 5, 1883; *New York World*, Apr. 5, 1883; Pletcher, *Rails, Mines, and Progress* (see Chap. 22, n. 41), 68–69, 151–71, 178–81; Don M. Coerver, *The Porfirian Interregnum: The Presidency of Manuel González of Mexico, 1880–1884* (Fort Worth, 1979), 200–210.

Chapter 25: Siege

1. The best political cartoons of Gould are found in the American edition of *Puck* for the 1880s. I am indebted to Kingdon Gould, Jr., for lending me two albums of enlarged reproductions of these cartoons.

2. *Public* 20:313, 409; *New York World*, Oct. 29, Oct. 30, Nov. 2–6, Nov. 10–12, Nov. 16, Nov. 24, Nov. 28, Nov. 29, Dec. 16, Dec. 18, Dec. 20–25, and Dec. 28, 1881; *New York Tribune*, Nov. 23 and Dec. 8, 1881, Dec, 12, 1883; *New York Times*, Nov. 23–25, Dec. 22, and Dec. 24, 1881; *Chronicle* 33:502, 527–28, 560, 642, 700, 717, 744, 34:86, 115. The term "stamped" (or "assented") refers to stock certificates stamped with the lower rate of return provided in the agreement. Holders of stamped certificates thereby assented to cancellation of the 10 percent rate stipulated in the original lease.

3. *New York Tribune*, Apr. 13, May 6, May 10, May 14, May 23, June 1, and June 13, 1882; *New York Times*, Dec. 27–30, 1881; *New York World*, Dec. 28–31, 1881. The Westbrook inquiry is covered in the *New York Tribune*, May 7–June 1, 1882; *New York Sun*, Apr. 13, Apr. 19, Apr. 20, Apr. 22, Apr. 29–May 14, May 24, May 31, and June 1, 1882.

4. *Chronicle* 35:50–51, 132, 347–48; *New York Times*, July 23, Nov. 5, and Nov. 9, 1882; *Public* 22:233, 296; *New York Herald*, Oct. 7, 1882; *New York World*, Oct. 7, 1882; *New York Sun*, July 12, Sept. 20, Oct. 20, and Nov. 9, 1882.

5. *New York Times*, Apr. 12, Apr. 20, Nov. 28, and Dec. 11–13, 1883; *Chronicle* 35:443–44, 456, 545, 575, 658, 36:169, 453–54, 37:48, 534, 719; *Public* 23:118, 132, 201; *New York Tribune*, Nov. 22–24, Nov. 28, Dec. 4, Dec. 8, and Dec. 12, 1883; *New York Sun*, Nov. 23 and Nov. 28, 1883.

6. *New York Times*, Dec. 10–14 and Dec. 21–22, 1883; *New York Sun*, Dec. 8, 1883.

7. *Chronicle* 32:288, 33:154, 201, 34:230; *Public* 20:71; *New York Times*, Apr. 26, 1881; *New York Sun*, July 29, 1881; Reid, *Telegraph in America* (see Chap. 17, n. 3), 601–4, 772–84.

8. *Public* 20:151, 21:119; *New York Sun*, Mar. 30, 1882; *New York Times*, Nov. 7, 1882;

New York Tribune, Nov. 30, 1882; Grodinsky, *Jay Gould* (see Prologue, n. 5), 449–51; *Chronicle* 34:336, 357, 637, 35:10, 331–32; *Bradstreet's*, May 27, 1882.

9. *Public* 22:87, 327, 403; *Chronicle* 35:172, 189, 393, 516–17, 545, 576, 637–38, 658, 720, 737, 764; *New York World*, Nov. 2, Nov. 6, Nov. 10, Nov. 17, Nov. 22, Nov. 30, Dec. 10, Dec. 17, Dec. 19, and Dec. 30, 1882, Jan. 3 and Jan. 10, 1883; *New York Sun*, Nov. 17, Nov. 30, and Dec. 25, 1882; *New York Times*, Nov. 7, 1882; *New York Tribune*, Nov. 30, 1882, Jan. 3, 1883; *New York Herald*, Nov. 2, Nov. 23, Dec. 1, Dec. 11, and Dec. 19, 1882; *Bradstreet's*, Jan. 6, 1883.

10. *Labor and Capital* (see Chap. 1, n. 7), 917–18; *New York Herald*, Dec. 14, 1882; *Chronicle* 36:30, 163, 181, 188, 197, 358; *Public*, 23:52, 100, 103; Reid, *Telegraph in America*, 604–5; *New York Times*, Feb. 12, 1883.

11. Perkins to Forbes, Mar. 11, 1882, BA; Forbes to Perkins, Dec. 9 and Dec. 13, 1882, BA.

12. Forbes to J. W. Garrett, Dec. 15, 1882, BA; Forbes to George Jones, Dec. 16 and Dec. 20, 1882, BA; Forbes to J. J. Higginson, Dec. 15, 1882, BA; Forbes to Richard Olney, Dec. 15, 1882, BA; Olney to Forbes, Dec. 20, 1882, BA; W. J. Ladd to Forbes, Dec. 28, 1882, BA.

13. *Chronicle* 34:336, 662; *Public*, 21:375; Forbes to Perkins, Dec. 13, 1882, BA; Forbes to J. W. Garrett, Dec. 15, 1882, BA.

14. *DAB*, 2:199–202; Mott, *American Journalism* (see Chap. 12, n. 32), 415–21. Bennett lacks a good biography. The relevant chapters in Don C. Seitz, *The James Gordon Bennetts* (Indianapolis, 1928), are anecdotal.

15. *New York Herald*, Nov. 22, 1868, Aug. 1, 1869. For the *Herald*'s use of cable dispatches see Whitelaw Reid, "A Decade of American Journalism," *Westminster Review* 128 (October 1887): 855–56.

16. J. W. Garrett to John King, Jr., July 4, 1882, GP; R. Garrett to J. W. Garrett, July 21, Aug. 24, and Sept. 29, 1882, GP; *Chronicle* 35:445, 496.

17. *New York Herald*, May 1, 1881, Nov. 7, Dec. 2, Dec. 11, and Dec. 19, 1882; *New York Sun*, Oct. 17, 1882.

18. W. H. Osborn to William Ackerman, May 16, 1883, *RL*; *Postal Telegraph* (see Chap. 17, n. 4), 163–64; *Public* 24:7, 116; *Chronicle* 37:151, 175; Forbes to Perkins, Sept. 4, 1883, BA; Perkins to Forbes, Sept. 7, 1883, BA; *New York World*, Aug. 15 and Aug. 16, 1883. The United Press of 1882 is not to be confused with the modern organization; see Mott, *American Journalism*, 492, and *Postal Telegraph*, 165–84.

19. *Public* 23:387, 24:85, 102, 149–50; *New York Sun*, Aug. 1 and Aug. 18, 1883; *New York World*, Aug. 16, Aug. 18, and Aug. 19, 1883; *New York Times*, Jan. 14, 1887; Reid, *Telegraph in America*, 695–712; *Bradstreet's*, July 21, 1883; *Chronicle* 37:85; *Labor and Capital*, 1084–85.

20. *Labor and Capital*, 1071–77, 1085–94; *Public* 24:85–86, 149; *New York Herald*, July 26 and Sept. 6, 1883.

21. *Public* 24:196; *New York Tribune*, Oct. 3, 1883; *Chronicle* 37:397. Grodinsky, *Jay Gould*, 448, observes that Gould's acquisition of Western Union, "far from being accepted as a masterpiece of corporate strategy, was almost universally regarded as a serious blunder." One exception was *Public* 22:38.

22. *New York Times*, June 10, 1883; *Public* 23:371, 24:116, 264; Pender to J. W. Garrett, Oct. 3, 1883, GP; *New York Herald*, Apr. 10, 1883.

23. *Chronicle* 34:337; RGD, NYC, 422:695; *New York Times*, Feb. 15, 1883; *New York Sun*, Oct. 3, 1882; *Public* 23:19.

24. *Public* 23:19.

25. Grodinsky, *Jay Gould*, 479–88, argues that Gould had resumed a heavy specula-

tive line, misread the market, and suffered heavy losses from the miscalculation. Although this interpretation is a possibility, I don't think the evidence sustains it.

26. Stedman, *Stock Exchange* (see Chap. 7, n. 5), 301; *New York Times*, Jan. 14, 1881. The tendency of financial writers to personify the struggles on Wall Street, while not surprising, is worth a study in itself. Like other areas of journalism, financial writing evolved a distinct and colorful style during these years.

27. *New York Herald*, Feb. 28, 1881. See also *New York World*, Mar. 1, 1881.

28. *Public* 20:388, 21:148; *Chronicle* 33:702, 709; Perkins to Forbes, Feb. 4, 1882, BA; *Philadelphia North American*, Dec. 21 and Dec. 23, 1881.

29. *Bradstreet's*, Feb. 4, 1882; *Public* 21:115, 131; *Chronicle* 34:223; Grodinsky, *Jay Gould*, 484.

30. *New York Times*, Mar. 14 and Mar. 15, 1882; *New York World*, Mar. 14, Mar. 21, and Mar. 22, 1884; *New York Sun*, Mar. 14 and Mar. 15, 1884; *Public* 21:164; *Chronicle* 34:301, 308; *Bradstreet's*, Mar. 18, 1882.

31. *Public* 21:228; *Bradstreet's*, Mar. 18, 1882.

32. Perkins to Forbes, Mar. 18, 1882, BA; *New York Times*, Mar. 14 and Mar. 19, 1882; *Chronicle* 34:301–2; *Public* 21:164, 211, 228. Within days the exhibit was widely referred to as Gould's "spring opening" or "spring showing."

33. *Chronicle* 34:387–88, 418, 428, 35:340, 389, 416, 422, 449, 501, 591–92, 596; *Public* 21:213, 244, 276, 324, 22:75, 132, 179, 267, 292, 315, 339, 372–73; *New York Herald*, Nov. 21 and Dec. 16, 1882; *New York Sun*, Apr. 3, 1883. Stock prices for 1882 are summarized in *Chronicle* 36:11–13.

34. *New York World*, Mar. 4 and Apr. 8, 1883; *New York Sun*, Apr. 8, 1883; *New York Times*, Apr. 8, 1883; Snow, *Helen Gould* (see Chap. 1, n. 2), 149–50, 152.

35. Snow, *Helen Gould*, 149; *New York Times*, Apr. 8, Apr. 15, Apr. 16, and Apr. 25, 1883; *Public* 23:243.

36. *New York Herald*, May 5 and May 6, 1883; *New York Times*, May 6, May 13, and June 10, 1883; Stedman, *Stock Exchange*, 305; *Public* 23:86, 219, 291, 299, 24:3; *Bradstreet's*, Apr. 28 and July 14, 1883; *Chronicle* 37:12, 435.

37. *Public* 23:83, 86–87; *New York Herald*, Apr. 7, May 13, May 18, and May 21, 1883; *New York Times*, June 10, 1883. The *Herald* of May 21 insisted Gould was selling "with a liberality . . . only limited by the ability of the market to absorb them."

38. *New York Times*, May 12–14, 1883; *New York Herald*, May 12, 1883; *Public* 23:291, 339, 371, 388, 404; *Chronicle* 37:42, 69; *Bradstreet's*, Apr. 21, Apr. 28, July 14, July 21, and July 28, 1883.

39. This interpretation disagrees with Grodinsky, *Jay Gould*, 486–87. I find his view inconsistent with what is known about Gould's behavior during 1882–83.

40. *Chronicle* 37:168; *Public* 24:117; *Bradstreet's*, Aug. 4, Aug. 11, Aug. 18, and Aug. 25, 1883.

41. Stedman, *Stock Exchange*, 307; *Stockholder* 24:500, 511; RGD, NYC, 416:100 A22, A83, A143.

42. Stedman, *Stock Exchange*, 306–8; RGD, NYC, 416:100 A158; *Public* 24:163, 179, 187, 203; *Chronicle* 37:314, *Bradstreet's*, Sept. 1, Sept. 15, Sept. 22, and Oct. 6, 1883.

43. *New York Times*, Sept. 5 and Sept. 6, 1883; *New York Sun*, Sept. 6, 1883; *Chronicle* 37:440; *Stockholder* 22:23, *Public* 24:163, 230, 243; *Bradstreet's*, Sept. 29, Oct. 20, and Oct. 27, 1883; *New York Tribune*, Oct. 15, 1883.

44. *Bradstreet's*, Nov. 24, Dec. 8, Dec. 15, and Dec. 29, 1883; *Public* 24:294; *Stockholder* 22:147, 178; *New York Times*, Dec. 16 and Dec. 31, 1883.

45. *Stockholder* 22:50, 148, 178; *Bradstreet's*, Sept. 29, Oct. 13, and Dec. 29, 1883;

Forbes to Frederick Billings, Sept. 19, 1882, BA; Forbes to Perkins, Oct. 22, 1883, BA; *Public* 24:147, 339; *Chronicle* 37:660; *New York Tribune*, Dec. 30, 1883.

46. *Bradstreet's*, Dec. 8, 1883.

Chapter 26: Immolation

1. *New York Herald*, Feb. 28, 1881.

2. Gould to Miller, Jan. 9, 188[0], HGS.

3. Gould to Clark, Jan. 27, 1884, KG; *New York Tribune*, Jan. 18, Jan. 29, and Feb. 5, 1884; *New York Herald*, Jan. 3 and Jan. 5, 1884; *Stockholder* 22:198; *Chronicle* 38:1–2, 7, 60; *Bradstreet's*, Dec. 8, 1883, Jan. 5 and Jan. 12, 1884.

4. Gould to Clark, Jan. 4, Jan. 9, and Jan. 27, 1884, KG; *Stockholder* 22:195, 211, 246, 247; *Bradstreet's*, Jan. 12 and Jan. 19, 1884; *New York Herald*, Jan. 8, Jan. 10, Jan. 18, Jan. 21, and Jan. 22, 1884; *New York Times*, Jan. 20, 1884. The Texas & Pacific pool is described in Dodge to W. T. Walters, Jan. 11, 1884, DR, 11:19.

5. *Chronicle* 38:148, 188; *Stockholder* 22:260, 262; *New York Tribune*, Jan. 29, 1884; *New York Herald*, Jan. 27 and Jan. 29, 1884; *Bradstreet's*, Feb. 2, 1884.

6. *Chronicle* 38:97, 108; *Bradstreet's*, Jan. 26 and Feb. 9, 1884; *Stockholder* 22:259, 276, 294; *New York Herald*, Jan. 23, 1884; *New York Tribune*, Jan. 23, 1884; Grodinsky, *Jay Gould* (see Prologue, n. 5), 489–90.

7. *Chronicle* 38:146–47, 251, 261; *Bradstreet's*, Mar. 1 and Mar. 8, 1884; *New York Times*, Feb. 2, Feb. 5, and Feb. 7, 1884; *Stockholder* 22:278, 326, 337–39, 341, 343, 371, 407; *New York Herald*, Feb. 2, Feb. 6, Feb. 10, Feb. 13, Feb. 21, and Feb. 28, 1884; *New York Tribune*, Mar. 28, Apr. 29, May 1, and May 23, 1884; *New York Sun*, Mar. 28, 1884.

8. Vining to V. M. Caime, Mar. 8, 1884, BA; Perkins to Forbes, Mar. 17, 1884 (two items), BA; *New York Herald*, Mar. 12, 1884; *New York Times*, Mar. 14, 1884; *Bradstreet's*, Apr. 12, 1884; *Chronicle* 38:468, 479.

9. *Stockholder* 22:263, 324, 339–40, 354, 356, 359, 386, 403, 438, 455; *Chronicle* 38:197, 295, 510; Trottman, *Union Pacific* (see Chap. 13, n. 1), 203–9; *New York Tribune*, Apr. 27, 1884; Gould to Clark, Jan. 4, 1884, KG.

10. *Bradstreet's*, Mar. 29, Apr. 5, Apr. 19, and Apr. 26, 1884; *Stockholder* 22:453; *Chronicle* 38:398, 424, 472, 479; *New York Herald*, Apr. 21 and Apr. 25, 1884; *New York Times*, Mar. 27, 1884; *New York Tribune*, Apr. 2 and Apr. 22, 1884.

11. *Stockholder* 22:467–69; *Chronicle* 38:510; *New York Herald*, Apr. 30, 1884; *Bradstreet's*, May 3, 1884; *New York Times*, May 4, 1884.

12. *Chronicle* 38:530, 549–53; *Stockholder* 22:482, 530, 599; *Bradstreet's*, May 3 and May 10, 1884; *New York Herald*, May 7 and Dec. 28, 1884; *New York Tribune*, May 1, May 2, and May 7, 1884; *New York Times*, May 7, 1884; *Boston Herald*, May 1 and May 4, 1884; Stedman, *Stock Exchange*, 309–15; McFeely, *Grant* (see Chap. 22, n. 42), 491–92.

13. *Bradstreet's*, May 17, 1884; *New York Herald*, May 15 and May 16, 1884; *New York Tribune*, May 14–17, 1884; Stedman, *Stock Exchange* (see Chap. 7, n. 5), 316–19; *Chronicle* 38:589, 663.

14. *New York Tribune*, May 15, 1884; *New York Sun*, May 15, 1884; *Bradstreet's*, May 17, 1884; *Chronicle* 40:11.

15. *Bradstreet's*, May 17 and May 24, 1884; *New York Tribune*, May 15–18, 1884; *New York Times*, May 15–18, 1884; *New York Herald*, May 17 and May 20, 1884; Stedman, *Stock Exchange*, 319–20. The Gould-Vanderbilt cables are reprinted in the *New York Tribune*, May 23, 1884, and *Stockholder* 22:534.

16. *Bradstreet's*, May 24, 1884; *New York Times*, May 21, 1884; *New York Herald*, May 20–22, 1884; *Philadelphia Press*, Nov. 30, 1885; *New York Tribune*, May 20–23, 1884.

17. *New York Herald*, May 22–24 and May 27–28, 1884; *New York Times*, May 25, 1884; *Stockholder* 22:533; Grodinsky, *Jay Gould*, 493–94.

18. The origins of this story remain an intriguing mystery. The earliest full-blown version I have found is in the *New York Herald*, Dec. 4, 1892, although there are vague references to it earlier. See for example *Stockholder* 22:693. Gould's early biographers clearly plucked it intact from the newspapers and handed it down to later writers. See White, *Wizard* (Prologue, n. 4), 175–76, 289–90; Northrop, *Life and Achievements* (Prologue, n. 2), 177–79; Halstead and Beale, *Jay Gould* (Prologue, n. 2), 84–86; Ogilvie, *Life and Death* (Prologue, n. 4), 117–19; Grodinsky, *Jay Gould*, 495–96; O'Connor, *Gould's Millions* (Prologue, n. 1), 232–35.

19. *New York Herald*, Dec. 4, 1892; Northrop, *Life and Achievements*, 179. My assertion that nothing appeared in the press is based on examining several financial journals and newspapers issue by issue for the year.

20. *Stockholder* 22:533, 549, 583; *Chronicle* 38:646, 688, 699, 732; *New York Herald*, June 4, June 8, and June 15, 1884; *New York Times*, June 8, 1884; *Bradstreet's*, June 7 and June 14, 1884.

21. *Bradstreet's*, June 7 and June 14, 1884; *New York Herald*, June 3, June 6, and June 22, 1884; *New York Tribune*, May 20, May 21, June 6, June 10, and June 12, 1884; Maury Klein, *History of the Louisville & Nashville Railroad* (New York, 1972), 195–222.

22. *Chronicle* 38:178; *New York Times*, May 9, 1884; *New York Tribune*, Jan. 18, Apr. 4, May 14, and June 1, 1884. For more detail on this episode see Anon., "Chapter of Wabash" (Chap. 20, n. 4), 184–89, and *Federal Reporter* 29 (1886):162–74.

23. *New York Tribune*, Apr. 15, May 30, and June 7, 1884; "Chapter of Wabash," 185–87; 14 *U.S.* 95, 6. Swayne, the Wabash counsel, noted correctly at the time of application that the bondholders did not have rights in court because no default had yet occurred and the lease allowed a grace period of six months before the bondholders could apply for a receiver.

24. Anon., "Chapter of Wabash," 187–89; *Federal Reporter* 29:165–66; *Stockholder* 22:583; *Chronicle* 38:639, 731, 756, 39:23; *New York Herald*, May 31, June 2, June 17, June 23, July 19, and July 22, 1884; *New York Tribune*, July 4, 1884; *Bradstreet's*, Aug. 2, 1884.

25. *New York Herald*, June 5, 1884; *New York Times*, June 6, 1884; *Chronicle* 38:680, 39:97–98, 325; *Stockholder* 22:595.

26. *New York Tribune*, Apr. 16, 1884; *New York Sun*, May 7 and May 15, 1884; *Bradstreet's*, May 10, 1884; *Chronicle* 38:550, 571.

27. *Bradstreet's*, Aug. 2, 1884; *New York Tribune*, May 21, June 6, June 7, June 15, and June 17, 1884; *Chronicle* 38:679, 39:128; *Stockholder* 22:563, 566, 698, 740, 23:101; *New York Times*, June 8, 1884; *New York Herald*, June 11, Nov. 13, and Dec. 7, 1884. The final agreement merged the three companies by exchange of stock on the following basis:

Company	*Existing Stock*	*Exch. Rate*	*New Stock*
Manhattan	$13,000,000	85	$11,050,000
New York Elevated	6,500,000	120	7,800,000
Metropolitan	6,500,000	110	7,150,000
	26,000,000		26,000,000

28. *Chronicle* 38:530, 541; *New York Tribune*, May 3, 1884; *Stockholder* 22:484; *New York Times*, July 31, 1884.

29. Charles Francis Adams, Jr., "Memorabilia 1888–1893," 39–40, CFA; *New York Times*, Aug. 20, 1884; Edward C. Kirkland, *Charles Francis Adams, Jr.: The Patrician at Bay* (Cambridge, Mass., 1965), 88; G. W. Holdrege to Perkins, May 17 and May 22, 1884, BA; Perkins to Forbes, May 23 and May 24, 1884, BA; Forbes to Perkins, May 23, 1884, BA.

30. Perkins to Forbes, May 23, 1884, BA; *New York Times*, July 31, 1884; Grodinsky, *Jay Gould*, 421–22.

31. For details see *New York Times*, July 31, 1884.

32. Ibid.; *New York Tribune*, June 15–19, 1884, Dec. 5, 1890; *New York Herald*, June 19 and July 2, 1884; UP Directors' Minutes, June 18, 1884, UP; *Stockholder* 22:485; *Chronicle* 38:639, 707, 39:23; *Bradstreet's*, June 21, 1884; Grodinsky, *Jay Gould*, 422.

33. Perkins to Forbes, May 23, 1884, BA.

34. This material is taken from New York papers for the summer months.

35. *New York Herald*, June 11 and Aug. 13, 1884; *Stockholder* 22:241; *Omaha Bee*, July 14, 1884.

36. *Stockholder* 22:611, 662; *New York Times*, July 15, 1884; *New York Herald*, July 15 and July 18, 1884.

37. *New York Herald*, July 18, 1884. The Mercantile board included Sage, Dillon, Field, Eckert, Humphreys, Henry Marquand, John T. Terry, Fred Ames, and Norvin Green.

38. *New York Herald*, June 24, 1884; *Stockholder* 22:599, 631; *Bradstreet's*, July 5, 1884; *Chronicle* 39:12; *New York Tribune*, June 20–22, 1884; Gould to Clark, June 11 and June 29, 1884, KG.

39. *Stockholder* 22:630, 675, 723; *Chronicle* 39:118, 196; *Bradstreet's*, June 21, July 26, Aug. 2, Aug. 9, and Aug. 29, 1884; *New York Herald*, July 19, Aug. 3, Aug. 22, Aug. 29, and Sept. 1, 1884; *New York Tribune*, Aug. 14, Aug. 15, Aug. 28, and Sept. 28, 1884.

40. *Chronicle* 39:150; *New York Herald*, Aug. 29 and Sept. 6, 1884.

41. *Chronicle* 38:635, 707, 39:71, 263–64; *Stockholder* 22:567; *New York Herald*, June 7, June 8, July 18, and July 19, 1884; *New York Times*, July 18, 1884; *New York Tribune*, June 8, 1884. For more detail see Grodinsky, *Jay Gould*, 464–66.

42. *Chronicle* 39:296, 348; *Stockholder* 22:790; *New York Herald*, Sept. 27, 1884; *New York Tribune*, Sept. 8 and Sept. 27, 1884. For examples of Bennett's campaign see the *New York Herald*, May 17, June 12, June 18, Aug. 28, Sept. 17, Oct. 1, and Dec. 16, 1884.

43. *Chronicle* 39:98, 117, 183, 381, 408, 462; *Bradstreet's*, Aug. 2 and Oct. 25, 1884; *New York Tribune*, June 5 and Sept. 28, 1884; Herapath's *Railway Journal*, Aug. 30, 1884; *New York Herald*, Aug. 7, Aug. 31, Oct. 9, Oct. 11, and Dec. 24, 1884; Grodinsky, *Jay Gould*, 428–29.

44. *Chronicle* 39:235; Adams, "Memorabilia 1888–1893," 64, 193, 199, CFA; Gould to Clark, Sept. 26 and Oct. 18, 1884, KG.

45. James S. Clarkson to Gould, Oct. 6, 1884, and notation Gould to Whitelaw Reid, Oct. 6, 1884, Reid Papers, Library of Congress. For the political background see Muzzey, *Blaine* (Chap. 15, n. 45), 256–325, and Robert D. Marcus, *Grand Old Party* (New York, 1971), 3–100.

46. Muzzey, *Blaine*, 316. Details on the banquet, complete with the roster of guests, are in the *New York Tribune*, Oct. 30, 1884.

47. *New York World*, Oct. 30, 1884. For the earlier attack on Gould's support of Blaine see *New York Herald*, May 22, June 12, July 13, Aug. 8, Sept. 30, and Oct. 15, 1884. The name "Jay Gould Blaine" first appeared in the *Herald* of Nov. 1, 1884. See especially ibid., Nov. 2, 1884.

48. *New York Herald*, Oct. 30–Nov. 4, 1884. Winslow was president of the Frisco.

49. Ibid., Nov. 5–8 and Dec. 28, 1884; *New York Tribune*, Oct. 23 and Nov. 5–18, 1884; *Boston Transcript*, Nov. 28 and Dec. 17, 1884; Muzzey, *Blaine*, 322–25. Muzzey quotes the text of Gould's telegram to Cleveland and claims to have seen the original.

50. *New York Times*, Dec. 8, 1884; *New York Herald*, Nov. 7, Nov. 8, and Nov. 26, 1884; Muzzey, *Blaine*, 323–24. For the election results see William G. Rice and Francis L. Stetson, "Was New York's Vote Stolen?" *North American Review* 199 (January 1914):79–92.

51. Mott, *American Journalism* (see Chap. 12, n. 32), 420, 434–36. For examples of Gould's alleged meddling in state politics see *Public* 22:163, 196–98, 294.

52. Gould to Clark, Oct. 18 and Nov. 17, 1884, Jan. 5, 1885, KG; *Bradstreet's*, Sept. 20, Oct. 4, Oct. 25, Nov. 8, and Nov. 22, 1884; *Stockholder* 23:6, 21, 22, 39, 84, 87; *New York Tribune*, Nov. 26, 1884; UP Stockholders' Minutes, Mar. 25, 1885, UP; *Chronicle* 39:515, 575; *New York Herald*, Nov. 19, Nov. 20, Nov. 23, and Dec. 4, 1884.

53. *Stockholder* 23:3; *Bradstreet's*, Oct. 25 and Nov. 8, 1884.

Chapter 27: Realignment

1. Gould's correspondence abounds with requests for and references to maps. For some examples see Gould to Clark, Aug. 17, Aug. 19, Dec. 6, and Dec. 14, 1885, KG; Gould to William Kerrigan, Dec. 31, 1886, Gould letterbooks in possession of Kingdon Gould, Jr. (hereafter GLB).

2. DR, Dec. 3, 1892, 13:921.

3. Wabash opened the year at 5, Wabash preferred at 13¾, Katy at 16, Texas & Pacific at 13⅜. *Bradstreet's*, Jan. 3, 1885.

4. For an explanation of these practices see Grodinsky, *Jay Gould* (Prologue, n. 5), 402–3. See also Masterson, *Katy Railroad* (Chap. 21, n. 2), 225. Much of the evidence on these practices is hearsay.

5. Hughitt to Joy, Aug. 29, 1885, JFJ; Adams to Perkins, Oct. 23, 1885, BA. For details on efforts to reach rate agreements see Grodinsky, *Transcontinental* (Chap. 13, n. 13), 262–68.

6. Perkins to Adams, Mar. 19, 1885, BA; Perkins to Potter, Mar. 27, 1885, BA; Gould to Clark, July 21, 1885, KG; *Bradstreet's*, Feb. 28, 1885.

7. *Annual Report to the Stockholders of the Missouri Pacific Railway Company*, 1885, 17–18; *Boston Transcript*, May 12, 1883; *Statistics and Directories*, 48, 63, 64, 69; MP Directors' Minutes, Nov. 13, 1883, MP; *Stockholder* 24:550; Perkins to Potter, Dec. 26, 1884, BA.

8. *Chronicle* 40:28, 41:356; Gould to Clark, July 21 and Sept. 21, 1885, KG; *New York Tribune*, Jan. 18, 1884; Perkins to Adams, Feb. 25 and May 20, 1885, BA; *Bradstreet's*, May 23, 1885. For details on the new projects see *MP Report*, 1885, 15–18, and *Statistics and Directories*, 53, 54, 57, 61, 62, 66, 67, 73.

9. Adams to Perkins, Sept. 19, Oct. 2, Oct. 6, and Nov. 9, 1885, BA; *Bradstreet's*, Oct. 17, 1885.

10. Adams to Perkins, Oct. 2, Oct. 14, Oct. 23, and Nov. 9, 1885, BA; Central Branch Exec. Comm. Minutes, Sept. 30, 1885, MP; MP Directors' Minutes, Oct. 1, 1885, MP; Forbes to T. S. Howland, Oct. 19, 1885, BA; Perkins to Forbes, Oct. 25 and Nov. 4, 1885, BA; Burton, "Missouri Pacific" (see Chap. 22, n. 2), 691–93.

11. The Omaha Belt episode has been entirely neglected by historians. Burton, "Missouri Pacific," 724, referred to its origins as "somewhat obscure." About forty of the letters from Gould to Clark, KG, discuss the project in sufficient detail to clarify Gould's

role and intentions. See also Omaha Belt Stockholders' and Directors' Minutes, July 25, Sept. 17, and Dec. 2, 1885, May 1, 1886, MP.

12. MP Exec. Comm. Minutes, Jan. 14, 1886, MP; *Chronicle* 42:93, 125; *Bradstreet's*, Jan. 23, 1886; Grodinsky, *Transcontinental*, 276–79; Perkins to Potter, Jan. 6, 1886, BA.

13. Forbes to Perkins, Oct. 30, 1885, BA; Perkins to Adams, Dec. 18, Dec. 28, and Dec. 31, 1885, BA; Perkins to Potter, Dec. 31, 1885, Jan. 5, 1886, BA. The best account of the expansion mania of 1886–87 is in Grodinsky, *Transcontinental*, 270–308. Gould shared the prevailing view that construction costs would never be cheaper.

14. Grodinsky, *Jay Gould*, 525–32. Gould's construction work is itemized in Burton, "Missouri Pacific," 681–729. His close attention to details of the program in Kansas can be followed in the correspondence in GLB for the period December 1885–March 1887.

15. Grodinsky, *Jay Gould*, 527, 529–30; Grodinsky, *Transcontinental*, 281–83; Burton, "Missouri Pacific," 691–700; *Chronicle* 44:564–65, 590–91; Perkins to Geddes, May 11, 1886, BA; Perkins to Strong, June 5, 1886, BA; Potter to Perkins, Sept. 2, 1886, BA; Forbes to Perkins, Oct. 24 and Nov. 6, 1886, BA; Forbes to Geddes, Oct. 26 and Nov. 6, 1886, BA; Forbes to Griswold, Oct. 27, 1886, BA; *Railroad Gazette* 18:748.

16. Perkins to W. J. Palmer, Dec. 31, 1885, BA; Potter to Perkins, May 27, 1886, BA; Perkins to Potter, Jan. 4 and Jan. 6, 1887, BA; *New York Times*, Sept. 14, 1886, Mar. 7, 1887; Dodge to Morgan Jones, May 18, 1886, Feb. 19, 1887, DR, 11:393, 685; Gould to George C. Smith, Sept. 1, Sept. 11, and Nov. 9, 1886, GLB; Gould to Clark, Dec. 28, 1886, Feb. 14 (two items), Feb. 16, Feb. 23, and Mar. 2, 1887, GLB, Aug. 14 and Aug. 20, 1887, KG; Clark to Gould, Aug. 3, 1887, KG; Gould to E. O. Walcott, Feb. 11, 1887, GLB; Gould to M. C. Gould, Dec. 8, 1886, GLB; Gould to D. H. Moffat, Feb. 14 and Feb. 26, 1887, GLB; Gould to Henry McLaughlin, Dec. 20, 1886, GLB; MP Exec. Comm. Minutes, Feb. 3 and Aug. 18, 1887, MP; Dodge to N. R. Gibson, Mar. 12, 1887, DP; *Chronicle* 44:586, 45:243, 538; Athearn, *Rebel of the Rockies* (see Chap. 19, n. 24), 150–70; Overton, *Gulf to Rockies* (see Chap. 15, n. 13), 131–91. The intercepted telegram suggests that, for all the indignant sputtering by Forbes and Perkins about Gould's reading messages sent over Western Union, they had no qualms about indulging in the same practice.

17. Perkins to Holdrege, Dec. 19, 1885, BA; Gould to George Gould, July 11, 1886, KG; Gould to Hoxie, June 25 and July 8, 1886, GLB; Gould to Clark, June 20, July 3, July 10, July 18, July 25, Aug. 29, Sept. 25, and Oct. 3, 1886, Aug. 20, 1887, KG; Adams to Dodge, Sept. 2, 1886, DP; Gould to Adams, Nov. 8, 1886, GLB; Gould to Clark, Feb. 15 and Feb. 16, 1887, GLB.

18. Gould to Kerrigan, Dec. 6, Dec. 9, Dec. 10, Dec. 13, and Dec. 21, 1886, GLB; Gould to Henry Wood, Dec. 10 and Dec. 28, 1886, GLB; Gould to C. T. Walker, Dec. 13 and Dec. 31, 1886, GLB; Gould to J. M. Row, Dec. 21, 1886, GLB; Gould to Clark, Dec. 27, 1886, GLB; *Chronicle* 44:59. For background on the Arkansas roads see Burton, "Missouri Pacific," 688–91, 715–17, 718–22.

19. Kerrigan to Gould, Jan. 3, 1887, GLB; Gould to Hoxie, July 6 and July 7, 1886, GLB; Gould to George C. Smith, July 23, 1886, GLB; Gould to Henry McLaughlin, Sept. 22, 1886, GLB; Gould to Kerrigan, Nov. 8, 1886, GLB; Gould to Clark, Dec. 28, 1886, Feb. 18, 1887 (two items), GLB; MP Exec. Comm. Minutes, Jan. 4, 1886, MP.

20. Gould to George R. Brown, Dec. 14, 1886, GLB; Gould to Edwin Atkins, Jan. 8, 1887, GLB; Gould to Kerrigan, Dec. 31, 1886, GLB; Gould to Clark, Jan. 5, Jan. 31, Feb. 10, Feb. 19, and Feb. 21, 1887, GLB, Feb. 22, 1887, KG; *New York Times*, Jan. 14, 1887; *Chronicle* 44:551, 45:642; Gould to Henry Wood, Feb. 5 and Mar. 3, 1887, GLB. My interpretation differs sharply from that in Grodinsky, *Jay Gould*, 536–37.

21. On Missouri construction see Gould to Kerrigan, Dec. 8, 1886, GLB; Gould to Clark, Feb. 16, Feb. 18, and Mar. 3, 1887, GLB. For Texas see Gould to Hoxie, July 6 and July 8, 1886, GLB; Gould to Kerrigan, Oct. 5, Oct. 30, Nov. 8, Nov. 9, Nov. 10 (two items), Nov. 19, Nov. 20, Nov. 22, Nov. 24, Dec. 1 (three items), Dec. 2, Dec. 11 (two items), Dec. 14, and Dec. 18, 1886, GLB. The Louisiana venture is detailed in Gould to Clark, Feb. 1, Feb. 9, Feb. 12, and Feb. 14 (three items), GLB.

22. Poor, *Manual 1885,* 793, *Manual 1888,* 791; *MP Report*, 1887, 6–15, 24–26; Gould to D. S. Smith, May 26, 1886, GLB; Gould to J. F. Dillon, Dec. 22, 1886, GLB; Gould to J. L. Cadwalader, Dec. 22, 1886, GLB; *Chronicle* 43:459, 746, 44:370, 621, 45:369; *Stockholder* 25:23, 42, 755.

23. *New York Times*, July 6, 1884; Adams to Perkins, Oct. 8, 1885, BA; Grodinsky, *Transcontinental*, 281, 298–302; Bryant, *Santa Fe* (see Chap. 21, n. 5), 123–41.

24. Perkins to W. W. Baldwin, Dec. 9, 1884, BA. The first Gould proposal is admirably treated in Grodinsky, *Jay Gould*, 428–30.

25. *Federal Reporter* 23 (1885):866; *Stockholder* 23:456; *Bradstreet's*, May 2 and June 20, 1885; Perkins to Potter, June 29, 1885, BA; Perkins to George B. Roberts, Aug. 18, 1885, BA.

26. *Chronicle* 40:570–71, 41:746; *Bradstreet's*, May 2, May 9, and June 20, 1885. The modified plan is outlined in "Memorandum," July 14, 1885, JFJ, and details on the negotiations are in JFJ for July–August 1885.

27. *Railway World* 30:409; *Federal Reporter* 29 (1886):167–68; *Chronicle* 42:217, 695; *Stockholder* 24:563; *New York Tribune*, June 7, 1886; Grodinsky, *Jay Gould*, 432–35.

28. Ibid.; Gould to James Hickson, Nov. 27, 1886, GLB.

29. *Federal Reporter* 29 (1886):169–70; *New York Tribune*, Dec. 8, 1886; *Bradstreet's*, Dec. 11, 1886. The latter called the decision "probably one of the most important decisions ever rendered in a railroad case."

30. *Federal Reporter* 29 (1886):170–74; Wabash General Agent to Joy, Jan. 15, 1886, JFJ.

31. *Bradstreet's*, Dec. 11, 1886; *Chronicle* 43:719, 44:10, 119, 344, 435, 466, 621, 654, 682, 782; *New York Tribune*, Dec. 8, Dec. 17, Dec. 18, and Dec. 31, 1886, Apr. 7, 1887; *New York Times*, Mar. 8, 1887; Gould to A. A. Talmage, Dec. 14 and Dec. 27, 1886, GLB; Gould to Charles Ridgeley, Dec. 17, 1886, GLB; Gould to Isaac H. Knox, Dec. 23, 1885, GLB. The division of the Wabash occurred in part because the different groups of bondholders separated along those lines and in part because of the jurisdictional limitations of the two federal judges. Gresham's district lay in Illinois, Brewer's in Missouri.

32. Gould's role in the Texas & Pacific is richly documented, particularly by the wealth of material in GLB and TPHR. For more detail on the underlying financial structure of the three divisions see Grodinsky, *Jay Gould*, 440–41; *Stockholder* 23:723.

33. The 1884 fight may be followed in the *Chronicle* 39:325, 565, 581, 607, 40:302; *Stockholder* 22:804–5; *Railroad Gazette* 16:860, 908; *New York Herald*, Dec. 11, 1884; *Bradstreet's*, Mar. 7, 1885. The report, dated Feb. 5, 1885, is in TPHR. See Simmonds, "Texas and Pacific" (Chap. 21, n. 4), 56–62.

34. R. S. Hayes to Dodge, Apr. 11, 1885, DR, 11:157; A. H. Calef to Dodge, Apr. 21, 1885, DR, 11:161–62; *Chronicle* 41:24, 77, 42:464; *Report of the Reorganization Committee to the Directors of the Texas and Pacific Railway Company*, Dec. 1, 1888, 1–4, TPHR. The receivers appointed were John C. Brown and Lionel A. Sheldon.

35. Wistar to C. E. Satterlee, Jan. 4, 1886, TPHR; Gould to Wistar, Jan. 5, 1886, TPHR; Wistar to Gould, Jan. 5 and Jan. 6, 1886, TPHR; Gould to Satterlee, Jan. 10, 1886 (two items), TPHR; *Chronicle* 42:61. See also Isaac J. Wistar, *The Autobiography of Isaac Jones Wistar* (New York, 1937), 492–94.

36. Gould to Wistar, Apr. 10 and Apr. 26, 1886, GLB; Wistar to Gould, May 6, 1886, TPHR; *Chronicle* 42:519, 552, 575–76, 604.

37. *Chronicle* 42:61, 519, 632, 664; *Bradstreet's*, June 12, 1886. For a specific example of this favoritism see *Chronicle* 42:207.

38. *Chronicle* 42:695, 774–75, 783; *Bradstreet's*, June 12, 1886; Gould to Wistar, June 17, 1886, GLB; Lionel A. Sheldon to Satterlee, Mar. 17, 1886, TPHR; Gould to Hoxie, June 29, June 30, July 1, and July 7, 1886, GLB; Wistar to Satterlee, July 26, 1886, TPHR.

39. Gould to Wistar, July 26 and July 27, 1886, GLB.

40. Ibid., July 14, 1886, GLB; *Chronicle* 43:12, 41, 103; *Stockholder* 24:683; *Bradstreet's*, July 31, 1886.

41. Gould to Fleming, July 20, 1886, GLB; Gould to Satterlee, July 31, Aug. 1–3, 1886, TPHR; Memo, Aug. 9, 1886, GLB; Gould to Wistar, July 27, Aug. 10 (three items), and Aug. 13 (two items), 1886, GLB; Gould to Drexel, Morgan and Kuhn, Loeb, Aug. 11, 1886, GLB; Gould to John G. Moore, Aug. 17, 1886, GLB; Gould to Wistar, Aug. 10, 1886, TPHR; *Chronicle* 43:133, 163–64, 191–92; *New York Tribune, Aug.* 11, 1886. The account in Grodinsky, *Jay Gould*, 443–44, does not describe the modified terms correctly and errs in emphasizing Morgan's role. The evidence reveals clearly that Schiff was the banker most actively involved in the negotiations.

42. *New York Tribune*, Aug. 11, 1886; *Chronicle* 43:191, 218, 275, 400; *Stockholder* 24:699, 764; *Reorg. Comm. Report*, 4–5.

43. Wistar to Satterlee, Jan. 13 and Jan. 22, 1887, TPHR; George Gould to Satterlee, Jan. 21, 1887, TPHR; Satterlee to J. H. Schiff, Jan. 24, 1887, TPHR; Gould to Wistar, Feb. 24 (three items) and Feb. 25, 1887, GLB; Wistar to Gould, Feb. 24, 1887, TPHR.

44. Wistar to Gould, Mar. 17 and Apr. 26, 1887, TPHR; Wistar to Satterlee, Dec. 16, 1887, TPHR; Wistar to John C. Brown, May 4, 1887, TPHR; *Reorg. Comm. Report*, 6; *Chronicle* 44:119, 309, 782, 45:401, 643, 821. For the Dos Passos episode see Gould to Dos Passos, May 1, 1886, with attached endorsements, TPHR.

45. Gould to Kerrigan, Oct. 2, Nov. 5, and Nov. 8, 1886, GLB; Gould to Clark, July 10, July 18, Aug. 29, and Sept. 25, 1886, Aug. 7, 1887, KG; Gould to George Gould, July 11, 1886, KG.

Chapter 28: Workday

1. *New York Herald*, Sept. 20, 1886.

2. George Gould, interview in the *New York Herald*, Sept. 20, 1886; Clara Morris, *Life on the Stage* (New York, 1905). Actress Morris recalled once meeting Jay backstage and was astonished to receive not the usual proposition but words of encouragement and an offer of help if she encountered trouble. "I thought of the gentle voice, the piercing eyes that had grown so kind, the friendly promise. . . . I am forced to believe Mr. Jay Gould was perfectly honest and sincere in his offer of assistance" (305).

3. *New York Herald*, Sept. 20, 1886; *New York Times*, Sept. 15, 1886.

4. Snow, *Helen Gould* (see Chap. 1, n. 2), 160; *New York Herald*, Sept. 10, Sept. 11, Sept. 18, Sept. 20, and Sept. 21, 1884, Sept. 20, 1886; *New York Tribune*, Sept. 11, Sept. 12, Sept. 16, Sept. 17, Sept. 19, and Sept. 20, 1884; *Stockholder* 23: 131; *New York Times*, Sept. 15, 1886.

5. *New York Times*, Sept. 15, 1886; Gould to Clark, June 20, 1886, KG. For Victoria's disappearance see *New York Herald*, Sept. 2–13, 1886.

6. Snow, *Helen Gould*, 161.

7. *New York Times*, Sept. 15, 1886; *New York Sun*, Sept. 15 and Sept. 16, 1886; *New York Herald*, Sept. 16 and Sept. 20, 1886; *New York Tribune*, Sept. 17, 1886; Snow, *Helen Gould*, 161–62.

8. Snow, *Helen Gould*, 162; Gould to Edith Gould, Oct. 10, 1887, KG.

9. *Stockholder* 23:357; *Bradstreet's*, Mar. 14 and Mar. 21, 1885; *New York Sun*, Mar. 7 and Mar. 11–15, 1885; Bureau of Labor Statistics and Inspection of Missouri, *The Official History of the Great Strike of 1886 on the Southwestern Railway System* (Jefferson City, Mo., 1887), 5–6; Foster Rhea Dulles, *Labor in America* (New York, 1966), 138–40; Thomas R. Brooks, *Toil and Trouble* (New York, 1964), 63–65. A copy of the *Official History* is in MP.

10. *New York Sun*, Mar. 26, 1885; *Labor Troubles in the South and West*, House Reports, 49 Cong., 2d sess., 1886 (Serial 2502), no. 4174, pt. 1, 29–30; Brooks, *Toil and Trouble*, 65; Dulles, *Labor in America*, 140; Norman J. Ware, *The Labor Movement in the United States, 1860–1895* (New York, 1929), 139–45.

11. Ware, *Labor Movement*, 145; Dulles, *Labor in America*, 141–42; Brooks, *Toil and Trouble*, 65; *Official History*, 7. Adams noted that "news of breakdown of strike by giving in of Mo. Pacific came on us this morning like a wet blanket." Charles Francis Adams, Jr., Diary, Mar. 16, 1885, CFA.

12. *Official History*, 7–10; Ware, *Labor Movement*, 147. The fullest sources for the strike are the *Official History* and the testimony in *Labor Troubles*. Most of the essential correspondence and documents are reproduced in both places, especially *Labor Troubles*, 331–86.

13. *Official History*, 12–17; Dulles, *Labor in America*, 143.

14. *New York Herald*, Sept. 20, 1886; MP Exec. Comm. Minutes, Mar. 8, 1886, MP. For typical versions of Gould's role see Ware, *Labor Movement*, 144–49; Dulles, *Labor in America*, 142–45; Brooks, *Toil and Trouble*, 66–67.

15. *New York Times*, Nov. 24, 1886; Frank W. Taussig, "The South-Western Strike of 1886," *Quarterly Journal of Economics* 1 (January 1887):216–17; Burton, "Missouri Pacific" (see Chap. 22, n. 2), 740; *Official History*, 117.

16. *Official History*, 17–28; *New York Sun*, Mar. 10, 1886.

17. *Official History*, 25, 29–50; *New York Sun*, Mar. 11, 1886; *Bradstreet's*, Mar. 14 and Mar. 21, 1885; *Chronicle* 42:380; Ware, *Labor Movement*, 147; *Labor Troubles*, 35–36.

18. *New York Sun*, Mar. 13, Mar. 17, and Mar. 20, 1886; *Labor Troubles*, 35–36. A convenient summary is in Taussig, "South-Western Strike," 187–89; *Official History*, 51–75; *Bradstreet's*, Mar. 27, 1886. The secret circular is reprinted in *Official History*, 69–75.

19. *Official History*, 64–68; *Labor Troubles*, 36–37; *New York Sun*, Mar. 28, 1886.

20. *Official History*, 68; *Labor Troubles*, 9–13, 37, 87.

21. *Official History*, 75–77; *Labor Troubles*, *13*, 37–39; *New York Sun*, Mar. 29 and Mar. 30, 1886.

22. The official stenographic report of the meeting is reprinted verbatim in *Labor Troubles*, 48–58. See also *Labor Troubles*, 87; *Official History*, 76–79; *New York Tribune*, Apr. 16 and Apr. 17, 1886. Hopkins affirmed that Gould had placed the arbitration question in Hoxie's hands and "could not well interfere. It would have destroyed all discipline on the road. Mr. Hoxie would have resigned at once." *Labor Troubles*, 75.

23. *Labor Troubles*, 22, 40–41; *Official History*, 79–80; *New York Times*, Apr. 1 and Apr. 3, 1886; *New York Tribune*, Apr. 1–3, 1886; *Chronicle* 42:409.

24. *Official History*, 80–88; *New York Sun*, Apr. 1, Apr. 4, Apr. 6, and Apr. 10, 1886; *New York Times*, Apr. 4 and Apr. 5, 1886; *New York Herald*, Apr. 4 and Apr. 5, 1886; *New York Tribune*, Apr. 4 and Apr. 5, 1886; Dulles, *Labor in America*, 144; Taussig, "South-Western Strike," 215.

25. *Official History*, 88–90; *Labor Troubles*, 41; *New York Times*, Apr. 6–9, 1886; *New York Herald*, Apr. 6, 1886; *New York Tribune*, Apr. 6–8, 1886.

26. The letters are reprinted in *Official History*, 90–100; *Labor Troubles*, 42–46; *New York Herald*, Apr. 16, 1886. An abridged version is in the *New York Tribune*, Apr. 15, 1886.

27. *Official History*, 100–108; *Labor Troubles*, 61–65, 379; *New York Times*, Apr. 7–10, 1886; *New York Herald*, Apr. 8–26, 1886; *New York Tribune*, Apr. 10–29, 1886; *New York Sun*, Apr. 16 and Apr. 17, 1886; *Stockholder* 24:463; Taussig, "South-Western Strike," 206–10.

28. *Official History*, 108–17; *New York Times*, Apr. 12 and Apr. 13, 1886; *New York Herald*, Apr. 13, Apr. 29, Apr. 30, and May 1–4, 1886; *New York Tribune*, May 2, May 4, and May 5, 1886.

29. *New York Tribune*, May 5, 1886; *New York Herald*, May 5 and May 6, 1886. For examples of other strikes see *New York Tribune*, Apr. 1 and Apr. 6, 1886.

30. *New York Times*, Nov. 24, 1886; *New York Tribune*, Aug. 8 and Nov. 24, 1886; Gould to Clark, Aug. 9, 1886, KG; Gould to Helen D. Gould, Aug. 20, 1886, GLB; Gould to United States Hotel, Aug. 20, 1886, GLB; Gould to A. G. Curtin, Sept. 1, 1886, GLB; Gould to Kerrigan, Oct. 28, 1886, GLB; MP Directors' Minutes, Nov. 26, 1886, MP; DR, 11:578; Gould to J. W. Midgley, Nov. 30, 1886, GLB.

31. These generalizations about Gould's work methods and habits are based on close study of his business correspondence. Individual citations would be so extensive as to tax the stamina of reader and writer alike. For an example of the Gould legend with regard to service see S. G. Reed, *The History of the Texas Railroads* (Houston, 1941), 355.

32. Gould to Kerrigan, Nov. 24, 1886, GLB. The italics are mine.

33. This description of a tour is a composite drawn from a wide variety of sources. For the 1887 tour see *New York Times*, Apr. 27, 1887.

34. *Angell* (see Chap 1, n. 13), 542–43; *New York Tribune*, Oct. 27, 1886; Gould to Kerrigan, Nov. 12 and Nov. 18, 1886, GLB; Gould to Pullman, Feb. 19, 1887, GLB.

35. Gould to Clark, Dec. 23, 1886, GLB.

36. *New York Tribune*, May 7, 1882; *New York Times*, May 7, 1882, Apr. 27, 1887, Jan. 3, 1892.

37. The correspondence on these matters in GLB is too extensive to cite fully. For examples see Gould to Hoxie, July 1, July 6, and July 12, 1886; Gould to Kerrigan, Sept. 11, Sept. 13, Oct. 30, Nov. 6, Nov. 20, Nov. 22, Dec. 8, Dec. 10, and Dec. 29, 1886; Gould to Abram Gould, Nov. 3 and Nov. 9, 1886, Feb. 28, 1887; Gould to Clark, Dec. 21 and Dec. 22, 1886, Jan. 8 and Feb. 16, 1887.

38. Gould to Clark, Feb. 10, 1887, GLB.

39. Gould to Kerrigan, Dec. 9, 1886, Jan. 3, 1887, GLB.

40. For examples of financial and legal work see Gould to Louis Fitzgerald, Sept. 16 and Sept. 23, 1886, GLB; Gould to R. M. McDowell, Sept. 23, 1886, GLB; Guy Phillips to Gould, Oct. 18, 1886, GLB; Gould to H. C. Deming, Oct. 22 and Oct. 28, 1886, GLB; Gould to Loomis L. White, Nov. 16, 1886, GLB; Gould to I. & S. Wormser, Nov. 17, 1886, GLB; Gould to Clark, Nov. 24, 1886, Feb. 16, 1887, GLB; Gould to S. B. French, Nov. 30, 1886, GLB; Gould to Kerrigan, Dec. 15, 1886, GLB; Gould to George C. Smith, Dec. 17, 1886, GLB. Earle, *Mr. Shearman and Mr. Sterling* (see Chap. 7, n. 4), 87, notes that Gould's business with that firm "continued to be very substantial (perhaps its largest fee account) until 1890."

41. Gould's attention to operating detail is reflected most obviously in his constant requests for data. See for example Gould to E. M. Morseman, Aug. 14, 1886, GLB; Gould to C. G. Warner, Aug. 31 and Oct. 30, 1886, GLB; Gould to Kerrigan, Sept. 10 and Nov. 30, 1886, GLB.

42. Gould to Clark, Apr. 3, 1887, KG.

43. For overviews see Chandler, *Visible Hand* (Prologue, n. 6), 145–87, and Grodinsky, *Transcontinental* (Chap. 13, n. 13), 312–35. Ironically, Adams was wrestling with similar problems on the Union Pacific. See his letters to Dodge in DP; Adams, "Memorabilia 1888–1893," 62–67, 95, 106–34.

44. Gould to Clark, Aug. 5, 1885, KG; Gould to George C. Smith, Nov. 26, 1886, GLB; Gould to Clark, Nov. 26, 1886, GLB; *New York Tribune*, Dec. 4, 1886; DR, 11:578, 589. Gould pushed Talmage for a vice-presidency of the Wabash. See Gould to O. D. Ashley, Mar. 1, 1887, GLB; Gould to Talmage, Mar. 1, 1887, GLB.

45. Adams, "Memorabilia 1888–1893," 199, CFA; Burton, "Missouri Pacific," 738–39. Unfortunately, Clark left no written account of his career. Approached once by a friend interested in doing a biography of him, Clark snapped, "Who in the devil wants to know anything about me?" The operating ratio is the ratio of expenses to earnings. Although its value has often been disputed, railroad executives considered it an important indicator.

46. Gould to Hoxie, July 10, 1886, GLB; Gould to Clark, Feb. 6 and May 4, 1887, KG; *New York Tribune*, Oct. 28, 1886.

47. Gould to Clark, Feb. 6, 1887, KG.

48. Gould to Hoxie, July 10, 1886, GLB; Gould to Clark, Dec. 20, 1886, Feb. 1, 1887, GLB; Gould to Kerrigan, Jan. 6, 1887, GLB; Gould to Clark, Feb. 6, 1887, KG; *New York Tribune*, Dec. 4, 1886.

49. Gould to Clark, May 10, 1887, KG; *New York Times*, July 25, 1887. In December Fleming raised a similar complaint about tie purchases on the Texas & Pacific. For this controversy see Fleming to Wistar, Dec. 12, 1887, TPHR; John C. Brown to Wistar, Dec. 26 and Dec. 31, 1887, TPHR; Brown to Schiff, Dec. 31, 1887, TPHR.

50. MP Exec. Comm. Minutes, Apr. 28, 1887, MP; Gould to Clark, Feb. 8, May 4, and May 28, 1887, KG.

51. Helen D. Gould to Sarah Northrop, May 25, [1887], JGL; Gould to Clark, May 28, 1887, KG. For accounts of Gould's testimony see *New York Times*, May 18, 1887, and *New York Tribune*, May 18–20, 1887. The fullest description of Gould and the commissioners is in the *New York Tribune*, May 22, 1887.

52. Isaac H. Bromley to Adams, May 17–19, 1887, UP; *New York Tribune*, May 22, 1887. The whole of Gould's testimony is in *PRC*, 446–592.

53. *New York Tribune*, June 3, 1887; Memorandum from John Rhodehamel to Christine Meadows, June 5, 1979, The Mount Vernon Ladies' Association of the Union; Elswyth Thane, *Mount Vernon: The Legacy* (Philadelphia, 1967), 175–78; Lily L. Macalester Laughton to Gould, June 14, 1887, MVLA; Gould to Laughton, June 17, 1887, MVLA; Laughton to Gould, June 29, 1887, MVLA. The deed for the land, dated July 23, 1887, is in MVLA. Gould paid $2,500 for the tract of 33 1/2 acres.

54. *New York Tribune*, June 14, 1887; *New York Times*, July 25 and Aug. 2, 1887; *New York World*, July 25, 1887.

55. Gould to Clark, July 15, 1887, KG.

56. Ibid., July 15, July 27, July 29, Aug. 7, and Aug. 20, 1887, KG. The switch from wood to coal was no small item. "Fuel per train mile run on the Iron M in June was .0869 agt .0466 on MoP," Gould observed in the Aug. 20 letter. "Using coal will reduce cost to not greater than MoP or a saving of $250,000 per year."

57. Ibid., July 27, July 29, July 31, and Aug. 7, 1887, KG.

58. Gould's plan, marked *"In strictest confidence,"* is in ibid., July 31, 1887. For Newman see MP Exec. Comm. Minutes, May 12, 1887, MP.

59. Ibid., July 27, Aug. 7, Aug. 20, and Sept. 17, 1887, KG; *MP Report*, 1887, 8–15; Poor, *Manual 1886*, 836, and *Manual 1887*, 773.

60. Gould to Clark, Sept. 17, 1887, KG.

61. *New York Tribune*, Dec. 9 and Dec. 10, 1886; Gould to Clark, Sept. 18 and Nov. 15, 1885, KG; *Federal Reporter* 29:171–72. The figures given are calculated from tables in *MP Report*, 1884, 15–16, and 1887, 27.

62. The correspondence on Gould's coal activities is too extensive to cite in full. For some examples see Gould to McDowell, May 12, June 23, June 25, June 28–30, Oct. 4, Nov. 6, Nov. 10, Nov. 15, Dec. 2, and Dec. 18, 1886, GLB; Gould to Clark, Dec. 16, 1886, Feb. 2, 1887, GLB; Edwin Gould to McDowell, June 29, 1886, GLB; Gould to Barney & Smith, Dec. 29, 1886, GLB; Gould to Kerrigan, Sept. 13, Sept. 21, Nov. 10 (two letters), and Nov. 15, 1886, GLB; Gould to Clark, Aug. 7, 1887, KG. The letter from Edwin Gould reaffirmed his father's belief that "it is more important to acquire more coal lands than to pay dividends now."

63. Gould to B. C. Brown, Nov. 30, 1886, GLB; Gould to Clark, Jan. 31, Feb. 1, Feb. 4, Feb. 17, Feb. 22, and Feb. 23, 1887, GLB; Gould to McDowell, Feb. 9, Feb. 17, Feb. 26, and Mar. 1, 1887, GLB.

64. *MP Report*, 1887, 29; Gould to E. B. Wheelock, May 12, May 13, May 24, and July 1, 1886, Feb. 1, 1887, GLB. The land grant belonged originally to the New Orleans, Baton Rouge & Vicksburg Railroad, also known as the Backbone. See *New York Times*, Jan. 10, 1885, and *Bradstreet's*, Mar. 21, 1885.

65. Gould to L. Phillips, Nov. 30, 1886, GLB.

66. *New York Tribune*, Oct. 10, 1886; *New York World*, Aug. 1, 1888; *New York Times*, Mar. 11, 1889; Grodinsky, *Jay Gould* (see Prologue, n. 5), 338–40.

67. Gould to Kerrigan, Oct. 2, Nov. 10, and Nov. 12, 1886, GLB; Gould to Taussig, Nov. 16, 1886 (two letters), Feb. 23, Feb. 26, and Feb. 28, 1887, GLB; Memorandum of Agreement, Dec. 10, 1886, GLB; Gould to Clark, Jan. 5, 1887, KG; Gould to Roberts, Feb. 21, 1887, GLB.

68. Burton, "Missouri Pacific," 722–23; Gould to Hoxie, June 29 and July 7, 1886, GLB; Gould to George C. Smith, July 19, July 22, July 23, Aug. 10, Aug. 17, and Sept. 25, 1886, GLB; Gould to Kerrigan, Oct. 28 and Oct. 30, 1886, GLB; *MP Report*, 1887, 26.

69. Gould to Kerrigan, Dec. 1, Dec. 21, and Dec. 22, 1886, GLB; Gould to George C. Smith, July 23 and Nov. 24, 1886, GLB; Gould to James Hill, July 8, 1886, GLB; Gould to George Gould, July 11, 1886, KG; Gould to Hoxie, June 29, June 30, July 2, July 7, July 9, and July 10, 1886, GLB; Gould to E. A. Smith, June 28, 1886, GLB; *New York Tribune*, Oct. 17 and Oct. 18, 1886.

70. Gould to George C. Smith, Dec. 1, 1886, GLB; Gould to Kerrigan, Dec. 8, Dec. 10, and Dec. 14, 1886, GLB; Burton, "Missouri Pacific," 723; Gould to Clark, Jan. 5, Aug. 20, and Oct. 16, 1887, KG; *New York Tribune*, May 17, 1887.

Chapter 29: Casualties

1. *New York Herald*, July 1 and Sept. 18, 1887; *New York Times*, Oct. 7, 1887; Snow, *Helen Gould* (see Chap. 1, n. 2), 140. One of Jay's powers of attorney to Edwin, dated Oct. 26, 1887, is in material in possession of Edwin Gould (hereafter EG).

2. Gould to George Gould, July 11, 1886, KG.

3. *New York Times*, July 13, 1933; *New York Tribune*, May 12, 1887; *Stockholder* 25:550.

4. Snow, *Helen Gould*, 153, 172–74. Nellie's obituary in the *New York Times*, Dec. 21, 1938, is filled with errors.

5. Snow, *Helen Gould*, 147, 169; *New York World*, Aug. 18, 1887; *New York Sun*, Aug. 18, 1887.

6. Snow, *Helen Gould*, 126–27, 141–43, 146–47, 175–76, 179; Guy Phillips to G. E. Palen, Nov. 12, 1886, GLB.

7. Gould to George Gould, July 11, 1886, KG.

8. *New York Times*, Oct. 18 and Dec. 1, 1885; *New York Sun*, Nov. 29, 1885; *Stockholder* 24:143; *Bradstreet's*, Dec. 5, 1885; *Economist* (London), Dec. 26, 1885. The partnership included Connor, Morosini, and George Gould as general partners and Jay as special partner. It was formed in January 1883 to last for two years, and was renewed in December 1885 for another year. See RGD, NYC, 422:695, and *New York Times*, Oct. 18, 1885.

9. RGD, NYC, 348:900 A98; *New York Times*, Nov. 12, 1885; *Stockholder* 24:500, 511, 26:1; *Chronicle* 41:665, 667–68, 42:588; *Bradstreet's*, Dec. 12, 1885, May 15, 1886; *DAB*, 19:176; *New York World*, Mar. 16, 1891; Stedman, *Stock Exchange* (See Chap. 7, n. 5), 328.

10. *Bradstreet's*, Oct. 3, Oct. 10, Dec. 12, and Dec. 19, 1885, Jan. 2 and May 22, 1886; *Stockholder* 23:424, 24:179, 25:97; *Chronicle* 41:382; RGD, NYC, 418:294, 300 A43. Heath owed Gould $260,000 at the time of his failure. See Stedman, *Stock Exchange*, 325.

11. *Chronicle* 40:172, 224, 233, 501, 41:13; *Bradstreet's*, Feb. 28, Apr. 25, and July 4, 1885; *New York Times*, Apr. 26, 1885; *Stockholder* 23:504, 611; *New York Tribune*, May 24, 1885; Grodinsky, *Jay Gould* (See Prologue, n. 5), 501–7.

12. *Chronicle* 41:123, 414, 439, 649; *Bradstreet's*, July 4 and Aug. 8, 1885; Stedman, *Stock Exchange*, 324–26; Grodinsky, *Jay Gould*, 508–14. Grodinsky asserts that Gould resented Morgan's intrusion and was contemptuous of his efforts. "To think that Morgan had thus succeeded where he had failed must have been galling, and his sense of personal frustration warped his business judgment." I find no evidence to substantiate this view, and that offered by Grodinsky is unconvincing.

13. *Bradstreet's*, Aug. 1, Aug. 15, and Oct. 3, 1885; *New York Tribune*, July 5, 1886; *Labor Troubles* (see Chap. 28, n. 10), 64; Gould to A. E. Bateman, Nov. 26, 1886, GLB; *New York World*, July 2, 1887.

14. *Bradstreet's*, Dec. 18, 1886; *New York Tribune*, Dec. 16, 1886; Grodinsky, *Jay Gould*, 512–14.

15. *Chronicle* 44:38, 802; *Stockholder*, 25:68; Gould to Clark, Feb. 1, 1887, GLB; T. J. Coolidge to Perkins, Sept. 16, 1887, BA.

16. Gould to George Gould, July 11, 1886, KG.

17. *Chronicle* 40:423, 626, 40:651, 685; *Stockholder* 23:536. See also Chap. 26.

18. *New York Herald*, July 14, 1885; *New York Times*, July 11–13, 1885; Reid, *Telegraph in America* (see Chap. 17, n. 3), 781. This episode is drawn from the lengthy accounts given in the *Herald*, *Times*, and *Tribune* for July 1885 and May–July 1886. The 1886 material consists largely of arguments and testimony from the Stokes suit, rich in background detail but, like all testimony, filled with contradictions and disputed points.

19. *New York Times*, July 11 and July 12, 1885; *New York Herald*, May 25–27 and June 26, 1886. Eckert's testimony in the Stokes suit leaves little doubt as to the orders given him.

20. *New York Herald*, Apr. 16, Apr. 27, May 2, May 26, and May 28, 1886; *New York Times*, July 11–16 and Dec. 28, 1885, May 25, June 30, and July 9–11, 1886; *Bradstreet's*, Aug. 15, 1885; *Chronicle* 41:75, 330. The Ingersoll quotation is in the *New York Herald*, May 22, 1886.

21. *Bradstreet's*, Aug. 15, 1885; Reid, *Telegraph in America*, 781, 785–90.

22. *Bradstreet's*, Aug. 15, 1885; *Railroad Gazette* 17:590; *Chronicle* 43:125; *New York*

Tribune, Dec. 14, 1884; *New York Herald*, Oct. 28, 1887. For data on Western Union see *Chronicle* 41:444, 43:458.

23. *Stockholder* 24:35; *Chronicle* 42:339, 729, 43:309; *New York Times*, Dec. 28, 1885; *New York Tribune*, Apr. 20, 1886; *New York Herald*, May 2, May 5, and Sept. 9, 1886; *Bradstreet's*, Apr. 24, 1886. The Bennett crusade can be followed in the *Herald* for May 1886. The quotation is taken from the May 25 issue. For the congressional investigation see *Military and Postal Telegraph*, House Reports, 50 Cong., 1st sess., 1888 (Serial 2598), no. 178, iii.

24. *New York Herald*, Oct. 28, 1887; Grodinsky, *Jay Gould*, 471.

25. Grodinsky offers a good case in point. His otherwise impressive study suffers from a steadfast refusal even to consider the effect of Gould's personality, motives, needs, and other personal factors on his career. To cite but one example, his analysis of Gould's later career takes no account of the influence exerted by his declining health.

26. *New York Tribune*, Oct. 27, 1887; *New York Times*, Mar. 12, 1887; *New York Herald*, Oct. 28, 1887; Grodinsky, *Jay Gould*, 472; Stedman, *Stock Exchange*, 329–30.

27. *Bradstreet's*, Aug. 1, Aug. 8, and Aug. 15, 1885; *Chronicle* 41:145; *New York Tribune*, July 4, July 5, Oct. 28, Dec. 10, Dec. 11, and Dec. 18, 1886.

28. *New York Tribune*, Oct. 25, 1887; *New York World*, Feb. 3, 1889. See also Garrett's obituary in the *New York Times*, July 30, 1896. The merger attempt is described in the *New York Herald*, Oct. 9, 1887.

29. *New York Herald*, Oct. 9, 1887; Stedman, *Stock Exchange*, 329–30; *New York Times*, Mar. 9, 1887. For Sully see Klein, *Great Richmond Terminal* (Chap. 9, n. 18), 49–50.

30. *New York Times*, Mar. 11 and Mar. 13, 1887; *New York Tribune*, Mar. 11, 1887; *New York Sun*, Mar. 12, Mar. 16, and Mar. 25, 1887; *Chronicle* 44:343; Klein, *Great Richmond Terminal*, 173–95. The saga of the B & O sale can be followed in detail in the *New York Times*, Mar. 9–25, 1887. Edward Hungerford, *The Story of the Baltimore & Ohio Railroad, 1827–1927* (New York, 1928), 2:161–64, makes no mention of this episode. With the delicacy of those who prefer to substitute discretion for history, Hungerford devotes only two uninformative paragraphs to Robert Garrett's administration.

31. The Garretts' animosity toward Gould is mentioned in the *New York Times*, Oct. 11, 1887. For Sage's assertion see ibid., Mar. 11, 1887. Gould's view that the telegraph was losing heavily is sprinkled through various issues of the *Times* during March 1887.

32. Stedman, *Stock Exchange*, 330–33; *New York Times*, Mar. 15–25, 1887; *New York World*, May 10 and June 7, 1887. The fullest sketch of Ives is in the *New York World*, Aug. 14, 1887.

33. *Philadelphia Press*, Mar. 13, 1887; *New York Times*, Mar. 21, 1887; *New York Tribune*, May 15, 1887; *New York Herald*, Sept. 4, 1887; Grodinsky, *Jay Gould*, 473–74.

34. *New York Tribune*, June 25, 1887; *New York Times*, July 13, July 17, July 21–24, Aug. 12–14, and Aug. 20–25, 1887; *New York Herald*, June 30, July 1, July 13, July 14, July 18, July 22, and Aug. 12, 1887; *New York Sun*, July 16, July 22, and July 23, 1887; *New York World*, July 13, Aug. 12, and Aug. 28, 1887; *Chronicle* 45:14, 78, 96, 106, 112, 136; Stedman, *Stock Exchange*, 333. Ives's later troubles may be followed in the *New York Times*, July 13–26, Sept. 19–26, and Oct. 11, 1888; *New York World*, June 13, 1890.

35. *New York Times*, July 24, Sept. 3, Sept. 4, Sept. 9, and Sept. 10, 1887; *New York World*, Aug. 6, 1889; *New York Sun*, Sept. 1 and Sept. 2, 1887; *Bradstreet's*, Oct. 15, 1887; *New York Herald*, Sept. 2, Sept. 4, Sept. 10, Sept. 18, Sept. 20, and Sept. 22, 1887; *Chronicle* 45:285, 293.

36. *New York Times*, Oct. 3–8, 1887; *New York Tribune*, Oct. 4–9, 1887; *New York*

Sun, Oct. 5–8, 1887; *New York Herald*, Oct. 7, 1887; *Chronicle* 45:334, 453, 465, 473. Western Union also agreed to assume the annual rent of $60,000 paid by B & O Telegraph to the parent company for the railroad's wires.

37. *New York Times*, Oct. 8–11, 1887; *New York Tribune*, Oct. 8–11, 1887; *New York Herald*, Oct. 9, 1887; *New York Sun*, Oct. 9–11, 1887.

38. *Chronicle* 45:509, 47:439–40; *Stockholder* 26:12; *New York Herald*, Oct. 12, Oct. 18, Oct. 19, and Oct. 28, 1887; *New York World*, July 23 and Aug. 9, 1887; *New York Sun*, Oct. 12, Oct. 13, and Oct. 28, 1887; *New York Tribune*, Oct. 28, 1887.

39. *Chronicle* 45:453; *Bradstreet's*, Oct. 15, 1887; *New York Herald*, Oct. 11, Oct. 12, Oct. 21, and Oct. 24, 1887, and Feb. 10, 1888. Grodinsky, *Jay Gould*, 476, called Western Union "the greatest success of his career." I agree with that appraisal.

40. *New York Times*, Oct. 11–13 and Oct. 24–25, 1887; *New York Herald*, Oct. 12, and Oct. 24, 1887; *New York Tribune*, Oct. 11–13, Oct. 16, Oct. 18, Oct. 23, Oct. 25, and Dec. 18, 1887.

41. *New York Tribune*, June 9, June 13, July 8, July 16, Aug. 6–8, Aug. 11, and Sept. 2, 1888; *New York Sun*, June 10, July 16, Aug. 6–11, and Aug. 27, 1888; *New York Times*, Sept. 4 and Oct. 29, 1887, Aug. 6, Aug. 7, and Aug. 12, 1888, July 30, 1896; RGD, Maryland, 12:459; *New York World*, June 10, Aug. 7, and Aug. 10, 1888, Feb. 3, 1889.

42. Gould to George Gould, July 11, 1886, KG.

43. Carter, *Cyrus Field* (see Chap. 20, n. 9), 1–338; "How Mr. Terry came to write his article on the Field & Manhattan as told by Mr. Sage," 4, HGS. This article, dated Jan. 27, 1893, bears the initials "J. P. M." It appears to have been taken down by Dr. John P. Munn. The conversation with Sage took place Dec. 22, 1892.

44. Carter, *Cyrus Field*, 340–42.

45. Ibid., 338; *Stockholder* 23:501, 776; *Chronicle* 41:556, 43:578; *Bradstreet's*, Nov. 14, 1885; *New York Tribune*, July 5, 1886. The largest increase in passengers came on the Second Avenue line, where the fare experiment had been implemented.

46. *New York Times*, Oct. 24, 1886; "How Mr. Terry," 1–2; *Chronicle* 44:173. For the stock prices see *Chronicle* 44:49.

47. Carter, *Cyrus Field*, 342; *New York Herald*, June 12, June 15, and June 16, 1887; *New York Times*, June 24, 1887; *New York World*, June 24, 1887; Stedman, *Stock Exchange*, 334; "How Mr. Terry," 2.

48. *New York Herald*, June 25 and June 26, 1887; *New York Times*, June 24 and June 25, 1887; *New York Tribune*, June 25, 1887; *New York World*, June 25, 1887; *Chronicle* 44:802–3.

49. *New York Times*, June 26–30, 1887; *New York Tribune*, June 26–30, 1887; *New York Herald*, June 26 and June 30, 1887; *New York World*, June 27, June 29, and June 30, 1887; *New York Sun*, June 29, 1887.

50. Ibid. The *Times* of June 30, 1887, favored the lamb-lion parable on the front page and the lamb-wolf version on the editorial page.

51. Standard versions of the Field episode are in Carter, *Cyrus Field*, 342–44; Josephson, *Robber Barons* (see Chap. 4, n. 15), 211; White, *Wizard* (see Prologue, n. 4), 168–69; Northrop, *Life and Achievements* (see Prologue, n. 2), 158–59; Warshow, *Jay Gould* (see Chap. 5, n. 1), 176–78; Myers, *American Fortunes* (see Chap. 5, n. 1), 489–90; Ogilvie, *Life and Death* (see Prologue, n. 4), 103–4; O'Connor, *Gould's Millions* (see Prologue, n. 1), 183–85. The most notable exception is Grodinsky, *Jay Gould*, 312–14, which offers the fullest and most balanced of the existing accounts.

52. *Stockholder* 24:687; *New York Tribune*, June 29, 1887. Two participants in the

Field episode left accounts of it. Sage's is found in "How Mr. Terry," Terry's in his letter to Helen Gould, Jan. 18, 1893, HGS. Both men were admittedly partial to Gould, but they did have personal knowledge of the facts and were outraged by the popular version of what happened. Carter's account is suspect in the extreme, marred by gaping holes, a vague chronology, some inaccurate details, and a limited understanding of both events and business history in general. These weaknesses are compounded by a complete lack of documentation, which he justifies with the most baffling rationale I have yet encountered: "I have deliberately omitted footnotes or numbered references . . . in the belief that a reader should not have to look elsewhere in a book for facts essential to the story." Ibid., 7.

53. "How Mr. Terry," 1–2; Terry to Helen Gould, Jan. 18, 1893, HGS. Western Union was then selling at about 75.

54. Terry to Helen Gould, Jan 18, 1893, HGS; "How Mr. Terry," 1; *New York Herald*, Sept. 1, 1887; *New York Times*, June 25 and July 1, 1887; *NCAB*, 12:227.

55. "How Mr. Terry," 2; Snow, *Helen Gould*, 210.

56. Terry to Helen Gould, Jan 18, 1893, HGS. Gould's explanation of the called loans is in the *New York Tribune*, June 29, 1887.

57. *New York Times*, July 3, 1887. The same columnist quotes T. W. Pearsall's opinion that Gould "could have forced Mr. Field to accept much harsher terms; but he paid a just market price, such as made it possible to sell again at a profit." Gould, of course, kept most of the stock.

58. "How Mr. Terry," 3–4; *New York Times*, July 1 and July 2, 1887; *New York World*, June 30 and July 1, 1887. Sage said he read the article in the *World*, but more likely he saw it in the *Times*. The *World*'s account of June 30 was the closest in many details to the version given here.

59. *New York Times*, July 8 and July 9, 1887; Snow, *Helen Gould*, 211. Field's statement is in the July 9 issue; see also the editorial.

60. Snow, *Helen Gould*, 211; Terry to Helen Gould, Jan. 18, 1893, HGS; "How Mr. Terry," 4–5. The column is in the *New York Times*, July 17, 1892. Grodinsky, *Jay Gould*, 314, 316, is the only writer I've found who uses this column as evidence. Unfortunately, he gives it the incorrect date of June 17, 1892.

61. Carter, *Cyrus Field*, 346–59; *New York Herald*, Dec. 15, 1888, Jan. 17, 1889, Dec. 1, Dec. 2, Dec. 16–20, Dec. 23, and Dec. 24, 1891; *New York Times*, Jan. 17, 1889. The quotation is in the *New York Evening Mail*, June 3, 1881. I have taken it from Trottman, *Union Pacific* (see Chap. 13, n. 1), 200.

62. Gould to Clark, Jan. 5, 1885, KG.

63. Ibid., Jan. 5, 1885, Sept. 6, 1886, KG; Gould to Daniel H. Lamont, July 25, 1887, Grover Cleveland Papers, Library of Congress; Cleveland to Gould, July 30, 1887, Cleveland Papers; Gould to Miller, Oct. 19, 1887, Cleveland Papers.

64. Gould to George C. Smith, Sept. 10, 1886, GLB; *Public* 22:163, 196–98, 294.

65. Gould's views on the pending interstate commerce bill are expressed in Gould to Clark, Dec. 23, 1886, GLB.

66. Z. L. White, "A Decade of American Journalism," *Westminster Review* 128 (October 1887):858. The new journalism and Pulitzer's role are summarized in Mott, *American Journalism* (see Chap. 12, n. 32), 430–39.

67. I am not certain of the precise date when the *Herald* first used this name. The earliest appearance I have seen is Sept. 19, 1887. Thereafter it was used regularly until Gould's death.

68. *New York Times*, Aug. 18, 1883. Financial journals joined the chorus of publica-

tions accusing the *Tribune* of being Gould's organ. See for example *Stockholder* 13:664–65. Dana's *Sun* called Reid Gould's "stool pigeon" and "hireling," and referred to the Tribune Building as "Gould's office." Duncan, *Whitelaw Reid* (see Chap. 12, n. 26), 48–49; H. W. Baehr, *The New York Tribune since the Civil War* (New York, 1936), 122.

69. *New York Times*, July 23, 1878, Feb. 27, 1879, Feb. 11 and Feb. 23, 1881; Perkins to Forbes, Feb. 18, 1889, BA. For a similar attack against the *World* see *New York Herald*, Nov. 15, Nov. 17, and Nov. 18, 1882.

70. Baehr, *New York Tribune*, 122; Duncan, *Whitelaw Reid*, 48–49; Mott, *American Journalism*, 423; J. S. Clarkson to Dodge, Aug. 10, 1887, DP.

71. *PRC*, 742–44; Gould to Clark, Feb. 7, 1878, KG. The *Mail and Express* was purchased by Elliott F. Shepard, a lawyer who was also William H. Vanderbilt's son-in-law. See Mott, *American Journalism*, 449.

72. Mott, *American Journalism*, 426–27; *Stockholder* 25:577. Keep's name is in Gould's address book, HGS, with the notation that he is a *Times* reporter.

73. Forbes to Perkins, Dec. 29, 1888, BA; *New York Herald*, Nov. 24, 1882. Earlier Forbes confided that "the Penn. RR folks told me they knew that the Western Union Board often had important railroad dispatches laid before them." Forbes to Griswold, July 4, 1879, BA. Gould did not then control Western Union.

74. *Postal Telegraph* (see Chap. 17, n. 4), 17, 179; *New York Herald*, Nov. 9, 1882; *Hearing before Committee on Railroads* (see Chap. 17, n. 5), 4–5; *New York Herald*, Nov. 9, 1882. For background detail see *Postal Telegraph*, 165–84, 287–316; *Hearing before Committee on Railroads*, 38–61; Reid, *Telegraph in America*, 793–803; Mott, *American Journalism*, 491–93. The contract between Western Union and Associated Press is reprinted in *Postal Telegraph*, 317–22.

75. *New York Times*, July 1, 1886; Snow, *Helen Gould*, 182–83. For a sample of Gould's press image see *New York Times*, Jan. 27, 1886. A selection of newspaper articles on the acquisition of Western Union by the government is in *Postal Telegraph*, 71–118.

76. Snow, *Helen Gould*, 184–85.

77. *Stockholder* 23:517.

78. Gould to John Main, Oct. 6, 1886, GLB; *New York Sun*, Dec. 20, 1886.

79. *New York Times*, June 30, 1887. The obituary does not make clear the cause of Talmage's death. Hayes married Hoxie's widow in 1889 and lived until 1905. See ibid., Mar. 3, 1905.

80. *New York World*, Feb. 9, 1889; *Who Was Who in America* (Chicago, 1942–50), 4:459. For Hopkins's family and father see the *New York Tribune*, June 18, 1887, and *DAB*, 9:215–17.

81. *New York World*, Feb. 9, 1889.

82. Ibid. Whitney was then secretary of the navy.

83. *New York Times*, Oct. 28, 1887; *New York Herald*, Oct. 26, 1887; *New York Sun*, Sept. 24, 1887; Snow, *Helen Gould*, 164.

84. *New York Times*, Sept. 25 and Oct. 28, 1887; *New York Herald*, Oct. 19 and Oct. 23, 1887; *New York Tribune*, Oct. 19, Oct. 20, Oct. 26, and Oct. 29, 1887.

85. The departure scene is taken from the *New York Times*, Oct. 30, 1887; *New York Sun*, Oct. 30, 1887; *New York Tribune*, Oct. 30, 1887; *New York Herald*, Oct. 30, 1887. A passenger list for the *Umbria* is in JGL.

86. For the yacht race see Gould to George Gould, July 11, 1886, KG; *New York Tribune*, July 16 and July 22, 1886.

87. *New York Times*, Nov. 6, 1887; *New York Sun*, Nov. 6, 1887; *New York Herald*,

Mar. 1, 1888; Helen D. Gould to Sarah Northrop, Dec. 28, 1887, JGL; Helen D. Gould to Anna Hough, Apr. 19, 1888, JGL; George Gould to Helen D. Gould, Dec. 15, 1887, KG; Snow, *Helen Gould*, 164; *New York Tribune*, Mar. 25, 1888.

Chapter 30: Deceptions

1. This timing is suggested by the fact that Gould was dreadfully ill during the summer of 1888. Snow, *Helen Gould* (see Chap. 1, n. 2), 168, reaches this same conclusion. A copy of Gould's death certificate is in HGS. Dr. Munn listed the chief cause, pulmonary consumption, as being of three years' duration, which would put the date at 1889. I do not think this figure need be taken precisely.

2. Gould to Charles Francis Adams, Jr., Aug. 19, 1886, GLB.

3. *New York Herald*, Jan. 1, Jan. 20, and Mar. 26, 1888; *Chronicle* 45:601, 665, 46:19–21, 94, 126, 345, 372–73; Kuhn, Loeb & Co. to Exec. Comm., Nov. 9, 1887, MP Exec. Comm. Minutes, Nov. 10, 1887, MP; MP Exec. Comm. Minutes, Nov. 1887–Mar. 1888, MP; *New York Times*, Jan. 22 and Mar. 20, 1888; *New York Tribune*, Mar. 21 and Mar. 22, 1888.

4. *MP Report* (see Chap. 27, n. 7), 1888, 1–13; *Stockholder*, Mar. 20, 1888; *Chronicle* 46:367–68, 372–73, 378–80; *New York Herald*, Mar. 21 and Mar. 22, 1888.

5. *Chronicle* 46:405; *New York Times*, Mar. 25, 1888; Perkins to Albert Fink, Feb. 9, 1888, BA; Perkins to Hughitt, Feb. 17, 1888, BA; Perkins to E. L. Godkin, Feb. 25, 1888, BA. A general reduction of dividends did occur. See *Chronicle* 47:5.

6. MP Exec. Comm. Minutes, Mar. 26, 1888, MP; *New York Herald*, Mar. 24 and Mar. 25, 1888; *New York Tribune*, Mar. 23–25, 1888; *New York Times*, Mar. 27, 1888; *Chronicle* 46:405.

7. *New York Times*, Mar. 3, May 17, May 20, June 13, June 14, and Dec. 11, 1885, July 9, Oct. 19, and Oct. 26–29, 1887; *New York Herald*, July 9, Oct. 19, and Oct. 30, 1887; *New York Tribune*, Oct. 19, Oct. 20, Oct. 26, and Oct. 29, 1887; *New York Sun*, Oct. 29, 1887; J. F. Dillon and Artemas Holmes to Adams, Oct. 17, 1887, UP; *Stockholder* 23:531.

8. *New York Herald*, Jan. 1, Jan. 2, Jan. 8, Jan. 10, Jan. 11, Jan. 18–22, and Jan. 31, 1888; *New York Tribune*, Dec. 31, 1887; *New York Sun*, Dec. 31, 1887.

9. *New York Herald*, Jan. 1, Feb. 14–17, and Mar. 1–8, 1888; *New York Tribune*, Jan. 14 and Feb. 29, 1888; *New York Sun*, Feb. 15, Feb. 29, Mar. 1, Mar. 5, and Mar. 6, 1888; *New York Times*, Mar. 1–3, 1888; *Stockholder* 26:6–7.

10. *New York Herald*, Feb. 22, Feb. 24, Mar. 6, Mar. 20–26, and Apr. 1, 1888; *New York Tribune*, Feb. 23, 1888.

11. *New York Tribune*, Mar. 25, 1888. The *Herald* reprinted this interview twice, on Mar. 26 and Mar. 30, 1888. The former included the heading, "Wasn't It Cruel about George, 'My Son George!' "

12. *New York Tribune*, Mar. 25 and Mar. 27, 1888; *New York Sun*, Mar. 26–28, 1888.

13. Ibid. For conflicting accounts of the divorce see the *New York Herald*, Mar. 1 and Mar. 29, 1888; *New York Sun*, Mar. 2, 1888; *New York World*, Feb. 9, 1889.

14. *New York Tribune*, Mar. 27–29, 1888; *New York Sun*, Mar. 27–29, 1888.

15. The only evidence I have found that bears on Bennett's position is a memoir written thirty-four years later by Julius Chambers, the managing editor of the *Herald*. He claimed that Gould did not write the famous letter to Bennett; that he alone was responsible for the *Herald*'s response to Gould's "brutal, malicious and fallacious attack"; that Bennett only approved his response afterward; and that it was the "most heroic act of my career." Little of what Chambers says coincides with the known facts, and his omissions

are glaring. For example, he omits all mention of the long enmity between Gould and Bennett and of the prolonged campaign in the *Herald* against Gould that preceded the letter. It is, in short, the selective recollection of an old man peering through the haze of an unreliable memory. Julius Chambers, *News Hunting on Three Continents* (New York, 1921), 306–7.

16. *New York Herald*, Mar. 27–31, 1888.

17. Ibid., Apr. 1, 1888; *New York Sun*, Apr. 1, 1888; *New York Tribune*, Apr. 1, 1888; *New York World*, Apr. 1, 1888. Chambers's assertion that Gould did not write this letter is without foundation. He wrote virtually all his own letters and would hardly allow someone else to compose so important and personal a document.

18. *New York Times*, Apr. 2, 1888; *New York Sun*, Apr. 1–3, 1888; *New York Herald*, Apr. 2, 1888.

19. *New York Herald*, Apr. 1–15 and Apr. 26, 1888.

20. Ibid., Apr. 3–7, Apr. 10, Apr. 12, and Apr. 13, 1888; *New York Times*, Mar. 28, Mar. 29, Apr. 4, and Apr. 12, 1888; *New York Tribune*, Apr. 12, 1888; *New York Sun*, Apr. 12, 1888; *Stockholder* 26:7; Adams, "Memorabilia 1888–1893," 310–11, CFA.

21. Gould and Sage to Adams, May 5, 1888, UP; *New York Herald*, Apr. 9, Apr. 11, and Apr. 24, 1888; *New York Times*, Apr. 20 and July 11, 1888; *New York World*, July 13, 1888; *New York Tribune*, July 20, July 21, and July 31, 1888; *Chronicle* 47:43, 133.

22. Gould to Clark, Apr. 12, 1888, KG.

23. *Chronicle* 46:87, 145, 472; Kuhn, Loeb to Gould, Mar. 30, 1888, in MP Directors' Minutes, Apr. 2, 1888, MP; *New York Herald*, Apr. 1, 1888.

24. *New York Times*, Jan. 2, 1888; Gould to Clark, Apr. 15 and May 19, 1888, KG; *Stockholder*, May 22, 1888. For the Wabash plan see *New York Times*, Sept. 4, 1887; *Chronicle* 46:621, 51:385.

25. Gould to John C. Brown, undated but probably Apr. 1888, TPHR; Wistar to C. E. Satterlee, Apr. 4, Apr. 11 (two items), Apr. 12, Apr. 13, Apr. 14 (two letters), Apr. 16, Apr. 17 (telegram and note appended to copy of agreement), and Apr. 25, 1888, TPHR; MP Exec. Comm. Minutes, Apr. 12, 1888, MP; Schiff to Wistar, Apr. 12, 1888, TPHR; S. A. Caldwell to Wistar, Apr. 13, 1888, TPHR; Schiff to Satterlee, Apr. 17, 1888, TPHR; Gould to L. S. Wolf, Apr. 24, 1888, TPHR; Kuhn, Loeb-Drexel, Morgan to Gould, Apr. 25, 1888, TPHR; *Chronicle* 46:134, 539, 621; Wistar to Satterlee, Apr. 4, 1888, TPHR; *Philadelphia Press*, Apr. 13, 1888; MP Directors' Minutes, May 3, 1888, MP; Gould to Clark, Feb. 13, 1889, KG.

26. *Chronicle* 46:148; *Railroad Gazette* 19:478. For background detail see Chap. 27 and Grodinsky, *Jay Gould* (Prologue, n. 5), 534–36.

27. *Chronicle* 40:120, 44:211, 267, 276, 45:565, 613; *Stockholder*, Oct. 29, 1887; Dodge to Clark, Dec. 9, 1887, DR, 11:1015. Masterson, *Katy Railroad* (see Chap. 21, n. 2), 222–25, recounts the Gould years under the title "The Rape of the Katy." His account of the violation is more suggestive than accurate, a titillation of tall tales and hearsay that fails to catch Gould in flagrante delicto.

28. Dodge to Clark, Dec. 9, 1887, DR, 11:1015. The Panhandle Road was the Dodge-Evans line between Pueblo and Fort Worth, the combined Denver, Texas & Fort Worth and Fort Worth & Denver City roads.

29. *Chronicle* 45:820, 46:149, 371; MP Exec. Comm. Minutes, Feb. 2, 1888, MP; *New York Tribune*, Mar. 27, 1888; *New York Herald*, Jan. 20, Mar. 27, Apr. 15, and Apr. 19, 1888; *New York Sun*, Mar. 27, 1888; *New York Times*, Apr. 8, 1888.

30. Grodinsky, *Jay Gould*, 538, 541. Gould was obliged to deny reports of impending receivership for the International as early as May 1887. See *Stockholder*, May 3, 1887.

31. Gould to Board of Directors, Apr. 18, 1888, MP Directors' Minutes, Apr. 19, 1888, MP; *New York Sun*, Apr. 20, 1888; *New York Times*, Apr. 20, 1888; *New York Tribune*, Apr. 20, 1888; *Chronicle* 46:511.

32. *New York Herald*, Apr. 21, Apr. 22, and Apr. 25, 1888; *New York Times*, Apr. 21, Apr. 25, and May 18, 1888; *New York Sun*, May 4, 1888; *New York Tribune*, Apr. 21, Apr. 25, May 10, May 11, and May 18, 1888; *Chronicle* 46:538, 601, 609, 650; *Stockholder*, Apr. 24, May 1, and Aug. 21, 1888.

33. MP Directors' Minutes, Apr. 19 and May 3, 1888, MP; MP Exec. Comm. Minutes, Apr. 26 and May 31, 1888, MP; IGN Exec. Comm. Minutes, May 2, 1888, MP; *Chronicle* 46:564, 573; *New York Times*, May 5 and Sept. 25, 1888; *New York Tribune*, May 8, 1888.

34. Gould to Clark, Apr. 25, 1888, MP Exec. Comm. Minutes, Apr. 26, 1888, MP; *New York Times*, May 5, May 6, June 10, and June 11, 1888; *New York Tribune*, May 30, June 1, and June 16, 1888; *Chronicle* 46:739; *Stockholder*, Aug. 21, 1888.

35. Gould to George Gould, Sept. 10, 1888, KG.

36. *New York Times*, June 2, June 3, and June 5, 1888; *New York World*, June 3–5, 1888; *Chronicle* 46:700. The illness rumors began even before the trip. See *New York Tribune* and *New York Sun*, May 27, 1888.

37. *New York Times*, June 10, 1888; *New York Tribune*, June 3 and June 4, 1888; *New York World*, June 3–5, 1888.

38. *New York Times*, June 3, June 11, and June 23, 1888; *New York World*, June 12, 1888; *Chronicle* 46:771. The *Times* noted that "Mr. Gould has his doctor with him. He does not travel nowadays without such an attendant."

39. *New York Tribune*, June 12, 1888; *New York Times*, June 12–14 and June 16, 1888; *New York World*, June 13, 1888; MP Exec. Comm. Minutes, June 14, 1888, MP; *Chronicle* 46:759.

40. *Stockholder*, July 9, 1888; *New York Times*, July 13–15, 1888; *New York World*, July 14, 1888; *New York Tribune*, July 10 and July 24, 1888.

41. *New York World*, July 17, 1888. The interview appeared first in the *Philadelphia Times*.

42. *New York Times*, July 28 and Aug. 20, 1888; Snow, *Helen Gould*, 166–68. Gould's sojourn at Saratoga is best followed in the *Times* and *World* between July 29 and Aug. 28, 1888. The *Tribune* gave little coverage, the *Herald* and *Sun* virtually none.

43. *New York Times*, July 29 and Aug. 1, 1888; *New York World*, Aug. 16, 1888; *New York Tribune*, Aug. 5 and Aug. 19, 1888; Snow, *Helen Gould*, 167–68.

44. *New York Times*, July 29 and Aug. 1–5, 1888; *New York World*, Aug. 19, 1888; MP Exec. Comm. Minutes, Aug. 7, 1888, MP.

45. *New York Times*, July 28 and Aug. 18, 1888; *New York World*, Aug. 16 and Aug. 27, 1888; *New York Tribune*, Aug. 5, 1888; *Stockholder*, Aug. 28, 1888.

46. *New York Times*, Aug. 5 and Aug. 13, 1888; *New York Herald*, Aug. 5, 1888. The events mentioned are taken from the *Times*, *Tribune*, and *World* for August 1888.

47. Snow, *Helen Gould*, 166, 168–69; Helen Gould to her mother, July 28, 1888, JGL.

48. *New York Times*, Aug. 29, Aug. 31, and Sept. 1, 1888; *New York World*, Aug. 31, 1888; Helen Gould to her mother, Aug. 31, 1888, HGS. Furlow Lodge remains in the Gould family and is presently owned by Kingdon Gould, Jr., who is George's grandson. It was at this lodge that the letters used in this biography were discovered. Mr. Gould also has papers relating to the construction of the lodge.

49. Helen Gould to her mother, Sept. 3[2] and Sept. 3, 1888, HGS; *Angell* (see Chap. 1, n. 13), 967–68; *New York Times*, Sept. 2, 1888; Snow, *Helen Gould*, 165–66.

50. Snow, *Helen Gould*, 165–66; *Angell*, 968.

51. Helen Gould to her mother, Aug. 31, Sept. 3[2], and Sept. 3, 1888, HGS; *New York Times*, Sept. 3–5, 1888; *New York World*, Sept. 5, 1888.

52. *New York World*, Sept. 5, 1888. For detail on the rate wars see Chap. 31.

53. Gould to Clark, Dec. 30, 1888, KG; *Chronicle* 48:22, 365–66; Gould to Clark, Sept. 30, Oct. 14, Oct. 24, Oct. 29, Dec. 28, and Dec. 30, 1888, KG.

54. *New York Tribune*, Oct. 31, 1888; Gould to Clark, Feb. 13, 1889, KG; *Chronicle* 48:324–25, 327; Dodge to Gould, Feb. 12, 1889, DR, 12:495–96.

55. *New York Tribune*, Nov. 29, 1887, Feb. 5, 1889; *Chronicle* 45:573, 46:621, 47:473, 48:190, 325–26. Grodinsky called this decision "the most sweeping victory which any group of bondholders had ever secured against Gould." Grodinsky, *Jay Gould*, 438.

56. MP Directors' Minutes, June 28, 1888, MP; *Stockholder*, Aug. 21, 1888; *Chronicle* 46:828, 47:140, 257–60; *New York Times*, Sept. 17, 1888; *New York Tribune*, Oct. 20, 1888.

57. For Gould's equipment purchases and bond negotiations on the Katy's behalf see Gould to George C. Smith, Sept. 22 and Oct. 5, 1886, GLB; Gould to Kerrigan, Oct. 30, 1886, GLB; Gould to James H. Smith, July 2, 1886, GLB.

58. *New York Tribune*, June 10, June 22, Sept. 27, Sept. 28, and Oct. 7, 1888; *New York World*, Sept. 17, 1888; *New York Times*, Sept. 25–30 and Oct. 7, 1888; *Chronicle* 47:490.

59. *Chronicle* 45:540, 46:574, 47:402, 403; *New York World*, Sept. 29, 1888; *New York Tribune*, Sept. 29, 1888; *New York Times*, Sept. 29 and Oct. 3, 1888; Gould to Clark, Oct. 29, 1888, KG.

60. *New York Times*, Sept. 29, 1888; *Chronicle* 47:515; Gould to Clark, Oct. 24 and Dec. 21, 1888, KG.

61. *New York Times*, Nov. 28, 1888; *Chronicle* 47:664, 708; IGN Exec. Comm. Minutes, Oct. 17 and Oct. 31, 1888, MP; Gould to Clark, Oct. 14, 1888, KG; Grodinsky, *Jay Gould*, 541.

62. Gould to Clark, Jan. 9, 1889, KG.

63. *New York Tribune*, Nov. 9–12, 1888.

Chapter 31: Losses

1. *New York World*, Jan. 14, Jan. 15, and Jan. 17, 1889, Dec. 3, 1892; *New York Sun*, Jan. 14 and Jan. 15, 1889; *New York Times*, Jan. 14, Jan. 15, and Jan. 17, 1889; *New York Herald*, Jan. 14 and Jan. 17, 1889; Snow, *Helen Gould* (see Chap. 1, n. 2), 169.

2. Gould to George Gould, May 1, 1892, KG; Snow, *Helen Gould*, 169–70.

3. *New York Times*, Feb. 8, 1889; *New York World*, Feb. 9, 1889. An obituary of Hopkins's father is in the *New York Tribune*, June 18, 1887.

4. *New York Times*, Feb. 8 and Feb. 9, 1889; *New York World*, Feb. 9–11, 1889. The Sloan story is in the *New York Times*, Dec. 6, 1888.

5. Gould to Clark, undated but clearly December 1888, KG; MP Exec. Comm. Minutes, Jan. 9, 1889, MP. The minutes contain copies of two telegrams from Clark on implementing the reductions. Hopkins was on the executive committee of the Missouri Pacific in 1890; see MP Exec. Comm. Minutes, Feb. 7, 1890, MP. He later served as president of the New York, Susquehanna & Western and as receiver for the Northwestern.

6. Gould to Clark, Jan. 3, 1889, KG; Clark to Dodge, Jan. 27, 1889, DP.

7. *Chronicle* 48:414. The Missouri Pacific's first mortgage bonds, issued at 6 percent, were refunded at 4 percent. The *Chronicle* estimated that if all other Missouri Pacific and Iron Mountain issues could be refunded at the same rate, the savings in interest would amount to $747,680 per year.

8. Gould to Clark, Dec. 31, 1888, Jan. 5 and Feb. 23, 1889, and undated but clearly December 1888, KG.

9. Ibid., Dec. 31, 1888, Jan. 3, Jan. 4, Jan. 5, Jan. 6, Jan. 7, Jan. 31, Feb. 23, and June 16, 1889, KG; *Chronicle* 48:355, 369; MP Exec. Comm. Minutes, May 8, 1889, MP. A mixed train consists of freight and passenger cars put together in the same train.

10. Gould to Clark, Dec. 31, 1888, Jan. 3, Feb. 13, Feb. 23, Mar. 3, May 28, June 16, June 27, and Sept. 28, 1889, KG; MP Exec. Comm. Minutes, May 2, 1889, MP; C. F. Meek to Dodge, May 14, 1889, DP.

11. Gould to Clark, Feb. 13, Sept. 28, Nov. 1, Nov. 7, and Dec. 31, 1889, and undated, KG; *Chronicle* 49:262. The undated note discusses the traffic and freight department problems. The internal evidence is insufficient to assign it a date with confidence. There is mention of a recent trip which, combined with other factors, leads me to believe it was written in November or December of 1889.

12. Gould to Clark, Dec. 12, 1889, GLB; Gould to George Smith, Dec. 9 and Dec. 24, 1889, GLB; Gould to Clark, Dec. 30 and Dec. 31, 1889, KG.

13. Gould to Frederic P. Olcott, Dec. 26, 1889, GLB. The letter refers to the Cotton Belt reorganization.

14. *New York Times*, May 16, 1889; *Chronicle* 48:663, 49:236, 51:385–86; Memorandum, Nov. 2, 1889, GLB. Some figures on Gould loans to the Wabash, dated Jan. 13 and Jan. 15, 1890, are in GLB.

15. IGN Exec. Comm. Minutes, Feb. 5, 1889, MP; Gould to Clark, Jan. 28 and Feb. 13, 1889, KG; *Chronicle* 48:251; *New York Times*, Feb. 5, Feb. 10, Feb. 15, and Feb. 18, 1889; *New York World*, Feb. 15, 1889.

16. Gould to Clark, Mar. 3, 1889, KG; *Chronicle* 48:453, 462; Reed, *Texas Railroads* (see Chap. 28, n. 31), 383; Masterson, *Katy Railroad* (see Chap. 21, n. 2), 239; Grodinsky, *Jay Gould* (see Prologue, n. 5), 543. A copy of the bill is in TPHR.

17. *Chronicle* 49:114, 206, 544–46, 702–3, 50:107, 275, 313, 423; *Bradstreet's*, Aug. 17, 1889; *New York Times*, Oct. 24 and Oct. 27, 1889, Jan. 18, Jan. 19, and Jan. 22, 1890; *New York Tribune*, Dec. 20, 1889; *Wall Street Journal* (hereafter *WSJ*), Aug. 9, Aug. 19, Oct. 24, and Dec. 5, 1889; *New York Sun*, Feb. 1 and Nov. 28, 1889; *Stockholder*, Dec. 17, 1889; *New York Times*, Jan. 18, Jan. 19, and Feb. 22, 1890; Memorandum, Dec. 9, 1889, GLB; Gould to Clark, Feb. 21, 1889, GLB; Gould to H. K. Enos, Feb. 24, 1889, GLB. The account of this episode in Grodinsky, *Jay Gould*, 543–47, is flawed by errors of chronology and by his emphasis on the Katy fight as a clash between Gould and Rockefeller.

18. *Chronicle* 48:601–2, 610, 634. Gould's hold on the Cotton Belt rested on a provision of the 1886 reorganization plan vesting control of the road in the hands of five trustees for a period of five years. Gould bought the securities mentioned on condition that he be allowed to name three of the five trustees.

19. Ibid., 48:601–2, 663, 49:301, 781; *Railway Review*, Dec. 7, 1889; Gould to James Speyer, Oct. 30, 1889, GLB; Gould to George S. Ellis, Nov. 8, 1889, GLB; Gould to Clark, Nov. 12, 1889, GLB; Gould to Olcott, Nov. 12, Nov. 13, and Dec. 26, 1889, GLB. The committee proposed taking the surplus securities at 37½; Gould wanted them left in the company's treasury until its credit had improved and they would fetch a higher price.

20. *Chronicle* 48:663, 50:139–40; *Stockholder*, Jan. 28, 1890; Gould to Olcott, Jan. 20, 1890, GLB.

21. Gould to Ellis, Nov. 8, 1889, June 28, 1890, GLB; Gould to Louis Fitzgerald, Feb. 5 and Feb. 15, 1890, GLB; Gould to W. B. Doddridge, Aug. 22, 1890, GLB; Gould to Fordyce, Oct. 10 and Oct. 13, 1890, GLB; Gould to Olcott, Dec. 4 and Dec. 6, 1890, GLB; *Chronicle* 50:498, 51:21, 114, 207, 570–71, 830, 52:121, 322, 643, 796; Poor, *Manual 1891*, 493–96. Gould's

continued hold on the Cotton Belt seems to have escaped the notice of contemporary observers and historians alike. Grodinsky, *Jay Gould*, 551, states that Gould was "unsuccessful in his efforts to dominate the reorganization" and that he "did not acquire control" of the new company.

22. See the map in Poor, *Manual 1891*, 785. The Panhandle's general manager declared that the Texas & Pacific "can get nothing for Colorado via their Missouri Pacific connections; they cannot make the time by their route." C. F. Meek to Dodge, May 14, 1889, DP.

23. Gould to his son [George], undated, KG; Gould to Clark, Jan. 5, Feb. 18, Feb. 27, and Sept. 21, 1889, KG, Sept. 28, Oct. 3, and Nov. 6, 1889, GLB; Gould to George C. Smith, Sept. 9 and Nov. 25, 1889, GLB; *Chronicle* 49:690; *New York Sun*, Feb. 22, 1890; Burton, "Missouri Pacific" (see Chap. 22, n. 2), 715–17.

24. Gould to R. M. McDowell, Sept. 6, 1889, GLB; Gould to George C. Smith, Dec. 9, Dec. 17, and Dec. 23, 1889, GLB; Gould to Clark, Feb. 18, June 16, and Sept. 21, 1889, KG, Sept. 5, Sept. 9 (two letters), Sept. 12, Sept. 14 (two letters), Sept. 20, Nov. 27, Nov. 29, Nov. 30 (two letters), Dec. 3, and Dec. 6, 1889, GLB; Gould to John A. Grant, Sept. 5 and Dec. 24, 1889, GLB; Gould to W. A. Bright, Nov. 11, 1889, GLB; Gould to E. B. Wheelock, Dec. 11, 1889, GLB; *Chronicle* 49:690; Burton, "Missouri Pacific," 719–22; Poor, *Manual 1891*, 788; Gould to Milton H. Smith, Dec. 9 and Dec. 18, 1889, GLB.

25. Gould to Clark, Sept. 24, Nov. 8, Nov. 18, Nov. 19, and Nov. 29, 1889, GLB, Nov. 15, 1889, KG; Clark to Gould, Sept. 12, 1889, MP; *New York Herald*, Nov. 25 and Dec. 14, 1888, Feb. 11–13 and Feb. 21, 1889; Klein, *Great Richmond Terminal* (see Chap. 9, n. 18), 219; Gould to Adams, Nov. 30, 1889, GLB.

26. Gould to Clark, Dec. 6 and Dec. 24, 1889, GLB; Adams to Gould, Dec. 14 and Dec. 17, 1889, UP; Gould to Adams, Dec. 17 and Dec. 24, 1889, GLB; Gould to George C. Smith, Dec. 23, 1889, GLB. For background on the bridge monopoly see *New York World*, Aug. 1, 1888, *New York Times*, Mar. 11, 1889, and Chap. 28.

27. Gould to William Taussig, Sept. 25, 1889, GLB; Gould to Clark, Nov. 20, Dec. 5, and Dec. 8, 1889, GLB, Dec. 6, 1889, KG.

28. Gould to Clark, Dec. 6, 1889, KG, Dec. 8, 1889, GLB; Gould to Roberts, Sept. 16, Oct. 3, and Nov. 1, 1889, GLB; Gould to J. P. Morgan, Sept. 25, 1889, GLB; Gould to William Taussig, Sept. 24, Sept. 25, Nov. 6, Nov. 13, Nov. 14, and Nov. 29, 1889, GLB; *Chronicle* 49:435.

29. Snow, *Helen Gould*, 170; Gould to Clark, Nov. 18 (two letters) and Nov. 29, 1889, GLB; Gould to George R. Kunz, Nov. 21, 1889, GLB; Gould to Mr. Sayng, Dec. 16, 1889, GLB; Gould to Mrs. Adele Eibeling, Sept. 18, 1889, GLB; Gould to Reid Northrop, Nov. 21, Nov. 25, and Dec. 23, 1889, GLB.

30. Gould to Clark, Oct. 1, 1889, GLB; Gould to Church Howe, Dec. 2, 1889, GLB; Gould to Kuhn, Loeb & Co., Dec. 18, 1889, GLB; Gould to W. S. Herndon, Dec. 21, 1889, GLB; Gould to Robert Fleming, Dec. 23, 1889, GLB; *New York Times*, Nov. 4, 1889; *Chronicle* 48:453, 49:262, 480, 744, 804; *Stockholder*, Nov. 12, Dec. 2, and Dec. 24, 1889; Gould to Marquand, Dec. 21, 1889, GLB.

31. Gould to Clark, Jan. 30, 1890, GLB.

32. Anyone who deals with the railroad history of this period knows the extent to which rail officers utilized the images, metaphors, and language of war and diplomacy in their correspondence. For a recent (and long overdue) study on this point see James A. Ward, "Image and Reality: The Railway Corporate-State Metaphor," *Business History Review* 55 (Winter 1981): 491–516.

33. Perkins to Geddes, May 25, 1886, BA.

34. Geddes to Perkins, Oct. 26, 1887, BA; Grodinsky, *Transcontinental* (see Chap. 13, n. 13), 312–35. For the Transcontinental Pool see *Stockholder* 23:484, 24:325, 372, 25:451; *Chronicle* 42:243; Gould to J. W. Midgley, Sept. 22, 1886, GLB. Details on the successors to the Iowa Pool are in Perkins to W. J. Ladd, May 11, 1885, BA; Potter to Perkins, July 16, 1886, BA; *Chronicle* 40:465, 43:34; *New York Tribune*, Sept. 11 and Sept. 16, 1886. On the Southwestern Pool see *Bradstreet's*, Aug. 22, 1885; *Railroad Gazette* 18:83; *Chronicle* 43:321; Gould to W. H. Newman, Nov. 16 and Nov. 17, 1886, GLB. The Colorado-Utah Pool is mentioned in the *New York Tribune*, July 10, 1886, and G. W. Holdrege to Henry B. Stone, Nov. 28, 1888, BA. Obviously the list is not all-inclusive.

35. *Chronicle* 46:145, 47:368–70, 514–16, 48:49; *New York Tribune*, Feb. 7 and Nov. 11, 1888; *New York Herald*, Nov. 14 and Nov. 25, 1888; *New York Sun*, Dec. 4, 1888.

36. *Stockholder*, Oct. 22, 1888; Adams to Perkins, Nov. 9, 1885, BA; *Chronicle* 47:368.

37. Perkins to Fink, Feb. 9, 1888, BA; Geddes to Perkins, Feb. 25, 1888, BA; T. J. Coolidge to Perkins, Mar. 12, 1888, BA.

38. *Chronicle* 47:515.

39. *New York Herald*, Nov. 24 and Nov. 25, 1888. The clearing house plan is in UP and reprinted in the *New York Times*, Nov. 28, 1888, and *Railroad Gazette* 20:796. There has long been confusion over its authorship. Winslow enclosed his version in a letter to Perkins, Nov. 14, 1888, BA. Grodinsky, *Jay Gould*, 560–61, and *Transcontinental*, 342–43, accepts the contemporary view that Gould originated the clearing house plan, then "saw the light and tried to evade responsibility for its failure." For decisive evidence of the version given here see Winslow to Adams, Nov. 23, 1888, UP; J. W. Midgley to Adams, Dec. 6, 1888, UP; Henry B. Stone to Perkins, Nov. 27 and Nov. 30, 1887, BA. The first distinguishes between the clearing house plan and "more complete plans, viz., an Operating Company (Gould's plan), or an Owning Company (Huntington's plan)." Midgley resolves the question by noting that "Mr. Gould had another scheme which we called 'The Operating Plan.' The Clearing House was an alternative one which I outlined as being a milder method, and one we could put into immediate effect."

40. *Railroad Gazette* 20:796; Perkins to Forbes, Jan. 19, 1888, BA; *New York Herald*, Dec. 16, 1888.

41. Grodinsky, *Jay Gould*, 561; Winslow to Adams, Nov. 23, 1888, UP; Stone to Perkins, Nov. 27 and Nov. 30, 1888, BA; *New York Herald*, Dec. 8, 1888.

42. Perkins to Forbes, Nov. 30, 1888, BA; Stone to Perkins, Nov. 30, 1888, BA; Winslow to Perkins, Dec. 4, 1888, BA; *New York Herald*, Nov. 28 and Nov. 29, 1888; *New York Times*, Dec. 4, 1888. Strong offered a plan of his own in a letter to Perkins, Dec. 8, 1888, BA. It was more informal and rested on pledges of good faith. See also *New York Tribune*, Dec. 15, 1888.

43. Stone to Perkins, Dec. 18, 1888, BA; *Chicago Tribune*, Dec. 15 and Dec. 16, 1888; *New York Herald*, Dec. 16, 1888; *New York Tribune*, Dec. 16 and Dec. 17, 1888. The western association was a loose amalgam of three regional associations, the Northwestern, Western, and Southwestern.

44. Winslow to Perkins, Dec. 4, 1888, BA; Perkins to Forbes, Nov. 30, 1888, BA; Adams, "Memorabilia 1888–1893," Dec. 16 and Dec. 23, 1888, 2–4, CFA; *New York Herald*, Dec. 16 and Dec. 17, 1888; *New York Tribune*, Dec. 19, 1888; *Railroad Gazette* 20:841.

45. MP Exec. Comm. Minutes, Dec. 5, 1888, MP; *New York Herald*, Dec. 2, Dec. 6, Dec. 17, and Dec. 21, 1888; *New York Tribune*, Dec. 6, Dec. 9, and Dec. 13, 1888.

46. Adams, "Memorabilia 1888–1893," Dec. 23, 1888, 4–5, CFA; *New York Tribune*, Dec. 22, 1888. To my knowledge, Adams left the only firsthand account of these meetings.

47. Adams, "Memorabilia 1888–1893," Dec. 23, 1888, 5–6, CFA. Grodinsky, *Jay Gould*, 561–62, describes these meetings much as he did the 1885 trunk line crisis, using the premise that "Morgan ignored Gould" and Gould resented being overshadowed. Apart from completely ignoring Gould's physical and emotional state at the time, Grodinsky offers no evidence of any tension between Gould and Morgan, who in fact had often worked closely together.

48. Adams, "Memorabilia 1888–1893," Dec. 23, 1888, 6–8, CFA; *New York Herald*, Dec. 22, 1888; *New York Tribune*, Dec. 22, 1888; *Chronicle* 47:768.

49. *New York World*, Jan. 5 and Jan. 7, 1889; *Chronicle* 48:2; *New York Herald*, Jan. 8, 1889; Adams, "Memorabilia 1888–1893," Jan. 13, 1889, 9, CFA.

50. Adams, "Memorabilia 1888–1893," Jan. 13, 1888, 9–15, CFA.

51. *New York Herald*, Jan. 9, 1889; *Chronicle* 48:60, 67.

52. Adams, "Memorabilia 1888–1893," Jan. 13, 1889, 15–23, CFA; *New York Herald*, Jan. 11, 1889; *Chronicle* 48:49–50, 60, 67–68, 80–81.

53. *New York Sun*, Jan. 4 and Jan. 16, 1889; *New York Herald*, Jan. 13, Jan. 22, Jan. 25, Jan. 29–31, Feb. 11–13, Feb. 21, and Feb. 23, 1889; *New York Times*, Feb. 17 and Feb. 21, 1889; *Chronicle* 48:210; Perkins to Forbes, Feb. 18, 1889, BA.

54. Adams, "Memorabilia 1888–1893," Feb. 24, 1889, 43–50, CFA; *New York Herald*, Feb. 22 and Feb. 23, 1889; *New York Times*, Feb. 25, 1889; *Chronicle* 48:251. Technically, neither Union Pacific nor Missouri Pacific signed. Hughitt had Adams's proxy to vote for the plan but could not sign for him. W. H. Newman voted for the plan but refused to sign for Gould.

55. *Bradstreet's*, Mar. 2, 1889; *Philadelphia Press*, Mar. 14 and Sept. 14, 1889; *Boston Herald*, July 20, 1889; *Chronicle* 48:828, 49:44, 561; *WSJ*, July 24 and Oct. 8, 1889; Adams to Dodge, Oct. 23, 1889, DP. For more detail on these episodes see Grodinsky, *Jay Gould*, 563–65, and *Transcontinental*, 345–49.

56. *Chronicle* 50:107, 174; Gould to Clark, Jan. 30, 1890, GLB.

57. *Chronicle* 50:174; Gould to Clark, Feb. 6, 1890, GLB.

Chapter 32: Last Hurrah

1. *New York Times*, Mar. 9, Apr. 28, and Apr. 30, 1890; *Chronicle* 50:220; *Railroad Gazette* 21:120, 136; *New York World*, May 4, 1890; Gould to Clark, May 13 and May 15 (two letters), 1890, GLB; Gould to George C. Smith, May 17, 1890, GLB. These letters refute the assertion in Grodinsky, *Jay Gould* (see Prologue, n. 5), 566, that "there is no gainsaying the fact that Gould's influence was exerted against the side of stability."

2. Gould to Walker, May 6, 1890, GLB; *New York Times*, May 23, 1890; *Chronicle* 50:783–86; Johnson and Supple, *Boston Capitalists* (see Chap. 13, n. 1), 323–28; Bryant, *Santa Fe* (see Chap. 21, n. 5), 150–52.

3. *New York Times*, May 28, 1890; Gould to Clark, May 24, 1890, GLB; *New York World*, May 29, 1890; *Stockholder*, June 3, 1890.

4. Gould to Walker, Aug. 7 and Aug. 19, 1890, GLB; *New York Times*, Aug. 7, 1890; *New York Tribune*, Sept. 19 and Sept. 30, 1890; *New York World*, Sept. 21, Sept. 28, and Oct. 5, 1890; *Bradstreet's*, Oct. 18, 1890.

5. Gould to John A. Grant, July 25, 1890, GLB.

6. Gould to Kerrigan, Nov. 18, 1886, GLB. The *Atalanta* is on display in Jefferson, Texas, where it has been lovingly restored and redecorated by the Jessie Allen Wise Garden Club. The interior decor has been changed somewhat by later users, notably George Gould, who had Edith's monogram carved on one of the beds. I am grateful to Mrs. G. W. Carpenter of the Excelsior House in Jefferson for her help during my stay. Visitors to the car should be advised, however, that much of the information given about it and Gould is inaccurate. The restoration of the car triggered a fresh round of blather about Gould in the papers. See for example *Dallas Morning News*, July 11, 1954.

7. *Stockholder*, Feb. 18, 1890; Gould to Clark, Feb. 21, 1890, GLB; George Gould to Edith Gould, Mar. 5, 1890, KG. The series of letters from George to Edith provides considerable detail on the trip.

8. George Gould to Edith Gould, Mar. 22, 1890, KG; Gould to George Gould, undated, KG; John C. Brown to Satterlee, Aug. 12, 1889, TPHR; Edward Taylor to Gould, May 23, 1890, TPHR; Gould to Grant, May 3, 1890, GLB; Riegel, *Western Railroads* (see Chap. 13, n. 1), 291.

9. Gould to Grant, Jan. 29 and May 12, 1890, GLB; *Chronicle* 50:590, 771, 783–86, 801, 812; *New York Times*, Apr. 23 and June 7, 1890; *Stockholder*, May 6, 1890; Gould to Olcott, Dec. 4, 1890, GLB.

10. Gould to Clark, Feb. 13 and Feb. 26, 1890, GLB, June 28, 1890, KG; Gould to J. A. Roosevelt, Feb. 12, Sept. 17 (two letters), and Sept. 29, 1890, GLB; Gould to Hallgarten & Co., May 28, 1890, GLB; Gould to Enos, Feb. 24 and Sept. 29, 1890, GLB; Gould to Cochran & Pierce, Sept. 30, 1890, GLB.

11. Gould to Grant, May 12, 1890, GLB; Gould to Clark, June 28, 1890, KG; Adams to Gould, Aug. 21 and Aug. 30, 1890, UP and TPHR; Gould to Adams, Aug. 22, 1890, GLB and UP; Gould to Adams, Sept. 3, 1890, UP; Gould to Huntington, Sept. 6, 1890, GLB; *Chronicle* 51:457, 52:957–58.

12. Gould to Clark, July 9, 1890, GLB.

13. MP Directors' Minutes, Jan. 24 and May 8, 1890, MP; MP Exec. Comm. Minutes, Feb. 7 and June 18, 1890, MP; *Chronicle* 50:519, 52:26; *Bradstreet's*, May 3, 1890; *Stockholder*, May 6, 1890; *New York Times*, May 10, 1890; *New York World*, May 25, 1890; *WSJ*, Dec. 23 and Dec. 24, 1889, Jan. 4, Feb. 13, Feb. 15, Feb. 18, Feb. 19, Apr. 5, and May 10, 1890; Gould to Mr. Parmley, June 9 and June 10, 1890, GLB; Gould and Sage to A. H. Calef, June 28, 1890, GLB.

14. Gould to Enos, Feb. 24, 1890, GLB; Gould to W. B. Doddridge, July 14, 1890, GLB; Gould to Newman Erb, July 15, 1890, GLB; Gould to George W. Ely, July 25, 1890, GLB; *Chronicle* 50:904.

15. MP Exec. Comm. Minutes, June 18, 1890, MP; *New York Times*, July 16, 1890; Gould to Clark, July 9, 1890, GLB, July 15, 1890, KG and GLB; Gould to Magoun, July 15, 1890, GLB; *New York World*, Sept. 18, 1890.

16. Gould to McDowell, Jan. 15, 1890, GLB; Gould to Clark, July 4, 1890, KG. For examples see Gould to Clark, June 27, July 7, and Nov. 18, 1890, GLB; Gould to McDowell, July 22, 1890, GLB; Gould to Doddridge, Oct. 7, 1890, GLB.

17. Gould to Clark, Jan. 15, 1890, KG, May 6, May 7, May 8, May 13, July 7, Sept. 25, Sept. 30, and Nov. 10, 1890, GLB; Gould to Grant, May 9, 1890, GLB; Gould to Northrop, Oct. 1, 1890, GLB; Gould to Clark, June 22, 1890, KG and GLB.

18. Fritz Redlich, *Steeped in Two Cultures* (New York, 1971), 33–64. The term "creative destruction" was coined by Joseph Schumpeter to characterize the process by which change occurs in the capitalist economy. See Joseph Schumpeter, *Capitalism,*

Socialism, and Democracy (New York, 1937). Obviously I disagree with the conclusion in Grodinsky, *Jay Gould*, 567, that Gould "found it difficult to operate in the spirit of harmony, in which mutual concessions were imperative."

19. Gould to Clark, July 15 and Sept. 28, 1890, KG. For Gould's pine and sawmill activities see Gould to E. B. Wheelock, Feb. 18, 1890, GLB; Gould to Clark, May 12, May 13, and July 2, 1890, GLB; Gould to R. W. Bringhurst, May 12, 1890, GLB; Gould to G. D. Woodbridge, June 2, 1890, GLB; Gould to C. W. Goodlander, Sept. 3, 1890, GLB.

20. Gould to Clark, Jan. 29, Jan. 31, Feb. 5 (two letters), Feb. 17, Feb. 18, Feb. 19, May 7, May 14, May 17, May 20, June 16 (two telegrams), June 21, June 23, June 24, and June 28, GLB; Gould to Grant, June 11, June 21, and Nov. 25, 1890, GLB; Gould to Grant and Wheelock, June 24, 1890, GLB; Gould to Milton H. Smith, June 10, 1890, GLB; Gould to D. A. Boody, May 20, 1890, GLB; Clark to Gould, May 18, 1890, GLB; Gould to S. A. Trefant (?), June 23, 1890, GLB; Gould to Huntington, July 29, 1890, GLB; Gould to Stuyvesant Fish, June 21, 1890, GLB; Fish to Gould, Nov. 28, 1890, TPHR; Poor, *Manual 1891*, 788. The cutoff, chartered as the Houston, Central Arkansas & Northern, was completed in December 1891. The Iron Mountain absorbed the company in December 1893. See Burton, "Missouri Pacific" (Chap. 22, n. 2), 720–22.

21. Gould to Clark, Jan. 15, 1890, KG, May 5 (two letters), May 6, May 7, May 10, May 12, June 13, June 14, June 19, June 28, July 9, July 15 (two telegrams), July 16, Sept. 24, and Sept. 25 (two telegrams), 1890, GLB; Gould to W. M. Murdock, May 10, 1890, GLB; Poor, *Manual 1891*, 786, 788; *New York Times*, Dec. 23, 1890.

22. Gould to Clark, Jan. 15, June 19, and undated, KG, May 15, June 7, June 11, June 13, June 25, June 26, July 28, Oct. 6, and Oct. 7, 1890, GLB; MP Exec. Comm. Minutes, Apr. 30, 1900, File 108, MP; Poor, *Manual 1890*, 521–22. The memorandum of agreement is described in the June 11 letter.

23. Gould to Clark, May 7, July 2, and July 16, 1890, GLB, June 30, 1890, KG and GLB.

24. Gould to Mr. Lanier, Nov. 24, 1890, GLB.

25. Snow, *Helen Gould* (see Chap. 1, n. 2), 201–2; *New York Tribune*, Dec. 3, 1892; *New York Times*, Dec. 3, 1892; *New York World*, Dec. 3, 1892. For examples of donations see Gould to Samuel Sloan, Jr., Nov. 30, 1886, GLB; Gould to John Crosby Brown, Sept. 10, 1886, GLB; Gould to Edward M. Field, Sept. 24, 1889, GLB; Gould to Mr. McFarland, Dec. 12, 1886, GLB.

26. *New York World*, Dec. 3, 1892; *New York Times*, July 8, 1890; Snow, *Helen Gould*, 203.

27. Gould to Anna Gould, Feb. 16, 1892, JGL; Snow, *Helen Gould*, 180–81, 203–4; *New York World*, Sept. 2, 1890; *New York Times*, Mar. 9 and Mar. 10, 1892.

28. Snow, *Helen Gould*, 181, 204–5; *New York Times*, Mar. 10, 1892; *New York Tribune*, Dec. 3, 1892. "Those were his actual words," Snow added.

29. See for example *New York World*, July 10, 1887, Dec. 1, 1890. An article on Brady is in the *New York World*, Aug. 27, 1890.

30. Gould to Clark, Dec. 24, 1890, KG.

31. For accounts of Koch's work see *New York Herald*, Nov. 10, Nov. 15–17, Nov. 19, Nov. 21, Dec. 12, and Dec. 19, 1890, Feb. 23, 1891; *New York World*, Nov. 19, Nov. 22, Nov. 23, and Nov. 30, 1890, Jan. 30, 1891. Gould's interest in Koch's discoveries is mentioned in Snow, *Helen Gould*, 193. Although Koch did not find a cure, his work was crucial. In 1882 he demonstrated the bacterial cause of several diseases, including tuberculosis, and received the Nobel Prize in 1905 for his work in developing tuberculin as a test for tuberculosis.

32. Details on the Union Pacific's expansion program are in Trottman, *Union*

Pacific (see Chap. 13, n. 1), 236–38; Athearn, *Union Pacific Country* (see Chap. 13, n. 4), 311–29; Overton, *Gulf to Rockies* (see Chap. 15, n. 13), 217–57; Adams, "Memorabilia 1888–1893," Feb. 24, May 16, and Oct. 6, 1889, 50, 58–61, 72–77, 97–98, CFA. The Short Line, the Utah & Northern, the Utah Central, and some smaller roads were merged and the Panhandle was combined with some Colorado branches into the Union Pacific, Denver & Gulf Railway.

33. Terms of the contract are in Gould to Fred Ames, Dec. 24, 1890, GLB. The Rock Island had a line between southern Nebraska and Colorado Springs but lacked a connection between this road and Council Bluffs.

34. *Chronicle* 40:481, 650, 653, 44:212; *Stockholder*, May 31, 1887; *New York Herald*, Sept. 27 and Oct. 3, 1887. Gould, Sage, and Dillon occupied three of the nine seats on the Pacific Mail board.

35. *Stockholder*, May 31, 1887; *New York Times*, June 28, June 29, Aug. 24, and Aug. 25, 1887; *New York Herald*, Sept. 22 and Oct. 3, 1887; *Chronicle* 45:426, 46:21. Huntington had also left the board in May 1887.

36. *Chronicle* 40:481, 46:706–7, 48:727–28, 50:768–69.

37. Ibid., 50:769, 771; *New York Times*, May 28, 1890; *New York World*, May 28, 1890; *WSJ*, Apr. 18, Apr. 22, May 22, and May 27, 1890. For background on Brice and Thomas see Klein, *Great Richmond Terminal* (Chap. 9, n. 18), 30–31, 50–52.

38. Adams, "Memorabilia 1888–1893," Dec. 23, 1888, 10, CFA; Adams to Dodge, Apr. 19, 1890, DP.

39. Adams to Gould, Apr. 2, 1890, UP; Dodge to Adams, Apr. 7, 1890, DR, 13:157; Gould to Clark, June 11, 1890 (two letters), GLB; Gould to C. G. Warner, June 16, 1890, GLB; Gould to Adams, Aug. 27 and Sept. 24, 1890, UP, Sept. 6, 1890, GLB; Adams to Dodge, Apr. 4 and Apr. 8, 1890, DP.

40. Adams, Diary, July 19 and Aug. 22, 1890, CFA; Adams, "Memorabilia 1888–1893," June 30, 1889, 84, CFA; *Stockholder*, Jan. 21, Mar. 18, and Mar. 25, 1890; *New York World*, June 9, June 16, June 23, and Sept. 25, 1890.

41. Trottman, *Union Pacific*, 236–37; Kirkland, *Adams* (see Chap. 26, n. 29), 123–24; Grodinsky, *Jay Gould*, 577; Adams to Dodge, May 1 and May 20, 1889, DP; Adams, "Memorabilia 1888–1893," Oct. 6, 1889, Nov. 23 and Nov. 30, 1890, 96–99, 154–55, 174, CFA.

42. Adams, "Memorabilia 1888–1893," June 18, 1889, July 4 and Nov. 23, 1890, 62–67, 117, 127–29, 169, and passim, CFA.

43. Adams, Diary, Sept. 8, 1890, CFA; Adams, "Memorabilia 1888–1893," Oct. 6, 1889, Apr. 27 and July 4, 1890, 98, 106–34, and passim, CFA; Adams to Dodge, May 1 and June 11, 1890, DP; R. S. Grant to G. M. Lane, Oct. 28, 1890, UP; Perkins to E. P. Ripley, Mar. 26, 1890, BA; *Stockholder*, Apr. 29, June 10, July 22, and Sept. 16, 1890.

44. *Chronicle* 51:348, 538; Gould to Newman Erb, Oct. 8, 1890, GLB; *New York World*, Oct. 12 and Oct. 19, 1890; Adams, Diary, Oct. 1 and Oct. 7, 1890, CFA. Details of Adams's western trip are in the diary, Oct. 1–30, CFA, and "Memorabilia 1888–1893," 141–53, CFA.

45. Adams, Diary, Nov. 6, 1890, CFA; *New York Herald*, Nov. 2, Nov. 7, Nov. 8, and Dec. 21, 1890; *New York World*, Nov. 9, 1890; Grant to Lane, Oct. 28, 1890, UP. The Succi saga may be followed in the *Herald* for November and December.

46. *New York World*, Nov. 9, 1890; *New York Herald*, Nov. 8, 1890; *New York Times*, June 10, 1890.

47. *New York Herald*, Nov. 11, 1890; *New York World*, Nov. 9 and Nov. 11, 1890; Adams, Diary, Nov. 7–11, 1890, CFA.

48. Adams to H. P. Frothingham, Nov. 10, 1890, UP; Adams to Blake, Boissevain &

Co., Nov. 11, 1890, UP; Adams to Hughitt, Nov. 10 and Nov. 12, 1890, UP; Adams, Diary, Nov. 11–13, 1890, CFA; *New York Herald*, Nov. 12 and Nov. 13, 1890; *New York World*, Nov. 12 and Nov. 13, 1890; Adams to C. S. Mellen, Nov. 12, 1890, UP; Adams to Dodge, Sept. 15, 1890, DP; Adams to J. T. Granger, Nov. 13, 1890, DP, UP; Adams to Grant, Nov. 13, 1890, UP.

49. Adams, Diary, Nov. 15–18, 1890, CFA; Adams, "Memorabilia 1888–1893," Nov. 23, 1890, 146–47, 153–63, CFA; Adams to J. S. Tebbets, Nov. 15, 1890, UP; *Chronicle* 51:681; *New York Herald*, Nov. 15–18, 1890; *New York World*, Nov. 14–21, 1890.

50. Gould to Clark, Nov. 17, 1890 (two letters), GLB; *New York Herald*, Nov. 19–21 and Dec. 10, 1890; *New York Times*, Nov. 18 and Nov. 19, 1890; *New York World*, Nov. 18–21, 1890.

51. *New York World*, Nov. 22, Nov. 23, and Nov. 26, 1890; *New York Herald*, Nov. 22, 1890.

52. *New York Herald*, Nov. 21 and Dec. 1, 1890; *New York World*, Nov. 20–26 and Nov. 29, 1890.

53. *New York World*, Nov. 22, Nov. 23, and Nov. 27, 1890; *New York Herald*, Nov. 21 and Nov. 22, 1890.

54. Adams, "Memorabilia 1888–1893," Nov. 23, 1890, 160–62, CFA; Adams, Diary, Nov. 15–17, 1890, CFA.

55. Adams, "Memorabilia 1888–1893," Nov. 23, 1890, 162–65, CFA; Adams, Diary, Nov. 18 and Nov. 19, 1890, CFA; *New York World*, Nov. 20 and Nov. 23, 1890; Charles Francis Adams, The *Autobiography of Charles Francis Adams*, ed. W. C. Ford (Boston, 1916), 190.

56. This scene is taken from Adams, "Memorabilia 1888–1893," 165–67, and Adams, Diary, Nov. 19 and Nov. 20, 1890, CFA. For a good example of the press accounts see *New York World*, Nov. 20–27, 1890.

57. Adams, "Memorabilia 1888–1893," 165–67, CFA; Adams, Diary, Nov. 19 and Nov. 20, 1890, CFA. The quotation is from the "Memorabilia," but the diary conveys the same sentiments: "With Ames to see Gould,—last word well said and got out of that with a sense of superiority to the poor little cad, who didn't look me in the eye."

58. Adams, Diary, Nov. 26, 1890, CFA; UP Directors' Minutes, Nov. 26, 1890, UP; *New York World*, Nov. 27, 1890; *New York Herald*, Nov. 27, 1890; *Chronicle* 51:748. Adams's belief that Gould wanted revenge is revealed in "Memorabilia 1888–1893," Nov. 23, 1890, Jan. 25 and Nov. 13, 1891, 147, 159, 171, 199, 325, CFA; Adams to Dodge, Nov. 22 and Dec. 2, 1890, DP; Adams, Diary, Oct. 24, 1894, CFA.

59. *New York World*, Nov. 27, 1890.

60. Ibid., Nov. 20, 1890.

61. Notice from Sidney Dillon, Dec. 8, 1890, UP; *New York Herald*, Dec. 10, 1890; Forbes to Perkins, Nov. 22, 1890, BA.

62. Adams, "Memorabilia 1888–1893," Nov. [Dec.] 7, 1890, 176, CFA; *New York Herald*, Dec. 5, Dec. 7, and Dec. 31, 1890; *New York World*, Nov. 29, Dec. 7–9, 1890; *Chronicle* 51:830; Gould to Clark, Dec. 24, 1890, KG; Dillon to Holcomb, Nov. 26 and Dec. 2, 1890, UP; Dillon to Clark, Dec. 24, 1890, UP; Dillon to Ames, Dec. 12, 1890, UP; Dillon to Dexter, Atkins, and Lane, Dec. 23, 1890, UP; Dillon to Lane, Dec. 27, 1890, UP; Guy Phillips to W. J. Quinlan, Jr., Dec. 31, 1890, GLB; Phillips to Lane, Dec. 31, 1890, GLB.

63. *Chronicle* 51:157–58, 361–62, 762, 768–69; *Bradstreet's*, Oct. 18, 1890; *Fourth Annual Report of the Interstate Commerce Commission* (Washington, D.C., 1890), 21–27; *New York Sun*, Nov. 15, 1890; Gould to C. G. Warner, Dec. 10, 1890, GLB.

64. *Chronicle* 50:783–86, 51:731; *Bradstreet's*, Oct. 18, 1890; *New York World*, Oct. 19 and Nov. 2, 1890.

65. Joseph Schumpeter, *The Theory of Economic Development* (Cambridge, Mass., 1961), 6.

66. *New York World*, Nov. 21 and Nov. 25, 1890; *New York Times*, Nov. 22, 1890.

67. UP Directors' Minutes, Nov. 26, 1890, UP; Gould to Frank Bond, Nov. 25, 1890, GLB; *New York World*, Nov. 22, 1890; *New York Herald*, Nov. 22 and Nov. 23, 1890. The complete text of Gould's memorandum is in the UP minutes.

68. C. F. Meek to Dodge, Dec. 1 and Dec. 3, 1890, DP; *New York World*, Nov. 21, 1890.

69. Gould to Morgan, Dec. 3, 1890, GLB; Meek to Dodge, Dec. 3, 1890, DP; Adams to S. Endicott Peabody, Nov. 15, 1890, UP; Perkins to Forbes, Dec. 4, 1890, BA; Forbes to Perkins, Dec. 3, 1890, BA; *New York Herald*, Nov. 26, 1890; *New York World*, Nov. 28, Dec. 5, and Dec. 7, 1890.

70. *Proceedings of the Meeting of the Presidents of Railway Companies West of Chicago and St. Louis at the House of Mr. J. Pierpont Morgan, New York City, on Dec. 15, 1890*, MP. My version is taken from this pamphlet, which was apparently printed as the official record of the meeting. It makes clear, as the newspaper accounts do not, that the presidents and Morgan did not ignore Gould's plan but rejected it before turning to that offered by Morgan.

71. Ibid. For press accounts of the agreement see *New York Herald*, Dec. 16 and Dec. 17, 1890; *New York World*, Dec. 16 and Dec. 17, 1890; *Chronicle* 51:877.

72. *New York Herald*, Dec. 16, 1890; *New York World*, Jan. 10, 1891; *New York Times*, Dec. 19 and Dec. 21, 1890; *Chronicle* 51:877.

73. *New York Herald*, Dec. 16 and Dec. 17, 1890; *New York Times*, Dec. 21, 1890; *New York World*, Dec. 17, 1890; *Chronicle* 51:877; Gould to Morgan, Dec. 23, 1890, GLB; Gould to Miller, Dec. 23 and Dec. 27, 1890, GLB.

74. Stickney's remarks and Gould's story are both in the *New York Herald*, Dec. 16, 1890.

Chapter 33: Surrender

1. *New York World*, Jan. 9–12, 1891; *New York Herald*, Dec. 30, 1890, Jan. 9 and Jan. 10, 1891; Dillon to Clark, Jan. 13, 1891, UP; *Chronicle* 52:82, 121, 125–26; Gould to Magoun, Jan. 21, 1891, GLB; Gould to Hughitt, Jan. 21 and Jan. 23, 1891, GLB; *New York Times*, Jan. 24, 1891; *New York Tribune*, Feb. 7, 1891.

2. *New York World*, Nov. 26, 1890; Gould to Clark, Dec. 23 and Dec. 24, 1890, GLB; Gould to Ames, Dec. 24, 1890, GLB.

3. Gould to Ames, Dec. 26, 1890, GLB; Gould to Clark, Dec. 29 and Dec. 30, 1890, GLB; Gould to Miller, Dec. 29, 1890, GLB. A summary of the two leases is in GLB, dated Dec. 30, 1890. In the letter to Miller Gould claimed he did not know of the leases' existence "till last evening." This was obviously not true.

4. *New York Herald*, Dec. 30, 1890, Jan. 3, 1891; *New York Times*, Jan. 4, 1891; *Chronicle* 52:41; Ames to Dodge, Jan. 3 and Jan. 14, 1891, DP; Gould to Clark, Jan. 23, 1891, GLB; UP Exec. Comm. Minutes, Jan. 22, 1891, UP; Dillon to Clark, Jan. 24, 1891, UP; J. F. Dillon to J. M. Thurston, Jan. 3, 1891, GLB.

5. Gould to Clark, Jan. 7, Jan. 13, Jan. 16, and Jan. 21, 1891, GLB; *New York Herald*, Jan. 31, 1891; *New York World*, June 14, July 28, and July 29, 1891; *New York Times*, July 30, 1891; Gould to Clark, Nov. 5, 1891, KG; *Chronicle* 53:157.

6. Gould to G. M. Lane, Jan. 5, 1891, UP; Guy Phillips to W. J. Quinlan, Jr., Jan. 7, 1891, GLB; Gould to Clark, Jan. 14, 1891, GLB; George Gould to Lane, Jan. 15 and Jan. 17, 1891, UP; Dillon to Clark, Jan. 24, Jan. 26, and Jan. 31, 1891, UP; *New York World*, Jan. 10, Jan. 25, and Feb. 11–13, 1891; *New York Herald*, Jan. 30, Feb. 1, Feb. 3, and Feb. 5, 1891; *WSJ*, Feb. 12, 1891; Geddes to Perkins, Mar. 4, 1891, BA.

7. *New York World*, Feb. 10–14, Feb. 16, and Feb. 24, 1891; *New York Herald*, Feb. 14, 1891; *Chronicle* 52:280.

8. Dillon to Clark, Feb. 14, 1891, UP; Helen Gould to Frank Gould, Feb. 7, 1891, HGS. Dillon described the trip as "a continued ovation, which was not altogether pleasant to any of us, as you are aware that neither Mr. Gould nor myself like display."

9. Ibid.; *New York World*, Feb. 9, Feb. 11, and Feb. 14, 1891; *New York Herald*, Feb. 14, 1891.

10. *New York Times*, July 30, 1891; Geddes to Perkins, Mar. 4, 1891, BA; *New York World*, Mar. 4, and Mar. 5, 1891. For the Terminal's downfall see Klein, *Great Richmond Terminal* (Chap. 9, n. 18), 235–58. Gould's only obligation was a $500,000 guarantee for the Terminal's floating debt. Gould to George Gould, Apr. 6, 1892, KG.

11. I am indebted to Dr. Peter M. Small for information on tuberculosis.

12. *Chronicle* 52:534; Helen Gould to her father, Apr. 7 and Apr. 9, 1891, HGS; Gould to Dillon, Apr. 17, 1891, UP; Perkins to Forbes, Apr. 19, 1891, BA; *Bradstreet's*, Apr. 18, 1891; *New York World*, Apr. 5, Apr. 16, and Apr. 24, 1891; *New York Tribune*, Apr. 16 and Apr. 18, 1891; *WSJ*, Apr. 18, 1891.

13. *Bradstreet's*, Apr. 18, 1891; *New York World*, Apr. 16 and Apr. 18, 1891; *New York Tribune*, Apr. 16, 1891; Gould to Clark, May 12, 1891, KG; *New York Times*, May 7, 1891; *WSJ*, Apr. 7, Apr. 23, and May 5–7, 1891.

14. Gould to Clark, May 7 and May 12, 1890, KG; *New York Times*, May 7, 1891; *New York World*, May 8, 1891; *Chronicle* 52:718.

15. Ibid.; *New York World*, June 17, July 4, July 14–16, and July 18, 1891; *New York Times*, July 4, 1891; *WSJ*, July 21, 1891; *Chronicle* 53:71; *Railroad Gazette* 23:522.

16. Gould to Dillon, undated [July 1891], UP.

17. *New York World*, May 2, May 31, June 12, June 18, June 19, June 23, June 24, Sept. 23, Sept. 24, and Oct. 17, 1891, Jan. 17 and Mar. 26, 1892.

18. *Chronicle* 52:659–60, 681, 53:76; *Stockholder*, Apr. 30, 1891; George Gould to Lane, Feb. 11 (three times), May 2, and May 26, 1891 (two items), UP; Gould to Dillon, Apr. 16, 1891, UP; Dillon to Gould, Apr. 17, 1891, UP; Gould to Lane, May 26, May 28, May 29, June 1, and June 11 (three items), 1891, UP; Edward Canfield to Gould, June 26, 1891, UP; Gould to Canfield, undated [June 1891], UP; "Statement of notes held by Jay Gould, President," June 19, 1891, UP; Canfield to Gould, June 27, 1891, UP; H. P. Frothingham to Lane, June 30, 1891, UP; *New York Times*, June 25, 1891. The file for Incoming Correspondence, Office of the President, UP, is crowded with letters and telegrams pertaining to the hunt for money during the spring and summer of 1892.

19. *New York Herald*, Nov. 30, 1890; *New York Times*, July 30, 1891; Hyde to Dillon, Aug. 23, 1891, UP; Gould to Dillon, Aug. 24 and Aug. 25, 1891, UP.

20. *New York World*, Sept. 9, 1891; *New York Tribune*, Sept. 9, 1891; Dillon to Canfield, July 31, 1891, UP; George Gould to Canfield, July 31, 1891, UP; exchange of telegrams between George Gould and Canfield, July 31, 1891, UP; Atkins to Ames, Aug. 6, 1891, UP; Dexter and Atkins to Gould, Aug. ?, 1891, UP; *Chronicle* 53:178–79.

21. Gould to Dodge, Aug. 6, 1891, UP; Gould to Atkins, Aug. 7, 1891, UP; Dillon to Gould, Aug. 7, 1891, UP; *New York World*, Aug. 9, 1891; *WSJ*, Sept. 8, 1891.

22. *Chronicle* 53:178–79, 187, 210–11, *New York World*, Aug. 12, 1891; Gould to Dillon, Aug. 13, 1891, UP; Dillon to Gould, Aug. 7 and Aug. 14, 1891, UP.

23. *Chronicle* 53:210, 257, 290, 368; *New York World*, Aug. 9, Aug. 12, and Aug. 19, 1891; *WSJ*, Aug. 15–21, Aug. 24, and Aug. 25, 1891; George Gould to Lane, Sept. 14, 1891 (two items), UP. My version differs substantially from that in Grodinsky, *Jay Gould* (see Prologue, n. 5), 585–87.

24. Dodge to W. T. Walters, Sept. 21, 1891, DP; *WSJ*, Sept. 23, Sept. 26–29, 1891; *New York World*, Sept. 4, Sept. 9, and Sept. 25–27, 1891; *New York Times*, Sept. 25–27, 1891; *Chronicle* 53:436–37, 475; O. W. Mink to Ames, Oct. 9, 1891, UP.

25. *New York Sun*, Sept. 26, 1891; *New York World*, Sept. 27, 1891. In the crazy quilt world of newspaper opinion it is worth recalling that for years the *Sun* was accused of being under Gould's thumb. Admitting that the *Sun* had long been friendly to Gould, the *World* still insisted that Morgan "has for several years past dictated the financial opinions of that paper."

26. *New York World*, July 27, 1891; *Bradstreet's*, Sept. 19, 1891; *New York Times*, Sept. 20, 1891.

27. *Bradstreet's*, Sept. 19, 1891.

28. *New York Times*, Sept. 24 and Sept. 25, 1891; *New York World*, Sept. 25, 1891.

29. *New York World*, Sept. 25 and Sept. 26, 1891; *Chronicle* 53:418, 427.

30. *WSJ*, Sept. 26, 1891.

31. Ibid., Sept. 26, 1891; *New York Times*, Sept. 27 and Sept. 29, 1891; *Chronicle*, 53:475; MP Directors' Minutes, Sept. 30, 1891, MP. Gould's report is reprinted in the *New York Times*, Oct. 1, 1891, and *Chronicle* 53:474–75. Gould's action had some defenders; see *WSJ*, Sept. 25–30, 1891.

32. *New York Times*, Oct. 2, 1891; *New York World*, Oct. 2, 1891; *Chronicle* 53:464; *WSJ*, Oct. 2, 1891. Periodic bouts of confusion of the sort described here are a common symptom of tuberculosis.

33. *New York Times*, Oct. 2, 1891; *New York World*, Oct. 2 and Oct.3, 1891; *WSJ*, Oct. 2, 1891.

34. *New York World*, Oct. 3 and Oct. 13, 1891; *New York Times*, Oct. 6, 1891.

35. *New York World*, Oct. 3, 1891; *New York Herald*, Dec. 13, 1891. The strength of this belief that better times awaited can be appreciated only by reading extensively in the sources, including many of those cited above.

36. *Chronicle* 53:593; *New York World*, Oct. 2, 1891.

37. Gould to Clark, Nov. 5, 1891, KG; *New York Times*, Nov. 7, 1891; Adams, "Memorabilia 1888–1893," Nov. 13, 1891, 324, CFA.

38. Gould to Clark, Jan. 14, 1891, GLB.

39. Ibid., Jan. 15 (two letters), Jan. 19 (two letters), and Jan. 20, 1891, GLB; Gould to George C. Smith, Jan. 14, 1891, GLB; Gould to Clark, Jan. 18 and Feb. 28 (two letters), KG; *New York Times*, Sept. 30, 1891; *New York World*, Aug. 14 and Sept. 27, 1891.

40. *Chronicle* 52:608–9, 957–58, 53:21, 257, 569, 604, 54:203, 329, 366–67, 483; R. S. Watkins to Dodge, Sept. 6, 1891, DR, 13:655; *New York World*, July 3, 1891; *New York Times*, June 9, 1891, Jan. 23, Feb. 19, Feb. 24, and Nov. 27, 1892; *New York Tribune*, Feb. 19, 1892.

41. *Chronicle* 48:530, 53:567, 55:724, 806; *Stockholder*, Apr. 22, July 29, and Oct. 8, 1890; *New York World*, May 30, July 22, Nov. 24, and Dec. 1, 1890, Sept. 7, 1891; *New York Herald*, Dec. 22, 1891.

42. *Chronicle* 45:641, 727, 47:592, 48:292, 368–69, 651–52, 50:313, 51:206, 414, 53:519, 713; *Stockholder*, Nov. 15, 1887, Nov. 19, 1889; *New York Tribune*, Jan. 11, 1888.

43. *Chronicle* 52:238; *New York Herald*, Feb. 24, 1889; *New York World*, May 13, June 12, June 16, and Oct. 8, 1891, Mar. 18, 1892; Gould to Robert A. Johnston, Jan. 24, 1891, GLB.

44. *New York Herald*, Feb. 24, 1889; *New York Times*, Jan. 29, 1891; *New York World*, Sept. 18, Oct. 5, and Dec. 24, 1890, May 13, June 9, and July 4, 1891. See also the cartoon in the *New York World*, Jan. 11, 1891.

45. *New York World*, Apr. 23, May 13, May 15, May 19, May 26, June 1, June 9, July 4, July 7, July 9, July 15, and July 23, 1891; *New York Times*, Dec. 12, 1891, Oct. 27, 1892; *Chronicle* 53:713.

46. Quoted in Snow, *Helen Gould* (see Chap. 1, n. 2), 193.

47. *New York World*, Feb. 15 and Apr. 5, 1891; Gould to [Anna Gould], undated, JGL; Helen Gould to her father, Apr. 9, 1891, HGS.

48. *New York Herald*, Dec. 27, 1891; *New York World*, Dec. 27, 1891; Howard Gould to Carl Biringer, Nov. 17, 1891, HGS; Snow, *Helen Gould*, 189–91. The lists of guests in the newspapers have the usual errors and contradictions.

49. *New York Herald*, Dec. 27, 1891; *New York World*, Dec. 27, 1891, Jan. 13, 1892.

50. *New York Times*, May 7 and May 9, 1891; *New York World*, July 18, July 19, and Dec. 5, 1891, Dec. 5, 1892. The account of the bombing in Snow, *Helen Gould*, 186–87, is dubious because of obvious errors of fact.

51. *New York World*, Dec. 30, 1891, Jan. 4, Jan. 17, Jan. 18, and Feb. 4, 1892; *New York Times*, Jan. 3, 1892.

52. Snow, *Helen Gould*, 170–71. Bezique is a card game similar to pinochle.

53. *New York World*, Feb. 25–27, 1892. For the gift episode see Chap. 32.

54. *New York World*, Feb. 27, Feb. 28, Mar. 6, Mar. 9, Mar. 10, Apr. 2, Apr. 3, Apr. 5, and Apr. 29, 1892; C. E. Satterlee to George Gould, Mar. 16, 1891, TPHR; John A. Grant to Satterlee, Oct. 12, 1891, TPHR; A. N. Towne to Gould, Apr. 1, 1892, JGL; Gould to George Gould, Apr. 6 and Apr. 29, 1892, KG.

55. Gould to George Gould, Apr. 6, 1892, KG; Edward Canfield to Clark, Mar. 23 and Apr. 6, 1892, UP; *Chronicle* 54:526, 711, 725; UP Directors' Minutes, Apr. 27, 1892, UP; *New York World*, Jan. 4, Apr. 7, and Apr. 28, 1892.

56. Gould to George Gould, Apr. 29, 1892, KG. This letter refutes the statement in Grodinsky, *Jay Gould*, 590, that "despite the outcome Gould was not elated."

57. UP Directors' Minutes, Apr. 27, 1892, UP; UP Stockholders' Minutes, Apr. 27, 1892, UP; UP Directors' Minutes, June 16, 1892, UP; MP Directors' Minutes, June 22, 1892, MP; *New York World*, Apr. 24, 1892; Canfield to Clark, June 6, 1892, UP; DR, 13:835; *New York Times*, July 13, 1892; *New York Tribune*, July 12–14, 1892; *Chronicle* 55:76.

58. Gould to George Gould, May 1, 1892, KG; *Chronicle* 54:704–6, 722–24, 801, 55:895; Gould to Clark, Sept. 28, 1892, KG.

59. *Chronicle* 55:101, 147, 235–37, 463, 504, 545; MP Directors' Minutes, Sept. 27, 1892,MP; *WSJ*, July 12, July 13, and Sept. 27, 1892; *New York Tribune*, July 13, July 15, July 23, and Nov. 8, 1892; *New York Times*, June 10, 1892.

60. *New York Herald*, Oct. 4, Oct. 8, and Oct. 12, 1892; *New York Tribune*, Nov. 11, Nov. 12, and Nov. 24, 1892; *Chronicle* 55:639; Grodinsky, *Transcontinental* (see Chap. 13, n. 13) 360–61.

61. MP Directors' Minutes, Oct. 11, 1892, MP; *New York Herald*, Oct. 26 and Oct. 27, 1892; *New York Times*, Oct. 27, 1892; *New York World*, Jan. 30, 1891.

62. *New York Tribune*, Dec. 2, 1892; Snow, *Helen Gould*, 193.

63. Snow, *Helen Gould*, 194; *New York Herald*, Dec. 1 and Dec. 2, 1892; *New York Tribune*, Dec. 1 and Dec. 2, 1892; *New York World*, Dec. 1 and Dec. 2, 1892. The death

certificate lists miliary tuberculosis as the immediate cause of death and states that it was contracted two weeks earlier. A copy is in HGS.

64. *New York Tribune*, Dec. 3, 1892.

65. The usual romanticized accounts of Gould's last moments are in all the New York papers for Dec. 3, 1892. All give the time of death as 9:15 A.M., but the death certificate gives it as 9:10.

66. Unless otherwise indicated, details of the preparations and funeral are taken from the *New York Herald*, Dec. 4–7, 1892, *New York Times*, Dec. 4–7, 1892, *New York Tribune*, Dec. 4–7, 1892, and *New York World*, Dec. 4–7, 1892.

67. Snow, *Helen Gould*, 196.

68. Thomas Bulfinch, *Bulfinch's Mythology* (New York, n.d.), 109. This is the Modern Library edition.

Epilogue: The Legend and the Legacy

1. For this discussion I have used a copy of Gould's will in possession of Kingdon Gould, Jr. For newspaper accounts see *New York Tribune*, Dec. 8, Dec. 10, and Dec. 13, 1892; *New York Herald*, Dec. 8 and Dec. 13, 1892; *New York World*, Dec. 8–10 and Dec. 13, 1892. The original will was made in December 1885. Codicils were added in February 1889, shortly after Helen's death, and on Nov. 21, 1892. It required two years of calculating to arrive at a definitive figure for the value of the estate, which was roughly $81 million gross and $73.2 million residuary. The precise amounts, including a list of all the securities held by Gould at his death, are in the *Wall Street Daily News*, Jan. 9, 1895.

2. There is no satisfactory study of the later Gould generations. Snow, *Helen Gould* (see Chap. 1, n. 2), views events from Helen's perspective and is delicate to the point of vast omissions. Two popular accounts, O'Connor, *Gould's Millions* (see Prologue, n. 1), and Hoyt, *The Goulds* (see Prologue, n. 6), concentrate on the sensational and the scandalous, both of which exist in abundance. Hoyt offers the fullest version but seldom gets beneath the surface of the headlines. Unless otherwise indicated, my account of Gould's children is drawn from these sources.

3. Some details on George's railroad ventures and the fight with Harriman can be found in Ernest Howard, *Wall Street Fifty Years after Erie* (Boston, 1923). See also Athearn, *Rebel of the Rockies* (Chap. 19, n. 24), 189–235.

4. Glimpses of Helen's life can be found in Snow, *Helen Gould*, 217–340. A less charitable view from the younger generation is in Celeste Andrews Seton, *Helen Gould Was My Mother-in-Law* (New York, 1953).

5. Snow, *Helen Gould*, 222–23, 232–48; Griffin, *Roxbury* (See Chap. 1, n. 4), 99–101, 124–25, 202–8.

6. Marquis Boni de Castellane, *How I Discovered America* (New York, 1924), 14–15. Boni's memoir also gives delightful if slanted views of George and Edith Gould and Helen.

7. Hoyt, *The Goulds*, 320–22, 327–28.

8. These quotations and several more in the same vein can be found in the *New York Times*, Dec. 3–8, 1892; *New York Tribune*, Dec. 3–8, 1892, *New York World*, Dec. 3–8, 1892; *New York Herald*, Dec. 3–8, 1892. For memorials to Gould see MP Directors' Minutes, Dec. 3, 1892, MP; UP Directors' Minutes, Dec. 16, 1892, UP. The Western Union and Manhattan resolutions, along with some other tributes, are reprinted in Snow, *Helen Gould*, 197–202. In this section I have also drawn extensively on my article, "In Search of Jay Gould" (see Prologue, n. 1).

9. *New York Times*, Dec. 3, 1892; Dodge to George Gould, Dec. 2, 1892, DR, 13:923.

10. For the market reaction see *New York Herald*, Dec. 3, 1892; *New York Times*, Dec. 3, 1892; *Chronicle* 55:928, 56:11; *New York Tribune*, Dec. 10, 1892. The Hendrick observation appeared in *American Illustrated Magazine* and is quoted in both O'Connor, *Gould's Millions*, 285, and Hoyt, *The Goulds*, 127. Hoyt at least takes issue with it.

11. A good introduction to this subject is Irvin G. Wyllie, *The Self-made Man in America* (New York, 1954). See especially chap. 7.

12. I have discussed this subject at length in Maury Klein and Harvey A. Kantor, *Prisoners of Progress: American Industrial Cities, 1850–1920* (New York, 1976).

13. Frank Norris, *The Pit* (New York, 1903), 63.

14. Quoted in Halstead and Beale, *Jay Gould* (see Prologue, n. 2), 218.

15. *New York Sun*, Aug. 25, 1877.

16. *New York Times*, Dec. 8, 1884. For references to Gould as effeminate see Morris, *Life on the Stage* (Chap. 28, n. 2), 304–5; Halstead and Beale, *Jay Gould*, 211–12, 290–92, 295; O'Connor, *Gould's Millions*, 244; White, *Wizard* (Prologue, n. 4), 216–17.

17. To refresh the reader's memory, these biographies are Halstead and Beale, *Jay Gould;* Northrop, *Life and Achievements* (see Prologue, n. 2); White, *Wizard;* Ogilvie, *Life and Death* (see Prologue, n. 4).

18. Justin Kaplan, Introduction to Mark Twain and Charles Dudley Warner, *The Gilded Age: A Tale of Today* (Seattle, 1968), vi.

19. Glenn Porter, *The Rise of Big Business*, 1860–1910 (Arlington Heights, Ill., 1973), 4. For the historiographical debate see Thomas C. Cochran, "The Legend of the Robber Barons," *The Pennsylvania Magazine of History and Biography* 74 (July 1950):307–21; Louis Galambos, "The Emerging Organizational Synthesis in Modern American History," *Business History Review* 44 (Autumn 1970):279–90.

20. Myers, *American Fortunes* (see Chap. 5, n. 1), 7. To this the publisher added, "There was no denunciation, no loose editorializing. . . . At no time did he indulge in tirades against personal traits, dispositions or temperaments. He was not concerned with the good or bad qualities of the individual founders and perpetuators of great fortunes." Anyone who reads Myers's work may reasonably conclude that the publisher read some other volume.

21. Josephson's remarks are in his foreword to the Harvest paperback edition of *The Robber Barons*, vi.

Index

INDEX

INDEX

Maury Klein, professor of history at the University of Rhode Island, has written half a dozen books on American railroads and business history, including a two-volume history of the Union Pacific Railroad, and several score scholarly and popular articles. He has written novels, a film script, and a play, and has consulted for public television. His interests range across America from the Civil War through the Vietnam era and beyond; they encompass railroads and newspapers; film and the theater; the lives of Scotch-Irish and Jewish immigrants. He lives in Narragansett, Rhode Island, with his wife and family.

The Evolution of Strategy

Is there a 'Western way of war' which pursues battles of annihilation and single-minded military victory? Is warfare on a path to ever greater destructive force? This magisterial new account answers these questions by tracing the history of Western thinking about strategy – the employment of military force as a political instrument – from antiquity to the present day. Assessing sources from Vegetius to contemporary America, and with a particular focus on strategy since the Napoleonic Wars, Beatrice Heuser explores the evolution of strategic thought, the social institutions, norms and patterns of behaviour within which it operates, the policies that guide it and the culture that influences it. Ranging across technology and warfare, total warfare and small wars as well as land, sea, air and nuclear warfare, she demonstrates that warfare and strategic thinking have fluctuated wildly in their aims, intensity, limitations and excesses over the past two millennia.

BEATRICE HEUSER holds the Chair of International History at the School of Politics and International Relations, University of Reading. Her publications include *Reading Clausewitz* (2002); *Nuclear Mentalities?* (1998) and *Nuclear Strategies and Forces for Europe, 1949-2000* (1997), both on nuclear issues in NATO as a whole, and Britain, France, and Germany in particular.

The Evolution of Strategy

Thinking War from Antiquity to the Present

BEATRICE HEUSER

CAMBRIDGE
UNIVERSITY PRESS

CAMBRIDGE
UNIVERSITY PRESS

University Printing House, Cambridge CB2 8BS, United Kingdom

Cambridge University Press is part of the University of Cambridge.

It furthers the University's mission by disseminating knowledge in the pursuit of education, learning and research at the highest international levels of excellence.

www.cambridge.org
Information on this title: www.cambridge.org/9780521155243

First published 2010

A catalogue record for this publication is available from the British Library

Library of Congress Cataloguing in Publication data
Heuser, Beatrice, 1961–
The evolution of strategy : thinking war from antiquity to the present / Beatrice Heuser.
p. cm.
Includes bibliographical references and index.
ISBN 978-0-521-19968-1 (hardback) – ISBN 978-0-521-15524-3 (pbk.)
1. Strategy–History. 2. War–History. I. Title.
U162.H48 2010
355.4–dc22
2010024605

ISBN 978-0-521-15524-3 Paperback

In memory of Julian Chrysostomides
1928–2008
scholar, teacher, friend

Contents

Acknowledgements

I want to express my gratitude to Michael Watson of Cambridge University Press for having helped me get this book accepted for publication, and to Dr Joe Maiolo for having put me in touch with him. I also want to draw attention to the edited volume with early strategists' texts and a commentary (*The Strategy Makers: Thoughts on War and Society from Machiavelli to Clausewitz*) that I am preparing for Greenwood Publishers, which will hopefully make up for the briefness with which I had to deal with some of the wonderful early texts on strategy that I have cited in this volume (Heuser 2010b).

My thanks for support go to the staff of the following libraries: the library of the Military History Research Office of the Bundeswehr in Potsdam (which owns the Bleckwenn collection of rare books on military subjects), especially the extremely competent chief librarian, Dr Annette Penkert, and Ralf Schöttler; the library of the University of the Bundeswehr in Neubiberg; the librarian of the Führungsakademie of the Bundeswehr, Dr des. Karen Schäfer; the Codrington Library of All Souls College and the Bodleian Library in Oxford; the British Library in London; the Spanish National Library in Madrid; the Staatsbibliothek in Berlin; and the Library of the University of Reading.

Special thanks go to the President of the University of the Bundeswehr in Neubiberg, Professor Merit Niehuss, who allowed me to do extensive research while teaching at that university from 2006 to 2007, to my current Head of School, Dr Philip Giddings, and our School Administrator, Patricia Hicks, with both of whom it was a delight to work.

Warmest thanks go to the following friends and colleagues who have sacrificed much time by reading chapters and giving invaluable advice and pointed out many errors: Dr Frank Tallett, for the chapters on early modern warfare; Professor Andrew Lambert, Professor Mike MccGwire and Professor Michael Salewski for their trenchant comments on my chapter on maritime warfare; and for equally helpful comments on the chapters on land and air warfare, Dr Dale Walton, Mr John Salmon, Wing Commander Chris Luck, Dr Robert Foley, Dr Simon Anglim and Colonel Dr John Olsen. Dr Jeremy Lester commented helpfully on my chapter on asymmetric warfare. Andrew Lambert kindly made available to me some of his extremely helpful unpublished manuscripts, which are given due recognition in the notes. Professor Martin van Creveld with his admirably comprehensive erudition about warfare across the centuries read the whole script, and I am especially grateful for the detailed and constructive comments he gave me at very short notice indeed. I count myself very lucky to have had their expertise to draw upon, especially as I needed it in a very short time, during what should have been their vacation. Thanks for individual references also go to Professor Hans van Wees for classical sources, Professor Lindy Grant on medieval warfare, Dr James Green for checking my legal passages, Drs Christian Hartmann and Dieter Pohl for figures of Soviet losses in the Second World War and Professor Sir Lawrence Freedman for various references. Dr Geoffrey Best kindly gave me guidance on issues of international law. All remaining errors are emphatically my own!

I also want to acknowledge the help I received from Susan Dickson with the Spanish texts, from Professor James Tang, who helped me choose a good translation for Sun Tzu, and from Stefano Damiani with Machiavelli, and to him and Sara Pesatori for their help with the conversion of the footnotes, and to Peter Randall for checking my text for style and language. And last and definitely not least, the forbearance of my husband and child must be mentioned, without whose constant support I could not have researched or written this book. My parents have, as always, been a great source of moral support.

This book is dedicated to my teacher and friend Julian Chrysostomides, born in Constantinople in 1928, one of the last real Byzantines. In her adopted country, Britain, she contributed to the flowering of scholarship on Byzantine history, and she educated countless young scholars. She died in late 2008, just before I could prevail upon her to read my manuscript. I hope my humble treatment of the subject of Byzantine warfare here will not be taken adversely to reflect upon her as my teacher.

A note on referencing

For the purposes of this book, I needed to adapt Harvard-style referencing, as it was not designed with archival or internet sources in mind, nor for ancient or medieval manuscripts that were first printed centuries after they were written, and translated later still. My references serve as a shorthand for indicating the original date when a text was written; any additional date refers to the translation I have used, so that for example (Anon. 6th c./1985: 10–135) refers to the sixth-century anonymous text *Peri Strategias* translated and edited by George Dennis in *Three Byzantine Military Treatises* (Washington, D.C.: Dumbarton Oaks, 1985), pp. 10–135; this full information can be found in the bibliography. Where only a later edition was available to me, this is indicated in the bibliography; nevertheless, the earlier date is given as the in-text references in order keep these short, even where the page reference applies to the later edition. With key texts which have been reprinted many times, however, I have adopted the classicists' convention of referring to book, chapter and verse rather than to page numbers. Finally, 'q.i.' stands for 'quoted in'; 't.i.' stands for 'text printed in'.

PART I

Introduction

1 *What is strategy?*

Man made War in his own image.
(Willmott 2002: 14)

The way in which a society makes war is a projection of that society itself.
(Sidebottom 2004: 35)

Art of war or science of war, and technical definitions of 'strategy'

'Thinking war': this is how the French sociologist Raymond Aron characterised Clausewitz's work (Aron 1976). It is a conceptual challenge to write about the evolution of Strategy, especially with an emphasis on the social institutions, norms and patterns of behaviour within which it operates, the policies that guide it and the culture that influences it. For, as we shall see presently, the use of the word 'strategy' has changed very considerably over time. This book's *main* purpose is *not* to provide a history of the word 'strategy' and all that it denoted over time. Instead, it will examine how people thought about the link between political aims and the use of force, or its threat, which we will refer to as Strategy with a capital 'S'. This definition will be applied retrospectively to find out how strategists – writers on the conduct of war – thought about this issue in the past, whether or not they employed the actual term 'strategy', which after classical antiquity only came into use again around 1800.[1]

[1] To use the terminology of linguistics, I am using an onomasiological approach to the evolution of the discourse on Strategy as defined above, not the semasiological approach, which would be a history of the use of the word 'strategy' (Penth 2006: 5–18).

Nevertheless, the evolution of the term 'strategy' itself must be our starting point, not least in order to understand why there is so little agreement on the use of the term, and why it has changed so much over time. The Greek word 'strategy' (either as *strategía* or *strategiké*) was used in antiquity for the art or skills of the general (the *strategós*) – 'the general is the one who practises strategy'. By the sixth century at the latest, however, at the time of Emperor Justinian, in Byzantine usage, a difference was made between 'strategy' – 'strategy is the means by which the general may defend his own lands and defeat his enemy's' – and, hierarchically subordinated to it, 'tactics' (*taktiké*), the 'science [*epistéme*] which enables one to organize and maneuver a body of armed men in an orderly manner' (Anon. 6th c./1985: 10–135). It is possible that such definitions had already found their place in earlier works, such as the lost parts of Aeneas Tacticus (*c.* 357 BCE) or Frontinus (*c.* 35–103 or 104 CE). In either case, Frontinus in his Latin work on stratagems or ruses used the Greek words both for stratagem (*strategémon*) and for strategy (*strategía*), as neither word had a proper Latin equivalent (Frontinus *c.* 1st c. CE: I). Nor did Greek texts of the following centuries distinguish systematically between strategy and tactics. Maurice (539–602), the East Roman (Byzantine) emperor (from 582) wrote a work known a *Strategikón*, which dealt mainly with technical aspects of the conduct of war. A similar subject matter was discussed in a book in Greek called *Taktiké Theoría* dating from the second century CE, written by Aelianus Tacticus.

Emperor Leo VI ('the Wise', 865–912, emperor from 886) drew extensively on Aelianus in his own work, which later became known, not entirely appropriately, as *Taktiká* (Leo *c.* 900/1917), as Leo used the terms *strategía* and *taktiké* in the same hierarchical way as the sixth-century work referred to above. It would be Leo's work that would bring this greater meaning of 'strategy' to the West. Count John of Nassau-Siegen (1561–1623) in his *Book of War* drew on Maurice's *Stratégikon* and on Leo's *Taktiká*. John did not adopt the Greek term 'strategy', circumscribing it with the general's (*Feldher*) tasks. The word '*tactic*' he actually used (John 'the Middle' 1610/1973: 17, 516, 642). John thus built on Leo's analytical framework, which resonated in the literature, even though the word *strategía* had not yet become integrated into the Western languages.

The majority of authors before the French Revolution wrote neither about 'strategy' nor 'tactics' but about military matters in the tradition of the Roman author Publius Flavius Vegetius Renatus, Vegetius for short, who lived in the late fourth century CE; or else they wrote 'military instructions' (Puységur 1690), or about the 'art of war' (Machiavelli 1521). In the Western world, the French Count Jacques Antoine Hippolyte Guibert (1743–91) was probably the first, in his *General Essay on Tactics*, to define higher and subordinate levels of the conduct of war, speaking of 'tactics' and 'grand tactics' when talking about war aims, the configurations of armed forces in relation to the political aims and several such dimensions which we would today regard as Strategy. Without ever using the word 'strategy', Guibert wrote about both what we would today call Strategy and Tactics, dwelling primarily on the relationship between the nature of a society, its internal values and foreign-policy objectives, with an overall Strategy derived from these values and objectives, the armed forces that match these and the way these should be employed, down to battlefield Tactics (Guibert 1772/1781). Just as Monsieur Jourdain had been speaking 'in prose' all his life without knowing the expression, Guibert was what today we would call a Strategic Theorist without thinking of himself in these terms.

Shortly after the publication of Guibert's *General Essay*, the Byzantine use of the terms which pertains even today was introduced in the West. In 1771 Paul-Gédéon Joly de Maizeroy (1719–80) translated Leo's *Taktiká* into French. He still hesitated to translate Leo's term '*strategía*' into French, and used 'the art of the general' in his translation itself, and '*stratégique*' in his commentary (Leo *c.* 900/1771: 5–7). But here, for the first time in the West, the two terms 'strategy' and 'tactics' were used in a hierarchical sense, strategy denoting the higher level, tactical the lower, of warfare. In 1777 Johann von Bourscheid in Vienna published a translation of Leo into German, more appropriately under the title *Emperor Leo the Wise's Strategy and Tactics* [*sic*]. From then onwards, the use of both terms in the Byzantine sense spread throughout the West.

Whether or not they used the term 'strategy', writers since antiquity posited that Strategy should be formulated on the basis of practical experience or theoretical reflections before being applied in war. Authors on war were divided as to whether they were writing about the art or the science of war, a debate that has not been settled to this day, and which from 1800 largely overlapped with the question whether 'strategy' concerned only theoretical reflection or also practical applicability. This question can probably be found first in the writing of Archduke Charles (1771–1847), the Habsburg commander in the wars against Napoleon, who in 1806 defined 'strategy' as 'the science of war: it designs the plan, circumscribes and determines the development of military operations; it is the particular science of the supreme commander'. 'Tactics', by contrast, he defined as 'the art of war. It teaches the way in which strategic designs are to be executed; it is the necessary skill of each leader of troops' (Waldtstätten 1882: 57; Anon. 1814: vii, 3).

In contrast to all these, the Prussian philosopher-general Carl von Clausewitz (1780–1831) in his masterpiece on war spoke out against this categorisation of warfare as either an art or a science. Instead, he wrote,

> we could more accurately compare it to commerce, which is also a conflict of human interests and activities, and it is *still* closer to politics, which in turn may be considered as a kind of commerce on a larger scale. Politics, moreover, is the womb in which war develops.

This is where we encounter the idea about the relationship between politics and war for which Clausewitz is most famous, namely that 'war is an act of policy' (Clausewitz 1832/1976, I: 1, 24).

Surprisingly, in view of his theoretical ideas on war expressed in other parts of his work, Clausewitz used very narrow definitions. In Book II of *On War* he defined 'strategy' merely as 'the use of engagements for the object of the war' (Clausewitz 1832/1976, II: 1; III: 1). It was not Clausewitz's narrow definition of 'strategy', but his definition of war that would impress future thinkers: war as 'an act of force to compel our enemy to do our will' (Clausewitz 1832/1976, I: 1, 2). This view would resonate through the strategic writing of the following centuries, to the point where it became a commonplace to define the aim in war, and thus victory, as the successful imposition of one's will upon the enemy, and to see all Strategy as a pursuit of that aim.

The narrow Clausewitzian and Jominian definition of 'strategy' would live on until the end of the nineteenth century. In 1845 French Marshal Marmont defined 'strategy' as 'the general movements which are made beyond the enemy's range of sight and before the battle', while 'tactics is the science of the application of manoeuvres' (Marmont 1845: 17–25). Writing in 1853, the French naval officer

Louis-Édouard, Count Bouët-Willaumez defined 'strategy' as 'the art of determining the decisive points of the theatre of war and the general lines and routes along which armies have to move to get there' (Taillemite 1999: 50). Indeed, this unimaginative definition would be echoed well into the twentieth century (Mordacq 1921: 15), albeit mainly outside France, where the words 'tactics' and 'strategy' were apparently rarely uttered until after France's crushing defeat at the hands of Prussia in 1870/1 (Mayer 1916: 7).

One of the echoes came from Prussian Field Marshal Helmuth von Moltke the Elder who saw the 'essence' [*Wesen*] of 'strategy' in the preparations needed to get troops to the battlefield simultaneously (q.i. Schlichting 1897: II: 11). Elsewhere he proclaimed more originally that 'strategy is a system of expediencies' which defied general principles that could be taught (Großer Generalstab 1911: 1). His Russian contemporary, General Mikhail Ivanovich Dragomirov, dismissed the concept of a 'science' of war out of hand, instead endorsing the concept of a 'theory of war' (q.i. Foch 1900/1918: 8). Other very technical definitions abounded, such as this by Clausewitz's contemporary Wilhelm von Willisen: 'Strategy is the doctrine of making connections ... the doctrine of battling [*Schlagen*] is tactics' (Willisen 1840: 26). Or take another, that of the Britons Sir Edward Hamley, General J.F. Maurice (1891:7; 1929: 3) and G.F.R. Henderson (1905: 39), who by 'strategy' understood 'the art of rightly directing the masses of troops towards the objects of the campaign'. 'The theatre of war is the province of strategy, the field of battle is the province of tactics.' French General Bonnal, lecturing at the Ecole de Guerre in 1892–3, told his students that '[s]trategy is the art of conceiving; tactics the science of execution' (Castex 1937: 6). In the Cold War, Marxist-Leninist definitions continued to follow narrow definitions of 'strategy' and 'tactics', adding the intermediary level of operation (Leebaert 1981: 14f.).

Clearly, these technical definitions did not make allowance for the political directives under which Strategy operated. Wider concepts were needed. The British military historian Henry Spenser Wilkinson, in discussing naval operations in 1894, gave this definition: 'A policy is national action directed to an end or purpose. The object set up must be one that the nation values and appreciates, or else the Government will have no support in its efforts to attain it. And the means must be suitable to the end' (Wilkinson 1894: 21). A decade later, Lt.-Col. Walter James, while using narrow definitions of 'strategy' and 'tactics', dwelt on the political aims of warfare:

> Strategy is largely affected by moral considerations. Of two different courses – one of which might give important political, the other more purely military results – it will sometimes be more advantageous to choose the former, because of the greater effect it will have on the course of the war. (James 1904: 17f.)

We see how gradually, the line between policy and 'strategy', especially 'grand strategy', was becoming blurred. The emphasis of the link between policy and military execution becomes particularly strong in the writings of Captain (later Sir) Basil Henry Liddell Hart, whose most important works stem from the 1920s, 1930s and 1940s. He dismissed earlier definitions as too narrow, instead developing the concept further again. For Liddell Hart, 'strategy' was 'the art of distributing and applying military means to fulfil the ends of policy' (Liddell Hart 1944: 229). This definition, which has great merits, is so broad, however, that Richards Betts would be justified in criticising it for making 'strategy' synonymous with foreign (or indeed any) policy (Betts 2001/2: 23).

This had already been recognised by French General André Beaufre (1902–75) and French sociologist Raymond Aron (1905–83). Aron suggested fusing the terms 'policy' and 'strategy' in the neologism 'praxeology'. Beaufre, however, decided to stick with 'strategy', using 'total strategy' as equivalent to the British term 'grand strategy'. Hence Beaufre argued that all warfare is 'total', by which he meant 'carried on in all fields of action', political, economic, military, cultural, and so forth (Beaufre 1966/1967: 19–23, 29). This, however, lends itself to considerable terminological confusion in view of other usage of the term 'total war' (as we shall see in chapter 7).

While Aron's term 'praxeology' failed to catch on, agreement on his insistence on the link between Strategy and practice spread. His American contemporary Bernard Brodie wrote in the middle of the Cold War that 'Strategic thinking, or "theory" if one prefers, is nothing if not pragmatic. Strategy is a "how to do it" ... guide to accomplishing something and doing it efficiently ... Above all, strategic theory is a theory for action' (Brodie 1973: 452f.). From this, Colin Gray developed the idea of 'strategic theory' which 'helps educate the strategist so that he can conceive of, plan, and execute strategy by his command performance' (Gray 2010).

With the introduction of the concept of 'grand strategy' in the Second World War, something closely akin to overall state policy on foreign and military affairs, new variations appear in our list of definitions. The US Joint Chiefs of Staff in their *Dictionary of the U.S. Military Terms for Joint Usage* of 1964 defined 'strategy' as the development and use of

> political, economic, psychological and military forces as necessary during peace and war, to afford the maximum support to policies, in order to increase the probabilities and favourable consequences of victory and to lessen the chances of defeat. (q.i. Luttwak 1987: 239–41)

The British political scientist Robert Neild in 1990 defined 'strategy' in an even wider way, as the pursuit of

> political aims by the use or possession of military means. In formulating strategy, the first step is to decide on political aims. Without political aims, war is mindless destruction and the possession of military means in peacetime is mindless waste. Once political aims are specified, the military means must be selected and tailored to fit those aims. (Neild 1990: 1)

Thus the link between policy at the highest level and the use of military force as its tool, postulated by Clausewitz but not yet coupled by him to the word 'strategy', gradually became a matter of universal consensus. And yet there was scope for further refinement of the concept, which, as we shall see, brought further essential dimensions of strategy into focus.

The articulation of different dimensions of Strategy

War as an instrument of politics

The rediscovery of the great political philosophers of antiquity and their ideas about the *polis*, the body politic, the state and its relation to its armed forces, made thinkers of the modern age write about the link between Strategy and politics. A crucial place in the translation of these classical philosophical concepts into modern times is held by Niccolò

Machiavelli (1469–1527), who besides writing on the *Art of War* (structured much like Vegetius' classical handbook) also wrote about politics, in his more famous work *The Prince* and in the *Discourses*. Other philosophers on the state, politics, justice and law, such as Matthew Sutcliffe (1546 or 1547–1629) in England, Justus Lipsius (1547–1606) and Hugo Grotius (1583–1645) in the Netherlands repeatedly touched on war in their works. Just as Roman law had developed concepts of a justifiable use of war, set in stone for the Christian world by Augustine of Hippo and after him Thomas Aquinas, they were mainly concerned with the legality and legitimacy of warfare.

A few exceptional writers in the tradition of Machiavelli brought these strands of thought together. The most prominent are Sutcliffe, dean of Exeter; the Spanish aristocrat, officer and diplomat Don Alvaro of Navia Osorio and Vigil, Viscount of Puerto, Marquis of Santa Cruz de Marcenado (1684–1732); and Guibert. Coming from the classical Roman and then Catholic just-war tradition (see chapter 2), they assumed that the end state of war should be peace, but a more just peace than that preceding the war (e.g. Saillans 1589/1591: ch. 5). For Sutcliffe, Lipsius and Grotius it was taken for granted that peace had to be the end state of war. In the eighteenth century, the Swiss philosopher Emerich de Vattel by contrast reflected on the consequences of the imposition of an *unjust* peace which would lead to renewed war (Vattel 1758/1834: Book IV). At the close of the eighteenth century, Dietrich Heinrich von Bülow in Prussia had no such qualms: he defined the purpose of all operations in war as bringing about 'peace, which one tries to force upon the enemy through the harm done to him, to be advantageous to oneself, and disadvantageous to him' (Bülow 1799: 12). Nevertheless, there was thus consensus from Cicero to the French Revolution that the only sensible aim of war could be a durable peace. Napoleon's insatiable expansionism, however, changed this perception.

The nexus between political war aims and the conduct of war was commonplace by the time Clausewitz put his pen to paper – it was so widely accepted that few saw the need to spell it out. One who did spell it out was August Wagner, who opined that no commander would be greatly successful unless he knew

> what is generally true about all wars; why each war ... has been started; which means are to be applied, not alone to win, but to achieve the aims, for the purpose of which one has taken up arms; in short, who has not

> reflected on his profession and is not able quickly to apply the fruits of his reflections to actual cases. (Wagner 1809: viii)

Another was Clausewitz's colleague at the War Academy in Berlin, Otto August Rühle von Lilienstern (1780–1847), next to Clausewitz the most outstandingly original German-speaking writer of that generation. In his *Handbook for Officers* (published 1817–18), he argued that every war had a cause and a purpose which would

> determine the character and the direction of all activity ... The individual operations have military purposes; the war as a whole always has a final political purpose, which means that war is undertaken ... in order to realise the political purpose upon which the State has decided.

According to Rühle, then, 'Every war and every [military] operation is based on a Wherefore? and Why?, a purpose and a cause, which will give a specific character and a definite direction to each of its actions' (Rühle 1818: 8). In the light of the Napoleonic Wars, and reflecting Vattel's concerns about the consequences of an unjust peace, Rühle concluded that

> victory is not always the necessary condition of conquest or of peace, and peace is not always the necessary result of victory and conquest ... Each war has ... a main purpose, which, however, is not always ... peace. Peace can be seen merely as the termination of the state of war. The obstacle which in war obstructs the attainment of the main purpose is the enemy, and it has to be cleared out of the way. In the best case this may lead to victory, but for this reason alone, victory is not the main purpose of the war, but only a subordinate purpose within war. If somebody concludes a peace without attaining the main purpose ... he can be called the defeated party, however many battles he may have won, even if he has won all of them.

Writing with the Napoleonic conquests in mind, he added:

> To the contrary, victory and conquest are often causes of the continuation, the renewal and the multiplication of war. Often, peace comes because none of the warring parties was able to defeat the other, and often war is not made in order to establish peace. (1818: 8f.)

Rühle pointed out the ambiguity of the term 'peace': is it merely the absence of war or the 'lasting friendly agreement of states among each other'? He drew attention to instances in history when peace was concluded so as better to prepare for the next war, and to instances where war was continued and drawn out because at least one side sought to achieve some gains other than victory. There were wars which were fought to further the personal interests of individual 'officers and state officials, or the army; in short, of some subordinate interest, but not for the sake of the common well-being of the state' (Rühle 1818: 8f.). Victory – if defined as the attainment of such particular interests – cannot therefore be the main purpose, but must be subordinated to much greater aims, such as the aim of turning one's enemy into a friendly power (Rühle 1818: 11).

> Moreover, in view of the political links and networks which all civilised [*kultivierte*] states entertain with one another, in all wars it is almost as important what impression the conduct and the results of the war have on the public opinion and the interest of the other temporarily neutral states, as what relationship the two warring parties have on account of the war [between them]. A temporary advantage, the early humiliation of the enemy, a conquest – however brilliant – are of little value for the state whose existence has to be calculated and secured for hundreds of years, if there is not the hope of keeping this advantage and the conquests for a long time, or if it creates the fear of a new, greater danger ... These concerns about public opinion and the political community of states are so important in determining the legal basis of war, and its essential usefulness, and explain why even very powerful states try at least to package their feuds in an acceptable way and accept limitations on their behaviour even in victory. (Rühle 1818: 12)

While many wars are fought for gains (*Nutzen*) or honour, as Rühle conceded, war *ought* to be only the means of states to obtain justice; 'according to the principles of morality, war should never be waged for any other purpose'. Unfortunately, who is in the right is not always clear, and

> war is thus the way in which states settle their legal quarrels, in one word: their [clashing] political aims, against each other with the use of force. It is the attainment of these political aims, which are the true final war aims, not victory, peace or conquest, if these are not perchance in line with the political intention. The army is merely the acting organ, the executive of the higher will. The army's and its leaders' entire mental activity

should aim to tailor the individual operations, to combine and execute them in a way that their success may deflect any danger from their state, or give it political advantages. (Rühle 1818: 13)

In view of Rühle's far-sighted observations quoted above, which Clausewitz, as his colleague, must have been familiar with, it is ironic that it is usually Clausewitz who gets all the credit for articulating this link between politics and warfare, especially as in *On War* he deliberately desisted from spelling out the implications. It was Clausewitz's posthumous rival, Jomini, who devoted a considerable part of his *Summary of the Art of War* of 1837 to what he calls the 'politics of war', which he uses in a way that comes very close to my definition of Strategy. Jomini compares favourably with Clausewitz in that Jomini reflected explicitly on the relationship between politics and war, especially on the political motives that would lead to war. 'A government goes to war', he wrote,

To reclaim certain rights or to defend them; to protect and maintain the great interests of the state, [such] as commerce, manufactures, or agriculture; to uphold neighbouring states whose existence is necessary either for the safety of the government or the balance of power; to fulfil the obligations of offensive and defensive alliances; to propagate political or religious theories, to crush them ... or to defend them; to increase the influence and power of the state by acquisitions of territory; to defend the threatened independence of the state; to avenge insulted honor; or, from a mania of conquest. (Jomini 1837/1868: 14)

Jomini divided wars into several categories, explaining that these different types of war required different ways of waging them. The categories were:

- 'Offensive wars to reclaim rights', which he regarded as 'the most just war[s]', even though they would normally be waged on territory at that stage held by the enemy (hence 'offensive', involving the invasion of somebody else's territory).
- Wars that were politically defensive, but 'offensive in a military point of view'. This would include pre-emptive wars, wars in which one attacked an enemy anticipating an attack by him. Jomini was convinced, however, that a defensive war carried out on one's own territories held great advantages, as it would have the support of the

population, a well-known theatre of operations and help from all the local authorities (Jomini 1837/1868: 17).

- 'Wars of expediency', to snatch something from an adversary who happened to be going through a time of weakness or disarray. What he had in mind was Frederick the Great's seizure of Silesia (Jomini 1837/1868: 18).
- 'Wars with or without allies'.
- 'Wars of intervention' in the 'internal affairs of a neighbouring state'.
- 'Aggressive war for Conquest and other Reasons' à la Genghis Khan, which could be 'a crime against humanity' [*sic*], even though Jomini thought that 'it is better to attack than to be invaded' (Jomini 1837/1868: 23).
- 'Wars of opinion' or what we would call ideological wars (such as the war between Revolutionary France and its adversaries, and, looking beyond Jomini's own times, Nazi Germany and the Soviet Union, or many conflicts of the Cold War pitting communists against non-communists).
- 'National wars', by which he meant wars of resistance against foreign invasion involving the mobilisation of the entire people, with the wars of resistance and liberation against Napoleon's forces in mind. He had personal experience of the Spanish War of Liberation of 1808–12, which he had experienced as particularly dreadful, using the expression 'wars of extermination' to describe them, when Spanish insurgents wiped out – exterminated – whole French units in ambushes and night attacks (Jomini 1837/1868: 29–35).
- 'Civil Wars, and Wars of Religion'.

He stressed that each of these wars had to be waged differently – in 'wars of opinion', 'national wars' and 'civil wars/wars of religion' the rabble was involved in a way in which it was not in 'wars of expediency'. His categories overlap in places; a defensive war might not be distinguishable from what he called 'national wars' as he saw these as defensive. Nevertheless, Jomini's categorisation goes a long way to take political aims into account as chief variable determining the character of any Strategy.

That one's conduct of war should be governed by politics was a disputed concept, however. Lossau in his handbook stated apodictically that where politics 'ceases to have its effects, war starts'. Politics only decides the moment when peace yields to war. At least he conceded that the politics – he should have said ideology – of a state determined its defensive or offensive disposition (Lossau 1815: 7). With this he founded the Prussian tradition of those who opposed the interference of political decision-makers in the conduct of war, and in the words of Colonel (later Field Marshal) von Manteuffel to Prussian Prince Frederic Charles in 1857, warning him to keep his nose out of what was the military's business: 'when the sword has been drawn, war ... steps into the foreground, becomes fully independent, and politics becomes its servant' (quoted by Hahlweg in Clausewitz 1832/1976: 67). Moltke would famously go even further in his resistance to Bismarck in the context of the Wars of German Unification (Carr 1991).

The political role of Strategy, even as applied within war, gradually won out against this attempt to cut politicians out of the conduct of war. This did not wipe out the tensions between political leaders and the executing military which this division of labour necessarily entailed. The technical approach which we sketched in the previous section still reverberates in the 1989 definitions of 'strategy' by the US military, but it simultaneously acknowledges the political aims of warfare (Handel 1996: 36). But by the late twentieth century, John Garnett's definition of 'strategy' as 'the way in which military power' is or might be 'used by governments in pursuit of their interests' would have found universal recognition (Garnett 1975: 3). The late Michael Handel (1942–2001), venerated teacher of generations of US officers, put it more simply and trenchantly: 'strategy is the development and use of all resources in peace and war in support of national policies to secure victory' (Handel 1996: 36). We see in both American definitions a much wider understanding of Strategy, which takes on board the nexus between policy and war as its instrument.

Dialectics of will

All the definitions of 'strategy' we have encountered so far fall short of taking into account that war has two sides: how can a definition of Strategy take into consideration the interaction of one's own side with the enemy? Clausewitz had been a trailblazer here – his comparison of war with wrestling incited later strategic thinkers to build this dimension into the relationship between power and the use of force, so as to take account of the *dialectics* of the use of force. The supreme commander of the Prussian forces in the wars of Wars of German Unification, Helmuth von Moltke the Elder (1800–91), was one of them. Like others before him, he described 'strategy' as positioned on a level between the higher sphere of politics and the lower plane of military operations.

Politics uses war to reach its purposes, it influences decisively the beginning and the end [of war], and retains the right to increase its demands during its course or to settle for lesser aims. Given this uncertainty, strategy can only try to obtain the highest possible aims which could conceivably lie within its reach in view of the available means. It is thus that [strategy] best serves politics, by working for the purpose of politics, but quite independently from [politics] in its actions.

He went on to explain that the next task of Strategy, after that of serving politics, is to prepare the means of waging war. This task had to be fulfilled as a function of given resources, geography, logistics and so forth.

Matters are different concerning the subsequent task of strategy: the military use of available resources, that is, in operations. This is where our will soon encounters the independent will of our adversary. Although we can impose limits on it, we can only break it by the means of tactics, [i.e.] through battle. (Moltke 1960: 316)

More importantly, however, Moltke produced the famous dictum that a battle plan does not survive the actual encounter with the enemy:

It is a ... delusion if to believe that one can determine a campaign plan far in advance and carry it out until the end. The first clash with the enemy's main forces creates a new situation, depending on its outcome. Much of what one had intended to do becomes impossible to carry out, some things become possible which could not have been expected earlier. The only thing the army command can do is correctly appreciate the changed circumstances, and then to give instructions to do what is appropriate for the foreseeable next phase. (q.i. Rohrschneider 1999: 157)

After the First World War, Johannes Kromayer developed this strategy further. In the middle of the great German 'Strategy debate' surrounding Delbrück (see chapters 4 and 7), he – rightly – argued that Clausewitz's idea that policy determines a firm set of war aims at the outset of war was deficient, as one's war aims, and policy itself, and with it Strategy, must surely change throughout any war as a function of the success or failure of one's operations (Kromayer 1925a: 401f.). In the middle of the Cold War, André Beaufre developed this idea further. In Clausewitzian terms he saw 'strategy' as 'the art of the dialectic of force, or more precisely, the dialectic of opposing wills, which use force for the settlement of their disputes' (Beaufre 1963/1965: 22). The American military historians Williamson Murray and Mark Grimsley came to similar conclusions: 'strategy is a process, a constant adaptation to shifting conditions and circumstances in a world where chance, uncertainty, and ambiguity dominate' (Murray and Grimsley 1994: 1; Murray 1999a: 33). This in turn was echoed by their British colleague Hew Strachan, who argues that 'strategy in war is a process' that requires continuous adjustment in the light of enemy action and a continuous reconsideration of policy and new policy-making, involving political leaders, military leaders and other experts (Strachan 2006: 59–82).

The realisation that any given Strategy must not be static (if it is to be successful), but must react to and be re-formed according to the interaction with an enemy, is another huge step forward in our understanding. It has not, however, penetrated Strategy-making sufficiently in practice, nor has the concession that should logically flow from it, namely, that the achievement of a stable peace will require concessions and a commitment to it from both sides (Heuser 2007a). Instead, the century and a half from the French Revolutionary Wars to the Second World War was dominated by the quest for the enemy's unconditional surrender, and thus de facto a replacement of the enemy regime. This would prove fatal if the 'hearts and minds' of the populations of the adversarial country(ies) could not be won by persuasion to embrace the post-war settlement.

Another crucial realisation is that Strategy is a function of variables – such as one's own political aims and the enemy's political aims – but of partly interconnected variables, which makes the whole equation even more complicated.

War as a function of multiple interdependent variables

The first step on this intellectual exploration of Strategy was Jomini's categorisation of wars, which implicitly shows that one's own war aims vary, and are thus variables. But further variables could be identified.

It was a discovery of Clausewitz's, in my view his most original and insightful one, that war is a function of variables some of which in turn are *interconnected*, that is they are functions of each other. Particularly famous is Clausewitz's 'remarkable trinity' of variables:

- 'Primordial violence, hatred, and enmity, which are to be regarded as a blind natural force', and these he correlated to the passions of the people as a whole: the more the people were involved in a war, the more they identified with it, the more violent the war would be.
- 'The play of chance and probability', together with 'the interplay of courage and talent' that depended on the peculiarities of the military commander and the army, the commander's *coup d'oeil*, the morale of the troops and so on.
- Policy, using war as its instrument, subjecting war 'to reason alone'. This he identified with the intentions of the government (in other words, its political war aims).

Clausewitz surmised that any war is a function of all three sets of variables and, crucially, recognised that they affect each other: for example, an upwards trend of violence, hatred and enmity might force governments to extend formerly modest war aims. Or a population's lack of emotional engagement in a war might undermine the morale of the armed forces committed. The Clausewitzian notion of war as a function of *inter*dependent variables was taken up by Marshal Ferdinand Foch (1851–1929) and others who studied him assiduously in the late nineteenth century in order to find out why the Prussians had defeated France so thoroughly in 1870/1 (Derrécagaix 1885; Foch 1900).

Clausewitz had identified further variables, strewn here and there in his text: in a particularly poignant chapter in Book VIII, he noted that every age has its own way of warfare, thus identifying what today we would call 'culture' as one crucial variable. Material variables, especially the terrain of the battlefield, the ability or inability to communicate fast and gather intelligence during battle (the 'fog of war'), and a number of other technical and circumstantial factors were also emphasised by him.

Clausewitz was not the first to have identified the cultural variable. Classical writers had commented on the very different 'ways of war' of individual culturally very different groups (Scythians, Persians, Huns, Saracens, Turks, etc). In modern times, the central theme in Guibert's *Essai général de tactique* was the nexus that he saw between a society's values (and thus, culture) and internal political system and its way of war (Guibert 1772/1781). Henry Humphrey Evans Lloyd (1718–83), a Welsh mercenary who in his life fought for Louis XV of France, for Empress Maria Theresia of Austria and Empress Elizabeth of Russia, identified political culture as a variable in the waging of war: drawing on the usual examples from classical antiquity but also from his own experiences, he differentiated between the ways despotic, monarchical and republican governments used force in interstate affairs. To him, a democracy (republic) was clearly predestined to have a defensive overall Strategy, and was ill equipped to wage long wars or wars far from home; Lloyd also assumed that democracies would have neither a standing army nor mercenaries, but that its population would rise up as militia to defend its own state (Lloyd 1781/2005: 458–78).

Guibert and Clausewitz noted, and the Clausewitz pupil (and critic) Martin van Creveld brought to our attention again in the late twentieth century, that different cultures perceive war differently. As we have seen, Rühle noted that not all pursue peace. There are cultures (and sub-cultures, thinking of sections of the military even of fairly peaceful civilisations) where the warrior is admired, and there are age-groups (particularly adolescents and young men) in which the excitement of the adventure of 'war' outweighs other cultural counterbalances. As van Creveld put it, there are people who like to make war, and political factors may simply be an excuse for doing so (Creveld 1991, ch. 6). One is well advised to take these factors into account before espousing any theory which sees any violent conflict as guided by realistic political aims from its beginning to its end. Clausewitz underlined the many forms a war can take by likening it to a chameleon: he described war as infinitely variable, depending on a multitude of contextual factors, the many variables alluded to above. In some conflicts between large groups of people (such as tribes, warring factions) or states (with or without the use of force), political aims can be fairly well established. They may show a conscious use of force or the threat of its use in support of these political aims, to change the will of the adversary and to settle the dispute to one's own advantage.

In other wars such aims are less easily discernible, if at all, which can be for a host of reasons: the lack of a strong, co-ordinated leadership and the resulting multitude of unconcerted, divergent interests, but also the previously invoked cultural factors.

This particular wheel has been reinvented by anthropologists towards the end of the twentieth century, albeit in a more systematic way, and with an emphasis on different variables. Employing case studies from different cultures, they have identified three sets of variables that dominate war, apart from situational constraints that are special for each case:

1. material variables, such as the natural environment (geographic features, climate ...), technology, the economy;
2. social institutions (anything from clan loyalty or kingship to statehood), including patterns of behaviour that are largely taken for granted and seen as norm;[2]
3. culture (mainly collectively shared belief clusters, images, symbols and myths), defined sometimes to include norms and patterns of behaviour (Snyder 2002: 14f.).

While some, like the anthropologist Raymond Kelly, have looked at these variables as independent from one another, the anthropologist Bronislaw Malinowski, pioneer in this area, but also the philosopher Emile Durkheim, the historian Ernest Gellner and the anthropologist Brian Ferguson postulated a hierarchy of factors ('a nested hierarchy') in which material factors were the most important, influencing institutions, which in turn influenced or even determined culture. Others, like the anthropologist Simon Harrison, reversed the order of nesting, insisting that it is culture that determines patterns of social behaviour and institutions, and even determines how people deal with their environment. In the light of the historical evidence of the evolution of warfare, especially in the Western world, one cannot but agree with the anthropologist Jack Snyder, who postulates reciprocal influences and causality among all three sets of variables. Snyder underlined the effects of their interaction and 'complex feedback relationships' in distinct circumstances (Snyder 2002: 12, 32f.). To sum up, then, anthropologists like some of the early strategists

[2] According to certain definitions I prefer, social institutions, and especially norms and patterns of behaviour, are subsumed under 'culture'.

before them analyse war as a function of material factors, social institutions and culture.

'Social institutions' and 'culture' overlap, according to many definitions. For this reason and for the purposes of this book, it is more useful to redefine these second and third levels of variables. Subsuming both categories into a single category of 'culture', we shall examine the conduct of war as a function of passive and active aspects of culture. Passive aspects are mindsets or beliefs about the world: for example with hostile groups confronting us, with lessons of past wars to guide us, with rules and conventions of behaviour which would be dangerous or immoral (or both) to ignore or challenge. As John Hattendorf put it, 'strategists think in the context of the prevailing cultural and national attitudes that surround them' (Hattendorf 2000: 1, 21, 127).

Here we should include existing social structures and institutions, and also beliefs and myths about oneself, one's own group, and one's enemies, beliefs about the working of the world, beliefs about moral obligations, existing customs or traditions that have to be upheld. Active aspects of culture are those where freedom of choice is more pronounced: these include the prioritisation of certain values and principles over others, the definition of political aims, the changes that may be brought about through the agency of war (or the threat of the use of armed force), the institutions, norms of behaviour yet to be created and prescribed as desirable. Active aspects of culture tend to be subordinate to its passive aspects: few people can escape, even in their imagination, the world in which they live and which many assume to be immutable. Yet the 'complex feedback relationship' which Jack Snyder postulates exists here too, as all innovation, once realised, in turn affects culture as it is passively perceived.

To return to Jomini, he named further variables, which included the degree of the passions aroused by either side in a war (which correctly implies that what for one side is a 'war of opinion' – fought out over ideological differences worth dying for – might merely be a war for a limited political aim for the other side, mobilising much less public support). Additional variables for Jomini were the military systems of both sides, that is how and whom they recruited as soldiers, with what reserves, financial resources and weapons, and the degree of their loyalty to their military and political commanders.

The 'character' of the head of state or government, the talents of the military leaders, the relationship of political decision-makers and military leadership and of course enduring physical factors such as geography and a state's wealth in human beings, in natural resources, industry and social structure, all were identified by Jomini as variables that would influence the conduct and outcome of a war (Jomini 1837/1868: 38–65). In the following chapters, this very useful list of variables will be applied to war in different ages.

Towards the end of the nineteenth century, Mahan produced another list of variables which he thought influenced Strategy. He listed geography, with its sub-aspects of sea lanes, harbours and territory (*Hinterland*), the size of the population, the character of the people and the character of the government, the political system and within it the ideology and political orientation of the political party/parties in government (Mahan 1890: 25, 57). The 'character of the people' was a *topos* going back to antiquity, of course, had often been invoked throughout history, but acquired a new appeal in the era of Social Darwinism. Writing a little later than Mahan, Sir Herbert Richmond (1871–1946), a Royal Navy admiral turned Cambridge don, wrote a study of Britain's war of 1739–48 by identifying the following factors that came into play: geography, and the requirements that different parts of the British Empire had, the ships available to the belligerents, the manpower, greater political and diplomatic aims, parliamentary politics and pressures and the wrangling among the main decision-makers and players on either side of the war within the opposing governments (Schurman 1965: 140f.).

Almost a century later, another naval specialist, the Briton Geoffrey Till, established a new list of (arguably interconnected) variables influencing Strategy, very much in the tradition of Mahan:

a. a maritime community;
b. resources (and the economic basis for a big navy);
c. styles of government;
d. geography and geopolitics;
e. shipping;
f. naval bases;
g. the fighting instrument (Till 1982: 75–90).

The Belgian teacher of Strategy Henri Bernard argued that military history had as its proper subject of study the evolution of the art of war over the centuries, as a function of the multiple variables he discussed, applying a Clausewitzian template, in his three volumes called *Total War and Revolutionary War* (essentially a course on military history up to 1945). His variables included demographic, social, political, economic, ideological, technological and institutional factors such as command structures within the armed forces (Bernard 1965: 5).

Again we see that this list is interconnected – no strong economy without the demographic basis to keep it going, no free society without a related economic structure, no totalitarian ideology without attempts to centralise control over all economic activities within the state. Large, standing armies evolved with the creation of centralised states with the infrastructure required to raise the taxes to maintain such an army. Nationalism arose first out of the rhetoric of the 'nation' which began to be used during the French Revolution, underpinned by the use of the *levée en masse*, which linked citizenship with the obligation to defend the nation. Subsequently, the growth of national printing presses and literacy helped further the growth of nationalism.

A refreshingly original categorisation was produced by T.E. Lawrence (1888–1935) on the basis of his reading of the eighteenth- and nineteenth-century texts. Lawrence defined three 'elements, one algebraical, one biological, and one psychological'. The first he liked to call *hecastics*, and it was

> pure science, subject to the laws of mathematics, without humanity. It dealt with known invariables, fixed conditions, space and time, inorganic things like hills and climates and railways, with mankind in type-masses too great for individual variety ... It was essentially formulable ... The second factor was biological, the breaking-point, life and death, or better, wear and tear. Bionomics seemed a good name for it. The war-philosophers had properly made it an art, and had elevated one item in it, 'effusion of blood', to the height of a principle. It became humanity in battle, an art touching every side of our corporal being, and very war. There was a line of variability (man) running through all its estimates. Its components were sensitive and illogical.

The third factor Lawrence described as psychological, 'of which propaganda is a stained and ignoble part', but linked it to what Xenophon called 'diathetic', from a Greek word for 'order'.

> Some of it concerns the crowd, and adjustment of spirit to the point where it becomes fit to exploit in action, the prearrangement of a changing opinion to a certain end ... It considers the capacity of mood of our men, their complexities and mutability, and the cultivation of what in them profits the intention. We had to arrange their minds in order for battle ... and through [our own men] ... the minds of the enemy ... and thirdly, the mind of the nation supporting us behind the firing line, and the mind of the hostile nation [a]waiting the verdict, and the neutrals looking on. (Lawrence 1920, 266f.)

We see echoes here of Clausewitz's trinity and a very trenchant and helpful perception of what in the evolution of warfare is variable, what are immutable constants.

New technical developments spawn ideologically driven fantasies and speculations about how to use armies, and about political consequences. Historical experiences of wars, especially traumatic ones, determine the subsequent preoccupations of survivors. Decisions made about war are a function of the structures of the societies that wage the wars, and decision-makers' ideas and views are conditioned by the mindset particular to their culture. War aims are dictated by the concepts of the world, of society, of friend and foe, and of notions of what one can achieve through military manpower and technology. And Strategy is a function of all these variables and many more. This analytical approach will be used to formulate a series of guiding questions to examine the evolution of Strategy in the following chapters.

Strategy in peace and war

But does force always have to be used in order to settle disputes? Can one not change the enemy's will by the threat of the use of force alone? Is Strategy not something that is part of peace as well as war? These questions were particularly acute when in the twentieth century authors realised that the absence of war did not amount to the absence of interstate strife, with the threat of another world war overshadowing peace.

As we have seen, some writers of the nineteenth century included the *preparation* for war in peacetime among the tasks of Strategy. In the age of total war, the realisation dawned on several writers on the subject that this was not the only role of Strategy outside periods of declared war. The political extremes of communism and fascism met not least in their common perception of peace as the continuation of war by other means. Western liberal thinkers, by contrast, developed a different perception. They began to see the use of force as only one instrument of state policy, alongside many others, like diplomacy, trade policies and so on. In turn, the latent threat of the use of force could be an instrument of state policy in times of peace. Western liberal thinkers like Norman Angell (1872–1967) did not doubt that conflicts of interest occurred in times of peace, but they tended to think more of these conflicts as resolvable without actual fighting. Even among them, few would doubt that politics, including of course the relations between states, would always include conflict and strife and a struggle of wills. But if military force could be a latent instrument of policy, Strategy needed to be redefined. In the interwar period, Admiral Castex noted:

> [S]trategy is ... the general conduct of operations ... Strategy prepares combat, makes the effort to carry it out in the best conditions, and to produce the best results ... [Strategy] contains [*détient*] the general idea to which the campaign is dedicated ... It guides tactics, leaving it room for action when its hour has arrived. Strategy is on each side of combat, tactics during combat. (Castex 1937: 9)

But Castex also adopted the contemporary idea that Strategy was operational in times of war and also in times of peace, and the view that naval conferences and diplomacy were just as much tools of Strategy as the use of force in war (Castex 1937: 17f.).

In 1943, while the Second World War was being waged and the public did not know the secrets of the Manhattan Project, the American scholar Edward Mead Earle edited his famous volume *The Makers of Modern Strategy*. Here he wrote, much as Moltke the Elder had done: 'Strategy deals with war, preparation for war, and the waging of war. Narrowly defined, it is the art of military command, of projecting and directing a campaign.'

He noted that the word was used more narrowly for the 'art of military command' which the general had mastered in order to 'win victory' in battle until the end of the eighteenth century.

> But as war and society have become more complicated – and war ... is an inherent part of society – strategy has of necessity required increasing consideration of nonmilitary factors, economic, psychological, moral,

political, and technological. Strategy, therefore, is not merely a concept of wartime, but is an inherent element of statecraft at all times ... In the present-day world, then, strategy is the art of controlling and utilizing the resources of a nation – or a coalition of nations – including its armed forces, to the end that its vital interests shall be effectively promoted and secured against enemies, actual, potential, or merely presumed. The highest type of strategy – sometimes called grand strategy – is that which so integrates the policies and armaments of the nation that the resort to war is either rendered unnecessary or is undertaken with the maximum chance of victory.

And it is in this broader sense that Earle used the term in his path-breaking volume (Earle 1943: viii).

After the Second World War, well into the Cold War, the study not only of international relations (really, interstate relations) but also of Strategy took off, with many new definitions being added, some coming from new disciplines. Thomas Schelling (1921–) came to strategic studies from economics and, specifically, games theory, introducing his own term, the 'strategy of conflict', in which various contestants (who might be state regimes) might try to reach outcomes to their conflicts that were mutually advantageous. In the nuclear age, neither side could find much comfort in the notion that Strategy was all about imposing one's will upon an enemy who could make the price for such a success exorbitant. With 'strategy of conflict', Schelling emphasised the 'interdependence of the adversaries' decisions', in conflicts which were equalled to 'variable-sum games', in which 'the sum of the gains of the participants involved is not fixed'. While victory seen as the imposition of one protagonist's will upon the other is a fixed-sum game in which one's gain is the other's loss, Schelling introduced the idea that – especially in the nuclear age – this is only one of many possible outcomes. Crucially, the avoidance of nuclear war might be a mutually advantageous outcome that would persuade warring parties to settle for less than an all-out imposition of their will upon the enemy (Schelling 1960: 3–5).

Robert Osgood, another leading American expert on security issues, drew similar consequences for the meaning of Strategy:

[M]ilitary strategy must now be understood as nothing less than the overall plan for utilizing the capacity for armed coercion – in conjunction with the economic, diplomatic, and psychological instruments of power – to support foreign policy most effectively by overt, covert, and tacit means. (Osgood 1962: 5)

The definition proposed by John Hattendorf implies a similar approach:

Strategy reaches for an overarching idea, reflecting an understanding of an entire war. While strategy involves this conceptual dimension, it is, at the same time, the comprehensive and actual direction of national power, including armed force, to achieve some measure of control over an opponent, and, by that control, to achieve specific practical and political ends. (Hattendorf 2000: 122)

Sir Lawrence Freedman, with his political science background, put it perhaps most elegantly: 'Strategy is about the relationship between (political) ends and (military, economic, political etc.) means. It is the art of creating power' (Freedman 2008: 32).

Oxford historian Hew Strachan rightly concluded that the term 'strategy' has thus undergone a considerable shift in meaning and usage since Clausewitz was writing. Until the First World War, 'strategy' was used by most writers to mean something below politics in a hierarchy of determinants. Since then, terms like 'grand strategy' or 'major strategy' (as opposed to 'pure strategy' or 'minor strategy') have been coined, embracing the pursuit of political ends (primarily in international relations) not only with military tools, but also with diplomatic, economic or even cultural instruments. The Cold War with its blurred distinction between war and peace finally pushed 'strategy' over the fence up to the level of politics, leading to a 'conflation of strategy and politics' (Strachan 2003). One attempt to bring clarity to this area is the introduction particularly in Britain of the term 'grand strategy', referring to the way political aims are translated into the use of different available tools of state politics (Cabinet Office, Historical Section 1956–76). But the expansion of the word 'strategy' in contemporary usage continues.

All in all, the word 'Strategy' is hard to press into one universally accepted definition valid through the ages. Nevertheless, important insights that have been gained by successive strategists building on previous generations include the following. Strategy is a comprehensive way to try to pursue political ends, including the threat or actual use of force, in a dialectic of wills – there have to be at least two sides to a conflict. These sides interact, and thus a Strategy will rarely be successful if it shows no adaptability. Before the French Revolution, it was not spelled out but tacitly assumed that the antagonists – mainly princes – had common aims and that Strategy was not a 'zero-sum game', as no side wanted the total destruction of the social order of the other; in most contexts, both sides assumed each other's survival. From the time of the French Revolution until 1945, by contrast, the assumption that Strategy was a 'zero-sum game' prevailed (although these words were introduced by Schelling only in the subsequent period). To win, one side had to impose its will upon the enemy; or at least this is what the vast majority of strategists assumed. The nuclear age made strategists aware again that conflicting sides might have common interests – especially, the avoidance of nuclear apocalypse. Strategy once again was opened up to the highly political dimensions of bargaining and the quest for 'win-win' solutions that might satisfy – at least to some extent – both sides and avoid worst outcomes (especially major or nuclear war).

In today's popular usage, the term 'strategy' is applied to many realms of life outside politics proper. Its merger with the jargon of economics and management is particularly prominent (Strachan 2003). Today, the advertisement of a vacant 'chair in strategy' is as likely to refer to a branch of business management as to anything military. Meanwhile, governments try to develop 'strategies' for dealing with unemployment, housing shortages, education and so on, and every business has a business plan or 'strategy'. In business-speak, 'strategy' is defined as 'the direction and scope of an organisation over the long term, which achieves advantages in a changing environment through its configuration of resources and competences with the aim of fulfilling stakeholder expectations' (Johnson *et al.* 2005: 9). The conflation of 'strategy' and 'politics' or even 'long-term economic aims and planning' and the vaguely synonymous use of the term 'strategy' with that of 'policy' (and thus the inflation of the term 'strategy') can be deplored or criticised as unhelpful, or taken as a matter of fact and worked around (Strachan 2003). In this book, however, I shall try not to use the terms as synonyms, but to keep them apart as far as possible. I shall be using the terminology of my sources where at all reasonable, unless this leads to excessive confusion, and given the nature and concerns of these sources, problems will not arise too often.

What is this book examining?

The evolution of the literature on Strategy is of interest for three reasons. First, because writings on Strategy influence the actual conduct of wars, sometimes directly, if the writer was at once a practitioner, or the Strategic concepts were applied or where the literature, staff college textbooks or field manuals prescribing standard operational procedures. More often, literature on Strategy would influence the practice of warfare indirectly, perhaps with a long delay in time, sometimes through ideas that were misrepresented, distorted and misapplied. Either way, the concepts expressed in the literature on Strategy help us better to understand the operational Strategies applied in actual wars.

Secondly, when implicit assumptions in the literature on Strategy are rendered explicit through textual and contextual analysis, we find revealing indicators of social institutions and norms, of the writers' perceptions and understanding of politics and relations between political entities (often in the light of 'lessons learned' from particular historical events), of their values, ideologies and passive and active culture more generally.

Thirdly, once this literature is analysed in its historical context, we can identify more timeless generalisations their authors have made with a claim to truth beyond the immediate material, cultural and other circumstantial factors in which it was created, which take us back to the quest for a better understanding of the phenomenon of war itself. Strategy, from this point of view, is the process by which human, material and cultural factors are brought to bear upon war.

We have noted that before the twentieth century the term 'strategy' was not used in the sense generally accepted today. But even without the actual terms, people thought – not always systematically – about some of the questions that arise from this definition, which will be used to interrogate the texts about warfare that we will be passing in review. They are:

- In their 'passive culture' or mindsets, do they see war as an inevitable part of the world, a legitimate activity, or something to be contained, avoided, or even ultimately banned? Is war seen as the 'normal' solution to such conflicts? If alternatives to waging war are seen, are they dismissed, and if so, why?

- What is the nature of any particular war, what aims are being fought for?
- For whom or which audience is the author writing, who is the presumed 'we' in the work, and what are 'we' fighting for? Who are the combatants? Are they part of the presumed 'we' as citizen-soldiers, or are they hirelings, mercenaries, economically precious but politically unimportant, to be used as the general saw fit?
- Who is the adversary, and how is the adversary seen – is it a hated enemy or an equal, is it an individual prince or a regime, a group, a nation or a race? Is he/she, are they seen as legitimate contestants, or rebels, heretics, indeed sub-humans?
- What are the parts played by particular parts of warfare – preparation, logistics, siege and, above all, battle?
- How is the adversarial population to be treated – punished for disobedience/heresy, annihilated in a Social Darwinian or racist contest for resources and *Lebensraum*, or spared if they are non-combatants?
- How is the other side to be defeated? Through victory on the battlefield, through occupation of the capital, the entire country, through starvation by blockade, destruction of infrastructure and resources, through social reform/revolution or regime change?
- What role do geography and technology play?
- How are the means of warfare – such as the armed forces, militias or professional armies, navies or air forces, conventional or nuclear weapons – and how are the basic postures – defensive, offensive – determined by the ideologies of those who wish to employ them?
- How did ideology, culture and political aims translate into the use of force, planned on paper or implemented in the field?

And once these points are established for particular writers in their historical context, we can ask,

- Was there a 'Western way of *thinking* about war', one recognisable overall set of answers to these questions, or were there always several?
- Did different approaches to Strategy exist side by side, or was there a succession of different ways of thinking? If the latter, were there distinct watersheds or turning points?
- Did they evolve from each other in a unidirectional fashion? That is, did Western thinking – against the background of practice – evolve

steadily from a limited towards an ever more total form of war, or did it meander backwards and forwards between limited and absolute or total war? Were there just these two poles?
- How distinct and original was thinking about naval and maritime Strategy, as opposed to the older strategies for land warfare? How little or much of a departure was thinking on air power and then on nuclear weapons?
- And on a more detailed level, where did certain ideas originate, who picked them up in later context, how did they apply them and perhaps pass them on in turn?

Far from all the authors considered here wrote about all these dimensions, and if they did, rarely in a coherent fashion. Many authors from Machiavelli to Clausewitz did not see a book on war as the proper place to reflect on the relationship between states, the world order, political aims and norms of conduct – for such reflections, one has to turn to Machiavelli's *Prince* or his *Discourses*, not his *Art of War.* Authors of manuals did not see politics as falling within their remit. In turn, the great lawyers like Hugo Grotius with his *Law of War and Peace* rarely reflected on how to pursue political aims on the battlefield.

Before the mid nineteenth century, the two sides only came together in the minds of a handful of exceptional individuals, such as Machiavelli, Sutcliffe, Santa Cruz de Marcenado and Guibert, several of whom had experience both in military command and in policy-making. Even after 1850, few wrote about the use of force in a political context, however evident it was to them that the use of force had a political aim. The numbers of those who consciously brought in the political dimension only truly expanded from the eve of the First World War onwards. In turn, 'Strategic Studies', with its heavy emphasis on this dimension, only became recognised in the Anglosphere from the mid twentieth century, and outside the Anglosphere few people know even today what to make of that word. Nevertheless, flashes of insight into their views of the world order, of potential or actual adversaries, of moral or customary constraints on war, on the political purposes of war in their age can be gleaned from the writings even of some of the field manualists, and there are parts on the conduct of war which are replete with political considerations even in manuals for 'the perfect captain'.

The literature considered here

Most writing on war falls into one of the following categories: historiography, in which causalities of events might be discussed and which thus produce comments on strategies and tactics chosen (and their appropriateness); manuals with practical instructions for generals or higher officers in the field; political philosophers discussing war or the structure and ethos of armed forces in the context of state and politics; theologians, lawyers and moral philosophers writing about the ethics of war; and finally, analyses of war, a genre that was largely invented by Clausewitz (Heuser 2007b).

Ancient historians writing about wars whose works were used time and again by subsequent generations, apart from Thucydides (*c.* 460–395 BCE), the one great favourite even today, included especially Xenophon (430–355 BCE) and Polybius (*c.* 200–*c.* 120 BCE). Some classical authors actually wrote manuals on how to conduct war, including Aeneas the Tactician (4th c. BCE), Flavius Arrianus (after 85–after 160 CE), and his contemporary Aelianus Tacticus. All these were printed for the first time in the sixteenth and seventeenth centuries. But none of them had the widespread readership of Vegetius. The Middle Ages produced little beyond translations or at best updates of Vegetius.

After the invention of the printing press, from 1470 to 1642, Geoffrey Parker has counted 162 English-language and 460 foreign-language books published on warfare (Parker 1976: 197 n. 7). These numbers increased steadily in the following centuries. In the seventeenth century, at least 100 books were published dealing with the 'art of war' and similar subjects. From 1700 to 1756, a further fifty-six such books appeared, with a significant increase from 1748. From the outbreak of the Seven Years' War in 1756 until the outbreak of the French Revolution in 1789, over 100 books on this subject area were published, and publication numbers remained high during and just after the French Revolutionary and Napoleonic Wars (Gat 1989: 25). There was a slump in the publication of monographs in the mid nineteenth century, although by now a number of professional journals were established in several Western countries, adding to the literature through articles.

Until the end of the nineteenth century, writing on war was almost exclusively the domain of soldiers or ex-soldiers on the one hand and international lawyers on the other. There were a few exceptions to this rule, including Friedrich Engels, Herbert Spencer and Jan Bloch, but even these belong to the second half of the nineteenth century. It was only after the First World War that more civilian thinkers took an interest in war. Accordingly, with rare exceptions, most writers about war until the early twentieth century tended to focus very much on tactics and what Peter Paret has called military 'specifics' (Paret 1983: 164–7). Strategic studies, analytical writing about war as an instrument of policy, increased exponentially after the Second World War, and the discipline of International Relations with its focus from its inception on conflict has established itself in many universities worldwide. Space makes it impossible to provide a fully comprehensive analysis of what has been written about Strategy, even in the Western world, since 1945. Sub-themes such as nuclear Strategy, or counterinsurgency alone, have produced libraries that can compete with those on Shakespeare. The need for a selective study of this literature is therefore blatant. The selection principles for this book will include the celebrity of the argument made (that is, its subsequent prominence in the debate and influence on others), its focus on something largely resembling or overlapping with our definition of Strategy and/or the degree to which it seems to me typical of evolving expert opinion on the subject.

This book is *not* about actual, applied government strategies, but mainly about the thinking of strategists who published their works. While I shall occasionally quote from some government documents, on the whole they will remain in the background, as will all writing on tactics (except where the word is used in the modern sense of Strategy, for example in Guibert's *General Essay on Tactics*). Manuals on the art (or science, as the authors prefer) of war will be used where they contain ruminations about the political purposes of war or other connections between politics and military force. Works written to educate princes or to help generals will be used, provided they tackle such a political dimension. The bulk of the literature considered in the present volume will be monographs (sometimes multi-volume works) on warfare. Some aimed to capture the nature of war and its relations to political ends, society and culture, even though none of them saw their conclusions on this level as entirely divorced from practical applicability; theirs was rarely a quest for understanding for the mere sake of understanding, but for more informed policy-making and conduct of war. Overall, my analysis will be 'front-loaded', in the sense that the earliest expressions of any idea dealt with here will be given more attention than those which, coming later, merely adopted and reiterated the concept.

A final note on the interaction of ideas is required. Before the French Revolution, knowledge travelled slowly, but finally had a wide effect. When Jacques François de Chastenet de Puységur, Marshal of France, was writing down his principles and rules on the art of war (between 1693 and 1743), he claimed to have no knowledge of any modern work on war except for the memoirs of Montecuccoli (1609–80) and Turenne (1611–75). He claimed that there had been no structured works on the principles on the art of war since Vegetius (Puységur 1748: 3). His book indicates that he had, after all, read Vauban on siege warfare and constructions. But otherwise, he had clearly missed out on the entire Italian, Spanish and English literature that had been produced until that date on the art of war, and he had even missed out on the books of his countrymen Henri, duc de Rohan (1636) and François de la Vallière (1666). The circulation of literature was already increasing in the second half of the eighteenth century, but by the end of the nineteenth century, on the eve of the First World War, European writers read each other's work quickly and avidly. While at the dawn of the twenty-first century electronic journals make it possible for articles to reach a worldwide readership on the very day of publication, there is in some areas of the debate about Strategy a greater parochialism now than from the Renaissance to the eve of the First World War, when the 'strategic community' was both smaller and more international.

The structure of the book

The British military historian Jeremy Black has most persuasively argued against the artificial imposition of cut-off dates, stressing the overlap of different eras and mindsets (Black 2005). I have found this amply confirmed in the literature on Strategy. There are long-term continuities in certain areas – especially the quest for the identification of eternal principles guiding warfare – from Vegetius to the present, and this is reflected in the chapters of Part II. Part III maps tendencies of thinking which originated at the latest in the Age of Reason, the eighteenth century, and found their apogee in the Second World

War. In shorthand, these come under the captions of the Napoleonic paradigm of war and the drive towards total war. Part IV discusses maritime strategy, which took off in the mid nineteenth century, strongly under the influence of prevailing fashions in thinking about land warfare. In turn, some maritime strategists developed concepts which inspired early air power theorists, discussed in Part V, together with the child of air power, nuclear strategy. While the twin atomic explosions of Hiroshima and Nagasaki shocked many strategists into reconsidering and indeed rejecting the Napoleonic paradigm, the turning point for many of them was arguably earlier, with the First World War, and it is this return to 'limited war' thinking that is analysed in Part VII. Small wars have always existed, and some of their key considerations have come to the forefront even in the context of major war, as we shall see in Parts VI and VII. In the Epilogue, we shall consider why there is such a gap between strategic theory and practical application.

PART II

Long-term constants

2 *Warfare and mindsets from Antiquity to the Middle Ages*

And the prince must ponder that victory comes from God and that his kingdom and his rule depend upon Him.

(Robert de Balsac 1502: g.ii)

Part II of the present book will deal, first, in chapters 2 and 3, with the changes in attitudes to war, the theory and practice of strategy from Antiquity to the French Revolutionary and Napoleonic Wars. Then, in chapter 4, it will deal with enduring themes and debates, some of which can be traced back one and a half millennia, some continuing to the present.

Technology and warfare

For the two hundred or so millennia of human existence for which we have archaeological evidence, and during the roughly six millennia for which we have any form of written records, until the nineteenth century, the human mastery of technology showed *no* steady progress. Related to this, Western warfare showed *no* steady development from primitive to ever more sophisticated, warfare did *not* become steadily more deadly or humane or limited or unlimited. The same is true for military technology of all sorts. Martin van Creveld has rightly noted that there were fluctuations in warfare, but no real quantum leaps ahead, between 750 BCE and 1500 CE, with many factors remaining 'unaltered well into the age of gunpowder', or even until well into the nineteenth century (Creveld 1989: 34f.). Indeed, today we must add that many forms of old technology have survived right into the nuclear age. The sword, used for nothing but warfare (as opposed to hunting or agricultural pursuits) since the Bronze Age, seemed to have ceased to have anything beyond symbolic functions in the nineteenth century. Yet machetes – originally cutting tools for use on vegetation – were used as swords in the Rwandan genocide of 1994. The last cavalry charge in history was thought to have occurred in the Second World War, but at least one took place in Afghanistan in the early twenty-first century.

Other technologies came and went. The crossbow had existed in ancient times, but was 'forgotten' in the Latin West along with many other inventions and technologies of classical Antiquity such as Greek Fire. The Greek *phalanx* or close formation of soldiers carrying spears and shields dropped out of use, but was 'reinvented' in several places in the high Middle Ages and in early modern times. At the end of the West Roman Empire, even at the time of Vegetius in the late fourth century CE, there was a decline in siegecraft. The sophisticated Roman siege engines were all but forgotten in the early Middle Ages; it was only in the high and late Middle Ages that people turned to Vegetius for inspiration for the invention of systems like the trebuchet with similar functions to engines the Romans had.

Like warfare, the development of military technology was thus not one-directional: both technological and organisational inventions could be forgotten or given up, and warfare in the 'Dark Ages' was certainly conducted in ways less sophisticated, but arguably at times no less effective, than those of the Roman Empire at its height. Over these two millennia, we observe a fluctuating pattern with different strands – disciplined infantry, agile horsemen, artillery, heavy horsemen, fortifications, siege engine and so forth – interweaving and dominating at different times, with few if any major technological innovations over the millennia for which we have archaeological evidence, until early modern times. Fashions in warfare, one might say, fluctuated, and certain tactics were used, forgotten and reintroduced from the outside or reinvented. The conduct of war has rarely if ever been static over much more than a century, as hostile groups encountering each other always sought to maximise the advantage they could draw from any particularly successful ways of fighting they had developed or were developing. Differences of tactics, or weapons technology, then usually became subjects of great interest to the inferior party, and technology transfer was always seen as desirable: the assertion that it is in no way dishonourable to learn from the enemy goes back at least to Ovid ('*fas est et ab hoste doceri*', Meta. 4.428).

The swift and highly mobile horsemen of the peoples of the eastern steppes with their bows and arrows inflicted many defeats on the Roman infantry as the disciplined and well-armoured Roman armies from the late second century CE found themselves at a disadvantage faced with the wily Barbarian horsemen. They managed to turn some of the horsemen into auxiliary forces of the Roman Empire (Luttwak 1976b). Bows and arrows, used by relatively lightly clad Arab horsemen against the Byzantines, in the hands of Saracens and Turks were still a menace for the crusaders in their heavy armour, and in the hands of Pandurs, Vlachs, Cossacks or Tatars, for the regular armies of the seventeenth, eighteenth and early nineteenth centuries.

The West in the meantime had developed the stirrup and the heavily armoured knight, a combination which seemed the height of military technology within the West from *c.* 800 to 1500, the 'cavalry interregnum' (Neill 1998: 496). From around 1200 the social dominance of the knight was no longer paralleled by physical inviolability, when the reinvention of the crossbow and the importing by the English of the Welsh longbow exposed him to death inflicted at a distance by mere commoners. Around 1300, the Scots, and the peoples of the Rhine, from Flanders to Switzerland, rediscovered what one could do with a tightly grouped infantry formation armed with shields and weapons that kept the attacking knights' weapons out of reach: pikes. The Swiss, the Flemings and the Scots developed the tightly packed formations of pikemen (the Swiss and German *Haufen*, the Scottish *schiltron*), a reinvention or rediscovery, one might say, of the Greek *phalanx*. Battles followed where such pikemen defeated knights.

Army sizes fluctuated considerably from antiquity until the late eighteenth century, and rarely reached six figures in battle on either side before the wars of the French Revolution (Hall 1997: 202–35). Technology hardly changed from the Thirty Years War to the mid nineteenth century. Gunpowder, which first appeared in Europe in the 1320s, was used in various forms – first for siege warfare, later fired by cannon in battle and even later still by hand-held guns. It necessitated changes in the conduct of warfare, which were introduced surprisingly slowly – new forms of fortifications to withstand cannon fire in sieges, new forms of drill for the artillerists who fired hand-held weapons, lest they get in each other's way. But the greatest leap forward in the way of waging war, the French Revolutionary and Napoleonic Wars, did not use technology much different from that of the Thirty Years War a good century and a half earlier, or of the Seven Years War half a century earlier. Technology only began to matter in a big way once the Industrial Revolution and the exponential take-off in the growth of the population of Europe from the eighteenth century, and from the mid nineteenth century, new scale-amplifying technology affected warfare.

Strikingly, authors writing on war were unaware of any major technological change since antiquity, or indeed any 'Military Revolution' brought on by gunpowder until well into the modern period. Even then, for about two centuries, there was a debate between the supporters of the 'ancients' (authorities from Antiquity) and the 'moderns' as to whether new generals, strategists and writers could be as good as the ancients, and whether new technology had transformed war to the extent that one could no longer learn much from the ancients. This debate was only settled largely in favour of the 'moderns' on the eve of the French Revolution (Heuser forthcoming).

Causes, aims and ethics of war from the Roman Empire to the late Middle Ages

The West embattled

Causes of wars also fluctuated wildly before the French Revolution. But literature on warfare rarely dwelt on this point. Vegetius at the end of the fourth century and Maurice around 600 reflect the aims in peace and war of an empire that had long peaked in its expansion and was now on the defensive (Luttwak 1976b). Neither wrote much about entering enemy territory, and if so, in the case of Maurice, only to intercept a retreating enemy army, on its way home from looting on the territory of the Byzantine Empire. The weakness of the late Roman and, after the fall of the West Roman Empire in 476, the East Roman (Byzantine) Empire is also reflected in their recurrent advice to avoid a set-piece battle, and to deal with the enemy through ruses, ambushes and surprise (Vegetius *c.* 387/1996). As the Emperor Maurice elaborated:

> To try simply to overpower the enemy in the open, hand to hand and face to face, even though you might appear to win, is an enterprise which is very risky and can result in serious harm. Apart from extreme emergency, it is ridiculous to try to gain a victory which is so costly and brings only empty glory. (Maurice *c.* 600/1984: 65; see also pp. 83, 96f., 107)

There was clearly at the latest from Vegetius' times – when the Roman Empire was in decline and beset by 'barbarians' – no 'Western way of war' that craved battles of annihilation, which did not even live on as an ideal or normative idea in the literature of the time (Sidebottom 2004: 1–15).

Byzantine rulers had long given up the notion that an enemy might be wiped out or turned into a constant ally (or that allies could be trusted indefinitely). Maurice cautioned against integrating allied forces too closely into the Byzantine forces, lest they undermine their cohesion and learn too much about the Byzantine way of waging war. They could obtain dangerous intelligence, should they one day turn into enemies, as Roman allies and clients proved repeatedly since Arminius turned on his former master in the battle of the Teutoburg Forest in 9 CE. Yet Maurice advised openness to proposals 'for peace on [mutually] advantageous terms coming from an enemy', even if that enemy had been defeated (Maurice *c.* 600/1984: 84, 87). This view reflects hundreds of years of Roman attempts to assimilate former enemies, which in the final two centuries of the West Roman Empire brought more and more men of 'barbarian' origin into the highest ranks of the armed forces and even onto the imperial throne. The same applied later to the East Roman Empire. The Goths, who had inflicted the crushing defeat of Adrianople on the Romans in 378, were later integrated into the imperial forces, changing their culture and ethos considerably; Maurice himself was overthrown and killed in 602 by a soldiers' revolt. His successor, Phokas, who had risen through ranks of the army, was half barbarian himself.

Phokas' descendant Nikephoros Phokas or Nikephoros II was thoroughly Romanised when he took up the fight against invaders from the south, especially the Arab Hamdanides of Aleppo and Cilicia in the mid-950s, approximately when he must have written or co-authored his *Peri Paradromes*, best translated as *About Skirmishes*. Nikephoros (the name meaning 'the one who carries victory') was emperor from 963 until his assassination in 969. His writings reflect a weak and overextended armed state beset by raids from external enemies hoping for booty; the armed forces of the Byzantine Empire were too overextended to be able to concentrate quickly in response to such surprise attacks in unexpected areas, and Nikephoros' manual discusses various expedients through which one might seek to drive back the invaders while avoiding a pitched battle which one's own forces might well lose (Nikephoros mid-950s). No 'Western way of war' of seeking decisive annihilation battles here: it was Byzantium's Persian, Arab, Saracen and Turkish enemies, who, 'in contrast the Byzantines, sought decisive victory on the battlefield. They, not the Byzantines, wanted battle' (Kaegi 1992: 128; see also Haldon 1999: 7f.).

With the exception of the short-lived reconquest under Justinian (r. 527–65) and Heraclius (r. 610–41) of areas of the Roman Empire in the west and south that had been lost to various invaders, the Byzantines were entirely on the defensive, and even Justinian's and Heraclius' reconquests were an attempted return to a status quo ante. They were an immediate part of the long tradition started in the West Roman Empire in the late fourth and fifth centuries of defending Christendom against heathen invaders. Thus, without any need for a specifically formulated holy war doctrine, but fighting under the protection of cross and icons carried into battle, they defended good against evil, cosmos against chaos (Haldon 1999: 7f., 13–33). With very few exceptions, the Byzantines preferred peace to war, in the words of Princess Anna Comnena, 'for peace is the end of all wars. Invariably to prefer war instead of peace, always to disregard the good end, is typical of foolish commanders and foolish political leaders, the mark of men who work for the destruction of their own state' (Anna Comnena *c.* 1148/1969: 381). Despite its warrior-saints, despite the centuries of fighting against invaders and despite the important role of the army in the empire, for the millennium of its rule, Byzantium was much less warlike than the other peoples of Europe, North Africa or the Middle East.

The origins of just war ethics

The classical bequest of Greece and Rome to the subsequent centuries dominated by the barbarians is one of the most debated, complex and intriguing question of European history. The bequest in terms of how to wage war, as we shall see, was the simplest part of it, for a thousand years largely confined to the work of Vegetius. It was another matter for the question of when it was just to wage war; here the reality differed largely from the Roman theory, and yet the Roman concepts would persuade generations until our own time (Rich 2001).

The ethical stipulation that war must be fought for the sake of (re-) establishing peace can be found as far back as Aristotle (*Nikomachean*

Ethics X.7), but does not seem to have affected pre-Roman Greek practice much. The most important conditions for a war to be *just* can be traced to Cicero, who stated that the pursuit of peace must be the foremost aim of any war; without the pursuit of peace a war cannot be just. War must not be undertaken if the same aim can be reached otherwise; here we have the condition that recourse to war, to be just, must be the last resort. If one resorts to war, it must be carried out with moderation (that is, the violence used must not be disproportionate to what the enemy had done) and after a formal declaration (Cicero 44 BCE: I.34–8; Tooke 1965: 10).

While it was the Romans more than the Greeks before them who cared about the concept of a 'just war', Greeks adopted it under their influence, as the example of Onosander shows. This first-century philosopher in *Strategikos* (*The General*), echoed Cicero, with views that would influence the later Byzantine writers on the subject:

> The causes of war ... should be marshalled with the greatest care; it should be evident to all that one fights on the side of justice. For then the gods also, kindly disposed, become comrades in arms to the soldiers, and men are more eager to take their stand against the foe. For with the knowledge that they are not fighting an aggressive but a defensive war, with consciences free from evil designs, they contribute a courage that is complete; while those who believe an unjust war is displeasing to heaven ... enter the war with fear. (Onosander 1st c./1923: IV.1–2)

The just-war theory that survived the fall of Rome was a surprisingly neat fusion of Roman just-war concepts, found especially in the writings of Cicero, and dicta from the Hebrew Bible and the Christian New Testament. This fusion was effected above all by Augustine of Hippo (354–430), one of the Western Church Fathers. Augustine did not use Greek literature, and accordingly did not build on earlier Greek thinking on war, but only on Roman concepts, especially the concept of a just war, that we find in Cicero (Mattox 2006: 14–18). Scattered over a variety of works including books and letters, Augustine had listed a series of conditions for a war to be just: the existence of a just cause for war (that is, war was just for a party that had been wronged); only God (or a rightful ruler empowered by God) could declare such a war; the war had to be waged with the intention of righting the wrong (for example, restoring land to its rightful owner) and bringing about peace, not the lust for revenge or booty (Mattox 2006: 44–91). Augustine's views would lay the foundations of the Western just-war tradition (Russell 1975). Similar views, probably without any direct link to Augustine's, can be found in the writings of East Roman Emperor Leo VI (865–912):

> It was not the destiny of men to wage war against each other ... As creatures that have been formed in the image of God ... [it is their destiny] to enjoy peace with each other, to love each other ... Since, however, since the origins of the world, the Enemy of our race [i.e. the Devil] has aimed to fight humans through their own nature, it is necessary for them to preempt the attacks against them, which He conducts through their equals [i.e. other humans]. They must not subject themselves to other warring nations, but seek their salvation in the science of the arts of war, to protect themselves through these against inimical attacks, and to throw back upon the Adversary the evil which He, upsetting peace, deserves to suffer, until all evil of the impious is expunged. And after the happy restitution of the common wealth, peace will be beloved and accepted by all. (Leo VI *c.* 900: Preface)

The practice and the theory

The waves of pagan invaders of the Roman Empire were quite unaffected by such thinking. From the middle of the third century, Germanic tribes, Huns, Arabs and Moors made raids into the Roman Empire, aiming above all to come away with booty. Gradually, as and when they encountered little and, as time went on, even less resistance, they turned to the conquest of territory and its exploitation as their main war aims (Schmitt 2005: 417–44; Whitby 2005: 355–85). Only when they settled down and their leaders developed a taste for the Roman-Christian lifestyle (and the benefits of the support of the Church for their standing within their own society) did they grudgingly allow themselves to become acquainted with just-war thinking. Christianity was a useful asset in this context: with the espousal of Christianity, tribal chiefs from the Goths settling in the Iberian Peninsula, to Clovis in Gaul and Charlemagne some generations later, from the Saxon Ottos to Polish Mieszko I and Bolisław or Stephen I of Hungary could obtain the Church's blessings for expansionist campaigns against pagan or 'heretical' tribes – ostensibly with the aim of bringing them into the fold of the Church. Annihilation, if they refused to give up their ways, was part of the legitimised programme, for example in Charlemagne's massacres of the Saxons to the east of his expanding realm. Nevertheless, in their martial ambitions, the rulers of medieval Europe stood firmly in the tradition of those migrant tribes that had brought about the fall of the Roman Empire in the West (Zeller 2001: 102–12).

Indeed, the medieval knight's very *raison d'être* was to fight, even if the purpose was to 'conserver, defendre et garder le peuple en transquillité', to protect the common people. It was practically their mission statement to go out and seek honour and show boldness, as the French nobleman and soldier Jean de Bueil (1406–77) put it in his novel *Le Jouvencel*. It was written in 1466 to instruct young knights in their obligations, and in the conduct of war (Bueil 1466/1887: 13f.). The *Book of the Order of Chivalry*, which was circulated throughout Western Europe not only in Latin but also in Catalan, French, Scots and English and can perhaps be traced back to the twelfth-century Majorca-born Ramón Lull or Llull, similarly defined the task of the knight to serve God and his master, and to protect women, widows, orphans, clerics, the sick and all men weaker than himself (Anon *c.* 1483: 38f.).

The Crusades were a fusion of the warrior cults of the Germanic tribes who had come from the east and the north, now sedentary throughout western and southern Europe, and the Augustinian definitions of a just cause given by the Catholic Church: the crusaders were sent out to aid their Christian brethren in the east, established there since Roman times, who were being prevented from visiting Christianity's places of pilgrimage in the Holy Land, or even from celebrating mass. But from this it followed that one could legitimately set out to reconquer from the infidel any territory that had once been in Christian possession – whether this was the Iberian Peninsula, or indeed anywhere within the confines of the former Roman Empire, including Palestine or North Africa. In general, war aims here were to overthrow the rulers (what today would be called 'regime change') and to break the dominance of the ruling Islamic elites, which could include massacres: if they had troubled the Christian population, local non-Christian populations were seen by the Christian Europeans as barbaric and as oppressors of the Christians and thus evil also in the eyes of God (Riley-Smith 2001: 127–40).

It has to be said, however, that Christian Europeans were in part responding to the way they themselves were being treated by the further waves of migrant people who continued to come crashing into their lands. Goths of both sorts, Avars and Huns, Magyars and Mongols, Vikings and Turks tended to loot and murder without restraint. Only when they settled, or began to see the advantages of commercial arrangements (the levying of Danegeld or other taxes or 'protection money') did they confine themselves to a less destructive form of exploitation, interrupted from time to time by punitive massacres or other reminders that the levies were worth paying. The general rule gradually became that Christians, at least from the late tenth century onwards, treated each other with relative respect and sometimes spared each other's civilians, as demanded by the Church. Nevertheless, massacres of non-combatants continued: particularly in the British Isles, Christian Englishmen fought mercilessly against Christian Welshmen, Scots and Irish, and in France against the local population if it sided with the Valois dynasty in the Hundred Years War. To be fair, massacres of civilians were inflicted by both sides (Strickland 2006: 107–40). But since the siege of Troy, there was a persistent pattern of bloody massacre of the burghers of a town that had refused to yield, once a siege had been successful. It was Roman, medieval, and indeed modern practice to treat citizens of towns thus invested as rebels and not as protected by customary law. Even though the *Book of the Order of Chivalry* that circulated in many copies and languages from the early thirteenth century expressly condemned the destruction of castles, cities, towns, the burning of houses, the cutting down of trees and the slaying of livestock (Anon. *c.* 1483: 42), Jean de Bueil, with his direct experience of the Hundred Years War, commented that it was the common people who suffered the most in war (Bueil 1466/1887: 15). This is thus not merely a twentieth century phenomenon.

After territorial expansion and exploitation, the next aim for medieval tribal chiefs was to provide for their sons. The fortunes of the male members of a dynasty were seen as more important than those of the collectivity (tribe, people or nation), and no concept comparable to that of statehood – of the motherland, polis or Rome – came above the pursuit of the best fortune of the ruling family. Consequently, the division of the patrimony among (male) heirs was the rule until well into the mid-second millennium CE, and with it the in-built cause of war in which rival kinsmen and -women tried to recreate for themselves, from the possessions of their brothers and nephews, the kingdom their fathers had held (Honig 2001: 113–26).

Rivalling the provision of lands and inheritance for all sons (and of dowries for daughters), by and by, the concept of passing on possessions en bloc to one sole heir came into being, leaving many other members of the family – from uncles and cousins to other sons – high and dry. The logical consequence was the creation of the predominant cause of war even until the French Revolution: dynastic contests over legitimate succession, prominent examples being the Hundred Years War between the Plantagenet and Valois dynasties and the War of the Roses in England (Curry 2001: 141–54). As historian Jeremy Black so aptly put it, 'War was the continuation of litigation by other means' (Black 1998: 47). Primogeniture – the inheritance of undivided lands by the first-born male – established itself in Europe from the twelfth century, but other practices – inheritance by younger brothers rather than sons of princes, or the division of lands among male relatives – persisted in many places. For example, it led to the division of the Habsburg possessions between what became the Spanish and Austrian branches of the family even in the sixteenth century.

Churchmen since Augustine of Hippo had struggled with the problem of reconciling the commandment not to kill with the need to punish offenders and to protect others from them. Those offending against God's will deserved punishment, particularly if they refused to relent, despite being given the opportunity to do so. Offences against God's will included uprisings against the rightful ruler, a key theme in all dynastic quarrels, but also heresy. Medieval clerics taking an interest in these matters were particularly exercised by the long-running confrontation, on the one hand, of the emperors of the Holy Roman Empire – seen as the successors of (Saint) Constantine the Great, who had created the Christian Roman Empire – and the papacy on the other. Much of their writing only makes sense in the context of this conflict, which challenged the world order itself as they understood it; examples include Dante Alighieri's *De Monarchia* of 1313 and Marsilius of Padua's *Defensor Pacis* of 1324, both of whom could only conceive of peace within a strong state or empire. Given these other preoccupations, there is little medieval literature of relevance to this book, beyond the Augustinian-Thomasian writing on just war, or historiography in the forms of chronicles of wars or crusades or the reigns of monarchs punctuated by battles and military campaigns.

The earliest medieval text on warfare that is no longer either a historical treatise or a legal one, nor a copy of Vegetius, is that of

Honoré Bonet or Bouvet. Writing during the Hundred Years War, this Provençal monk and theologian in his *Tree of Battles* (*c.* 1385) could not imagine a world without war, and in the tradition of Augustine actually proclaimed the rightful place of war in the divine order:

> [T]he truth is that war is not an evil thing, but is good and virtuous; for war, by its very nature, seeks nothing other than to set wrong right, and to turn dissension to peace, in accordance with Scripture. And if in war many evil things are done, they never come from the nature of war, but from false usage ... for all good things, and all virtue, come from God ... [W]ar comes from God, and not merely that He permits war, but that He has ordained it.

Bonet then put forward examples drawn from the Hebrew Bible in which God commanded the people of Israel to take up arms against His enemies (Bonet *c.* 1385/1493: 125). As a cleric and scholar, however, Bonet was concerned about how to limit war to 'proper usage' through a due observance of existing laws and customs of war, its legitimacy (the *ius ad bellum*), its legitimate conduct (the *ius in bello*), the proper treatment of non-combatants, prisoners and so forth.

Drawing heavily on Bonet's *Tree of Battles*, the Venetian-born Christine de Pisan (or Pizan) in her *Livre des faits des armes et de la chevalerie* of 1408–10 covered the question of the justice of war (Saux 2004). Both Bonet and Christine added this dimension to the subject-matter of Vegetius – on whom Christine also drew. Bonet and Christine differentiated between the three conditions listed by Augustine – just cause, legitimate authority, right intention – and motivations that could not be approved of – the desire for revenge or the desire to conquer somebody else's lands. Christine, who with her experience of politics at the French court and in Italian city states did not separate warfare from politics, proposed a political alternative to waging war, if a prince should feel wronged and offended by another: to assemble 'a great council of wise men in his parliament ... and not only will he assemble those of his own realm, but in order there be no suspicion of failure, he will also call upon some from foreign countries that are known not to take sides, elder statesmen as well as legal advisors and others', to whom both sides should put their view of the situation. Only if such a council could not bring about a settlement between the two parties should they be allowed to proceed to war (Christine 1410/1999: 13–17).

Christine's work was popular enough to warrant printing in French in 1488, but her name was omitted as that of the author – probably, as she had put it herself, because it struck contemporaries as 'unusual for women' to write on this subject matter (Christine 1410/1999: 12). The book's success and Franco-English cultural proximity of this age is illustrated by Henry VII of England only one year later commissioning an English translation of Christine's popular work. This and the printing of the English version was undertaken by the pioneer of printing in England, William Caxton, 'so every gentleman born to arms and all manner of men of war, captains, soldiers, victuallers and all others would know how they ought to behave in the feats of wars and battles'. It is perhaps the first instance of a ruler imposing the reading of a field manual; the twenty surviving printed copies testify to the success of this endeavour (Nicholson 2004: 14–19).

Despite Honoré Bonet's and Christine de Pisan's concerns about the rules of war, and although the Church tried hard to protect Christian non-combatants from the ravages of warfare, civilians were rarely spared, particularly not in the wars of religion, and warfare depended on any state's ability to pay for it. Onosander (1st c./1923: XXXV.1) had thought 'Plundering should not be permitted after *every* battle', implying that after certain battles it was acceptable, and Vegetius had emphasised the need to ensure that one's own forces had sufficient supplies, while using famine against the enemy. Besieged fortresses and towns would be forced by famine to surrender, and cutting the enemy's forces off from supplies was also advantageous: 'By this strategy, if the enemy collect together, suffer famine, and if they disperse, are easily beaten by frequent surprise attacks' (Vegetius *c.* 387/1996: 69). His advice was certainly heeded throughout the Middle Ages and in the early modern period. While medieval armies shied away from battles – these were generally seen as a divine ordeal, and the outcome could cost a king his crown – they pillaged and plundered happily, and engaged in scorched-earth policy to deny areas and food to the adversary and to punish populations for siding with opponents (Honig 2001; Goosens 2001).

The use of force against heretics – dissenting Christians among one's *own* population – had existed already in the Byzantine Empire. Heretics challenged the orthodoxy of the majority and with it the whole edifice of a God-given state founded on the claim to have the only right opinion about what God's will might be. Heretics thus were hated more and treated worse even than the infidel, and religious wars – whether they were the wars against the Paulicians or Bogomils in the east, the Albigensian Crusade, the Hussite wars in Bohemia or the persecutions of the Lollards in England – tended towards greater cruelty, more massacres of civilians, the killing of prisoners and the shedding of the restraints on war by now normally practised between Christians (Strickland 2006). Massacres continued to be common when the occupants of fortresses or other fortified places were successfully besieged and failed to surrender in time.

With all this warfare going on, why was there relatively little writing on the way to wage war in the Middle Ages? One – and perhaps the most conclusive – reason is that it was widely thought that God decided the outcome of war, and man only had a limited ability to influence it, mainly by following a cause that was just in the eyes of the Lord, preferably combined with an upright lifestyle. This conviction is found throughout the Middle Ages in East and West and into the early Renaissance (Haldon 1999: 22f.). Even at the very beginning of the sixteenth century, Robert de Balsac in his *Ship of Princes and Battles* reminded kings that victory was a gift of God (see the quotation at the beginning of this chapter), and a little later, the famous French Chevalier de Bayard, knight without fear or reproach, when confronted with superior enemy numbers, chose to attack, saying 'victory comes from God and not from the number of men' (Champier 1525/1992: 164). As late as 1588, his countryman François de Saillans, writing under the pseudonym of Bertrand de Loque, in his *Instructions for the Warres* asserted that 'the victory dependeth not of the multitude of fighting men, but of the grace and fauour of God' (Saillans 1589/1591: 21). Saillans was still typical of the thinking of his time. Almost a century later, Paul Hay du Chastelet still saw just war as God-willed, and its outcome directly decided upon by God (Hay du Chastelet 1668/1757: 5–8).

With such an independent variable overriding all other factors in war, what was the use of speculating much about the others? The fact that Saillans and Hay, like Vegetius the Christian in the fourth century CE but unlike Western authors in the Middle Ages, wrote so extensively about the other variables suggests, however, that their trust in divine providence was tempered by the conviction that He expected humans to do their part in making a successful outcome of a just war possible. The 'rebirth' of classical attitudes to arts and sciences, going hand in hand with the rediscovery of classical works but also the discovery of Byzantine authors, can thus be seen very clearly in treatises on war in early modern Europe, to which we shall now turn.

3 *Warfare and mindsets in early modern Europe*

It seems to me to be imprudence to commit the fortune of a state to the event of a general battle where the fate of arms is always uncertain and dangerous.

(Hay du Chastelet 1668/1757: 112)

Causes, aims, and practice of war in early modern Europe

Religion would ultimately recede in the thinking of early modern authors, but not before divisions within Christendom had been a major cause for two centuries of wars. This element was present, as we have seen, in the medieval persecutions of heretics. In the sixteenth century, it would tear the Holy Roman Empire and the Habsburg lands apart, as well as France and, on a lesser scale, England, and it would be a major factor in the Thirty Years War. But other drivers for war known to the Middle Ages equally persisted.

The culture of early modern European elites, like that of their medieval predecessors, saw warfare primarily as an opportunity to win glory and renown, and the means to claim and to defend property where courts did not exist or would not arbitrate in their favour. The entire political system of early modern Europe consisted of volatile alliances between thoroughly selfish dynasties, who by marriage, inheritance or war sought to amass for themselves as much territory, wealth and subjects as possible (Black 1987: 20–2). The utterly self-serving nature of these dynasties' politics was mitigated by religious factors, which at the time occupied the place filled by ideology in the twentieth century. Speaking not only of princely dynasties, Niccolò Machiavelli and, echoing him, Raymond Beccarie de Pavie, sieur de Fourquevaux (1508–74) opined that 'men do sooner forget the deaths of their father, then the loss of theyr patrimony' (Machiavelli 1532b/1961: 97; Fourquevaux 1549: 260).

Slightly differently from principalities, in some Italian city states, collectivities could own territory or wage war over its possession. Either way, the wars between these cities aimed not at the overthrow of a system, let alone the annihilation of a people. On the contrary, the prince (or oligarchy) waging such war wanted to step into the shoes of the prince (or oligarchy) ousted through the defeat of the enemy, to enjoy the full fruits of the possessions. Contenders on both sides tended to see each other as equals, as bound by many rules and obligations, rules supported by the Church and its lawyers, except when adversaries could be seen as rebels or heretics, in which case all clemency tended to be suspended. Post-conflict reconstruction and the imperative of clemency practised by the victor play an important part in the military writings of early modern Europe. Paul Hay du Chastelet, who held public offices under Louis XIV, wrote at great length about the obligations of the victor to behave magnanimously towards the defeated, to rebuild towns and cities that had been damaged in sieges and to win the allegiance of the former enemy populations (Hay du Chastelet 1668/1757: 148–51, 155).

As in the Middle Ages when the warrior cult of the knight characterised Europe, early modern Europe was essentially a bellicose society. In the light of this continuity, we need other factors to explain why there were so many wars between 'great powers' ('states which play a major role in international politics with respect to security-related issues') in the sixteenth and seventeenth centuries, compared with following centuries when 'great-power wars' became less frequent but produced more casualties (Levy 1982: 278–300). One such factor was that the religious inhibitions that had prevented some medieval men from bringing upon themselves a divine judgement through battle receded. Battles were increasingly seen as something that human action (other than praying and repenting of one's sins) could influence, even if unpredictable divine intentions were soon enough replaced by equally unpredictable factors such as Clausewitz's 'friction' and 'chance', neither of which had to do with morality. Instead, battles were now a decision-making mechanism that could settle disputes, but not a last moral tribunal. Famously, 'just causes' in the Catholic tradition of Augustine and Thomas Aquinas were ignored by Machiavelli, who wrote in his *History of Florence* (composed between 1521 and 1525):

> Those who make war have always and very naturally designed to enrich themselves and impoverish the enemy; neither is victory sought or conquest

desirable, excepting to strengthen themselves and weaken the enemy. Hence it follows, that those who are impoverished by victory or debilitated by conquest, must either have gone beyond, or fallen short of, the end for which wars are made. (Machiavelli 1532a/1847: 254f.)

The aim of territorial conquest, considered quite amorally, was also taken for granted as the prime mover in war by the Spaniard Bernardino de Mendoza (1541–1604), a nobleman from Castile who as a cavalry officer accompanied the Duke of Alba on his campaign to subdue the rebellious Netherlands. Mendoza called the desire to conquer natural in all men, and 'how much more so in kings' (Mendoza 1595/1597: 10). Two hundred years later, Kant described monarchs as resolving 'on war as on a party of pleasure, for reasons most frivolous', and one of his prescriptions for eternal peace was to legislate that '[a]ny state, of whatever extent, shall never pass under the dominion of another state whether by inheritance, exchange, purchase or donation' (Kant 1795/1796: 3, 17).

Causes and thus aims of wars in early modern times, as given in justifications (*claregationes et manifesta*) published by princes, include the following:

- the attempt to reassert or impose a universal monarchy (the Holy Roman Empire or a similar notion) or the fight against such predominance (Cardinal Richelieu's favourite reason for war);
- fight against a rebellion;
- hereditary/dynastic claims;
- attempts to (re)create an equilibrium of powers or a balance of powers;
- economic interests;
- the call for a crusade in the form of a fight against Turks (who from the ninth century onwards were steadily pushing westwards, invading and occupying parts of Europe after their conquest of the Byzantine Empire in 1453, enlarging their areas of conquest until their second unsuccessful siege of Vienna in 1683);
- preventive action against imminent threats;
- religious issues (*religionis necessitas*);
- defence of one's subjects against a bellicose attack;
- defence of liberties tied to membership of estates;
- treaty commitments;
- redressing of grievances (*iniuria vindicta*). (Repgen 1985: 21)

Critical contemporary observers noted, however, that these justifications were not always the real or main reasons for going to war. Indeed, unlike in the writings of the Middle Ages, we find recurrent from Niccolò Machiavelli to Hugo Grotius and Kant a cynicism about rulers' and governments' professed justifications for their actions. For Machiavelli, the ultimate cause of war was 'man's nature', his ambitions, his greed and the eternal dynamics of the relations between states (Machiavelli 1531: I.37; II.9). Emeric Crucé (1590–1648) in his *Nouveau Cynée* of 1623 noted that foreign wars were undertaken 'for honour, for profit, or for the reparation of some wrong, or for the exercise'. Religion he rather saw as a pretext. Maximilien de Béthune, duc de Sully (1559–1641), adviser of Henri IV of France, listed 'jealousy, avarice, ambition and vanity', while to him as a Huguenot, religious diversity was a serious factor in war (q.i. Hartmann 1995: 39, 56f.). Grotius, too, pondered whether the reasons given for going to war are pretexts or not. Under the heading 'On the Unjust Causes of Wars' he wrote:

> Some wars were founded upon real motives and others only upon colourful pretexts ... Though most powers, when engaging in war, are desirous to colour over their real motives with justifiable pretexts, yet some, totally disregarding such methods of vindication, seem able to give no better reason for their conduct, than what is told by the Roman Lawyers of a robber, who being asked, what right he had to a thing which he had seized, replied, it was his own, because he had taken it into his possession ... Others make use of pretexts, which though plausible at first sight, will not bear the examination and test of moral rectitude, and, when stripped of their disguise, such pretexts will be found fraught with injustice. (Grotius 1625/1901: 267f.)

Like Grotius, the Strasbourg historian and lawyer Johann Heinrich Boecler (1611–72) commented that explanations such as the list above, as given in public documents, were issued by princes or the leaders of republics for public consumption. Often, however, these contained only pretexts designed to dissimulate the more private and truer causes. Among these he listed 'the love for ruining others, the immense desire for one's own lucre, avarice and ambition' (q.i. Repgen 1985: 24f.). Neither Grotius nor other contemporaries seem to have seen this behaviour on the part of rulers as new to their age, but saw it as an enduring, age-old phenomenon, citing classical examples to confirm this (Grotius 1625/1901: 267).

With a good number of princes falling into the category of ambitious as opposed to peaceful (as Feuquières put it), the obligation of princes to pursue any dynastic claim to territory they felt militarily able to sustain, from the time of Charlemagne if not earlier, created a structural driver for war. Even in the high Middle Ages, a pattern could be observed whereby different princes tended to gang up on an overly ambitious peer set upon territorial aggrandisement, witness for example the alignments of the Holy Roman emperor and the English monarch against the French monarch at Bouvines in 1214. This balance of power pattern or, to be more accurate, the increasingly automatic tendency to counterbalance an expansionist power, was developed to a fine art in the early modern period. Henri, duc de Rohan in his *Perfect Captain* of 1636 observed that small states only existed because the jealousies of their (bigger) neighbours stopped those from swallowing the small ones, and that for their defence, they had to rely mainly on this mechanism (Rohan 1636: 352f.).

The Swiss diplomat and lawyer Emerich de Vattel in his *The Law of Nations* of 1758 reflected the preoccupation of the princes of his age with a distribution of power among several states (as opposed to the unitary world-view which was such an important point of reference from Constantine I's Roman Empire until the Westphalian Peace Treaties of 1648 or even the Peace of Utrecht in 1713). In Vattel's age, there was no one state which could justly claim hegemony in Europe, not even Maria Theresia's Holy Roman Empire, which by now was seen as the House (or dynasty) of Austria and its possessions. Vattel's contemporaries were obsessed with the danger that one prince might upset this distribution of power, and – like Habsburg emperor Charles V in the sixteenth century or Swedish king Gustavus Adolphus in the Thirty Years War, or Louis XIV of France – might aspire to dominate Europe. In this context, Vattel struggled with the moral dilemma posed by preventive war: if a neighbour increases his power (and lands) by whatever means (lawful or unlawful), and we are afraid that the neighbour will next move against us, is it justifiable to attack that state to prevent this aggrandisement? From experience he thought that the stronger a state grows, the more it is likely to bully its neighbours, however just the reasons were for which it grew in the first place. But Vattel thought prudence dictated that one should not use unlawful means for the attainment of just and laudable ends (Vattel 1758/1834: 307f.). Instead, Vattel described his contemporaries' favourite counter-measure to a bid for aggrandisement and domination:

> Europe forms a political system, an integral body, closely connected by the relations and different interests of the nations inhabiting this part of the world. It is not, as formerly, a confused heap of detached pieces, each of which thought itself very little concerned in the fate of the others ... The continual attention of sovereigns to every occurrence ... and the perpetual negotiations, make of modern Europe a kind of republic, of which the members – each independent, but all linked together by the ties of common interest – unite for the maintenance of order and liberty. Hence arose that famous scheme of the political balance, or the equilibrium of power; by which is understood such a disposition of things, as that no one potentate be able absolutely to predominate, and prescribe laws to the others. The surest means of preserving that equilibrium would be, that no power should be much superior to the others, that all, or at least the greater part, should be nearly equal in force. Such a project has been attributed to Henry IV [of France].

This counterbalancing might even lead to the formation of outright confederacies to oppose a predominant power (Vattel 1758/1834: 312).

Several sets of motives could of course come together. Gustavus Adolphus commented about the Thirty Years War that 'all the wars of Europe are now blended into one' (q.i. Roberts 1995: 18); the same could be said for the confluence of the multiple causes of this war.

As the Catholic Church could by definition not impose universal rules on contesting factions belonging to different religious denominations, the previous near-monopoly of ecclesiastic lawyers in the formulation of the rules of (that is, restraints on) war gave way to the writings of secular lawyers. In part they tried to describe common practice, in part derive rules logically from first principles. Political writings of this period, such as Jean Bodin's, Hugo Grotius' and Thomas Hobbes's views on sovereignty and the prince's or state's monopoly of the use of force, give us insights into the perceptions, what we have called 'passive culture', of the times, the beliefs forming the conceptual framework within which war aims could be formulated. Thus while we still saw multiple references to Christianity in Saillans's writings, Hugo Grotius, writing only a few decades later, under the impression of the Eighty Years War between the Dutch and the Spanish (which fused with the wider European Thirty Years

War) famously tried to found both *ius ad bellum* and *ius in bello* on a secular reasoning, not exclusively on Christianity or its different confessions. Grotius, following the Catholic tradition of Francisco de Vitoria (1483–1546) and Alberico Gentile (1552–1608), which in turn built on Thomas Aquinas and Augustine, and Cicero, insisted on the need for a just cause for a war. Grotius established rules for the conduct of war: first, that only what was necessary to conduct and complete a just war might be done, and nothing more. Any action beyond that which is necessary would be illegitimate. Secondly, that the legitimate position one had might yet be changed in the course of a war, so that, for example, one might have the right to fight against parties that joined an adversary's side. Finally, one might have to accept certain wrongs in order to achieve the redressing of the primary grievance. As an example he gave the conquest of a province rightfully belonging to the enemy, in order to exchange it later for one's own territory which the enemy had unjustly seized. The enemy's women and children could be killed according to the *ius gentium*, he argued with reference to examples from Homer and Thucydides. He personally thought this unjust, as they were unarmed and thus killing them was unnecessary (Grotius 1625/1901: III.4).

By and by, with the waning of the religious cause of war, and the centralised states' monopoly on armed forces, the dynastic driver for war became the most potent and by far the most dominant among the causes of war listed above. The division between medieval wars of succession (which tended to have a civil war element and which tended to weaken, not strengthen, the crown) and the modern wars of succession (which strengthened the state monopoly on the use of force), rather than being a neat line, is more of a century of overlap between old and new. Bernardino de Mendoza, writing in 1595, saw crises of succession as most dangerous for any kingdom, as it gave the vassals a choice of whom to back, weakening the kingdom and being a prime cause for civil war – an observation which fitted well for English experiences of the twelfth to fifteenth centuries and French experiences from the Hundred Years War until the assassination of Henri IV in Paris (Mendoza 1595/1957: 14). The tendency of early modern states to fission into civil wars declined as the centralised monarchies gained strength and monopolised power within the state. The *ancien régime* of the late seventeenth and of the eighteenth century fought wars almost exclusively over dynastic claims to lands, but now largely without competing claims from among its own aristocracy. Claims were made when the territory was of strategic or economic value, and legal claims might be fabricated; good legal claims by contrast were no longer pursued just for their own sake (Tallett and Felix 2008). In the eighteenth century, civil wars resulting from rival dynastic claims were seen by contemporaries as 'almost impossible ... in Europe' (Lloyd 1781/2005: 476). The nobility, who had throughout the high and late Middle Ages and in early modern times had a proclivity to challenge royal successions and fight against each other, had calmed down in all European countries, but this went along with the need to keep it occupied and away from intrigues and feuds among themselves or against the crown. The nobility's self-perception and claim to its rank was entwined with its descent from the warrior caste of the medieval knights, and martial values were still intrinsic to its identity. What better way to keep them out of trouble domestically than to give them high offices in the royal armies, directing their dynamism against external foes and harnessing it to the furtherance of the dynastic interest of the ruling family (Kunisch 1987)?

This phenomenon was particularly pronounced in absolute monarchies. In the same measure in which the state was identified with the person of the monarch, his or her dynastic claims became the supreme *raison d'état*. Even though he was admonished by authors of his era never to undertake anything but wars in a just cause (Hay du Chastelet 1668/1757: 4f., 54f.), Louis XIV with his wars gave Boecler every reason to suspect that legal 'just causes' for going to war were pretexts. John Lynn has shown very persuasively how the Sun King, in the cultural context of his age, cornered himself into a logic in which he staged one preventive war after another. He had locked himself into a web of dynastic obligations, a loudly proclaimed quest for glory in war (which included the seizure of any occasion to increase the territory of France, including territory that could not be claimed dynastically), and fear of attacks from all those increasingly fearful of him, who, convinced that he would strike sooner rather than later, wanted to catch him before he was fully prepared. The result was what the great French military historian André Corvisier has called '*la défense aggressive*', which made Louis look like a 'relentless, insatiable conqueror' (John Lynn), winding up the spiral of preventive wars ever further (q.i. Lynn 1994: 178–204). Louis XIV's aphorism that he *was* the state could be reformulated as '*la guerre, c'est moi*'; war was the essence of his rule (Kunisch 1987).

The writings of Frederick II (the Great) of Prussia, especially his political testaments, fully confirm the concerns of the encyclopedist. As a young man of 27, Frederick in his *Anti-Machiavell* (1739) had extolled the virtues of peace and of a stable equilibrium of powers in Europe, noting the dangers to all of Europe that would result from the upsetting of this delicate balance by a new parvenu monarchy set upon expansion (Frederick II 1739). In 1748, after the First and Second Silesian Wars against Empress Maria Theresia of Austria (1740–2 and 1744–5) he could still postulate in his military writings that wars should be short and sharp because long wars would undermine discipline and would be too draining of the resources of a kingdom (Frederick II 1882: 86). Soon the Seven Years War proved that shortness was less easily attained than propagated.

A good dozen years after writing his *Anti-Machiavell*, when Frederick wrote his first political testament, he found himself in agreement with Machiavelli, who had posited that a disinterested power surrounded by ambitious neighbours, refusing to play their game, would soon go under (Frederick II 1752: 366). Just cause no longer features in Frederick's discussion of wars: he now saw but two causes, 'vanity' and 'self-interest', of which he dismissed the first and fully endorsed the second, elsewhere stressing the importance of 'glory' (Frederick II 1752: 348, 396). A decade later, when Frederick wrote his second political testament, he came even closer to seeing the world like Louis XIV before him:

> A prince who makes war because he is worried, frivolous, [and] has upsetting [*désordonnée*] ambition, is just as worthy of being condemned as a judge who uses the sword of the law to pierce an innocent. War is good if it is made to support the interests [*considerations*] of a State, to maintain its safety or to contain the projects of an ambitious prince consisting of conquests harmful to your interests ... Honour, the yearning for glory and the good of the fatherland must animate those who dedicate themselves to arms, without vile passions sullying such noble sentiments. (Frederick II 1768: 556)

The Hohenzollerns' interest was to expand their territorial possessions, until they would become 'one of the most considerable powers of Europe' (Frederick II 1752: 376). In 1768 he wrote, 'The first preoccupation of a prince must be to protect what he has [*de se soutenir*], the second his aggrandisement', aims which required the utmost adroitness of manoeuvre (Frederick II 1768: 650). Prussia was rapidly becoming the expansionist upstart of which, as heir to the throne, the Hohenzollern prince himself had warned in his *Anti-Machiavell*. It was indeed upsetting the status quo, based on the delicate balances of interests, territorial ambitions and powers in Europe. Frederick further contributed to this confusion and upheaval by deliberately attempting to defy predictability and advocating that the successful prince should always vary his conduct so as not to become a calculable entity in the other powers' plans (Frederick II 1752: 396). This also meant that one could not be a dependable constant – at least in the medium or long term – for one's own allies, and, *mutatis mutandis*, meant that no power could rely on the assistance of allies: Frederick warned his heir to count less upon the help of allies than upon his own forces. 'You never make conquests except by yourself' (Frederick II 1768: 652).

For Frederick, alliance politics should be entirely free of sentiments or other personal preferences (such as, for example, religious affiliation or family ties). Alliances should be formed for short periods of time, and entirely as a function of one's own short-term interests (Frederick II 1752: 344). At least, he advised against breaking treaties without a very good reason (Frederick II 1752: 396). In practice, Frederick did not always follow this maxim, as his family ties with the Hanoverians and Brunswick and his 'irreconcilable enmity' with the Habsburgs whose lands he coveted were the two recurrent patterns in his alliance policies. But Frederick's articulation of the value- and sentiment-free pattern of alliances is the clearest representation of the post-Westphalian (and thus post-religious), *ancien régime* counterbalancing-of-powers game which played such a prevalent role in passive culture and assumptions about interstate relations until at least the end of the twentieth century.

Examples of this prevailing culture are the respective war aims of the belligerents in the Seven Years War. For Frederick, as we have already noted, they were the increase of territory ruled by the Hohenzollern dynasty. For Empress Catherine II of Russia, Great Chancellor Aleksey Petrovich Bestuzhnev-Rjumin defined the aim of reducing the Hohenzollern's possessions (by dividing them up with the Habsburgs) to the point where they returned to being minor princelings with a negligibly small power base. In a letter of 24 July 1759 to her Field Marshal Daun, Empress Maria Theresia claimed that nothing but 'the weakening of the king of Prussia' would eliminate a host of evils, and that this was thus

> the true aim of the present war, not merely the reconquest of Silesia and Glatz, but the happiness of the human species and the maintenance of our holy religion, of which I constitute almost the only support in Germany. (q.i. Kunisch 1975: 220)

In this context, war was the norm, peace only the time in which one prepared and trained for war, as Frederick mused with a reference to Vegetius (Frederick II 1752: 406). Nevertheless, princes sought to seek justifications – just causes – for their action, finding which they might well leave to their lawyers or diplomats, as Kant suggested (Kant 1795/1796: 17). That they did so shows that, however callous they were, they recognised the moral imperative of presenting their war as fought in a just cause.

The ethics of war in early modern Europe

The legitimacy of warfare: ius ad bellum

Early modern history saw the numbers of battles and sieges increase dramatically compared with the second half of the Middle Ages (Black 1987: 4–11; Tallett 1992: 13). Bloody conflicts between organised entities with numbers of participants well into four- or five-figure numbers did not take place solely between sovereign princes, oligarchies or what can now accurately be called states. There are many examples of large-scale insurrections, and wars of state fission, in which the people of certain parts of a political entity sought to break away and form their own entity. Examples include the failed attempts to establish independent statehood by Bohemia in the 1620s (the origins and early part of the Thirty Years War), Catalonia in the 1640s and later Transylvania, Hungary and Ukraine in the early part of the eighteenth century (Black 1987: 11).

In the sixteenth century, medieval scholastic arguments along the lines of Bonet's were still marshalled for the legitimacy of warfare. Writing at the end of the sixteenth century, Saillans introduced his

Discourse of War with the question of 'Whether it bee lawfull for a Christian Prince, or Magistrate, to make warre', conceding that 'All men are not of opinion, that the Christian Prince, or ciuill Magistrate may warre with a safe conscience against the enemies of his estate, or of the Church.' He set out to argue the contrary, drawing on examples from the Bible:

> The first reason is, because God hath so expresly commaunded ... The second reason is, because God himselfe hath at sundry times counselled, and taught his seruantes, how they ought to gouerne themselues in warre ... The third reason is, because God himselfe in holy writ is called 'the God of Hostes' and 'the Lord God of battels'. Likewise, that iust warres are called 'the battles of God'. The fourth reason is, because many Kinges and Princes are highly commended in holy writt, for that they had warred valiantly against their enemies ... The fifth reason is, because Jesus Christ and his Disciples haue allowed the warre ... The sixt reason is grounded on the law of armes, and the lawfull vse of the sword: for the Prince beareth not the sword in vaine, saith S. Paul, because he is the seruant of God [Rom 13.4], for the safegard of the good, and the punishment of the wicked. (Saillans 1589/1591: 1–3)

Unlike Machiavelli, Saillans still peppered his reflections on what might constitute a just cause for resort to war with Christian considerations – he was, after all, writing for 'Christian Princes'. He wrote, 'The causes are two that may moue vnto warre. One, that respecteth the goods of this world, and the naturall life of man: and this is of two sortes. First when it commeth to the point of repulsing the violence and iniury of tyrants that giue the onset', because self-defence of a prince cannot be denied.

> Secondly, when it commeth to the pinch of relieuing those that are allyed and haue concluded some league of amity with vs being wrongfully oppressed: For Christian Princes may conclude allyance with their next neighbour Nations, so that nothing bee done against the honor of God. Some will say, that the bornes and iurisdictions are distinct, and that it is against all right to inuade another. I answere that indeed it is ill done to incroch one uppon another, when iuasion is pretended to each others territories ... to appropriate himselfe that, which is yet in controuersie. But here the case standeth otherwise, when it commeth to the push ... to defend the right of those that are allianced and confederate vnto vs.

Thirdly, any Christian Prince was obliged to fight in defence of the Church. Finally, 'a Prince may chastice by warre, or otherwise, his heretical subiects, but with some moderation: and yet not before he haue conuinced them to be heretiques by the word of God'. Even given these conditions, wrote Saillans, 'because of the great inconueniences and mischiefs that accompany warre, princes ought not to attempt it, except in time of great extremity' (Saillans 1589/1591: 4–14). Almost a century later, Paul Hay du Chastelet (1619–?70 or 82) admonished Louis XIV to initiate war only as a last resort (Hay du Chastelet 1668/1757: 57–9).

But by and large, military writers did not see it as their responsibility to reflect on the causes and aims of war. The French aristocrat Antoine de Pas, Marquis of Feuquières (1648–1711), the son of a great French general of the Thirty Years War, himself lieutenant-general under Louis XIV, felt no need to tell his son, for whom he was writing, and thus his readers, more about the circumstances of war than that Europe was divided into states, some of which were ruled by princes, others being republics; and that some princes were peaceful and others ambitious. In the latter case, he wrote in a Machiavellian vein, the prince had to 'benefit from the divisions which he will have adroitly sown among his neighbours', and had to 'use any pretexts which they furnished, in order to quarrel with them' (Feuquières 1731: 9).

By contrast, 'D.J.' (possibly the Chevalier de Jaucourt), writing in Diderot's and d'Alembert's *Encyclopedia*, had similar views to Boecler's on just causes of wars: they were 'the defence of one's religion, one's fatherland, one's possessions, and one's person, against tyrants and unjust aggressors'. However, he thought that '[a]s the sovereigns feel the force of this truth, they take great trouble to circulate manifestos to justify the war they undertake, while carefully hiding from the public or from themselves the true motives which guide them' (D.J. 1757: 966). Tactfully, the ensuing example was one from antiquity, but the express dismissal of 'glory ... aggrandisement, and utility' as well as 'personal interest', if the other had not committed any wrong, is indirect criticism of the reasons so often given by princes at the time. The author went on to argue that even a legal reason which was technically correct could not truly serve as justification for warfare which in reality stemmed from different motives – such as the quest for 'vain glory', the desire to become feared (*redoubtable*), 'to exercise one's armed forces, to increase one's domination', or to acquire the more fertile territory belonging to others, motivations which he asserted were 'very common'. With a resonance of the Spanish theologians like Vitoria, the author declared unjust the conquest of any territory belonging to another people, such as the Native Americans – whose lands the *conquistadores* had seized by force, under the pretext that the natives practised 'idolatry'. Following the tradition of Augustine, Aquinas and Grotius, the *Encyclopedia* thus argued that only such a war was just that had a defensive aim or aimed at the reparation of damages caused by the other side, and that it could only be just if its aim was to establish a solid and lasting peace. The author explained, however, that an offensive campaign could result from just such a set of causes, and that a defensive campaign might aim to protect unjustly acquired possessions. The cause itself might be restitutive, the campaign offensive. In view of the great suffering brought about by war, war must in any case be the last resort, even if one had a just cause. This cause could be just, but still had to be important enough to warrant the recourse to arms – for example self-preservation or the public good. Moreover, there needed to be a fair chance of success (D.J. 1757: 966f.).

The encyclopedist considered it to be part of 'European military laws' that one had to spare the life of prisoners of war as well as that of anybody surrendering and asking for quarter, and of course all non-combatants, that is, old people, women, children and those who because of age or profession could not carry arms. Any dishonourable act against women was particularly unlawful (D.J. 1757: 966f.). D.J.'s text highlights the remarkable continuity in the Western just-war tradition from Roman times to the Enlightenment, notwithstanding the frequent divergences from it in practice.

Writing at the same time, Vattel was mostly concerned with the laws governing a declaration of war and then conduct in war. Vattel's writing reflects the multitude of aims in waging war either in his time or in bygone times, which included self-defence,

> the prosecution of some rights, or ... safety. We attack a nation with a view either to obtain something to which we lay claim, to punish it for an injury it has done to us, or to prevent one which it is preparing to do, and thus avert a danger with which it seems to threaten us. I do not here speak of the justice of war ... all I here propose is, to indicate, in general, the various

objects for which a nation takes up arms, – objects which may furnish lawful pretences, or unjust pretences, but which are at least susceptible of a colour of right.

He added that offensive wars were, of course, also waged for 'conquest, or the desire of invading the property of others', but these were not 'regular warfare, but ... robbery' (Vattel 1758/1834: 293). While Vattel still came from the Augustinian just-war tradition, in which 'War cannot be just on both sides', he conceded that 'since nations are equal and independent, and cannot claim a right of judgment over each other, it follows, that, in every case susceptible of doubt, the arms of the two parties at war are to be accounted equally lawful, at least as to external effects, and until the decision of the cause' (Vattel 1758/1834: 306).

An even more secular approach formed the basis of the views of international law of Vattel's contemporary, Montesquieu (1689–1755):

The law of nations is naturally founded on this principle, that different nations ought in time of peace to do one another all the good they can, and in time of war as little injury as possible, without prejudicing their real interests. The object of war is victory; that of victory is conquest; and that of conquest, preservation. From this and the preceding principle all those rules are derived which constitute the law of nations. (q.i. Hobbs 1979: 4)

Here we see a total acceptance of 'aggrandisement' as the natural right of nations (no longer princes) to set out to conquer territory, even if Montesquieu gave this a defensive interpretation.

Crucially, however, in all these contexts, what was absent before and after the wars of religion, and then again until the French Revolution, was the war aim of social revolution. No prince, republic or other state waged war to destroy the social and political structure of another country and to export a revolutionary ideology. As Hay du Chastelet put it succinctly: the prince 'only overthrows thrones in order to strengthen them' (Hay du Chastelet 1668/1757: 310).

Restraints on the conduct of war: ius in bello

Rules on how to conduct war (*ius in bello*) were honoured almost as much in their breach (although this invariably led to indignation and complaints) as in their observance. The invention of printing made it possible for rulers and military commanders to issue all their soldiers with 'Articles of War' which invariably listed looting, theft and rape as punishable offences, carrying the death penalty (see for example Sutcliffe 1593). Soldiers could not plead ignorance, even if they could not read, as the Articles were also read out at weekly army assemblies (Tallett 2006: 22). Nevertheless, irregular payment of troops and unreliable food supplies pushed soldiers to disregard these rules even at the risk of their own lives. There are even examples where one side used force prior to declaring war, especially in naval and colonial contexts, far from a controlling centre. By contrast, wars did not generally start by accident, but hostile attitudes were long established and grievances well publicised. Only major drawn-out conflicts were rarely intended – usually, at least in conventional state-*versus*-state wars, just as in previous or following centuries, both sides hoped for swift victories (Black 1987: 14, 16f.).

As previously in the Byzantine Empire, the religious wars of the sixteenth and seventeenth centuries were remarkable for their cruelty and their large numbers of casualties. From the English Civil Wars onwards, religious beliefs first blended with, and were gradually replaced by, ideological struggles, which in equal measure challenged the legitimacy and inner logic of existing states, and were therefore fought more fiercely.

Even though there are examples of campaigns in which the commanding princes or generals consciously sought to cause little collateral damage to the local populations, there are countless examples of atrocities committed against non-combatants, both individually and collectively, and many instances of mass starvation through artificially created famines to deprive the enemy's armies of vital food and fodder for their horses. In the twelfth century, Henry II of England had practised scorched-earth tactics in his fight to subdue Ireland. Henry VIII four centuries later did much the same, and so did his daughter Elizabeth I, followed in turn by Oliver Cromwell in the mid seventeenth century. In the sixteenth century, Maurice of Nassau destroyed more than 200 villages in the Dutch Eighty Years War against Spanish overlordship. Religious wars and civil wars in particular abounded with such examples; apparently such acts were easier to justify to oneself when one defined the victims of one's actions as heretics, unbelievers or devils.

Casualty figures for seventeenth-century warfare as a proportion of the population of the areas affected were topped only by figures of the twentieth century. The sufferings of the non-combatants were not much different from those of medieval times in quality, but increased populations led to increased numbers of people suffering the 'miseries and sorrows of war', title of the famous cycle of etchings by Jacques Callot (1592–1635). During the Thirty Years War, it was regular practice to burn down villages, but also to burn harvests, mills and bakehouses to deny the enemy (and any neutral or inimical population) their benefit (Tallett 1992: 59f.). This scorched-earth policy, coupled with the forced requisitioning of food and fodder from the areas through which armies passed, continued well into the eighteenth century and led, directly or indirectly through famine in the subsequent months, to the high casualty figures among the non-combatants. Rulers tried to spare their own populations – one of the reasons why they and military authors were so keen to carry warfare into the enemy's lands. Even though time and again military authors warned against pillaging and the use of force against local populations, it is in fact wrong to argue that these were less affected by early modern warfare than the populations of the twentieth century. (Only the eighteenth century in some parts of Europe, and the nineteenth century after Napoleon, saw some protection extended explicitly to civilian populations, and then only in western parts of Europe, and to white civilians in the American Civil War.) While special rights and immunities of non-combatants were often disregarded in practice, such behaviour was not advocated by soldiers writing on the conduct of war; indeed, several soldiers or ex-soldiers warned against it, while in the Middle Ages it had been only clerics and Christine de Pisan who had counselled against such actions on moral grounds (Rocca 1582, especially books 2, 3 and 4; Fourquevaux 1549: 222–57).

Rules of war had been developed in the Middle Ages – both for the circumstances in which a Christian prince could declare war on another, and for the conduct of such a war. Saillans formulated 'Certain rules and lawes to be practised of those that vndertake to wage warre, to the end that God may blesse and prosper their actions':

1. That warres be not taken in hand, but for some great and important affaires ...

2. That the Prince be not troubled vpon euery light occasion, but before hee make warre, that he assay by all meanes possible to maintaine peace and concord: for if the end of war be to seeke to conclude peace againe (as Cicero saith [*De Officiis* 1]) without all doubt the Christian Prince ought to procure and seeke all honest and lawfull meanes, whereof hee can bethinke himselfe, to pacify and accord all differences, before he procede to armes and to warre ...
3. That no man warre, to the end to vsurpe the goods and inheritance of other men ...
4. That the Prince that will vndertake to warre against others, first let him knowe how to rule and ouercome his owne passions and affections ...
5. That the prince vrged of necessity to warre, chose out wise Captaines and guides well experienced and valiant men ...
6. That he choose also for his men at armes and souldiers, men that feare God ...
7. That good and holy lawes haue course and vigoure in Christian armies, and not onley in townes and Cities, but also abroade in the middest of the campaigne ...
8. That the leaders and Captaines bee valiant and couragious in all such difficulties, as of custome fall out in exploits of warre ...
9. That Captaines be diligent and painfull, and in so wise ydle and sluggish ...
10. That nothing be enterprised or attempted without asking aduise at the mouth of God ...
11. That the armies put their trust in God alone, and not in the helpe of man ...
12. That the armies above all thinges be Godly and holy, if wee will haue them worke good effect (Saillans 1589/1591: 17–31).

Hay du Chastelet admonished princes and their generals to be generous in victory, to look after the enemy's wounded as well as their own, to protect the citizens of towns that had surrendered from pillage and worse, and to woo the 'hearts' of people in occupied territories, so that they would become one's own loyal subjects (Hay du Chastelet 1668/1757: 148–57, 310). A century later, Vattel urged his readers to abide by the general rule of sparing non-combatants, as far as possible, and as long as one was not dealing 'with a savage nation, who observe no rules, and never give quarter' (Vattel 1758/1834: 347). A separate treatment of Hellenes and of barbarians, of Romans and uncivilized peoples, of Christians and the infidel, of Europeans and of natives of other continents runs throughout this period, to be taken up again and vastly amplified by the Europeans of the late nineteenth century in their colonial wars. By then, Social Darwinism and explicit racism injected a new viciousness into European warfare, but the continuity of this pattern of thought whereby operational restraints on war did not apply to certain 'others' runs uninterrupted from Antiquity to the present.

As religious wars tended to be particularly cruel, they spurred humanist endeavours to limit such excesses, most famously associated with Erasmus of Rotterdam's *Antipolemos* of 1521. The perceived need for mutual respect for other Christian confessions was first codified in the famous Confessional Peace of Augsburg of 1555, but it took almost another century until, after the Thirty Years War, this became more generally accepted in Europe, leading to a secularisation of states' foreign policy pursuits, making them subject (again, as in the times of the *condottieri*) above all to financial considerations. By the second half of the seventeenth century, Louis XIV's France fought Catholic and Protestant countries indiscriminately. This did not mean, however, that the Vegetian and medieval practice of creating artificial famines and thus causing the suffering and deaths of large numbers of non-combatants ceased entirely. Under Louis XIV, the Palatinate was subjected to a scorched-earth policy, and its towns and cities sacked and burned (Lynn 2002), contradicting the notion of writers of the following generations that warfare between 1648 and 1792 was particularly civilised (see below).

Eternal war or eternal peace?

Many sixteenth-century men were still under the influence of the earlier Christian teaching which saw war as either God-given (like Honoré Bonet) or saw in war yet another innate vice of man – going back to the murder of Abel by Cain – or indeed a reflection of the 'war in heaven' which had occurred in the very beginning when God expelled Lucifer from his celestial realm. As we noted at the end of the previous chapter, the notion that war was God-given and God-willed lived on well into early modern times. But as the Age of Enlightenment dawned and purely faith-based explanations of the world became less acceptable, the notion that war was an inevitable part of God's plans for man receded. Biblical precedents gradually paled as models for modern warfare, just as religious pretexts for going to war were seen increasingly cynically by secular writers, of whom Machiavelli, writing a century and a half before Hay, was probably the first. Most early modern authors still took it for granted that war could not possibly be eliminated, and sometimes even took up the complaint of the ancients that long periods of peace were an evil too, because they made society weak and armed forces undisciplined (Guibert 1772/1781: vff.). The bewailing one's own decadence in comparison with the strength of the simpler lifestyle of one's enemies goes back to Herodotus' admiration for the frugal life of the Persians (Herodotus I.69f.), Xenophon's worries about the collapse of the previously equally frugal Spartan lifestyle in his *Lakedaimonion Politeia*, Roman Republican fears of their society's decadence leading to the legislation of 187 BCE (Livy XXXIX.6) and Tacitus' praise for the Germanic peoples and his indirect complaints about his fellow citizens' life of luxury in his *Germania*. The following millennium seemed to furnish more proof of the weakness of sophistication and luxury compared with barbarian simplicity and martial prowess. Both the West and then the East Roman Empires had indeed succumbed to much less civilised invading tribes, and the theme continued to preoccupy writers in early modern Europe, and not only in France. The third Marquis Santa Cruz de Marcenado, for example, pondered both the benefits and the dangers of peace, even though he insisted on the Augustinian criteria for just wars and the condemnation in principle of wars between Christian princes, and included Augustine's stipulation that it is unethical to start a war one is too weak to win (Santa Cruz de Marcenado 1724–30 [1726]/1885: II, 54–72). Some apologists for war argued that war channelled passions outward and made use of unruly elements (for example Hay du Chastelet 1668/1757: 2f.).

Lawyers and moral philosophers increasingly likened warfare to crime, stressing the lawlessness of the state of war and the many transgressions in civilian society committed by soldiers, even outside war; they saw war as brutalising those involved in it, who in turn imported brutal, lawless and unrestrained modes of behaviour into civil society. Either way, from early modern times onwards, violence was increasingly seen as a function of human passions rather than of

God's design, and sometimes as caused by culture or geography. Some sixteenth- and early seventeenth-century writers worried as much about violence in the form of duelling as about effects of war such as the slaughtering, starving or burning of civilians, rape and pillaging, and saw no need for reflecting on the political origins of wars (Noue 1587; Digges 1604; Hale 1971: 3–26).

The English philosopher Thomas Hobbes (1588–1679), writing as civil war loomed in England, is most popularly remembered for his view that the state of nature is a state of war of all against all, and that this is mitigated by the creation of the 'Commonwealth', his word for what we today usually term the state (Hobbes 1642: ch. I.12). From this many have drawn the conclusion that war is endemic to the human condition, and that interstate relations necessarily assume the form of war of all (states) against all others. Even Kant still contended that '[w]ith men, the state of nature is not a state of peace, but of war' (Kant 1795/1796: 11). This was not, however, a point of consensus among writers in early modern Europe, even though military men writing about war often took this position. The Saxon Samuel Pufendorf (1632–94), a Lutheran minister's son, argued that the first humans – Adam and Eve – had loved each other, and that the state of nature was thus one of love, not enmity. Nor did he accept Hobbes's reasons for arguing that once human societies had formed up into states ('commonwealths'), these must necessarily be inclined to go to war with each other (Pufendorf 1672/1749: 106–13).

In late Antiquity and in the Middle Ages, the only sphere of lasting peace that could be imagined was within the *Pax Romana*, the Roman – and then Christian Roman – Empire. This claim was inherited by the Holy Roman Empire, but it did not include all Christian states within its confines, and wars between Christian entities were notoriously present in the Middle Ages, in defiance of the papacy's views on this, which presented the alternative (but never realised) model of a peaceful Christian universe. The Grand Design of Henri IV of France and his adviser Sully merely replaced the Holy Roman Empire (of which France was crucially not a part) with the project for the constitution of a European super-state including not only the Holy Roman Empire but also France, England and Scotland, significantly to be called *Respublica Christiana* (Sully 1632/1921: 38). This term had been claimed a century earlier by political philosophers for the Holy Roman Empire alone, with the political programme for Emperor Charles V to unite the entire Christian world within it. Such an idea of a super-state still underlay the project of 1712 of a European Union [*sic*], published by Charles-Irénée Castel, Abbé de Saint-Pierre (1658–1743) in his *Project for Rendering Peace in Europe Everlasting*. The abbé's concept was derided by some contemporaries, but not by all. Jean-Jacques Rousseau liked it so much that he invested his own time to cut it from two volumes to one longish pamphlet, thus ensuring a much larger readership; the product was then occasionally wrongly attributed to Rousseau himself (Rousseau 1761/1761). Even Voltaire, who wrote a sarcastic treatise in response to the abbé's plans, later admitted that he preferred it to the nicest strategies for war (Voltaire 1769).

A crucial new departure in thinking about war and peace came from the pen of Count Jacques Antoine Hippolyte de Guibert, in his *General Essay on Tactics*, which he wrote shortly after the end of the Seven Years War, although he delayed its publication by several years for fear of censorship. There he mapped out a community of states which would all be republics and therefore by his definition not expansionist, because government would be exclusively in the interest of the people, who would only be prepared to wage war in defence of their own territory (Guibert 1772/1781: xxxiiif.). Guibert's idea predated Immanuel Kant's *On Eternal Peace* (1795) with its plan for a confederation of peaceful republics, without central enforcing authority, by twenty-five years. Guibert later became more sceptical, and his youthful ideas have been unjustly all but forgotten. It is Kant's prescriptions for a universal peace treaty that are mainly remembered, which included the abolition of all standing armies and a pledge of mutual non-intervention (Kant 1795/1796: 2–10).

The French Revolutionary and Napoleonic Wars could not entirely quell this hope that war might be abolished through human arrangements, which inspired the Holy Alliance of 1815 formed by the victor powers after Napoleon's defeat. They were submerged, however, in the 'realism', militant, xenophobic nationalism and Social Darwinism of the nineteenth century, which we shall discuss in chapter 5. Kant's ideas on the peaceful coexistence of republics would remain famous and gain renewed popularity in the twentieth century, particularly after the end of the Cold War.

4 *Themes in early thinking about Strategy*

The profusion of variables in war has never discouraged the search for foolproof systems.

(Galula 1964/2005: xii)

There is universality even amidst great diversity. Strategy and strategic effect do not differ from war to war, among geographic environments, or among culturally distinct belligerent polities.

(Gray 1999a: 362)

Chapters 2 and 3 have dealt with attitudes to war and its purpose in late Antiquity, the Middle Ages and early modern history, highlighting both changes and continuities. This chapter will dwell on aspects which have partly or entirely continued to the present: the basic pattern of siege warfare; debates about the recruitment and status of troops; the question of the importance of battle as well as defensive and offensive strategies; the spectrum of warfare from limited to major; and the ceaseless quest for eternal principles guiding warfare and ruling war.

Sieges and static defences from Troy to Basra

From Antiquity to the present, territorial conquest has never been achieved by a total physical occupation of all land. Instead, the symbolic seizure of certain settlements and strategically crucial points like mountain passes or bridges or crossroads (frequently coincident with settlements) had and has to make do. In turn, such places would be defended, best done by fortifying them.

Settlements fortified by stone walls or at least by mud walls, wooden fences and/or ditches had existed since the Neolithic Age. Greek cities had elaborate fortifications; the pattern of a heavily defended enclosure around the royal precinct (the *acropolis*), surrounded by the dwellings of the lesser townsfolk, themselves protected by another wall, can be found in most layers of Troy, as well as in the earliest layout of many other ancient towns and cities, including Rome, where the old stronghold was the Capitol Hill. The Romans had fortified some of their frontiers. In a less sophisticated way, such land walls for the defence of entire regions continued to be built in the early Middle Ages – Offa's Dyke in Britain, Gunfreds Danevirke under Danish rule and the Götavirke in Sweden, but the practice was then discontinued until the earthworks of the Enlightenment and the Maginot and Siegfried Lines of the twentieth century.

The medieval castle was a development either of Roman fortresses or of the fortified Roman town or the rulers' settlement at the heart of villages, towns or cities along the model of ancient settlements. It seems that after the fall of the West Roman Empire, Occidentals for centuries forgot how to build stone castles and other stone fortifications, notwithstanding the fact that most of them must have lived within a few days' marches at most of surviving examples of Roman fortifications or at least stone walls (Sander 1934: 457–76). It was only in the late tenth century that the stone castle was reinvented in France. The relatively fast spread of castles around Europe from around 1000 is less a technological innovation than a return to a more sophisticated past, a rediscovery or early renaissance of Roman castle architecture (as the name *castellum* itself suggests) after a decline in technology in the previous 600 years.

For the civilians, 'fortified places' – which soon included entire towns, as in Roman times – were the main places of refuge when hostile armies broke into their lands. For governments, they were the places from which the land was administered, including of course the raising of troops or money for war. Armed forces, too, would seek refuge in these places, both from enemy armies and from trying weather conditions. From there, they would sally forth to attack adversarial armies in defensive campaigns. Food and munitions were stored in such places, which limited the range of movement of all armies who relied upon supplies from the authorities for whom they were fighting – as opposed to living off the land.

By contrast, invading and conquering armies necessarily had to move away from their own country's sources of supply and thus had longer supply lines; consequently, they sought to rely on allied (fortified) towns (like the Holy Roman emperors in their campaigns in

Italy), build new fortifications (such as Edward I's castles in Wales) or conquer existing strong points both to deprive their adversaries of them and to put them to their own use. The latter explains the many sieges in the Hundred Years War and in the Franco-Burgundian wars, or the campaigns of the Thirty Years War (Croxton 1998; DeVries 1998). Either way, armies operating outside their own lands were usually forced to live off the land simply because they could not master the logistical challenge of bringing all their victuals and ammunition with them. If the warring factions had sufficient money, they could buy victuals, but until the late sixteenth century warring monarchies tended to run out of money within a few months if not weeks; where soldiers' pay was in arrears, these quickly resorted to loot and plunder and forced requisitions, invariably thereby increasing if not engendering local hostility to these occupying forces.

Sieges occurred more frequently in the high and late Middle Ages than battles, and sieges continued to be frequent in early modern European history, even while the frequency of battles increased significantly. Siege warfare was, however, transformed into a vastly more expensive affair for both sides by the advent of gunpowder, and in response, the *trace italienne*, defensive fortifications, which reached their architectural apogee in the age of Louis XIV of France. Much was written about siege warfare – considered as a distinct part of the art of war for example by the French encyclopedists (Tressan 1757: 992) – in all the literature analysed here until the end of the twentieth century, but it tended to be primarily of a technical nature. Only Machiavelli made points of a more political nature: he advised his readers not to impose harsh conditions on besieged towns, but to tempt them into surrender by promises – including mendacious ones – of good terms. For if the besieged feared that they would be treated as rebels and slaughtered once their town was taken, it would make them only more determined to hold out (Machiavelli 1531/1998: 334–59, 442f.). In any case, Machiavelli's and many other early modern authors' persistent admonitions not to allow one's victorious soldiers to run amok after taking a town was often disregarded in practice: there are many horrifying examples of massacres and mass looting following the fall of towns in early modern times – most notoriously the rape of Magdeburg in 1631 – and in more recent history, from the fall Saragossa in 1809 in the Peninsular War to the genocide in Srebrenica in the Yugoslav War in 1995. It varies from case to case as to whether such massacres ensued from poor discipline or were part of a deliberately adopted strategy.

Defensive fortifications, in any case, were a strong element of strategy not only in classical times but also in the later Middle Ages and then in modern times, up to the French Revolution. By the end of the eighteenth century, many parts of mainland Europe from the Channel to the Vistula were studded with stand-alone fortifications built in the tradition of the *trace italienne*, and fortified towns of all sizes had huge star-shaped earthworks and walls all around them. Until the French Revolution, sieges of fortified places and towns were central to warfare and as common in Europe as pitched battles. The most famous defender of such fortifications was Louis XIV's adviser and chief engineer, Sébastien le Prestre de Vauban (1633–1707), whose elaborate scheme, moving away from individual castles and fortified towns scattered throughout the country to two to three lines of new fortifications along France's frontiers, was largely implemented in the reign of his patron (Vauban 1685, 1704, 1737–42).

Also on the other side of the Atlantic, in the French and Indian Wars, as in the American War of Independence, sieges were prominent in warfare. While many encounters were hardly more than skirmishes of a 'small war' calibre, the bigger encounters were either straightforward sieges of towns or forts, as the names of the encounters show – Fort Beauséjour, Fort Bull, Fort Oswego, Fort William Henry, Louisburg, Fort Carillon, Fort Frontenac, Fort Duquesne, Fort Ligonier, Fort Niagara – or battles which took place near towns and forts, such as the battle of the Plains of Abraham outside Quebec, following the siege of Quebec, in 1759.

This explains the growing preoccupation of eighteenth-century writers with the organisation of supplies and with the distances between fortified storage facilities and the troops. These distances, and the roads between these two points, or lines, became something of a fetish for a number or writers, first among whom was Adam Dietrich Heinrich von Bülow (1757–1808). He has remained famous only for his mathematical approach to war, a typical example of the mathematical interest in warfare shown by many of his contemporaries, from Guibert in France to the Hanoverian Gerhard von Scharnhorst (1755–1813). Bülow was particularly interested in the geometry of supply lines, 'inner lines' for most defending armies fighting on their own territory and angles of attack. Scharnhorst and other artillerists concentrated on the mathematics of calculating artillery trajectories for aiming canon and angles for defensive and offensive siege trenches. The equally eccentric Prussian captain-turned-writer Georg Heinrich von Berenhorst (1733–1814) found that since Vauban, fortifications of towns and places had expanded enormously, as had the effort to besiege them in time and space (Berenhorst 1797: 42).

It was the abandonment of reliance on fortified places for supplies that allowed the French Revolutionary and Napoleonic armies to range much farther and faster than most of their predecessors since Roman times. The Austrian Archduke Charles, writing retrospectively about the Napoleonic Wars, explained the waning of the importance of fortresses:

> The deployment of larger masses [of troops] than in previous times, paired with a lightness and acceleration of their movement that we had not been used to, necessarily had to have a changing impact on warfare in its entirety, and thus also on the effects of fortifications. During the time of calm that followed the Seven Years War one saw fortifications as perfect defensive protection of the hinterland without any regard for ownership, circumference or manning, and without demanding more than the passive defence of their ramparts. It is little wonder that one preferred small fortifications to bigger ones, convinced that they could achieve more with lesser means. This delusion dissipated as the armies grew in numbers. [Since then] fortifications in unimportant locations, of small size and manning are only of importance if the adversary's forces are so small that they have to limit their enterprise to the conquest of small areas, being unable to spare the smallest number of troops ... With large-scale [armies], the influence of such fortifications disappears altogether. (Charles 1838: 209f.)

The only exception that he saw in the continued relevance of smaller fortifications was their role in people's war (generally planned insurgencies): arms and supplies for the population could be kept at such fortresses which took on the role of arms depots (Charles 1838: 209f.).

But even in the Napoleonic Wars, in which the armies initially swept around fortifications, second-line soldiers were often detailed to deal with them. And in the Peninsular War, the long stretch of free-standing fortifications (in the tradition of the great walls rather than fortified places) at las Torres Vedras gave Wellington a crucial advantage in his defence against the French in 1810. Furthermore, the fact that a number of capitals were not fortified crucially influenced the Napoleonic Wars. The lack of well-maintained, modern fortifications around Vienna, Berlin, Madrid, Moscow and Paris ruled out the option for the defeated Austrian, Prussian, Spanish, Russian and French armies in 1805, 1806, 1808, 1812 and 1814 respectively to withdraw to their capitals and attempt to defend themselves against Napoleon's armies or the Coalition (Brialmont 1895: 11f.).

In the wars from the second half of the nineteenth century, sieges of important cities – Sevastopol (1854–5 in the Crimean War), Atlanta and Petersburg (1864 and 1864–5 in the American Civil War), Plevna/ Pleven (1877–8 in the Russo-Turkish War) – came to the fore again. In 1860 Wilhelm von Willisen, author of a three-volume work on major war, produced a monograph on fortifications, inspired by debates on the need to fortify Berlin. Willisen argued that it was impossible to erect fortifications that could fully protect the – quickly expanding – Prussian capital in a palpable, physical sense, and reasoned instead that it had to be complemented by an indirect, strategic defence in depth through army movements far to the east and south-east of the capital. While the land to Berlin's east (and throughout Prussia along the Baltic) is notoriously bare of physical barriers and inviting to invading forces, Willisen argued this more generally, as few places, to his mind, could be properly defended by fortifications alone (Willisen 1860).

Further examples can be found in the Franco-Prussian War, where some fortifications were found wanting and in dire need of modernisation, and fell to the German armies early on. After the war, this realisation triggered a flurry of constructions of (quickly outdated) fortifications in France. But it was Belgium that produced one of the most prominent theoreticians on the construction of fortifications, Henri-Alexis de Brialmont (1821–1903), who rose to become the Belgian Inspector-General of Fortifications and of the Engineer Corps in 1875. He initially conceived of a defensive Strategy for Belgium that – against much understandable opposition – planned for the need to abandon the capital, Brussels, and to withdraw to Belgium's second-largest city, Antwerp. Taking advantage of its estuary location, from 1864 the city was turned into the fortified national redoubt at his instigation (Brialmont 1895).

After the Franco-Prussian War, Brialmont saw the danger that lay in a renewed conflict between the Germans and the French; in view of the French fortifications that were now under construction, he predicted that the Germans would move through Belgium in order to attack France. He therefore urged the construction of fortifications along the Meuse. From 1887 to 1892, the cities of Liège and Namur were thus heavily fortified, albeit not heavily enough to withstand the new German 305 and 420 mm howitzers brought into action against them in 1914. Liège fell after only two days, Namur was even less of an obstacle to the German army. Christopher Duffy, in his biographical sketch of Brialmont, regretted 'that he devoted so much of his talent to permanent fortification, a cause that was beyond saving' (Duffy 1967: xliii). And yet, the First World War on the Western Front could be said to have degenerated into one big siege. Nor was the belief in the utility of great walls a thing of the past, as the construction of the Maginot Line showed subsequently.

It seemed at first that the arrival of aircraft and later of long-range missiles made sieges of towns or cities (in the form of ground forces surrounding the built-up areas until they surrendered) obsolete, yet cities like Leningrad and Stalingrad were not only bombed but besieged in the Second World War, and sieges occasionally occurred in the smaller wars of the following decades, until the sieges of Iraqi towns in the war begun by the US and its coalition partners in 2003.

Sieges have thus not become obsolete, but the great bulk of what has been written about them has been overtaken by technical changes. Most technical aspects in the literature are therefore omitted here. The principle of fortifications – the augmentation of human strength for defence by static defences infinitely more resistant – continues to play its role in Strategy, both where it is entirely defensive or where it is aiming to hold an enemy's onslaught in one place the better to be able to turn his flank and counter-attack in another. By and large, defences in history have been a function of the former – a mainly defensive Strategy – whether this was in the Roman Empire when it had stopped expanding and began to be subject to invasions, the Iron Curtain during the Cold War with its more or less elaborate physical barriers throughout Europe or Israel's security fence.

Feudal levies, mercenaries or militia?

Another key theme linked to the purposes of war and thus Strategy, which again goes back to Antiquity, is the question of who should fight, and how these soldiers should be recruited, with all the political, social and administrative implications. Rome had started with the ideal of the free farmer-citizen who defended his land and Rome. But Rome had then adopted many different systems including long-term service in a professional army (in return for land and early retirement) and also the integration – from auxiliary forces to full members of a professional army – of 'barbarian' tribes originating outside Italy. This led to a change of the basic culture and spirit of the Roman army, and was seen as having contributed to the fall of the West Roman Empire to barbarian tribes and their chiefs. The Byzantines were therefore wary of such mercenaries, but the theme would also be picked up in the West in the Renaissance. In what historians used to call 'feudal' medieval Europe, lords had the right to ask their retainers to provide annual military service. Getting soldiers who were doing their military service as a part of a feudal obligation to stay armed beyond the previously agreed annual period (perhaps one or two months) could be very difficult.

As the monetary economy flourished again in the twelfth century, military service began to be replaced here and there by payments which the lord could translate into arms procurement, food and fodder, or use for the hire of soldiers who became increasingly professional. Individual leaders started turning these mercenaries into what today would be called private military companies, for hire to anybody willing to pay their wages and able to pay regularly.

Neither the pike nor the crossbow nor the longbow constituted a technological quantum leap in the way gunpowder did *in the long run* – spears, bows and arrows were all among the oldest weapons known to man. Yet all of these came together to transform later medieval warfare very profoundly, making lowly infantrymen more important because they became dangerous even to a well-protected knight, amounting to what has been called an infantry revolution in the fourteenth century. Infantrymen were much cheaper to equip than mounted 'men-at-arms'. From the eleventh century until the Hundred Years War, European armies had been dominated by knights ('feudal warrior-aristocrats') on horseback, fighting to capture and exchange for ransom rather than to kill. By contrast the armies after the Hundred Years War

> differed from this description on every single count. They were drawn from the common population (albeit often led by aristocrats); they served

for pay; they fought primarily on foot, in close-order linear formations which relied more on missile fire than shock action; and they fought to kill. (Rogers 1995: 56)

The Hundred Years War had already been fought to a great extent with mercenaries. The result of this was that during peaceful spells within the period of the Hundred Years War, and after its end, there were bands of soldiers who could not find a way to reintegrate into the civilian labour force and became a plague throughout western Europe. Charles VII of France was the first western European prince to reinstate the Roman concept of the state monopoly of commanding armies, simultaneously setting up the first standing (professional) army in the West since the fall of Rome. This would serve as a model to all of Europe, and become the external symbol of sovereignty.

With the Renaissance came the politicisation of the issue of who the soldiers defending the polity should be. Machiavelli, inspired by the Roman model, strongly opposed the use of mercenaries, as they felt no loyalty to the polity they were engaged to defend. The Roman citizen-solider, by contrast, had defended what was his and the *urbs* (Rome) and its possessions. In *The Prince*, Machiavelli wrote:

[T]he arms on which a prince bases the defence of his state are either his own, or mercenary, or auxiliary, or composite. Mercenaries and auxiliaries are useless and dangerous. If a prince bases the defence of his state on mercenaries he will never achieve stability or security. For mercenaries are disunited, thirsty for power, undisciplined, and disloyal; they are brave among their friends and cowards before their enemy ... they avoid defeat just so long as they avoid battle; in peacetime you are despoiled by them, and in wartime by the enemy ... a republic which has its own citizen army is far less likely to be subjugated by one of its own citizens than a republic whose forces are not its own. [Thus] Rome and Sparta endured for many centuries, armed and free. The Swiss are strongly armed and completely free. (Machiavelli 1532b/1961: 77–9)

Yet in the two centuries after Machiavelli mercenaries became increasingly useful to warring factions as entrepreneurs took charge of recruiting, paying, training, equipping, housing, feeding and moving them (Mallett 1974; Parrott 2009). Wallenstein for one was such an entrepreneur on the biggest scale. Even so, for ideological reasons, the theme of the native or citizen-soldier or a militia became important from the seventeenth century (Sweden in the Thirty Years War, England in the Civil Wars): henceforth, polities seeing themselves as republics or commonwealths either implemented or toyed with the idea of an obligation for all men to perform military service.

Spanish authors of the sixteenth, seventeenth and eighteenth centuries also tended to come out against mercenaries, and – earlier perhaps than writers from other countries – against multinational armies. Jerónimo Jiménez de Urrea (1510–74) was concerned mainly with deflecting the Spanish nobles' overdeveloped quest for personal honour away from duelling to fighting for their sovereign, while claiming that one could not gain honour from serving a foreign prince. Yet he postulated that even as a mercenary, any officer or soldier was bound by duty to serve that prince loyally, just as the soldier was duty-bound to serve his captain (Urrea 1556: 163–5). Four decades later, Urrea's countryman Mendoza argued that it would be preferable to have culturally homogeneous armies – armies whose soldiers hailed only from one *nación*. He claimed to see patterns of mutiny and conspiracy within armies composed of different nationalities with leading captains also hailing from different countries, but acknowledged that it would be difficult – at any rate for Spain – to raise enough soldiers from its native population to form a nationally 'pure' army (Mendoza 1595: 64f.).

While piketeers or pikemen decreased in number and were replaced by musketeers, the infantry remained the bulk of all armies, outnumbering cavalry. Sir James Turner in *Pallas Armata* (1683) wrote:

> the ancient distinction between the Cavalry and Infantry, as to their birth and breeding, is wholly taken away, men's qualities and extractions being little ... regarded ... most of the Horsemen, as well as of the Foot, being composed of the Scum of the Commons. (q.i. Roberts 1995: 23)

Gustavus Adolphus' army in the Thirty Years War, hailed as 'the first truly national European army' (Michael Roberts), in reality consisted not only of Swedish militiamen but also of large numbers of mercenaries. In the seventeenth century most armies including the Swedish continued to contain mercenary forces, drawn above all from Scotland, Switzerland and the Balkans, countries with a surplus of poor, yet hardy males, with a strong fighting culture. Only by the end of the seventeenth century did the monarchs of Europe gain full control of their armies, doing away with military enterprisers: this 'was a significant development; for once the armies became royal (as the navies already were), the way was open for their eventually becoming national' (Roberts 1995: 16f., 22). In the reign of the Sun King in France, Hay du Chastelet advocated recruiting one's army above all from natives (Hay du Chastelet 1668/1757: 22f.), even though foreign mercenaries would continue to serve in the French armies for a good century yet.

In the mid eighteenth century, Emerich de Vattel thought it the norm that only the sovereign had the right to raise troops (Vattel 1758/1834: 293). 'Every citizen is bound to serve and defend the state as far as he is capable.' Only those should be exempted 'who are incapable of handling arms, or supporting the fatigues of war', these being old men, children and women.

> Although there be some women who are equal to men in strength and courage, yet such instances are not usual; and rules must necessarily be general, and derived from the ordinary course of things. Besides, women are necessary for other services in society; and, in short, the mixture of both sexes in armies would be attended with too many inconveniences. (Vattel 1758/1834: 294)

Vattel recognised that in earlier communities, all males had been called upon to fight once there was war. Yet he also observed that societies had gradually moved towards a division of labour, and in Vattel's day societies everywhere tended to have 'regular troops' (professional armies). Vattel, however, thought in principle all citizens should be involved in the war effort (Vattel 1758/1834: 294f.). If mercenaries there were – foreign soldiers who do not owe military service but perform it in return for pay – they should have the right to equal treatment as natives, but it was the latter whom Vattel clearly preferred (Vattel 1758/1834: 297f.). As we shall see in chapter 6, a similar argument underlay Guibert's plea in his *General Essay on Tactics* for the constitution of a militia or citizen-army (Guibert 1772/1781: viiif.).

Vattel's and Guibert's contemporary, the Welshman Henry Humphrey Evans Lloyd, who sold his own services to several European monarchs, retained his fundamental admiration of the

British political system in which he had been raised, which, as a constitutional monarchy, he termed almost a republic as the monarch could not do as he pleased (Lloyd 1781/2005: 470–4). He divided states into despotic states, (absolute) monarchies, constitutional monarchies and republics. The despotic states to him included the Ottoman Empire; among constitutional monarchies, oddly enough, he included China and India besides Britain. Absolute monarchies he identified with professional, mercenary armies, which harmed both the nation and its ruler (Lloyd 1781/2005: 464–9). Republics – a term he also used interchangeably with democracies – he argued, were defended by their citizens, which made democracies fundamentally defensive, but very committed to the cause of the defence of their state. Lloyd, as had Guibert before him, postulated that different forms of government fought wars differently, an idea that Clausewitz developed into a cultural variable in his work (Lloyd 1781/2005, 458–78). As earlier with Guibert and later with Kant, and another century later with Jean Jaurès (see chapter 7), we see again a notion linking republics with defensive warfare.

Drill, discipline, mathematical war and the abhorrence of chance

What role did chance play when the purpose of all teaching and most writing about war was to influence the outcome of war in one's own favour? Chance was clearly an upsetting concept for military planners. Paul Hay du Chastelet in 1668 explained to his readers that the monarch should fear war or at least never assume its positive outcome, because God's will – Christian destiny – might be different from his, and was always incalculable. It was not really chance that influenced battle, but God who decided everything, down to the trajectory of the canon-ball or bullet which killed, 'as there is nothing in the universe that happens except through His immutable and eternal order' (Hay du Chastelet 1668/1757, 56f.). This was small comfort, and in early modern history, the art of war aimed, where humanly possible, to eliminate chance.

What soldiers from about 1300 did need, if they were using pikes, arrows or firearms, was tight tactical co-ordination, effected by long training, drill and discipline. Not surprisingly, in view of what we have learned about the denial of change, this was justified primarily in terms unrelated to the technological requirements. It was a new *Renaissance* mentality which led Justus Lipsius to advocate drill and exercises, which he thought would inculcate stoicism in the soldiers; Maurice of Nassau was very keen on Aelianus' description of Roman close-order (infantry) drill, which was soon practised by the army of the Dutch Republic.

The need not only for practice but for never-ending drill characterised the armies of early modern Europe that relied increasingly on hand-held, single-shot guns. The strictly choreographed movements of lines of soldiers which had been developed from the sixteenth century onwards were still very much in place in the eighteenth, by which time soldiers were clad in colourful uniforms; this form of war is picturesquely referred to by the French as the 'lace war', *la guerre en dentelles*. Such discipline was vital, as was movement in neat formations, columns and lines, as the danger of being shot by one's own comrades was great if their movements and one's own were not highly predictable and precise. Discipline was famously the leitmotif of Frederick II of Prussia's military style and, unsurprisingly, forms a prominent part of his military writings. He called for nothing less than perfection in the discipline and drill of his soldiers, well aware of all too human tendencies towards laziness, and postulated that the officers had to keep after their soldiers incessantly to drive them to give their best in all circumstances, and to keep them busy with hard physical work and exercises, in war and peace. He also listed many measures to be taken to minimize desertion which was at a high rate in eighteenth century armies, precisely because of the hard drill that was imposed on soldiers (Frederick II 1748: 174–6).

Consequently, the calculability of actions was foremost in the mind of writers on war. They called for maximum planning in all areas – from the crucial supplies and logistics mentioned above to exercises and movements in actual battle. Maurice de Saxe wrote that 'war can be made without leaving anything to chance. And this is the highest point of perfection and skill in a general' (Saxe 1732: 298f.). Chance, as several writers recognised, could throw the spanner into these delicate works at any moment, and chance was the enemy of good planning. Chance was thus seen as inimical and something that again had to be hedged against, but also as the central explanation why the best plans might come to naught (Kessel 1933: 248–76). Accordingly, several writers saw the defence of a country as easier than an offensive war: the defence was more easily planned, within one's own, well-known territory, with one's own resources, on inner lines, and less vulnerable to the vicissitudes of fortune or to a breakdown in supplies.

Battle avoidance or decisive battles?

Classicist Victor David Hanson has claimed that there was a Greek commitment to decisive pitched battles (Hanson 1989), a claim that has been contested by fellow classicists. Hans van Wees has shown that surprise attack, siege warfare and skirmishes were, even in ancient Greek culture, more frequent than major battles (Wees 2004: 131–50), and Harry Sidebottom argues that the pitched battle may have been an ideal in some periods and places in classical antiquity, but it was not the practice (Sidebottom 2004). In this book, we have demonstrated that Hanson is certainly wrong to elevate any Greek ideal of major battles into a 'Western way of war', for which he claims continuity throughout the ages. Whatever may have been the prevailing thinking and practice in the eight centuries between the age of Pericles and the late Roman Empire, in the fourteen centuries between Vegetius and the French Revolution, few believed either in the inevitability or the unconditional desirability of battle. Nor were battles always decisive. By contrast, ruses, diplomacy and other indirect approaches were seen as highly desirable by all. Views were balanced among those who thought the defensive stronger, and those favouring the offensive. The verdict on these points of the different writers we have quoted was based on empirical findings, not on ideological preconception – something that would change radically in the second half of the nineteenth century.

We have noted that in the high and late Middle Ages, far fewer battles were fought in Europe than in the early modern period. In Italy, the trend towards an increase in numbers of battles was noticeably among the city states in the fifteenth century, even though strife among them had been consistent throughout the earlier centuries. The frequency of battles increased towards the late seventeenth and eighteenth centuries, but did not necessarily make battles decisive for the outcome of war.

Attitudes towards battles and campaigns changed over time. Sun Tzu famously favoured the avoidance of battle, where other means to bring down the enemy were available, but he had no impact on Western military thinking until the twentieth century, when the West began to take note of Sun Tzu's disciple, Mao.

In the Occident, one can identify different phases – in the Middle Ages, battles were seen as divine ordeals, best avoided (lest God might favour the enemy in view of everybody's inherent sinfulness). The majority of European writers on war, and many generals, from the fourth to the late eighteenth centuries can be counted among the 'Fabians' (Liddell Hart 1930), an expression inspired by the Roman general Fabius Maximus (*c.* 280–203 BCE), nicknamed *Cunctator*, the hesitator, who time and again avoided a full confrontation and pitched battle during the Carthaginian invasion of Italy in 218–203 BCE; frequent direct references were made to him. While there were strategists who asserted the centrality of battles and their central role in all warfare, the Fabians, equally experienced in military affairs, counselled avoidance of decisive battles (especially where a favourable outcome for one's own side could not be predicted confidently).

Vegetius preferred bloodless victories through ruses, or through starving or surprising the enemy to bloody battles, and emphasised the need to keep one's own plans secret from the enemy for as long as possible (Vegetius *c.* 387/1996: 116–19). Most of Byzantine warfare, with the exception of the brief reconquests of Justinian and Heraclius, was marked by the defensive (Chrysostomides 2001: 91–101). Unsurprisingly, Maurice's *Strategikon* clearly puts him among the Fabians as he wrote: 'It is safer and more advantageous to overcome the enemy by planning and generalship than by sheer force; in the one case the result is achieved without loss to oneself, while in the other some price has to be paid.' And '[i]t is well to hurt the enemy by deceit, by raids or by hunger, and never be enticed to a pitched battle, which is a demonstration more of luck than of bravery' (Maurice 600/1984: 82f.). We find the same caution about battles – to be avoided if conditions are not entirely favourable – in the anonymous Greek treatise on 'strategy' of the sixth century (Anon. 6th c./1985: 102f.).

It was 750 years later that the Western cleric Honoré Bonet by contrast contended that it was more 'virtuous' to 'attack one's enemies than to await attack'. He reconciled this with Christianity, as he assumed that the good Christian knight would be fighting the enemies of God, and he also conceded – drawing on Aristotle – that in some contexts one should patiently await the right moment for an attack, and not rush it (Bonet *c.* 1385/1493: 120f.). He was not followed in this point by his disciple, Christine de Pisan. Drawing on several classical writers, Christine collected quotations from them counselling prudence: it was better 'to use diet and hunger before steel' (Caesar), 'good military leaders' should 'be old men in character' (Lucius Aemilius Paulus Macedonus) and Scipio Africanus, she noted, had emphasised that it was prudent 'not only to offer one's enemy a way to escape, but also help him to find it' (Christine 1410/1999: 99f.).

Machiavelli in his *Discourses* gave a very balanced view of when the offensive was more appropriate than a defensive posture, based on Livy's historical examples but also on examples from more recent times. In short, he thought that a well-defended (fortified, organised) country should risk a defensive stance, while a country with poor defences would be better off trying to carry a war into the adversary's territory (Machiavelli 1531/1998: 306–8). Writing at much the same time, the sieur de Fourquevaux cautioned that a general should never give battle unless he was convinced that it would turn to his advantage (Fourquevaux 1549). Lazarus Schwendi (1522–84) urged his readers to err on the side of caution even if chances looked good: 'Whoever could defeat the enemy by waiting or starving him is foolish if he decides on a battle instead.' But Schwendi was not a partisan of a defensive Strategy in all circumstances, as he thought there was much to lose and little to win by adopting it (q.i. Delbrück 1920a: 4. 396f.). Mendoza warned against initiating a battle, as even the victorious battle was expensive and led to the death of many; he told the Spanish Crown Prince, to whom he dedicated his book, to approach battle with a 'leaden foot' (Mendoza 1595/1597: 109).

In the opposite camp, views were similar. The Dutch opponents of the Spaniards were equally reluctant to give battle. John 'the Middle' of Nassau-Siegen, one of the Dutch military leaders in their wars against the Spanish Habsburgs, advised against skirmishes which might escalate into an unplanned battle. Yet he concluded from the Dutch experience that many small battles could achieve more than few encounters, even if the latter were 'big and strong' (John 'the Middle' 1597/1960: 129f.). William Louis of Orange wrote to his cousin Maurice in 1607, with a direct reference to Fabius Cunctator: 'we have to conduct our affairs in a way that will not subject them to the chances of battle', and urged him 'not to proceed to give battle, except out of extreme necessity'. In the same year, William Dilich published his *Warbook* in Hesse, expressing his preference for 'winning nothing' rather than suffering a battle and its consequent losses (both q.i. Delbrück 1920a: 397).

These views influenced the French in turn. Jean de Billon in his treatise on the Dutch military reforms actually outlined tactics of exhausting the enemy, by moving one's troops about, keeping the enemy guessing as to whether or when one might give battle. He described these tactics as preferable to the 'uncertain fortune' of giving battle (Billon 1613: x).

Critics of this cautious approach (who at the same time confirm its prevalence) include Henri, duc de Rohan, writing during the Thirty Years War:

> Of all actions of war the most glorious and the most important is to give battle; winning one or two leads to the acquisition or the overthrow of entire empires. In antiquity all wars were decided by battles, which led to such quick conquests. Now war is made more in the fashion of the fox than of the lion, and it is based more on sieges than on combat. And yet there are even today nations which mostly decide their wars by battles, like the Turks and the Persian; and even among the Christians we have seen recently that several battles took place in Germany. (Rohan 1636: 257f.)

Like Rohan, Raimondo de Montecuccoli, one of the Habsburgs' generals in the last part of the Thirty Years War, urged his readers, 'Do not avoid combat, but seek to give it where you have the advantage' (q.i. Chaliand 1994b: 567).

And yet even the Thirty Years War saw fewer battles per year than other protracted wars. Of the two famous imperial generals of this war, Johann Tserklaes Tilly favoured battles and by and large emerged successfully from the majority of the forty he fought as a general. Albrecht von Wallenstein (or Waldstein), by contrast, hesitated to give battle, even though he performed well when he was not laid up with one of his periodic illnesses (Domes 1995: 148–62).

Twenty years after the Westphalian peace treaties, Hay du Chastelet in France thought of battle as crucial to any decision in war, and urged his reader to put it at the centre of all efforts. Interestingly, he was a civilian, and never saw battle himself (Hay du

Chastelet 1668/1757: 127–9). By contrast, Feuquières, as an experienced general, was among the Fabians. Only extreme necessity or good reasons, he argued, should drive one to give battle. Among the good reasons he counted one's own numeric superiority; fear of mass desertions among one's own ranks; or a window of opportunity before the enemy received his own reinforcements. He urged that one should weigh the advantages one might gain from a victory against the disadvantages of a defeat – if the latter were greater, one should not run the risk (Feuquières 1731: 124). Feuquières was succeeded as leader of the sceptics by Maurice de Saxe, who led the French forces in the War of Spanish Succession (1701–14), in the Austro-Turkish Wars under Prince Eugene (1716–18) and the Wars of the Austrian Succession (1740–8). In his *Dreams about the Art of War*, he wrote,

> I do not favour pitched battles, especially at the beginning of war, and I am convinced that a skilful general could make war all his life without being forced into one. Nothing so reduces the enemy to absurdity as this method; nothing advances affairs better. Frequent small engagements will dissipate the enemy until he is forced to hide from you. I do not mean to say by this that when an opportunity occurs to crush the enemy that he should not be attacked, nor that advantage should not be taken of his mistakes ... But when a battle is joined under favourable circumstances, one should know how to profit from victory and, above all, should be contented to have won the field of battle in accordance with the present ... custom ... On the contrary, the pursuit should be pushed to the limit. And the retreat ... will be turned into a rout. (Saxe 1732: 298f.)

The pursuit of enemies after a battle had been won and lost was exceptional in his era, and reserved for armed forces who could be defined as traitors and insurgents against legitimate authorities, who were thus treated as criminals, like the Jacobite supporters after their lost battles against the Hanoverian kings of Britain (especially after Culloden in 1746). No such post-battle slaughter thus took place when Maurice de Saxe defeated the Duke of Cumberland (son of George II) a year earlier at the battle of Tournai in 1745, as this was seen as a contest between two legitimate opponents.

In Spain, Santa Cruz de Marcenado was among those urging caution against giving battle unless one could be very confident of its positive outcome for one's own side (Santa Cruz de Marcenado 1724–30 [1727]/1885: X). His Austrian Habsburg contemporary, Field Marshal Count Khevenhüller (1683–1744) elaborated the pros and cons of the decision to give battle, developing decision-making criteria for when a battle was to be avoided, when sought (q.i. Delbrück 1920a: 400f.). As late as 1815, Johann Konstantin von Lossau cautioned against accepting a battle which the enemy sought to impose, as this would give the initiative – and the leading role – to the adversary (Lossau 1815: 40f.).

Even Frederick II of Prussia, while favouring the offensive, had cautioned against battles that did not usefully further one's war aims. Commanders should avoid being forced into battle in adverse circumstances, and if the enemy sought to impose a battle, he automatically imposed the 'laws' of that battle – the battlefield, the positions, the time and weather conditions. A commander, he wrote, must have a good reason to seek battle; otherwise, as we have seen, his general rule (which he himself was unable to apply) was that wars should be short and lively (Frederick II 1748: 180–2).

There has been a considerable debate about whether Frederick really aimed for short, sharp wars. Did he instead, as the Prussian-German historian Hans Delbrück (1848–1929) argued both before and after the First World War, favour manoeuvres and last-minute decisions on whether or not to give battle? In a comparative study of Pericles and Frederick II of Prussia, Delbrück opposed a 'strategy of throwing down' or 'crushing' the enemy in decisive battles, to an 'exhaustion strategy' or 'tiring strategy'. In the latter, big battles would be avoided; the enemy would be worn out not only by pinpricks and skirmishes, but also by the cost of keeping his own forces in being for a long time, to wear the enemy down through constant manoeuvring requiring constant marches in response, which would lead to an exhaustion of supplies, demoralisation of the forces, desertion and indirect success (Delbrück 1890; Hobohm 1920: 203ff.). Delbrück saw two categories of generals, the ones who primarily sought to crush the enemy, among whom he counted Alexander the Great, Caesar, Napoleon, Gneisenau, Moltke, and those who sought to wear him out until he came to terms: Pericles, Hannibal, Gustavus Adolphus, Eugene, Marlborough, Frederick II and Wellington. Majority opinion today tends to be that Frederick, like Tilly or Marlborough, might have wished for quick wars with decisive battles, but as his adversaries, especially Daun, the field marshal of Empress Maria Theresia, tended not to, and other circumstances mitigated against them, he was forced to manoeuvre and avoid battle just as much as the other side.

Guibert was torn on the subject of the performance of the admired Frederick II of Prussia. Guibert had as a youth witnessed the Seven Years War, and especially the ignominious defeat of the French at Rossbach (1757), where French losses had been almost twenty-five times greater than Prussian losses, thanks to Frederick's use of the oblique order of battle he liked to apply. While this was not Frederick's only victory, it was an exceptional one. It was far more common even for him to be unable to bring about decisive battles. Either the defeated side managed to reconstitute armies fairly soon after the defeat, or refused to give battle and manoeuvred around him, or he himself chose not to engage in battle, or the victor did not manage to exploit the victory in battle diplomatically and politically, as the other side formed new coalitions or had other reasons to hope that the situation could still be reversed.

Frederick II was not the only one to find that 'decisive' battles eluded him: the same applied before his time, to other military commanders of the *ancien régime* as much as to the commanders of the Thirty Years War. In the intervening period, the Duke of Marlborough, who had won several battles in the War of the Spanish Succession, had found it equally difficult to bring about battles and to create the conditions for them to be truly decisive diplomatically (Ostwald 2000: 649–78).

Both contemporaries and later historians saw the Thirty Years War as well as several of the wars of the *ancien régime*, especially the Great Northern War or the Seven Years War, as notable for their indecisive campaigns and battles. Bülow wrote

> A defeat is of much less consequence in the modern system of war, than it was among the ancients ... The number of the killed is never very great; that of the wounded is considerable, but they recover ... Modern battles never weaken an army to such a degree, but that it may be ready for a fresh attack a few days after; and even that is not necessary, as every victory may be rendered fruitless by Strategic manoeuvres on the flanks and rear of the enemy. (Bülow 1799/1825: 107f.)

There were notable exceptions in that period. Gustavus Adolphus of Sweden and Frederick II of Prussia would be remembered for their keenness on battle, especially by writers of the nineteenth century. But confronted with the battle-eager Huns, Arabs, Magyars and Turks, over a millennium or more, the Byzantines and west Europeans were not only on the defensive, but also on the hesitant side, even if there were some crucial and 'decisive battles' in which, pushed into a corner, they managed to hold their own.

Offensive and defence

On average, writers before the second half of the nineteenth century were divided as to whether the defensive or the offensive was preferable. For Feuquières, there were five different sorts of war: defensive wars (to be waged, he thought, mainly by a weaker party, or if otherwise attacked), offensive wars (which one should only wage if much stronger than one's adversary), wars in which both sides were perfectly equally balanced, wars fought as a result of treaty obligations to allies and civil war (which alone he regarded as a very unfortunate business). Feuquières expressed no preference, but simply commented on the problems and opportunities arising from each form of war (Feuquières 1731: 113–57). Frederick II of Prussia, who was willing to take risks and to bluff in his wars, dismissed any purely defensive plan. Instead, he argued in favour of an offensive defensive, that is, a defensive Strategy that would turn into counter-attack at the earliest opportunity. He much preferred aggression, however, unbalancing the European state system in the process of his opportunistic campaigns, coming close to engendering an alliance against Prussia which might have brought down his dynasty (Frederick II 1748: 176). Henry Lloyd, who fought against Frederick in several campaigns, recognised only two forms of war, defensive and offensive. For Lloyd, neither form was by definition superior (Lloyd 1781/2005: 485–99).

The comte de Tressan, writing on war in Diderot's and d'Alembert's *Encyclopedia*, still took a completely technical view of war. According to Tressan, the 'State of War' – by which he meant the totality of all preparations for war, including the making available of human and other resources, their training and the war plan – should incline to the offensive or to defence in view of the balance of forces – one's own side vs. the opponent's. He made the barest reference to differing war aims as independent of resources, such as a potential opponent's wish to conquer or reappropriate a disputed province. Otherwise, he described offensive or defensive wars primarily as functions of the forces available on one's own side. If these were weaker or smaller than those of the enemy, then one should choose a defensive posture; if one's own forces were larger or stronger, one could choose the offensive. He thought that a defensive war was more difficult to conduct than an offensive war, where one could feed one's forces from the enemy's territory, which had the added advantage of denying these resources to the enemy (Tressan 1757: 988–91). Famously, Carl von Clausewitz after his experience of Russian defence in depth against Napoleon's *grande armée* in the campaign of 1812, in which only a small fraction of Napoleon's forces ever returned, would also proclaim defence the stronger form of war (Clausewitz 1832/1976, VI: 1).

The wars of the French Revolution, and those of Napoleon, proved a watershed in this context. Thenceforth and until the end of the Second World War, the 'decisive battle' became the be-all and end-all of warfare, indeed, the very aim of all Strategy.

Limited and unlimited wars

An important but mistaken notion spread among writers on Strategy even before the French Revolutionary and Napoleonic Wars, and that was the assumption that wars until then had been 'limited' in aims and means, and that they became 'unlimited' in aims and means under Napoleon. We find it with Vattel (1758/1834: 347) and Guibert, who regarded the warfare of his age – with the exception of that of Frederick II of Prussia – with distaste: 'Today, all of Europe is civilised', he wrote disparagingly.

> Wars have become less cruel. Outside combat, blood is no longer shed. Towns are no longer destroyed. The countryside is no longer ravaged. The vanquished people are only asked to pay some form of tribute, often less exacting than the taxes that they pay to their sovereign. Spared by their conqueror, their fate does not become worse [after a defeat]. All the States of Europe govern themselves, more or less, according to the same laws and according to the same principles. As a result, necessarily, the nations take less interest in wars. The quarrel, whatever it is, isn't theirs. They regard it simply as that of the government. Therefore, the support for this quarrel is left to mercenaries, and the military is regarded as a cumbersome group of

people and cannot count itself among the other groups within society. As a result, patriotism is extinct, and bravery is weakening as if by an epidemic. (Guibert 1772/1781: 187f.)

Describing the state of affairs before the French Revolution, the Hanoverian Johann Friedrich von Decken (1769–1840) thought that 'wars have become less frequent and less devastating', thanks to the state monopoly on standing armies (Decken 1800: 134). Clausewitz, who had not witnessed the 'limited' wars of the *ancien régime* himself, writing more than half a century after Guibert, uncritically paraphrased Guibert's impression of them (Clausewitz 1832/1976, VIII: 3). From this Clausewitz formed his much-quoted paradigm of 'limited' (especially *ancien régime*) war vs. absolute (especially Napoleonic) war, which was at the centre of his major revisions of *On War* which remained incomplete at his death (Aron 1976: I).

There was some truth in this with regard to wars immediately preceding the French Revolutionary Wars, but even among the wars of the *ancien régime* more generally, some were barely limited by rules of conduct in war. If we go back further in time, the Thirty Years War was anything but limited in nature, in any sense that both Guibert and Clausewitz would have recognised: relative to the total population of the areas where the war was waged, greater losses of life were caused (with up to two-thirds of the population dead in places) than in the First or even the Second World War.

Nevertheless, the perception of the eighteenth century as an age of limited war was passed on to the following generations and became commonplace. To give just two examples, both dating from 1933: the British strategist Basil Liddell Hart in his appropriately entitled *The Ghost of Napoleon* claimed that 'The features of the eighteenth century that first strike the historical observer are the rareness of battles and the indecisiveness of campaigns' (Liddell Hart 1933: 20). The Italian historian Guglielmo Ferrero (1871–1942) wrote similarly:

Restricted [i.e. limited] warfare was one of the loftiest achievements of the eighteenth century. It belongs to a class of hot-house plants which can only thrive in an aristocratic and qualitative civilisation. We are no longer capable of it. It is one of the fine things we have lost as a result of the French Revolution. In losing it we won many other kinds of progress, but it might in the end wipe out all that we have achieved. That is the great danger. (Ferrero 1933: 63f.)

Few would, like Russell Weigley, point out that the early modern age had been the *Age of* (frequent) *Battles*.

Particularly when seen from a long-term perspective, the notion that wars before 1792 were limited and thereafter became unlimited in scope, decisiveness and war aims is a misconception. This is because, as we have seen, war aims, and the strategies employed to achieve them, have fluctuated widely in European history and the history of war in other parts of the world. In fact almost all major categories of warfare, and Strategy, have existed before the French Revolution, from terrorism and irregular warfare and insurgency to genocide, from limited, small-scale skirmishes to massive battles with large-scale casualties. Technological change did not play the direct and dominant role as the simple, all-conditioning variable which determined the intensity of wars. Instead, throughout history, the use that was made of technology has depended on social, cultural, political and ideological variables. The evolution of warfare has not been one-directional. The notion of a one-way development of war from limited to unlimited (and ever more bloody) has to be rejected as false, just as the Whig interpretation of history in general as one of steady progress was exposed as false.

Frank Tallett has argued that it was only logistics that restricted the options available to military commanders in early modern Europe. War, he argued, was rarely limited in its effects on civilians, or in its war aims: these were grandiose, aiming (before the Peace of Augsburg 1555 and again in the early decades of the seventeenth century), to change the religious face of Europe. Or else they aimed to change the order of the known world (as did several parties in the Thirty Years War), or to swallow up sovereign states as in the Seven Years War (Tallett 1992: 67f.).

While this sounds persuasive, the fact remains that contemporaries saw the French Revolutionary Wars and the Napoleonic Wars as a watershed, and it changed their thinking on warfare, finally prising it away from the denial of change. But if change there was, did all warfare change? Were there no eternal constants, no eternal laws governing war the way they governed physics or the human body? If one acknowledged change in technology and other circumstances of war, could one still seek to identify eternal patterns of warfare? Were there eternal maxims applicable through the ages? Could one still distil these from history?

The enduring quest for eternal principles governing warfare

During the Vietnam War, the French strategist David Galula remarked:

> Ever since men have thought and fought (sometimes in the reverse order), attempts have been made to study war – philosophically, because the human mind loves, and needs to lean on, a frame of reference; practically, with the object of drawing useful lessons for the next war. Such studies have led, in extreme cases, to the denial that any lesson at all can be inferred from past wars, if it is asserted that the conduct of war is only a matter of inspiration and circumstances; or conversely, they have led to the construction of doctrines and their retention as rigid articles of faith, regardless of facts and situation. (Galula 1964/2005: xii)

Indeed, most works on warfare since antiquity started from the firm assumption that one could learn lessons from past examples, just as theologians or lawyers deduced ethical tenets or laws from the writings of the ancients and the Bible. This was of course the central assumption on which Vegetius and others had constructed their field manuals, and this assumption survives until the present as the foundation of empirical research in all the social sciences. Moreover, authors until well into the eighteenth century denied that changes in military technology, especially what has retrospectively been called the military revolution (Roberts 1956) flowing from the introduction of gunpowder, transformed war to the extent that historical examples drawn especially from classical antiquity and the works of the ancients should be abandoned as guidance to its conduct (Heuser forthcoming).

Since antiquity, such empiricism was the basis of all sciences. Addressing himself to his patron, Lorenzo de' Medici, Machiavelli wrote:

> As for intellectual training, the prince should read history, studying the actions of eminent men to see how they conducted themselves during war and to discover the reasons for their victories or their defeats, so that he can avoid the latter and imitate the former. Above all else, he should read

history so that he can do what eminent men have done before him ... In this way, it is said, Alexander the Great imitated Achilles, Caesar imitated Achilles; and Scipio, Cyrus. (Machiavelli 1532b/1961: 89f.)

This is an attitude that Machiavelli shared with medieval writers on Strategy. Even another century and a half later, with gunpowder firmly established on the battlefield not only in the form of cannon but also of light artillery, most writers on Strategy agreed that (in the words of historian Azar Gat) 'while the forms [*Formen*] of war may change with time, its spirit [*Geist*], or essence [*Wesen*], remains unchanged' (Gat 1992: 67). Writers in the seventeenth and eighteenth centuries, like Vegetius and Machiavelli before them, wanted to extract the 'best principles' of fighting from the study of past wars, in the words of the baron de Traverse (1752: v).

The reflections and recommendations made by the early modern writers on military matters tended to focus, in the tradition of Vegetius, on the craft of war, and generally mention few political considerations. Good examples of this are the works of Diego de Alava y Viamont, Montecuccoli, Puységur and Feuquières. All four concentrated to the exclusion of almost anything else on what the 'perfect captain' or general needed to prepare for war and conduct his campaigns and battles successfully. They and other military professionals saw themselves as highly skilled craftsmen, or artisan/artists, or other specialists in their field. They hardly if ever touched on political issues.

It was logical that they all sought principles easy to articulate, to teach and to learn, as what they all had in common is that they wrote instruction manuals. An early example of such rules can be found in Henri, duc de Rohan's seven rules on how to win battles:

1. Never to let oneself be forced to give battle against one's will.
2. To choose a battlefield fitting to the quality and the number of soldiers.
3. To deploy one's army on the battlefield in accordance with the quality of one's soldiers so that it is to one's own advantage: put the cavalry before infantry or infantry before cavalry, depending on which is stronger ... and have a reserve, like the Romans.
4. To have several good generals, not just one.
5. Keep good distances between forces in the order of battle.

6. The bravest soldiers should fight on the wings, not in the centre.
7. Do not allow pursuit of enemy forces or pillage. (Rohan 1636: 259ff.)

Such sets of rules were drawn up, as in the writings of Machiavelli, Bernardino Rocca or Bertrand de Loque, for the prince as supreme commander of his armed forces, or like the treatises of the sieur de Fourquevaux for the general, or even, like the works of Thomas and Dudley Digges, or Henri, duc de Rohan, for an officer at any level, the 'perfect captain'. All of them sought to formulate doctrine as defined by Barry Posen, the part of Strategy that concerned only the armed forces, that they were to learn and apply to their warfare (Posen 1984: 13f.). This tradition of writing instruction manuals continued unabated until the end of the eighteenth century. After the early example of Henry VII of England distributing Christine de Pisan's *Boke of the fayt of armes and of Chyualrye* to his knights in the late fifteenth century, it was mainly from the eighteenth century that specially prepared instruction manuals or field manuals were issued to officers as compulsory reading. Among them we find Frederick II of Prussia's field manual of 1761, written by himself for his officers. By 1783 at the latest Britain had issued similar regulations (Anon. 1783). Napoleon himself opined that

> [s]trategy ... rests on principles which no one must forget. All the generals of old and all those following in their footsteps with dignity have only done great deeds because they have internalised the natural rules and principles of the art of war, which are the good fit [*justesse*] of combinations [of forces], the exact concept of the means to be applied, resistance to difficulties. (q.i. Foch 1909: 3)

Books of instruction also continued to originate from the pens of private authors. One such was Johann Friedrich Konstantin von Lossau, who stipulated that the 'spirit of war' was always the same, even if it was difficult to try to theorize about it – theories had the catch of being either 'too general and thus subject to endless modifications', or 'actually have their eyes fixed on very particular circumstances' which they pretend to be worthy of universal generalisation (Lossau 1815: 35).

Scharnhorst in 1792 wrote the main German-language field manual in the classical style that can be seen as exemplary for its kind, in the tradition of Vegetius. It can be taken as a prime example of *ancien régime* thinking, reflecting no interest in political or ethical dimensions. Instead, what interested him as a mathematician and artillerist were the technicalities of siege warfare and targeting (Scharnhorst 1793: 390–442).

In the middle of the Napoleonic Wars Archduke Charles postulated both the existence of immutable principles of war and a constant change in the way they had to be applied, as a result of constant evolutions in weaponry and military technology (Charles 1806: 88f.; Waldtstätten 1882: 225). Shortly after their end, the Bavarian engineer Joseph Ritter von Xylander (1794–1854) professed scepticism about the 'science of strategy' being able to yield any such rules: 'The science [*Wissenschaft*] of Strategy does not yield rules the direct application of which must lead to certain events; it only yields general principles of wars, from which rules can be abstracted for the infinite number of different cases that result from reality.' He did, however, settle for one, that of the offensive as key to success (Xylander 1818: 20, 34).

Clausewitz can also be subsumed in this generalisation, as his writings express both the belief in eternal, intrinsic characteristics of war, while emphasising the infinite variety of its manifestation in reality (Clausewitz 1832/1976, VIII). In a note in a paper of 1827 Clausewitz enumerated a list of principles of warfare:

- that defence is the stronger form [of warfare] with the negative purpose, attack the weaker form with the positive purpose;
- that the great successes also determine the smaller ones;
- that one can thus trace back strategic effects to certain centres of gravity;
- that a demonstration is a weaker use of force than a real attack, that it must therefore be based on particular conditions;
- that victory does not merely consist of the conquest of the battlefield but in the destruction of the physical and moral fighting forces, and that this is mostly only achieved in the follow-up to the battle that has been won;
- that success is always the greatest where the victory was achieved by fighting;
- that therefore the change from one line and direction [of attack] to another can only be regarded as a necessary evil;

- that an attack on the rear [*Umgehung*] is justified only by [one's own] superiority in general or by the superiority of one's own line of communications and retreat over that of the adversary;
- that flanking positions are conditional upon the same circumstances [as spelled out in the previous point];
- that every attack weakens as it progresses (t.i. Clausewitz 1832/1976: 182f.; Gat 1989: 255–63).

Attempts to try to boil down the rules guiding successful warfare had become fashionable. Clausewitz' contemporary and rival Antoine Henri Jomini is most commonly associated with such a list of principles, writing in an American edition of his works:

> There exist a small number of fundamental principles of war, and if they are found sometimes modified according to circumstances, they can nevertheless serve in general as a compass to the chief of any army ... Natural genius will doubtless know how, by happy inspirations, to apply [these] principles as well as the best-studied theory could do it ... The fundamental principles upon which rest all good combinations of war have always existed ... These principles are unchangeable; they are independent of the nature of the arms employed [!], of times and places ... For thirty centuries there have lived generals who have been more or less happy in their application. (q.i. Gat 1989: 112)

Jomini's recipe for success, deduced from Napoleon, was 'the employment of masses upon the decisive points' (Jomini 1811/1865: 149). In his *Summary of the Art of War*, he wrote that 'Strategy, particularly, may be regulated by fixed laws resembling those of the positive sciences', but conceded that this was 'not true for war as a whole' (Jomini 1837/1868: 321).

Shortly after the Franco-Prussian War, French officers were taught rules of war, with opinion differing only on whether there were twenty-four or forty-one. There were opponents of such reductionism, but the rules dominated the teaching methods (Mayer 1916: 7). Opponents included General Berthaud, but also General Lewal (Lewal 1892: 49ff.). Writing a good decade later, the French General Victor Derrécagaix related 'strategy' to a small number of immutable principles, while tactics were eternally subject to mutation under the influence of new weapons. But when pressed to pronounce on what such an eternal principle might be, the only one he articulated was 'to be the strongest at the decisive point' (Derrécagaix 1885: I: 4, 375).

In 1895 the first French field manual containing a set of rules was produced. The essential rules were:

1. To preserve one's own freedom of action and to limit that of the enemy.
2. To impose one's will upon the adversary and to avoid submitting to his [an echo of Clausewitz].
3. To economise one's forces in favour of mass and to apply that to a conveniently chosen point and to a favourable movement, in order to produce a decisive development.

Everything else, it was explained, flowed from these principles (q.i. Bonnal 1920: 18f.).

On the eve of the First World War, Clausewitz's disciple Sir Julian Corbett produced a more sophisticated argument centred on human nature. He held that theoretical study

> can at least determine the normal. By careful collation of past events it becomes clear that certain lines of conduct tend normally to produce certain effects; that wars tend to take certain forms each with a marked idiosyncrasy, that these forms are normally related to the object of the war and to its value to one or both belligerents; that a system of operations which suits one form may not be that best suited to another. We can even go further. By pursuing an historical and comparative method we can detect that even the human factor is not quite indeterminable. We can assert that certain situations will normally produce, whether in ourselves or in our adversaries, certain moral states on which we may calculate. (Corbett 1911: 9)

Alfred Thayer Mahan was characteristically more interested in rules and maxims in the Jominian tradition:

> Maxims of war ... are not so much positive rules as they are the developments and applications of a few general principles. They resemble the ever varying, yet essentially like, forms that spring from living seeds, rather than the rigid framework to which the free growth of a plant is sometimes forced to bend itself. But it does not therefore follow that there can be no such maxims, or that they have little certainty or little value. Jomini well says, 'When the application of a rule and the consequent maneuver have procured victory a hundred times for skilful generals, shall their occasional failure be

> a sufficient reason for entirely denying their value and for distrusting the effect of the study of the art? Shall a theory be pronounced absurd because it has only three-fourths of the whole number of chances in its favor?' Not so; the maxim, rooting itself in a principle, formulates a rule generally correct under the conditions; but the teacher must admit that each case has its own features ... which modify the application of the rule, and may even make it at times wholly inapplicable. It is for the skill of the artist in war rightly to apply the principles and rules in each case. (Mahan 1918: 229f.)

In the light of technological change, however, the French General Colin, writing shortly before the Great War, was sceptical about what in his time was the commonplace of 'immutable grand principles of war'. He opined that many among them were defunct, while others that were in fact 'eternal' had been neglected and might make a reappearance (Colin 1911: 233).

Marshal Foch by contrast tried to reinforce the teaching of Jomini's list of principles or variables: Foch invoked Napoleon (who said that 'The principles of war are those which have directed the great Commanders whose great deeds have been handed down to us by History') and Jomini, who extolled the virtues of 'sound theories founded on principles both true and justified by facts', which, in addition to history, were the true training for commanders. Foch summarised: 'There is, then such a thing as a theory of war. That theory starts from a number of principles[.] The principle of economy of forces. The principle of freedom of action. The principle of free disposal of forces. The principle of security, etc.' (Foch 1900: 8f.). Foch complained that in the past, the variable of morale had been neglected by his countrymen in writing on the subject (Foch 1900: 3).

The first listing of such principles in the USA occurred in the *Training and Regulations 10-5* field manual of 1921 (q.i. Brodie 1959: 24). In Britain it was J.F.C. Fuller who introduced the idea of such rules, following Foch, after his countrymen – like Sir Edward Hamley (Hamley 1872; J.F. Maurice 1891: 8) – had long rejected the notion that warfare could be guided by eternal principles.

Fuller boldly condensed all his wisdom into 'eight principles of war, and they constitute the laws of every scientifically fought boxing match as of every battle'. Echoing Foch, these principles are:

1st Principle: The principle of the objective.
2nd Principle: The principle of the offensive.

3rd Principle: The principle of security.
4th Principle: The principle of concentration.
5th Principle: The principle of economy of force.
6th Principle: The principle of movement.
7th Principle: The principle of surprise.
8th Principle: The principle of co-operation (Fuller 1923: 27f.).

Fuller's principles made their way into the British *Field Service Regulations* of 1924 (vol. II). They were:

- maintenance of the objective
- offensive action
- surprise
- concentration ('of superior force at the decisive time and place')
- economy of force ('involves the correct distribution and employment of all resources in order to develop their striking power to the utmost')
- security ('of a force and of its communications')
- mobility ('implies flexibility and the power to manoeuvre and act with rapidity')
- co-operation

and Basil Liddell Hart wrote extensive comments upon them (Liddell Hart 1944: 179f.).

The 'economy of forces' is an often misunderstood concept which goes back at least to Clausewitz. Admiral Castex explained it as really meaning the management (in the original Greek meaning of the word, husbandry) of forces: distribution, organisation, deployment, provisioning, with an underlying notion of a harmonious order underpinning all this (Castex 1934, IV: 54–61). In another volume of his work, however, he defined the economy of forces as prioritising a principal aim and giving fewer resources to secondary aims (Castex 1939, II: 7). As Bernard Brodie observed:

> The term 'economy of force' ... derives from an interpretation governed by the nineteenth century connotation of the word 'economy', meaning judicious management but not necessarily limited use. Thus, the violation of the indicated principle is suggested most flagrantly by a *failure to use* to good military purpose forces that are available ... Of late, however, the term has often been interpreted as though it demanded 'economizing' of forces, that is, a withholding of use. (Brodie 1959: 26)

Liddell Hart subsequently simplified his recipe for British military success in a chapter headed 'The concentrated essence of war':

DO
1. Choose the line (or course) of least expectation.
2. Exploit the line of least resistance.
3. Take a line of operation which offers alternative objectives.
4. Ensure that both plan and dispositions are elastic, or adaptable.

DON'T
5. Don't lunge whilst your opponent can parry.
6. Don't renew an attack along the same line (or in the same form) after it has once failed. (Liddell Hart 1932: 310–14)

In 1928, Fuller modified his principles, with the most important four remaining:

> The offensive should be assumed on all occasions when circumstances permit.
> Offensive action should spring from a strong defensive basis.
> The defensive should be so organised as to permit of it changing into an offensive at the shortest possible notice.
> The offensive cannot be too strong in reserves, therefore the defensive should not employ a weapon beyond the minimum necessary to establish security. (Fuller 1928: 1)

These Fullerian principles still inspire British field regulations.

Several British and American naval strategists of the late nineteenth and early twentieth century – the historical school, one might call them – were particularly convinced that they could draw lessons for the present from their countries' naval history (see chapter 9). Yet their deductions from past experiences were constantly challenged by changing technology. No lesser man than Admiral Sir John Fisher, British First Sea Lord shortly before and again during the First World War, and patron of several naval historians, thought that 'Whatever service the past may be to other professions, it can be categorically stated in regard to the Navy that history is a record of exploded ideas. Every condition of the past is altered' (q.i. Kennedy 1989: 183).

By contrast, Charles Callwell, writing in 1905, argued that 'While naval tactics and military tactics are constantly going through a process of evolution as the science of producing arms of destruction progresses, the broad principles of strategy ashore and afloat remain unchanged from century to century.' He conceded, however, that

> it is none the less true that the principles of maritime strategy have in the course of years undergone appreciable modifications, in conformity to a certain extent with the advances which have taken place in the craft of the shipwright. Then, again, developments and discoveries in the science of electrical communication are exerting no small influence over the principles governing the applications of strategy to modern conditions at sea. Progress in civilisation, moreover, has tended to more clearly define the relations which should exist between belligerents and neutrals, than [before]. (Callwell 1905: 23f.)

Continuities of a different sort were at the heart of an analysis of war that stood more in the Clausewitzian tradition, and this we find in the writings of Hans Delbrück that we have already alluded to. Even before the First World War, Delbrück thought that he had identified two basic strategies, one of which he called the Strategy of exhausting the enemy through constant manoeuvring, pinpricks and the avoidance of battle where possible, thus drawing out the war (*Ermattungsstrategie*), which, as we have seen, he pinned on Frederick II. The other Strategy he identified was that of seeking to crush his army in a decisive battle (*Niederwerfungsstrategie*; see also chapter 7). Delbrück thought he could see this dual pattern right through history, from Pericles to Napoleon. One of his critics, the historian Otto Hinze, baulked at this:

> Concrete historical systems do not repeat themselves in world history. What repeats itself in the flight of manifestations are only the elements of such systems which can be identified analytically, i.e. abstract principles or tendencies, such as the principles of crushing or exhausting the adversary ... But those, I argue, are present in all strategies, only in very different mixtures and in infinite gradations. These differences constitute the individual, concrete historical system of strategy in different eras and with different military leaders. (Hintze 1920–1: 145f.)

Regaining a superior Clausewitzian level of analysis and turning away from mere Do's and Don'ts for armed forces' instruction manuals, the Anglo-American strategist Colin Gray sought to identify eternal truths about warfare that could be seen to apply whatever the technology, culture and other context of a war. Gray asserts that:

> Notwithstanding the uniqueness of every historical context, only a mind monumentally closed to evidence and reason could fail to acknowledge the generic continuities in strategic history ... [W]ar has an unchanging nature ... [while] warfare takes many forms, not infrequently even within the same war. In the broadest terms, warfare comes in two varieties, regular and irregular.

Factors that Gray regards as ever present he enumerates, following Clausewitz, as the 'trinity of passion, chance and probability, and reason' and the restrictions that can be imposed upon the conduct of war by rational actors. These will seek to control war because of considerations of cost, policy, Strategy, of fear and of culture. Some may prefer not to use all their power in 'the interest of a civilised order' (Gray 2005: 16–19). In his *Fighting Talk* he sought to identify tenets about war that are of lasting applicability, indeed, truth (Gray 2007b). Unlike some of the early modern writers, however, Gray in no way denies change: rather, he argues that 'the strategic thread is the most significant of the several engines of historical change', albeit not 'the sole driver' (Gray 2007a: 280).

The quest for eternal principles determining success and failure in war, the wise application of which might produce a favourable outcome for one's own side, thus continues to this day. And yet crucial parameters were subject to change, such as the aims of war and the means to conduct it.

PART III

The Napoleonic paradigm and Total War

5 *The age and mindset of the Napoleonic paradigm*

Napoleon was not a personality, but a principle.
(The Duke of Wellington in 1837)[1]

Causes of wars, world-views and war aims 1792–1914

Chapters 2 and 3 charted the fluctuations and the constants in warfare prior to the French Revolution. From Cicero until the French Revolution, there was agreement in the Romano-Christian world that war could only be justified if it led to peace, and a better peace than previously. There were varied opinions about the importance of battle in warfare, and many counselled extreme restraint in seeking and giving battle. While military men wrote about victory in battle, writers of all professions were aware that a lost battle did not necessarily mean the end of war, and that victory in battle needed to be followed up in certain ways to lead to peace. Even though thinkers from Machiavelli, Urrea and Bernardino de Mendoza to Vattel favoured militias drawn from local populations, armies were more often than not multinational, composed of local levies and often foreign mercenaries.

Most of this changed dramatically with the French Revolutionary and Napoleonic Wars, known in Britain as the 'Great War' until the First World War claimed that epithet. They were a watershed not only in the conduct of war itself, but also in thinking about war (Ford 1963/4: 18–29). Without any technological revolution, the French fought differently, with much larger armies than Europe had seen since antiquity, reaching six-figure numbers under Napoleon. His organisation of armies into columns was one element of his success,

[1] Quoted by Philip Henry, fifth Earl of Stanhope, *Notes of Conversations with the Duke of Wellington, 1831–1851* (London: John Murray, 1888): 102.

but the new tactics applied by the French were aptly summarised by the Prussian General Valentini who saw it as the conduct of small (or irregular) war, but on a much larger scale (see Part VI of this book): in battle, France's soldiers relinquished drilled formations and sought out the enemy individually, whether by careful targeting of their muskets, or with sabre or bayonet (Valentini 1779/1820: 1–3). When challenged by the royalist armies who sought to put King Louis XVI back on the throne, the French Revolutionaries fought for themselves and for their newly won liberties. They were inspired by new political ideals, and fought for their own cause, something which Count Jacques Antoine Hippolyte de Guibert had predicted would give them a very special force (Guibert 1772/1781: 138). The Napoleonic armies – composed of an increasing percentage of foreigners – fought for France, but above all for their charismatic leader, for honour and glory.

Contemporaries unsympathetic to the French Republic or to Napoleon struggled with the supposed secrets of French success. If they were monarchists like the Prussians, they tried to ignore the appeal of democracy, talking condescendingly of the French rabble. Still, they gradually conceded that the nation in arms as a concept revived if not engendered by the French Revolution had transformed warfare in scale and intensity, and introduced a force and a passion into the warfare of the subsequent 150 years that had been absent since the last 'wars of opinion' (Jomini), the religious wars of previous centuries.

By contrast Napoleon's admirers, like Jomini, created a Napoleonic paradigm for success: they emphasised his monumental victories, his fast moves and surprising attacks, his clever concentration of his troops on one point in the adversaries' forces, the 'centre of gravity', and his relentless pursuit of battles with casualty figures on both sides entirely unprecedented in living memory. This latter interpretation in particular, popularised by both Jomini and Clausewitz, would seize the imagination of contemporary commentators and later strategists, who up to the present have tended to write admiringly about Napoleon's 'annihilation battles', his reckless pursuit of victory. Very few realised, like the French socialist Jean Jaurès (1859–1914), that Napoleon had ultimately failed both at Leipzig and finally at Waterloo, and nothing of his empire survived. Even along the way, he had engendered more enmity than victories, to the point where ultimately he had most of Europe united against him (Jaurès 1911: 53–5).

Those who resisted the French discovered xenophobic nationalism for themselves, a tremendously important change of European passive culture, by and by to spread to most parts of the globe. They were mostly opposed to (French) republicanism, (French) democracy, (French) secularism, in fact, all things French (including Napoleon's relatives as imposed monarchs), and they upheld their own old royal families. They also discovered the cult of their own language, culture and people, aspiring to statehood for their own nation. Tyroleans, Spaniards and Prussians fought not only for their princes but also for their 'nation', their culture. From the loving rehabilitation of vernacular languages and peasant lore that had been ignored in the great century of French cultural predominance in Europe grew angry self-assertion, amply fuelled by the inferiority complex which the defeats at the hands of the French had engendered. This complex turned into an aggressive need to prove one's supposed superiority. Logically, it took the shock of a similar defeat for France in 1870/1 to turn 'peasants into Frenchmen' and to engender, there too, a xenophobic nationalism of a quality not previously known in France (Weber 1977; Audoin-Rouzeau 1989).

Ironically, it had been Napoleon and the French who had given other Europeans the concept of national unity – as for example in Northern Italy, where Napoleon established the Italian Republic (later transformed into the Kingdom of Italy) and first united principalities that had been separate entities for centuries. Then there was Poland, temporarily liberated by Napoleon from Prussian and Russian encroachments, where his memory is sacred. The seeds of nationalism were thus sown by the Napoleonic occupations (Llobera 1994: 179–210).

Ethnic or xenophobic nationalism, discovered almost overnight in the first decade and a half of the nineteenth century, initially seemed to run out of steam after the defeat of Napoleon and the collapse of his empire, but it continued to gestate throughout Europe. The contest between *ancien régime* and republicanism continued, culminating in the European revolutions of 1830 and 1848. Ethnic nationalism was the 'third way': it could be harnessed to monarchies who had lost their absolutist claims to power, or it could be married to democratic tendencies, or it could occupy an ambiguous space between the two, where the 'people' were glorified without gaining the vote. As the republican and democratic forces failed to triumph over monarchies in 1830 and 1848, democratic urges in on the European continent were increasingly channelled into ever-growing national pride and concomitant popular hatred of supposed enemy 'nations'. Xenophobia was turned into poisonous ideas of one's nation's superiority and irreconcilable, hateful competition with others through the admixture of the popular reading of Malthusian and Darwinian teachings about the scarcity of resources in the world, and a competition of nations or 'races' for survival. Nationalism had blossomed earlier in some places – in Britain, there had been English antecedents in the sixteenth century, and the Napoleonic Wars forged British nationalism; the French had since the Middle Ages seen themselves as having succeeded the Israelites as God's new chosen people. Elsewhere nationalism took until the First World War to develop its full force. Either way, nationalism transformed warfare.

The deliberate mass killing of civilians simply because they belonged to one group rather than another had been extensively practised in classical antiquity, in the Middle Ages, in the religious wars of early modern Europe and in colonial warfare. In the nineteenth century, it reappeared in Europe, spreading from the Balkans. 'It is the moods of the peoples, lust for annexation and revenge, the desire to draw close to peoples of the same tribe ... which endanger peace' wrote the Prussian Field Marshal Helmuth von Moltke the Elder late in his life (Großer Generalstab 1911: 4). He knew what he was talking about: he had spent several years as adviser to the Ottoman Empire, a pressure-cooker heated up by the force of nationalism that was boiling from Greece to Armenia, from Bulgaria to Egypt. The 1870s in particular saw horrible atrocities committed by both sides in the Balkan territories of the Ottoman Empire. With an exponentially grown world population, the following century would see genocide on scales unprecedented in history.

The American Civil War (1861–5) is an exceptional case in this context. Obviously, the opposition between both sides was not primarily based on xenophobic nationalism, but on sentiments not unrelated. White Southerners believed in their own inherent racial supremacy over their slaves, with the issue of the abolition of slavery dominating the debates between North and South in the decade before the outbreak of the war and providing a key cause. Northerners generally had a comparable faith in American exceptionalism, in being God's chosen people. The extreme religious convictions of William Tecumseh Sherman (1820–91), one of the chief generals of the Union, that God's hand and His divine providence could be seen throughout the war are not untypical of the culture or religious revival of his age (McPherson 1997; Bower 2000). Nor was Sherman's or General Philip H. Sheridan's attitude to enemy civilians, and their respective campaigns for the destruction of civilian morale in Georgia, the Carolinas and in the Shenandoah Valley of Virginia, merely the product of the passions of the moment. During the Franco-Prussian War, Sheridan joined the Prussian side as an observer, urging Bismarck to cause 'the inhabitants [of France] so much suffering that they must long for peace, and for their government to demand it. The people must be left nothing but their eyes to weep with over the war' (q.i. Degler 1997: 68). Indeed, the American Civil War has been called 'the first of the modern total wars' (Williams 1954: 3), although in view of the definition of the term which we will discuss in the next chapter, it can at best be regarded as such in terms of the large-scale mobilisation of men and resources (Neely 1997 29–52).

The Franco-Prussian War of 1870/1 was carefully orchestrated by Bismarck with two ends in mind: one, the seizure of Alsace and Lorraine, which interested him not only because the populations spoke a German dialect as remote from modern German as Dutch, and could therefore be declared ethnic Germans who should be integrated into the new German nation-state. That Bismarck was capitalising on nationalist sentiments rather than on deeply held national convictions is apparent from his lack of interest in the inclusion of Austria, where modern German was spoken, but which rivalled Prussia for German leadership. Alsace and Lorraine were also interesting because of their industry and natural resources, and their strategic location on the west bank of the Rhine. Which led to Bismarck's second aim: to override the deep historical rivalries among the many German principalities through confronting them with an outside threat, France: it was arguably his cleverest manoeuvre to have manipulated the French Emperor Napoleon III into declaring war on Germany, mainly out of hurt pride. This was a war with limited aims, but national feelings were mobilised on both sides to the extent that the limited territorial losses suffered by France – and the huge punitive payments imposed by Germany – led to smouldering irredentism and lasting animosity.

Both the Franco-Prussian War and the Russo-Japanese War of 1904/5 followed the new pattern set by the American Civil War in terms of how mass-produced weapons, quantum leaps in the yield of firepower, railroads, the telegraph and other innovations could increase both the sizes of armies involved and the numbers of casualties inflicted. The First World War would trump all three in terms of casualties.

The Russo-Japanese War, like the Franco-Prussian War, was fought over the control of a limited albeit strategically important piece of territory in Russia's Far East. By contrast, the war aims of the First World War were less clear. The war notoriously started, much like the Franco-Prussian War, over an issue of pride and personal injury to the imperial family: after the Austrian crown prince and his wife had been assassinated by a Serb terrorist, Serbia refused to do Austria's bidding with regard to the treatment of the terrorist himself and Serb army officers he had implicated. In the absence of an international criminal court, or any international organisation that could have contained the situation and effected a mutually satisfactory handling of the situation, Austria went to war with Serbia, Russia as Serbia's ally went to war with Austria, Germany as Austria's ally went to war with Russia, France as Russia's ally went to war with Germany, and Britain followed suit as Germany, in its infamous Schlieffen Plan pre-emptive attack against France, crossed through neutral Belgium, guaranteed by Britain in turn. What war aims were for any party from this point was far from clear. True, France wanted to repossess its provinces and to humiliate the Germans with a peace treaty as smarting as the one imposed by Bismarck in 1871, Serbia wanted Austrian possessions in the Balkans to form a greater Serb kingdom and Britain wanted to liberate Belgium, the neutrality of which seemed important to it. But in the context of this, much else was achieved, especially the destruction and the parcelling out of the German, the Austro-Hungarian and the Ottoman Empires. Nor would the Germans and the Austrians have done otherwise had they won – witness the territorial concessions exacted by Germany from Russia at the Peace of Brest-Litovsk of March 1918, even if they had been dreamt up by clever diplomats at the green table and had not been part of any design in 1914. None of this had been part of a real 'grand strategy' prior to 1914, and all sides made it up as they went along. The only clear plans which had existed in 1914 – the German army's Schlieffen Plan for a sickle attack on northern France through Belgium and the German navy's plans for attacking America or at least Scandinavia and preferably Britain – were at best, as Stig Förster has convincingly shown, an attempt to pre-empt all others in what was seen to be a window of opportunity (Förster 1995, 1999). But even there, the absence of properly articulated demands that could have been the basis of quick armistice negotiations made nonsense of it politically (Kennedy 1976, 1979; Wilson 1995). While Clausewitz's view that no one in his right mind should ever start a war without a clear idea of what he wants to achieve is praiseworthy, the reality of the First World War was different. But then, this was not an age of enlightenment and rationality, but one which prided itself in its irrational passions and emotions, its xenophobic nationalism and its glorification of war.

Militarism

For another novelty in the passive culture of the nineteenth century was 'militarism', a word coined by Madame de Chastenay just after the Napoleonic Wars, but taken up only in 1861 by Pierre Proudhon writing about *War and Peace*, soon used widely in Germany, France and Britain (Berghahn 1981: 6). The phenomenon, which also spread to the United States and indeed to other parts of the globe like Japan, mixing well with the old Ottoman warrior-culture, had several aspects. One was that of a new and qualitatively different admiration for the military, even though the cult of the warrior can be observed in many societies past and present. There are cultures in which all adult men are under pressure to prove themselves as warriors; there are other cultures, found in Europe from the Dark Ages until the *ancien régime*, in which an elite claims its position of superiority as a warrior caste. The militarism of the nineteenth century differed from this in that through various forms of conscription, entire classes of men spent anything from one to five years in a purely military environment, with its culture of shouted orders and unquestioning obedience, and the inescapable respect for the higher ranks which flowed from their power to punish or pardon and order all the others about. Military behaviour spread into civilian spheres, and no profession, no occupation was generally held in greater esteem than that of the military officer (or, for a young man, his time in military service). The lawyer Hermann Kantorowicz (1877–1940) noted about the spirit of that age that 'the officer became the male role model for primary school teachers, postal employees and greengrocers' (q.i. Picht 1954: 461). Even children were put into miniature imitation naval uniforms, the famous sailor suits, throughout North America and Europe.

Whole nations were militarised. Worse still, from the point of view of policy-making, civilian leaders tended to be ignored unless they donned uniforms, as Prince Otto von Bismarck noted with dismay, and monarchs with their waning powers vis-à-vis parliaments or oligarchies who increasingly encroached upon regal prerogatives still liked to assume the posture of military commanders-in-chief; to this day monarchs and members of royal families incongruously like to be portrayed in uniforms. The posturing of Emperor William II of Germany as supreme military commander and his meddling in military matters contributed to the lack of co-ordination of German military planning and political leadership on the eve of and during the First World War (Ritter 1954: 21–48).

Accordingly, those writing on war and Strategy between the death of Clausewitz and the outbreak of the First World War throughout the West were mostly military officers, usually lieutenant-colonels or colonels at the time of writing, several of whom attained the level of general or even marshal. They had all experienced major military action first-hand, but also knew colonial policing, such as G.F.R. Henderson (1853–1904), J.F. Maurice (1841–1912) and C.E. Callwell (1859–1928). Most of them had plenty of international experience – either in the respective empires, or, in the case of Moltke the Elder and Colmar von der Goltz (1843–1916), in the Ottoman Empire. Several were military instructors in staff academies, which accounts for the sprinkling of civilians like the later Oxford don Henry Spenser Wilkinson (1853–1937), Chichele Professor of Military History at All Souls College, Oxford, and Sir Julian Corbett (1854–1922), teacher of British naval officers.

The others were not only very similar in background, but they were also all avidly reading each other, in what was perhaps a more international debate than any since. Literature took much less time to travel from one country to another, and translators were busy. Curiously, nationalism, while so crucial to the animosity between these countries which was to drive them into the Great War, did not prevent strategic theorists from reading very closely and respecting their enemy's works. Of course they also used much the same case studies from recent history, and drew very similar lessons from them.

Militarism was also an accepted cultural feature of their general readership. While it is generally thought today that the populations of Britain, France and Germany were less universally happy about the outbreak of the First World War than had long been claimed, there was widespread jubilation (Liddell Hart 1944: 17), and at any rate they were well prepared for heroic self-sacrifice (Bruendel 2003). The British had long been led to think of themselves as a warrior nation (Paris 2000), but this applied also to the French and to the self-appointed 'heirs' of Frederick the Great. Militarism – the special, very powerful role in society of the military, whose leaders were still to a large extent recruited from the same social strata as 100 years earlier – was weaker in Britain. It was fairly strong – especially on the right and among the more affluent classes – in France and Germany. Only a handful of critical minds doubted the inherent superiority of the moustachioed or bearded caste of professional officers and the hierarchy of uniformed human machines created to serve and obey them.

Opinions on civil–military relations were not divided according to nationality, but right across national boundaries. One central issue was whether the civilian leadership – the politicians – should have the final say on the conduct of military operations even *during* war, or the supreme military leader – the famous Bismarck vs. Moltke controversy in the war of 1870/1. In Prussia, General Wilhelm von Blume (1835–1919) had a complex prescription for civil–military relations, which could suit even full-fledged democracies today, with the stark exception of his claim, reflecting the realities of the Second German Empire (1871–1918), that the military commander was not subservient to the politician, but to the emperor, who appointed both without parliamentary checks (Blume 1882: 16). Blume emphasised the importance of constant discussion between politicians and the military, in which the former would explain their aims, the latter explain what they could do or what they could not do, and the former would have to adapt their aims accordingly. Moreover, a constant adaptation of war aims might become necessary during the course of a war (Blume 1882: 19–21). General Rudolf von Caemmerer (1845–1911) sided with Bismarck in the controversy over whether the military leader or the political leader should have the final say on the major lines to follow during war time (Caemmerer 1904/1905: 84–7). By contrast French General Jules Louis Lewal (1828–1903) and his colleague Commandant (later General) Jean Colin (1864–1917) thought, like Moltke, that the politicians should keep out of the conduct of war altogether. They could and should, both conceded, have a say in the declaration of war, in the war aims and in the armistice and peace negotiations. Other than that, they should respect the military's sole decisions on how to conduct the war (Lewal 1892: 16; Colin 1911: 341–6). The German General Friedrich von Bernhardi (1849–1930) in *On War of Today* conceded that political aims, and thus political decision-makers, had to dominate war:

> War is only ever a means to reach an aim which is altogether outside its sphere. Therefore, war cannot decide on this aim for itself, by determining the military target according to its own judgement. If one conceded this privilege to war, one would always incur the danger that war, freed from all bonds, would be ... fought for its own sake, or that its achievements would fall short of what is politically necessary.

But in barring politicians from interfering in the conduct of war, Bernhardi spoke for the majority of his colleagues in the West: if politics, that is, political decision-makers

> define their aims in relation to the means of power available to the state, and in co-operation with the military leader determine the military targets which are to be reached, it must nevertheless *never interfere in the conduct of war itself* and *attempt to order it to take a particular course* to actually reach the military targets ... For war is the continuation of politics with *other* means and will be unfaithful to its innermost essence, as soon as it makes use of political means ... If politics [i.e. decision-makers] wanted to conduct the military action onto a political course, and use war not as an independent means of politics, but as a political means, politics would put at risk military success in all cases. (Bernhardi 1912b, II: 205, 207f.)

Another aspect of militarism was the related world-view in which violent solutions to the political disagreements of states were the best, most noble and most honourable. Negotiations with another state – always seen as adversary – let alone compromises were seen as cowardly, unmanly, weak. In the spirit of militarism, political decisions were thus taken with a proclivity towards violent solutions.

The cultural differences between Britain, France and Germany on the eve of the First World War were very small compared with the differences between today's Britain and the pre-1914 Britain, between today's France and pre-1914 France, between today's Germany and pre-1914 Germany. If anything, they understood each other far too well, as they all were out to be top nation and to prevent the others from starving them out of colonies, *Lebensraum* and existence in the deeply Hobbesian international environment of the times. In defiance of any theory postulating that people go to war with each other because of *differences* in ideology, in the case of the First World War, it was because they were so similar to each other, and understood each other so well, that they all sought to get their retaliation in first.

The influence of Social Darwinism and racism

Even during the Franco-Prussian War, the term 'racial war' (*Rassenkrieg*) cropped up in soldiers' letters (Stonemann 2001: 288). Social Darwinism was inappropriately named after Charles Darwin, owing much more to Herbert Spencer's *Principles of Biology* of 1864, where he coined the famous notion that only the fittest race/nation would survive a colossal struggle for resources against all others in a world that was getting too small for all of them. This notion was only a little more prominent (if that) in Germany than in France and Britain.

Helmuth von Moltke the Elder thought of the world as a constant fight of 'that what will be against that what is', in an almost biological analogy presupposing limited resources fought over by competing races. Moltke's and Bismarck's Wars of German Unification in themselves stand for the distinctiveness of the spirit of the age. Bismarck built a feeling of common destiny among the peoples from several German-speaking states (traditional adversaries and rivals) by manipulating them into waging war against common constructed enemies. Absurdly, the victory over France in the last of these wars was used to proclaim a united German Empire, or *Reich*, absurd, as Austria, the major German-speaking power, as homeland of the Habsburgs the true heir to the Holy Roman Empire, was deliberately left out; like the American Civil War, the Wars of German Unification were thus in part a secessionist war. Bismarck believed that only the shock of wars and victories, of 'blood and iron' could bring the other

German states to surrender their distinct historical and regional identities to the united Germany under Prussian tutelage. War – the military solution – was preferred over any alternatives. Bismarck's Strategy had its equivalent in the wars of Italian unification. Although Moltke and Bismarck were competitors for power, their views in this matter are in agreement and stand for the international dimension of militarism. 'Eternal peace is a dream', wrote Moltke disparagingly,

> *and not even a beautiful one*, while war is an element of God's world order. In war, the most noble virtues of man unfold, which would otherwise slumber and become extinct: courage and abstention, loyalty to one's duty and willingness to make the sacrifice of one's life. The experiences of war are a lasting influence, strengthening a man's ability for all future.

Conceding that war brought suffering for the families, Moltke described this as an inescapable part of 'this world', the *condition humaine*, 'according to the will of God'. He harked back to the poet Schiller's *Sturm und Drang* phase in which he glorified war as 'good', as it made people more resourceful, forced them to think and take action. As a result of 'improving morality', there might be fewer wars in the future, he speculated,

> but no State can do totally without war ... As long as distinct nations exist, there will be conflicts which can only be sorted out through the use of force. I regard war as the *last*, but a fully justified means, to defend the existence, the independence, and *the honour* of a State. (Großer Generalstab 1911: 1–3)

Deterrence was not a concept congenial to this culture. Moltke's contemporary Wilhelm von Willisen argued that the sole purpose of armies was war (Willisen 1860: 66). Similarly, Lorenz von Stein, writing in 1872, saw war as the only external purpose of armed forces, the internal purpose being the protection of the independence of the state. War, in his words, was the benchmark of all peoples, their 'highest life-task' (Stein 1872: 4f., 33).

The widely read Wilhelm von Blume grudgingly conceded that 'every war destroys much happiness in life, annihilates the fruits of much industry, demands great sacrifices, and remains an evil, albeit one that is at times necessary and salutary in its consequences'. He believed in the 'invigorating' effects of war on a population involved in it. Deploying the purple prose so popular in his age, he admired the way in which a *Volk* (people) might engage

all its strength, its possessions and blood in order to secure the life conditions of the State, for its honour, its cultural mission. Spirits are forcefully stimulated and directed towards ideals, slumbering forces are awakened. Bravery and manly strength, sense of duty and self-denial, triumph over caution, sloth, indolence, selfishness; the feeling of solidarity, the self-consciousness of the State are given life, what is true and real in people and State seizes its rights, while lies, incompetence and pretence are unmasked. Such a war is like a thunderstorm which after hot summer days sweeps across the land, fearsome and destructive, but at the same time cleansing, impregnating, awakening weakened forces to new life ... thus States and peoples intermittently need war in order to retain their vigour ...

The dangers of an excessively long peace grow for the people just as its luxury grows. A people that amasses wealth and becomes soft with an agreeably good life, will fall victim to the changes of borders and states through war so much the faster, as its riches will instil desires abroad. (Blume 1882: 8f.)

This morbid quest for a purge, a purification of the nation from its corruption and social problems had developed not only in Germany, but throughout Europe; as we have seen in chapter 2, it goes back to classical antiquity. Nevertheless, Social Darwinism added a new dimension to it, when the military writers of the late nineteenth century almost to a man worried about the non-martial *mentalité* of their fellow nationals, which might give the edge to less civilised, more robustly barbarian peoples once the ultimate showdown was upon them all (see, for example, Blume 1882: 41–6). Time and again, the strategic theorists stressed that it was counterproductive for the higher aim of the survival of the nation if the individual soldiers valued their own lives – the currency through which the glory, independence and liberty of the nation had to be bought. For they all agreed that the ultimate showdown would come and was inevitable (Cairns 1953: 283, 285). Sacrifice instead of comfort and indulgence was seen as the panacea for society's ills. Here is David Lloyd George speaking on 19 September 1914:

We have been living in a sheltered valley for generations. We have been too comfortable and too indulgent ... and the stern hand of Fate has scourged us

> to an elevation where we can see the great everlasting things that matter for a nation – the great peaks we had forgotten of Honour, Duty, Patriotism, and, clad in glittering white, the great pinnacle of Sacrifice pointing like a rugged finger to Heaven. (q.i. Eksteins 1990: 189)

Social Darwinism also inspired visions of states and empires as organic systems with a childhood, adulthood and old age or decline, followed by demise. The German geographer and geopolitician Friedrich Ratzel (1844–1904) described states as organisms. In his writing he created the basis for the German cult linking 'blood and soil' ('*Blut und Boden*') by harping on about the relationship of people and the soil and geography, and the related theme of people and space ('*Volk und Raum*'). This soon turned into a complaint about the overpopulation of Germany, leading to the logic of colonial expansion either overseas or within Europe, a battle cry taken up by the German colonial movement and later by the National Socialists (Ratzel 1897: 65f.).

Peoples, cultures and societies were described along with species as less developed or more developed. Ratzel projected a similar hierarchy into forms of war: the *lowest* (!) form to him was 'war of annihilation', 'where the eradication of the adversarial people becomes the sole purpose, as one disdains its territory; a part [of the people] is killed, a part abducted into slavery, the land is left uncultivated or falls to the victor as a collateral gains'. The word 'annihilation' here clearly stands for genocide. After this, and only slightly less barbaric according to Ratzel's categorisation, came closely related forms of war: wars for booty and wars of conquest of territory. Most of the wars of the previous two millennia, he thought, had been wars for territory, which in his view even applied to the national wars of the nineteenth century, where peoples wanted to reconquer lost territory, or make their national territory more compact and less patchy. Trade wars, another one of his categories, he thought were never just about trade, but usually reflected long-standing conflicts. They were similar to colonial wars, but both, unlike the other categories, were ill suited to arouse great popular passions (Ratzel 1897: 65f.). Where the French invoked the 'civilising mission' of France (*mission civilisatrice*) and Bernhardi spoke of Germany's 'highest cultural tasks' (Bernhardi 1912b, I: 7), Ratzel fantasised about the 'heroic traits of the colonists'. While Ratzel emphasised a hierarchy of advanced and primitive cultures rather than supposed biological (racial) differences between populations (Ratzel 1897: 65f.), his works easily lent themselves to racist interpretations.

The most extreme protagonist of Social Darwinist views of the struggle for the survival of the fittest must be Bernhardi. He was the faithful offspring of a line of sadistic and misanthropic German high-priests of the state and adulators of war as purifier of society from Frederick the Great, Goethe and Schiller in their bad moments, to Hegel and Fichte, to August von Schlegel, Treitschke and the German Social Darwinists like Claus Wagner, all quoted by Bernhardi:

> The natural law, to which all laws of Nature can be reduced, is the law of struggle. All ... property, all thoughts, inventions, and institutions, as, indeed, the social system itself, are a result of the intrasocial struggle in which one survives and another is cast out. The extrasocial ... struggle which guides the external development of societies, nations, and races, is war ... In what does the creative power of this struggle consist? In growth and decay, in the victory of the one factor and in the defeat of the other! This struggle is a creator, in that it eliminates. (Bernhardi 1912a: 11f.)

From all this Bernhardi deduced a 'duty to make war' (Bernhardi 1912a: 38–55). Bringing in the arguments of Social Darwinism, Bernhardi succinctly summed up the dilemma of the ever-expanding world population as many saw it since Malthus:

> War is a biological necessity of the first importance, a regulative element in the life of mankind which cannot be dispensed with, since without it an unhealthy development will follow, which excludes every advancement of the race, and therefore all real civilization. 'War is the father of all things.' [Heraclitus] The sages of antiquity long before Darwin recognized this. The struggle for existence is, in the life of Nature, the basis of all healthy development ... The law of the stronger holds good everywhere ... The weaker succumbs ... Strong, healthy, and flourishing nations increase in numbers. From a given moment they require a continual expansion of their frontiers, they require new territory for the accommodation of their surplus population. Since almost every part of the globe is inhabited, new territory must, as a rule, be obtained ... by conquest, which thus becomes a law of necessity ... [Eventually], might gives the right to occupy or to conquer. Might is at once the supreme right, and the dispute as to what is right is decided by the arbitrament [*sic*] of war. War gives a biologically just

decision, since its decisions rest on the very nature of things. (Bernhardi 1912a: 10, 15)

Hitler's argument about the need for '*Lebensraum*' was fully fledged well before the First World War (Puschner 2001).

Bernhardi also argued that the qualities of armies depended on the spirit of the nations, their discipline, strength or moral weakness, whether they are 'infected by revolutions' or by 'increasing opulence'. Germany had a (growing) population of 65 million on a territory the same size as France with a population of only 40 million; Germany needed more food, and therefore more colonies, which could only be at the expense of other colonial powers. In Bernhardi's view, Germany faced a 'European war against superior enemies', and Germany's freedom of action was hampered to an intolerable degree. Germany's position in the world – 'purchased ... so dearly with the blood of our best' – was being threatened. Germans should ensure that the needs of their country were met, as befitted its greatness, as 'the very existence of our people as a civilized nation' was endangered. Germans should set out on their 'civilizing tasks' with a greater colonial empire, increasing the 'influence of Teutonic culture in all parts of the globe'. France, Russia and Britain were Germany's enemies (Bernhardi 1912b, I: 7).

Writing almost thirty years before Bernhardi, Wilhelm von Blume did not sound quite as brutal, but his writing, too, echoed a world in which all were at war against all. It reflected the spirit of the Bismarckian world when no traditional friendships between states could be counted upon any longer after Bismarck had used all and deceived all in order to promote the rise of the Second German Empire at the expense of all its neighbours – 'The friendship of other states today might even tomorrow turn into enmity' – and therefore he advised against relying on the support of any other state (Blume 1882: 49). The Prussian Field Marshal Colmar von der Goltz got his ideas on the influence of discipline from Darwin – it makes the armed forces so much more powerful than an armed mob (Maurice 1891: 43). Von der Goltz claimed that high culture and great prowess in war went together, while decadence sometimes went along with a continued knowledge of warfare, absent bravery and self-sacrifice (Goltz 1883/1906: 8ff.).

Similarly, von der Goltz quoted Clausewitz's dismissal of naïf views of humanity, which played straight into the hands of the enemy who would gain ascendancy (Goltz 1883/1906: 8ff.). 'The matter thus lies in a nutshell', he wrote.

> All projects of disarmament are framed in misconception of our present political life, which proceeds from the tribal consanguinity of nations. Owing to the community of interests, which today prevails in every nation, the various peoples confront each other like persons among whom a natural inherent selfishness becomes the source of disputes, individual goodwill notwithstanding. National egotism is inseparable from our ideas of national greatness. This egotism will always appeal to arms when other means fail, and where would a tribunal of arbitration be found which were capable of dictating peace? Only a world-empire could do this. But world-empires owe their being to, and are inseparable from, wars. (Goltz 1883/1906: 10f.)

Echoing Moltke the Elder in his famous exchange of open letters with the Swiss professor Bluntschli, von der Goltz wrote: 'We must accept what the gods send. What is ... absolutely certain, is that *wars are the fate of mankind, the inevitable destiny of nations; and that eternal peace is not the lot of mortals in this world*' (Goltz 1883/1906: 570).

This passive culture, this world-view was not alien to Britons. Lt.-Col. F.N. Maude, writing in 1905, thought that

> Clausewitz was the first to define War as an extreme form of human competition. In other words, he did for the nation what Darwin subsequently did for individuals, viz. he showed that War was nothing more or less than the 'struggle for the survival of the fittest' on the national plan. (Maude 1905: v f.)

Biological concepts were employed all round, to argue that along with the new weaponry, warfare now required even stronger and even hardier human beings to conduct it – precisely a scarce commodity unless a physical (and moral) reform could be effected among the masses of young men destined to go and fight (Altham 1914, I: 1–5). The French *Jeune École* of maritime Strategy discussed in chapter 9 was also thoroughly taken by such biologically deterministic arguments (Bueb 1971: 19). Ratzel praised 'war' as the extreme manifestation of the 'masculine traits of ... the will to dominate', while peace 'fettered the man to his wife and descendants' and was the expression of the 'feminine-conservative sexual life', which in the context of his times was not meant positively (Ratzel 1897: 65f.). Historian Antulio Echevarria sees in this fear of decadence and decline the origins of the cult of the offensive, which all sides hoped would stem the tide of moral disintegration (Echevarria 2002a).

Exceptionally among the nineteenth-century strategic thinkers, Blume understood that modern weapons were a great equalizer – no longer would it be important whether an individual was brave, healthy, robust, muscular and sporty: the bullet or bayonet pierced his chest as easily as that of the weakling, the coward, the pale miner or the malnourished industrial worker (Blume 1882: 42). This realisation, much as the introduction of the crossbow in the twelfth century, threatened any system of internal social hierarchy posited on the inherent, physical, hereditary superiority of some families over others, and thus of the old system of class rule which despite growing democracy was alive and well not only in the imperial Germany of the Junkers, but also in Britain and France (Cairns 1953: 273–85; Gordon 1974: 191–226). How could classes expect to rule others if they were not stronger, fitter and less vulnerable than their social inferiors?

Of the belligerent, nationalist ethos that was spreading through Europe like a fatal, contagious disease towards the end of the nineteenth century, the Spanish-born American philosopher George Santayana wrote:

> The spirit in which parties and nations beyond the pale of English liberty confront one another is not motherly nor brotherly nor Christian. Their valorousness and morality consists in their indomitable egotism ... a desire which is quite primitive. (q.i. Eksteins 1990: 128)

But the disease touched not only people 'beyond the pale of English liberty'. Social Darwinism spread widely among these military authors on all sides. Its argument ran along the following lines: only the fittest nation/race (both terms still used fairly synonymously) would survive in a struggle pitting nation/race against nation/race. The ruling classes of all three countries – possibly more in Britain and France than in the aggressively upwardly mobile German Empire – also succumbed to cultural pessimism (Gordon 1974: 199). They became conscious that their industrialised societies had large working classes living mostly in poor conditions, who were physically anything but fit. All the military writers considered here were convinced of the high civilisation and culture attained by their own nation, but saw with concern the lack of martial lust and prowess, alongside the disintegration of old social structures, customs and class structures (Blume 1882: 46f.). Such cultural 'superiority complexes', accompanied by the fear of one's own cultural decline in relation to vigorous barbarian foes outlived the First World War not only in Germany, where it turned into the diabolical racism of National Socialism. To give but one example of such language, the British military historian, later General, J.F.C. Fuller (1878–1966) in 1923 explained his view that 'Though the desire of man is peace, the law of life is war; the fittest, mentally or bodily, survive, and the less fit supply them with food, labour and service' (Fuller 1923: 7).

Morale

The importance of 'morale' fitted in with the belief that a Darwinian struggle for the survival of the fittest race was looming, for 'morale' was seen as a function of national or racial superiority. The recognition of the importance of morale, as an independent factor alongside physical forces, goes back at least to the French naval specialist Audibert Ramatuelle, writing at the time of the French Revolution and of Napoleon (Depeyre 1990b: 85). This is a particularly French theme, and late in the nineteenth century came to substitute for numbers, when the French realized that the German population was 'outbreeding' the French, as the French had already developed the civilised trend towards smaller families, often with only one son, the famous *fils unique*.

The particular French answer to this, paired with the doctrine of the *offensive à outrance*, culminating in the views of Marshal Ferdinand Foch, was that the French simply had to force themselves to be mentally, spiritually and emotionally superior to the numerically superior and militarily obviously very impressive Germans (Foch 1900). It was the *élan* and the will that had to make the French superior to the Germans. Over and over, French military writers had to tell themselves that they were superior – and blot out the memory of the ignominious defeat of 1870/1 (Cairns 1953: 280–2).

Similar ideas, emphasizing the morale over *matériel*, can also be found in British writing: Admiral Sir Herbert Richmond called 'the human factor ... the most influential of all the elements in war, transcending that material factor which tends to dominate the minds of superficial men' (Richmond 1931: 298). Writing in mid-1921, Captain Basil Liddell Hart (1895–1970) agreed on the importance of morale. 'The strength of an army depends on its moral foundation more than its numbers; the strength of an armed nation depends on the morale of its citizens. If this crumbles the resistance of their armies will also crumble, as an inevitable sequel.' Liddell Hart paraphrased Foch (without acknowledgement):

> Man in war is not beaten, and cannot be beaten, until he [has] acknowledged himself to be beaten ... So long as war persists as an instrument of policy, the objects of that policy can never be attained until the opponent admits his defeat ... Hence it can be seen that victory is a moral, rather than a material effect. To conquer, one has to make one's foes feel the force of a moral superiority which will shake their faith in their own power to win. This demoralization is achieved, however, by a concrete effect. The infliction of casualties will not alone produce it. The survivors, who alone retain the power to admit defeat, must themselves feel the superiority of the opponent. The concrete proof of this to them comes clearest from the fact of being driven back – not a few yards only, for their morale will survive this; but being hurled back in confusion, from which there seems no hope of recovery. The demoralization which begets a general conviction of inferiority comes from retreat and the break-up of organization. (Liddell Hart 1944: 20)

Morale henceforth became something all military doctrines considered as a crucial factor.

The original 'realists'

Against the widely held belief that universal struggles for national survival were afoot, amid the growing tide of nationalist selfishness that matched such a zero-sum Darwinian world-view, the nineteenth century gave birth to what would in the twentieth century become known as 'realist' interpretations of international relations. This saw interstate relations in terms of Hobbesian anarchy, in which each state would fend for itself, totally selfishly, without any loyalty to any other party, and with at best a regard to the balance of power logic inherited from early modern Europe, practised so ruthlessly by Bismarck.

Where selfishness was previously the dominating motivation of princes and their dynasties, it was now the narrow selfishness of nations that was widely accepted as the key determinant of interstate relations. This supposed a total distrust of international law and interstate conventions which, nevertheless, were increasing in number and scope during the same period (Honig 1995/6). The 'realist' approach also fitted well with the Napoleonic paradigm, as it was mutating into the pursuit of total war (see the following chapter).

Moltke himself had, surprisingly, held rather pacific tendencies in his youth. In 1840 he professed his faith in 'the much derided concept of general European peace', and in 1842 he explained his belief that social and economic developments in Europe, especially trade and industry, would increase the hankering of its peoples for peace. By the mid-1850s, as his military career flourished, his view had changed profoundly, and he wrote disparagingly of legal and diplomatic attempts to settle disputes between states. At the time of the Franco-Prussian War he wrote that in view of the fluctuating goals of diplomacy, 'strategy has no choice but to strive for the highest goal attainable with the means given' (q.i. Gat 1992: 72).

Blume recognised the restraints of international law, humanitarian concerns and world opinion on warfare. He conceded that certain laws were followed by civilised nations, but only their conscience, and their concern about the opinions of third parties, could make them uphold these laws. There was no point, he thought, in formulating any conventions which go beyond existing laws as they would be disregarded in war (Blume 1882: 1f.).

Typically for the 'realists' of this period, von der Goltz totally distrusted any proposals for disarmament. 'Only that nation is secure which is prepared at any moment to defend its independence sword in hand' (Goltz 1883/1906: 8–11). Von Bernhardi thought that 'The enmities surrounding us cannot be exorcised by diplomacy' (Bernhardi 1912b, I: 12). Writing on the eve of the First World War, he opined:

> The arbitration courts, which the contracting powers engage to obey, are meant not only to lessen the dangers of war, but to remove them altogether. This is the publicly avowed object of such politics. In reality, it is hardly caused by an ideal love of peace, but is evidently meant to serve quite different political purposes, [namely their particular advantages] ... We

Germans therefore must not be deceived by such official efforts to maintain the peace. (Bernhardi 1912b, I: 10f.)

Bernhardi concluded: 'If we wish to gain the position in the world that is due to us, we must rely on our sword, renounce all weakly visions of peace, and eye the dangers surrounding us with resolute and unflinching courage' (Bernhardi 1912b, I: 11).

Britain and the US also tended to distrust international agreements, either not ratifying them, in the case of the US, or, in the case of Britain, secretly (or indeed openly) basing their planning on the assumption that they would not be respected by either party in war time. Lord Goschen, First Lord of the Admiralty, declared in 1898 with regard to the danger of the interruption of commerce by an adversary:

I know that it is said that we could secure ourselves from this particular danger by accepting an alteration in the rules of International War which would exempt merchant ships from capture in war; but the life of a people must rest in something stronger than the adhesion of a belligerent to a technical rule made in peace. We know too well that if the decisive issue of victory depended on their breach, plausible reasons would be found, probably on the pretext of reprisals, for breaking a law which had no other sanction than the authority of a paper agreement. (q.i. Gretton 1965: 9)

It is no accident that the most prominent writer on air power, Giulio Douhet, writing after the First World War, also belonged to the 'treaties-are-pointless' school:

All the restrictions, all the international agreements made during peacetime are fated to be swept away like dried leaves on the winds of war. A man who is fighting a life-and-death fight – as all wars are nowadays – has the right to use any means to keep his life ... The limitations applied to the so-called inhuman and atrocious means of war are nothing but international demagogic hypocrisies.

He therefore predicted that poison gas would be used in a future war, despite its prohibition (Douhet 1928/1983: 181f.).

Owing to extreme necessity, all contenders must use all means without hesitation, whether or not they are forbidden by treaties, which after all are nothing but scraps of paper compared to the tragedy which would follow. (Douhet 1928/1983: 189)

A common ploy in the context of such an argument was always to profess one's devotion to principles, but to declare them unworkable, as the enemy would not adhere to them, as in the following passage of a memorandum which was circulated by Royal Air Force Chief of Staff Hugh Trenchard in 1928 – at a time when Trenchard still suspected that the next war by Britain would be fought *against France*: 'If this restriction' on bombing cities

were feasible, I should be the last to quarrel with it; but it is not feasible. In a vital struggle all available weapons always have been used and always will be used ... [F]oreign thinkers on war have already shown beyond all doubt that our enemies will exploit their advantage over us in this respect and will thus force us to conform and to counter their attacks in kind. (q.i. Chaliand 1994: 910)

Opponents of this 'realist' approach – who were in a small minority – included the socialists of Europe, like Bertha von Suttner, August Bebel and, on the more extreme end, Karl Liebknecht. A particularly prominent opponent was Jaurès in France, who wanted for his country a defensive Strategy, on the basis of which, if attacked, France could submit her cause 'to the arbitration of civilised humanity' (Jaurès 1911: 72). But even a self-styled pacifist like Liddell Hart, writing in 1925, was sceptical of the value of treaties, as opposed to a wide-ranging change of heart:

The humanization of war rests not in 'scraps of paper', which nations will always tear up if they feel that their national life is endangered by them, but in an enlightened realization that the unlimited spread of death and destruction endangers a victor's own future prosperity and reputation. (Liddell Hart 1944: 36)

The consequence of this distrust of international law and engagements was horrendous. Here just one example from naval warfare: 1909 saw the Declaration of London, which regulated the circumstances in which neutral shipping could be treated as hostile (if it carried contraband or enemy personnel; contributed to the enemy's intelligence-gathering efforts) and might even be destroyed if its conduct to a port 'would involve danger to the safety of the [capturing] warship or to the success of the operations in which she is engaged at the time'.[2] Historian Bryan Ranft commented:

> When war came in 1914, Britain successfully resisted pressures from neutrals, led by the United States, to abide by the Declaration and gradually established, using heavy pressure on European neutrals, measures designed to cut off Germany from all seaborne supplies, measures incidentally accepted by the United States when she became a belligerent. In both [world] wars Germany used these measures as a justification for waging unrestricted submarine warfare as an exercise of the recognised right of reprisal.

Limitations on submarine operations against merchant shipping were imposed at the Washington Naval Conference (1921–2), the London Conference (1930) and the Submarine Protocol of 1936, which was ratified by over forty states by 1939, including Germany, the USA, Britain, France and Japan. Nevertheless, unrestricted attacks on merchant shipping began early in the Second World War (Ranft 1982: 170). The 'realist' mind frame of crucial individuals thus ensured that the 'scraps of paper', intended to tame war and render it more humane, were indeed blown aside when the storm broke.

So much for the passive culture and the world-views of the era of the Napoleonic paradigm. In the following chapter we shall explore the different dimensions of this paradigm, and watch its transformation from the pursuit of victory in battle to the pursuit of total war.

[2] Declaration concerning the Laws of Naval War 208 Consol. T.S. 338 (1909), Art. 49, www1.umn.edu/humanrts/instree/1909b.htm, accessed 13 October 2009.

6 *The Napoleonic paradigm transformed*

From total mobilisation to Total War

[T]he absolute form of war is deduced, not from any of the changes in weapons or in the organizations of armies, but from the entrance of nations into the arena which was before occupied by 'sovereigns and statesmen'.

(Henry Spenser Wilkinson 1910: 181)

This chapter will examine the predominating elements of Strategy in the long nineteenth century and map the evolution of the Napoleonic paradigm into one of Total War along two strands. The first was rooted in the perceived need to make warfare – defensive warfare – more effective by harnessing the entire nation's strength to the cause of the defence of the country, including all its members, all its productivity, all its wealth. In truth this was more an ideal of the French Revolution than a Napoleonic measure, but the two were often merged into one in the minds of men. This ideal originated even before the Revolution, in the Enlightenment, with Count Guibert. It is what would in the First World War be called 'total war' in the sense of 'total mobilisation of one's national resources' by the French politician Léon Daudet, who claimed to have invented the term (Daudet 1918: 8f.).

The second is the quest for a war-deciding battle, which is truly the core of the Napoleonic paradigm. Clausewitz saw the 'battle of annihilation' as central to Napoleon's success (Clausewitz 1832/1976, IV: 11[1]). The sense of the term 'annihilation' would evolve over time in a terrible way, under the shadow of Social Darwinism and racism, as we shall see.

[1] Howard and Paret (in their translation of Clausewitz 1832/1976: 258) translate '*Vernichtung*' as 'destruction' rather than 'annihilation', but in view of the subsequent impact of Clausewitz's writing and the use of his ideas by others in the nineteenth and early twentieth centuries, 'annihilation' is the more appropriate translation.

As we shall see in the following chapter, three different responses to the horrors of the First World War shared the quest for an alternative Strategy to that of attrition. One response was to embrace the defensive in the most static possible way, with the Maginot Line. A second response was to adopt a more indirect approach to confronting the enemy. Of these, one approach sought a new way of mobile war to lead to a more decisive form of battle than what had turned into the attrition warfare of the Western Front. The other concentrated on targeting a different part of the enemy – his civilian population where it was undefended. This last response overlapped greatly with another war aim which had its roots elsewhere and aimed *not* at the targeting of civilians to bring war to a quicker end, but at annihilating an enemy people down to the last man, woman and child. This was a war aim that was not unknown to classical Antiquity, nor to the medieval world or to the wars of religion, and that was present in the ruthless fight against the aristocracy and royalist insurgents in the French Revolution: the annihilation of the enemy, a concept that from the second half of the nineteenth century was interpreted in an increasingly racist and comprehensive way. This is 'total war' in the sense in which the most famous National Socialist strategist, First World War leader General Erich Ludendorff, elaborated it in his eponymous book of 1935. He defined the aims of a future war as 'the *annihilation* of the enemy Army *and of the enemy nation*' (Ludendorff 1935/1936: 168; my emphasis).

From this literature, and contemporary usage, Stig Förster and Jörg Nagler have developed four criteria for the use of the term 'Total War':

- Total mobilisation: going back to definitions used by Daudet and others, it includes total mobilisation of human and material resources for the war.
- The pitting not of prince against prince, or army against army, but of nation against nation, until one or the other is totally annihilated, enslaved, subdued. Förster and Nagler home in on Ludendorff's claim that Total War should be fought 'for the purpose of fighting a war for the life or death of a nation' but included in their definition also total control by the military of one's own country for this purpose, so that peace becomes nothing but a prelude to war.
- According to Förster and Nagler, Total War is organised by modern states with gigantic bureaucratic and military machines: the state,

originally justified to protect its citizens, 'now turned into a murderous fighting machine that used its citizens as "human material" [*Menschenmaterial*]'.

- Förster and Nagler emphasised unlimited war aims as an essential characteristic of Total War. Unconditional surrender, an American term which goes back to the American Civil War (!), was demanded by one side. The other side – the Germans in the Second World War – aimed for physical annihilation or enslavement of enemy peoples, soldiers and civilians alike.

Förster and Nagler summed up their definition thus:

> Total War, at least theoretically, consists of total mobilization of all the nation's resources by a highly organized and centralized state for a military conflict with unlimited war aims (such as complete conquest and subjugation of the enemy) and unrestricted use of force (against the enemy's armies and civil population alike, going as far as complete destruction of the home front, extermination, and genocide). (Förster and Nagler 1997: 10f.)

As we have noted, the killing of unarmed civilians had existed since time immemorial. Massacres of unarmed populations are the first proven form of warfare (Guilaine and Zammit 2001: 120–2). The question arises: is all genocide Total War? Förster and Nagler acknowledge earlier forms of deliberate mass killings of civilians, but see the presence of industrialised mass societies as a prerequisite for 'modern' total war, and therefore discourage the use of the term 'Total War' for earlier cases of genocidal warfare, even though these meet other criteria of their definition (Förster and Nagler 1997: 10f.). This last point is debatable, but we should make these definitions our starting point for what is discussed in this chapter: the transformation of a quest for military victory into a war of national survival or even genocide.

The quest for total victory

In the light of the experience of the Napoleonic Wars, Clausewitz's colleague Rühle von Lilienstern noted in 1818 that war was always conducted for political aims which might transcend battlefield victory or even the establishment of (a short-term) peace (see chapter 1). And Clausewitz, as Napoleon's interpreter, is most famously remembered today for having articulated the link between political purpose and war. The era that began a couple of decades after Clausewitz's death, however, and ended with the two World Wars was marked, above all, by the dismissal of political constraints by most writers on Strategy. Limited war became inconceivable to most. Instead, they ceaselessly emphasised the need for a decisive victory, modelled on Napoleon's battles of annihilation, in which the enemy's armies had to be 'ruined', 'wiped out', 'crushed'. This could only be achieved, they thought, in a commitment to the offensive, as the more 'vigorous', active, indeed 'virile' form of war (Blume 1882: 8f.).

What were the strategic-political aims of such a war? For the Bavarian Joseph Ritter von Xylander, shortly after the Napoleonic Wars, the aim of war was the occupation of and conquest of the territory and possessions of the enemy (Xylander 1818: 17f.). Half a century later, the German university professor Lorenz von Stein from Schleswig-Holstein defined as aims in war the maintenance of the independence of one's own state, and 'victory over the enemy abroad' (Stein 1872: 33). Victory as an absolute concept – not linked to the achievement of political aims – became the be-all and end-all of all strategic writing, and it had to be sought on the battlefield, as the Italian naval historian and Dominican priest Alberto Guglielmotti put it, as just one of many articulating this ill-defined idea (Ferrante 1993: 106f.). Nor was Goltz's war aim of 'complete subjection of the enemy' defined any further (Goltz 1883/1906: 9f., 13). Echoes of Clausewitz's general definition of the aim of war of 'imposing our will on the enemy' were found everywhere. At the time of the outbreak of the Great War, British Major General E.A. Altham defined military success as the ability 'to compel' one's 'adversary to conform to the nation's will' (Altham 1914, I: 3).

Victory was sought for its own sake. As Liddell Hart correctly observed, '[a] "decisive victory" is apt in military language to have a mystical sense which is by no means synonymous with its actual effect'. It did not matter that '[t]he statement hardly accords with the last two or three thousand years of experience' (Liddell Hart 1944: 43). And he mused critically, '[i]t is extraordinary how many victories have been gained in wars without the effect that in normal language is meant by the word "decisive"' (Liddell Hart 1944: 50). The strategic theorists of the era dominated by the Napoleonic paradigm saw it. They wanted *la victoire pour la victoire.*

The Romano-Christian idea that only just wars should be fought, with their prerequisite of a just cause, now rarely featured. If any nexus between the conduct of war and political cause was ever spelled out, it would sound like Scottish Lt.-Col. Walter H. James's elaboration of the law of the bully:

> The object of war is peace on the victor's terms. To obtain it, it is necessary to bring home to the hostile nation the futility of resistance, and the right method to pursue is that which soonest demonstrates the uselessness of further opposition. Campaigns, therefore, should be conducted simply with this end in view. Hence during the operations, while military reasons will usually play the chief part, moral or political considerations must not be neglected. (James 1904: 10)

Nor was any objective framework of justice invoked by Major-General F.B. Maurice, who defined the 'ultimate object of war' as 'either the complete conquest of the enemy; or to make him sue for peace on satisfactory terms; or to cause him to abandon the purpose for which he went to war' (Maurice 1929: 68, 86). After the First World War, his colleague J.F.C. Fuller served us evidence that Social Darwinism was not dead, even in Britain: 'Nature knows nothing of morality, unless morality be defined as race survival' (Fuller 1923: 15). If one's world-view was one in which irreconcilable 'races' were pitted against each other, then they must of necessity come to blows, and each side had an interest that such a clash should be decisive.

Such all-encompassing, that is, total war obviously called for total victory, brought about by the enemy's unconditional surrender. The genesis of this concept can be traced at least to the American Civil War; the victory of the German coalition over France in 1871 proclaimed in the mirrored hall at Versailles was translated into an imposed peace treaty which was quite as punishing as the Treaty of Versailles imposed by a revanchist France and her allies on Germany in 1918, in the same hall of mirrors. Both engaged the defeated party – an entire nation and as yet unborn generations – not only to payment of reparations for decades to come, but also deprived it of territory, and in each case went along with regime change (even though the latter was not imposed by the victors). Unconditional surrender would again be espoused as a war aim by the allies fighting Germany, and this time Japan, in the Second World War, making it arguably more difficult than absolutely necessary for the Japanese government to terminate the war. In the end, the Japanese were shocked into unconditional surrender by the ultimate weapon of Total War, the atom bomb.

The centrality of the battle

Rühle von Lilienstern had remarked soberly after the Napoleonic Wars that a party in war should be seen as defeated if it had not attained its main political purpose in that war, 'however many battles he may have won' (Rühle 1818: 8f.; see chapter 1). This view would be abandoned by almost everybody in the following century and a half, only Sir Julian Corbett dissenting, as we shall see in the next chapter. Basil Liddell Hart rightly observed that '[s]ince the late eighteenth century the nations have been enslaved by a military doctrine – the doctrine of a fight to a finish – from which, once committed to war, they were helpless to shake free' (Liddell Hart 1944: 43, 50). To Moltke, writing in 'Instructions for Superior Commanders of Troops',

> The victory in the decision by arms is the most important moment in war. Only victory breaks the enemy will and compels him to submit to our own. Neither the occupation of territory nor the capturing of fortified places, but only the destruction of the enemy fighting-power will, as a rule, decide. This is the primary object of operations. (q.i. Gat 1992: 68)

For Wilhelm von Willisen, also, the art of war had but one purpose, and that is to achieve victory in battle (Willisen 1840: 26). Colmar von der Goltz argued that one could not vanquish an enemy by defeating his minor armies: the victory has to be achieved sooner or later by confronting his main army on the battlefield. Instead, one must 'unite all the available forces for the decisive moment' to achieve that (Goltz 1895/1899: 9–21).

Von der Goltz elaborated on the effect of mass politics on war, and on the great national interests that were at stake: 'War nowadays generally appears in its natural form, i.e. as a bloody encounter of nations, in which each contending side seeks the complete defeat, or, if possible, the destruction of the enemy.' Anything less than that would not do, as Napoleon's wars had purportedly shown. Von der Goltz could not imagine wars with limited aims, short of the 'overthrow of the enemy', engaged in with merely a portion of one's force.

He thought that trivial causes would not generally culminate in wars (the reactions to the assassination of Sarajevo in 1914 must have come as a surprise to him). Any government attempting to cut short a war would be prevented from doing so by the 'the inflamed national feeling of the people' who

> will at once interpose and not allow those in authority to make the fate of the whole war dependent on a misfortune sustained by a small portion of its fighting resources. Public opinion will insist upon their reinforcement; the enemy will do likewise, and thus, little by little, contrary to all original intentions, the whole strength of both the combatants will become engaged. Now that states and nationalities are in most cases almost identical, they resemble persons who would rather lose their lives than their honour. (Goltz 1895/1899: 5–7)

Even in a small controversy, he thought it unlikely that a large army would be held in readiness with only a small section striking blows. One would try to get the enemy to accept one's will as quickly as possible, which would be done by as hard a blow to start with as possible. One could not, he argued, defeat an enemy by defeating his minor armies: the victory had to be achieved sooner or later by confronting his main army on the battlefield. One should 'unite all the available forces for the decisive moment' to achieve that (Goltz 1895/1899: 8–20). Furthermore, he wrote,

> [w]here such forces set the great machinery of war into motion, wars can only end with the entire annihilation of one party, or the complete exhaustion of both ... [T]he growing national sentiment and the political realisation of the principles of nationality have increased to a marvellous extent the powers of resistance of States.
>
> Future war
>
> will be waged with a destructive force such as has hitherto never been displayed. War is now an exodus of nations, and no longer a mere conflict between armies. All moral energy will be gathered for a life and death struggle, the whole sum of the intelligence residing in either people will be employed for their mutual destruction ... The growth of national motives of jealousy and national enmity entails a corresponding display of force. (Goltz 1895/1899: 463–69)

The British military held similar views: in the words of Henderson, the aim of 'strategy' had to be to bring 'the enemy to battle, while tactics are the methods by which a commander seeks to overwhelm him when battle is joined' (Henderson 1905: 39). The 1920s British *Field Service Regulations* proclaimed 'battle' to be 'the decisive act in war' (Maurice 1929: 68); the early 1930s edition of the *Regulations* stated that 'Battle is usually the decisive act in land warfare' (Liddell Hart 1944: 50).

Writing in 1900, Henry Spenser Wilkinson embraced von der Goltz's views, which he thought were 'held by all Continental military men and by all Continental Governments'. Spenser Wilkinson, with his widely read publications, ensured that the British, too, became convinced that

> war in our time is bound to be a struggle for national existence, in which everything is risked, and in preparation for which, therefore, no conceivable exertion must be spared ... the absolute form of war is deduced, not from any of the changes in weapons or in the organizations of armies, but from the entrance of nations into the arena which was before occupied by 'sovereigns and statesmen'. The dynastic form of war was a courtly duel, which went on until the first wound, whereupon honour was satisfied, reparation made, and the episode closed. The national form is a bitter quarrel, and a fight which ends only when one or the other combatant lies prostrate and helpless at the mercy of his foe, whose first anxiety will be to prevent the beaten enemy from ever recovering sufficient power to be able to renew the quarrel with hope of success. (Wilkinson 1910: 180)

His countryman General Maude underlined the need to get the whole country (including business) organised for a major war, because a defeat would be so catastrophic (Maude 1905: 125).

What France had to do to win was explained in similar terms: echoing Moltke, General J.L. Lewal noted that there was but one objective in warfare, 'to win, overwhelm the adversary materially and morally, to oblige him to ask for mercy' (Lewal 1892: 17). General Messimy emphasised that 'Victory is not achieved through the possession of a town or territory, but through the destruction of the adversarial forces'; Lieutenant Colonel Rousset argued, 'One has to think exclusively of battle. All efforts, all thoughts, all preparations have to pertain to the assurance of its success'; and General Langlois spelled out as the main aim 'to ensure that one wins the battle' (Langlois 1906: 74). The rest would come by itself.

Annihilation of the enemy

A word that was frequently used in this context, and which would grow in meaning, was to 'annihilate' (in German: *vernichten*). The term was frequently linked with the aim of any battle from the age of Napoleon until the First World War. Writing just slightly before Clausewitz, Xylander in Bavaria wrote that 'to annihilate the inimical army must be the aim of the commander in the largest battles as much as in the smallest engagement'. Xylander added, however, that the war aim must be to occupy the enemy's territory (Xylander 1818: 8, 17f.). Clausewitz himself wrote: '[T] he annihilation of the inimical armed forces is … always the means to reach the purpose of the engagement' (Clausewitz 1832/1976, I: 2). This term, originally used to describe the destruction of enemy forces by Napoleon in his big battles, over time took on a life of its own.

Some strategists rightly shunned the ambiguity of the term 'annihilation'. Clausewitz himself had used 'destruction' as well as annihilation, and had written, 'The fighting forces must be destroyed: that is, they must be put in such a condition that they can no longer carry on the fight. Whenever we use the phrase "destruction of the enemy's forces" this alone is what we mean' (Clausewitz 1832/1976, I: 2). Colmar von der Goltz still echoed Clausewitz when he wrote:

> [W]e mean that, by the annihilation of a portion of his fighting power, we make him despair altogether of any subsequent favourable turn in the hostilities; – and by 'destruction', we imply that we reduce him to such a physical and moral state that he feels himself incapable of continuing the struggle.

While claiming that 'war … must aim at the complete subjection of the enemy', von der Goltz noted that armies did not have to be completely wiped out to achieve that. Once they lost 20 per cent of their battle strength, or at most half, the battle would normally be decided (Goltz 1883/1906: 7, 9f., 13).

Colonel J.F. Maurice was also more precise:

> [T]he aims of strategy directed against the actual condition of the armies of our time are twofold – first, to break up the organic force of the opposing

army by dealing in concentrated force with fractions of the enemy, and secondly, to threaten, and if possible to destroy, the enemy's connection with the sources from which he draws his supplies. (Maurice 1891: 32)

Spenser Wilkinson explicitly developed Clausewitz's and von der Goltz's ideas of what the logical aim in battle must be, that

each side would, from beginning to end, aim at the total destruction of the enemy's forces, by which is meant, not necessarily killing all his men, but making an end of all his armies. The armies who surrendered at Sedan and Metz were in this sense destroyed, though only a percentage of the men were killed or hurt. (Wilkinson 1910: 180)

Nevertheless, the military sense of this term soon receded. We have seen above that with Ratzel the word 'annihilation' was used synonymously with genocide, and this meaning would become predominant in the thinking and later actions of the National Socialists.

In practice, for the looming First World War, this meant the need for mass armies which would be given the task of 'annihilating' each other. This resulted in a Strategy of attrition, of *Materialschlacht*, and of 'bleeding the other side dry' in the hope that one's own human resources would not run out, as generals from Falkenhayn to Foch and Hague did on both sides in the First World War. 'Annihilation' would fully unfold its horror also for civilians in the Second World War, at the hands of Hitler's eager supporters.

The universal cult of the offensive

Another characteristic of this age was the cult of the offensive. 'Military technology', wrote Jack Snyder, an expert on the First World War, 'should have made the European strategic balance in July 1914 a model of stability'. In his view, both the Boer War and the Russo-Japanese War had given witness to the enhanced strength of the defensive through innovations in firepower and inner lines or interior logistics improved by railways (Snyder 1984: 108). And yet the overwhelming majority of military writers of the time all favoured the offensive, their judgement coloured by a shared cultural and ideological predilection.

We have seen that in the centuries up to Napoleon, there was no consensus as to whether defence or offensive was preferable – most writers took for granted that both options had to exist, and a few, especially Clausewitz, actually thought that the defence had the better cards (see chapter 3). This view was not shared by his contemporaries, and became anathema in the nineteenth century, until the First World War. Xylander had written as early as 1818, 'The attack is the actual basic principle of war; only through the offensive can a result be arrived at' (Xylander 1818: 34). A century later, Bernhardi expressed the conviction, that especially in modern mass warfare, the offensive is the by far superior form of warlike operation [*Verfahren*]', thus explicitly contradicting 'the greatest of military authorities', Clausewitz (Bernhardi 1911: 399). He concluded that

> an offensive operation holds out greater chances of success than a defensive one, and that it commends itself especially to the weaker party, as long as the balance of forces [between the adversaries] holds any possibility at all of a favourable decision. A certain degree of superiority cannot be levelled even by the most ingenious offensive; but in that case the defensive is completely unable to bring about a favourable decision. (Bernhardi 1911: 411)

Clausewitz's rival Jomini is most famous for his writing on Napoleon's big, often campaign-deciding battles, and his praise for them as central to warfare. Even so, he had seen advantages in waging a purely defensive war on one's own territory where shorter supply lines favoured the defence (Jomini 1837/1868: 17, 72). Jomini's, not Clausewitz's, views on defence and offensive would be the lodestar of strategic writing for the rest of the century. General Jean Auguste Berthaut (French War Minister 1876–7) praised Jomini's commitment to the offensive as it gave the commander 'freedom of action' and the ability to dominate the movements of his adversary, who was forced into a reactive mode of conduct (Berthaut 1881: 19f.). General Victor Bernard Derrécagaix (1833–1915) equally echoed Jomini when spelling out the only law that he saw fit to guide French Strategy, namely 'to be the strongest at the decisive point', as 'strategy has but one goal, victory'. The offensive alone would enable France to reach her aim, he wrote, with a reference to von der Goltz's agreement on the subject (Derrécagaix 1885, I: 375–80, 617f.).

Berthaut also followed Jomini's views on the strategic offensive: 'The offensive ... gives [the general] all his freedom of actions and ... subjects his adversary's movements to his.' The offensive has the advantage of surprise, choice of place, gives the possibility of taking the war outside of one's own territory. Berthaut quoted Marshal Marmont, who had said 'One dominates the thinking of the adversary, and a first success sometimes gives an ascendant that one will not lose again' (Berthaut 1881: 19f.). Berthaut backed this up with many historical examples, and of course the clinching argument that Napoleon had always chosen the offensive. After all, mused Berthaut, just like his other European colleagues: 'Only the strategic offensive ... gives important and decisive results: one therefore has to make all efforts to seize it while one is stronger and even, in some circumstances, when one is the weaker [party].' On the down side, this meant that one had to be able to be the first to take action, before the adversary, which was not always politically advisable (Berthaut 1881: 28).

Captain Georges Gilbert, incarnation of the cult of the offensive, was but one who called for a revival of Napoleonic warfare. The argument ran roughly as follows: France had invented the modern (Napoleonic) form of major warfare, she was the first to apply it in its most perfect form. But then France lost her *élan*, neglected her own traditions and through misgovernment forgot her Napoleonic military heritage, no longer listened to her special genius and was punished for it in 1870, when France's enemies used the Napoleonic way of war against her. The only prescription for success would therefore be for France to recover her own genius: then she would find her way to victory again (Gilbert 1890, 1892).

Also in France, Commandant J.J.H. Mordacq saw a link between diplomacy and war aims: 'An energetic, courageous diplomacy almost always corresponds to an offensive war, that is carried out most vigorously; and reciprocally, a timid foreign policy, that is short-sighted, is usually followed by a defensive war, conducted without *esprit*, without a clear guiding idea' (Mordacq 1912: 2).

Even the mere discussion of a defensive Strategy became dishonourable. General Lanrezac, commander-in-chief of the French Fifth Army at the outbreak of the First World War, addressed a group of high-ranking French officers; at the end of his talk, he supposedly said: 'The doors are well shut? Well, in that case I can speak to you about the defensive' (Castex 1934, IV: 137). The strategy expert Douglas Porch judges that the 'doctrine of the offensive' in France seemed to obviate the need for 'serious reform' of the armed forces and it compensated for real innovation in arms technology and for cohesion within the armed forces, whose officers, like the country itself, were still very torn between republicanism and left-wing sentiments on the one hand and reactionary-Catholic values on the other (Porch 2006: 142).

East of the Rhine, Willisen in his *Theory of Major War* seemed reluctant to tackle 'the Teaching of the Defence' at all, as he saw it as passive, weak and incapable of proving decisive in the course of a campaign (Willisen 1840: 115ff.). Moltke the Elder conceded that the defensive party could choose the places from which to defend itself, forcing the attackers to take certain routes and to expose themselves to the defender's fire. But he praised the offensive for imposing the 'law' of action upon both sides, by choosing the way and means and aim of the war, with the connotations of strength, confidence, decisiveness, while the defender had to wait in [demoralising] uncertainty (Großer Generalstab 1911: 141). Colonel Blume, writing some years after the Franco-Prussian War, noted that 'an energetic conduct of war strengthens the national character; by contrast, wars conducted over a long time and lamely entail the greatest material losses and moral damages' (Blume 1882: 1f.). And the geographer Ratzel opined, 'It is always the first *aim* of war to penetrate into the territory of the adversary' (Ratzel 1897: 65 f.).

Von der Goltz for one was not blind to the renaissance of defensive fortifications. Since 1870/1, France had taken the lead and started building vast fortifications for a future theatre of war, with many other countries following suit, including Germany, Poland, Lithuania, Austria, Switzerland, Italy, Belgium and Rumania. There was a big shift towards recognition of the need for a 'closer union between siege operations – hitherto regarded as a thing distinctly apart – and action in the open field, and, furthermore, of making more general use of entrenchments in the latter phase of warfare' (Goltz 1883/1906: 2f.). Nevertheless, von der Goltz insisted that the distinctive characteristics of the modern method of conducting war would be:

1. Calling up the military resources of the country to such an extent that, victory being gained, one could proceed to enforce a favourable peace with the least possible delay.
2. Placing all one's forces in readiness at the very commencement of the war.

3. A ceaseless and untiring prosecution of the campaign until the organized resistance of the enemy is broken in decisive battle; after which, until the conclusion of peace, a less strenuous action and one more sparing of the 'instrument of war' would ensue.

'The defence is ... strictly speaking, not a complete form of the art of war with which we can reach the object of the war, namely, the overthrow of the enemy.' This had to be the object of every war, Goltz opined (Goltz 1895/1899: 21f.).

If France wanted to be offensive, the decision to go ahead had to be taken as soon as possible. Accordingly, mobilisation had to take place as soon and quickly as possible (Derrécagaix 1885, I: 375ff.). In his section on defensive operations, Derrécagaix emphasised all the disadvantages of the defensive, again quoting a suitable passage from von der Goltz. This included the concession that the defence could have its advantages – simpler, shorter inner lines and so forth. Derrécagaix commented:

> The defensive is thus a mode of operations that can lead an army to victory ... a nation can be dragged forcibly into a war or be threatened with aggression by its neighbours. What should one do in that case, other than proudly to accept this provocation, with the determination not to lay down one's arms until one's fatherland has been freed? In the conditions of modern warfare, the circumstances that define the need to [resort to] the defensive are, generally speaking: 1. The inferiority of [one's own] forces; 2. Delays in the mobilisation of [one's] armed forces. Numeric inferiority obliges nations to espouse a particular way of war and to prepare all their efforts in times of peace in the perspective of the sole aim of defending one's territories. This is the situation of second-rate powers ... which leads us away of the main subject to be treated here ...

namely France's defence, as he did not see France as a second-rate power (Derrécagaix 1885, I: 615–18).

Even the Belgian General Brialmont, greatest advocate of defensive fortifications, succumbed to the temptation to write enthusiastically about the offensive: 'The offensive increases courage, augments the confidence of the soldier, and gives a freedom of movement which multiplies possible force configurations and favours success' (Brialmont 1895: xxix).

The Russo-Japanese War gave an insight into the devastating effects of the new firepower on the aggressor. And yet its lessons were ambiguous: the victor, Japan, had seized the initiative from the beginning to pre-empt its opponents, and achieved success before Russia was ready. Instead of cautioning about the new strength of any defensive position well-equipped with firepower, it thus also furnished arguments for those favouring the offensive at all costs. As W.H. James wrote, '[t]he offensive, moreover, carries the war into the enemy's country, raises the enthusiasm of the troops and saves the nation, which takes it, from the evils of war in its own land ... The offensive, then, is the resource of the strong' (James 1904: 37, 39). He reflected on Clausewitz's theory of the diminishing returns of the offensive. As the offensive army advances, it has to detail troops to garrison the conquered area, leading to a diminution of its own strength. He conceded that one needed large numbers and good organisation to start an offensive.

> But given these, there can be no doubt of its advantage. The moral gain is great, the soldier feels he is superior to his adversary when led with determination against him; and this mental attitude leads more than half-way along the road to victory ... Whatever the advantages of defensive strategy may be, no sane being would propose to exemplify them by allowing the enemy to invade this country before defeating him. (James 1904: 40f.)

James knew that this was not merely a British perspective, but that it was shared by 'every Continental nation'. He predicted the race to mobilise and pre-empt an attack from the other side: every party, he thought prophetically,

> seeks by the most careful organisation in time of peace to ensure that, in the event of a conflict with a neighbour, it may be able to gather its forces with the utmost rapidity on the frontier and anticipate the assembly of the hostile armies. The object then of the first battles would be to render a further offensive possible, and so enable the victor to continue the war in the territories of his defeated opponent ... the quicker a nation can put its troops on a war footing, the greater will be its chances of success, other things being equal. (James 1904: 42f.)

To confirm that this tenet was just as firmly entrenched by then in the British military mind as it was in France or Germany, it suffices to quote General C.E. Callwell, the great British guru of small war, who writing in 1899 contrasted it to the prevailing offensive doctrine. He explained well why it was so firmly believed that the offensive was necessary, and had to be implemented pre-emptively, as soon as possible upon the outset of war:

> In regular warfare between two nations possessing highly organized military systems, this taking the lead at the outset is an object which both seek to obtain. Each has its plan of campaign. The one that gets the start can thereby dislocate the whole scheme of operations which has been elaborated, in theory, by the other. (Callwell 1896/1906: 72)

Pre-emption was thus at the centre of the cult of the offensive.

Writing with hindsight in 1916, the French General Bonnal noted: 'General Bernhardi, and all those who have studied war, have claimed that the offensive has all virtues' but also 'the genius of the German people'. What they had not foreseen, however, was the stalemate that came into being in the First World War resulting from each side applying the same overall Strategy (Bonnal 1916: 1–5). By then, Émile Mayer, standing for many, had fully recognised the defensive power of the new weaponry: 'The most important transformations in military technology have augmented the armies' capacity for defence' (Mayer 1916: 83f.).

Total mobilisation or professional military elites?

Dreaming about the perfect state, the perfect polity, its perfect internal order and its perfect external behaviour, Guibert, writing his *Essai général de tactique* in the late 1760s, made an impassioned plea for a citizen army (a militia) to replace any paid professional armed forces. The advantages he saw as manifold: a lower cost of defence resulting from an (unpaid) duty for each male citizen to defend his polity, while still pursuing his normal job in civilian life. If attacked, however, such a citizen army would fight with vigour unmatched by mercenary armies who were not defending their own land, family or prosperity. Citizens would be defending their own cause, not that of a feeble government or a dynasty. Such a reform would make France resistant against all attacks from the outside and prosperous and strong internally (Guibert 1772/1781).

Such an army would be defensive: if would pose no threat to its neighbours, as it would only fight if attacked. Guibert likened such an army, once provoked to fight, to a fierce northerly wind bending the slender reeds that stood up against it. Nothing could stop such an armed nation before it had completely subdued all its enemies (Guibert 1772/1781: 134–51).

Guibert's ideas contained contradictions. Like so many before him, Guibert took up the ancient refrain of one's own nation's decadence and excess of luxury and the need to restore its ancient simplicity, austerity, hardiness and warlike spirit. But a warlike culture tends to become a warring culture, and it is difficult to extol martial virtues without succumbing to the temptation to practise them on one's neighbours if the opportunity presents itself. On the one hand, Guibert extolled a citizen army and a nation in arms as not posing a threat to its neighbours and thus becoming the building-block of a peaceful world which ideally should consist of nothing but states similarly configured. On the other hand, he dreamt of more decisive battles than those he had witnessed, of a war that would truly change the political situation, and the 'fierce wind from the north' he imagined might carry 'a fixed plan of *agrandissement*', expansion, which would then result in the subjugation of all peoples around (Guibert 1772/1781: 137).

Guibert changed his views on the composition of armies as he grew older. As the French Revolution was breaking out, in a treatise called *On Public Force*, he extolled the virtues of a professional army, which alone could be sent to fight far away from its home in defence of the state's interests (Guibert 1790). He still favoured a militia that would ensure the defence of its country if attacked, but it should be framed – *encadré* – by a professional army. The contradiction between the warlike, austere spirit which he wished for the entire nation in his earlier work and a peaceful society with a spirit of law-abidance and conflict-avoidance had dawned on him. He now began to see armed civilians, and above all men whose profession is war, as a potential threat to a peaceful society. In his late work, therefore, he postulated that professional soldiers should be denied the citizen rights of voting for a parliament or standing for elections, and the militia should decidedly not have the right to bear arms in peacetime (Guibert 1790: 567–638).

It was Guibert's earlier ideal of a peaceful and defensive army of citizen-soldiers lived on: in February 1793, the French Revolutionary Isnard, in order to justify the *levée* of 300,000 men, exclaimed 'May the victory of the Revolution be the victory of universal liberty, but also the victory of universal peace!' (q. i. Jaurès 1911: 133). Guibert came to be regarded as prophet of the *levée en masse*, proclaimed in August 1793, to defend France against the invading forces of Habsburg emperor Francis II and other supporters of his kinsman by marriage, Louis XVI. The rural populations of northern and eastern France rose up to fight the invading coalition forces, with men and women, old and young, resisting alongside the regular forces. The myth of the *levée en masse* was born (Heuser 2009).

While the Revolutionary National Assembly in 1789 had abolished compulsory military service, it was soon brought back. The 'nation in arms', already encountered in the writings of Machiavelli and Vattel, now became French. Every (male) citizen, pronounced General Edmond Dubois-Crancé (1747–1814), War Minister of the Directorate in 1799, must be a soldier, a view that was enshrined in Article 109 of the French Constitution of 1791. As Deputy Barère said on 23 August 1793: 'France's military contingent, fighting for her freedom, comprises her entire population, all her industry, all her building works, all her engineering works' (Serman and Bertaud 1998: 79). The *levée en masse* thus already contained the concept of total mobilisation of one's citizenry, but also one's industrial and economic resources, characteristics of 'Total War', as we have seen.

Professional armies, conscripts and militias

Napoleon followed Guibert's line of argument and stated his belief that a nation in arms – 'a nation defended by the people' – would be 'invincible'. An emperor, he also postulated, should rely on 'national soldiers', not upon mercenaries – a maxim he himself soon abandoned in practice (Napoleon 1906: 158, 178). Nevertheless, the six-figure armies for which he managed to recruit the men throughout Europe swept through the Continent from west to east, bringing down one regime after another through devastating victories in battle. To counter Napoleon's armies, Scharnhorst and his protégé Clausewitz grudgingly favoured the establishment of a militia to back up a regular professional army, recruited purely locally – this fitted nascent German nationalism (Usczeck and Gudzent 1986: 219–25).

Other monarchists in Europe saw things differently. Thus Archduke Charles had a decided preference for professional forces. 'If the soldier's physical adroitness alone was sufficient, then there would be nothing better than militias [*Bürgerwachen*].' But in his view, and despite his experience with the French Revolutionary forces, the fighting of militias was nothing but 'puppet play' compared with that of professional soldiers (Waldtstätten 1882: 82). Gerhard Scharnhorst's friend and fellow Hanoverian Johann Friedrich von Decken also argued for a standing (professional) army, as he believed war to be an endemic problem, requiring the need for the best possible armed forces (Decken 1800: 68). What he had in mind were warriors like the Scots, who since the Jacobite defeat at and exile after Culloden in 1746 could be found in mercenary armies throughout Europe (Decken 1800: 11, 15). To fight with an all-volunteer force: (1) society must be differentiated into different occupations; (2) the theatre of war must be nearby; (3) war must be over quickly; (4) armies must not be too large – he thought it impossible to maintain an all volunteer army of 100 000; (5) weapons must be easy to learn to use and handle (Decken 1800: 31f., 34). While sophisticated cultures were capable of sophisticated ways of waging war, they also tended to decline – here again the old *topos* of decadence (Decken 1800: 39, 49). His conditions for a state to delegate responsibility for defence to a part of the population were:

1. these soldiers must be particularly loyal and self-sacrificing;
2. they must not be employed for other tasks, such as in agriculture;
3. neighbouring states must not have standing armies;
4. there must be no fortifications that have to be defended permanently;
5. there must be no fighting far away, no colonial ambitions;
6. they must not be employed in wars of aggression (Decken 1800: 40).

Decken realised that there was an argument for a voluntary force, a militia serving to defend their own country. He feared, however, that such a militia would of necessity become forced to fight and would lose the will to do so, if their concerns about what was happening at home began to outweigh their commitment to the war. Free men, he thought, were more likely to go to war if they had slaves tending to their fields at home. As a nobleman, he connected this with the ideal of a warrior class, while simultaneously expressing concerns about the potential consequences of arming the poor. Nor did he want to follow Guibert's advice and have two forces, a militia for homeland defence and a professional force for expeditions abroad (Decken 1800: 84).

Decken was aware of criticism that could be levelled against standing armies, for example by Edmund Burke, who had told the British parliament that all standing armies could become dangerous to the constitution, or Kant's plea for the gradual dissolution of all armies or the danger of Caesarism with the example of Oliver Cromwell in mind, who had commanded more loyalty among his troops than did the king. Had not Montesquieu warned that the soldier's duty of obedience was incompatible with his republican rights as a citizens? Would he not therefore hate all those who had those rights? And yet Decken preferred to hedge and prepare against such dangers, for instance by rotating officers through command posts frequently to prevent their troops from becoming too attached to them, in order to have the advantages of a standing professional army (Decken 1800: 92, 118–20, 122, 134).

These advantages according to Decken included above all the 'taming' of war. Standing armies he thought had a stabilising effect because the adversary knew what he was up against: this is a deterrence argument. A regularly paid [i.e. standing] army would not plunder, and thus war would affect civilians less (Decken 1800: 134–40). As Decken followed Montesquieu's and the mature Guibert's reasoning that a soldier, who had to obey, could not be a citizen with citizen's rights, he could not see militias as compatible with republics either (Decken 1800: xii, 44, 64, 81, 118–20).

In contrast with this aristocratic plea for a professional army, the 'nation in arms' in France became an ideal associated not only with nationalist movements, but also with Republicanism. Jomini had written:

> National wars ... are the most formidable of all. This name can only be applied to such as are waged against a united people, or a great majority of them, filled with a noble ardour and determined to sustain their

independence: then every step is disputed, the army holds only its camp-ground, its supplies can only be obtained at the point of the sword, and its convoys are everywhere threatened or captured. This spectacle of the spontaneous uprising of a nation is rarely seen, and though there be in it something grand and noble which commands our admiration, the consequences are so terrible that, for the sake of humanity, we ought to hope never to see it. (Jomini 1837/1868: 29)

For several reasons, this idea was difficult to accept for the elites of the countries under French occupation. In Prussia, Scharnhorst's and Clausewitz's plans for resisting the Napoleonic conquest were the mobilisation of the entire adult male population as a militia in a people's war. But was this a temporary measure, or did they indeed want to establish armed forces based on general conscription in peacetime too? The commission instituted by King Frederick William III of Prussia in 1807 produced a majority report, written by General Scharnhorst, and a minority report by Colonel Boguslawski. Scharnhorst's majority report proposed the immediate introduction of conscription, in other words universal military service for all Prussian men, explicitly on the model of the law of conscription of the French Revolution. It counselled the immediate abolition of the interdiction on the bearing of arms by civilians, which had been imposed to stop illegal hunting. It also counselled a better treatment of all soldiers so that they might take pride in the defence of their country.

Boguslawski, by contrast, pleaded with the king not to expect townsfolk and soft noblemen to be as hardy and well prepared for the hardships of military life as peasant boys, and therefore to allow for the French practice, introduced soon after conscription, of *remplacement*: paying for somebody else to do your military service for you. Further, Boguslawski saw no point in introducing universal conscription as long as Prussia was not directly under attack, or if the state's finances and political circumstances allowed this, which, by implication, he doubted (Miszellen 1889). The Prussian king, worried about this attempt to 'turn everything into soldiers', put the report to one side for a while, and at the end of a year asked for a new one, the one which finally produced the military reforms of 1808, in which Scharnhorst's egalitarian ambitions were toned down. Nevertheless, the king had it proclaimed on 3 August 1808 that henceforth every male subject of his, of whatever birth, was liable to military service (Lehmann 1922: 436–57).

But how much could one expect from a conscript army? Like the mature Guibert, Scharnhorst, one of the chief Prussian reformers together with General Count August Wilhelm Neidhardt von Gneisenau (1760–1831), never ceased to believe that a conscript force had to be complemented by and integrated with a professional force, which was to be promoted on the basis of merit, not of seniority or descent (Wollstein 1978: 325–52; Greiselis 1999: 102–17). The problem that Guibert had recognised in part was that the spirit of such an army was different from that of a militia, in which the nation had a stake in its own defence. Guibert had felt concern about their assertiveness and their martial, uncivic spirit. He did not fully articulate what this meant, but we can sense the mature Guibert's fear of the armed masses.

There was also another side to this issue. If men were defending their nation, their status had to rise, and less emphasis could be put on their subject-status vis-à-vis the monarch. For the Prussian reformers, both elements were still quite compatible, while they abhorred French concepts of democracy or people's sovereignty: Karl vom Stein saw 'Germany's nobility and culture firmly and indissolubly chained to the happiness of the Prussian monarchy' (Haussherr 1960: 271). Scharnhorst argued against the continued practice of corporal punishment in such conscription-based armed forces (Demeter 1960: 226). There was more to the issue of a citizen army than just the need to treat citizen–soldiers better in the armed forces. The biggest issue of all was how to deny them what for the French Revolution was the rationale at the heart of a citizen army: namely, that the citizenry, the nation, was the new sovereign, and that the citizen-soldiers defended their own rule in a democracy. The reactionary societies who aimed to free themselves not only from the Napoleonic occupation but also from everything that France stood for – most notably, democracy – hated this as much as they hated the French. No way would Frederick William III of Prussia allow himself to be turned into a figurehead of a constitutional monarchy of a quasi-democracy, giving his people the vote. It needed considerable persuasion on the part of Gneisenau to get him to appeal to 'my people' to initiate the Prussian resistance in 1812, and even the reformers were nationalists, but not democrats. Indeed, Clausewitz, the most famous disciple and protégé of both

Scharnhorst and Gneisenau, distrusted the people, whom he saw as the source of 'the primordial violence of [the] nature [of war], the hatred and enmity which must be regarded as a blind natural instinct' (Clausewitz 1832/1976, I: 1, 28), and he hated and feared the French model of democracy. Instead – under the influence of his half-British wife – he liked the oligarchic system of Britain. Other reformers equally cited the British militia system as a desirable model (Lehmann 1922: 438f.).

Feelings were no different among the Spanish elite. They were happy to have various partisans fighting for Spain against the French, but the *juntas* – the royally appointed or self-appointed provisional governments in different regions – early on tried everything to regain complete control over the self-organised rebels, and to integrate them into the regular armed forces, lest they become in every way too independent from their control (see Part VI). Democracy, like French secularism and anti-clericalism, was anathema also in Spain.

Whether one was afraid of them or not, the nations – numerically larger than ever before – were now a factor in warfare, as warfare had become a national business. Mass warfare and the nation in arms went hand in hand, one conditioning the other.

In Prussia, the issue of conscription versus professional army remained central in the nineteenth century. Classically for a nationalist, Colonel Blume, writing in the 1880s, thought war beneficial only if the people themselves fought for their state; by contrast he thought a war fought by mercenaries extremely noxious for the body politic (Blume 1882: 8f.). Bernhardi spoke for all when he thought that mass armies would be the most significant feature of the next war – 'armies of millions ... the like of which have not been seen before in history' (Bernhardi 1912b: 62). On this point there was agreement between the left and the right in Germany (Bebel 1892).

To simplify, from the French Revolution to the end of the twentieth century, two antagonistic tendencies in French politics engaged in a tug of war over the nature of French armed forces: one was the more conservative, reactionary, royalist tendency, which wanted a professional army (or at least a system whereby rich individuals could buy themselves out of military service) and feared the population, and the other a more republican, egalitarian tendency, in the tradition of the young Guibert and Carnot, which demanded universal conscription and was deeply suspicious of the professional army, a tendency that was embodied in the revolutions of 1830 and 1848. The prevailing view of the military chiefs was that in order to have a good soldier, it was indispensable that he should serve for a long time and become thoroughly disciplined, and thoroughly different from a civilian. Conservatives accused the Republicans of wanting to weaken the professional French army through universal conscription, depriving it of its military spirit. The result was, as it was remarked, that 'The army is a nation in the nation, that is the vice of our time ... It asks itself incessantly whether it is slave or ruler of the state; this body searches everywhere for its spirit and cannot find it' (Montheilhet 1932: 10). The French middle and upper classes and important members of the generality, in the words of General Antoine Chanzy (1823–83), saw in universal military service the 'danger of arming the nation' (q.i. Montheilhet 1932: 13). The professional army that had been created by the restoration survived without much change the revolutions of 1830 and 1848. For in France, the army had two functions, an external one and an internal one: to restore order there in all the unrest which had troubled France since the French Revolution. As a contemporary parliamentarian, Francisque Bouvet, commented in 1851, 'the armed forces are not so much designed for international war, but to serve as an instrument of power domestically and to underpin ... strong governments' (q.i. Montheilhet 1932: 31).

The Republicans continually criticized this. In 1848 the Republican-minded General Christophe de Lamoricière demanded

> that the Army should be constituted in the image of the nation, that it is animated by its spirit, its thinking. Against obligatory military service, the argument is made that France is not used to so much rigour, our values are not Republican enough to comply with [conscription], we are not Spartans. Well, I say: alas for France, if today she has recognised that the Republic is her only anchor of salvation, but she still does not have enough faith in the value of the institutions that she has created for herself [i.e. the Republic] to dare hope that they will function according to these values and will harmonize them with the conditions of its new existence! Alas for her, for she can only found an indestructible Republic once she has eradicated the old monarchist values. (q.i. Montheilhet 1932: 27)

Another Republican, Etienne Vacherot, in his book on *Democracy* argued that:

Democracy does not go along with a real [professional] army, however national-minded it may be. It goes along with the nation in arms. If this system is more appropriate to defence than to conquest, this is a happy constraint in which civilization should rejoice. With such military institutions [universal military service], a government is not tempted to become involved in adventures by the ambition of sovereigns or the temptations of the military spirit. If an external war is recognised as just and necessary, it will be an entire nation that will rise to wage it ... [With a militia], a people is invincible on its own territory; and, if justice and civilization oblige it to carry its forces abroad, it becomes irresistible through the strength of its national *élan*. The military strength of standing armies is more brilliant but more fragile, quicker at achieving startling victories and great conquests, but more subject to irreparable defeats, and above all more feeble in the face of invasions. A people without militia forgets the military arts and finds itself reduced to relying only on its professional army in supreme danger. Once this army is destroyed by a superior force, such a people without discipline and military initiative will suffer foreign invasion ... A military people, on the other hand, is an irrepressible force; there is no attacking army of which it could not get the better. (Vacherot 1860: 314f.)

The military operations in which France was involved between 1815 and 1870 did not involve any vital interests. With strange echoes of the Prussian Boguslawski half a century earlier, the French politician Adolphe Thiers echoed Guibert's differentiation between homeland defence by militias and the need for professional armies for action far away from home, when he argued that in the face of large-scale aggression, 'the nation in arms may be adequate; but, when we have to fight what I shall call political wars [i.e. limited wars], in which enthusiasm does not play the principal part, such an army would be inadequate' (q.i. Montheilhet 1932: 35).

When the Prussians won over the Austrians at the battle of Sadowa/Königgrätz in 1866, Emperor Napoleon III tried to reintroduce conscription, as Prussia was now seen as a threat to France. Republicans in his country formed a curious alliance with the emperor to bring about a return to Revolutionary concepts of the military, and to recreate a nation in arms. But there was too much opposition: all proposals were watered down into inefficiency in the law of 1868. The politician Adolphe Thiers summarised the majority view that:

The peasant, put into the ranks of the army, finds there conditions that are better than at home ... But military service is an intolerable tyranny for the man who is destined to a civilian career ... The middle-class people who have a taste for the military go to the military schools.

Thiers repeated his views almost literally in 1868 and 1872: 'The society in which everybody is a soldier is a barbarian society ... In the countries in which everybody is a soldier, everybody is badly off ...', and obligatory military 'service would inflame all heads and would put a gun on the shoulder of every Socialist' (q.i. Montheilhet 1932: 30, 65). After France's defeat of 1870/1, Thiers became the first President of the French Third Republic. Although conscription was reintroduced, the Third Republic continued with the old system of buying oneself out.

French society in the restoration period, as in the Second Republic and the Second Empire, was thus a thoroughly non-militarized one. Its dominant classes had a strong aversion to all things martial. As Prosper Mérimée wrote:

By preaching that money is the sovereign good, one has profoundly changed the belligerent sentiments of France, I won't say among the people in general, but among the higher classes. The idea of risking one's life has become very repugnant, and those who regard themselves as honest folk would describe it as base and crude. (q.i. Montheilhet 1932: 47)

Incarnating the Republican tradition of the French Revolution, and also imbued with the defensive spirit of Guibert's militia, the French socialist leader Jaurès was impressed by the Swiss model of reserves, which offered an example of total mobilisation for defence that could turn into an irresistible offensive power after an initial defensive phase. Such a defensive force could choose and prepare the battlefield to confront the invader, and could confront him with the 'armed nation' itself (Jaurès 1911: 95, 163–95). Jaurès therefore pleaded for the reliance on such a reserve, well trained and well equipped: including the twelve classes of reserves, France might thus muster 2 million men, plus territorial forces. With these she would confront the 900,000 attacking first-line German soldiers, according to calculations. France should be able to deploy her forces in a way that suited her and would be well placed for a Napoleon-style victory over the Germans (Jaurès 1911: 104f.). We find echoes of Guibert's *Essai* (Jaurès 1911: 122):

If we want for France a defensive strategy, it is because we want for her a defensive policy, a policy of peace and equity ... A people that, wanting peace, gives proof of this to the world ... which until the eve of war proposes the litigation to the arbitration of civilised humanity, a people which, even when the storm has broken still asks humankind to settle the conflict, such a people has such an awareness of being in the right that it will be prepared for all sacrifices to save its honour and its life. It is determined to put up an untameable and prolonged resistance. By contrast, in a nation that has been thrown into war by a movement of pride and robbery, malaise will grow from hour to hour. (Jaurès 1911: 106)

In order to restructure France's armed forces for an optimal defensive posture, Jaurès turned to the revival of militias, a term which was scorned at the time. Von der Goltz, for example, opposed militias, and Moltke had derided France's armies (Jaurès 1911: 109). But for France, Jaurès argued, mass armies had their roots in the Revolution, with Carnot and the law of 1 August 1792 (Jaurès 1911: 116). Jaurès consciously followed this tradition:

It is thus absolutely necessary that, in the free countries, every citizen should be a soldier, or nobody should be one. But France, encircled with ambitious and warlike nations, obviously cannot make do without an army; it is thus necessary, to take up the expression of Jean-Jacques Rousseau, that every citizen should be soldier by duty, and no one by profession. (Jaurès 1911: 117)

By the time Jaurès was writing, consensus in France had shifted after the stunning defeat of 1870/1. While Jaurès's defensive approach, which we will examine in detail in the following chapter, was anathema to the majority, they agreed with him on the need for a mass army to compete with the numerically superior Germans. One advocate of a conscript army in France, General Lewal, had emphasised the nexus between the need for mass armies and industrialisation, which was coming to dominate all spheres of social activity (Lewal 1892: 87).

As a consequence of technological innovation, armies now could occupy larger spaces, and thus larger armies were called for, as J.F. Maurice observed in Britain before the Great War (Maurice 1891: 14–18). General Maude also underscored the importance of mass. The new stress on land warfare in such large numbers was a departure from British traditions, yet Maude thought armies had much more to offer than navies:

> The land forces of this Empire exist for the purpose of compelling any possible enemy or enemies to submit to our will. The fleet alone cannot accomplish this object, because 'ironclads cannot climb hills' [as the Ottoman Sultan commented on the occasion of a British naval demonstration at Ducigno 1880]. Hence we need an organisation capable of very great expansion, and regulations – tactical, logistical, etc. – capable of enabling the Army, this organisation, to meet any European army or part of it in its own country on equal terms. (Maude 1905: 125)

In the late nineteenth century, Colmar von der Goltz was alone in thinking that in a more distant future there might be

> the coming of a time when the armed millions of the present will have played out their part. A new Alexander will arise who, with a small body of well-equipped and skilled warriors, will drive the impotent hordes before him, when, in their eagerness to multiply, they shall have overstepped all proper bounds, have lost internal cohesion, and, like the greenbanner army of China, have become transformed into a numberless, but effete, host of Philistines.

But he agreed with the others that war in the near future would be conducted by mass armies, and his writing is based on that assumption (Goltz 1883: 5).

Again the odd one out, Sir Julian Corbett criticised his contemporaries' obsession with mass armies. They had this

> idea of making war not merely with a professional standing army, but with the whole armed nation – a conception which of course was not really Napoleon's. It was but a revival of the universal practice which obtained in the barbaric stages of social development, and which every civilisation in turn had abandoned as economically unsound and subversive of specialisation in citizenship. (Corbett 1911: 20–2)

Corbett pointed to the mistaken notion that it was 'the armed forces of the enemy and not his territory' that was one's main objective – mistaken, as the victory over an army did not mean that the enemy state was ready to surrender (Corbett 1911: 20–2).

Following Corbett, after the First World War, it was Basil Liddell who became the principal British critic of conscription and mass armies, in his view a crass fallacy which had proved so disastrous for the millions of young men killed and maimed in this war. He explained:

The theory of human mass dominated the military mind from Waterloo to the World War. This monster was the child of the French Revolution by Napoleon ... The wave of democracy in the 1840s fostered the growth of this mass-theory. The idea of the nation in arms appealed to the democratic mind. Other minds, also short-sighted, were quick to take advantage of it. Equality of service was a convenient substitute for the reality [of democracy]. The victories of the Prussians in 1866 and 1870 were regarded by the world as confirming the truth of this theory.

Prescriptions for mass warfare, he realised, had become

fixed on the mind of Europe. Proclaimed by soldiers everywhere, not least in vanquished France, as indisputable truths, they were submissively accepted by a generation of statesmen dangerously ignorant of war. The ... threefold consequences were: to make war more difficult to avoid, more difficult to conduct successfully, and more difficult to terminate save by sheer exhaustion. (Liddell Hart 1944: 33, 133)

He mused after the First World War that once mobilised, conscript mass armies of Germany, France and Russia had made it hard if not impossible to call off a war. 'Moreover, these armies were so cumbrous, their movements so complex, that even direction could not be modified' (Liddell Hart 1944: 18). At the time, in words reminiscent of Thiers's, J.F.C. Fuller also explicitly linked conscription to the world's return 'to tribal barbarism', caused by the French and Industrial Revolutions (Fuller 1961: 33).

Having been raised in the Clemenceau tradition of seeing his own country as world policeman, Charles de Gaulle wanted to see a world order which would only be completed after the founding of the United Nations after the following Great War. At the Geneva Disarmament Conference in the early 1930s France had proposed the creation of an international police force composed of contingents from countries willing to contribute them. 'And of what could this force be composed, except of professional soldiers? One cannot imagine Governments calling upon conscripts and reserves to go and separate Japan and China ... or to eject the racist militias from Austria or the Saar ... So the professional soldier becomes the necessary guarantee of all great human hopes' (Gaulle 1934/1940: 71). De Gaulle's conclusion was that 'the tendencies of the world, the conditions of an international organization of peace, at all events our own duty of assisting the weak and maintaining order in the Empire, all combine to compel us to create professional troops' (Gaulle 1934/1940: 74).

Writing in the early 1930s, de Gaulle obviously saw Germany as France's main adversary, past and future, although until 'recently' Germany had been limited in arms and weaker than France. Nevertheless, unlike most French strategists of 1871–1914, and like Basil Liddell Hart, de Gaulle had had a horror of mass armies before the Second World War, as the wars fought with them – Moltke's wars, the American Civil War, the First World War – had been so terrible. If evolution moved away from mass armies, 'it would be a priceless boon for the human race'. While he thought war an inevitable part of human existence, 'its horrors depend, very largely, on the dimensions one gives to it. On the whole, no form of battle is more sanguinary than that of nations-in-arms' (Gaulle 1934/1940: 76f.). As for France, with a numerically smaller population compared with Germany, only mechanization could make 'Destiny ... once more ... serve the fortunes of France' (Gaulle 1934/1940: 83). After the Second World War, de Gaulle would change his views on this – he would embrace conscription as something French society needed as a socially integrating, nation-building factor. Even then, he would not entrust France's vital defence to a professional army. The French sanctuary would be guarded, instead, by the magic of nuclear weapons (Heuser 1998a: ch. 2).

Others who put their faith in new technology also saw the option of moving from mass army to a small professional force, for example the US military air pioneer General William (Billy) Mitchell, who thought it 'probable that future wars again will be conducted by a special class, the air force, as it was by the armoured knights in the Middle Ages' (Mitchell 1925: 19).

Nevertheless, mass armies fought the First World War, and conscript mass armies fought the Second World War. After 1945, even the Korean War, France's Indochina War and America's Vietnam War were fought with conscript soldiers (see chapter 18).

Fear of the 'excited masses'

Guibert had foreseen and, after the French Revolutionary and Napoleonic Wars, Jomini and Clausewitz had recognised the enormous force of populations emotionally and physically mobilised for a war. 'If England has proved that money will procure soldiers and auxiliaries, France has proved that love of country and honour are equally productive', as Jomini wrote (Jomini 1837/1868: 51). The conservative writers of the second half of the nineteenth century found this daunting: there was a widespread fear among the French right and the more affluent parts of French society that social inferiors might rise against their masters as the French had done in 1789, 1830 and 1848, and again 1871. With the great exception of Jaurès, the French military writers considered here stood in awe of the rabble and potential uprisings, which they felt the strong need to suppress (Faivre 1985: 63–91).

In France the military establishment continued to be a nation within the nation, dominated by particularly Conservative, reactionary, authoritarian sentiments, and they were fairly anti-intellectual. Caesarism – the rise of popular military leaders to the top – was rampant. In 1873, after Adolphe Thiers, General Marie Edmé MacMahon was elected second President of the Third Republic; he attempted to rule against the National Assembly and was forced to resign in 1879. In 1889, Georges Boulanger, Minister of War, who was strongly supported by the army, was forced to go into exile at the eleventh hour by Republican elements which had uncovered his plans for a putsch.

The Dreyfus affair exposed the differences between the spirit of the Republic and the spirit of the army and showed the enormous polarization in French society between the republican, egalitarian forces (who continued to want a nation in arms) and the Conservative reactionary forces, particularly in the army, who did not share republican values but were anti-Semitic and allied to the Catholic Church. The two sides feared each other more than they feared the external enemy. Again, the armed forces and their Strategy were deeply influenced not by the external threat but by the internal dynamics of French society, and the two deeply conflicting sets of values which divided French society so profoundly.

It was only in the last years before the First World War that the French army was really integrated into the nation, and that conscription really began to effect the population in general. Even going into the war, the conservative part of French society was still fearful of an insurrection of the left against the state. In fact the opposite was produced by this most nationalist of all wars: the 'sacred union' (*union sacrée*). The army, with its universal military service, was the quintessential expression of this 'sacred union' (Gorce 1963: 131).

In Britain and Germany also, by the second half of the nineteenth century, the 'have-nots' were beginning to get organised, causing the conservative 'haves' nightmares. They seemed to fear their own lower-class compatriots more than the enemy across the frontiers. Meanwhile, the post-Dreyfussard *Kulturkampf* in France made the left hesitate until about the time of the Second Moroccan Crisis (1911) to decide that the Germans were more hateful than the repressive French armed forces (Porch 2006).

On the one hand, then, were those military officers and writers in France, Germany and Britain who feared the physical weakness, the lack of moral fibre and the pacifism of (at least large segments of) their own populations. On the other were those who saw the masses as a factor escalating violence, whether as potential social revolutionaries, or as a blood-thirsty rabble who would not accept an armistice until an enemy was utterly defeated. Writing after the Franco-Prussian War, which had seen the most extensive mobilisation of conscripts in German-speaking states yet, Moltke the Elder saw 'the introduction of general military service' as important in integrating 'the educated classes of society into the armies. Admittedly, the rougher and more brutal elements remained, but they are no longer the sole components of the military.' He nevertheless continued to fear 'popular passions pushing for war'. He cautioned: 'We must not allow the inner quality of the army to be weakened, otherwise we will end up with militias. Wars conducted by militias characteristically last much longer, and for that reason demand more sacrifices of money and lives than all other wars.' He listed the French 'experiments' with militias of 1792 and in the Franco-Prussian War, as well as the American Civil War, as 'horrors' which 'nobody would wish to transplant onto European soil'.

An armed crowd is far from being an army, and it is barbarian to lead them into combat ... savagery and cruelty inevitably accompany the arming of the people ... What happens when the government loses the reins of power, when power is assumed by the masses, is demonstrated by the history of the Paris *commune.*

Moltke's was thus a passionate plea for lengthy military service to turn even conscripts into disciplined, drilled professionals through an intensive education (Großer Generalstab 1911: 2–11). Von der Goltz equally thought that 'the excited masses are more prone to clamour for war than Cabinets' (Goltz 1895/1899: 3). Nevertheless, he was convinced, using very Guibertian–Clausewitzian terminology, that 'The day of Cabinet wars is over ... Wars have become solely the concern of the nations engaged' (Goltz 1883/1906: 9).

Such feelings were not unique to the German upper classes. Lt.-Col. W.H. James also emphasised the role of popular sentiment:

> In democratic nations the people are supposed to be the arbiters of their own destinies, but they are, as history shows us, often swept along by a sudden wave of sentiment which may hurry them into a position where war becomes unavoidable. Countries in which public opinion has great power are much more liable to be thus affected than those in which the Government is in the hand of a few individuals placed above such influence. (James 1904: 11)

He had little faith in the military policies of democracies:

> The more autocratically a nation is governed the more consistent its policy will usually be. An autocratic ruler, although not independent of his environment, is less affected by it, and is more likely to influence the current of politics by impressing his will on his statesmen and even on his successors. (James 1904: 12)

Spenser Wilkinson was fascinated by the writings of von der Goltz, who 'expresses the conviction, *shared by all contemporary military thinkers*, that the identification of Governments with nations has made permanent what Clausewitz called the absolute type of war' (Wilkinson 1910: 181; my emphasis). Wilkinson therefore pleaded with his countrymen to cease thinking of future war as limited, and to think of throwing the nation's entire resources into it instead.

To sum up, writers on warfare with a republican, democratic frame of mind tended to favour militias, all the way up to Jaurès writing in early twentieth-century France and the spiritual fathers of the West German Bundeswehr. Monarchists, or others favouring oligarchic forms of state and governance, preferred professional armies, distrusting the (potentially revolutionary) masses. Nationalists of all hues liked mass armies of conscripts. Only the British had the geographic luxury of not having to think about such a major contingency until 1914.

Mass armies were indeed the forces that would clash in the First World War. And to some extent, the fears of the conservative strategists were vindicated. After several classes of soldiers allowed themselves to be slaughtered, there were uprisings in 1917/18 not only in Russia, but also in Germany and France. While the socialist revolution was contained in Western and Central Europe and limited to Russia, the tensions between left and right, between social classes, the haves and have-nots, continued to haunt those who thought about Strategy. But these were not the only traumas brought on by the war that was sold to the public as the Last of the Last (*La Der des Ders*), and yet became the first of *two* World Wars, indeed the first of two *Total* wars.

It brought together the strands of *Total War* discussed at the beginning of this chapter: the well-nigh total mobilisation of the populations of the countries at war, and the annihilation of army after army on the Western Front. Moreover there was the genocidal annihilation of the Armenians carried out by the Ottoman Turks. Again, these strands fused in the Second World War, with its societies put on war footing, its mobilisation of civilians and industry, its air attacks on cities, its battles of annihilation at Stalingrad and Kursk and, finally, the largest genocide in war that the world has seen inflicted on Russians and Jews by the Germans.

7 Challenges to the Napoleonic paradigm versus the culmination of Total War

Do you want Total War? If necessary, do you want it more total and more radical than we can imagine it today?

(Goebbels's speech at the Palace of Sports, Berlin, 18 February 1943)

Mars mechanised: the Napoleonic paradigm versus technological innovation

We have seen that Americans on both sides of the Civil War rendered homage to the Napoleonic paradigm, which they knew through Jomini's writings; the Prussians developed their own version of it, ostensibly through their selective reading of Clausewitz but mainly under the increasingly 'realist' military leadership of Moltke; the French felt the need to rediscover Napoleonic warfare after their defeat at the hands of the German coalition in 1870/1 which they attributed to the Prussian officers' studies of Napoleon, again through the medium of Clausewitz. Just as Renaissance men had felt the need to resurrect the wisdom of the ancients and to model their strategies on those of great generals of Antiquity, the emulation of the Napoleonic way of war became the lodestar of strategy from *c.* 1860 to 1918, as we have seen in the previous chapter.

Nevertheless, there were strong voices who doubted that the Napoleonic paradigm could be applied without significant modifications, pointing to the great technological innovations affecting warfare in this period. Unlike in early modern history, these innovations were both fast in coming and their significance was not denied; however, what the consequences would be was open to debate. In addition to belief in Social Darwinism and biology, and in the right of might, there was the changing factor of technology. The Wars of German Unification and the American Civil War especially had seen revolutionary changes there. Both used railways, the telegraph and industrial mass-production of weapons, ammunition and other equipment, and both harnessed mass armies to a popular cause (Degler 1997: 53–74), which made them 'industrialised people's wars' (Förster and Nagler 1997: 9). Liddell Hart gave a succinct summary of the 'revolution in warfare' caused by the new technology from the introduction of the railways and the interwar period (he was writing in 1934):

> [The evolution of war since the mid eighteenth] century has been marked by four main trends. First, was the growth of size. From France under the Revolution and Napoleon, to America in the Civil War and Prussia under Moltke, the armies swelled to the millions of 1914–1918. Second, came the growth of fire-power, beginning with the adoption of rifles and breech-loading weapons. This, imposed on size, conduced [*sic*] to a growing paralysis of warfare on land and sea. Third was the growth of industrialisation ... And fourth was the revolutionary growth of mobility, due in turn to the steam engine and the motor.

This revolution of technology had 'mechanised Mars' (Liddell Hart 1944: 24f., 69).

Completely unlike in the previous centuries, technology became a hotly debated subject in the mid nineteenth century, and remains one. Not only in debates on naval/maritime Strategy, but also those on land warfare, and later air power, was there from now on always a '*matériel*' school which focused on this aspect at the expense of all others, and we can see a continuity here lasting to the works of the ardent admirers of the 'revolution in military affairs' at the end of the twentieth century. But if so much had changed in technology, was the French Revolution still the great watershed in view of the technological changes in the late nineteenth century (Echevarria 2002a: 199–214, 2002b: 84–91)?

It is significant that the ideology-driven Bernhardi, while writing at length on the importance of new technology from railways to firepower, still dwelt on the importance of cavalry (Bernhardi 1912b: 190). By contrast in France, General Derrécagaix wrote that after the German victory 1870/1:

> We have understood that since the day when the Prussian example has forced the European powers to transform their military institutions, and to arm all their able-bodied men, since the perfecting of firearms, since the development of the railway network and electrical communications,

the rules of the military art have undergone changes which every day are affirming themselves more, and which put armies in a true period of transition. (Derrécagaix 1885, I: 1)

At the very end of the nineteenth century, the German General Sigismund von Schlichting started his work on tactical and strategic principles of the present by echoing Clausewitz (*On War*, III.8), 'times have changed, and ... war has changed profoundly and is inspired by very different forces' from those which obtained during the wars of Frederick II and Napoleon (Schlichting 1897: 9).

In Germany, Britain and France there were strategic theorists who bravely contended that the next war would have to be short as no long war could be sustained because of the lethality of the new weapons or because of the speed with which campaigns would now be carried out (Cairns 1953: 282f.; Farrar 1972: 39–52). A few well-informed individuals, however, usually in the armed forces, predicted rightly that it would be long and drawn-out. Even the German military leadership did not really believe in the possibility of a short war, as Stig Förster has shown; instead, they embarked on a desperate gamble, at best hoping to be able to seize the initiative and in a sharp, short *coup de main* against France to give Germany a good position from which to win a longer war of attrition (Förster 1995: 61–95; Showalter 2000: 679–710; Herwig 2002: 681–93). While General Joffre in France also proclaimed that the war would be short, Douglas Porch shows that military opinion in France doubted that the war could go on for long because they had little faith in the power of resistance of their own nation – 'undermined' and 'weakened' morally by the disease of pacifism (Porch 2006: 121).

The pre-First World War debate about the length of the next war reads much like the post-First World War debate about the effects of aerial bombing on any future war (see chapter 12). Von der Goltz, for example, thought that the stronger the weapons, the more fear they caused ('deterrent effect'), the sooner the battle would be over, and is 'generally ... less bloody in proportion as the engines of destruction have attained greater perfection'. A single modern shrapnel bomb could kill ten to twenty people, with consequently great moral effect; therefore, he extrapolated, war should be over more quickly.

In spite of ... the enhanced effect of the weapons of war, no battle of modern times has produced such carnage as did those of Eylau or Borodino.

> But most bloody of all were the battles of ancient times, in which the attack was made with a club or the short Roman sword.

Nowadays, battles would simply be more rapidly decided, war concluded faster; 'that is certainly an end to be desired, since, war, in these modern times, by displaying itself in its natural and violent form, convulses all creation' (Goltz 1883/1906: 13).

In the United Kingdom, in 1891, Colonel Esdaile had argued that the new weapons had brought on 'an absolute revolution in all our present systems of tactics and strategy' and that the new balance between offensive and defence inclined towards the defence. But in the same year, Spenser Wilkinson pronounced this to be 'erroneous', arguing that there had been no revolution on the battlefield, only 'progress. Weapons produce their old effects at longer distances and more speedily. But the effects are not new in character' (Wilkinson 1910: 156f., 162). Looking backwards rather than forwards, he took a leaf from the writing of Moltke the Elder, who in his 1865 work *Remarks upon the Influence of Improved Firearms upon Fighting* had said:

> Generally speaking, the consequence of the peculiarities of the improved firearms will be to strengthen the defence as against the attack. The defender can choose his position in such a way that the enemy must advance over an open plain ... The attacker has, from the fact of attacking, certain evident advantages, which he will always retain. (q.i. Wilkinson 1910: 163)

These included the choice of the object, and the way by which to reach it. Spenser Wilkinson agreed with Moltke that

> in future the defence must seek the open plain, the attack, broken ground ... The less the chance of success for a frontal attack, the more surely will the enemy turn against our flanks, and the more important it becomes to secure them ... The smaller the force and the shorter its front, the easier it is to turn its defensive position. (Wilkinson 1910: 156f.)

Flanking movements would be necessary, even in what would seem to be a frontal assault. Little did Spenser Wilkinson foresee that in this desperate attempt to outflank the enemy, the mass armies on the Western Front would spread themselves out halfway across Europe, along a front entirely lined by trenches.

Others did foresee this 'crisis of fortifications'. In France, General Hippolyte Langlois (1839–1912) was very sceptical of the value of the fortifications which France constructed after the 1870/1 war, with Belfort, Épinal, Toul and Verdun in the front line, but also fortresses further inland along inroads. He thought they would hamper France's flexibility (Langlois 1906: 15–36). He argued that after assuring the defence of the French territory against an initial German assault, France would have to prepare the offensive into enemy territory (Langlois 1906: 83–6).

Notwithstanding the enormous innovations in military technology, Corbett rightly noted that his contemporaries continued to be obsessed with Napoleonic warfare, which they saw as the only way, and 'the bare suggestion that there may be other ways' was branded by them as heresy (Corbett 1911: 20–2). Liddell Hart, writing in the mid-1920s, found this still to be the dominant paradigm. 'The textbook principle of aiming "to destroy the enemy's main forces" came to be visualized in the [First] World War as a destruction of the enemy's troops, rather than the disruption of his military organization.' This was reflected in

> the memoirs, and letters, of those who directed our strategy – with the constant emphasis upon 'killing Germans'. Compared with this purpose there was small concentration of thought upon the far deeper effects of paralysing the enemy's brain, physically and psychologically: his nerve systems and arteries. (Liddell Hart 1944: 57)

While consensus on the Napoleonic paradigm prevailed, the historical school more generally, that is, those authors stressing continuity in warfare and harking back to historical examples as inspirations for their Strategy – respectively, Hannibal's successful indirect approach at Cannae and Napoleon's division of the enemy's forces at Austerlitz (Groote 1990: 33–55) – did not go unchallenged in the light of new technology. Von der Goltz commented:

> Although the leading principles of warfare are said to be eternal, yet the phenomena which have to be dealt and reckoned with are liable to continuous change. War, as an act of human intercourse, is, in its eternal form, subject to all the same changes which affect the latter. Railways and telegraphs, which show new ways to trade, have also opened hitherto closed paths to military science ... Military precepts are thus continually changing

in their application, and it may rightly be said, that every age has its own peculiar mode of warfare. Hence, methods which, in 1870, led to triumph, cannot now be regarded as the absolute standard for the future; new conditions, ushered in by the present, compel us to devise fresh ways and means. (Goltz 1883/1906: 1)

Yet his prescriptions for the following war were the application of the Napoleonic paradigm with massively added firepower; and few writers on Strategy had as much influence on the thinking of their time – both inside and outside their own country – as von der Goltz and Bernhardi.

The dissenters: Corbett's limited wars and Jaurès's defensive army

Before the First World War, challenges to the Napoleonic paradigm came from other quarters, too, even though in this atmosphere of rabid nationalism and militarism it was difficult to dissent. It took not only exceptional intelligence and character, but also originality of thought to do so, and it helped if one was not brainwashed by military service. Thus it was two highly educated civilians who were the main dissenters, Sir Julian Corbett and Jean Jaurès.

Unlike almost all his contemporary strategic thinkers, Corbett thought that limited war was by no means dead. He was well aware of opinion around him. He distinguished 'between the German or Continental School of Strategy and the British or Maritime School – that is, our own traditional School, which too many writers both at home and abroad quietly assume to have no existence'. The latter school's views, in his opinion, was familiar with a distinction between what Clausewitz called unlimited and limited war, a distinction also found in Jomini (1837/1868: 330–1), who had written: 'There are two different kinds [of war], one which may be called territorial or geographical ... the other on the contrary consists exclusively in the destruction or disorganisation of the enemy's forces without concerning yourself with geographical points of any kind.' The first Jomini had called 'offensive wars to assert rights', such as Frederick's conquest of Silesia. By and large, this coincided with Clausewitz's own idea of limited wars, which were seen as very different from the unlimited aims of Napoleon. As Corbett remarked and most of

Jomini's disciples chose to ignore, Jomini had criticised Napoleon for his unlimited aims in his campaign against Russia in 1812 – with a limited territorial aim Napoleon might have got out of it again (Corbett 1911: 46).

Corbett pointed out that at the time of Clausewitz, the German states' territories were scattered all over central Europe, but by the time Corbett was writing, consolidated states had formed with contiguous territory. When Clausewitz 'conceived the idea, the only kind of limited object he had in his mind was ... "some conquests on the frontiers of the enemy's country", such as Silesia and Saxony for Frederick the Great, Belgium in his own war plan, and Alsace-Lorraine in that of Moltke'. Corbett observed that this would be unworkable in reality, when nations would see these provinces as 'organic parts' of their country, and when little would stop a nation from stepping up its efforts to defend them. In his view, Clausewitz's concept, while no longer applicable to wars between nationalist states in Europe, might still apply to at least one side in a colonial conflict overseas.

German analysts agreed with Corbett only on the unfeasibility of limited war in Europe. The rise of nationalism with all its symbolism of unity of nation, territory and state outdated Clausewitz's idea that war might take the form of the conquest of a hamlet for the mere purpose of acquiring a negotiating chip for diplomacy (Hintze 1920–1: 131–77). A nation, especially if ethnically or racially defined, could not allow any of its members to fall (or remain) under foreign domination – a creed which turned the ethnic patchwork of Central and Eastern Europe into a field of landmines, all of which duly exploded in the First and Second World Wars.

As we shall see in chapters 9 and 10, Corbett would be either admired or hated by his contemporaries; he had many more unorthodox views for which he was finally disowned by the British Admiralty, which arguably speeded up his death (Grove, Introduction to Corbett 1911/1988).

On the French side, almost the only voice crying in the wilderness and arguing against the cult of the offensive was that of Jean Jaurès, France's leading socialist. A philosopher and historian by training, Jaurès significantly broached strategic questions coming from the classic French *topos* of composition and recruitment of the armed force. Jaurès in particular challenged Captain George Gilbert's and General Langlois's call for a revival of Napoleonic war, as Jaurès thought this had been a 'mortal period ... for the military institutions of France and for [our] national defence'. As we have seen Jaurès had understood the significance of the ultimate defeat of Napoleon, but he was in a small minority; most of his compatriots still saw Napoleon as France's genius. Napoleon, Jaurès argued, had gone for vainglory, causing destruction and sterility, when French Revolutionary warfare contained so many more promises and possibilities that could have been to the benefit of mankind more generally (Jaurès 1911: 53–5). If anything, he argued that one should rehabilitate the warfare of Henri de Turenne, Louis XIV's general, who fought mainly limited, defensive wars against the Holy Roman Empire's leading generals. This was of course the sort of limited, defensive warfare despised by Jaurès's contemporaries (Jaurès 1911: 57f., 61f.). At the same time Jaurès was sceptical of all historical models and 'servile imitation'. Moreover, he pointed out a contradiction: if those like Captain Gilbert cautioned against 'Prusso-mania' among French militaries, that is, an admiration of the Prussian military because of its victory against France in 1870, this Prussian style was actually a derivative of Napoleon's way of war, and Bismarck's imperial policies had more than a little in common with Napoleon's: neither was founded primarily on the rule of law and justice, but on crude force (Jaurès 1911: 61f.).

Jaurès then went on to draw attention to Clausewitz's conclusion from the Napoleonic Wars that *defence*, not the offensive, was the stronger form of war. By then Clausewitz was greatly, albeit grudgingly, admired in France as a Napoleon analyst and as the intellectual vehicle carrying his secret of success – 'the rapid, concentrated and bold offensive' – from France to Prussia (Jaurès 1911: 75).

Like Mao three decades later, the Socialist Jaurès was particularly inspired by Clausewitz's recipe of for defensive 'people's war'. This, he argued, should be the inspiration for France (Jaurès 1911: 76f.).

> For if tomorrow, war breaks out between France and Germany ... Germany will certainly resort to the offensive ... [S]he will brusquely invade French territory and ... will seek to strike at the main French forces, only just assembled, with one of those formidable strikes which will knock out the adversary or leave him at least so shaken that even in a prolonged fight, he will not be able to recover full combat strength and the élan of victory. (Jaurès 1911: 81)

Jaurès drew on the writings of French Captain Louis Nathaniel Rossel (1844–71), who like Jaurès had advocated a defensive Strategy for France with her demographic inferiority to Germany, and who died in the siege of Paris (Jaurès 1911: 86). Properly organised to defend France, 'total application of her forces' could lead to success. 'By contrast, France risked losing everything by launching herself into a premature offensive, uncovering her own hinterland' (Jaurès 1911: 89–92).

There is a parallel between Jaurès's thinking and the much less developed plea, made almost twenty years earlier, by a German Socialist, August Bebel (1840–1913), for an egalitarian system of conscription for his country. Like Jaurès, Bebel was immune to the militarism of his age, and in 1892 incurred the anger of his fellow parliamentarians in Germany's powerless Reichstag when he pleaded for a return to France of Alsace and Lorraine, as he saw Germany's annexation of these two French provinces as the prime cause for Europe's insecurity. Bebel deliberately quoted the French republican tradition of conscription as a model for Germany, which again did not endear him to the rabid nationalists by whom he was surrounded (Bebel 1892). Bebel and Jaurès disagreed famously on some interpretations of Marxism, but both, with their Socialist internationalism, sought in vain to stem the tide of bellicosity which led to the deluge of 1914–18. Jaurès would be one of its first victims: on 31 July 1914, he was assassinated for opposing the brewing war.

Lessons of the First World War

What united all participants of the Great War on the Western Front after this war was the utter determination never again to wage a war in the way in which this one had been waged, at least between 1914–17. Even those who were convinced that 'the normal condition of mankind is not peace but war', and that Britain, at any rate, before 1914 had benefited from war, after 1918 thought that 'a new era had begun in which civilization, as a whole, cannot benefit, but must suffer, from warfare between its integral parts' (Groves 1934: 30f.). As the British army officer J.F.C. Fuller – no shrinking violet he – noted:

I can understand the Quaker spirit, the Crusader spirit, and the Mercenary spirit. I can understand a man detesting war, glorying in war, looking upon it as a good business. But I cannot understand anyone wishing to repeat the last war'. (Fuller 1928: v)

The strategic and tactical stalemate, the slaughter at the Western Front, the feeling that the politicians and indeed the population back home had betrayed the soldiers were universal, and all strategists drawing lessons from this war emphasised that next time round everything had to be done differently. The next war had to be brought to a conclusion quickly, and any – *any* – means, any new technology that might make this possible, was considered. The subliminal hatred for one's own government and civilian population which had accepted the suffering of millions in the trenches and even retrospectively did not understand what the survivors had gone through was usually sublimated in a call for carrying the next war far into the territory of the enemy, to make the (enemy) population, and not just the soldiers at the front, feel the scourge of war.

Indeed, the general cultural reaction especially in Britain, France and Belgium was a turn away from jingoism and towards pacificism (Martin Ceadle), that is, the proclivity always to give negotiations and peace a greater chance than militarism and war. Charles de Gaulle described the change of mentality in his book of 1932:

Everything in the climate of opinion generated by the ... peace combines to disturb the mind of the professional soldier. The masses, after having been exposed for so long to the horrors of violence, violently react against them. A sort of *mystique* spreads rapidly which not only calls down curses upon war in general, but leads men to believe that it is an outmoded activity, for no better reason than that they want it to be so, and this fervour breeds its own form of exorcism. The world is noisy with the condemnation of battle, murder and sudden death. To inspire a sense of guilt, the visual arts are widely employed to make men familiar with the ravages of war. A veil is drawn over the achievements and the heroism of those who did the fighting. No longer is that sense of glory evoked in which, throughout the centuries, nations have found consolation for their sufferings, but only the memory of blood and tears and death. History is distorted so that the battles of the past shall be forgotten, and the profession of arms attacked root and branch. (Gaulle 1932/1960: 12)

To paraphrase Oswald Spengler, the First World War was the traumatic Myth of the Twentieth Century for France and Britain. France lost one in six of her mobilised men, dead or missing in the war. In addition, over 3 million Frenchmen were wounded, of whom over a million remained disabled for life. Germany lost 1.8 million dead or missing, with 4.2 million wounded. Britain lost just under a million dead or missing, with over 2 million wounded. Austria-Hungary lost 1.2 million with 3.6 million wounded; Russia lost 1.7 million, the largest figure after Germany, with the greatest number of wounded, just short of 5 million (Doughty 1985: 72).

The war had seen far-reaching mobilisation, but was still far from realising the ideal of the French National Assembly of 1793 or of total war in terms of mobilisation. A fifth of the total population of France had been involved in the war effort, that is, in the armed forces or military industry. In Germany it was 18%, in Italy 15%, in Austria-Hungary 14%, in Britain 13% and in Russia 10%. But all sides drew the lesson that in any future war, it had to be less a matter of 'business as usual', and the civilian economy and population had to be mobilised more totally still.

Strategy responses to the First World War

The German military historian Hans Delbrück made himself very unpopular with the establishment and the Reichswehr by criticising the German military commanders of the First World War. Looking at Falkenhayn's Strategy in his attack on Verdun, he commented that it

> was no attempt to break through; it was not a battle; it did not aim to bring about a great tactical decision. Had we finally taken Verdun, it would of course have been of great importance for morale, but in Falkenhayn's view, such a success was not entirely necessary. The point of this enterprise ... was the exploitation of the advantage of our ... positions, in order to inflict on the enemy much greater losses than we were suffering ourselves. The aim was not to beat the French, but to bleed them to death. As great as the pain was that they had to suffer, they had to hold Verdun, as ... its surrender would have been an unbearable loss of prestige ... This bleeding [the enemy] is a form of ... exhaustion-strategy. It is neither merely passive waiting, nor manoeuvring without bloodshed. (Delbrück 1920b: 49)

As we have seen in chapter 3, Delbrück had developed an analogous dualist interpretation of warfare through the ages, claiming particularly that Frederick the Great had pursued a Strategy of attrition, not of decisive battle. This was tantamount to an attack on Germany's sacred national symbols. In the age of the cult of the offensive, nationalist Prussians and Germans found it impossible to accept such a verdict on 'their' Frederick, and a heated debate ensued (Kromayer 1925a: 394). Reinhold Koser was among Delbrück's angry critics. Misrepresenting the 'exhaustion strategists' as those who refused to give battle altogether, he could haughtily claim that this could not possible apply to Frederick. Koser rightly noted that the Strategy of 'crushing' the adversary varied as a function of the different ages and cultures in which it was applied. The princes and generals of the *ancien régime* may have aimed to crush the enemy's armies, but not overthrow a regime or a state with all its structures as Napoleon would. At best, a prince of the *ancien régime* wanted to replace another prince as ruler of a state, but social or political revolution was not on the agenda. With the rise of the nation(alist) states of the nineteenth century, however, the aim of crushing the enemy nation (and not merely its armies, or even revolutionising its social and state structures) was conceived and spread throughout Europe (Koser 1904).

Delbrück would achieve more lasting fame abroad. His interpretational pattern was taken up by the twentieth-century American historian Russell Weigley, who adopted it in his famous classic on *The American Way of War*. As a weak and, in terms of manpower and ships, initially small state, the young USA, according to Weigley, had to adopt a Strategy of attrition or exhaustion of the adversary in any conflict. As the USA's population and economic and especially industrial power grew, it could transform its way of war into war with the aim of the annihilation of the adversary's armed forces in battle (very much Napoleon-style), a Strategy it practised from 1865 until 1945 and favours to this day (Weigley 1976). Critics of Weigley have shown that while this might have been the preferred Strategy, necessity forced the USA to choose deterrence or indeed piecemeal operations more akin to attrition (as defined by Delbrück and Weigley) on several occasions between 1865 and 1941, and especially after 1945 (Linn 2002). Nevertheless, the Delbrückian distinction proved a useful analytical tool.

While Delbrück himself did not really put forward an alternative account of how to deal with warfare in the context of all this new technology, other German critics of the conduct of the First World War concentrated on individual military campaigns, especially the Ludendorff offensive of 1917/18. This was found deficient in the light of Napoleonic imperative of concentrating forces on one point, as opposed to the broad offensive along the whole front from Ypres to Reims (Mayr 1925). This led to the call for a revival of manoeuvre warfare, and the development of the *Blitzkrieg* operations of the Wehrmacht, practised so successfully in the early stages of the Second World War.

French Strategy: total defence

The First World War purged France of her militarism and any adulation of war stemming from previous centuries. French Strategy became entirely defensive. Jaurès, first French victim of the Great War, was vindicated posthumously. General André Beaufre commented forty years later: '[T]he victors of 1918 lost interest in strategy because they had been taught, not Strategy as such, but *a* strategy which was held up as the be-all and end-all of the art. This particular strategy' – the pursuit of annihilation battle – 'proved false. The idol [Strategy] was therefore torn down' (Beaufre 1963/1965: 13).

In concrete terms, France had not thought about the consequences of the loss of her north-eastern frontier regions, which produced three-quarters of the total iron ore of France, and between two-thirds and three-quarters of her coal. But these regions were lost at the beginning of the war in 1914, opening the eyes of the French leadership to the need to defend the area where so much of France's economic resources were. The French learned from this how vulnerable and crucial at once their eastern industry was, whence the emphasis on the need for better defences in a future war. Hence their opting for a predominantly defensive posture, with the construction of the Maginot Line, promoted in particular by the Generals Maxime Weygand and Maurice Gamelin (Doughty 1985: 126f.). Originally plans had foreseen the movement into Belgium almost to the border between Belgium and Germany. But Belgium's declaration of neutrality in 1939 rendered this plan politically impossible. Plan D, operational from 1933–5, would have meant going forward to the Namur–Dyle River–Antwerp line ('Dyle Plan'). There was also a Plan E or Escaut Plan, to move forward to Tournai, the Escaut River, and then Antwerp (Doughty 1985: 66). With the Maginot Line in place, it was hoped that French forces could concentrate on the defence of the north alone. But sitting behind the Maginot Line, unable to pass through neutral Belgium because a democracy abiding by international law did not do such a thing, France was quite unable to take any offensive action, come to the aid of Czechoslovakia or Poland or even take effective action when Hitler remilitarised the Rhineland in 1936. France's static defences were at odds with what Jaurès had called for, and proved unfortunate.

While the lessons the Germans had drawn from their experience in the East and the West during the First World War led them to insist on the need for mobility and thus the tank, the generals and marshals of the right in France, including Pétain himself, but also Generals Weygand, Debeney and Gamelin, were sceptical of the tank and the aeroplane. Instead, their lesson from the First World War was that firepower was the most important acquisition to make, and that well-fortified trenches and defence systems could not be overrun, if they were protected with ample firepower.

There was a minority view, based on movement, pre-First World War Plan XVII, which advocated mass movement forwards. Foch and Joffre were in favour of this immediately after the First World War, and later de Gaulle took an interest in the mobility provided by tank warfare. But among the French leadership, they were in the minority.

The indirect approach I: manoeuvre warfare

The distrust and ignorance of technology of many military leaders had proved fatal for millions in the First World War. Germany and Russia drew the lesson that mobility had to be the key to future operations. Both started planning for mobile warfare, what would later become famous as German '*Blitzkrieg*' operations. There were no great theorists of this outside the armed forces. The brains behind this manoeuvre warfare committed little of their thinking to published writing, or if so, merely retrospectively.

The commander-in-chief of the American Expeditionary Forces (AEF) in the First World War, General John Pershing, had espoused an all-out Strategy of going for the annihilation of the adversary's armed forces, confirming Russell Weigley's model of an American way of war. But the reality was the incremental build-up of US forces in Europe as a result of delays caused by training and transport by sea imposed even on the AEF campaigns, which historians critical of Weigley's interpretation have characterised as attritional rather than decisive (Grotelueschen 2000; Linn 2002: 521–3).

The USA initially returned to the preferred 'American way of war', with the *Field Service Regulations* of 1923 fully embracing again the annihilation of the enemy's armed forces in battle as the unlimited objective of any war (Linn 2002: 525). By 1934, in contrast, General Douglas MacArthur as army Chief of Staff described future war as being fought by small, highly mobile elites. '[T]he American defense system envisions gradual rather than a simultaneous commitment of forces to action whenever an emergency may arise.' The entire young male population of the USA would, if necessary, be mobilised for a larger-scale conflict (Linn 2002: 527). Innovation in the US military was stymied in the interwar years, however; David E. Johnson concludes that there was a lack of innovative writers of the calibre of a Fuller, a Liddell Hart, a Charles de Gaulle or a Guderian, and the prevailing US mentality was apathetic to military pursuits, resisting the expenditure that would have been needed to effect any major transformations of the sort that MacArthur proposed. By the outbreak of the Second World War, the US military was thus neither on the ground nor in the air transformed into the decisive weapon MacArthur or Billy Mitchell had dreamed of (see chapter 12), and both became once again, as in the previous war, above all instruments of attrition, precisely what the few innovators had hoped to avoid (Johnson 1998: 218, 225).

The need to revive manoeuvre warfare was the main lesson drawn from the trench warfare that had characterised the First World War on the Western Front. In a future war, units should advance rapidly to defeat enemy armies partly by the surprise of timing, partly by the surprise of the angle of attack, or by encirclement, cutting them off from their supplies or other units. The argument has been made that the long-range cavalry raids into enemy territory which formed an important part of the Confederacy's Strategy in the American Civil War were seen as a model by European strategists of the interwar period. But they had no need to look across the Atlantic for models (Dwyer 1999). Here, too, the Napoleonic Wars and the Franco-Prussian War provided model enough, and the revival of manoeuvre warfare can be seen as yet another tribute paid to the Napoleonic paradigm.

Arguments for speed and surprise were anything but new (Holleindre forthcoming). Because of the perennial nature of war, von der Decken had argued during the Napoleonic Wars that pre-emptive war had to be an option: he wanted a standing army capable of attacking first, at short notice, preferably to fight one's own defensive war on the enemy's territory. Underlying was a balance of power argument: even a smaller country might have to rise against a bigger one that was growing too powerful and threatening (Decken 1800: 173f., 181). Surprise was of course one of the agreed 'principles of war' of the field manuals of the early twentieth century. But one strategist took the concept to new dimensions, and this was Captain Basil Liddell Hart with his book *Strategy: The Indirect Approach*, frequently reprinted after it was first published as *Decisive Wars of History* in 1929, arguably Liddell Hart's greatest work (Danchev 1999: 314). By his own definition, Liddell Hart's 'strategy of indirect approach' was 'the highest and widest fulfilment of the principle of surprise' (Liddell Hart 1944: 238). With this binary concept, Liddell Hart had much in common with Delbrück's binary theory of strategies of attrition and strategies of direct confrontation, and Liddell Hart's would become as famous. Writing at the beginning of the 1930s, Liddell Hart like Delbrück before him criticised the frontal assaults of the First World War:

> Foch's idea of surprise was, as he said, guided by 'mechanics' and consisted 'in applying superior forces at one point', the reserve being 'hurled as one block'. This conception was built on a fallacy; it was already undermined by the improvement of weapons to which Foch and his contemporaries gave so little heed. The theory of the Greek phalanx, with its reliance on mass, is nullified by the machine-gun. The more ranks, the more swathes of dead – that is all. In face of this hard reality, the mechanistic theory of surprise broke down in the [First] World War. (Liddell Hart 1944: 204)

Foreshadowing the thinking of the American theorists of the 1960s and 1970s, he wrote in 1926, 'The real target in war is the mind of the enemy command, not the bodies of his troops. If we operate against his troops it is fundamentally for the effect that action will produce on the mind and will of the commander' (Liddell Hart 1944: 48). At the outset of the Second World War, Liddell Hart contrasted the indirect approach with the doctrine of mass: 'Although war is contrary to reason, since it is a means of deciding issues by force when discussion fails to produce an agreed solution, the conduct of war must be controlled by reason if its object is to be fulfilled' (Liddell Hart 1944: 177).

Liddell Hart claimed to have derived the concept of the 'Strategy of Indirect Approach' from his historical studies in the 1920s, summing it up as follows:

> [T]he fact [has] emerged that a direct approach to the object or objective along the 'line of natural expectation' has ever tended to negative results. The reason being that the strength of an enemy country or force lies far less in its numbers or resources than in its stability or equilibrium – of control, morale and supply ... To move along the 'line of natural expectation' is to consolidate the enemy's equilibrium. And by stiffening it to augment its resisting power ... In contrast, the decisive victories in military history have come from the strategy of indirect approach, wherein the dislocation of the enemy's moral, mental or material balance is the vital prelude to an attempt at his overthrow. (Liddell Hart 1944: 238)

Liddell Hart's emphasis lay on manoeuvre warfare, and he later claimed to have influenced the German Wehrmacht to invent '*Blitzkrieg*' – a myth created by Liddell Hart himself that has since been discredited (Searle 1998). Another strategist interested in the possibilities offered by new technology was J.F.C. Fuller, whose works concerned mainly mechanisation, admittedly with the possibilities in manoeuvre warfare which this entailed (Fuller 1920).

In France, too, there were a few strategic thinkers who were very keen on a mobile way of war. There was an early precedent in France for interest in manoeuvre. One of the concerns in France before the First World War had been that because France was a republic, its mobilisation would take longer than in Germany, a monarchy, where the emperor could simply order it. The horror scenario on the French side had been that of a brusque German advance which might at best be slowed down by the forces deployed along the frontier. Langlois had hoped that France could compensate for the numeric superiority of the Germans by an 'essentially French quality: subtlety', plus mobility, *vitesse*, initiative, 'in short, manoeuvre' (Langlois 1906: 1–13).

As noted above, the young Charles de Gaulle was among this unfashionable interwar minority who saw the tank as bringing an evolution to warfare which, 'insofar as one owes it to mechanization, gives back to quality, as opposed to quantity, the importance which it had at one time lost' (Gaulle 1934/1940: 59). De Gaulle thought it was a 'technical necessity which is driving the military system towards a professional army'. But there was also another factor, namely, that

> the exclusive system of the nation-in-arms is only suitable to conflicts in which the stake is unlimited. In order to justify a call to arms of all men capable of bearing them, the death of millions, the loss of vast riches and the social and moral confusion which are the characteristics of mass warfare, we must first of all have violent quarrels, the class of frenzied hatred and ambition, and threatened bondage. That, at any rate, was the state of mind of the nations of Europe before 1914, convinced that war would offer them the alternatives of death or victory.

But writing his book just before Hitler came to power, de Gaulle thought that 'the conditions from which "total war" [*sic*] grew', with an obvious albeit implicit reference to the First World War, were 'gradually disappearing and making way for others. There are good reasons for believing that a war starting tomorrow would only be remotely connected, at the beginning, with the premature attack of mobilized masses.' Like Jan Bloch before him, he even mistakenly believed that 'the ubiquity of wealth, the overlapping of interests and the infiltration of ideas has created among the nations an interdependence which compels them to limit their [military-expansionist] ambitions' (Gaulle 1934/1940: 63–5).

The indirect approach II: target the population

The alternative to harnessing new technology to one's cause was the targeting of the enemy's weakest spot, described variously as his 'soft underbelly', or a 'centre of gravity' of a different sort from that of the enemy's armed forces that had previously been seen as such. In 1903 Prince Louis of Battenberg, director of naval intelligence in the United Kingdom, speaking to the Royal Commission on Food Supply, said how awful it would be if any of the other powers would use a naval blockade against the United Kingdom: 'there is the larger question of humanity. You cannot condemn forty millions [Britons] to starvation on the ground that they assist in defending their country, because you include women and children.' By the same token, it became clear to the British military leaders that Britain could turn the tables on the continental powers that might actually be as vulnerable to a naval blockade as the United Kingdom. Moral reservations were laid aside, and by 1908 high-ranking naval officers wanted the naval blockade to be adopted as a Strategy option. It was not adopted until well into the First World War, as it was ineffective as long as Germany could obtain supplies via Belgium or the Netherlands. The assurance of the latter's neutrality had of course been the crucial reason for the truncation of the Schlieffen Plan by Moltke the Younger (Gooch 1994: 292f.).

Bernhardi understood the new possibilities offered by naval blockade: 'The course of events at sea may mean starvation for the population' (Bernhardi 1912b: 16). Sir Julian Corbett was in favour of blockading the Germans:

> In order to increase the pressure on the enemy and to strike at him financially it is necessary to undertake secondary operations against his trade. We must be careful [however] to keep this in its proper place, and to remember that it is not the primary object of the war. (q.i. Haggie 1985: 122)

The blockade of Germany in the First World War did not bring the war to an end any earlier, but it led to increased civilian mortality, especially among the very young and very old and the sick.

After the First World War, the Berlin history professor Otto Hintze, the bitter adversary of Hans Delbrück, pleaded for an 'indirect' [*sic*] approach of aiming to exhaust the enemy in a future war. These included tiring out his armed forces through manoeuvres, forcing them to undertake long marches without giving battle, or through making their upkeep excessively expensive by drawing out a war, or through ruining his economy in other ways, such as a continental blockade, or by cutting off food supplies to his population through a blockade, or by exploiting the social and political tensions within the enemy nation to foster revolutionary tendencies or civil wars. Hintze identified such measures in the continental blockade of Napoleon, or in the Strategy of Moltke in the second phase of the Franco-Prussian War. Strategy, he argued, is a mixture of these tendencies, but the mix shows infinite variation in its proportions. Any military commander would prefer to crush the adversary in a decisive annihilation battle, but this opportunity rarely presented itself, and rarely would it decide the war (Hintze 1920–1: 131–77).

Writing in 1925, Liddell Hart had not as yet developed moral revulsion against blockade, but saw other disadvantages.

> First, it can only be successful where the enemy country is not self-supporting, and can be entirely surrounded – or at any rate its supplies from outside effectively intercepted. Second, it is slow to take effect, and so imposes a strain on the resources of the blockading country.

In his outrage over the dying and suffering of the soldiers in the First World War, he for a while was a fervent advocate of turning any war into 'a duel between two nations', with all the consequences this implied, namely the explicit targeting of civilians (Liddell Hart 1944: 49, 52).

For 'terror' was the weapon, other than manoeuvre and static defences, that strategists often named that might change the future of war. The thought that 'terror' might lead to a quicker end to a war, and that its sum total might thus be less horrifying in terms of human suffering than a long-drawn-out war was not altogether new, nor can the possibility be denied that the advocates of such 'terror' campaigns had some naïf belief in this logic. Even in 1569, Thomas Churchyard, a pamphleteer who accompanied the English general Sir Humphrey Gilbert on his campaign to subdue Munster in Ireland, wrote in defence of the atrocities that Gilbert and his men committed: 'through the terror which the people conceived thereby, it made short wars' (q.i. Selesky 1994: 61). Three hundred years later, the American Lieber Code, standing at the beginning of the era of establishing multilaterally agreed laws of war, stated, 'the more vigorously wars are pursued, the better it is for humanity. Sharp wars are brief' (q.i. Biddle 1994: 147). Optimists even in the late nineteenth century thought that while military technology was making war bloodier, it would become shorter and more decisive and might even lead to long-term peace and the abandonment of war. This included men like Hiram Maxim, inventor of the Maxim gun, who thought that making war 'appalling to contemplate' would make 'nations pause'. He thought that 'men who are good students of human nature', among whom he clearly included himself, had to realise 'that the best way to preserve peace is to make war as terrible as possible – terrible in its toll of blood and money, terrible in its widespread ravages, and terrible in its uncertainty' (Budiansky 2003: 10). In 1899 an international movement tried to ban hot-air balloons as precursors of aircraft that might be used for bombardment. At the time, the American delegate Captain William Crozier voiced the following considerations:

> Who can say that such an invention will not be of a kind to make its use possible at a critical point on the field of battle, at a critical moment of the conflict, under conditions so defined and concentrated that it would decide the victory ... localizing at important points the destruction of life and property, and ... sparing the sufferings of all who are not at the precise spot where the result is decided? Such use tends to diminish the evils of war. (Budiansky 2003: 11)

After the First World War this desperate belief became widespread. To get away from the bloodshed of the trenches, J.F.C. Fuller advocated the use of 'terror', always in the hope that it would terminate a future war more quickly, with less overall bloodshed. The destruction of the enemy's possessions might not be the best way to proceed: 'A nation which destroys the economic resources of its enemy, destroys its eventual markets, and thus wounds itself.' He hoped that the principle of destruction could be replaced by this new principle of imposing one's 'will at the least possible general loss'. In order to achieve this,

> the means of warfare must be changed, for the present means are means of killing, means of blood; they must be replaced by terrifying means, means of mind. The present implements of war must be scrapped and these bloody tools must be replaced by weapons the moral effect of which is so terrific that a nation attacked by them will lose its mental balance and will compel its government to accept the hostile policy without further demur. (Fuller 1923: 28–30)

The third indirect approach, then, would be the use of air power against the enemy's industrial centres and populations (see chapter 13). Writing in the interwar period, General F.B. Maurice reflected the debate on the indirect approach. If the war aim was 'the restoration of peace on satisfactory terms' and if this was 'achieved by overcoming the opponent's will to continue the struggle', how was this to be done? '[T]here has never been any question until recently, that once a state of war has arisen, the way to overcome the opponent's will was to defeat his armed forces.' In the British *Field Service Regulations* it was now 'suggested that it is within the power of aircraft to terrify the civil population of an enemy country to an extent which will overcome the opponent's will even if his armed forces are undefeated' (Maurice 1929: 68).

Total War and genocide: Ludendorff and the National Socialists

Nevertheless, it was a substantial moral and conceptual step away from targeting the population in order to break the enemy's will to resist, to practise genocide. The genocide of the Armenians by the Ottoman Empire during the First World War stood in a long tradition of smaller-scale, mutually inflicted Balkan atrocities. It was the first large-scale forerunner of the two great state-perpetrated genocides or democides: the Holodomor of 1932–3, Stalin's organised famines in the Ukraine, and the Holocaust, the German persecution above all of Jews. To this must be added the radical ill-treatment of the Slavs by the Germans during the Second World War, which resulted in several times as many millions of deaths among Soviet citizens by deliberately engineered starvation, much on the lines of the Holodomor (Davies and Wheatcroft 1994: 57–80).

The key strategic thinker and practitioner who actually advocated such a Strategy is Erich Ludendorff, one of the German supreme commanders in the First World War. He has the dubious fame of having been Adolf Hitler's rival for the leadership of the German National Socialist Workers' Party founded soon after the war; in 1925 he was this party's candidate for the presidency of Weimar Republic Germany. In 1935, frustrated that the little Austrian had beaten him to becoming *Führer* of Germany, and under the influence of his racist wife, he published his *Total War*. His wife, who believed in the application of breeding principles for domesticated animals to human beings, had added a special touch to Ludendorff's views, rooted in the Social Darwinist thinking of the previous century. But in his hands, they were transformed into a book that, next to Hitler's *Mein Kampf*, comes the closest to defining what the Germans would do in the Second World War, namely wage a Total War which added to total mobilisation the dimension of genocide.

Some passages in *Total War* are little different from the writings of those who even before the First World War had advocated the application of naval blockades to get at the enemy's nation as the true centre of enemy gravity, or Douhet or Trenchard (see chapters 9, 12 and 13). But none of the others had gone to the explicit length to which Ludendorff went in spelling out the war aims; nor did the others aim for genocide, as he and the National Socialists did.

He agreed with those such as Bernhardi and Colin that limited wars as Clausewitz had described, the 'cabinet wars' which he had still thought possible, were now utterly outdated, and like some after him at the end of the twentieth century consigned Clausewitz to the waste-paper basket of history. The fighting in the First World War affected great swathes of territory and the civilian populations living there:

> Not only the armies, but the populations themselves are now indirectly subjected to the operations of war ... Nations are now directly involved in a war through blockades and propaganda. Total warfare is thus directed not only against fighting forces, but indirectly also against the nations themselves ... The nature of a Total War postulates that it can be waged only when the existence of the entire nation is actually being threatened, and the latter is really determined to wage such a war. (Ludendorff 1935: 15f.)

War he described as 'the highest test of a nation for the preservation of its existence', and the German nation was being challenged, in his view, just in this way.

> The more nations are regaining their racial consciousness ... the stronger the knowledge is growing of the world-destroying activities of super and international powers, of the Jewish people and the Roman Church ... who are striving for universal power above and over the nations. (Ludendorff 1935: 24)

And he concluded, '[f]or a morally strong people, the war decision lies solely in the victory on the battlefield and in the annihilation of the enemy Army and of the enemy nation' (Ludendorff 1935: 168). These lines would seem like hyperbole, if it were not for the fact that this is precisely the policy that the National Socialists embraced.

The Second World War: culmination of Total War

The National Socialists infamously developed Social Darwinism to its racist extreme. A stunningly obvious but neglected element of continuity between the First World War and the war aims of Hitler and his supporters lies in the applied Strategy and the strategic writing of Ludendorff. Peace, to Ludendorff as to Hitler, and indeed to Lenin and Stalin, was the continuation of war by other means. To Ludendorff in theory and Hitler in practice, Auschwitz was a proper battlefield of this war of annihilation, along with Stalingrad, besieged Leningrad or Kursk. Hitler's government did, consciously, aim at the total annihilation of the Jewish people, and indeed of the enslavement or annihilation of the Slavs. Besides the killing of Europe's Jews in the Holocaust, Hitler's government consciously implemented policies directly or indirectly causing the starvation of Russian civilians. In addition to Red Army losses of 8.7–13 million (Krivosheev 1997: 79) and to other acts of war such as the bombing of Leningrad, the Germans thus deliberately caused the deaths of between 7 and 10 million civilian Soviet citizens, total Soviet losses being estimated at between 25 and 26 million (Davies and Wheatcroft 1994: 58, 78f.).

The Second World War, more so even than its great predecessor, was by every definition Total War. It is irrelevant whether Hitler had hoped to get away with salami-tactics of aggrandisement, or whether the Japanese hoped the Americans would not react as they did to the attack on Pearl Harbor. Both the Germans and the Japanese made as great an effort as is humanly and collectively possible to win a world war which they had partly inadvertently started. Their adversaries had to retaliate in kind. As President Roosevelt told Congress on 11 January 1943:

> We wage total war because our very existence is threatened. Without that supreme effort we cannot hope to retain the freedom and self-respect which give life its value. Total war is a grim reality. It means the dedication of our lives and resources to a single objective: victory. Total war in a democracy is a violent conflict in which everyone must anticipate that both lives and possessions will be assigned to their most effective use in the common effort – the effort for community survival – National survival. In total war we are all soldiers, whether in uniform, overalls, or shirt sleeves. Total war requires nothing less than organizing all the human and material resources of the Nation. (q.i. Hobbs 1979: 61f.)

The Second World War produced genocide, but also the greatest tank battles ever, with the Battle of Kursk an outstanding example, and the greatest sieges ever, with the siege of Leningrad haunting memories still. It produced the apogee of city bombing, with 'conventional' ordnance and the first and so far only use of atomic bombs in war producing firestorms with comparable effects.

The age of Total, genocidal War did not come to an end in 1945. Genocide and democide occurred also in the Cold War. It was arguably practised by Tito in Yugoslavia, by Mao during the Cultural Revolution and by Pol Pot in Cambodia (Rummel 1992). After the end of the Cold War, the Hutus practised it against the Tutsis in Rwanda with more primitive weapons, and on a smaller scale it was carried out by Milošević's followers in Bosnia. Again, we find that the boundaries between different eras cannot be neatly drawn.

Barbarism in warfare

At this point, an excursion into the progressive barbarisation of warfare from the nineteenth to the twentieth century, which contemporary observers commented upon, is necessary. Changes in the practice of warfare have much in common with fashion, and this applies also, perhaps mainly, to their effects and to how these are perceived. The early twentieth-century cultural historian Max von Boehn noted that fashion in clothes is in large part driven by the fact that the novelty of the exposure of a part of the body previously covered seems particularly shocking, no matter how much flesh used to be revealed, and how little now (Boehn 1904). A similar observation can be made about the perception of barbarism in warfare: it was often what was unusual that shocked in many contexts.

No matter how few deaths were caused in total by, say, Native American practices in warfare relative to the settlers' use of shotguns, or by the *francs tireurs* on the French side of the Franco-Prussian War relative to the massive number of casualties caused in battle by the needle-point gun or by gas in the First World War compared with the machine gun, it was scalping, *francs tireurs* and gas that were homed in on as particularly barbaric, not the objectively greater lethality of other weapons and practices. The bombing of Guernica in 1937, which may at worst have caused over 1,000 deaths, shook world opinion, while the bombing of Warsaw by the Germans two years later, causing probably twenty to forty times as many deaths, is merely one of many such events of the Second World War. Thus the perception of barbarism in warfare is dependent on cultural norms, and the perceptions rooted in these, and is not something that would stand up to objective benchmarks, such as overall number of casualties resulting from it, relative to casualties resulting from other measures.

Moreover, the conduct of war since Roman times, as we have seen, did not develop in a one-way fashion from very simple warfare towards ever greater professionalism or the ever greater observation of restraints on warfare, or warfare on an ever larger scale. Instead, the evolution of warfare fluctuated quite strongly. What we *can* see after the chaos that spread in Europe with the end of the West Roman Empire and the gradual reconstruction of a wider civilisation with accepted norms is an initially slow but, from the nineteenth century onwards, constant progress with the *ius in bello* to incorporate more and more humanitarian considerations designed to ward off actions seen as particularly cruel. Nevertheless, unarmed people of all ages and sexes were the casualties of war, by direct killing, maiming, wounding and rape, or indirectly through famine and disease, throughout the two millennia covered in this book.

There are, however, objective criteria for barbarism. One concerns the killing of the unarmed, those unable to defend themselves. Relative to the horrors of previous wars, the total numbers of non-combatants killed or wounded rose to completely unprecedented numbers from the mid nineteenth to the twentieth century, culminating in the democide inflicted by Stalin and Hitler.

There are at least three main reasons for this, one, quite simply, the exponential growth of populations around the world that explains the difference of scale. The twentieth century thus saw a quantum leap; its mass atrocities were not unprecedented in *intention*, but entirely unprecedented in the *numbers* of victims. While even after the Second World War, the terrible record of the Thirty Years War of having led to the deaths – directly, but more often indirectly – of one third of the population of the theatre of war remains unparalleled in *relative* terms, in *absolute* numbers, the casualties of the Second World War of course far surpassed it. More Europeans died in the Second World War than populated the continent at any one time in the seventeenth century.

The second reason is that only an advanced state administration, composed of 'willing executioners', obedient and without compassion, and possessing the technological means to keep track of the population (what the German bureaucrats called *Erfassen*) can practise genocide on the scale of the Ottoman Empire against the Armenians, Stalin's Soviet Union in the Holodomor, Germany with the Holocaust and its mass-starvation of Russians or even Rwanda's Hutus against the Tutsis. Only these two factors jointly made possible the extent of mobilisation and the industrialised warfare of the American Civil War, the German Wars of Unification and then the two World Wars. Equally, they made possible the geno- or democides of the twentieth century, including even the low-tech Rwandan genocide, with their unprecedented total numbers.

But the third is an ideology that proclaimed the killing of certain categories of people as absolutely necessary for one's own 'race's' or 'class's' survival. It was needed as the motivation to overcome what Bismarck reportedly called a 'culpable laziness in killing' (q.i. Degler 1997: 68), which from time to time seized even Wehrmacht soldiers and SS men in the Second World War. And such ideologies flourished particularly in the 130 years after Napoleon's defeat at Waterloo. In the following chapters we shall see how the interplay of new technical possibilities and dangers, values, ideals and ideology shaped thinking, particularly with regard to maritime Strategy.

PART IV

Naval and maritime Strategy

8 *Long-term trends and early maritime Strategy*

The rule of the sea is indeed a great matter.
(Pericles, q.i. Thucydides I.143, 20)

Those nations and cities that have the command of the sea, even if they are foiled on land, they can never be thoroughly vanquished, before they are beaten from the sea.
(Matthew Sutcliffe 1593: 273)

Strategy on land, at sea and in the air

Terminology for land warfare and sea warfare is partly different. Where we talk of Strategy, Tactics, sometimes of operational art, with relation to the sea, it has been within the wider context of 'command of the sea' or 'seapower', a term traced to Thucydides' 'rule of the sea' (Starr 1978). Admiral Sir Herbert Richmond, who upon retiring from the Royal Navy became a Cambridge don, defined it thus just after the Second World War:

> Sea power is that form of national strength which enables its possessor to send his armies and commerce across stretches of sea and ocean which lie between his country or the country of his allies, and those territories to which he needs access in war; and to prevent his enemy from doing the same. (Richmond 1947: ix)

The term 'strategy' appeared even later in a naval context than in Western writing on war in general. An entirely unsurprising definition in the Clausewitzian and Jominian tradition (see chapter 1) is that of Giovanni Sechi in his *Elementi di arte militare marittima* of 1906:

> The conduct of naval warfare belongs to naval strategy; the execution of the strategic concepts is attributed to logistics if one is not in the presence

of the enemy, to tactics when the cannon thunders ... Consequently, we can say that strategy is the spirit which thinks, logistics and tactics are the arms [*brazzi*] that act. (q.i. Castex 1937: 6)

Here, as in general definitions of Strategy, much value was added by bringing in the political dimension. Sir Julian Corbett very helpfully introduced 'maritime strategy' as a generic term which subsumes the narrower concept of 'naval strategy' for the use of force at sea. Taking a wider, holistic perspective on Strategy, Corbett understood that 'maritime strategy' necessarily had to involve land warfare, but might – and in the case of island states like Great Britain usually did – involve the use of navies. If both land and naval warfare were involved, Corbett recommended the use of the term 'maritime strategy', with the purely sea-bound element of this Strategy being naval warfare:

By *maritime strategy* we mean the principles which govern a war in which the sea is a substantial factor. Naval strategy is but that part of it which determines the movements of the fleet when maritime strategy has determined what part the fleet must play in relation to the action of the land forces ... it is almost impossible that a war can be decided by naval action alone.

On their own, navies could but exhaust the enemy, and would simultaneously exhaust themselves. For a 'firm decision', co-operation of land and naval forces was required.

Since men live upon the land and not upon the sea, great issues between nations at war have always been decided – except in the rarest cases – either by what your army can do against your enemy's territory and national life or else by the fear of what the fleet makes it possible for your army to do.

In Corbett's system of thinking, 'maritime strategy' incorporated 'naval strategy', which was its derivative (Corbett 1911: 15f.). This definition subsequently became widely accepted and will be employed here, as far as it does not misrepresent views (Gretton 1965: 3).

US navy historian John Hattendorf built on Corbett:

[G]rand strategy is the comprehensive direction of power to achieve particular national goals, within it maritime strategy is the comprehensive direction of all aspects of national power that relate to a nation's interests at sea. The Navy serves this purpose, but maritime strategy is not purely a naval preserve. Maritime strategy ... include[s] diplomacy, the safety and defense of merchant trade at sea, fishing, the exploitation, conservation, regulation and defense of the exclusive economic zone at sea, coastal defense, security of national borders, the protection of offshore islands as well as participation in regional and worldwide concerns relating to the use of oceans, the skies over the oceans and the land under the sea. (Hattendorf 2000: 256)

As with the definitions of Strategy in general which we discussed in chapter 1, Hattendorf's broad definition would by no means have been generally understood or accepted in the eighteenth, nineteenth or even early twentieth centuries. Indeed, many who accepted Corbett's point continued to write about 'naval strategy', 'naval thinking' or 'naval warfare', so we can only follow the current convention, noting that when 'naval' is used here, it will be interpreted in the Corbettian sense of 'maritime', unless stated otherwise (Hill 2006: 160).

Yet another – typically French – approach is given by the historian Jean Pagès (1990: 15):

Naval thinking is the discourse in which strategic doctrine is born, accompanied by essential tactical rules to which the leaders of the naval forces of the country have to refer in preparing to carry out all their operations. It takes account of political, military and economic imperatives, and of geographical constraints. It is susceptible to evolve as a function of changes within the country or outside it. Naval thinking can be profoundly influenced by an ideology. In one word, it is at the outset an intellectual step which leads to decisions and concrete actions.

This usefully recalls some of our main tools in the analysis of the evolution of Strategy attempted in this book.

Eternal principles, enduring variables or unique geographic situations?

There are eternal principles governing naval warfare, of which the importance of land is perhaps the greatest: John Hattendorf laconically reminds his readers that 'Throughout naval history, naval battles have mainly taken place close to land' (Hattendorf 2000: 258). As with writing about land warfare, there were those who sought to identify eternal principles, or at least enduring variables, in maritime warfare. For example Admiral Sir Cyprian Bridge (1839–1924), who had sailed all parts of the globe before becoming director of naval intelligence and Britain's chief strategic planner in the 1890s, complained about the tendency to focus excessively on new weapons. Writing in 1907, he argued that 'From age to age, from one stage of culture to another, men remain essentially the same. They can add to their acquired knowledge; but in boldness, fortitude, wariness, energy, persistence, the savage of Guadalcanal is essentially on an equality with the graduate of Oxford or the General Staff Officer of Berlin' (Bridge 1907: 3, 5).

While acknowledging human nature as a constant, many historians of the 'all-history-is-unique-and-incomparable' persuasion believe that no generalisations can be drawn from the history of warfare, as every instance is so unique as not to lend itself to any analysis of recurrent patterns. There have been very few naval or maritime powers in world history. At sea, interstate war is, even more so than most land warfare, determined by geography. The number of 'island states' with overseas empires but without land frontiers to worry about – all told, throughout human history – at best reached two figures. By the time writers tackled the principles of naval warfare, there were only three states that could qualify: Britain, Japan and, stretching the concept, the USA. Generalisations about the need (and even remote possibility) of achieving naval dominance, or 'command of the sea', could perhaps apply to these states, but hardly to any state with finite resources needing to worry about its land frontiers at the same time. To state the obvious, land-locked states have to worry about armies and indeed about their vulnerability to aircraft and missiles, but not at all about attacks from the sea.

There are many substantial differences between land warfare and warfare at sea. One which Corbett identified is that 'lines of communication', or 'lines of operation', can be switched in the open seas. 'Seeking out the enemy's fleet' on the open oceans is thus more difficult than on land (Corbett 1911: 158f.). The concept of concentration of effort also differed on land and at sea, because fleets also need to protect commerce and not just win battles. Corbett noted that there was nothing comparable in land warfare.

> Nor is it more profitable to declare that the only sound way to protect your commerce is to destroy the enemy's fleet ... What are you to do if the enemy refuses to permit you to destroy his fleets? ... the more you concentrate your force and efforts to secure the desired decision, the more you will expose your trade to sporadic attack. (Corbett 1911: 160f.)

To a greater extent than in writing on land and air warfare, writing about naval warfare was therefore country-specific. If Jomini's and the young Clausewitz's fixation with decisive or 'annihilation' battles could be applied to the sea, even as an ideal concept, it only made sense as a Strategy for the world's half-dozen biggest navies in conflicts with each other. And in practice, what was a dream worth pursuing for Britain and the USA was a chimera for France, Italy, Germany and perhaps even the largely ice-bound fleets of the Soviet Union. Indeed only a small number of states could afford a navy with worldwide reach and major ships (Howard 1980: 79); this has been true since ships first travelled around the globe.

In addition technological change affected maritime warfare more than land warfare, and made debates about war at sea more focused on the latest technology and at an earlier point, as we shall see. Paul Hay du Chastelet, writing soon after the creation of the French navy in the mid seventeenth century, thought that:

> Whatever knowledge we have of ancient maritime wars conducted by the [ancients] ... we can barely draw any rule from them for what we do in our own times: our usages are too different from practice in antiquity, and the invention of artillery has made all machines used then useless. One could hardly form precepts on what our own ancestors have done, unless one drew lessons from what has been done long since the invention of cannon. Thus the principles [of naval warfare] are quite new ... The size of the ships, the types of weapons, the equipment and the fire ships are changing almost every year, and rarely does one use the same methods in the same battles. (Hay du Chastelet 1668/1757: 160)

Writing about naval/maritime warfare, power or Strategy tends to revolve around two themes: powers with a potential for predominating an entire ocean or even on a global scale – which basically means, in early modern history, Spain, the Netherlands and Britain, in the early modern period also briefly France, later just Britain and the USA, with regional bids for domination made by Japan and Russia;

or second-tier powers who had to work out how to protect themselves and pursue their interests in the face of a stronger adversary, and the extent to which they might challenge and equal (or even outclass) this adversary. A third permutation exists, that of the second-rank naval power that opts for a permanent alliance with a first-rank power (or one or more equals), which enables it to concentrate its forces on other adversaries (single or multiple). Short of this, third-rank or small naval powers could never do more than see to their coastal defences, the defence of their territorial waters and perhaps engage in piracy, if they were far enough removed from the great powers to be able to get away with it.

As we shall see, the first theme, that of the first-tier powers and those aspiring to join them, tended to go along with a focus on big battles annihilating enemy fleets. It should have carried the health warning 'for superpowers only' (Hill 1986: 79), but in fact did not, as it flourished in the ugly age of the universal Social Darwinist scramble for 'world power'. The second theme, pursued by the second-tier powers, is perhaps the most interesting, as it revolves around ways of maximising one's own particular strengths (geographic, possibly in terms of ships and weapons), avoiding annihilation by a numerically superior enemy and defending one's own interests (mainly commerce and protection from invasion). The third permutation was explored primarily by Britain and France in the twentieth century, albeit very reluctantly, as dependence on an ally, however much strength it brought, always reduced their freedom of decision-making in other ways.

Admiral Sir Reginald Custance (1847–1935) observed that since about the 1860s there were two schools of naval teaching: first, the 'historical school', which sought 'eternal truths' (Lucien Poirier) about naval warfare in historical examples; secondly, the '*matériel* school', which focused on new developments in technology, arguing that the past cannot teach us much. We have seen that this school had also existed for, and indeed dominated writing about, land warfare since the middle of the nineteenth century, and a similar school would later hold its own in air power and nuclear debates. Custance by contrast was on the side of the 'eternal truths in history' school and resented the domination of the 'technology changes everything' or *matériel* school (Custance 1907). Interestingly, as we shall see, there was little general agreement within these two schools on some of the most central themes, such as the desirability of battle or the value of commerce raiding or brick-and-mortar coastal defences.

Writing in the age of oar and sail

Writing at the end of the nineteenth century, one of Britain's earliest naval educators, Admiral Philip Howard Colomb (1831–99), claimed that 'Naval warfare is of comparative modern origin. Sea fights there were, no doubt, in very ancient times, but sea fights do not of themselves constitute naval warfare' (Colomb 1891: 1). Admittedly, there is little that shows a connection in the minds of early writers between political aims and military Strategy. Yet gradually, we see the following debates emerge:

- a quest by some for the 'command of the sea' as opposed to others who say it cannot be obtained;
- naval battle, with some saying it must be sought and others that it should be avoided if a favourable outcome is uncertain;
- the advocacy of a *guerre de course*, that is, privateering and the attack on enemy trade, as alternative;
- and finally, blockades, and how to defend against them.

Writing about naval warfare can, like writing about war more generally, be traced back to antiquity: Vegetius (Book IV, chs. 31–43), Leo (XIX) and in the Middle Ages Christine de Pisan and Jean de Bueil all wrote about it, but from a very tactical perspective. In the Renaissance, Thucydides' 'rule of the sea' was rediscovered, and from this Francis Bacon (1561–1626) derived his claim that 'He that commands the sea is at great liberty and may take as much and as little of the war as he will' (q.i. Lambert n.d. b). The claim was exaggerated further by Sir Walter Raleigh (1555–1618): 'Whosoever commandeth the sea commandeth trade; whosoever commandeth trade commandeth the riches of the world', and thus the world itself. The Italian Tomaso Campanella later put it even more concisely, '*Che e signore del mare e signore della terra*' (q.i. Olivier 2004: 38).

In contrast to Bacon and Raleigh, Hugo Grotius in the Netherlands was the first to dismiss the 'command of the sea' as a chimera. In his *About the Freedom of the Sea* of 1609 and subsequently his *On the Laws of War and Peace*, Grotius argued that the ocean 'cannot be reduced to a state of private property' (Grotius 1625/1901: 103–8).

By implication this meant that 'command of the sea' could never be more than an ideal – a realisation that later writers had to labour to rediscover.

Few authors of the early modern era showed an interest in the strategic dimensions of naval warfare. One has to agree with the French Admiral Raoul Castex (1878–1968), who claimed that there was an 'almost complete void in the writing on naval Strategy before the French Revolution, which stands in utter contrast to the work of authors writing about the army in the same period' (Castex 1937: 31). We do, however, find one parallel to writings on land Strategy, and that concerns the desirability of battle.

Naval battle versus privateering and commerce warfare

The Frenchman Charles de la Rouvraye (1783–1836 or 1850) was an early fan of naval battles of annihilation: 'Combat does not cease until one of the two [sides] is totally destroyed', later adding, 'You cannot abandon the field of battle until there are no enemies left to fight, or until one cannot fight any longer' (Rouvraye 1815: 8, 35). By contrast his Swiss contemporary Victor Emmanuel Thellung de Courtelary (1760–1842) was still imbued with the *ancien régime*'s caution with regard to giving battle, and warned against it (q.i. Langendorff 1999: 87–9). While up to the Napoleonic Wars, the age of oar and sail saw many naval battles, including major ones, in the subsequent period until Tsushima (1905) there were very few; arguably, only three wars were strongly affected by naval operations, but none was a battle between two hostile fleets. The first of these was the war between Germany and Denmark (1848–9), when the Danish fleet stopped a German general from crossing to the island of Alsen, where the Danish army was. The second was the American Civil War, where the North had not only superior numbers in armies but also the command of the sea, the coast and the major rivers, the last being the key to their victory in this very major war, while the South had no fleet (Wilkinson 1894: 3). The third was the Battle of Lissa (1866), in which Italy attempted to come to the aid of Prussia in its war against Austria, but was defeated.

Nevertheless, the age of sail was replete with smaller-scale naval activity. Piracy, which goes back to antiquity, had been exploited politically since the high Middle Ages, when princes began to issue 'letters of marque and reprisals' to give such entrepreneurs state protection – and secure a cut of the prize for their state. This is intimately linked to war, as in times of war, states sought to intercept each other's merchant shipping, particularly where the claim could be made that cargo carried by enemy merchantmen served their war effort.

From the seventeenth century, wars between state-owned navies (the Anglo-Dutch Wars) proved very expensive, while commerce raiding was cheap and profitable. The Anglo-Dutch Wars were perhaps the 'purest' naval wars in modern times, and were settled by economic exhaustion. The following Anglo-French Wars, by contrast, had a land dimension centring on the Low Countries and allowed France to use alternative naval strategies. The *guerre de course* thus played an important part in Franco-British conflicts between 1689 and 1815 (the 'Second Hundred Years War') because the Royal Navy was much superior in numbers, putting France at a disadvantage in pitched battles. Attacks were also made by all sides on vessels with neutral flags, creating pressure for legislation. From this it has been argued that there were two different traditions, the supposedly French tradition of commerce raiding and a British tradition of seeking battle (summarised in the myth of 'Trafalgar'). The reality, on both sides, was a mixture, with a significant amount of bluff thrown in on the British side.

The Declaration of Paris of 1856 outlawed privateering. Henceforth states' navies monopolised commerce war, which thus stopped being a lucrative business practice and became a standard element of naval Strategy. But it now entailed the destruction of cargo, ships and often crew as an 'acceptable alternative to capture', because in the American Civil War, the South had little opportunity to sell prizes or to return them to their own harbours (Olivier 2004: 8).

A French 'tradition'?

Modern French naval history can be traced back to Cardinal Richelieu's (1584–1642) forceful programme of naval construction and centralised administration, undertaken in the knowledge that Europe now depended greatly on the import of gold and silver, as well as other merchandise, from overseas, and in the conviction that the French monarchy's sovereignty had to be defended and demonstrated on sea as much as on land (Granier 1992: 37–53). Richelieu's convictions about the importance of commerce and the centrality of its protection to France's naval tasks in every way foreshadow similar American and British thinking towards the end of the nineteenth century.

It has been argued primarily by Britons since John Clerk of Eldin (1728–82) and most famously by the American Alfred Thayer Mahan (1840–1914) that the French traditionally shunned battles and, instead, inclined towards the harassment of the enemy's shipping and protection of one's own (Mahan 1890, 1892; Depeyre 1992: 70). Indeed, a few French writers took this line, most famously Sébastien Le Prestre de Vauban, Louis XIV's minister and architect of France's ring of fortifications, who in advocating the avoidance of major battles was an important precursor of the later French *Jeune École*. In 1695, he wrote in favour of the abandonment of large-scale naval warfare in favour of *la petite guerre navale*, the naval equivalent of the irregular warfare that was known at the time as 'small war'. He wrote: 'War at sea conducted by big fleets [*en corps d'armée*] has never yielded what the King had hoped from it, and will never do so ... because [the great countries of Europe] will probably always be stronger at sea than us.' As the French could not hope to become 'masters of the sea', he recommended that they harass Anglo-Dutch commercial shipping, a Strategy he thought 'less expensive, less hazardous, and less costly to the state' while 'enriching the kingdom'. This course of action he praised as simple, yet subtle and advantageous: it would cost the adversaries much to protect their shipping adequately, while they could not retaliate in kind, as France relied so little on overseas commerce. Moreover, France had geography on her side: 'France has all the advantages of the *course* which surpass in every way those of her neighbours, as all the commerce of its enemies goes back and forth within reach of her coasts and her most important harbours.' France could thus cut the 'nerves' of her enemies, suing for peace on much better conditions than by following any other Strategy (q.i. Motte 2003: 93).

By contrast, other French naval theorists of the seventeenth and eighteenth centuries, from Paul Hoste (1652–1700) to Jacques Raymond vicomte de Grenier (1736–1803), were just as keen on battle as British practitioners, possibly more so than contemporary writers on land warfare. As Captain Audibert Ramatuelle (1759–1840) wrote in 1802, 'The French navy has always preferred the glory of assuring or protecting a conquest to the perhaps more brilliant, but effectively less real, seizing of some ships; and in this it has been closer to the actual aim ... of war': conquest (q.i. Castex 1937: 33). Moreover, the French official naval doctrinal manuals of the nineteenth century (formulated in 1832, 1857, 1864 and 1878) had the naval battle at their centre (Darrieus 1993: 209–31). So the French 'tradition' of battle-avoidance and preference for the *guerre de course* is something of a myth. It was really only in the second half of the nineteenth century that several French naval strategists advocated it strongly, in view of the double menace of Britain and its empire on one side, and the growing and aggressive Germany on the other. This combination made it impossible for France to meet both threats on an equal level.

A British 'tradition'?

In British writing, it is hard to tease out any principles before the second half of the eighteenth century. It is by and large a British national characteristic not to articulate things that are a shared tacit assumption – the need is not perceived, there is enough continuity among people and ideas to pass them on by osmosis, 'learning by doing'. This also applies to British naval Strategy, passed on from one generation of sailors and military commanders to the next, as naval historian Andrew Lambert has rightly observed (2003b).

Notwithstanding Raleigh's and Bacon's great words about the 'command of the sea', England initially practised the *guerre de course*, particularly when it faced the much stronger Spanish in the sixteenth century and then the redoubtable Dutch. The English were seen by their sixteenth- and early seventeenth-century contemporaries as dangerous pirates. As Britain became richer and its navy grew in size, the – tacit – British Strategy was mainly defensive and deterrent, even if Britain deployed ships with advanced offensive potential, and occasionally realised its latent threat of destroying an adversary's strategic port through bombardment from the sea, especially in far-flung parts of the empire. This can be subsumed under 'gunboat diplomacy', which could function both as what twentieth-century strategists would call 'deterrence' ('don't do this to me or I shall do that to you') and 'coercion' ('do this or I shall do that to you'). On the whole, Britain's leaders preferred 'deterrence, arms racing, negotiation, or concession, usually in a combination to meet the circumstances' to large-scale naval battles that would be costly to the victor and vanquished alike (Lambert 2003b: 165). Britain ran its empire on perpetual 'imperial overstretch', its ships constantly tasked to police an area far larger than could be reasonably expected of them, but this fact seems to have remained unrecognised by many of its adversaries. Even as a first-rank power, Britain had much more strategic weight than actual military force. Where possible, Britain offset other powers' forces indirectly, through intercepting their commerce where it was conducted across the sea. Britain's main rivals – permanently or temporarily France, Spain, the Netherlands, from the nineteenth century Russia, and finally Germany, Japan and Italy in the twentieth century – all needed to muster armies as well as fleets, while Britain could afford not to raise armies as large as those of these adversaries.

It is true that in the seventeenth and eighteenth centuries, when dealing with the numerically inferior fleets of France, and during the French Revolutionary and Napoleonic Wars, the British navy was configured to give battle, while its continental adversaries, especially towards the end of this period, had little to gain and much to lose from a direct confrontation. Thus the concept of the 'fleet in being' came into existence. It is usually traced back to Admiral Arthur Herbert, Earl of Torrington (1647–1716), who developed it in the context of war against France at the end of the seventeenth century. This was incidentally at the same time as Frederick William I of Prussia (1688–1740) created his 'army in being', one might say, which he used to deter other powers from attacking his country, while avoiding battle where possible. Similarly, Torrington in an encounter with the French at Beachy Head in 1690 tried to avoid a showdown. Instead, he thought,

> [w]hilst we observe the French, they can make no attempt [to attack] either on sea or shore, but with great disadvantage ... Most men were in fear that the French would invade; but I was always of another opinion: for I always said, that whilst we had a fleet in being, they would not dare to make an attempt. (q.i. Till 1982: 113)

Unfortunately, the Anglo-Dutch leadership did not agree and forced the English fleet to give battle at Beachy Head after all, where it was squarely defeated; what saved Britain from a French invasion was not Torrington's posture, but the failure of France to exploit her victory.

In the nineteenth century, Britain made much of 'gunboat diplomacy', the ability to threaten any port with bombardment, which in the case of threatening a power of equal or greater strength amounted to deterrence; interestingly, this is when the use of the term 'deterrence' spread in English (Freedman 2004). Thus Britain carried out its latent threat in the War of 1812 against the USA (famously burning Washington), and again in the *Trent* crisis of 1861, forcing Americans, French and Russians to invest heavily in fortifications. In the age of steam, armoured vessels and long-range artillery, naval bombardment could do considerable damage to forts and fortified harbours situated within river estuaries (Lambert 2003b: 164–95). The Crimean War had its share of it with the British bombardment of Sweaborg near Helsingfors (Helsinki) in Finland in August 1855, and in October 1855, Fort Kinburn on the Kinburn Peninsula, dominating the estuary of the Dnieper in the Black Sea. Finally in April 1856, the British threatened Kronstadt on the island of Kotlin, which dominated the estuary of the Neva River and thus the access to St Petersburg, persuading Russia to sue for peace. In this context, Britain did not fight major naval battles, as Russia refused to give them and resorted to a fleet-in-being Strategy. So instead, Britain blockaded or attacked enemy harbours; it sought the strategic fruits of victory through applying pressure to the shore. Again, one can see parallels with eighteenth-century 'manoeuvre tactics' as employed by land strategists at the time, parading armies for show, in the hope that their bluff would not be called, but sometimes achieving the desired effect on potential adversaries (Salewski 2002, I: ch. 2).

At the same time, Britain posed a threat to any adversary's commercial or military fleets through unpredictable harassment. In some ways, 'gunboat diplomacy' resembles guerrilla or partisan warfare (see chapters 15 and 16): a relatively small, very mobile navy can keep all adversaries and all colonies around the globe on their toes, forcing them to invest in costly coastal defences, while it is the navy that decides where to strike and where to catch enemies or insurgents against colonial power by surprise (Esdaile 1995). Bases all over the globe were crucial to Britain's empire – for the purpose of provisioning and refitting its ships, but also for protecting its shipping and thus its commerce – and to its Strategy of harassing adversaries' fleets far from home waters. At the end of the Napoleonic Wars, Britain had naval bases right around the world. By creating constant insecurity for Britain's adversaries, Britain could be said to 'command the sea', or at least large parts of the world's oceans. For other powers, it paid to be on good terms with the United Kingdom. In short, Britain's 'tradition' was not so much one of seeking decisive battles at all cost, but the flexible adaptation to circumstances, derived from instinct and long experience rather than principle and strategic reflection.

Mahan and Basil Liddell Hart were the main creators of the master narrative of a particular 'British way of war'. Britain's overwhelming naval penchant is undeniable, unsurprisingly given its geography. But Mahan's Jominian emphasis on decisive naval battles sought by Britain in its history is not. Slightly more realistic is Liddell Hart's characterisation of 'the historic Strategy of Britain', where he argued that since the Elizabethan age, the proven successful Strategy for Britain has been to keep off the continent and to offset land power by 'sea pressure on the enemy' and 'financial support to all possible allies' (Liddell Hart 1932: 36f.). Such generalisations always depend on the time-span examined. If British history since Roman times is considered, the period when England/Britain did not have continental possessions to defend is very short, beginning with Queen Victoria's accession to the throne, and even she, with her German husband and her European grandchildren, was hardly disinterested in the fate of the Continent. Given the same long-term perspective, Britain's history as a naval power with a sphere of rule beyond its coastal waters amounts only to a quarter of its recorded history.

While two myths have been exposed as such in this chapter, we have also encountered some maritime Strategies which were to be of lasting importance. These were, first, the Strategy of seeking a degree of predominance at sea, which would oblige the state adopting it to try to chase away or give battle to any hostile navies. Secondly, and as one alternative to this, there was the opportunist Strategy of enriching oneself by raiding the enemy's commerce while avoiding large-scale encounters. Thirdly, as another alternative to the first, there was the Strategy of maintaining a 'fleet in being', to be a direct or indirect deterrent to other fleets, while not seeking battle, if it could be avoided.

In the following chapters, we shall see permutations of these Strategies being developed, adopted or abandoned, partly for practical, but also in part for ideological reasons.

9 *The age of steam to the First World War*

In order to determine the way in which one has to fight, it is important to define first the aim one pursues in fighting. This aim is the annihilation of the enemy.

(René Daveluy 1902: 19)

At sea, the old Strategy is obsolete: the objective is no longer to seek out in the principal navy of the enemy to put it out of operation ... A navy must serve (1) to protect the nation's coasts and the maritime commerce; (2) to attack the coasts and the commerce of the nation's enemies ... [The aim is to] attack the enemy either when he leaves his base of operations, or en route, before his arrival in sight of the threatened coast; to pursue him and to fall upon him during his retreat once he has succeeded in executing his attack on the littoral; to strike at him when he thinks he has escaped pursuit and wants to return to his port ... One can see that our strategy is simple, unitary, scientific, modern.

(Z and Montechant 1893: 407–9)

The 'Anglo-Saxon' writers in the age of steam

As late as the end of the nineteenth century, Philip Colomb still complained that no one had written seriously about naval warfare. Thenceforth, he thought most naval writers were obsessed with '[t]he struggle ... for the mastery at sea, whether territorial conquest was or was not to follow success in this respect'. Philip Colomb maintained against this the argument that commerce across the oceans had grown, and intercepting it and appropriating the wares was advantageous in itself (Colomb 1891: iiif., 32). Despite the actual rarity of naval battles, the naval orthodoxy of the late nineteenth century read something like this: 'The primary object of our battle-fleet is to seek out and destroy that of the enemy.' The parallel to the contemporary prevalence of the Napoleonic paradigm is obvious. As Corbett remarked, however, the enemy might not oblige – he might simply 'remove his fleet from the board altogether. He may withdraw it into a defended port, where it is absolutely out of your reach without the assistance of an army' (Corbett 1911: 156).

The matériel *and the historical schools*

We have seen in writing on warfare in general how the mid nineteenth century ushered in an understandable obsession with new technology and the changes it brought to warfare, which brought challenges to the Napoleonic paradigm and to the view that lessons of history could be applied to Strategy in a very simple way (see chapter 7). This pattern of a clash of a 'historical' and a 'technological' or '*matériel* school' is paralleled in the naval warfare literature of the same period. The writers who focused on technology are the least interesting in the long run, as the technology that fascinated them itself became quickly outdated, unless they reflected also on how technology would allow them to attain political aims. So we shall confine ourselves here to only one prominent example of this '*matériel*' thinking: Admiral Sir John Fisher, who was to direct Britain's naval war effort between 1914 and April 1915, wrote on the eve of that war:

> The submarine is the coming type of war vessel for sea fighting ... It means that the whole foundation of our traditional naval strategy ... has broken down! The foundation of that strategy was blockade. The Fleet did not exist merely to win battles – that was the means not the end. The ultimate purpose of the Fleet was to make blockade possible for us and impossible for the enemy ... Surface ships can no longer maintain or prevent blockade ... All our old ideas of strategy are simmering in the melting pot! (q.i. Till 2006: 62)

Such shocks were produced by several successive technological innovations – steam, the torpedo, the ironclad, aviation, radar, missiles, to name but some. These did indeed introduce major changes, providing new dangers and new opportunities, not necessarily changing the strategic purpose of the use of navies. Therefore, more interesting than the *matériel* school, for our purposes, are members of the historical school. As we shall see, they diverged in their findings, and there were sub-groupings, such as the blue water school (or the British

Mahanians), or the bricks-and-mortar school, mainly soldiers writing on naval matters. Their interest in the past as a data bank for them to draw upon, analyse and interpret was not different from the methodology used by writers on warfare in general since Vegetius, but in the context of writing about naval Strategy, it seemed new.

In the last third of the nineteenth century a small number of men, no more than a dozen, on either side of the Atlantic began to see the need to articulate enduring principles of naval Strategy, that could be passed on from generation to generation, notwithstanding changes in technology, primarily with the aim of 'educating the navy' (Schurman 1965). The institutions they helped found, or in which they taught, have contributed greatly to their fame, although their number is disproportionally small compared with those writing on land warfare. They included a mathematician by training, the Briton John Knox Laughton (1830–1915), instructor at the Royal Naval College from 1866, later Professor of Modern History at King's College, University of London, where he created the antecedents of what today is a large department of war studies. Then there was the American Admiral Stephen Bleecker Luce (1827–1917), founder of the United States Naval War College in 1885. He recruited the naval officer Alfred Thayer Mahan, himself the son of a military instructor at the US Military Academy at West Point. Mahan's major writings were only published from 1890 onwards and earned him the reputation of America's 'naval schoolmaster' (Schurman 1965). What these men had in common was the positivist conviction going back to antiquity, that eternal truths about or lasting principles of warfare could be discovered through the study of history.

In Britain, the 'historical school' included the blue water school. It was led by the Colomb brothers, Philip (1831–99), who rose to the rank of Vice Admiral in the Royal Navy, and John (1838–1909), originally an artillerist with the Royal Marines, later an MP, who tended to lecture the House of Commons on naval matters. None of the early members of the 'historical school' had studied history in any formal educational context, but they saw historical evidence as a database for early political-science-type reasoning about military policy (Heuser 2007b). In the words of one of the writers of this school, Sir Cyprian Bridge, their purpose was 'to show the value – indeed the necessity – of a knowledge of naval history, which, it is held, ought to be studied not as a mere gratification of antiquarian predilection, but as a record of the lessons of naval warfare' from which he sought to deduce 'sound principles' to guide the officer in future wars. He did, however, warn of false lessons from history (Bridge 1907: viif., 8). The historical school also includes the by far cleverest and most original of these men, a lawyer by background, Sir Julian Corbett.

The purpose and value of historiography was much debated at the time. Mahan and Philip Colomb saw past historical events – experience of warfare – much like data in scientific experiments, and sought to deduce – largely 'immutable' – principles from them to provide guidance for the present (Schurman 1965: 37–59, 72f.). Corbett was more prudent in his formulations, but the general principle of using historical 'data' in the quest for iterative patterns was common to them all.

Occasionally, specialists on land warfare turned to writing on naval warfare, such as Henry Spenser Wilkinson, like Corbett a lawyer by training, later Chichele Professor of the History of War at Oxford, and General Sir Charles Edward Callwell, better known for his writings on counterinsurgency.

In the next sections, we shall focus on the main subjects of their debates.

The importance of commerce

The importance of commerce is seen by Britons and Americans as a predominantly Anglo-Saxon theme, even though the naval writers of other countries equally focused on it. There was general agreement among all those analysing Britain's position in the world in the eighteenth and nineteenth centuries that trade by sea was the key to her preeminence, and that this applied to any power aspiring to increase its wealth through overseas commerce. John Hattendorf has identified the great commonality between British and American approaches, which put utility and commercial gains for their own country at the centre of all naval policies (Hattendorf 2000: 109–20). Yet Mahan, Laughton and initially Richmond gave trade protection a relatively lower priority than the need to crush the enemy in a decisive battle (Schurman 1965: 136); Richmond later reversed his position (Richmond 1930). This reflects their emphasis on the need to acquire 'command of the sea' as the prior condition for one's own prevalence.

The Colombs emphasised the great importance of her commerce to Britain's wealth and position in the world. They extrapolated from this the need for a strong navy to protect Britain's shipping. The brothers came to stand for a policy of prioritising naval expenditure at the time when the primacy of the army was an accepted tenet in British military spending (Schurman 1965: 16–59).

Mahan is exceptional in deliberately playing on the link between commerce and navy, especially in democracies: 'where merchant shipping exists, it tends logically to develop the form of protection which is called naval; but it has become perfectly evident ... that a navy may be necessary where there is no shipping'. He thought that American President Monroe's Doctrine of 1823, the commitment to defend the Western hemisphere against any interference from other powers, especially Spain, had commercial roots, but that it was also

> partly military, defensive against European aggressions and dangerous propinquity; partly political, in sympathy with communities struggling for freedom. A broad basis of mercantile maritime interests and shipping will doubtless conduce to naval efficiency, by supplying a reserve of material and personnel. Also, in representative governments, military interests cannot without loss dispense with the backing which is supplied by a widely spread, deeply rooted, civil interest, such as merchant shipping would afford us.

He concluded, '[t]o prepare for war in time of peace is impracticable to commercial representative nations, because the people in general will not give sufficient heed to military necessities, or to international problems, to feel the pressure which induces readiness' (Mahan 1911: 446f.).

Spenser Wilkinson saw a similar nexus for Britain: 'if British trade is to be kept up, it is absolutely necessary to be ready for ... prompt and complete victory at sea; and this course, evidently the best if war must come, is also the surest way of averting war altogether' (Wilkinson 1894: 63). Admiral Sir Reginald Custance echoed Mahan:

> In the conduct of a war are to be distinguished three influences – the political, the economic, and the military – which react on each other and tend to strengthen or weaken the national effort according as they pull in the same or opposite directions. The connecting links between the three influences are the controlling minds – whatever be the form of Government – and

public opinion, which are moved first by the political object, and later by economic necessities; then both react on the military effort. Each influence possesses a spiritual, mental, or moral as well as a material side. (Custance 1919: 1)

Command of the sea

John Colomb used Thucydides' and Raleigh's term 'command of the sea' without further definition (Olivier 2004: 17). He was accused of 'virtually declar[ing] that naval war was an end in itself' and not the means to an end. Navy supporters, by the same token, saw him as the author of the 'New Testament' of the blue water school, that is, those advocating naval supremacy in defence spending over spending on the army (Schurman 1965: 56f.). Colomb was convinced that he had deduced his principles of eternal value systematically from his studies of British naval operations since the Anglo-Dutch Wars. Mahan, whose series of extraordinarily influential monographs began to be published in 1890, reached similar conclusions. He, too, stood accused of presenting navies as able to win wars on their own without substantial support from land forces.

Mahan's ambition for his own country to dominate the sea had an interesting limitation, however: he thought that the USA could only achieve this jointly with another power, not on its own. This was not so much for want of resources: in contrast to the European writers on the eve of the First World War who deeply feared the bellicosity of their own populations (see chapter 6), Mahan, as we have seen, thought that democracies ('popular governments') were unwilling to foot the bill necessary for such a military posture.

Is it true that, as the French claimed, 'the sea brooks only one mistress'? Mahan answered his own rhetorical question: 'The control of the sea, even in general, and still more in particular restricted districts, has at times and for long periods remained in doubt; the balance inclining now to this side, now to that. Contending navies have ranged its waters in mutual defiance' (Mahan 1911: 256). For Mahan, the obvious partner to help the USA achieve naval supremacy was clearly Britain. Anglo-American relations continued to be rocky well beyond the First World War, when US and British military planners still thought they might in the foreseeable future have to fight a war against each other (Hattendorf *et al.* 1993: Doc. 442), but Mahan with his admiration for English naval history yearned for the 'special relationship' from the 1890s, if not earlier.

At the time, Britain had a 'two-power standard' for its navy, that is, the postulate (formulated in 1893) that the Royal Navy should be the size of her two main potential adversaries' navies taken together (Gretton 1965: 6). As the Anglo-German naval race made itself felt in the first decade of the twentieth century, the First Lord of the Admiralty notified the British Prime Minister of his fears that the two-power standard might be inappropriate in dealing with Germany; instead, a mere margin of superiority over the German High Seas Fleet should be Britain's goal (Hattendorf 1993: 754f.). This became concrete policy soon: in the following years, the Admiralty tried to develop and hold a '60% superiority in vessels of the dreadnought type over the German navy' (Hattendorf 1993: 757).

Even then, the mantra of 'command of the sea' continued to haunt British minds. Bridge also placed the greatest emphasis on the command of the sea as the aim in naval warfare, explaining that it meant 'control of maritime communications ... The power that obtains this control can attack its enemy where it pleases, and evidently the control must be obtained before a great military expedition can be sent across the sea.' He thought that command of the sea could at times – but not usually – be obtained without fighting – for example, by Britain and France in the Crimean War, and Britain in the Boer War. Sometimes it could be obtained by intimidating the enemy. He conceded that command of the sea was never complete – small enemy raids would always remain possible. The command of the sea could be won, for example, by making the enemy fear one more than one feared him, with him thinking one would do unto him what he wanted to do unto one, and before he could do it. Surprise was thus useful, but Bridge also cautioned against rash attacks (Bridge 1907: 123, 131, 139).

Yet Bridge belonged to the decisive-battle school: 'To gain command of the sea we must defeat the enemy's navy.' While one might just drive it into a port and keep it there, mindful that it would not break out, 'The enemy's force remains the objective until its destruction or the end of the war' (Bridge 1907: 141). Henry Spenser Wilkinson agreed: 'The command of the sea is to be had only by destroying or crippling the hostile navy. Until this has been done the transport of troops by sea is a dangerous operation' (Wilkinson 1894: 33).

Callwell preferred the term 'maritime preponderance' rather than 'command of the sea' (the latter being more complete, the former more usual). It is rare that nothing is left of the beaten fleet (the implication of 'command of the sea'). Even Trafalgar, he noted, had not resulted in total interdiction of enemy shipping (Callwell 1905: 1ff.).

Corbett was more modest in his ambitions for 'command of the sea' and the function of naval battles. He approved of the idea of 'seeking out' the enemy only in order to boost morale, or to strike 'before the enemy's mobilization is complete' (Corbett 1911: 174). Corbett explained how one might secure command of the sea, although he rarely used the expression and spoke more of a 'working command':

1. Methods of securing command:
   (a) By obtaining a decision.
   (b) By blockade.
2. Methods of disputing command:
   (a) Principle of 'the fleet in being'.
   (b) Minor counter-attacks.
3. Methods of exercising command:
   (a) Defence against invasion.
   (b) Attack and defence of commerce.
   (c) Attack, defence, and support of military expeditions (Corbett 1911: 165f.).

Depending on one's geographic location and preoccupations, however, one could have more modest ambitions. Napoleon reportedly said in 1804, 'Let us be masters of the Straits [the English Channel] for six hours and we shall be masters of the world' (q.i. Till 1982: 130).

Attack, defence, and annihilation battles

Three other major debates concerning naval Strategy will be summarised only briefly here, as they so closely paralleled those conducted over land warfare. The first concerned defence and the offensive and, most famously, Mahan and his British followers, in the Jominian tradition, strongly favoured the offensive. Notably, there were also two dissenters: Admiral Hyacinthe Laurent Théophile Aube (1826–90) of the *Jeune École* held strongly that in naval warfare 'the defence is the superior objective, in contrast to what happens on land, where the offensive appears as the surest means of keeping the national soil inviolate by invading the enemy's soil' (q.i. Ceillier 1990: 206). The other was Corbett, who saw the advantages and disadvantages of both a defensive and an offensive Strategy.

The second, closely related theme is that of the decisive or annihilation battle. If Mahan's work on the navy can be characterised by one idea, it is the need for a large fleet ideally to crush an enemy in a decisive battle, as Mahan's hero Nelson had done at Trafalgar. Mahan thus applied the Napoleonic paradigm to naval Strategy (see chapters 5 and 6). This interpretation of the blue water school's and Mahan's teaching thrived in the contemporary cultural context of the naval equivalent of militarism, sometimes referred to as 'navalism'. For 'navalists' of all countries, the great naval battle became the absolute aim without much justification, as '*l'art pour l'art*', as Michael Salewski has noted (2007: 90).

For Corbett, by contrast, battle was a means to an end, perhaps not always a necessary means, and certainly not an end in itself. As he used to point out, even confronted with naval personnel who worshipped Nelson as the English god of war, it was not Trafalgar that decisively defeated Napoleon (Schurman 1965: 147–84). The Nelson admirers were represented in Britain by Admiral Sir Reginald Custance and his advocacy of naval battle and the *offensive à outrance* (Custance 1919: 3). This they regarded as 'the true spirit ... and traditions of the navy' (Custance 1907: 189, 228). The British Admiralty proclaimed at the 1902 Colonial Conference:

> The primary object of the British Navy is not to defend anything, but to attack the fleets of the enemy, and by defeating them to afford protection to British Dominions, supplies and commerce. This is the ultimate aim ... The traditional role of the Royal Navy is not to act on the defensive, but to prepare to attack the force which threatens – in other words to assume the offensive. (q.i. Till 2006: 80)

Like him, Spenser Wilkinson and General Callwell argued relentlessly that the destruction of the enemy's fleet was crucial to the successful outcome of a war (Wilkinson 1894: 50f.; Callwell 1905: 51). This view was widely held – in harmony with the 'primacy-of-battle fetish' that reigned throughout the Western world (Marder 1961: 306). This conviction crossed the boundaries between the *matériel* school and the historical school.

Equally related was the third theme, that of the concentration of forces. Naval writers on this point sounded just like their khaki comrades. Take Admiral Louis Edouard Bouët-Willaumez, writing in the mid-century: 'The tactical combinations of a good general or admiral must above all aim to operate with superior forces and combined effort on a decisive point' (q.i. Taillemite 1999: 50). Echoing Jomini, Mahan himself wrote that 'strategic advantage' lay 'in concentration, in central position, and in interior lines' (Mahan 1911: 76).

Fortifications, naval defences and bases

More particular to maritime Strategy are several other themes. Fears of invasions of Britain obviously go back to Roman times, and came to the fore whenever there was a crisis, especially in Anglo–French or later British–French relations. In the British wars against France of the late eighteenth and early nineteenth century, this fear agitated Britons, but the navy was deemed sufficient to ward off such foes. There was Admiral John Jervis, first Earl St Vincent's (1735–1823) apocryphal quip that he never claimed that Napoleon couldn't invade Britain, but that he couldn't come by sea. Indeed, Napoleon's great invasion plan with army, navy, and hot air balloons, as illustrated on the front cover of this book, never materialised. From the mid nineteenth century, however, invasion scares increased with the introduction of steamships which were supposed to have 'thrown a steam bridge across the Channel' for any invader, as the long-time British Foreign Secretary and then Prime Minister Lord Palmerston put it (Till 2006: 61). This gave rise to a 'bricks-and-mortars school' of advocates of coastal fortifications. The second half of the nineteenth century therefore saw the erection of a string of red-brick defences along the southern coast of England, just as coastal fortifications had become a feature of the Atlantic seaboard of the United States of America.

John Colomb, in an essay of 1867 entitled 'The Protection of our Commerce and Distribution of Our Naval Forces Considered', wrote:

> [O]ur policy should be this: in times of peace to provide for the safety and welfare of our merchant fleets on the high sea, and at trading ports *not* in our possession, nor in that of any European power; and in time of war with any European power or America, our object should be, to ensure the safety of those ports *in* our possession, and to afford protection, not only to them, but to as great an area as possible around them. (q.i Olivier 2004: 17)

Both Colombs centred their campaign for naval reforms on the call for secure naval bases along Britain's sea shores but also along the lines of Britain's commercial shipping routes throughout the empire and Britain's overseas bases, a view shared by Richmond (Schurman 1965: 144) and Callwell (1905: 65–89). They rightly recognised that in view of the regular need for refuelling with coal which distinguished steamships from sailing-ships, well-provisioned overseas bases were more important than ever before. Moreover, compared with wooden ships, metal ships needed more professional refurbishing at regular intervals to fight the effects of salty sea water on their hulls, and this meant the need for dry docks along shipping routes. The Colombs, Richmond and Callwell succeeded in raising public and governmental awareness of this issue and influenced policy-making (Schurman 1965: 21–4, 33, 42–6, 50).

In 1875, Colonel Sir William Jervois, writing for the Admiralty, called upon the British Secretary of State for War to improve the defences of Britain's many bases, only some of which were adequately fortified. Jervois argued for fortifications, as opposed to 'floating defences' in the form of ships protecting the bases, as the former seemed to him significantly less expensive (Hattendorf 1993: 595–8).

The 'bricks-and-mortar school' were derided by the blue water school and the Mahanians. Yet it was clearly a British strategic aim in case of war where possible to capture enemy's naval bases, naval stations or maritime fortifications – terms used synonymously (Callwell 1905: 94ff.). But for this, naval action alone might not be enough, as the Russo-Japanese War demonstrated: 'The reduction [i.e. destruction] of a maritime stronghold must generally be effected by attack from the land side.' This war reinforced the insight that in general, the defeat of an enemy navy has to be followed up with operations on land – a point which even Mahan conceded (Mahan 1911: 435).

Although himself a great advocate of drawing lessons from history, Admiral Sir Cyprian Bridge, a typical example of the blue water school, warned that 'Naval history down to our own day is filled with instances of wrong deductions from observed occurrences.' One of the false lessons in his view was the policy of putting fortifications everywhere in the British Empire, a lesson deduced from the siege of Sevastopol in the Crimean War. 'Not only were most of these useless, the scheme of defence in accordance with which they were constructed was also so radically unsound that it brought about a great relative decline in our naval strength, thus exposing the country to serious peril' (Bridge 1907: 10f.). Instead, Bridge advocated strategic defence: What he called the three divisions of naval warfare were 'coast defence, colonial defence, and defence of commerce'. Coastal defence he used in the largest sense, not merely the defence of a particular port or stretch of coast – that would be 'local defence'. Bridge thought it was a waste of money to spend much on particular points, rather than on a flexible sea-going fleet which could reconfigure to protect particular points (strategic defence). In his view, it made sense to have as big a fleet as possible, even if this led to an arms race. To concentrate his fleet for an invasion, the enemy had to give away his plans, and would then be easier to catch. Raids would occur either way, thought Bridge, and could not be excluded totally, but they would have 'little influence on the course of a war as a whole'. For local defence, he thought one might need special smaller craft. Passive defences – fortifications – he dismissed as too cumbersome and limited in their effectiveness. For colonial defence also the best answer, according to Bridge, was 'an adequate sea-going force'. He thought that the same applied to defence of commerce (Bridge 1907: 91–116). In war, maritime trade should be secured by putting merchant vessels in convoys protected by an escort of men-of-war, or keep cruisers around trade routes to ward off attackers. The latter choice, however, required too many cruisers to be affordable. Typical of the 'realists' who dominated his generation (see chapter 5), Bridge was dismissive of international treaties and conventions, claiming it was against man's nature to respect them (Bridge 1907: 162). Such scepticism was and is the luxury of citizens of great powers who have the option of effecting their own security in alternative ways to trying to uphold general norms.

In general, the blue water school thus argued that one should not invest in land fortresses – which would be purely defensive – but in a navy instead, which could be defensive and offensive. Mahan proved exceptional here when he warned against the view that 'military force can always, under all circumstances, dispense with secure bases of operations'. He rejected 'the opinion that the navy is the proper instrument, generally speaking, for coast defense in the narrow sense of the expression, which limits it to the defense of ports' (Mahan 1911: 153). But fortresses had other limitations: 'Fortresses, coast or other, defend only in virtue of the offensive power contained behind their walls. A coast fortress defends the nation to which it belongs chiefly by the fleet it shelters.' This in his view applied also to fortresses along land frontiers, which above all shelter the garrisons which will attack invading armies, and thus be offensive, even if they defend their country. 'Strategically, coast fortresses are not for defense, but for offense, by sheltering and sustaining that force which against an invader is the offensive arm; that is, the navy' (Mahan 1911: 432–5).

The army's bricks-and-mortar school in turn were dismissive of the blue water school and the Mahanians, Britain's General Wolseley cautioning in 1896:

> I know nothing that is more liable to disaster and danger than something which floats on the water. We often find in peace and in the calmest weather our best ironclads running into each other. We find great storms dispersing and almost destroying some of the finest fleets that ever sailed. Therefore, it is essentially necessary for this country that it should always have a powerful army, at least sufficiently strong to defend our own shore. (q.i. Gretton 1965: 12)

Interservice rivalry was clearly an important factor in the different naval perspectives (on this, see also the Epilogue to this book).

Naval blockade

Raids on enemy ports must have been among the earliest forms of naval warfare: actual blockades featured even in the Peloponnesian War. The pinnacle of blockades during the age of sail was the 'continental system' of Napoleon, an attempt to strangulate Britain's economy through a Continent-wide embargo on trade with Britain. In turn, the Royal Navy arrested any ship sailing towards any continental port, no matter what flag it was flying (Wilkinson 1894: 12f.). In the American Civil War, the blockade of the South by the North was very effective. In turn, Southern (Confederate Navy) ships pursued the *guerre de course* with some success.

British authors in particular wrote about blockades, which before the First World War aimed primarily at undermining an adversary's war effort. The authors in question included Philip Colomb, Corbett,

Callwell and the official historian Captain Stephen Roskill (1903–82). They divided blockades into the sub-categories of:

- the 'close blockade', where the enemy fleet was 'sealed up' in one or several ports, and associated with Lord St Vincent during the Napoleonic Wars;
- the 'open' or 'distant' blockade, associated with Admiral Richard Howe, the first Earl Howe (1726–99), in the naval campaigns against revolutionary France (Schurman 1965: 45f.).

In practice, the difference between them could be blurred, and navies might need extra help from armies to succeed, as Callwell thought in the light of the Russo-Japanese War (Callwell 1905: 120ff.). Callwell also warned that an adversary's navy could re-emerge if it was only bottled up in the harbours and had not been destroyed in battle (Callwell 1905: 127).

In view of the threat posed by torpedoes to capital ships, after 1888 the United Kingdom moved naval exercises from close blockades to distant blockades (Olivier 2004: 15). By 1911, the United Kingdom planned almost exclusively for distant or 'open' blockades. Blockade was both feared by and seen as promising for Britain. Sir John Fisher famously said in 1904, that 'It is not *invasion* we have to fear if our Navy is beaten, it's *starvation*!' (q.i. Till 2006: 75).

Amphibious/maritime operations

Britain had had a pioneer in the thinking on joint navy and army operations in the person of Thomas More Molyneux, writing in the mid eighteenth century (Molyneux 1759). The subject was by and large neglected by writers until well over a century later, when Philip Colomb wrote that 'the dividing line between attempts to gain the command of the sea in order to facilitate a descent upon the land, and descents upon the land with an admitted want of command of the sea, is an exceedingly fine one' (Colomb 1891: 203). He emphasised the mutually reinforcing factors in land and naval warfare, where coastal targets were concerned: 'certain conditions – command of the sea, sufficient and well-handled land forces, landings either away from the batteries, or after their fire has been temporarily silenced ... small vessels – have always been necessary to secure the success of territorial attack' (Colomb 1891: 430).

Perhaps not surprisingly it is a general, Callwell, who is called 'the father of Joint thinking' (Colin Gray), in that he developed the famous concept of co-operation between army and navy. Callwell liked to quote Foreign Secretary Sir Edward Grey's dictum that 'The British Army is a projectile to be fired by the Navy' (Callwell 1905: xiii). Callwell wrote his book on *Military Operations and Maritime Preponderance* during the Russo-Japanese War, which illustrated what could be achieved through co-operation between army and navy. Callwell raised the question whether sailors should be trained to go ashore to fight. His answer was that one should rather take soldiers, transported by sea. He argued that only a joint ('amphibious') force could cause the adversary extensive damage, not a navy acting on its own (Callwell 1905: 148–55, 170, 182f.). Sir Cyprian Bridge, Callwell's contemporary, although a naval man to boot, advocated 'jointness' on the one hand on the grounds of similarities and connections between naval and land warfare. Both on land and at sea, he opined, 'the primary objective is the enemy's armed force'. On the other hand, differences made for complementarity:

> [N]aval campaigns by themselves are not likely to end a war or cause the complete surrender of one side. A purely naval contest may wear out one belligerent; but the process will be long, and if one side is quite worn out, the other will almost certainly have begun to feel the effects of fatigue. Consequently, as a rule naval strategy should aim at enabling a land army to give the finishing stroke. (Bridge 1907: 22–4, 163–83)

Several other thinkers took a similar line, such as Colonel George Furse, who had a particular interest in logistics, and David Hannay (Furse 1897). Writing on the eve of the First World War, George Grey Aston of the Royal Marines was one of the first to integrate aircraft into his vision of joint operations of the future (Aston 1914). In theory, then, the value of jointness was well recognised, even if in practice it often foundered on interservice rivalries. The British Admiralty conceded, at the end of the Great War, that 'The war has been fought, and the final decision reached, on land; but the land campaign was rendered possible only by reinforcements and supplies from oversea' (Hattendorf *et al.* 1993: Doc.441).

Coercion and deterrence before the nuclear age: the fleet in being and gunboat diplomacy

Two centuries after Torrington invented the concept of the 'fleet in being', Sir Philip Colomb took up the view that a naval force 'in being', even if numerically inferior to that of an adversary, might deter an invasion of the British Isles. He described the navy as a 'shield or guard', while the army was the 'spear to strike' an adversary (Till 2006: 82f.). Since then naval strategies for second-tier navies have been subsumed under this heading.

The British Empire was based on a Strategy of deterrence, coercion and bluff, although this was codified nowhere – one future British Secretary of State for War commented in 1900 that it was true 'to say that our existing military organisation is based upon no known and accepted principle' (q.i. McDermott 1985: 101). If deterrence or coercion were applied, it was done so without any conscious articulation of this concept. British Strategy was seen more as 'muddling through'. In the words of one War Office official, 'we are attempting to maintain the largest Empire the world has ever seen with armaments and reserves that would be insufficient for a third class military power'. Until the Boer War, Britain's Strategy when confronting Asian and African countries had largely exploited British technological superiority, which would inspire terror and induce the local population to co-operate, in ignorance of the fact that British manpower resources were severely limited. In most cases, the British were bluffing and could have done little to increase their pressure. Coercion and deterrence, making much of a threat that could barely be realised but was occasionally implemented to drive the message home, but otherwise signalled through the parading of gunboats and the deployment of small, well-equipped units, was the one recognisable pattern in British military dealings with potential and actual colonies (McDermott 1985: 101).

The dividing line between peace and war was thus fluid in British maritime operations even in the nineteenth century. Bridge for one acknowledged by 1907 that 'There is a strategy of peace as well as of war.' With a large fleet, any power would have a special weight also in peacetime (Bridge 1907: 24f.). Bridge argued that 'In every detail the strategy of peace should conduce to furthering the object which we are likely to have in war' (Bridge 1907: 24f.). Corbett characteristically had much sympathy for the fleet in being as a deterrent, in the hope that one need not give battle every time, and in certain circumstances could retreat and avoid it. In auspicious circumstances, however, such a fleet might be used to attack (Corbett 1911: 26). Mahan, equally characteristically, dismissed the 'fleet in being' school (Mahan 1911: 398–400).

Gunboat diplomacy was a practice uniquely developed by naval warfare, unless one admits parallels with police patrolling on the highways. The nineteenth century for Britain and France was the era of imperial policing, which heavily involved their respective navies. William Ewart Gladstone (1809–98), the British Liberal statesman and Prime Minister, strongly advocated Britain's assumption of a world police role, most famously remembered in his Midlothian electoral campaign speeches in 1879. Gladstone was defeated in 1893 over the issue of naval spending. There had never been universal consensus on this: one of Gladstone's Foreign Secretaries, later Liberal Prime Minister, Lord Rosebery (1847–1929), told Queen Victoria: 'we cannot afford to be the Knight Errant of the World, careering about to redress grievances and help the weak' (q.i. Crewe-Milnes 1931, II: 426).

The subject of gunboat diplomacy was most convincingly analysed by a diplomat, James Cable:

> Gunboat diplomacy is the use or threat of limited naval force, otherwise than as an act of war, in order to secure advantage, or to avert loss, either in the furtherance of an international dispute or else against foreign nationals within the territory or the jurisdiction of their own state. (Cable 1971: 39)

Cable identified essentially different sorts of gunboat diplomacy since the twentieth century: what he calls 'island grabbing' – a sudden *coup de main* on a long-coveted neighbouring island – and the deployment of American carriers near an adversary's or a threatened ally's coast, which can include the imposition of a blockade (Cable 1971: 24, 131–3). What was needed in times of peace to resist gunboat diplomacy 'is not a surface force capable of winning a naval battle on its own, but one which cannot simply be shouldered out of the way' – echoes of the fleet in being!

> The inherent advantages of coastal defence are so great that even an inferior fleet can deprive an assailant's threat of credibility, let alone prevent him

from achieving a *fait accompli*, whereas bombers and submarines cannot be presented as a counter without at least an implicit threat of escalation. But the fleet must exist and be ready to sail. (Cable 1971: 187)

While the label was invented in Britain, gunboat diplomacy was clearly practised by other powers as well. As we shall see in chapters 12 and 14, the similarities between this thinking and some nuclear Strategies show the roots of the latter in naval deterrence.

French naval theorists in the late nineteenth and early twentieth centuries

In France, too, there appeared a historical and a *matériel* school. The *matériel* school, for whom the development of steamships, mines, stronger and more precise cannon, ironclads, torpedoes, submarines and later the air forces were the central points of concern, dominated around the turn of the century, but then the pendulum swung back towards the historical school, strongly inspired by the writings of Mahan. Unlike the 'Anglo-Saxons', adherents to the French historical school did not necessarily claim to find eternal invariables; much to the contrary, big changes, resulting from new technology, were stressed by them. The difference between the (politically more left-wing) *matériel* school and the (politically more conservative) historical school lay more in the policies they prescribed.

The French matériel *school and the* Jeune École

Politically, like the *Jeune École*, the French *matériel* school as a whole was associated with the left and with social-climbers from more humble backgrounds with a technological education and engineering mindset. They looked down upon by the old naval elite with their *littéraire* and anti-scientific tradition (Motte 1990: 127f.).

The pure *matériel* school was particularly strong in France. Ever since the Enlightenment, the French have been welcoming new technology, eager to try to apply it to solve problems, and to compensate for what from the late nineteenth century was seen as France's weakness in relation to her enemies. Unlike the 'Anglo-Saxons', the French tended to see new technology as an opportunity more than a threat.

A precursor of the *matériel* school in France, and particularly of its radical sub-group, the *Jeune École* ('new school'), from whom they clearly drew some key ideas, was Henri-Joseph Paixhans (1783–1854), an army artillerist who promoted the R & D of the exploding shells named after him. Most of his writing concerned artillery innovations, and besides urging their adoption, he raised the question of the moral justification of their introduction. He did not see how any of the missiles he advocated made killing more horrible than those already existing; it was not these arms, he argued, that were inhumane, but war itself, caused by human deficiencies and ambitions. Optimistically, writing in the middle of the nineteenth century, he thought humanity was becoming ever less ferocious. At the same time, he saw his country using war to promote the freedom of the seas for all, and not just for the British and the Americans. (Funnily enough, the British and the Americans also claimed to be fighting for freedom for all.) Paixhans anticipated the *Jeune École* argument that, armed with the new missiles, smaller ships could destroy the biggest ships of the line and thus make them obsolete, thus pleading for the acquisition of more numerous, smaller ships, such as swift frigates and corvettes. His emphasis on speed, again, would be taken up by the *Jeune École* (Taillemite 1994: 106–32).

Unlike the Anglosphere, where late nineteenth-century writers on naval matters were mainly civilians of various backgrounds, in France the writers were mainly professional naval officers, like Admiral Jean-Baptiste Grivel (1778–1869), who had risen to the defence of the French navy against a campaign to abolish it in view of its dramatic lack of success over centuries (Grivel 1837: 165–94), his son, Baron Louis Antoine Richild Grivel (1827–82), who also rose to the rank of Admiral, René Daveluy and Ambroise Baudry, both of whom rose to high ranks in the French navy.

The two Grivels, father and son, both writing in the first half of the nineteenth century, incarnated the French preference for the *guerre de course*, commerce-raiding. The father came out strongly in favour of distributing the captured vessel's goods among one's own sailors, and practically advocated the *guerre de course à outrance*:

> Let us use frankly the resources remaining to us [against a stronger enemy]; let us use them without scruples, and in a way that leads us to our aim ... even though this form of war ... does not bring us any positive benefit,

> granted that we burn and sink anything we get our hands on, we shall have gained enough on the day on which we will have proved to our enemy that we can cut off his sinews of war [literally: nerve] by attacking his commerce. (Grivel 1837: 190)

Where in the classical adage money was described as the 'sinews of war,' commerce was here substituted for it.

The son, Richild Grivel, spilled much ink on the great move from sail to steam, and other technological innovations (Grivel 1869). He identified three Strategy options for any power:

- coastal defence and attack;
- *la grande guerre*;
- cruiser warfare (the new form of commerce raiding) as most suited to France against the United Kingdom because of the Royal Navy's superior size.

Richild Grivel thought one should use bigger ships to fight smaller vessels, but shun battle with the United Kingdom, anticipating the tenet of the *Jeune École*, which was: 'shamelessly flee from the strong; shamelessly attack the weak'. France, he emphasised, needed different strategies depending on whether she would encounter 'an *inferior* or *superior* power (speaking navally)'. On this depended, in his view, whether France should opt for a fleet Strategy to crush an inferior enemy, or a cruiser Strategy to harass a superior foe (Olivier 2004: 281f.). The latter allowed a state 'to hit directly commerce and industry, that is to say, the very sources of the enemy's prosperity'. This should be France's Strategy against Britain, or the Strategy of a continental power towards an island power (of which, of course, there were not that many in the world).

By contrast, if an island power wanted to affect the continent, then mastery of the sea – and Richild Grivel went along with the Elizabethan tenet that this meant mastery of the world's commerce and thus the world – was desirable: one would combine naval and army action, attacks on enemy ports, impose blockades (Taillemite 1992: 91). Her geography alone, the predicament of having to face both a possible land war against the German states and the naval power Britain, necessarily deprived France of the option of matching both sides on a completely equal footing. It would only become clear after 1871 how true this was. But Richild Grivel may have been the first to have introduced the concept of France having to counter her enemies as David countered Goliath – with the weapons (and the Strategy) of the weaker against the stronger, '*le faible contre le fort*' (Taillemite 1992: 94). This is a subject that would become the leitmotiv of French thinking in the nuclear age, but the perceived need to compensate for the innate quantitative superiority of adversaries is all-persistent in French military thinking from about this time, deeply reinforced by the humiliating and surprising defeat of 1870/1.

The recognition that France could not aspire to a first-class navy by British standards was at the heart of the ideas of the *Jeune École*. Admiral Louis Edouard, Count Bouët-Willaumez (1808–71), summarised the rationale for his country's navy in the mid nineteenth century as follows:

> Our navy protects our maritime commerce and our colonies: leading the second-class navies [of the world], she has for centuries depended on the great principles of the freedom of the seas ... In case of war, the navy serves as a force multiplier for the armies, transporting them across great distances, recruiting and supplying them; it makes it possible to attack the enemy wherever he is vulnerable. (q.i. Taillemite 1999: 61)

French naval officers believed France had to stand up to Britain, the foremost naval power of the world in the eighteenth and nineteenth centuries; the question was: how?

The group of French naval thinkers who became known as the 'young school' tried to find an answer to this. Conscious of France's limited capabilities compared with Britain, they proposed a package of coastal defence, anti-blockade warfare and attacks on the enemy's commerce in the tradition of the *guerre de course* (Bueb 1971).

The actual birth of *Jeune École* thinking has been dated to the French government's budget report of 1879. In the French National Assembly, Deputy Ernest Lamy articulated an alternative Strategy for France as 'second naval power' in a changing Europe. He implied that France now had to worry also about the German navy, so France should aspire to have a navy that 'could hold out against the combined efforts of two fleets other than' the Royal Navy. But the costs of a navy were now so high that he thought France could not construct many ironclads to fight a war of major battles with many ships of the line (*guerre d'escadre*). Instead, France should ensure the defence of her coasts, produce strong ships with strong motors and strong guns, capable of staying at sea for a long time, and with them pursue a *guerre de course* (Ceillier 1990: 198).

The Strategy advocated by Lamy was to pose a real threat to maritime commerce in the form of amphibious assaults and close blockade. Admiral Jean-Pierre Edmond Jurien de la Gravière (1812–92), later president of the French Academy of Sciences and member of the Académie française, put it as follows:

> It is up to the nations for whom the navy only comes second in national ranking [after the army], and who are too poor to imitate the luxury of England, to work out what the future is. A good idea of this future can give their navy a menacing step forward. Every invention which threatens the giants and tends to emancipate the flies is a progress which the French navy could not welcome more, as it would no longer have to double its forces and its strength in a few years. (q.i. Ceillier 1990: 203)

A young journalist, Gabriel Charmes (1850–86), took up the idea and started to campaign for the necessary military reforms and the ruthless – and indeed lawless – pursuit of attacks on the enemy's commerce, with audacity and courage (the famous French *élan*). Charmes brought to the *Jeune École* fashionable concepts, focusing on the image of France as the weaker power having to tackle a stronger power:

> Battles between big fleets [*la guerre d'escadre*], blockade warfare, wars aimed at diverting forces on the continent, will in future only be memories of the past. There are only two forms of war left: the *guerre de course* in the open seas and coastal warfare against undefended towns ... The *guerre de course* is the weapon of the weak against the strong. It has to be [pursued] without limit and without pity, but it remains a moral[ly acceptable] endeavour: even though the means for reaching it are terrible and savage, it is certainly not a result which is contrary to the progress of human society, as this new power of the weak which one day will assure the entire freedom of the seas will prize away predominance from some nations that are luckier than others. (q.i. Monaque 1997: 60)

Gabriel Charmes thus cast France in the role of the defender of the rights of the weaker nations against the stronger ones – foreshadowing the key feature of French nuclear Strategy. This fitted well with France's self-perception, since the French Revolution, as 'soldier of humanity'

(Chuter 1996). In defence procurement terms, Charmes argued for more role specialisation and thus the construction of a range of different types of ships, gun-boats, torpedo-boats, cruisers. Concomitantly, he opposed the trend towards ever larger, heavier ironclads that was prevalent in France at the time (Ceillier 1990: 207f., 211).

These ideas came close to becoming government policy when Admiral Aube, with experience in the colonies, became Navy Minister in Charles de Freycinet's short-lived cabinet of 1886–7 and suspended the battleship construction programme. Instead, he supported the construction of cruisers and torpedo-boats. He argued for French missions of coastal defence (including in France's colonies) and commerce raiding with the help of cruisers specially built for this purpose. He was a new prophet of total economic war, fully aware of the effects this would have on the enemy nation (Monaque 1994: 133–43). In 1885, Aube wrote in starkly Social Darwinist terms:

> War is the negation of law. It ... is the recourse to force – which rules the world – of an entire people in the incessant and universal struggle for existence. Everything is therefore not only permissible but legitimate against the enemy ... Therefore the torpedo-boat will follow from afar, invisible [to] the liner it has met; and, once night has fallen, perfectly silently ... it will send into the abyss liner, cargo, crew, passengers; and, his soul not only at rest but fully satisfied, the captain of the torpedo-boat will continue his cruise. (q.i. Olivier 2004: 134)

Aube thus totally disregarded the legal obligations under the Declaration of Paris of 1856. He also recommended shelling ports to put pressure on the population, thus fully anticipating the argument for city bombardment put forward by the early air power strategists (see chapter 12). Aube hoped that economic and financial chaos would result for the enemy nation from French commerce raiding. In terms that Moltke, Bernhardi and indeed Ludendorff and Hitler would have endorsed, Aube described the conflicts of the future as 'the decisive wars, the wars for survival ... veritable struggles for life' (Olivier 2004: 134f.). Although Aube thought that the French navy should adopt a defensive Strategy, his concept of how to conduct it foreshadowed the total war against the enemy population that would be put into practice by the British naval blockade in the First World War:

> – What does one want from war? – To be the stronger side to impose one's law [upon the other]. – How does one impose it? – Through exhaustion [*en accablant*]. – Whom does one exhaust? – The members of the nation in their persons, their interests, their possessions. One thus has to attack the most abundant mainsprings of life. To destroy the fleet of England [!] is to defeat its pride; to sink the ships which bring to the English bread, meat, cotton, the salaries for the workers, it is that to conduct war against England. The principle of strategy is henceforth: not to waste time on the enemy fleet, but to concern oneself only with its riches and its ruin. (q.i. Ceillier 1990: 206)

Here is a link that would fit well into Talmon's view of the connections between French Revolutionary democracy and totalitarianism (Talmon 1952): Admiral Aube was postulating a war against the entire enemy nation, civilians included, men, women, young and old. His disciples at least harboured the arcane hope that somehow wars could in this way be brought to a swifter end, foreshadowing the thinking of several of the air power theorists after the First World War (Motte 1990: 121f.). The members of the *Jeune École* were seen as republicans or even radicals (that is, on the left in the French political spectrum), standing out in the sea of Conservatism which the French navy otherwise represented – the *Marine* was also known as 'La Royale', not only because the Navy Ministry happened to be located in the rue Royale in Paris, but also because of its Catholic, Conservative and even royalist leanings (Jackson 2000: 131). The members of the *Jeune École* presented the torpedo-boat as a democratic vessel as opposed to the cruiser and capital ship, which they saw as fetishes of the elitist thinking of reactionaries (Motte 1994: 149, 151).

Other *Jeune École* writers included Commandant Paul Fontin and Lieutenant J.H. Vigot, who, writing under pseudonyms, wanted to show 'that our country will be invincible on the seas, on condition that we stop pursuing the ... chimera of wars of big fleets'. They accused their government of lacking a coherent Strategy. France should above all renounce the construction of ironclads, 'those ruinous mastodons', which would produce no good effect in war. Instead, she should put all her money into cruisers, gun-boats and torpedo-boats, nothing else, until the submarine was ready. An anonymous writer – French military officers were not allowed to publicise their views – accused France of stupidly copying foreign countries (that is, Britain). He and his colleagues saw little point in this, as both countries were so different in strategic terms – Britain had all to lose away from home, on the oceans of the world, while France had all to lose on her own home territory. 'The defensive system of France's coasts is the foundation of her offensive power' (Anon. preface to Z and Montechant 1893: ix–9).

Another officer, writing as 'Z', together with a civilian, H. Montechant, poured scorn on those who claimed that naval Strategy follows immutable principles, regardless of technological developments. The authors argued instead that naval Strategy is a function of new engineering and new weapons systems. What they saw as immutable by contrast was France's need, for geographic reasons, of a defensive naval Strategy.

While in Admiral Tourville's time (1642–1701), two fleets, coming within sight of one another, had to give battle, because of the speed of modern ships, this was no longer necessary, said Z and Montechant. In Tourville's era, the greatest immediate danger in war had been the disembarkation of enemy forces on one's territory; today, enemy navies would start by bombing ports. Then, sailing ships had to follow the wind, now ships could go anywhere. Then, one could hide in the Channel; with modern means of signalling and the speed of ships, this was no longer possible without being detected. For France, major battles 'in the open seas' were now less beneficial than the *guerre de course* and coastal defences. The members of the *Jeune École* concluded that France only stood to lose if she tried to match the enemy's ironclad fleets and if, as in previous centuries, she sought out the enemy's main fleet to give battle. Acknowledging that this quest for major battle dominated land warfare, they asserted that in this respect naval war was the antithesis to land warfare. This was the 'simple, unitary, scientific, modern' maritime Strategy that they advocated for France (Z and Montechant 1893: 407–9).

With the fall of the Freycinet cabinet in 1887, Admiral Aube's and the *Jeune École*'s influence was temporarily terminated, although their ideas lived on, and came to the fore again after the Great War. They also had some impact on thinking in other countries: Austria-Hungary in the 1880s adopted the *Jeune École*'s ideas (which ignored international law on prizes and so on) because they meant a 'fleet on the cheap' (Olivier 2004: 135). Moreover, the submarine warfare of both World Wars was conducted in the spirit of the *Jeune École*.

The French Mahanians

Unlike all the authors quoted above, one could subscribe to the *matériel* school and yet be a French Mahanian, a fan of the great battle. This can be seen in the *Naval Battle* (1912) of Ambroise Baudry, written in the staccato style of which his contemporary Foch was so fond, with memorable phrases that could be drilled into dull brains. Baudry did little besides translating into terms of naval warfare the prevailing views on future war among writers on Strategy in general, the Colins and Gilberts of his age, based on Clausewitzian terminology. Battle was all-important, whether for army or navy. Small wars had become unthinkable to Baudry; major wars would be all the more intensive for that reason. Only a decisive victory would enable that side to impose its will on the adversary. The *guerre de course* made no sense to Baudry, as the only aim was the all-out annihilation battle, without difference between land or sea warfare, both based on the total mobilisation of the nation. Blockades were a waste of effort. Building small ships meant squandering money; everybody (that is, France) had to strive to build nothing less than the largest possible fleet. Nothing less than a duel of the giants would settle scores. This was the *offensive à outrance* gone to sea (Baudry 1912: 1–47, 224–7).

The French historical school subscribed to this principle, which until the First World War predominated over the *matériel* school and the *Jeune École*. The French historical school were thus, to a man, entirely unoriginal Mahanians, and were in agreement with the 'decisive battle' fashion of the age, the spirit of the annihilation battle and the *offensive à outrance* advocated by the land strategists. While some French writers accused Mahan of Francophobia (the heroes of his histories were mostly the British), many of the 'historians' found a source of pride in his extolling of some feats of French naval history, such as those of Admiral Suffren (1729–88), who had preferred decisive battles over coastal defences and the *guerre de course*. At any rate, around the turn of the century Mahan's books were devoured in France much as in Britain, Germany, Japan, Italy and elsewhere (Motte 1994: 145–72).

Two writers typical of this conservative Mahanian thinking are Admiral René Daveluy (1863–1939) and Admiral Gabriel Darrieus (1859–1931), the latter, like Mahan, primarily a naval educator.

Besides Mahan, René Daveluy was influenced by Prussian land warfare writing, centring on the importance and centrality of the annihilation battle. In the tradition of Guibert and Clausewitz, he belittled the wars of the *ancien régime*:

> [T]here was an era when one pretended to make war without fighting; one did not seek battle, one took it on with regret. This period coincided with the American War of Independence, and the false ideas which spread at the time had their effect also on the navy ... As one did not admit the need to get rid of the enemy, to wipe him out [*supprimer*], the predominant aim was to get out of an engagement with the least damages. One only entered into an engagement half-heartedly, one fought outside the effective range of cannon, then one disengaged without getting hurt ... Indecisive battles are sterile. (Daveluy 1902: 19f.)

He accused the French navy of a tendency to want to preserve their ships and of staying on the defensive, which he condemned: 'The true protection of the hardware [*équipement*] is victory' (Daveluy 1902: 22f.).

> Moral victories which leave the adversary intact only have effects of little importance or duration. By the fact that one leaves to the enemy the ability to come back to the battlefield, one gives him the possibility of reconquering what one has lost ... The annihilation of the enemy is one of the general rules of war; it flows from its very essence, from its definition. It is a necessity which prevails even more at sea than on land. (Daveluy 1902: 20f.)

And he concluded: 'The aim of combat ... is the annihilation of the enemy' (Daveluy 1902: 19f.). 'There are no two ways of fighting, there is only one: one has to seek out and destroy the enemy' (Daveluy 1902: 26). And this had to be done with 'the mass or concentration of forces: that is the weapon of victory' (Daveluy 1902: 44). After battle should come the inexorable pursuit of the defeated party, which 'nothing must stop' (Daveluy 1902: 100f.). 'Military strategy teaches to attack first of all the [enemy's] principal army.' If you could strike at the heart of a nation, you should do so (Daveluy 1905: 23f.). For Daveluy, the decisive battle was key to all success in naval warfare: to him it was not a question of either protecting the coast or attacking the adversary's territory or his commerce. 'Destroy the enemy and you will have both these results simultaneously' (Daveluy 1905: 8).

And, just like Foch and the other French land warfare strategists, Daveluy raved about the importance of moral force: under the French navy's motto 'Honour and Fatherland', he postulated, one should write 'it is the most obstinate who will win' (Daveluy 1902: 127–33). As it was difficult to explain (after France's defeat at the hands of the Germans in 1870/1) how the French navy could be crucial in a future war to free Alsace and Lorraine, Daveluy pointed to the growth of the French navy's preferred enemy, the Royal Navy (Daveluy 1905: xiv, xvi). The constant pointing to 'England' as enemy by the French navy is noteworthy. Instead of using the term 'Britain' that stood for a liberal state set steadily on the path to democracy, it conjured up centuries of enmity, furnishing excuses to both sides for not uniting against oppressive states throughout the world.

Quite in keeping with these sentiments, Daveluy shared with many of his contemporaries a disdain for alliances (Daveluy 1905: 45). When discussing the advantages and disadvantages of offensive and defence, however, he showed himself more subtle than some. He admitted that a fully offensive role made little sense as the navy could not seize ground the way armies could on land. The navy's mission, besides, always had to be to protect their nation's 'floating riches', which presented a great part of the 'national capital', and had to police the seas to prevent an invasion of the '*metropole*' (homeland) and the colonies. At the same time, the navy had to resort to attacks on the enemy's 'interests'. The navy thus had multiple tasks, which often meant the need to prioritise. Concentrating only on the defence of the coastlines, however, would be the death knell to any colonial power (Daveluy 1905: 51–8). Equally, he dismissed the *guerre de course*, noting condescendingly that it had actually been started by the English, but later used by the French against them. This, he argued, should be left to slow vessels out in the colonies; one must not use expensive new ships for this diversionary purpose (Daveluy 1905: 240–76).

Admiral Gabriel Darrieus's writings, like Mahan's and Corbett's, stemmed from his lectures at the French higher naval staff college (Darrieus 1993: 209–31). His patron, Navy Minister M.E. Lockroy (1895–6), had founded the *École de Guerre de la Marine.* Unlike Jomini, but like Mahan, Laughton, the Colombs and Corbett, Darrieus did not claim that hard and fast rules ('code') could be found (or deduced from history) that could be applied to the conduct of war, leading dependably to victory. His aim, he wrote, was both 'more modest and more useful: it consists of seeking in the past general indications which are appropriate to give to a great military leader, which, all other things being equal, will orient him towards greater chances of success' (Darrieus 1907: 7).

Darrieus parted company with the Anglo-Saxon commitment to naval predominance, writing (before Corbett) that it was 'megalomania' to believe in the possibility of a total, real 'mastery of the sea', a dream that only the British might be forgiven for dreaming. Nevertheless, he shared the Mahanian desire for decisive and big naval battles (*guerre d'escadre*). The French tended to quote Montesquieu on this matter, who in his *Esprit des lois* had famously pronounced against defensive warfare as it was bad for morale, '*décourageant*', giving the enemy the advantage of showing courage and energy. Darrieus had also read his German contemporaries, and accepted the views of von der Goltz and the others that the attack, the offensive, was infinitely better than the defensive. Quoting Moltke and von der Goltz, he asserted that the destruction of the enemy fleet was the main aim of war (Darrieus 1907: 96, 287f., 299f., 358). Concomitantly, Darrieus like Daveluy dismissed the *guerre de course* as an aberration, claiming that historical examples showed its lack of success. The big battle was what counted for Darrieus, and submarines or cruisers could only have minor roles in this central element of naval war (q.i. Darrieus and Estival 1990: 100f., 111f.). It was with this thinking that a generation of naval officers entered the First World War, and found itself quite disappointed.

Mahan's thinking raised a central question for France's Strategy: how could France (or Germany, or Italy, or Spain or any other continental power), not an island empire, afford the military posture of Britain that need not truly worry about invading armies? This had of course been the central line of criticism put forward by the *Jeune École*, but there were also writers after the demise of the *Jeune École* in 1887 who raised this question. Commandant Abeille, writing in 1912, dismissed the deductions generally derived from Mahan's writings as 'simplifications bordering on the ridiculous and generalisations so outrageous that they cannot withstand examination'. Mahan himself he accused of ignoring variables such as the intellectual, religious and political evolution of humanity, simultaneous land warfare and geographic specificities: were not Mahan's works with their selective choice of historical examples written specifically as advocacy of a certain naval policy for his own country, America, and were they not in essence advocacy of a political ideology (Motte 1994: 170f.)?

An alternative option for France – one not discussed by the strategists – was a diplomatic solution, long overdue: in recognition of the convergence of the political values and ideologies of France and Britain (and the USA), to abandon attempts to match them quantitatively and to secure an alliance with Britain (and the USA in the First and Second World Wars and, from 1949, in the Atlantic Alliance). The turning point in French operational Strategy was thus the conclusion of the *Entente cordiale* in 1904, which kept France's western 'back' free for her to concentrate, in her land defence, on Germany, and with her navy, before 1918 on the Austro-Hungarian Empire, and on Italy after Mussolini's rise to power in 1922.

Germany before the First World War

There is not the space to do justice to the naval writers of all countries of the period. Italy, with its outstanding traditions in naval commerce and power in the Mediterranean, in the nineteenth and early twentieth centuries produced a vivid debate that would feed into early air power theory. Here, too, the cult of the offensive had its followers and some notable adversaries (Ferrante 1993).

Germany, by contrast, was a latecomer on the naval scene. It was only with the arrival of Admiral Alfred von Tirpitz at the head of the new German Empire's equally new navy that this changed. Tirpitz dismissed any *guerre de course* or cruiser warfare Strategy that might have fitted a colonial policy. He wanted the German navy concentrated on the North Sea, not the Baltic, where Germany might have played the role of a regional hegemon. In an internal memorandum (*Dienstschrift* IX) on Tactics and Strategy of 1894, the fight for command of the sea was clearly defined as the ultimate aim of naval warfare. Much in the offensive spirit of the age, a defensive Strategy was deprecated as inactive and thus as 'moral suicide', and the navy was tasked to take any war to the coasts of an enemy, beyond its own territorial waters. In the context of the strategic offensive, giving battle with all its capital ships at the earliest possible opportunity was the aim of all efforts. Commerce raids was the Strategy of the defeated and the weak, and thus not for

Germany. Initially German naval Strategy under Tirpitz, unsurprisingly, was Moltke and the Prussian school of army thinkers gone to sea (Berghahn 1971).

In 1899 Tirpitz formulated a second Strategy for Germany whereby the High Seas Fleet should aspire to represent a 'risk' or deterrent, but not a threat, to Britain. The logic resembled both the 'fleet-in-being' logic and that of France's future nuclear Strategy in an uncanny way: the High Seas Fleet should be of such importance that a battle with it would leave even a victorious world power (that is, Britain) significantly weakened vis-à-vis any third power, so that such a strong power would avoid a direct clash with Germany (Rosinski 1977: 54). This 'risk strategy' with its accompanying naval building programme had the effect, however, of provoking Britain into changing over from focusing its naval planning on war against France and Russia (or even America) to war with Germany. It ushered in the naval arms race which led to the *Dreadnought* and gave grist to the mills of the 'decisive battle' school (Salewski 1979), and ultimately ensured Germany would be weakened by a naval arms race. The *Entente cordiale* between Britain and France in 1904 allowed the United Kingdom to withdraw its ships from the Mediterranean, which was left to the French to defend, and to concentrate them on the Channel, the North Sea and the Atlantic – which put paid to Tirpitz's risk Strategy as it supposed a threat to Britain from a third party. Like the French navy, the German navy fancied itself as the great challenger of the British Empire, in a struggle between 'Teutonic' and 'Anglo-Saxon races'.

There were dissenting voices, however. German Vice Admiral Victor Valois considered the challenge to British power heroic but unprofessional. Valois was a cruiser man, arguing against the commitment to capital ships of the greatest tonnage. Captain von Maltzahn equally came out in a struggle *against* the fetish of command of the seas, recognising that the *Reich* was a second-rank power in this respect. Vice Admiral Karl Galster advocated a 'small war' against Britain's commerce instead of giving a decisive battle, and urged his superiors to see to the defences of the *Reich*'s coasts and harbours – accordingly, he was a torpedo man (Brezet 1990: 129f.). This view sat uncomfortably with the offensive spirit of the age, and with Emperor Wilhelm II's obsession with rivalry with his British relations.

Conclusions

The full range of thinking about maritime Strategy had unfolded in the nineteenth century. Across national boundaries, different Strategies were advocated and different approaches existed and fuelled debates within each major naval power. But approaches and Strategies did not necessarily go together. Some historically minded writers argued for big battles – Trafalgar resonated as an aspired-to model – or else for alternatives like commerce raiding or the deterrent Strategy of the fleet in being. *Matériel* schools with their concern about the impact of ever-changing technology could argue, one day, for big battles with ironclads, another for commerce raids with torpedoes fired from cruisers as stand-off weapons and a third day for the abandonment of both in favour of the invisible submarine.

Crucial concepts were developed here which would influence air power and nuclear thinking and later Strategy in general: deterrence, coercive Strategies derived from the subtle use of force as political instrument known as gunboat diplomacy, but also the role of blockades in attempting to bring an enemy state to its knees by hitting its 'soft underbelly', the civilian populations. There was an ideological element in many of these approaches, or else a practical commonsense argument, sometimes both. But the ideological element did not necessarily have to fit the overall disposition of the political culture in which it had grown. Thus one would hardly have expected 'liberal' countries with a developed concern about human rights like France or Britain to have developed the *Jeune École*'s or the blockade advocates' disdain for civilian lives, nor Tirpitz in Germany to have so eagerly embraced the at least partly defensive 'fleet in being' strategy. The reality is more complex and shows that within each of these countries, arguments of practicality might contend with ethical concerns, and traces of ideologies otherwise alien to that political culture could flourish even within it.

10 *The World Wars and their lessons for maritime Strategists*

The purpose of war at sea ... is to protect one's own naval traffic and to cut off the enemy's.

(Admiral Kurt Assmann, 1943: 29)

The First World War

However many strategists like Corbett had earlier emphasised the need for amphibious operations, on the eve of the Great War the British, American and indeed German navies wanted to win decisive battles by themselves, not plan jointly with their khaki comrades. Under the influence of Mahan, but also of strategists of land warfare, American official naval Strategy also favoured the offensive, even though when war came, it was not as offensive in execution (Kennedy 1976, 1989). The Royal Navy's preferred Strategy was a distant blockade and the economic war, but many within it had a soft spot for Mahanian teaching.

The First World War saw only two supremely important naval operations, if one does not count the transport of the British Expeditionary Force to the European mainland, the bottling up of German military vessels in the North Sea and Baltic after the naval battle of the Falkland Islands on 8 December 1914 or the fight against German submarines. These two operations were the unsuccessful attempt to seize the Turkish Straits with the amphibious landings on the Gallipoli (Gelibolu) Peninsula in 1915, and the battle off Jutland (Skagerrak) between the British Grand Fleet and the German High Seas Fleet on 31 May 1916.

The British naval blockade of Germany gradually greatly affected the German economy. In early 1915, food shortages had already become painful. The overall result, rather than undermining the enemy's war effort decisively, was starvation among the German and Austrian populations, with around 900,000 deaths attributed to its effects.

Admiral Fisher, the First Sea Lord, had originally hoped to use a (near) blockade to force the German High Seas Fleet to meet the Royal Navy in a major battle, while it was his long-time adviser Corbett who recognised in the blockade the tool to 'exert pressure on the citizens and their collective lives'. This was an idea Corbett had picked up from his German colleague, General Colmar von der Goltz, who even before the First World War had developed the *Jeune École*-style idea that ways should be found to force the other side to beg for peace by 'making the enemy's country feel the burdens of war with such weight that the desire for peace will prevail' (Corbett 1911: 97f., see also Lambert 2010).

US strategic planning for war with Japan in the interwar period goes back to the impact on Americans of the Japanese victory against Russia in 1904/5, and to anti-Asian riots in San Francisco in 1907. A series of interwar plans under the title 'War Plan Orange' (and its successors in the 'Rainbow' series) all assumed three phases, starting with a Japanese attack on American outposts in the Pacific, followed by a gradual reversal turning into a US counteroffensive, leading to phase three, a blockade of Japan and air bombardment of her industry, until Japan surrendered (Hattendorf 2000: 124). Prior to the Second World War, the concept of the blockade was thus well established in US military thinking, as was the concept of air bombardment of industry as a complement with similar aims.

As we shall see in chapters 12 and 13, this logic would connect the rationale of the blockade – on the basis of the arguments of Goltz and Corbett – with the rationale of city bombardment – as advocated by Douhet and Trenchard and implemented most memorably against Coventry, Dresden, Tokyo, Hiroshima and Nagasaki.

Both World Wars saw the triumph and then the defeat of the submarine. In the First World War, of every 12.5 tons of shipping sunk by the Germans, 11 tons were sunk by submarines. In the Second World War, two-thirds of the shipping sunk by the Germans was as a result of submarine attacks, 13 per cent was caused by German aircraft (Lindsey 1980: 32f.). The perception was widespread in Britain that it nearly lost the war at sea to German U-boat attacks, but for the belated adoption of the convoy system in 1917. This was profoundly counterintuitive. The Royal Navy initially opposed it for a host of reasons, such as their conviction that merchant vessels would be unable to keep station and would constantly collide with each other in convoys, that the ships would have to sail at the speed of the slowest among them and that ports would not be able to cope with the need to deal with a large influx of ships all at the same time (Hattendorf *et al.* 1993: Doc. 440). Britain had been almost entirely unprepared, at the outbreak of the Great War, for this menace, being – perhaps excessively – focused on the German cruiser challenge (Gordon 2006: 89–102).

British lessons

Just like strategists of pure land warfare, Western writers on maritime Strategy were convinced, after the Great War, that any future war must be conducted differently – quicker, more conclusively, with fewer military casualties, with stronger effects on the enemy population who were expected to put pressure on their governments to admit defeat sooner. Here, as in land warfare, but unlike air power thinking, the cult of the offensive had abated significantly in Britain after the Great War.

While Britons continued to hold on to the firm belief that being British meant 'naturally' being 'sailors and an amphibious race' (in the words of First World War Prime Minister David Lloyd George), the bellicosity of Social Darwinism had given way in Britain (and France) to a war-weary, defensive mindset, recasting self-perception and history, marvellously summed up by Lloyd George as a Liberal MP in 1930:

> The whole history of the British Navy has been one of *defence*, sometimes of our own coasts, more generally defence of world freedom from one tyrant or another. Its strength ... has always been employed in defence of liberal civilisation, the freedom of nations and individuals. It is today, and has for centuries been, the only effective and honest police force of the seas ... For centuries past the British Navy ... has been the chief agent for the suppression of gun-running, the chief liberator of slaves ... it will take centuries of effort by the League of Nations to achieve as much at Geneva for peace, freedom and honour as has been achieved by the British Navy on the high seas of the world. (Bell 2000: 180f.; my emphasis)

In 1921, British navy planners were still happily drawing up contingency plans for war with the USA (Hattendorf *et al.* 1993: Doc. 442). By 1922, however, this new defensive stance, against a background of imperial overstretch, rocketing costs of the 'post-Jutland' ships and steeply rising social expenditure by the United Kingdom government, led to the acceptance of parity with the US fleet in the Washington Naval Treaty, where the ratio for the navies of USA, Britain, Japan, France and Italy was fixed at 5:5:3:1.75:1.75. Britain accepted German naval rearmament with the Anglo-German Naval Agreement in 1935, and by 1940 it conceded an overall reduction of the role of the navy in British Strategy. Meanwhile, the new air force, holding out promises of defence against new threats from the air, gained exponentially in importance (Lambert 1994: 97). While imperial policing clearly continued to be a role for the navy – and continued to be seen as such well into the twenty-first century, in the new guise of world policeman – the central role that many naval thinkers had wished for the navy was on the wane, as even naval writers conceded (Kennedy 1936: 45–96, 221–8).

Liddell Hart was neither the first nor the last Briton who thought that in Britannia's hands, 'the Trident [is] a mightier weapon than the sword', as one British MP put it in 1910 (q.i. Till 2006: 119). He argued for British abstention from continental warfare, except by supporting proxies fighting on the side where British interests lay. This argument received renewed attention in the Cold War, when Michael Howard, later long-time doyen among British strategists and known for his staunch commitment to NATO and European defence, temporarily used it to plead for British disengagement from the Continent (Howard 1958). This nostalgia was anachronistic, however, in an age of aircraft and missiles.

The waning of decisive battle

The naval battle of Jutland in 1916 had been the only big naval battle in which the Royal Navy's Grand Fleet fought the German High Seas Fleet in the First World War, and – arguably – ended in a draw: while the German fleet was smaller, both fleets were approximately equally equipped technologically, and the Germans owed their ability to fend off defeat to training, not to the quality of their technology (Hattendorf 2000: 44f.). Different lessons were drawn from this. Extreme supporters of victory through decisive battle, such as Commander John Creswell, Custance, Captain Bernard Acworth and Commander Russell Grenfell, thought Admiral Jellicoe had been too cautious during this battle. They thought the consequent development of the war had proved that it was more expensive for Britain to try to secure command of the sea by methods short of annihilating the main enemy fleet. Had this been achieved, patrolling trade routes would have been much easier and cheaper. Richmond remained sitting on the fence. He saw the pros and cons of attempting to impose a large-scale naval battle on the enemy's principal fleet. On the one hand, he conceded that its pursuit at all costs might detract attention from more urgent and vital operations, depending on circumstances, as the overall aim always had to be the protection and defence of British and Commonwealth trade routes, followed by the ability to support the army with amphibious operations. On the other hand, much could be done by wiping out the enemy's capital ships: once that had been achieved, this would liberate one's own (remaining) ships to act as escorts for convoys, to impose blockades or to attack the enemy's overseas possessions.

> The dangers of invasion are removed and ships and men and material are set free for protection of trade, or attack upon trade. The whole experience of war tells the same tale – a great victory is followed by a dispersion of the ships that had concentrated for it. (q.i. Till 2006: 115)

On the other end of the spectrum, Richmond's patron Corbett, the official historian of the Great War at sea, displeased the Admiralty with his unenthusiastic analysis of the importance of battle when dealing with Jutland. It famously led to the official disavowal of his conclusions in a preface, in which the Admiralty complained especially about 'the tendency to minimise the importance of seeking battle and of forcing it to a conclusion' (q.i. Till 2006: 114).

Corbett was not alone with his views, however. Sir Archibald Hurd, perhaps the earliest retrospective critic of the performance of the Royal Navy in the Great War, wrote in 1918,

> If the Battle of Jutland had resulted in the annihilation of the [German] High Seas Fleet our position would not have been greatly altered: Germany would still have possessed in her destroyers, submarines and minelayers the only active element of her naval power; her coast defences ... would have remained. The great ships would have gone, and to that extent our great ships would have been set free. For what purpose could they have been used after the German High Seas Fleet had been destroyed? (Hurd 1918: 126)

Many opinion- and policy-makers in the navy were persuaded by Corbett's and Hurd's position, and lost any obsession with decisive battles à la Jutland, if ever they had had one. They returned to the old British naval mission of protecting Britain's coasts and commerce, and threatening any enemy's. In the interwar years, the Royal Navy planned and exercised for a long struggle of this sort, not for big naval battles. Meanwhile, the decline of the British navy itself was precipitous in the face of a rising triple threat from Germany, Italy and Japan (Bell 2000: 1–47).

John Creswell and Russell Grenfell, writing in 1936 and 1937 respectively, still thought that Britain needed a battlefield supremacy which alone could enable her to reach a decisive decision in battle – blockade, attack on enemy trade and anything else he regarded as second best. Both were still steeped in the pre-First World War spirit of the offensive (Creswell 1936). Grenfell deplored the waning of the Nelsonian heritage in the First World War: Jutland should have been more decisive, should have been key to a British presence in the Baltic. 'In addition to the glamour of victory and the appeal it would have made to the imagination of the world, it would, I believe, have convinced all nations that the final victory was bound to be with the Allies, and it might easily have shortened the war by a year.' Grenfell believed there were two primary objectives in naval warfare, 'of which the destruction of the enemy's armed forces is one and the control of sea communications the other, both being interrelated' (q.i. Till 1982: 103). This was nostalgic thinking; a more realistic appraisal of the inconclusive battle of Jutland led to a spreading concern on both sides of the Atlantic that Mahan's belief in the possibilities of sea power had been exaggerated (Sumida 1999: 40). Yet in the interwar period, the US navy's preferred exercise scenarios tended to revolve around decisive, ever larger-scale naval battles.

After years of cuts in spending, the dire straits in which all three branches of the armed forces found themselves, coming on top of war-weariness, greatly contributed to the British government's penchant for appeasement. The Royal Navy's leadership were among the strongest advocates of this policy (Gordon 1994: 63–85).

While Mahan had still envisaged Britain as the senior partner in a special Anglo-American naval consortium, Britain was rapidly sinking to being the USA's equal and then, after the Second World War, junior naval partner. The stages of this decline were the US Naval and Shipping Acts of 1916, shortly before US entry into the First World War, which initiated a significant expansion of the US fleets, commercial and military. The next step was the already mentioned Washington Naval Treaty of 1922. In the Second World War, the US domination of the Pacific was acknowledged by Britain, Australia and New Zealand. Confirming this, in 1951 Australia and New Zealand concluded the ANZUS (Australia, New Zealand, US) mutual defence treaty with the US which left out the United Kingdom (Hattendorf 2000: 109–20). Britain, too, continued to place her reliance on the US, in her case in the Atlantic: the Royal Navy was subsequently configured to play a role almost exclusively as the USA's ally in NATO, with a clear complementarity of roles – Britain's navy was reoriented primarily towards blocking Soviet naval access to the North Sea and the North Atlantic. Britain now played the role of junior partner that Mahan half a century earlier had envisaged for the USA.

As the official historian of British naval operations in the Second World War, Captain Stephen Roskill also underlined the importance of the offensive tradition, but in a new context of jointness:

> Though the exercise of maritime power in defence of trade is essential to the nation's war economy, and it alone can provide the conditions from which the final decisive offensive will be launched, it is by exercising this same heritage in the despatch of great military expeditions overseas that a maritime strategy can be crowned by final victory. (Roskill 1954–61: 12)

Roskill cautioned against 'indecisive battles' with unnecessary loss of shipping and lives (Gardner 2006: 151, 157).

Amphibious operations, convoys and blockades

As the major British and Commonwealth amphibious operation of the First World War, the ill-fated Gallipoli campaign of 1915 received equally contradictory reviews as the Battle of Jutland. Neither Corbett, Richmond nor Callwell thought that the outcome of the unsuccessful and casualty-heavy campaign in the Dardanelles had invalidated the concept of amphibious operations altogether. Indeed, consensus remained that amphibious operations were good in principle and might be applied in the future (Till 2006: 110–13). Churchill in the Second World War proposed a similar angle of strategic attack on the spreading German sphere of power to an unenthusiastic Stalin, but then Turkey's neutrality was in the end more valuable than the potential gains from such an assault. The growing roles of air power, however, made other British strategists more sceptical about the future possibilities of seaborne invasions: while the Second World War was looming, Liddell Hart, for example, thought another Dardanelles-style amphibious operation would lead to an even greater catastrophe (Liddell Hart 1939: 131).

The persistent attacks on Allied merchant shipping by German submarines during the First World War put paid to the British and American tendency to dismiss the *guerre de course* as outdated or irrelevant. Importantly, the U-boat menace could not be dealt with offensively in a decisive naval battle, but could only be met defensively, and through the application of the convoy system (Kennedy 1989: 174f.). Against French opposition, Britain failed to impose the elimination of submarines on all sides in the Washington Naval Conference of 1921–2; the Royal Navy feared them precisely because they posed a new challenge to its battle fleets, while regarding them, in Admiral Wilson's words, as 'underhand, unfair, and Damned unEnglish' (Kennedy 1989: 182) – much in the same spirit as the Second Lateran Council in 1139 had banned the crossbow as unchristian (and no more successfully).

Few writers had foreseen the importance of convoys. This was one of the points which Corbett, otherwise extremely clear-sighted and indeed prophetic, had got thoroughly wrong in his *Principles of Maritime Strategy*. Corbett, a lawyer by training, equated blockade with the occupation of enemy territory as a weapon of war, thereby foreshadowing an important element of Total War (Corbett 1911: 245). One of the great lessons of the First World War was that it was pointless to try to protect lines of communication as though these were railways, every inch of which had to be intact for a train to be able to move on it. It was infinitely more sensible to move convoys of ships about under the protection of naval vessels (Till 2006: 77).

As far as blockades were concerned, the British ban on German (and Austrian) overseas trade in the First World War had had effects beyond those of any historical precedent. Richmond, Hurd and other supporters of the navy claimed that the blockade of Germany had made a vital contribution to Western victory. Few if any of the British strategists lost any sleep over the fact that, technically successful as it was, the blockade claimed victims mainly among non-combatants.

French lessons

Even though the French navy had ranked fourth among the world's navies on the outbreak of the First World War, it had fought no major naval battles in this war, but concentrated its efforts on the Mediterranean, more to counter the influence of its allies Britain and Italy than to combat the Austro-Hungarian navy. For the defences of her imperial lines of communication, France had to rely to a larger extent on Britain than imagined even by the architects of the *Entente cordiale*. The need for the alliance with the United Kingdom was strongly confirmed, much to the annoyance of the French navy leadership, who had been raised on rivalry with Britain.

Although the French navy joined in the blockade against Germany, remarkably few Frenchmen wrote about naval or maritime Strategy in the interwar period. Nevertheless, some lessons seemed to stand out. The most important concerned the paucity of big, decisive naval battles (or their complete absence, in the case of France). But like so many historical events, one could see it in two different lights: either a vindication of the usefulness of a mere fleet in being (the French, in this case, which checked and paralysed the Austro-Hungarian navy), or the call for an avoidance of the slaughter in the Western trenches through the shifting of the main theatre of war from land to sea, where a decisive battle might be fought less bloodily. Even before the Great War, its bloodiness on land had been prophesied, and a less painful naval alternative had been proposed by Ambroise Baudry (1912: 10). After the war, most French authors abandoned any hope of France being able to confront a first-rate power (such as Britain or the USA) in a *guerre d'escadre*, a major battle between capital ships, or to secure, on her own, 'command of the sea'. They confirmed the need for an alliance with the United Kingdom, which alone would allow France at least to predominate (against Italy) in the Mediterranean, where it might seek 'command of the sea' on a local scale (Motte 1990: 123–7).

Commandant C.C. Richard resuscitated the *Jeune École*: his lesson drawn from the Battle of Jutland was that the concept of the decisive naval battle of annihilation was inappropriately imported into naval

Strategy from land warfare. In his view, the attack on and defence of naval communications and the blockade were the true objects of naval warfare:

> Unlike on land, at sea, the destruction of the organised forces of the enemy does not constitute the unique, necessary and sufficient means of imposing one's will upon the adversary. Battle, while always desirable, is only useful where this destruction is needed to reach the double objective of the true mode of action of naval power: the maintenance of the integrity of its communications, and the rupture of those of the enemy.

And in the light of the French experience in the Mediterranean in the First World War, the concept of a fleet in being, to Richard, was attractive. Such a Strategy could be appropriate for a second-rank navy (like that of France) 'on condition that their fleet truly demonstrates its existence through an incessant activity, jointly with that of its *corsairs*', the modern-day *corsairs* par excellence being the submarines (q.i. Motte 1990: 149f.).

In the light of the submarines' *guerre de course* and the few big naval engagements otherwise, one might have expected the *Jeune École* to come out of the First World War triumphant. Yet the 'Royal Way' of the Mahanians continued to predominate in the French navy's educational establishments, albeit not completely immune to revision, reconstruction and updating. Among those musing on the lessons of the First World War, one stands out: the reconstructed or neo-Mahanian Admiral Raoul Castex, whose main work, the six-volume *Théories stratégiques*, was first printed between 1927 and 1935.

Castex stood in the tradition of the Mahanians Daveluy and Darrieus, and indeed of the Prussians and Germans since Moltke.

> The defeat of the organised force [of the enemy] is ... the first objective [of war]. Concomitantly, the quest for combat ... imposes itself. All aspirations, all dispositions have to be oriented towards this ... Until we have achieved this crucial defeat of the organised forces of the enemy, we have to abandon all other preoccupations, [such as] the defence of our coasts or communications. That would harm our central objective, lead us to disperse our forces and would expose us to a dissipation of efforts ... What matters is to win in the naval theatre itself ... We shall practice the offensive which alone assures the full flowering of our concept, without worries about the risks which await us on this path. (Castex 1937: 220f.)

He retracted somewhat from this absolute statement when he rehabilitated other components of naval warfare, declaring, for example, that the creation of coastal defences in peacetime was desirable (Castex 1934: 167–76). The *guerre de course*, however, he thought had brought only failure to France, despite the heroic memories of France's *corsaires* (Castex 1934: 111, 166). He conceded that victory could be achieved only through 'an integrated offensive' of all forces, surface fleet, submarines, air forces. But submarine and air offensives were secondary, the surface fleet played the central role, and the main enemy was the enemy's surface fleet. It made sense for the air force to attack enemy military and commercial harbours, 'treating them just like other enemy towns'. But this was ancillary to the main fleet battle (Castex 1934: 116f., 142–9).

Castex dwelt on the disadvantages of defence on land, above all, demoralisation through 'waiting, expecting, anxiety, uncertainty, which depressed courage'; as usual, Montesquieu was invoked. Castex conceded reluctantly that there might also be merits in the defence of land. By contrast, he saw no advantages in naval defence (Castex 1934: 106, 150–65).

He admitted that a thoroughly offensive naval Strategy was more difficult to carry out than an offensive in land warfare, as the immense width of the oceans could never be entirely dominated, and the enemy could withdraw into his ports (Castex 1934: 126). Castex conceded a point to Corbett, who had warned against a blind rush into an offensive at all costs, as an offensive Strategy requires the means both in quality and quantity to carry them out (Castex 1934: 121). It was thus all-important that France acquire these, and that she begin to favour an offensive posture if at all possible (Castex 1934: 136).

As a true Moltkean, Castex emphatically denied that there was any innate link between the overall political war aims of a country and its choice of an offensive or defensive Strategy. Thus he had difficulties explaining why 'pacific' France had in 1914 adopted an offensive Strategy (Castex 1934: 133–6). He turned a blind eye to the alternative explanation, namely, that France's militarism offset its defensive and pacific spirit, and that nationalistically infected democracies might do terrible things.

Castex, like Liddell Hart, de Gaulle, Guderian and other contemporaries, showed a great interest in the potential of manoeuvre, intelligent movement to create a favourable situation for oneself. Despite his devotion to the offensive, he did not favour a mindless, frontal offensive à la Verdun or the Somme (Vascotto 2001: 81–8). Instead, he, like several key thinkers of the interwar period, devoted a good deal of thought to ways of outflanking the enemy fleet, catching it by surprise:

> Instead of a homogenous, symmetrical, unimaginative ... distribution of forces we choose an asymmetric arrangement, with an eccentric point of gravity [*désaxé*], oriented towards a previously determined direction, intended by the spirit which moves the forces and the material. (Castex 1939: 10, 16)

Castex's views were untypical for war-weary, defensive France. In terms of overall Strategy, French defence spending in the interwar years was concentrated on the army and the fledgling air force, and funds were taken away from the French navy, especially during and after the Great Depression. The French Admiralty resented the growing French dependence on co-operation with Britain, and the political leadership's acceptance in 1922 in the Washington Naval Treaty of having a navy only little more than a third the size of that of Britain and the USA, and equal to that of Italy. Realising French fears, Italy soon cast off the shackles of the Washington Treaty, and like Germany engaged in a major naval construction programme, both of which soon troubled the French navy. British and French strategies for dealing with these growing 'threats in being' differed diametrically. While France tried to defend the status quo fixed in the Versailles and Washington Treaties at all costs, Britain's overall Strategy during the interwar period was to release pressure from fascist, nationalistically charged Italy and Germany by making limited concessions to what was seen as their understandable ambitions (Scammell 1997: 92–118). The *Royale*, but also the civilian French government, saw the Anglo-German Naval Agreement of 1935 as a perfidious betrayal by Britain. And yet France's strategic dependence on the United Kingdom and the empire was inescapable. The French Admiralty did not cease to stress that this spelled disaster for France; they almost anticipated the ultimate act of what the French saw as 'Albion's perfidy' at Mers el Kebir, where the largest part of the French fleet was sunk in mid-1940 without prior warning to France, resulting in the death of 1,300 French crew, in a British surprise attack to prevent it from falling into the hands of the Germans (Heuser 1998c). This overshadowed a naval co-operation which was dictated by the otherwise closest possible convergence of values and political ideology, and by strategic necessities (Jackson 2000: 130–59).

The second-tier powers

Germany

As Britain failed to fulfil the German naval leadership's expectations in terms of battle-mindedness, a small number of German officers under Rear-Admiral Hipper proposed more offensive action against British lines of communication. The German submarine or '*U-Boot*' provided the perfect tool for this. Yet this was pursued at once in defiance of the laws of war (provisions for saving sailors and passengers from attacked ships were soon abandoned) and only half-heartedly from a strategic point of view. For the naval leadership continued to have its eyes on its capital ships in the North Sea, and craved the big battle.

From 1915, a young naval captain, Wolfgang Wegener (1875–1956), saw this dilemma and pleaded for a transfer of Germany's naval centre of gravity to the Baltic Sea, where the High Seas Fleet had a real chance of cutting off all allied supplies to Russia, while abandoning the North Sea and with it the quest for a decisive battle. Wegener produced a memorandum to this effect, which was rejected by Tirpitz as 'poison for the fleet' (Rahn 1990: 138f.). After the war, Wegener, by now Vice Admiral, produced a critique of German naval action in the First World War, in which he accused the High Seas Fleet of having wasted its strength on the Jutland campaign. Instead, he thought it should have tried 'to reach the Atlantic at all costs', with an amphibious campaign that should have secured southern Norway. Wegener noted that the German navy's leaders at the beginning of the First World War were convinced that Britain would attack Germany's northern shores – in the belief that the stronger always attacks. The German Admiralty was so blinded by the prevailing doctrines of land warfare, he argued, that they did not understand British inaction.

But in fact, argued Wegener, the Royal Navy's behaviour made eminent sense, as Britain's strategic position was almost unassailable – her supplies came from the Atlantic, her lines of communication were out of reach for the High Seas Fleet, while German trade was easy to cut off in the English Channel or near Scotland. The British could thus turn the North Sea into a 'dead ocean' with little effort, and had no reason to change their Strategy (Wegener 1941: 4f.). In all of this, Britain's position was strategically defensive. Only the German submarine attacks on British shipping across the Atlantic eventually affected the United Kingdom (Wegener 1941: 8).

Instead, Germany should have taken a different approach: 'Command of the sea', he explained, 'means controlling commercial lines of communication. Such a line was within our reach', namely, that to Denmark, Sweden and Norway, and thence to the Atlantic, to break through Britain's blockade. But Germany had failed to reach out for it (Wegener 1941: 11). Such a Strategy would have made sense, as it would have taken account of actual commercial needs and geostrategic givens. Instead, the High Seas Fleet was yearning for 'the battle "as such"', 'a battle, which does not exist' (Wegener 1941: 20–3, 46). Battle should be seen as means, not as an aim in itself (Wegener 1941: 32). Instead, 'naval warfare is a struggle about the lines of communication ... nothing else' (Wegener 1941: 49).

Wegener argued that 'for the army ... strategy only begins with the outbreak of war, while for the navy, it is detached from tactics and thus begins before war, ideally in peacetime'. Groping towards the realisation that the greatest German mistake before and during the First World War had been the total absence of political direction for military planning (but also the lack of co-ordination between the two services), he stated, 'naval strategy is not a mere military task but a joint task for the military man and the politician, in war and peace' (Wegener 1941: 64). 'World politics ... is naval power' (Wegener 1941: 80).

Another assault on the prevailing Tirpitzian orthodoxy came from Captain Otto Groos (1882–1970), a contemporary of Wegener's (although with a less brilliant career), who was Germany's foremost disciple of Corbett. Like Wegener, Groos saw the need to point to differences between naval and land Strategy: while in the latter, the destruction of the enemy forces and the occupation of land might be the necessary aim, this was quite different for the navy. He quoted the geographer Ratzel, who had stressed that war at sea tended to take place in much greater spaces, and could not be localised, the way war naturally is on land. Consequently he argued, like Wegener, that battles, especially naval battles, could not be aims in themselves. Nor were they the only way to achieve one's aims in war at sea. Only an understanding of the overall strategic situation – on land, at sea – could lead to sensible choices in Strategy (Groos 1929: 44f., 57ff., 76f.). Like Wegener, Groos emphasised the peculiarities of every country's geographic situation. Options such as the command of the Atlantic were simply not available to Germany, while Germany might achieve much with its cruisers and submarines through commerce raiding (Groos 1929: 181–95).

While Groos sympathised with a 'fleet in being' Strategy, he also saw its problems. He gave his own definition of its aims:

> To deny the command of the sea to a numerically superior enemy who needs ... it for the realisation of his plans. This should be done through a decisive battle, while one's own fleet acts defensively, but takes advantage of every opportunity to injure the enemy through counterstrokes.

The limits of a 'fleet in being' Strategy was that it would never allow one to gain command of the sea oneself (Groos 1929: 121).

A contemporary observer detected a trend in interwar German thinking putting increasing emphasis on the interception of enemy trade and protection of one's own (Rosinski 1977: 61–5). It culminated in the writings of Admiral Kurt Assmann (1883–1962), which put economic warfare – *Wirtschaftskrieg* – at the centre of all naval warfare in the industrial age, and pointed to its centrality in British naval Strategy in both World Wars. This, not the pursuit of any decisive battle, in Assmann's view had to be the German navy's main preoccupation (Assmann 1943: 2–16, 29).

On the eve of the Second World War, Admiral Erich Raeder (1876–1960), commander-in-chief of the German *Kriegsmarine* (as the navy had been called since 1935), adopted something much akin to a fleet in being Strategy:

> Enemy naval forces, even if inferior in strength, are only to be attacked if this should be necessary to achieve the main objective. Frequent changes in the operational area will provide uncertainty and delays in the sailing of the enemy's shipping, even if no material success is achieved. The temporary disappearance of German warships in remote areas will add to the enemy's confusion. (q.i. Till 1982: 119)

Raeder, like most of his generation, saw Britain as Germany's main and traditional adversary and the USA as Britain's natural ally, however isolationist it seemed at the time. And yet in 1929, tasked with a definition of its objectives in times of war, the navy directorate, under Admiral Raeder, planned for naval operations against Poland and France. Raeder underscored that it was necessary for Germany to avoid any major conflict with Britain, simply too powerful an adversary. By contrast, interrupting the trade of France and Poland seemed a realistic option (Rahn 1976: 281–6).

Raeder and Wegener were realistic, recognising Germany's limitations even when they began to build a new fleet in breach of the peace treaty obligations. When in May 1938 Hitler issued a directive in which he listed France, the USSR and Britain as adversaries for the sake of military planning, the naval directorate's (*Seekriegsleitung*) section Skl 1 produced a 'Memorandum on Britain' in October 1938. It prioritised the *guerre de course* against Britain and her colonies; its motto was 'Naval warfare is the struggle about the economic and military lines of communication.' From this followed the principal task: to destroy the enemy's 'rule of the sea in areas of the sea which are needed by the adversary for his sea communications'. The document stated that this would only be possible if Germany dispatched units of cruisers to these areas in advance of any blockade Britain was likely to impose on Germany; moreover these units would have to be self-sufficient and must not rely upon supplies for three months. There were echoes here of the hope that the British might be persuaded to come to an agreement with their German fellow Aryans – as Hitler himself hoped fervently – as any alternative would be so unpalatable and costly (Salewski 1970: 45–58). This was, however, a considerable misperception of Chamberlain's motives for appeasing Germany, where racial sympathies played no part.

Once the Second World War was under way, the *Kriegsmarine* hardly played a role in the defeat of Poland and France. Raeder and the leadership of the High Seas Fleet were unenthusiastic about Hitler's two-front war against the 'Anglo-Saxon Powers' in the West and the USSR in the East. In true German (and French) naval tradition, they wanted to turn against the West: they saw the British Empire, plus the USA, almost as one sole entity. They urged a declaration of war against Washington even in 1939, and advocated unlimited submarine warfare against all shipping supplying the British Isles, in the full knowledge that such action had brought the USA into the First World War (Herwig 1971: 650). Hitler only declared war on the USA on 11 December 1941, in the wake of allied Japan's attack on Pearl Harbor, when the great German invasion of the USSR – 'case BARBAROSSA' – was well under way. Much to the chagrin of the naval leadership, BARBAROSSA continued to have total priority until the defeat of Germany.

In 1942 the naval leadership tried once more to develop a comprehensive Strategy, referred to as 'great plan' by Michael Salewski, in order to concert the entire Strategy of the Third Reich to defeat the German navy's favourite enemy, Britain, in a triple indirect approach. The army's campaign against Russia should turn south through the Caucasus to deprive Britain of supplies from the Persian oilfields and the Middle East; Rommel's land campaign in North Africa was to aim at seizing Egypt and the Suez Canal; and Germany's ally Japan was to be persuaded to advance to the Indian Ocean to deny Britain further supplies from there. Much to their dismay, however, Japan did not fully comply, as it continued to see the USA, not Britain, as its main enemy. Thus the 'great plan' was dead even in mid-1942 (Salewski 1975: 72–107).

Germany's defeat put an end to any independent German naval Strategy. The Bundeswehr, created in 1955, had to content itself with a small role in the North Sea and Baltic, always acting alongside allies.

Lessons elsewhere

While the Italian Admiralty had been fervent Mahanians on the eve of the First World War, the war itself did not at all follow their expectations. Capital ships were destroyed – not in battles, but in their harbours, both on the Italian and the Austrian side. Other than that, war in the Adriatic in particular was a stand-off, with mutual paralysis generated by two fleets acting, one might say, as 'fleets in being', with capital ships more important as political symbols than military assets.

Nevertheless, even after this war, there was a tendency towards Mahanian thinking and there continued to be fervent advocates of the construction of capital ships. The fascist victory in 1922 encouraged great dreams and great ideas, postulating that the Mediterranean should once again become 'our sea', *mare nostro*. Mussolini had his naval strategists to support these dreams, such as Oscar di Giamberardino (1881–1960), Edoardo Squadrilli and Vittorino Moccagatta, who preached the offensive, the cult of the big battle and the need to annihilate an enemy's fleet to obtain a decisive victory, all of which fitted perfectly into the fascist ideology (Giamberardino 1937; Squadrilli 1937). Alongside these, however, there were also followers of a more cautious approach: Alfredo Baistrocchi, convinced that the next war would again be a war of attrition, underscored the primary task for the Italian navy of defending Italy's maritime commerce; Giuseppe Fioravanzo (1891–1975) tended towards a 'fleet in being' Strategy, which he adapted to mean a fleet free to fight wherever and whenever it wished to, but defensively (Baistrocchi 1924; Fioravanzo 1930–1). Fioravanzo was the main opponent of Giulio Douhet, the air power theorist, in defending the continuing and equal importance of the navy: he argued that navies were still and would continue to be vital for transport, commerce, troop movements and supplies, a role even an air force with much larger aircraft could not hope to usurp in the foreseeable future. Moreover, Fioravanzo defended the need for an integrated use of all three forces in a future major war, 'an integrated war', as he called it. Douhet, by contrast, thought that navies need only be small, with fast ships and submarines to defend the coastlines from enemy attacks at sea, while the main action in future would be in the air (Ferrante 1997: 153–69).

Of the world's other naval powers, only Japan had remained unscathed by the First World War. The Japanese navy leadership were admirers of Mahan, and had his writings translated into Japanese even before the nineteenth century was out, so that his works could be read at all naval academies. The Japanese defeat of the Russian fleet in 1905 seemed to them ample confirmation that Mahan was right. They were convinced that in their rivalry with the USA, what was needed was a major naval battle, for the purpose of which they had to get the US navy to concentrate in one place in the western Pacific. Japanese aviation should then help bring about a defeat of the US navy. For the Japanese naval aviator and politician Minoru Genda (1904–89), this hinged on the creation of aircraft-carriers, from which their aircraft could launch their attack. Genda was not without opponents who disliked this particular form of marriage between air forces and navy, but he prevailed (Budiansky 2003: 257). Japan's grand Strategy in the

1930s and 1940s was a function of its interservice rivalry, the navy drawing it into a conflict with the USA, the army into land warfare in China; the resulting compromise that tried to do both was as spectacularly unsuccessful as the close co-operation between army and navy at Port Arthur and Tsushima in 1904–5 had been successful.

US lessons from the Second World War

As Eric Grove and Geoffrey Till have underscored, 'British and American experience in the Second World War had been quite dissimilar' and so were the images of the utility of navies in the Cold War. The USA had fought and won classic naval battles, while the Royal Navy had been much more focused on the protection of its own naval communications and the interception of enemy ships and trade. The US navy emerged from the Second World War with a Mahanian emphasis on ships on the largest end of the spectrum still going strong, and with decisive naval battles still at the centre of naval education. Aircraft-carriers in particular had been at the centre of the US navy since 1941, and had played a large role in the Pacific theatre. And even among those strategists who had previously believed that a purely naval Strategy could exist independently of land forces, the First World War had sown doubts. By contrast, the Second World War saw the transformation of the battleship into a useful albeit subsidiary weapon, in support of combined operations: this was shown in the successful British attack on the Italian fleet at anchor at Taranto on 11 November 1940, the Japanese attack on Pearl Harbor on 7 December 1941, the US victories at Midway on 4 June 1942 and near the Philippines at Leyte Gulf on 23–26 October 1944 (Grove and Till 1989: 281–83; Till 1982: 109).

The decisive naval battles of the Second World War had above all involved the US navy in the Far Eastern theatre. While some strategists acknowledged that these naval battles had to be seen in the context of submarine warfare against Japanese commercial and military shipping, and the indecisiveness of naval operations before air power brought to bear the ultimate weapon on the war in the Far East, these naval battles led some to insist that the nuclear technological revolution had not changed 'the fundamental principles of sea power' any more than had the arrival of the steamship (Gretton 1965: 4). Indeed, what one US Secretary of War said about the US navy during the

Second World War continued to hold well into the Cold War period, namely that its leaders and spokesmen 'frequently seemed to retire from the realm of logic into a dim religious world in which Neptune was God, Mahan his prophet, and the United States Navy the only true Church' – with Trafalgar its Epiphany, one might have added, replaced after the Second World War by the US navy's experience in the Pacific war (q.i. Till 1994b: 197).

The course of the Second World War in Europe, however, and particularly its gruesome air-power dominated finale in the Far East, reinforced the call for 'jointness': the conclusion for most was that it was pointless to think in terms of a 'naval' vs. a 'continental' Strategy. By the end of the Second World War, the battleship had clearly yielded its 'queen-of-the-naval-battlefield' position to the aircraft-carrier, the recognition of the end of an era where one could fantasise about decisions being brought about by navies alone, without land or air dimensions (Clark and Barnes 1966: 63).

Conclusions

While the USA came out of the Second World War still believing in naval battles, the faith in Mahanian ideas had been shaken in all other major players. While Castex's Mahanianism in France survived the First World War, German thinkers urged their superiors to reconsider after their experiences with the Battle of Jutland/Skagerak. By the end of the Second World War, the myth of Trafalgar had paled considerably, and naval thinkers on all sides found themselves confused as to what course to steer. The simplest, bureaucratically logical way was to cleave to earlier configurations where possible and to replace obsolete shipping with as many of the same types as still affordable, or, if one was a rich nation, to try to keep up with the Joneses in naval procurement (which particularly applied to France).

However, the full unfolding of air power that the Second World War had brought, and then the nuclear revolution, imposed new considerations, which only gradually made their way into the particularly long planning and procurement cycles of navies. These considerations we shall discuss in the following chapter.

11 Maritime Strategies in the nuclear age

Come to think of it, I would not put anything on the surface of the ocean – it's too good a target [for nuclear weapons].

(Edward Teller in the mid-1950s, q.i. Till 1982: 183)

The Cold War framework

The interwar period (especially the 1930s) and the Cold War had in common that they were at once characterised by the absence of major war and the ever-present fear of it. The French naval officer Hubert de Moineville rightly commented about the Cold War:

[I]n the state of the world today, and in that unprotected space that is the sea, peace does not really exist. The word is used, mainly as a linguistic convenience, to describe the permanent state of tension in which we live. (Moineville 1982/1983: 9)

What continued in the Cold War was the preoccupation of the *matériel* school with new technologies and the problems of overall strategic analysis in the context of ever new imponderables presented by technological innovations and revolutions. The focus on fleet action was lost in the preoccupation with all these new technological features (Martin 1967: 10). Similarly, specialists accused NATO and successive Cold War governments of lacking a grand Strategy with regard to the employment of naval power (Hattendorf 1982: 59). Sir Peter Gretton (1912–92), who rose to the rank of Admiral and Fifth Sea Lord and upon his retirement from the Royal Navy became a research fellow at Oxford, thought NATO and US Strategy incoherent:

Here we seem to have the Army and Navy preparing for a long war in which reserves could be used, while the Air Force stressed the Strategic Bomber attack ... [the scenario of] a long global war still survived [the

Korean War and the advent of thermo-nuclear weapons] as the main reason for which the expansion programme had been planned, though already the talk of broken-backed wars cast doubt on the whole concept ... What, in the meantime, had been the development of strategic thought in NATO naval circles and in particular at the Norfolk headquarters of SACLANT? ... none. (Gretton 1965: 34f., 44)

Gretton was not the only critic, as the 'great debate' about NATO's nuclear Strategy of those years showed (Aron 1963b). As one of those SACLANT planners commented, '[t]he "concept" of fighting a nuclear war was inherently incoherent ... Meanwhile the armed services had to strive to develop contingency plans for handling the results' of a nuclear exchange. They tried to do their best with a 'surreal' mission.[1]

In the Cold War, the very geography of the world seemed to change, but not just for technological reasons. Obviously, the advent of aircraft had transformed Strategy since the First World War, and in the Cold War, the development of aircraft with the capacity to reach any point on the globe and to remain airborne for more than a day thanks to in-air refuelling began to change the meaning of geographic distances. Missiles did the same.

But changes in the perception of geography occurred also thanks to political changes. In the interwar period the Mediterranean had been of unprecedented importance to Britain on account of the surge of British and imperial shipping through this sea and on through the Suez Canal to Asia (Hattendorf *et al.* 1993: Doc. 446). With the existence of the Suez Canal, the Mediterranean was Britain's lifeline to its empire until 1949, when the Indian subcontinent ceased to be a British colony. Yet the perception of the importance of this line of communication continued even thereafter, as Britain's attempt to secure the Suez Canal in 1956 proved. With the loss of empire, the Mediterranean's importance for Britain evaporated, although the realisation of this took long to sink in. Only in 1967–9 did Whitehall's Defence White Papers draw a line under the history of British Empire and military commitments east of Suez, to refocus on the North Sea and the Atlantic, with only occasional sorties into the Mediterranean, where British forces had their bases in Gibraltar and Cyprus.

[1] Michael MccGwire, letter to the author, 15 August 2008.

A similar reorientation with regard to the Far East took place for France after the loss of Indochina in 1954, while the Mediterranean emphatically remained central to France's maritime Strategy. Both Britain and France have retained individual island possessions and bases scattered throughout the world, the naval mobility to send ships to remote places for disaster relief and potential crises requiring the evacuation of their own nationals. But this is a far cry from their firm military commitments at the height of empire.

The Atlantic played varying roles in US Strategy. From the principal zone through which hostile European forces could navigate to pose a direct threat to the young United States, the Atlantic turned into a protected zone of influence under the Monroe Doctrine, to be left largely to Britain to defend in the two World Wars, while in the Second, the USA concentrated its naval efforts on the Pacific. In the Cold War, the situation changed again, and the USA assumed some of the burden of defending the Atlantic, but also a major role in the Mediterranean and the other open seas between South Atlantic and Pacific.

New interests in the sea developed globally with the discovery of off-shore mineral resources and led to legal, albeit rarely military, disputes. Certain islands remained flashpoints, especially in the waters between Turkey, still expanding in good Ottoman tradition with the invasion of Cyprus in 1974, and Greece, still on the defensive in the tradition of its Byzantine ancestors; between Japan and China; and between China and several other neighbours. A classic instance of island snatching, performed by Argentina with regard to the British-held Falklands Islands in 1982, led to what was perhaps the most significant naval encounter of the post-1945 era, although it would be an exaggeration to call it a '"fleet versus fleet" battle' (Grove 1990: 159). Michael Salewski has called this the last nineteenth-century-style naval war. British–Spanish wrangling over Gibraltar by contrast was kept to a civilised level by mutual membership in the European Union.

As the world population exploded in size, fish-stocks diminished as a result of over-fishing. This led to tensions between Chile, Ecuador and Peru, and at various times to near-war conflicts over fishery rights between Britain and Iceland (the 'Cod Wars' of 1958, 1972–3 and 1975–6, which escalated up to the ramming of ships and cutting of nets) and Canada and Spain (the 'Turbot War' of 1995). The distances travelled by fishing crews tended to be greater than in previous times. Whaling, too, led to international tensions. Navies assumed new roles in the protection or exclusion of fishing fleets.

Is decisive battle still possible?

Both the British and the US navies continued to prefer a Strategy which, in the context of an overall defensive scenario of Soviet or later Warsaw Treaty Organisation attack, was offensive and 'forward', focusing on areas as close to the enemy's bases as possible, and as far away from one's own coastline (Grove and Till 1989: 281–3).

There was a turning point in this respect in Soviet naval Strategy, when in 1979 the foremost Soviet naval strategist of the Cold War, Fleet Admiral Sergei Georgevich Gorshkov, commander in chief of the Soviet Red Fleet, published his *The Sea Power of the State.* He emphasised that decisive battles had not been frequent in the two World Wars, and that 'Most of the combat clashes of the major forces' in the Second were 'associated with operations against the shore ... or to ensure transoceanic or sea communications' (Gorshkov 1979: 11).

Like Gorshkov, Western experts increasingly doubted that a Third World War scenario including major naval battles was realistic. Hedley Bull, the Australian strategist and Oxford don, thought that

> between nuclear powers a major war at sea is difficult to envisage ... [If forming] part of a 'general war' involving a strategic nuclear exchange between the super-powers ... a slow-moving struggle for command of the seas is likely to be rendered pointless before it is underway. There have been speculations that a limited war might be fought at sea by nuclear powers, but the interests at stake in such a conflict would be so vital, especially for the Western powers, that it is hard to conceive that the strategic nuclear threshold would not be crossed. (Bull 1980: 7)

Thus in the nuclear age, the 'decisive battle' school among naval thinkers was soon dying out. It became ever clearer that the tasks of navies had to be seen in conjunction with the tasks of armies and air forces on land, with navies relegated to support roles, albeit crucial or even essential ones. British historians from Michael Howard to Paul Kennedy emphasised in their works the interconnectedness of naval and army actions, best subsumed under Corbett's term of 'maritime strategy'; they were unconvinced, by contrast, that the navy had much of a role to play in isolation in the 1970s.

The impact of nuclear weapons

When in mid-August 1945 leading members of the Royal Navy leadership put their ideas on paper about the effects of the invention of the nuclear bomb, their views on its impact on the navy were as follows:

- 'The best place to put British nuclear weapons might be in rockets, preferably launched from ships or', thus the extraordinarily foresightful thinking of Professor P.M.S. Blackett, director of naval operational research, 'on submarines, as they were least prone to detection and pre-emption'. Atomic-powered nuclear submarines would be a very desirable acquisition on account of their stealth.
- The possibility was great that if the next war were to begin with nuclear attacks on 'civilian morale' (the British euphemism for the bombing of population centres), then the navy would not have a great role to play.
- Ports, naval bases, naval convoys and large ships would be particularly vulnerable to atomic attacks (Hattendorf *et al.* 1993: Doc. 449).

If the aircraft-carrier had emerged from the Second World War as the new queen of the sea, how did nuclear weapons affect it? Edward Teller, a physicist who helped build the H-bomb, in the 1950s dismissed the aircraft-carrier as useless, being excessively vulnerable in the nuclear age.

> It looked to me like quite a good target. In fact, if I project my mind into a time, when not only we, but also a potential enemy, have plenty of atomic bombs, I would not put so many dollars and so many people into so good a target. (q.i. Till 1982: 183)

Teller's view on this matter became widely known and affected the thinking of – at least – Western thinkers about future configurations and missions of navies (Géré 1992: 206). At the same time British admirals, generals and field marshals quarrelled over this point – while Sir John Slessor of the RAF declared the carrier obsolete, the vice-chief of the naval staff, Vice Admiral W.W. Davis, tried to persuade the British commander-in-chief in the Mediterranean, Admiral the Earl Mountbatten of Burma, of their continuing value (Hattendorf *et al.* 1993: Doc. 453). France and eventually the USSR also persuaded themselves that carriers had important roles to play outside all-out nuclear war and that it was worth investing in them. In the early 2000s, Britain once more debated a costly two-carrier acquisition programme.

Initially, both the Soviet and Western militaries thought of nuclear weapons as a way of causing much more effective and powerful explosions than with conventional ordnance, and they integrated the new invention into their operational planning in this role. Then they came to see them as a profoundly different weapon, the use of which would make all other forms of war and all other weapons irrelevant. From around the 1960s at the latest, when knowledge of the less obvious effects of nuclear explosions spread among leading service personnel, both Western armies and navies concluded that 'nuclear weapons interfered with good fighting' (Hill 2006: 160). This accounts for conventional proclivities among military planners in the USA from around 1960, while the Soviets first tried to ignore the effects of nuclear explosions on their own personnel, and then, in the late 1970s, were also converted to conventional war alternatives. 'At least up to 1975 the consensus among [Western] naval officers appears to have been that strategic nuclear deterrence was not an element of seapower as such, but a politico-military capability that happened to be most conveniently placed at sea' (Hill 2006: 164). Initially, 'American military and naval thinking had been based on the notion that deterrence required the explicit threat of escalation to the nuclear level. That threat alone was ... considered to be sufficient to preclude warfare' until the Korean War came along in 1950. But even then, until the mid-1970s, American naval thinkers had difficulties in fully accepting that other forms of deterrence and limitations on war had to be developed, as wars occurred as frequently in the nuclear age as previously. Nuclear weapons had merely added to its complexity (Hattendorf 2004: 7; see also chapters 17 and 18).

From the 1980s until the early 1990s, Britain's and France's strategic nuclear deterrent became increasingly identified with their nuclear-powered submarines equipped with long-range nuclear missiles, so that the identification of sea power and nuclear weapons became somewhat stronger. British and French navies also had carrier-borne aircraft capable of firing nuclear missiles. Since the end of the Cold War, however, the British, French and American navies' nuclear dimension has diminished greatly, with the exception of the submarines carrying the 'strategic deterrent', now significantly reduced in numbers. Meanwhile, naval officers of all nuclear powers were happy to keep their ships 'nuclear-free zones', to make them employable in lesser – but more likely – situations.

Command of the sea in the nuclear age?

'Command of the sea' had finally been recognised as unobtainable in the nuclear age, four centuries after Hugo Grotius had put forward this view. Roskill, for one, wrote that it was unreasonable to expect 'complete control'; it would suffice to have 'an ability to pass ships safely across an area of water which may be quite small in extent or may cover many thousands of square miles of ocean ... it is far more common for control to be in dispute than undisputed ... sporadic attacks will remain a possibility' (q.i. Gardner 2006: 150). The US Admirals Stansfield Turner and Henry E. Eccles proposed a reformulation of the old concept of command of the sea. Turner, writing in 1974, postulated the substitution of the term 'sea control', explaining that:

> This change in terminology ... is a deliberate attempt to acknowledge the limitations on ocean control brought about by the development of the submarine and airplane ... The new term 'Sea Control' is intended to connote more realistic control in limited areas and for limited periods of time ... It is no longer conceivable, except in the most limited sense, to totally control the seas for one's own side or to totally deny them to an enemy.

The purpose of all of this was to 'ensure industrial supplies' for one's own side, to transport and deploy, and thereafter reinforce and resupply, one's military forces overseas, to supply allies in war, and to 'provide safety for naval forces in the Projection of Power Ashore role'. In turn, one sought to deny all these abilities to the enemy (q.i. Till 1982: 189f.). Admiral Eccles accordingly broke down 'sea control', as he put it, 'by area and by time':

Absolute control (command of the sea) Complete freedom to operate without interruption. Enemy cannot operate at all.

Working control General ability to operate with high degree of freedom. Enemy can only operate with high risk.
Control in dispute Each side operates with considerable risk. This then involves the need to establish working control for limited portions for limited times to conduct specific operations.
Enemy working control Position 2 reversed.
Enemy absolute control (command of the sea) Position 1 reversed. (q.i. Till 1982: 189)

A NATO study on naval power of March 1969, called the Brosio Study after the then Secretary-General, translated this into the mission of being able 'to control and use the seas throughout the entire conflict spectrum'. The defence of NATO's sea lanes against submarines stood at the centre of this mission (Sokolsky 1989: 310, 314). Yet throughout the Cold War this, and NATO's sea-lift capability in all imaginable circumstances where the theatre of operations would be sea-bound, were secondary to the mission of maintaining a credible nuclear deterrent, meaning nuclear weapons that could be launched and would reach their targets with some certainty, even after one or more NATO countries had been hit by enemy nuclear weapons. And this mission was increasingly conferred upon submarines carrying nuclear-charged missiles (SSBNs).

Increasingly, it was understood that 'command of the sea is relative and partial rather than absolute and general' (Hill 1986: 81). A maritime power would be measured by its ability to use the sea for its own purposes, to deny it to adversaries, and by its reach at sea – regional or global (MccGwire 1976: 15).

Along with command of the sea, the decisive battle fell into disrepute in the nuclear age. No serious naval thinker still imagined in a hyper-Mahanian way that his service could decide war more or less on its own, even though Mahan remained the point of reference of naval thinkers throughout the world, from Italy (Ferrante 1999) to China (Holmes and Yoshihara 2005) and beyond (Rosenberg 2000). As we shall see, this dream passed to some of the air force thinkers. Naval strategists all turned into maritime strategists, that is, to follow Corbett's definitions, they henceforth recognised the importance of jointness, of the closest possible co-operation between navy, air forces and land forces. As the Italian strategist Giamberardino put it in 1947: 'Any particularist and independent concept of each of the three services must be considered as overtaken and arbitrary' (q.i.

Ferrante 1999: 173). This was a jibe addressed not at old-fashioned naval thinkers, but at the air forces, where Douhetian ideas of the third service as a war-winning force lived on (see chapter 12). Subsequently, it was also applied by naval men in several NATO member states as a criticism of NATO's strategies, which until the late 1950s relied so heavily on the American Strategic Air Command's (SAC) ability to destroy the enemy's war effort by striking at his big cities and key infrastructure.

Multiple roles for navies

Nevertheless, Geoffrey Till asserted in 1982 that there was 'little evidence to support the contention that the general idea of a maritime strategy' in the Corbettian sense 'or of sea power itself are moribund matters of interest only to historians' (Till 1982: 239). NATO collectively had very strong navies throughout the Cold War, mainly composed of contingents above all from the USA, Britain and France, with the Netherlands making a disproportionally large contribution for the size of the country. Throughout the Cold War, NATO's naval forces were superior to those of the Warsaw Pact, creating an imbalance inverse to that of the conventional land warfare superiority of WTO forces. But were and are such navies of any use to the West in the nuclear age?

There was a creeping fear that as a whole service, navies might have lost their central role in major war. The Western war scenarios of the earliest years of NATO's existence assumed all-out aggression on the part of the Soviet Union and its satellites (as its allies in Eastern Europe were referred to in the early Cold War) against Western Europe, to be met with strategic nuclear bombardment of the USSR by SAC. Once the USSR had a nuclear arsenal, it was assumed it would equally use all its nuclear weapons. This Doomsday scenario assumed also that war would be short – where, then, could navies bring their weight to bear? As carriers of troops and provisions that would take weeks to arrive in Europe across the Atlantic? As the long-term protectors of commercial shipping or as defenders against repeated attempts of penetration into NATO waters by the enemy? The British navy first realised that in such a scenario it would be redundant, and in the early 1950s pushed the idea that after an initial mutual attack of unprecedented violence and destruction, there would still be a period of 'broken-backed warfare' in which the land and air forces might be enormously decimated, but in which the navies would come into their own (Baylis 1995: 126–77; Heuser 1997, chs. 1 and 2). This concept that Europe might be resupplied by sea a month and a half after the major nuclear exchange apparently also made sense to naval men in other NATO countries, such as Admiral di Giamberardino of Italy, whose navies might otherwise face the same predicament (Ferrante 1999: 179).

Vice Admiral Sir Peter Gretton commented: 'It seems to me to be hypocritical to talk of preparing to fight a long general war at sea, and to put forward claims for the exercise of maritime strategies in such a war seems absurd' (Gretton 1965: 79). On the other side of the East–West divide, Admiral Gorshkov equally deplored the views of some influential '"authorities" who considered that with the appearance of atomic weapons the Navy had completely lost its value as a branch of the armed forces. According to their views, all of the basic missions in a future war allegedly could be fully resolved without the participation of the Navy' (q.i. Grove and Till 1989: 293).

Even though in the early 1950s, navies desperately sought to retain their role in general war, most naval men were unhappy about thinking exclusively about this scenario, where after a short, intensive nuclear exchange, there might be little left for them to do. In contrast to the Soviet navy, the Royal Navy, the US navy and the French navy had traditions of naval use in limited-war scenarios, which they resurrected even in the nuclear age (Grove 1990: 189). These words of an admiral of the Royal Navy from the nineteenth century applied again from the late 1950s and into the twenty-first century:

> I don't think we thought very much about war with a big W. We looked on the Navy more as a World Police Force than as a warlike institution. We considered that our job was to safeguard law and order throughout the world – safeguard civilisation, put out fires on shore, and act as guide, philosopher and friend to the merchant ships of all nations. (q.i. Till 1982: 208)

This thinking came to the fore again in the Cold War (perhaps minus the 'philosophy', not something British military men are inclined towards).

Navies therefore needed new reasons for existing, and these were found in peace, or limited conflicts, rather than in major war. In the Cold War, tasks for navies were thus described by Gretton as:

[P]atrols of ships and maritime aircraft to prevent gun-running and the infiltration of saboteurs and guerrillas, of flag-showing visits to troubled areas, and of the ability to land seamen, marines and soldiers speedily and at short notice to deal with civil unrest and disturbance – police tasks which the sailor has performed with discretion for a hundred years and for which he is eminently suitable. (Gretton 1965: 80)

Gretton contrasted outdated and remaining tasks for the navies of NATO powers. Outdated principles included the two-nation standard in peace time, as well as the unique naval task of fighting against the enemy's fleet. Enduring tasks included those of protecting one's own land from invasion, and equally one's shipping (shared with the air force), amphibious operations, advanced bases and the ability to attack enemy shipping and bases (Gretton 1965: 24–8).

For a reasonable statement of what a second-tier power could aspire to as general and even geographically transferable principles, three French pieces furnish good examples. One is by the French Admiral Adolphe Auguste Marie Lepotier (1898–1978), prolific author on naval/maritime Strategy from the 1930s until the 1960s, here with a piece dating from 18 January 1949 on 'eternal missions' of the navy: to retain 'freedom of action' and strategic mobility for one's own forces, to secure supplies from overseas which of course included the need to protect one's own ports and coasts, the ability to engage in joint operations with or without allies, to carry 'out naval air operations against the territory or the sea communications of the enemy', and direct defence (q.i. Géré 1992: 188).

The second example is an anonymous essay published in 1955, which contrasts the eternal principles and the variables for the particular case of France. Permanent tasks are listed as:

– the protection of maritime communications; – French presence in overseas territories; – support at all times of the policies of the government; – in times of war, operations at sea and along the coasts in cooperation with the other services and with allied nations. Currently this last mission is carried out in the framework of the Atlantic Pact. There is no change that could question the need for the navy ... As long as there are other surface ships that carry supplies and fuel necessary for the armies and for populations, we need military surface ships alongside them to protect them.

The variables by contrast were identified as 'the proportions between different categories of vessels' within a navy 'and the proportions between naval and air forces. The technical progress of arms is the essential cause of these changes' (q.i. Géré 1992: 186f.). At the same time, Vice Admiral Pierre Barjot argued that naval power is subtler and more applicable in limited-war scenarios than air power.

Today, air forces are prisoners of weapons of mass destruction ... This will be increasingly so in the future ... Air warfare is 'all or nothing'. By contrast, the navy allows nuances. In limited wars, the navy can intervene where strategic air power runs the risk of setting fire to the world. If strategic air power is a weapon of intimidation, its effects cannot be calibrated, and it may be nailed to the ground while the navy retains the spectrum of its possibilities ... The navy offers the possibility ... to enter into a war or not, and to calibrate political interventions. The navy is a factor of peace. (Barjot 1955: 293f.)

Despite its Socialist Realism and scientific Marxism-Leninism, the Soviet Red Fleet in some aspects– such as the acquisition of aircraft-carriers – sought to rival the USA, suggesting that national pride and rivalry with America prevailed over any ideology-specific deductions about the uses of naval forces in the furtherance of international socialism. Decisive differences from US preferences persisted, as a result of the geographic-climatic dilemmas that have confronted Russian navies for centuries, revolving around the problems of access to the open sea and ice-free waters and the need to maintain four fleets. But the parallels between Gorshkov's limited war/Cold War use for the navy – the support of friendly regimes and revolutionary movements worldwide – and those put forward by, say, the UK's Admiral Gretton for limited/Cold War scenarios are striking (Gretton 1965; Gorshkov 1979; Till 1982: 68–74). Gorshkov's book introduced a new tone into Soviet strategic writing. His list of peacetime roles for the Soviet navy dwelt particularly on its role in limited wars (Gorshkov 1979). Hitherto, it had been dominated by the concept of East–West tensions as a continuous strife, in which even peaceful coexistence was different from peace in that it equalled ideological strife which, according to Marxist-Leninist historical necessity, had to be won ultimately by communism. The Soviet military and political leadership did not want a Third World War, on the contrary, ever since the Great Patriotic War of 1941–5, they had hoped fervently that it could be avoided. If it did come, however, the USSR must 'win', whatever that meant. But until the 1970s it seemed that Soviet military writers were not allowed to ponder the concept of limited war. From then on, Soviet fleet exercises, especially Okean 70, which included even Cuba, indicated that the USSR was developing a new rationale for its navy outside that of an all-out East–West conflict. Again, close parallels between Eastern and Western naval thinking are evident when Russian Admiral S.E. Zakharov wrote in the 1970s:

> Soviet military doctrine ... recognises the possibility of the outbreak of local wars which take place without the use of nuclear weapons. In this [kind of war] Soviet military doctrine allots an important place to armed battle on the sea, which could acquire an enormous strategic significance. (q.i. Till 1982: 184)

In the West, some experts questioned the role of navies in an age of air power with global reach. In the words of Rear Admiral Richard Hill, it was said in the Cold War, 'with some justification, that the naval staffs of the West should get down on their knees nightly to thank their Maker for the Soviet Navy' (Hill 2006: 166). Once the USSR had disappeared, one *enfant terrible* even argued for the abolition of the US navy, his tongue only partly in his cheek (q.i. Hattendorf 2006: 10).

But great-power navies could claim roles in more limited conflicts. Even without participation in fighting, their role might be considerable, said strategic theorist Eric Grove:

> Their presence can act to set the boundaries for a conflict by demonstrating the willingness of larger powers to intervene if things get out of control. They may be used to exert pressure on one or both sides to accept a ceasefire, especially in a protracted conflict in which neither side can clearly win. Similarly, the presence of one major power's naval forces may neutralize the activities of the other by the demonstration of the potential capacity to disrupt any planned intervention. (Grove 1990: 218)

Navies have the advantage over land forces that they can be brought to crisis zones – near the sea – fairly quickly, can intervene but transform themselves again from 'gladiators to helpful bystanders, and back again with little need to delay or reconfigure'. They can be 'extricated so much more easily than their land ... equivalents' if 'things go wrong' (Till 1994b: 180, 190). Navies could thus limit the risk of escalation. Paul Nitze, US Secretary of the Navy 1963–7, was a key proponent of this view (Grove and Till 1989: 298).

Navies were quick to claim their place in the 'low intensity conflicts' of the Cold War, if only to transport, supply and evacuate their countries' ground forces, but also in the range of operations well short of war that involve coastal patrolling to control smuggling and illegal immigration, to protect against piracy or guard offshore installations (Hill 1986: 111–31). After the end of the Cold War, John Hattendorf aptly summarised the multiple roles of navies, in a 'joint' context. In wartime, the focus of 'the military element in maritime strategy' would be to establish control of strategically important parts of the sea, and use them for one's own purposes. Such 'control' can take a range of forms: it can be 'general or limited, absolute or merely governing, widespread or local, permanent or temporary'. Once some degree of control is achieved, the navy's tasks in war are:

- Protecting and facilitating one's own and allied merchant shipping and military supplies at sea.
- Maintaining safe passage for shipping through restricted waters and access to ports and harbours.
- Denying commercial shipping to an enemy.
- Protecting the coast and offshore resources.
- Acquiring advanced bases.
- Moving and supporting troops and advanced bases.
- Gaining and maintaining local air and sea control in support of air and land operations.

In the tradition of the 'fleet in being' thinkers, Hattendorf acknowledged the latent importance of navies in peace and war, as 'the foundation upon which the diplomatic and policing role rest' (Hattendorf 2000: 236–8f.).

At any rate, after the arrival of aircraft, navies could no longer be seen as a country's 'first line of defence', as the US navy was described in Mahan's age. Nor could they be presented as the only instruments of maritime power and the ability to project it. A case thus had to be made each time for the use of naval assets as opposed to alternative means equally capable of achieving any particular mission.

Modern 'gunboat diplomacy', power projection and conventional deterrence

Another buzzword was that of 'power projection' to denote naval peacetime roles, especially the naval transport of troops and aircraft and anything else around the world, and diplomacy underpinned by naval deployments, up to and including the threat of limited force (United Kingdom, Ministry of Defence 1957: para. 37), formerly subsumed under 'gunboat diplomacy'. As Captain Stephen Roskill noted:

> The functions of our maritime forces in an amphibious expedition [in which the navy carries troops to another shore] differ considerably from those of the forces employed on mercantile convoy work. In the latter case their duties end with the safe arrival of the convoy in port; but in the former case they must continue to support and assist the army after it has landed, and continue to maintain the maritime control on which success on land hinges. (q.i. Gardner 2006: 154)

In the 1930s, deterrence became linked to air power by the British, and after 1945 it became the foundation of British, US and then NATO nuclear Strategy. Nuclear deterrence was initially a concept alien to strategists from other NATO member states. The Italian ex-naval officer Romeo Bernotti (1877–1974), for example, in his ruminations of 1954 on the form that a Third World War would take, could not conceive of this role for nuclear weapons. To him, nuclear weapons were a means of destruction, and would be used as such in a future conflict, in a way that deprived 'winning' of any real sense (Bernotti 1954; Ferrante 1999: 175). Admittedly, both declaratory and operational NATO Strategy at the time had little but destructive aims. 'Massive retaliation' emphasised the use of nuclear weapons both on the battlefield, where battlefield nuclear weapons were becoming available, both for naval and land use, and for strategic bombardment as an immediate response to aggression. However, the corresponding NATO Strategy document put deterrence first, and there was an underlying conviction on the part of key NATO and member state decision-makers on this subject that this emphasis on massive and ubiquitous use of nuclear weapons would enhance deterrence (Heuser 1997, ch. 2).

In the mid-1960s, Sir Peter Gretton was emphasising the role of the navy in military interventions in limited wars outside Europe, a line of argument which declined in Britain after the government decision at the end of that decade to cut her commitments to areas east of Suez. In 1980, Hedley Bull developed James Cable's theme of 'gunboat diplomacy' in situations short of war, even though Bull rejected the term in view of its historical association with the 'coercion of weak states by strong ones'. Navies were used in the Cold War to support friends and clients, coerce enemies, neutralise 'similar activities by other naval powers, [to exert] a more diffuse influence in politically ambiguous situations in which even one's own objectives may be uncertain, or merely [to advertise] one's own sea power or "showing the flag"'.

Moreover, sea power could serve '[a]s an instrument of diplomacy', where it had 'certain classical advantages vis-à-vis land power and, more recently, air power'. These included, first, flexibility: 'a naval force can be sent and withdrawn, and its size and activities varied, with a higher expectation that it will remain subject to control than is possible when ground forces are committed'. After several extended wars of decolonisation and the American experience of Vietnam, Western powers had become reluctant to use 'ground forces in contested areas of the Third World' for fear that these could 'lead to uncontrollable involvements' (Bull 1980: 8).

The second advantage was that of visibility (Luttwak 1974: 14). The third Bull called 'universality and pervasiveness', the ability of navies to get to any part of the oceans even in the absence of bases. This was something aircraft could do in principle, but even with in-air refuelling, they could not visibly circle an area for days, let alone weeks or months.

The flexibility of navies is valued because of the fear of nuclear war and the belief that there is less risk of unintended expansion of a war when only naval forces are involved than where ground or air forces are used – partly for reasons of command and control, and partly because destruction confined to the sea is bound to be less than destruction which spreads to the land.

Navies could be visibly deployed without actually using force – exploiting the latent threat of the use of force, an important instrument of Cold War diplomacy (Bull 1980: 8). This theme was developed especially by Edward Luttwak, who from the 1970s became a much-listened-to guru on Strategy in the USA. In *Political Uses of Sea Power* (1974), he developed his 'theory of suasion' in relation to naval activities. He divided 'suasion' into active and latent. Active 'suasion', he argued, could be either supportive (of an ally) or coercive (of an adversary), the latter of which could be either positive (compellent) or negative (deterrent). Latent 'suasion' again could be deterrent or supportive (Luttwak 1974: 1ff.).

Luttwak made the argument, previously developed by Pierre M. Gallois (see chapters 12 and 13), that no force can [per]'suade' in the absence of a willingness to carry out the latent threat if necessary. This is the central problem with an incredible threat made in the context of deterrence: 'To speak softly while carrying a big stick may be less effective as a deterrent than to make a firm, overt commitment to use a rather smaller stick.' This problem is linked, he argued, to the fact that governments address multiple audiences, which he called the 'inherent indirection' of the 'suasive' measure: 'while the U.S. Sixth Fleet may be continuously deterring Russian and Arab moves against American interests as well as reinforcing the Alliance, it may also be giving unintended encouragement to Israeli activism in a manner inimical to the interests of the United States' (Luttwak 1974: 12f.).

Gunboat diplomacy – whether under this term or any other – was practised mainly by the USA in the Cold War; other states resorted to island snatching if these were close to their own territorial waters, in the hope of extending their geographic possessions. Hedley Bull rightly observed that this expansion of Soviet naval activity, including the expansion of its navy, in the 1970s led to a change in the opportunities available to the United States, which had truly dominated the oceans since the end of the Second World War. From the 1970s until the end of the Cold War, the USA had to face the possibility that the Soviet Union would neutralise American naval deployments by creating a naval counter-presence nearby; the USA could not hope to dominate situations of concern to the USSR as unilaterally as it had done in the Cuban Missile Crisis (Bull 1980: 9). Yet the USA continued to use its naval forces for military purposes, such as the seizing of the island of Grenada in 1983 and the bombing of Libya in 1986. The Soviet Union by contrast had rarely if ever used its navy quite in this way (Grove 1990: 158). While the Soviet navy had not been created as an instrument of peacetime policy, it began to operate as one in the 1980s, when it was used at times to deter US attempts to stifle anti-colonial movements.

Blockade and protection of 'lines of communication' in the nuclear age

One of Stephen Roskill's lessons of the Second World War was the affirmation of the utility of the blockade, 'one of the chief means whereby a nation which is stronger at sea may be able to impose its will on one which, though stronger on land, is not self-supporting in food and raw materials'. Despite the German civilian casualties produced by naval blockade in the First World War, Roskill thought this 'a relatively humane form of war' albeit slow to take effect (q.i. Gardner 2006: 153). Commercial blockade had been less effective in the Second than in the First World War. Germany had made comprehensive preparations for it in the interwar period, with immense efforts to substitute raw materials available for those only accessible through maritime imports. During the Second World War, land conquests provided it with new sources of raw materials (Gardner 2006: 156). The Cold War would see blockades applied in many instances – usually affecting civilians rather than the regimes with whom the blockading powers disagreed.

In the East–West conflict of the Cold War, blockades took on a new dimension, too: no longer was blockade a matter of preventing ships from leaving ports (close blockade); it became a matter of preventing Warsaw Pact fleets from accessing the Mediterranean and Atlantic (distant blockade). In turn, for the Soviet Union and its allies, Strategy revolved around preventing NATO ships from gaining a foothold in the Baltic which it had in peacetime through the navies of Denmark and the Federal Republic of Germany. Planning revolved around three areas: the Turkish Straits (Bosphorus and Dardanelles) which were the sole exit from the Black Sea to the Mediterranean; the waters north of Norway, south of Svalbard (Spitsbergen) that are largely ice-bound in winter; and the exit from the Baltic Sea to the North Sea through

Kattegat and Skagerrak. US Admiral Stansfield Turner preferred to call this not blockade, but 'sortie control':

> Bottling up an opponent in his ports or on his bases ... Today's blockade seeks destruction of individual units as they sortie. If we assume an opponent will be in control of the air near his ports, sortie control tactics must primarily depend on submarines and mines ... a most economical means of cutting off a nation's use of the seas or ability to interfere. (q.i. Till 1982: 187)

Concomitantly, the need for the protection of 'lines of communication' (LOC) was still a key concept of Cold War naval Strategy, even though one can hardly disagree with Sir Peter Gretton when he wrote:

> When I read on the naval recruiting posters 'Join the Navy and help to protect Britain's Sea Lanes', my heart sinks for another ignorant officer has been allowed to perpetuate the old aimless catchword ... It is *ships* which must be protected, not lines drawn across charts. The core of the problem of sailing a ship or ships from one port to another in war is to control the slice of water in which the ship floats, as well as the air above and the depths below. Any wider degree of control is welcome but not essential. (q.i. Gardner 2006: 156)

Strategies for second-tier powers

If even Americans had to concede that 'command of the sea' was little more than an ideal, this was clearer still to the second-tier powers, among which Britain had to count itself soon after the end of the Second World War. Most Britons failed to recognise this prior to the Suez Crisis of 1956 and pretended that Britain continued to be a global player until the late 1960s, when she relinquished her commitments east of Suez. Yet Britain and the British continued to have a disproportionate degree of influence in naval matters. After Suez, British naval capacity exceeded the NATO requirements for operations in the North Sea, the Atlantic and the Mediterranean, focusing increasingly on the defence of the Greenland-Iceland-United Kingdom gap in the North Sea (Hill 2006: 161). In NATO, the Royal Navy played a role second only to the (admittedly, much, much larger) US navy. It is also true for intellectual input: the majority of widely read writers on naval

Strategy of the twentieth century have been Britons. John Hattendorf, doyen of US naval educators, listed as the most famous (in the USA!) in the latter part of the twentieth century Laurence Martin, Edward Luttwak, Ken Booth and Sir James Cable – of these, only Luttwak is not British (Hattendorf 2000: 203).

Britain had become a medium power, as Admiral Hill rightly found, and he built on this realisation an explicit analysis of maritime Strategy for medium powers. Medium powers he defined as powers that lie between the self-sufficient superpowers and the insufficient small powers. Examples he had in mind included France, India, Japan, Australia, Israel and Britain, acknowledging special historical baggage, self-perception and other cultural peculiarities (Hill 1986).

In the tradition of political science and international relations, Hill used power in two senses, that of a state capable of taking action and that of 'the ability to influence events'. Such power in the latter sense, he argued, was generated by goods and money, by knowledge and ideas, and by arms.

> The first is intrinsically desirable, and provides the base for others forms of power, but is often operationally unusable; the second is powerful but slow-acting, not always controllable; the third dangerous and either menacing or violent. They may be thought an unholy trio, and they are certainly not the three Graces, but they are the means by which states seek to protect their vulnerabilities and promote their interests. (Hill 1986: 7)

As 'maritime power is the ability to use the sea' (Hill 1986: 48) he explained a state's Strategy as a balancing act between the means of power available, the degree of the state's control over them and over the situation in general (including influence on allies) and resulting from both the ability to initiate and sustain policies.

Having forged a permanent alliance with the USA in NATO, and having lost almost all her colonies, Britain gave up entirely her balance-of-power tradition. By 1945 at the earliest and 1991 at the latest, the greatest power on earth was the USA, and 'realist' as well as 'balance-of-power' logic should have led Britain to counterbalance America, as France attempted to do under de Gaulle in the late 1960s and again, briefly, after the end of the Cold War. At least Britain should, according to this 'realist' theory, have refrained from bandwaggoning. But Britain did no such thing. Instead, Britain remained the loyal ally of the USA.

Alliance Strategies

One difference between the Cold War and previous eras concerned alliances: they were structured and institutionalised as never before. In the Warsaw Treaty Organisation and in NATO, member states subjected their own Strategy-making to that of the alliance as a whole, albeit to varying degrees. An Italian commentator saw in NATO the 'end of the military-national identity' of Italy, and illustrated the extreme subordination of all Italian strategic debates, naval or otherwise, under the imperative of alliance cohesion (Ferrante 1999: 171–88). The Federal Republic of Germany had no Strategy of its own, to the point where its Defence White Books (Papers) were derivatives of NATO strategies. Japan, too, lost any Strategy of its own, and any offensive, aggressive edge, tied to the USA in a bilateral relationship. The disappearance of aggressive intentions and planning among those three countries was one of the best results of the Second World War, and it led to a substantial down-scaling of the roles and potential of their navies.

There was a certain degree of freeloading on the part of NATO member states in ostensibly and rhetorically putting the Alliance first in all they were doing, as Admiral J.R. Hill commented (Hill 1986: 81), but perhaps it was only France, and temporarily Greece, that truly managed to optimise their own selfish state interests while benefiting from Alliance membership and the coverage of the American nuclear umbrella, a commitment not actually spelled out in the Treaty itself (Heuser 1997: 15–25).

The French discovered early on that they had problems with the Atlantic Alliance. Even on the eve of the signing of the North Atlantic Treaty, Admiral Lepotier reminded an all-services French military audience of the disadvantages of alliances, very accurately foreseeing many of the problems that would bedevil NATO throughout the decades to come. States belonging to a geographically spread-out alliance would have divergent interests in any theatre of war. Each ally would put himself (and any colonial possessions) first *in extremis*. Each would think about his national prestige when command posts and headquarters were assigned, the states with great naval traditions being particularly keen to protect their reputations (Géré 1992: 188). Indeed, France's Strategy, going back to the eve of the First World War, of opting for an alliance with a Western naval great power in order to concentrate her own naval efforts on the Mediterranean, soon after the conclusion of the North Atlantic Alliance brought her into conflict with her two main naval allies, who had other views on that area. While for France the Mediterranean was above all a north–south line of communication with her North African possessions, but also a theatre she wanted to dominate as local great power, for Britain it was until the late 1960s an east–west line of communication to her empire beyond Suez. For the USA, it was a supply route to southern Europe in an East–West general war scenario; consequently, the USA made it increasingly clear that it intended to be 'top dog' in the Mediterranean.

As a Briton, Hill had every reason to criticise successive British governments for their extreme submissiveness in their alliance with the USA (so clearly in contrast with France's stance), and he rightly underscored the strategic cost of alliance. While it does indeed act as a force multiplier on the one side (keeping a state's back free), it comes at the price of independence in many matters of defence decision-making.

> Dependence represents a diminution of power not an accretion of it. The state's interests become less identifiable as they become less independently defensible. Freedom of action, restricted in the military sphere, becomes more and more constrained in diplomacy, and finally in economic matters. (Hill 1986: 222)

What this criticism of dependence on an alliance and especially on one major ally does not deal with is the very limited influence that medium powers, let alone small powers, can exercise in an anarchic world without permanent alliances. The same holds true for a world without generally respected norms, which on the one hand restrict the freedom of action of states, but which on the other clarify who is acting unlawfully, and who by contrast deserves the solidarity and protection of the other states. This last point, although only of limited reliability, contrasted the performance of the UN with that of the League of Nations in the interwar period, and in the eyes of others is the greatest *acquis* of the post-1945 period.

Geoffrey Till has underscored the utility of navies in servicing alliance systems. They are useful in joint exercises, and helpful for friendship visits and on-board receptions, the pretty side of the coin of gunboat diplomacy (Till 1994b: 191). But one must go further and recognise that alliances alone, permanent ones or ones at least for the 'foreseeable future', allowed Britain, France and Japan (not to mention Germany or Italy) to escape from the predicament of second-tier powers. And this is the assumption that they had to face several other powers alone, with two-nation standard navies, which either burdened them with intolerable military expenses or led Germany and Japan to such foolish acts of (supposedly preventive) aggression in the late nineteenth and early twentieth centuries.

Change of world-views and principles in conducting international affairs

Power projection has been widely seen as acceptable even after 1945. By contrast, the perception has disappeared from naval Strategy writing that it is legitimate to start wars, especially large-scale wars. It has been eclipsed by the concomitant development of international law since the Briand–Kellogg Pact of 1928 that outlawed the use of aggressive war as a legitimate instrument of politics.

Hedley Bull for one drew attention to the change not only of international law but also of (corresponding) mentality since Mahanian times, when the great naval strategists developed their theories. While Mahan – like Frederick II of Prussia – had had no hesitations about advocating aggression as a Strategy for the USA, with the aim of increasing its possessions, 'States today no longer use sea power to seize colonies; they experience inhibitions, unknown in Mahan's time, in overtly coercing weak states; and they do not describe their actions as aggression.' Indeed, he argued, towards the end of the Cold War:

> Today no international order can be viable that is not based on a wider consensus than can be found among the Western powers, or between these powers and the Soviet Union. A consensus will have to be sought that embraces the Third World, whose alienation from the present international order cannot be brought to an end without a radical redistribution – not merely ... of wealth and resources, but also of political and military power. The great powers' pursuit of the traditional form of global naval ascendancy, viewed by the Third World as the source of its historical subjection and a bulwark of its present status of dependence, flies in the face of this requirement. (Bull 1980: 11)

Bull, who ended his career as an Oxford don, was not simply an isolated example of views in the ivory towers of think tanks and universities. In the USA, the Brookings Institution argued in 1977 that military power rarely achieved lasting political results and such traditional forms of domination were outdated. Such arguments came to the fore periodically in debates about the structure of the US navy, especially the size of its carrier fleet (Pay 1994). Despite some overall force reductions, the US navy thus robustly survived the demise of the Mahanian 'decisive battle' scenario, still strong in the first half of the Cold War. The 'power projection' argument is underpinned variously with the need to pursue US interests (in the language of US Republicans) or world order (in the language of President George H.W. Bush, but also the US Democrats). The debate on the number of carriers the USA needs is thus a small mirror of debates about America's place in the world and the shape which America would like the world to have, just as Strategy is a mirror of views of the world and the use of force which the writer on Strategy sees as legitimate or at least justifiable.

Conclusions

Naval Strategy has almost always been intimately connected with land Strategy, while the opposite cannot be said. As Corbett correctly noted, naval Strategy was ultimately always about the access to and control of land, with the seas being means of transport of goods and men, defence of transport, attack and defence of littoral areas or points or sea passages.

Geography dominated naval Strategy, while it only affected tactics in land and air warfare, Strategy less so. But while anybody could have a use for armies, a limited number of states had access to the sea. Long – and not ice-bound – coasts are needed to make a navy a key element of overall Strategy, or even one equal or superior to the other services. As a direct function of geography, successive German states or indeed France could simply never aspire to 'command of the sea' in the way in which island states and states with many geographic characteristics in common with them could – and these are finite in number: in different ages, Spain, Portugal, Britain, the USA and Japan. Some island states, surprisingly for geo-strategists, never tried their hand at it – the Philippines, the Indonesian islands before

Dutch conquest, the Malay peoples, Madagascar or Sri Lanka might have played the role of a Venice or Genoa, but emphatically did not. Russia's access to the sea has been hampered by geography and concomitant climate to this day, and the Ottoman Empire could never dominate more than the eastern Mediterranean and the Black Sea. The constellations were finite, and so were the options.

The change from sail to steam was more of a revolution than the slow introduction of and adaptation to gunpowder – at any rate, it occurred in a far shorter span of time. This resulted in the emergence of two interpretational patterns in naval Strategy, the historical and *matériel* schools. The first tended to demonstrate a fundamental unchanging geographic pattern of naval warfare, while the latter emphasised change through technology and the need constantly to adapt Strategy. Both schools included believers in the centrality of naval battle ('Mahanians'), much in keeping with the offensive, battle-centric spirit of the age. However, both also had proponents of more indirect naval pressure on adversaries, such as the *guerre de course* (supposedly the French tradition) and blockade, or the fleet in being, even though throughout history the British and the French in particular had followed each of these strategies at different times, sometimes simultaneously.

In the late nineteenth century 'Mahanians' liked to see navies as able to win wars almost on their own, with ground forces relegated to secondary, albeit not wholly dispensable, importance. They equally believed in the centrality of decisive naval battles. In this, they foreshadowed the belief of later air power strategists that their service alone could be the war-winning weapon. A similar claim would later be made more legitimately for nuclear weapons, which each of the services would try to acquire to protect its claim to a major – if not the decisive – role in a Third World War.

Other nineteenth-century strategists saw the practical importance of great power deterrence and gunboat diplomacy to settle issues in one's favour without big battles. Navies, as von der Goltz and Corbett thought, might decide wars through the imposition of blockades that could leave the strangled societies begging for peace. In practice this did not happen, as such populations rarely have the means, or even will, directly to bring about a change in their governments' policies, as their governments are seen as only defence against the evil forces starving them (Chuter 1997).

The advent of nuclear weapons initially brought into question the very need for navies in a future all-out nuclear war, likely to be over too soon for navies to fully assume their protective or (trans-Atlantic) transport roles. In such a context it was really only the nuclear-powered submarine with nuclear missiles (SSBNs) that gave navies a role. From the second half of the Cold War, Navies reinvented themselves, so that their foremost *raison d'être* was no longer major war, but anything short of it, from gunboat diplomacy to all sorts of more limited contingencies. With the exception of the SSBN (and the role of denying Soviet SSBNs access to the North Sea and the Atlantic), the US, British and French navies became predominantly support services for air and ground forces, whom they transported to the theatres of operations and kept supplied and nursed if there were casualties. These navies substituted for the bases in colonies or on allied territory which could become unavailable over time. With the exception of roles in alliance contexts played by several navies – notably the Dutch, Canadian and Australian navies and the Japanese Self-Defence Forces – none of the other navies have seriously developed or maintained capacities going beyond their own regional waters, and roles beyond equally local, traditional ones.

If in our world – where globalisation began in 1492 – it is increasingly difficult politically to keep one's head down and pay attention only to regional issues, this has implications for alliance policies. From the time of the Washington Naval Treaty, or at the latest from the Second World War, Britain, which had come close to 'commanding the sea' in the nineteenth century, recognised the importance of close and continuing co-operation with the superpower USA. Successive British governments of the second half of the twentieth century thought – in the words of a particularly influential British diplomat, Sir William Strang (1951) – that 'We have in fact no alternative but to work with them' (q.i. Grove and Till 1989: 274). This is a view likely to dominate Britain's military policies in the twenty-first century, unless the USA one day decides to cut Britain loose. Other states aspired to a position like Britain's in the nineteenth century. America was one of them (succeeding in taking Britain's place in the twentieth century); Japan and the USSR were partially and temporarily successful. No others came near, trapped as they were in their respective geographic boundaries and concerns about land warfare on other fronts. France, like Britain, had to concede the need to form a lasting alliance, with the United Kingdom, or the USA or both.

At the beginning of the twenty-first century, 'Mahanian' big naval battles are difficult to imagine, while many of Corbett's views still stand up very well: all military action at sea has an element of force projection onto land, and necessarily requires some form of 'jointness', that is, co-operation between naval and other forces (whether these be marines, naval air arms or the separate services). Fleets in being – generally, smaller fleets, whether mere coastal defence navies or blue-water capable formations – have many times proved their worth as either as 'risk factors' in a Tirpitzian sense, that is, as forces that tie down a certain number of enemy forces, or as deterrents to piracy (Hattendorf 2000: 260).

The former can be seen as an enduring factor, while piracy as a factor in naval operation has fluctuated. It is on the rise again in the Pacific and Indian Oceans, so that policing, protection of shipping (but only against individual assaults, not on the scale on which Allied shipping had to be defended against German submarines in the two World Wars) and coastal defence tasks are eminently topical. Medium-sized and small navies are growing in number in Latin America, Africa and Asia, navies that would have little use for Mahanian ideas. However, the growing navies of India and China may prove exceptions. The navies of the Permanent Five in the UN Security Council, especially the self-designated 'world policemen' the USA, Britain and France, are likely to be involved more in the 'management of turmoil' (Geoffrey Till) than in the sorts of operations that were at the centre of their training and thinking (albeit not their operations) for most of the nineteenth and twentieth centuries. In general, Till early on predicted rightly that the post-Cold War world would see a shift of the tasks of Western navies away from 'blue water naval operations' (especially 'securing control of the sea' or the protection of shipping against large-scale and systematic commerce raiding by enemy military vessels), towards what the Americans began to call 'littoral warfare', the projection of power ashore (Till 1994a: 179). But, as the wide-angle perspective of this book suggests, that is not likely to be the final shift in the long and fluctuating history of maritime warfare.

PART V

Air Power and nuclear Strategy

12 *War in the third dimension*

Whoever is master of the air will be master of the world.
(General Jean-Henri Jauneaud, q.i. Artzet 1999: 47)

[F]light ... is still so new, so beautiful ... so swift, that its value in peace and war ... is bound to be somewhat magnified ... [T]he result is that we tend to ascribe to it boundless potentialities and to reject other implements of war as outmoded. We are under the sway of a dogma of innovation, just as blind and as dangerous as that [claiming] there is nothing essentially new in war.
(Bernard Brodie 1942: 193)

Child and grandchild of naval Strategy

We have seen that writing on naval warfare differed in many points from that on land warfare. Air Strategy emerged in most ways as the child of naval or maritime Strategy, and much less of thinking on land warfare. The main tenet inherited from the land Strategy thinkers was the Clausewitzian war aim of 'imposing one's will upon the enemy'.

The inheritance from naval thinking was both much larger and more detailed. The three crucial bequests from naval writing to air power theory were:

- the debate about whether the navy's (or as it might be the air force's) actions should be subordinated to the needs of land warfare, which was essentially the argument of Corbett;
- the claim that the navy (or then, the air force) could win a war more or less on its own, mainly in a central and decisive naval battle (or strategic air campaign);
- the perceived possibility of using the navy (or later, the air force) to circumvent the enemy's strongest or hardest part, his military forces, and carry the war directly to the enemy's vital but softest

> part, his civilian populations in his big cities. While the overall aim of advocates of this route – through blockade or city bombing – was to defeat the enemy state, not to annihilate his population in a genocidal fashion, a Social Darwinist element resonated in this thinking.

Some concepts – such as command of the sea/command of the air, sea power/air power – could be transposed one to one. Other translation of ideas from naval and maritime Strategy to air Strategy included cross-overs and mergers of the ideas of several 'schools' on maritime Strategy. At times, the genealogy of ideas can be explicitly traced, as was particularly true for the writers on general Strategy in the nineteenth and early twentieth centuries and their influence on naval/maritime Strategy thinkers. But with the growth of formal educational curricula and cheaply available specialist periodicals in the nineteenth and early twentieth centuries, ideas spread in ways that elude specific tracing. Some were 'in the air': when finding in somebody's writing on air Strategy traces of a previous author's or school's ideas, we often cannot prove that these were consciously assimilated. In some cases a certain idea may well have been reinvented, rediscovered. In others, two individuals, dealing with similar questions, having read similar things, may independently have hit upon very similar answers.

The beginnings of air power

It took centuries of writing about war before some writers began to focus on Strategy – as nexus of the use of force for political ends. Writing on maritime Strategy in relation to political purpose only took off with the age of steam. By contrast, in the case of air warfare, all these developments were squeezed into a very short time indeed, as most major ideas on air warfare were produced within four decades of the Wright Brothers' first heavier-than-air flight in 1903, only three years after the first *Zeppelin* airship or 'dirigible'. It has been claimed 'that almost every basic tactical and strategic application for air power had been tried out, at least experimentally, by the end of the 1914–18 War' (J.W.R. Taylor, q.i. Wells 2000: 23). This is almost true, with the exception of Soviet air-landing operations of the late 1920s (Sterrett 2007: 8). Something similar can be said for air power

Strategy: most of its possible variants had been articulated by the time the Second World War broke out. If some proved inadequate thanks to the still limited technology available, the ideas themselves would prove resilient to these misadventures and flourish once technology caught up, in turn paving the way for most of the key arguments of nuclear Strategy.

Even in 1893, a British Royal Engineer, Major J.D. Fullerton, told an international meeting of military engineers in Chicago that the future development of airborne vehicles meant as great a revolution as that brought about by gunpowder, that future war might start with battles in the air and end with 'the arrival of the aerial fleet over the enemy capital'; meanwhile, the 'command of the air' would be crucial to obtain, as without it, armies on the ground would be at the enemy's mercy (q.i. Wells 2000: 7). Essentially, he had put his finger on both the future 'strategic' and 'tactical' roles of air forces, which we shall discuss below. The 'tactical' use for the support of army or navy was the gathering of information, which the French, for example, had done even during the wars of the French Revolution with the help of hot-air balloons (see also the illustration on the front cover of this book). Yet this function took a while to gain acceptance: on the eve of the First World War, General Douglas Haig was still dismissive of the potential of aircraft, instead putting his faith in cavalry (q.i. Wells 2000: 9). Others by contrast – such as Friedrich von Bernhardi – quickly grasped the possibilities of air warfare: 'Even the air must be conquered; dirigible balloons and flying machines will form quite a new feature in the conduct of war' (Bernhardi 1912b, I: 15).

More important still was the fact that with air power, war would become the business of the people – the ordinary non-combatant citizens of the states involved in any war – in a way neither Guibert nor Clausewitz could have foreseen. As H.G. Wells later recalled in his thoughts on the eve of the First World War:

> Before any practical flying had occurred, I reasoned that air warfare, by making warfare three dimensional, would abolish the war front and with that the possibility of distinguishing between civilian and combatant or of bringing a war to a conclusive end. This I argued, must not only intensify but must alter the ordinary man's attitude to warfare. He can no longer regard it as we did the Boer War for example as a vivid spectacle in which his participation is that of a paying spectator at a cricket or baseball match. (q.i. Budiansky 2003: 8f.)

General concepts adopted from land and naval warfare

As early air strategists usually came originally from the other services, it is logical that concepts from ground and naval Strategy featured prominently in air Strategy; the wheel was not entirely reinvented. One inherited concept was that of the concentration of forces on one point, which we have encountered as part of the Napoleonic paradigm. Admiral Raoul Castex added 'command of the sea' to the mix: 'the interest of concentrating the largest possible number of aircraft on the decisive point is evident. Only from this point of view does it make sense to speak about "command of the air"' (Castex 1934: 34). In Britain, General (later Field Marshal) Jan Smuts wrote in 1917:

> It is important that we should not only secure air predominance, but secure it on a very large scale; and having secured it in this war, we should make every effort and sacrifice to maintain it for the future. Air supremacy may in the long run become as important a factor in the defence of the Empire as sea supremacy. (q.i. Tedder 1948: 84)

The most prominent thinker on air Strategy was the Italian, General Giulio Douhet (1869–1930). His advocacy of an air force began even before the First World War, and his starting points were key concepts of naval warfare which he transposed to the air:

> We are fully conscious today of the importance of having command of the seas, but soon the command of the air will be no less important because only by having such a command – and only then – can we make use of the advantages made possible by aerial observation and the ability to see targets clearly – advantages which we shall not be able fully to enjoy until we have the aerial power to keep the enemy grounded. (q.i. Chaliand 1994: 893)

In his most famous work, the *Domination of the Air* (translated as *Command of the Air*), Douhet seemed to echo H.G. Wells:

> By virtue of this new weapon, the repercussions of war are no longer limited by the farthest artillery range of surface guns, but can be directly felt for hundreds and hundreds of miles over all the lands and seas of nations at war. No longer can areas exist in which life can be lived in safety and

> tranquillity, nor can the battlefield any longer be limited to actual combatants. On the contrary, the battlefield will be limited only by the boundaries of the nations at war, and all of their citizens will become combatants, since all of them will be exposed to the aerial offensives of the enemy. There will be no distinction any longer between soldiers and civilians. (Douhet 1921/1983: 9–10)

And he formulated: 'To have command of the air means to be in a position to prevent the enemy from flying while retaining the ability to fly oneself.' Foreseeing the development of bomber fleets, he concluded:

> A nation which has command of the air is in a position to protect its own territory from enemy aerial attack and even to put a halt to the enemy's auxiliary actions in support of his land and sea operations, leaving him powerless to do much of anything. Such offensive actions can not only cut off an opponent's army and navy from their bases of operations, but can also bomb the interior of the enemy's country so devastatingly that the physical and moral resistance of the people would also collapse.

Thus 'to have command of the air is to have victory'. Concomitantly, 'to be beaten in the air means defeat and acceptance of whatever terms the enemy may be pleased to impose' (Douhet 1921/1983: 24f., 28).

But as with 'command of the sea', critics had pointed out that the ideal of 'command of the air' was difficult or impossible to implement. The Soviet strategists were aware of the benefits of the 'command of the air', but also of how ephemeral it was. N.A. Iatsuk, head of the tactics department of the Air Academy, wrote that 'Without air superiority, even having superiority on land, it is difficult to defeat the enemy, and if the strengths are even, it is impossible.' He concluded that however much the army and fleet were to beg the air force for close air support or for other action against military targets, 'the war should begin with a decisive strike' on the enemy air force (q.i. Sterrett 2007: 25). Aleksei Sergeevich Algazin, one of the generals later purged by Stalin, writing in 1928, noted that 'air superiority can only be gained through massive numerical superiority and for a very short time' (q.i. Sterrett 2007: 29). P.P. Ionov, appointed chief of staff of the Red Army airborne troops in 1941, echoed traditional land and naval warfare concepts about the concentration of forces:

> To gain air superiority in a given operation means to gain, for the entire period of the operation, freedom of action for one's own air force and to exclude or significantly constrain the activity of the enemy air forces in supporting the ground forces. Thus this superiority will be temporary (for the time of the conduct of the operation) and local (in the region of active operations). It can be achieved through massed air power, at the cost of weakening other less important axes. Falling short of the enemy in the battle for strategic air superiority, concentrated force on selected axes at the time of the operation makes it possible for the weaker side to gain air superiority. (q.i. Sterrett 2007: 43)

In France, Castex, who, exceptionally, agreed with Corbett on this point, thought it unrealistic to think about any permanent command of the air just as there was little point in thinking about a lasting command of the sea. With air power, the absurdity of the concept seemed to him even more blatant, as aircraft could stay in the air for a much shorter time than ships could stay at sea (Castex 1934: 34). In the British Royal Air Force (RAF), John Slessor (1897–1979) emphasised that one no longer spoke about 'command of the seas', but more modestly of 'control of sea communications', and equally, not of 'command of the air' but of 'air superiority' (Slessor 1936: 266). Half a century later, Air Marshal M.J. Armitage and Air Commodore R.A. (Tony) Mason updated the definition:

> Classically, air superiority was only a means by which freedom of action was gained so that bomber forces could achieve their primary aim. But once the bomber offensive became outdated, at least in the superpower confrontation, and once attention focused on a likely short war rather than a long war of attrition, then not only the notion of air superiority but even the use of the term itself lost currency. Air superiority was replaced by expressions such as 'local air superiority', 'favourable air situation', or even by the phrase a 'tolerable air situation'. (Armitage and Mason 1985: 266)

And yet only six years later, it was crucial in the First and Second Gulf Wars (1991 and 2003) for the success of the Western coalition against Iraq that they should have complete control of the air over and around Iraq; throughout the 1990s, the UN declared a large area of Iraq a 'no-fly zone' that was patrolled by US and British aircraft, and the same air control was exercised in the NATO bombing campaign of targets in Serbia and Kosovo in 1999 (Operation Allied Force). 'Command of the air', or at least 'local air superiority', is thus not only highly desirable, but, with contemporary technology, achievable against a technologically inferior adversary.

The obsession with the offensive that had dominated the long nineteenth century up to the First World War lived on in some minds. A crucial one was Hugh Trenchard (1873–1956), an early aviator who dominated the infancy and youth of the RAF. As commander of the Royal Flying Corps in France he wrote a memorandum for General Haig in September 1916:

> An aeroplane is an offensive and not a defensive weapon ... British aviation has been guided by a policy of relentless and incessant offensive. Our machines have continually attacked the enemy on his side of the line, bombed his aerodromes, and carried out attacks on places of importance far behind the lines ... [T]his has had the effect so far on the enemy of compelling him to keep back or to detail portions of his forces in the air for defensive purposes ... [T]he sound policy, then, which should guide all warfare in the air would seem to be this: to exploit the *moral effect* of the aeroplane on the enemy, but not let him exploit it on ourselves ... [T]his can only be done by attacking and by continuing to attack. (q.i. Wells 2000: 13; my emphasis)

Trenchard, the Bernhardi of Britain, supervised the establishment of the first British long-range bomber force in mid-1918 (Murray 1999b: 73). After the First World War, he fought against budget allocations on air defences, arguing instead that one should concentrate on the preparation of a bomber offensive against any adversary. With this he set the tone of RAF doctrine until the end of the Second World War (Biddle 2002: 70, 75). Even though, at the time, Trenchard was still uninterested in the use of the air force in a role independent of the war on the battlefield, we see here in embryo his later ideas on *morale bombing*, which were taken up not only in Britain, but also in France (Vauthier 1930: 50–4). Trenchard's thinking and Giulio Douhet's thinking were so close as to be indistinguishable in many points, yet they clearly developed their views in parallel – there is no firm evidence that one read the other (Golovine 1936: 7f.). For

Douhet, 'The air plane [sic] is the offensive par excellence' (Douhet 1921/1983: 15).

> The very magnitude of possible aerial offensives cries for an answer to the question, 'How can we defend ourselves against them?' To this I have always answered, 'By attacking.' ... Viewed in its true light, aerial warfare admits of no defense, only offense. *We must therefore resign ourselves to the offensives the enemy inflicts upon us, while striving to put all our resources to work to inflict even heavier ones upon him.* (Douhet 1921/1983: 52, 55)

So it was 'futile to divert aerial forces to defense' (Douhet 1921/1983: 60f.). As we have seen in chapter 10, Douhet got into a substantial debate on the matter with Captain Giuseppe Fioravanzo of the Italian navy, who by contrast argued that like naval superiority, air superiority could also be contested; the one who had won it had to defend it (Ferrante 1997: 156–64).

In France, the Douhet-follower Pierre Faure used Clausewitzian terminology of *negative* defence and *positive* offensive to demonstrate that for France, the ground forces behind the Maginot Line would take on purely negative roles if Germany attacked. Only the air force could take on the positive role of counteroffensive (Faure 1931: 25). Such Douhetian–Trenchardian thinking would resonate throughout the following century.

Deterrence through air power was a further important concept imported primarily from the navy (see chapter 9). It was already fully developed in the 1930s and became of crucial importance with nuclear Strategy, that special heir of air power. Initially, some dismissed it. Before the Second World War, Slessor could not see an air force on the ground as 'a fleet in being' in the same way as a navy (Slessor 1936: 34). Even after the war, Air Marshal Lord Tedder claimed it was nonsense to have an air force 'in being' in the way one can achieve political-military effects with a 'fleet in being': apart from the vulnerability of aircraft and airfields, an air force could not have the same visibility as a fleet (Tedder 1948: 84).

This view was soon revised, and, by any other name, 'fleet in being' thinking played an important role in air power and nuclear Strategy. The most famous occasion where an air force 'in being' had openly affected an adversary was on the occasion of the flyover of new Soviet bombers on Soviet Aviation Day in July 1955, attended by foreign observers. A small number of Soviet intercontinental bombers overflew the same terrain several times, leading the foreign observers to conclude that the Soviet Union had a much larger total number available than they had in reality. There followed US hysteria about a supposed 'bomber gap' between the Soviet and American arsenals.

From the 1960s, the main argument that would be made for US air force (USAF) bases in Europe, once the USAF had acquired intercontinental bombers, was their visibility and thus their political function of signalling US presence to foes and reassuring allies. The same argument of visibility was later made in 1979–83 for the deployment of the land-based intermediate-range nuclear forces (INF) in Europe. We shall see also that the 'fleet in being' concept in other ways informed French nuclear Strategy.

Service rivalry

Another inherited feature was that of interservice rivalry, in fights over the state's military budget. Where previously some followers of Mahan had dreamed that the navy might decide wars, from the First World War onwards, it was the air force's turn to nourish this dream. From the beginning, Douhet, and along with him some other air enthusiasts, postulated that aircraft and an air fleet should form a service in their own right, and not become subservient to the army or the navy. As 'air fleets will grow in importance as they get larger', Douhet wrote, '[t]he army and navy should not then see in the airplane merely an auxiliary arm of limited usefulness. They should rather see in the plane a third brother, younger of course, of the powerful family of war' (q.i. Chaliand 1994: 893). General Jan Smuts went further still, telling the British cabinet in the summer of 1917:

> [T]he day may not be far off when aerial operations with their devastation of enemy lands and destruction of industrial and populous centres on a vast scale may become the principal operations of war, to which the older forms of military and naval operations may become secondary and subordinate. (q.i. Jones 1973: 136)

Indeed, his views found general acceptance in the cabinet. In the following year, they thought, air bombardment of 'the chief industrial centres ... may form the determining factor in bringing about peace' (q.i. Budiansky 2003: 98). With its vital role for British Strategy thus recognised, the RAF became truly independent on 1 April 1918, while in the USA, air forces continued to be part of army and navy until after the Second World War. While the US navy retained its air assets, an independent US air force was created in 1947.

The French, who prior to the *Entente cordiale* saw themselves as doubly challenged by the numerically superior Royal Navy at sea and by the numerically superior German armies on land, pondered technological solutions to this in both naval and land warfare. With the arrival of air power, they developed an interest in yet another technological fix. As the aviation pioneer Clément Ader (1841–1925) put it, squeezed between Britain as the mistress of the seas and Germany with her superior land forces, France had no choice but to try to predominate, at least, in the air (Artzet 1999: 47).

Douhet developed the prototype of what after the Second World War would be NATO's sword–shield debates of the 1950s and 1960s: Douhet assigned the shield or holding role to the ground forces, that of the sword to the air force, which could plunge deep into the enemy nation's heart. His motto was: 'resist on the ground and create a mass in the air'. Not surprisingly, Douhet got into a quarrel with naval strategists on the subject of whether the navy was still needed. Douhet grudgingly conceded that the navy still had a role to play, but in view of the vulnerability of its bases to naval and to air attack, it would in future be quite secondary to that of the air force, which alone could strike not only the enemy's coastal production centres but those deep within his country (Ferrante 1997: 156–64).

The same pattern of an early air force-navy antagonism can be found in the USA, where General William (Billy) Mitchell (1879–1936), a First World War flying ace, posed an outright challenge to the grand Mahanian concept that the 'navy was America's first line of defence' with the argument that any ship could be sunk by a bomber. He went even further than Douhet in declaring the navy and army obsolete with the advent of air power, albeit modestly defined by him as 'the ability to do something in the air'. His argument, like Douhet's, was that the air force could be used against 'vital centers' of the enemy nation (q.i. Wells 2000: 59), a phrase which came to dominate Western air power and nuclear strategies for the rest of the twentieth century.

Not only will it insure peace and contentment throughout the nation because, in case of national emergency, air power, properly developed, can hold off any hostile air force which may seek to fly over and attack our country, but it can also hold off hostile shipping which seeks to ... menace our shores. (Mitchell 1925: ix)

Much to the ire of the sailors, he argued that:

The missions of armies and navies are very greatly changed from what they were ... Surface navies have entirely lost their mission of defending a coast because aircraft can destroy or sink any seacraft coming within their radius of operation ... aircraft today are the only effective means of coast protection. Consequently, navies have been pushed out on the high seas. The menace of submarines from below and aircraft from above constitutes such a condition that the surface ship as an element of war is disappearing. (Mitchell 1925: xvf., 56)

And he added:

[A]n entirely new method of conducting war at a distance will come into being ... superior air power will dominate all sea areas when they act from land bases and that no seacraft, whether carrying aircraft or not, is able to contest their aerial supremacy. (Mitchell 1925: 11f.)

On the one hand, Mitchell claimed that both aircraft and submarines were essentially defensive (Mitchell 1925: 120). Yet on the other he asserted that 'No defense from the ground is capable of stopping air raids over a country' (Mitchell 1925: 123). While antagonising the other services, Mitchell thus became the idol of the American military aviators.

Whereas Trenchard and Douhet probably developed their virtually identical views in parallel, the French Douhetians clearly took their views from the Italian general. His influence reached its height in the 1930s, when Colonel P. Vauthier published a synthesis of Douhet's ideas, issued with a preface by First World War hero Marshal Pétain. In 1931 Pierre Faure warned of another great French defeat à la Charleroi in August 1914 unless France developed a strong air force of bombers, which would at the same time not create a crushing economic weight for his country (Faure 1931: 9f.). Indeed, Faure hoped for a 'quick and cheap victory' through air power, should its deterrence force fail (Faure 1935: 43). Faure went further and argued that the occupation of the ground no longer had the same importance as before – at best, the army would occupy the ground as symbolic confirmation that the air force had won the war. Faure expressed his astonishment 'that the doctrine of the occupation of the terrain is still so widely accepted' (Faure 1931: 42, 46). This same argument, as we shall see, would be made in the context of the 1991 Gulf War. Faure spelled out the conclusion: 'aviation is likely to become the preponderant weapon ... The machine will have more importance than man ... [T]he nation whose aviation outclasses that of its adversaries has the greatest chances of victory' (Faure 1931: 44–6).

This interservice rivalry was observed at close quarters by J.M. Spaight, a lawyer and Whitehall civil servant. He noted that there were two sides in the struggle for a 'philosophy' of air power in its early decades: one side thought air power ineffective, the other argued that air power had relegated armies and fleets to 'the scrap heap'. The latter Spaight identified with Douhet and his followers – he was careful not to mention Trenchard – with their 'gospel of war and damnation' (Spaight 1938: 169). Air Commodore L.E.O. Charlton, for example, predicted that henceforth armies and navies would only serve as shields from behind which the air forces could attack the enemy and decide the outcome of the war; from this followed that the 'old ideas of offence and defence have undergone a complete change' with air power incarnating the offensive (Charlton 1935: 166). Two years later he even went further, bluntly declaring armies and sea power as 'obsolescent'; the next war would be conducted 'mainly in the air' (Charlton 1937: 70–82, 205).

Two former naval men took the same line, as the army and navy in their view were no longer forces that could take the offensive (Kennedy 1936: 58–79, 149–68). One, in a long diatribe about the British navy, its traditionalism, its lobbyists, its stick-in-the-mud supporters who distrusted any innovation, accused them of blindness to air forces being 'the decisive arm in war today' (Macmillan 1941: 25).

In the USA, Major Alexander de Seversky wrote that 'The most significant single fact about the war now in progress is the emergence of aviation as a paramount and *decisive factor* in war-making' (Seversky 1942: 3). He argued that aircraft had taken over the traditional tasks of the navy:

Navies have lost their function of strategic offensive ... The blockade of an enemy nation has become a function of air power ... Only air power can defeat air power ... Land-based aviation is always superior to ship-borne aviation.

He doubtless endeared himself to his navy comrades, particularly with one chapter headed 'The Twilight of Sea Power', where he wrote:

Navies are no longer lords of the seas. Their authority is being rapidly restricted and in some respects wholly wiped out. Certain naval units may be salvaged for auxiliary jobs under the protection of air power. A portion may make itself felt against backward nations lacking effective aircraft. But the rest – especially battleships – will be consigned to museums of outlived weapons along with the bow and arrow and the blunderbuss. (Seversky 1942: 155)

Seversky's most intelligent critic was probably the Czech Ferdinand Otto Miksche, who had served on the Republican side in the Spanish Civil War and argued that 'while the aircraft has a place in the war machine, it can never of itself *become* the war machine' (Miksche 1943: 9f.; my emphasis). Seversky had claimed that aviation was emerging as the '*decisive factor* in war-making' (Seversky 1942: 3). Miksche, by contrast, showed that as the Second World War had unfolded:

The Polish, French and Russian campaigns – not even the battle of Crete [*sic*] – were not decided in the air. They were decided on land, with the Luftwaffe playing an important role in these battles ... *support*[*ing*] the other two sections of the Wehrmacht in their operations – sometimes directly, sometimes indirectly, but generally combining both methods. (Miksche 1943: 13)

Miksche dismissed Seversky and Douhet as utopian in the light of the unfolding campaigns of the Second World War:

[A]n air concentration by itself is as incapable of achieving a decision as a concentration on land can if operating alone. It is also a mistaken idea that a decision must or could be reached in the air before it is reached on land. The two concentrations, on land and in the air, are integral. Both are necessary for victory; each supplements and reacts to the other. (Miksche 1943: 21)

Had not even arch-fascist Ludendorff dismissed the notion that a war could be decided by air bombardment alone (Miksche 1943: 26)?

Miksche's informed scepticism about the claims of the air force was widely shared even in the interwar years. Marion W. Ackworth, writing as 'Neon', held that aircraft were so utterly vulnerable and unreliable, and their reach so limited, that there was no point in the state continuing investing in them on a large scale. Air power would be 'utterly ineffectual so far as winning the war is concerned'. They could play a minor role in the support of navies and armies that would bear the brunt of any war, and bring about a decision (Neon 1927: 127–30, 190). While being overly pessimistic with regard to the future role of aircraft, 'Neon' was right in claiming, at least for the years of the Second World War, that 'Aerial bombing is from its very nature absurdly inaccurate, therefore indiscriminate. Successful attacks on specific and protected targets are operations beyond the scope of "air power".' And from this 'Neon' concluded that 'Bombing attacks on the non-military population are deliberately intended' and thus contrary to international law (Neon 1927: 178, 184). Similarly, Admiral Sir Herbert Richmond in 1923 and Admiral Sir Richard Custance in 1924 defended the importance of the navy and army against the air force or what Custance called the 'aery' (Spaight 1938: 50). Richmond scathingly summarised the views of the strategic or city bombing school: they aimed not to defeat the enemy armed forces, but 'to overcome the enemy people', who could now be attacked directly, 'without any need of overcoming a defending force'. Instead, 'Air forces will bombard the great centres of life and industry, the organisations of transport, water-supply and other internal national services, the administrative establishments and the civil population itself' (Richmond 1934: 100f.). In France, General Albert Niessel, translator of Clausewitz and Falkenhayn, also emphasised that it was the ground forces who in his view must decide the outcome of wars (Spaight 1938: 53). In the United Kingdom Admiral John Tovey, commander of the Home Fleet, protested to the Admiralty against the 'absolute priority' given to Bomber Command and its city raids in terms of defence budget and effort. In the spring of 1942 he wrote:

> The Navy could no longer carry out its much increased task without adequate air-support; that support had not been forthcoming ... Whatever

> the results of the bombing of cities might be, and this was the subject of keen controversy, it could not of itself win the war, whereas the failure of our sea communications would assuredly lose it ... It was difficult to believe that the population of Cologne would notice much difference between a raid of 1000 bombers and one by 750. (Budiansky 2003: 274–7)

While the critics did much to keep air force expenditures down, it is the strategic or city bombing that gained force on the Western side in the Second World War. Trenchard's successor in spirit, Field Marshal Sir Arthur 'Bomber' Harris (1892–1984) and USAAF General Carl Andrew Spaatz (1891–1974) both thought little of the performance of the other two services. The chief of the Imperial General Staff, General Alan Brooke, recalled: 'Harris told us how well he might have won the war had it not been for the handicap imposed by the existence of the other two services.' At the time, Spaatz had opined that the Normandy landing would be pointless and doomed to failure (Budiansky 2003: 299).

Spaatz felt vindicated in his belief in air power dominance with Hiroshima and Nagasaki. US General Frederick L. Anderson wrote to Spaatz: 'I wish to congratulate you upon proving to the world that a nation can be defeated by air power alone' (q.i. Wells 2000: 151). Douhet and the city bombing school seemed vindicated by the aircraft's new ordnance, the atomic bomb.

The rivalling claims to predominance of the different services became even if anything more pronounced in the USA after the Second World War, when initially the USA cut down its armed forces drastically, and the shares in the remaining pie became that much more contested. Rivalry remained prominent and disruptive when the USA began to rearm from 1950, resulting in intense competition for resources (Wells 2000: 149–74, and 175–204).

The air forces in turn had their comeuppance when the 'unmanned bomber' (cruise missiles) and ballistic missiles started to usurp their place. Would these replace the 'manned bomber' and do them out of business? Dwight D. Eisenhower as US President told the Joint Chiefs of Staff in 1960 that he no longer saw any need for the bomber (Trachtenberg 1986: 756).

The conventional wars of the 1950s to the 1980s neither excluded air forces nor allowed them to 'go it alone'. The one success story, that of LINEBACKER II, proved ephemeral (see below). The picture seemed to change, with Douhet once more in the ascendant (this time with conventional, but unprecedentedly precise air power) in the Gulf War of 1991. Former US Marine Corps General Bernard Trainor noted that it was here that the coalition ground campaign had, for the first time, supported the air campaign. The Gulf War Air Power Survey (GWAPS), conducted by a mix of military experts and leading independent academics, concluded that 'for the first time in history, air power had reached the expectations of its proponents' (GWAPS 1993, II.i: 327). For, as USAF Colonel Dennis Drew commented, in the Gulf War, the impact of air power 'was clearly overwhelming and decisive ... [T]he air victory ... symbolised the maturity of air power, the domination of air power and the need for a new paradigm of warfare' (q.i. Mason 1994: 235–78).

Moreover, proponents of strategic bombing claimed after the 1991 Gulf War 'that the Gulf War heralds a new age in which strategic bombing will be the strongest form of military power, dramatically enhancing America's coercive capabilities and options'. As the academic Robert Pape observed, in all subsequent crises, such as the Yugoslav Wars in the 1990s, 'increasingly the first question in debates over American intervention is becoming, Can air power alone do the job?' Pape's answer to this was 'no' (Pape 1996: 314, 318f.). As the GWAPS team explained, boots were still needed on the ground there (and subsequently in crises such as Kosovo, Afghanistan, again Iraq ...) to hedge against a 'stab-in-the-back' legend such as that which the German army had created after 1918, claiming it was not military defeat, but political betrayal that had led to the demise of Wilhelm II's *Reich* in the First World War (GWAPS 1993, II.i: 32).

These debates about the role of the air forces in the general context of war and other military operations, and especially debates about strategic bombing, continue to this day. The next chapter will examine generically all the Strategies proposed or air power, again tracing traditions and links to earlier thinking which we have identified in previous chapters.

13 *Four schools of air power*

The strategic/city bombing school was only one of four, of which three existed from the beginning of the history of aviation, but several of which would only approach realism with further developments of technology. The strategic/city bombing school at least initially was inspired by Social Darwinist concepts of 'national wars' in which nation was pitted against nation in the war effort and no target was out of bounds. It achieved its summit with the nuclear bombing of Hiroshima and Nagasaki.

The second is the military targets school, often referred to as tactical, loved best by the other two services. The main tasks of the air force under this heading would be to provide close air support (CAS) for the other two services, or, beyond the actual field of battle, reconnaissance and the targeting of enemy military forces (and those only) that were approaching the battlefield from afar.

The third school – overlapping with the first two – concentrates on leadership targeting, and depends on precision bombing, although its concepts also go back to early thinking about air Strategy. It is posited not on a more or less explicit tendency to kill enemy civilians indiscriminately, but on a concept of justice that would target the wicked leaders (rogue regimes, tyrants, oppressors) in 'decapitation strikes' which would bring the enemy war effort to an end. Often there is the assumption that this would lead to the liberation of their long-suffering, oppressed and perhaps misguided populations. This carries echoes of the war aims of the French Revolution, and of the USA in the First and Second World Wars.

The fourth school was the political signalling school, which had the shortest life-span, from the end of the 1950s to the end of the 1960s, with a few blips of revivals in the 1990s.

The strategic or city bombing school

The strategic or city bombing school has much in common with the 'blue water school' or Mahanian tradition of naval thinking, and was therefore dubbed the 'blue skies school' by John Spaight. Its claims for what air power could do, he noted, went far beyond what the blue water school ever claimed for itself (Spaight 1938: 37–9). Both liked to claim that their service could almost single-handedly bring about the decision in war, and that it was on the open seas – or by targeting the enemy's 'vitals' from the air – that a war should be fought and won. They differed greatly, of course, in the nature of their adversaries: the blue water thinkers saw the enemy navy as main opponent, while the city bombers sought to avoid the clash with enemy armed forces and to take the war to the undefended masses of the enemy's population instead. 'The classic air-power theorist had never quailed at the notion of killing civilians' (Budiansky 2003: 285). As the German Admiral Tirpitz put it during the First World War, 'If one could set fire to London in thirty places, then what in a small way [i.e. with a single bomb] is odious would retire before something fine and powerful.' And historian Williamson Murray showed that 'All the major powers involved in the air war' during the First World War 'pursued strategic bombing to some degree' (Murray 1999b: 72).

Yet from the beginning of the use of air power, the legitimacy of bombing civilian targets was known to be ethically problematic. The Hague Conventions on Land Warfare of 1899 and 1907 had outlawed the bombing or shelling of cities (but had failed to articulate the possibility that this could be done from the air).

The Hague Aerial Bombardment Rules of 1923 that had made this quite explicit failed to be ratified by the states that had participated in the conference. They included:

> Art. 22: Aerial bombardment for the purpose of terrorizing the civilian population, of destroying or damaging private property not of military character, or of injuring non-combatants is prohibited.
>
> Art. 24: Aerial bombardment is legitimate only when directed at a military objective, that is to say, an object of which the destruction or injury would constitute a distinct military advantage to the belligerent.

Military targets, according to Article 24, included factories producing military supplies and lines of communications. Article 25 listed targets to be avoided if at all possible: historic monuments, places of worship, buildings dedicated to art or science and hospitals. Nevertheless, Article 24 provided the loophole which could excuse any bombing as long as one protested that one was only in pursuit of a 'military objective', whatever the collateral damage. In the words of James Spaight, adviser to the British delegation:

> The doctrine of the 'military objective' ... will be no adequate protection. Any belligerent who chooses will be able to keep within the rules ... and yet use his air arms for a purpose quite distinct from the destruction of objects of military importance, namely, for the creation of a moral, political, or psychological effect within the enemy country. (Budiansky 2003: 188)

During the 1920s the hope was still alive that the 1923 rules might become binding, and in subsequent years there was some awareness that they might become observed in practice and thereby turn into customary law. In July 1939, Hermann Goering as chief of the German air force gave out instructions on the conduct of air warfare which neatly sums up the prevailing paradox:

> Any target is of military importance if it is important for the enemy's conduct of war. [Targets] of military importance are e.g. water, gas, and power supplies, food stores, large-scale bakeries, cool storage buildings, etc. ... Terror attacks on the civilian population are contrary to international law. Nevertheless, the context of the war may make even this form of aerial attack necessary. (Goering 1939)

And indeed, this is how Germany began the Second World War with its air attacks on Warsaw.

Deterrence through city targeting?

Some even thought optimistically, *before* the invention of nuclear weapons, that air power alone might deter war altogether. We have seen in chapter 10 that 'deterrence' – dissuading an adversary from a particular course of action by threatening intolerable punishment – was practised in British maritime Strategy long before the word was invented, and Alexander George and Richard Smoke rightly note that example of deterrence posturing can be found in Thucydides (George and Smoke 1974: 12). In the context of air power, it gained wide prominence in the interwar period, but especially in the 1930s. Pierre Faure spelled it out most plainly in France early in that decade. If France built a bomber force, he opined, ideally one bigger than that of her adversaries,

> [t]he fear of terrible retaliation [against their aggression], without precedent in history with its human hecatombs [of dead], would certainly make those on the other side of the Rhine think before they undertake a new war. If the German people knew that an attack by its army on France would signify the probable and immediate destruction of twenty of its largest cities and the population they contained, it would seem doubtful that the German government would take this risk ... 400 aircraft ... capable of annihilating twenty German cities, the bridges of the Rhine, important railway stations and the great industrial centres, are the best guarantee of peace that we could own. (Faure 1931: 9f., 13)

In his second book of 1935, however, Faure fully accepted the possibility that deterrence might fail; and, he argued, 'because war has only one aim: victory', as a result, it would be 'more important to destroy a city than an enemy air squadron'. So if war did break out, he believed it would be reduced to one phase only, 'an aerial offensive destined exclusively to take moral and material disorder to the enemy', much as NATO strategists from 1949 until the late 1950s thought that a Third World War would start with a massive nuclear exchange. Like the nuclear strategists later, Faure thought that the alternative would be a return to 'an abominable war of mutual attrition' on the battlefield (Faure 1935: 74, 81, 83). His great hope was that with a sizeable bomber force (in French possession), 'war would become an impossibility, and the aircraft will *have killed off war*' (Faure 1935: 202; my emphasis; the title of his book translates as *The Aeroplane Will Kill War*). This strikingly foreshadows the hope of the early nuclear age that war would now become an impossibility, that one had entered the age of the *non-guerre* (non-war); in the optimistic words of French strategists: '*la guerre est morte*', war is dead (Heuser 1998a: 84f.).

In the same years, this was also the British government's thinking in backing the RAF leadership's preference for a bomber force over air defences. British Prime Minister Stanley Baldwin (Prime Minister 1935–7) and his advisers put their money on the deterrent effect of the British bomber force on any adversary (Bialer 1980).

Deterrence theory was thus fully established well before the nuclear age, as George Quester has so brilliantly shown (1966). The effect of this deep commitment by the British to deterrence in the form of an offensive strategic posture resulted in a mirror imaging with regard to German intentions, and ultimately in *self*-deterrence: Chamberlain reached a settlement with Hitler in Munich because of his fear of a 'knock-out blow' of the Luftwaffe against Britain, and the ultimate commitment to the defence of Poland came after a false alarm about a planned German surprise attack on the West in early 1939 (Watt 1989).

Strategic bombing: the debate

Strategic bombing was the extension of the thinking behind the naval blockade. In the First World War, the British government had attempted to break German morale by starving the population through a naval blockade. That the idea was crucial, not the technology, can be shown by the fact that other ways of achieving the same aim were explored. Some British scientists during the First World War had already researched ways of starving the Germans into submission by burning Germany's crops from the air (Budiansky 2003: 110).

Before, and even after, a declaration of war, city bombing might be illegal, as Captain Pierre Yvon of the French navy noted, but he doubted that this made it less likely in the future. Yvon drew attention to the German bombing of arms factories in London suburbs in the First World War – these targets could hardly be distinguished from civilian targets, even if this bombing was arguably not carried out with the objective of terrorising civilians. Against this background, he elaborated on the parallels between naval blockade and air force-imposed blockades, stressing the similarities (Yvon 1924: 92, 97, 101–18). Indeed, Douhet even advocated pre-emptive war: 'the air arm ... may try for the destruction of the enemy's capital *even before war is declared*' (Douhet 1928/1983: 188; my emphasis).

This crucial link between naval blockade thinking and city bombing was spelled out particularly clearly by British Royal Flying Corps veteran Captain Norman Macmillan (1892–1976). Two years into the Second World War, he advocated imposing 'air blockades', which would stop vital supplies (food and raw materials) reaching the enemy's factories.

> Unlike a sea blockade, which is an external form of blockade, air blockade is at its most effective when applied against internal targets within the enemy country or zone of occupation. So the aeroplane ... is an instrument capable of bringing about blockade in a new form in which the weapon is directed straight at the citizen. To be applied with success, air blockade must have a wide application. Its effect is purely destructive ... (unlike sea blockade, which ... can seize prizes) ... It is a more cruel form of blockade, one which ignores all the canons of war, for it puts into the front line women and children, and aged and non-combatant men, as much as or even more than the armed forces. It is a weapon used not only to destroy the commodities before they reach the consumer, but impartially to destroy the consumer also, by fire as well as high explosive, and perhaps by gas. (Macmillan 1941: 109f.)

Macmillan was hopeful of socio-political effects that the bombing might have:

> Air blockade might be employed in such a way as to cause internal disruption in a country by the excitement of some of the human passions ... War and revolution stalk hand in hand. And it is no less than a German revolution that Britain has pledged herself to secure by the overthrow of the Nazis by force of arms. (Macmillan 1941: 113)

Blockade and air bombardment were usually euphemistically said to be designed to reduce the enemy nation's 'morale', a term, as we have seen, that in the nineteenth century was linked with the armed forces, not the nation as a whole. All sides seem to have hit on this 'intellectually mushy' (Budiansky 2003: 136) concept pretty much simultaneously. In May 1917 the Germans rolled out the new big Gotha bombers, with a range allowing them to strike London. The crews of these and other German bombers were told that their targets were 'the morale of the English people' (q.i. Budiansky 2003: 96). Hugh Trenchard wrote in a memorandum of November 1917: 'Actual experience goes to show that the moral effect of bombing industrial towns may be great, even though the material effect is, in fact, small.' In May 1918 he wrote: 'The anxiety as to whether an attack is likely to take place is probably just as demoralising to the industrial population as the attack itself.' He therefore wanted plans to stage bombing raids throughout Germany so that Germans everywhere would worry about whether they would be the next to be hit (Budiansky 2003: 101). It fits the Social Darwinist – or should we say racist? – mindset behind these statements on both sides perfectly that a Royal Flying Corps paper on targeting of 1917 argued that bombing German chemical works would produce a great effect on morale, as 'a majority of German chemists' were Jews who were 'not usually brave'. The paper also posited that German bombing of the East End of London was more likely to lead to panic because of the many 'Jews' who lived there (Budiansky 2003: 102f.).

In January 1918, the British Air Policy Committee proclaimed a new Strategy to be followed, 'to attack the important German towns systematically' to shake 'the morale of workmen' to the point where 'output is seriously interfered with' (Jones 1973: 161f.). While the Strategy was not carried out by the British air force during the First World War, the doctrine survived (Jones 1973: 188). In his first lecture, Air Vice Marshal Sir Robert Brooke-Popham, the first commandant of the Air Staff College, told his officer-students: 'it is now the will-power of the enemy nation that has to be broken, and to do this is the object of any country that goes to war' (q.i. Meilinger 2003: 52). He told the Royal United Services Institute in a lecture in December 1919:

> The material effect of bombs ... is small compared to its moral effect, and I think most people will agree that bombing from aircraft has considerably greater moral effect than ordinary shelling. This of course applies with still greater force to towns and places far behind the line which, at any rate previous to the war, considered themselves immune from attack. (q.i. Biddle 2002: 79)

When the RAF created 'home defence' forces after the First World War, they still put most of their eggs in the other, the offensive basket of the retaliatory bomber force. In a memorandum of 19 July 1923, Trenchard as the Chief of Air Staff advocated concentrating on bombers, still holding the widely shared assumption that in the next war the enemy would be France. If Britain bombed France, he felt 'that although there would be an outcry, the French in a bombing duel would probably squeal before we did. That was really the final thing. The nation that would stand being bombed longest would win in the end' (q.i. Jones 1987: 29). He personally was convinced

> that a high casualty rate would have a greater effect on the morale of the French pilots than it would on ours. Casualties affected the French more

than they did the British ... the policy of hitting the French nation and making them squeal before we did was a vital one – more vital than anything else ... The Army policy was to defeat the enemy Army – ours to defeat the enemy nation. (q.i. Jones 1987: 30)

Brooke-Popham's successor at the RAF Staff College in the early 1930s, Commandant Joubert de la Ferté, like Trenchard and Brooke-Popham was still thinking in such Social Darwinist terms:

The nation with the better fighting men, the stouter hearts, the better leaders and equipment will, in the end, outfight the weaker, who will be forced more and more on the defensive, until finally the successful offensive of the stronger side will have given it the security it requires. (q.i. Biddle 2002: 101)

Applicants to the RAF Staff College were made to sit an entrance exam, and as one participant recalled, not surprisingly the expected answer to one question was clearly 'that the only appropriate use for air power lay in the offensive against enemy morale' (Budiansky 2003: 134–6). None of this sounds very different from the German Luftwaffe officer Albert Kesselring's statement at roughly the same time that Germany should aim for a strategic bombing competition with its adversaries, convinced that an enemy nation would withstand it less well than the German *Volksgemeinschaft*. Kesselring became Chief of Staff of the Luftwaffe in 1936, and rising to field marshal remained throughout the war one of Hitler's most trusted generals (Murray 1999b: 97–9).

By contrast, even in 1917, the French air service adopted a policy paper in which they warned that there was no point in bombing just for the sake of bombing: 'on ne bombarde pas pour bombarder' (Budiansky 2003: 101).

But even for most advocates of city bombing, it was not merely a case of bombing for the sake of bombing. We have seen that in the desperate quest for alternatives to the trench warfare of the First World War (see chapter 7), several strategists explored alternatives of all sorts, what Liddell Hart called indirect approaches. Spaight commented on how

The Great War ... left men's minds wearied, disillusioned, disgusted with the murderous slaughter, the slow, muddled, wasteful ineffectiveness of the tactics of attrition. The air seemed to offer a way of escape from the condition of deadlock, of stalemate to which war had been reduced. It was a new domain which could be called in, if not to redress the balance, at least to relieve the impasse that had been reached in the old. (Spaight 1938: 168)

One seeking for alternatives whom we have already encountered was J.F.C. Fuller. His vision of future war was that 'Fleets of aeroplanes will attack the enemy's great industrial and governing centres. All these attacks will be made against the civil population in order to compel it to accept the will of the attacker' (Fuller 1920: 42). He emphasised that 'the policy of a nation is founded on the will of its civil inhabitants and that the supreme military power of aircraft is their ability to "hop" over armies and fleets and attack what is in the rear of them' (Fuller 1923: 148). Basil Liddell Hart, who later liked to think of himself as a pacifist, in 1925 advocated a Strategy that sought to 'discover and exploit the Achilles' heel of the enemy nation' striking 'not against its strongest bulwark but against its most vulnerable spot', just as Paris, prince of Troy, had killed Achilles by striking at his vulnerable heel. The enemy's cities he saw as the modern equivalent of Achilles' heel in his ambiguously-named book *Paris and the Future of War*; Paris could stand both for the Trojan prince who had defeated Achilles and for the French capital, potential target for British air bombardment, if France was the future enemy (Liddell Hart 1925: 26f.).

The British government's former scientific adviser, Hardinge Giffard, second Earl of Halsbury (1880–1943), wrote in the *Daily Mail* on 8 July 1927: future war would 'not merely be armies engaging armies but whole nations mobilised against nations'. With the total involvement of one's own and one's enemy's nations in war, the distinction between combatant and non-combatant was disappearing (see chapter 7).

The girl filling a shell at a factory is just as much part of the machinery of war as the soldier who fires it. She is much more vulnerable and will certainly be attacked. It is impossible to say that such an attack would be unjustified. The matter does not end with mere munitions workers. The central organisations essential to modern warfare are carried on in 'open towns' and largely by civilians. An attempt to paralyse them would be perfectly legitimate. The first conclusion, therefore, that emerges is that an attack will be made upon the civilian population. (q.i. Jones 1987: 40)

Trenchard also listed economic and military-strategic factors as reasons for city bombing. In his famous memorandum of May 1928, he argued that in future, it would no longer be 'necessary for an air force, in order to defeat the enemy nation, to defeat its armed forces first. Air power ... can pass over the enemy navies and armies, and ... attack direct the centres of production, transportation and communication', which of course would be located in urban centres, but this he did not mention. He denied vehemently that this would be

> indiscriminate bombing of a city for the sole purpose of terrorising the civilian population. It is an entirely different matter to terrorise munitions workers (men and women) into absenting themselves from work ... through fear of an air attack upon the factory or dock concerned.

Invoking the right conceded to any belligerent in the Hague Convention of 1907 of destroying munitions destined to be used against him, Trenchard created the fiction that massive destruction of a city could be kept separate from well-targeted daytime attacks on munitions factories or harbours (Webster and Frankland 1961: 71–6). From here it was only a small step to Harris's 'dehousing' Strategy, which sought to convey the impression that the houses would be destroyed by night-time bombing raids, while the workers and their families were not inside them.

Until the introduction of precision-guided munitions (PGMs) in the 1970s, however, technology was still far from fulfilling the dreams of early air strategists. Bombing from great altitudes proved much less effective in practice than in artificial test situations. The experience of the Spanish Civil War, used by the Germans, Italians and Soviets as a test for their air forces and doctrine, proved to them, and to other observers like the French, that ground support and low-level bombing of undefended targets were the more feasible options.

The RAF's experience early in the Second World War showed that daylight bombing (let alone battlefield support!) resulted in large losses of planes, so they embraced night-time (and thus of necessity big-target) bombing. At the time it was still technically impossible, in high-altitude night-time raids, to differentiate between targeting factories and surrounding residential areas. When the USA came into the war, its air forces, to the annoyance of the RAF, initially opted for daytime precision bombing. But the Americans, too, soon abandoned it because of the heavy losses they incurred. The turning point was the American Eighth Air Force's attack on Schweinfurt in August and October 1943, when almost a fifth and then a quarter of the bombers were lost, many more being damaged and beyond repair. From this point onwards, American military targeting gave way to city bombing, whatever General Arnold later claimed. American bombers resorted to 'drenching' targets in order to increase the likelihood of direct hits (Budiansky 2003: 286, 320, 323).

Where, then, were the restrictions of the laws of war in that context? 'Realists' shrugged off such restrictions as mere scraps of paper (see chapter 5). The Conservative MP and aeronautical engineer Sir Charles Dennistoun 'Dennis' Burney believed like Trenchard that international law would fall by the wayside when every country had become nothing but 'one huge arsenal'. Everybody would use their weapons without restraint. 'In any future war', he opined, 'all the vital parts of the city [of London] could be heavily bombarded within twenty-four hours, and all our defensive preparations would be powerless to prevent it' (Burney 1934: 74).

Many authors foresaw accurately how war would evolve and understood the moral dilemma of facing ruthless 'realists' like Hitler and Mussolini, who, like his protégé Douhet, believed that the role of air forces was 'to break up enemy formations, to command the air, and to weaken the morale of the enemy's civilian population'. Mussolini added with insouciance: 'Centres of population on both sides are bound to suffer' (q.i. Spaight 1938: 161).

Acknowledging the moral dilemma that city bombing posed, Brigadier P.R.C. Groves in the United Kingdom nevertheless thought those pacifists misguided who, in order to prevent such a future development of war, supported 'the reduction or even the abolition of our already inadequate Air Force'.

> For ... if a pacifist people, owing to an objection in principle to the use of force, are not prepared to oppose force by force, they will in the long run strengthen the rule of violence, that is the rule of their militarist opponents. The cause of peace cannot be promoted by leaving Right weaponless. (Groves 1934: 33)

Lord Thomson, who in the interwar years was twice Secretary of State for Air, foresaw ruthless bombing of densely populated areas, leaving 'both victors and vanquished ... with ruined cities ... hospitals filled with the maimed and mutilated of all ages and of both sexes, asylums crowded with unfortunate human beings whom terror has made insane'. Thomson saw great possibilities for air power to underpin international peace, but also warned of the dangers of its abuse in war. He rightly feared that technology had outrun the moral development of humanity (Thomson 1927: 26f., 190–2).

Alexander Seversky, writing in the USA in the 1930s, put his finger on a crucial question: 'Does the attacker aim at the possession of the enemy country or at its elimination as an economic and political factor?' The difference mattered, because '[t]he war of possession calls for ground forces, for aviation co-ordinated with those forces, for air power on a tight leash in order to avoid unnecessary destruction' (q.i. Hobbs 1979: 90). Wars before the spread of Social Darwinism, racism and 'totalitarian democracy' (Talmon 1952) had aimed at conquest and exploitation, not at annihilation. In the political-ideological climate of the interwar years, annihilation of the enemy became an end in itself, as the Germans demonstrated with their extermination camps.

Some American airmen also thought city targeting desirable well before practical difficulties put paid to the prevailing US Strategy of bombing military targets. General H.H. Arnold and Lt.-Col. Ira Eaker thought of bombardment of a city as the best way to 'break the will' of the enemy people – note the Clausewitzian term (Arnold and Eaker 1938: 129). 'Hap' Arnold would later command the US Army Air Forces during the entire Second World War, as American bombing degenerated from unsuccessful attempts at precision strikes into city busting. In August 1944, General Carl Spaatz informed him of British pleas to 'join with them in morale bombing'. Spaatz was worried that 'any deviation from our present policy' of precision bombing would result in the US being accused of unethical behaviour. Yet the USAAF slid into this bombing pattern, mainly because technically it had few alternatives if US bombers were to have a reasonable chance of survival. Even after the infamous raid on Dresden in the night of 13/14 February 1945, Arnold and Spaatz denied publicly that the USAAF had changed its targeting away from 'military objectives' (Budiansky 2003: 321f.). At the same time, the USA completely switched to a city-flattening bombing Strategy in Japan – with the first massive use of incendiary bombs being made under Curtis LeMay against Tokyo on 9 March 1945. The joint target plan that advocated this showed awareness of the magnitude of civilian casualties it would entail, but justified it in terms of the targets being civilian 'workers', whose killing would lead to 'losses in labor' (Gentile 2001: 87) – an echo of the Trenchard memorandum of 1928.

Punishing the enemy nation – on a small scale

During the Conference at the Hague in 1922–3, the British delegation asked for a clause to be inserted in the Convention to allow indiscriminate bombing 'for police purposes in certain outlying regions', that is, for the sort of operations they had been conducting in Iraq, Somalia and Afghanistan (Budiansky 2003: 188). Colonial warfare 1897–1923, according to RAF estimates, had cost the lives of thousands of British soldiers and far fewer tribesmen. Since the RAF was brought in, the relation was inverted. The 1920 British Army Directive regarding Mesopotamia read like a prescription for scorched-earth measures in the Hundred Years War or in Cromwell's campaigns against Ireland:

> Villages will be razed to the ground and all woodwork removed. Pressure will be brought on the inhabitants by cutting off water power and by destroying water lifts; efforts to carry out cultivation will be interfered with, and the systematic collection of supplies of all kinds beyond our actual requirements will be carried out, the area being cleared of the necessities of life. (q.i. Meilinger 2003: 50)

In this context, a discussion between General Sir Claud Graham, Chief of the General Staff of the Indian army, and Air Marshal John Salmond, Commanding Air Officer in India, with regard to British military action against Afghanistan is exemplary. Salmond protested against aerial bombing of Kabul and native villages. Graham replied that Afghanistan was 'not a signatory to the Hague or Geneva Conventions, the Afghans mutilated and ill-treated wounded opponents, they were not a civilised nation and he assumed, therefore, that there were no reservations.' He added, 'Personally, if I were doing the bombing, I should not care what the restrictions were ... In the past they had not used the Air Force ruthlessly enough' (q.i. Meilinger 2003: 50). (In a similar way, but on a much larger scale, the fact that the USSR had not ratified the Geneva Conventions would be invoked in the Second World War by the Germans to excuse their horrendous treatment of Soviet prisoners.) Financial restraints also affected the decision for aerial bombing. It was used against Afghan rebels in 1919 with bombers which had been completed just too late to be used against Berlin in the First World War. When in 1920 the 'Mad Mullah' Mohammed bin Abdulla Hassan in British Somaliland staged a revolt against the British occupying forces, the forecasted costs of subduing him with ground forces were off-putting. By contrast, the RAF promised to do it much more cheaply, through bombardment from the air (which they did effectively, as Hassan did not have any anti-air defences). These bombing raids, under the heading of 'air control', were applied also to Iraq in 1922, again under the command of John Salmond, with the aim of establishing 'a tradition' that would teach the 'natives' what to expect in return for insubordination. In the words of one RAF officer:

> One objective must be selected – preferably the most inaccessible village of the most prominent tribe which it is desired to *punish* ... The attack with bombs and machine guns must be relentless and unremitting and carried on continuously by day and night, on houses, inhabitants, crops and cattle ... The news of the punishment will spread like wildfire ... This sounds brutal, I know, but it must be made brutal to start with. The threat alone in the future will prove efficacious if the lesson is once properly learnt. (q.i. Budiansky 2003: 143; my emphasis)

Again, we can identify a visible link between the condescension with which the imperial policing was undertaken, and morale bombing. In 1924, Wing Commander C.H.K. Edmonds explained it at a lecture on 'Air Strategy' at the Royal United Services Institution:

> Just as in the small war our continuous bombing made the tribe's life intolerable and brought [the enemy] to heel, so in the case of the big war our object is to destroy the enemy's morale – we must make him feel that life has become so impossible that he prefers to accept peace on our terms. (q.i. Budiansky 2003: 144)

In a paper of 1922 Trenchard even went as far as to suggest that air power might be used to quell 'industrial disturbances' in 'settled countries' including India, Egypt and Ireland, or even England.

Churchill as Secretary of State for War and Air told him never again to mention the last two in such a context (Budiansky 2003: 146f.). This reflects not only the overlap between class thinking and Social Darwinism on the part of Trenchard (and many of his contemporaries even in liberal countries), but also the more racist thinking of, in this case, imputed by Churchill to his audience, for whom the treatment meted out to the 'coloured' peoples was one thing, that used against whites – one's own people – another.

But punishment was not merely a European idea, inspired by notions of imperial policing. In the USA, Billy Mitchell prophesied in 1925 that in future, states would use air power as their first resort, as 'first punitive element', in any conflict with another state, because it was so easy to call upon (Mitchell 1925: 131).

The lesson of the German bombing of London in the First World War and in the Spanish Civil War, as anybody could see, was that the bombing of cities reinforces solidarity among the victim populations and the determination to hold out (Pape 1990, 1992, 1996; Budiansky 2003: 8f., 207–9). But this was not what the RAF wanted to hear (Chuter 1996). As the War Cabinet directed on 14 February 1942, even before Harris took over, the air offensive should have as its 'primary objective ... the morale of the enemy civil population and, in particular, industrial workers'. Aiming points would be specifically 'built-up areas' of cities, and soon Harris issued the instructions for 'dehousing' workers (Biddle 1994: 152). On 30 March 1942, in a memorandum to the Prime Minister, Harris, too, advocated that the bombing should be directed against German cities so that a third of the German population could be 'turned out of their house and home', claiming that 'dehousing' had a worse effect on morale than having 'friends and relatives killed': 'There seems to be little doubt this would break the spirit of the people.' In this he put forward the same argument as Machiavelli had four centuries earlier (see chapter 3). Churchill was persuaded by this argument, and told Stalin in the late spring of 1942, 'Morale is a military target ... We shall seek no mercy and we shall show none' (Budiansky 2003: 284).

In general, the British bombing campaigns against German cities were welcomed by Britons and allies alike. A Gallup poll of 1938 found that 91 per cent of the Americans asked thought that 'all nations should agree not to bomb civilians in cities in wartime'. But three days after Pearl Harbor, 67 per cent of Americans asked were in favour of indiscriminate bombing of cities, with only 10 per cent firmly opposed. Only a few had qualms about city bombing, including the British government's scientific advisers, Sir Henry Tizard and Lord (Solly) Zuckerman. The latter pointed out that technological options were driving Strategy, not the other way round, as it should be. There were also objections from religious quarters. On 11 March 1944 *The New York Times* reported that 28 clergymen, professionals and others protested against bombing of civilians, calling upon 'Christian people ... to examine themselves and their participation in this carnival of death'. But this only provoked a public outcry against 'softheartedness' towards the 'rapacious Germans' (Budiansky 2003: 282–5, 320f.).

Robert Pape concluded in the 1990s from this and subsequent air operations aiming at punishment and coercion that 'conventional punishment' – that is, military operations without the use or threatened use of nuclear weapons – 'rarely succeeds'. In 'conventional' war contexts, 'denial strategies work best' – that is, the use of all forces to deny the enemy a military objective. Robert Pape looked at thirty-three cases and, among these, at five in detail, Japan 1945, Germany 1945, Korea 1953, Vietnam 1965–8 and 1972 and Iraq 1991, and concluded that:

> The evidence shows that it is the threat of military failure, which I call denial, and not threats to civilians, which we may call punishment, which provides the critical leverage in conventional coercion. Although nuclear weapons can make punishment the critical factor, in conventional conflicts even highly capable assailants often cannot threaten or inflict enough pain to coerce successfully. Conventional munitions have limited destructive power, and the modern nation-state is not a delicate mechanism that can easily be brought to the point of collapse. Moreover, governments are often willing to countenance considerable civilian punishment to achieve important territorial aims. Consequently, coercion based on punishing civilians rarely succeeds. (Pape 1996: 10)

In a nuclear context, by contrast, denial strategies did not seem useful to him, while risk strategies were most likely to be effective, and 'nuclear punishment' – which he conceded 'should be ... rare' – could in his view be 'effective'. His empirical database for the latter is of course limited to just Hiroshima and Nagasaki, plus the Cuban Missile Crisis and one or two other Cold War crises, making his conclusions about nuclear coercion difficult to back up with a significant base of evidence. Where 'conventional' campaigns are concerned, however, his findings are very persuasive. Most importantly, he, like British Ministry of Defence official David Chuter, exposed the false expectation that populations subject to bombing would rise up to overthrow their governments or at least coerce their governments to submit to the will of the enemy. An arguable exception to this was the 1999 air campaign to free Kosovo from Serb domination, but it followed many years of conflict over the fission of Yugoslavia and mounting Western pressure on the Serb leader (Lake 2009).

Will terror shorten the war?

In chapters 6 and 7, we saw that in the late nineteenth century, and especially after the Great War of 1914–18, several people – from Hieram Maxim and Alfred Nobel to J.F.C. Fuller and Liddell Hart – had hoped that weapons inspiring ever greater terror would shorten if not prevent any future war. This was also put forward as one justification of the terrible drift towards city bombing that started with the First World War. Even in 1909, R.P. Hearne wrote that:

> I agree that airships will make war more terrible, but I have endeavoured to show that its very terror will delay an outbreak. Airships will also render warfare more localised in its destruction (that is to say, more humane), more decisive, and more rapid ... [T]he destruction will be more closely restricted to the combatants, and there will be far less of that cruel slaughter of non-combatants and that widespread and useless destruction of property which are likely to result from ordinary methods of warfare. (Hearne 1909: xxix)

'Strategic' air attacks – on enemy cities – did indeed occur in the First World War: the Germans attacked London and Paris, and the Allies staged airborne reprisals against Karlsruhe, Trier and Saarbrücken. Some saw this as repulsive even then (Wells 2000: 40). By contrast, the British magazine *Flight*, reflecting the desire of military men to bring the experience of this ghastly war home to the civilians, in 1917 called for the bombing of Berlin in retaliation for the attacks on Britain: 'Until we can bring home to the originators of bombing raids on towns not ordinarily in the military zone the horror and injustice of it, these raids will continue' (q.i. Budiansky 2003: 97).

While in reality the record of British bombing effectiveness during the First World War had been poor, on 1 January 1919 Trenchard published the official record of the British air force's performance in the war, claiming that morale bombing had been effective: 'At present the moral effect of bombing stands undoubtedly to the material effect in a proportion of 20 to 1.' By what reckoning he got to this equation, he did not explain (q.i. Budiansky 2003: 103, 131).

Even Douhet belonged to those who after the First World War thought that terror would shorten the next war and cut the losses. 'Mercifully', he wrote,

the decision will be quick in this kind of war, since the decisive blows will be directed at civilians, that element of the countries at war least able to sustain them. These future wars may yet prove to be more humane than wars in the past in spite of all, because they may in the long run shed less blood. But there is no doubt that nations who find themselves unprepared to sustain them will be lost. (Douhet 1921/1983: 60f.)

Douhet later noted his disdain for

that peculiar traditional notion which makes people weep to hear of a few women and children killed in an air raid, and leaves them unmoved to hear of thousands of soldiers killed in action. All human lives are equally valuable; but because tradition holds that the soldier is fated to die in battle, his death does not upset them much, despite the fact that a soldier, a robust young man, should be considered to have the maximum individual value in the general economy of humanity ...

Any distinction between belligerents and nonbelligerents is no longer admissible today either in fact or theory. Not in theory because when nations are at war, everyone takes a part in it: the soldier carrying his gun, the woman loading shells in a factory, the farmer growing wheat, the scientist experimenting in his laboratory. Not in fact because nowadays the offensive may reach anyone; and it begins to look as though the safest place may be the trenches. (Douhet 1928/1983: 195f.)

Billy Mitchell also subscribed to the view that city bombing would have benign overall effects:

No longer will the tedious and expensive process of wearing down the enemy's land forces by continuous attacks be resorted to. The air forces will strike immediately at the enemy's manufacturing and food centers, railways, bridges, canals and harbors. The saving of lives, man power and expenditure will be tremendous to the winning side. The losing side will have to accept without question the dominating conditions of its adversary, as he will stop entirely the manufacture of aircraft by the vanquished.

He therefore thought victories might be 'sharp and decisive' (Mitchell 1925: xv f., 15). Similarly, Liddell Hart in his *Paris, or the Future of War* dismissed the 'ethical objection' to 'the seeming brutality of an attack on the civilian population', adding:

The events of the last war have, however, in some measure acclimatized the world to the idea that in a war between nations the damage cannot be restricted merely to the paid gladiators. When, moreover, the truth is realized that a swift and sudden blow of this nature inflicts a total of injury far less than when spread over a number of years, the common sense of mankind will show that the ethical objection to this form of war is at least not greater than to the cannon-fodder wars of the past. (Liddell Hart 1925: 50–2)

In 1928, Trenchard sought to nip any criticism of his Strategy in the bud:

I emphatically do not advocate indiscriminate bombing, and I think that air action will be far less indiscriminate and far less brutal and will obtain its end with far less casualties than either naval blockade, a naval bombardment, or sieges, or when military formations are hurled against the enemy's strongest points protected by barbed wire and covered by mass artillery and machine guns. (q.i. Jones 1987: 45)

We see here foreshadowed the expectation of the half a million lives saved on one's own side with which Truman would justify the use of the atom bomb on Hiroshima and Nagasaki. In Britain, Brigadier Groves explained that:

In Europe, warfare hitherto primarily an affair of fronts will be henceforth primarily an affair of areas. In this 'War of Areas' the aim of each belligerent will be to bring such pressure to bear upon the enemy people as to force them to oblige their government to sue for peace. The method of applying this pressure will be by aerial bombardment of national nerve-centres, chief among which are the great cities. (Groves 1934: 32)

But the reality proved different: if anybody among the Second World War Allies actually harboured the illusion that a victory over Hitler's regime could be effected without a ground invasion, by bombing the German population into an insurgency against their own government, to force Hitler to sue for peace, they were gravely mistaken. The same illusion was entertained by some in the First Gulf War, and equally exposed as such (GWAPS 1993, II.ii: 15). As we have seen, this assumed mechanism with which bombing a country would trigger an uprising to depose the leadership and sue for peace has so far not materialised.

Criticism of strategic bombing

Several authors showed great concern about the possible effects of city bombing even before it was so widely practised in the Second World War. 'Neon' described air bombardment of towns and villages, which had already been carried out many times in British and French colonial warfare in the 1920s, as 'alien and abhorrent to the instincts of the British nation, as it must be to all civilised and Christian people' (Neon 1927: 176, 189). Lord Beatty in a letter to *The Times* (2 May 1930) criticised 'the adoption, as a basic war aim, of attack on enemy non-combatants' as 'un-English' and unacceptable to the majority of Britons (q.i. Spaight 1938: 51). Admiral Sir Herbert Richmond condemned the prevalent views on the use of air power, involving the attacking of civilian populations, on ethical grounds: 'Frightfulness, expressly repudiated recently in the case of sea warfare, appears to be a fundamental principle in the air' (Richmond was referring to the Submarine Rules of the Washington Treaty of 1922). There were equally Americans who thought it unacceptable as a Strategy for their country, for example Major-General C.P. Summerall of the 2nd Corps and Captain W.S. Pye of the US navy, testifying to the Swight Morrow Aircraft Boards in 1925. Summerall saw bombing non-combatants as contrary to the laws of war: 'If that falls everything falls.' Pye was afraid the US would become known as '"the baby killers" and the "Boches" of the future' (all q.i. Spaight 1938: 51f.).

There was opposition to the Trenchard memorandum of May 1928 from the chief of the Imperial General Staff, Field Marshal George Francis Milne, and the Chief of the Naval Staff, Admiral Charles Madden. Madden criticised the assumption that Trenchard took 'for granted that direct air attack on the centres of production, transportation and communication must succeed in paralysing the life and effort of the community and therefore [in] winning the war'. But 'No evidence has so far been produced that such bombing in the face of counter attack will have such a result.' Both Milne and Madden feared rightly that air attacks of this sort might stiffen, rather than undermine, enemy resistance. Both criticised Trenchard's priorities as amounting to indiscriminate attacks on civilians, which would be unethical. Madden feared the conflict with international law, because, as Milne pointed out, Trenchard's doctrine 'amounts to one which advocates unrestricted warfare against the civil population of one's enemy' (Webster and Frankland 1961: 76–83).

At the Geneva Disarmament Conference 1932–4, Germany, Denmark, Sweden, Spain, the USSR and the Hedjaz proposed the elimination of military aircraft; others proposed a ban on bombing from the air, but no resolution was adopted (Spaight 1938: 62–5). Germany walked out of the Disarmament Conference in 1933, and that was the end of that. Nevertheless, long before the invention of nuclear weapons, the conventional free-fall 'bomb was, for most men, the Beast of the Revelation whose number they could not calculate', with the consequence that 'The civilised world took alarm' and tried to ban it like the dumdum bullet or chemical weapons before it (Spaight 1938: 169).

Strong criticism of the Trenchard–Portal–Harris Strategy came from within the RAF, even during the war itself. In 1941, Sir Wilfrid Freeman, Vice Chief of the Air Staff, disagreed with Trenchard's assessment of the weakness of German stamina in standing up to the British bombing: 'Lord Trenchard's theory ... depends on a basis which is fundamentally unsound. Material damage would be negligible and the enemy's morale, if not stimulated, will certainly be strengthened in a very short time' (Biddle 2002: 196). After the war, inhibitions about putting forward ethical arguments against city bombing decreased. In 1949 an anonymous author published a pamphlet called *The Strategic Bombing Myth* which not only emphasised the importance of bombing German transportation for the collapse of the German army, but also described the strategic bombing of cities as immoral and ineffective (Gentile 2001: 160).

While the debate about the ethics of British and American air power Strategy in the Second World War is still going on (Markusen and

Kopf 1995), few today would fail to see where the moral problem lies, and this applies equally to views on what targets air forces or nuclear weapons can legitimately hold at risk. From the Second World War onwards, in US military parlance, 'strategic' in the context of air warfare meant (a) nuclear means and (b) the targeting of cities, including where possible the enemy leadership hiding in bunkers that could be cracked only by nuclear weapons.

Panacea targets

We have seen that an alternative to population targeting was championed by the Americans, but in the context of overly optimistic assessments of possible precision. The alternative revolved around the claim that a society could be paralysed by striking some vital points in its complex economy, points usually identified as power supplies (especially oil and electricity), transport nodes (key railways or railway or road intersections, key bridges), or particular industries. All of these were dismissed by adversaries of this approach as 'panacea targets' that would not work; unsurprisingly, the main champions of terror bombing, especially Air Chief Marshal Sir Arthur Harris, who headed the British strategic air command from 1942, were among these critics (Webster and Frankland 1961: 290). From 1943, the USA particularly targeted synthetic oil production in Germany; Britain targeted Romanian oilfields in sorties from Italy. Attempts were made to take out ball bearing factories, but also fighter aircraft factories, a particularly famous example being the US daylight precision bombing campaign of February 1944, 'Big Week', which proved very costly in terms of the USA's own losses. Yet German fighter aircraft production rose – while the factory buildings had suffered extensive damage, much of the production machinery had remained intact or had been dispersed in time to avoid destruction (GWAPS 1993, II.ii: 29f., 59). The lack of success of these sacrifices did not become clear immediately: in November 1945, General Arnold still clung to special targets:

> The Strategic Theory, as applied to the United States air warfare concept, postulates that air attack on internal enemy vitals can so deplete specific industrial and economic resources, and on occasion the will to resist, as to make continued resistance by the enemy impossible. To accomplish the strategic purpose, it is necessary to destroy only a small proportion of

industry, probably not more than a fraction of the total required to conduct modern warfare on a large scale. Indiscriminate widespread destruction of enemy industry is simply a waste of effort. Examination of any national economy will disclose several specific industries or other national activities without which the nation cannot effectively carry on modern warfare. It is conceivable that there will aways be one industry, such as the oil industry in Germany, so necessary to all phases of the national war-making ability that its destruction would be fatal to the nation. (q.i. GWAPS 1993, II.ii: 9f.)

The strategic bombing surveys conducted by the USA and Britain after the Second World War, however, confirmed Zuckerman's scepticism, showing that the results of the bombing were most unsatisfactory. They had not been the single most important cause of the defeat of Germany, either through lowering morale or through fatally injuring the German war machine. Nor had the bombing of Japan's cities with conventional ordnance, even the firebombing, pushed the Japanese into surrender; it had taken the atomic bombs to do this (Chuter 1997).

On the basis of the findings of the United Kingdom strategic bombing survey, Admiral Sir Gerald Dickens concluded that the bombing strengthened German morale, did not lead to a significant fall in productivity and was uncivilised. Dickens argued that the bombing made reconstruction work exceedingly difficult, but that this reconstruction was needed at the end of the war (Dickens 1946: 39). Often cited were the retrospective statements of Albert Speer, German Minister of Armaments and War, that a focused attack on the chemical industry would have brought Germany to its knees (Webster and Frankland 1961: 284). Wing Commander Hubert Raymond Allen was very critical of Trenchard and his pupils Sir Charles Portal and Sir Arthur Harris. Writing about 'The Scandal of Hamburg' and 'The Rape of Dresden', he argued that 'it was all unnecessary!'. After a careful examination of arguments made by them and the strategic bombing survey evidence (and indeed the perceptions of Albert Speer), Allen concluded

that the power of Bomber Command was totally misapplied in the Second World War. Area bombing was forced on Sir Arthur Harris because his aircrews could hit nothing smaller than a city except in special conditions. But the position changed quite soon after he assumed command and he failed to recognize this startling fact ... Bomber Command should have sought

> out the several weak points in the German economy and concentrated on them, in the same way that the Germans were hoping to sever Britain's sea communications by submarine warfare and put her out of the war. (Allen 1972: 131, 177, 181)

There is little to suggest, however, that the air forces of the Second World War would have been capable, against enemy defences, of carrying out such a precision air campaign. Attempts to follow such priorities in the Korean War with the 'Air Pressure Strategy', where especially hydroelectric power stations were targeted, had some, but not decisive success (O'Neill 1985: 379). Nor did selective targeting of this sort really succeed in the Vietnam War, but it has been argued convincingly that without taking the bombing raids into Chinese territory, whence North Korean and North Vietnamese supplies came, it was unlikely to paralyse either war effort anyway. The authors of the GWAPS have modestly argued that one reason for the success of Desert Storm, which included a large element of panacea targeting (especially of power supply and communication), was that Iraq did not rely on an external power for its supplies (GWAPS 1993, II.ii: 365). By then, Warden's 'centres of gravity' had replaced 'panacea targeting' (see below).

The military targets school: denial

Tactical support or strategic weapon?

France emerged from the First World War with purely military missions for its air force. Here is how the first instruction book for 'large units' of 1921 described them:

> Reconnaissance aircraft act for the benefit of the command, of artillery and infantry ... Bomber aircraft attack targets on the battlefield, extend the range of action of the artillery and intervene against the [lines of] communication and the installations of the enemy, up to the limits of their range of action. They operate night and day, searching for [opportunities] of massive action that alone are capable of producing appreciable results. Fighter aircraft grant the reconnaissance and bomber aircraft their freedom of action on the battlefield, giving them cover against the actions of the enemy air force; they seek out the enemy aircraft to fight and destroy them. (q.i. Marill 1997: 26)

Oddly, the time of the planning (from 1926) and construction (from 1930) of the Maginot line coincided with heavy French flirtation with a Douhetian Strategy. Those charmed by it included First World War hero Marshal Philippe Pétain (1856–1951), the French air force leadership under General Victor Denain (1880–1952) and at least one of the Air Ministers, Pierre Cot (1895–1977) and his adviser General Jean-Henri Jauneaud (1892–1976). From the official institution of the French air force as an independent service in 1933 until the outbreak of the Spanish Civil War, they pushed for the construction of 200 bombers (Marill 1997: 28f.). Then France switched to a different approach, pushed by General Paul François Maurice Armengaud (1879–1970) and inspired by the thinking of the Italian Mecozzi, who like the Germans in the 1930s preferred close air support for battlefield operations. Gradually Cot and his 'brains trust', Generals Denain and Jauneaud, fell in with this line, supported by the lessons drawn from the Spanish Civil War. At the time when Cot left office in 1938, a specific form of close air support predominated, the *aviation d'arrêt*, whose mission was to stop an enemy invasion on the ground. The overall result of these several successive shifts within a few years has been described as 'chaos', one plausible explanation of France's poor performance in 1940 (Vivier 1997).

The USA in the 1920s and 1930s had the luxury of not being too concerned about invasions and, in the predominant spirit of isolationism, had little intention of getting involved in wars abroad. As the US Army Air Corps continued as an entity subordinated to the army until the establishment in 1947 of the USAF, its proclivity for military targeting in the interwar period was assured.

Even in the RAF, not everybody favoured the emphasis on bombers that the Trenchard–Portal–Harris school imposed. Slessor wrote in 1936 that air forces had the advantage of particular flexibility – they could be committed to this mission one day, and to another, far away, the next, whereas it took much longer to move armies and navies. The main air offensive, he suggested, should be directed against the enemy's lines of communication, and other targets defended by enemy aircraft. In this context one's own air forces might be able to avoid the enemy's ground forces and aircraft. The 'supplementary offensive' should be directed against the enemy's own air force as such. '[A]ir forces are less able to decline actual combat than are armies or navies.' One should, however, use one's own aircraft for the strategic and tactical defensive only occasionally, and avoid it if possible (Slessor 1936: 34, 37). Consequently, the RAF entered the Second World War with a pronounced disdain for co-operation with either of the other two services, even the coastal command. Its leaders showed persistent reluctance to have their bombers accompanied by fighter aircraft on their missions, wanting to keep them back for the defence of the United Kingdom only. In 1940, the RAF balked at giving close air support to the retreating British and French ground forces and barely covered the evacuation from Dunkirk (Murray 1999b: 87–9, 120f.).

The Soviet Union had developed a strategic bomber force in the mid-1930s, but by the time it was invaded by Hitler's Wehrmacht in 1941, it had, in the words of air historian Edward Homze, 'the most tactically oriented air force in Europe' (q.i. Wells 2000: 27). The Soviet Union's air force lagged behind the others in modernisation: by contrast, the Luftwaffe was ahead of the others, producing aircraft of the latest technology from 1937–8, making it 'the world's most advanced airforce' by 1939. The RAF only made the transition to this new generation of aircraft in 1938–9; France from late 1939; and the USSR from 1942 (Murray 1999b: 105, 109).

Soviet strategists had from the Civil War onwards concentrated on the use of aviation to support the actions of ground forces, albeit not in what would later be called 'close air support'. Early on, the Soviets tried their luck in somewhat deeper strikes behind the enemy rear, against enemy aviation, but also against ground targets to cut off supplies and destroy forces not yet applied to the battle. Despite some debate about alternatives, this would remain the hallmark of Soviet air power use until and into the Second World War (Sterrett 2007: 54f., 86–126). The Soviet Union, and even more so Germany and Italy, had learnt from the Spanish Civil War: despite the notoriety acquired by the bombing of Guernica, the most effective use of air power there had been in battlefield support, especially at the fortnight-long Battle of Guadalajara in March 1937. Here the Republicans fought with the support of Soviet fighter aircraft against the Nationalists with their Italian backing. On this occasion, fighter aircraft for the first time posed a great challenge to bombers, which had to be defended by fighters against attacking fighters (Budiansky 2003: 206f.; Sterrett 2007: 60–4).

Unlike the air forces of its future adversaries, the Luftwaffe paid attention to the entire spectrum of possible air force missions, while the RAF and the American air force only caught up with fully developing all these missions by 1943 (Murray 1999b:116). German Strategy included targeting military installations and enemy forces, entirely in keeping with international law (Corum 1995b, 1996, 1997). While the first field manual of the German Luftwaffe was written in 1935 without the benefit of its experiences of Spain, it already emphasised military targeting and close air support. The Luftwaffe from the very outset of the Second World War used the full panoply of possible air operations, from CAS to the strategic bombing of Warsaw, Rotterdam and then Belgrade (Boog 1992b). The Luftwaffe targeted airfields and ground forces' installations in Poland, France, Belgium and the Netherlands, with greater success in the west than in Poland, as the French, Belgians and Dutch had not dispersed their aircraft.

Subsequently, both the French and the British air forces did very poorly in close air battlefield support. In the summer of 1940, when the Luftwaffe tried to grind Britain down with its relentless bomber attacks on the British mainland, the Battle of Britain proved all those who had claimed that air power had no defensive role wrong: this operation was won by British fighter aircraft (Budiansky 2003: 225f.).

In contrast to the attacks on Britain, in 1941 the Luftwaffe achieved an easy victory over the Soviet air force similar to those achieved over the Poles, French, Belgians and Dutch, with surprise on their side. While the Soviet air force was not altogether wiped out on the ground and attempted a defence, it performed poorly. Disorganised after the Stalinist purges, it even had to resort to using long-range 'strategic' bomber aircraft for close air support of ground forces. The Soviet air force only began to recover at around the time of the Battle of Stalingrad in late 1942, and was numerically far superior to the Luftwaffe by the end of the war. Even so, in 1943 the Luftwaffe managed to bomb Soviet industry – its medium-range bombers were just capable of this – while the Soviet bomber-building programme got on its way too late to have a decisive impact on the Great Patriotic War (Wells 2000: 85–110).

The Japanese air force was largely oriented towards a Strategy of targeting military installations. In May–September 1939 over Mongolia, air force-to-air force battles took place between Japanese and Soviet air forces, won by the Japanese, after the Khalkhin-Gol incident. Japanese aircraft sank the Royal Navy's two capital ships off Singapore in late 1941, spelling out the end of British naval involvement in the war in the Pacific and of its protective role vis-à-vis Australia and New Zealand. The attack on the US fleet anchoring at Pearl Harbor in December 1941 also fits the pattern of military targeting. The Japanese naval air force proved particularly important in this war, and the Japanese air force saw the support of the other two services as its main function, although it also bombed Chinese cities well into the mainland, such as Chunking (Wells 2000: 73–84).

Military targeting was also very important in the Mediterranean theatre. Allied action against German lines of communication across the Mediterranean was successful, not least because of Allied control of the air over North Africa; Rommel's operations in North Africa were handicapped by this (Budiansky 2003: 293). Of course, submarines played a huge part here. Yet German submarines also fell prey to Allied aircraft: the turn of the tide came in 1941 from which point a third, and increasingly more, attacks on submarines were staged by aircraft.

Despite these successes, the RAF was very unenthusiastic about convoy-protection. This is particularly true for Bomber Harris, who disliked such 'purely defensive' operations. The US aviators protested about being sent to 'look for a needle in a haystack' and preferred to bomb submarines 'where they are built and launched', in the ports. Attempts to do just that – bombing French ports like Saint-Nazaire in which the Germans were building their submarines – resulted only in vast destruction of the harbour towns, while the hardened shelters under which the submarines were being constructed escaped unscathed. By contrast, General Dwight D. Eisenhower, commander of the Allied forces, emphasised the use of aircraft against the German U-boats in the Mediterranean as he was preparing the invasion of North Africa (Budiansky 2003: 274–8f.).

We have already seen that the US in principle preferred military targeting but found it technically unsustainable in the Second World War. Only one year after its end, the Greek Civil War created the need for renewed targeting discussions which could not possibly revolve around 'city busting', as the rebels were hiding in mountainous regions and did not have the support of entire cities. Then in 1950 the Korean War erupted, and again simply targeting North Korean cities was not the solution. While Thomas K. Finletter, US Secretary of the Air Force (1950–3), thought that Korea was a one-off and need not influence US Strategy or force procurement, General Otto P. Weyland,

US commander of the Far Eastern air force (1951–5), thought the Korean War something of a model for the future. So did Sir John Slessor in the United Kingdom, who had by now risen to command the RAF:

> We must expect to be faced with other Koreas ... The idea that superior air power can in some way be a substitute for hard slogging and professional skill on the ground in this sort of war is beguiling but illusory ... all this is cold comfort for anyone who hopes that air power will provide some kind of short cut to victory. (q.i. Armitage and Mason 1985: 45)

The civil wars and anti-colonial insurgencies of the Cold War also evoked traditional reactions and can be seen as continuities in the use of air forces from the imperial policing of the 1920s and 1930s. Helicopters, first produced on a large scale late in the Second World War, made an important contribution to French counterinsurgency campaigns in Algeria from 1955. They became the quintessential air weapon to be used in support of ground forces, rising to particular prominence in the Vietnam War, despite their vulnerability to fire from the ground.

In turning away from the painful experience of Vietnam, the American military from 1975 to 1983 developed the new 'preferred' Strategy of AirLand Battle, and, as a derivative, NATO's Strategy of follow-on-forces attacks (FOFA). Both foresaw the use of conventional air power (and, as the last resort, nuclear weapons) mainly against military targets, largely in support of the operations of the ground forces, but preferably in deep strikes way beyond the 'forward edge of the battle area' (FEBA), launched from air bases on the ground or from ships. The emphasis of this Strategy was on jointness – hence the quaint spelling of AirLand Battle. Strategic (nuclear) targeting was only part of AirLand Battle as the ultimate threat of escalation. If the air forces were going to interfere with the enemy's supplies they preferred to cut them off at the source ('strategic interdiction') rather than in the immediate vicinity of the battlefield ('battlefield air interdiction'). All of this was made possible by the infinitely greater precision in bombing that came with precision-guided munitions (PGM) technology from the 1970s onwards. Meanwhile, also as a result of progress in technology since the beginning of the air power age, the survival rate of bomber planes carrying out attacks had risen exponentially, giving Western governments more confidence in their use without courting the risk of unacceptable levels of casualties among their own air crews

(Budiansky 2003: 421). All of these measures, with military targets bombed or held at risk, amounted to what Robert Pape has called the use of air power with the purpose of 'denying' the enemy victory (Pape 1996); the bombing is carried out not to punish the enemy, but to withstand aggression.

The leadership targeting school: decapitation

PGMs also provided the crucial prerequisite for the 'leadership targeting school', which aimed to inflict punishment on the chief evil doers, avoiding collateral damage when possible, a notion that existed also in nuclear Strategy in the 1980s. Leadership targeting is arguably a sub-category of panacea targeting, but it has a special moral dimension. Underlying this approach to air power is a notion of divine justice, which can now finally be administered by humans in war, from afar, in all places, without too much cost and pain to one's own side. The aircraft or the precision missile unleashes the lightning of Zeus that strikes down the tyrant, and ideally not the innocent people surrounding him.

Deterrence might be preferable, but would it always work? Professor Frederick Lindemann, later first Viscount Cherwell, a physicist from Oxford and one of the British government's scientific advisers on the eve of and during the Second World War, did not think deterrence, trying to keep the other side in check through threats of reprisals, would work against 'the heads of gangster governments' – what we today call 'rogue regimes' (Budiansky 2003: 189f.). Once the time of deterrence was gone and the Second World War was well under way, bombing from the air proved quite incapable of punishing the enemy leaders: Hitler survived the massive bombardment of Berlin in his bunker to end his own life.

During the nuclear age, as we shall see, the concept of leadership targeting – then referred to as 'decapitation' – was revived. This entailed the development of special 'earth penetrating weapons' or 'bunker busters' against 'very hard targets', which could only be tackled with nuclear weapons before the development of GBU 28 'bunkerbuster ordnance' with depleted uranium BLU-113 penetrators. While PGM fans uphold 'decapitation strikes' as an option, the GWAPS showed that the inability to 'land a bomb on Saddam's lap' (or later, Milošević's of Yugoslavia, or Osama bin Laden's) was due to insufficient intelligence, not technical ineffectiveness (GWAPS 1993, II.ii: 367; Pape 1996: 322f.).

But other ways of bringing an enemy state to its knees were considered. Even before Iraq invaded Kuwait, in 1988 Colonel John Warden of the USAF had developed the theme of 'strategic air warfare' as *the* best way to wage war in a paper, 'Centers of Gravity – the Key to Success in War' (Olsen 2003, 82; Olsen 2007, 112–14). Warden deliberately tried to exclude the targeting of civilians and aimed to limit collateral damage. Instead, he tried to target the leadership directly from the opening of an air campaign, with the aim of achieving 'strategic paralysis'. While this might include 'decapitation' of the enemy leadership, the crucial result sought was the paralysing of its decision-making and executive abilities, by cutting it off from its military forces and from incoming information. Beyond this, wrote Warden, 'Military objectives and campaign plans must be tied to political objectives *as seen through the enemy's eyes, not one's own.*' Warden made the intended end-state the criterion for target selection. Warden's concepts rejected any ground-forces-support role for the air force, advocating instead deep strikes into enemy territory, with particular care given to target selection. Warden claimed that, blinkered by the fear of a Third World War, the other services were pursuing war-fighting, not war-winning strategies (Olsen 2003: 84f.).

Inspired by Clausewitz, Warden's 'centres of gravity' had the following objectives:

> [A] state realizes its political objectives when the enemy command structure (i.e., the enemy leader or leaders) is forced by direct or indirect action to make concessions. Control of the enemy command structure, civil and military, must be the ultimate aim of all military operations. At the strategic and operational levels, inducing the enemy to make the desired concessions requires identification and attack of those parts of the enemy state and military structure which are most essential to his ability and desire to wage war.

Concessions the enemy might be asked to make can range from 'his right to existence' to 'his desire to destroy his opponent. In between these extremes, he can concede a province, a trade right, or his intention to conduct a military offensive' (Warden 1992: 63). INSTANT THUNDER was a plan for strikes against eighty-four targets in Iraq over six to nine days (Olsen 2003: 64; Olsen 2007, 140–82). Warden believed that this air campaign, on its own, could not only force Saddam to withdraw his troops from Kuwait, but also bring down Saddam's regime, as the population (including the armed forces) would rise up against him, once his tyrannical power over them was undermined by a paralysis of government through the destruction of its brain, heart and nervous system, in short, its vital centres.

Eventually Warden's ideas strongly influenced the four-phase concept of DESERT STORM as actually implemented:

Phase I: Strategic Air Campaign against Iraq;
Phase II: Air Campaign against Iraqi air forces in Kuwait;
Phase III: Ground combat power attrition to neutralize the Republican Guard and isolate the Kuwait battlefield;
Phase IV: Ground attack to eject Iraqi forces from Kuwait (Olsen 2003: 131).

The strategic air campaign included the leadership itself, electricity supplies and oil refineries, air defences and known sites connected with Iraq's weapons of mass destruction programmes. Effectively, this was a mix of leadership, counter-military, counter-force and panacea targeting.

Warden saw the ground forces as a fall-back force: a 'cocked fist' to be 'held in reserve'. This annoyed army leaders (Budiansky 2003: 414–18). To a large extent, Warden's quarrel with the army leadership was a rerun of the sword–shield debates in nuclear Strategy of the 1950s and 1960s.

The Gulf War Air Power Survey concluded, however, that the idea of fomenting a revolt against Saddam Hussein or a coup against him through a few days of strategic bombing had been unrealistic (GWAPS 1993, II.ii: 15). As Budiansky noted, '[t]he political strength of Saddam's regime was such that only a campaign aimed at breaking Iraq and probably involving tens of thousands of casualties could have toppled the dictator' (Budiansky 2003: 427f.). Critics like Pape and Chuter pointed out afterwards that in this assumption about the population's willingness to rise up and overthrow a tyrannical regime, Warden followed the erroneous assumption of the city bombing school's morale bombing arguments. In fact no top Iraqi leader was killed in the air strikes, despite many strikes on government buildings and presidential residences. Nor did communications between Saddam and his commanders break down, as Warden had hoped.

Eliot Cohen, a lecturer at Johns Hopkins University and director of the Gulf War Air Power Survey, poured scorn on the 'passive collection of targets' as excessively mechanistic and in no way responsive to the enemy's moves, culture, psychology and social vulnerabilities (Budiansky 2003: 427f.).

Eight years later, when Operation Allied Force over Kosovo was carried out, there were calls for leadership decapitation (that is, the specific targeting of Serb leader Milosevic), but the intelligence was lacking to find Milosevic, as it had been in the 1991 Gulf War with regard to Saddam Hussein. Among the shortcomings of Desert Storm and of Operation Allied Force was that little thought was given to what to do once the enemy leader(s) was/were overthrown and how to turn this into a lasting victory (Olsen 2003: 288–90). The Supreme Allied Commander in Europe (SACEUR) decreed that the first requirement was not to have any NATO casualties or even losses of aircraft. The Kosovo campaign was a spectacular success, with fewer NATO casualties than in any air operation before. No resistance was put up by the Serbs on the ground. Christopher Coker of the London School of Economics commented: 'Humanitarianism, it would seem, is not just an objective. Western societies can now fight wars which minimise human suffering, that of their enemies as well as their own.' He remained sceptical, however, of attempts to 'sanitise' war and render it more 'humane', driven by the 'liberal conscience' (Coker 2001: 2f.). While it is true that damage to civilians and even to the environment were relatively limited in the Kosovo campaign, in this and the following conflicts in which the USAF and its allies participated, 'collateral damage' has continued to result in the unintended deaths of non-combatants. This is difficult to reconcile with the increasingly humanitarian aims that Western liberal democracies list among their aims in using force.

The political signalling school: games theories

The political signalling school of air targeting, the last and least to be discussed here, had its roots on the one hand in operations research or analysis thinking introduced into the policy-making process by mathematicians and economists, and on the other in games theory derived from economics. It dominated the early air power application by the USA in the Vietnam War.

Four days after becoming President, Lyndon Johnson approved National Security Action Memorandum NSAM 273, which defined America's war aims in Vietnam as: 'to assist the people and government of [South Vietnam] to win their contest against the externally directed and supported communist conspiracy'. All US 'decisions and actions' should be geared to ensure 'the effectiveness of their contribution to this purpose' (q.i. Clodfelter 1989: 40). After the Tonkin Gulf incident in 1964 (a North Vietnamese attack on a US cruiser), President Johnson's advisers urged him to bomb North Vietnam from the air. Of three options – (A), stepping up of covert activities; (B), all-out bombing of North Vietnam; (C), slow squeeze, gradually escalating – the third (C) was chosen, with a bit of (A) mixed in. The Joint Chiefs of Staff preferred option (B), or at least a heavier bombing than option (C) as approved by Johnson.

The bombing campaign – ROLLING THUNDER – started on 2 March 1965 and went on for three and a half years, a length of time not originally planned. The public was told that the campaign was designed to punish the North for its intervention in the South, to undercut this support and to raise the morale of the people in the South. ROLLING THUNDER can be described as tactically offensive, even though the overall American Strategy was defensive and aimed at the restoration of the status quo ante in Vietnam. Over time, the bombing was stepped up both geographically and quantitatively to cover almost the entire North; Haiphong harbour was a target, but Hanoi, the North Vietnamese capital, was spared deliberately. In 1968 the bombing rates were wound down. Paradoxically, the North Vietnamese government gained kudos from withstanding ROLLING THUNDER, and managed to motivate its population to support the war. North Vietnam had espoused a 'long war' Strategy of seeking to exhaust the patience of their adversary (Smith 1994: 48f., 211–24, 240f.).

The concept underlying ROLLING THUNDER stemmed from early 1960s strategic thinking that can be traced to political scientists Robert E. Osgood (1916–91) and Thomas C. Schelling (b. 1921), labelled 'graduated and reciprocated initiatives in tension' or GRIT, a variation of what they also called 'conflict management'. This was based on economic 'games theories' which assumed that conflicts, including bloody wars, could be played dispassionately with one side signalling its willingness to play with certain self-limitations in the hope that the other would make the same concession, or that one side might make conciliatory gestures, while threatening to up the ante if the enemy took advantage of such a self-limitation and refused to make an analogous 'gentlemanly' concession. McGeorge Bundy (1919–96), for one, wanted to apply such targeting principles as developed by the RAND think tank, including categories of targets deliberately *not* attacked, with the idea that the North Vietnamese should realise that 'there is always a prospect of worse to come'. While Curtis LeMay (1906–90) claimed to want to 'bomb them back into the Stone Age', the actual air strikes were very discriminating, deliberately avoiding whole categories of targets, especially big conurbations. President Johnson and his collaborators retained the closest control on the targeting, approving 'packages' of targets for the weekly bombing at their Tuesday White House meetings (Budiansky 2003: 379).

But the 'pauses' in the bombing which the US government introduced in the vain hope that they would lead the North Vietnamese to seek negotiations and an armistice merely allowed the latter to repair damage, bring in new supplies and open up new lines of communication. Threatening to up the ante, escalating to a level of violence thus far not present in the game, like raising the stakes in a game of poker, while signalling willingness to come to an understanding and to de-escalate proved dramatically unsuccessful in Vietnam. The American side had much less to lose than the Vietnamese communists, for whom everything was at stake – not only their lives, but their society, their ideals and their people. Thus one 'player' was trying to fight an economic signalling 'game', using bombers instead of light signals or Morse code, while the other was fighting an irregular war, dismissing any game rules the first was trying to impose. One 'player' played with limited stakes only, the other invested all his stakes, willing to fight to the death. It turned out that it was this commitment and engagement which was decisive, not the objective strength in terms of quality or quantity of military hardware.

Arguably, South Vietnamese morale was initially boosted a little by ROLLING THUNDER, but not in the following years (Smith 1994: 211–24). Crucially, as Robert Osgood had to concede,

> Hanoi was not persuaded. After three years the experiment in punitive bargaining was abandoned as a failure. Perhaps the signals were not clear. Indeed, in response to public protests throughout the world, the

US government stressed the purely military objectives of the bombing as though to deny their punitive function. (Osgood 1979a: 115)

The Tet Offensive in 1968 showed that ROLLING THUNDER had failed to break the will of the North Vietnamese regime. ROLLING THUNDER came to stand for failure.

The Nixon administration, determined to bring the war to an end, adopted a very different approach with its air strikes which for the first time used PGMs in great quantity. The LINEBACKER I operation of May to November 1972, and even more so the LINEBACKER II December 1972 'Christmas bombings', were qualitatively different from previous bombings, as by now North Vietnam was fighting a regular war. The aim, which was attained in January 1973, was to get the North Vietnamese to the conference table in Paris. LINEBACKER II comes closest in US practice to DESERT STORM in the First Gulf War, as it was a round-the-clock operation, with the use of laser-guided munitions, targeting also Hanoi and Haiphong; while targets were mainly military facilities and transport nodes, up to 2,000 civilians were killed unintentionally (GWAPS 1993, II.ii: 63–5). While more PGMs were used in the First Gulf War, the difference was not an order of magnitude.

The USAF air power theories as applied to Vietnam, as one analyst opined,

derive their coherence less from how they interacted technically in the events of the 1960s and 1970s than from their common origin in the thinking done between World Wars I and II. First, air power's proponents ... have typically stressed the essential novelty of the air age and the consequent irrelevance of historical experience. The new principles and practices of air power supposedly superseded old military lessons and dogmas. (Mrozek 1989: 5–7)

The theme of LINEBACKER, however, was an old one, with its overwhelming focus on strategic air offensive (strategic bombing and autonomous bombing offensive), with roots not only in Billy Mitchell's writings but also in Mahan's (Mrozek 1989: 183). Clodfelter summarized: ROLLING THUNDER on the one hand and LINEBACKER I and II on the other 'differed in their effectiveness as political instruments, and the political objectives guiding them contributed to the disparity of results'. With ROLLING THUNDER,

President Lyndon Johnson turned to air power to help achieve his positive goal of an independent, stable, non-Communist South Vietnam. At the same time, his negative objectives – to prevent a third world war and to keep both domestic and world public attention focused away from Vietnam – limited *Rolling Thunder.*

By contrast, with LINEBACKER,

President Richard Nixon's ... positive political goal was an American withdrawal that did not abandon the South to an imminent Communist takeover ... Even after he decided to court President Nguyen Van Thieu, Nixon's positive goals remained more limited than Johnson's ... Negative goals had a marginal impact on Nixon's application of air power. His détente with the Chinese and Soviets removed the threat of an expanded conflict ... The lack of negative objectives allowed Nixon to expand the bombing until it threatened to wreck Hanoi's capability to fight by rendering its army impotent.

Before North Vietnam's 1968 Tet Offensive against the South and its allies,

the Southern war was a guerrilla conflict ... *Rolling Thunder* could have affected northern war-making capacity only by attacking two targets: people and food. The destruction of either the North's population centres or its agricultural system would have had a minimal impact on the war in the South, however ... the prospect of North Vietnam's ruin did not guarantee a South Vietnamese victory ... The cessation of Northern support was no guarantee that Saigon could survive against the Viet Cong.

Again by contrast,

Nixon's *Linebacker* campaigns were effective political instruments ... [in part] because the war's nature changed in 1972. After the decimation of the Viet Cong in the 1968 Tet Offensive, the North Vietnamese Army was the only military force capable of achieving ... unification. (Clodfelter 1989: 204–8)

While ROLLING THUNDER remains until this day an example of how not to, LINEBACKER II was seen by several strategists, including Richard Shultz, as a good example of the use of air forces or 'compellence' – forcing an enemy to do something he did not want to do (Shultz 1992: 171–91). However, Robert Pape thinks that

> PGMs have done nothing to enhance the coercive strength of strategic air power. Punishment, risk, and decapitation strategies had little merit before PGMs, and they have little merit now. Denial remains the most effective coercive air Strategy, and PGMs have further increased the superiority of theatre air power over strategic bombing. (Pape 1996: 326f.)

There was an element of signalling in the First Gulf War's Strategy, too, and a way in which strikes that were supposed to have purely tactical or operational consequences suddenly rose to strategic importance. The most famous example was the destruction during DESERT STORM of the Al Firdos district bunker in Baghdad, which unbeknownst to US intelligence was being used as a shelter for civilians. The ensuing international furore caused by these casualties resulted in a significant reduction of strikes against Baghdad and Basra in the subsequent campaign, as it became a priority to avoid a repetition of this unintended outcome. The 'signalling' does not seem to have worked. After the war, Iraqi officials said that they had taken US strategic air strikes as evidence that there would be no extensive bombardment of Baghdad and Basra and that therefore they should 'hold tight and the Americans will go away'. Saddam Hussein's Interior Minister commented: 'We would have understood carpet bombing, but we didn't understand this other'; he thought the United States was 'just being spiteful' (Pape 1996: 320f.)

Conclusions

These four schools of air power are still with us today, although what we have called the 'political signalling school' has in some way changed beyond recognition, and games theory is no longer fashionable. The dividing lines between them have become blurred at times. For example, the denial to the enemy of control over his military apparatus by striking key communications centres could be seen either as decapitation or as military targeting. But by and large, these different approaches are still here. Moreover, they can be found transposed into nuclear Strategies, which are the subject of the next chapter.

14 *Nuclear Strategy*

Science has brought forth this danger; but the real problem is in the minds and hearts of men.

(Albert Einstein on nuclear weapons, q.i. Sokol 1961: 71)

Targets

'A history of air Strategy is really a history of targeting', wrote Philip Meilinger (2003: 170), and this applies equally to nuclear Strategy. This is also a practical way to point out the direct continuity from air power targeting to nuclear targeting. At the same time, the choice of targeting allows us to cut through rhetorical-ideological packaging, pointing to real, or in the case of plans, potential physical effects. Given the lack of precision that could be achieved with the aircraft technology of the 1940s, it is a matter of complete indifference to those killed in the bombing whether the intention in hitting them was one of deliberate indiscriminate bombing, dehousing them or targeting the factory two miles up the road. This of course also applies to whatever intentions and principles guided nuclear Strategy. Equally, it allows us to get to some hard facts, frequently obscured through rhetoric about 'knock-out blows', 'annihilation', 'surgical strikes' with 'collateral damage' or 'dehousing'.

City targeting from the First to the Second World War was partly a function of ideology, but definitely dictated by technological limitations which virtually ruled out precision in long-distance bombing. After Hiroshima and Nagasaki, in practice there was a clear preference for air attacks on targets directly connected to the enemy war effort, whether these were enemy armed forces themselves (land, naval, air) or logistic in the widest sense (transport-, communication- or supplies-related).

There was a remarkable continuity between the preferences of the pre-1939 military target school and the more general preferences of

Table 1 *Targets for conventional strategic bombing*

Targets ➔ 🡻 Operation	Military installations, including airfields, arms industry	Ports, transport nodes, railway lines	Civilian industries, power stations, gas works, oil refineries	Residential areas, civilians	Elites, leadership
Gen. Sykes (UK) 1911		✓			
Adm. Paul Behncke (Germ.) Aug. 1914		✓		(✓)	✓
Kaiser Wilhelm II Feb./July 1915		✓		⦸/✓	⦸
British air attacks against Germany 1918		✓		✓	
Gen. Douhet (It.) 1921	(✓)	✓	✓	✓	
Marshal Brooke-Popham (UK) 1926		✓	✓		
Pierre Faure (Fr.) 1931		✓	✓		
Macmillan (UK) 1938, 1941		✓	✓	✓	
RAF Western Air Plans 1939		✓	✓		
Germans in WW II	✓	✓	✓	✓	
UK in WW II	✓	✓	✓	✓	
US in WW II	✓	✓	✓	⦸/✓	
US in Korea 1950–3	✓	✓	✓	(⦸)	⦸

Vietnam (USA) Rolling Thunder	✓	✓	✓	⊘	⊘
Vietnam (USA) Linebacker I and II	✓ (I and II)	✓ (I and II)	✓ (I and II)	✓ (II)	✓ (II)
Gulf War (USA) Desert Storm 1991	✓	✓	✓	⊘	✓
Kosovo Allied Force 1999	✓	✓	✓	⊘	✓
Afghanistan 2001	✓	✓		⊘	✓
Gulf War II 2003	✓	✓	(✓)	⊘	✓

✓ targeted or recommended as target
⊘ withhold target

Sources:

Report by the British General Frederick Sykes, 1911 on the future of aircraft; German Rear Admiral Paul Behncke, Deputy Chief of the Naval Staff, memorandum of 20 August 1914 on the usefulness of air attacks on cities; views of Emperor William II of February 1915 and May 1915 (Budiansky 2003: 45–8).

Giulio Douhet and Pierre Faure (Douhet 1921/1927: 28, 41; Faure, 1931: 13, 31f., 43). Brooke-Popham 1926 (Spaight 1938: 45).

The RAF's 'Western Air Plans' for bombing of Germany on the eve of the Second World War; several of these were applied by mid-1941 (Budiansky 2003: 282).

Captain Norman Macmillan, air correspondent of the *Daily Mail* (Macmillan 1938: 55–7, 1941: 113).

General 'Hap' Arnold and Ira Eaker (Arnold and Eaker 1938: 128).

Summary of air power use in Korean War (Armitage and Mason 1985: 32, 36f., 38f.; O'Neill 1985).

Rolling Thunder (Smith 1994: 211–24; Budiansky 2003: 380f.).

Linebacker I and II (GWAPS 1993, II.ii: 63–5; Michel 2001).

Desert Storm in 1991 (GWAPS 1993, II.ii; Budiansky 2003: 414–18).

Allied Force in 1999 (Budiansky 2003: 430).

Second Gulf War of 2003 (Budiansky 2003: 437).

Afghanistan campaign (2001–) (Batschelet 2002).

Table 2 *Targets for nuclear bombing*

Nuclear weapons Targets ➔ ➔ Plans	Military installations, including airfields, arms industry, missile bases, other nuclear installations	Counter-force	Ports, transport nodes, railway lines	Civilian industries, power stations, gas works, oil refineries	Residential areas, civilians	Elites, leadership
US and NATO until 1954	✓	✓	✓		✓	✓
US SIOP 1960		✓		✓	✓	✓
British preferred		✓				
McNamara (USA) 1962	✓					
NATO 1969–first and follow-on use	✓		(✓)		⊘	⊘
NATO 1969–general nuclear response				✓	✓	✓
NSDM242 (USA) 1974	✓		✓	✓		

Colin Gray (1979)	✓			⊘	✓
WTO until 1987	✓	✓	✓		

✓ targeted or recommended as target
⊘ withhold target

Sources:

British nuclear targeting preferences (Heuser 1997: 66–9).
US and NATO Strategy in the 1950s, SIOP 1960 (Ball 1986: 61, 64–9, 72–83).
McNamara in his Ann Arbor speech, 1962 (Olsen 2003: 72).
Limited Nuclear Options (Davis 1975/6).
President Richard Nixon on 17 January 1974 in a National Security Decision Memorandum NSDM 242; SIOP of 1983 (Ball 1986: 72–83).
Colin Gray (Gray 1979: 65–7, 1984: 66n).
SIOP 6F of 1989 (Ball and Toth 1990).Warsaw Treaty Organisation (Heuser 1993).

'conventional' (that is, non-nuclear) air strategists after 1945, when the 'strategic or city bombing school' emigrated in toto to the nuclear domain. With increasing technical precision, 'conventional' air Strategy largely abandoned city bombing, although this was treated as a 'worst case' or 'withhold option', applied for example at the end of the Vietnam War in LINEBACKER II, and even then not in the form of indiscriminate bombing of residential areas. Instead, 'conventional' air Strategy focused on targets of the enemy's war machine, to the extent that air technology and intelligence available permitted in each context. With the increasing precision given to nuclear weapons also (as these moved from the oldest nuclear technology of free-fall bombs to missiles and then precision targeting guided by satellite intelligence), the military targeting school was also able to become prominent in the nuclear context, as we shall see.

Nuclear Strategy is the child of air Strategy, in that the concepts governing it were mostly derived from air power theory. Some concepts were owed to naval or maritime thinking, but these were usually imported into nuclear thinking via air Strategy. Again, as with the influence of naval/maritime thinking on air thinking, the transfer was not directly one to one. Air Strategy was dominated by four schools, the strategic or city bombing school, the military target school and the two less important schools advocating leadership targeting and political signalling. In the 1930s, as we have seen, official British and some French air power thinking clung to the hope that a British or French strategic bomber force would deter the next war's opponents from using their air forces to bomb British and French cities, but this hope proved unfounded. A conventional 'deterrence school' was thus not viable in the pre-nuclear age. By contrast, the deterrence concept came to dominate nuclear Strategy. In fact, the strategic or city bombing school of the pre-nuclear age became the nuclear deterrence school, while the military targeting preferences of conventional air strategy underpinned nuclear war-fighting strategy.

Political signalling, which was applied conventionally in the Vietnam War, came to inspire NATO's compromise nuclear Strategy MC 14/3 and its derivative documents from 1967 (Heuser 1997: ch. 3). Leadership targeting also popped up from time to time in a nuclear context for deeply buried (leadership) bunkers that could not be taken out with deep-penetrating missiles. But even these concepts were usually harnessed to the overall sought-after effect of deterrence.

Deterrence

The British 'fleet in being' heritage

The history of nuclear weapons began with the aim of deterrence. Just as the British had hoped in the 1930s that the RAF's bomber force would have a *deterrent* effect on its enemies in a future war, deterrence was at the root of the collaborative Western nuclear programme from which America's first atom bombs stemmed. Two physicists at the University of Birmingham, who had recently escaped from the Nazi *Reich*, Otto Frisch and Rudolf Peierls, in a memorandum submitted to the British government in February 1940 urged it to commission the research and development of nuclear weapons. They themselves had worked out the basics of what was required, and added:

> We have no information that the same idea has also occurred to other scientists but since all the theoretical data bearing on this problem are published, it is quite conceivable that Germany is, in fact, developing this weapon ... If one works on [this] assumption ... it must be realized that no shelters are available that would be effective and that could be used on a large scale. The most effective reply would be a counter-threat with a similar bomb. Therefore it seems to us important to start production as soon and as rapidly as possible, even if it is not intended to use the bomb as a means of attack. (t. i. Arnold 2003: 111–26)

Even in mid-August 1945, the British navy leadership, in pooling their thoughts and reactions to Hiroshima and Nagasaki, came up with the following ideas, as expressed by Rear Admiral R.D. Oliver:

- The only hope of countering an atomic threat was deterrence, i.e. the possession of atomic bombs by Britain.
- 'So far in history, the appearance of a new weapon has always been followed by suitable countermeasures.' While '[t]he likelihood of this occurring in the case of atomic missiles' was 'not yet known', there was not much optimism that this would be the case.
- If Britain were attacked by atomic bombs, this small country, albeit the head of the empire, would 'be rendered ineffective from the war-making point of view and the survival of the "British Empire idea" would then hang upon the ability of the Dominions (and the

United States of America) to bring forth a greater counter-blow with rapidity'.
- All concentrations of military, industrial, or governmental power would be particularly vulnerable to atomic attacks, i.e. cities, military bases of all sorts (ports, any other naval bases, air bases, army bases), naval convoys ...
- The net effect was 'that the price worth paying for peace is now very much higher, and that the main function of our armed forces should be the *prevention of major* war, rather than the ability to fight it on purely military grounds' (Hattendorf *et al.* 1993: Doc. 449; my emphasis).

At the same time in America, the academic Bernard Brodie made the case for nuclear deterrence:

> [T]he first and most vital step in any American security program for the age of atomic bombs is to take measures to guarantee to ourselves in case of attack the possibility of retaliation in kind. The writer in making that statement is not for the moment concerned about who will *win* the next war in which atomic bombs are used. Thus far the chief purpose of our military establishment has been to win wars. From now on its chief purpose must be to avert them. It can have almost no other useful purpose. (Brodie 1946: 76)

As Brodie said later, 'nuclear weapons do by their very existence in large numbers make obsolete the use of and hence need for conventional forces on anything like the scale of either world war' (Brodie 1973: 412). Had nuclear weapons now usurped the role of the navy and then the air force as self-proclaimed war-winning – or indeed war-averting – services, making the other services redundant?

Deterrence would be seen by many as the only rationale nuclear weapons could ever have, and the desire for 'deterrence' was thus at the heart of British government policy to develop nuclear weapons (Gowing 1974). As British Admiral J.R. Hill commented:

> Inevitably, over the next 40 years, complications crept in. First, great confusion was caused by the application of the noun deterrent to the means of nuclear bombardment alone. Frequently the atomic weapon and its means of delivery were referred to as *the* deterrent, and this led to many distortions and errors of emphasis ... Theories such as massive

retaliation ... gave nuclear bombardment a deterrent function over too wide a range of issues, a function it could not credibly sustain. The misconception is inclined to survive in the more simplistic French attitudes to nuclear retaliation in response to any attack on metropolitan France. (Hill 1986: 80)

Very quickly, the possession of nuclear weapons by several parties led many strategists to the conviction that only deterrence made sense, because a nuclear war was unwinnable. Contrary to all we now know about Soviet and Warsaw Pact Strategy until Mikhail Gorbachev took it in hand in the late 1980s (Heuser 1993), Westerners, applying culturally blind mirror-imaging, began to assume in the 1950s that the Soviet leadership must share this view. While in 1949 and 1950 there was still the fear of a Third World War by Stalin's design, by the late 1950s, Admiral the Earl Mountbatten of Burma, the First Sea Lord of Britain, wrote:

We believe that so long as we remain united with our allies, maintain the nuclear deterrent and sufficient conventional forces to show our ability and determination to protect our vital interests in all circumstances, direct war with Russia is unlikely to break out except by miscalculation. (Hattendorf *et al.* 1993: Doc. 455)

NATO documents usually added the possibility of war by 'accident' (Heuser 1997: 1–14).

Those who argued for international control of nuclear forces with the Baruch Plan of 1946, or since then for a drastic reduction in nuclear arms while taking care not to bring back conventional war in the absence of a nuclear deterrent, tended to explain the continued need for nuclear weapons in terms similar to the concept of a 'fleet in being'. They usually coupled it to the argument for the need to forestall a rogue regime becoming the only nuclear power on earth. In this context, numbers only mattered up the minimum needed to make a nuclear force 'sufficient' to be credible – a British discovery. With the technology available since the 1960s, this can be expressed in the concrete terms of three or four nuclear-powered submarines (to ensure that in a crisis, at least two could be active) carrying no more than fifty independently targetable and launchable nuclear missiles each.

Closely connected to the concept of existential deterrence is the concept of accepted mutual vulnerability. As noted by the British planners even in 1945 (see above), until nuclear weapons were invented, new offensive technology had always tended to find its defensive answer. Before 1945, Admiral Sir Herbert Richmond's belief was shared widely: 'Man is so ingenious an animal that he finds a way of providing an antidote to most inventions. The accuracy of fire with the bomb and the torpedo has increased, and, simultaneously, the means of resisting the attack have improved. Finality is far distant' (Richmond 1934: 112). The advent of nuclear weapons had changed this. In France, the aviator General Pierre Marie Gallois argued that nuclear weapons have put an end, at least for the time being, to the old technological struggle between 'sword' and 'shield'. No reliable 'shield' was in sight to counter a nuclear attack, as even an anti-ballistic missile defence system would be fallible enough to let a small percentage of attacking missiles (or aircraft) through, sufficient to cause unbearable damage to the other side. He used this to back up his argument that a nuclear 'battle' could no longer take place (Gallois 1977: 220–2). Since the dawn of the nuclear age, scientists have worked on responses to a nuclear attack, and several were developed, including, in the late 1960s and early 1970s, such anti-ballistic missile (ABM) systems, and the Strategic Defense Initiative ('Star Wars') of the early 1980s, with space-based technology. Unfortunately, no such system has as yet been able to achieve 100 per cent effectiveness, and anything less than a 100 per cent is unsatisfactory in a context where of, say, 1,000 incoming nuclear weapons, a penetration rate of even 0.5 per cent could spell the end of five major cities.

This led strategists and key decision-makers grudgingly to concede that it was paradoxically better to live with mutually assured destruction – MAD – than with missile defences. This point was implicitly conceded even by the Soviets when they agreed to sign, with the USA, the ABM Treaty of 1972, limiting ABM systems to the protection of two sites only, one the country's capital, one a site containing one's own strategic nuclear weapons.

The paradox of this mutually agreed perpetuation of vulnerability by enemies otherwise continuing to entertain operational war plans against each other has worried strategists ever since. Colin Gray accused US governments of having allowed a misplaced balance-of-power thinking to creep in (Gray 1984: 7). Other critics included President Ronald

Reagan, who with 'Star Wars' temporarily threatened to undermine the ABM Treaty, and George W. Bush, who after the USA's unilateral withdrawal from the treaty in 2002 pushed for a limited ABM system too small to undermine any deterrence relationship with nationalist Russia, but large enough to deal with perhaps a small, two-digit number of missiles fired by a nuclear proliferator elsewhere in the world.

Notwithstanding this de facto agreement by the Soviets to what in the West would be described as a state of nuclear deterrence, the Soviets did not really embrace it. They described their own Strategy towards the West as one of restraining it (*sderzhivaniye*). Western Strategy they described as intimidation (*ustrasheniye*) (Betts 1987: 5). This different perception on the matter led them to fail to grasp or deliberately ignore NATO concepts of 'signalling' resolve with restraint, and to abstain from developing limited nuclear options on a Western model (see below).

Deterrence by threat of punishment and French nuclear Strategy

While the United Kingdom adopted a counter-force preference, as we shall see, its declaratory nuclear Strategy and the characteristics of its 'strategic' nuclear system always hinged on the ability to strike 'key aspects of Soviet state power', *vulgo*, the 'Moscow criterion', that is, the ability to penetrate ABM defences around Moscow (Heuser 1997: 76–8). For a small nuclear force, any counter-force option, let alone tactical nuclear war-fighting in support of ground forces, was limited by its capacity and could be implemented only in an alliance context.

French nuclear Strategy was to a large extent a successor of the British 'fleet in being' Strategy, with a probably unconscious admixture of Tirpitz's 'risk strategy' (with its deterrence of a stronger force by a weaker force) and a conscious borrowing from the *Jeune École*. Even in 1945, Admiral Castex had hit on the formula that the atom was 'a great leveller', it equalised the standing of stronger and weaker powers, as long as the weaker power also had a nuclear weapon (Castex 1945). This idea was avidly taken up by General Gallois (1960: 4), the main thinker behind the *dissuasion pure et dure* (pure and hard deterrence) school which under de Gaulle became French government orthodoxy (Heuser 1997: 93–123). Just like Tirpitz,

Gallois argued that his country needed only a smallish nuclear force, as the Soviets did not merely have to take into consideration this force, but also how weakened they themselves would stand in relation to the USA, if France had inflicted upon the Soviets all the damage she could do (Gallois 1960: 185). Gallois was partly inspired in this argument by the British debate, where British nuclear forces were sometimes invoked as a trigger to bring in the USA on a conflict (Heuser 1998a: 34). Gallois and de Gaulle, however, firmly believed that this link with another superpower could not be relied upon in a formal alliance: unlike Raymond Aron, his main adversary in the *Grand Débat*, Gallois believed that the atomic bomb had killed alliances, a theme he developed in his belief that the Americans and the Soviets would sell out the Europeans in the Anti-Ballistic Missile and Strategic Arms Limitation treaties of the early 1970s (Gallois 1960: 187, 1975).

This was of course the key difference between France and her NATO allies who – willy-nilly in the case of those that did not or could not develop nuclear weapons, willingly in the case of successive British governments – preferred to put their money on the alliance with America as their nuclear guarantor. Robert Osgood later recalled that 'Outside Europe the credibility … of [the USA threatening to initiate] the use of nuclear weapons under any circumstances seem[ed] to reach its high point in 1954, during the Quemoy and Matsu crisis and the fall of Dien Bien Phu, and has steadily declined ever since' (Osgood 1979a: 104). Not just outside Europe: many Europeans, too, came to think of the American 'guarantee' – which was never written down anywhere, not even in the North Atlantic Treaty's Article V – as a bluff. Even Americans shared France's doubts: Bernard Brodie thought that the USA 'may not forever be willing to incur' the risk of nuclear attack on it 'in order to defend the nations of western Europe. It is therefore time we began to think of some alternatives to total war as a means of defending' Western Europe (Brodie 1959: 336). He therefore approved of the deployment of US and indigenous British and French nuclear weapons in Western Europe, but also substantial ground forces as a 'trip wire', to force the USSR to make any aggression equally substantial (Brodie 1959: 346). Where in naval Strategy the benefits of alliances as force multipliers and their credibility are uncontested, in the extreme case in which nuclear weapons would play an explicit role, reliance on an ally is much less so. Let us look at this paradox from a different perspective.

In the context of an obsession with mathematical-economic formulae that had spread to France from the other side of the Atlantic, Gallois developed a formula for deterrence which has by far outlived any of its time. He described deterrence as the product of the technical performance of weapons or military means in one's possession, and of the subjective perception by the side to be deterred of the will of the threatened nation (or rather its decision-makers) to use these means (Gallois 1960: 151f., 209).

This formula, perhaps even more helpfully rendered as 'the product of the force of the means available and the will to use them', provides an insightful key into the whole phenomenon of deterrence – and self-deterrence. For example, anybody but a pacifist would be very likely to have the will to use conventional weapons to defend themselves against aggression. An aggressor might not, however, be deterred by conventional weapons, despite harbouring no doubt that his adversary would indeed use them. The limited means would weaken the deterrence effect considerably. Yet a state would be more likely to use its soldiers or its navy than its nuclear weapons in support of an ally in distress. The matter is different with nuclear weapons. Here, the willingness of key decision-makers to risk a nuclear defence against a aggressor who himself owns nuclear weapons might indeed be doubted, as these decision-makers might be 'self-deterred' by the danger of nuclear escalation. Though the means available might be terrifying, the product of means and perceived willingness to use them might tend towards zero, depending on the personalities of the decision-makers one was dealing with. There is no reason to doubt that Hitler would have used nuclear weapons, had he owned them. In the Cold War, no Soviet leader was willing to bring on nuclear war deliberately to speed up the spread of socialism to the rest of the globe. But then, no Soviet leader of the nuclear age was ever faced with an invasion of his country, and in such a case the perception would be different. It might be different again from that of European leaders with more humanist convictions.

For alliances this means that it is the best option to commit firmly to mutual *conventional* defence; this at least is credible, faced with anything short of a full-fledged nuclear adversary, and to make all necessary arrangements (peacetime command structures, organisation, exercises, logistics networks and so on) to make them credible. In the Cold War, the problem with this otherwise sensible proposition, favoured by the USA, was that the Europeans rejected the course of preparing for conventional war in Europe. For them, the devastation caused by such a 'merely' conventional war, the Second World War, had been too great to contemplate a repetition. Europeans thus preferred *nuclear* deterrence, however lacking in credibility this might be.

Gallois's deterrence *pure et dure* was the essence of this European option, and it was uncompromisingly tied to city targeting. He described it as one of the central paradoxes of nuclear peace that with the 'resolutely pacific intentions' of using force only if it was itself attacked, the state had to commit itself to targeting the enemy's population. By the same logic, it would be the attacker that would target his victim's military means (Gallois 1960: 180). Refusal to give battle – the *non-bataille* (Gallois 1976: 22) – was central to France's posture and especially to de Gaulle's withdrawal from NATO's integrated military command just as other alliance members wanted to come to an agreement on what to do if deterrence failed.

The descendant of Faure's unrealistic dreams of deterrence in the pre-nuclear age, French official nuclear Strategy is wedded particularly strongly to the notion that nuclear weapons have abolished at least major war. Dissenting voices, including those of prominent retired military officers, have been accused of being lackeys of the Americans, and have never been able to make much of an impact. The beauty of French Strategy is that it can satisfy both pacifists (Gallois 1976: 113), as atomic war 'will not take place', and the most uncompromising anti-American nationalists.

Both British and French strategies essentially hold enemy cities 'at risk', and as we have noted, this continued to be the ultimate fall-back option of US nuclear Strategy. This is the essence of any deterrence by threat of punishment, where targets especially valuable to the adversary must be held at risk. It does not rule out sparing them deliberately during a war, and for as long as possible hitting only targets that contribute to the enemy's fighting strength, but sparing cities, as Robert McNamara proposed to his NATO colleagues in Athens in 1962 (McNamara 1962).

This raised the question, however, of what the Soviet leadership really valued most, its population – of which it seemed willing to sacrifice a considerable proportion – or its political control over the society as a whole and its military in particular (Albert Wohlstetter, q.i. Gray 1984: 18). We have seen that the USA had adopted in the 1970s, inter alia, a 'counter-recovery' targeting list, to destroy Soviet industry that would be vital to the recovery of the USSR after a Third World War. Colin Gray took issue with that, arguing that it should be the Soviet regime that should be eliminated through conscious targeting choices. He disapproved of population targeting on moral grounds, and thought that even counter-recovery targeting would indirectly punish the surviving population after a nuclear war (Gray 1979, 1984). His arguments on this point remain persuasive.

Morality, legality and credibility

This brings us to the legality of population targeting, the implied threat of deterrence. In 1961 the UN General Assembly passed a Resolution on the Prohibition on the Use of Nuclear and Thermonuclear Weapons, saying 'any state using nuclear and thermonuclear weapons is to be considered as violating the Charter of the United Nations, as acting contrary to the laws of humanity and as committing a crime against mankind and civilization'. It was introduced by African and Asian states, supported by the USSR, and passed with fifty-five votes to twenty, the USA, the United Kingdom, France and seven other NATO members among those voting against, and twenty-six abstentions. The USA tried to introduce an amendment about the use of nuclear weapons in individual or collective self-defence, but the amendment was defeated by fifty votes to twenty-eight with twenty abstentions (Rosenberg 1994: 165).

In 1965, the Second Vatican Council in its *Pastoral Constitution of the Church in the Modern World* addressed but failed to resolve the dilemma of the morality of nuclear deterrence. Essentially, the Catholic Church's stand was that deterrence was ethically justifiable, but not nuclear use (Osgood and Tucker 1967: 195–322). This makes little sense, if nuclear deterrence depends on the credibility of the implementation of the nuclear threat, as summed up in Gallois's formula above.

On 6 July 1996 the International Court of Justice gave an Advisory Opinion that even 'the *threat* or use of nuclear weapons would generally be contrary to the rules of international law' (my emphasis).

Nuclear weapons states are under obligation 'to bring to a conclusion negotiations leading to nuclear disarmament in all its aspects under strict and effective international control'.

> In the long run, international law, and with it the stability of the international order which it is intended to govern, are bound to suffer from the continuing difference of view with regard to the legal status of weapons as deadly as nuclear weapons. It is consequently important to put an end to this debate of affairs: the long-promised complete nuclear disarmament appears to be the most appropriate means of achieving that result. (ICJ 1996)

A good decade earlier, during the last Cold War crisis, when American intermediate range nuclear forces were deployed in Europe, American Catholic bishops in a pastoral letter of 1983 gave their approval of the concept of deterrence for the sake of peace, but denounced any nuclear use in war (US National Conference of Catholic Bishops 1983). The illogic of condoning a nuclear deterrence stand but denying it any credibility in the form of executable nuclear plans was most famously exposed by Albert Wohlstetter (1983). In the United Kingdom, the Permanent Undersecretary and thus the highest-ranking civil servant in the Ministry of Defence and the brains behind official British – and much of NATO – nuclear Strategy at the time, Michael Quinlan, equally denounced the illogic of such a stance that would deprive any deterrence posture of credibility (Quinlan 1986). The same point was made by Colin Gray (Gray 1984). This criticism – rejected by the French for their own Strategy – provides the nexus between a nuclear deterrence posture and a nuclear war-fighting Strategy.

Nuclear war-fighting Strategy

The apogee of strategic bombardment

As we have seen, the effects of the atomic bombs in 1945 were claimed as the triumph of Douhet's, Trenchard's and Mitchell's ideas, which previously had not been realisable with mere conventional ordnance. The Japanese emperor felt forced to surrender, and the Allies had clearly 'imposed their will upon the enemy', to use the Clausewitzian expression. This is the common perception even now; whether or to what extent this was truly the sole effect of the bombing of Hiroshima and Nagasaki is the subject of considerable retrospective debate among historians (Heuser 1999: 8–34).

Simultaneously, it was one extreme of 'annihilation', and this strategy lived on into the Cold War. In 1949, Vice Admiral Arthur W. Radford testified to Congress that the USAF's atomic Strategy aimed at a 'war of annihilation'. But could this be reconciled with Western values, upon which the principles of the UN Charter were founded? Shortly after the war, President Truman, who had defended his decision to use atomic bombs against Hiroshima and Nagasaki with the argument that it had 'saved half a million American lives' and cut short the war, began to reason in a more complex way. During the Berlin blockade of 1948–9, Stuart Symington, Truman's Secretary for the Air Force, wanted control of US nuclear weaponry to be given to the air force and taken away from the civilian Atomic Energy Commission, so that they could train while 'handling it'. Truman refused, explaining 'that this is not a military weapon ... It is used to wipe out women and children and unarmed people, and not for military use. So we have got to treat this differently from rifles and cannon and ordinary things like that' (q.i. Budiansky 2003: 351, 353).

Late in 1947, the USAF directorate of intelligence prepared a study of targets for atomic bombing within the USSR, with the horrific title 'To Kill a Nation', and the targets were above all industry located in and near seventy Soviet cities. A sub-committee of the US Joint Chiefs of Staff, the Joint Strategic Plans Committee, concluded in a report of 15 March 1948:

> Atomic bombs will be used by the U.S. Strategic concept: To destroy the capacity and will of the enemy to continue hostilities. Initially to launch attacks designed to exploit the power of atomic weapons against the war-making capacity of the enemy. (q.i. Budiansky 2003: 349)

This thinking continued right through the Cold War. When General Thomas S. Power, Chief of the US Air Staff, was briefed by RAND in December 1960 on the concept of 'counterforce' doctrine and the virtues of restraint and controlled escalation with 'withhold targets', he exclaimed: 'Why do we want to restrain ourselves? Restraint! Why are you so concerned with saving *their* lives? The whole *idea* is to kill the bastards! Look, at the end of the war, if there are two

Americans and one Russian, we win!' (q.i. Budiansky 2003: 366). This was not unlike the Soviet definition of victory prevalent until the late 1980s, when Gorbachev rightly put an end to such thinking (Heuser 1998b).

There was even, on the Western side, a short flirtation with preventive nuclear war, at a time when the Soviet arsenal was still so small and unprotected by means of hardened shelters as to be incapable of surviving a large-scale counter-force attack by the US even with conventional bombers. Admiral Ralph A. Ofstie, who had previously been a director of the US strategic bombing survey, in 1948 advocated preventive war; while the US would give its plans away if it made any conventional military preparations, such a preventive war could be prepared by stockpiling nuclear weapons and building bombers (Gentile 2001: 155). General Orvil A. Anderson, commandant of the Air War College, in September 1950 (that is, after the outbreak of the Korean War) made the mistake of recommending such a course of action in an interview with an Alabama journalist, which cost him his job; others shared his view, but General Hoyt Vandenberg, Air Force Chief of Staff, would not tolerate such public statements. Even Bernard Brodie seems to have flirted with the idea at the time (Gentile 2001: 159–61), as did the British philosopher Bertrand Russell before turning into a disarmer.

On the Western side, there were critics of city targeting from the moment such practices were aired, and not only among the anti-nuclear movements which formed among the public, especially in Western Europe in the 1950s. George Kennan in the late 1940s, when briefed about USAF strike plans against the USSR, noted: '[i]f you drop atomic bombs on Moscow, Leningrad, and the rest, you will simply convince the Russians that you are barbarians trying to destroy their very society and they will rise up and wage an indeterminate guerrilla war against the West'. Perhaps under his influence, a review of plans of the American Strategic Air Command (SAC) in April 1949 argued that bombing Soviet cities with nuclear weapons would 'not per se bring about capitulation, destroy the roots of Communism or critically weaken the power of Soviet leadership to dominate people'. Instead, it might well 'validate Soviet propaganda ... unify the people ... and increase their will to fight' (q.i. Budiansky 2003: 353).

Other critics of SAC's Strategy tended to find themselves in the US navy. Admiral Ralph A. Ofstie, who in 1948 was still advocating that the US should 'knock hell out of Moscow with atomic bombs', by October 1949 argued that the killing of civilians in the bombing of enemy cities had been immoral, and that it would be similarly immoral to kill Soviet civilians, and that such 'military methods' could not be endorsed by the US people as they were 'so contrary to our fundamental ideals'. Admiral Radford took the same position about the 'massive killing of noncombatants'. Admiral Thomas Kinkaid condemned the city bombing of the Second World War as 'terrorizing bombardment' violating the laws of war. Admiral W.H.P. Blandy thought that 'no sane man would derive any satisfaction from killing women and children' as it was nothing but the 'slaughter of innocent people' (Gentile 2001: 154).

Nevertheless, as we have seen, city targeting continued to dominate Western nuclear employment planning. As Stanford professor Anthony Sokol noted, the prevailing idea even in the 1960s was that 'strategic bombing' of the enemy's 'vital centers' (Billy Mitchell) would be central to any future East–West war, which would be even more 'total' than the Second World War had been.

> It is the basis of most of the currently accepted theories of 'total war' and of the assumption that a revolution has occurred in the method of conducting wars in general. It is also the foundation of the present strict distinction between 'strategic' weapons and forms of fighting – those which affect the enemy nation directly – and 'tactical' ones which are used in the literal fighting between the opposing armed forces. (Sokol 1961: 30)

This was not just the public assumption, but operational war planning. In his survey of US contingency plans, Jeffrey Richelson found continuity of population targeting from 1946 to the late 1980s. Soviet cities always led the target list. Richelson pointed out that this was incompatible with the US ratification of Protocol I of 1977 of the 1949 Geneva Convention, which in Article 85 prohibits 'making the civilian population or individual civilians the object of attack'. This protocol tried to avoid the loophole of the 1907 Hague Convention, which had made military considerations possible excuses for targeting built-up areas. We find in the 1977 protocol that prohibitions covered attacks that 'may be expected to cause incidental loss of civilian life, injury to civilians, damage to civilian objects, or a combination thereof, which would be excessive in relation to the concrete and direct military advantage anticipated'. While even this formulation is wide open to interpretation (who decides what is excessive?), it is indeed difficult to reconcile it with hundreds of thousands or indeed millions of non-combatant casualties (Richelson 1986: 244).

Counter-force targeting

In the Warsaw Treaty Organisation (WTO), the Marxist-Leninist faith in the inevitable triumph of socialism, taken to mean the necessary survival of a socialist society of sorts, dominated Strategy, but was repeatedly assailed by some who had a clearer picture of the physical consequences of nuclear war. There was much speculation among Western experts as to what operational Soviet and WTO Strategy was, and we still have little access to archives that would give us a definitive answer. Much, however, can be surmised from the exercises conducted by the WTO, some on the ground, some at a boardroom planning level.

Targeting was counter-force and counter-military: the WTO exercised the destruction of NATO nuclear forces (missiles and aircraft) before these were launched, on receiving warning that they would be launched, fictitiously launching WTO weapons at the very moment of Western launch (which makes no sense logically). With these impracticable simultaneous-launch scenarios, WTO exercises tried to make allowances for Brezhnev's 1977 oral commitment to 'no first use'. The numbers of missiles fired in the exercises were limited only by the available launchers; firing would therefore have taken place in several salvos. There is no indication of any attempt at 'symbolic' or signalling use, nor is there any indication that the WTO would have reacted with differentiated nuclear use (i.e. with a selective rather than an all-out nuclear response) to NATO limited nuclear use packages. The WTO exercises indicate above all an operational Strategy of trying to knock out – by conventional or nuclear means, both options were exercised – enemy nuclear weapons, but also command centres, air defences and other military installations directly connected with NATO's ability to use nuclear weapons in the European theatre. Surviving WTO nuclear strike plans for these exercises never included the direct targeting of population centres. But the collateral damage to big cities – for example Hamburg – from nuclear hits on military installations in suburbs would have been lethal for a large part of the population, depending on the direction of the wind (Heuser 1993). The exercises give us no indication, however, how such 'theatre' use of WTO nuclear weapons would have related to 'strategic' use for targets in the USA, or even targets in the United Kingdom.

If any sense can be made of what was an ambiguous mishmash of postures (Baylis 1995), in the case of deterrence failure, preferred official British nuclear Strategy in the 1950s and 1960s would have been to use British strategic forces in counter-force strikes, which presupposed scenarios of joint action with the USA (Heuser 1997: 63–92). The issue of credibility arose here, too. At the end of the 1950s, the NATO powers abandoned their most-likely war scenario inherited from Douhetian thinking of the 1920s and British air power thinking of the 1930s, in which a major war would start with an all-out air attack by the other side, possibly as a 'bolt from the blue'. Thereafter, the scenario planned for was more one of 'accident or miscalculation', with a conventional attack by the WTO which NATO would try to stop, but which might get out of hand. The question was, how should such an attack be met, if conventional defences proved insufficient? There had to be options other than the incineration of Soviet cities as the next-best response. Could nuclear weapons be used 'tactically', in support of military operations, to deny the enemy a successful invasion?

Tactical nuclear use: deterrence by denial

At the beginning of the Cold War, and in the case of the USSR until almost the end of the Cold War, some military men, but also civilians, saw in nuclear weapons – at least the smaller, low-yield weapons that were developed from the 1950s – weapons of military use, just like any previous form of weapon. NATO Strategy MC 48 of 1954 ('Massive Retaliation') planned for their use on the battlefield in large numbers, but in rhetoric and operational planning, this was mixed up with massive strikes against cities in communist countries, all assumed to be part of the 'communist bloc' at the time.

In 1956 the British retired naval officer Anthony Buzzard, pleading for a more sophisticated Strategy than 'Massive Retaliation', made the suggestion that the West should declare that it would henceforth make 'distinctions' in terminology but also in planning and possible use, borrowing from air force terminology: 'The *tactical* use of nuclear weapons ... is to be confined to 'atomic [i.e. low-yield] weapons, and is to exclude even these from use against towns and cities. Their *strategic* use ... is to include hydrogen weapons and the mass destruction of targets in towns and cities' (q.i. Baylis 1995: 198). In fact, this differentiation was henceforth made in popular parlance. Despite the fact that many other terms were used by insiders in the coming decades for a host of reasons related to Strategy and technology (such as battlefield weapons or short-range nuclear forces (SNF), theatre nuclear forces (TNF), intermediate range nuclear forces (INF) and so forth), several leading strategists – including key figures like Sir John Slessor and Pentagon planners – ultimately regarded such a distinction as impracticable (Baylis 1991: 143–6). While Buzzard's aim in recommending this differentiation had not been to make nuclear weapons more useable but to make escalation to city busting less inevitable – he wanted a more 'graduated deterrence', his linguistic differentiation coincided with the period in which military planners envisaged nuclear use in all circumstances (Heuser 1997: chs. 2 and 3).

This applied even to 'limited war' scenarios. Since the expulsion of Yugoslavia from Cominform in 1948 and the subsequent Soviet planning for an invasion of Yugoslavia which the West felt it should counter, US government planners and key decision-makers had developed the concept of limited war (see chapter 17), a concept which rose to some public prominence with the Korean War, and also that of limited nuclear war. In 1951, the American Joint Chiefs of Staff (JCS) developed options for the use of nuclear weapons against Soviet satellite states that might invade Yugoslavia to return it forcibly to the communist fold. They deliberately excluded Soviet targets, in the hope that they might be able to contain aggression against Yugoslavia without escalation to world war (Heuser 1989: 167). From its inception, 'containment' was not a passive Strategy, but one aiming, like its more honestly named successor Strategy, to 'roll back' Soviet control of Eastern Europe and the Far East, but always in reaction to a strongly perceived, centralised communist military threat (Heuser 1989).

In 1951, the targets for such a 'limited' nuclear war as envisaged by the JCS would have been cities for want of the ability to hit anything else with bombers loaded with free-fall atomic bombs – the hydrogen bomb was first tested in the following year. Even in the late 1950s,

Bernard Brodie still thought that 'Limited war might conceivably include strategic bombing carried on in a selective or otherwise limited manner, for example bombing with nuclear weapons on selected targets such as airfields while being as careful as possible not to hit cities' (Brodie 1959: 310).

The question remained unanswered as to whether such a use would have inevitably resulted in escalation. The hope of keeping a nuclear war limited grew with the development of low-yield 'battlefield' or 'tactical' nuclear weapons, undertaken since 1951, with usable prototypes ready by 1953. Thomas Schelling at the RAND corporation wrote six years later:

> With the development of small-size, small-yield nuclear weapons suitable for local use by ground troops with modest equipment, and with the development of nuclear depth charges and nuclear rockets for air-to-air combat, the technical characteristics of nuclear weapons have ceased to provide much basis, if any, for treating nuclear weapons as peculiarly different from other weapons in the conduct of limited war.

But some argued that

> there are political disadvantages in our using nuclear weapons in limited war, particularly in our using them first, and that those who consider a fireball as as moral as napalm for burning a man to death must nevertheless recognize a worldwide revulsion against nuclear weapons as a political fact. In other words, if there is no longer a technical military dividing line between the effects of nuclear weapons and the effects of other weapons, there is a political distinction, a restraint exercised by the reactions of allies and neutrals. (Schelling 1959: 1)

It was such battlefield nuclear weapons which led Henry Kissinger to embrace the concept of limited nuclear war, which he thought possible, if combined with the tactics of 'small, highly mobile, self-contained units, relying largely on air transport even within the combat zone'. He was convinced that both sides would refrain from escalating such a conflict to a strategic level (Kissinger 1957: 174–202, esp. 180).

> War between nuclear powers has to be planned on the assumption that it is likely to be a nuclear war. Nuclear war should be fought as something less than an all-out war. Limited nuclear war represents our most effective

strategy against nuclear powers or against a major power which is capable of substituting manpower for technology. (Kissinger 1957: 199)

But it would work only in certain conditions, namely if strategic exchanges were avoided (that is, escalation to all-out war), if the enemy was not expected to surrender unconditionally, if one came across as determined and yet created a framework for a settlement of the conflict that was acceptable to the adversary and if one employed force in a discriminating or 'graduated' way (Kissinger 1957: 201).

By the early 1960s, however, official US thinking was that nuclear war should not be viewed in the category of 'limited' war, a term henceforth reserved for warfare involving the use exclusively of conventional weapons (Trachtenberg 1986: 763). Robert Osgood, an American strategist who devoted much thought to the concept of 'limited war', soon concluded it was impossible for Europe:

Most official studies and war games indicated that, even if it could be limited geographically, a tactical nuclear war in Europe would probably ... devastate the European allies, and that it would require more rather than less manpower. Moreover, given the growing Soviet tactical nuclear weapon force, NATO's ports, airfields, supplies, and logistics seemed particularly vulnerable in a tactical nuclear war. (Osgood 1979a: 106)

He added:

One trouble with all strategies of local war in Europe was that the Soviet Union showed virtually no inclination to be a partner to them. Rather, Soviet doctrine, published sources, and war manoeuvres, seemed rigidly geared to a strategy of blitzkrieg – a sudden offensive strike with conventional and nuclear (both battlefield and strategic) weapons intended to defeat and disorganize the NATO powers. Although Soviet military writings in the late 1960s envisaged the possibility of limited non-nuclear exchanges, they remained as hostile as ever to ideas of controlled escalation and intra-war bargaining by limited options of any kind, and especially by nuclear options. (Osgood 1979a: 107)

Air power thinking also underlay the question of what role nuclear weapons would play in relation to conventional weapons. The reversal of scenarios of the end of the 1950s from massive initial attack to gradual escalation from an initially conventional war (Heuser

1997: 3–14) went along with a debate about 'sword and shield'. Douhet saw the ground forces as a shield, there to hold up an enemy's aggression on the ground, while the air forces as sword would attack his vital parts, his cities, in strategic strikes. This view dominated NATO Strategy in the early 1950s, until the introduction of battlefield nuclear weapons could help NATO ground forces 'deny' enemy ground forces any conquest on the ground. If the enemy – the WTO – did not persist in its attack and did not start employing nuclear weapons on any level, a war might be terminated there without any need to resort to 'strategic' nuclear bombing by SAC.

Alternatively, with hugely increased conventional ground forces, the preferred US strategic option of the Kennedy administration would have been to try to repel a WTO attack with conventional forces only, using them as 'sword' to defeat aggression, while the nuclear forces (and assumed fear of nuclear escalation on all sides) would have acted as 'shield', or perhaps more accurately, a lid on escalation to a nuclear level. As noted above, the Europeans could not live with this Strategy, as in their experience even a conventional war on their territory was unbearable. This US–European debate marked the parting of ways of France and NATO where nuclear Strategy was concerned.

But for the other Europeans, too, deterrence by conventional or battlefield nuclear 'denial' was unacceptable. They insisted that NATO Strategy had to put an end to aggression as quickly as possible by bringing in the threat of nuclear escalation. There was here, always, a divergence of interests for Europeans and Americans; NATO always had a problem with the credibility of America's nuclear protection promise to its allies. Albert Wohlstetter thought that no nation would commit suicide for another; indeed, '[s]uicidal threats are *in general* not a reliable means' of deterrence (Wohlstetter 1983: 30). Once Soviet bombers and missiles could reach the USA, NATO always had to struggle with the credibility of any threat of nuclear use.

Political signalling for war termination, and the danger of escalation

Even in 1954, somebody had hit on the possible need to restore deterrence, once it had broken down. Not just recognising an either-or of deterrence or, in the absence of deterrence, war fighting, Warren

Amster, an American operations analyst, noted that there had to be three levels to deterrence:

1. deterrence to prevent the outbreak of war
2. 'to stop such a war, should it start' and
3. 'failing this, to carry out ... destruction of the attacker's country' (q.i. Sherwin 1956: 135).

The US–European debate of the early 1960s was resolved by NATO – minus France – through a compromise between deterrence by denial, tactical use and deterrence by political signalling for the purpose of war termination. This was expressed in the political guidelines for first or early follow-on use of nuclear weapons by NATO, in which strategic nuclear weapons were indeed seen as a shield or lid, not to be resorted to except in extremis, in the great hope that war could be terminated through restoration of 'intra-war deterrence' through limited use of nuclear weapons by NATO. Targets would have been military, specifically not population centres. There was some debate as to where these targets should be. John P. Craven thought that the first use of nuclear weapons – as tactical use – was more likely to occur on sea against enemy ships, as collateral damage there would have been minimal (Craven 1980: 82). Thought was given to other options for such 'tactical' use with the possibility of avoiding escalation, although many strategists doubted that such a scenario was realistic (Bull 1980: 8). NATO finally agreed that as the purpose of such a first use of nuclear weapons would have been to *signal* to the USSR NATO's resolve, to expose the Soviet leadership's *miscalculations* about any lack of determination on the part of NATO to defend itself, the targets had better be of significant importance for the WTO's war effort, on land (Heuser 1997: 52–7).

The French, no longer included in NATO nuclear planning, felt compelled to make some concessions concerning the problem of how to get from nothing to all-out nuclear retaliation. Guy Brossollet, a French defence official, wrote a study on the *Non-Battle* (refusal of battle) which largely reflected official Strategy: taking up an idea developed by General Fourquet in the late 1960s, Brossollet argued that France needed to 'test' the adversary's intentions with conventional forces, but not in order to deny them conquest physically, only in order to show that they could not invade Western Europe without resistance. The threat of nuclear weapons was held to prevent escalation after such a 'test', as surely the enemy would recoil from risking nuclear war. The 150 air-mobile units charged with conducting such 'tests' imagined by Guy Brossollet (which he called 'modules') would be deployed in depth and would aim to disrupt the enemy's advance, without seeking to defeat him in a major battle; they knew they were not capable of doing so. If these conventionally armed 'modules' proved unable to stop the enemy (that is, the enemy would accept the losses they inflicted and not believe that France was prepared to escalate to nuclear war), Brossollet (in keeping with operational French Strategy) counselled the use of short-range, low-yield nuclear weapons (the *Pluton* ground-launched missiles), again with the aim of 'testing' the adversary's resolve, not with the aim of defeating his onslaught. Only if the enemy proved determined to carry on would France use her strategic nuclear arsenal, held back until such a time as her last trump (Brossollet 1975). The 'test' is the only French concession to the need for an actual nuclear use plan to bolster the credibility of what is otherwise an all-or-nothing nuclear stance.

If initial nuclear use did not persuade the enemy that aggression promised no success, then NATO Strategy would have foreseen follow-on nuclear use. The concept of an escalation to several successive levels is associated with the work of the US strategist Herman Kahn. Famously, he elaborated on 'rungs' on an 'escalation ladder', emphasising that one could move up and down on this ladder, in the hope that the final (forty-fourth) rung of 'Spasm or insensate war' would never be reached (Kahn 1965). Kahn's concept owed something to Amster's 1950s ideas of calibrating nuclear responses (to minimise the danger of unintended escalation) by, for example, using just as many nuclear carriers in retaliation as the enemy had sent out, or hitting as many targets as the enemy (Sherwin 1956: 135).

The crucial objection to this approach, and indeed to any political signalling intention, which still holds today, was formulated in the same year by Thomas Schelling, ironically one of the thinkers with whom it originated. He cast doubt on the ability to signal such intentions to the adversary in the midst of the confusion invariably caused by any nuclear strikes in a war-time situation that was by definition confused (Schelling 1965: 228). Indeed there is no indication that the WTO would have agreed to 'play this game' rooted in economists' games theory. WTO Strategy did not differentiate in its responses to Western nuclear use – as far as we can see, its Strategy was completely independent from Western action except for launch-on-warning, to which the response would have been pre-planned launches in large numbers, not differentiated responses (Heuser 1993).

Central to all of NATO Strategy was also the danger of self-deterrence. This was the reason why the early NATO Strategy of 'Massive Retaliation' was abandoned by all but France. As Bernard Brodie commented:

> [P]rotagonists of the massive-retaliation doctrine [support] that position not because they preferred big wars to little ones but because they have been convinced that such a policy minimizes the danger of *any* war. It makes sense, they imply, to threaten a possibly suicidal reaction to aggression as long as the chance of aggression is thereby reduced almost to zero.

The problem is, however, 'its conspicuous lack of credibility' (Brodie 1959: 349f.). Others, too, doubted this credibility. Lawrence Freedman saw NATO Strategy in the early 1980s as a 'myth', a 'mixture of confusion and uncertainty'; in his magisterial work on the evolution of nuclear Strategy he came away with the 'sense of an enormous, and somewhat transparent, bluff' (Freedman 1989: 78–80, 157, 428). Paradoxically, the only logical way out of this dilemma was to introduce an element of irrationality, in Thomas Schelling's words, 'a threat that leaves something to chance'. Decision-makers might find it in their interest to convey the impression that war might escape their control (Schelling 1960: 83–118). We have here a third attitude to chance: the strategists of early modern Europe had tried to contain it through drill, planning and mathematics; Frederick II and Napoleon had tried to take advantage of its opportunities, and here it was introduced to try to render something credible which otherwise had little plausibility. Lawrence Freedman concluded his study on nuclear Strategy with the observation that the uncertainties as to whether nuclear deterrence would work in all future scenarios, and whether in political contexts as yet unforeseeable they would not lead to major disaster, could only leave untroubled those with an 'optimism unjustified by any historical or political perspective'.

> [T]he position we have reached is one where stability depends on something that is more the antithesis of strategy than its apotheosis – on threats that things will get out of hand, that we might act irrationally, that

possibly through an inadvertence we could set in motion a process that in its development and conclusion would be beyond human control and comprehension ... *C'est magnifique, mais ce n'est pas la stratégie.* (Freedman 1989: 432f.)

The Soviets seem to have gone one step further, at least eliminating the paradox: reportedly, they introduced a 'Dead Hand' nuclear retaliation system that would have launched an all-out nuclear counteroffensive in case of a nuclear attack on the USSR, even if all Soviet command centres had been annihilated by the enemy, unstoppable by any human agency. This eliminated chance and the human element from the equation. But such a retaliatory strike was no meaningful Strategy in the sense in which we have used the term in this book, as no meaningful political ends, except crudest revenge, could have been achieved in this way. If the world had been immersed in radioactive nuclear winter for years after the explosion of tens of thousands of nuclear weapons, as some scientists predicted, there could also be no meaningful victory.

Can nuclear war be won?

We have seen that nuclear weapons were from their first use linked by Brodie and other foresightful thinkers with the future need to deter, rather than fight, any major war. Sir John Slessor, shortly after his retirement from the RAF in 1952, wrote in *Air Force* magazine: 'I assure you that in my opinion everybody is going to lose the next war' (q.i. Budiansky 2003: 364). Caspar Weinberger, US Secretary of Defense, merely reiterated what by then was a mantra when in 1983 he told Congress that 'We, for our part, are under no illusions about the dangers of a nuclear war between major powers; we believe that neither side could win such a war' (q.i. Gray 1984: 7). It had taken some time for this realisation to spread in the West, however, and even today, there is no way of knowing whether this conclusion is universally accepted.

Between 1957 and the adoption of MC 14/3 in 1967, the Strategy document which would remain operational for NATO until the end of the Cold War, NATO gradually backed away from any hope of 'winning' a war against the Soviet Union, let alone 'defeating' it. At best, the war aim in a (it was *always* assumed) defensive war against the

Warsaw Pact was to bring aggression to a stop and, ideally, restore the status quo ante. But even the latter was a pious wish. By contrast, the WTO backed away from an offensive posture only with its Strategy of 1987, and only the draft Strategy of 1990 espoused as strategic aims, first, the prevention of all war, and second, should this fail, war termination. All previous WTO strategies had seen the ('historically inevitable') victory of socialism – even if defined as the survival of just a few socialists – as the war aim (Heuser 1998b).

In the late 1970s and early 1980s, the United States espoused a 'Countervailing Strategy' central to which was the aim of denying any aggressor victory – the fruits of aggression – through adroit force employment, including nuclear targeting (Makins 1981). One of its architects, Walter Slocombe, Deputy Undersecretary for Defense for Policy Planning, explained that the Strategy did 'not assume that the United States can "win" a limited nuclear war', but that it would 'ensure as best we can that the *Soviets* do not believe they could win such a war' (Slocombe 1981: 24).

There were Western strategists, however, who were impatient with the notion that nuclear war would be 'unwinnable', in particular Reagan advisers Colin Gray and Keith Payne. Admittedly, Gray had a modest definition of 'victory' as 'no more and certainly no less than that the United States achieves its political objectives (whatever they may be – and they may be quite modest)' (Gray 1984: 6). This is a very acceptable definition. If the conditions were met that any political objectives must be reasonable in relation to means, technology, risks and the avoidance of intolerable developments, then even NATO's Strategy aiming at 'war termination' (or the re-establishment of deterrence 'intra-war') would constitute a 'victory', thus defined. This does not, however, seem to have been enough for Gray and Payne, who challengingly claimed that 'Victory is Possible' in nuclear war, calling for the adoption of targeting principles to correspond to a 'Theory of Victory' (Gray 1979; Gray and Payne 1980).

Notwithstanding these two catchy titles, their argument had some plausibility, and took Catholic teaching on 'just war' as its key criteria. They rightly criticised the lack of coherence in the formulation of US Strategy in terms of military planning, weapons development and procurement and arms control, and the lack of planning beyond limited nuclear options for early use in war, as opposed to an integrated plan for what to do if further escalation became necessary (short of, again, 'general nuclear response'). On the one hand, they advocated giving greater protection to the United States through civil defence (which was all but non-existent, compared with that of the USSR), and ballistic and air defences, thus paving the way for a welcoming of Reagan's Strategic Defense Initiative. They never ceased to stress the implausibility of any nuclear use that would result in hundreds of millions of American deaths through Soviet retaliation. On the other hand they advocated a shift away from the targeting of Soviet industry, especially its recovery ability, and the corresponding collateral damage among civilians, and towards the targeting of the relocation bunkers of the Soviet political and military leadership, communication centres, means of communication and other instruments of domestic control (Gray and Payne 1980; Gray 1984). But in the heightened sensitivity of this Second Cold War, the aggressive packaging of their recommendations in terms of 'victory' and 'war-fighting' made many fearful of their views.

By contrast, there is a different tendency in Western thinking, rooted both in the anti-nuclear fears of the disarmers and in the political realism of Michael Howard and others who recognised even in the Cold War that nuclear disarmament might only make the world 'safe for conventional war'. It is the quest for ways to delegitimise and marginalise nuclear weapons, to reduce them to 'fleets in being' that will dampen the bellicosity of actors on the world scene, without themselves remaining a major risk to security (Yost 1990). The end of the Cold War was both caused by and made possible by massive nuclear arms reductions above all in the USA and the Soviet Union, but also in Britain and France, with enormous reductions also of US nuclear weapons on the territory of non-nuclear NATO members, and corresponding withdrawal of Soviet weapons from the states of the defunct WTO (Smith 2002). Tactical (or non-strategic) nuclear weapons of all sorts disappeared from Europe virtually overnight, with France persuaded to follow suit; those strategists who had worked hard to bring about the strategies of the early 1980s briefly shed tears over the elimination of 'the wrong' weapons, while very quickly all sides were delighted with their removal (Garrett 1987; Davis 1988).

New initiatives were put forward of an ever more practical and decreasingly ideological nature on how to reduce nuclear arsenals further, until they might be little more than 'virtual', 'post-existential', perhaps even unassembled. Surprising individuals were among those now seriously interested in the study of how nuclear weapons could be abolished, including Robert McNamara and the late Sir Michael Quinlan. The two or three 'realists' who immediately after the end of the Cold War proclaimed that nuclear proliferation would make the world safer, and that one might best start with Germany (Mearsheimer 1990), were booed down overwhelmingly, not least by Germans.

Russia in the 1990s moved to a deterrence by denial Strategy resembling that of the USA in the mid-1950s, when battlefield nuclear weapons were to be used to make up for conventional weapons and personnel deficiencies. By contrast, and with the exception of Israel, nuclear proliferation outside Europe has in each case gone together with a weighty political agenda in which the nuclear weapons, real or still aspired to, are a political weapon and not crucial to any military Strategy. Nevertheless, here as in Strategies for conventional wars that we shall discuss in the following chapters, the West may be very much out of line with the thinking in some other cultures of the world.

War taken to its absurd extreme

The possibilities of nuclear warfare have taken war to its most absurd extreme. For anybody valuing human life as much as liberty, which after all is nothing if there are no humans to enjoy it, or if humans can enjoy it only amid death and rubble, the Armageddon scenarios of the Cold War would have made no sense, if implemented.

We have seen, however, that 'merely conventional' weapons, with precision guidance, whether fired from land, sea or from aircraft, have reawakened old dreams of meting out divine justice upon tyrants, rogues and villains, ideally sparing non-combatants, and all preferably at no cost to oneself (called by the French *la guerre zéro morts*, war with zero deaths, naturally only on one's own side). A rough realisation of this dream was made possible by 'smart' missiles which America has put into use since the 1970s for precision targeting.

And yet it is difficult to accept the argument that 'limited nuclear use' with 'small-yield' nuclear weapons is possible merely because, so the argument goes, it would not amount to much more than the use of high-yield 'conventional' weapons such as fuel-air explosives. Both are in fact a quantum leap up from the 'conventional' weapons used in the World Wars and even in Korea and Vietnam, and this should not commend them to us. The Maxim gun, dynamite, air bombardment and even nuclear weapons have not banned *all* war, let alone eliminated all the causes for wars. The British planners of 1945 were right, in that so far no *technological* defence has been found to counter nuclear weapons fully. Yet there is an answer, an asymmetric one, to enormously strong powers, even those operating with ultra-sophisticated technology, even to those owning nuclear weapons. This we shall explore in the next chapter.

PART VI

Asymmetric or 'small' wars

15 *From partisan warfare to people's war*

'The Way' means inducing the people to have the same aim as the leadership, so that they will share death and share life, without fear of danger.

(Sun Tzu *c.* 500–300 BCE: 1)

I prove to the conquering peoples the folly of wanting to invade a country where the threatened nation is disposed to ... rise up spontaneously, *en masse*.

(Le Mière de Corvey 1823: xivf.)

Two meanings of 'small war'

Armies with state-of-the-art weaponry, fighting pitched battles against armies of similar size, similar organisation and similar equipment, both representing recognised states and governments, fighting according to unspoken or even legally encoded rules – these are the characteristics of what is thought of as conventional, symmetric warfare. At the high end of the quantitative spectrum, this is termed major war, and whole nations might be mobilised in its support. At the low end it may be a limited engagement like those of the 'cabinet wars' of the eighteenth century, often indecisive and perhaps even relatively clement on populations in general.

By contrast, this chapter deals with wars fought between parties that are fundamentally unequal, though not in all respects, as we shall see. Crucially – and increasingly over the centuries – they have been unequal in that one side has authority, a recognised claim to a monopoly of power and increasingly a state apparatus in some form, mostly including 'regular' armed forces, usually with more powerful weapons. The other side tends to lack all or most of those attributes, but by their very insurgence challenge the authority and position of their adversaries. The British officer Charles E. Callwell, who has many valuable insights to offer despite his sickeningly racist terminology, distinguished three sub-categories: 'campaigns of conquest or annexation' against an enemy inferior usually in equipment and army size; 'campaigns for the suppression of insurrection or lawlessness or for the settlement of conquered or annexed territory'; and finally 'campaigns undertaken to wipe out an insult, to avenge a wrong, or to overthrow a dangerous enemy' (Callwell 1896/1906: 25). Of these, the second category is probably encountered most frequently today. While for the Western powers it usually means interference in the internal matters of a foreign country, for that country, 'an insurgency is a civil war', as the French theorist and practitioner David Galula (1919–67) realised (1964/2005: 2).

Since the early nineteenth century, many have used the term 'small war' synonymously for any asymmetric war. Using Callwell's word, small (or asymmetric) wars thus

> include the partisan warfare which usually arises when trained soldiers are employed in the quelling of sedition and of insurrections in [developed] countries; they include campaigns of conquest when a Great Power adds the territory of [underdeveloped countries] to its possessions; and they include punitive expeditions against tribes bordering upon distant colonies. (Callwell 1896/1906: 24)

Asymmetry can of course come in all forms and variations, and exists also in many forms of 'regular' warfare: when Germany invaded Belgium in 1914, for example, there was clearly an asymmetry of numbers. The asymmetry examined in this chapter, however, concerns above all the claims to authority, and in most cases the wielding of this authority through the command of the entity's central 'regular' forces. Callwell explained this well: 'The expression "small war" has in reality no particular connection with the scale on which any campaign may be carried out; it is simply used to denote, in default of a better [term], operations of regular armies against irregular, or comparatively speaking irregular, forces' (Callwell 1896/1906: 21). At the outset of an insurgency, as Galula observed, the

> overwhelming superiority in tangible assets favours the powers against whom it is staged: they have diplomatic recognition; legitimate power in the executive, legislative, and judicial branches; control of the administration and police; financial resources; industrial and agricultural

resources ... transport and communications facilities; use and control of the information and propaganda media; command of the armed forces

and the police. And yet the insurgents, with their blatant inferiority in all those dimensions, have moral factors on their side, which might ultimately bring them victory. In view of this, 'the insurgent needs so little to achieve so much whereas the counterinsurgent needs so much to achieve so little' (Galula 1964/2005: 3f., 46).

There is more semantic confusion in this subject area than anywhere else, as terminology has changed so often, even though the phenomenon of asymmetric war – in our limited sense of the organised uprising of 'the weak' against 'the strong' – is perennial. We must therefore start with definitions.

What we are dealing with in this chapter is sometimes claimed to be a 'new' form of war that has come to the fore since the end of the Cold War in 1989–91 (Kaldor 1999). Certainly, some technological aspects of these wars are new – the use of mobile phones and the Internet, for example. But in its essentials, the wars concerning us here are at least as old as any form of war and predate state formation. They are in themselves challenges to state formation, with its crucial aspect of establishing an internal monopoly on the use of armed force and a single central government. This is why they were so particularly widespread in Europe whenever there were no clearly recognised and functioning states, wherever state-forming processes were underway and whenever larger entities – empires – were disintegrating and fighting occurred over the future structures and borders of the remainder. There is not the space here to deal extensively with historical examples, but these go back at least to the insurgency of David against the Hebrew king Saul (around 1000 BCE?), to the rebellions by conquered peoples against Rome and to the many incursions into the Roman Empire by mounted nomads, which ultimately brought down both the West and a thousand years later the East Roman Empires (for historical examples, see Ellis 1975). Western culture was greatly influenced, for example, by the ultimately successful – and thus supposedly God-willed – insurgency of the Maccabee brothers and their supporters against the Hellenistic king Antiochus IV in Israel, and the nine-year civil war this triggered. In Western cultures, no opportunity would be lost to cite this example to justify insurgencies.

Today most military officers will think in this context of the Chinese sage Sun Tzu, who lived sometime between 500 and 300 BCE, as the starting point of all writing on asymmetric warfare. He particularly developed ideas of catching an adversary off guard, unbalancing him, avoiding central battle, while seeking to overcome him by indirect means:

> A military operation involves deception ... when [your enemies] are strong, avoid them ... When their military leadership is obstreperous, you should irritate them to make them angry – then they will become impetuous and ignore their original strategy. Use humility to make them haughty. Tire them by flight. Cause division among them. Attack when they are unprepared, make your move when they do not expect it. Strike at their gaps, attack when they are lax, don't let them figure out how to prepare ... The military has no constant form, just as water has no constant shape – adapt as you face the enemy, without letting them know beforehand what you are going to do. (Sun Tzu *c.* 500–300 BCE: 6–9)

Sun Tzu's ideas have been taken to form the basic pattern of an 'Eastern' style of war contrasted with the supposed 'Western' tradition of seeking major battle that Victor Davis Hanson has perhaps done most to publicise (see chapters 1–4).

Sun Tzu remained largely unknown in the West until the twentieth century, although he was first translated by a French missionary shortly before the French Revolution (Amiot 1772), and there are apocryphal stories that Napoleon read him. It is mainly after the Second World War that interest in him was engendered in the West through the confrontation with Mao's success against, first, the Japanese and then the Chinese Nationalist forces, and then the spread of Mao's ideas for the conduct of 'revolutionary war', which blended Sun Tzu and other Chinese sages with Clausewitz's 'people's war' (McNeilly 2001).

While we can find examples of asymmetric warfare in the West throughout the period covered in this book, it seems that until the end of the eighteenth century there is no specific Western literature on how to conduct such an uprising; the relevant literature comes from the side of those challenged by it. Machiavelli and later sixteenth-century writers such as Raymond Beccarie de Pavie, sieur de Fourquevaux, had already written about how to deal with occupied territories and with actual or potential 'rebels', as we shall see presently; the literature on what would much later be called 'counterinsurgency' thus predates any Western writing – albeit not practice – on how to conduct an uprising and fight as weaker, irregular forces against a superior power.

Since its introduction during the Spanish–Dutch Eighty Years War, the term 'small war' has provided considerable terminological confusion for later generations. Sebastián de Covarrubias already uses it in his dictionary of 1611, re-edited in 1674, but the term to him meant feud or civil war (Covarrubias 1611/1943: 666). By the following century, the term was in wide usage, especially in French, where it was translated as '*la petite guerre*'. But now, and up to the French Revolution, it was used exclusively to denote special operations conducted by special forces acting in wars alongside regular armies. These were rooted in the practices of the Roman, the Byzantine and the Habsburg Empires of recruiting such special units from tribes who fought much like the Easterners encountered by the crusaders: they were lightly armed, employed hit-and-run tactics and could be used for reconnaissance, espionage, sabotage and ambushes. By the seventeenth and then the eighteenth centuries, the Habsburg practice of employing units made up entirely of men from such tribes was being emulated in other parts of Europe (Selig and Skaggs 1991: 12–14). These units were called *Partheyen* in German, *parties* in French, and their leader alone was called a *Partheygänger* or *partisan*. To repeat, 'partisan warfare', until well into the nineteenth century, thus meant what we now call the conduct of 'special operations'. Until the American War of Independence and then the Napoleonic Wars, partisan warfare was entirely devoid of ideology or other political content, but was professional warfare carried out by professional units specialising not in regular war and pitched battles but in irregular activities (Rink 1999). These professionals were also called *Jäger* (huntsmen or foresters) in German and, later, *chasseurs* in French; the terms survive today, where units of *Jäger* are still trained and earmarked for particular territories such as mountain warfare.

The classical texts on partisan warfare start with a *Treatise on Small War for Free Companies* by Armand François de la Croix (d. 1743), published by his son in 1752. It was followed in 1756 by *Small War* by Thomas-Auguste le Roy de Grandmaison (1715–1801) and in 1759 by *The Partisan* by the presumably Hungarian Captain de Jeney, who seems to have fought on the Austrian side in the Seven Years War. They were still being reprinted when further works were published, such as those by the Prussians Georg Wilhelm Freiherr von Valentini and Friedrich Leopold Klipstein, both of 1799, which differ little in their prescriptions for the use of light troops, both cavalry and infantry, for special operations from those published before the French Revolution. The same can be said for Clausewitz's lectures on small war, held at the General War School in 1810–12, and for several of the works written well after the Napoleonic Wars but based on their experience (for example Decker 1822; Chrzanowski 1839; Davidov 1841). From the point of view of our enquiry into the political purpose of the use of force, these manuals reveal little about their authors' mindsets, but are primarily tactical-technical in nature, and follow in almost every way the narrow manual tradition that we have traced back to Vegetius and beyond.

Two wars or periods of war cumulatively changed 'small wars', in the sense of special operations conducted by professional special forces, to ideologically driven 'peoples' wars' (Clausewitz's term) or insurgencies, including operations by small bands of ideologically driven fighters with some degree of support from the local populations (Heuser 2010a). The first was the American War of Independence (1775–83), the second the French Revolutionary and Napoleonic Wars (1792–1815). In the first of these, irregular forces fought alongside regular armies on both sides, just as in wars of the seventeenth or eighteenth centuries. However, on the American side, the irregulars were ideologically motivated to fight for their own cause – freedom from British tutelage. Several American partisan leaders such as Daniel Morgan and Andrew Pickens had gained experience in the previous French and Indian Wars, and in many ways were copying tactics of the native American tribes (Starkey 1998; Schmidt 2003: 169). General Nathaniel Greene organised partisan bands to trouble the British redcoats in the swamps of Eastern Carolina, led among others by 'Swamp Fox' Francis Marion. Greene described his tactics in a terminology which sounds very pertinent to partisan warfare more than two centuries on:

> I have been obliged to practice that by finesse which I dared not attempt by force ... There are few generals that has [*sic*] run oftener or more lustily than I have done ... But I have taken care not to run too far, and commonly I have run as fast forward as backward, to convince our Enemy that we were like a Crab, that could run either way ... We fight, get beat, rise and fight again. (q.i. Weigley 1976: 36)

The question has been raised whether this early American experience with ideologically motivated irregular warfare had any effect on subsequent European changes in warfare: after all, as we have seen, French Revolutionary warfare was described by contemporaries as a large-scale version of small (or irregular) war (see chapter 5). Historian Peter Paret found little evidence for such a transatlantic transfer (Paret 1964). By contrast, the historians Selig and Skaggs have argued that with his emphasis on the combination of independent action and corporate discipline, the Hessian mercenary Johannes Ewald (1744–1813) brought back from his experience of fighting alongside the British in the American War of Independence ideas that would have an impact on the transformation of war in Europe (Selig and Skaggs 1991: 26–9; see also Schmidt 2003). Yet in their manuals on small war, neither Ewald nor his compatriot Andreas Emmerich (1737–1809) with his similar experiences in America addressed the strategic dimensions of warfare that characterised the works of a Machiavelli or a Guibert.

The second, even bigger laboratory of Strategy was that of the French Revolutionary and Napoleonic Wars. In the royalist Vendée uprisings against the French Revolution and in the insurgencies in the Tyrol, in Spain, Russia and finally Prussia against the invasion and occupation by Napoleon and his allies, the insurgents were, by later standards, 'reactionary' – they rejected the innovations brought by the French – and they discovered nationalism for themselves. The most important among these was arguably the Spanish War of Independence (1808–12), more commonly known as the *Guerrilla*, which gave the Spanish term for 'small war' a different, politically loaded meaning that has lasted to this day. Historians have, however, underscored the complexity of this particular *Guerrilla* with a capital 'G'. It should be stressed again that it was primarily reactionary, Roman Catholic, aiming at the restoration of the status quo ante and especially the Bourbon royal family. The forces that rose up to fight the French comprised remainders of the former Spanish regular army, *miquelets* or *somatens* (home guards or reserves that had long existed), *partidas*, *cuadrillas*, *cuerpos francos* (free corps) or special forces that had served the crown before Napoleon's invasion, augmented by many volunteers. From this point onwards, a further semantic confusion arose – the Spanish noun '*guerrilla*' would mutate, via the intermediate expression 'parties of guerrilla', to denote, not 'small war', but those waging it on the side of the rebels (Esdaile 2002; Moliner Prada 2005).

There were some politically 'progressive' insurgencies in the nineteenth century, such as the insurgencies culminating in revolutions in France in 1840, 1848 and 1871, and in combination with nationalism, the anti-Ottoman insurgencies (which became wars of independence) throughout the Balkans. Among peoples for whom the creation of a nation state (in either the ethnic or the political sense) was an aim, insurgencies against the (foreign) state power could be progressive, as in the nineteenth century in Latin America, but it could also have distinctly anti-democratic values, or anything in between, such as the Garibaldi-led insurgencies that culminated in the Italian Wars of Unification. The association of people's war and left-wing movements became strong during the Second World War and the Cold War, when communists tended to support any other movement directed, first, against the Axis powers, and later against Western colonial powers or the USA, even if the latter sought to export democracy. But even during the Cold War, communist insurgencies tended to have a strong element of nationalism – ceaselessly stressed by Vo Nguyen Giap (1911–) in Vietnam, for example, to the point of claiming Vietnamese antecedents in strategic thinking and rejecting support from foreign soldiers (Giap 1970a: 107, 1971: 25, 95).

After the end of the Cold War this identification of people's war and 'progressive' left-wing causes receded: it is still found in Latin America, but in other parts of the world – Afghanistan, Iraq, some African states – insurgencies against governments are extremely 'reactionary', especially where old tribal warlordism and/or Islam come into play.

There was a remarkable degree of personal continuity among those experiencing small wars or people's wars in different countries and contexts. As noted, Ewald and Emmerich had fought under British command in the American War of Independence; roughly thirty years later, Emmerich was executed at the age of 72 for attempting to stage an anti-Napoleonic insurgency back in his native Hesse. Jean Frédéric Auguste Le Mière de Corvey (1770–1832) had fought on the Revolutionaries' side against the insurgents of the Vendée. Carl von Clausewitz (1780–1831) had experienced the harassment of the invading Napoleonic army by Russian irregulars in the campaign of 1812 and the *grande armée*'s terrible losses (Clausewitz 1823), an experience which resonates throughout his chapter on 'People's War' in *On War*. The Russian General Denis Vasilevich Davidov (1784–1839) also drew on this campaign for his own otherwise fairly conventional theories on partisan war (Davidov 1841). The Pole Wojciech Chrzanowski (1793–1861) fought alongside Napoleon in this campaign and in the battles of Leipzig and Waterloo. Heinrich von Brandt (1789–1868), Le Mière de Corvey and Thomas Robert Bugeaud de la Piconnerie, Duke of Isly (1784–1849) had all fought – on opposite sides – in the Spanish War of Independence and drew their respective lessons from this experience, which Bugeaud de la Piconnerie subsequently applied to French colonial warfare. Several Spanish partisans of the *Guerrilla* of 1808–12 fought in the subsequent Carlist Wars in Spain, or left Europe and as *gauchos* joined wars of independence in Latin America against the Spanish and the Portuguese. In turn, prior to organising the Italian national movement, Giuseppe Garibaldi (1807–82) had in 1837–48 fought as *corsaire* and *gaucho* for desperate causes in Latin America: for the Republica del Rio Grande Do Sul, for a rebellion against the Empire of Brazil, for the defence of the Republic of Uruguay, against the Argentine dictator Juan Manuel de Rosas. From this time he kept a red shirt (which his men also would later wear in Italy and France), a poncho from the *pampa*, a large *gaucho* hat. Such continuity of experience and application must have existed previously, and is reflected in the writings of educated officers on how to repress revolts but difficult to trace in any writings of rebel leaders before the late eighteenth century.

With the works of Auguste Blanqui (1805–81) and Garibaldi, works on how to conduct insurgencies themselves became ideologically laden. By the time of Ernesto 'Che' Guevara (1928–67), well after the Second World War, the *guerrillero* became co-terminous with the social revolutionary, fighting for social justice (Che 1960/2003: 38–41). Che admitted that seen from the outside, *guerrilleros* might not look different from bandit gangs, who 'have all the characteristics of a guerrilla army, homogeneity, respect for the leader, valor, knowledge of the ground, and often, even good understanding of the tactics to be employed. The only thing missing is support of the people' (Che 1960/2003: 10). The support of the people, however, linked to the *guerrilleros*' fight for the people's interests, was the defining characteristic of revolutionary warfare (Galula 1964/2005: 4).

While in the nineteenth century, Britain and France had harshly repressed insurgencies in their own colonies, they had sympathised with national uprisings in Europe against the Ottoman, Russian and even Habsburg Empires. The First World War ushered in a period in which the Western liberal powers, especially Britain, France and the United States, were generally sympathetic to guerrilla uprisings, only one example of which was the British encouragement of the Arab revolt against the Ottoman Empire in the Great War itself. Then in the Second World War, the British and American governments and the French government in exile supported the resistance movements fighting against the Germans and Italians. The assumption that partisans were good and on 'our' side is still strong in C.N.M. Blair's work written for the British Ministry of Defence in 1957. Then NATO had plans, in case of a defeat of its regular forces at the hands of Warsaw Pact forces, to deny the enemy the fruits of his victory by supporting Europe-wide resistance movements, much as the American Office of Strategic Services and the British Special Operations Executive had done during the Second World War. While the examples discussed by Blair included operations that were hostile to the West – most notably in Malaya – the general gist of the book concerns lessons to be learned on how to stage uprisings.

It is only with the French and American experiences in Indochina and Algeria that this liberal perception of insurgencies as being worthy of support turned, and that most Western authors once again concentrated on how to counter such insurgencies, now almost always communist-led or supported. The insurgent side – waging 'revolutionary war', as both sides insisted (Galula 1964/2005: xiii) – was henceforth represented by communist authors, especially Mao Tse-tung (1893–1976), Giap and 'Che' Guevara. For Giap as for other communist writers, only a 'war of liberation' could be a 'just war', and it had to link national liberation and world proletarian revolution (Giap 1970b: 16, 1971: 24f.).

Other terms used in connection with asymmetric warfare also shifted their meaning. In the nineteenth century, the word 'partisan' had graduated from referring solely to the leader of a detachment of irregular forces to referring to all its members. Through the pens of those who did not know Spanish, the *guerrillero* (partisan in this wider sense) turned into a *guerrilla*, in the process of this even often losing an 'r'; the same authors tend to talk tautologically about 'guerrilla warfare'. During the twentieth century, 'guerrilla' often came to carry ideological weight through its association with 'revolutionary war'. Other words crept in, often linked to particular technology and organisation, such as the riflemen of the American Wars of the eighteenth century, the German *Freischütz* or sniper of the early nineteenth century (immortalised in the ghastly eponymous opera), translated exactly into the *francs-tireurs* of the Franco-Prussian War 1870/1 and of subsequent wars. Those fighting against *guerrilleros* or partisans liked to call them bandits or *banditi*, a word that made the rounds in the Italian *Risorgimento* (the various insurgencies and wars of national unity, 1815–61), *brigands* (as the French called Algerian insurgents in the 1830s) or *Banden*, the favourite word of the Wehrmacht in repressing local resistance in the Second World War. Thus some terms – such as 'insurgency' and 'counterinsurgency' (often abbreviated as COIN, '*Aufstandsbekämpfung*' in German, '*contreinsurrection*' in French) – remained neutral and descriptive, others – especially '*Bandenbekämpfung*' – acquired much ideological weight.

The mosquito and the lion: tactics

As noted above, prior to the nineteenth century there was little of political, let alone strategic dimension in the writings about insurgencies and irregular war listed above. Those of a 'special operations' character before the French Revolution always relied upon regular forces to bring about the decision in the war on the battlefield. Partisan tactics are well described in the anonymous eye-witness account of Richard the Lionheart's part in the Third Crusade, in which the crusaders, crossing Byzantine territory in Asia Minor on their way to the Holy Land, clashed with Turkish invaders. The account describes the hit-and-run tactics of the lightly armed Turkish horsemen. Whenever the crusaders pursued the Turks, these would flee, but as soon as their pursuers were out of breath, the Turks would turn about and set upon them. They were compared in this anonymous account with flies buzzing around their prey, flying off when they are shooed away, and returning to the attack as soon as one stopped going after them (Anon. *c.* 1200?/1997: 234f.). Two characteristics of these twelfth-century Turkish tactics became recurrent themes in partisan warfare, as we shall see: the use of light cavalry and the avoidance of pitched battles. The partisan detachments were by definition those units who avoided a regular battle, but might rejoin the regular forces after their special duties had been carried out (Jeney 1759/1760: 104–6). Heinrich von Brandt likened the Napoleonic army on the Iberian Peninsula to the lion of Aesop's fable, who was plagued by the mosquito until he surrendered (Brandt 1823: 64). Only after the watershed of the Spanish War of Independence, when partisan warfare merged with insurgency, and Napoleon's doomed campaign in Russia, was it thought necessary to spell out that pitched battles had to be avoided by irregular fighters. Now the ranks of professionals were swelled with untrained volunteers. Le Mière de Corvey thus noted that:

> Parties of *guerrillas* have to avoid clashes in the open, against disciplined troops, except if they are in the majority or can take the enemy by surprise. For in limiting oneself to conducting this partisan war in the mountains and countryside covered [in trees], one inflicts an enormous pain upon him without much risk. (Mière 1823: 100)

Clausewitz's people's war aimed to deny control of the land to an invader if the regular army had been defeated: 'Militia and bands of armed civilians cannot and should not be employed against the main enemy force – or indeed against any sizable enemy force. They are not supposed to pulverize the core but nibble at the shell and around the edges' (Clausewitz 1832/1976, VI: 26). According to his concepts, the militia (*Landsturm*) would wear the enemy down, but it would eventually be left to a re-formed regular army to bring about the decisive victory over the enemy forces (Clausewitz 1812/1966: 716–20). Chrzanowski's views were similar: his ideal scenario of a defensive war had partisans (referring now to all irregular fighters) fight alongside a regular army, if their country had not yet been completely occupied by enemy armed forces; the partisans would seek to fight in the enemy's rear or on his flanks. Chrzanowski assumed that the partisans would be organised in advance by the central government, from local levies. If the regular forces were defeated, the defence would become the task of the partisans. The longer a war dragged on, the better it would be for the partisans, and the more the enemy army would be weakened. Partisans must still avoid direct confrontation with the enemy's army. Instead, they should try to destroy their lines of communication, intercept high-ranking officers or officials travelling on their own and kill individual soldiers, which would force the enemy to go about only in big units. Little actions such as this, he thought, would finally wear down even the strongest army (Chrzanowski 1839/1846: 5f., 10–12).

Enrico Gentilini (1806–?) in his pamphlet on partisan warfare of 1848 also gave a well-rounded list of tasks of partisans or *stracorridori*:

> The principal objective of this warfare is to fall upon the enemy in those places which he cannot garrison with a considerable number of men; to torment him continually, tire him and deny supplies, without exposing oneself to serious risks ... The main aim is to surprise the enemy's posts, throw his camp into disarray, assault him at night in his tents, attack him from above whilst he is on the march, surprise his convoys, cut his communications, be the first to occupy those places which one is almost sure the enemy wishes to take; destroy his arms and powder workshops and other military establishments ... eliminate unsympathetic generals, suspect authorities, and hold to ransom that part of the population known to support the enemy, intercept the couriers, burn the magazines. (q.i. Ellis 1975: 82)

Garibaldi's model greatly resembled those of Le Mière de Corvey, Clausewitz and Chrzanowski, although there seems no evidence that he had read them. His prescriptions were extremely classical: don't allow the enemy to catch you, confuse him about your real intentions and with that purpose in mind, change camp sites, never move at the same time on successive days, march with as little baggage as possible, camp in well-hidden places, make a great show of setting off on main roads and then climb through bushes and over rocks to get to secondary roads, announce that you are going to X when in reality you set off for Y, make yourself entirely familiar with the terrain, observe the adversary. Garibaldi noted that the partisan leader had to take great care to look after his men, their food supplies, keep up their morale, recruit them with great discernment of their qualities. Garibaldi wanted more irregular troops – light infantrymen (*bersaglieri*) – for the main body of the Italian army, indeed, he wanted the whole army to be trained in the style of these light, easily deployable infantrymen. These could attack the enemy everywhere, masque the columns as they advanced and protect them against fire from the enemy's marksmen, allowing the columns to reach their aims and to spread out and approach the enemy. Each company in his view thus needed a unit of *bersaglieri*. Still, the main fighting in such a regular army would be done by the column of regular soldiers (Garibaldi 1848/1984, 1866/1886).

Garibaldi's actual fighters were all volunteers: he asked very much of his dedicated followers, but never forced anybody to join up. In 1849, when they had to give up the besieged city of Rimini to the French, Garibaldi wanted to withdraw to Rome and appealed to his soldiers to come along, saying, 'Soldiers! What I offer those who want to follow me is this: hunger, the cold, thirst, no money, no roof, no ammunitions, but constant skirmishes, forced marches.' The only reward he offered the patriotic volunteers was 'glory' (q.i. Heyries 1998: 237f.). About half of Garibaldi's followers in the Italian Wars of Independence were of middle-class origins, with university degrees and highly idealistic. Conservative Italians of all social backgrounds were afraid of their revolutionary potential. Later in life, however, Garibaldi became keener to fight classical wars with little units of well-trained soldiers, resorting to bands only when no regular soldiers were available (Heyries 1998: 239).

At the same time as Garibaldi, Auguste Blanqui in France developed concepts of urban guerrilla, in time for the Commune of 1871 to implement some of his ideas. Obviously, urban insurgents would not have horses, but again, Blanqui emphasised attrition and the sparing use of force (Blanqui 1868). T.E. Lawrence in turn picked up the baton of theoretical writing about how to stage an insurgency and guerrilla warfare on the back of his own experience in the First World War, rightly noting that insurgencies had never previously been written about in English from the point of view of the insurgents (Lawrence 1920/2005: 272). From his experience, Lawrence drew the following generalisations about the 'Science of Guerrilla Warfare':

> It must have a sophisticated alien enemy, in the form of a disciplined army of occupation too small to fulfil the doctrine of acreage: too few to adjust number to space, in order to dominate the whole area effectively from fortified posts. It must have a friendly population not actively friendly, but sympathetic to the point of not betraying rebel movements to the enemy. Rebellions can be made by 2 per cent active in a striking force, and 98 per cent passively sympathetic. The few active rebels must have the qualities of speed and endurance, ubiquity and independence of arteries of supply. They must have the technical equipment to destroy or paralyse the enemy's organised communications, for irregular war is fairly [Wilhelm von] Willisen's definition of strategy, 'the study of communication' in its

extreme degree, of attack where the enemy is not. In fifty words: granted mobility, security (in the form of denying targets to the enemy), time, and doctrine (the idea to convert every subject to friendliness), victory will rest with the insurgents, for the algebraic factors are in the end decisive, and against them perfections of means and spirit struggle quite in vain. (Lawrence 1920/2005: 273)

There are several aspects of guerrilla which Lawrence was the first to articulate. These include the use of speed, mobility and space in depth, rather than clinging to strong points, however great their symbolism might be (in his case, Medina); patience and time, rather than the frantic quest for a quick, decisive victory so characteristic of Strategy on all sides on the eve of the First World War.

Similar descriptions, emphasising the need for fluidity, surprise, ambushes and so on are also found in the communists' writings. Sun Tzu's recommendations are clearly reflected in the following passage of Mao's *On Guerrilla Warfare*:

In guerrilla warfare, select the tactic of seeming to come from the east and attacking from the west; avoid the solid, attack the hollow; attack; withdraw; deliver a lightning blow, seek a lightning decision. When guerrillas engage a stronger enemy, they withdraw when he advances; harass him when he stops; strike him when he is weary; pursue him when he withdraws. In guerrilla strategy, the enemy's rear, flanks, and other vulnerable spots are his vital points, and there he must be harassed, attacked, dispersed, exhausted and annihilated. (Mao 1937/2000: 46)

In Vietnam, Vo Nguyen Giap explained that the Communist Party had decreed the following Strategy: 'disperse and then regroup', 'mount ambushes', 'attack surprisingly' (Giap 1970a: 124). Mao famously compared guerrilla units

to innumerable gnats, which, by biting a giant both in front and in rear, ultimately exhaust him. They make themselves as unendurable as a group of cruel and hateful devils, and as they grow and attain gigantic proportions, they will find that their victim is not only exhausted but practically perishing. (Mao 1937/2000: 54)

Galula identified the force multiplier which made the gnats so powerful: in insurgencies and counterinsurgency warfare, every military action can have political effects that far outweigh its purely military implications; a tactical move can carry strategic weight (Galula 1964/2005: 5). Tactical mistakes can therefore be difficult to compensate.

Bases and base areas

Again it was Lawrence who first articulated what in practice had been long known: 'rebellion must have an unassailable base, something guarded not merely from attack' (Lawrence 1920/2005: 273). Lawrence's concepts of bases and the need for impenetrable, inviolable rear areas was taken up – or reinvented – by Mao (Mao 1937/2000: 52f.) and Giap. In the late 1940s, Mao's Red Army, now opposing the Nationalist Chinese forces, could again move from its hit-and-run tactics to the behaviour of a regular army, establishing itself firmly in 'base areas' which they began to defend against the regular-army enemy. The Red Army and the communist forces had in the meantime grown in strength. After the Second World War, the communists therefore changed their Strategy, and began to defend the area over which they held command against enemy attacks.

In Vietnam, the guerrilla fighters started by establishing the liberated region of Viet Bac, then guerrilla bases in other regions. Rear bases were established in the mountains, then in the plains and springboards for attacks on the enemy's neuralgic points. To complement the rear bases in rural areas, Giap formulated the need to create revolutionary bases in the cities, because workers were, according to classical Marxist-Leninist teaching, the most politicised population. Giap never ceased to emphasise the importance of consolidating the rear, which depended entirely on the support of the local populations (Giap 1970a: 67f., 73–5, 1970b, 42–55, 1971: 59–76). Fusing all these areas together, the Vietnamese communists could proceed to found the Democratic Republic of Vietnam. They attempted to transfer their dual approach of creating guerrilla-dominated zones and actual guerrilla-held rear bases to South Vietnam ('liberated regions': Giap 1971: 62f.). The most important 'great rear area' for the 'liberation' of South Vietnam was of course, in Giap's terminology, Vietnam's socialist North. Indeed the socialist camp as a whole was Vietnam's vast rear area. This had to be consolidated, and the enemy's areas of domination had to be ceaselessly attacked (Giap 1971: 71, 75).

A repeated pattern in such insurgencies was that the rebels could hold out indefinitely when they had safe bases in remote and inaccessible areas within their own country. As air power and visual intelligence from the air grew in precision, however, the alternative grew in importance, namely to have safe havens outside one's own country, politically inaccessible to the counterinsurgency forces. Thus in the Greek Civil War, the rebels found refuge in Yugoslavia, until their retreat was barred when Tito closed the border in 1949. Upon this they were bombed into submission by the USAF in the Grammos and Vitsi mountains. In both Afghanistan wars of 1979–89 and 2001–, the Taliban retreated across the border into Pakistan when under pressure from the Soviets and NATO forces respectively. Defence bases are thus a strategic dimension of insurgencies.

Annihilating the enemy?

Lawrence consciously developed guerrilla tactics against the prevailing fashion of the day, which he associated with Foch's definition of modern war as 'to seek for the enemy's army, his centre of power, and destroy it in battle' (Lawrence 1920/2005: 264). Since the anti-Napoleonic wars, this terminology had haunted also writings on insurgency. Le Mière de Corvey wrote:

> When undertaking a national war in order to preserve one's independence and to defend against a foreign invasion, one needs a war of extermination [sic], and the enemy must needs be expelled, or else the nation that defends itself has to be invaded, and the winner has the right to treat it as a conquered country. (Mière 1823: vii)

Lawrence by contrast first formulated the aim of pinning down, not killing (let alone *annihilating*) enemy forces, by letting them hold certain garrisons and harassing them and their supply lines in unexpected places, never confronting them in numbers. Lawrence's Arab allies were content, under his influence, not to prioritise killing Turks:

> If they would go quietly, our war would end. If not, we would try to drive them out: in the last resort we would be compelled to the desperate course of blood, on the maxim of 'murder' war, but as cheaply as possible for

ourselves, since the Arabs were fighting for freedom, a pleasure only to be tasted by a man alive. (Lawrence 1920/2005: 264)

After Lawrence, however, the old hot-blooded quest for annihilation of the enemy returned in the literature on guerrilla. For Giap, for example, the strategic aim would always be to win against larger numbers, better equipment, to *annihilate* as many enemies as possible (Giap 1970a: 62, 102). Che Guevara was convinced that 'war is always a struggle in which each contender tries to annihilate the other' (Che 1960/2003). The radical terminology he used seems to have exceeded his actions in practice; in his essay of 1966 which was read out at the Tricontinental Conference in Havana in the same year, he appealed to the communist rebels of the world to take the war into the homes of the enemy, to make war 'total' for him (Che 1968/2008: 173).

In reality insurgents can rarely hope to *annihilate* superior enemy forces, let alone wage 'Total War' according to the definition we have given in chapter 6. They can use terrorism, as was done so widely in the Algerian War and which saps the support of the population of a colonial power for its own colonial wars. However, terrorism, if practised against civilians, tends to alienate 'world opinion', and ultimately works against insurgents, unless the repression on the part of the counterinsurgents is even more spectacularly brutal.

Can guerrilla lead to victory? Phased guerrilla

Could partisans, could an insurgency win against regular forces by themselves? Lawrence had flirted with the notion that his Arabs might have achieved strategic victory by themselves, without external help (Lawrence 1920/2005: 272). This is a hotly debated subject. Can insurgents secure victory on their own, without regular forces engaged in major war bringing about a decision elsewhere, and if so, how?

August Rühle von Lilienstern, who himself had been a member of a partisan unit in the anti-Napoleonic wars, realised 'that war is *generally* composed of both forms of war ... and the operations and endeavours that occur in campaigns partly belong to the realm of small war, partly to that of major war' (Rühle 1818: 1). Partisans before the French Revolution were always auxiliaries of larger, conventional forces. The Spanish War of Independence against Napoleon was brought to an end by the victory of Wellington's regular armies. Enrico Gentilini emphasised the complementarity between partisans and regular forces. The partisans' missions included keeping intact their own regular army's communications, 'to watch over the enemy so that he does not take us [our own army] by surprise, and finally to reconnoitre ... his positions'. He did not think partisan detachments could aim to do more than achieve 'minor successes' (q.i. Ellis 1975: 82). As we have seen, Le Mière de Corvey and the other writers of the early nineteenth century agreed with him.

When Lawrence was sent out to help Arabs organise resistance against the Ottoman Empire in the First World War, the Arab forces on their own could not have defeated the Ottoman Turks; the big decisions in the First World War were brought about by the clashes of the regular armies elsewhere. Notwithstanding their leaders' optimism about them, Lawrence's Arabs and Tito's Partisans would hardly have prevailed in the two World Wars respectively if it had not been for the enemy's defeat by regular forces. The Yugoslavs liked to claim that they had liberated themselves, which the Soviets countered with the argument that the Balkans would not have been freed without the Red Army's overall victory over the Wehrmacht and the Soviet seizure of Berlin. A similar pattern applies even in the Cold War, even though the great East–West Third World War famously did not take place. Arguably the French defeat in Algeria and that of the French and the USA in Indochina were due to the larger context of the Cold War, which prevented both from lashing out as hard as they might have done, for fear of escalation to global – and even nuclear – war. As Galula put it, in a context of global ideological competition, 'revolutionary war' always had an international dimension (Galula 1964/2005: 3).

Callwell had paved the way to a different take on this problem by conceiving of different 'stages' or phases in the conquest of an underdeveloped country:

> In the first stage the [superior, invading forces] overthrow the armies and levies which the rulers and chieftains in the invaded country gather for its defence ...; in the second stage organized resistance has ceased, and is replaced by the war of ambushes and surprises, of murdered straggler and of stern reprisals ...

that is, the insurgency phase (Callwell 1896/1906: 26).

The concept of phases could also be used to explain the swelling of an insurgent movement to the point where it could turn into an army and defeat the regular forces. In the 1930s and 1940s, people's war, plus the communist ingredient, became a potent mix as people's revolutionary war in the hands of Mao Tse-tung in China and the two Indochinese Marxists Giap and Truong Chinh. In both cases, there was a foreign occupying force in the country, Japan during the Second World War and in Indochina the French, who tried to re-establish themselves as colonial powers after the defeat of Japan. But in both countries, the communist insurgency ended up being directed just as much against indigenous powers, in the case of China, Chiang Kai-shek's Nationalists, in the case of Indochina the Emperor Bao Dai, and later the Diem regime backed by the USA. Writing in 1937, after the Japanese had brutally occupied all of China, Mao explained that: 'ours is the resistance of a semicolonial country against imperialism'.

The ideological problem with Mao's revolutionary war was that China fulfilled the revolutionary model of Karl Marx even less than did the Russian Revolution. China in the 1920s, 1930s and 1940s consisted mainly of an agricultural economy, and not at all an industrialised society in which proletarian masses were impatient to overthrow a capitalist bourgeoisie. Moreover, as Mao noted, the rivals of the communist forces, under the Kuomintang's and Chiang Kai-shek's leadership, was strong: the Chinese Nationalists themselves exploited the potential for anti-colonialist sentiments in the population. The Red Army in turn was initially weak confronted with the regular forces of the Nationalist government.

According to classical Marxist-Leninist teaching, the country was thus in no way ripe for revolution. Moreover, the precedent of the Russian revolutionary war could not be applied in all details. But could there be two different models of revolutionary warfare? This in itself posed a serious problem for communist orthodoxy. Mao, however, decided that this doctrinal problem could be overcome, and devised his own doctrine for communist warfare. Mao defined all guerrilla as war 'fought by every class of men against invaders and oppressors' (Mao 1937/2000: 42, 58). He codified his own views of how to conduct an insurgency by describing seven progressive steps through which the movement had to pass in fighting against a foreign occupier:

1. Arousing and organizing the people.
2. Achieving internal unification politically.
3. Establishing bases.
4. Equipping forces.
5. Recovering national strength.
6. Destroying enemy's national strength.
7. Regaining lost territories.

The policy or Strategy was one of creating a united anti-Japanese front, the goal to achieve the complete emancipation of the Chinese people from foreign oppression, which in itself was seen as revolutionary by Mao. The support for this goal by the mass of the Chinese population was seen by him as critical (Mao 1937/2000: 42f.). In keeping with most European writers before him, Mao wrote:

There is in guerrilla warfare no such thing as a decisive battle; there is nothing comparable to the fixed, passive defence that characterizes orthodox war. In guerrilla warfare, the transformation of a moving situation into a positional defensive situation never arises. The general features of reconnaissance, partial deployment, general deployment, and development of the attack that are usual in mobile warfare are not common in guerrilla war ... When we discuss the terms 'front' and 'rear' it must be remembered, that while guerrillas do have bases, their primary field of activity is in the enemy's rear areas. They themselves have no rear.

These guerrilla fighters' tasks were

to exterminate small forces of the enemy; to harass and weaken large forces; to attack enemy lines of communications; to establish bases capable of supporting independent operations in the enemy's rear, to force the enemy to disperse his strength; and to co-ordinate all these activities with those of the regular armies on distant battle fronts. (Mao 1937/2000: 52f.)

Mao did not prescribe a one-directional progress of revolutionary guerrilla warfare from one step to the next, but saw the possibility for the need to regress to an earlier stage, if circumstances required: 'guerrilla units formed from the people may gradually develop into regular units and, when operating as such, employ the tactics of orthodox mobile war'. Indeed, he even recognised that 'orthodox armies may, due to changes in the situation, temporarily function as guerrilla [units]' (Mao 1937/2000: 54).

Writing a year later on the need to prepare for 'Protracted War', Mao no longer spoke of seven steps but of three stages of revolutionary warfare. He described China as being engaged in a first stage of resistance against the foreign occupier.

> In this stage the form of fighting we should adopt is primarily mobile warfare, supplemented by guerrilla and positional warfare ... At the tail end of the first stage, the enemy will be forced to fix certain terminal points to his strategic offensive owing to his shortage of troops and our firm resistance, and upon reaching them he will stop his strategic offensive and enter the stage of safeguarding his occupied areas.

The crucial points of the first stage or phase of the Chinese revolutionary warfare, when the Red Army was weak and the enemy strong, was not to insist unduly on defending land at all cost. The Red Army in this first stage rejected any position warfare – fundamentally unfavourable to them – in favour of a flexible war of movement.

The second stage Mao defined as 'one of strategic stalemate':

> [T]he enemy will attempt to safeguard the occupied areas and to make them his own by the fraudulent method of setting up puppet governments ... Taking advantage of the fact that the enemy's rear is unguarded, our guerrilla warfare will develop extensively in the first stage, and many base areas will be established, seriously threatening the enemy's consolidation of the occupied areas, and so in the second stage there will still be widespread fighting. In this stage, our form of fighting will be primarily guerrilla warfare, supplemented by mobile warfare.

He anticipated that the Chinese would still 'retain a large regular army', but that they would not be able to launch a strategic counter-offensive at this stage, because the enemy – the Japanese, later the Nationalists – would in all likelihood 'adopt a strategically defensive position in the big cities and along the main lines of communication under [their] occupation' while the Chinese would have inferior equipment to theirs. At this stage, some Chinese communist forces would have to be deployed along the front line, but most would be dispersed behind the enemy's rear. From there they could launch a guerrilla offensive against areas occupied by the enemy. Mao envisaged that the fighting would be fierce and that 'the country will suffer serious devastation'. If the struggle went well, the guerrilla forces might hope to hold two-thirds of the territory, and even the final third remaining in enemy hands would be in part contested by both sides. Mao anticipated that this second stage would last quite a long time, something for which one had to prepare psychologically. Finally the third stage would see the great counteroffensive to recover the territories one had lost.

> In the third stage, our war will no longer be one of strategic defensive, but will turn into a strategic counter-offensive manifesting itself in strategic offensives; and it will no longer be fought on strategically interior lines, but will shift gradually to strategically exterior lines ... The third stage will be the last in the protracted war ... Our primary form of fighting will still be mobile warfare, but positional warfare will rise to importance. While positional defence cannot be regarded as important in the first stage because of the prevailing circumstances, positional attack will become quite important in the third stage ... In the third stage guerrilla warfare will again provide strategic support by supplementing mobile and positional warfare, but it will not be the primary form as in the second stage. (Mao 1938)

Mao's concept of stages or phases would greatly impress Western counterinsurgency writers from Galula (1964/2005: 29ff.) and Roger Trinquier (1961/2006 and 1968) to the authors of Field Manual 3-24 of 2006 (US Army and US Marine Corps 2006). Not all insurgents, however, plan their war on this pattern. Mao himself acknowledged this: he always went along with Clausewitz's (and, taken from Clausewitz, Lenin's) tenet that the wars of all eras were distinct and that no single recipe for their conduct could be prescribed for all of them (Mao 1937/2000: 49). Rather than involving himself in the debate between Stalin on the one hand and Tito and his Greek followers on the other in 1948–9 on how to prosecute the Greek Civil War, he sagely stuck to his line.

Mao also developed a concept of different speeds of operation which were designed to unbalance a stronger enemy. Strategically, his Red Army was clearly weaker than the Nationalist forces to begin with. Therefore, strategic encounters had to be ruled out, the large, decisive battle had to be avoided. Where the enemy was keen to stage a major encounter, the revolutionary forces had to seek to delay it as much as possible: they had to be patient, seeking to change the time scale on a strategic level to one which would be to their advantage. On an operational and tactical level, however, they needed to transform the strategic superiority of the enemy into an operational and tactical inferiority. Here, quick, short operations were more advantageous to the enemy, so the aim of the communist forces was to harass him, and withdraw quickly, before the cumbersome large forces of the enemy could react properly to a surprise attack and could beat the communist aggressor by virtue of the regular forces' superior size and strength. The communist insurgents had the advantage, as in the First World War, the Arabs against the Turks, of choosing the time and place of their harassing attack: the regular army of the opponent thus had to be on its guard at all times, which sapped morale and created a permanent state of insecurity and tension.

The Vietnamese were the other practitioners and thinkers who contributed to the theoretical literature on how to conduct guerrilla, mainly through the doctrines of Ho Chi Minh (1890–1969) and Giap, who rose to become Minister of National Defence of the Democratic Republic of Vietnam, Vice Prime Minister and the commander-in-chief of the armed forces of North Vietnam. Unlike Mao, Giap did not believe in phases and saw guerrilla and regular war as going hand in hand, being carried on simultaneously and fluctuating from one to the other. He spoke of guerrilla and regular war as being 'enmeshed' like hair in a comb. As appropriate, one should form clandestine forces, groups of *guerrilleros* or larger guerrilla detachments (Giap 1970a: 120, 122, 125).

Che Guevara acknowledged that there were two different forms of guerrilla: one complementary to the operations of a big regular army (he cited Ukrainian partisans in the Second World War, fighting against the Germans, as an example), the other 'the case of an armed group engaged in struggle against the constituted power, whether colonial or not, which establishes itself as the only base and which builds itself up in rural areas'. It was this second sort that he was concerned with (Che 1960/2003: 11). He drew the lessons from the Cuban Revolution that the second form of insurgency, consisting of 'popular forces', could actually win a war against regular armed forces. However, he roundly dismissed the possibility of a victory brought about by guerrilla tactics alone, while adopting Mao's concept of phases: the last would simply not be guerrilla, but regular warfare (Che 1960/2003: 7–9, 13). The guerrilla, following Mao's concept of phases, had to gather strength. Initially, it would be the fight of the *guerrilleros*, the 'armed nucleus, the fighting vanguard of the people'. But as the whole guerrilla was being fought to promote the interests of the people, the ranks of the *guerrilleros* would swell, until, ultimately, they could defeat the opposing regular army (Che 1960/2003: 10). Che Guevara conceived of the first phase as that in which the *guerrilleros* would harass the enemy, while concentrating on the avoidance of destruction by the enemy, operating from impenetrable redoubts and bases. Only in a second phase, when they had grown considerably in number, could they begin campaigns of attrition against the enemy. They would now strike at the enemy ceaselessly, not letting his soldiers sleep, attacking his lines of communications. For this the *guerrilleros* needed total support of the population, who had to be convinced that the partisans were invincible. In a final phase, the enemy's central bases could be attacked, and the enemy could be completely destroyed (Che 1960/2003: 14–16).

Che Guevara had the same problem as Mao, in that the Latin countries for which he was envisaging insurgencies did not comply with the Marxist-Leninist ideal of revolution arising in the cities of advanced industrial societies. Cuba had convinced him that 'In underdeveloped America', as in Russia and China, 'the countryside is the basic area for armed fighting.' But he also elaborated on concepts of urban guerrilla, probably as the first since Blanqui (Che 1960/2003: 8: 35ff.). If society as a whole was not yet ripe for a revolution, he argued that the *guerrilleros* could form the hearths (*focos*) of a movement, which could spread its fire over the entire country at any time (Che 1963/2003: 142–58). The practice did not bear out this hope, but insurgencies of one form or the other continue to characterise the political landscape of Latin America. Interestingly, however, it shows the possibility even for a Marxist of rethinking the concept of phases; despite US counterinsurgents' obsession with a three-phase guerrilla, it should not be taken for granted that all insurgents conceive of their enterprise in this way.

Force structures

While Lawrence's fighters were all volunteers, the Vietnamese after the Second World War resorted to as extensive a mobilisation of the population as possible. Giap, educated at the French grammar school at Hue and at the University of Hanoi with its French ethos, was clearly disingenuous when claiming that the motto he adopted – 'each habitant is a soldier' – was an 'old saying' rather than a direct derivative of the French Revolutionary ideal that 'each citizen is a soldier' (Giap 1971: 38f.). Rather than referring to eighteenth-century French antecedents (see chapter 6), Giap preferred to refer to the Lenin quotation: 'To conduct war, it is necessary to mobilize all people's forces, to turn the entire country into a revolutionary bastion, to devote everything to the war, and to use all forces and national resources to protect the revolution.' Giap extrapolated from this that 'The entire country has fought the aggressor under the vanguard Party's leadership'; 'even the women must fight' (Giap 1970b, 14, 26, 1971: 38f.). In December 1946, Ho Chi Minh proclaimed:

> Men and women, old and young, regardless of religious creed, political affiliation and nationality, all Vietnamese must stand up to fight the French colonialists to save the Fatherland! Those who have rifles will use their rifles, those who have swords will use their swords! Those who have no swords will use spades, hoes or sticks! (q.i. Giap 1971: 100)

Giap stressed the importance of loyal cadres, and the unity of the army and the people, among whom the fighters must be able to move freely (Giap 1970a: 52).

For Giap, there were three categories of forces: regular forces, regional forces and the popular militia.

- *The militia*: guerrilla forces, self-defence forces, organised village by village, factory, street, fighting with prudence or sophistication. Giap declared that being engaged in production also counted as fighting.
- *Regional forces*: these provided the hard core of defence of the region. They should work closely with the *guerrilleros*, partisans and regular troops. Their tasks: to annihilate the enemy, to promote the guerrilla, to defend the population.
- Finally *regular forces*: these should be mobile, fight where needed and where possible. They should include navy and air force. Their aim should be to annihilate [*sic*] the enemy forces (Giap 1970a: 53f.).

Che Guevara similarly stressed the identity of interests between *guerrilleros* and the oppressed classes for whom they were fighting and, consequently, the lack of firm boundaries between them – as noted, he saw the fighters merely as the vanguard of the army that would by and by constitute itself (Che 1963/2003: 142–58).

Guerrilleros: Jesuits or bandits?

Indeed, a rigid categorisation and division between crack forces, main body of fighters and non-combatants is antithetical to an insurgency. Standard operational procedures, drill and automated masses make an adversary predictable. All of this is alien to irregular forces. Already de la Croix and his son had noted that one could not carry out partisan warfare with regular soldiers. The right men for the job had a tendency towards insubordination, and they would never fight in quite the same way as ordinary disciplined soldiers (Croix 1752: 44f.)

This lack of mechanical obedience and discipline, however, was at the same time an advantage: Clausewitz in his lectures on partisan warfare had stressed the need for exceptionally intelligent and flexible officers to lead partisan detachments (Clausewitz 1810–11/1966). Callwell, too, had recognised this:

> The more irregular and the less organized the forces of the [insurgents] are, the more independent do they become of strategical rules. An army which disperses if it is beaten cannot be treated in the same way as an army which under such circumstances retreats in as compact and regular formation as the case admits of towards its base. (Callwell 1896/1906: 52)

Lawrence underscored this point even more, fully understanding the advantages of military 'indiscipline', ad-hocery and initiative: this made for subtleness: 'Guerrillas must be allowed liberal work-room … Guerrilla war is far more intellectual than a bayonet charge.' Lack of drill and robot-like discipline, lack of robot-like behaviour all contributed to the incalculability of the guerrilla operation, and were actually assets (Lawrence 1929/2005: 283).

Others were less confident that the *guerrilleros*' free spirit and tendency towards taking the initiative, often bordering on insubordination, was a good thing. From the Spanish War of Independence itself, the juntas were concerned about how to reintegrate the partisan units into their regular armies and struggled to keep control over them. This was the leitmotif of Jose Maria Carvajal's regulations for partisans published on 11 June 1812, when the war against Napoleon was barely won (Carvajal 1812). Karl Marx, in studying the Spanish *Guerrilla*, had commented:

> As to the guerrillas, it is evident that, having for some years figured upon the theatre of sanguinary contests, taken to roving habits, freely indulged all of their passions of hatred, revenge, and love of plunder, they must, in times of peace, form a most dangerous mob, always ready at a nod in the name of any party or principle, to step forward for him who is able to give them good pay or to afford them a pretext for plundering excursions. (q.i. Gerber 1988: 156)

The Prussian establishment, like the Spanish government, was very wary of the Schills and Lützows and other self-appointed freedom fighters in the anti-Napoleonic wars. Even those for whom the *guerrilleros* were fighting might by Conservative inclination turn against them at the smallest transgression, let alone when noble social revolutionaries lapsed into bandit ways. The communist writers on guerrilla therefore never ceased to stress the need for their fighters to behave well and to avoid giving credence to the enemy's inevitable accusations that they were merely bandits. Playing on the popular connotations of intelligence, sly machinations, but purity of spirit, personal lifestyle and intention, Che Guevara famously described the *guerrillero* as 'the Jesuit of war' (Che 1960/2003: 12). Only by living thus could they gain the loyal support of the population for whose good they were fighting.

Hearts and minds I

The need to operate within a benign population had already been commented upon by Emmerich and Ewald, and was standard fare even for the irregular partisans of the eighteenth century. From the Spanish War of Independence, guerrilla forces were fighting for what they saw as the population's cause. The need to treat the population for whom one was after all fighting well was thus spelled out over and over, from Garibaldi to the communist writers and practitioners of the twentieth century.

Like the early modern writers dealing with regular war before them (see chapters 2–4), writers on guerrilla emphasised the need to treat local populations well to secure their support. Like Davidov seven years before him, Gentilini urged good behaviour vis-à-vis locals:

> It must be made a very serious offence for any [partisan] to enter their homes and disturb the peace of the occupants, show rudeness or shamelessly steal anything ... By committing such an infamous act you will bring down upon yourselves the hatred of the inhabitants, and you will have many more enemies to contend with; and as well as being unable to carry out your own plans, you will put yourself into the very difficulties which should afflict the enemy. (q.i. Ellis 1975: 82)

In 1870 Garibaldi wrote instructions for the volunteers of the army of the Vosges, which he wanted to be the nucleus of a national army, and his objective was to transform this guerrilla into a popular insurrection. About the relationship between *guerrilleros* and population, he noted: 'The volunteers and partisans have to make themselves respected and loved by the population at all cost.' Mutual respect he saw as important everywhere, also within the armed forces (q.i. Heyries 1998: 242f.).

Mao defined the people as central to the guerrilla. Only with the full support of the population could a partisan move among it like a 'fish in water', without being caught (Mao 1937/2000: 93). Accordingly, his army had to put into practice 'Three Rules' and 'Eight Remarks' which proscribed stealing, selfish or unjust behaviour, dishonesty, immodest and inconsiderate behaviour (Mao 1937: 92). The struggle for the support of the population is also central to Ho Chi Minh's very similar 'Twelve Recommendations' of 5 April 1948, which added the prohibition of forced requisitions, perjury and contemptuous behaviour to the list, and urged fighters 'not to give offence to people's faith and customs (such as to lie down before the altar, to raise feet over the hearth, to play music in the house, etc.)', and as the basis of this, 'to study the customs of each region so as to be acquainted with them in order to create an atmosphere of sympathy, then gradually to explain to the people to abate their superstitions', to 'help the people in their daily work (harvesting, fetching firewood, carrying wood, sewing, etc.)', to educate them, and in turn to behave in a self-disciplined and respectful way (q.i. Giap 1974: vf.).

Che Guevara similarly described many ways in which the *guerrilleros* should gain the confidence and support of the peasants. An important part of this was to begin the social revolution – usually turning on the redistribution of land – even while the fighting was still going on. In terms reminiscent of writings of the sixteenth to the eighteenth centuries, the former medical student admonished his fighters to show mercy to wounded enemies, up to and including the administration of first aid (Che 1960/2003: 25, 38–42). This contradicts his emphasis on the need to 'annihilate' the enemy's forces. The reality, however, both in Algeria and in Indochina/Vietnam, was one of brutal terrorisation of the population by both belligerent sides to secure their co-operation.

Brutality is a feature of insurgencies, especially if it is directed against an 'other' population perceived as 'colonisers', oppressors, ethnically or religiously different groups, whether foreign or locally born. No attempts are then usually made to win this 'other' population's hearts and minds, but terrorist attacks are particularly common here – witness Tamil Tiger terrorism practised in particular against the Singhalese in Sri Lanka from about 1980 to their final – equally brutal – repression in 2009, or the Palestinian suicide bombers in Israel in the early 2000s. As we shall see, 'hearts-and-minds' campaigns on both sides presuppose a perception of the populations thus wooed as 'part of us', and not as an intrinsic part of the repressive enemy group.

Defence in depth

Guerrilla Strategy could also be employed to defend a weak country against large-scale aggression. The earliest example of a full development of this subject, and one without direct influence on later Western traditions, can be found in East Roman (Byzantine) Emperor Nikephoros II Phokas's *Peri Paradrómes*.

Nikephoros' prescriptions placed heavy emphasis on intelligence – from the deployment of frontier guards who were not expected to be able to fend off the invaders, but to summon help, to the constant tracking of Arab raiding parties once they had penetrated into Byzantine territory. Apart from that, his book mainly prescribed what would later be regarded as irregular warfare against the intruders: ambushes, surprise attacks, bottling up the enemy forces in narrow passages when they are trying to get out of the Byzantine realm, attacks on their rearguard and baggage train, keeping as closely in touch with them as possible while trying to elude enemy surprise attacks by going as far as changing one's own camp site several times during the night. This is a rare example of what one might call *symmetric* irregular warfare, even though both sides used what to them, at the time, were regular soldiers: from the eighth to tenth centuries, both the Arabs and the Byzantines were clearly using what would later be thought of as partisan warfare (Nikephoros mid-950s).

Analogous defensive concepts for states with weak armies were reinvented in the light of the heritage of the Spanish *Guerrilla*, and the Russian overall Strategy against the French in 1812, revolving around popular defence of a country in all its depth, to make occupation difficult for any invader. We have mentioned this above for the writings of Clausewitz, Le Mière de Corvey and Chrzanowski. We have seen in chapter 4 that Wilhelm von Willisen also contributed to the literature about defence in depth. The defensive aspect of this concept – a *levée en masse* to defend one's homeland against an invasion – was fully developed by Jean Jaurès for France as an alternative to the prevailing creed of the *offensive à outrance* (see chapter 7).

Similar concepts were developed by Sweden and Switzerland in the twentieth century, where the adult male population would, after a shorter period of active military service and training, remain a militia or territorial army, ready to organise themselves to form thousands of small resistance units to deny an occupying army full control of the country. Elements of this were present in the self-constituted resistance movements of the Second World War that formed against German and Italian occupation from the Channel to Russia. The Yugoslav and Greek partisans could rightly pride themselves, perhaps not on having defeated the Germans, but on having tied down many German divisions, thus weakening them in relation to the Red Army and the Western Allies (Roberts 1976).

A general mobilisation of the population through the Home Guard to resist a possible invasion of the British homeland during the Second World War was the subject of the writing of Tom Wintringham, a former commander of the British battalion in the International Brigade in the Spanish Civil War. He was authorised by the British government to teach Local Defence Volunteers 'ungentlemanly warfare'. One of his works, *People's War* (1942), lays out his ideas about active resistance to potential occupation. But by June 1942 the British government ruled out any guerrilla activities for the Home Guard;

it seems that Wintringham's revolutionary ideas about homeland defence were matched, in the eyes of British authorities, by all-too revolutionary ambitions for British society, and again we see the pattern of suspicions about unconventional warfare that might not be sufficiently controllable by the powers that be (Purcell 2004).

The NATO member states were not blind to the possibilities that might be offered by such concepts of total defence. In concrete terms, NATO itself planned 'stay-behind' operations, of which unfairly perhaps the most famous is GLADIO, for the formation of resistance movements in many forms and on many levels, should a Third World War like its predecessor have led to partial surrender and the evacuation of Western and Southern Europe by US and British forces (Nuti and Riste 2007). This also contextualises C.N.M. Blair's study of resistance movements, written in the 1950s. Since then, however, the NATO powers have lost interest in this option. Alternative defence options to NATO's nuclear-cum-conventional deterrence strategies were put forward in several countries, relying heavily on concepts derived from guerrilla. They would not, however, become official policy in any NATO member state (see chapter 17). There is a nexus between this and the legal problems facing irregular fighters even today, as we shall see presently.

16 *Counterinsurgency*

As a general rule the quelling of rebellion in distant colonies means protracted, thankless, invertebrate war ... It is a singular feature of small wars that from the point of view of strategy the regular forces are upon the whole at a distinct disadvantage as compared to their antagonists.

(Callwell 1896/1906: 27, 85)

This is not the sort of struggle where you take a hill, plant the flag and go home with a victory parade ... It's not war with a simple slogan.

(General David Petraeus, 11 September 2008[1])

The legal status of insurgents

Lucifer, the rebel angel, was expelled from Heaven by an angry God, and throughout its history, Christianity has treated rebellion against legitimate authority not only as an offence but as a capital crime. Insurgents threaten the very power structure of society. This is why medieval monarchs resorted to such brutal repression, and why insurgents were customarily treated so much more harshly if captured than regular soldiers. Grotius took great pains to emphasise that subjects had to accept injury inflicted upon them by rulers for the sake of not bringing instability to the society as a whole. Yet he argued that a prince issuing a command directly contrary to the law of God could be disobeyed, likewise a prince encroaching upon the rights of other parts of the government, such as a parliament ('Senate'), or the territory of another (Grotius 1625, I: 4). This could be seen as a legitimisation of the 'people's war' of resistance against an invasion force,

[1] http://news.bbc.co.uk/1/hi/world/middle-east/7610405.stm, accessed 13 September 2008.

once a regular army is defeated by the invaders, but had no lasting impact on international law.

Few civilisations, confronted with insurgents rebelling against what was perceived on the other side as legitimate authority, took them prisoner; most killed them out of hand. An attempt to change this was made in the American Civil War, where both sides had irregulars fighting for them. In 1863, Order 100 or the 'Lieber Code' was created among other things to stop summary executions of partisans upon their arrest. Henceforth, there was – in theory – to be no legal distinction between regular and irregular soldiers, and irregulars were to be given the same protection from martial law as regulars – including the right to a military trial (Beckett 2001: 30). But no international consensus on this point could be found, not least as by definition, irregulars also tended to break the *ius in bello* themselves: generally, they would not wear recognisable uniforms or other outward signs of being combatants, precisely so that they could pass as civilians and be able to hide more easily among the civilian population. Central to insurgencies is their illegality or semi-legality. The Hague Convention of 1899 struggled with this issue, but without a satisfactory result. So guerrilla fighters continued to be hunted down and, more often than not, killed immediately or after tortuous interrogation.

Until well into the second half of the twentieth century, international law posited that the population of a state occupied by (foreign) military forces after their victory in battle had a duty to obey the occupying powers, which by virtue of their victory became the sole legitimate authority in the state. This meant that anybody resisting the occupants was considered as an outlaw, and consequently was not protected by the laws of war which protect regular combatants: the latter were to be taken prisoner, not summarily executed, while irregular fighters had no such protection. The Hague Convention of 1907 allowed the recognition of militias and volunteer corps as protected by the 'laws, rights, and duties of war' if they:

1. were commanded by a person responsible for his subordinates;
2. had a fixed distinctive emblem recognizable at a distance;
3. carried arms openly; and
4. conducted their operations in accordance with the laws and customs of war.

The laudable aims here were clearly to protect civilians against reprisals aimed at combatants. But while militias and volunteer corps since the nineteenth century at the latest generally were distinctive armies and carried arms openly, this does not apply to partisans or insurgents, whose way of war turns on stealth, surprise and on hiding rather than facing an open confrontation – and this is done usually by blending into the non-combatant population. International law was formulated both to protect combatants after their surrender or capture and to protect civilians against the use of force aimed at combatants who were still fighting, which in turn is only possible if the differences between combatants and non-combatants are blatant. The Hague Convention of 1907 thus does not cover clandestine fighters or insurgents. This dilemma was recognised at the time, which is why a vague preamble was agreed upon with regard to 'cases not included in the Regulations [of the Hague Convention], the inhabitants and the belligerents remain under the protection and the rule of the principles of the law of nations, as they result from the usages established among civilized peoples, from the laws of humanity, and the dictates of the public conscience' (t.i. Roberts and Guelff 2000: 70–81). The Geneva Convention of 1949, while introducing many improvements in the protection of non-combatants and prisoners of war (that is, captured regular combatants), did little to enhance the rights of clandestine fighters or insurgents. While in principle the rights and privileges of protection of life are due even to spies and saboteurs, where granting such rights and privileges would be 'prejudicial to the security' of the state against whom the espionage or sabotage is directed, the rights and privileges can be ignored (Convention IV, Article 5: Roberts and Guelff 2000: 303).

Article 37 of the 1977 Additional Protocol I to the Geneva Conventions (to which crucial states, such as the USA and Israel, are not signatories) expressly prohibits 'perfidy', including 'the feigning of civilian, non-combatant status', and ruses such as 'the use of camouflage, decoys, mock operations and misinformation', all standard tools of guerrilla (Roberts and Guelff 2000: 442). Attempts were made to widen the application of the protection of the laws of war, to include armed forces responsible to 'a government or an authority not recognised by an adverse Party', paramilitaries or other 'persons who have taken part in hostilities' who claim the status of prisoners of war. By contrast, a spy – captured while engaging in espionage – has no protection, nor a mercenary.

The problem with clandestine operations was again tackled:

In order to promote the protection of the civilian population from the effects of hostilities, combatants are obliged to distinguish themselves from the civilian population while they are engaged in an attack or in a military operation preparatory to an attack. Recognizing, however, that there are situations in armed conflicts where, owing to the nature of the hostilities an armed combatant cannot so distinguish himself, he shall retain his status of combatant, provided that, in such situations, he carries his arms openly: (a) during each military engagement, and (b) during such time as he is visible to the adversary while he is engaged in a military deployment preceding the launching of an attack in which he is to participate. (Geneva Protocol I, 1977, Articles 43–7: Roberts and Guelff 2000: 444–7)

The great change introduced with the 1977 Additional Protocol I is thus that insurgents or factions in a civil war – if acting under and answerable to some higher, albeit self-proclaimed authority – can receive the same protection as combatants, even if the authority leading them is not recognised by the adversary. It is thus not until the late 1970s that international law accorded some protection to insurgents, provided the insurgency is recognised as an armed conflict (and provided the insurgents themselves operated within the above constraints).

What of the application of this protocol? Even at the outset of the twenty-first century, insurgents can still in practice be categorised as criminals by the occupants, and treated accordingly (for example, tried for murder rather than for a war crime, the latter of course excluding the open killing of an enemy combatant). The equation of insurgents with outlaws by their opponents thus enjoys not only great continuity and continuing popularity, but is also the underlying attitude explaining the particularly repressive treatment inflicted on captured insurgents, and even on non-combatants supporting them or merely not betraying them to the occupying power.

Brutal repression

There are many examples of counterinsurgency operations (COIN) in history, and they obviously go as far back as examples of insurgencies. The Romans repressed uprisings against them mercilessly, from the destruction of the town of Numantia on the Iberic Peninsula (153

BCE) to the suppression of the revolts of Vercingetorix by Caesar and of Boudicca by the early empire. These revolts would in the early nineteenth century be commemorated by the new nationalists of all these countries as shining examples for them to be emulated and avenged in their opposition against Napoleon.

British and Irish history in turn is particularly rich in terrible punishments wrought by the English upon populations for harbouring insurgents, matched by the medieval Scots and the Welsh, by their destructive raids on English border regions. The entire Eighty Years Dutch War of Independence against Spain was such an insurgency, and Spanish attempts to quell it with their cruelty fed the great Black Legend (*leyenda negra*) of Spanish brutality.

Extremes of cruelty were reached in the French Republic's treatment of the rebels of the Vendée. It consisted not only of a scorched-earth policy, but has been called genocide by some, as rebel villages were not only razed to the ground but their inhabitants were slaughtered indiscriminately, much the way burghers of towns and cities were butchered from antiquity to the early modern period if they refused to surrender to a besieging army. It is thought that about 150,000 Vendéeans were killed (Secher 2003).

The Vendée, in turn, provides crucial evidence for those who argue that Europeans actually exported to their colonies their home-grown penchant for massacring civilians as a counterinsurgency measure (Wagner 1995). Alexis de Tocqueville wrote in 1841: 'I have brought back from Africa the terrible impression that currently we are waging war in a far more barbarous way than the Arabs themselves' (q.i. Frémeaux 2003: 10f.). The commanding general in Algeria, Duvivier, in fighting the insurgents under Abd el-Kader, summed up French measures: 'for the past eleven years, we have knocked down the buildings, burnt the harvests, destroyed the trees, massacred men, women and children, with an ever growing fury'. In French colonial policies in northern Africa, pillage and the ever-present threat of atrocities were aimed at terrorising the tribes not only as a punishment for rebellion, but also to deter them from resorting to any form of insubordination (q.i. Frémeaux 2003: 11f.). Clearly, a form of 'state terrorism' was being practised. It is therefore not surprising to read Callwell praising the conduct of French General Lazare Hoche (1768–97) in the Vendée as a model to be emulated (Callwell 1896/1906: 41).

One of the most infamous examples of this practice are the orders given by the Wehrmacht Field Marshal Wilhelm Keitel (1882–1946) in September 1941 to deal with insurgents:

> We must undertake without delay the most rigorous measures to affirm the authority of the occupying power and to avoid an extension of these attacks. We must never lose sight of the fact that, in occupied countries, a human life is worth less than nothing and that intimidation is only possible through extraordinarily harsh measures. When taking reprisals for the death of a German soldier, the execution of 50 to 100 Communists is essential ... The method of execution should reinforce yet more the impact of the punishment. (q.i. Ellis 1975: 149)

The Wehrmacht also burned down villages as reprisals, and in an inverted variation of Greek and Roman practice, slaughtered women, children and old people, taking off the men as slave labourers (Jungfer and Meyer 1997; Arnold 2005). German commanders in the East expressly decided that the justifications for resistance which the Hague Convention did accord to people who had not had the time to organise a classic defence were to be ignored (HIS 1996: 138). Keitel was sentenced to death by the Nuremberg tribunal as a major war criminal.

Having lost Indochina, in the subsequent fight to keep Algeria, the French responded with brutality: they resorted to torture and swift executions of suspects (Trinquier 1961/2006: 20). During the Indochina War, in the French school for counterinsurgency in northern Indochina, there was an inscription on a wall: 'Never forget – the enemy does not fight this war in accordance with French Army Rules' (Fall 1965: 13).

It has become a well-established tradition to contrast this with the success of the British in Malaya, where the British managed to isolate the insurgents from the passive population, protect the latter from the insurgents and implement social and economic reforms that addressed most existing grievances (Thompson 1966; Cable 1986; Stubbs 1989; Nagl 2002). It helped immensely that the rebels in Malaya formed an ethnically visibly distinct group, as they were nearly all Chinese. Almost identical methods were applied by the British in Kenya against the Mau Mau (1952–6), who again came from a distinct ethnic group, the Kikuyu, with a total of 1.5 million members. The Mau Mau themselves numbered at most 20,000. This time, it was the Kikuyu who were moved to fortified villages which they had to build themselves under very difficult conditions; a further 70,000 Kikuyu were moved into detention camps, and the hardships that came with the move and the construction of the villages in the middle of nowhere with no infrastructure led to an estimated 100,000 deaths among them (Markel 2006; Anderson 2005).

High-casualty repression did not end here, however. It became a particular hallmark of COIN in Latin America, Africa and Asia. In 1982 General Efraín Montt of Guatemala promised in a memorable expression that to deal with the insurgency in his country, he would 'dry up the human sea in which the guerrilla fish swim', echoing Mao's metaphor (q.i. Valentino *et al.* 2004: 399). Next to more humane approaches, this theme also resonates in the very considerable literature which has been published since the beginning of the 'war against terror', with its specific targeting of Afghanistan since 2001 and the beginning of the Iraqi insurgency in 2004. The metaphor employed was that of 'draining the swamp', in direct reference to earlier British exploits, despite the concession that this form of 'collective punishment' is morally repulsive (Markel 2006: 47). As we shall see, however, it is the non-punitive alternative based on persuasion and incentives that has become the basis for consensus.

Hunting the insurgents and becoming like them?

The most immediate pressing problem of counterinsurgency has always been how to find and capture the insurgents. To hunt down insurgents, as we have seen, regular armies were mostly useless. One of the secrets of the success of counterinsurgency has been seen by many writers in the transformation of the hunters into quasi-*guerrilleros* themselves, sharing the insurgents' same rough lifestyle. In the mid-1890s, Spanish General Valeriano Weyler (1838–1930) brought new ideas and methods to bear against the long-successful insurgents in Cuba, by transforming the operating patterns of the regular forces to mirror those of the guerrilla fighters (Ellis 1975: 102f.).

In the Second World War, the German Wehrmacht issued a manual *On Fighting Bands* (1944). It set down the need to hunt down insurgents with special 'hunting commandos' (*Jagdkommandos*) that were to chase the guerrilla fighters until they could round them up on convenient killing grounds surrounded by the commandos (*Kessel*). The qualities this required in the commandos come very close to those of the guerrilla fighters themselves:

> The fighting of bands is no second-rate fight. It requires soldiers who are particularly agile, cunning, fighting like hunters, hardened and frugal. Only continual vigilance protects troops from serious casualties ... A hunting section should not be smaller than a platoon nor stronger than a company. Fights with superior forces should be avoided ... Hunting is an intensified, indefatigable kind of pursuit, its aim is outrunning, bringing to a stand and crushing or capturing the prey. (q.i. Ellis 1975: 149)

On another level also, counterinsurgent forces are faced with the question as to whether they need to become like those they are fighting in order to be able to subdue them. We have already noted that many rules of war are not obeyed by guerrilla fighters. In the nineteenth and early twentieth centuries, commanders of counterinsurgency forces often felt this gave them the right to respond in kind. The British Field Services Regulations in 1912 stated:

> In campaigns against savages, the armaments, tactics, and characteristics of the enemy, and the nature of theatre of operations demand that the normal application of the principles of regular warfare be considerably modified. (q.i. Moreman 1996: 109)

As human rights and humanitarian conventions became more explicitly codified in the twentieth century, especially after the Second World War, counterinsurgency forces felt that they were obliged to fight with one hand tied behind their backs if in contrast to the insurgents they were obliged to follow proper rules of engagement, uphold the laws of war and all associated conventions; otherwise they would be criticised as no better morally than those they were fighting. This problem was always particularly pronounced when the guerrilla fighters resorted to terrorist acts against civilians associated with those they were fighting against, such as the terrorist acts committed by the Algerian Liberation Movement (FLN) against French *colons* in the 'white' parts of Algiers. In Algeria the French decided to abandon 'civilised rules of conduct'. This clearly worked against them, both in the medium and in the long term. Tactically, it was successful: the French made significant strides towards controlling Algeria again militarily – they even liked to think that they had almost won the war and were betrayed by de Gaulle (Galula 1964/2005: 69). Strategically, in the sense of achieving the political aims, it was extremely counterproductive, as it consolidated public opinion against the French in Algeria, but also increased sympathy worldwide for the Algerian struggle to achieve independence, while such sympathy had been undermined originally by the FLN's use of terrorism against French civilians.

It is thus a general pattern that guerrilla warfare asymmetrically pits a weak force circumventing the rules and limitations of regular war against a much stronger opponent whose *forte* is precisely that of regular warfare. The Goliath will always have the tendency to exploit his natural strengths to counter the guerrilla fighters, to 'capitalize on our asymmetric advantages' (Ancker and Burke 2003: 25). In the case of the US armed forces, this would mean, *inter alia*, capitalising on superior firepower – the heavy-handed approach which proved so counterproductive both in Vietnam and Iraq. In fact, the Goliath has to try to become more like the nimble-footed and agile David whom he is fighting, if he does not want to risk losing the main prize: the population of the country in which the guerrilla is waged.

Hearts and minds II

Concerns about winning over the sympathy of the populations in fought-over territories and towns goes back to Livy, whose views on the subject were taken up by Machiavelli in his discussion of 'The Management of Conquered Peoples': 'subject peoples should be generously treated or wiped out' (Machiavelli 1531/1998: 349). While practice often differed, the authors of the Renaissance, including Machiavelli himself, advocated the first of these options, recoiling from the latter. Thus Machiavelli, listing historical examples and their consequences, pronounced against the constructions of fortresses with the purpose of holding down the local population:

> [W]hen a prince or a republic is afraid of its subjects and fears they may rebel, the root cause of this fear must lie in the hatred which such subjects have for their rulers ... due to their misbehaviour; a misbehaviour which is due to their fancying they can hold them by force, or to their foolish way of governing them.

In dealing with any rebellion thus caused by poor governance, he conceded that brutality might bring short-term rewards for the oppressors, but warned that even those utterly reduced to poverty and disarmed could rise again, driven by the fury over this treatment; moreover, 'If you kill their leaders and suppress all other signs of insurrection, like the heads of the Hydra other leaders will arise' (Machiavelli 1531/1998: 353–356).

A very different approach to fighting down insurgents can be found in writings on the art of war that devote space to the pacification of a conquered country at the end of hostilities. Several works of the sixteenth century deal with this issue. All writing on *The Art of War*, the Italian Bernardino Rocca (1582) and the two Frenchmen Fourquevaux (1548/1549), and François de Saillans (1589/1591) insisted on the great need for discipline among soldiers, especially at the end of and after hostilities. Fourquevaux in his career had been both a general and a French ambassador to Spain. This helps explain why he was one of the rare military men who discussed the art of war not merely in terms of winning battles, but also of securing the political aims of the war itself. He prescribed the banishment of 'certain citizens' from occupied towns so as to prevent them from organising an uprising. Like most contemporary authors, he was extremely concerned to prevent any pillaging and other misconduct in the countryside, and to prevent atrocities after the capture of a town. He proscribed the destruction of the possessions of townsfolk, and any misconduct against women, 'for there is no crime that dooth so much offend the hearts of the people, as to see their wiues seduced or forced' (Fourquevaux 1548/1589: 212: 260).

A contemporary of theirs was the Spaniard Bernardino de Mendoza, who in his *Theory and Practice of War* dealt specifically with how to fight insurgencies. He gave very reasonable advice on a number of points. An insurgency had to be fought as quickly as possible, he argued, lest it spread throughout the country. He assumed, incidentally, that among the insurgents would be found a prince's vassals, that is, leading members of the aristocracy. Accordingly, he cautioned against fighting them in open battle, unless the prince was very certain that he would prevail in such a contest of armed force. In any other situation, he strongly urged that the prince should negotiate a peace with the vassals. In case the prince was victorious, however, he recommended that the leaders of the insurgency should be dealt with severely, but to make an example of them, not out of vengeance. The pacification of the land he thought possible only if the prince paired justice and clemency, castigating the heads of the insurgent movement, but pardoning the mass of insurgents, for, he argued, 'it is not possible to punish a multitude' (Mendoza 1595/1597: 30). We see here and in the writings of the authors considered below an underlying understanding for and sympathy with insurgents, who, in this humanist view, might well have good reason to rise up in protest against poor government and injustice.

In the second half of the seventeenth century, Paul Hay du Chastelet put his mind to dealing with insurgencies of the type which he described as civil war. France had seen its measure of these, and Hay thought it was up to princes and magistrates to prevent them through good governance and through preventing resentments caused by encroachments upon the rights of citizens. If a civil war broke out, the government should react flexibly, according to the laws of medicine, which administered cures according to the overall state of the patient. Different countries, he noted, had different customs; the Spanish never forgave rebels, while the French always pardoned them. Hay recommended to his king incessant negotiations with the rebel leaders to find mutually acceptable compromises (Hay du Chastelet 1668/1757: 265–71).

Mendoza influenced the earliest truly great author of comprehensive ideas on counterinsurgency: Santa Cruz de Marcenado. A leading military officer, diplomat and finally governor of one of the Spanish possessions in North Africa, he was another one of the very rare writers on Strategy to consider military measures in the light of their political outcome. One of his eleven volumes of *Military and Political Reflections* was dedicated entirely to the subject of 'Rebellions' or insurgencies from the point of view of the government trying to suppress them (Santa Cruz de Marcenado 1724–30 [1727]/2004, VIII).

Following Livy, he stated his conviction that insurgencies are unlikely to succeed or are less dangerous if they are not brought about by poor government; therefore, government should be 'gentle and just' (Santa Cruz de Marcenado 1724–30 [1727]/2004, VIII: ch. 1). The conflict might not be resolved unless at least a fair part of the grievances was first addressed. Like Hay, Santa Cruz emphasised the need to act quickly to stamp out a rebellion, as the tide would turn in favour of the rebels the longer it took the counterinsurgency forces to gain the upper hand (ch. 45). This phenomenon was rediscovered in the early twenty-first century, when analysts of the Iraq fiasco would speak of 'occupation fatigue' which would set in unless quick, positive results were produced by the benign occupiers (Ford 2005: 54). In dealing with an occupied population, Santa Cruz, like Mao and Ho after him, advocated respect for traditional customs, laws, religion and privileges of the region concerned (chs. 2, 10). He spoke out firmly against punishment of rebels without a regular trial, as that would undermine the rule of law (ch. 4). If people complained about the treatment they received in an area occupied by the military, he advocated changing the garrison, even the military commander (chs. 5, 6). He recommended entertaining the people in some way – clearly thinking of the Roman *panem et circenses* to turn their minds away from insurrection (ch. 13). To get potential rebels off the streets, he advocated recruiting unemployed youngsters into the army (ch. 33). To avoid the instrumentalisation of malcontentment by foreign powers, Santa Cruz urged controlling borders and roads leading abroad (ch. 34). Dangerous individuals, he thought, should be taken hostage and removed from their normal environment in which they could help plot the insurgency; they should, however, be treated well in their isolation. Rebels, where they could be seized, should be taken prisoner, but other rebels should be offered pardon if they went home within a given number of days (chs. 42, 53, 59, 62). He underlined that the physical extermination of the rebels was not the aim, but the curbing of the rebellion: today's rebels would ideally be tomorrow's productive citizens (ch. 43). Accordingly, he advised avoiding bloody reprisals, as they merely served to create martyrs for the cause and increased support for the insurgency (ch. 50). Santa Cruz proposed the promulgation of articles aimed mainly at disarming the insurgents and the population in general (who might or might not decide to support the insurgents), while issuing passports to members of the public moving about on normal business to distinguish them from the rebels, who should be intercepted in their movements. Secret meetings should be prohibited, but for those observing the rules, a guarantee of their lives and property should be given (ch. 55).

Crucially, the population should be given a stake in the order assaulted by the rebels, by looking to the economic welfare of the region: trade, manufacture and all other production should be protected and encouraged to improve the economic well-being of the area, tax reliefs should be granted and to bring the potentially disaffected population over to the side of enlightenment and truth – assuming, of course, that the legitimate government represented these values – Santa Cruz recommended the establishment of schools and universities to enlighten the populations of the areas concerned (ch. 61).

Santa Cruz thus recognised that in combating insurgents, a reign of terror would not ultimately lead to a change of heart of the locals. But, as Clausewitz would say a century later, prevailing in war means imposing one's will upon the enemy, and how much easier to impose one's will if the enemy can be convinced that he can live with, perhaps even thrive in, the new situation? What Santa Cruz was clearly advocating was giving the population of the area concerned a stake in a reformed state which the counterinsurgency tried to establish (state-building!) and defend.

This is perhaps the most important line of thinking on counterinsurgency, but there is no evidence that Santa Cruz de Marcenado's writings were much read. For one, they have not been translated into English to this day. There is an isolated recognition in the work of Henry Humphry Evans Lloyd that insurgencies ('revolts') are caused by 'injuries, real and supposed', and that these injuries have been inflicted upon the insurgents by a sovereign authority whom they therefore aim to destroy. '[I]t is during this contest that the greatest efforts are made, because there is no alternative; freedom or slavery is the result of it.' Accordingly,

> if you persevere in attacking them, and treat them with cruelty and contempt, they will not be satisfied with any conditions you may grant them, but, moved by revenge, which is a very active principle, pursue their oppressors to destruction. (Lloyd 1781/2005: 453)

After Lloyd, Le Mière de Corvey opined: 'if one wants to preserve one's conquest, one has to treat the defeated gently, that is the only way to gain their loyalty. Unfortunately, all the wars in which the *levée en masse* has been used, are fed by some fanaticism, be it a political spirit, a religious one, etc.; without this [ideologisation], there would not have been wars of extermination the results of which are horrible.' In a footnote, Le Mière gave the example of the atrocities with which Charlemagne had dealt with the Saxons during their forced conversion; he clearly did not approve of this king to whom posterity gave the title 'the Great' (Mière 1823: vii–ix).

Clausewitz had already hit on the importance of winning the population's support. He thus also recognised it as a war aim 'to win over public opinion'. Initially, he proposed to do this through crude force, through crushing victories and the seizure of the capital (Clausewitz 1810–11/1966: 1070). Later Clausewitz commented on

> what an enormous contribution the heart and temper ['*Gesinnung*': disposition, opinion] of a nation can make to the sum total of its politics, war potential and fighting strength. Now that governments have been conscious of these resources, we cannot expect them to remain unused in the future, whether the war is fought in self-defence or in order to satisfy intense ambitions. (Clausewitz 1832/1976: 220)

Cast more in the humanist mould of thinking of Santa Cruz, but without knowledge of his writing, some of the most interesting writers on this subject came from France in the late nineteenth and early twentieth century. In 1895 General François Jacques André Duchemin described French operations in Indochina in flowery but enlightened language:

> The pirate [insurgent] is a plant which grows only on certain soil. The most efficient method is to render the ground unsuitable to him ... There are no pirates in completely organised countries. To pluck wild plants is not sufficient: one must plough the conquered soil, enclose it and then sow it with the good grain, which is the only way to make it unsuitable to the tares. The same happens on the land desolated by piracy: armed occupation, with or without armed combat, ploughs it; the establishment of a military belt encloses and isolates it; finally the reconstitution and equipment of the population, the installation of markets and cultures, the construction of roads, sow the good grain and make the conquered region unsuitable to the pirate; if it is not the latter himself who, transformed, co-operates in this evolutionary process. (q.i. Gottman 1941: 242)

The same spirit of enlightened self-interest is reflected in the writings of General Joseph Galliéni (1849–1916) and Marshal Hubert Lyautey (1854–1934), although not necessarily always put into practice by the French. Lyautey advocated 'peaceful penetration' of the hostile populations or occupied areas, writing an article in

1900 on 'The Colonial Role of the Army', in which he argued about Morocco:

> The best means for achieving pacification in our new colony is provided by combined application of force and politics. It must be remembered that, in the course of our colonial struggle, we should turn to destruction only as a last resort, and only as a preliminary to better reconstruction. We must always treat the country and its inhabitants with consideration, since the former is destined to receive our future colonial enterprises and the latter will be our main agents and collaborators in the development or our enterprises. Every time that the necessities of war force one of our colonial officers to take action against a village or inhabited centre, his first concern, once submission of the inhabitants has been achieved, should be reconstruction of the village, creation of a market, and establishment of a school. It is by combined use of politics and force that pacification of a country and its future organisation will be achieved. Political action is by far the most important. It derives its greater power from organisation of the country and its inhabitants. (q.i. Gottman 1941: 243)

At much the same time, an American commentator on US operations in the Philippines wrote that '[I]f army officers and the army have had to know something of the art of war, they have had to know and use far more of the art of pacification. In the Philippines their work was four-fifths peace and one fifth war-making' (q.i. Linn 2002: 519).

Despite the much earlier antecedents, the term struggle for 'the hearts and minds' is said to have been coined by the British high commissioner in Malaya, Field Marshal Sir Gerald Templer, during the 'emergency' of 1948–57. He spoke of the need to 'win the hearts and minds of the population' (q.i. Lapping 1985: 224). The phrase subsequently became hackneyed, to the point where Americans talked about 'WHAMing' the peasantry in Vietnam (Johnson 1973: 41). In practice, however, this happened rarely: the Briton Robert Thompson and the Australian Robert O'Neill criticised precisely the failure of the Americans, in the Vietnam War, to engage earlier in the battle for the support of the Vietnamese population itself (O'Neill 1968; Thompson 1970). Thompson pointed to the futility of attempting to crush an insurrection, if the resulting situation or state was politically and economically not feasible. He emphasised moreover that the government, however outraged its own forces might be by atrocities committed by partisans, had to stay on the side of legality at all times: if one was trying to set up a state of law, one could not give it credibility if one acted oneself unlawfully in the process of setting it up or restoring it. Thus he saw trials to which the public was not admitted or executions under martial law as signs only of the weakness of a government and of the danger that it would soon collapse completely. No government, he felt, could impose itself and win the sympathy of the local population if it engaged in torture or execution of terrorists, as both remained criminal offences in a state of law. Moreover, Thompson thought that to be successful, the leaders of a counterinsurgency (particularly the foreign leaders) had to have a comprehensive master-plan which would help them keep focused on what sort of society and economy they were seeking to create: you could never, he argued, merely go back to the status quo ante (Thompson 1970).

Galula and his compatriot Roger Trinquier (1908–86) pointed out, however, that in 'revolutionary war', there may be a battle for the support of the population about whom these wars were ostensibly being fought, but the population would more often than not be terrorised by the partisans into giving their support. Both Galula and Trinquier emphasised the need to protect the population from the bullying and terror of the insurgent leaders in their quest for support (Trinquier 1961/2006: 27; Galula 1964/2005: 52f.). What the French had sought to do, in Indochina and Algeria, had been to turn the insurgents' own weapons against them: terror, illegitimacy, torture (Trinquier 1961/2006: 20, 68ff., 1968), something Galula strongly condemned, pointing at the asymmetry of moral judgement regarding insurgent and counterinsurgency actions.

For two decades after the fall of Vietnam, the whole subject area of counterinsurgency felt too sore for American official strategists to give it serious thought. A turning point came in the early 2000s. The American Field Manual Interim 3-07.22 of 2004 already rightly reflected a new awareness of cultural specificity and its enormous importance in warfare, especially counterinsurgency. Starting in a very Clausewitzian vein, it argued that:

> The center of gravity in counterinsurgency operations is the population. Therefore, understanding the local society and gaining its support is critical to success. For U.S. forces to operate effectively among a local population and gain and maintain their support, it is important to develop a thorough understanding of the society and its culture, including its history,

tribal/family/social structure, values, religions, customs, and needs. (US Army 2004)

A mental gate opened with the experiences of Iraq in 2004 for a flood of publications on counterinsurgency. May it suffice here to home in on the centre of the shift in thinking which led to the adoption in the USA in December 2006 of a new counterinsurgency field manual (FM 3-24; US Army and US Marine Corps 2006). The team who had produced it had been led by Generals David Petraeus and James Mattis. Petraeus himself had experienced the insurgency and civil war in Iraq in 2004 and formulated fourteen recommendations, prophesying the tenor of the field manual. They converge closely with the emphasis on respect for local customs we have already found in Santa Cruz, but also in Mao and Ho. Besides drawing on his own experience, Petraeus, who himself holds a Ph D, was influenced both by historians of past COIN campaigns and by the anthropologist Montgomery McFate:

1. Delegate as much as possible to the locals, so that they feel ownership of what is being done – an idea that can be traced to T.E. Lawrence.
2. Visible improvements in the country concerned have to be effected quickly, otherwise the population will conclude that things were better before the counterinsurgency operations got under way.
3. Financial investment in reconstruction is of prime importance.
4. 'Increasing the number of stakeholders' in the new order 'is critical to success.'
5. A military operation must only be carried out if it is clear that the political (and psychological) costs (for example, through collateral damage, civilian casualties and so forth) do not outweigh the military gains.
6. Intelligence on the identity, movements and whereabouts of the insurgent fighters is crucial, so that they can be targeted individually and discriminately, without accidentally hurting civilians among whom they are hiding.
7. The military as much as NGOs or specialised units must engage in state-building ('nation-building', in US terminology) or civic action, that is, in activities designed to improve the infrastructure and the living conditions of the population in general and to mend the social fabric, and regenerate or strengthen trust in the forces of law.

8. It is not enough to create new police forces – to back them, the ministries of justice, the interior and so forth have to be reformed, as well as the administrations serving them.
9. Cultural awareness is crucial in order not to alienate the local population.
10. Counterinsurgency is only successful when economic and social recovery and the improvement of living conditions go along with military success.
11. For a military force engaging in counterinsurgency operations in a foreign country to help a friendly government, the ultimate success depends on the behaviour and acceptance of the local leaders.
12. The implementation of all the above points depends on the awareness and adroitness of junior commanders.
13. On all levels, especially junior levels, the military leaders have to show initiative, flexibility, imagination and ability to innovate.
14. 'A leader's most important task is to set the right tone', to emphasise the importance of non-fighting tasks over purely military missions and to avoid over-reactions to especially trying situations (Petraeus 2006).

Conclusions

The flood of literature on this topic has by no means dried up, although a consensus seems to have been established that in counterinsurgency, it is crucial to focus on the needs and interest of the population of that area. 'Hearts and minds' may sound too soppy for some – 'providing security' may be more appropriate (Lynn 2005: 24). David Edelstein in his study of military occupations from 1815 to 2003 also put it prosaically: COIN will only be successful if the occupied population recognises the need for occupation, feels solidarity with the occupiers in the face of a threat common to both sides and has 'a credible guarantee' that the occupiers will 'withdraw and return control to an indigenous government in a timely manner' (Edelstein 2004: 51).

Or, in the words of McFate and her fellow anthropologist Andrea Jackson:

> To win support counterinsurgents must be able to selectively provide security – or take it away. [COIN] forces must become the arbiter of economic

> well-being by providing goods, services, and income – or by taking them away. [COIN] forces must develop and disseminate narratives, symbols, and messages that resonate with the population's pre-existing cultural system or counter those of the opposition. And finally, counterinsurgents must co-opt existing traditional leaders whose authority can augment the legitimacy of the government or prevent the opposition from co-opting them. (McFate and Jackson 2006: 13)

An unresolved tension exists between playing on local customs and traditions on the one hand and the aims on the other to build states with democratic institutions practising good governance and ensuring human rights – these in themselves may be incompatible with the local customs and traditions.

But the central consensus is there: only the side that puts the population's interests first will ultimately win them over – terror will bring about temporary conformity and obedience, but not long-term success. When COIN forces are seen as occupiers, or when insurgents are seen as brutal oppressors, people may be cowed into co-operation, but it will not become a self-sustaining force in that society (McFate and Jackson 2006: 18). COIN cannot be seen as a zero-sum game, and above all, the winners must be the local population, not the foreign states intervening in their country.

In the context of small wars, writers realised early on that the key to lasting success had to be the pacification of a country or region, not by the brute imposition of force, but by holding out a better deal to the population. The people had to be persuaded that they did not need to support rebels in order to see an improvement of their lives and an addressing of their grievances. Persuasion, not the imposition of one's will (which in chapter 1 we have seen to be by consensus the essence of most definitions of military Strategy these days) is at the heart of success in asymmetric warfare. As we shall see in the following chapter, after 1945, this realisation would spread to liberal thinking about other warfare, too.

PART VII

The quest for new paradigms after the World Wars

17 *Wars without victories, victories without peace*

[T]he field of battle is the only tribunal before which states plead their cause; but victory, by gaining the suit, does not decide in favour of the cause. Though the treaty of peace puts an end to the present war, it does not abolish a state of war (a state where continually new pretences for war are found).

(Immanuel Kant 1795/1796: 24)

Indirect strategy seeks to obtain a result by methods other than military victory.

(General André Beaufre 1963/1965: 108)

The First World War as turning point?

The major theme of this chapter does not start with Hiroshima in 1945. We could trace it back to Jaurès's and August Bebel's reactions to the Franco-Prussian War of 1870/1. But it is even more clearly present in reactions to the war of which contemporaries had ardently hoped that it would be '*la der des ders*', the last of its kind, and which despite hecatombs of casualties failed to produce a lasting peace. As J.F.C. Fuller concluded from this experience, '[e]xacted peace can be no more than an armistice' (Fuller 1961: 153). The strong perception of the futility of the sacrifices of that Great War by those who had fought in it and later witnessed Hitler's attempts to reverse all its outcomes greatly influenced strategic thinkers.

Another line of thinking concludes from the ravages caused by war that little can be worth such destruction and misery. As wars mobilised entire nations whose populations were growing exponentially, and entire industrialised state economies, the degree of destruction grew in turn. Polish-born intellectual Jan Bloch (1899) thus argued that the cost of war and destruction for advanced societies was so high that few causes could warrant it. Yet the Great

War soon showed that governments did not necessarily follow this logic.

After Bloch, and especially after the watershed of the First World War, Norman Angell and others tried to think about how to limit or even ban war. There was agreement, expressed for example in the foundation of the Royal Institute of International Affairs at Chatham House in London, that a better understanding of interstate relations was needed to ward off the repetition of this disaster (Howard 1979). In a reawakening of the eighteenth-century quest for institutionalised peace, which we have seen particularly in the proposals of Saint Pierre and Kant, but as a concrete after-effect of the Great War, in 1928 French Foreign Minister Aristide Briand and US Secretary of State Frank Kellogg had signed a 'pact' named after them, outlawing war. At the time, there was an important strand of opinion among foreign policy makers and international lawyers that this was the right way to go, and many other countries adhered to the pact – without any legally binding consequences, however. The war-weariness which this reflected in the following decade greatly contributed to the appeasement of Hitler in the 1930s.

As all that had been won by the liberal states through the sacrifices of the First World War was turning to ashes with the outbreak of the Second World War, Liddell Hart wrote even in 1939, echoing Machiavelli, that the more violence one used, the greater resistance would become. Violence tended

> to consolidate the enemy's troops and people behind their leaders ... The more intent you appear to impose a peace entirely of your own choosing, by conquest, the stiffer the obstacle you will raise in your path ... if and when you reach your military goal, the more you ask of the defeated side the more trouble you will have, and the more cause you will provide for an ultimate attempt to reverse the settlement achieved by war.

A military victory, even one achieved at greatest cost, did not automatically mean lasting achievement of one's war aims: most notably, the Versailles Peace Treaty had not brought a lasting peace. 'Force is a vicious spiral – unless its application is controlled by the most carefully reasoned calculation. Thus war, which begins by denying reason, comes to vindicate it – throughout all phases of the struggle' (Liddell Hart 1944: 177). Nuclear weapons were thus not the only factor cautioning against major war. They were in some respects the manifestation of something greater, but they brought the message home to those who had not yet learnt it by 1939.

After the Second World War came the foundation of the International Institute for Strategic Studies in London, the RAND Corporation in California (the name comes from 'Research and Development') and several other institutes or university departments from Scandinavia and the Netherlands to Australia devoted to peace research or polemology. The latter was the term which Gaston Bouthoul used for his institute in France, meaning science of war or war studies. Their aims were the same, whatever their names (peace studies or war studies), as expressed by Liddell Hart and Bouthoul: 'If you want peace, study war' (Liddell Hart 1954; Bouthoul 1976).

A new dichotomy went along with this disillusionment with major war. As Liddell Hart put it, if one's own

> object of war is to obtain a better peace ... it is essential to conduct war with constant regard to the peace you desire ... [Therefore] if you concentrate exclusively on victory, with no thought for the after-effect ... it is almost certain that the peace will be a bad one, containing the germs of another war.

Therefore, '[v]ictory in the true sense implies that the state of peace, and of one's people, is better after the war than before' (q.i. Kennedy 1991: 2f.). Augustine of Hippo had said much the same one and a half millennia earlier.

Several authors in the Augustinian tradition had in the past stressed that wars should not be fought unless they were just. But they tended to focus on the just cause, and less on the right intention, or the reasonable expectation of achieving an end-state of peace that was better than the initial situation, without incurring disproportionate cost in terms of suffering, death and destruction in getting from the beginning of war to its end (see chapters 2 and 3). Later, casting aside any obligation to justify their actions to a greater moral authority, many princes of the modern age, and later 'realists' and statesmen of other persuasions, felt little need to articulate more than their unilateral, subjective interest to justify going to war. In the late Cold War, the 'Weinberger Doctrine' formulated by the then US Secretary of Defense in 1986 as guidance to Strategy, while sensible in many ways, did not dwell on any just cause, legitimate authority or right intentions. It mentioned only the principle of 'last resort', that is, that war should be undertaken only if all other options had failed, a principle derived from Roman legal practice (Russell 1975: 4f.).

By contrast, the movement to outlaw war after the Second World War led to the Charter of the United Nations outlawing the use of force as a furthering of one's own state's interests, except in self-defence against aggression, which by extension branded the aggressor as an outlaw (UN Charter 1945, Chapter VII, Article 51). And yet it was only at the turn to the new millennium that the concept of 'grand strategy', as Paul Kennedy rightly remarked, was linked with peace as much as with war (Kennedy 1991: 4). The concept of a 'better peace' was clarified by the UN's 'Secure World' report of 2004, commissioned by the Secretary-General: the authors declared that the use of force might be condoned by the UN Security Council under Chapter VII, if all of five criteria were met. These are that the threat must be serious enough, that the purpose of the use of force must be only to halt or prevent the threat in question, that it is the last resort, that proportionate means are applied (that is, not more than necessary) and that the consequences of action are likely to address the threat, and will not be worse than the consequences of inaction (UN 2004: 58). This sets crucial parameters for Strategy: the use of force cannot be seen as legitimate if its consequences are worse than the wrong it was supposed to undo. The document reflects a remarkable continuity in 'Western' – now purportedly global – ethics and thinking about war since Roman times, notwithstanding Western deviations in the age of total war.

Causes, conduct and ethics of wars since 1945

Only citizens of the most cozened states of the world will associate the 'Cold War' with a 'long peace' that has supposedly given way to a period of 'new wars'. There were dozens of hot wars during the Cold War, and tens of millions of casualties, more still counting intra-state wars and persecutions. Some, such as the Korean War, the several Arab–Israeli wars, the Iran–Iraq War 1980–8, and since the end of the Cold War, the Ethiopia–Eritrea War 1998–2000 and the Gulf Wars of 1991 and 2003, easily fall into the same category as the major wars of previous centuries with regular armies clashing in major battles. Measured by nineteenth-century standards, they can bear comparison with Moltke's wars against Austria and Denmark or the Crimean War, and they were major wars for the countries on whose territories they were fought. From the perspective of at least one side, they had aspects in common with the two World Wars. They were 'limited' only in that, while taking place in the nuclear age, they were not fought with nuclear weapons. A simple either–or categorisation of 'limited'–'unlimited' hardly does them justice. Genocidal (Ludendorffian 'total') war aims have shown themselves to be perfectly possible in local wars waged with limited manpower without major battles, and limited territorial war aims were perfectly compatible with major tank battles in the Arab–Israeli wars or the Ethiopia–Eritrea War.

What was distinctive during the Cold War, notwithstanding their 'containment' and the more honestly named 'roll-back' Strategies (Heuser 1989), was that the Western states tended to be on the reactive side. The wars that did occur were interpreted by the West 'as a part of a worldwide struggle between communism' and the West. At a second glance, however, as British Field Marshal Lord Carver noted rightly, most of the wars of the Cold War

> were not directly concerned with the struggle between these opposing systems or ideologies, unless the fight against colonialism or imperialism is considered as having been instigated, as opposed to merely supported, by communism. The principal causes of these wars were either the contest between imperial authorities, attempting to maintain that authority, and the nationalist movements which challenged it, or the struggle for power between rival elements of the population as the imperial authority withdrew ... These rivalries were at times based on differences of political outlook, but almost all were heavily influenced by racial, historical and cultural differences. (Carver 1990: 290)

This characterisation of wars post-1945 also fits the wars since 1991, especially if one slightly widens the definition by integrating the fact that perceived racial (or better: ethnic), historical and cultural (including religious) conflicts between 'rival elements of the population' usually flared up whenever an authoritarian or other strong regime collapsed. This accounts for the explosion of Yugoslavia after the death of Tito, or the periodic outbursts of violence in African countries at times of presidential elections. Such wars occurred both before and since the end of the Cold War. It is misleading to call them 'new wars', even if the Internet or mobile phones have added a new touch here and there. What has changed with the end of the Cold War is not the nature of these wars, but the fact that Western powers could intervene in local conflicts, even, as in the case of Kosovo, against the wish of Russia, without having to fear that this might escalate into a nuclear apocalypse. But curiously, self-limitations in the conduct of such operations would now replace what in the Cold War had been the fear of nuclear war. And this has its roots in the ethical shift back to pre-nineteenth-century values that had taken place.

The consequence of the outlawing of aggressive war by the UN Charter was that after 1945, few wars have been properly declared (Brodie 1973: 229). Thus, as James Cable observed, '[t]he foreign offices of the world, or their legal advisers, were so impressed by the Nuremberg Judgment' on the German war crimes, and indeed the pronouncement of the Tokyo war crimes tribunal, that henceforth they eschewed entering into a legal situation through the declaration of war in which *ius in bello* applied, by which they might later be judged (Cable 1971: 12). Many wars since 1945 were neither formally begun nor formally ended; they were described as 'emergency', 'troubles', 'crisis', 'affair'. With casualty figures comparable to losses in major wars, however, they were clearly wars, if defined as the use of violence by one organised group against another (Heuser 2008). This applied particularly to violence used by governments against unwanted segments of their own populations, in democide as defined by R.J. Rummel, that thrived in extremist ideologies, be they racist, hyper-nationalist or communist. They tended to occur in contexts seen by the perpetrators as war threatening the survival of their own narrowly defined group, even if the war was undeclared or was a 'Cold War' ideological struggle. As the most important battleground in the Cold War, just as in the wars of religion in early modern times, was 'in our minds' (Rees 1964: xvi), it seemed all the more reason for fanatics to suppress opposition by annihilating it.

While genocide and democide were still taking place in communist and other authoritarian states in numbers too terrible to contemplate (Rummel 1992), in liberal democratic states, much thinking about war developed in the opposite direction, that indicated by the Briand–Kellogg Pact and the UN Charter. US strategist Bernard Brodie observed that while attitudes to war 'in our western civilization' had changed remarkably little over the millennia, they have changed ... rapidly in the half century since World War I, and with especial force since the coming of nuclear weapons. Where war was once accepted as inevitably a part of the human conditions, regrettable ... but offering valued compensations in opportunities for valor and human greatness – or, more recently, in opportunities for the ascendancy of superior peoples – the modern attitude has moved towards rejection of the concept of war as a means of resolving international or other disputes.

Instead, he found that after 1945, 'justifications of war and of preparations for war appear to be confined largely to self-defense – expanded by the superpowers to include defense of client states – or, in a very few instances, correction of what is conceived to be the most blatant injustice' (Brodie 1973: 274). Brodie also remarked that, in the Cold War, the words 'glory' and 'honour' all but disappeared from debates about Strategy (Brodie 1973: 229). Indeed, when towards the end of the Cold War, a British field marshal would write in defence of the military that 'Their profession is an honourable one' (Carver 1990: 300), the word is used almost apologetically, not as the rallying cry to arms that it was from antiquity until the Second World War. Instead, the term 'security' had ubiquitously replaced the term 'honour', and, as Brodie found, could be harnessed to just about any cause (Brodie 1973: 274, 345–55), what the 'critical security studies' school would later refer to as the 'securitization' of subjects (Buzan *et al.* 1998). Richard Betts has recently made the same observation with regard to the buzzword 'credibility', especially in relation to NATO, used to justify the threat or use of force where previously honour or loss of face might have been invoked (Betts 2000: 49). But the context was here that unless the firm positions which NATO took and its implicit threats were credible in small contingencies, it could not deter major war.

Side by side with this there continued to be the Napoleonic paradigm of the 'pursuit of victory' in war as the guiding principle, which Brian Bond rightly traces 'from Napoleon to Saddam Hussein' (Bond 1996). On the whole, militaries in the Western world continued to plan for 'victory', but the list of exceptions to this grew. In 1954, the American *Field Service Regulations* FM 100–5 introduced the concept of 'wars of limited objective', and relinquished the term 'victory' as a war aim: 'Victory alone as an aim of war cannot be justified, since in itself victory does not always assure the realization of national objectives' (q.i. Summers 1982: 67). It was also in the mid-1950s that NATO dropped the concept of victory from its war aims in case of a Warsaw Pact attack (Heuser 1998b: 311–28). In the final stages of its campaign in Algeria, France decided consciously not to use all its might to retain this colony. Equally the USA in Vietnam arguably did not aim for 'victory' (Gray 1979: 71), and of course any notion of a military 'victory' is alien to the peace-keeping operations which came into being since 1945 and then became ever more frequent, and ever longer.

Even in regular war, the means available for the pursuit of victory were capped. In the 1950s planners still assumed that nuclear weapons would be available – especially in an East–West war, but not only there – like any other munitions. By the early 1960s, however, perceptions and terminology had changed: 'limited war' came to mean 'conventional war' in American parlance. What is more, the assumption was henceforth that all war other than a Third World War would be, had to be limited. And a Third World War, it was fervently hoped, would 'not take place' (Gallois 1976: 113). It also became clear by then at the latest that it was difficult for the military to plan and organise simultaneously for conventional and nuclear operations, as NATO did for its defence throughout the Cold War. It meant that during the first phase of conventional defence – which everybody hoped would not escalate to nuclear war – NATO forces could not fight even with all conventional means available: for example, a number of aircraft had to be kept back from conventional missions to be ready for nuclear missions. This led many air officers to resent the nuclear dimension, as they felt they had to fight the perhaps all-important conventional phase with one hand tied behind their back (Trachtenberg 1986: 760). Similar feelings were expressed by naval officers (see chapter 11).

The West did not renounce the use of force altogether – the chief lesson of the fight against Hitler's Germany, at least for Britain, France and the USA, balanced the pacifist sentiments that had firmly established themselves in the trenches of Flanders in 1914–18 and, before that, in the American Civil War. This lesson was that there were contexts in which war was just and warranted, and the freedom of one's own civilisations was worth defending even at the cost of millions of lives (although perhaps not tens of millions). As Field Marshal Lord Carver put it,

> It is as well to remind those who quote the injunction in the Sermon on the Mount to 'turn the other cheek ...' that ... if evil is not resisted, it will prevail. That is the justification for the use of force to deter, and if necessary, to defeat those who turn to it to further their own ends. (Carver 1990: 300)

The Soviet Union had retained a similar lesson from its Great Patriotic War, except that its leadership was willing to accept much higher casualties than the West. By contrast, this lesson did not come naturally to the Germans, Austrians or Italians, who only after the Second World War internalised a profoundly pacifist stance with concessions only to the right to self-defence (Heuser 1999: 154–91).

Even after the Second World War, police actions, wars against aggressors, rogue states or insurgencies perceived as illegitimate were undertaken quite readily by the colonial powers in retreat, or by the self-appointed world policemen, the USA, Britain, France and indeed the USSR (in Afghanistan). Yet the assumption was shared on all sides that these actions must not turn into a major war, and especially not a nuclear war. There was a growing consensus since the early 1950s that only the ultimate issue of Western freedom and survival – and on the Eastern side, the very survival of socialism – could warrant the risk of nuclear war. Dissenters questioned that even this was a cause worth risking the end of civilisation for.

This is the solution to the 'riddle' which Liddell Hart set out to solve, when he wondered why it was that:

> On this side of the [Iron] curtain, the language used about Soviet policy is stronger than was ever addressed to Hitler before war came, but what even Americans say is mild compared with the abusive terms the Russians hurl at American policy. In the light of history, it hardly seems that such intense friction and mutual suspicion can go on indefinitely without resulting in an explosion. (Liddell Hart 1950: 90)

In this context, there was a new, odd employment of *ancien régime* balance-of-power thinking. In the 'realist' mode of thinking, ideology is discounted and all states are thought equally selfish, but capable, out of pure self-interest, of co-operating temporarily with other states to pull down another power who is poised to bid for regional or world domination. To such a mindset, a 'balance of powers' even with an ideological enemy, is an acceptable world order. In the Cold

War, 'balance of power' thinking took a new twist (Betts 1987: 14f., 20, 91, 175–9). It was neatly summed up in Tom Lehrer's 'Nuclear Proliferation Song': after the USA, 'Russia got the bomb, and that was OK, / 'cause the balance of powers was preserved that way.' A variant of this perception is that of 'spheres of influence', on the agreement of which Churchill and Stalin constructed the post-Second World War order, in Moscow in December 1944. In the middle of one of the greatest ideological conflicts ever, many on the liberal side would rather concede moral equivalence to the other than risk blowing up the world.

The return of spectator-sport warfare

What, if anything, has changed since the end of the Cold War? Spheres of influence still seem accepted to a large extent, although the Russian sphere has contracted significantly. America is in many ways the world hegemon, but not to the extent that it would or could risk a major war to defend Georgia against Russia or Taiwan against the People's Republic of China. The spectre of a Third World War has paled. By contrast, the surprise betrayed by many at renewed manifestations of more limited forms of war that had receded somewhat during the Cold War is unjustified. They had clearly been unduly mesmerised by the spectre of the apocalypse that did not happen. Neither the Gulf War of 1990/1, nor the Wars of Secession in Yugoslavia in the 1990s, nor the genocide in Rwanda, nor the short campaign and ensuing quagmire in Iraq 2003 was essentially 'new'. Least new is the phenomenon of terrorism, well known long before 9/11, the attacks with hijacked aircraft against buildings in New York and Washington in 2001. 'Anarchists' famously used terrorism in Moscow, Paris and London before the First World War, and from the 1950s to the 1990s, European states had repeatedly suffered terrorist attacks on a smaller scale. This was usually a mix of the same fervent anti-Western reactionary nationalism or religious fanaticism that characterised 9/11 and now takes the guise of Islamism throughout Africa and Asia, with ultra-left political splinter groups like the Red Army Faction in West Germany or the Red Brigades in Italy, who were usually supported with training, weapons or finance from within the communist world. Even the USA had experienced terrorist attacks, such as the bombing of buildings in the 1970s by the radical left 'Weathermen', the bank robberies, murders and VIP kidnappings in the same decade carried out in the US by the 'Symbionese Liberation Army', the Oklahoma City bombing of 1995 and an earlier attempt by Al Qaeda to blow up the World Trade Center.

All these types of conflict had more dimensions than merely the conflict between communism and liberalism. Thus their principal patterns have not changed with the end of the Cold War, except for the subtraction of communist or American financing which used to fuel most local conflicts in the Cold War. The Rwandan civil war at the end of the twentieth century proved that the Nazi genocide was not made possible only by good railway logistics, by hard-working civil servants, obedient SS or Wehrmacht units and superior German industrial technology. The war aim of genocide, we had to learn, can be pursued successfully with primitive machetes.

As Edward Rice has shown, it is *not* true that 'Wars of the Third Kind' – wars within underdeveloped states – have spread only since the end of the Cold War (Rice 1990). They have been a very long-term phenomenon, at best disrupted by periods of colonialism. It *is* true that most wars of the present are 'Wars of the Third Kind', and that in those wars, '[t]here are no set strategies and tactics. Innovation, surprise, and unpredictability are necessities and virtues'. Modern weapons technology of a very expensive sort rarely plays a role, the emphasis is on improvised explosive devices, small arms and Kalashnikovs. The distinction between civilians and soldiers disappears, not least in civilian casualty figures of 90 per cent in intra-state wars (Holsti 1996: 37).

What has changed for the West is that it is now *less directly* implicated in most wars of the present. These 'Wars of the Third Kind' take place locally, for local reasons and aims, between local factions, they are rarely directed against former colonial powers; they do not threaten to bring on the end of the world and the three terrorist attacks of 9/11, the Madrid bombing in March 2004 and 7/7 (on London's public transport in 2005) are sad but negligible occurrences compared quantitatively with the casualties suffered in any war. That they occupy so much political space in the West is a matter of Western choice, not of necessity. Indeed, the West can 'choose' to become involved in these wars or stand aloof, a notion going back to the great medieval Jewish theologian Moses Maimonides (12th century/1940: 320). The West can look on, almost as 'spectators' to war, just as H.G. Wells described Europe looking on during the Boer War (see chapter 12). According to Colin McInnes the 'spectator-sport warfare' of the present has the following characteristics:

1. The localization of war, leading to expeditionary strategies and force requirements.
2. The targeting in war of an enemy leadership or regime, not society.
3. The requirement to minimize collateral damage.
4. The requirement to minimize risk and exposure for both Western society and its armed forces.
5. The belief that the enemy armed forces are no longer the target in war or necessarily the means to victory. (McInnes 2002: 143)

From a Western point of view, argues General Sir Rupert Smith,

> War no longer exists. Confrontation, conflict and combat undoubtedly exist all around the world ... and states still have armed forces which they use as a symbol of power. None the less ... war as battle in a field between men and machinery, war as a massive deciding event in a dispute in international affairs: such war no longer exists ... [A] paradigm shift in war has undoubtedly occurred: from armies with comparable forces doing battle on a field to strategic confrontation between a range of combatants, not all of which are armies, and using different types of weapons, often improvised. The old paradigm was that of interstate industrial war. The new one is the paradigm of war amongst the people. (Smith 2005: 1, 3)

We see here again the dominance of the nineteenth- and early twentieth-century model of warfare, against which current wars are measured, therefore seeming new to some observers. Taking a larger perspective, however, Martin van Creveld first noted that the current wars look more like a return to much *older* forms of war, perhaps to the most original forms of war, the sub-state, pre-state wars, with warring groups vying for power, much as in Europe between the fall of the West Roman Empire and the end of the Thirty Years War (Creveld 1991). What *is* relatively new is the detachment that the Western powers have from these wars, which for them are wars of choice, as Lawrence Freedman put it (1993). Only if oil or narcotics are involved is the West directly affected, only if the war destabilises a volatile region with weapons of mass destruction can a case for direct concern be constructed. But held up against the previous colonial withdrawal wars or the communist threat in the Cold War, the actual effects on and potential threats to the West are significantly smaller since 1991, to the point where it is difficult to explain to a Belgian or Canadian or US Mid-Western electorate why they should feel concerned.

The relinquishment of the Napoleonic paradigm

'Violent conflict aiming at military victory' was the Strategy characteristic of the Napoleonic era, wrote General André Beaufre in the early 1960s. With 'Wagnerian romanticism', strategists of the nineteenth and the first half of the twentieth centuries had distorted Clausewitz's analysis of Napoleon's recipe for success into the mistaken notion that this was 'the only orthodox strategy' which 'gave birth to the two great World Wars ... both of which showed up the limitations of the Clausewitz-Napoleon strategy'. But then, 'haunted by the two fruitless catastrophes of 1914–18 and 1939–45 and now armed with all the resources of modern science', he thought humanity might 'have found a method of preventing a repetition of these catastrophes': nuclear deterrence to ban major, albeit not small wars (Beaufre 1963/1965: 28, 104).

As we saw in the previous chapters, the time since Napoleon had seen three ideas come to dominate strategic writing: one was that of the total mobilisation of one's own nation and national resources for war and the second was the quest for a decisive battle leading to the imposition of one's will on the crushed enemy. The third, which grew through Social Darwinist and racist notions and combined in the West with the desire to sidestep the enemy's military, was to target the enemy's civilians, or in its extreme form, at the hands of Hitler, to enslave and annihilate entire populations.

With its roots going at least as deep as the First World War, but shaken into explosive growth by Hiroshima and Nagasaki, a new guiding idea came to dominate strategic thinking, even though many strategists did not immediately realise how all-pervasive its implications were in the global communist–liberal confrontation. It was, in short, the idea that in a world of nuclear powers, the aim could no longer be the annihilation of the enemy in any form, nor even the uncompromising imposition of one's will, as the enemy – or his bigger ally – would have the means either to prevent this or to wreak an unbearable nuclear revenge. Liddell Hart wrote after the Second World War that, 'old concepts and old definitions of Strategy have become not only obsolete but non-sensical with the development of nuclear weapons. To aim at winning a war, to take victory as your object, is no more than a state of lunacy' (Liddell Hart 1960: 66). Martin van Creveld in 1999 even thought that the Western abandonment of the quest for victory was due entirely to the invention of nuclear weapons:

> From the beginning of history, political organizations going to war against each other could hope to preserve themselves by defeating the enemy and gaining a victory; but now, assuming only that the vanquished side will retain a handful of weapons ready for use, the link between victory and self-preservation has been cut. (Creveld 1999: 337)

The new dominating theme became the search for outcomes of mutual advantage (Schelling 1960), or for outcomes with which both sides could live. In the view of US strategist Morton Halperin, the existence of thermonuclear weapons meant that the two superpowers now had a vital interest to avoid either an 'explosion' (meaning 'the sudden transformation of a local war into a central war by the unleashing of strategic nuclear forces') or 'expansion' (meaning 'a gradual increase in the level of military force employed'); 'explosion' and 'expansion' together he defined as 'escalation' (Halperin 1963: 3), which *must* be avoided.

The superpowers could no longer aspire to clear-cut victories, in the sense of a unilateral imposition of one's will upon a disarmed enemy against his will, or a clear-cut 'zero-sum game': the realisation dawned on strategic thinkers that wars did not necessarily promise lasting victory. The outcomes of wars could still include a temporary armistice, hailed as a victory by one side or the other. But as neither side could totally 'disarm' the other, or 'bomb' the other 'back into the Stone Age' for fear of retaliation (whatever the rhetoric), even a temporarily defeated enemy would come back to the charge within a matter of years. Demonstrations of this were the Vietnam War after the Tet Offensive and even after the LINEBACKER bombings, the Arab–Israeli Wars and the conflicts between India and Pakistan or Ethiopia and Eritrea. The arrival of nuclear weapons thus changed many people's thinking, here sooner, there later, from Brodie in

1945 to Reagan and Gorbachev in the early 1980s. Philosophers and strategists pondered life under the perpetual threat of the nuclear destruction of the world, peace movements formed as never before, anti-nuclear demonstrations marked the outlook and passive culture of whole generations.

Even in 1952, the British nuclear-scientist-turned-strategist P.M.S. Blackett pronounced that total war had been abolished with the invention of thermonuclear weapons (Blackett 1956: 5). Slessor in 1958 cautiously rephrased Blackett's verdict: '*major all-out* war has abolished itself with the advent of nuclear weapons' (Slessor 1958: 320; my emphasis). French thinkers went further. After the loss of France's colonial empire removed the direct obligation to respond to smaller contingencies except in wars of choice, several French strategists professed their belief right up to the end of the Cold War that 'war is dead' (Le Borgne 1987). They only gradually gave up their Europe-focused optimism at the end of the Cold War, re-engaging in warfare in the hope that it could be a war with zero casualties (*la guerre zéro morts*), until we now find French military authors returning to the consensus that wars are again quite 'probable' (Desportes 2008).

Back in the late 1950s, Harvard-historian-turned-strategist Henry Kissinger had begun to suspect that nuclear weapons were irrelevant in the conflicts between a nuclear power and a non-nuclear power, especially if the latter was underdeveloped (Kissinger 1957: 127). Perhaps the West had put too much emphasis on nuclear deterrence, resulting in its defeats on a conventional level. This was due, in the words of Captain Stephen Roskill,

> to the fact that the Russian leaders know perfectly well that, unless they provoke us to an altogether insupportable extent, it is inconceivable that we should initiate the use of nuclear weapons. Thus, while we have ... a deterrent which ... can never be used to support our policy on any secondary issue, they are able to ... extend their zones of political and economic domination with little hindrance ... in the era of nuclear parity, in a period of 'cold war', or in a conflict for limited purpose, a strategy based mainly on nuclear power cannot be effective. (Roskill 1962: 251)

As global peace was difficult to achieve, the first aim was to limit wars that *were* taking place, to stop them from becoming nuclear.

The return of limited wars

We owe the differentiation between 'limited' and 'unlimited' war to Clausewitz, or rather to Guibert (1772/1977: 149f.). Much credit must go to Sir Julian Corbett for having imported this concept into the strategic thinking of the twentieth century (see chapter 1). It had been dismissed as completely outdated, however, by the 'realist' strategists writing on the eve of the First World War and in the interwar period, notably by Jean Colin (1911/1912: 322–46), von der Goltz (1895/1899: 5–7) and of course Erich Ludendorff (1935: 15f.). The 'realist' belief in the prevalence of Total War was initially reinforced, not questioned, by the First World War. Bernard Brodie rightly observed that 'Following World War I it became axiomatic that modern war means total war.' And this 'seemed to be confirmed and reinforced by World War II' (Brodie 1959: 307f.). Even on the eve of the twenty-first century, American strategists would see 'modern war' as coterminous with world war (Murray 1999: 75).

As Americans are culturally prone to a Manichaean view of a perpetual fight of good against evil, they assumed, in the words of historian David Rees,

> that all states are equally interested in peace, that force and 'power politics' are always to be detected in international relations, and that the conflicting policies of countries and power groups can usually be harmonised by the same means that govern internal domestic differences – due process, reason, common sense, elementary morality and institutions such as the United Nations.

If another state would not abide by these rules, the American proclivity to see it as 'evil' came to the fore again, and it had to be fought accordingly. War, as seen by Americans,

> can only be justified when fought as a crusade against tyrants in a mood of righteous indignation. Then, maximum force must be used to end the conflict as quickly as possible, and so total wars fought by democracies quickly take on an ideological character as witness the two great wars of the [twentieth] century. (Rees 1964: xi)

In similar terms, US strategic analyst Alexander George noted that

> conditioned by World Wars I and II, many Americans viewed war ... as a moral crusade against an aggressor. There was little room in such an image of war for limiting one's objectives or limiting the means one employed to achieve them ... The image of war as a moral crusade encouraged the belief that when war broke out, military factors and military judgment should have great weight, if indeed not dominate, in shaping policies for the conduct of the war. (George 1971: 1f.)

Confronted with enemies representing absolute evil, the use of 'maximum force' with the aim to annihilate the enemy's armies and to overthrow his regime became the 'American way of war'. It had crystallised since the American Civil War and was very much a manifestation of the Napoleonic paradigm. The US military had a pronounced preference for all-out war aiming at total victory, regardless of political aims (Kissinger 1957: 90; Brodie 1959: 315; Osgood 1979a: 99).

Yet the Spanish Civil War (1936–9), and soon after it the Second World War, the Greek Civil War (1946–9), the Berlin Airlift (1948–9), the French Indochina War (1946–54) and the Malay Emergency (1948–60) demonstrated that 'limited war' existed, forms of war that were 'modern' in many respects and yet could not be put in the category of the World Wars. Shortly before the outbreak of the Korean War, Basil Liddell Hart asked, almost paraphrasing Clausewitz's thinking on absolute and real war:

> Can war be limited? *Logic* says – 'No. War is the sphere of violence, and it would be illogical to hesitate in using any extreme of violence that can help you to win the war.' *History* replies – 'Such logic makes nonsense. You go to war to win the peace, not just for the sake of fighting. Extremes of violence may frustrate your purpose, so that victory becomes a boomerang. Moreover it is a matter of historical fact that war has been limited in many ways'. (Liddell Hart 1950: 366)

Initially, as the Korean War broke out, policy analysts, and political and military leaders throughout NATO feared that this might be merely the first in a series of Moscow-led moves which would culminate in an invasion of North Atlantic Treaty territory (Osgood 1979a: 96). But when the outbreak of war in the Far Eastern peninsula remained an isolated case, strategic thinkers turned to Clausewitz's writing about 'limited war' for an explanation. Rees claimed that 'the

Korean war was the first important war in American history that was not a crusade'. At any rate, it was not, as previous wars, fought until the total defeat of the enemy (Rees 1964: xi).

Korea raised great problems for American Strategy. Total war always aims for 'overthrowing the enemy leadership', what has since become known as 'regime change'. It invariably strengthens resistance on the part of the adversary; he is still in control of the armed forces and internal security forces. It may well introduce an element of civil war. For Kissinger, this was irreconcilable with superpower conflict in a nuclear world (Kissinger 1957: 127). Presidents Truman and Eisenhower thought on the same lines. In Korea, contrary to the American way of war, the United States government settled for only a partial victory. It deliberately restricted the nature and scope of its intervention in order to avoid a direct armed encounter with the Soviet Union or a protracted war with China (Osgood 1979a: 96).

American strategic theorists were well aware that limited wars had occurred before the nuclear age. Charles E. Osgood noted that '[l]imited wars are as old as the history of mankind, as ubiquitous as armed conflict'. Indeed, upon reflection Osgood realised that wars 'fought to annihilate, to completely defeat or dominate, the adversary' had actually been 'momentous and rare'. What was distinct since 1945, however, was that 'strategies of limited war ... are derived particularly from the fear of nuclear destruction and the exigencies of the Cold War' (Osgood 1979a: 93). This was why Korea was the 'first modern limited war' (Brodie 1973: 63; see also Halperin 1963: 2).

Osgood's widely accepted definition of 'limited war' was that it had to be limited 'in both means and ends'. Initially, American planners did not discount entirely the use of nuclear weapons in limited war (Kissinger 1957: 137–45; Osgood 1979a: 95). As we have seen in chapter 14, this idea was abandoned in the early 1960s, and henceforth 'limited war' was taken to mean a war fought without nuclear weapons.

Even before Vietnam, American strategists identified the difficulties of 'winning' such a war that was limited from a US perspective, if 'victory' was defined in the Clausewitzian sense of imposing one's will upon the enemy. Henry Kissinger wrote:

> A limited war ... is fought for specific political objectives which, by their very existence, tend to establish a relationship between the force to be

employed and the goal to be attained. It reflects an attempt to *affect* the opponents' will, not to *crush* it, to make the conditions to be imposed seem more attractive than continued resistance, to strike for specific goals and not for complete annihilation. (Kissinger 1957: 120)

In other words, limited wars had to be wars of *persuasion*, where arguments and incentives complemented limited use of force.

Rees explained that, 'Limited war is ... political war par excellence, in that "purely military" considerations are excluded' (Rees 1964: xvi). A limited war, according to Osgood,

is fought for ends far short of the complete subordination of one state's will to another's, by means which involve far less than the total military resources of the belligerents and which leave the civilians life and the armed forces of the belligerents largely intact ... limited war is not only a matter of degree but also a matter of national perspective, since a local war that is limited from the standpoint of external participants might be total from the standpoint of the local belligerents, as in the Korean and Vietnam wars. Clearly, the Western definition of limited war ... reflects not some universal reality but the interests of the Western allies, and especially of the United States, in a particular period of international conflict. (Osgood 1979a: 94)

It thus became clear quickly to a few strategists that what was limited war – occasionally referred to as 'local war' (Halperin 1963: 2) – for the USA was hardly limited for Korea and other countries on whose territories the war was fought. In the words of Bernard Brodie: 'What is a limited war to us, limited in terms of emotional as well as material commitments, may be total war to our opponents and to one or more of our allies; thus, their demonstration of "resolve" may well exceed ours' (Brodie 1959: 315). This foreshadowed the configuration that would characterise the Vietnam War, where the Americans, notwithstanding their superior weaponry and forces, had a limited commitment and were ultimately defeated by an inferior adversary with a greater commitment, an adversary who tried to mobilise totally for this war (see chapter 15).

But short of achieving a victory in the traditional way of winning a war militarily, could one side or the other further its aims? Western strategists thought that this might well be the case, but for the communist side rather than the American, which was so wedded to the

Napoleonic paradigm. Writing just after the end of the Korean War, while France was still up to its neck in colonial withdrawal wars, French Vice Admiral Pierre Barjot asked:

> [M]ay not the hydrogen bomb incite Communism to take recourse exclusively to conventional weapons, speculating on the psychological self-deterrence of the Western powers ... [as] to unleash the H-bomb which would lead to the suicide of humanity? ... The risk of setting the world ablaze has brought about a return to limited wars. (Barjot 1955: 229)

Limited war was thus not to the advantage of the West, which would in such a context be self-deterred from using its nuclear weapons. Osgood explained this in terms of two strands of 'limited war' thinking. One was the Western strand, using force with restraint to deter and contain the other side, in the hope of avoiding general war. 'The other strand, inspired by Mao Tse-tung and Third-World nationalism and propounded by revolutionary nationalists, sought to use guerrilla warfare' to fight for independence against Western colonialism. It uses 'limited means – the Strategy of insurgency – to achieve total political conquest' (Osgood 1979a: 93). This strand was proving remarkably successful, even in the face of Western counterinsurgency. In the Cold War it was seen, perhaps excessively, in the context of the ideological strife which made each side support any state, faction or group as long as that was also locked in conflict with the other side. France and the USA emerged particularly scarred. Overall, the Western track record was not good, from Indonesia and Indochina/Vietnam to Algeria, and the wars of Angola and Mozambique against Portugal.

In view of the traumatic experiences of Indochina and Algeria, it was a French strategist who applied a particularly systematic analysis to the problem of Strategy in the nuclear age, and to limited wars.

General Beaufre and the systematic reconsideration of Strategy

General André Beaufre was a friend of Liddell Hart's, and his work was influenced in part by Liddell Hart's thinking about the 'indirect approach' in Strategy (see chapter 7). Beaufre's *Strategy of Action* of 1966 complements his earlier book on the 'strategy of deterrence'; he saw these as two opposite Strategies. His 'strategy of action' included coercion through the use of armed force, or merely its threat. As Beaufre explained, 'action is always based upon a dialectic between the possible gain and the possible loss, a balance sheet between the expectation of success and the fear of the risks involved' (Beaufre 1966/1967: 28, 32)

Beaufre distinguished 'five levels of action':

1. Complete peace.
2. The cold war levels:
   (a) the level of insidious intervention
   (b) the level of overt intervention
3. The levels involving the use of armed force:
   (a) the conventional war level (ranging from limited military intervention to full-scale military war)
   (b) the nuclear war level, ranging from warning shots to the nuclear paroxysm. (Beaufre 1966/1967: 62)

Beaufre drew attention to complications to Strategy-making arising from conflicting political objectives (Beaufre 1966/1967: 70). Following Clausewitz, he prescribed a systematic approach, starting with the diagnosis of the political situation, followed by setting out one's political and strategic objectives, drawing up a balance sheet of one's own objectives and means and the adversary's objectives and means and finishing with a plan of action that would allow for different phases of operation. These had to include the whole range of actions from the preparation of domestic public opinion and the appeal to international opinion in order to isolate the enemy, and then diplomatic and military action, ideally achieving concessions without the shedding of blood (Beaufre 1966/1967: 83–101). Sun Tzu and all the Western Fabians would have approved (see chapter 4).

Using his earlier definition of 'modes', Beaufre described 'direct and indirect "modes" of action', as part of a 'total strategy' comprising all 'modes'. One mode of action for him was the 'total strategic manoeuvre' consisting of 'actions which may be carried out in fields other than military in order to render maximum assistance to the military action' (Beaufre 1966/1967: 102–27). As Beaufre explained:

The concept of total strategy in the indirect 'mode' is extremely logical but it differs considerably from the habits formed in the nineteenth century when direct strategy was the normal method of action ... [It means] using the minimum of force and military resources, the art of maximum exploitation of the narrow margin of freedom of action left to us by nuclear or political deterrence in order to achieve the desired, frequently large-scale, results ... [I]ndirect strategy amounts to a tough form of negotiation. (Beaufre 1966/1967: 112f., 115)

Taking up Liddell Hart's idea of the indirect approach, Beaufre explained that: 'the new fundamental principle is that limited war, on the material level of forces ... is unlimited on the psychological and moral level ... The new strategy is a very indirect military strategy where the psychological decision is sought directly by the preponderant application of psychological means' (Beaufre 1972: 42). After an overview of warfare in previous centuries, Beaufre opined that:

[O]nly by a pseudo-Clausewitzian aberration have we been led to believe that cases existed where one aim was clearly predominant over all the others. 'Winning the war' is not a political aim; as Liddell Hart has clearly demonstrated, the real political aim is the type of peace to follow the war. (Beaufre 1966/1967: 24)

Here again we see the relinquishment of the belief that 'victory' in a military sense alone was the be-all and end-all of war.

Indeed, Beaufre thought that war 'under the sign of a Clausewitzian strategy, is today impossible' in view of the extremes to which nuclear weapons would take it: 'In conventional warfare, the defeat of armed forces and the occupation of territory were the signal of capitulation. This convention has lost a good part of its validity.' In the Second World War, Yugoslavs, Greeks, Chinese, all occupied by enemy forces, 'resisted for years ... waging guerrilla war'. After the end of the worldwide conflagration, the Vietnamese with much inferior weapons managed to hold their own successively against the French and the Americans, and the Algerians against the French. The outcomes of these conflicts were determined less and less by 'military victory. In such a situation, one sees that the strategy of limited war is essentially different from conventional war: the conventional war aims to win, *limited war aims to convince*' (Beaufre 1972: 41). Here Beaufre was echoing the Spanish philosopher and writer Miguel de Unamuno (1864–1936), who famously told General Franco's nationalists, 'you may be winning, but you have not convinced [me]'. True victory in an ideological context, one might deduce from this, would only be one in which the other side is not militarily beaten, but persuaded to accept a new situation.

Coercion

If persuasion in an entirely non-violent political sense failed, how could force be brought to bear on any situation in the nuclear age? This is the subject of much literature, especially in the Anglosphere, on 'coercion', introduced above all by Thomas Schelling (1966) and Alexander George, who wrote:

> Traditionally force and threats of force have been used in several ways in order to influence the calculations and behaviour of opponents in world politics. We can identify at least four general Strategies:
>
> 1. The quick, decisive military strategy
> 2. Coercive diplomacy
> 3. Strategy of attrition
> 4. Test of capabilities within very restrictive ground rules. (George 1971: 16–21)

George did not claim that this was a new invention, but on the contrary pointed out that the essentials of the Strategy of coercive diplomacy

> have long been known, although the coercive diplomacy of an earlier era ... was not articulated systematically. Rather, it was part of the conventional wisdom of those who engaged in statecraft and diplomacy, implicit in what they did and occasionally discussed in memoirs rather than explicated in scholarly works. (George 1971: 21)

Coercive diplomacy in his terminology could be both offensive and defensive. A defensive use of coercive diplomacy would be the attempt

> to persuade an opponent to stop doing something he is already doing ... or to undo what he has already accomplished. In contrast to this defensive use of the strategy, coercion may also be employed offensively to get the opponent to do something he has not done and does not want to do – to make him pay a price, give up territory – in order to avoid the threatened sanctions. An analogy here is the robber who persuades his victim to turn over his money peacefully.

In other words, this was 'diplomatic blackmail' (George 1971: 24); where nuclear weapons are involved in this process, 'diplomatic blackmail' would become 'nuclear blackmail' (Betts 1987: 13f.). Thomas Schelling explained:

> The bargaining power that comes from the physical harm a nation can do to another nation is reflected in notions like deterrence, retaliation and reprisal, terrorism and wars of nerve, nuclear blackmail, armistice and surrender, as well as in reciprocal efforts to restrain that harm in the treatment of prisoners, in the limitation of war, and in the regulation of armaments. Military force can sometimes be used to achieve an objective forcibly, without persuasion or intimidation; usually, though ... military *potential* is used to influence other[s] ... by the harm it *could* do to them [my emphasis]. It ... can be used for evil or in self-protection, even in the pursuit of peace; but used as bargaining power it is part of diplomacy – the uglier, more negative, less civilized part of diplomacy – nevertheless, diplomacy.

The fear of nuclear escalation of any local war did not prevent powers from 'trying it on', from rattling their sabres, or putting pressure on smaller powers.

> There is no traditional name for this kind of diplomacy. It is not 'military strategy' which has usually meant the art or science of military victory, and while the object of victory has traditionally been described as 'imposing one's will on the enemy', how to do that has typically received less attention than the conduct of campaigns and wars. It is a part of diplomacy that, at least in [the USA], was abnormal and episodic, not central and continuous, and that was often abdicated to the military when war was imminent or in progress.

Since 1945, however, 'this part of diplomacy' had become 'central and continuous; in the United States there has been a revolution in the relations of military to foreign policy'. In his book Schelling tried to identify 'principles that underlie this diplomacy of violence', trying to ascertain

> how countries do use their capacity for violence as a bargaining power, or at least how they try to use it, what the difficulties and dangers are and some of the causes of success or failure. Success to some extent, failure even more, is not an exclusively competitive notion; when violence is involved,

the interests even of adversaries overlap. Without the overlap there would be no bargaining, just a tug-of-war. (Schelling 1966: vf.)

Here again, we find the idea that it is necessary to part company with any unilateral imposition of one's will upon the enemy, and to build on interests held in common with the adversary.

Semantic confusion persists in view of different definitions: many authors use 'coercion', 'coercive diplomacy' and 'compellence' interchangeably. According to Lawrence Freedman's comprehensive synthesis on the subject, 'coercion' should be 'defined as the deliberate and purposive use of overt threats to influence another's strategic choices. The term was chosen to distinguish it from both compellence, which is but one variant of coercion, and coercive diplomacy.' Building on the definitions by George and Schelling, Freedman suggests the following use of the terms *compellence* and *deterrence*, as the aggressive and defensive sub-modes of *coercion*: 'compellence' should be used for a '*forcible* offence' such as 'taking something, occupying a place or disarming an enemy or a territory, by some direct action that the enemy is unable to block'. This might include 'his withdrawal or his acquiescence, or his collaboration by an action that threatens to hurt, often one that could not forcibly accomplish its aim but that, nevertheless, can hurt enough to induce compliance'. Thus:

Deterrence involves a demand of inaction and *compellence* a demand of action. Deterrence involves explaining what must not be done and the consequence if it is – and then waiting. The overt act is up to the opponent. Compellence, in contrast, usually involves *initiating* an action (or an irrevocable commitment to action) that can cease, or become harmless, only if the opponent responds. (Freedman 1998: 18f.)

Freedman himself diverged from Alexander George by arguing that 'coercive activity is not inherently defensive or offensive (for this depends essentially on the objectives being pursued) nor does it necessarily involve modest force'. Freedman put more emphasis on 'the importance of counter-coercion and the influence of specific acts of coercion on strategic relationships over the long term' (Freedman 1998: 3f.).

Analysing the considerable literature on coercion, Robert Pape usefully grouped it into four broad schools or 'families'. The first of these focused on the comparable strength of resolve and reputation for determination. This 'family' included Fred Iklé (1964) and Schelling himself (1966). The second of Pape's 'families' emphasised 'the balance of interests' and indeed stakes in a dispute, which serves to explain how a militarily stronger power might not prevail when pitted against a weaker power that has little to lose and is completely committed to a cause. Pape included Osgood and Tucker (1967), with their comprehensive study of 'force' or 'military power' in its practical, but also legal and moral dimension, answering in the negative the question of whether 'force' and 'war' were obsolete in the nuclear age. Here Pape also listed George *et al.*'s work on the *Limits of Coercive Diplomacy* (1971). Pape's third 'family' of strategists had pondered particularly 'the vulnerability of the adversary's civilian population to air attack' and the leverage that might be gained from threatening it. These were the early air theorists who urged upon their governments a city bombing Strategy, including, of course, Douhet (1921), J.F.C. Fuller (1923), with many thinkers of this era well summed up by Quester (1966). It also included Liddell Hart at his worst, when he advocated city bombing, with *Paris, or the Future of War* (1925). Pape's fourth 'family' 'points to the balance of forces: striking military targets in the adversary's homeland shifts the military balance, thereby compelling the victim to modify its behaviour'. Here Pape named Edward Luttwak and Paul Nitze (Pape 1996: 7), but Richard Betts's work (1987) should also be included in this category. Pape then examined the relative success in using 'threats to civilians ("punishment") and threats of military failure ("denial")' for the purposes of deterrence or coercion.

Punishment threatens to inflict costs heavier than the value of anything the challenger could gain, and denial threatens to defeat the adventure, so that the challenger gains nothing but must still suffer the cost of the conflict. (Pape 1996: 7)

Pape examined historical cases, concluding,

Leaders are often drawn to military coercion because it is perceived as a quick and cheap solution to otherwise difficult and expensive international problems. Nonetheless, statesmen very often overestimate the prospects for successful coercion and underestimate the costs. Coercive attempts often fail, even when assailants have superior capability and inflict great punishment on the target state. Famous failures include German attempts to coerce Britain in 1917 and again in 1940, the French occupation of the Rhineland in 1923–1924, Italian efforts to coerce Ethiopia in [1935], the American embargo against Japan in 1941, Allied bombing of Germany in World War II, American efforts against North Vietnam from 1965 to 1968, and Soviet operations against the Afghan rebels from 1979 to 1988.

By contrast, Pape found 'virtually no instances in which coercion was not tried when a credible claim can be made that it would likely have succeeded' (Pape 1996: 2).

Analytically, it is rarely justifiable to reduce complex negotiations to the one sub-category of coercion. As Freedman noted, it is difficult to distinguish between, say, Iraq in the late 1980s trying to persuade Kuwait to write off Iraqi debts by diplomatic means alone – when Iraq's superior armed forces always loomed in the background – and actual threats of invasion (Freedman 1998: 18). In both cases, force plays a role; the psychological impact of latent threats, however, may vary depending on the perception of the target of that threat. Compellence, like deterrence, may to some extent lie in the eye of the beholder.

A purely defensive posture in the parlance of the US political scientists could also be termed 'suasion' or 'dissuasion' (the latter being also the French word for 'deterrence'). With such an aim in mind, a different way to achieve it was taken by those Europeans studying the issue, a more practical, *matériel*-orientated approach.

Defensive defence and the relinquishment of victory

The nuclear age had brought forth the disarmers, advocating the elimination of nuclear weapons and the reduction – ideally elimination – of armaments, following the precedents of arms limitations treaties of the interwar period. Soon after the Second World War, Liddell Hart, for example, advocated the abolition of all 'weapons of mass destruction', but also of heavy tanks and heavy guns, admitting that this was difficult for him to write as he was a great fan of mechanised armies:

As a military theorist, one had seen in armoured mobility the solution of the military problem of reviving the power of the offensive. But when one came to look at the problem from the wider point of view – of peaceful nations wishing to check aggression – the sensible solution was to nullify such a

> rebirth by abolishing the means of it ... The core of the whole disarmament problem lies in *convincing* the aggressive that victory is unattainable from the start. The most effective means is to annul the chances of successful attack. To sterilise offensive potency is to sterilise war itself. (Liddell Hart 1950: 354, 358; my emphasis)

Also in the 1950s, a number of strategists like Liddell Hart tried to think of ways in which East–West tensions could be structurally contained through force structures, especially along the Central Front, that might not lend themselves to scenarios of provocation and escalation. Upon retiring from the RAF, Sir John Slessor put forward alternative proposals for the structure of the German Bundeswehr created in 1955 (Dean 1987/8: 65). British military historian Michael Howard developed ideas of Western disengagement from the Continent, in reaction to Polish proposals, known as the Rapacky Plan (Howard 1958). In the 1970s the retired Brigadier Kenneth Hunt, under the aegis of the IISS, worked on force structures which might bring more stability to the Central Front areas without looking like an open invitation for the Warsaw Pact to invade. Military historians John Keegan and Trevor Dupuy, much like Hunt, thought about hardware solutions, in their case focusing on fortifications and obstacles (Dean 1987/8: 65, 77). Even at the very end of the Cold War, some of the finest British strategists, Ken Booth and John Baylis from the University of Wales, examined alternative defence plans to those developed by NATO (Booth and Baylis 1989).

Sparked off by the last peak in East–West tensions, there was a further increase in publications on options of defending against a major (that is, Warsaw Pact) attack with conventional means only. In Britain, the Alternative Defence Commission undertook serious critical studies of how to remodel British defence to make it work without nuclear weapons; they concluded that it meant the need for a very much larger conventional force, and indeed the reintroduction of conscription (Alternative Defence Commission 1983, 1985). Other authors came from neutral countries (Austria, Sweden), from peace research institutes or from parties on the left and peace movements in NATO countries. General Emil Spannocchini, Chief of the Austrian Army, very seriously examined low-intensity, guerrilla warfare options for his country in the light of the writings and applied Strategies of Mao, Giap and Tito, turning essentially on a double denial: denying the enemy a major, decisive battle (which he would win, in view of his superiority), but also denying the enemy the successful occupation of the country. We have seen how these concepts have their roots in the Spanish and Russian resistance against Napoleon (see chapter 15). Any small country with a mountain redoubt or otherwise inaccessible areas had a chance of retreating to it and holding out against even a major adversary by denying him total control of the country, and by irritating him, guerrilla-fashion, with continual skirmishes; if the Swiss had espoused this concept, why not Austria (Spannocchini and Brossollet 1976)? At the end of the Cold War, authors from Scandinavian countries both within and outside NATO examined such options (Boserup 1990; Crenzien 1990; Nokkala 1991).

In Germany, a number of ex-Bundeswehr officers such as General Gerd Schmückle, Brigadier Eckart Afheldt and Siegfried Freundl; peace researchers from universities and think tanks, such as the nuclear-physicist-turned-peace-activist Carl Friedrich von Weizsäcker, the retired Brigadier Schmückle and Albrecht von Müller; the university lecturers Theodor Ebert, Götz Neuneck and Dieter Lutz; and staff from the Social Sciences Research Institute of the Bundeswehr, tried to think through similar concepts. They all sought to get away from NATO's reliance on nuclear deterrence, built on the threat to use them if necessary (Hannig 1984). The alternatives put forward usually involved asymmetric responses: guerrilla tactics, small units inflicting pin-pricks on the enemy, denying him the control of the country rather than giving battle, setting up early versions of electronically connected networks that could operate independently, if necessary, while consciously avoiding any military configurations, deployments and weapons systems which might be perceived as aggressive (Afheldt 1976, 1983). Some of the writing was inspired by guerrilla and counter-guerrilla manuals. Like guerrilla fighters, such small defensive units would have relied heavily upon the indirect support of the indigenous population and administration, who were, however, barred by international law from offering active resistance themselves – quite a complicated equation (Weizsäcker 1984; Schmückle and von Müller 1988; Møller 1991).

'Defensive defence' Strategies tended not, however, given any serious consideration by governments of NATO member states, as they were so utterly at variance with nuclear-deterrence-based NATO Strategy. Moreover, in pointing out the differences between European and

American interests (Afheldt 1983: 164–208, 289–306), they were seen as undermining Alliance solidarity. Finally, they would have favoured the Soviet Union: until 1987 the Warsaw Treaty Organisation had a war-winning Strategy in which it would have used all its forces, possibly merely in a conventional mode, but nonetheless with enormous firepower, to crush any opposition to its supposedly defensive invasion of Western European NATO territory (Heuser 1993, 1998b). By contrast, these concepts would have been of extreme interest in reconstructing European defences after the end of the Cold War, especially in the former Warsaw Pact countries that in several waves joined NATO. Non-provocative forces, 'structurally unable to attack' (Schmückle and von Müller), configured in such a way might have made it considerably easier for Russia to accept the transformation of European security in the 1990s and the first decade of the twenty-first century. But even in these entirely changed circumstances, they were not given much consideration.

What both Cold War NATO Strategies from the mid-1950s and defensive defence had in common was that neither aimed at *victory* as a war aim (Heuser 1998b). US Field Manual FM 100-5 of 1954 had already omitted the term, and FM 100–5 of 1962 replaced the concept of limited war with that of limited means. The manual postulated that: 'The essential objective of United States military forces will be to terminate the conflict rapidly and decisively in a manner best calculated to prevent its spread to general (nuclear) war' (q.i. Summers 1982: 69). As Western strategists focused increasingly on persuasion, deterrence, non-offensive defence and even coercive diplomacy, the classical, medieval and modern paradigm of war as a legitimate instrument of state politics was indeed receding. Most Western theoreticians and policy-makers came to agree that victory could not be a meaningful concept in all-out nuclear war (Kissinger 1957: 86, 90). The same realisation crept into the Soviet Strategy debate, although it did not become dominant until the late 1980s. In January 1964, two authors writing in Soviet military journals, Major General N. Sushko and Major Kondratkov, challenged the orthodox doctrine that war continued to be the continuation of politics by rational means: they argued that 'on account of its clear character of destruction and annihilation', nuclear war had 'ceased to be a reliable means to reach political aims, whatever these may be' (q.i. Dahm 1970: 394). Both were subsequently disavowed, but the ensuing debates about the maxim that 'war is a rational tool of politics' was a touchstone for Soviet orthodoxy, and the abandonment of this maxim marked the collapse of this ideology in Europe (Heuser 2002: 143–50). Russian Strategy rediscovered victory in 1993, when their 'Basic Provisions of the Military Doctrine of the Russian Federation' once more spelled out the defeat of an enemy as a war aim. By contrast, for most Western liberals in the early twenty-first century, *victory* (like 'glory') seems of little value as a thing in itself, as the price at which it might come seems disproportionate to the gains and as it does not guarantee peace.

18 *No end of history: the dialectic continues*

In war, there is no substitute for victory.
(General Douglas MacArthur 1951)

Do you want to win, or do you want peace?
(Martti Ahtisaari, q.i. Lyall 2008)

The Napoleonic paradigm strikes back: Summers's Clausewitzian critique

At the end of the Cold War, the euphoric hope spread that the collapse of Soviet-led communism might mean the end of ideological strife in the world. Assuming that no new (or old) ideology (or religion) would rise up to challenge liberal democracy as 'best of all possible' systems, to paraphrase Voltaire, and assuming that wars originated mainly from ideological disagreement, one could argue that this was, in a Hegelian or Marxist sense, the end of dialectical history (Fukuyama 1989). But the dialectical debate between those who favour the abandonment of the Napoleonic paradigm and those to whom it is the only thing that makes sense in military terms continues.

Not even all Western strategists found this cultural shift in the West easy to accept, as we have seen in the context of the evolution of air power after 1945. In 1982 Edward Luttwak deplored the fact that

> the West has become comfortably habituated to defeat, Victory is viewed with great suspicion, if not outright hostility. After all, if the right-thinking are to achieve their great aim of abolishing war they must first persuade us that victory is futile, or, better still, actually harmful. To use Stalinist language, one might say that the struggle against war requires the prior destruction of the very idea of victory. (Luttwak 1982: 17)

His criticism, crucially, highlights the fact that this Western shift of perceptions is unparalleled by thinking in many other countries and cultures. Indeed, Robert Cooper argues that it is not even shared widely within the USA, which is still operating very much as a 'modern' state that fiercely defends its right to use force as it sees fit, with or without a mandate from the world's higher authority, the UN, which has now taken the place of God in the just war theology of old. For Cooper, it is only Europeans that have acquired a 'post-modern' mindset leading them to renounce the use of force to settle their own quarrels by force. Even he rightly points out that the peaceful European Union's central problem is its coexistence on one planet with 'modern' states that have not renounced the use of force, and with 'pre-modern' entities aspiring to statehood and a state monopoly of the use of force, but in reality torn by rivalling claims to power (Cooper 2000).

Especially before the end of the Cold War, it was not only strategic theorists like Edward Luttwak, Colin Gray and Keith Payne who found the culture shift in the West, or in the EU plus perhaps Japan, Australia, New Zealand and some elements of the American intelligentsia, problematic. Above all the American military found it trying to live with wars without victory, more so than, for example, the United Kingdom with its long history of imperial policing. In 1982, US Colonel Harry G. Summers published 'A Critical Analysis of the Vietnam War', which he called *On Strategy*. This publicised the US Army War College's own critical analysis of the Vietnam War and drew heavily on Clausewitz, implicitly dismissing the sophisticated theories of the political use of the military instrument for 'coercion' discussed above.

What were the reasons of American failure in Vietnam? 'One of the most frustrating aspects of the Vietnam War from the Army's point of view is that as far as logistics and tactics were concerned we succeeded in everything we set out to do ... How could we have succeeded so well, yet failed so miserably?' (Summers 1982: 1). Colonel Summers drew up a long list of reasons, each time correlating them to passages from *On War* for enlightenment. First, Summers pointed to the confusion of what Clausewitz called the mere 'preparation for war, and war proper', the latter requiring the intelligent application of fighting forces, equipment and so forth in the war itself – the employment of the armed forces to secure the objectives of policy. Secondly, Summers underlined that the political objectives themselves were flawed, as they did not meet Clausewitz's criterion for what constitutes appropriate ends of Strategy: the choice of 'objectives that will finally lead to peace' (Summers 1982: 4).

A third mistake was identified by Summers as the failure of the US government to mobilise the American people's support for the war effort, pointing at Clausewitz's analysis of the difference between the eighteenth-century cabinet wars and the French Revolutionary and Napoleonic Wars, which in his view had been so successful because war had become 'the business of the people'. Americans thus saw Vietnam as '[President] Johnson's war' or 'Nixon's war' (Summers 1982: 6). This criticism was not just Summers's: British journalist Henry Brandon had talked to leading US civilian and military men in Saigon in late 1967 and asked them about US public opinion; all answered that that was not for them to consider, but for the US President. Brandon objected:

> But foreign policy begins at home. The fact that President Johnson ignored American public opinion and disregarded the mounting opposition to the war as he fought it ... aroused deep mistrust in government and in the military. The war was seen by many as unnecessary; by others as immoral. It inflamed the already restless youth and created an atmosphere conducive to violence whether on the university campuses or in the black ghettos. (q.i. Brodie 1973: 439f.)

Bernard Brodie commented that 'Brandon was pointing to what was obviously a failure in strategy', to the point where 'American performance in Vietnam was adversely affected. The fighting morale of American troops began to decline conspicuously even before their numbers were reduced' (Brodie 1973: 440). This failure to mobilize popular support for the American war effort in Summers's view lay in the fact that the key military figures, '[w]ith World War II fresh in their minds ... equated mobilizing national will with total war, and they believed total war unthinkable in a nuclear age'. They disregarded the fact that the USA had fought several limited wars in the nineteenth century, and that each had involved the mobilisation of public support (Summers 1982: 13). In Vietnam the USA had been much less committed to the war than were its opponents.

With regard to the formulation of Strategy, its application in war and the conduct of the war itself, Summers again identified several reasons for America's failure. Instead of concentrating on classical,

Clausewitzian Strategy, US leaders had concentrated on technology – *matériel* –, 'the internal organization of the army, and the mechanism of its movements' (Summers 1982: 3). The Strategy formulated by the civilians, according to Summers, failed to define war aims which were realistically achievable. He accused the Johnson and Nixon administrations of having been overly concerned with the danger of an escalation of the conflict through Chinese intervention (as had happened in the Korean War) or even Soviet intervention, either of which might have led to nuclear war. As Henry Kissinger observed on this point:

> The paradox we never solved was that we had entered the Korean War because we were afraid that to fail to do so would produce a much graver danger to Europe in the near future. But then the very reluctance to face an all-out onslaught on Europe severely circumscribed the risks we were prepared to run to prevail in Korea ... Ten years later we encountered the same dilemmas in Vietnam. Once more we became involved because we considered the warfare in Indochina the manifestation of a coordinated global Communist strategy. Again we sought to limit our risks because the very global challenge of which Indochina seemed to be a part also made Vietnam appear as an unprofitable place for a showdown. (Kissinger 1979: 64)

Central to Harry Summers's critique of US Strategy in Vietnam was, of course, the relinquishment of the concept of 'victory'. He argued that the USA from the beginning declared the limits of its involvement in Vietnam, something on which its adversaries could base their own Strategy.

Summers concluded that the resulting Strategy sought excessively to localise US military action to South Vietnam with initially only severely circumscribed air action against North Vietnam. Instead of concentrating on North Vietnam as the enemy's centre of gravity, then, the successive US governments focused on 'counterinsurgency', limited to the territory of South Vietnam (Summers 1982: 174, 177). Summers saw analogies in the Korean War, where the US leadership had been extremely cautious about air attacks on Chinese territory – reflecting the general 'limited war' theme of escalation avoidance. Until the early 1970s, the USA did little to fight against the true centre of its opponent's gravity, North Vietnam itself, the 'great rear area', as Giap called it (see chapter 15). Actions against the North were limited to air strikes and naval operations. General Maxwell Taylor told the

US Senate in 1966 that it was not the US objective to defeat North Vietnam, but merely 'to cause them to mend their ways'. Dismissing the danger of a massive escalation of the war at least within the region, Summers complained that victory against the principal adversary was thus excluded as an objective (Summers 1982: 103). For fear of spreading the war to other parts of Asia, the USA regarded China as taboo, and only used air strikes against Laos and Cambodia, avoiding a ground invasion, even though both countries were sources of supplies to the Vietnamese communists. Summers criticised the fact that the USA thus focused fighting on the 'minor centre of gravity' – the Viet Cong and the North Vietnamese armed forces, while sanctuarising the 'major centre of gravity', thus failing to adopt a Strategy that could lead to US strategic success. 'Ironically, our tactical successes [in South Vietnam] did not prevent our strategic failure and North Vietnam's tactical failures did not prevent their strategic success' (Summers 1982: 88–91, 129).

Finally, Summers underlined the conclusions of a former commander of the US forces in South Vietnam and subsequent Chief of Staff, General Frederick C. Weyand: in Vietnam, the military had been 'called upon to perform political, economic and social tasks beyond its capability while at the same time it was limited in its authority to accomplish those military tasks of which it was capable' (q.i. Summers 1982: 79). In Summers's view, only the South Vietnamese themselves might have succeeded in the state-building tasks which the Americans decided to shoulder in the South, as only they had a credible long-term stake in the country.

Summers's lecture tours on this subject were well received – the US military itself went through a major rethink about US Strategy, and with Summers's Clausewitzian analysis reached the conclusion that in Vietnam, it had been asked to do the impossible and stopped from doing what it did best – namely fight large-scale wars with massive technological superiorities, in the quest for victory, not for political signalling. The outcome of this revision was the AirLand Battle doctrine, developed in the late 1970s and early 1980s, and given official endorsement with Field Manual 100-5 *Operations* of 1982. It implicitly reintroduced the concept of victory:

This doctrine is based upon securing or retaining the initiative and exercising it aggressively to *defeat* the enemy. Destruction of the opposing force is achieved by throwing the enemy off balance with powerful initial blows struck against critical units and areas whose loss will degrade the coherence of enemy operations. (Batschelet 2002; my emphasis)

When the US tried to turn AirLand Battle into the basic operational doctrine for NATO, it had to be substantially modified for several reasons. Victory and defeat were no more acceptable to America's European Allies during this period – the last big crisis of the Cold War – than in the previous two decades. Moreover, European allies objected to the deep (counter-) strike element by land forces. While AirLand Battle included the option of counterattacking on land with deep enveloping movements to get behind the first echelon of attacking enemy forces, this was removed from the NATO concept, follow-on-forces-attack (FOFA), as NATO members did not want to be accused of aggressive planning for NATO forces to cross the Iron Curtain and move into Warsaw Treaty Organisation (WTO) territory (Gouré and Cooper 1984).

Either way, fortunately no opportunity presented itself to the US or NATO to apply these new doctrines before the Cold War drew to a close in Europe, in what was clearly a 'win-win situation' for all sides, as the Third World War had indeed not taken place. Just after the end of the Cold War, however, the opportunity would present itself to put major aspects of AirLand Battle to the test in a different context in Iraq.

Major war since 1945

As noted above, the Korean War was a major war in all respects other than the abstention by the West from using atomic bombs. For the Vietnamese, France's Indochina War of nine years was major, and the Tet Offensive of 1968 surely qualifies the eight-year war with US involvement as a 'major war', even when measured with the yardstick of Moltke's wars in the 1860s. So were the Arab–Israeli Wars, even though they lasted days rather than months or years; in 1956 and in 1967, the Israelis made impressive demonstrations of what they had learnt from the *Blitzkrieg* operations of the Wehrmacht. Then in 1991, something much akin to AirLand Battle was applied by the USA in the First Gulf War against Iraq. While the air bombing campaign had partly divergent priorities from the 1982 concepts of AirLand Battle (see chapter 13), here again was a Strategy that made the most of the American *forte*: overwhelming firepower, superior technology, mass, all in pursuit of a crushing military victory. Summers seemed amply vindicated: this resounding military success was celebrated as the post-1945 apogee of the 'American way of war', a comeback of the Napoleonic–Clausewitzian paradigm. America's huge military and technological superiority had been brought to bear on an enemy, the forces had acted together, a huge air campaign had prepared the way for a ground invasion which met hardly any resistance and the American military felt it could shake off the galling experience of failure associated with their Vietnam experience and walk tall again. This allowed US strategic thinkers to focus for a while on their preferred way of war – based on technological solutions to politico-military problems.

This phase coincided, not accidentally, with a wave of military-technological innovations, another 'revolution in military affairs'. Strategists threw themselves into technological topics, much like the *matériel* school in maritime Strategy in the late nineteenth century (see chapter 9), or the technology experts in air and nuclear Strategy, and the buzzwords were 'network-centric warfare' and the like. Yet in studying major US wars in the twentieth century, analyst Stephen Biddle reminded his readers of the many other predictions of revolutionary change as a result of technological innovation in the past, which had proved wrong or exaggerated: from Hieram Maxim, who thought that his gun would make war too horrible to envisage, and Alfred Nobel, who thought dynamite would prohibit war because it would make it too terrible, to Jan Bloch, who thought major war obsolete, to the *Jeune École* of French naval Strategy, who thought the torpedo would make the capital ship obsolete and would revolutionise naval warfare, the post-First World War air power thinkers like Douhet, who thought warfare on the ground would be reduced to almost no importance and who believed that the Second World War could be won in weeks with a 'knockout blow' from the air, and to those who thought major war would be impossible once the atom bomb had been invented (Biddle 2004: 198f.). This is not to say that 'network-centric warfare' was an insignificant innovation – it simply did not provide the key breakthrough solution to US warfare at the turn of the millennium.

Rather than being an amazing departure in warfare, the First Gulf War was at best the perfection of a long-standing tradition (Rosen forthcoming). The casualties on the coalition side were around 200 dead and 750 wounded, a low record to be beaten only by NATO performance in the Kosovo conflict, with no combatant deaths on the allied side; but Kosovo was not by any stretch of the imagination the sort of major war in which the US military specialised. What was still very traditional about the Gulf War of 1991 was the attitude of the US military to enemy casualties. As in the air bombing of the Second World War or in Korea, and like the military of any country in any pre-1945 war, the US military still gloated about enemy casualties.

In Vietnam, the enemy was treated as a statistic in the 'body counts' that were supposed to feed figures into the computers of operations analysis. In Summers's view, it was courting moral outrage to bureaucratise the language used by the military to describe their actions: 'We did not kill the enemy, we "inflicted casualties"; we did not destroy things, we "neutralized targets".' What was new was that the bloody truth behind these euphemisms (many of which were neither neologisms nor unique to the Vietnam War) was brought home to millions of television spectators every evening. The incongruence between the language of government and military pronouncements and the visible horror of the war eroded what little public support there had ever been (Summers 1982: 35–7).

In the First Gulf War, the US armed forces thought it wise to go to the opposite extreme in not publicising enemy casualty numbers. But this went hand in hand with at best indifference to them, which revolted sensitive elements of Western public opinion. The pictures of the burnt-out retreating Iraqi vehicles and charred bodies on the road to Basra, let alone civilians killed as 'collateral damage', challenged this one-sided approach among the publics of liberal states.

More critically still, once more victory did not bring lasting peace. In the years 1991–2003, in the absence of a proper post-war settlement, Iraq could not be given a stake in any 'peace'. It was a period in which the UN continued to seek to exact compliance from Iraq on a range of points. The UN's Strategy was to lay aside military means – with the exception of the patrols of the no-fly zone, and the enforcement of sanctions, which were supposed to change Saddam Hussein's policies through economic pressure. The consequences were no less drastic than military action, however, and on a smaller scale had effects similar to the allied blockade of Germany in the First World War. According to estimates by the International Red Cross, the sanctions imposed between 1989 and 1999 led to significant increases in infant mortality rates (ICRC 1999), of which Iraqis themselves were very aware, staging frequent *anti-Western* protests about them. Arguably this was due more to how the regime chose to use the UN's 'Oil-for-Food Programme' than to a deliberate UN policy to withhold vital medicines and food. Nevertheless, how could the results be tolerated given that the UN Security Council's measures aimed to coerce Iraq's oppressive regime, not punish the population?

When at the time of the Second Gulf War, ostensibly fought (like Western involvement in the Yugoslav Wars and Afghanistan from 2001) for the benefit of an oppressed people, US Secretary of Defense Donald Rumsfeld shrugged off the looting unleashed by the US-led coalition's invasion with the words 'stuff happens', public revulsion in the West ensued. Memories resurfaced of the infamous US officer's justification from the Vietnam War – 'we had to destroy the village to save it'. Was this 'effects-based' Western Strategy?

Moreover, if 'war in the age of globalisation is conducted on behalf of a universalised notion of mankind, rather than on behalf of a culturally specific community that exists among others', then, as Reading academic Andreas Behncke put it, 'the enemy [regime] has become a criminalised and morally inferior agent' (Behncke 2006: 937). If by implication Western powers intervene as police forces, they have to prevent collateral damage among the public they are there to protect. There can be no 'enemy nation' in this model; the loss of even one 'innocent bystander' is a catastrophe. In addition, the 'world police forces' must not create unnecessary obstacles to the most important goal – peace, in the form of post-war reconstruction, state building and the rehabilitation of the people for whom the intervention was effected.

The return of small wars

Small wars had never disappeared, least of all in the Cold War (see chapter 15), but with the waning of the fear of a Third World War, they once again claimed the liberal states' full attention. With the

Yugoslav Wars of the 1990s, an old debate that had existed during the Vietnam War came to the fore again: what was the West doing in intervening, especially as the intervention was justified exclusively on 'humanitarian' grounds? Whose interests was it serving? Western powers were no longer directly involved as colonial powers. Instead, their intervention was so clearly defined as being *in support of* the local populations, not against them, that 'collateral damage' of any sort became intolerable, and incompatible with the fundamental aims of intervention. Unlike the First Gulf War, Yugoslavia was not, as one US general had infamously explained to CNN, a matter of 'us against the In'juns'. There were no battlefield victories to be won, massive firepower mattered little (except in the air campaign to end the crisis over Kosovo), and AirLand Battle-type concepts had no applicability. This had little to do with high tech, and much to do with history and violent politics, more so than Western powers were accustomed to in their own 'post-modern' societies. And, as Stephen Biddle notes, low-intensity wars cannot be won by high tech alone (Biddle 2004: 199–202).

The next round of the US and British debates about Strategy, in the early 2000s, thus took place with a different focus – it concerned not the FM 100–5 series of field manuals for major war, but FM3-24 on *Counterinsurgency* (see chapter 16). So with Bosnia, Kosovo, Afghanistan and Iraq, against the background of Islamic fundamentalists' hatred of the West, the pendulum swung away again from *matériel* (or new technology) and back towards the 'soft' approach to Strategy – historical analysis, analysis of the background of the conflict, anthropological keys to the understanding of societies. The focus shifted from military victory to gaining the support – if not the hearts and minds – of the majority of the populations. Military victory, in such a context, is meaningless, as General Sir Mike Jackson, heading NATO's Kosovo force, told a journalist in 2000, when asked when he expected 'victory' in Kosovo. Post-conflict reconstruction, reconciliation, peace-building, peace-keeping and, above all, the transformation of bellicose cultures into peaceful ones takes decades, not weeks or months. By contrast, 'exit strategies' tend to use the wrong time scales, corresponding to terms in office of the Western leaders who initiated such campaigns, not to the time frames needed for cultures to be changed.

Conscripts or professional armies?

The new concept of Western armies acting as world police forces calls for soldiers different from those of the high firepower mass armies that confronted the Warsaw Pact under the threat of nuclear war. But what forces does one need for policing the world? Even before the Second World War, Basil Liddle Hart, J.F.C. Fuller and Charles de Gaulle had called for professional armies as a means of *limiting* warfare. The two Britons at any rate returned to this call in the 1950s, just as NATO doctrine first embraced the creed that the defence of Western Europe in the nuclear age required mass armies. Even in 1950, Liddell Hart was passionately opposed to the 'prehistoric mammoth', the 'antiquated idea' of a conscript army, on which he blamed the horrors of the major wars of 1864–1945 (Liddell Hart 1950: 324–330). He also thought it incompatible with liberty and the individualism of civilised societies, arguing that:

> Conscription immensely increases the power of the State over the individual. It has been of great service to dictators as a means of enslaving the people to their own purposes. Liberty-loving peoples are foolish if they help to preserve such a system as a natural and proper custom. For conscription has been the cancer of civilisation. (Liddell Hart 1950: 338)

Two countries that saw themselves as world policemen and that with their seats on the UN Security Council were given authority to play this role, the United Kingdom, with its long tradition in this matter, and the USA had already abandoned conscription during the Cold War – one after Korea, the other after Vietnam. They did so for exactly the reason that Guibert had identified 200 years earlier, namely, that one cannot easily fight expeditionary wars overseas with conscript armies.

The debate about professional versus conscript forces had raged in France throughout the nineteenth century. During the Cold War, France seemed firmly wedded to mass armies – to a large extent for philosophical and domestic political reasons: de Gaulle as creator of the Fifth Republic in 1958 abandoned his earlier views and espoused the dogma of the citizen-soldier to place his state firmly in the Republican tradition (see chapter 6). France thenceforth tried to do both – be a world policeman and retain conscription – and ran into serious problems. Indochina was fought with professional forces; Algeria with conscripts who pushed French public opinion to favour giving up. The First Gulf War of 1991, however, demonstrated to France after the end of the Cold War how much better British (let alone the remaining superpower's) professional forces were adapted to this sort of operation, and was crucial in moving the country to give up its hallowed Republican tradition of conscription a decade later. It is probably a sign of France's confidence in its republicanism that at the turn of the millennium, she could move to a professional army without massive debate, soul-searching or protests.

In Germany, one of the (debatable) explanations of the catastrophe of the Third Reich was that the (professional) Reichswehr had submitted all too meekly to the rule of the National Socialists, when a citizen-soldier army, imbued with the ideals of democracy, might have stood up against this regime and defended democratic values and law against orders from above. This notion – and the US and NATO Strategy at the time which called for more conventional forces – led to the creation of the Bundeswehr in the mid-1950s as a mass, conscript army. Most European NATO members – all those geared mainly towards the defence of their own territory against Warsaw Pact invasion – had conscript armies until the end of the Cold War to offset the WTO's mass armies. That something other than this rationale was involved in the case of the Bundeswehr, however, can be illustrated by the fact that since 1991, most European armies have abandoned conscription and moved towards professional armies, while only Germany, together with Greece and Turkey, who eye each other wearily across the Aegean, refuse to budge.

But there is a third option. Taking up ideas like those of de Gaulle in the early 1930s, Liddell Hart in 1950 wrote about the feasibility of an 'international force', effectively a standing army at the command of the UN. Its forces, he thought, could use just one single language – the French Foreign Legion had shown this was possible – and ideally should have standardised weapons, first perhaps with different countries supplying different types of weapons, one tanks, another aircraft, or, later, different countries co-operating to provide components of weapons systems. 'This would pave the way for the international authority itself to take over the manufacture of the major weapons – which would be the safest system.' He did not think that, being multi-national, it would suffer from a lack of nationalist fervour.

They could well develop their own *esprit de corps* and pride, like the French Foreign Legion. Moreover,

> the importance of pugnacity is decreasing in the more modern types of force, and replaced by the dispassionate determination that accompanies pride in craftsmanship. Pugnacity matters less in a tankman than in an infantryman. It is not required in the artilleryman ... Patriotic ardour will hardly improve [the] aim [of a bomber crew]. The more mechanical that weapons become, the less will the absence of a national spirit be felt.

Liddell Hart wanted this international force ideally to take charge of the world's 'more powerful offensive weapons, especially the longer-range ones', so that this force could impose the decisions 'of the international authority'. He opined 'that there can be no security for any nation without a partial surrender of national rights to a higher authority in which each shares' (Liddell Hart 1950: 345–8).

The debate about a UN standing force has resurfaced periodically, most recently after the end of the Cold War with UN Secretary-General Boutros Boutros-Ghali's Agenda for Peace. It foundered, however, on the sovereigntist resistance of major state governments (especially the Permanent Five Members of the Security Council) which want to control their own armed forces and keep the UN subordinate to their decision-making.

De facto most of the armed forces of the Western countries today are professional – even the Bundeswehr, which has token conscripts, but for its operations abroad relies on professionals in all-volunteer units. International forces like the ones which de Gaulle and Liddell Hart dreamed of exist, except that they are today still under national command, operating in coalitions, or else they are private military companies, usually in the pay of states just like mercenaries of old. Their tasks include some dangerous patrolling, and even raiding missions and actual battles, such as that in Helmand Province in Afghanistan in mid-2006. There is little left in such contexts of the militarism which made the Bismarck–Moltke wars, the American Civil War and the First World War possible. Gone is the lure of great gains to be made by war that made it a lucrative business for its princes. We see a return, however, of war entrepreneurs much like those who had made profitable business from Antiquity to the nineteenth century.

Something that has not changed – notwithstanding the UN's attempts to create a UN standing force – is the close attachment of armed forces to states, making them a particularly easy instrument for states to control and use – much easier than the economy, exchange rates or other instruments of foreign policy. There is thus still, as General Sir Michael Rose noted critically, a tendency to turn intractable foreign policy problems into something to be tackled by military force. The former commander of UNPROFOR in Bosnia agreed with Liddell Hart, who had said: 'The more I reflect on the experience of history, the more I come to see the instability of solutions achieved by force' (Rose 1999: 10). But in view of the recession of the state in many other areas, states clearly cling to their monopoly on the use of force – and to armed force as an instrument of statecraft – more determinedly than to just about any other prerequisite of statehood.

This must, however, be offset against the overall cultural shift in the West, where attitudes to armed forces are concerned. A decade after the end of the Cold War, the British military historian Jeremy Black turned his gaze to the present and future of war, and rather than a 'revolution in military affairs', he saw a 'revolution in attitudes towards the military'. The end of the Cold War had come with a decline of willingness to serve in the military, a gradual move from conscript armies to professional armies with few exceptions throughout the Western world – including, as we have seen, the abandonment of the French Revolutionary ideal of the 'nation in arms' and every citizen as a solider – women gaining access to the military and new policing and social-worker roles for soldiers, all of which profoundly challenge the old role model of the military. Although old ideas and stereotypes of male heroism in soldiering are still reinforced through films, novels and electronic games, attitudes are shifting (Black 2004: 13–20). The military today in liberal societies is seen and respected more as a professional fire brigade or a police force, there to protect security and fight rogue states, than as warriors or a special social caste that was once supposedly the hallmark of all Indo-European societies. Christian Churches are embarrassed today by the medieval ideal of the crusader which Pope Urban II created. Chivalry remains part of the cultural memory in terms of protecting the weak, not primarily of killing others. The pretty coats which rendered soldiery attractive in the eighteenth and nineteenth centuries are replaced by practical but dreary camouflage outfits; and the annual Trooping of the Colour in

London, the Edinburgh Tattoo or the 14 July parade in France are about all that remains of this colourful pageantry. Indeed, there are European countries that stage no military parades. Militarism has subsided together with the Social Darwinist rage of the nineteenth and early twentieth centuries.

Future developments

To reiterate, all this is probably true only for liberal states, not for many other cultures in the world, where new ideals of suicide warriors-of-God or martyrs for one's own ethnic group/race seem to flourish along with militarism, cultural intolerance and bellicose mentalities. It is useful to remind ourselves, as did Emperor Charles V, with his global empire on which the sun did not set, that 'Peace ... is beautiful to utter, but difficult to have, for everyone knows that one cannot have it without *the enemy's consent*. [To gain this consent] one must perform great labours, which is easy to say but difficult to do' (q.i. Brandi 1937: 190; my emphasis).

Nor have we reached the end of history in the sense of universal consensus on the best system of government and order of society within the West. The evidence examined in this book underscores the multi-directional development of thinking about war and Strategy over time. If the past saw no one-way development, it is unlikely that the future will. All we can do is to conclude that at present, the trend in Western thinking is one that, starting with the First World War but strongly reinforced by Hiroshima, has turned its back on the cult of the offensive or of major war and decisive battles as the solution for long-standing political problems. Instead, we see the use of force, comparable to police action against violent criminals, as a necessary evil for the protection of world society and local communities, and aim to limit casualties on all sides, struggling to establish a viable peace. This could still be the result simply of 'considerations of expediency and utility' rather than a more deeply rooted transformation of attitudes in the West, as Osgood and Tucker warned (1967: 358f.).

In that wondrous trinity of victory, peace and justice, which any Strategy that stands in the just war tradition should aim to bring together, liberal societies are in danger today of forgetting that peace, too, is not an end in itself. Peace without justice is no more likely to endure than victory without a mutually acceptable peace, seen by all sides as built on some degree of justice. The Western peace movements tried to banish this thought, and were partly taken by surprise at the commitment with which peoples in the communist world rose up against the status quo, risking violent repression, in 1989–91. Equally, many in the 'post-modern states' of Europe (Cooper 2000) that had left behind the nationalism of the 1789–1945 era, were surprised to find people willing to sacrifice what they perceived as an unacceptable peace, go to war, kill and die for nationalism, as was demonstrated in Yugoslavia in 1991–9. Nor could Americans understand the Islamic opposition to the American-dominated peaceful new world order that came into being in 1991 with the disappearance of the Soviet Union as superpower. Worthy of our full support as it is, the Western change of mentality in the last hundred years is a delicate plant. It will require great efforts to protect and nurture it, so that it can flower and spread its seeds to create a more benign world.

19 *Epilogue: Strategy-making versus bureaucratic politics*

[I]ntramural struggles over policy can consume so much time and attention that dealing with external realities can become secondary.

(Clifford 1991: 93)

Where you stand depends on where you sit.

(Miles 1978: 399)

Policy and Strategy in practice

If we review the recorded wars of human history, it is rarely, if ever, possible to point to any precise dividing line between politics and the use of armed force as its tool. Much writing about Strategy presupposes that there is none: the ideal for Guibert, Clausewitz and many others was a Strategy-making process in which the supreme political decision-maker was identical with the supreme military commander, in the tradition of Alexander the Great, Gustavus Adolphus, Frederick II of Prussia or Napoleon. Once it became clear that the norm was that the two were not identical, many nineteenth- and early twentieth-century military men, led by Prussian military thinkers, postulated that politics (and diplomacy) had no role to play once war had been declared, and could only resume its functions once war had been ended by victory/defeat or armistice (Lossau 1815: 7). This assumption (which originated in Prussia, but was not confined to it) was largely a function of the contempt in which the age of militarism held politics, politicians, ministers, diplomats and civilians in general. The Second World War would prove a watershed in this respect. Since then, the demands of politics and technical possibilities and limitations of warfare have had too many mutual implications to be seen as separable, particularly when with the spread of democracy, public opinion matters more than ever before.

The one-time strategic adviser and US Secretary of Defense Henry Kissinger, himself a historian of the nineteenth century, wrote in the Cold War:

> A separation of strategy and policy can only be achieved to the detriment of both. It causes military power to become identified with the most absolute applications of power and it tempts diplomacy into an over-concern with finesse. Since the difficult problems of national policy are in the area where political, economic, psychological and military factors overlap we should give up the fiction that there is such a thing as 'purely' military advice.

Nor did he believe in the existence of 'purely' political considerations (Kissinger 1957: 422).

The vast literature that is the product of research on decision-making processes in international relations (with particular attention paid to decisions for war and peace) indicates that matters are more complex still. The Clausewitzian model according to which governments make decisions about the use of military force with clear political objects in mind is in itself a crude over-simplification: any one decision-maker will have several objectives, some mutually exclusive or conflicting, and any group of decision-makers will have even more heterogeneous objectives ranging from the promotion of their own career to the interests of the state they lead or the protection of international order and peace. They may have very mistaken views about the possibilities offered by the military means available to them, or about the convictions, ideology, interests, intentions and capabilities of their adversaries, and a host of other factors. Their perceptions and thus their objectives may change over time, and this is likely to be a function of their appreciation of the evolution of the military conflict itself. Particularly at the outset of war, and during critical situations, decision-makers, not blessed with the historian's benefit of hindsight, tend to stumble blindly in the dark, above all concerned with surviving the crisis and preventing the worst from happening. It is thus rarely (if ever) the case that the decision to go to war is a pure function of threat assessments, assessments of one's own and of the adversary's available means and of clear political war aims, with clear military aims derived from them that can still be clearly recognised at the end of hostilities (Bull 1961: 49).

John Hattendorf rightly noted that Strategy throughout history, especially in the centuries before bureaucracy came to dominate, when applied in practice, was rarely fully thought through on an analytical level. 'The practitioners of strategy are those at the highest levels, who, in the crises of the moment, often do not explicitly record why they take the actions they do.' What scholars with their penchant for patterns and logically coherent ideas may identify as strategic concepts may just be traditions, passed on almost as if by osmosis rather than articulated principle, or what is perceived in a particular culture as the 'obvious' or 'commonsensical' thing to do – where passive culture dominates active culture. The inspiration of the moment upon which a supreme commander draws is usually due to 'long and conscientious self-culture in the line of his profession' of which he may not be fully conscious (Hattendorf 2000: 1, 21, 127).

This explains why Colin Gray sees Strategy as 'the bridge that relates military power to political purpose; it is neither military power *per se* nor political purpose ... [S]trategy ... [is] the use that is made of force and the threat of force for the ends of policy' (Gray 1999a: 17). Richard Betts argues that it cannot really be more than such a bridge, as any greater degree of 'integration of policy and operations is a prescription for civil-military tension' (Betts 2001/2: 28). In view of the different cultures of bureaucratic decision-making in government and international organisations, and the decision-making by politicians and the military, greater integration would be like the unstable mix of oil and water.

Strategic concepts formally adopted by governments rarely existed in a formal written form before the late nineteenth century. Since their usage has become widespread, they have been characterised more by contradictions and compromises resulting from bureaucratic politics than by a logical application of explicit principles; indeed, this is why Richard Betts has found that often there is no Strategy in any way like that defined in chapter 1 and that 'Strategy' is 'an illusion' (Betts 2000). One therefore has to be aware of the corporate character of any government document that is taken to reflect or define government 'strategy'. Such documents, as the British Admiral Richard Hill observed astutely, are 'a distillation of compromises' (Hill 2006: 161); it accounts for the ambiguity John Baylis identified as central to British strategy (Baylis 1995).

It is a deficiency of most definitions of 'strategy' that they assume that the (mainly) two entities contending with each other are monolithic.

In looking at the making of Strategy, to use Newton's expression, the definitions assume a black box: decisions are made within it, but the principles and rules by which they are made are subject to the logic of political ends, in a particular geographic setting, with particular means available (armed forces, on the basis of a particular economy, population, wealth ...). If this were so, we would have reason to be surprised at the incoherence of so many Strategy decisions. We have already mentioned the evolution of economies, states, administrative apparatuses of states (civil service, tax collection, investment in infrastructure ...), and technology as variables. But there is a further variable, which applied as much to Gustavus Adolphus' decision-making as to Frederick II's, as to the coalitions fighting against Napoleon: the multitude of actors. This variable has an internal and an external dimension. The external dimension of this variable is the older one: it is other entities (principalities, kingdoms, republics, empires ...) interacting with the belligerents. It is potential and actual allies of either side, their attempts to draw political, economic or other advantage from the situation, by becoming belligerents themselves or by blackmailing one or both the belligerents with the threat of joining the other side – with the whole spectrum of alliance politics, interstate relations and diplomacy. Machiavelli, in discussing historical examples, found that an alliance or 'league is governed by a council, which must needs be slower in arriving at any decision than are those who dwell within one and the same' country (Machiavelli 1531/1998: 268). In the seventeenth century Paul Hay du Chastelet was advising Louis XIV:

Alliances with monarchies are preferable to the friendship of republics as the latter act with excessive slowness, their prudence is timid, they are excessively cautious before committing themselves, their council-meetings are never secret, there are always hidden desires and interests which trouble the governments, and it is rarely the case that all those who have authority in [government] affairs have the same inclinations and sentiments, to the point that at every turn there are new difficulties. By contrast, in monarchies, all is decided by the behaviour and will of one single person ..., so that everything is done with greater promptness and more authority than in republics, except when a prince thinks that his advantage in protecting his allies is not sufficiently at stake to stand by his alliance treaties. Instead, the spirit of republics only has as a principle [governing] all its movements nothing but utility, which one prefers on all occasion to that which would give more glory and greater esteem to the state. (Hay du Chastelet 1668/1757: 185f.)

After Napoleon's defeat, the Concert of Europe as a formal mechanism of concertation between several states introduced a new dimension to alliance politics in the nineteenth century, and since 1918 the League of Nations and since 1945 the United Nations and a growing number of international organisations have given some framework to interstate relations. In short, these are what the British naval specialist Fred Jane called 'international complications' (Jane 1906: 163).

Such 'international complications' arising from alliance relations have a crucial effect on the conduct of military operations, where so much depends on command structures – which ally's commanders can overrule commanders from other allied states? – which in turn tend to depend on the mix of alliance forces – which ally contributes most, and to whom are smaller numbers subordinated? In the light of the Elizabethan wars on the Continent, Matthew Sutcliffe noted:

> Whoever ... meaneth to winne honor in the succour of his friends abroad, let him as much as he can endeuour to cary with him a sufficient force. Small numbers are neither esteemed of enemies, nor friends. Into the field they dare not come, for they are too fewe and too weake; being penned up in cities they famish. If our friends be stronger than our ayde, then are they commaunded by them. If any of their leaders want gouernement, our men that are put to all hard services, pay the prise of their folly. If any calamitie happen to their army, our people first feele it. They shift to them selves, being in their owne countrey; ours are slayne both of enemies and friendes, and if victuals ware scant they sterve first. (Sutcliffe 1593: 104f.)

Three and a half centuries later, Charles de Gaulle, with his scepticism about alliances derived from his experience in the Second World War, would not have expressed this very differently.

The internal manifestation of this variable is collective policy-making, in which even a President of the US, France, Russia or Finland, or a Prime Minister of Britain, Norway or Australia cannot overrule internal opposition – from opposition or the government's own ministers, officials or military – in the way in which a tyrant, a dictator and to a lesser extent any monarch, appointed by the Grace of God, could (and even then, tyrannicide and regicide existed). Today there may be a supreme decision-maker with whom 'the buck stops', and yet decision-making in democracies (or even dictatorial oligarchies) is a bargaining process which adds entirely new rationales to the making of Strategy. And here, as political scientists have long emphasised, political and personal battles enter into the picture and, equally importantly, bureaucratic politics.

From the seventeenth century, but especially from the mid nineteenth century, until the present, as bureaucracies grew in all states and technology became ever more complex, and with it procurement decisions, government decision-making became correspondingly more complex. Prince Eugene of Savoy (1663–1736) and Frederick II of Prussia, who had the great advantage of being able to make decisions all by himself, thought that if a general wanted to do nothing, all he needed to do was summon a council of war, where all the opinions voiced by those consulted would be sure to cancel each other out and create stalemate (Frederick II 1882: 89). Guibert, with his own extensive experience of government decision-making, wrote in the mid eighteenth century:

In almost all states of Europe, the different branches of administration are governed by particular ministers, whose interests and views jar, and are reciprocally detrimental to each other; each of them is occupied exclusively with his object. One might imagine the other departments belonged to a different nation. Happy, indeed, are those States where the ministers, jealous of each other, do not act as open enemies. (Guibert 1772/1781: xiii)

Napoleon himself is credited with the maxim that:

When war is begun, the presence of a deliberative body is a nuisance and often fatal. The turbulent, the ambitious, those who want to be talked about, and to be popular, and dominant, raise themselves up by their own authority to be advocates of the people. As counsellors of the prince, they defend those who are not attacked, they claim to know all, want to determine all, direct all. Successively, they become censors, creators of factions, and rebels. (Napoleon 1906: 142–4)

In addition to the competition between rivalling ministers, interservice rivalry in societies with armies *and* navies may well have existed in some modest form for centuries. All of this, taken together under the heading of bureaucratic politics, has become ever more pronounced and is crucial to any understanding of the configuration and equipment of armed forces, leaving reasons of Strategy, let alone any grand strategic concept, well behind.

The fight over money and government spending, especially on wars, was already at the centre of the tense relations between Tudor and Stuart monarchs and their parliaments, culminating in the

English Civil War. The fight for money and thus resources between the institutions and organisations concerned with military policy took its place centre stage from the end of the nineteenth century. Henceforth, Strategy was never again just about relations with the enemy. Henceforth, increasingly, Strategy was made, especially in peacetime, to further the interests of one's state vis-à-vis others (allies included), government vs. opposition, one minister against another, one section of a ministry against another, the different branches of the armed forces among themselves, navy vs. army, paratroopers vs. marines, tank units vs. lightly armed infantry, and so on. In democracies, Strategy – often reflected in White Papers or Blue Books – was increasingly the product of committee work, with the agencies concerned represented, the final document being much less a work of logical coherence than of compromise, balancing the vested interests represented at the drafting table. Meanwhile, there were ever-fluctuating and evolving collective views on Strategy, sub-issues and related issues, often communicated by osmosis rather than reasoning, through 'group think' (Janis 1982) engendered by frequent meetings in camera, by commonly read newspapers and the odd pertinent lecture remembered dimly from staff college or listened to in one of the think tanks. As Admiral Hill reflected, strategic principles that were supposed to answer perceived and ever-evolving strategic problems are usually the final outcome of such committee work, usually supplemented by 'late-night gossip' in the Ministry of Defence, 'and philosophising over cheap sherry, with nearly all the ideas discarded next day but a nugget or two remaining. Nothing was ever written down; what was worthwhile stuck in the mind' (Hill 1986: 5).

Add to this the proclivity of armed forces to fight wars not in ways that are most appropriate to reach the desired end state with regard to the adversary. Given any say in the matter, they prefer to fight the wars they have prepared for, for which they have acquired equipment, for which they have configured and which they want to play out in reality. Similarly, diplomats want to prove the usefulness of the treaty or the international organisation they have promoted, so it *must* be the answer to the security problem of the day. Add to this 'public opinion' – another black box in need of opening up. The media chip in, furnishing fresh or old but reinforced arguments for one decision or another. A myriad individual and institutional interests come into play, each eloquently defended in the ever more complex decision-making process within states and among states. Different 'black boxes' in which Strategy is made produce such varied results that they can no longer be explained in the simple terms of old definitions of Strategy. Strategy-making becomes absorbed by all forms of political interplay, which can be distinguished from normal politics only in that the use of force or the threat of the use of force play a central role.

John Hattendorf has summarised the way in which Strategy is made in the leading world hegemon today, the USA. This is little more than an example of any democratic policy-making progress, albeit on a larger scale than in Britain, Poland or Italy, let alone Estonia or Slovenia. Hattendorf sees strategic decision-making as taking place in four areas:

1. The state's grand Strategy on the highest political level is formulated by the President and modified or supported by Congress.
2. Arms procurement, along with statements on the need for these weapons and their role in various conflict scenarios, is undertaken by each of the services, within a budget agreed within the government and in general terms with Congress, and coordinated by the US Secretary of Defense.
3. Contingency plans for a spectrum of imaginable scenarios of war or military intervention are made under the auspices of and according to the terms of reference defined by the Joint Chiefs of Staff.
4. 'Operational planning, the preparation of precise plans for wartime operations, is done by ... various ... commanders in chief' and by specially designated teams under their supervision.

As Hattendorf has rightly remarked, in theory, these areas 'should directly complement one another' (Hattendorf 2000: 201f.). In practice, however, there is any amount of friction (Hattendorf 2004). This arises from the interaction of diverging bureaucratic interest, from alliance politics, from the moving of the goalposts over time (planning for grand Strategy should in theory dominate the three other areas, but while planning there is being carried on within the parameters defined by grand Strategy, grand Strategy itself might change in reaction to new external or internal events) or the different time frames particular to these different areas (a presidency in a democracy may last five years, whereas a planning-to-deployment-to-scrapping cycle for large weapons systems, tanks, aircraft or submarines and surface vessels must be measured in terms of several decades). In practice, one might thus see a Strategy that was agreed upon in year x being implemented by some countries and services in years $x+4$ to $x+6$, while, in the meantime, by year $x+2$ Strategy had shifted. To give a concrete example, NATO's Strategy of symmetric response to the Soviet-cum-satellite threat was agreed at Lisbon in 1952. On the basis of this, the Federal Republic of Germany was admitted to NATO in 1954, and its army, the Bundeswehr, was set up between 1955 and 1958. In 1954, however, NATO adopted 'Massive Retaliation', opting to meet Soviet conventional aggression asymmetrically with nuclear answers, shifting the emphasis away from mass armies to high tech forces, equipped with nuclear weapons which the Bundeswehr was not to have. What use was the Bundeswehr as a mass army in such a Strategy? Intra-alliance friction ensued (Heuser 1997).

Turning to procurement, the implementation of decisions in this area made in year x might only be fully completed by year $x+30$, if they are not cancelled or modified along the way; where major weapons systems are concerned, there is usually a point when large-scale modifications cannot be made any longer and where the cancellation costs would be absurdly high. So any Strategy has to operate on the basis of procurement decisions connected with a grand Strategy several times removed. Thus NATO and the Warsaw Pact were deeply committed to a grand Strategy of détente while the US and Soviet missile stocks were still being built up to unprecedented numbers, undermining the very policies prevailing at the time.

This also means that military planners never think they have the quantity and quality of arms they need. Since complex weapons systems take several years to get from the drawing board to deployment, military planners have a tendency to feel unready for war, which they prefer to imagine as taking place in five or ten years' time, after the next Five Year Plan, after the next procurement spree. The replacement of one type of tank, frigate, fighter aircraft and so on by another never takes place in a short period of, say, months, but (especially in peacetime) more usually over years, which means there are always some old weapons systems operating alongside the new ones, logistical and interoperability complications arising from this and some military men inexperienced in the use of either type of equipment. Ironically, while Strategy-making has come to be dominated by debates about the implications of ever-newer technologies and the appropriateness of weapons systems, the implementation of any Strategy is hugely hampered by the time gap between planning and execution.

In the process of government Strategy-making, one would expect a logical ('Cartesian') deduction from first principles, as sketched in André Beaufre's work (see chapter 17): starting with a definition of grand Strategy, then moving on to analyse what part in pursuing it military power might play, how exercised and with what means – and what part of those means would be affordable. Instead Strategy-makers have a tendency to 'enter that sequence halfway through' (Richard Hill), not questioning many unspoken assumptions and 'gut instincts' (Geoffrey Till) – about means, the way to use military power and so forth – which are all too often taken for granted, often anachronistically so (Till 1994a: 184; Hill 2006: 162).

Peacetime alliance politics and Strategy conducted through international organisations, a novelty of the post-Second World War era, have created a new and very powerful dimension of bureaucratic politics. In times of peace before 1945, politics had been a game to be played by the two or three services, plus forceful politicians in government opposing each other (a particularly strong dimension in democracies with coalition governments like France), plus the military-industrial complex. In the post-1945 world there has been the additional, higher playing field on the international level, where national prestige is a huge consideration in alliance relations, where thousands of little deals are made, linking procurement, bases and wrangling over the assignment of command posts and the delimitations of their geographic and other competences (Maloney 1995).

To give an example, a state may decide to join in a UN-led military intervention or peace-keeping operation, not primarily in order to help resolve a conflict somewhere or to stop bloodshed and stabilise a precarious situation. While such aims will be somewhere on the list, other considerations may well predominate, such as the influence this state may wield in this operation and beyond this operation (precisely *because* of its involvement in this operation) on other international issues, in the UN and other international forums. The state may wish to strengthen the UN or another collective security organisation, or indeed contribute to the effectiveness of one as opposed to another (Till 1994a: 192). It may wish to benefit from training together with other, well- or better-equipped powers; it may hope to benefit from sharing information, or even technology, in the process. Its navy or air force or army may want to prove that it is needed and can perform in earnest. Many considerations of this kind may serve to put the resolution of the conflict itself and the fate of its direct participants and victims further down on any list of priorities, usually at their cost. While such reasoning would have been entirely familiar to Frederick II of Prussia, some of the motivations seem quite 'frivolous' to those like Kant primarily concerned with limiting the miseries brought by war. This need not merely be the victims themselves, or the Red Cross that tries to keep the survivors alive, but 'public opinion' or 'international opinion' in general. Appeals to such a public in turn can have an impact, in democracies with an uncontrolled press, on government decision-making.

The frailty of human logic

In his 'theory of interdependent decision', Thomas Schelling assumes 'cool-headed' behaviour underlying threats and responses, reprisals, brinkmanship, cheating, trusting and so forth as that is more helpful for his theorising (albeit less realistic). He calls this 'rational behaviour'.

> Rationality is a collection of attributes, and departures from complete rationality may be in many different directions. Irrationality can imply a disorderly and inconsistent value system, faulty calculation, an inability to receive messages or to communicate efficiently, it can imply random or haphazard influences in the reaching of decisions or the transmission of them, or in the receipt or conveyance of information; and it sometimes merely reflects the collective nature of a decision among individuals who do not have identical value systems and whose organizational arrangements and communication systems do not cause them to act like a single entity. (Schelling 1960: 16)

There are several problems with these assumptions about rationality. No value system is entirely orderly and consistent. No communication can ever be guaranteed to be efficient, there can never be certainty that in complex negotiations one has been understood fully or that one has understood fully. No group of decision-makers will ever act entirely like a single entity. By contrast, conflict-engendering values – such as national pride – may well be *shared* by opposed nations, and yet produce only conflict. Individual decision-makers who are supposed to be allies may well *share* the goal of self-promotion at the expense of everyone else. It is thus not always, as is so often claimed, *differences* in culture, values or in the quite rational conclusions drawn from these that account for their eagerness to wage war; instead, this may be due to *shared* bellicose values, a *shared* emphasis on 'honour' rather than peace, and *shared* simple selfishness and disregard for the suffering this will bring upon others.

What wonder, therefore, that Strategy in practice rarely follows the precepts laid out by strategic thinkers. In fact, their writings – even if the authors are major establishment figures or responsible for the education of future decision-makers – exercise at best an indirect influence on the making of Strategy. Even Frederick II of Prussia, who was in the enviable position of being strategic thinker, supreme decision maker and commander-in-chief in one, could not implement the Strategy of short, sharp wars that he himself thought most desirable. Instead, the major contribution of strategic thinkers is perhaps that they create a framework for a (partly public, partly internal governmental) debate which in turn plays a part – but only a part – in the processes of decision-making on Strategy. It is other factors, such as financial limits or the narrow interests of all the entities involved in Strategy-making, that have by far the greater impact in each concrete case. One could only change that, as Kant suggested, if decision-makers were forced at least to listen to the moral philosophers (Kant 1795/1797: 43–5) – we might add, the strategists and international lawyers – which of course they are not. (A cynic might say that on the occasions that they do, decision-makers are just as likely to misunderstand or misapply arguments made by strategic thinkers than to be constructively influenced by them).

Either way, those *thinking war* will always hope for it to be neater, more coherent, purposeful, goal-oriented than it turns out to be in practice. In reality, even if the most logical, coherent and humane Strategy were applied throughout – and, to switch register, with the most just cause, authority and intention – in view of the suffering it causes, there is no such thing as a good war, to paraphrase Benjamin Franklin.

20 *Summaries and conclusions*

The first chapter of this book listed a series of questions which have been addressed along the way, but overall answers must be given now, based on the findings of this book.

From antiquity to the Thirty Years War, and again in the age dominated by the Napoleonic paradigm, most writers on war assumed that it is an inevitable part of the world. Many sought ways to limit it, and a few, writing especially in the Enlightenment and again in the twentieth century, even hoped that it could be abolished altogether through binding legal arrangements and restructurings of societies. The abolitionists and the limiters met in making efforts to reduce the pain and suffering caused by war. Some merely sought ways to offset their own side's weakness, like some maritime strategists of secondary powers. Several Roman Catholic thinkers like Bonet in the fourteenth century and Hay du Chastelet in the seventeenth, as well as Moltke and the other 'realists' and Social Darwinists from the mid nineteenth century onwards, not only dismissed alternative solutions to war but positively welcomed it as part of God's or Nature's great design. They tended to favour battles and the offensive. As we might expect, the limiters largely advocated defence, deterrence and indirect approaches to reaching one's goal. Complicating the picture, however, some limiters advocated seizing the initiative by instigating offensive and decisive battles, or using terror, including the bombing of cities, in the hope of offsetting their own perceived weakness or shortening the war. Both sides were driven by their view of the world and their ideologies. Technology could generally be used to defend arguments on either side.

Although there is a solid tradition stretching from Cicero to the UN that urges princes or other governments not to wage wars that do not fulfil criteria for being 'just', wars were fought for all conceivable reasons, ranging from the narrow family interests of princely houses to perceived 'national' or even 'racial' causes. The more limited the group whose interests were being fought for, the less the emotional engagement of the population, the greater the need for a professional, paid army to fight such wars. The more the masses had been mobilised by religion or other ideologies to think of a cause as their own, the greater the tendency towards total war. Atrocities against civilians and mass slaughter on battlefields (as opposed to the observance of an armistice once one side surrendered) thus generally went along with ideologically motivated wars, from religious wars (Kortüm 2001) to nationalist and 'racial' wars – for atrocities to be committed, it was usually sufficient if only one side saw them thus (Foley 2001). Convincing oneself that one was dealing with 'sub-humans', eternal enemy species or heathens/heretics and so forth hated by someone's own enraged deity always whipped up the fighting spirit, and so did the belief that one's own defeat would lead to the rule of some Empire of Evil.

The treatment of 'enemy' populations varied accordingly. If they were seen as valuable assets for one's own economy, or potential good subjects, they were sometimes spared or, in the worst case, enslaved. If they were seen as inherently evil, or as competitors for limited resources, this could quickly lead to atrocities or even genocide. In the Christian tradition, atrocities were always legislated against except if the victims were heretics (refusing to see the 'errors of their ways'), rebels or had themselves (really or supposedly) oppressed Christians. Most early modern writers admonished their readers to treat enemy civilians with clemency. Notwithstanding such admonitions, atrocities occurred frequently as soldiers got out of hand, explanations for which must be sought in the realms of psychology or biology, not reason. Atrocities on a genocidal scale were also at times ordered by princes and other legitimate governments, or at least condoned. There was a relative lull in atrocities in Europe (outside the Ottoman Empire) from 1648 until 1914. As the drift to Total War in its racist dimension did away with the taboo of maltreating civilians, sea power made it possible to starve them through blockades, and then air power to bomb them. There were parallel developments in ground warfare. Ideology here clearly drove technology, not vice versa, and technology gave racists and nationalist Social Darwinists new ideas of how they could pursue their aims.

There was a correlation also between ideology and who should do the fighting. Those who wanted to limit war might opt for either a militia – the citizen-soldier or conscript – in the hope of adopting a clearly defensive Strategy, in the tradition of the young Guibert and Jaurès. Or they might want a professional army, properly paid and well disciplined and thus not prone to get out of hand, who could be sent abroad to carry out the policies of their government or even an international organisation working for peace to oppose aggression, in the tradition of the older Guibert, Liddell Hart and the young de Gaulle. Curiously, this meant that the very defensive thinkers' commitment to citizen armies could coincide with the views of ultra-offensive, nationalistic and expansionist thinkers.

They diverged greatly on the importance of battle. Until the French Revolution, opinion on the desirability and centrality of battle was divided. Most writers counselled prudence. Either way, battles and battlefield victory were seen as instrumental to something else, not as an aim in themselves. By contrast, from Napoleon until Castex and Guderian, a paradigm predominated in which battle was seen as central, inevitable and the only way to bring about a decision, with all too little thought given to how a subsequent peace should be secured. This, if any, was the age of a supposed 'Western way of war', but it was a short period in the few millennia for which we have historical evidence (Weltman 1995). By contrast, the indirect, political use of latent military power, the perpetual preparation of battles which one hoped would never have to take place, in the form of deterrence, coercion, suasion and so on, while used in previous centuries, came to dominate from 1945, hand in hand with a widening of the concept of Strategy explicitly to include policies in times of peace and any policies with military dimensions.

Small wars of the insurgency, civil war and people's war type existed throughout, side by side with major war, sometimes as the more common form, especially prior to the re-establishment of state structures in Europe after the Middle Ages, receding at other times, as in Europe – but not on other continents – between 1918 and 1991. Even in this period, small wars in the form of insurgencies against superior state military power troubled major Western powers in other theatres. In dealing with such insurgencies, lines of thought with regard to the need for persuasion, and indeed mutually advantageous or 'win-win' outcomes – above all, peace settlements in the interest of the populations affected – were developed that in turn influenced Western thinking about war in general after the watershed of the two World Wars.

Thinking about Strategy in relation to the peace one wanted to achieve and could realistically expect should have come part and parcel with the Romano-Christian just war tradition. There were traces of it in the works of writers of the sixteenth to the eighteenth centuries, but it was not as central as it should have been. Military men in particular concentrated on warfare rather than on its effects and what was to be achieved by it politically. Concerns about long-term stability were more central to writing on insurgencies (small wars of the people's war type) before they were extended, mainly after 1945, to wars more generally. By contrast, most strategic theory concerning major war between 1792 and 1945 was dominated by the Napoleonic paradigm: it was focused on decisive battle and sought to 'impose one's will on the enemy' in a 'zero-sum' fashion (you win, I lose; I win, you lose). The post-1945 shift back to concerns to limit war is usually attributed to the invention of nuclear weapons, but has its roots in an earlier war-weariness, the waning of Western nationalism and militarism and the growth of humanitarian awareness. In view of the long-standing compartmentalisation of thinking about the conduct of war – traditionally relegated to military professionals – and politics or international law – traditionally engaged in by civilians – there is much that remains to be explored in the relationship between the conduct of war and the achievement of a just and therefore lasting peace.

In the context of the need to remain within international law, the division of war into four levels – political, strategic, operational, tactical – as borrowed from Soviet doctrine may be difficult to maintain. As we have seen in chapter 13, in air strikes, a tactical mishap can acquire strategic dimensions, and the same can be seen in counterinsurgency, where accidental collateral damage, or the misbehaviour of counterinsurgency forces, can do damage on a strategic or even political level when the benevolence of the population is so central to success. So the question arises in this context, too, whether Strategy must not be rethought and its levels redefined.

Finally, we have tracked down the origins and traced the history of some ideas in Strategy, surfacing and disappearing again, to resurface in another area, like colours in a woven carpet. Naval and maritime Strategy was in part derived from more general, or land-focused writing about war, but at the same time influenced to a much greater extent by geography and the peculiarities of states concerned. Air power and nuclear Strategy in turn were strongly derived from previous naval theory, adapting primarily the concepts of blockade, deterrence and the 'fleet in being'. The Napoleonic paradigm resonated through both naval and air power thinking, although there were important dissenting views, especially concerning the uses of sea power.

We have found, perhaps with some surprise, that there is not always a neat correlation between a basically humanitarian and ethical approach to war in a particular culture and the Strategies advocated or even adopted by its armed forces. We found examples of this oddity in the *Jeune École*'s cynical advocacy of targeting even vessels with civilians on board, which admittedly never seriously dominated French operational Strategy, and the French and British practice of naval blockades to starve the enemy's civilian populations. We saw it in the surprisingly restrained and – initially – reasonable Strategies of Tirpitz with his deterrent fleet, which he admittedly soon abandoned, but also in the Strategies advocated in the interwar years by Wegener, Assmann and Groos. We saw it in the Social Darwinist tenor of Trenchard's and the RAF's city bombing Strategy, also during that period, which contrasted with the restrained and international law abiding doctrine of the Wehrmacht, admittedly discarded at the first shot of the Second World War. We saw them in the initially quite restrained development of the Soviet Red Fleet and indeed the Chinese fleet in the Cold War, and the former's conversion to an explicit statement of its roles outside major war contexts. These examples caution against any a-priori assumptions that a liberal power would necessarily pursue a 'liberal way of war', which may be another myth worth exploding, or that authoritarian powers, even of the expansionistic sort, are necessarily as streamlined in their thinking and planning and as unitary in action as their ideology suggests, even if aspiring to be totalitarian. Each case deserves a closer look, and the overall ideology of the society out of which a Strategy is born is only one factor that will determine this child's features.

The broad picture that emerges is one of fluctuation, not constants. Today's relatively humanitarian and ethical approach to Strategy, in keeping with ethical concepts going back to Antiquity which were present but often neglected in the intervening centuries, is thus unlikely to last, unless we work hard, generation after generation, to make it last. Nor is it shared in all parts of the world. We can make every effort to spread it, but we will be up against great odds, which is not an excuse for not trying. Ultimately, as it took many so long to recognise, a stable world order in which the use of force is as severely circumscribed and regulated as police action, and in which peace can be maintained while legitimate needs can be satisfied and grievances can be addressed through mechanisms not involving bloodshed, is beneficial to all sides. As Michael Howard has argued, it took many generations to hit upon 'the invention of peace' and to develop the rudimentary underpinnings of a world peace order. Its future depends on the successful establishment and survival of a 'world community' sharing the same 'cultural norms' and above all the quest for peace to maintain it. Peace as such, he argues, 'is not an order natural to mankind: it is artificial, intricate and highly volatile' (Howard 2001: 104f.).

If we wish to nurture it and persuade others to share this value, we would do well to discard the notion derived from Clausewitz that the aims of Strategy should be 'to impose our will' unilaterally upon the enemy. Even if force is used, what ultimately is needed is the enemy's willingness to be persuaded to accept a new situation, which they will only do if they have a stake in it, and the prospect of a better life. We can only win lastingly through convincing, as Unamuno implied, and we can only win ultimately if the other side benefits as well.

Bibliography

Aelianus, Claudius Tacticus *c.* 106: *Tactica sive De Aciebus constituendis.* Leiden: Ludovius Elzevirius, 1613; trans. John Bingham: *The Tactics of Aelian.* London: Laurence Lisle, 1616.

Aeneas Tacticus *c.* 357 BCE: *Poliorkētika*, ed. and trans. William A. Oldfather, in *Aeneas Tacticus, Asclepiodotus, Onosander.* Loeb Classical Library. London: Heinemann, 1923: 1–55.

Afheldt, Horst 1976: *Verteidigung und Frieden: Politik mit militärischen Mitteln.* Munich: Carl Hanser.

1983: *Defensive Verteidigung.* Reinbeck near Hamburg: Rororo.

Alava [Alaba] y Viamont, Diego de 1590: *El Perfecto Capitán instruido en la Disciplina Militar y nueva ciencia de la Artillería.* Madrid: Pedro Madrigal; modern Spanish edn, Madrid: Ministerio de Defensa, 1994.

Alford, Jonathan (ed.) 1980: *Sea Power and Influence: Old Issues and New Challenges.* Montclair, NJ: Allanheld, Osmun & Co; Farnborough: Gower for the IISS.

Allen, Hubert Raymond 1972: *The Legacy of Lord Trenchard.* London: Cassell.

Alternative Defence Commission 1983: *Defence without the Bomb.* London: Taylor and Francis.

1985: *Without the Bomb: Non-Nuclear Defence Policies for Britain.* London: Paladin/Granada.

Altham, E.A. 1914: *The Principles of War Historically Illustrated*, 2 vols. London: Macmillan.

Amblimont, Marquis Fuchsamberg d' 1788: *Tactique navale, ou traité sur les évolutions, sur les signaux et sur les mouvements de guerre.* Paris: P.-F.Didot le Jeune.

Amiot, Père Joseph-Maire and Joseph de Guignes (trans. and ed.) 1772: *Art militaire des Chinois.* Paris: Didot l'Ainé. Containing 'Les Treize articles sur l'art militaire, ouvrage composé en chinois par Suntse ...'.

Ancker, Col. Clinton J. and Lt.- Col. Michael D. Burke 2003: 'Doctrine for Asymmetric Warfare', *Military Review* (July–Aug.): 18–25.

Anderson, David 2005: *Histories of the Hanged: The Dirty War in Kenya and the End of Empire.* New York: Norton.

Andreopoulos, George 1994: 'The Age of National Liberation Movements', in Michael Howard, George Andreopoulos and Mark Shulman (eds.): *The Laws of War: Constraints on Warfare in the Western World.* New Haven: Yale University Press: 191–213.

Angell, Norman 1958: *Defence and the English-Speaking World.* London: Pall Mall Press.

Anna Comnena *c.* 1148: *The Alexiad*, trans. E.R.A. Sewter. Harmondsworth: Penguin, 1969.

Anon. 6th century: *Peri Strategias*, ed. and trans. George Dennis: *Three Byzantine Military Treatises.* Washington, D.C.: Dumbarton Oaks, 1985.

c. 1200?: *Itinerarium peregrinorum et gesta regis Ricardi*, trans. Helen J. Nicholson: *Chronicle of the Third Crusade.* Aldershot: Ashgate, 1997.

c. 1483: *The Book of the Order of Chivalry.* London: William Caxton; ed. Alfred T.P. Byles, London: Kegan Paul, 2004.

1699: *Le parfait homme de guerre, ou idée d'un heros accompli.* Amsterdam: Jacques Desbordes.

[Paul-Gédéon Joly de Maizeroy] 1777: *Théorie de la guerre, où l'on expose la constitution et formation de l'Infanterie et de la Cavalerie ... avec l'application des principes de la grande Tactique, suivie de démonstration sur la Stratégique.* Lausanne: s.n.

1783: *A Collection of Regulations, Orders, and Instructions formed and issued for the use of the Army in consequence of the Pay-Office-Act and other occasions subsequent to 24th of December 1783.* London: War Office.

[a Prussian officer] 1799: *Abhandlung über den kleinen Krieg und über den Gebrauch der leichten Truppen, mit Rücksicht auf den französischen Krieg mit Anmerkungen von L.S. von Brenkenhoff.* Berlin: Christian Friedrich Himburg.

1802: *Nothwendige Randglossen zu den Betrachtungen über einige Unrichtigkeiten in den Betrachtungen über die Kriegskunst, über ihre Fortschritte, ihre Widersprüche und ihre Zuverlässigkeit* [von Berenhorst]. Leipzig: Gerhard Fleicher the Younger.

1814: *Grundsätze der Strategie erläutert durch die Darstellung des Feldzugs von 1796 in Deutschland*, 3 vols. Vienna: Anton Strauss.

Armitage, M.J. and R.A. Mason 1985: *Airpower in the Nuclear Age, 1945–1984: Theory and Practice*, 2nd edn. London: Macmillan (1st edn 1983).

Arnold, H.H. and Ira Eaker 1938: *This Flying Game*, 2nd edn. New York: Funk and Wagnallis.

Arnold, Klaus J. 2005: *Die Wehrmacht und die Besatzungspolitik in den besetzten Gebieten der Sowjetunion. Kriegführung und Radikalisierung im 'Unternehmen Barbarossa'*. Berlin: Ducker und Humblot.

Arnold, Lorna (ed.) 2003: 'The History of Nuclear Weapons: The Frisch-Peierls Memorandum on the Possible Construction of Atomic Bombs of February 1940', *Cold War History* **3/3**: 111–26.

Aron, Raymond 1963a: *Douze Leçons d'introduction à la stratégie atomique*. Paris: Institut d'Études Politiques de l'Université de Paris.

1963b: *Le Grand Débat*. Paris: Calmann-Lévy; trans. Ernst Pawel: *The Great Debate*. Westport, CT: Greenwood Press, 1965.

1976: *Penser la Guerre: Clausewitz*, 2 vols. Paris: Gallimard.

Artzet, Hervé 1999: 'Le général Jean-Henri Jauneaud: une personnalité aéronautique de l'Entre-deux-guerres', *Revue Historique des Armées* **4**: 46–56.

Asclepiodotus before *c*. 100 BCE: *Technē Taktikē*, ed. and trans. W.A. and C.H. Oldfather. Loeb Classical Library. London: William Heinemann, 1923.

Assmann, Kurt 1943: *Wandlungen in der Seekriegsführung*. Berlin: Mittler.

Aston, George Grey 1914: *Sea, Land and Air Strategy*. s.l.: Murray.

Audoin-Rouzeau, Stéphane 1989: *1870: la France dans la guerre*. Paris: Armand Colin.

Backerra, Manfred 1983: 'Zur sowjetischen Militärdoktrin seit 1945', *Beiträge zur Konfliktforschung* **1**: 35–55.

Bahr, Egon and Dieter Lutz 1988 (eds.): *Gemeinsame Sicherheit, Konventionelle Stabilität*. Baden-Baden: Nomos.

Bailey, Sydney 1987: *War and Conscience in the Nuclear Age*. Basingstoke: Macmillan.

Baistrocchi, Alfredo 1924: *Per l'Efficienza dell'Italia*. Livorno: Belforte.

Ball, Desmond 1986: 'The Development of the SIOP, 1960–1983', in Desmond Ball and Jeffrey Richelson (eds.): *Strategic Nuclear Targeting*. Ithaca, NY: Cornell University Press: 57–83.

Ball, Desmond and Robert Toth 1990: 'Revising the SIOP. Taking War-Fighting to Dangerous Extremes', *International Security* **14/4**: 65–92.

Balsac, Robert de 1502: *La Nef des princes et des batailles*. Lyons: Guillaume Balsarin.

Banse, Ewald 1932: *Raum und Volk im Weltkriege*. Oldenbourg: Stalling; trans. Alan Harris: *Germany, Prepare for War!* London: Lovat Dickson, 1934.

Barjot, Pierre 1955: *Vers la Marine de l'âge atomique*. Paris: Amiot Dumont.

1961: *Histoire de la guerre aéro-navale.* Paris: Flammarion.

Barjot, Pierre and Jean Savant 1961: *Histoire mondiale de la Marine.* Paris: Flammarion.

Barker, Thomas 1975: *The Military Intellectual and Battle: Raimondo Montecuccoli and the Thirty Years' War.* Albany, NY: State University of New York Press.

Barker, Thomas, Jeremy Black and Weston Cook. 1997: 'Geoffrey Parker's Military Revolution: Three Reviews of the Second Edition', *Journal of Military History* **61/2**: 347–54.

Barret, Robert 1598: *The Theorike and Practike of Moderne Warres, Discoursed in Dialogue wise.* London: William Ponsonby.

Batschelet, Allen 2002: 'Effects-Based Operations: A New Operational Model?' Carlisle, PA: US Army War College, online at www.iwar.org.uk/militaryresources/effect-based-ops/ebo.pdf, accessed 27 July 2008.

Baudry, Ambroise 1912: *La Bataille navale.* Paris: Berger-Levrault.

Baylis, John 1991: 'Anthony Buzzard', in John Baylis and John Garnett (eds.): *Makers of Nuclear Strategy.* London: Pinter: 136–55.

1995: *Ambiguity and Deterrence: British Nuclear Strategy, 1945–1964.* Oxford: Clarendon Press.

Beaufre, André 1963: *L'Introduction à la stratégie.* Paris: A. Colin; trans. R.H. Barry : *An Introduction to Strategy.* London: Faber and Faber, 1965.

1964: *Dissuasion et stratégie.* Paris: A. Colin; trans. R.H. Barry: *Deterrence and Strategy.* London: Faber and Faber, 1965.

1966: *Stratégie de l'action.* Paris: A. Colin; trans. R.H. Barry: *Strategy of Action.* London: Faber and Faber, 1967.

1972: *La Guerre révolutionnaire: les Formes nouvelles de la guerre.* Paris: Fayard.

Bebel, August 1892: *Stehendes Heer oder Volkswehr? Rede ... zur gegenwärtigen Militärlage gehalten in der Reichstagssitzung vom 13. Dezember 1892.* Nuremberg: Wörlein, 1893.

Beckett, Ian Frederick William 2001: *Modern Insurgencies and Counter-Insurgencies: Guerrillas and their Opponents since 1750.* London: Routledge.

Behncke, Andreas 2006: 'The Re-Enchantement of War in Popular Culture', *Millennium* **34/3**: 937–49.

Belissa, Marc and Patrice Leclerq 2001: 'The Revolutionary Period, 1789–1802', in Anja V. Hartmann and Beatrice Heuser (eds.): *War, Peace and World Orders in European History.* London: Routledge: 203–13.

Bell, Christopher 2000: *The Royal Navy, Strategy and Seapower between the Wars.* Basingstoke: Palgrave Macmillan.

Benzinger, Wolfram 2006: 'Zwischen *bellum iustum* und modernem Völkerrecht: Überlegungen zum Denken über Krieg und Frieden am Ende des Mittelalters', *Militärgeschichtliche Zeitschrift* **65/1**: 131–51.

Berenhorst, Georg Heinrich von 1797: *Betrachtungen über die Kriegskunst, über ihre Fortschritte, ihre Widersprüche und ihre Zuverlässigkeit*; enlarged 3rd edn, Leipzig: 1827; repr. Osnabrück: Biblio Verlag, 1978.

Berghahn, Volker 1971: *Der Tirpitz-Plan: Genesis und Verfall einer innenpolitischen Krisenstrategie unter Wilhelm II*. Düsseldorf: Droste.

1981: *Militarism: The History of an International Debate, 1861–1979*. Leamington Spa: Berg.

Bernand, André 1999: *Guerre et violence dans la Grèce antique*. Paris: Hachette Littératures.

Bernard, Henri 1965: *Guerre totale et Guerre révolutionnaire*, vol. I. Brussels and Paris: Brepols.

Bernhardi, Friedrich von 1911: 'Clausewitz über Angriff und Verteidigung: Versuch einer Widerlegung', *Beiheft zum Militär-Wochenblatt* **12**: 399–412.

1912a: *Deutschland und der nächste Krieg*. Stuttgart: Cotta; trans. Allen Powles: *Germany and the Next War*. London: Edward Arnold, 1912.

1912b: *Vom heutigen Kriege*. Berlin: Mittler; trans. Karl von Donat: *On War of Today*, 2 vols. London: Hugh Rees, 1912.

Bernotti, Romeo 1954: *Questa crisi mondiale*. Livorno: Tirrena.

Bertaud, Jean-Paul 1988: *The Army of the French Revolution: From Citizen-Soldiers to Instrument of Power*. Princeton University Press.

Berthaut, Jean Auguste 1881: *Principes de stratégie: Étude sur la conduite des armées*. Paris: Librairie Militaire de DuMayne.

Betts, Richard 1983: 'Conventional Strategy: New Critics, Old Choices', *International Security* **7/4**: 140–62.

1987: *Nuclear Blackmail and Nuclear Balance*. Washington, D.C.: The Brookings Institution.

2000: 'Is Strategy an Illusion?' *International Security* **25/2**: 5–50.

2001/2: 'The Trouble with Strategy', *Joint Force Quarterly* (Autumn/Winter): 23–30.

Bialer, Uri 1980: *The Shadow of the Bomber: The Fear of Air Attack in British Politics, 1932–1939*. London: Royal Historical Society.

Biddle, Stephen 1996: 'Victory Misunderstood. What the Gulf War Tells Us about the Future of Conflict', *International Security* **21/2**: 139–79.

2004: *Military Power: Explaining Victory and Defeat in Modern Battle*. Princeton University Press.

Biddle, Tami Davis 1994: 'Air Power', in Michael Howard, George J. Andreopoulos and Mark R. Shulman (eds.): *The Laws of War: Constraints on Warfare in the Western World*. New Haven: Yale University Press: 140–59.

1995: 'British and American Approaches to Strategic Bombing: Their Origins and Implementation in the World War II Combined Bomber Offensive', *Journal of Strategic Studies* **18/1**: 91–144.

2002: *Rhetoric and Reality in Air Warfare: The Evolution of British and American Ideas about Strategic Bombing, 1914–1945*. Princeton University Press.

Bijker, Wiebe, Thomas Hughes and Trevor Pinch 1987: *The Social Construction of Technological Systems*. Cambridge, MA: MIT Press.

Billon, Jean de 1613: *Les Principes de l'art militaire. Lyon s*; repr. Rouen: Berthelin, 1641.

Black, Jeremy 1994: 'Eighteenth-Century Warfare Reconsidered', *War in History* **1/2**: 215–32.

1998: *Why Wars Happen*. London: Reaktion Books.

1999: *Warfare in the Eighteenth Century*. London: Cassell.

2001: *Western Warfare, 1775–1882*. Chesham: Acumen.

2004: *War and the New Disorder in the 21st Century*. New York: Continuum.

2005: *Introduction to Global Military History, 1750 to the Present Day*. London: Routledge.

Black, Jeremy (ed.) 1987: *The Origins of War in Early Modern Europe*. Edinburgh: John Donald Publishers.

Blackett, P.M.S. 1956: *Atomic Weapons and East–West Relations*. Cambridge University Press.

1962: *Studies of War Nuclear and Conventional*. Edinburgh: Oliver Boyd.

Blair, C.N.M. 1957: *Guerrilla Warfare*. London: Ministry of Defence.

Blanqui, Auguste 1868: *Instruction pour une prise d'armes*. Paris: Sens et Tonka, 2000.

Bloch, Jan 1899: *Is War Now Impossible?*, trans. and ed. W.T. Stead. London: Grant Richards.

Blume, Wilhelm von 1882: *Strategie: eine Studie*. Berlin: Mittler.

1912: *Strategie ihre Aufgaben und Mittel*. Berlin: Mittler.

Blunt, V.E.R. 1942: *The Use of Air Power*; repr. Harrisburg, PA: Military Services Publishing, 1943.

Bluth, Christoph 1995: 'Offensive Defence in the Warsaw Pact: Reinterpreting Military Doctrine', *Journal of Strategic Studies* **18/4**: 55–77.

Boehn, Max von der 1907: *Die Mode*, 8 vols. Munich: Bruckmann, 1907–1910.

Bond, Brian 1977: 'Liddell Hart and the German Generals', *Military Affairs* **41/1**: 16–20.

1996: *The Pursuit of Victory: From Napoleon to Saddam Hussein.* Oxford University Press.

Bonet [or Bouvet], Honoré *c.* 1385: *L'Arbre des batailles nouuellement imprimé et corrigé.* Paris: Jehan du Pré, 1493; trans. G.W. Coopland: *Tree of Battles.* Liverpool University Press, 1949.

Bonnal, G.A. 1916: *Les Conditions de la guerre moderne.* Paris: de Boccard.

1920: *De la Méthode dans les hautes études militaires en Allemagne et en France.* Paris: Albert Frontemoing.

Bonnet, Gabriel 1968: *La Guerre révolutionnaire du Vietnam.* Paris: Payot.

Bonte, Florimond 1968: *Qu'est-ce que la guerre atomique?* Paris: Éditions Sociales.

Boog, Horst 1988: 'Das Problem der Selbständigkeit der Luftstreitkräfte in Deutschland 1908–1945', *Militärgeschichtliche Mitteilungen* **1**: 31–60.

1992a: 'Der strategische Bombenkrieg. Lufwaffe, Royal Air Force und US Army Air Forces im Vergleich bis 1945', *Militärgeschichte* **2**: 20–30.

1992b: 'Der strategische Bombenkrieg Luftwaffe. Royal Air Force und US Army Air Forces im Vergleich bis 1945', *Militärgeschichtliche Beiträge* **6**: 18–28.

1995a: '"Baedeker-Angriffe" und Fernstflugzeugprojekte 1942. Die strategische Ohnmacht der Luftwaffe', *Militärgeschichtliche Beiträge* **4**: 93–111.

1995b: 'Bombenkriegslegenden', *Militärgeschichtliche Beiträge* **9**: 22–9.

Booth, Ken and John Baylis 1989: *Britain, NATO and Nuclear Weapons, Alternative Defence versus Alliance Reform.* London: Macmillan.

Borgert, Heinz-Ludger 1999: 'Die Planung "Barbarossa" in den Akten der Seekriegsleitung 1941', *Mars* **5**: 202–29.

Boserup, Anders (ed.) 1990: *The Foundations of Defensive Defence* New York: St Martin's Press.

Bourdé de Villehuet, Jacques 1765: *Le Manœuvrier ou essai sur la théorie et pratique des mouvements du navire et des évolutions navales.* Paris: Guérin et de la Tour.

Bouthoul, Gaston and Carrère, R. 1976: *Le Défi de la guerre, 1740–1974.* Paris: Presses Universitaires de France.

Bower, Stephen 2000: 'The Theology of the Battlefield: William Tecumseh Sherman and the U.S. Civil War', *Journal of Military History* **64**: 1005–34.

Brandi, Karl 1937: *Kaiser Karl V: Werden und Schicksal einer Persönlichkeit und eines Weltreiches*. Munich: Bruckmann.

Brandt, Heinrich von 1823: *Ueber Spanien mit besonderer Hinsicht auf einen etwaigen Krieg*. Berlin: Schüppel'sche Buchhandlung.

1829: *Handbuch für den ersten Unterricht in der höhern Kriegskunst*. Berlin: Schüppel; repr. Wiesbaden: LTR Verlag, 1981.

Brezet, François-Emmanuel 1990: 'La Pensée navale allemande des origins à 1914', in Hervé Coutau-Bégarie (ed.): *L'Évolution de la pensée navale*, vol. I. Paris: Economica: 119–33.

Brialmont, Henri-Alexis 1895: *La Défense des états et la fortification à la fin du XIXe siècle*, 2 vols.; repr. Osnabrück: Biblio Verlag, 1967.

Bridge, Cyprian Arthur George 1907: *Art of Naval Warfare*. London: Smith, Elder and Co.

1910: *Sea-Power, and Other Studies*. London: Smith, Elder and Co.; repr. Teddington: Echo Library, 2006.

Brodie, Bernard 1942: *A Guide to Naval Strategy*; repr. New York: Praeger, 1965.

1946: *The Absolute Weapons*. New York: Harcourt, Brace.

1959: *Strategy in the Missile Age*. Princeton University Press.

1973: *War and Politics*. New York: Macmillan; repr. London: Cassell, 1974.

1976: 'The Continuing Relevance of *On War*', in Carl von Clausewitz: *On War*, ed. and trans. Michael Howard and Peter Paret. Princeton University Press: 45–58.

Brossollet, Guy 1975: *Essai sur la non-bataille*. Paris: Belin.

Brown, Frederik 1991: 'AirLand Battle Future: The Other Side of the Coin', *Military Review* **71/2**: 13–24.

Brown, Harold 1980: 'Documentation. American Nuclear Doctrine', *Survival* **22/5**: 267–9.

Brown, Neville 1963: *Strategic Mobility*. London: Chatto and Windus for the ISS.

1986: *The Future of Air Power*. London: Croom Helm.

1992: *The Strategic Revolution*. London: Brassey's.

Bruendel, Steffen 2003: *Volksgemeinschaft oder Volksstaat: Die 'Ideen von 1914' und die Neuordnung Deutschlands im Ersten Weltkrieg*. Berlin: Akademie Verlag.

Buchan, Alastair (ed.) 1966: *A World of Nuclear Powers*. Englewood Cliffs: Prentice Hall.

Buchfinck, Ernst 1938: 'Moltke und Schlieffen', *Historische Zeitschrift* **158**: 308–22.

Budiansky, Stephen 2003: *Air Power from Kitty Hawk to Gulf War II*. London: Viking Penguin.

Bueb, Volckmar 1971: *Die 'Junge Schule' der französischen Marine Strategie und Politik 1875–1900*. Boppard am Rhein: Boldt.

Bueil, Jean de 1466: *Le Jouvencel*. Paris: Antoine Verard, 1493; ed. Léon Lecestre: *Le Jouvencel par Jean de Bueil*. Paris: Renouard, 1887.

Buffet, Cyril and Beatrice Heuser 1998: *Haunted by History: Myths in International Relations*. Oxford: Berghahn.

Bugeaud de la Piconnerie, duc d'Isly, Thomas Robert 1846: *Aperçus sur quelques détails de la guerre*, 3rd edn. Paris: Leneuve & Dumaine.

1850: *De la Stratégie, de la Tactique, des Retraites et du Passage des Défilés dans les montagnes des Kabyles*. Algiers: Imprimerie du Gouvernement.

1883: *Œuvres militaires*. Paris: L. Baudoin.

Bull, Hedley 1961: *The Control of the Arms Race*. London: Weidenfeld and Nicolson.

1980: 'Sea Power and Political Influence', in Jonathan Alford (ed.): *Sea Power and Influence: Old Issues and New Challenges*. Farnborough: Gower for the IISS: 3–11.

1987: *Hedley Bull on Arms Control*, ed. Robert O'Neill and David N. Schwartz. Basingstoke: Macmillan for the IISS.

Bülow, Dietrich Heinrich Freiherr von 1799: *Geist des neuern Kriegssystems hergeleitet aus dem Grundsatze einer Basis der Operationen*. Hamburg: Benjamin Gottlieb Hoffmann; 3rd edn, Hamburg: August Campe, 1835; trans. Gen. Charles Malortie: *The Spirit of the Modern System of War*. London: Egerton, 1806; 3rd edn, 1825.

1805: *Leitsätze des neuern Krieges oder reine und angewandte Strategie*. Berlin: Heinrich Frölich.

Burney, C.D. 1934: *The World, the Air and the Future*. London: Alfred A. Knopf.

Buzan, Barry, Ole Waever and Jaap de Wilde 1998: *Security: A New Framework for Analysis*. Boulder, CO: Lynne Rienner.

C–, General (Chrzanowskli) 1852: *Kleines Kriegshandbuch für Offiziere: Abriss der angewandten Taktik aller Waffen, der Generalstab und der Parteigängerkrieg*. Halle: C.A. Schwetschke.

Cabinet Office, Historical Section 1956–76: *Grand Strategy: The History of the Second World War*, 5 vols. London: HMSO.

Cable, James 1971: *Gunboat Diplomacy 1919–1991: Political Applications of Limited Naval Force*. London: Chatto & Windus for the IISS; 3rd edn, Basingstoke: Macmillan, 1994.

Cable, Larry E. 1986: *Conflict of Myths: Development of Counterinsurgency Doctrine and the Vietnam War*. Albany, NY: State University of New York Press.

Caemmerer, Rudolf von 1904: *Die Entwicklung der Strategischen Wissenschaft im 19. Jahrhundert.* Berlin: Baensch; trans. Karl von Donat: *The Development of Strategical Science*, London: Hugh Rees, 1905.

Cairns, John 1953: 'International Politics and the Military Mind: The Case of the French Republic, 1911–1914', *Journal of Modern History* **25** (Dec.): 273–85.

Caldwell, Cyril 1943: *Air Power in Total War.* New York: Coward-McCann.

Calingaert, Daniel 1988: 'Nuclear Weapons and the Korean War', *Journal of Strategic Studies* **11/2**: 177–202.

Callwell, Charles E. 1896: *Small Wars, Their Principles and Practice.* London: HMSO; 3rd edn, London: HMSO, 1906.

1905: *Military Operations and Maritime Preponderance: Their Relations and Interdependence*; repr. Annapolis, MD: Naval Institute Press, 1996.

Campbell, John 1989: 'The Soviet Union, the United States, and the Twin Crises of Hungary and Suez', in W.R. Louis and Roger Owen (eds.): *Suez 1956 – The Crisis and Its Consequences.* Oxford: Clarendon Press: 233–53.

Carlton, Eric 1996: 'Massacres: An Historical Perspective', *Journal of Military History* **60/3**: 540–1.

Carpenter, Galen Ted 2000: 'NATO's New Strategic Concept: Coherent Blueprint or Conceptual Muddle?', *Journal of Strategic Studies* **23/3**: 7–28.

Carr, William 1991: *The Origins of the Wars of German Unification.* London: Longman.

Carvajal, José Maria 1812: *Reglamento para las Partidas de Guerrilla.* Cádiz: Don Nicolas Gomez de Requena.

Carver, Michael 1990: *War since 1945*, 2nd edn. Atlantic Highlands, NJ: Ashfield.

Castex, Raoul 1934: *Théories stratégiques*, vol. IV. Paris: SEGMC.

1937: *Théories stratégiques*, vol. I, 2nd edn. Paris: SEGMC (1st edn 1929).

1939: *Théories stratégiques*, vol. II, 2nd edn. Paris: SEGMC (1st edn 1930).

1945: 'Aperçus sur la bombe atomique', *Revue Défense Nationale* 2; reprinted in Dominique David (ed.): *La Politique de défense de la France: textes et documents.* Paris: Fondation pour les Études de Défense Nationale, 1989: 65–70.

Ceillier, Capitaine 1990: 'Les Idées stratégiques en France de 1870 à 1914: la Jeune Ecole', in Hervé Coutau-Bégarie (ed.): *L'Évolution de la pensée navale*, vol. I. Paris: Economica: 195–231.

Chaliand, Gérard (ed.) 1990: *Anthologie Mondiale de la Stratégie*. Paris: Robert Laffont.
1994a: *Stratégies de la guérilla: de la Longue Marche à nos jours*. Paris: Payot & Rivages.
1994b: *The Art of War in World History*. Berkeley, CA: University of California Press.
Champier, Symphorien 1525: *Les Gestes ensemble la vie du preulx Chevalier Bayard*. Lyons: Gilbert de Villiers; ed. Denis Crouzet, Paris: Imprimerie Nationale, 1992.
Charles, Archduke of Austria 1806: *Grundsätze der höhern Kriegskunst für die Generäle der österreichischen Armee*. Vienna: Imperial and Royal Printing Office.
1838: 'Das Kriegswesen in Folge der franzöischen Revolutionskriege', in Freiherr von Waldtstätten (ed.): *Erzherzog Karl: Ausgewählte militärische Schriften*. Berlin: Richard Wilhelmi, 1882: 199–216.
1882: *Ausgewählte militärische Schriften*. Berlin: R. Wilhelmi.
Charlton, L.E.O. 1935: *War from the Air: Past, Present, Future*. London: Nelson.
1937: *The Menace of the Clouds*. London: Hodge.
Charnay, Jean-Paul, Jean Chagniot and Léo Hamon 1981: *Guibert ou le soldat philosophe*. Château de Vincennes: CERSC.
Che Guevara, Ernesto 1960: *La Guerra de guerrillas*. Havana: Minfar; trans. J.P. Morray, in Marc Becker (ed.): *Che Guevara: Guerrilla Warfare*. London: Souvenir Press, 2003: 1–141.
1963: 'Guerra de guerrillas: un método', *Cuba Socialista* **3/25** (Sept.); anon. trans., in Becker (ed.): *Che Guevara*: 142–60.
1968: 'Mensaje a la Tricontinental'. Havana: May, online at www.filosofia.org/hem/dep/cri/ri12094.htm, accessed 20 September 2009; anon. trans., in Becker (ed.): *Che Guevara*: 161–75.
Chesney, George Tomkyns 1871: *The Battle of Dorking*. Edinburgh: W. Blackwood.
Cheynet, Jean-Claude 2003: 'La guérilla à Byzance. Byzance sous le choc des invasions', *Revue Historique des Armées* **231**: 13–23.
Chickering, Roger and Stig Förster (eds.) 2000: *Great War, Total War: Combat and Mobilisation on the Western Front, 1914–1918*. Cambridge University Press.
Chickering, Roger, Stig Förster and Bernd Greiner (eds.) 2005: *A World at Total War: Global Conflict and the Politics of Destruction, 1940–1945*. Cambridge University Press.
Christine de Pisan 1410: *Livre des fais d'armes et de Chevalerie, later published as L'Art de la chevalerie selon Végèce*. Paris: Antoine Verard, 1488; trans. as *Boke of the fayt of armes and of Chyualrye*.

London: William Caxton, 1489; trans. Sumner Willard and ed. Charity Cannon Willard: *The Book of Deeds of Arms and of Chivalry*. University Park, PA: Penn State University Press, 1999.

Chrysostomides, Julian 2001: 'Byzantine Concepts of War and Peace', in Anja Hartmann and Beatrice Heuser (eds.): *War, Peace and World Orders in European History*. London: Routledge: 91–101.

Chrzanowski, General Wojciech 1839: *Über den Partheigängerkrieg: eine Skizze*, trans. A Prussian Officer; repr. Berlin: Stuhrsche Buchhandlung, 1846.

Chuter, David 1996: *Humanity's Soldier: France and International Security, 1919–2001*. Providence, RI: Berg.

1997: 'Triumph of the Will? Or, Why Surrender Is Not Always Inevitable', *Review of International Studies* **23**: 381–400.

Cicero : 44 BCE: *De Officiis*, trans. and ed. M.T. Griffin and E.M. Atkins: *On Duties*. Cambridge University Press, 1991.

Citino, Robert 1999: *The Path to Blitzkrieg: Doctrine and Training in the German Army, 1920–1939*. Boulder, CO: Lynne Rienner.

2004: *Blitzkrieg to Desert Storm: Evolution of Operational Warfare*. Lawrence, KS: University Press of Kansas.

Clark, J.J. and Dwight Barnes 1966: *Sea Power and Its Meaning*; 2nd edn, New York: Franklin Watts, 1968.

Clausewitz, Carl von 1810–11: 'Lectures on Small War', in Werner Hahlweg (ed.): *Carl von Clausewitz: Schriften – Aufsätze – Studien – Briefe*, vol. I. Göttingen: Vandenhoeck & Ruprecht, 1966.

1810–12: 'Übersicht des Sr. königl. Hoheit dem Kronprinzen in den Jahren 1810, 1811 und 1812 vom Verfasser erteilten militärischen Unterrichts', in *Vom Kriege*, 19th edn, ed. Werner Hahlweg. Bonn: Ferd. Dümmlers Verlag, 1991: 1041–86.

1812: 'Bekenntnisdenkschrift', in Hahlweg (ed.): *Carl von Clausewitz: Schriften – Aufsätze – Studien – Briefe*, vol. I: 682–750.

c. 1823: 'Feldzug von 1812 in Russland', in Hahlweg (ed.): *Carl von Clausewitz: Schriften – Aufsätze – Studien – Briefe*, vol. II: 717–924.

1832: *Vom Kriege*, 19th edn, ed. Werner Hahlweg. Bonn: Ferd. Dümmlers Verlag, 1991; trans. and ed. Michael Howard and Peter Paret: *On War*. Princeton University Press, 1976.

Clayton, Anthony 1994: *The Wars of French Decolonisation*. London: Longman.

Clerk, John, of Eldin [the Elder] 1790: *An Essay on Naval Tactics, Systematical and Historical*. London: T. Cadell.

Clifford, J. Gary 1991: 'Bureaucratic Politics', in Michael Hogan and Thomas Paterson (eds.): *Explaining the History of American Foreign Relations*; 2nd edn, Cambridge University Press, 2004: 91–102.

Clodfelter, Mark 1989: *The Limits of Air Power: The American Bombing of North Vietnam.* New York: Free Press; London: Collier Macmillan.

Cohen, Eliot 1991: 'Churchill and Coalition Strategy in World War II', in Paul Kennedy (ed.): *Grand Strategies in War and Peace.* New Haven: Yale University Press: 43–70.

Coker, Christopher 1986: *A Nation in Retreat? Britain's Defence Commitment.* London: Brassey's.

2001: *Humane Warfare.* London: Routledge.

2004: *The Future of War: The Re-Enchantment of War in the 21st Century.* Oxford: Blackwell Publishing.

Colin, Jean 1911: *Les Transformations de la guerre.* Paris: Flammarion; repr. Paris: Economica, 1989; trans. L.H.R. Pope-Hennessy: *The Transformation of War.* London: Hugh Rees, 1912.

Collier, Basil 1974: *A History of Air Power.* London: Weidenfeld and Nicolson.

Collins, Joseph 1992: 'Desert Storm and the Lessons of Learning', *Parameters* (Autumn): 83–95.

Colomb, John Charles Ready 1880: *The Defence of Great and Greater Britain: Sketches of Its Naval, Military, and Political Aspects.* London: Edward Stanford.

Colomb, Philip Howard 1891: *Naval Warfare: Its Ruling Principles and Practice Historically Treated.* London: W.H. Allen.

Colson, Bruno 1991: 'La Culture stratégique Americaine et le Golfe', *Stratégique* **3** and 4: 51–2 and 113–32.

Colson, Bruno and Jean-Pierre Colson 1994: 'Les Penseurs navals hollandais', in Hervé Coutau-Bégarie (ed.): *L'Évolution de la pensée navale*, vol. IV. Paris: Economica: 173–90.

Cookson, J.E. 1998: 'The British Armed Nation, 1793–1815', *Journal of Military History* **62/3**: 619–20.

Cooper, Robert 2000: *The Postmodern State and the World Order.* London: Demos.

Copel, Étienne 1984: *Vaincre la guerre.* Paris: Lieu Commun.

Corbett, Julian 1911: *Some Principles of Maritime Strategy.* London: Longman's, Green and Co.; repr. Annapolis, MD: Naval Institute Press, 1988.

Corum, James 1994: *The Roots of Blitzkrieg.* Lawrence, KS: University Press of Kansas.

1995a: 'The Luftwaffe and the Coalition Air War in Spain, 1936–1939', *Journal of Strategic Studies* **18/1**: 68–90.

1995b: 'The Luftwaffe's Army Support Doctrine, 1918–1941', *Journal of Military History* **59/1**: 53–77.

1996: 'From Biplanes to Blitzkrieg: The Development of German Air Doctrine between the Wars', *War in History* **3/1**: 85–101.

1997: 'The Development of Strategic Air War Concepts in Interwar Germany, 1919–1939', *Air Power History*: 18–35.

1998: 'The Spanish Civil War: Lessons Learned and Not Learned by the Great Powers' *Journal of Military History* **62/2**: 313–34.

Corum, James and Wray Johnson 2003: *Airpower in Small Wars: Fighting Insurgents and Terrorists*. Lawrence, KS: University Press of Kansas.

Covarrubias, Sebastián de 1611/74: *Tesoro de la Lengua Castellana o Española* (1611 with additions from 1674), ed. Martín Riquer and repr. Barcelona: S.A. Horta, 1943.

Crane, Conrad 2000: 'To Avert Impending Disaster: American Military Plans to Use Atomic Weapons during the Korean War', *Journal of Strategic Studies* **23/2**: 72–88.

Craven, John 1980: 'The Future of the Sea-Based Deterrent', in Alford (ed.): *Sea Power and Influence*: 80–5.

Crenzien, Bodo 1990: *Stabilitet og sikkerhed: militaere aspekter af konventionel nedrustning*. Copenhagen: Sikkerheds- og Nedrustningspolitiske Udvalg.

Creswell, John 1936: *Naval Warfare: An Introductory Study*. London: Sampson Low.

Creveld, Martin van 1989: *Technology and War from 2000 BC to the Present*. New York: Free Press.

1991: *The Transformation of War*. New York: Free Press; London: Brassey's (published as *On Future War*).

1999: *The Rise and Decline of the State*. Cambridge University Press.

2004: *Supplying War*, 2nd edn. Cambridge University Press.

Creveld, Martin van, Steven L. Canby and Kenneth S. Brower 1994: *Air Power and Maneuver Warfare*. Honolulu, HI: University Press of the Pacific; repr. 2002.

Crewe-Milnes, R.O.A. 1931: *Lord Rosebery*, 2 vols. s.l.: Murray.

Croix, Armand François de la 1752: *Traité de la petite guerre pour les compagnies franches*. Paris: A. Boudet.

Croxton, Derek 1998: 'A Territorial Imperative? The Military Revolution, Strategy and Peacemaking in the Thirty Years War', *War in History* **5/3**: 253–79.

Cruso, John 1644: *Military Instructions for the Cavallrie or Rules and Directions for the Service of Horse, Collected out of Divers Forrein Authors, Ancient and Modern*. Cambridge: Roger Daniel.

Curry, Anne 2001: 'War, Peace and National Identity in the Hundred Years' War', in Anja V. Hartmann and Beatrice Heuser (eds.): *War, Peace and World Orders in European History*. London: Routledge: 141–54.

Custance, Reginald 1919: *War at Sea*. Edinburgh and London: William Blackwood.
1924: *A Study of War*. London: Constable and Co.
[as 'Barfleur'] 1907: *Naval Policy*. Edinburgh and London: William Blackwood.
D.J. (Chevalier de Jaucourt?) 1757: 'Guerre, droit naturel et politique', in Denis Diderot and Jean le Rond d'Alembert: *Encyclopédie ou Dictionnaire raisonné des Sciences, des Arts et des Métiers*, vol. VII. Paris: Briasson, David, Le Breton et Durand.
Dahm, Helmut 1970: 'Die sowjetische Militärdoktrin', *Osteuropa* **6**: 384–98.
Danchev, Alex 1998: *Alchemist of War: The Life of Basil Liddell Hart*. London: Weidenfeld and Nicolson.
1999: 'Liddell Hart and the Indirect Approach', *Journal of Military History* **63/2**: 313–37.
Danel, Raymond and Jean Cuny 1980: L'Aviation française de bombardement et de renseignement: 1918–1940. Paris: Larivière.
Darrieus, Gabriel 1907: *La Guerre sur mer*. Paris: A. Challamel.
1993: 'Les Livres de Tactique français du Père Hoste au XIXe siècle', in Hervé Coutau-Bégarie (ed.): *L'Évolution de la pensée navale*, vol. III. Paris: Economica: 209–31.
Darrieus, Henri and Bernard Estival 1990: 'Darrieus et la renaissance d'une pensée maritime en France avant la Première Guerre Mondiale', in Hervé Coutau-Bégarie (ed.): *L'Évolution de la pensée navale*, vol. I. Paris: Economica: 89–117.
Daudet, Léon 1918: *La Guerre totale*. Paris: Nouvelle Librairie Nationale.
Daveluy, René 1902: *Étude sur le combat naval*. Paris: Berger-Levrault.
1905: Étude sur la stratégie navale. Paris: Berger-Levrault.
Davidov [or Davidoff], Denis Vasilevich 1841: *Essai sur la guerre de partisans*, trans. from the Russian by Count Héraclius de Polignac. Paris: J. Corréard.
Davies, R.W. and S.G. Wheatcroft 1994: 'Population', in R.W. Davies, Mark Harrison and S.G. Wheatcroft (eds.): *The Economic Transformation of the Soviet Union, 1913–1945*. Cambridge University Press: 57–80.
Davis, Lance and Stanley Engerman 2006: *Naval Blockades in Peace and War: An Economic History since 1750*. Cambridge University Press.
Davis, Lynn 1975/6: *Limited Nuclear Options: Deterrence and the New American Doctrine*, Adelphi Paper 121. London: IISS: 1–22.
1988: 'Lessons of the INF Treaty', *Foreign Affairs* **66/4**: 720–34.

Dean, Jonathan 1987/8: 'Alternative Defence: Answer to NATO's Central Front Problems?', *International Affairs* **63/4**: 61–82.

Debray, Régis 1972: *Révolution dans la Révolution?* Paris: Maspéro.

1974a: *La Critique des armes*, 2 vols. Paris: Seuil.

1974b: *La Guérilla du Che*. Paris: Seuil.

Decken, Johann Friedrich von 1800: *Betrachtungen über das Verhältnis des Kriegsstandes zu dem Zwecke der Staaten*. Hanover: Helwig'sche Hofbuchhandlung; repr. Osnabrück: Biblio Verlag, 1982.

Decker, Major Carl von 1822: *Der kleine Krieg im Geist der neueren Kriegsführung oder Abhandlung über die Verwendung und den Gebrauch aller drei Waffen im kleinen Kriege*; 3rd edn, Berlin: Ernst Siegfried Mittler, 1828.

Degler, Carl 1997: 'The American Civil War and the German Wars of Unification', in Stig Förster and Jörg Nagler (eds.): *On the Road to Total War: The American Civil War and the German Wars of Unification, 1861–1871*. Cambridge University Press: 53–73.

Deist, Wilhelm and Herbert Schottelius (eds.) 1981: *Marine und Marinepolitik im Kaiserlichen Deutschland*, 2nd edn. Düsseldorf: Droste.

Delbrück, Hans 1886: *Historische und politische Aufsätze*; repr. Berlin: von Walther und Apolat, 1887.

1890: *Die Strategie des Perikles erläutert durch die Strategie Friedrichs des Großen*. Berlin: Reimer.

1900: *Geschichte der Kriegskunst*, vol. I, *Das Altertum von den Perserkriegen bis Caesar*. Berlin: Stilke; 3rd edn, 1920; repr. Hamburg: Nikol, 2003.

1901a: *Geschichte der Kriegskunst*, vol. II, *Die Germanen*; 3rd edn, 1921; repr. Hamburg, Nikol, 2003.

1901b: *Geschichte der Kriegskunst*, vol. III, *Das Mittelalter*; 3rd edn, 1921, repr. Hamburg, Nikol, 2003.

1920a: *Geschichte der Kriegskunst*, vol. IV, *Die Neuzeit: Vom Kriegswesen der Renaissance bis zu Napoleon*. Berlin: Stilke; repr. Hamburg: Nikol, 2003.

1920b: 'Falkenhayn und Ludendorff', *Preußische Jahrbücher* **180/2**: 35–50.

1925: 'Zu Kromayers neuer Theorie der Strategie', *Historische Zeitschrift* **132**: 447–9.

Delmas, Claude 1974: *Le Second Âge nucléaire*. Paris: Presses Universitaires de France.

Delmas, Jean 1992: *Histoire Militaire de la France*, André Corvisier (gen. ed.), vol. II, *De 1715 à 1871*. Paris: Presses Universitaires de France.

Demeter, Karl 1960: 'Scharnhorst', in Werner Hahlweg (ed.): *Klassiker der Kriegskunst*. Darmstadt: Wehr- und Wissen-Verlag: 211–27.

Depeyre, Michel 1990a: 'Le Père Paul Hoste, fondateur de la pensée navale moderne', in Hervé Coutau-Bégarie (ed.): *L'Évolution de la pensée navale*, vol. I. Paris: Economica: 79–88.

1990b: 'Audibert Ramatuelle ou des Enseignements perdus', in Hervé Coutau-Bégarie (ed.): *L'Évolution de la pensée navale*, vol. I. Paris: Economica: 89–98.

1992: 'John Clerk of Eldin, un penseur naval contesté', in Hervé Coutau-Bégarie (ed.): *L'Évolution de la pensée navale*, vol. II. Paris: Economica: 57–78.

1998: *Tactique et stratégies navales de la France et du Royaume-Uni de 1690 à 1815*. Paris: Economica.

Derrécagaix, Victor 1885: *La Guerre moderne*, 2 vols. Paris: Librairie militaire de L. Baudoin; 2nd edn, Paris: Librairie militaire de L. Baudoin, vol. I, 1890; vol. II, 1891.

Desportes, Vincent 2008: *La Guerre probable*, 2nd edn. Paris: Economica.

Devine, A.M. 1993: 'Arrian's "Tactica"', in Wolfgang Haase (ed.): *Rise and Decline of the Roman World*. Berlin: de Gruyter: 312–38.

DeVries, Kelly 1998: 'Gunpowder Weaponry and the Rise of the Early Modern State', *War in History* **5/2**: 127–45.

Dickens, Gerald 1946: *Bombing and Strategy: The Fallacy of Total War*. London: Sampson Low, Marston and Co.

Digges, Thomas and Dudley Digges 1604: *Four Paradoxes, or Politique Discourses Concerning Militarie Discipline*. London: Lownes.

Dingman, Roger 1988/9: 'Atomic Diplomacy during the Korean War', *International Security* **13/3**: 50–89.

Dodge, Toby 2006/7: 'How Iraq Was Lost', *Survival* **48/4**: 157–72.

Domes, Wolfgang 1995: 'Tilly und Wallenstein. Ihr Aufstieg zum Heerführer – ein Vergleich', *Mars* **1**: 148–62.

Donagan, Barbara 1988: 'Codes and Conduct in the English Civil War', *Past and Present* **118**: 65–95.

Dosal, Paul 2003: *Commandante Che: Guerrilla Soldier, Commander, and Strategist, 1956–1967*. University Park, PA: Penn State University Press.

Doughty, Robert Allan 1985: *The Seeds of Disaster: The Development of the French Army Doctrine, 1919–1939*. Hamden, CT: Archon Books.

Douglas, Howard 1858: *On Naval Warfare with Steam* London: n.p; 2nd edn, London: John Murray, 1860.

Douhet, Giulio 1921: *Il Dominio dell'Aria* Rome: L'Amministrazione della Guerra; 2nd edn, Rome: s.n., 1927; trans. Dino Ferrari: *The Command of the Air*. Washington, D.C.: Office of Air Force History, 1983.

1928: *Probabili Aspetti della Guerra futura*. Palermo: Sandron; trans. Dino Ferrari in: *The Command of the Air*. Washington, D.C.: Office of Air Force History, 1983: 143–207.

Downes, Alexander 2006: 'Desperate Times, Desperate Measures: The Causes of Civilian Victimization in War', *International Security* **30/4**: 152–95.

Dragon, Gilbert (ed.) 1986: *Traité sur la guérilla de l'Empereur Nicéphore Phocas (963–969)*. Paris: Éditions du CNRS.

Duby, Georges 1993: *France in the Middle Ages, 987–1460: From Hugh Capet to Joan of Arc*, trans. from the French by Juliet Vale. Oxford: Blackwell.

Duchhardt, Heinz 2001: 'Interstate War and Peace in Early Modern Europe', in Anja V. Hartmann and Beatrice Heuser (eds.): *War, Peace and World Orders in European History*. London: Routledge: 185–96.

Duffy, Christopher 1967: 'Introduction', in General A.H. Brialmont: *La Défense des états et la fortification à la fin du XIXe siècle*. Osnabrück: Biblio Verlag, 1967 (first published Brussels, 1895): v–xii.

Dwyer, Christopher 1999: 'Raiding Strategy: As Applied by the Western Confederate Cavalry in the American Civil War', *Journal of Military History* **63**: 263–81.

Dwyer, Philip 1994: 'The Politics of Prussian Neutrality 1795–1805', *German History* **12/3**: 351–73.

Earle, Edward Mead (ed.) 1943: *Makers of Modern Strategy from Machiavelli to Hitler*. Princeton University Press.

Echevarria, Antulio 1996: 'General Staff Historian Hugo Freiherr von Freytag-Loringhoven and the Dialectics of German Military Thought', *Journal of Military History* **60/3**: 471–94.

2000: *After Clausewitz: German Military Thinkers before the Great War*. Lawrence, KS: University of Kansas Press.

2002a: 'The "Cult of the Offensive" Revisited: Confronting Technological Change before the Great War', *Journal of Strategic Studies* **25/1**: 199–214.

2002b: 'Combining Firepower and Versatility: Remaking the "Arm of Decision" before the Great War', *RUSI Journal* **147/3**: 84–91.

Edelstein, David 2004: 'Occupational Hazards: Why Military Occupations Succeed or Fail', *International Security* **29/1**: 49–91.

Eksteins, Modris 1990: *The Rites of Spring: The Great War and the Birth of the Modern Age*. London: Black Swan.

Ellis, John 1975: *A Short History of Guerrilla Warfare*. London: Ian Allan.

Emmerich, Andreas 1789: *The Partisan in War or the Use of a Corps of Light Troops to an Army*. London: H. Reynell.

Epstein, Robert 1995: 'Napoleon's Last Victory and the Emergence of Modern War', *Journal of Strategic Studies* **18/1**: 269–70.

Erasmus, Desiderius, van Rotterdam 1521: *Antipolemus*, trans. V. Knox: *Antipolemus: Or the Plea of Reason, Religion, and Humanity against War*. London: Dilly, 1794.

Esdaile, Charles 1995: *The Wars of Napoleon*. London: Longman.

2002: *The Peninsular War*. London: Penguin.

2008: 'De-constructing the French Wars: Napoleon as Anti-Strategist', *Journal of Strategic Studies* **31/4**: 512–52.

Eshel, David 1991: 'Desert Storm: A Textbook Victory', *Military Technology* **4**: 28–34.

Eshel, Tamir 1991: 'The Most Successful Air Campaign Ever?', *Military Technology* **4**: 36–44.

Evans, Nicholas 2000: 'Boer War Tactics Re-examined', *RUSI Journal* **145/4**: 71–6.

Evera, Stephen Van 1984: 'The Cult of the Offensive and the Origins of the First World War', *International Security* **9/1**: 58–107.

1998: 'Offense, Defense, and the Causes of War', *International Security* **22/4**: 5–43.

Ewald, Johann 1785: *Abhandlung über den kleinen Krieg*. Cassel: Johann Jacob Cramer; trans. Robert A. Selig and David Curtis Skaggs: *Treatise on Partisan Warfare*. New York: Greenwood Press, 1991.

1790: *Abhandlung von dem Dienst der leichten Truppe*. Flensburg, Schleswig and Leipzig: s.n.

1798: *Belehrungen über den Krieg, besonders über den kleinen Krieg, durch Beispiele großer Helden und kluger und tapferer Männer*. Schleswig: J.G. Röhß.

[Ewald, Johann] 1774: *Gedanken eines Hessischen Officiers über das, was man bey Führung eines Detaschements im Felde zu thun hat*. Cassel: Johann Jacob Cramer.

Faivre, Maurice 1985: 'La Pensée militaire de Jaurès', *Stratégique* **25**: 63–120.

Fall, Bernard 1961: *Street without Joy*. Harrisburg, PA: Strackpole Books; revised 1989 by Dorothy Fall; repr. Mechanicsburg, PA: Stackpole Books, 1994.

1965: 'The Theory and Practice of Counterinsurgency', *Naval War College Review* (Winter), online at http://smallwarsjournal.com/documents/fall.pdf, accessed 29 April 2010.

Farrar, L.L. 1972: 'The Short-War Illusion. The Syndrome of German Strategy, Aug.–December, 1914', *Militärgeschichtliche Mitteilungen* **2**: 39–52.

1976: 'Separate Peace – General Peace – Total War. The Crisis in German Policy during the Spring of 1917', *Militärgeschichtliche Mitteilungen* **2**: 51–80.

Farrar-Hockley, Anthony 1998: '"The Shock of Battle" 1914–1916', *RUSI Journal* **143/1**: 56–7.

Faure, Pierre 1931: *Vers un nouveau Charleroi*. Paris: Alexis Redier.

1935: *L'Avion tuera la guerre*. Paris: Gallimard.

Ferguson, Niall 1992: 'Germany and the Origins of the First World War: New Perspectives', *Historical Journal* **35/3**: 725–52.

Ferrante, Enzio 1993: 'La Pensée navale italienne: II. De Lissa à la Grande Guerre', in Hervé Coutau-Bégarie (ed.): *L'Évolution de la pensée navale*, vol. III. Paris: Economica: 97–122.

1997: 'La Pensée navale italienne entre les deux guerres mondiales', in Hervé Coutau-Bégarie (ed.): *L'Évolution de la Pensée navale*, vol. VI. Paris: Economica: 153–69.

1999: 'La Pensée navale italienne après 1945', in Hervé Coutau-Bégarie (ed.): *L'Évolution de la pensée navale*, vol. VII. Paris: Economica: 171–88.

Ferrero, Guglielmo 1933: *Peace and War*, trans. Bertha Pritchard. London: Macmillan and Ayer Publishing.

Ferrill, Arther 1991: 'The Grand Strategy of the Roman Empire', in Paul Kennedy (ed.): *Grand Strategies in War and Peace*. New Haven: Yale University Press: 71–85.

Feuquières, Antoine de Pas, Marquis de 1731: *Mémoires sur la Guerre où l'on a rassemblé les Maximes les plus nécessaires dans les operations de l'Art Militaire*. Amsterdam: François Changuion.

Fioravanzo, Giuseppe 1930–1: *La guerra sul mare e la guerra integrale*. Turin: Schioppo.

Fischer, Fritz 1959: 'Deutsche Kriegsziele Revolutionierung und Separatfrieden im Osten 1914–1918', *Historische Zeitschrift* **188**: 249–310.

Fissel, Mark Charles 2001: *English Warfare, 1511–1642*. London: Routledge.

Flavius Arrianus *c.* 136: *Tactica* [*Techne Taktike*]; Greek text ed. and trans. into Latin by Johannes Scheffer as *Arriani* Tactica *et Mauricii* Ars Militaris. Uppsala: Henricus Curio, 1664; facsimile repr., Osnabrück: Biblio Verlag, 1967.

Foch, Ferdinand 1900: *Principes de la guerre: Conférences faites en 1900 à l'École supérieure de Guerre*; 5th edn, Paris: Berger-Levrault, 1918; trans. Hilaire Belloc: *The Principles of War*. London: Chapman and Hall, 1918.

1909: *De la Conduite de la guerre: la Manœuvre pour la bataille.* Paris: Berger-Levrault.

Folard, Jean Charles Chevalier de 1724: *Nouvelles Découvertes sur la guerre, dans une dissertation sur Polybe.* Paris: Jean-François Josse and Claude Labottière.

1729: *Traité de la colonne et de l'ordre profond.* Paris: s.n.

Foley, Robert 2001: 'From Volkskrieg to Vernichtungskrieg: German Concepts of Warfare, 1871–1935', in Anja V. Hartmann and Beatrice Heuser (eds.): *War, Peace and World Orders in European History.* London: Routledge: 214–25.

2002a: 'Institutionalized Innovation: The German Army and the Changing Nature of War 1871–1914', *RUSI Journal* **147/2**: 84–90.

Foley, Robert (ed. and trans.) 2002b: *Alfred von Schlieffen's Military Writings.* London: Frank Cass.

Foot, R.J. 1985: *The Wrong War – American Policy and the Dimensions of the Korean Conflict 1950–1953.* Ithaca, NY: Cornell University Press.

1988/9: 'Nuclear Coercion and the Ending of the Korean Conflict', *International Security* **13/3**: 92–109.

Forage, Paul 2002: 'Bombs for Peace: A Comparative Study of the Use of Air Power in the Balkans', *Armed Forces* and *Society* **28/2**: 211–32.

Ford, Christopher M. 2005: 'Speak No Evil: Targeting a Population's Neutrality to Defeat Insurgency', *Parameters* (Summer): 51–66.

Ford, Franklin 1963/4: 'The Revolutionary-Napoleonic Era: How Much of a Watershed?', *American Historical Review* **69**: 18–29.

Förster, Stig 1995: 'Der deutsche Generalstab und die Illusion des kurzen Krieges, 1871–1914. Metakritik eines Mythos', *Militärgeschichtliche Mitteilungen* **54**: 61–95.

1999: 'Dreams and Nightmares: German Military Leadership and the Images of Future Warfare, 1871–1914', in Manfred Boemeke, Roger Chickering and Stig Förster (eds.): *Anticipating Total War: The German and American Experiences, 1871–1914.* Cambridge University Press: 343–76.

Förster, Stig and Jörg Nagler 1997: 'Introduction', in Stig Förster and Jörg Nagler (eds.): *On the Road to Total War: The American Civil War and the German Wars of Unification, 1861–1871.* Cambridge University Press: 1–28.

Fourquevaux, Raymond Beccarie de Pavie, sieur de 1549: *Instructions sur le faict de la guerre.* Paris: M. Vascosan; trans. Paul Iue [with wrong attribution of original authorship to General William de Bellay]: *Instructions for the Warres.* London: Thomas Man and Tobie Cooke, 1589.

Francesco Maria II, Duke of Urbino 1583: *Discorsi militari*. Ferrara: Dominico Mammarelli.

Frederick II of Prussia 1739: *Der Antimachiavell*, in Gustav Berthold Volz (ed.): *Die Werke Friedrichs des Großen*, vol. VII, *Antimachiavell und Testamente*. Berlin: Reimar Hobbing, 1912.

1748: 'Die Generalprinzipien des Krieges und ihre Anwendung auf die Taktik und Disziplin der preussischen Truppen', in Werner Hahlweg (ed.): *Klassiker der Kriegskunst*. Darmstadt: Wehr- und Wissen-Verlag, 1960: 174–82.

1752: 'Political Testament' of 1752, in Richard Dietrich (ed.): *Die politischen Testamente der Hohenzollern*. Cologne and Vienna: Böhlau Publishers, 1986: 328–461.

1761: *Des Koenigs von Preussen Majestaet Unterricht von der Kriegskunst an seine Generalen*. Frankfurt and Leipzig: s.n.

1768: 'Political Testament' of 1768, in Richard Dietrich (ed.): *Die politischen Testamente der Hohenzollern*. Cologne and Vienna: Böhlau Publishers, 1986: 462–697.

1882: *Militärische Schriften*, Adalbert von Taysen (ed.). Berlin: Richard Wilhelmi.

1899: *Anschauungen vom Kriege*, in Großer Generalstab (ed.): *Zwischenzum Generalstabswerk*; repr. Bad Honnef: LTR Verlag, 1982.

Freedman, Lawrence 1986: *The Price of Peace: Living with the Nuclear Dilemma*. London: Firethorn Press.

1989: *The Evolution of Nuclear Strategy*, 2nd edn. Basingstoke: Macmillan.

1991: 'Strategy in the Gulf War', *International Security* **16/2**: 5–41.

1993: 'The Politics of Military Intervention within Europe', in N. Gnesotto (ed.): *War and Peace: European Conflict Prevention*, Chaillot Paper no. 11. Paris: Institute of Security Studies: 29–38.

1996: 'Vietnam and the Disillusioned Strategist', *International Affairs* **72**: 133–51.

2004: *Deterrence*. Cambridge: Polity Press.

2008: 'Strategic Studies and the Problem of Power', in Thomas Mahnken and Joseph Maiolo (eds.): *Strategic Studies: A Reader*. Abingdon: Routledge: 22–33.

Freedman, Lawrence (ed.) 1998: *Strategic Coercion: Concepts and Cases*. Oxford University Press.

Frémeaux, Jacques 2003: 'Guerre et violence en Algérie: 1830–1870; 1954–1964', *Revue Historique des Armées*, **231**: 9–19.

Friedberg, Aaron 1982: 'The Evolution of U.S. Strategic "Doctrine", 1945–1981', in Samuel P. Huntington (ed.): *The Strategic Imperative: New Policies for American Security*. Cambridge, MA: Ballinger: 53–96.

Friedman, Norman 2001: *Seapower as Strategy: Navies and National Interests*. Annapolis, MD: Naval Institute Press.
Frontinus, Sextus Julius *c.* 1st century CE: *Libri quatuor Strategematicon*. Paris: s.n., 1690; trans. C.E. Bennett: *The Stratagems*. London: Heinemann, 1925.
Frost, Robert 2000: *The Northern Wars, 1558–1721*. Harlow: Pearson Education.
Fukuyama, Francis 1989: 'Have We Reached the End of History?' *The National Interest* **16** (Summer): 3–18.
Fuller, J.F.C. 1920: *Tanks in the Great War, 1914–1918*. London: John Murray.
1923: *The Reformation of War*. London: Hutchinson.
1925: *The Foundations of the Science of War*. London: Hutchinson.
1928: *On Future Warfare*. London: Sifton Praed.
1931: *Lectures on F[ield] S[ervices] R[egulations] II*. London: Sifton Praed.
1932a: *The Dragon's Teeth: A Study of War and Peace*. London: Constable.
1932b: *War and Western Civilisation, 1832–1932*. London: Duckworth.
1937: *Towards Armageddon: The Defence Problem and Its Solution*. London: Lovat Dickson.
1939: *Decisive Battles: Their Influence upon History and Civilisation*, 2 vols. London: Eyre and Spottiswoode.
1951: *How to Defeat Russia*. London: Eyre and Spottiswoode.
1961: *The Conduct of War, 1789–1961*. London: Eyre and Spottiswoode.
Furet, Marie-Françoise 1973: *Le Désarmement nucléaire*. Paris: Pédone.
Furse, George Armand 1897: *Military Expeditions beyond the Seas*, 2 vols. London: Clowes and Son.
Gallois, Pierre 1960: *Stratégie de l'âge nucléaire*. Paris: Calmann-Lévy.
1967: *Paradoxes de la paix*. Paris: Presses du temps présent.
1972: *L'Europe change de maître*. Paris: L'Herne.
1975: *La Grande Berne: l'Atome et les négociations Est–Ouest*. Paris: Plon.
1976: *L'Adieu aux armées*. Paris: Albin Michel.
1977: *Le Renoncement: de la France défendue à l'Europe protégée*. Paris: Plon.
1994: *Livre noir sur la défense*. Paris: Payot & Rivages.
Galloway, Archie 1986: 'FM 100–5: Who Influenced Whom?', *Military Review* **66/3**: 46–51.
Galula, David 1964: *Counterinsurgency Warfare, Theory and Practice*. New York: Praeger; 3rd edn, New York: Praeger, 2005.

Gardner, Jock 2006: 'All Sorts of War: British Naval Thinking and Technology in the Second World War', in Geoffrey Till (ed.): *The Development of British Naval Thinking*. London: Routledge: 134–59.

Gardner, W.J.R. 1996: 'The First World War at Sea and Elsewhere', *Historical Journal* **39/3**: 839–43.

Garibaldi, Giuseppe 1848: 'Decalogo tattico', in Oreste Bovio: 'L'arte militare de Giuseppe Garibaldi', in *Garibaldi, generale della Libertà*. Rome: Ufficio Storico SME, 1984: 21–44.

1866: *Consigli tatticidel Generale Garibaldi a' suoi volontari*; 2nd edn, Rome: Biblioteca della Camera dei Deputati, 1886.

Garnett, John 1975: 'Strategic Studies and Its Assumptions', in John Baylis, Ken Booth, John Garnett and Phil Williams: *Contemporary Strategy: Theories and Policies*. London: Croom Helm: 1–8.

Garrard, William 1591: *The Arte of Warre*. London: Roger Warde.

Garraud, Philippe 2001: 'L'Aviation de bombardement française pendant la campagne 1939–1940: capacités et emploi', *Revue Historique des Armées* **224**: 89–100.

Garrett, James 1987: 'Nuclear Weapons for the Battlefield: Deterrent or Fantasy?', *Journal of Strategic Studies* **10/2**: 168–88.

Gat, Azar 1989: *The Origins of Military Thought from the Enlightenment to Clausewitz*. Oxford University Press.

1992: *Military Thought in the Nineteenth Century*. Oxford University Press.

1998: *Fascist and Liberal Visions of War: Fuller, Liddell Hart, Douhet, and Other Modernists*. Oxford: Clarendon Press.

2001: 'Ideology, National Policy, Technology and Strategic Doctrine between the World Wars', *Journal of Strategic Studies* **24/3**: 1–18.

Gaulle, Charles de 1932: *Le Fil de l'epée*. Paris: Berger-Levrault; trans. Gerard Hopkins: *The Edge of the Sword*. London: Faber and Faber, 1960.

1934: *Vers l'Armée de métier*. Paris: Berger-Levrault; anon. trans.: *The Army of the Future*. London: Hutchinson, 1940.

1938: *La France et son armée*. Paris: Plon; repr. London: Hutchinson, 1948.

1940: 'Mémorandum du 26 janvier 1940', in Gérard Chaliand (ed.): *Anthologie mondiale de la stratégie*. Paris: Robert Laffont, 1990: 1208–13.

Gentile, Gian 2000: 'Shaping the Past Battlefield, "For the Future": The United States Strategic Bombing Survey's Evaluation of the American Air War against Japan', *Journal of Military History* **64**: 1085–112.

2001: *How Effective Is Strategic Bombing? Lessons Learned from World War II to Kosovo*. New York University Press.

George, Alexander and Richard Smoke 1974: *Deterrence in American Foreign Policy: Theory and Practice*. New York: Columbia University Press.

George, Alexander, William Simons and David Hall 1971: *The Limits of Coercive Diplomacy*. Boston: Little, Brown.

Géré, François 1992: 'La Pensée navale française sous la IVe République', in Hervé Coutau-Bégarie (ed.): *L'Évolution de la pensée navale*, vol. II. Paris: Economica: 181–210.

Giamberardino, Oscar di 1937: *L'Arte della Guerra in mare*. Rome: Ministry of the Navy.

Giap, General Vo Nguyen 1970a: *Guerre de libération: Politique – stratégie – tactique*. Paris: Éditions Sociales.

1970b: *Banner of People's War, the Party's Military Line*. London: Pall Mall Press.

1971: *National Liberation War in Vietnam*. Hanoi: Foreign Languages Publishing House.

1974: *People's War, People's Army: The Viet Cong Insurrection Manual*. Dehra Dun, India: Natraj Publishers.

Gibson, Irving 1945: 'The Maginot Line', *Journal of Modern History* **17**: 130–46.

Gilbert, Georges 1890: *Essais de critique militaire*. Paris: Librairie de la Nouvelle Revue; repr. Paris: Hachette, 1978.

1892: *Sept Études militaires*. Paris: Librairie de la Nouvelle Revue.

Girardet, Raoul 1989: *Problèmes militaires et stratégiques contemporaines*. Paris: Dalloz.

Girling, J.L.S. 1970: 'The Guam Doctrine', *International Affairs* **46/1**: 48–62.

Goering, Hermann 1939: Directive RW 5/v. 336a of 20 July 1939; repr. 30 September 1939. Bundesarchiv Militärarchiv (Freiburg).

Golovine, N[icholas?] 1936: *Air Strategy*. London: Gale and Polden.

Goltz, Colmar Freiherr v.d. 1883: *Das Volk in Waffen*. Berlin: Decker; 2nd edn, Berlin: Decker, 1899; trans. Philip A. Ashworth: *The Nation in Arms*. London: Hugh Rees, 1906.

1895: *Kriegführung: Kurze Lehre ihrer wichtigsten Grundsätze und Formen*. Berlin: Decker.

1895: *Kriegführung: Kurze Lehre ihrer Grundsätze und Formen*. Berlin: R. v. Deckers Verlag; trans. Major G.F. Leverson: *The Conduct of War: A Short Treatise on Its Most Important Branches and Guiding Rules*. London: Kegan Paul, 1899.

Gooch, John 1994: 'The Weary Titan: Strategy and Policy in Great Britain, 1890–1918', in Williamson Murray, Macgregor Knox and Alvin

Bernstein (eds.): *The Making of Strategy: Rulers, States and War.* Cambridge University Press: 278–306.

Goosens, Aline 2001: 'Wars of Religion: The Examples of France, Spain and the Low Countries in the 16th century', in Anja V. Hartmann and Beatrice Heuser (eds.): *War, Peace and World Orders in European History.* London: Routledge: 160–74.

Gorce, Paul Marie de la 1963: *La République et son Armée.* Paris: Fayard.

Gordon, Andrew 1994: 'The Admiralty and Imperial Overstretch, 1902–1941', *Journal of Strategic Studies* **17/1**: 63–85.

2006: '1914–18: The Proof of the Pudding', in Geoffrey Till (ed.): *The Development of British Naval Thinking.* London: Routledge: 89–102.

Gordon, Michael 1974: 'Domestic Conflict and the Origins of the First World War: The British and the German Cases', *Journal of Modern History* **46**: 191–226.

2006/7: 'Break Point? Iraq and America's Military Forces', *Survival* **48/4**: 67–82.

Gorshkov, Sergei Georgievich 1979: *The Sea Power of the State*, anon. trans. from the Russian. Oxford: Pergamon.

Gottman, J. : 'Bugeaud, Galliéni, Lyautey 1941: The Development of French Colonial Warfare', in Edward Meade Earle (ed.): *Makers of Modern Strategy.* Princeton University Press: 234–59.

Gouré, Daniel and Jeffrey Cooper 1984: 'Conventional Deep Strike: A Critical Look', *Comparative Strategy* **4/3**: 215–48.

Gowing, Margaret, with Lorna Arnold 1974: *Independence and Deterrence: Britain and Atomic Energy 1945–1952.* London: Macmillan.

Grandmaison, [Capitaine Thomas-Antoine le Roy] 1756: *La Petite Guerre, ou traité du service des troupes légères en campagne.* s.l: s.n.

Granier, Hubert 1992: 'La Pensée navale française dans la première moitié du XVIIe siècle 1600–1661', in Hervé Coutau-Bégarie (ed.): *L'Évolution de la pensée navale*, vol. II. Paris: Economica: 37–53.

Gray, Colin 1979: 'Nuclear Strategy: The Case for a Theory of Victory', *International Security* **4/1**: 54–87.

1984: 'War-Fighting for Deterrence', *Journal of Strategic Studies* **7/1**: 5–28.

1990: *War, Peace, and Victory: Strategy and Statecraft for the Next Century.* New York: Simon & Schuster.

1992: *The Leverage of Sea Power: The Strategic Advantage of Navies in War.* New York: Free Press.

1996: *Explorations in Strategy.* Westport, CT: Greenwood.

1999a: *Modern Strategy*. Oxford University Press.
1999b: *The Second Nuclear Age*. Boulder, CO: Lynne Rienner.
2005: 'Future Warfare, or, the Triumph of History', *RUSI Journal* **150/5**: 16–19.
2007a: *War, Peace and International Relations*. London: Routledge.
2007b: *Fighting Talk*. Westport, CT: Praeger.
2010: *The Strategy Bridge: Theory for Practice*. Oxford University Press.
Gray, Colin and Keith Payne 1980: 'Victory is Possible', *Foreign Policy* (Summer): 14–27; repr. in Philip Bobbit, Lawrence Freedman and Gregory Treverton (eds.): *US Nuclear Strategy – A Reader*. London: Macmillan, 1989: 466–76.
Greiselis, Waldis 1999: 'Scharnhorst 1755–1813 und die Wehrpflicht', *Mars* **5**: 102–17.
Grenfell, Commander Russell 1937: *The Art of the Admiral*. London: Faber and Faber.
Grenier, Jacques Raymond, vicomte de 1787: *L'Art de guerre sur mer, ou tactique navale, assujettie à de nouveaux principes et à un nouvel ordre de bataille*. Paris: Fermin Didot.
Gretton, Peter 1965: *Maritime Strategy*. London: Cassell.
Grimsley, Mark 1998: 'In Not So Dubious Battle: The Motivations of American Civil War Soldiers', *Journal of Military History* **63**: 175–88.
Grivel, Jean 1837: *Considérations navales en réponse à la brochure de M. de Pradt*, repr. in Hervé Coutau-Bégarie (ed.): *L'Évolution de la pensée navale*, vol. I. Paris: Fondation pour les Études de Défense Nationale, 1990: 65–94.
Grivel, Richild Louis Antoine 1869: *De la Guerre maritime avant et depuis les nouvelles inventions*. Paris: Bertrand.
Groos, Otto 1929: *Seekriegslehren im Lichte des Weltkrieges*. Berlin: Mittler und Sohn.
Groote, Wolfgang von 1990: 'Historische Vorbilder des Feldzugs 1914 im Westen', *Militärgeschichtliche Mitteilungen* **1**: 33–55.
Großer Generalstab (ed.) 1911: *Moltkes Militärische Werke*, vol. IV, *Kriegslehren*, Part 1. Berlin: Ernst Siegfried Mittler und Sohn.
Großer Generalstab, Abtheilung für Kriegsgeschichte 1899: *Friedrich des Großen Anschauungen vom Kriege*, Kriegsgeschichtliche Einzelschriften nos. 27–30. Berlin: Großer Generalstab; repr. Bad Honnef: LIT Verlag, 1982.
Grotelueschen, Mark 2000: *Doctrine under Trial: American Artillery Employment in World War I*. Westport, CT: Greenwood.
Grotius, Hugo (Hugo de Groot) 1625: *De iure Belli ac Pacis*. Paris: Nicolas Boun; trans. several anon. authors: *Of the Rights of War and Peace*,

3 vols. London: D. Brown, T. Ward and W. Meares, 1715; trans. A.C. Campbell: *The Rights of War and Peace.* Washington, D.C. and London: M. Walter Dunne, 1901.

Grove, Eric 1990: *The Future of Sea Power.* London: Routledge.

Grove, Eric and Geoffrey Till 1989: 'Anglo-American Maritime Strategy in the Era of Massive Retaliation, 1945–60', in John B. Hattendorf and Robert S. Jordan (eds.): *Maritime Strategy and the Balance of Power.* Basingstoke: Macmillan: 271–303.

Groves, P.R.C. 1934: *Behind the Smoke Screen.* London: Faber and Faber.

Gruber, Ira 1993: 'The American Civil War: Engaging the Passions of a People', *Historical Journal* **36/4**: 987–91.

Guertner, Gary (ed.) 1993: *The Search for Strategy.* Westport, CT: Greenwood Press.

Guibert, Jacques Antoine Hippolyte, comte de 1772: *Essai général de la* tactique. London: chez les libraries associés; repr. in Guibert: *Stratégiques.* Paris: Herne, 1977; trans. Lieut. Douglas: *A General Essay on Tactics.* Whitehall: Millar, 1781.

1790: *De la Force publique.* Paris: Didot l'aîné; repr. in Guibert: *Stratégiques.* Paris: Herne, 1977.

Guilaine, Jean and Jean Zammit 2001: *Le Sentier de la guerre: Visage de la violence préhistorique.* Paris: Seuil.

GWAPS 1993: *Gulf War Air Power Survey 1993*, ed. Eliot Cohen. Washington, D.C.: US Government Printing Office.

Hacker, Barton 2005: 'The Machines of War: Western Military Technology, 1850–2000', *History and Technology* **21/3**: 255–300.

Hackett, John, John Barraclough, Bernard Burrows, Kenneth Hunt, Ian McGeoch, Norman Macrae and John Strawson 1978: *The Third World War.* London: Sidgwick and Jackson; repr. London: Sphere Books, 1983.

Hagemann, Ernst 1940: *Die deutsche Lehre vom Kriege: Von Berenhorst zu Clauseswitz.* Berlin: Mittler.

Haggie, P. 1985: 'The Royal Navy and War Planning in the Fisher Era', in Paul M. Kennedy (ed.): *The War Plans of the Great Powers 1880–1914.* London: Hyman: 118–32.

Hahlweg, Werner 1957: *Carl von Clausewitz: Soldat, Politiker, Denker.* Göttingen: Musterschmidt Verlag.

Hahlweg, Werner (ed.) 1966: *Carl von Clausewitz: Schriften – Aufsätze – Studien – Briefe*, vol. I. Göttingen: Vandenhoeck and Ruprecht.

Hahlweg, Werner (ed.) 1979: *Carl von Clausewitz: Verstreute kleine Schriften.* Osnabrück: Biblio Verlag.

Hahlweg, Werner (ed.) 1960: *Klassiker der Kriegskunst.* Darmstadt: Wehr- und Wissen-Verlag.

Haldon, John 1999: *Warfare, State and Society in the Byzantine World, 565–1204*. London: UCL Press.

Hale, John 1962: 'War and Public Opinion in the Fifteenth and Sixteenth Centuries', *Past and Present* **22**: 18–35.

1971: 'Sixteenth-Century Explanations of War and Violence', *Past and Present* **51**: 3–26.

Halévy, Dominique 1960: *Contre la Bombe*. Paris: Éditions du Minuit.

Hall, Bert 1997: *Weapons and Warfare in Renaissance Europe: Gunpowder, Technology, and Tactics*. Baltimore, MD: Johns Hopkins University Press.

Halliday, Jon 1993: 'A Secret War. US and Soviet Air Forces Clashed Directly in Korea', *Far Eastern Economic Review* **22**: 32–6.

Hallion, Richard 1992: *Storm over Iraq: Air Power and the Gulf War*. Washington, D.C.: Smithsonian Institute Press.

Halperin, Morton 1963: *Limited War in the Nuclear Age*. New York: John Wiley and Sons.

1967: *Contemporary Military Strategy*. Boston: Little, Brown; 2nd edn, London: Faber and Faber, 1972.

Hamer, David 1998: *Bombers vs. Battleships: The Struggle between Ships and Aircraft for the Control of the Surface of the Seas*. Annapolis, MD: Naval Institute Press.

Hamley, Edward Bruce 1872: *Operations of War Explained and Illustrated*. Edinburgh: William Blackwood and Sons.

Hamm, Manfred 1988: 'The AirLand Battle Doctrine: NATO Strategy and Arms Control in Europe', *Comparative Strategy* **7/3**: 183–211.

Hamon, Léo 1966: *La Stratégie contre la guerre*. Paris: Grasset.

Handel, Michael 1996: *Masters of War: Classical Strategic Thought*, 2nd edn. London: Frank Cass.

Hannay, David 1913: *The Navy and Sea Power*. London: Williams and Norgate.

Hannig, Norbert 1984: *Abschreckung durch konventionelle Waffen: Das David–Goliath Prinzip*. Berlin: Verlag Spitz.

Hanson, Victor David 1989: *The Western Way of War: Infantry Battle in Classical Greece*. New York: Alfred Knopf.

Harris, J.P. 1995: 'The Myth of Blitzkrieg', *War in History* **2/3**: 335–52.

Harris, Paul 1998 'The British Expeditionary Force in the Hundred Days' Campaign', *RUSI Journal* **143/6**: 72–4.

Hartmann, Anja 1995: *Rêveurs de Paix? Friedenspläne bei Crucé, Richelieu und Sully*. Hamburg: Krämer.

2001: 'Identities and Mentalities in the Thirty Years' War', in Anja Hartmann and Beatrice Heuser (eds.): *War, Peace and World Orders in European History*. London: Routledge: 174–85.

Hattendorf, John 1982: 'American Thinking on Naval Strategy', in Geoffrey Till: *Maritime Strategy in the Nuclear Age*. London: Macmillan: 58–67.
2000: *Naval History and Maritime Strategy*. Malabar, FL: Krieger.
2004: *The Evolution of the U.S. Navy's Maritime Strategy, 1977–1986*. Newport, RI: Naval War College Press.
2006: *U.S. Naval Strategy in the 1990s*. Newport, RI: Naval War College Press.
Hattendorf, John (ed.) 2003: *War at Sea in the Middle Ages and the Renaissance*. Woodbridge: Boydell.
Hattendorf, John and Robert Jordan (eds.) 1989: *Maritime Strategy and the Balance of Power: Britain and America in the Twentieth Century*. Basingstoke: Macmillan for St Antony's College.
Hattendorf, John, R.J.B. Knight, A.W.H. Pearsall, N.A.M. Rodger and Geoffrey Till 1993 (eds.): *British Naval Documents, 1204–1960*. Aldershot: Scolar Press.
Hatton, P.H.S. 1967: 'The First World War, Britain and Germany in 1914: The July Crisis and War Aims', *Past and Present* **36**–8: 138–43.
Haussherr, Hans 1960: 'Stein und Hardenberg', *Historische Zeitschrift* **190**: 271.
Hay du Chastelet, Paul 1668: *Traité de la guerre, ou politique militaire*. Paris: Iean Gvignard, 1668; repr. Paris: Jombert, 1757.
Healey, Denis 1960: *The Race against the H-Bomb*. London: Fabian Society.
Hearne, R.P. 1908: *Airships in Peace and War*; 2nd edn, London: John Lane, 1909.
1909: *Aerial Warfare*. London: John Lane, The Bodley Head.
Hecht, Rudolf 1986: 'USA, Europa, Sowjetunion: Politik und Strategie 1950 bis 1960', *Österreichische Militärische Zeitschrift* **5**: 411–33.
Heckmann, Marie-Luise 2006: 'Römische Kriegsdeutungen und römische Kriegführung im frühen Mittelalter', *Militärgeschichtliche Zeitschrift* **65/1**: 19–62.
Henderson, David 1995: 'Shaping the United Kingdom's Minimum Nuclear Deterrent for the Turn of the Century', *RUSI Journal* **140/3**: 43–7.
Henderson, G.F.R. 1905: *The Science of War: A Collection of Essays and Lectures, 1892–1903*. London: Longmans, Green.
Herberg-Rothe, Andreas 2004: 'Staatenkrieg und nicht-staatliche Kriege in Clausewitz' *Vom Kriege*', in Thomas Jäger, Gerhard Kümmel and Marika Lerch (eds.): *Sicherheit und Freiheit*. Baden-Baden: Nomos: 31–46.
Herwig, Holger 1971: 'Prelude to Weltblitzkrieg: Germany's Naval Policy toward the United States of America, 1939–41', *Journal of Modern History* **43/4**: 649–68.

2002: 'Germany and the "Short-War" Illusion: Toward a New Interpretation?', *Journal of Military History* **66/3**: 681–93.

Heuser, Beatrice 1989: *Western 'Containment' Policies in the Cold War: The Yugoslav Case, 1948–1953*. London: Routledge.

1993: 'Warsaw Pact Military Doctrines in the 70s and 80s: Findings in the East German Archives', *Comparative Strategy* **12/4**: 437–57.

1997: *NATO, Britain, France and the FRG: Nuclear Strategies and Forces for Europe, 1949–2000*. Basingstoke: Macmillan.

1998a: *Nuclear Mentalities? Strategies and Belief Systems in Britain, France and the FRG*. Basingstoke: Macmillan.

1998b: 'Victory in a Nuclear War? A Comparison of NATO and WTO War Aims and Strategies', *Contemporary European History* **7/3**: 311–28.

1998c: 'Dunkirk, Dien Bien Phu, and Suez, or Why France Doesn't Trust Allies and Has Learned to Love the Bomb', in Cyril Buffet and Beatrice Heuser (eds.): *Haunted by History: Myths in International Relations*. Oxford: Berghahn: 157–74.

1999: *The Bomb: Nuclear Weapons in Their Historical, Strategic and Ethical Context*. London: Longman.

2002: *Reading Clausewitz*. London: Pimlico.

2006: 'Frédéric le Grand et le Comte de Guibert: une dialectique stratégique', *Revue Militaire Suisse*, special issue *Histoire*, **151/6–7**: 26–32.

2007a: 'Clausewitz' Ideas of Strategy and Victory', in Andreas Herberg-Rothe and Hew Strachan (eds.): *Clausewitz in the Twenty-First Century*. Oxford University Press: 132–63.

2007b: 'Clausewitz' Methodology and the Traditions of British Writing on War', in Major-General Mungo Melvin (ed.): *The Blenheim Conference Papers: The Silent Alliance – 300 Years of Anglo-German Cooperation*. Occasional Paper 52. Shrivenham: UK Strategic and Combat Studies Institute: 46–58.

2008: 'Misleading Paradigms of War: States and Non-State Actors, Combatants and Non-Combatants', *War and Society* **27/2**: 1–24.

2009: 'Guibert: Prophet of Total War?', in Stig Förster and Roger Chickering (eds.): *War in an Age of Revolution: The Wars of American Independence and French Revolution, 1775–1815*. Cambridge University Press: 49–67.

2010a: 'Small Wars in the Age of Clausewitz: The Watershed between Partisan War and People's War', *Journal of Strategic Studies* **33/1**: 139–62.

2010b: *The Strategy Makers*. Westport, CT: Praeger-Greenwood.

Hexham, Henry 1637: *The Principles of the Art Militarie, Practised in the Warres of the Vnited Netherlands*, Part I. London: M.P.; 2nd edn, Delft, 1642; Part II, Delft: Antony of Heusden, 1642.

Heyries, Hubert 1998: 'La Conception Garibaldienne de la guérilla: Armée de la Nation ou Nation armée?', in Centre d'histoire et de prospective militaires (ed.): *Guerre civile – guérilla – terrorisme: entre hier et avenir*. Actes du Symposium, vol. II. Pully, Switzerland: Centre d'histoire et de prospective militaires: 231–44.

Hickey, Donald R. 2001: 'The War of 1812: Still a Forgotten Conflict?', *Journal of Military History* **65**: 741–69.

Hill, J.R. 1986: *Maritime Strategy for Medium Powers*. London Croom Helm.

Hill, Richard 2006: 'British Naval Thinking in the Nuclear Age', in Geoffrey Till (ed.): *The Development of British Naval Thinking*. London: Routledge: 160–81.

Hintze, Otto 1920–1: 'Delbrück, Clausewitz und die Strategie Friedrichs des Großen', *Forschung zur Brandenburgisch-Preußischen Geschichte* **33/1**: 131–78.

HIS (Hamburger Institut für Sozialforschung) (ed.) 1996: *Vernichtungskrieg: Verbrechen der Wehrmacht 1941–1944*. Hamburg: Hamburger Edition.

H[obbes], T[homas] 1642: *Elementorum philosophiae sectio tertia de cive*. Paris: s.n.

Hobbs, Richard 1979: *The Myth of Victory: What Is Victory in War?* Boulder, CO: Westview Press.

Hobohm, Martin 1920: 'Delbrück, Clausewitz und die Kritik des Weltkrieges', *Preußische Jahrbücher* **181**: 203–32.

Hoffmann, Joachim 1968: 'Wandlungen im Kriegsbild der preußischen Armee zur Zeit der nationalen Einigungskriege', *Militärgeschichtliche Mitteilungen* **1**: 5–33.

Hohenlohe-Ingelfingen, Prinz Kraft zu 1898: *Strategische Briefe*, 2 vols. Berlin: Ernst Siegfried Mittler und Sohn; trans. Capt. Walter H. James: *Letters on Strategy*, 2 vols. London: Kegan Paul, Trench, Trübner and Co.

Holleindre, Jean-Vincent forthcoming: *La Surprise comme concepte stratégique depuis l'Antiquité*. Paris: Presses Universitaires de France.

Holmes, James and Toshi Yoshihara 2005: 'The Influence of Mahan upon China's Maritime Strategy', *Comparative Strategy* **24**: 23–51.

2008: *Chinese Naval Strategy in the 21st Century: The Turn to Mahan*. London: Routledge.

Holsti, Kalevi 1996: *The State, War, and the State of War*. Cambridge University Press.

Honig, Jan-Willem 1995/6: 'Totalitarianism and Realism: Hans Morgenthau's German Years', *Security Studies* 5: 283–313.

1997: 'Strategy in a Post-Clausewitzian Setting', in Gert de Nooy (ed.): *The Clausewitzian Dictum and the Future of Western Military Strategy*. The Hague: Kluwer Law International: 109–21.

2001: 'Warfare in the Middle Ages', in Anja Hartmann and Beatrice Heuser (eds.): *War, Peace and World Orders in European History*. London: Routledge: 113–26.

Hornblower, Simon 2001: 'Greeks and Persians: West against East', in Anja V. Hartmann and Beatrice Heuser (eds.): *War, Peace and World Orders in European History*. London: Routledge: 48–61.

Horne, Alastair 1977: *A Savage War of Peace: Algeria 1954–62*. London: Macmillan.

Hoste, Paul, S.J. 1697: *L'Art des armées navales, ou Traité des évolutions navales*. Lyons: Anisson et Posuel.

Howard, Michael 1958: *Disengagement in Europe*. Harmondsworth: Penguin.

1965: 'Jomini and the Classical Tradition', in Michael Howard (ed.): *The Theory and Practice of War*. London: Cassell: 5–20.

1970: *Studies in War and Peace*. London: Temple Smith.

1972: *The Continental Commitment: The Dilemma of British Defence Policy in the Era of Two World Wars*. London: Temple Smith.

1980: 'Order and Conflict at Sea in the 1980s', in Jonathan Alford (ed.): *Sea Power and Influence: Old Issues and New Challenges*. Farnborough: Gower for the IISS: 70–9.

1991a: 'British Grand Strategy in World War I', in Paul Kennedy (ed.): *Grand Strategies in War and Peace*. New Haven: Yale University Press: 31–42.

1991b: *The Lessons of History*. Oxford University Press.

2001: *The Invention of Peace*. London: Profile Books.

2008: *War and the Liberal Conscience*, 2nd edn. London: Hurst (1st edn, London: Temple Smith, 1978).

Howard, Michael (ed.) 1979: *Restraints on War: Studies in the Limitation of Armed Conflict*. Oxford University Press.

Hoyer, Johann Gottfried von 1829: *Handbuch für Offiziere in den angewandten Theilen der Krieges-Wissenschaften*, vol. IV, *Von der Strategie*. Hannover: Helwigsche Hof-Buchhandlung.

Hurd, Archibald Spicer 1918: *The British Fleet in the Great War*. London: Constable.

ICJ (International Court of Justice) 1996: Advisory Opinion of 6 July 1996, online at www.icj-cij.org/docket/files/95/7495.pdf?PHPSESSID=979b7ff9658f23ba15d951e476c42213, accessed 19 July 2008.

ICRC (International Committee of the Red Cross) 1999: 'Iraq, 1989–1999: A Decade of Sanctions'. Report of 14 December 1999, online at www.icrc.org/Web/eng/siteeng0.nsf/iwpList74/4BBFCEC7FF4B7A3CC1256B66005E0FB6, accessed 11 August 2008.

Iklé, Fred Charles 1964: *How Nations Negotiate*. New York: Praeger.

Iue, Paul 1589: *The Practise of Fortification*. London: Thomas Orwin.

Ivanov, Leonid 1989: 'Entretien avec le Maréchal Sergue? Akhromeiev. Plaidoyer pour une autre doctrine militaire', *Polique Internationale* **43**: 297–304.

Jackson, Peter 2000: 'Naval Policy and National Strategy in France, 1933–1937', *Journal of Strategic Studies* **23/4**: 130–59.

Jacobi von Wallhausen, Johann 1616: *Manuale Militare, Oder Kriegss Manual*. Frankfurt: Paul Jacobi.

Jähns, Max 1971: *Militärgeschichtliche Aufsätze 1866–1883*, ed. Ursula von Gersdorff. Osnabrück: Biblio Verlag.

James, Harold and James Marzenna 1994: 'The Origins of the Cold War: Some New Documents', *Historical Journal* **37/3**: 615–22.

James, W.H. 1904: *Modern Strategy*, 2nd edn. Edinburgh: William Blackwood and Sons.

Jane, Frederick T. 1906: *Heresies of Sea Power*. London: Longmans, Green.

Janis, Irving L. 1982: *Groupthink: Psychological Studies of Policy Decisions and Fiascoes*. Boston: Houghton Mifflin.

Jauffret, Jean-Charles and Maurice Vaisse (eds.) 2001: *Militaires et guérillas dans la Guerre d'Algérie*. Brussels: Complexe.

Jauneaud, Commandant Marcel 1923: *L'Évolution de l'aéronautique*. Paris: Flammarion.

Jaurès, Jean 1911: *L'Armée nouvelle: l'Organisation socialiste de la France*. Paris: J. Rouff; repr. Paris: Éditions Sociales, 1978.

Jeney, Capitaine de 1759: *Le Partisan, ou l'art de faire la petite-guerre avec succes selon le génie de nos jours*. The Hague: s.n.; trans. by an officer in the Army [Roger Stevenson]: *The Partisan, or, the Art of Making War in Detachments*. London: R. Griffiths, 1760.

John 'the Middle' of Nassau-Siegen 1597: 'Observationes', in J.W. Wijn: 'Johann der Mittlere von Nassau-Siegen', in Werner Hahlweg (ed.): *Klassiker der Kriegskunst*. Darmstadt: Wehr- und Wissen-Verlag, 1960: 119–33.

1610: *Das Kriegsbuch*, in Werner Hahlweg (ed.): *Die Heeresreform der Oranier: Das Kriegsbuch des Grafen Johann von Nassau-Siegen*. Wiesbaden: Historische Kommission, 1973.

Johnson, Chalmers 1973: *Autopsy on People's War*. London: University of California Press.

Johnson, David 1998: *Fast Tanks and Heavy Bombers: Innovation in the US Army, 1917–1945*. Ithaca, NY: Cornell University Press.

2007: *Learning Large Lessons: The Evolving Roles of Ground Power and Air Power in the Post-Cold War Era*. Santa Monica, CA:RAND.

Johnson, Gerry, Kevan Scholes and Richard Whittington 2005: *Exploring Corporate Strategy*, 2nd edn. Harlow: Pearson Education.

Johnson, J.H. 1997: 'Stalemate! The Great Trench Warfare Battles of 1915–17', *Journal of Strategic Studies* **20/1** 204–5.

Joliot-Curie, Frédéric 1959: *La Paix, le désarmement et la coopération internationale*. Paris: Éditions Défense de la Paix.

Jomini, Antoine Henri Baron de 1805: *Traité de grande tactique*. Paris: n.p, An 13.

1811: *Traité des grandes opérations militaires*, 4 vols.; 2nd edn. Paris: Magimel; trans. Col. S.B. Holabird: *Treatise on Grand Military Operations, or a Critical and Military History of the Wars of Frederick the Great*, 2 vols. New York: D. Van Nostrand, 1865.

1837: *Précis de l'art de la guerre*. Paris: s.n.; repr. Paris: Éditions Ivréa, 1994; trans. Capt. G.H. Mendell and Capt. W.P. Craighill: *The Art of War*. Philadelphia: J.B. Lippincott, 1868; facsimile repr. London: Greenhill; Novato, CA: Presidio Press, 1992; repr. London: Greenhill; Mechanicsburg, PA: Stackpole Books, 1996.

Jones, Neville 1973: *The Origins of Strategic Bombing*. London: William Kimber.

1987: *The Beginnings of Strategic Air Power: A History of the British Bomber Force, 1923–39*. London: Frank Cass.

Jungfer, Eberhard and Ahlrich Meyer (eds.) 1997: *Repression und Kriegsverbrechen. Die Bekämpfung von Widerstands- und Partisanenbewegungen gegen die deutsche Besatzung in West- und Südeuropa*. Berlin and Göttingen: Verlag der Buchläden.

Kaegi, Walter Emil 1983: *Some Thoughts on Byzantine Military Strategy*. Brookline, MA: Hellenic College Press.

1992: *Byzantium and the Early Islamic Conquests*. Cambridge University Press.

Kahn, Herman 1960: *On Thermonuclear War*. Princeton University Press.

1965: *On Escalation*. New York: Praeger.

Kaldor, Mary 1999: *New Wars and Old Wars*. Stanford, CA: University of California Press.

Kant, Immanuel 1795: *Zum Ewigen Frieden*. Koenigsberg: Friedrich Nicolovius; anon. trans.: *Project for a Perpetual Peace*. London: Vernor and Hood, 1796.

Kantorowicz, Ernst 1951: 'Pro Patria Mori in Medieval Political Thought', *American Historical Review* **56**: 473–93.

Kapranov, Nikolai 1992: 'A Russian Perspective on the Future of Nuclear Weapons', in Patrick Garrity and Steven Maaranen (eds.): *Nuclear Weapons in the Changing World: Perspectives from Europe, Asia, and North America*. New York: Plenum Press: 77–90.

Karsh, Efraim 1996: 'Reflections on the 1990–91 Gulf Conflict', *Journal of Strategic Studies* **19/3**: 303–20.

Kaufmann, William 1956: 'Limited Warfare', in Kaufmann (ed.): *Military Policy and National Security*. Princeton University Press.

Keen, Maurice (ed.) 1999: *Medieval Warfare: A History*. Oxford University Press.

Kennedy, J.R. 1936: *Modern War and Defence Reconstruction*. London: Hutchinson.

Kennedy, Paul 1974: 'Mahan versus Mackinder', *Militärgeschichtliche Mitteilungen* **2**: 39–66.

1976: 'The Operations Plan of the Great Powers, 1880–1914', *Militärgeschichtliche Mitteilungen* **1**: 188–206.

1989: 'The Relevance of the Pre-war British and American Maritime Strategies to the First World War and Its Aftermath, 1898–1920', in John B. Hattendorf and Robert S. Jordan (eds.): *Maritime Strategy and the Balance of Power*. Basingstoke: Macmillan: 165–88.

1991: 'Grand Strategy in War and Peace: Toward a Broader Definition', in Paul Kennedy (ed.): *Grand Strategies in War and Peace*. New Haven: Yale University Press: 1–10.

Kennedy, Paul (ed.) 1979: *The War Plans of the Great Powers, 1880–1914*. London: Hyman.

Kennedy, Paul and George Andreopoulos 1994: 'The Laws of War: Some Concluding Reflections', in Michael Howard, George Andreopoulos and Mark Shulman (eds.): *The Laws of War: Constraints on Warfare in the Western World*. New Haven: Yale University Press: 214–25.

Kessel, Eberhard 1933: 'Die Wandlung der Kriegskunst im Zeitalter der Französischen Revolution', *Historische Zeitschrift* **148**: 248–76.

1953: 'Zur Genesis der modernen Kriegslehre', *Wehrwissenschaftliche Rundschau* **3/9**: 405–23.

1987: *Militärgeschichte und Kreigstheorie in neuerer Zeit: Ausgewählte Aufsätze*. Berlin: Duncker und Humblot.

Killigrew, John 1996: 'Chinese Military Writings', *War in History* **3/3**: 353–7.

King-Hall, Stephen 1958: *Defence in the Nuclear Age*. London: Victor Gollancz.

Kingston-McCloughry, E.J. 1960: *Defence-Policy and Strategy*. London: Stevens.

Kissinger, Henry 1957: *Nuclear Weapons and Foreign Policy*. New York: Harper and Bros. for the Council on Foreign Relations.
1979: *The White House Years*. London: Weidenfeld and Nicolson.
Kiszely, John 1996: 'The British Army and Approaches to Warfare since 1945', *Journal of Strategic Studies* **19/4**: 179–206.
Kitson, Frank 1971: *Low Intensity Operations: Subversion, Insurgency and Peacekeeping*. London: Faber and Faber.
Klipstein, Friedrich Leopold 1799: *Versuch einer Theorie des Dienstes der leichten Truppen besonders in Bezug auf leichte Infanterie*. Darmstadt: G.F. Heyer.
Kortüm, Hans-Henning 2001: 'Der Krieg im Mittelater als Gegenstand der Historischen Kulturwissenschaften', in Hans-Henning Kortüm (ed.): *Der Krieg im Mittelalter*. Berlin: Akademie-Verlag: 13–43.
Kortunov, Sergei 1992: 'The Deterrent Forces of the Commonwealth of Independent States', in Patrick J. Garrity and Steven A. Maaranen (eds.): *Nuclear Weapons in the Changing World: Perspectives from Europe, Asia, and North America*. New York: Plenum Press: 65–76.
Koser, Reinhold 1904: 'Die preußische Kriegsführung im Siebenjährigen Kriege', *Historische Zeitschrift* **92**: 239–73.
Kraus, Karl 1958: 'Vom Werden, Wesen und Wirken des Preußischen Generalstabes', *Geschichte in Wissenschaft und Unterricht* **9**: 199–217 and 256–76.
Krivosheev, Grigorij (ed.) 1997: *Soviet Casualties and Combat Losses in the Twentieth Century*. London: Greenhill Books.
Kroener, Bernhard 1987: 'Friedrich der Große und die Grundzüge der europäischen Kriegsführung seiner Zeit', in Militärgeschichtliches Forschungsamt (ed.): *Militärgeschichtliche Beiträge*, vol. I. Herford: E.S. Mittler: 15–26.
Kromayer, Johannes 1925a: 'Waren Hannibal und Friedrich der Große wirklich Ermüdungsstrategen?', *Historische Zeitschrift* **131**: 393–408.
1925b: 'Antwort', *Historische Zeitschrift* **132**: 449–50.
Kunisch, Johannes 1975: 'Der Ausgang des Siebenjährigen Krieges', *Zeitschrift für Historische Forschung* **2**: 173–222.
1987: 'La Guerre c'est moi: zum Problem der Staatenkonflikte im Zeitalter des Absolutismus', *Zeitschrift für Historische Forschung* **14**: 407–38.
Lake, Daniel 2009: 'The Limits of Coercive Airpower: NATO's "Victory" in Kosovo Revisited', *International Security* **34/1**: 83–112.
Lambert, Andrew 1994: 'Seapower 1939–1940: Churchill and the Strategic Origins of the Battle of the Atlantic', *Journal of Strategic Studies* **17/1**: 86–108.

2003a: 'The Future of Air Power', *RUSI Journal* **148/3**: 46–53.
2003b: 'Winning without Fighting: British Grand Strategy and Its Application to the United States, 1815–65', in Bradford A. Lee and Karl F. Walling (eds.): *Strategic Logic and Political Rationality: Essays in Honor of Michael I. Handel*. London: Frank Cass: 164–95.
2009: 'The Possibility of Ultimate Action in the Baltic: The Royal Navy at War, 1914–1916', in Michael Epkenhaus, Jörg Hillman and Frank Nagler (eds.), *Skagerrakschlacht*. Munich: Oldenbourg.
2010: 'Sea Power', in George Kassimeris and John Buckley (eds.), *The Ashgate Research Companion to Modern Warfare*. Farnham: Ashgate, 73–88.
Lambeth, Benjamin 1999: 'Air Power, Space Power and Geography', *Journal of Strategic Studies* **22**: 63–82.
Langendorff, Jean-Jacques 1999: 'Victor Emmanuel Thellung de Courtelary: un théoricien maritime suisse du début du XIXe siècle', in Hervé Coutau-Bégarie (ed.): *L'Évolution de la pensée navale*, vol. VII. Paris: Economica: 79–98.
Langford, David 1979: *War in 2080: The Future of Military Technology*. Stanmore, NSW: Cassell Australia.
Langlois, Hippolyte 1906: *Questions de défense nationale*. Paris and Nancy: Berger-Levrault.
Lapp, Ralph 1963: *Kill and Overkill: The Strategy of Annihilation*. London: Weidenfeld and Nicolson.
Lapping, Brian 1985: *End of Empire*. London: Granada.
Laurent, Jacques 1990: 'Projet de "Doctrine Militaire de l'URSS": Texte fondamental ou exercice de style?'. Paris: Centre de Sociologie de la Défense Nationale. Programme d'Études des Armées Socialistes.
1992: 'La Russie en quête de doctrine militaire', *Stratégique* **56**: 47–79.
Laury, Georges 1991: *Le Seuil nucléaire: Stratégie, prolifération, contrôle*. Campin: Éditions Universitaires.
Laver, Harry 2005: 'Preemption and the Evolution of America's Strategic Defense', *Parameters* (Summer): 107–20.
Lawrence, T.E. 1920: 'The Evolution of a Revolt', *Army Quarterly* (October): 55–69; repr. in Malcolm Brown (ed.): *T.E. Lawrence in War and Peace*. Mechanicsburg, PA: Stackpole Books; London: Greenhill Books, 2005: 261–73.
1929: 'Science of Guerrilla Warfare', *Encyclopedia Britannica*, 14th edn, vol. X. Chicago; repr. in Malcolm Brown (ed.): *T.E. Lawrence in War and Peace*. Pennsylvania: Stackpole Books; London: Greenhill Books, 2005: 274–84.
Le Borgne, Claude 1987: *La Guerre est morte*. Paris: Grasset.

Lecoq, Anne-Marie (ed.) 2001: *La Querelle des Anciens et des Modernes, XVIIe–XVIIIe siècles*. Paris: Gallimard.

Leebaert, Derek (ed.) 1981: *Soviet Military Thinking*. London: George Allen and Unwin.

Lefebvre, Georges 1962: *The French Revolution from Its Origins to 1793*, 2 vols., trans. E.M. Evanson. London: Routledge, Kegan and Paul.

Legault, Albert and Lindsey, Georges 1973: *Le Feu nucléaire*. Paris: Éditions du Seuil.

Lehmann, Max 1922: 'Zur Geschichte der preußischen Heeresreform von 1808', *Historische Zeitschrift* **126**: 436–57.

Leo VI the Wise, Emperor *c.* 900: *Taktike Theoria*, ed. Robert Vári: *Leonis imperatoris Tactica*, 2 vols. Budapest: University Press, 1917; Latin trans. John Checo: *De bellico apparatu liber*. Basle: Michael Isengrinium, 1554; French trans. M. Joly de Maizeroy: *Institutions militaires de l'Empereur Léon le Philosophe*, 2 vols. Paris: Claude-Antoine Jombert, 1771; English trans. George Dennis: *The Taktika of Leo*. Washington, D.C.: Dumbarton Oaks, 2010.

Lessard, Pierre 2005: 'Campaign Design for Winning the War ... and the Peace', *Parameters* (Summer): 36–49.

Levine, Robert 1987: 'The Strategic Nuclear Debate'. RAND Report R – 3565 – FF/CC/RC. Santa Monica, CA: The RAND Corporation.

Levy, Jack 1982: 'Historical Trends in Great Power War, 1495–1975', *International Studies Quarterly* **26/2**: 278–300.

Lewal, Jules Louis 1892: *Introduction à la partie positive de la stratégie*. Paris: Librairie Militaire L. Baudoin.

1893: *Stratégie de marche*. Paris: Librairie Militaire de L. Baudoin.

1895: *Stratégie de combat*. Paris: Librairie Militaire de L. Baudoin.

Liaropoulos, Andrew 2006: 'Revolutions in Warfare: Theoretical Paradigms and Historical Evidence – The Napoleonic and First World War Revolutions in Military Affairs', *Journal of Military History* **70**: 363–84.

Liddell Hart, Basil 1925: *Paris, or the Future of War*. London: Kegan Paul.

1930: *The Real War, 1914–1918*. London: Faber and Faber.

1932: *The British Way in Warfare, 1895–1970*. London: Faber and Faber.

1933: *The Ghost of Napoleon*. London: Faber and Faber.

1937: *Europe in Arms*. London: Faber and Faber.

1939: *Defence of Britain*. London: Faber and Faber.

1941: *The Strategy of Indirect Approach* [new edition of *The Decisive Wars of History* of 1929]. London: Faber and Faber.

1944: *Thoughts on War*. London: Faber and Faber.

1946: *Revolution in Warfare*. London: Faber and Faber.

1950: *Defence of the West*. London: Cassell.
1954: *Strategy*. London: Faber and Faber; 2nd edn, London: Faber and Faber, 1967.
1960: *Deterrent or Defence: A Fresh Look at the West's Military Position*. London: Stevens and Son.
1967: *Strategy, the Indirect Approach*, 4th rev. and enlarged edn of *The Strategy of Indirect Approach*. London: Faber and Faber.
Lindsey, George 1980: 'Tactical Anti-submarine Warfare: The Past and the Future', in Jonathan Alford (ed.): *Sea Power and Influence: Old Issues and New Challenges*. Montclair, NJ: Allanheld, Osmun & Co; Farnborough: Gower for the IISS.
Linke, Horst Günther 1982: 'Rußlands Weg in den Ersten Weltkrieg und seine Kriegsziele 1914–1917', *Militärgeschichtliche Mitteilungen* **2**: 9–34.
Linn, Brian 2002: 'The American Way of War Revisited', *Journal of Military History* **66**: 501–30.
Lipsius, Justus 1637: *Justi Lipsii opera omnia*. Antwerp: Platin.
Llobera, Josep 1994: *The God of Modernity: The Development of Nationalism in Western Europe*. Oxford and Providence, RI: Berg.
Lloyd, Henry Humphry Evans 1781: *Continuation of the History of the Late War in Germany, between the King of Prussia, and the Empress of Germany and her Allies*, Part II, in Patrick Speelman (ed.): *War, Society and Enlightenment: The Works of General Lloyd*. Leiden and Boston: Brill, 2005: 375–478.
Lobov, V.N. 1992: 'Le Général Svechin et l'Évolution de l'art militaire: ses Idées face à l'épreuve du temps', *Stratégique* **56**: 31–46.
Londoño, Sancho de 1594?: *Discurso sobre la forma de reducir la Disciplina Militar a mejor y antiguo estado*. Madrid: Ministry of Defence, 1992.
Loque, Bertrand de, *see* Saillans, François de.
Lossau, Johann Friedrich Konstantin von Lossau [Anon.] 1815: *Der Krieg: Für wahre Krieger*. Leipzig: Engelmann.
Ludendorff, Erich 1935: *Der Totale Krieg*. Munich: Ludendorff Publishing; trans. A.S. Rapoport: *The Nation at War*. London: Hutchinson, 1936.
Luttwak, Edward 1974: *The Political Uses of Sea Power*. Baltimore, MD: Johns Hopkins University Press.
1976a: *Strategic Power: Military Capabilities and Political Utility*. Washington Papers 38. London: Sage.
1976b: *The Grand Strategy of the Roman Empire: From the First Century AD to the Third*. London: Weidenfeld and Nicolson.
1982: 'On the Meaning of Victory', *Washington Quarterly* **5/4**: 17–24.
1983: *Grand Strategy of the Soviet Union*. London: Weidenfeld and Nicolson.

1985: *Strategy and History*. New Brunswick: Transaction Books.

1987: *Strategy: The Logic of War and Peace*. Cambridge, MA: Belknap Press of Harvard University Press; 2nd edn, Cambridge, MA: Belknap Press of Harvard University Press, 2001.

1995: 'Towards Post-Heroic Warfare', *Foreign Affairs* **74/3**: 109–22.

1996: 'A Post-Heroic Military Policy', *Foreign Affairs* **75/4**: 33–44.

Lyall, Sarah 2008: 'Former Finnish President Wins Nobel Peace Prize', *The New York Times*, 8 October.

Lyautey, Marshal Louis Hubert Gonzalve 1891: 'Le Rôle social de l'officier', *Revue des Deux Mondes*, 15 March 1891; repr. in *Rayonnement*, ed. Patrick Heidsieck. Montreal: Gallimard/Éditions Variétés, 1944: 105–18.

Lynn, John 1984: *The Bayonets of the Republic: Motivation and Tactics in the Army of Revolutionary France, 1791–94*. Urbana and Chicago: University of Illinois Press.

1994: 'A Quest for Glory: The Formation of Strategy under Louis XIV, 1661–1715', in Williamson Murray, MacGregor Knox and Alvin Bernstein (eds.): *The Makers of Strategy: Rulers, States, and War*. Cambridge University Press: 178–204.

2002: *The French Wars 1667–1714: The Sun King at War*. Oxford: Osprey.

2005: 'Patterns of Insurgency and Counterinsurgency', *Military Review* (July–Aug.), 22–7.

MacArthur, Douglas 1951: 'Farewell Address to the US Congress', online at www.americanrhetoric.com/speeches/douglasmacarthurfarewelladdress.htm, accessed 16 December 2009.

Machiavelli, Niccolò 1521: *Libro della Arte della Guerra*. Florence: li Heredi di Philippo di Giunta [originally published as Niecoluci, Amadeo: *Arte de la Guerra*]; trans. Ellis Farnworth: *The Art of War*. New York: Da Capo Press, 1965.

1531: *I Discorsi*. Florence: Bernardo di Giunta; trans. L.J. Walker SJ and Brian Richardson: *The Discourses*. London: Penguin, 1998.

1532a: *Historie Fiorentine*. Florence: Bernardo di Giunta; anon. trans.: *The History of Florence and of the Affairs of Italy from the Earliest Times* ... London: Henry G. Bohn, 1847.

1532b: *Il Principe*. Florence: Bernardo di Giunta; trans. George Bull: *The Prince*. Harmondsworth: Penguin, 1961.

Mackinlay, John 1999: 'What is New about New Wars? *RUSI Journal* **144/4**: 84–6.

Macmillan, Norman 1938: *The Chosen Instrument*. London: John Lane, The Bodley Head.

1941: *Air Strategy*. London: Hutchinson.

Mahan, Alfred Thayer 1890: *The Influence of Sea Power upon History, 1660–1783*. Boston: Little, Brown.

1892: *The Influence of Sea Power upon the French Revolution and Empire, 1793–1812*, 2 vols. Boston: Little, Brown and Co.

1897: *The Life of Nelson: The Embodiment of the Sea Power of Great Britain*, 2 vols. Boston: Little, Brown and Co.

1905: *Sea Power in Its Relations to the War of 1812*, 2 vols Boston: Little, Brown and Co.

1911: *Naval Strategy Compared and Contrasted with the Principles and Practice of Military Operations on Land*. Boston: Little, Brown and Co.

1918: *Naval Strategy Compared and Contrasted with the Principles and Practice of Military Operations on Land*. Boston: Little, Brown and Co. (reprint of 1911 edn).

Mahar, Michael 2000: 'Air Power to Ensure Victory', *RUSI Journal* **145/5**: 61–2.

Maimonides, Moses 12th century: *The Book of Divine Commandments (Sefer Ha-Mitzvoth)*, vol. I, trans. Charles Chavel. London: Soncino Press, 1940.

Makarov, S.O. 1993: 'Discussion sur des questions de tactique navale', with a commentary by David R. Jones, in Hervé Coutau-Bégarie (ed.): *L'Évolution de la pensée navale*, vol. III. Paris: Economica: 149–70.

Makins, Christopher 1981: 'TNF Modernization and "Countervailing Strategy"', *Survival* **23/4**: 157–64.

Mallet, Allain Manesson 1684: *Les travaux de Mars, ou l'art de la guerre*. Paris: Denys Thierry.

Mallett, Michael Edward 1974: *Mercenaries and Their Masters: Warfare in Renaissance Italy*. London: The Bodley Head.

Maloney, Sean 1995: *Securing Command of the Sea: NATO Naval Planning, 1948–54*. Annapolis, MD: Naval Institute Press.

Maltzahn, Curt von 1906: *Der Seekrieg*. Leipzig: B.G. Teubner.

Mao Tse-tung 1937: *Yu Chi Chan*, trans. Samuel B. Griffith: *On Guerrilla Warfare*. New York: Praeger, 1961; reprinted Champaign, IL: University of Illinois Press, 2000.

1938: 'On Protracted War', May, online at www.marx2mao.com/Mao/PW38.html#s1, accessed 19 September 2009.

Marder, Arthur Jacob 1941: *British Naval Policy, 1880–1905: The Anatomy of British Sea Power*. London and Norwood, MA: Putnam's; reprinted as *The Anatomy of British Sea Power*. London: Frank Cass, 1964.

1961: *From the Dreadnought to Scapa Flow*. Oxford University Press.

Marill, J-M. 1997: 'La Doctrine militaire française entre les deux guerres', *Revue Historique des Armées* **206/1**: 24–34.

Markel, Wade 2006: 'Draining the Swamp: The British Strategy of Population Control', *Parameters* (Spring), 35–48.
Markham, J. David 1999: 'Wellington's Lost Soldiers: British POWs', *RUSI Journal* **144/6**: 83–9.
Markusen, Eric and David Kopf 1995: *The Holocaust and Strategic Bombing: Genocide and Total War in the Twentieth Century*. Boulder, CO: Westview Press.
Marmont, Marshal, duc de Raguse 1845: *L'Esprit des institutions militaires*. Paris: Librairie Militaire J Dumaine.
Martin, Jean-Clément 2006: *Violence et révolution: Essai sur la naissance d'un mythe national*. Paris: Seuil.
Martin, Laurence W. 1967: *The Sea in Modern Strategy*. London: Chatto and Windus for the IISS.
Martin, Laurence W. (ed.) 1979: *Strategic Thought in the Nuclear Age*. London: Heinemann.
Mason, R. Anthony 1994: *Airpower: A Centennial Appraisal*. London: Brassey's.
1995: 'Air Power: A Centennial Appraisal', *Journal of Strategic Studies* **18/4**: 141–2.
Mattox, J.M. 2006: *Saint Augustine and the Theory of Just War*. London: Continuum.
Maude, F.N. 1905: *Evolution of Modern Strategy from the 18th Century to the Present Time*. London: William Clowes.
Maurice, Emperor *c*. 600: *Ars Militaris*, printed in: *Arriani* Tactica *et Mauricii* Ars Militaris. Uppsala: Henricus Curio, 1664; trans. George Dennis: *Strategikon: Handbook of Byzantine Military Strategy*. Philadelphia: University of Pennsylvania Press, 1984.
Maurice, F.B. 1929: *British Strategy*. London: Constable.
Maurice, J.F. 1891: *War*. London: Macmillan.
1895: *How Far the Lessons of the Franco-German War Are Now out of date*. Aldershot lecture 22 January 1895. London: Gale and Polden.
Mayer, Emile 1916: *Autour de la Guerre actuelle: Essai de psychologie militaire*. s.l.: s.n.
Mayers, David 1986: 'Containment and the Primacy of Diplomacy. George Kennan's Views, 1947–1948', *International Security* **11/1**: 124–62.
Mayr, K. 1925: 'Clausewitz und der Zusammenbruch 1918', *Archiv für Politik und Geschichte* **5/2**: 389–416.
MccGwire, Michael 1976: 'Maritime Strategy and the Super-powers', in IISS (ed.): *Power at Sea*, Part II, *Superpowers and Navies*. London: IISS: 15–24.
MccGwire, Michael (ed.) 1975: *Soviet Naval Policy*. New York: Praeger.
McDermott, J. 1985: 'The Revolution in British Military Thinking from the Boer War to the Moroccan Crisis', in Paul M. Kennedy (ed.):

The War Plans of the Great Powers, 1880–1914. London: Hyman: 99–117.

McFate, Montgomery 2005: 'Anthropology and Counterinsurgency', *Military Review* (Mar.–April): 24–38.

McFate, Montgomery and Andrea Jackson 2006: 'Counterinsurgency and the Four Tools of Political Competition', *Military Review* (Jan.–Feb.): 13–26.

McInnes, Colin 2002: *Spectator-Sport War: The West and Contemporary Conflict*. Boulder, CO: Lynne Rienner.

McMahon, Robert 1991: 'Credibility and World Power: Exploring the Psychological Dimension in Postwar American Diplomacy', *Diplomatic History* **15/4**: 455–71.

McNamara, Robert 1962: 'Speech to NATO Council, Athens', repr. in Philip Bobbit, Lawrence Freedman and Gregory Treverton (eds.): *US Nuclear Strategy – A Reader*. London: Macmillan, 1989: 205–22.

McNeilly, Mark R. 2001: *Sun Tzu and the Art of Modern Warfare*. Oxford University Press.

McPherson, James 1997: *For Cause and Comrades: Why Men Fought in the Civil War*. New York: Oxford University Press.

Mearsheimer, John 1990: 'Back to the Future: Instability in Europe after the Cold War', *International Security* **15/1**: 5–56.

Meilinger, Phillip 1995a: 'Proselytiser and Prophet: Alexander P. de Seversky and American Airpower', *Journal of Strategic Studies* **18/1**: 8–35.

1995b: 'John C Slessor and the Genesis of Air Interdiction', *RUSI Journal* **140/4**: 43–8.

1996: 'Trenchard and "Morale Bombing": The Evolution of the Royal Air Force Doctrine before World War II', *Journal of Military History* **60/1**: 243–70.

2000: 'The Historiography of Airpower: Theory and Doctrine', *Journal of Military History* **64/2**: 467–502.

2003: *Airwar: Theory and Practice*. London: Frank Cass.

Mendoza, D. Bernardino de 1595: *Teórica y práctica de la guerra*. Madrid: Pedro Madrigal; repr. Madrid: Ministerio de Defensa, 1998; trans. Sir Edwarde Hoby: *Theorique and Practise of Warre*. Middelburg: Richard Schilders, 1597.

Menedez Valdes, Francisco de 1559: *Espejo y deceplina militar*. Brussels: s.n.; repr. Madrid: Ministero de Defensa, 1989.

Mercer, Donald 1984: 'Targeting Soviet Forces', *Military Review* **64/5**: 23–38.

Messerschmidt, Manfred 1980: 'Preußens Militär in seinem gesellschaftlichen Umfeld', *Geschichte und Gesellschaft*, Sonderheft 6, *Preußen im Rückblick*: 43–88.

Metz, Steven 2001: 'Strategic Asymmetry', *Military Review* (July–August): 23–31.

Meyer, Berthold 1985: *Atomwaffenfreie Zonen und Vertrauensbildung in Europa*. Frankfurt: Campus.

Meynert, Hermann 1973: *Geschichte des Kriegswesens und der Heeresverfassung in Europa*, Part I. Graz: Akademische Druck- u. Verlagsanstalt.

Michel, Marshall 2001: *The Eleven Days of Christmas: America's Last Vietnam Battle*. San Francisco: Encounter.

Mière de Corvey, J.F.A. Le 1823: *Des Partisans et des corps irréguliers*. Paris: Anselin & Pochard *et al.*

Miksche, F.O. 1941: *Blitzkrieg*, trans. Tom Wintringham. London: Faber and Faber.

1943: *Is Bombing Decisive? A Study of the Organisation and Tactical Employment of Modern Air Fleets*. London: Allen and Unwin.

1958: *La Faillite de la stratégie atomique*. Paris: Le Livre Contemporain.

1959: *The Failure of Atomic Strategy and a New Proposal for the Defence of the West*. London: Faber and Faber.

Miles, Rufus 1978: 'The Origin and Meaning of Miles' Law', *Public Administration Review* **38/5**: 399–403.

Mills, Richard 1990: 'Follow-on Forces Attack: The Need for a Critical Reassessment', in *RUSI and Brassey's Defence Yearbook*. London: Brassey's: 125–37.

Miszellen 1889: *Historische Zeitschrift* 61: 98–109.

Mitchell, William 1925: *Winged Defense: The Development and Possibilities of Modern Air Power, Economic and Military*. New York: G.P. Putnam's Sons.

Moineville, Hubert 1982: *La Guerre navale*. Paris: Presses Universitaires de France; trans. P.R. Compton-Hall: *Naval Warfare Today and Tomorrow*. Oxford: Blackwell, 1983.

Møller, Bjørn 1991: *Resolving the Security Dilemma in Europe: The German Debate on Non-Offensive Defence*. London: Brassey's.

Moliner Prada, Antonio 2005: 'Popular Resistance in Catalonia: *Somatens* and *Miquelets*, 1808–14', in Charles Esdaile (ed.): *Popular Resistance in the French Wars: Patriots, Partisans and Land Pirates*. Basingstoke: Palgrave Macmillan: 91–114.

Moltke the Elder, Helmut von 1960: 'Über Strategie: Aufsatz aus dem Jahre 1871', in Werner Hahlweg (ed.): *Klassiker der Kriegskunst*. Darmstadt: Wehr- und Wissen-Verlag: 316–18.

Molyneux, Thomas More 1759: *Conjunct Expeditions; or expeditions that have been carried on jointly by the fleet and army, with a commentary on littoral War*, vol. II. London: s.n.

Monaque, Rémi 1994: 'L'Amiral Aube, ses Idées, son Action', in Hervé Coutau-Bégarie (ed.): *L'Évolution de la pensée navale*, vol. IV. Paris: Economica: 133–43.

1997: 'Gabriel Charmes: Propagandist enthousiaste, mais parfois infidèle, des idées de l'Amiral Aube', in Hervé Coutau-Bégarie (ed.): *L'Évolution de la pensée navale*, vol. VI. Paris: Economica: 57–66.

Montecuccoli, Raimondo 1704: *Memorie*. Colonia: Henricus de Huyssen; later edition known as *Arte universal de la Guerra*. Lisbon: s.n., 1708.

Montgomery, John 2001: 'Using Limited Air Power in the 20th Century', *RUSI Journal* **146**:6: 59–62.

Montheilhet, J. 1932: *Les Institutions militaires de la France 1814–1932*. Paris: Félix Alcan.

Montluc, Blaise de 1592: *Comentaires de messire Blaise de Montluc maréchal de France, où sont descrits les combats etc. 1521–1575*. Bordeaux: Simon Millanges.

Mordacq, J.J.H. 1912: *Politique et stratégie dans une démocratie*. Paris: Plon.

1921: *La Stratégie: Historique, évolution*, 3rd edn. Paris: L. Fournier.

Moreman, T.R. 1996: '"Small Wars" and "Imperial Policing": The British Army and the Theory and Practice of Colonial Warfare in the British Empire, 1919–1939', *Journal of Strategic Studies* **19/4** (Dec.): 105–31.

Motte, Martin 1990: 'L'après-grande guerre dans la *Revue Maritime*, 1920–1923', in Hervé Coutau-Bégarie (ed.): *L'Évolution de la pensée navale*, vol. I. Paris: Economica: 117–51.

1994: 'L'Influence de Mahan sur la Marine française', in Hervé Coutau-Bégarie (ed.): *L'Évolution de la pensée navale*, vol. IV. Paris: Economica: 145–72.

2003: 'Cosaques de la mer: Aperçus sur la guérilla navale, de Vauban à Ben Landen', *Revue Historique des Armées* **2**: 91–103.

Mrozek, Donald 1989: *Air Power and the Ground War in Vietnam*. Washington, D.C.: Pergamon-Brassey's.

Mufti, Malik 2007: 'The Art of Jihad', *History of Political Thought* **28/2**: 189–207.

Muller, Jean 1973: *La Bombe en question*. Neulise: 'Combat non-violent'.

Murray, Stewart 1909: *The Reality of War: An Introduction to Clausewitz*. London: Hugh Rees.

Murray, Williamson 1981: 'The Luftwaffe before the Second World War: A Mission, a Strategy?', *Journal of Strategic Studies* **4/3**: 261–70.

1983: 'A Tale of Two Doctrines: The Luftwaffe's "Conduct of the Air War" and the USAF's Manual 1–1', *Journal of Strategic Studies* **6/4**: 84–93.

1992: 'Reflections on the Combined Bomber Offensive', *Militärgeschichtliche Mitteilungen* **51**: 73–94.

1993: 'Operations', in Eliot Cohen (ed.): *Gulf War Air Power Survey*, vol. II.i.Washington, D.C.: US Government Printing Office: 1–313.

1999a: 'Military Culture Does Matter', *Strategic Review* **2**: 33.

1999b: *War in the Air, 1914–1945*. London: Cassell.

Murray, Williamson, MacGregor Knox and Alvin Bernstein 1994: *The Making of Strategy: Rulers, States, and War*. New York: Cambridge University Press.

Nagl, John 2002: *Learning to Eat Soup with a Knife: Experiences of Malaya and Vietnam*. University of Chicago Press.

Napoleon 1906: *Maximes*. London: Arthur L. Humphreys.

Neely, Mark 1997: 'Was the Civil War a Total War?', in Stig Förster and Jörg Nagler (eds.): *On the Road to Total War: The American Civil War and the German Wars of Unification, 1861–1871*. Cambridge University Press: 29–52.

Neild, Robert 1990: *An Essay on Strategy As It Affects the Achievement of Peace in a Nuclear Setting*. Basingstoke: Macmillan.

Neill, Donald 1998: 'Ancestral Voices: The Influence of the Ancients on the Military Thought of the Seventeenth and Eighteenth Centuries', *Journal of Military History* **62/3**: 487–520.

Neon 1927: *The Great Delusion*. London: Benn.

Nicholson, Helen 2004: *Medieval Warfare*. Basingstoke: Palgrave Macmillan.

Nikephoros Phokas (mid-950s): *Peri Paradromes* [On Skirmishing], known in Latin as *De Velitatione*; trans. and ed. George Dennis: *Three Byzantine Military Treatises*. Washington, D.C.: Dumbarton Oaks, 1985: 146–239.

Noel-Baker, Philip 1958: *The Arms Race: A Programme for World Disarmament*. London: John Calder.

Nokkala, Arto 1991: *Non-Offensive Defence*. Helsinki: War College.

Noue, François de la [called Bras de Fer] 1587: *Discovrs politiqves et militaires du Seigneur de la Noue*, 2nd edn. Basle: François Forest.

Nuti, Leopoldo and Olav Riste 2007: 'Preparing for a Soviet Occupation: The Strategy of "Stay Behind"', *Journal of Strategic Studies* **30/6** (Dec.): 937–54.

O'Neill, Robert 1968: *Indo-China Tragedy 1945–1954*. London and New York: Frederick Warne.

1969a: *General Giap: Politician and Strategist*. Melbourne: Cassell Australia.

1969b: *The Strategy of General Giap since 1964*. Canberra Papers on Strategy and Defence no. 6. Canberra: Australian National University Press.

1985: *Australia in the Korean War, 1950–53*, vol. II, *Combat Operations*. Canberra: Australian Government Publishing Service.

Olivier, David 2004: *German Naval Strategy 1856–1888*. London: Frank Cass.

Olsen, John Andreas 2003: *Strategic Air Power in Desert Storm*. London: Frank Cass.

2007: *John Warden and the Renaissance of American Air Power*. Washington, D.C.: Potomac Books.

2009: *A History of Air Warfare*. Washington, D.C.: Potomac Books.

Onosander 1st century: *Onosandri strategeticus sive de imperatoris institutione*. Paris: Saugrain, 1599; trans. Illinois Greek Club: *Aeneas Tacitus, Asclepiodotus, and Onasander*. Loeb Classical Library. New York: G.P. Putnam's Sons, 1923.

Osgood, Robert 1957: *Limited War: The Challenge to American Strategy*. University of Chicago Press; repr. University of Chicago Press, 1965.

1962: *NATO: The Entangling Alliance*. University of Chicago Press.

1979a: 'The Post-War Strategy of Limited War: Before, during and after Vietnam', in Laurence Martin (ed.): *Strategic Thought in the Nuclear Age*. London: Heinemann: 93–130.

1979b: *Limited War Revisited*. Boulder, CO: Westview Press.

Osgood, Robert and Robert Tucker 1967: *Force, Order, and Justice*. Baltimore, MD: Johns Hopkins University Press.

Osinga, Frans 2007: *Science, Strategy and War: The Strategic Theory of John Boyd*. London: Routledge.

Ostwald, Jamel 2000: 'The "Decisive" Battle of Ramillies, 1706: Prerequisites for Decisiveness in Early Modern Warfare', *Journal of Military History* **64/3**: 649–77.

Overy, Richard 1992: 'Air Power and the Origins of Deterrence Theory before 1939', *Journal of Strategic Studies* **15/1**: 73–101.

Owen, David 1972: *The Politics of Defence*. London: Jonathan Cape.

Pagès, Jean 1990: 'La Pensée navale athénienne aux Ve et IVe siècles avant J.C.', in Hervé Coutau-Bégarie (ed.): *L'Évolution de la pensée navale*, vol. I. Paris: Economica: 50–78.

1994: 'Un Traité de tactique navale du XVIe siècle: le Libre III *Della Milizia Marittima* de Cristoforo da Canal', in Hervé Coutau-Bégarie (ed.): *L'Évolution de la pensée navale*, vol. IV. Paris: Economica: 37–64.

Palmer, Gregory 1978: *The McNamara Strategy and the Vietnam War*. London: Greenwood Press.

Pape, Robert 1990: 'Coercive Air Power in the Vietnam War', *International Security* **15/2**: 103–46.

1992: 'Coercion and Military Strategy: Why Denial Works and Punishment Doesn't', *Journal of Strategic Studies* **15/4**: 423–75.

1996: *Bombing to Win: Air Power and Coercion in War*. Ithaca, NY: Cornell University Press.

Paret, Peter 1964: 'Colonial Experience and European Military Reform at the End of the Eighteenth Century', *Historical Research* **37/95**: 47–59.

1983: 'Revolutions in Warfare: An Earlier Generation of Interpreters', in Bernard Brodie, Michael D. Intriligator and Roman Kolkowicz (eds.): *National Security and International Stability*. Cambridge, MA: Oelschlager, Gunn and Hain: 157–69.

Paret, Peter (ed.) 1986: *Makers of Modern Strategy*. Princeton University Press.

Paris, Michael 2000: *Warrior Nation: Images of War in British Popular Culture, 1850–2000*. London: Reaktion Books.

Parker, Geoffrey 1973: 'Mutiny and Discontent in the Spanish Army of Flanders 1572–1607', *Past and Present* **5**: 38–52.

1976: 'The "Military Revolution", 1560–1660 – A Myth?', *Journal of Modern History* **48**: 195–214.

1988: *The Military Revolution: Military Innovation and the Rise of the West 1500–1800*. Cambridge University Press.

1994: 'Early Modern Europe', in Michael Howard, George J. Andreopoulos and Mark R. Shulman (eds.): *The Laws of War: Constraints on Warfare in the Western World*. New Haven: Yale University Press: 40–58.

Parks, W. Hays 1995: '"Precision" and "Area" Bombing: Who Did Which, and When?', *Journal of Strategic Studies* **18/1**: 145–74.

Parquin, Charles 1843: *Souvenirs et campagne d'un vieux soldat de l'Empire*. Paris: ND des Victoires; ed. and trans. B.T. Jones: *Napoleon's Army: The Military Memoirs of Charles Parquin*. London: Greenhill Books, 1987.

Parrot, David 1985: 'Strategy and Tactics in the Thirty Years' War: The "Military Revolution"', *Militärgeschichtliche Mitteilungen* **2**: 7–25.

2009: 'From Military Enterprise to Standing Armies: War, State and Society in Western Europe, 1600–1700', in Frank Tallett and David Trim (eds.): *European Warfare, 1350–1750*. Cambridge University Press: 74–95.

Pay, John 1994: 'Full Circle: The US Navy and Its Carriers, 1974–1993', *Journal of Strategic Studies* **17/1**: 124–47.

Peddie, John 1994: *The Roman War Machine*. Stroud: Alan Sutton.

Penth, Sabine 2006: 'Konzeptionen Historischer Semantik am Beispiel der Begriffe "Friede" und "Krieg"', *Militärgeschichtliche Zeitschrift* **65/1**: 5–18.

Peters, John 1987: 'Evaluating FOFA As a Deterrent', *RUSI Journal* **132/4**: 39–44.

Petraeus, David 2006: 'Learning Counterinsurgency: Observations from Soldiering in Iraq', *Military Review* (Jan.–Feb.): 2–12.

Phillips, Gervase 1999: 'Strategy and Its Limitations: The Anglo-Scots Wars, 1480–1550', *War in History* **6/4**: 396–416.

Picht, Werner 1954: 'Der Begriff "Militarismus"', *Geschichte in Wissenschaft und Unterricht* **5**: 455–69.

Pierce, Watson O'Dell 1937: *Air War*. London: Watts.

Pierron, Edouard 1887–96: *Stratégie et grande tactique*, 4 vols. Paris: Berger-Levrault.

Ploetz, Michael 2001: 'Enemy Image and Identity in the Warsaw Pact', in Anja V. Hartmann and Beatrice Heuser (eds.): *War, Peace and World Orders in European History*. London: Routledge: 226–36.

Po, Guido 1929: *Il Potere marittimo e irifornimenti dell'Italia in guerra*. Rome: Societa Anonima Poligrafica Italiana.

Poirer, Lucien 1977: *Des Stratégies nucléaires*. Paris: Hachette.

1982: *Essais de stratégie théorique*. Paris: FEDN; supplement to *Stratégique* 13/1.

1985: *Les Voix de la stratégie*. Paris: Fayard.

Polišenský, J.V. 1968: 'The Thirty Years' War and the Crises and Revolutions of Seventeenth-Century Europe', *Past and Present* **39**: 34–43.

Porch, Douglas 1991: 'Arms and Alliances: French Grand Strategy and Policy in 1914 and 1940', in Paul Kennedy (ed.): *Grand Strategies in War and Peace*. New Haven: Yale University Press: 125–44.

2006: 'French War Plans, 1914: The "Balance of Power Paradox"', *Journal of Strategic Studies* **29/1**: 117–44.

Porroni, Marques Annibale 1676: *Tratado universal militar moderno*. Venice: Francesco Nicolini.

Posen, Barry R. 1984: *The Sources of Military Doctrine: France, Britain, and Germany between the World Wars*. Ithaca, NY and London: Cornell University Press.

Praissac, Sieur du 1625: *Les discovrs militaires*. Rouen: Jean Bouley.

Prete, Roy 1985: 'French Military War Aims, 1914–1916', *Historical Journal* **28/4**: 887–99.

Pufendorf, Samuel von 1672: *De jure naturae et gentium*. Londini Scanorum [Lund]: s.n.; trans. Basil Kennet: *The Law of Nature and Nations*. London: Bonwicke and others, 1749.

Purcell, Hugh 2004: 'Tom Wintringham: Revolutionary Patriot', *History Today* **54/10**: 11–17.

Puschner, Uwe 2001: *Die völkische Bewegung im wilhelminischen Kaiserreich: Sprache, Rasse, Religion*. Darmstadt: Wissenschaftliche Buchgesellschaft.

Puységur, Jacques de Chastenet, Seigneur de 1690: *Les Mémoires ... avec les Instructions Militaires*. Paris: J. Morel; repr. Amsterdam: Abraham Wolfgang, 1690.

Puységur, Maréchal Jacques François de Chastenet, Marquis de 1748: *Art de la guerre, par principes et par règles*. Paris: Charles-Antoine Jombert; repr. Paris: Charles-Antoine Jombert, 1768.

Quester, George 1966: *Deterrence before Hiroshima*. New York: John Wiley.

1977: *Offense and Defense in the International System*. New York: John Wiley and Sons; repr. with new introduction, New Brunswick: Transaction, 2003.

Quinlan, Michael 1986: 'Nuclear Weapons – The Basic Issues', *Ampleforth Journal* **91/2**: 26–7.

Rahn, Werner 1976: *Reichsmarine und Landesverteidigung 1919–1928*. Munich: Oldenbourg.

1990: 'La Réflexion stratégique dans la Marine allemande de 1914 à 1945', in Hervé Coutau-Bégarie (ed.): *L'Évolution de la pensée navale*, vol. I. Paris: Economica: 135–62.

Raleigh, Sir Walter 1650: *Excellent Observations and Notes, Concerning the Royall Navy and Sea Services*. London: T. Warren for Humphrey Moseley; also published as *Judicious and Select Essayes and Observations*.

Ralston, Joseph 1999: 'The Revolution in US Air Power', *RUSI Journal* **144/6**: 54–62.

Ramatuelle, Audibert 1802: *Cours élémentaire de tactique navale*. Paris: Baudouin, Year X (includes volume of illustrations).

Ranft, Bryan 1982: 'Soviet Strategy', in Geoffrey Till (ed.): *Maritime Strategy in the Nuclear Age*. London: Macmillan: 68–74.

Ratzel, Friedrich 1897: *Politische Geographie*. Leipzig; 3rd edn, Munich: R. Oldenbourg, 1923.

Rees, David 1964: *Korea: The Limited War*. London: Macmillan.

Reid, Brian Holden 1987: *J.F.C. Fuller: Military Thinker*. London: Macmillan.

1998: *Studies in British Military Thought: Debates with Fuller and Liddell Hart*. Lincoln: University of Nebraska Press.

Repgen, Konrad 1985: 'Kriegslegitimationen in Alteuropa. Entwurf einer Historischen Typologie', *Historische Zeitschrift* **241**: 27–49.

Rice, Condoleezza 1991: 'The Evolution of Soviet Grand Strategy', in Paul Kennedy (ed.): *Grand Strategies in War and Peace*. New Haven: Yale University Press: 145–65.

Rice, Edward 1990: *Wars of the Third Kind: Conflict in Underdeveloped Countries*. Berkeley: University of California Press.

Rich, John 2001: 'Warfare and External Relations in the Middle Roman Republic', in Anja V. Hartmann and Beatrice Heuser (eds.):

War, Peace and World Orders in European History. London: Routledge: 62–71.

Richardot, Philippe 1998a: 'La Datation *du De Re Militari* de Végèce', *Latomus* **57/1**: 136–47.

1998b: *Végèce et la Culture militaire au Moyen Âge (Ve–XVe siècles)*. Paris: Economica.

1999: 'Y a-t-il une pensée navale dans l'Occident médiéval?', in Hervé Coutau-Bégarie (ed.): *L'Évolution de la pensée navale*, vol. VII. Paris: Economica: 13–23.

Richardson, William 1986: 'FM 100–5: The AirLand Battle in 1986', *Military Review* **66/3**: 4–11.

Richelson, Jeffrey 1983: 'PD-59, NSDD-13 and the Reagan Strategic Modernization Program', *Journal of Strategic Studies* **6/2**: 125–46.

1986: 'Population Targeting and U.S. Strategic Doctrine', in Desmond Ball and Jeffrey Richelson (eds.): *Strategic Nuclear Targeting*. Ithaca, NY: Cornell University Press: 234–343.

Richmond, Herbert 1930: *Naval Warfare*. London: Ernest Benn.

1931: *The Navy in India, 1763–1783*. London: Ernest Benn.

1934: *Seapower in the Modern World*. London: Bell.

1947: *Statesmen and Sea Power*. Oxford: Clarendon Press; repr. Westport, CT: Greenwood, 1974.

Riley-Smith, Jonathan 2001: 'The Crusading Movement', in Anja V. Hartmann and Beatrice Heuser (eds.): *War, Peace and World Orders in European History*. London: Routledge: 127–40.

Rink, Martin 1999: *Vom Partheygänger zum Partisanen: die Konzeption des kleinen Krieges in Preussen, 1740–1813*. Frankfurt: Peter Lang.

Ritter, Gerhard 1954: 'Das Problem des Militarismus in Deutschland', *Historische Zeitschrift* **177**: 21–48.

1961: 'Der Anteil der Militärs an der Kriegskatastrophe von 1914', *Historische Zeitschrift* **193**: 72–91.

Roberts, Adam 1976: *Nations in Arms: The Theory and Practice of Territorial Defence*. London: Chatto and Windus for the IISS.

Roberts, Adam and Richard Guelff (eds.) 2000: *Documents on the Laws of War*, 3rd edn. Oxford University Press.

Roberts, Michael 1956: *The Military Revolution, 1560–1660*. Belfast: Marjory Boyd.

1995: 'The Military Revolution 1560–1660', in Clifford J. Rogers (ed.): *The Military Revolution Debate*. Boulder, CO: Westview: 13–35.

Robertson, Scot 1998: 'The Development of Royal Air Force Strategic Bombing Doctrine between the Wars. A Revolution in Military Affairs?', *Airpower: Theory and Practice* **12/1**: 37–52.

Rocca Piacentino, Bernardino 1582: *De discorsi di gverra Libri quattro*. Venice: Damiano Denaro.

Rogers, Bernard 1983: 'Sword and Shield: ACE Attack of Warsaw Pact Follow-on-Forces', *NATO's Sixteen Nations* **28/1**: 16–26.

Rogers, Clifford 1995: 'The Military Revolutions of the Hundred Years' War', in Rogers (ed.): *The Military Revolution Debate*. Boulder, CO: Westview: 55–94.

1998: 'The Efficacy of the English Longbow: A Reply to Kelly DeVries', *War in History* **5/2**: 233–42.

Rohan, Henri duc de 1636: *Le Parfaict Capitaine, Autrement l'abrégé des guerres de Gaule des commentaires de César*. Paris: Iean Hovzé; repr. with an introduction by Werner Hahlweg, Osnabrück: Biblio Verlag, 1972.

Rohrschneider, Kai 1999: 'Die Schlacht von Beaumont, 30. Aug. 1870: die operative Vorbereitung der Schlacht von Sedan', *Mars* 5: 125–61.

Rose, François de 1995: *La Troisième Guerre mondiale n'a pas eu lieu: l'Alliance atlantique et la paix*. Paris: Desclée de Brouwer.

Rose, Michael 1999: 'Are We Creating a Universal Culture of Violence?', *RUSI Journal* **144/2**: 9–13.

Rosen, Stephen Peter 1982: 'Vietnam and the American Theory of Limited War', *International Security* **7/2**: 83–113.

forthcoming: 'The Impact of the Office of Net Assessment on the American Military in the Matter of the Revolution of Military Affairs', *Journal of Strategic Studies*, Special Issue: *Military Theory in the Post-Cold War Era*.

Rosenberg, David Alan 1981/82: 'A Smoking Radiating Ruin at the End of Two Hours', *International Security* **6/3**: 3–38.

1986: 'U.S. Nuclear War Planning, 1945–1960', in Desmond Ball and Jeffrey Richelson (eds.): *Strategic Nuclear Targeting*. Ithaca, NY: Cornell University Press: 35–322.

1994: 'Nuclear War Planning', in Michael Howard, George J. Andreopoulos and Mark R. Shulman (eds.): *The Laws of War: Constraints on Warfare in the Western World*. New Haven: Yale University Press: 160–90.

2000: 'American Postwar Air Doctrine and Organization: The Navy Experience', in Mark K. Wells (ed.): *Air Power: Promise and Reality*. Chicago: Imprint: 175–204.

Rosinski, Herbert 1977: *The Development of Naval Thought: Essays*. Newport, RI: Naval War College Press.

Roskill, S.W. 1954–61: *The War at Sea, 1939–1945*, 5 vols. London: HMSO.

1962: *The Strategy of Sea Power*. London: Collins.

Rothenberg, Gunther 1989: 'Soldiers and the Revolution: The French Army, Society, and the State, 1788–99', *Historical Journal* **32**: 981–95.

1994: 'The Age of Napoleon', in Michael Howard, George Andreopoulos and Mark Shulman (eds.): *The Laws of War: Constraints on Warfare in the Western World*. New Haven: Yale University Press: 86–97.

Rougeron, Camille 1970: 'Les Enseignements de la Guerre au Vietnam', *Forces Aèriennes Françaises* **265**: 39–54.

Rousseau, Jean Jacques 1761: *Extrait du projet de paix perpétuelle de M. l'abbé de Saint-Pierre*. s.l.: s.n.; anon. trans.: *A Project for Perpetual Peace*. London: Cooper, 1761.

Rouvraye, Charles chevalier de la 1815: *Traité sur l'art des combats de mer*. Paris: Bachelier.

Rühle von Lilienstern, Otto August 1813: *Apologie des Krieges*; 2nd expanded edn, 1818; repr. Vienna: Age d'homme-Karolinger, 1984.

1817: *Handbuch für den Offizier zur Belehrung im Frieden und zum Gebrauch im Felde*, vol. I. Berlin: G. Reimer.

1818: *Handbuch für den Offizier zur Belehrung im Frieden und zum Gebrauch im Felde*, vol. II. Berlin: G. Reimer.

Rummel, R.J. 1992: *Democide*. Brunswick, NJ: Transaction.

Rupp, Daniel 2006: 'Am Gegner lernen? Zur Taktik französischer Heere in den Schlachten von Kortrijk, Arques und am Pevelenberg', *Militärgeschichtliche Zeitschrift* **65/1**: 89–112.

Russell, Frederick 1975: *The Just War in the Middle Ages*. Cambridge University Press.

Rüstow, Wilhelm 1872–4: *Strategie und Taktik der Neuesten Zeit*, 3 vols. Zurich: Friedrich Schultheß.

Sabrosky, Alan Ned and Robert Sloane 1988: 'The Weinberger Doctrine: The Six Criteria for the Use of Military Force', in Sabrosky and Sloane (eds.): *The Recourse to War: An Appraisal of the Weinberger Doctrine*. Carlisle Barracks, PA: US Army War College Strategic Studies Institute: 11–13.

Sagan, Scott 1978: 'SIOP-62: The Nuclear War Plan Briefing to President Kennedy', *International Security* **12/1**: 22–51.

Saillans, François de [published as Bertrand de Loque] 1589: *Deux Traités: l'un de la guerre, l'autre du duel*. Lyons: Iacob Rayoyre; trans. I. Eliot: *Discovrses of Warre and Single Combat*. London: Iohn Wolfe, 1591.

Saint-Pierre, Charles-Irénée Castel, Abbé de 1712: *Projet pour rendre perpetuelle la paix en Europe*, 2 vols. Utrecht: A. Schouten; anon. trans.: *A Project for Settling an Everlasting Peace in Europe*. London: Ferd. Burleigh, 1714.

Salazar, Diego de 1536: *Tratado de Re militari: Historia de todas las guerras civilas*. Alcala de Henares: Miguel de Eguia; repr. in Eva Botella Ordinares (ed.): *Tratado de Re militri di Diego de Salazar*. Madrid: Ministry of Defence, Secretario Technica, 2000.

Salewski, Michael 1970: *Die deutsche Seekriegsleitung, 1935–1945*, vol. I, *1935–1941*. Frankfurt: Bernard und Graefe.

1973: *Die deutsche Seekriegsleitung, 1935–1945*, vol. III, *Denkschriften und Lagebetrachtungen, 1938–1945*. Frankfurt: Bernard und Graefe.

1975: *Die deutsche Seekriegsleitung, 1935–1945*, vol. II, *1941–1945*. Frankfurt: Bernard und Graefe.

1979: *Tirpitz: Aufstieg, Macht, Scheitern*. Göttingen: Musterschmidt.

2002: *Die Deutschen und die See: Studien zur deutschen Marinegeschichte des 19. und 20. Jahrhunderts*, 2 vols. Stuttgart: Franz Steiner.

2007: 'Seeschlachten und Weltgeschichte', *Historische Mitteilungen der Rankegesellschaft* **20**: 74–92.

Samuels, Martin 1990: 'The Reality of Cannae', *Militärgeschichtliche Mitteilungen* **1**: 7–31.

Sander, Erich 1934: 'Der Verfall der römischen Belagerungskunst', *Historische Zeitschrift* **149**: 457–76.

Santa Cruz de Marcenado, Don Alvaro de Navia Osorio y Vigil, Vizconde de Puerto, Marques de 1724–30: *Reflexiones Militares*, 11 Books, of which Books 1–10, Turin: Juan Francisco Mairesse, 1724–7; Book 11, Paris: Simon Langlois, 1730; abridged edition, Barcelona: Revista Cientifico-Militar, 1885.

Saux, Françoise Le 2004: 'War and Knighthood in Christine de Pizan's *Livre des faits d'armes et de chevllerie*', in Corrinne Saunders, Françoise Le Saux and Neil Thomas (eds.) *Writing War: Medieval Literary Responses to Warfare*. Cambridge: D.S. Brewer: 93–106.

Saxe, Maurice de / Moritz von Sachsen 1732: *Rêveries sur l'art de la guerre*. The Hague: s.n., 1756; trans. US Army in Brig.-Gen. T.R. Phillips (ed.): *Roots of Strategy*, vol. V. Mechanicsburg, PA: Stackpole Books, 1985: 177–300.

Scammell, Clare 1997: 'The Royal Navy and the Strategic Origins of the Anglo-German Naval Agreement of 1935', *Journal of Strategic Studies* **20/2**: 92–118.

Scharnhorst, Gerhard 1793: *Militärisches Taschenbuch zum Gebrauch im Felde*, 3rd impression Hanover: Helwigsche Hofbuchhandlung, 1794; facsimile repr. Osnabrück: Biblio Verlag, 1980 with an introduction by Ulrich Marwedel.

Schelling, Thomas 1959: 'Nuclear Weapons and Limited War'. RAND Report P-1620 (20 Feb.). Santa Monica, CA: RAND Corporation.

1960: *The Strategy of Conflict*. Cambridge, MA: Harvard University Press.

1965: 'Controlled Response and Strategic Warfare', *Adelphi Papers* 5/19: 3–11; repr. in Philip Bobbit, Lawrence Freedman and Gregory

Treverton (eds.): *US Nuclear Strategy – A Reader.* London: Macmillan, 1989: 223–36.

1966: *Arms and Influence.* New Haven: Yale University Press.

Schelling, Thomas and Morton Halperin 1961: *Strategy and Arms Control.* New York: The Twentieth Century Fund.

Scherff, Walter von 1892: *Delbrück und Bernhardi.* Berlin: A. Bath.

Schilling, Warner 1981: 'U.S. Strategic Nuclear Concepts in the 1970s', *International Security* **6/2**: 48–79.

Schlichting, General Sigismund von 1897: *Taktische und Strategische Grundsätze der Gegenwart.* Ernst Siegfried Mittler and Son; 3rd edn, Berlin: Ernst Siegfried Mittler and Son, 1898.

Schmähling, Elmar 1990: *Der unmögliche Krieg: Sicherheit und Verteidigung vor der Jahrtausendwende.* Düsseldorf: ECON Verlag.

Schmidt, Peer 2003: 'Der Guerrillero: die Entstehung des Partisanen in der Sattelzeit der Moderne – eine atlantische Perspektive, 1776–1848', *Geschichte und Gesellschaft* **29**: 161–90.

Schmitt, Oliver 2005: 'Kriegführung und tribale Gesellschaft', in Burkhard Meißner, Oliver Schmitt and Michael Sommer (eds.): *Krieg, Gesellschaft, Institutionen: Beiträge zu einer vergleichenden Kriegsgeschichte.* Berlin: Akademie Verlag: 417–44.

Schmückle, Gerd and Albrecht von Müller 1988: *Das Konzept der 'Stabilen Abhaltung'.* Starnberg: Max Planck Gesellschaft.

Schurman, Donald 1965: *The Education of a Navy.* London: Cassell.

1981: *Julian S. Corbett, 1854–1922.* London: Royal Historical Society.

Searle, Alaric 1998: 'A Very Special Relationship: Basil Liddell Hart, Wehrmacht Generals and the Debate on West German Rearmament, 1945–1953', *War in History* **5/3**: 327–57.

Searle, Thomas 2002: '"It Made a Lot of Sense to Kill Skilled Workers": The Firebombing of Tokyo in March 1945', *Journal of Military History* **66/1**: 103–33.

Secher, Reynald 2003: *A French Genocide: The Vendée.* Notre Dame, IN: Notre Dame University Press.

Sechi, Giovanni 1905: *Elementi di arte militare marittima*, vol. I. Livorno: Guisti.

Selesky, Harold 1994: 'Colonial America', in Michael Howard, George J. Andreopoulos and Mark R. Shulman (eds.): *The Laws of War: Constraints on Warfare in the Western World.* New Haven: Yale University Press: 59–85.

Selig, Robert A and David Curtis Skaggs 1991: 'Introductory Essay', in Johann Ewald: *Treatise on Partisan Wafare.* New York: Greenwood Press: 1–38.

Serman, William and Jean-Paul Bertaud 1998: *Nouvelle Histoire militaire de la France, 1789–1919.* Paris: Fayard.

Seversky, Alexander de 1942: *Victory through Air Power.* New York: Simon & Schuster.

Sherwin, C.W. 1956: 'Securing Peace through Military Technology', *Bulletin of the Atomic Scientists* **12**: 159–64; repr. in Philip Bobbit, Lawrence Freedman and Gregory Treverton (eds.): *US Nuclear Strategy – A Reader.* London: Macmillan, 1989: 131–42.

Showalter, Dennis 1990: 'German Grand Strategy: A Contradiction in Terms?', *Militärgeschichtliche Mitteilungen* **2**: 65–102.

1991: 'Total War for Limited Objectives: An Interpretation of German Grand Strategy', in Paul Kennedy (ed.): *Grand Strategies in War and Peace.* New Haven: Yale University Press: 105–24.

1994: 'Hubertusburg to Auerstädt: The Prussian Army in Decline?', *German History* **12/3**: 308–33.

1996: *The Wars of Frederick the Great.* London: Longman.

2000: 'From Deterrence to Doomsday Machine: The German Way of War, 1890–1914', *Journal of Military History* **64/3**: 679–710.

Shultz, Richard 1992: 'Compellence and the Role of Air Power as a Political Instrument', in Richard Shultz and Robert Pfaltzgraff (eds.): *The Future of Air Power in the Aftermath of the Gulf War.* Maxwell, AL: Air University Press: 171–91.

Sidebottom, Harry 2004: *Ancient Warfare: A Very Short Introduction.* Oxford University Press.

Slessor, John 1936: *Air Power and Armies.* London: Oxford University Press.

1954: *Strategy for the West.* London: Cassell.

1957: *The Great Deterrent: A Collection of Lectures, Articles, and Broadcasts on the Development of Strategic Policy in the Nuclear Age.* London: Cassell.

1958: 'A New Look at Strategy from the West', *Orbis* **11**: 320–36.

1961: *What Price Coexistence? A Policy for the Western Alliance.* London: Cassell.

Sloan, Geoffrey 1999: 'Sir Halford J. Mackinder: The Heartland Theory Then and Now', *Journal of Strategic Studies* **22/2**: 15–38.

Slocombe, Walter 1981: 'The Countervailing Strategy', *International Security* **5/4**: 18–27.

Smith, John 1994: *Rolling Thunder: The American Strategic Bombing Campaign against North Vietnam, 1965–1968.* Walton on Thames: Air Research Publications.

1996: 'The Strategic Bombing Campaign: North Vietnam, 1965–1968', *Journal of Strategic Studies* **19/2**: 280–98.

Smith, Martin 2002: '"In a Box in the Corner"? NATO's Theatre Nuclear Weapons, 1989–99', *Journal of Strategic Studies* **25/1**: 1–20.

Smith, Merritt Roe and Leo Marx 1994 (eds.): *Does Technology Drive History? The Dilemma of Technological Determinism*. Cambridge, MA: MIT Press.

Smith, Rupert 2005: *On the Utility of Force: The Art of War in the Modern World*. London: Allen Lane; repr. Harmondsworth: Penguin, 2006.

Smythe, Sir John 1590: *Certain Discourses Military*. London: Richard Jones; ed. J.R. Hale, Ithaca, NY: Cornell University Press, 1964.

Snyder, Jack 1984: 'Civil-Military Relations and the Cult of the Offensive, 1914 and 1918', *International Security* **9/1**: 108–46.

2002: 'Anarchy and Culture: Insights from the Anthropology of War', *International Organization* **56/1**: 5–20.

Sokol, Anthony Eugene 1961: *Seapower in the Nuclear Age*. Washington, D.C.: Public Affairs Press.

Sokolovsky, V.D. 1963: *Military Strategy: Soviet Doctrine and Concepts*. London: Pall Mall Press.

Sokolsky, Joel 1989: 'Anglo-American Maritime Strategy, 1960–1980', in John Hattendorf and Robert Jordan (eds.): *Maritime Strategy and the Balance of Power*. Basingstoke: Macmillan: 304–29.

Sokov, Nikolai 2004: 'Russia's New National Security Concept: The Nuclear Angle', online at www.nti.org/db/nisprofs/over/concept.htm, accessed 7 August 2008.

Sondhaus, Lawrence 2001: *Naval Warfare 1815–1914*. New York: Routledge.

Spaight, James Molony 1932: *An International Air Force*. London: Gale and Polden.

1938: *Air Power in the Next War*. London: Geoffrey Bles.

Spannocchini, E. and Gérard Brossollet 1976: *Verteidigung ohne Selbstzerstörung: Das Ende der Schlacht*. Vienna and Munich: Carl Hanser.

Squadrilli, Edoardo 1937: *Politica marinara e Impero fascista*. Rome: Stabilimento tipografico del genio civile.

Stacey, Robert 1994: 'The Age of Chivalry', in Michael Howard, George Andreopoulos and Mark Shulman (eds.): *The Laws of War: Constraints on Warfare in the Western World*. New Haven: Yale University Press: 27–39.

Starkey, Armstrong 1998: *European and Native American Warfare, 1675–1815*. London: University College London Press.

Starr, C.G. 1978: 'Thucydides on Sea Power', *Mnemosyne* **31**: 343–55.

Steel, David 1784: *Elements and Practice of Riggings and Seamanship*, 2 vols. London: D. Steel.

Stein, Dr Lorenz von 1872: *Die Lehre vom Heerwesen als Theil der Staatswissenschaft*. Stuttgart: s.n., repr. Osnabrück: Biblio Verlag, 1967.

Sterrett, James 2007: *Soviet Air Force Theory, 1918–1945*. London: Routledge.

Stevenson, Roger ['An officer'] 1770: *Military Instructions for Officers detached in the Field containing a scheme for forming a corps of a partisan illustrated with plans of the manoeuvres necessary in carrying on the petite guerre*. London: s.n.; republished as *Military Instructions for Officers detached in the field containing a scheme for forming a corps of partisans* ... London: John Millan, Edward and Charles Dilly, 1779.

Stonemann, Mark R. 2001: 'The Bavarian Army and French Civilians in the War of 1870–1871: A Cultural Interpretation', *War in History* **8/1**: 280–99.

Strachan, Hew 1994: 'Germany in the First World War: The Problem of Strategy', *German History* **12/2**: 237–49.

1998: 'The Battle of the Somme and British Strategy', *Journal of Strategic Studies* **21/1**: 79–95.

2000: 'The First World War', *Historical Journal* **43/3**: 889–903.

2003: 'The Meaning of Strategy: Historical Reflections'. Inaugural lecture of 4 December 2003. Excerpts in *Newsletter of the University of Oxford*, 15 January 2004.

2005: 'The Lost Meaning of Strategy', *Survival* **47/3**: 35–54.

2006: 'Making Strategy: Civil-Military Relations after Iraq', *Survival* **48/3**: 59–82.

Strickland, Matthew 2006: 'Rules of War or War without Rules? Some Reflections on Conduct and the Treatment of Non-Combatants in Medieval Transcultural Wars', in Hans-Henning Kortüm (ed.): *Transcultural Wars from the Middle Ages to the 21st Century*. Berlin: Akademie Verlag: 107–40.

Stubbs, Richard 1989: *Hearts and Minds in Guerrilla Warfare: The Malayan Emergency 1948–1960*. Oxford University Press.

Sully, Maximilien de Béthune, duc de 1632: *Mémoires des sages et royalles oeconomies d'Estat, domestiques, politiques et militaires de Henry le Grand, l'exemplaire des roys, le prince des vertus, des armes et des loix et le père en effet de ses peuples françois*. Amstelredam [i.e. Sully-sur-Loire]: Aletinosgraphe de Clearetimelee et Graphexechon de Pistariste [i.e. Jacques Bouquet]; trans. Charlotte Lennox: *Memoirs of Maximilian de Béthune ... Prime Minister of Henry the Great*. London: J. Rivington and Sons, 1778; repr. with an introduction by David Ogg: *Sully's Grand Design of Henry IV*. London: Sweet and Maxwell, 1921.

Sumida, Jon Tetsuro 1999: 'Alfred Thayer Mahan, Geopolitician', *Journal of Strategic Studies* **22/2–3**: 39–62.

2000: *Inventing Grand Strategy, Teaching Command: The Classic Works of Alfred Thayer Mahan Reconsidered*. Washington, D.C.: Woodrow Wilson Press.

Summers, Harry G. 1985: 'Principles of War and Low Intensity Conflict', *Military Review* **65/3** (March), 43–49.

Summers Jr, Colonel Harry G. 1982: *On Strategy: A Critical Analysis of the Vietnam War*. Novato, CA: Presidio.

Sun Tzu *c*. 500–300 BCE: *The Art of War*, trans. Thomas Cleary. Boston: Shambala, 1991.

Sutcliffe, Matthew 1593: *The Practice, Proceedings and Lawes of Armes*. London: Deputies of C. Barker.

Svechin, Aleksandr A. 1992: *Strategy*. Minneapolis, MN: East View Publications.

Taillemite, Étienne 1992: 'Un théoricien méconnu de la guerre maritime: l'Amiral Richild Grivel', in Hervé Coutau-Bégarie (ed.): *L'Évolution de la pensée navale*, vol. II. Paris: Economica: 87–113.

1994: 'Henri-Joseph Paixhans et sa nouvelle Force maritime', in Hervé Coutau-Bégarie (ed.): *L'Évolution de la pensée navale*, vol. IV. Paris: Economica: 105–32.

1999: 'Deux pré-Mahaniens français: Bouët-Willaumez et Penoat', in Hervé Coutau-Bégarie (ed.): *L'Évolution de la pensée navale*, vol. VII. Paris: Economica: 45–77.

Tallett, Frank 1992: *War and Society in Early-Modern Europe, 1495–1715*. London: Routledge.

2006: 'Barbarism in War: Soldiers and Civilians in the British Isles, *c*. 1641–1652', in G. Kassimoris (ed.): *Warrior's Dishonour: Barbarity, Morality and Torture in Modern Warfare*. Aldershot: Ashgate: 19–39.

Tallett, Frank and Joël Félix 2008: 'The French Experience', in Christopher Storrs: *The Fiscal-Military State in Eighteenth-Century Europe*. Aldershot: Ashgate: 147–66.

Talmon, Jacob Laib 1952: *The Origins of Totalitarian Democracy*. London: Secker and Warburg.

Tedder, Arthur William 1948: *Air Power in War: The Lees Knowles Lectures*. London: Hodder and Stoughton.

Tertrais, Bruno 1994: *L'Arme nucléaire après la guerre froide*. Paris: Economica.

Theobald, Joseph von 1817: *Strategische Studien: Ein Beitrag zur Beförderung einer gründlichen Kenntniß der Strategie*. Stuttgart and Tübingen: J.O. Cotta'sche Buchhandlung.

Thierry, Hubert 1970: *Les Armes atomiques et la politique internationale*. Paris: Dunod.

Thompson, Robert 1966: *Defeating Communist Insurgency: Malaya and Vietnam*. London: Chatto and Windus.

1970: *Revolutionary War in a World Strategy, 1945–1969*. London: Secker & Warburg.

Thomson, Lord 1927: *Air Facts and Problems*. London: John Murray.

Thucydides 5th century BCE: *History of the Peloponnesian War*, trans. Rex Warner. Harmondsworth: Penguin, 1954.

Till, Geoffrey 1982: *Maritime Strategy in the Nuclear Age*. London: Macmillan.

1994a: 'Maritime Strategy and the Twenty-First Century', *Journal of Strategic Studies* **17/1**: 176–99.

Till, Geoffrey (ed.) 1994b: *Seapower: Theory and Practice*. London: Frank Cass.

2006: *The Development of British Naval Thinking*. London: Routledge.

Tilly, Charles 1964: *The Vendée*. London: Arnold.

Tooke, Joan 1965: *The Just War in Aquinas and Grotius*. London: SPCK.

Trachtenberg, Marc 1986: 'The Question of No-First-Use', *Orbis* **29/4**: 53–69.

Traverse, baron de (ed.) 1752: *Extrait de la première partie de l'Art de Guerre de M. le Maréchal de Puységur*. Paris: Charles-Ant. Jombert et Hochereau l'Ainé.

Tressan, comte de 1757: 'Guerre', in Denis Diderot and Jean le Rond d'Alembert : *Encyclopédie ou Dictionnaire raisonné des Sciences, des Arts et des Métiers*, vol. VII. Paris: Briasson, David, Le Breton and Durand.

Trinquier, Roger 1961: *La Guerre modern*. Paris: La Table Ronde; trans. Daniel Lee: *Modern Warfare: A French View of Counterinsurgency*. Westport, CT: Praeger, 2006.

1968: *Guerre – Subversion – Révolution*. Paris: Laffont.

Twigge, Stephen and Alan Macmillan 1996: 'Britain, the United States, and the Development of NATO Strategy, 1950–1964', *Journal of Strategic Studies* **19/2**: 260–81.

UN High Level Panel on Threats, Challenges and Change (2004): 'A More Secure World: Our Shared Responsiblity', online at www.un.org/secureworld/report.pdf, accessed 14 July 2009.

United Kingdom, Ministry of Defence 1957: *Defence: Outline of Future Policy*. Cmnd. 124. London: HMSO.

Urrea, Jerónimo Jiménez de 1556: *Diálogo de la Verdadera Honra Militar*. Venice: ?Marchio Sessa; repr. Madrid: Ministry of Defence, 1992.

US Army 2004: *Counterinsurgency Field Manual* FM 3–07.22 (interim), online at www.fas.org/irp/doddir/army/fmi3–07–22.pdf, accessed 26 October 2006.

US Army and US Marine Corps 2006: *Counterinsurgency Field Manual* FM 3–24, MCWP 3–33.5, online at www.fas.org/irp/doddir/army/fm3–24.pdf, accessed 29 April 2010.

US National Conference of Catholic Bishops 1983: 'A Challenge of Peace: God's Promise and our Response' (19 May 1983), online at www.americancatholic.org/Newsletters/CU/ac0883.asp, accessed 3 August 2008.

Usczeck, Hansjürgen and Christa Gudzent (eds.) 1986: *Gerhard von Scharnhorst: Ausgewählte militärische Schriften*. East Berlin: Militärverlag der DDR.

Vacherot, Étienne 1860: *La Démocratie*. Paris: Chamerot.

Valentini, Georg Wilhelm Freiherr von 1799: *Abhandlung über den kleinen Krieg und über den Gebrauch der leichten Truppen mit Rücksicht auf den französischen Krieg*; 4th edn, Berlin: J.W. Boicke, 1820.

Valentino, Benjamin, Paul Huth and Dylan Balch-Lindsay 2004: '"Draining the Sea": Mass Killing and Guerrilla Warfare', *International Organization* **58** (Spring): 375–407.

Vallière, François la Baume le Blanc, Chevalier de la 1666: *Pratiques et maximes de la guerre*. Paris: Estienne Loyson.

Valturius, Robertus 1472: *De Arte Militari*. Verona: Johannes ex Verona.

Vascotto, Vezio 2001: 'L'amiral Castex et ses théories stratégiques', *Revue Historique des Armées*, special issue Mélanges **224**: 81–8.

Vattel, Emerich de 1758: *Le Droit des gens, ou Principes de la loi naturelle*, Book III 'De la guerre'. Leyden: La Compagnie; trans. Joseph Chitty: *The Law of Nations; or, Principles of the Law of Nature*. London: S. Sweet and Stevens, 1834.

Vauban, Sébastien le Prestre de 1685: *Le Directeur général des fortifications*. s.l.: Van Bulderen.

1704: *Le Triomphe de la méthode: le Traité de l'attaque des places de Monsieur de Vauban, Ingénieur du Roi*. MS in the Direction Centrale du Génie, Paris, repr. Paris: Gallimard, 1992.

1737–42: *De l'Attaque et de la défense des places*. s.l.: De Hondt.

Vauthier, Arsène 1930: *Le Danger aérien et l'avenir du pays*. Paris: Berger-Levrault.

Vauthier, Colonel P. 1935: *La Doctrine de guerre du Général Douhet*. Paris: Berger-Levrault.

Vegetius, Flavius Renatus *c.* 387: *Epitoma de re militari*, trans. N.P. Milner: *Epitome of Military Science*, 2nd edn. University of Liverpool Press, 1996.

Verbruggen, J.F. 1994: 'De rol van de ruiterij in de Middeleeuwse oorlogvoering', *Revue Belge d'Histoire Militaire* **30**: 389–416.

Verbruggen, Jan Frans 1997: *The Art of Warfare in Western Europe during the Middle Ages: From the Eighth Century to 1340*, 2nd edn. Woodbridge: Boydell Press (first published 1977).

Vivier, Thierry 1997: 'Le Douhétisme français entre tradition et innovation, 1933–39', *Revue Historique des Armées* **206/1**: 89–99.

Voltaire, François Marie Arouet de 1769: *De la Paix perpetuelle, par le Docteur Goodheart*. Amsterdam: M.-M. Rey.

W., C. von 1808: *Marginalien zu den Grundsätzen der höhern Kriegskunst für die Generale der östreichischen Armee*. Weimar: Commission des Landes-Industrie-Comptoirs.

Wagner, August 1809: *Grundzüge der reinen Strategie*. Amsterdam: Kunst- und Industrie-Comptoir.

Wagner, Michael 1995: '"Normalkrieg" oder "Völkermord"? Neue Forschungen zur Niederwerfung des Aufstandes in der Vendée', *Francia* **22/2**: 177–85.

Waldstätten, Johann Freiherr von 1895: *Strategische Grundsätze in ihrer Anwendung auf den Feldzug in Italien 1866*. Vienna: L.W. Seidel.

Waldtstätten, Freiherr von (ed.) 1882: *Erzherzog Karl: Ausgewählte militärische Schriften*. Berlin: Richard Wilhelmi.

Warden, John 1992: 'Employing Air Power in the Twenty-First Century', in Richard H. Shultz, Jr and Robert L. Pfaltzgraff, Jr (eds.): *The Future of Air Power in the Aftermath of the Gulf War*. Maxwell Air Force Base, AL: Air University Press: 57–82.

Watt, Donald Cameron 1989: *How War Came*. London: Heinemann.

Weber, Eugene 1977: *Peasants into Frenchmen: The Modernisation of Rural France 1870–1914*. London: Chatto and Windus.

Webster, Charles and Nobel Frankland 1961: *The Strategic Air Offensive against Germany*, vol. IV. London: HMSO.

Wees, Hans van 2001: 'War and Peace in Ancient Greece', in Anja Hartmann and Beatrice Heuser (eds.): *War, Peace and World Orders in European History*. London: Routledge: 33–48.

2004: *Greek Warfare: Myths and Realities*. London: Duckworth.

Wegener, Wolfgang 1929: *Die Seestrategie des Weltkrieges*. Berlin: Mittler; 2nd rev. edn, Berlin: Mittler, 1941.

Weigley, Russell 1976: *The American Way of War: A History of United States Military Strategy and Policy*. New York: Macmillan.

1991: *The Age of Battles*. Bloomington, IN: Indiana University Press.

Weizsäcker, Carl Friedrich von 1984: *Die Praxis der defensiven Verteidigung*. Hameln: Spontholz.

Wells, H.G. 1908: *The War in the Air*, serial publication, *Pall Mall Magazine*, January–December; book form, London: Odhams, n.d. = 1921.

Wells, Mark (ed.) 2000: *Air Power: Primes and Reality*. Chicago: Imprint Publications.

Weltman, John (1995): *World Politics and the Evolution of War*. Baltimore, MD: Johns Hopkins University Press.

West, Philip 1989: 'Interpreting the Korean War', *American Historical Review* **94/1**: 80–96.

Whitby, Michael 2005: 'War and State in Late Antiquity: Some Economic and Political Connections', in Burkhard Meißner, Oliver Schmitt and Michael Sommer (eds.): *Krieg, Gesellschaft, Institutionen: Beiträge zu einer vergleichenden Kriegsgeschichte*. Berlin: Akademie Verlag: 355–86.

Wijn, J.W. 1960: 'Johann der Mittlere von Nassau-Siegen', in Werner Hahlweg (ed.): *Klassiker der Kriegskunst*. Darmstadt: Wehr- und Wissen-Verlag: 119–33.

Wilkinson, Henry Spenser 1894: *Command of the Sea and Brain of the Navy*. Westminster and London: Archibald Constable.

1910: *War and Policy*, 2nd edn: London: Constable.

William (Guilhelmus) Count of Nassau and Alain Claude de Mestre 1675: *Annibal et Scipion, ou les Grands Capitaines, avec les ordres et plans de batailles*. The Hague: s.n.; also published as Nassau, Comte G.L. de: *Annibal et Scipion ou les Grands Capitaines avec les ordres and plans des batailles*. Amsterdam: Marc-Michel Rez, 1768.

Williams, Al 1940: *Air Power*. New York: Coward-McCann.

Williams, Sir Roger 1590: *A Briefe Discourse of Warre*. London: Thomas Orwin.

Williams, T. H. 1954: *Lincoln and His Generals*. New York: McGraw-Hill.

Willisen, Wilhelm von 1840: *Die Theorie des großen Krieges*, 3 vols.; 2nd edn, Berlin: Duncker und Humblot, 1868.

1860: *Ueber große Landes-Vertheidigung, oder über Festungsbau und Heer-Bildung in Preußen*. Berlin: Duncker und Humblot.

Willmott, H.P. 2002: *When Men Lost Faith in Reason: Reflections on War and Society in the 20th Century*. Westport, CT: Praeger.

Wilson, A.J. 1973: 'The War in Vietnam – Reflections on Counter-Insurgency Operations', *RUSI Journal* **118/1**: 20–7.

Wilson, K.M. 1994: 'Understanding the "Misunderstanding" of 1 August 1914', *Historical Journal* **37/4**: 885–9.

Wilson Keith (ed.) 1995: *Decisions for War, 1914*. London: UCL Press.

Wilson, Trevor and Robin Prior 2001: 'Conflict, Technology, and the Impact of Industrialization: The Great War 1914–18', *Journal of Strategic Studies* **24/3**: 128–57.

Windsor, Philip 2002: *Strategic Thinking: An Introduction and Farewell*, ed. Mats Berdal and Spyros Economides. Boulder, CO: Lynne Rienner.

Wintringham, Tom 1940: *Armies of Freemen*. London: George Routledge & Sons.

Wohlstetter, Albert 1959: 'The Delicate Balance of Terror', *Foreign Affairs* **97/2**: 211–34.

1983: 'Bishops, Statesmen, and Other Strategists', *Commentary* **75/6**: 15–35.

Wollstein, Günter 1978: 'Scharnhorst und die Französische Revolution', *Historische Zeitschrift* **227**: 325–52.

Xylander, J. Ritter von 1818: *Die Strategie und ihre Anwendung*. Munich: Joseph Lindauer; 2nd rev. edn, Munich: Joseph Lindauer, 1821.

1831: *Untersuchungen über das Heer-Wesen unserer Zeit*. Munich: J.B. Cotta'sche Buchhandlungen.

1847: *Truppen-Lehre der Infantrie, Cavalerie und Artillerie*, 3rd edn. Munich: Joseph Lindbauer'sche Buchhandlung.

Yazov, D.T. 1989: 'On Soviet Military Doctrine', *RUSI Journal* **134/4**: 1–4.

Yost, David 1990: 'The Delegitimization of Nuclear Deterrence?', *Armed Forces and Society* **16/4**: 487–508.

Yvon, Pierre 1924: *La Guerre aérienne: Contribution à l'étude de ses lois*. Paris: Per orbem.

Z, Commandant and H. Montechant 1893: *Essai de stratégie navale*. Paris and Nancy, Berger-Levrault.

Zanthier, Friedrich Wilhelm von 1775: *Freyer Auszug aus des Herrn Marquis de Santa Cruz de Marzenado, Gedanken von Kriegs- und Staatsgeschäften, nebst einem Versuch über die Kunst den Krieg zu studieren*. Göttingen and Gotha: Johann Christian Dieterich.

Zeller, Bernhard 2001: 'Collective Identities, War and Integration in the Early Middle Ages', in Anja V. Hartmann and Beatrice Heuser (eds.): *War, Peace and World Orders in European History*. London: Routledge: 102–12

Zeppelin, Graf von 1908: *Die Eroberung der Luft*. Stuttgart: Deutsche Verlagsanstalt.

Index

CPSIA information can be obtained
at www.ICGtesting.com
Printed in the USA
LVHW081334270322
714519LV00009B/552